ELECTRON FLOW IN ORGANIC CHEMISTRY

ELECTRON FLOW IN ORGANIC CHEMISTRY

A DECISION-BASED GUIDE TO ORGANIC MECHANISMS

Third Edition

Paul H. Scudder

New College of Florida
Sarasota, Florida, USA

WILEY

Published by John Wiley & Sons, Inc., Hoboken, New Jersey.
Published simultaneously in Canada.

For general information on our other products and services or for technical support, please contact our Customer Care Department within the United States at (800) 762-2974, outside the United States at (317) 572-3993 or fax (317) 572-4002.

Wiley also publishes its books in a variety of electronic formats. Some content that appears in print may not be available in electronic formats. For more information about Wiley products, visit our web site at www.wiley.com.

Library of Congress Cataloging-in-Publication Data applied for

Paperback: 9781119718932; e-pdf: 9781119718970; e-pub: 9781119718994

Cover Design: Wiley

Cover Image: Courtesy of Paul H. Scudder

SKY10054597_090123

Contents in Brief

Contents

Contents

Preface

PREFACE TO THE STUDENT

Critical Thinking Approach

I want you to do well in organic chemistry. And even more, I would like you to understand why people become so interested in the field that they choose to make it their life's work. As a professor who taught Organic for 40 years, I recognized that probably less than 10% of the students in my Organic course were there of their own free will; almost all were taking Organic because it was required for something else. This makes for a tough audience. I wanted to engage my students and show them that Organic was not entirely dreary memorization of material out of a five-pound textbook. Organic can, not only be interesting, but also be important and relevant to their life, and can develop essential critical thinking skills that carry over to other fields.

It was more important for me to teach scientific critical thinking to my class of prospective biologists, biochemists, physicians, and chemists than to simply teach one more example of any reaction. Every student in my class needed to learn how to reason through complex problems. Organic Chemistry is full of these, and so the puzzles that my students would solve, in learning the scientific method, would be organic reactions. I asked my students to give me reasonable hypotheses for how a reaction works. As a puzzle to solve, organic reactions can teach critical thinking skills, which then can be applied to many other fields. I certainly want my physician to be great at critical thinking, for that is the essence of an accurate diagnosis.

Secrets to Success

This section gives you some of the secrets to success that have come from research on why some people succeed and others do not. Many books have been written on this topic, and ideas that will help you do well in Organic are summarized here. A long time ago, professor Lewis Terman of Stanford University developed the original IQ test and followed up on the youth that had top scores on that test. He found almost no correlation with how they turned out later in life. If smarts do not correlate with success, then what does? Recent research has turned up two character traits that can predict success, and the good news is that they are both learned and are not inherent. The learned traits are "mindset" and "grit," so let's look at each.

Mindset is the work of psychology professor Carol Dweck of Stanford University. She showed that people often had attitudes toward challenges that mostly fell into two categories. The "fixed mindset" group felt that their abilities were inherent, something they were born with, and not able to change. If they were "not good at math," then they would find another topic that came easily. If they failed, they took it personally, because it seemed they were not smart enough to succeed in that area. They often felt that working hard on a topic would not make up for a lack of an inherent gift. Therefore, they gave up easily.

Those with a "growth mindset" believed that ability was not inherent, but built through effort and practice. They realized that they could get better at something by working at it. For example, nobody becomes good a playing a musical instrument without a lot of practice. A growth mindset response to doing poorly at something was to work harder, not to quit. They just had not finished learning the topic or skill. If you look back at something you can do well, you will find that you spent a lot of time working at it. With a growth mindset, you can achieve much more than someone with a fixed mindset. Instead of saying, "I can't do it" just say, "I can't do it <u>yet</u>" and keep at it.

The second trait, "grit," is the work of psychology professor Angela Duckworth of the University of Pennsylvania. She defines grit as the perseverance and passion for long-term goals. Someone with the growth mindset needs grit to keep working at what they have not learned "yet." A good predictor of future success is the growth mindset, joined with the grit to not give up when things get difficult. Giving up is not a success strategy for anything.

Preface

The Challenge of Organic Chemistry

Organic Chemistry is a difficult topic; it is a cumulative year-long course, and is unlike any science course you have had before. Many other courses could be broken into separate modules with individual exams, and then it was off to a new topic. Organic Chemistry is one very long interconnected story. It is much more like a foreign language, where what you learn on day one is still needed at the end of the year. However, we tend to remember things that are recent and those things that we use a lot. The year-long nature of an Organic Chemistry course therefore requires much practice to retain the material for the length of the course. Cramming just does not work, but in the long run practice does.

This text uses 18 elemental mechanistic units, the electron flow pathways, to explain all organic reactions. These units are used over and over again and become the core of understanding how all the organic reactions work. The continual repetition of these basic units helps you retain the material and allows you to understand why these reactions occur. It is much easier to understand and retain information that makes conceptual sense.

An Expert Systems Approach to Organic Chemistry

Critical thinking requires us to ask, “Does this make sense?” If not, it's probably wrong. A believable scientific hypothesis has to be reasonable and pass cross-checks. We need to “generate and select” alternatives, which is the essence of a good critical thinking process. The map of all alternatives can be represented as a tree, what AI calls the “problem space.” This text extracts the essence of the field: the conceptual tools, the general rules, the trends, the modes of analysis, and everything that one would use to construct an expert system. We need an efficient way to navigate this problem space from start to goal. For that, we need a small set of essential principles, or “control knowledge,” to guide our route selection decisions toward an acceptable answer.

This text makes use of analysis tools common to expert systems, but rare in organic chemistry texts (such as flow charts as decision maps, correlation matrices to show all possible interactions, and simplified energy surfaces used as problem space maps). Good intuition comes from your automatic use of control knowledge to guide your decisions. If you can internalize this expert-system decision process, you will develop a chemical intuition and are well on your way to becoming an expert yourself. How to learn organic chemistry using this critical thinking approach is the essence of this book.

A Link to Biochemistry

For many students in a chemistry, biology, or premedical curriculum, the next course they will take after a year of Organic Chemistry is a year of Biochemistry, the organic chemistry of living cells. Since the cell is the world's best organic chemist, biochemistry has much to teach us about elegant reaction mechanisms. How does a cell manage to make all the compounds it needs from simple reactants, in water, at room temperature, and producing only the correct shape and not the useless mirror image? Organic Chemistry and Biochemistry are much more interesting when you think of them as an opportunity to learn the secrets of how life works. This text makes use of biochemical reaction mechanism examples to help you understand the magic and the elegance of the chemistry of living systems. It is just a taste of the things to come in a biochemistry course. You can see why physicians need to know organic and biochemistry, for if you can understand how a living system works, you might be able to repair it when things start to go wrong. I wish you success in this endeavor.

TO THE INSTRUCTOR

Unique Decision-Based Approach Expanded

This third edition provides students with something that they cannot get anywhere else: a chemical intuition based on learning and internalizing a cross-checked-decision process. An important part of the scientific method is the ability to postulate a reasonable hypothesis. This text teaches students how to write reasonable reaction mechanisms, and assumes only a general chemistry background. This text provides tools for handling large amounts of information. It emphasizes the "why?" of organic chemistry in order to help make sense of all the material.

To be able to teach students to make good decisions, we need to teach "control knowledge," which is the essence of a good intuition. These are checks of reasonability that include, among other things: stability trends, compatibility with the media pH, evaluation of energetics, and similarity to known processes. In addition to using flow charts to illustrate the problem analysis process, the third edition increases the use of energy surfaces as problem space maps to help with illustration of these concepts, while continuing the rigorous mechanistic approach to organic chemistry.

Originating with the first edition of this text was the concept of mechanisms being built from a limited number of elementary electron flow pathways, and that learning to assemble these pathways in a reasonable manner is all that is necessary to master mechanisms in organic chemistry. The impressive advantage that a decision-based approach has over memorization is that it engages the student. The instructor can ask questions like, "Why did it do this and not that?" New reactions become puzzles to solve, not simply another item to memorize.

This text uses several concepts and tools new to most undergraduate organic texts (but common in advanced texts) to aid in understanding the most difficult sections of the typical organic course. Hard–soft acid–base theory is used to guide decisions and to explain and predict the dual reactivity of many species. This text uses energy surfaces as maps of the terrain so that students have a physical model to help with the more complex decisions. Energy surfaces serve as bigger building blocks for multistep mechanisms.

Changes From The Second Edition

Besides the usual clarifications and modifications necessary to bring the text up to date, the text has been expanded to reinforce a decision-based approach. There are more flowcharts, correlation matrices, and algorithms that illustrate decision processes. Energy surfaces, normally the domain of graduate texts, serve as concept maps and allow students to visualize alternative routes.

All curly electron flow arrows are now in red to make those arrows stand out from the drawn structures. Color is also used to highlight and focus the reader's attention on parts of the molecule when deciding what parts have changed and to illustrate viable routes on the energy surfaces.

The text has been made more accessible to beginning students by adding more background discussion of how General Chemistry concepts can be used to show that structure determines reactivity. Proton transfer mechanisms and predictions of acid–base reactions are introduced early, in Chapter 3, setting up the discussion of organic reactions. To increase readability, all frontier molecular orbital theory reactivity explanations have been moved much later, to Chapter 13, "Qualitative Molecular Orbital Theory and Pericyclic Reactions."

Chapter 10, "Choosing the Most Probable Path," has been expanded with energy surfaces linked together like large building blocks to show the problem space terrain for each example and to show how the reactant structure and reaction conditions influence the route favored.

This chapter develops a general approach to all organic reactions by showing how to focus on the most reactive centers and choose the best route. The ΔpK_a cross-check developed for Prof. Jorgensen's CAMEO expert system for predicting organic reaction products is often used in this text for deciding reasonable reaction energetics.

New Chapters

A new Chapter 11, "Cross-checks and Decision Boundaries," focuses on the "control knowledge" of good mechanistic thinking. What makes for a reasonable mechanistic hypothesis? This chapter collects all of the more important mechanistic cross-checks introduced earlier in the text, and shows the critical thinking needed to avoid many of the common errors pulled from student exams and other sources. It presents what basic principles had been forgotten in the erroneous example and discusses more reasonable alternatives.

Chapter 14 also new, "Organic Structure Elucidation Strategies," closes out the text by summarizing useful tactics to solve spectral problems. In my years teaching a problems-based Structure Elucidation course, I found many books that were good collections of spectral problems, other texts that covered the instrumentation theory of how the spectra originated, but only a few books that detailed strategies on how to approach solving these problems.

A larger collection of important tools is gathered together in the appendix, including a new section, "A Bridge to Transition Metal Organometallics," designed to cover the gap between organic mechanisms and transition metal organometallics by retaining a few electron-pushing concepts to aid in the understanding of organometallic mechanisms.

Answers to Odd Numbered Problems Provided Within the Text

Since this text is often used independently of a classroom setting, having the answers to a group of problems is beneficial for testing the understanding of the material. Since this text is also used in courses, having assignable problems that do not have their answers in the back of the book is also desired for assigned homework. The answers to the odd numbered problems are in the back of the book. Each mechanistic answer comes with a detailed problem space map and discusses and evaluates alternate routes.

More Biochemical Examples

Biochemical examples give added relevance for the biology majors and premedical students who make up a significant portion of the undergraduate organic students. The elegance of biochemical processes in optimizing a low energy route can be appreciated and understood by looking at enzymatic mechanisms. These examples also provide a bridge if this text is to be used for review of organic before a biochemistry or enzymology course.

Online Aids

No matter what you hand out on the first day of class, your exams are your syllabus. Unfortunately, the students' universal test of importance of any material is, "Is this going to be on the exam?" Therefore, if you do not alter the way you test on the material, you have not significantly changed your course. In addition to the answers to all the exercises, supplemental exercises with detailed answers for each chapter are included in the solutions manual for this text to aid in implementing a decision-based approach to organic chemistry.

Applications

This textbook is designed to be flexible in its instructive role. It can be used in the major's undergraduate organic chemistry course as a short, highly mechanistic supplemental text. It can be used as the primary text in an advanced undergraduate or beginning graduate course in organic reaction mechanisms, or as a supplemental review text for graduate courses in physical organic chemistry, enzymatic reaction mechanisms, or biochemistry.

Preface

This text is the product of forty years of teaching organic chemistry at New College, the Honors College of the State of Florida. I am so grateful to have been given that opportunity.

Acknowledgments

I would like to thank all that have helped to bring this book to fruition, especially my father, Prof. Harvey I. Scudder, who helped me refine an algorithm-based teaching approach, and my Ph.D. mentor, Prof. Barry M. Trost. I am indebted to my students, who helped me work through the many versions of this text, to my colleagues at New College, and to the reviewers of this manuscript. I will maintain an errata list and encourage anyone to send me errors not on the list. I gratefully acknowledge the encouragement of my parents, and my wife, son, and daughter, who inspired me to keep writing in the face of an ever-growing project. Finally, I would like to thank all those at John Wiley & Sons who made the publication of this book possible.

This book is dedicated to my students, who have taught me to question everything.

1

BONDING AND ELECTRON DISTRIBUTION

Electron Flow in Organic Chemistry: A Decision-Based Guide to Organic Mechanisms,
Third Edition. Paul H. Scudder.

1.1 THE DECISION-BASED APPROACH TO ORGANIC CHEMISTRY

The Preface mentioned that this decision-based approach to organic chemistry is modeled after the scientific method. A good hypothesis is just a reasonable guess. You will learn how to recognize alternatives and how to judge which alternative is most reasonable. This is the essence of critical thinking, a crucial skill for scientists, physicians, and life in general. You will develop a good intuition, for intuition can be considered just an internalized decision process. We will use the artificial intelligence concepts of problem spaces and tree searches to help you develop this intuition for organic chemistry.

1.1.1 Introduction to Problem Spaces

If you were planning a road trip across the US, you would need a map of the highways. It would allow you to see all routes from your starting city to your goal city. You would then choose the best route for what you wanted to see and the time you had for the trip. This is exactly the process you want to go through for understanding organic chemistry. We need a map and the ability to choose the best route. Our maps of problems are called problem spaces and are often shown as trees, with a decision to be made at each branch point.

Figure 1.1A illustrates a generic problem space and some of the approaches to working from the start at the top of the tree at point S down to the expected answer. If that route is from S to A to D to I, some students may attempt to memorize "S goes to I" without understanding the process involved. In order to provide a greater understanding, instructors spend book and class time explaining the expected route to the answer. However, students may see the "lightning strike" to product as shown in Figure 1.1B but not understand the choices that were made along the way. Instructors may feel that, if shown enough times, students will be able to do it themselves. But we can't expect you to make good choices if you have not been taught how to make those decisions. When you work problems and depart from a reasonable route, you need to see your answer discussed to understand how better choices could have been made.

1.1.2 Introduction to Tree Searches

There are several ways to search a problem space tree. Computers often do a "breadth-first" search by considering every possible route and selecting the most efficient one. We might do this, if we planned our road trip with a map of all major roads. However, while working most science problems we are not provided with a complete problem space map, but rather must generate one on our own. We need to employ a different search strategy. New students often approach science problems "depth-first" by going with the first thing that occurs to them and hoping to get lucky (Figure 1.1C). If they don't reach the expected answer, they back up to the last branch and try another route. However, the more complex the problem space is, the less successful this depth-first search process is. Organic chemistry can have complex highly branched problem spaces, so we again need a different tree search strategy. The tree search strategy that excels in complex problem spaces is a "best-first" search, and that is the focus of this text.

A "best-first" search process is shown in Figure 1.1D, where alternatives are generated and the best alternative is picked at each branch point. Mastering and internalizing this "generate and select" process is crucial to becoming a good scientist or physician. After all, diagnosis is just a physician's search of a complex problem space. An internalized best-first search becomes the essence of a good intuition. If you can master this "generate and select" process and apply it to other fields, it may be the most valuable thing you can take away from a course in organic chemistry.

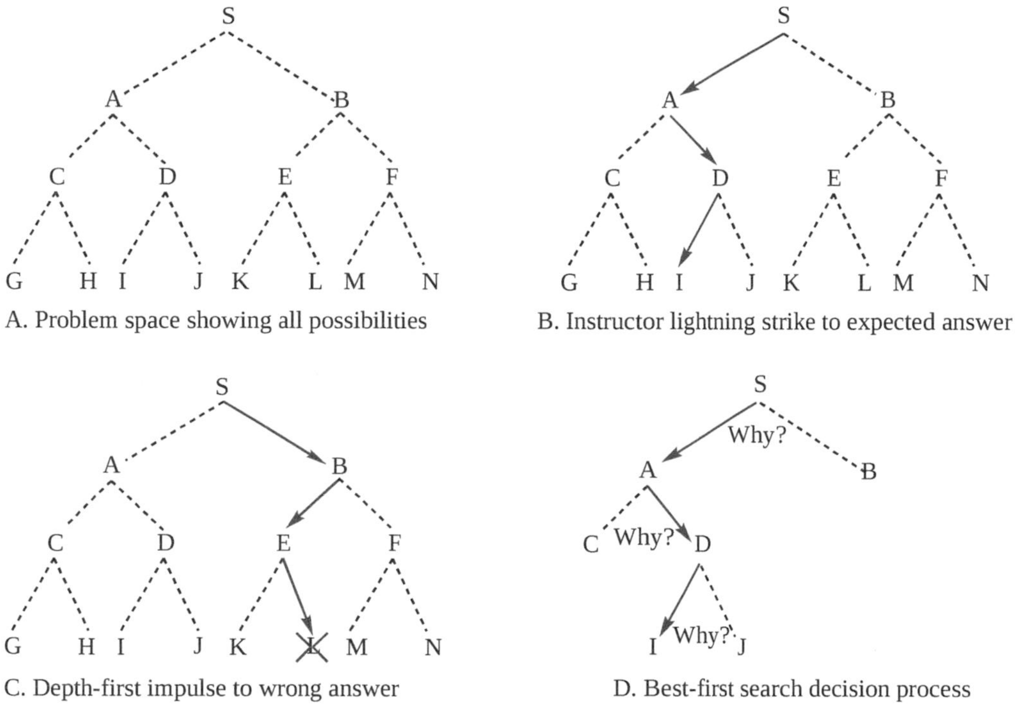

Figure 1.1 A generic problem space and some strategies for tree searches in the problem space.

1.1.3 Introduction to Control Knowledge

First, we need to recognize when we are at a branch point and that a decision needs to be made. Then, to be able to make good decisions in a best-first search, we need to master and utilize "control knowledge." These are checks of reasonableness to help you select the best route. For our road trip, this may be the quality of the highway, the season of year, and how much time we have. For organic problems, control knowledge includes the stability of intermediates, reactivity trends of reactants, whether the medium is acidic or basic, evaluation of energetics, and similarity to known processes. We need to understand the principles of stability that will determine which route is of lower energy. This text will use flowcharts and energy surfaces as problem space maps to illustrate what decisions need to be made. The good news is that the control knowledge that applies to thousands of organic reactions is limited and relatively easy to understand.

An important part of control knowledge is to recognize the boundaries of the problem. When we say an answer is in "the ballpark," we are just testing for reasonable bounds. If a friend bragged he or she could run a mile in a minute, you would have good reason not to believe, considering the fastest runners barely break a four-minute mile. You have tested incoming data with control knowledge before deciding, and that is the essence of critical thinking. Table 1.1 illustrates a few common values and upper bounds that are useful in organic chemistry.

Table 1.1 Common Values and Upper Bounds for Common Chemical Reactions

Measurement	Common for organic	Other examples	Upper bound
Density	Often <1 g/cm^3	1.0 g/cm^3 (water)	22.61 g/cm^3 (osmium)
Concentration	11.2 mol/L (benzene)	55.5 mol/L (water)	118.9 mol/L (osmium)
Temperature	–78 to +220°C	25°C (room temp.)	1400°C (glass melts)

Each of the dozen electron flow path alternatives that we build our explanations of more complex organic reactions has boundaries that need to be considered. These path limitations are included as each electron flow path is introduced in Chapter 4 and are summarized in Chapter 7. Another important aspect of control knowledge is using the reactivity trends of reactants to select the hottest site for reaction in a molecule. This allows us to focus on only the most important part of the molecule and not be distracted by differences in parts of the molecule that do not matter, like the unreactive hydrocarbon skeleton. In this way you won't "slip on the grease" when the hydrocarbon section changes, but the hot spot remains the same. Also, the stability trends of intermediates can be used to predict the lowest-energy route when two or more intermediates are possible. Since energy is often limited, the lowest-energy route is the fastest and often the predominant route. Stability trends of products determine the route in reversible systems, as the most stable product is the one formed. More complex decisions involve multiple factors, which contribute to a "decision boundary" between possible choices, as discussed in Chapter 9.

1.1.4 Preview of the Goals of Beginning Chapters

Chapter 1 – Bonding and Electron Distribution

Since favorable organic reactions usually break weak bonds and make stronger ones, you need to understand the different types of bonding. You must be able to represent the structure of molecules with Lewis structures and resonance forms, to be able to find the electron-rich and electron-poor "hot spots" of the molecule. Once the hot spots are identified, you need to be able to draw the process of bond breakage and bond formation with electron flow "curly arrows." The first chapter gives us the means to discuss the "why" of organic reactions using the common language of Lewis dot structures and electron flow arrows. So first we need to understand that "language" of the explanations, to be able to draw a Lewis structure correctly and understand how curly arrows move electrons.

Chapter 2 – The Process of Bond Formation

To be able to judge the most favorable route, we must understand qualitatively both thermodynamics and kinetics. Therefore we need to understand the process of bond making and breaking, what makes bonds strong or weak, and how the energetics of a process makes some routes more favorable than others.

Chapter 3 – Proton Transfer and the Principles of Stability

To ease into the discussion of organic reactions, it is best to start with a familiar reaction like the acid–base reaction of proton transfer and thoroughly understand how structure determines reactivity. If we can understand proton transfer and the energetics associated with various possibilities, we have come a long way toward understanding organic reactions. Organic reactions use the same principles and often have proton transfer as a first step.

Chapter 4 – Important Reaction Archetypes

Other than proton transfer, only about four mechanistic reaction types form the core of an organic course. It is important to recognize and understand these four important reaction archetypes. What does the problem space for each archetype look like, and how do the reactant structures and reaction conditions influence the most favored route?

Later Chapters

The later chapters develop a general approach to all organic reactions by showing how to focus on the most reactive centers and choose the best route. This book provides tools for handling large amounts of information. It will emphasize the grammar of organic chemistry that will help you make sense of all the material. Without this essential logic to guide you, organic chemistry can become a flood of disconnected reactions to be memorized that will challenge even those good at memorization.

This book groups reactions together that have similar processes. Reactants are put into generic groups that behave similarly. General Chemistry did this by grouping all the bases together and all the acids in a separate group. Looking at the chemical structure, we can predict the chemical behavior of HCl would be similar to HBr and the other acids.

A reaction mechanism is our best guess at the steps for how the reaction occurs. Even the most complex reactions can be broken down into a sequence of basic electron flow steps, called electron flow pathways, which become the building blocks of all the common mechanistic processes. The purpose of this text is to teach the understanding and proper combination of these building blocks. Just a dozen electron flow paths can explain almost all of the common reactions found in an undergraduate organic chemistry course. The mechanistic problem space illustrates the correct use of these pathways. The best advantage that a decision-based approach has over memorization is that it engages the mind. We can ask questions like, “Why did it do this and not that?” New reactions become puzzles to solve, not another thing to be memorized.

Our approach is to understand why and when a reaction will occur and to establish a command of organic chemistry based more upon understanding the basics than upon memorization. The most impressive result of organizing the material by mechanistic process is that you will develop a sound chemical intuition and will be capable of making good educated guesses. In addition, the how and why of organic chemistry is far more interesting than the memorization of its components. We will make use of trends and general rules as tools to aid us in understanding what has happened in an organic reaction. These tools have their exceptions and limitations, but they allow us to develop an overall “feel” for organic reactions. We can worry about the exceptions, if necessary, after the bulk of the material is mastered.

1.1.5 The Principle of Electron Flow

A reaction will occur when there is an energetically accessible path by which electrons can flow from an electron source to an electron sink. A reaction is a flow of electron density from an electron-rich region to an electron-deficient region (the obvious exceptions to this are free-radical reactions, treated separately in Chapter 12). To know which regions are electron rich or electron deficient, we must be able to predict the distribution of electron density over a molecule.

A *nucleophile* (“nucleus-loving,” $Nu:^-$) is a *Lewis base* (electron pair donor) that has an available electron pair for bonding. Nucleophiles act as electron sources and can be negatively charged like hydroxide anion or neutral like water. An *electrophile* (“electron-loving,” E^+) is a *Lewis acid* (electron pair acceptor) that can accept two electrons to form a bond. Electrophiles act as electron sinks and can be positively charged like hydronium cation or neutral like bromine.

Any bond formed is the combination of a nucleophile and an electrophile. The most probable product of a reaction results from the best electron source bonding with the best electron acceptor. The curved-arrow notation (Section 1.4) allows us to describe the flow of electrons from source to sink.

The concept of flow is very important. Just as water flows under the influence of gravity, electrons flow under the influence of charge: from electron-rich atoms to electron-deficient atoms. A poor electron source will not react with a poor electron sink within a useful length of time. One might consider this a case of no push and no pull, so no appreciable electron movement occurs. A poor electron sink requires combination with a good electron source for a reaction to occur.

We need to answer the questions: What properties distinguish a good electron source, a good electron acceptor, and a good pathway for electron flow? There are relatively few pathways through which the common electron sources and sinks react. The use of generic electron sources and sinks and generic electron flow pathways makes the similarities and interrelationships of the major reactions in organic chemistry become obvious. The electron flow pathways become the building blocks of even complex organic reaction mechanisms, so all the mechanisms seem to "flow" from first principles.

Before we explore the problem space for a simple proton transfer reaction, we need to understand the basics of bonding and define a consistent nomenclature. In order to use the electron flow paths, you first need to be able to keep track of atoms and electrons—write Lewis structures correctly and easily.

1.2 IONIC AND COVALENT BONDING

Before trying to understand stable molecules, let's try to understand stable atomic electron configurations. The outer electrons (valence electrons) are involved in bonding. Figure 1.2 shows the valence electrons of the atoms of elements common in organic chemistry. An atom might achieve the stable filled shell electron configuration of a noble gas (group VIIIA) in any of three ways: lose electrons, gain electrons, or share electrons.

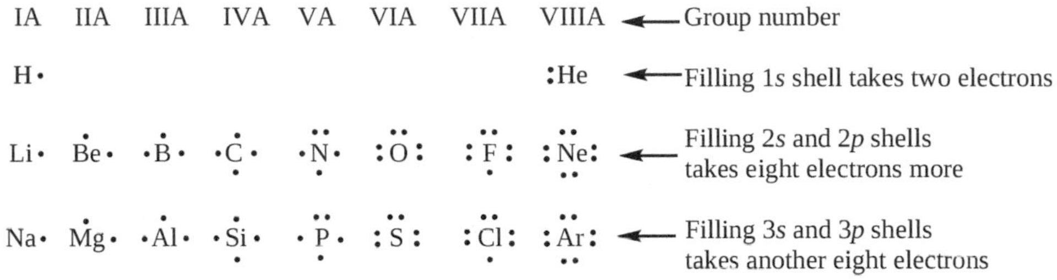

Figure 1.2 The valence electrons of elements in the first three rows of the periodic table.

In the first row of the periodic table, the two valence electrons of helium fill the 1*s* shell, making it very stable. If lithium loses one of its three total electrons, it now has the helium duet and is stable as the Li^+ cation. The hydrogen atom needs two electrons to form the helium duet, and it can achieve that either by gaining an electron outright to form hydride anion or by sharing an electron with another atom. The first row is special in needing only an electron duet for stability; higher rows require an octet of valence electrons for stability. Because hydrogen can add one electron to form hydride, it is sometimes drawn in two places on the periodic table, in Groups IA and in VIIA.

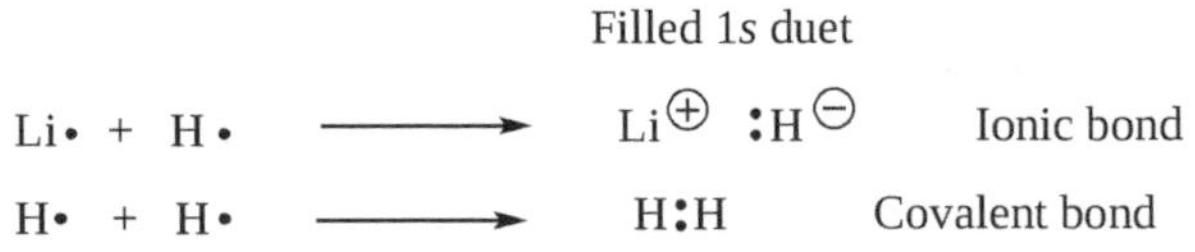

In the second row, neon has a complete valence shell with both the 2*s* and 2*p* orbitals filled (in addition to the filled 1*s* helium core). Sodium has one more electron than neon; losing that electron gives Na^+, which has neon's stable filled shell. If a fluorine atom gains one electron to become fluoride, it has the complete octet of neon and a filled shell.

To generalize, elements of Groups IA and IIA commonly lose their valence electrons to step back to the last noble gas electron configuration. For this reason, it is normal to write all compounds of elements in Groups IA and IIA as ionic. Halogens of Group VIIA can gain an electron to step forward to the next noble gas electron configuration of the halide ion. Groups IIIA through VIIA can gain stable valence octets by sharing electrons in covalent bonds.

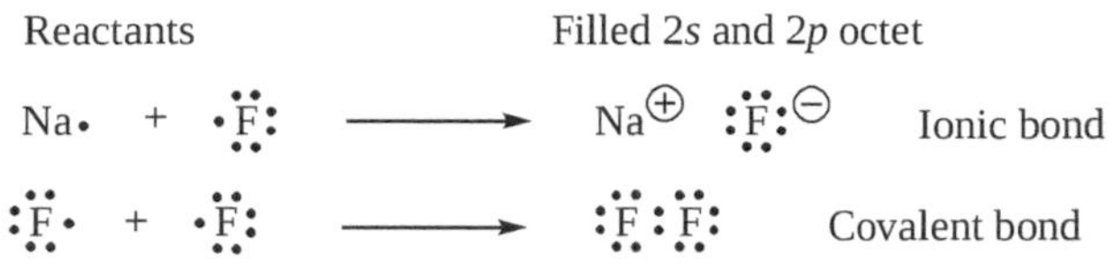

Ionic and covalent bonds are the two extremes of bonding. In a covalent bond, a pair of electrons is shared between two atoms. The shared pair of electrons is in a bonding orbital made from the overlap of an atomic orbital from each atom. For a strong covalent bond, these shared atomic orbitals must overlap well and must be similar in energy. In an ionic bond, that pair of electrons resides primarily on one of the two atoms, producing two oppositely charged ions that are attracted to each other. This attraction is expressed mathematically by Coulomb's law, which says the attractive force is directly proportional to the charge of each ion and inversely proportional to the square of the distance between them. A strong ionic bond has opposite charges very close to each other. Reactions tend to break weak bonds and make strong ones, so a reaction can be favored by the formation of strong ionic bonds or strong covalent bonds or both (the HSAB principle, Section 2.4).

Most bonds fall between these two extremes. Pauling defined *electronegativity* as the power of an atom in a molecule to attract electrons to itself. The electronegativity difference of two bonded atoms is an indication of how polarized, how ionic, the bond is. In homonuclear diatomic molecules like H_2 there is pure covalency, no polarization, for the electronegativity difference is zero. Common table salt, Na^+Cl^-, contains a good example of an ionic bond in which the electronegativity difference is large. Table 1.2 is a partial periodic table of the average electronegativities of commonly encountered elements. Elements more electronegative than carbon make up the upper right corner in the periodic table and are in boldface. Electronegativity is the result of increasing nuclear charge on the valence electrons. For example, the electrons in the 2*p* orbital get pulled closer to the nucleus in going from boron to fluorine as the nuclear charge increases. Likewise, Na^+ and F^- both have 10 electrons, but Na^+ is much smaller and its electrons are held tighter than fluoride due to sodium's greater nuclear charge, 11 protons in sodium vs. 9 protons in fluorine.

Table 1.2 Average Electronegativities of Selected Elements

H 2.20									
Li 0.98					B 2.04	C 2.55	**N 3.04**	**O 3.44**	**F 3.98**
Na 0.93	Mg 1.31				Al 1.61	Si 1.90	P 2.19	**S 2.58**	**Cl 3.16**
K 0.82			Cu (I) 1.90	Zn 1.65				Se 2.55	**Br 2.96**
				Cd 1.69					**I 2.66**

The electronegativity of an atom will vary depending on what else is bonded to it. For example, the carbon atom of a CF_3 group is more electronegative than the carbon atom of a CH_3 group. In the CF_3 group, the three highly electronegative fluorines withdraw electron density from the CF_3 carbon atom, which in turn will withdraw more electron density from whatever it's bonded to.

Both the degree and the direction of the polarization of a bond can be predicted by the electronegativity difference. The bonding electron pair is more likely to be found around the more electronegative atom. Carbon can be either partially plus, δ+, or partially minus, δ–, depending on the electronegativity of the group bonded to it. The polarization of the bonding electrons will become very important in understanding how two species interact, since like charges repel and unlike charges attract. A partially plus carbon will attract a negative ion; a partly negative carbon will attract a positive one.

1.3 LEWIS STRUCTURES AND RESONANCE FORMS

Electron flow paths are written in the language of Lewis dot structures and curved arrows. Lewis dot structures are used to keep track of all valence electrons, and curved arrows are used to symbolize electron movement. You must be able to draw a proper Lewis structure complete with formal charges accurately and quickly. Your command of curved arrows must also be automatic. These two points cannot be overemphasized, since all explanations of reactions will be expressed in the language of Lewis structures and curved arrows. A Lewis structure contains the proper number of valence electrons, the correct distribution of those electrons over the atoms, and the correct formal charge. We will show all valence electrons; lone pairs are shown as darkened dots and bonds by lines.

An atom in a molecule is most stable if it can achieve the electronic configuration of the nearest noble gas, thus having a completely filled valence shell. Hydrogen with two electrons around it, a duet, achieves the configuration of helium. Second-row elements achieve the configuration of neon with an octet of valence electrons. Third-row elements like sulfur achieve an octet but may also expand their valence shell; for example SF_6 is a stable molecule with six single bonds to sulfur (12 bonding electrons total).

1.3.1 Procedure for Drawing Lewis Structures

Use the periodic table to find the valence electrons contributed from each atom. Add an additional electron for a negative charge, or subtract one to account for a positive charge to get the **total number of valence electrons**. Then **draw single bonds between all connected atoms to establish a skeleton**, or preliminary structure. You need to know the pattern in which the atoms are connected. If you have to guess at the connectivity, the most symmetrical structure is often correct. Sulfuric acid, for example, has the one sulfur in the center surrounded by the four oxygens, two of which have attached hydrogens. Since hydrogen forms one covalent bond, it is always on the outside of the structure. **Place any additional bonds between adjacent atoms that both have incomplete octets to satisfy the following general bonding trends**. There may be more than one way to do this (see the next section). Do not exceed the octet for second-row elements. Higher-row elements like phosphorus or sulfur can exceed an octet.

General bonding trend	Example
One bond: H, F, Cl, Br, I	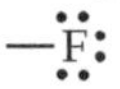
Two bonds: O, S	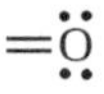
Three bonds: N, P	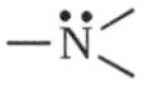
Four bonds: C, Si	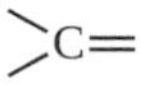

Use these trends to connect the atoms. Multiply the number of bonds by two to get the number of electrons shared in bonds. **Subtract the number of shared electrons from the number of valence electrons to get the number of unshared electrons** (if any). Place the unshared electrons as lone pairs on atoms that need them to make octets.

Assign formal charges. Formal charge is a comparison of the number of electrons an atom "owns" in the Lewis structure with the number it would have if it were free. The atom owns half of the electrons that it shares in a bond, but all of its unshared electrons.

Formal charge = free atom valence − (# of bonds + unshared # of electrons)

The sum of the formal charges must be equal to the total charge of the species. As a cross-check, usually if an atom has more than the number of bonds listed in the general bonding trends, it will have a positive formal charge; if it has less, it will bear a negative formal charge. Formal charge can help identify electron-rich and electron-deficient areas. However, a molecule may have more than one Lewis structure to consider (Section 1.3.2).

Example: Methoxide anion, CH_3O^-

The total number of valence electrons is 14; we get 6 from O, 4 from C, 1 from each of 3 H's, and 1 for the minus charge. It took four bonds to connect the atoms, so the number of shared electrons is 8. We have used eight electrons, and there are six electrons remaining to be added as lone pairs to complete oxygen's octet.

Valence 14
– Shared –8
Left over 6

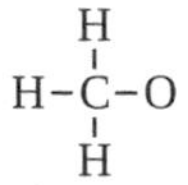

All that is left to do is assign formal charge. Oxygen started with six valence electrons, and in this structure has one bond to it and six unshared electrons, so –1 must be its formal charge. A check shows that the shells are correct for all the atoms, all the valence electrons have been used, and that the sum of the formal charges equals the total charge. The final structure is:

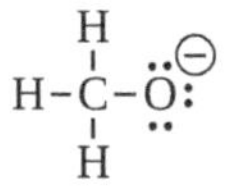

We can't have a bottle of just anions. All compounds are charge balanced, so the methoxide anion above would have a cation to balance the negative charge. This is often a group 1A cation like Li^+, Na^+, or K^+. These cations have the stable configuration of a noble gas, and therefore are just spectator ions. To simplify things, organic chemists often don't bother writing the spectator ions because they don't participate in the reaction.

Example: Acetaldehyde, CH_3CHO

The total number of valence electrons is 18; we get 6 from the O, 4 from each of 2 C's, 1 from each of 4 H's. The six-bond skeleton shares 12 electrons.

Valence 18
– Shared –12
Left over 6

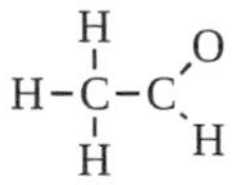

Another bond must be made because two adjacent atoms, C and O, have less than expected from the general bonding trends. This bond should go between those two atoms giving the structure:

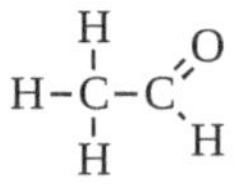

Seven bonds used 14 electrons; the 4 unshared valence electrons are added as lone pairs to complete the octet of oxygen. Formal charges are all zero; the final structure is:

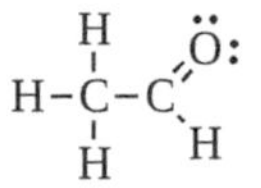

Cross-check: Valid Lewis structures:

- **have correct number of valence electrons,**
- **have correct formal charge (sums to total charge), and**
- **do not exceed octets for second-row elements.**

Common Errors

Throughout the text the incorrect example will appear on the left with the corrected version to the right of it, followed by an explanation on why the left example is incorrect.

Lewis Structure Errors

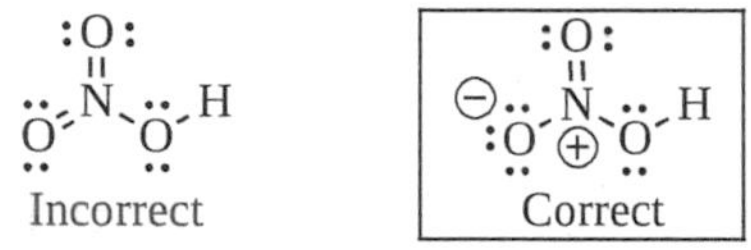

It is impossible to flow electrons correctly if you cannot keep track of them accurately. The structure on the left has exceeded the octet for nitrogen. Five bonds to N means 10 electrons around it. Since nitrogen is a second-row element, it cannot expand its valence shell.

1.3.2 Resonance Hybrids

Often, one Lewis structure is not sufficient to describe the electron distribution in a molecule. In many cases, the use of a *resonance hybrid* is necessary. All resonance forms are valid Lewis structures. In resonance forms, **only the electrons move and not the atomic nuclei.** Each resonance form does not have a separate existence but is a part of a hybrid whole. The use of a double-headed arrow, ↔, between the forms reinforces the notion of a hybrid representation of a single structure. The forms are not in equilibrium with each other (equilibrium is shown by two opposing arrows).

Only one hybrid exists, not individual equilibrating resonance forms. We are trying to describe a molecule's rather diffuse electron cloud with lines and dots; it is not surprising that often one Lewis structure is insufficient. The acetate anion, CH_3COO^-, requires two resonance forms to describe it. Both structures are of equal importance; the carbon–oxygen bond lengths are equal. Generally, an ion is more stable if it has more resonance forms of similar energy (implied by similar structure).

Frequently, the hybrid cannot be represented by equally weighted resonance forms. The hybrid may be more like one form than the other, and thus the resonance forms are denoted as major and minor. For formaldehyde, $H_2C{=}O$, the resonance forms represent the uneven electron distribution of a polarized multiple bond; the charge-separated minor resonance form places a negative charge on the more electronegative atom. Oxygen is winning the electronegativity tug-of-war for the bonding electrons, and mixing in this minor resonance form acknowledges that. The form on the far right does not contribute because it is contrary to the polarization of the bond. A good general rule is to avoid having an incomplete octet on a highly electronegative atom, like nitrogen or oxygen.

Resonance forms for acetamide, CH_3CONH_2, reveal the polarization of the amide group. The partially negative oxygen, not the partially positive nitrogen, will be the reactive site when an amide is used as an electron source.

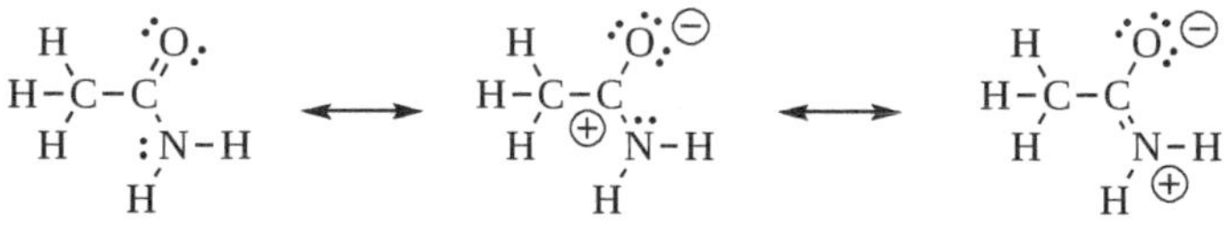

Major Minor Minor

The major resonance form will have the most covalent bonds and complete shells and the least amount of charge separation. It will seem to be the most stable of the possible resonance forms. Place any negative charges on electronegative atoms. Resonance structures with an incomplete shell are usually minor. Resonance forms having an electronegative atom with an incomplete octet are insignificant.

Occasionally "no-bond" resonance forms are needed to describe the electron distribution in a molecule and are usually associated with very acidic hydrogens. The polarized bond in hydrochloric acid can be represented by using resonance structures to show the partial ionic nature of the bond. The atoms do not move, only the electrons.

Major Minor

Cross-check: The major resonance contributor will have:
- **the most covalent bonds,**
- **the least amount of charge separation,**
- **the most complete octets and duets possible, and**
- **any negative charges on the most electronegative atoms.**

1.4 CURVED-ARROW NOTATION

A full-headed curved arrow indicates the movement of two electrons from the tail of the arrow to the head. A half-headed arrow (fishhook) indicates the shift of one electron likewise. The two ways that a bond can break are *heterolytic* (two electrons) or *homolytic* (one electron). Homolytic processes are unusual and will be treated separately in Chapter 12. For ease of identification, all electron flow arrows in this book will be red.

Heterolytic cleavage (full headed arrow)

Homolytic cleavage (fishhook arrows)

Arrows indicate a movement or flow of electrons that **must come from a site of electron density**, either a lone pair or a bond, and move to a site that can accept additional electron density.

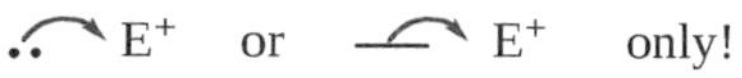

If an arrow comes from a bond, that bond is broken. If an arrow comes from a lone pair, the lone pair is removed and a new bond is formed at the head of the arrow. If the head of the arrow points between two atoms, it forms a new bond between them. If it points to an atom, it forms a new lone pair on that atom.

A source of confusion for beginning students is that for intermolecular bond-forming reactions, some authors will point the arrow between the two atoms, whereas others will point it directly at the second atom. We will try a compromise—when an arrow goes between two molecules, the head of the arrow is drawn close to the appropriate atom on the second molecule. The following are slightly different ways to show the formation of a bond between Nu:⁻ and E^+ to give the Nu–E bond; we will use the arrow notation on the right. A useful generalization is that an arrow that comes from a lone pair will always form a bond, not another lone pair.

The bond or lone pair from which the first arrow in an electron flow originates is called the *electron source*. The head of the last arrow in an electron flow points to the *electron sink*. **Arrows will always point away from negative charges and toward positive charges.** Sometimes it is useful to use arrows to interconvert resonance structures, but those arrows do not really indicate electron flow.

Some specific examples will help illustrate the correct use of arrows in reactions:

The first arrow on the left comes from the lone pair on the electron-rich hydroxide anion and makes a bond between the hydroxide oxygen and the hydrogen. The second arrow breaks the O–H bond and makes a new lone pair on oxygen. Note that with correct electron bookkeeping **the charge on one side of the transformation arrow will be the same as on the other side** (in this case one minus charge). Charge is conserved. If the electron movement signified by the curved arrows is correct, the products will also be valid Lewis structures.

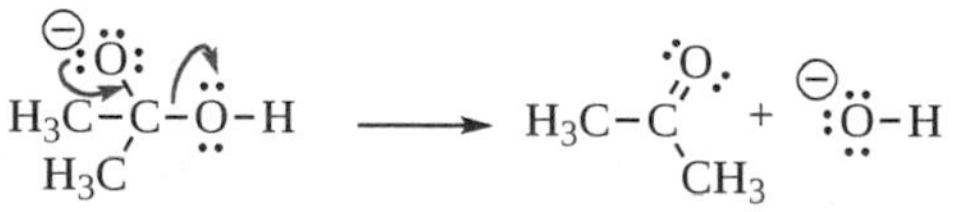

In this second example, the flow comes from the lone pair on the negative oxygen and forms a double bond. The flow continues by breaking the carbon–oxygen bond to form a new lone pair on the oxygen of the hydroxide ion. The electron source in this example changes from negative to neutral because the flow removes electrons from it; the sink becomes negative in accepting the electron flow.

In this last example, the flow starts with the electron-rich sulfur anion and forms a carbon–sulfur bond with the CH_2 group. The pi bond breaks and forms a new pi bond. The flow finishes by breaking the carbon–oxygen pi bond and forming a new lone pair on the electronegative oxygen. Every time an arrow forms a bond to an atom that already has a complete octet, another bond to that atom must break, so the octet is not exceeded.

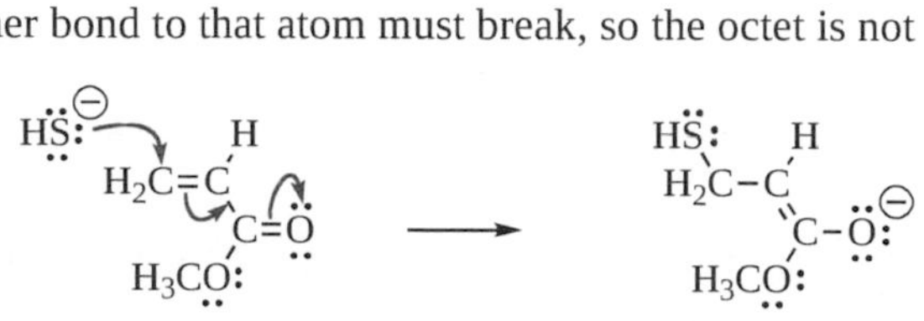

Exercise: Cover the right side of the previous reactions and draw the product.

A good way to see whether you have mastered arrows and the concept of electron flow is to provide the arrows given the reactants and products. In these one-step mechanism problems, you must decide which bonds were made and broken and in which direction the electron flow went. Here is an example:

Note the bonding changes: Oxygen has one less lone pair and has formed a new bond to the hydrogen from HBr; the H–Br bond is broken, and bromine now has another lone pair. Now look at the charges: Oxygen is now positive, and bromine is now negative. Electron flow must have come from oxygen (the source) and ended up on bromine (the sink) to account for the change in charge. Only one set of arrows could be correct: The first arrow must come from the oxygen lone pair and form a new O–H bond; the second arrow must break the H–Br bond and form a new lone pair on bromine.

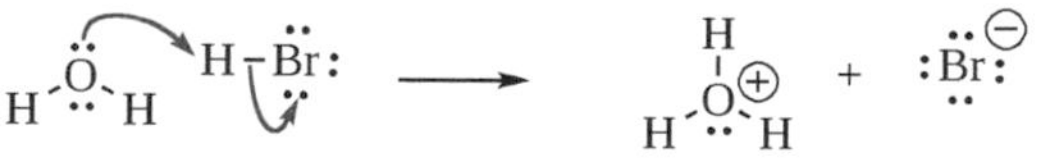

A more complex example might be useful.

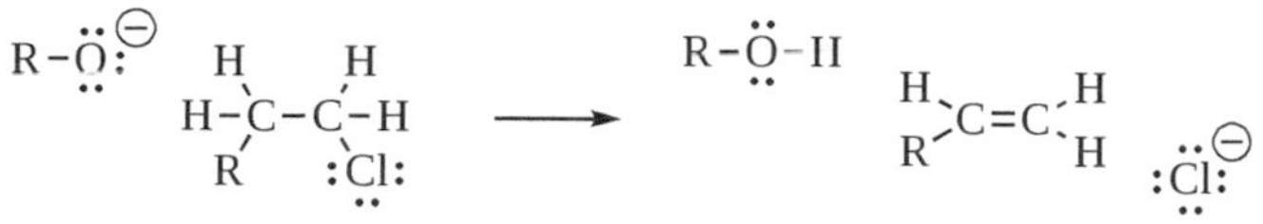

Oxygen again has lost a lone pair and formed a bond to hydrogen. A carbon–carbon double bond has formed, the carbon–chlorine bond is broken, and a new lone pair is on chlorine. The minus charge on oxygen in the reactants is now on chlorine in the products; the flow must have come from oxygen (the source) to chlorine (the sink). Again only one set of arrows could be correct: The first arrow must come from the lone pair on the negative oxygen and form an O–H bond; the second arrow must break the C–H bond and form a double bond; the third arrow must break the C–Cl bond and form a lone pair on chlorine.

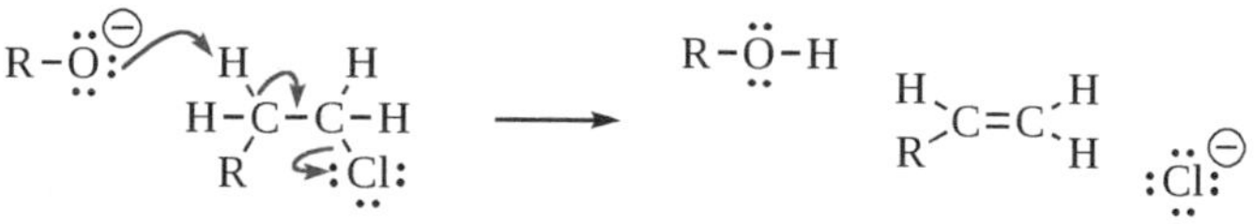

The flow of negative charge tells you the direction of electron flow in this last example.

The lone pair on nitrogen is gone and a new carbon–nitrogen bond has formed; the carbon–iodine bond is broken and iodine now has an extra lone pair. The charges indicate that electron density has been drained away from nitrogen (which has become positive in the product) and deposited on iodine to give the negative iodide anion. Only one set of arrows could be correct: The first arrow starts from the nitrogen lone pair and forms the N–C bond; the second arrow breaks the C–I bond and forms a new lone pair on iodine.

Exercise: Cover the answers in these previous examples and draw the arrows.

Precautions: It is very important that you pay strict attention to the Lewis structures and arrow positions. **Arrows always point in the same direction as the electron flow, never against it.** Arrows help us keep track of how the charges move. Because arrows move electrons from tail to head, the formal charge on the atom at the tail of the arrow increases by one (becomes more positive) and the formal charge on the atom at the head of the arrow decreases by one (becomes more negative). Never use a curved arrow to indicate the motion of atoms; curved arrows are reserved for electron flow only. Be forewarned that some texts may combine several steps on one structure to avoid redrawing a structure; others may show a partial set of arrows and expect you to fill in the rest mentally. In your study and practice always draw out each electron flow step completely, for errors that would otherwise be easy to find may become difficult to locate if several steps are jumbled together.

1.4.1 Good Arrow Pushing Habits

- Draw Lewis structures near any arrows (to keep track of atoms and electrons).
- Start every arrow from a **drawn** pair of electrons (not from a minus formal charge).
- Check Lewis structure octets and duets (the most important cross-check).
- Check formal charge and charge balance (or your next step goes astray).
- Use the **known** electron flow paths (introduced over the next few chapters).

1.4.2 Common Errors

Ligand-Rich vs. Electron-Deficient Errors

Groups like NH_4^+ and H_3O^+ are cationic but not electron deficient because both nitrogen and oxygen have complete octets. For example, the nitrogen in NH_4^+ has four groups bound to it, completing its octet and giving it a formal charge of +1.

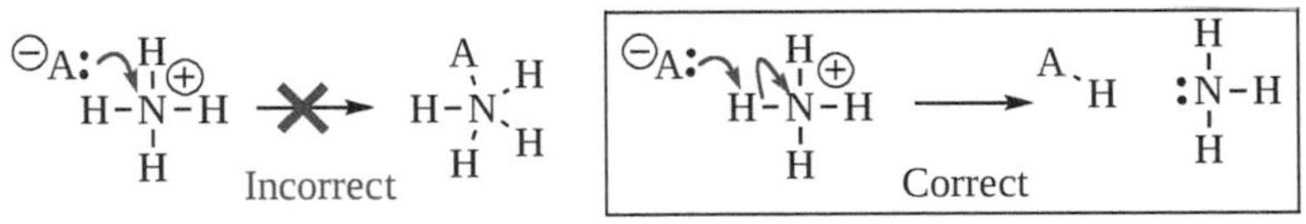

Any attack by an anion on the nitrogen would form a fifth bond and exceed its octet. Second-row elements, like C, N, and O, cannot exceed their octet. The resonance forms of the ammonium ion indicate that nitrogen is **winning** the electronegativity tug-of-war. The hydrogen atoms in the ammonium ion are partially positive and therefore acidic.

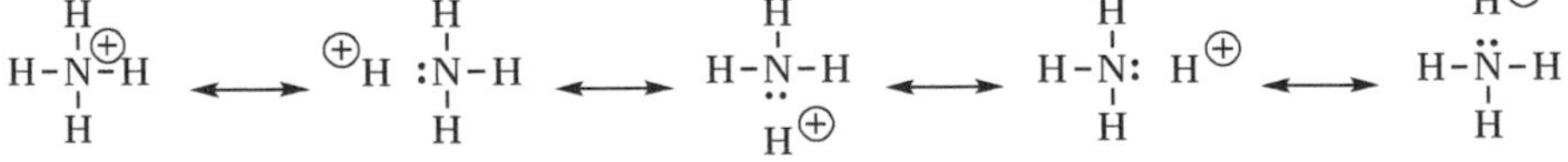

Similar errors are also seen with the even more acidic hydronium ion, H_3O^+, where again oxygen cannot exceed its octet and is much more electronegative than hydrogen.

Off-the-Path Errors

Until you have the principles of mechanistic organic chemistry thoroughly mastered, it is best to restrict your mechanistic proposals to simple combinations of the electron flow pathways, shown in Chapter 7. You may see a shortcut that with several arrows would allow you to transform the lines and dots of the Lewis structure of the reactant into the lines and dots of the product, but that is not the point of it. What you are trying to do with arrows is guess what is really going on in the reaction, and for that you should use processes that are actually known to exist. The pathways are a very powerful mechanistic vocabulary, and there are very few mechanistic processes that cannot be expressed as a simple combination of them.

Charge Errors

Typical organic solvents have difficulty stabilizing adjacent like charges. Therefore, avoid forming adjacent like charges unless the reaction solvent is water. **Very few organic intermediates have multiple charges.** Charge must be conserved; the total charge on both sides of the reaction arrow must be the same. Charge is part of your electron count and must balance through each step. Since arrows move electrons, the charges on your final structure must be consistent with the electron movement indicated by your arrows.

Becomes more positive or less negative. Formal charge increases by 1	→	Becomes more negative or less positive. Formal charge decreases by 1

Electron Flow Backwards Error

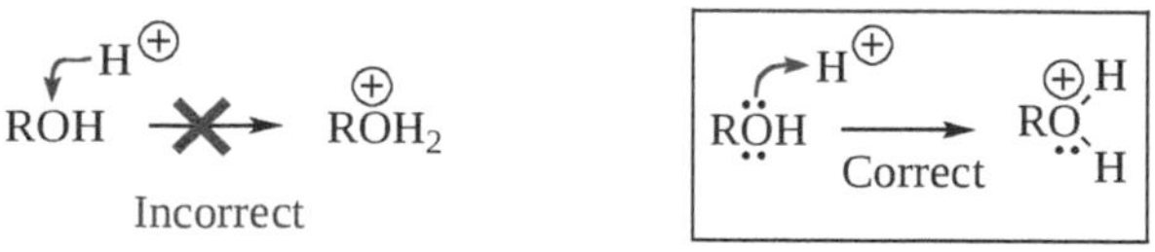

A proton has no electrons at all, so an arrow cannot start from it. This arrow error is seen in some biochemistry texts to mean, "put the proton there." Always use arrows only for electron movement, nothing else. **Arrows will always head toward positive centers and away from negative ones.**

Electron Flow Continuity Errors

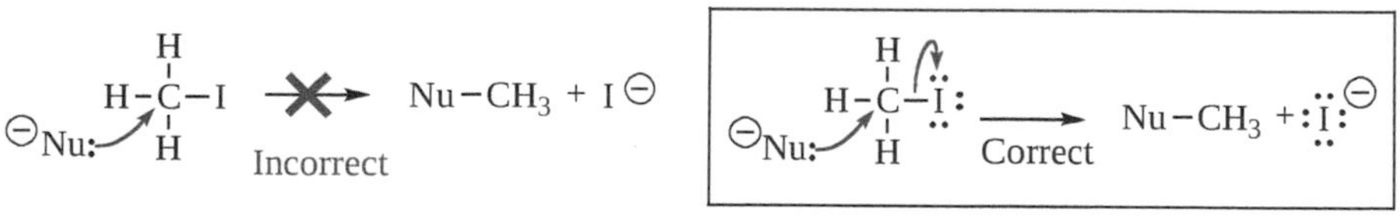

To be fair, some authors just draw the first arrow of an electron flow and expect the reader to supply the rest; it is best to see the entire flow drawn out.

Curved Arrows That Bounce or Miss Their Target

Beginners are often not careful in their drawing of arrows and may start an arrow from the wrong bond or put the head in between or on the wrong atoms. Since it is easy to spot, we won't give an example here. Always draw out all the electron pairs on any group that you are going to draw arrows near. Since an electron flow arrow moves electrons, start all arrows from a bond or electron pair, not from the formal charge. A formal charge does not represent or stand in for a drawn electron pair. For example, the anion BH_4^- has two electrons in each of the four boron–hydrogen single bonds. The formal charge just indicates that boron has one more electron than it started with.

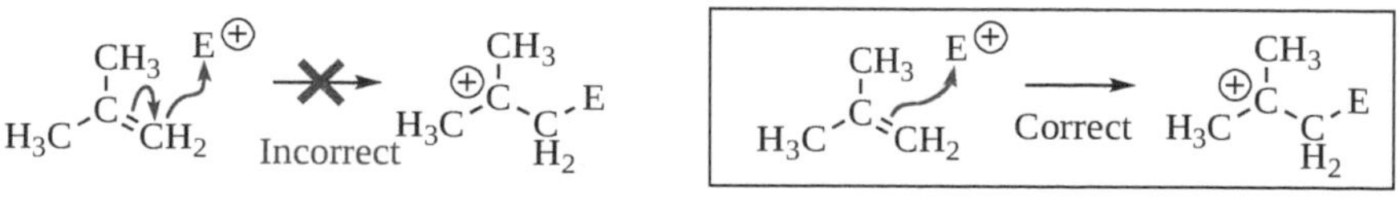

This "electron bounce" uses two arrows when only one is required. The second arrow is not starting from a drawn electron pair. **All** arrows must start from a **drawn** electron pair. It is much harder to detect errors when arrows are not used carefully.

1.5 NOMENCLATURE AND ABBREVIATIONS

1.5.1 Common Abbreviations

A complete table of abbreviations used in this text is in the Appendix, but a few are so common they need to be learned early.

b	Brønsted–Lowry base, accepts a proton from a Brønsted–Lowry acid
δ+	A partial positive charge
δ–	A partial negative charge
E	Electrophile or Lewis acid, an electron pair acceptor
ERG	Electron-releasing group
EWG	Electron-withdrawing group
Et	Ethyl, CH_3CH_2
HA	Brønsted–Lowry acid, donates a proton to a Brønsted–Lowry base
L	Leaving group, departs with its bonding electron pair
M	Metal
MO	Molecular orbital
Nu	Nucleophile or Lewis base, an electron pair donor
Ph	Phenyl group, C_6H_5, a monosubstituted benzene
R	Any alkyl chain
X	Chlorine, bromine, or iodine
Y, Z	Heteroatoms, commonly oxygen, nitrogen, or sulfur
......	Partially broken bond (or weak complexation)

1.5.2 Line Structure

A line structure or skeletal formula is a very easy way to represent organic structures. Each corner in the line corresponds to a carbon atom, and the hydrogen atoms are not drawn. Line structure is fast and convenient to use but can lead to difficult to locate errors. **Any portion of a molecule that is participating in the reaction must be drawn out, showing all the carbons and hydrogens.** It is much too easy to forget about the hydrogen atoms omitted in line structure. The parts of the molecule not participating in the reaction may be shown in line structure or abbreviated. Draw out functional groups (see Table 1.3) and show all H's on heteroatoms.

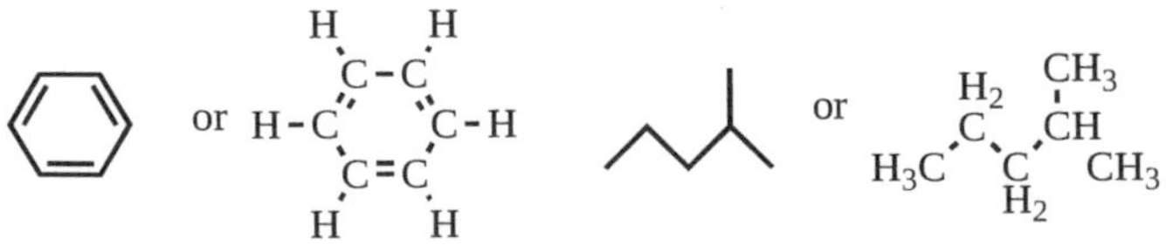

Organic chemistry is like a foreign language: It is cumulative and requires that vocabulary be learned in addition to grammar. Vocabulary is best learned as you need it, but there is so much to learn that a head start is helpful. You must be able to count in organic, know the common functional groups, and name simple compounds. Compounds are usually numbered from the end with the most oxidized functional group down the longest chain, giving other groups off the chain the lowest numbers if there is a choice. For detail, see Notes on Nomenclature in the Appendix. **Use Table 1.3 to learn the Lewis structures of the common functional groups, so you won't have to draw them from scratch every time they are needed.** A much larger functional group glossary is in the Appendix. In tables, lines off of a carbon are used to denote a bond to R or H.

Table 1.3 Lewis Structures of Common Functional Groups

Name	Functional group	Example	Typed version
Acyl halide	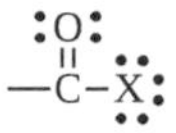	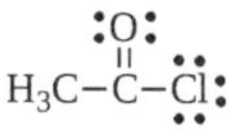	CH_3COCl
Alcohol	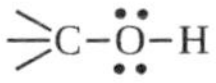	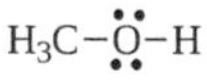	CH_3OH
Aldehyde	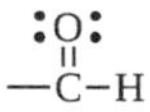	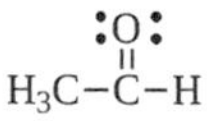	CH_3CHO
Alkane	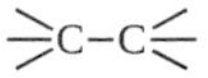	H_3C-CH_3	CH_3CH_3
Alkene	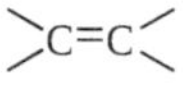	$H_2C=CH_2$	CH_2CH_2
Alkyl halide	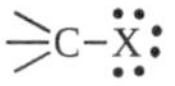	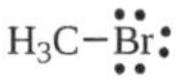	CH_3Br
Alkyne	$-C\equiv C-$	$HC\equiv CH$	HCCH
Amide	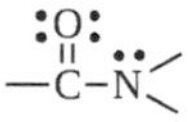	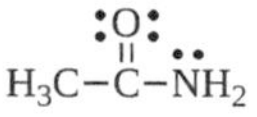	CH_3CONH_2
Amine	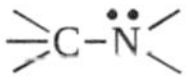	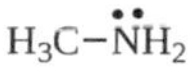	CH_3NH_2
Carboxylic acid	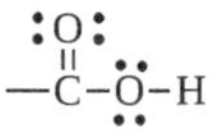	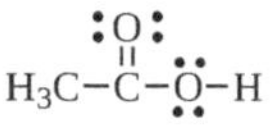	CH_3COOH
Ester	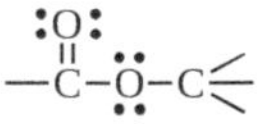	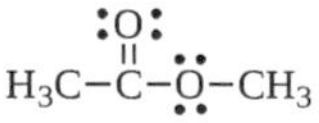	$CH_3CO_2CH_3$
Ether	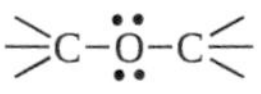	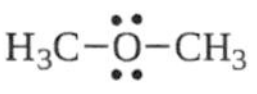	CH_3OCH_3
Ketone	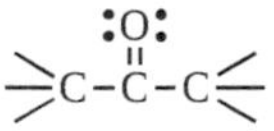	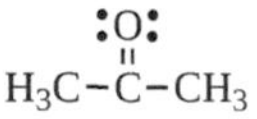	CH_3COCH_3
Nitrile	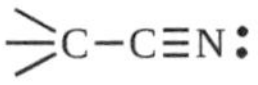	$H_3C-C\equiv N:$	CH_3CN

Exercise: Cover all but the far left side of the page and draw Lewis structures for all functional groups. Cover the left side and name all functional groups from the Lewis structures. Flash cards are useful in learning vocabulary like this.

1.5.3 List of the First 10 Alkanes

1	Methane	CH_4
2	Ethane	CH_3CH_3
3	Propane	$CH_3CH_2CH_3$
4	Butane	$CH_3CH_2CH_2CH_3$
5	Pentane	$CH_3CH_2CH_2CH_2CH_3$
6	Hexane	$CH_3CH_2CH_2CH_2CH_2CH_3$
7	Heptane	$CH_3CH_2CH_2CH_2CH_2CH_2CH_3$
8	Octane	$CH_3CH_2CH_2CH_2CH_2CH_2CH_2CH_3$
9	Nonane	$CH_3CH_2CH_2CH_2CH_2CH_2CH_2CH_2CH_3$
10	Decane	$CH_3CH_2CH_2CH_2CH_2CH_2CH_2CH_2CH_2CH_3$

1.6 THE SHAPES OF MOLECULES

Valence Shell Electron Pair Repulsion (VSEPR) Theory

Once you can draw a good Lewis structure of a molecule, you can use valence shell electron pair repulsion theory (VSEPR) to get its shape. VSEPR theory arises from the fact that repulsion between pairs of bonding and lone pair electrons determine the shape of the molecule. For the purpose of VSEPR, a double or triple bond is considered as a group, and a lone pair acts as if it is larger than a bonding pair. Figure 1.3 shows the shapes of molecules as a function of how many groups crowd around the center atom. The solid wedge in Figure 1.3 symbolizes that the bond is coming out of the paper, and the dashed wedge symbolizes that the bond is going back into the paper. You can demonstrate this VSEPR effect by tying together balloons of the same size, each corresponding to a group of electrons. Only third-row and higher atoms (not carbon) can bond five or six groups.

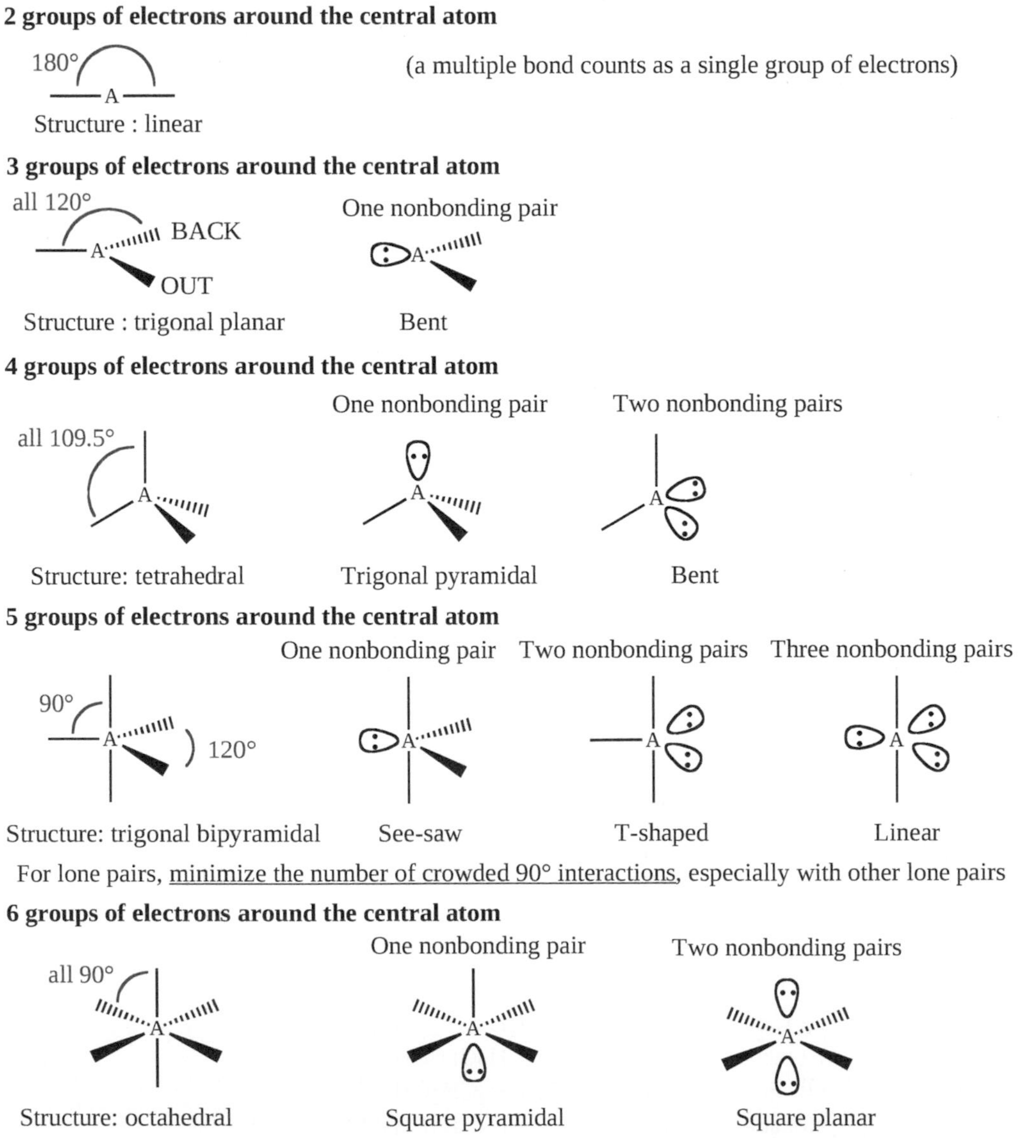

Figure 1.3 VSEPR-predicted shapes of molecules.

1.7 AN ORBITAL VIEW OF BONDING

1.7.1 Review of Orbitals Used in Bonding

You may remember from General Chemistry that electrons do not have orbits around an atom but have a cloud-like probability distribution around an atom called an orbital. Each atomic orbital holds two electrons and is described by a mathematical expression called a wave function. We can get the energy of the orbital and the electron's probability distribution from its wave function. The discussion of wave functions is above the level of this text, but if you are interested, a qualitative review and discussion of orbital theory and waves is in the beginning of Chapter 13. For now, we just need to understand qualitatively what the math of quantum mechanics tells us about wave functions.

Figure 1.4 shows several ways of representing a $2p$ orbital. Since a three-dimensional probability distribution cloud of varying electron density is difficult to draw like Figure 1.4a, we often draw orbital surfaces that enclose almost all of the electron cloud as in Figure 1.4b. We arbitrarily shade one lobe of the $2p$ orbital to show that its math sign is different from the other lobe. This math sign is a way of acknowledging the wave nature of the orbital, as in the peak and trough of a wave. There is a planar node in the center, because the math expression of the orbital goes through zero there. The most correct representation is shown in Figure 1.4b, but the commonly used simplified version, Figure 1.4c, is drawn much slimmer so it does not obscure other things in the drawing. When we draw two skinny p orbitals next to each other, it may look like they don't overlap, but they actually do.

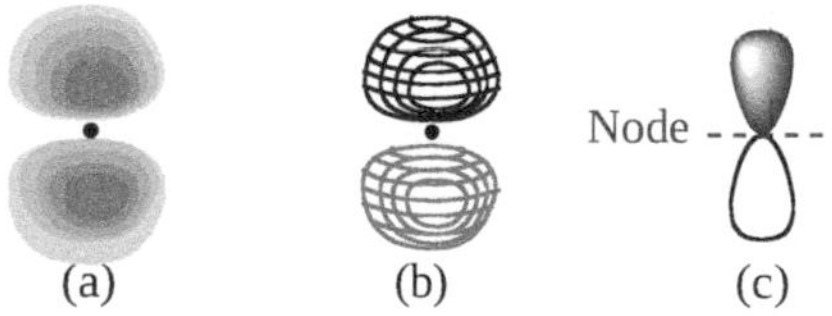

Figure 1.4 Representations of a $2p$ orbital. The nucleus is indicated by a heavy dot in the center. (*a*) Dot density cross-section. (*b*) Three-dimensional surface. (*c*) Simplified version.

The atomic orbitals used for bonding in organic compounds are the $1s$ (for hydrogen), the $2s$, $2p_x$, $2p_y$, $2p_z$ (for carbon), and occasionally the $3d$ (for third row elements). The s orbitals are ball shaped, whereas the p orbitals are dumbbell shaped and aligned along each of the three coordinate axes, x, y, and z. The four bonding orbitals for carbon are drawn below in Figure 1.5.

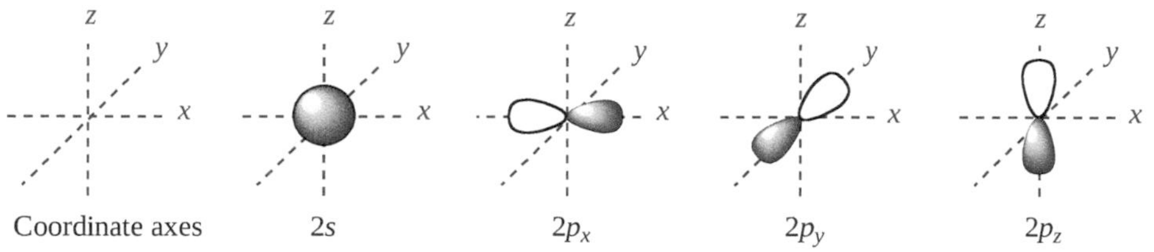

Figure 1.5 Representations of the bonding orbitals of carbon.

1.7.2 Using *p* Orbitals to Make the Sigma and Pi Bonds of Nitrogen

We can make all the bonding orbitals of molecular nitrogen using just the $2p_x$, $2p_y$, and $2p_z$ atomic orbitals of Figure 1.5. As we bring two nitrogen atoms together, each with their own set of p orbitals, the p orbitals will interact. If two orbitals are directly pointed at each other and overlap in phase, a sigma bond is formed. If the orbital overlap is side by side and in phase, a pi bond is formed. For every in-phase bonding combination of atomic orbitals, there is a higher energy out-of-phase antibonding combination. Figure 1.6 shows the bonding and antibonding molecular orbitals of nitrogen. From six atomic orbitals we built six molecular orbitals; no orbitals were gained or lost. In nitrogen only the three bonding molecular orbitals (not the antibonding) are filled with electrons, giving nitrogen a triple bond. The triple bond of nitrogen is made up of one sigma bond from the overlap of two $2p_x$ atomic orbitals, and there are also two perpendicular pi bonds, one from the overlap of two $2p_y$, and another from the overlap of two $2p_z$ atomic orbitals.

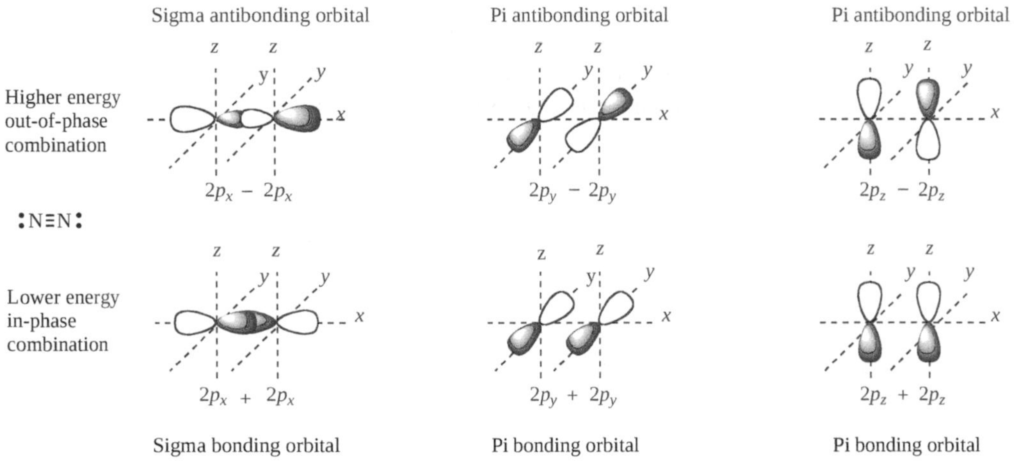

Figure 1.6 The in-phase and out-of-phase mixing of the $2p_x$, $2p_y$, and $2p_z$ atomic orbitals on two nitrogen atoms to give the bonding and antibonding molecular orbitals of molecular nitrogen.

1.7.3 Hybridization

It is convenient to have orbitals for bonding that point along the directions found from VSEPR. Toward this end, we can mathematically mix atomic orbitals on the same atom to get hybrid orbitals that point in the right directions and are useful for the description of bonding with other atoms. Figure 1.7 shows how a 2*s* orbital can be added to and subtracted from a 2*p* orbital to give the two *sp* hybrid orbitals. To subtract, we change the mathematical sign (shading) of the 2*s* orbital and then add it to the 2*p* orbital. The two orbital wave functions reinforce where the mathematical signs are the same (the lobe gets larger) and cancel out where the sign is different (the lobe gets smaller). The number of orbitals is conserved: When we combine two atomic orbitals, we get two hybrid orbitals. We now have two hybrid orbitals pointing 180° from each other able to be used for bonding to two other atoms in a linear arrangement.

Figure 1.7 The mixing of a 2*p* and a 2*s* orbital on an atom giving two *sp* hybrids.

Combining the 2*s* and one of the 2*p* orbitals to get two *sp* hybrids leaves two 2*p* orbitals remaining. This is shown in the orbital energy diagram in Figure 1.8 for *sp* hybridization. Because the 2*s* orbital is lower in energy and closer to the nucleus than a 2*p* orbital, **hybrid orbitals that contain a higher %*s* character will form bonds that will be shorter, stronger, and lower in energy.**

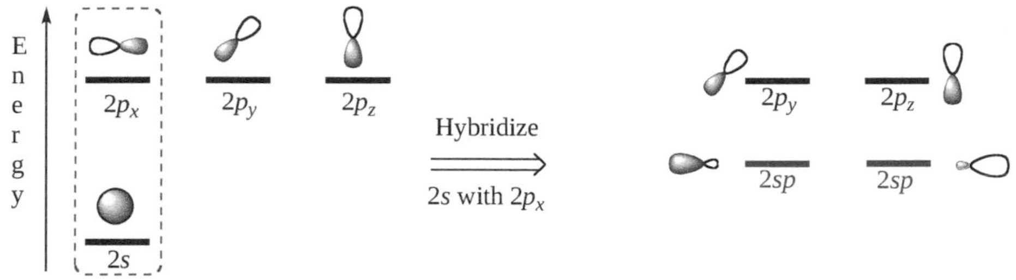

Figure 1.8 Orbital energy levels for mixing a 2*s* and a 2*p* orbital to form two *sp* hybrids.

As shown in Figure 1.5, carbon has one 2*s* and three 2*p* orbitals that can be used for hybridization. Hybridizing the 2*s* with the p_x and the p_y gives three sp^2 hybrid orbitals in the xy plane with a 120° angle between them, Figure 1.9. There is the one p_z orbital left.

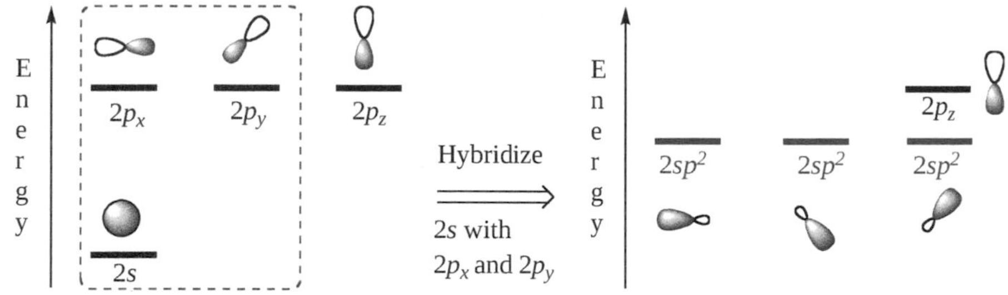

Figure 1.9 Orbital energy levels for mixing a 2*s* with two 2*p* orbitals to form three sp^2 hybrids.

Finally, if we hybridize 2*s* and all three 2*p* orbitals we get four sp^3 hybrid orbitals pointing to the corners of a tetrahedron, Figure 1.10.

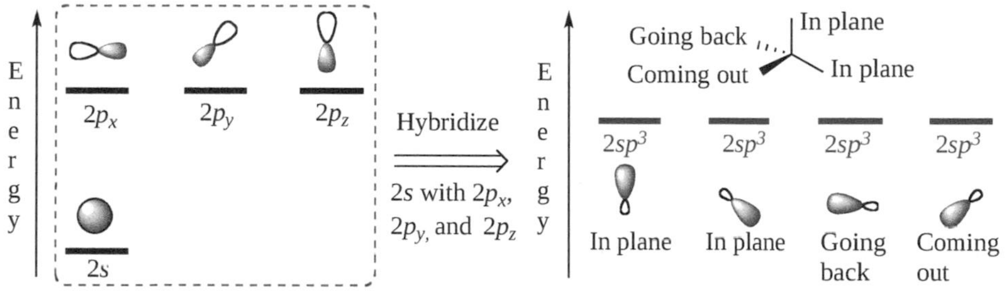

Figure 1.10 Orbital energy levels for mixing a 2*s* with three 2*p* orbitals to form four sp^3 hybrids.

The properties of all possible hybridizations of carbon are summarized in Table 1.4. To determine the hybridization of a carbon atom, just use VSEPR on the Lewis structure. A linear arrangement needs an *sp* carbon; a trigonal planar arrangement needs an sp^2 carbon. A tetrahedral arrangement usually needs sp^3 hybridization, but if a lone pair on the atom is involved in resonance, the lone pair is then in a *p* orbital, and the atom is trigonal planar, sp^2.

Table 1.4 Summary of the Properties of Hybrid Orbitals

Hybridization of carbon	*sp*	sp^2	sp^3
Number of hybrid orbitals	2	3	4
Interorbital angle	180°	120°	109.5°
%*s* character	50	33	25
%*p* character	50	67	75
Orientation	Linear	Trigonal	Tetrahedral
Electronegativity of C	3.29	2.75	2.48
Remaining *p* orbitals	2	1	0

1.7.4 Single Bonds

We can overlap two sp^3 orbitals to produce a sigma bond. The in-phase combination overlaps lobes of the same mathematical sign and is the lower-energy *bonding* orbital. The higher-energy out-of-phase combination is called the *antibonding* orbital and is usually empty. Sigma (σ) bond orbitals are cylindrical along the axis of the bond. Therefore, the ends can rotate about the axis of the bond without losing any overlap. **Strong bonds are due to good overlap of the bonding orbitals.**

To show that an sp^3 carbon is tetrahedral, organic molecules are usually drawn in three dimensions. The solid wedge in Figure 1.11 symbolizes the bond is coming out of the paper, and the dashed wedge symbolizes the bond is going back into the paper. The tetrahedral shape of sp^3 carbon has profound biochemical implications. *Stereochemistry* is the description of how identically connected atoms of a molecule are arranged in space (discussed in Section 2.3). The stereochemistry of a pharmaceutical can make all the difference between potent useful biological activity and undesired effects.

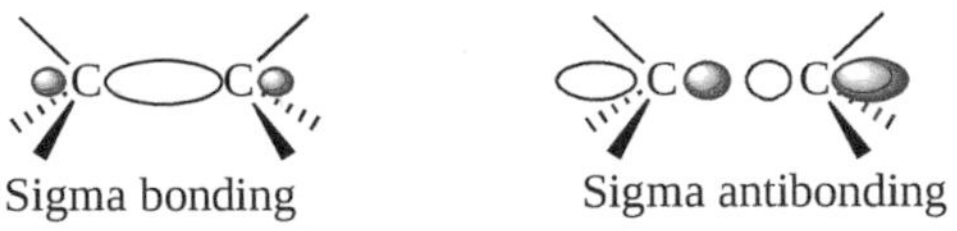

Figure 1.11 The bonding and antibonding orbitals of a carbon–carbon single bond are made from the combination of two sp^3 hybrid orbitals.

Conformational isomers differ only in rotations about single bonds. The eclipsed and staggered conformational isomers of ethane, CH_3CH_3, are shown in Figure 1.12. Although single bonds are drawn as if they were "frozen in space," rotation usually has no significant barrier at room temperature (free rotation at room temperature). Ethane's rotational barrier is tiny, only 3 kcal/mol (13 kJ/mol). One kilocalorie/mol is equal to 4.184 kilojoules/mol.

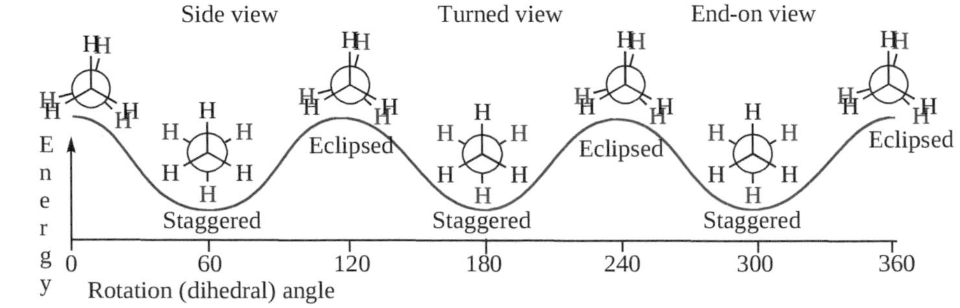

Figure 1.12 Views of two conformational isomers of ethane, staggered and eclipsed. The end-on view is called a Newman projection. Below them is a plot of the energy as the C–C bond rotates.

The staggered conformation is the lowest energy; the eclipsed conformation is highest since the C–H bonds are in the process of passing each other and their electron clouds repel. Rotation about the center C–C bond in butane, $CH_3CH_2CH_2CH_3$, requires passing two methyl groups with a slightly larger barrier of 4.5 kcal/mol (19 kJ/mol). The staggered form when the methyl groups are opposite (*anti*) is about 0.8 kcal/mol (3.4 kJ/mol) lower in energy than the staggered form, where the methyl groups are 60° to each other (*gauche*). A significant rotation barrier may arise if two very large groups would bump into one another when rotation about the single bond occurs. The best way to learn what organic compounds actually look like is by working with molecular models.

1.7.5 Double Bonds

Pi bonds are made by two *p* orbitals interacting side by side in the same plane and are relatively reactive due to less effective orbital overlap. The in-phase and out-of-phase combination of two 2*p* orbitals yield the pi molecular orbitals for ethene, CH_2=CH_2, shown in Figure 1.13. Double bonds result from a sigma bond and a pi bond between the two bonding atoms. Since that sigma bond lies in the nodal plane of the pi bond, the sigma and pi bonds of a double bond are considered independent.

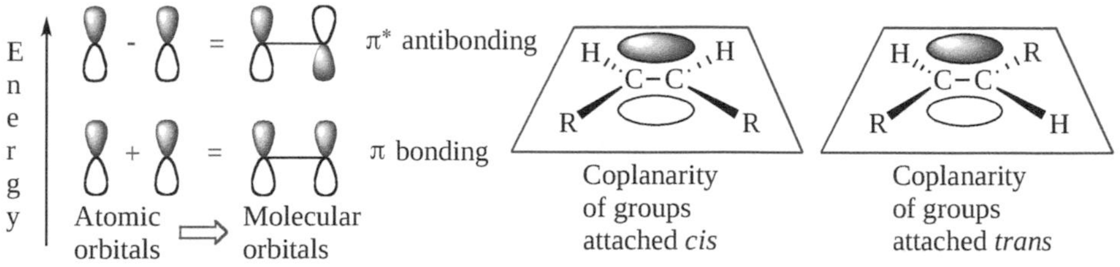

Figure 1.13 The bonding and antibonding orbitals of a carbon–carbon pi bond are made from the side-by-side combination of two 2*p* orbitals by subtraction and addition. Mixing two atomic orbitals gives two molecular orbitals. As shown, all atoms adjacent to the pi bond lie in a plane.

The requirement that the *p* orbitals overlap causes a very large barrier to rotation about the double bond, about 63 kcal/mol (264 kJ/mol). *Cis* (two groups on the same side) and *trans* (two groups on opposite sides) double-bond isomers do not interconvert at any reasonable temperatures. The *trans* isomer tends to be slightly more stable than the *cis* isomer because the groups may bump into one another when they are *cis*. More alkyl substitution on the double bond (replacing an H on the double bond by an R) makes the alkene slightly more stable. For example, an equilibrium mixture of butenes is found to contain 3% 1-butene, 23% *cis*-2-butene (R equals CH_3 above), and 74% *trans*-2-butene.

1.7.6 Triple Bonds

A triple bond is composed of one sigma bond and two pi bonds. Since the two pi bonds are perpendicular to each other, they are treated as two separate, noninteracting pi bonds. **Perpendicular orbitals do not interact.** The triple bond is linear, and atoms bonded to it lie in a straight line (Figure 1.14). Organic chemists tend to draw skinny *p* orbitals (see Figure 1.4); the triple bond is overall cylindrically symmetrical.

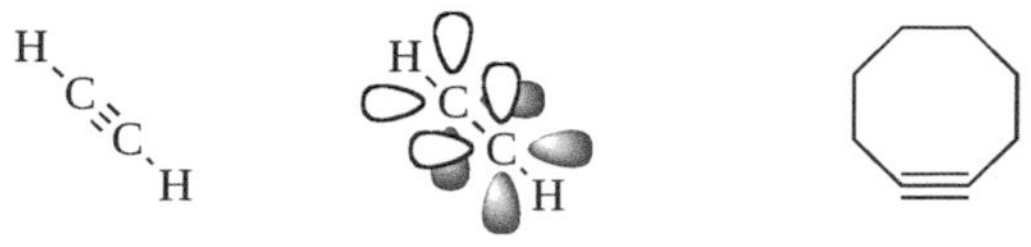

Figure 1.14 The pi orbitals of a triple bond. Cyclooctyne is shown on right.

The average bond strength of a triple bond is 200 kcal/mol (837 kJ/mol); therefore each pi bond is worth about 59 kcal/mol (247 kJ/mol). If placed in a ring, the triple bond can be bent from collinearity and become weaker and much more reactive. Cyclooctyne is the smallest ring that contains a triple bond and is still stable enough to be isolated.

1.7.7 Cumulenes

Cumulenes are compounds with two adjacent perpendicular double bonds (Figure 1.15). Each double bond can be considered separately. The central carbon is *sp* hybridized, and the two CH_2s lie in perpendicular planes. In heterocumulenes, heteroatoms have replaced one or more of the carbons, as in carbon dioxide, O=C=O.

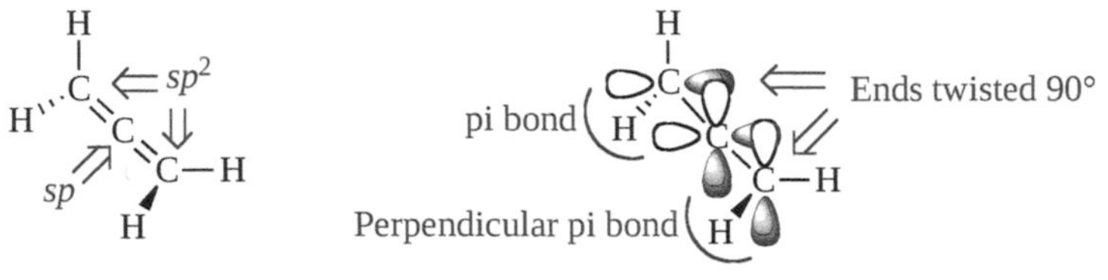

Figure 1.15 The pi orbitals of a cumulene.

1.8 MOLECULAR REPULSIONS, ATTRACTIONS, AND HYDROGEN BONDING

1.8.1 Nonbonded Repulsions

All groups on a molecule take up space, and this occupied space is called *steric bulk*. Interactions that attempt to force two groups to occupy the same space are very unfavorable. The like-charged electron clouds on the two groups repel each other, and as they are forced together the extra energy required is called *steric hindrance*, or nonbonded repulsion. If the path that the nucleophile must follow to satisfy orbital overlap requirements has too much steric hindrance, the reaction will not occur. The nucleophile will just bounce off the group that is in the way. Bulky nucleophiles need very open sites for attack, and hindered sites can be attacked only by small nucleophiles. The steric interactions between site and nucleophile are important in the competition between an anion acting as a nucleophile or acting as a base.

How close is too close? If the distance between two groups is less than the sum of their van der Waals radii, then strong repulsion begins. For most groups, this distance is between 3 and 4 Å, which is about twice the normal C–C single bond length of 1.54 Å. Models are very useful for comparing the relative sizes of groups. Shown below are some common groups ordered by increasing size: hydrogen, methyl, ethyl, isopropyl, phenyl, and *tert*-butyl.

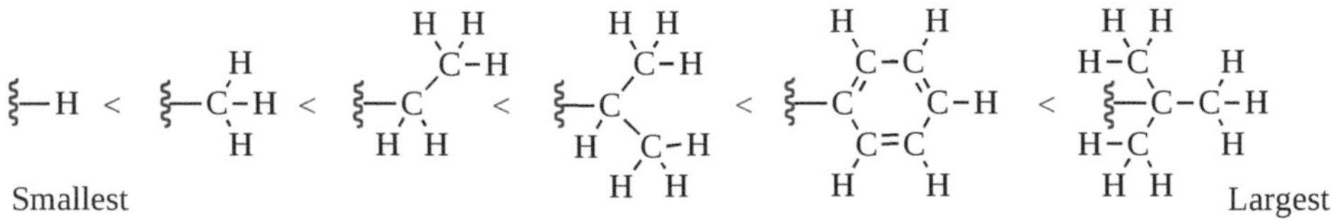

1.8.2 Dipole Attractions and Repulsions

Unlike charges attract, and like charges repel each other. The uneven distribution of electrons in a molecule can result in a pair of partial charges, represented by δ+ and δ− or by ⟼, a dipole moment arrow with the plus end indicated. Polar covalent bonds create bond dipoles. Another dipole arises from a lone pair of electrons in a hybrid orbital. If the center of positive charge (from the nuclei positions) is different from the center of negative charge (from the electron cloud) then the molecule will have a net dipole. Individual dipoles will vector sum to the net dipole. Water has both bond dipoles and lone pair dipoles, Figure 1.16.

Figure 1.16 The net dipole consists of summing components of bond dipoles and lone pair dipoles. With symmetrical compounds like O=C=O the dipoles will cancel resulting in no net dipole moment.

1.8.3 Hydrogen Bonding

Hydrogen bonding is the overlap of a lone pair orbital of a heteroatom with an H atom bonded to another heteroatom, commonly O, N, F, and sometimes S and Cl. With rare exceptions, the C–H bond is not significantly polarized and does not hydrogen bond. The attraction of the permanent δ+ of a polarized hydrogen–heteroatom bond with the δ− of a heteroatom lone pair provides an ionic component. Although the strength of a typical hydrogen bond to oxygen averages about 5 kcal/mol (21 kJ/mol), its importance should not be underestimated. Hydrogen bonding causes water to be a liquid rather than a gas at room temperature. It is calculated that the oxygen in water at room temperature is estimated to be bonding to 3.7 hydrogen atoms. Hydrogen bonding is dynamic and short lived. When water freezes it forms a hexagonal lattice with each oxygen bonded or hydrogen bonded to four hydrogen atoms. This open hexagonal lattice of ice is less dense than liquid water, and that is the reason that ice floats. Hydrogen bonds with a linear arrangement of all three atoms are most common. Often a very weak bond like a hydrogen bond or a complexation is indicated by a dotted line, ·······, which is also used to symbolize a partially broken or made bond. Don't start electron flow arrows from such a line; always use Lewis structures near any arrows to keep track of electrons.

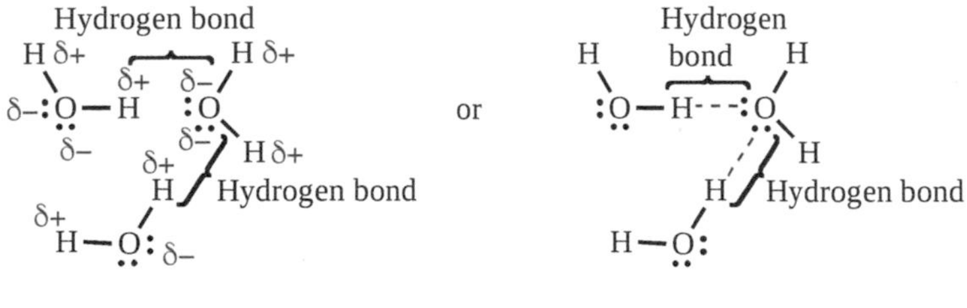

1.8.4 Cation–Pi Interactions

Pi-adducts (formerly called pi-complexes), also often called electron donor–acceptor interactions, are often a weak association of an electron-rich molecule with an electron-poor species. The donor is commonly the electron cloud of a pi bond or a group of pi bonds; the acceptor can be a metal cation, a halogen like Br_2, or an electron-poor organic compound. In the absence of solvent, as can occur in an enzyme cavity, the cation–pi interaction can be stronger than hydrogen bonding. An example is shown below, where a cation snuggles into the face of the aromatic pi cloud of benzene (see aromaticity, Sections 1.9.3 and 13.4).

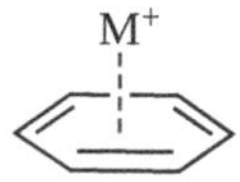

The gas-phase attraction between benzene and various cations decreases with increasing cation size (Li^+ at 38 kcal/mol, Na^+ at 27 kcal/mol, and K^+ at 18 kcal/mol). Figure 1.17 shows the cation–pi interaction between an alkene and silver ion.

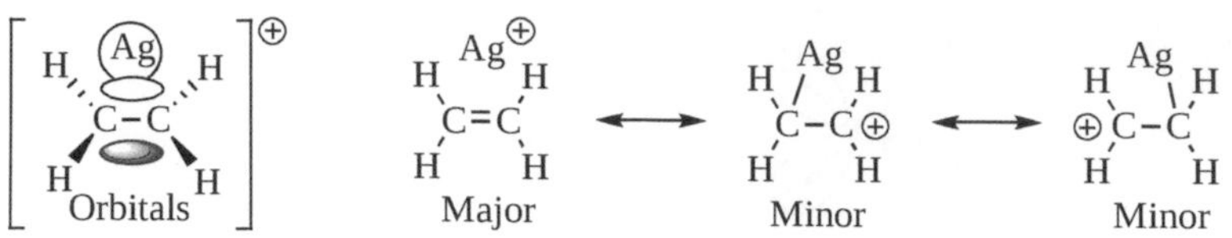

Figure 1.17 A silver ion pi-adduct. Resonance forms describe the orbital overlap shown at left.

1.9 CONJUGATION, VINYLOGY, AND AROMATICITY

1.9.1 Conjugation

Two pi systems connected by a single bond behave as one *conjugated* system because the *p* orbitals of the two pi systems are close enough to have good overlap and additional pi bonding. Figure 1.18 shows how a double bond and a carbonyl group can be placed in a molecule to form both a conjugated system and an unconjugated system. In the conjugated system, the partial plus of the carbonyl is delocalized into the second pi bond. The additional pi bonding makes a conjugated system more stable than the corresponding unconjugated system by about 3.5 kcal/mol (14.6 kJ/mol); any equilibrium between the two favors the conjugated system.

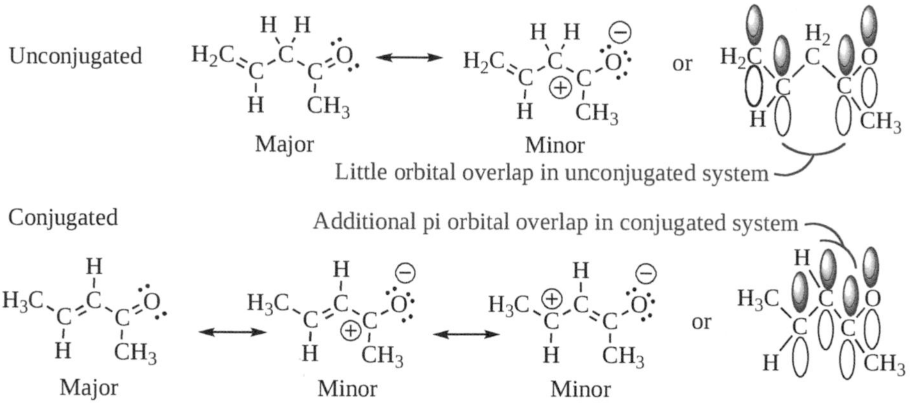

Figure 1.18 The resonance forms and pi overlap of an unconjugated system (top) and a conjugated system (bottom) of a carbon–carbon pi bond and a carbon–oxygen pi bond.

Another example of conjugation is the nitrogen lone pair joined by resonance with a carbonyl group shown in Figure 1.19 for amides. The nitrogen lone pair is not in an sp^3 orbital but in a *p* orbital that overlaps the *p* orbitals of the pi bond to make a three-*p* orbital pi system. Any three-*p* orbital pi system is called an *allylic system*. The conjugated system will have properties different from its individual parts because it now acts as a hybrid and not as two independent functional groups. The third resonance form requires the *p* orbital of the nitrogen lone pair to line up with the *p* orbital on carbon to get the proper overlap for the double bond. This required alignment gives amides a significant C–N rotational barrier of 15 to 20 kcal/mol (63 to 84 kJ/mol). The amide links in proteins do not rotate at room temperature and thus greatly contribute to the three-dimensional stability of proteins.

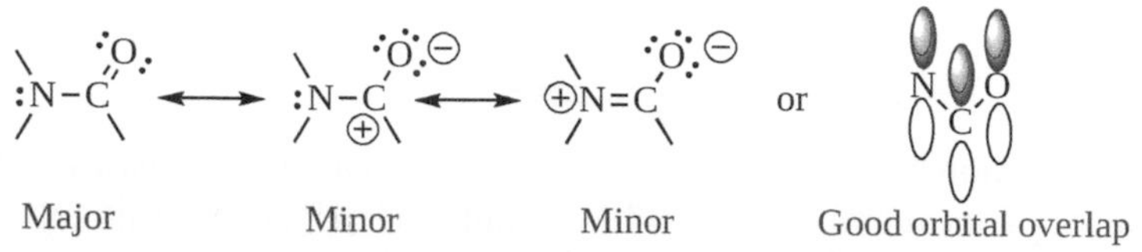

Figure 1.19 Conjugation of a nitrogen lone pair with a carbonyl to give an amide group.

Still another example of conjugation is the carboxylic acid group in which a hydroxyl is conjugated with a carbonyl group. The resonance forms of the carboxylic acid explain the increased acidity of the hydrogen (major form is on the left, with more minor forms to the right):

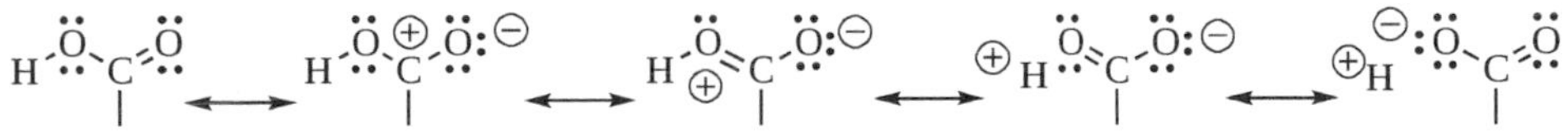

1.9.2 Vinylogy

Vinylogy is the extension of the properties of a system by the insertion of a carbon–carbon double bond. The conjugated system increases in length, but the physical properties and chemical reactivity remain approximately the same. A good example of this is the comparison of a carboxylic acid and a vinylogous carboxylic acid (Figure 1.20). The hydroxyl group is still conjugated with the carbonyl through the intervening pi bond.

Carboxylic acid Vinylogous carboxylic acid

Figure 1.20 The insertion of a carbon–carbon double bond between the OH and the C=O of the carboxylic acid creates the vinylogous carboxylic acid on the right.

The resonance forms for a vinylogous carboxylic acid shown below are very similar to the carboxylic acid ones shown above, and likewise can account for an acidic hydrogen.

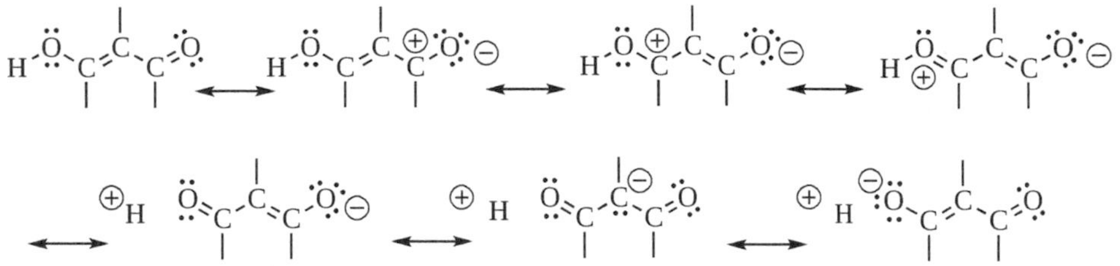

Ascorbic acid, vitamin C, is an important vinylogous carboxylic acid. The following structures compare several other vinylogous systems with their respective parent systems.

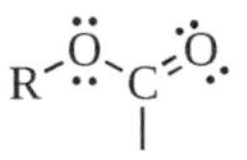

Ester

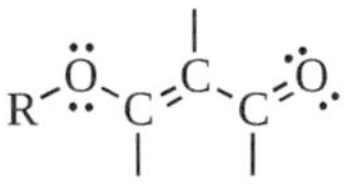

Vinylogous ester

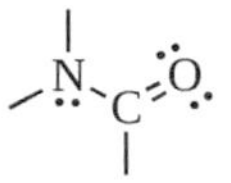

Amide

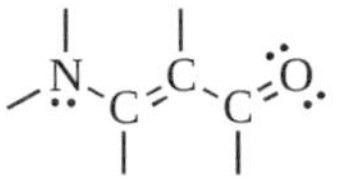

Vinylogous amide

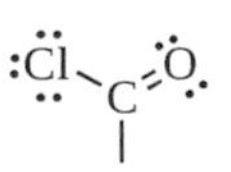

Acyl chloride

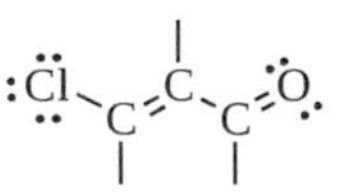

Vinylogous acyl chloride

1.9.3 Aromaticity

Benzene, C_6H_6, is much more stable and less reactive to electrophiles than simple alkenes, is described by two neutral resonance structures of equal importance, and is the prime example of an aromatic compound:

Any cyclic conjugated compound will be especially stable when the ring contains **4*n* + 2 pi electrons**, where *n* is any integer (*Hückel's rule*). This extra stability, "resonance stabilization," of the 2, 6, 10, 14, 18, ... pi electron cyclic conjugated systems is termed *aromaticity* and results from the pattern of molecular orbital energies of cyclic conjugated systems (Figure 1.21). Although the derivation of the pattern of these molecular orbitals is beyond the scope of this text, a qualitative explanation of them is given in Section 13.4. As with atomic orbital energy levels that are filled in each noble gas, filled molecular energy levels are more stable than partially filled levels. It takes two electrons to fill the bottom level, then four more to fill the next level, and four more for the next, giving the 2, 6, 10 pattern of stability.

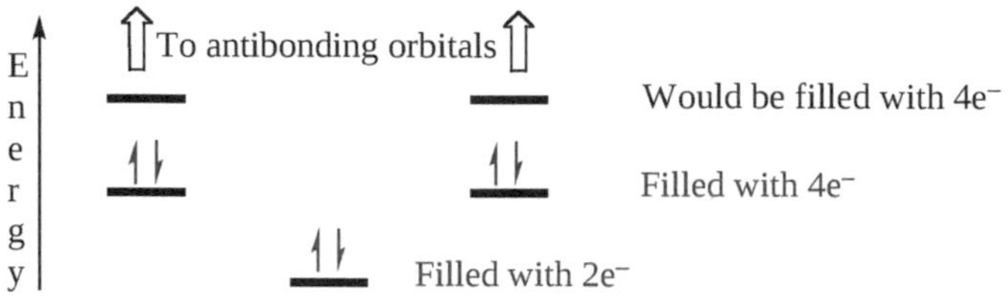

Figure 1.21 The general pattern of molecular orbital energy levels for cyclic conjugated systems. Six electrons are shown, filling the second MO level (stable). A total of four electrons would leave that MO half-filled (unstable).

The resonance stabilization of benzene is estimated to be about 36 kcal/mol (151 kJ/mol), but this is not an experimentally verifiable number. The 4*n* pi electron systems (4, 8, 12, 16, ...), with half-filled energy levels, not only are less stable than the 4*n* + 2 systems but are destabilized relative to their open-chain analogs and are called *antiaromatic*.

To summarize, **aromatic stabilization occurs in rings that have an unbroken conjugated loop of *p* orbitals (a *p* orbital on each atom in the loop with no big twists between) and 4*n* + 2 pi electrons in the loop.**

If the molecule has an unbroken loop of *p* orbitals, you will be able to draw at least two resonance structures that shift the double bonds around the loop, as we did for benzene. If any atom in the ring has four groups bound to it, then there can be no *p* orbital on that atom; the pi loop is broken, and the ring cannot be aromatic.

To determine possible aromaticity, first determine if the loop is intact, then determine how many electrons occupy the loop. Count each double bond as contributing two electrons. A lone pair on an atom in the loop is counted as contributing two electrons if there is no double bond to that atom. (If an atom is doubly bonded and contains a lone pair, the lone pair cannot be counted since it must be in an orbital that is perpendicular to the loop.) If an atom has two lone pairs, just one is counted (only one can align with the loop). A triple bond can contribute only two electrons (one of the two pi bonds must be perpendicular to the loop). A carbocation contributes an empty *p* orbital to the loop (but contributes no electrons).

Example problem

Which of the following compounds are stabilized by aromaticity?

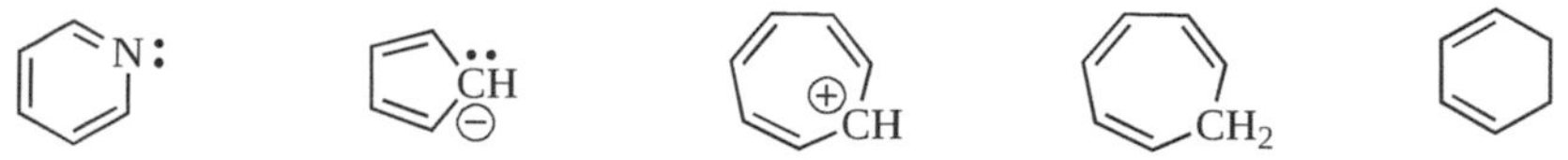

Answer: The first three on the left are aromatic. The far left compound has six electrons in the loop because the nitrogen lone pair is in an sp^2 orbital that is perpendicular to the loop of *p* orbitals, and so is not counted. The anion and the cation both have six electrons in the loop and a *p* orbital on every atom. Neither of the two compounds on the right has a complete loop of *p* orbitals because the CH_2 group(s) breaks the loop of pi conjugation.

1.10 SUMMARY

In this chapter, we learned that **structure determines reactivity**. We have seen that weak bonds are a source of reactivity. Strong bonds are made by good overlap of similar-sized orbitals (same row on periodic table). Bends or twists that decrease orbital overlap weaken bonds. Lewis structures and resonance forms along with electron flow arrows allow us to keep track of electrons and explain the changes that occur in reactions. VSEPR will help us predict the shape of molecules. Next we must review how bonds are made and broken, and what makes reactions favorable. Critical concepts and skills from this chapter are:

Best-First Search
- Generate all reasonable alternatives and select the best one.

Valid Lewis Structures:
- have correct number of valence electrons,
- have correct formal charges that sum to the total charge, and
- do not exceed octets for second-row elements.

The Major Resonance Contributor will have:
- the most covalent bonds,
- the least amount of charge separation,
- the most complete octets and duets as possible, and
- any negative charges on the most electronegative atoms.

Good Arrow Pushing Habits
- Draw full Lewis structures near all arrows (to keep track of electrons and atoms).
- Start every arrow from a **drawn** pair of electrons (not from a minus formal charge).
- Check Lewis structure octets and duets (the most important cross-check).
- Check formal charge and charge balance (or your next step goes astray).
- Use the **known** electron flow paths (introduced over the next few chapters).

VSEPR
- Groups of electrons, bonds and nonbonding pairs, repel each other and arrange themselves in space to minimize this repulsion. Lone pairs are given more space if possible (minimize 90° lone pair interactions).

Conjugation, Vinylogy, and Aromaticity
- Conjugation is the favorable overlap of adjacent p orbitals or pi systems to form a single larger pi system.
- Vinylogy is the extension of the properties of a pi system by the insertion of a pi bond in conjugation.
- Aromatic stabilization is found for cyclic conjugated rings with $4n + 2$ pi electrons.

Essential Skills
- Lewis structures—be able to rapidly draw any compound's major resonance form with the correct formal charges.
- Arrows—understand and be able to supply the arrows for converting between resonance forms, and for one-step reactions given the reactants and products.

CHAPTER 1 EXERCISES

1.1 Draw Lewis structures with resonance for the following neutral compounds. Decide the major resonance form, if any.

O–O–O $\quad$ H_2C–N–N $\quad$ O–C–O $\quad$ H–O–N–O $\quad$ H–O–NO_2

1.2 Draw Lewis structures with resonance for the following charged species. Decide the major resonance form, if any.

$H_2CNH_2^+$ $\quad$ OCN^- $\quad$ $HOCO_2^-$ $\quad$ H_2COH^+ $\quad$ $H_2CCHCH_2^+$

1.3. What is the polarization of the indicated bond?

Br–Br $\quad$ HO–Cl $\quad$ H_2B–H $\quad$ H_2C=O $\quad$ I–Cl

1.4 Circle the electrophiles and underline the nucleophiles in the following group.

BF_3 $\quad$ H^+ $\quad$ Ne $\quad$ NH_3 $\quad$ ^-CN

1.5 Draw the Lewis structure(s) that would be the product of the arrows.

(a)

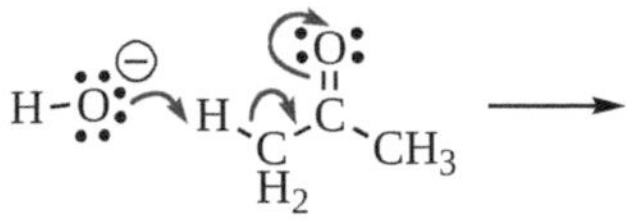

(b)

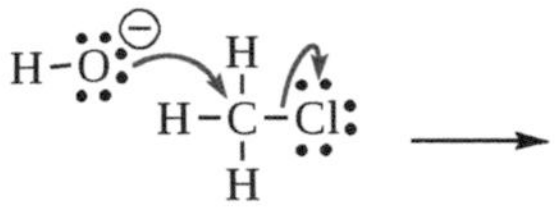

(c)

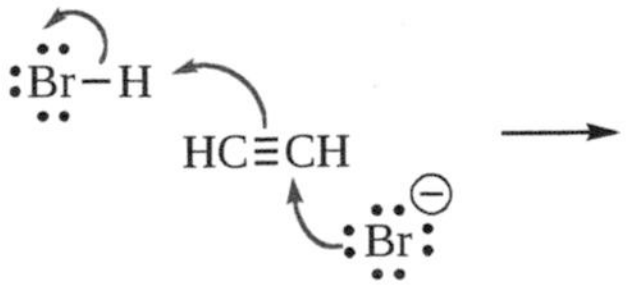

1.6 Draw the Lewis structure(s) that would be the product of the arrows.

(a)

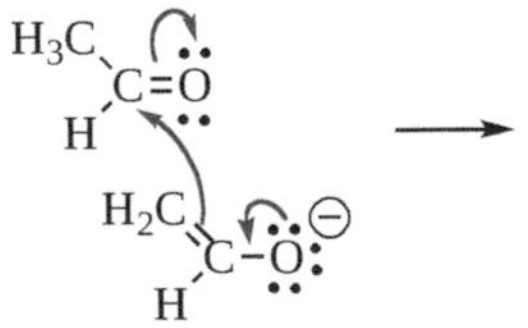

(b)

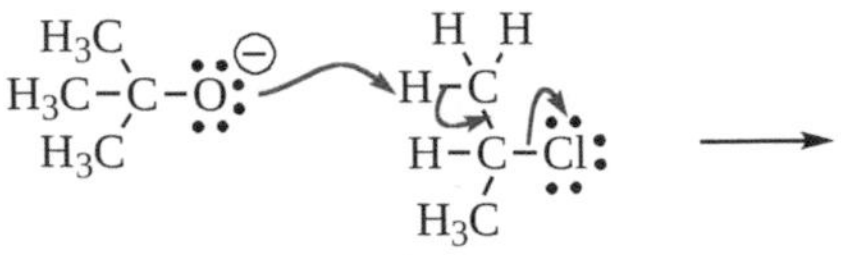

(c)

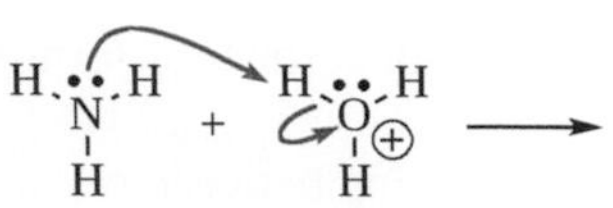

1.7 Give the curved arrows necessary for the following reactions.

(a)

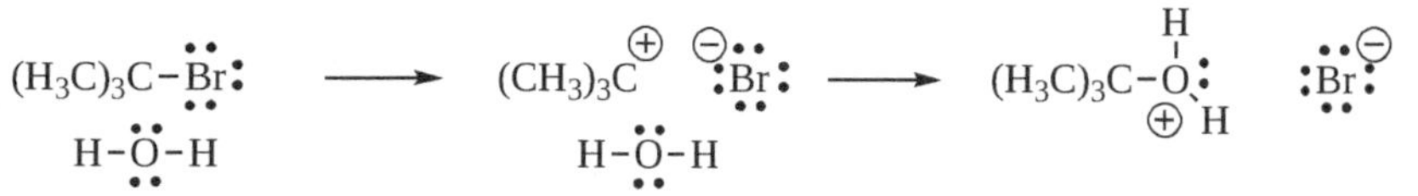

(b)

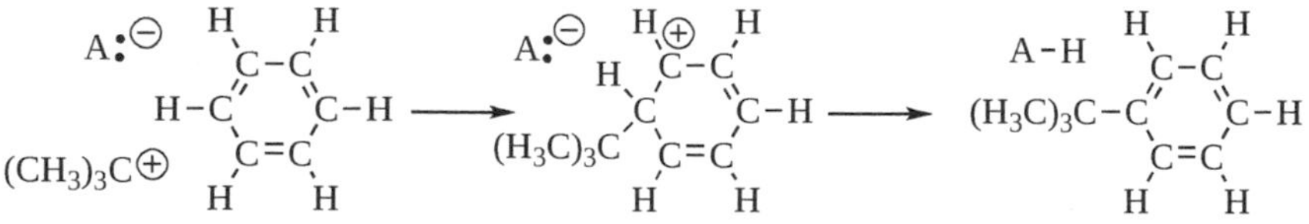

(c)

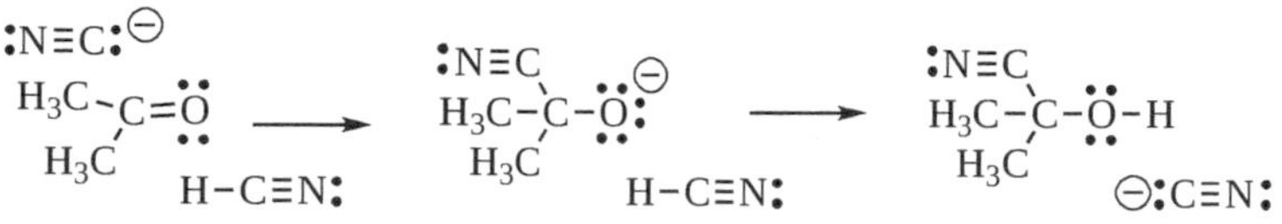

1.8 Give the curved arrows necessary for the following reactions.

(a)

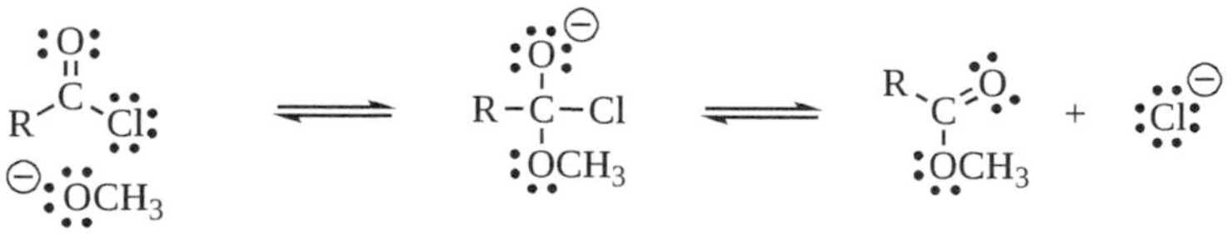

(b)

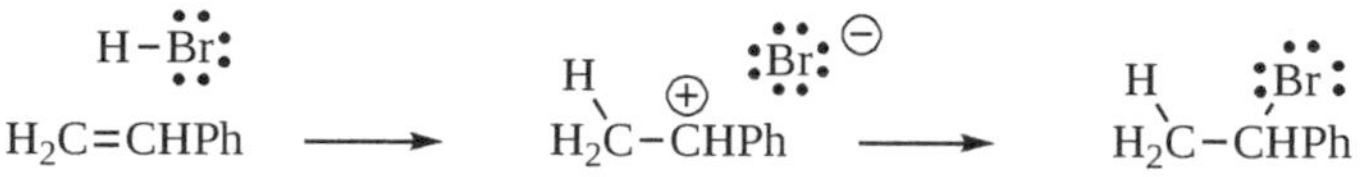

(c)

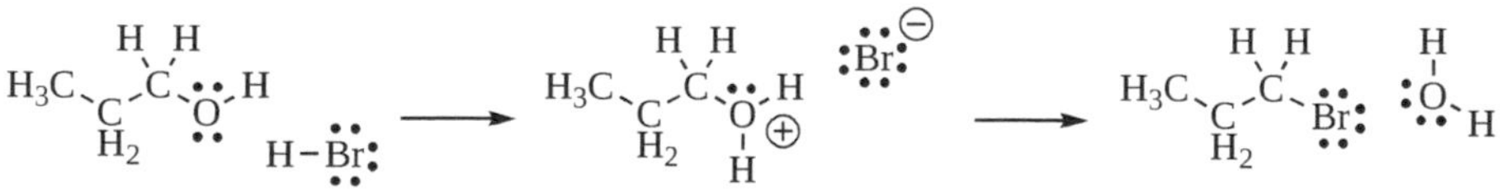

1.9 Draw full Lewis structures for the following line structures.

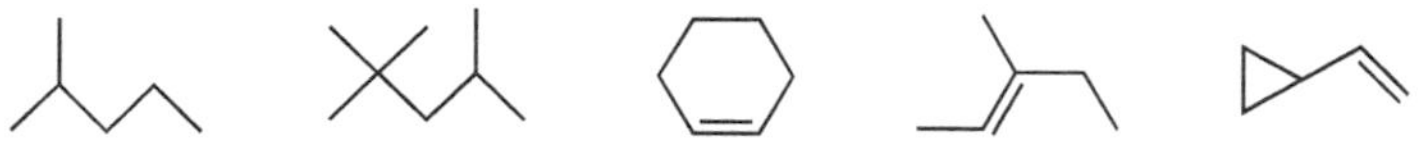

1.10 In the following structures circle any carbon atom that bears a significant partial positive charge. (Hint: look for electronegative atoms.)

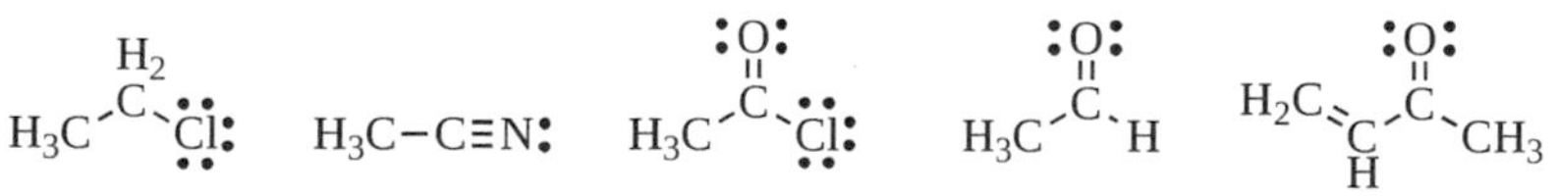

1.11 In the following structures circle any carbon atom that bears a negative or significant partial negative charge. (Hint: draw some resonance forms.)

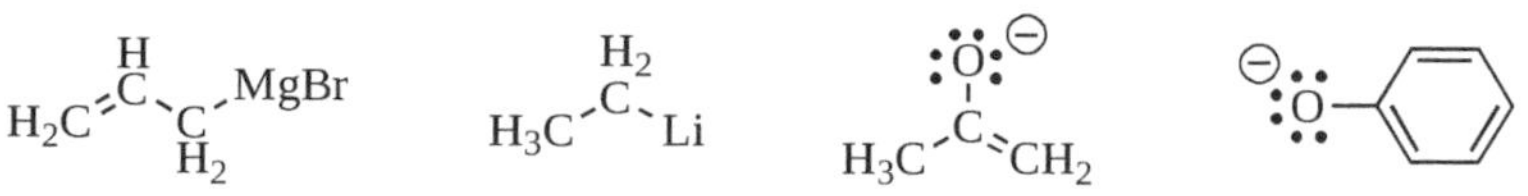

1.12 Give the hybridization of the carbons in these structures.

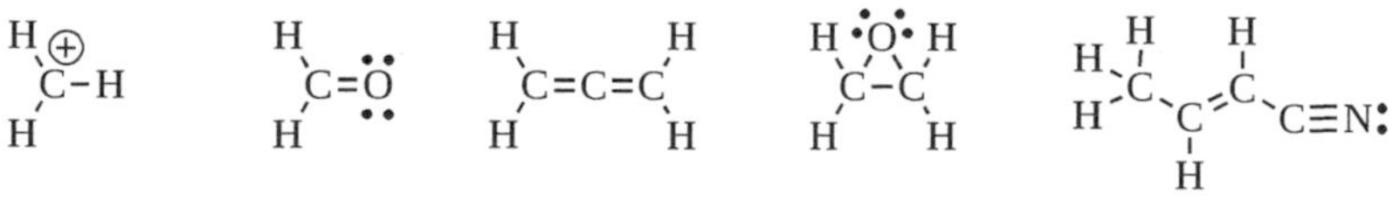

1.13 For each of the resonance pairs below, determine which is the major contributor.

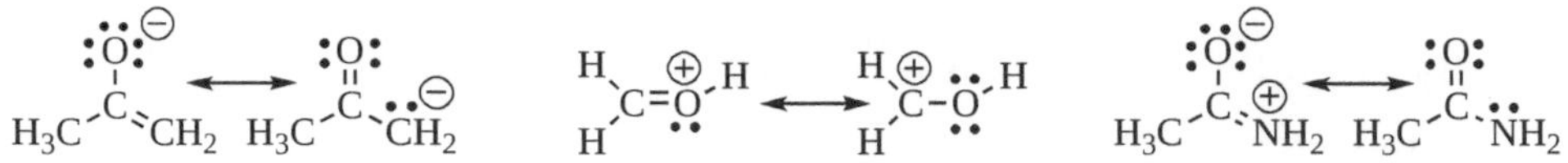

1.14 Draw the pi system for:

(a) an amide (b) carbon dioxide (c) an ester (d) vinylogous amide (e) nitrile

1.15 Circle the aromatic compounds in the following list.

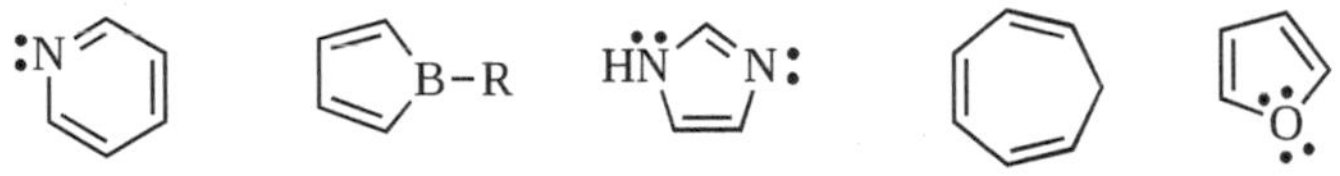

1.16 Circle the conjugated systems in the following list.

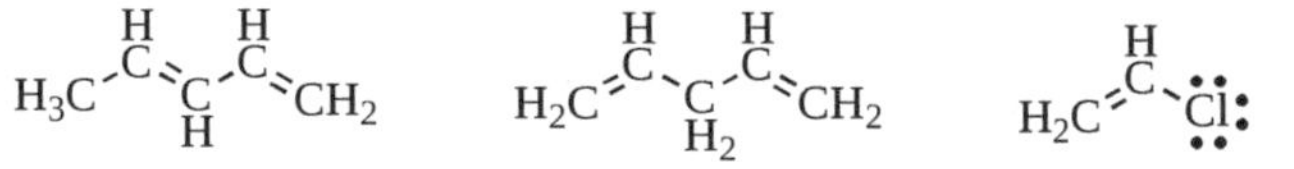

$H_3C-C\equiv C-CH_3$

1.17 First draw a good Lewis structure, then decide the structure of the following compounds, using VSEPR. For connectivity on these, put the unique atom in the center.

SF_4 CO_2 BF_3 H_2O NH_3

1.18 First draw a good Lewis structure, then decide the structure of the following ions, using VSEPR. For connectivity on these, put the unique atom in the center.

NO_2^- NO_2^+ NO_3^- SO_4^{-2} PO_4^{-3}

1.19 Rank the following on the size of the atom or ion (use #1 for largest).

Ne Na^+ F^- Mg^{2+} Al^{3+}

1.20 Circle the molecules that have a dipole moment.

NO_2^- NO_2^+ NO_3^- NH_4^+ $AlCl_4^-$

1.21 Circle the polar covalent and underline the mostly ionic compounds below.

Br_2 LiBr HCl KBr H_2O

1.22 Draw full Lewis structures for the following line structures of biomolecules.

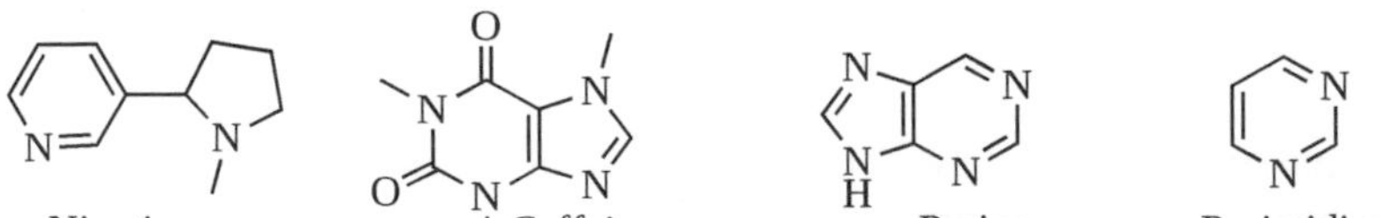

Nicotine Caffeine Purine Pyrimidine

1.23 Decide which of the following anions is the most stable, the least stable.

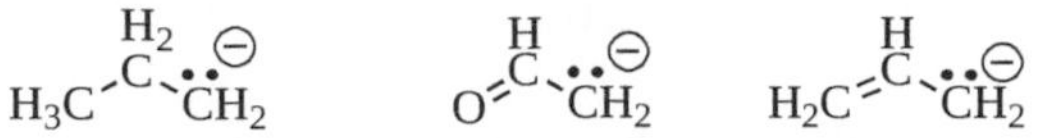

1.24 Decide which of the following cations is the most stable, the least stable.

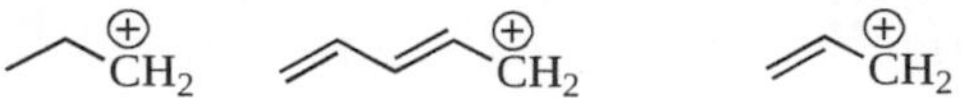

2

THE PROCESS OF BOND FORMATION

Electron Flow in Organic Chemistry: A Decision-Based Guide to Organic Mechanisms,
Third Edition. Paul H. Scudder.

2.1 ENERGETICS CONTROL KNOWLEDGE

Energetically favorable reactions usually break weak bonds and form strong ones. To be able to judge the most favorable route, we must understand qualitatively both thermodynamics, which determines the position of equilibrium, and kinetics, which determines the rate of reaction. The products of reversible reactions are controlled by thermodynamics; products of irreversible reactions are controlled by kinetics. Therefore, we need a quick way to tell if a reaction is expected to be reversible. Energetics control knowledge allows us to decide whether a reaction is too energetically uphill to occur (within bounds). If the reverse reaction is too uphill, then the process is considered irreversible. To predict reaction products, we need to understand the process of bond making and breaking and how the energy of the process makes some reactions more favorable than others.

2.2 ORBITAL OVERLAP IN COVALENT BOND FORMATION

2.2.1 Sigma Bonding

Strong bonds require good overlap of the bonding orbitals. **The formation of sigma bonds requires approximate colinearity of the reacting orbitals** because that produces the best overlap, thus the strongest bond. A weak bond is a site of reactivity. The average bond strength for a C–C single bond is 83 kcal/mol (347 kJ/mol). Deformed sigma bonds can be made, but at a cost. Rings containing three and four atoms are strained because the orbitals can no longer be directed along a line between the atoms, and their overlap and bond strength decreases (see Figure 2.1). This "*ring strain*" makes small rings easy to break. Strain energy destabilizes a three-membered ring by 27.5 kcal/mol (115 kJ/mol) and a four-membered ring by 26.3 kcal/mol (110 kJ/mol). Rings of five, six, and seven atoms are relatively strain free.

Orbitals of different sizes overlap poorly and therefore form weak bonds. In general, the overlap with a carbon-based orbital decreases as one goes down a column in the periodic table. For example, the strengths of a 2*p*, 3*p*, 4*p*, and 5*p* halogen to carbon bond are, respectively, C–F, 116 kcal/mol (485 kJ/mol); C–Cl, 81 kcal/mol (339 kJ/mol); C–Br, 68 kcal/mol (285 kJ/mol); and C–I, 51 kcal/mol (213 kJ/mol). Bonds between atoms that both contain lone pairs of electrons are weak because of lone pair–lone pair repulsion. Examples are the halogens: F_2 at 37 kcal/mol (155 kJ/mol), Cl_2 at 58 kcal/mol (243 kJ/mol), Br_2 at 46 kcal/mol (192 kJ/mol), and I_2 at 36 kcal/mol (151 kJ/mol).

Figure 2.1 The larger six-membered ring on the left does not distort the sigma bond; the three-membered ring on the right does. Bent sigma bonds are weaker and more reactive.

2.2.2 Pi Bonding

The formation of pi bonds requires approximate coplanarity of the reacting orbitals. The average bond strength of a double bond is 146 kcal/mol (611 kJ/mol); after subtraction of the sigma bond strength, the pi bond is worth about 63 kcal/mol (264 kJ/mol). For best overlap, the two *p* orbitals of the pi bond must lie in the same plane. The overlap of a pi bond is greatly diminished if one of the ends is twisted so that the two *p* orbitals are no longer coplanar. The orbital alignment requirements have some slack; an error of 10° off the proper angle appears to have little effect. Pi overlap falls off with the cosine of the twist angle. Pi-bond-forming reactions that involve deformations from coplanarity of up to 30° can occur in rare cases, but such compounds are reactive and difficult to make. The greater the amount of this twist, the less stable the pi bond. The extreme, a 90° twist, would have no pi bonding at all because perpendicular orbitals do not interact, with a loss of about 63 kcal/mol (264 kJ/mol) bonding energy stabilization.

A *trans* double bond in a ring is twisted as the ring size gets smaller (Figure 2.2). The smallest ring containing a *trans* double bond and is still stable enough to be isolated at room temperature is *trans*-cyclooctene, which is about 11 kcal/mol (46 kJ/mol) more strained than *cis*-cyclooctene. Cyclobutene, with an untwisted *cis* double bond, is stable well above room temperature.

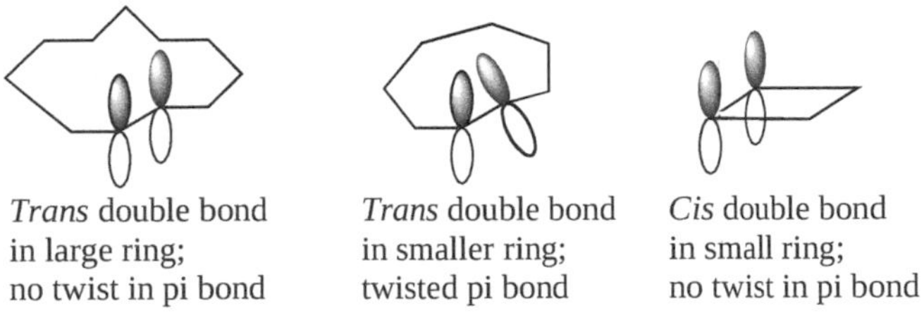

Figure 2.2 The distortion of a pi bond with varying ring size.

Pi bonds between orbitals of different sizes are very weak because of poor overlap. The 2*p*–2*p* pi bond of an untwisted carbon–carbon double bond has good overlap (Figure 2.3, left) because the orbitals are of the same size, but the mismatched sizes of the 2*p*–3*p* carbon–sulfur double bond does not (Figure 2.3, middle). Another weak pi bond because of poor overlap is the 2*p*–3*d* carbon–phosphorus pi bond (Figure 2.3, right).

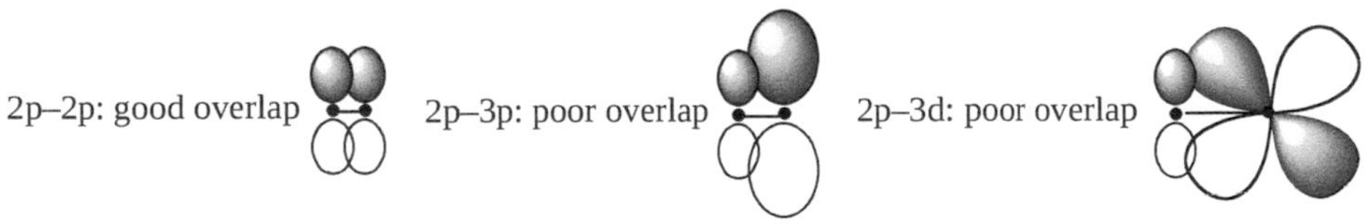

Figure 2.3 The better pi overlap in a 2*p*–2*p* pi bond vs. a 2*p*–3*p* pi bond or a 2*p*–3*d* pi bond.

Electron flow must occur through overlapping orbitals. Deformed sigma and pi bonds are less stable, and so the paths forming them are necessarily of higher energy; this is important since we will be attempting to predict the lowest-energy path on the energy surface. If the orbital overlap is poor, the path is less likely to occur. Good orbital overlap must occur in the product (thus strong bonds) and also all along the path of electron flow to the product. Reactions that occur in two or more steps have the opportunity between steps to rotate groups into orbital alignment for the next step. Reactions that occur in one step (concerted reactions) do not have this option; their pathways may be of higher energy because of rigid orientation requirements in the transition state. It is relatively easy to forget the three-dimensional nature of organic compounds when they are written flat on a page. Whenever you have the slightest suspicion that the orbital alignment may be poor, build the molecular model.

Stereoelectronic effects occur when the position of an orbital affects the course of a reaction. These orbital-position effects are a direct consequence of the need for the best orbital overlap in forming a new bond. Enzymes use the orbital alignment requirements of a reaction to achieve selectivity. Depending on the enzyme, the reactant will be held in such a way that the appropriate bonds are rotated into the alignment required for the desired reaction. In the following example, the aromatic ring conjugated to the C=N pi bond forms the pi system of the electron sink. The N–C single bond can rotate to bring either the C–R, C–H, or C–COO$^-$ bond into alignment with the *p* orbitals of the pi system of the electron sink. One enzyme holds the molecule in the active site such that only the C–COO⁻ bond, indicated in bold, aligns with the pi system of the electron sink and is selectively broken. With minor deviations, the entire molecule lies in the plane of the paper except the bond in bold to the carboxylate anion, which comes up out of the plane. In this conformation, only the bond in bold is properly aligned to form the N=C pi bond in the product.

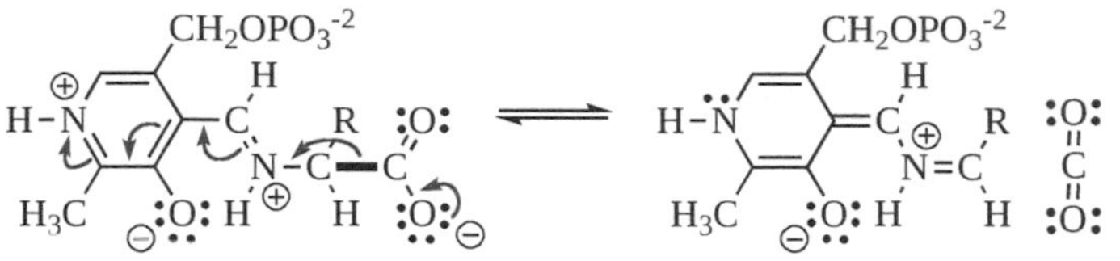

2.2.3 Microscopic Reversibility

The most basic rule of energetics is **the principle of microscopic reversibility: The forward and reverse reactions follow the same lowest-energy route but in opposite directions. There is only one energetically best pathway.** If A to B is the lowest-energy path forward, then B to C to A cannot be the lowest-energy path back. Remember to look in both directions when deciding on the lowest-energy path. The lowest-energy path for addition of a nucleophile is reversed to get the best path for kicking it back out.

Often the orbital alignment in one direction is clearer than in the opposite direction, so microscopic reversibility can help. For example, when a nucleophile attacks a three-*p* orbital (allylic) system such as an ester on the central carbon atom, it must approach the central *p* orbital along its axis, as shown in Figure 2.4.

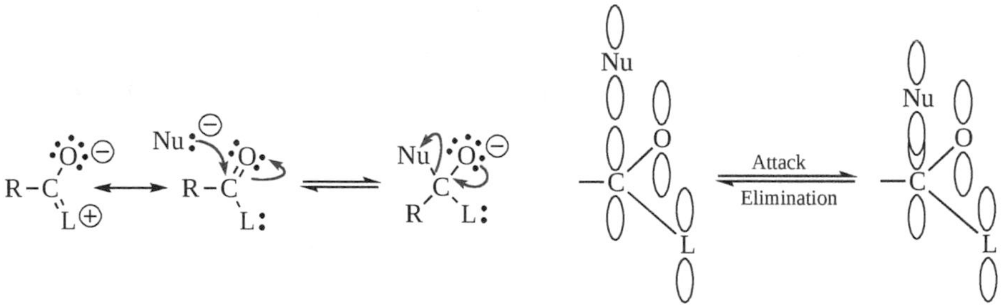

Figure 2.4 The illustration of microscopic reversibility for nucleophilic attack.

The remaining *p* orbitals in the product are parallel to the newly formed bond. It is less clear in the reverse reaction, which follows the same path but in the opposite direction, that the lone pair orbitals of O and L must be aligned parallel to each other and to the breaking bond at the transition state. The orbitals are then lined up so that they can easily become the allylic pi system of the ester. If one of the lone pairs were not lined up, the allylic system could not be established at the transition state, and that transition state would be much higher in energy. Allylic stabilization for radicals (Chapter 13) is worth 12.3 kcal/mol (51.4 kJ/mol).

Another example is the attack by an electrophile on the carboxylate anion, an allylic system bearing more than one lone pair (Figure 2.5). Since the allylic system stabilizes both starting material and product, the lowest-energy route would maintain that allylic system throughout the reaction path. In general, **the transition state of lowest energy maximizes the extent of bonding.** The loss of the electrophile (a proton, for example) would create a new lone pair in the plane of the carboxylate, perpendicular to the allylic pi system. The reverse reaction must be electrophilic attack on a lone pair that is not part of the allylic pi system. Because of the principle of microscopic reversibility, we can use either the forward or reverse reaction to decide the best route.

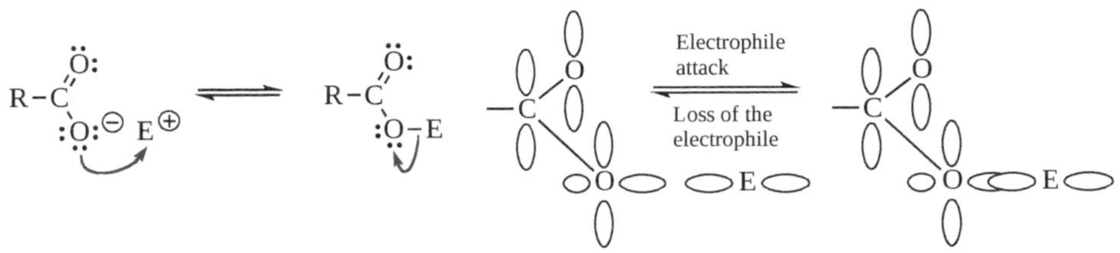

Figure 2.5 The illustration of microscopic reversibility for electrophilic attack.

2.3 THE ISOMER FAMILY TREE

Figure 2.6 shows the relationships between the different types of isomers. *Conformational isomers* just differ by single bond rotations, and therefore are interconvertible. *Stereoisomers* are isomers that have the same sequence of bonds and differ only in the arrangement of atoms in space. Mirror image pairs that are not superimposable are called *Enantiomers* (if the pair superimposes, then the two are the same). Enantiomers are discussed in more detail below. Stereoisomers that are not enantiomers are called *diastereomers*. For example, *cis* and *trans* double bond isomers are diastereomers. Diastereomers often have different physical and chemical properties.

Isomers
(Same molecular formula, but different structures or configuration in space)

Constitutional isomers
(different bonding order)

Same bonding order,
different arrangement in space

Stereoisomers
cannot be made equivalent by
rotations about single bonds

Conformational isomers
can be made equivalent by
rotations about single bonds

Enantiomers
(mirror images)

Diastereomers
(all others)

Figure 2.6 The isomer family tree.

A very important implication of the three-dimensional nature of organic compounds is that they can be *chiral*, occurring in left- and right-handed forms called enantiomers. A compound is chiral if it does not superimpose on its mirror image (Figure 2.7). A carbon atom with four different groups is called a chiral center.

Mirror–image pair

Flip left one over and try to superimpose

Figure 2.7 A chiral compound will not superimpose on its mirror image.

A molecule can be chiral without having a chiral center if it has a chiral shape such as a propeller or helix. A compound can contain chiral centers and not be chiral if there is an internal mirror plane. Whenever you are uncertain, build the model of the compound and try to superimpose it on a model of its mirror image. See the Appendix for how to name chiral compounds. Enantiomers have identical physical and chemical properties; they can be distinguished only in a chiral environment, such as in the active site of an enzyme. A 1:1 mix of two enantiomers is called a *racemic mixture* (symbolized by ±). Originally, most pharmaceuticals were synthesized as the racemic mixture because chiral synthesis was a real challenge. However, the enantiomers of many pharmaceuticals not only are often ineffective but also add to the toxicity, so great effort is now made to synthesize the effective enantiomer without its "evil twin."

Example problem

What is the relationship between each of the following pairs of compounds?

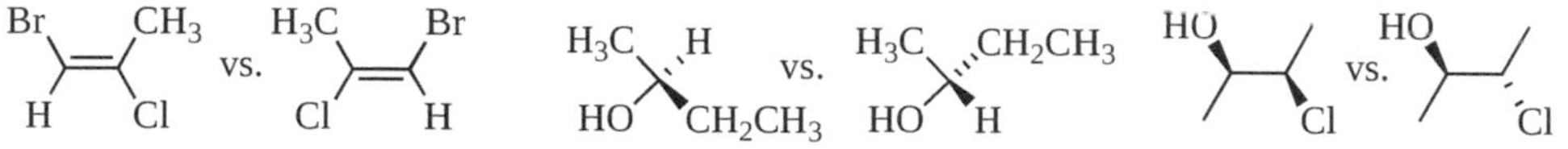

A B C D E F

Answer: Compounds A and B are the same (flip B over onto A; they are both flat). Compounds C and D are enantiomers (mirror image pairs; move C in front of D as if viewing C in a mirror). Compounds E and F are diastereomers (they are not mirror images, but have different orientations in space).

2.4 POLARIZABILITY AND HARD AND SOFT ACID–BASE THEORY

We have seen that strong covalent bonds are made by good overlap of two orbitals that are close in energy. Strong ionic bonds, on the other hand, are made when two oppositely charged ions get close together. Since favorable reactions tend to form the strongest bonds, either ionic or covalent, we can predict some reactions by checking whether the reactants or products have the stronger bonds. A good ionic bond partner has a small radius and a high charge and holds its electrons tightly and is said to be **hard**. A good covalent bond partner is the opposite, **soft**, able to share its valence electrons with valence orbitals that can be distorted toward the partner.

Hardness or *softness* is a property of a Lewis acid or base that is independent of its strength. An important aspect of electron availability is **polarizability, the ease of distortion of the valence electron shell of an atom by an adjacent charge**. For example, the valence electrons of iodide are shielded from the nuclear charge by all of the core electrons and thus are capable of being distorted toward a partially positive reactive site much more easily than the valence electrons of fluoride. **Polarizability increases going down a column in the periodic table.** For our purposes, it is more important to be able to compare the relative hardness of acids and bases; several trends are helpful.

Soft base—It is often a large, neutral species, or one with a diffuse charge of a weakly electronegative element with **high polarizability** that is easy to oxidize. **More resonance forms of an ion indicate a more diffuse charge.** The less charged or more diffuse the charge (delocalization softens), the softer the base is. The more polarizable the base, the softer it is:

$$\text{harder } F^- > Cl^- > Br^- > I^- \text{ softer}$$

Hard base—It is most often a small negatively charged ion of a strongly electronegative element with **low polarizability** that is difficult to oxidize. More electronegative elements hold their valence electrons tighter and are harder. Electronegativity increases going to the right on a row of the periodic table, so bases get harder going to the right on a row:

$$\text{softer } H_3C^- > H_2N^- > HO^- > F^- \text{ harder}$$

Hard acid—It is most often a small positively charged ion of an element with **low polarizability**. The more charge it has, the harder the acid is:

harder $Al^{3+} > Mg^{2+} > Na^{+}$ softer (but still reasonably hard)

Soft acid—It is often a large, neutral, or diffusely charged species or element with **high polarizability**. Again, the hard-to-soft trends for acids are more important than absolute values. The less charged or more diffuse the charge, the softer the acid is. The more polarizable acid is softer:

harder $Mg^{2+} > Cu^{2+} > Cd^{2+} > Hg^{2+}$ softer

The HSAB principle: Hard bases favor binding with hard acids; soft bases favor binding with soft acids.

The soft–soft interaction is a covalent bond, which is favored by good overlap of orbitals that are relatively close in energy. The hard–hard interaction is an ionic bond, which is favored by highly charged species that have small radii so that they can get close together to form a strong ionic bond. The soft–hard interaction is relatively weak. The HSAB principle simply expresses the tendency of reactions to form strong bonds, either ionic or covalent. **When a pair of molecules collides, two attractive forces lead to reaction: the hard–hard attraction (opposite charges attracting each other), and the soft–soft attraction (the interaction of filled orbitals with empty orbitals).**

An organometallic can react with a metal salt to produce a new organometallic and a new salt in the process called *transmetallation*. As expected from the HSAB principle, this equilibrium favors the formation of the more covalent organometallic from the softer pair and the more ionic salt from the harder pair. Transmetallation allows the conversion of a reactive organometallic into a more covalent, less reactive organometallic:

$$2\ RMgCl + CdCl_2 \rightarrow R_2Cd + 2\ MgCl_2$$

R soft–Mg hard + Cd soft–Cl hard → R soft–Cd soft + Mg hard–Cl hard

2.5 THERMODYNAMICS, POSITION OF EQUILIBRIUM

This text will use energy surfaces as problem space maps, and also will use the related energy diagrams to explain why a particular reaction may be favorable and how factors influence that favorability. We need to understand why and if a process is energetically downhill to predict whether it has a chance of happening.

2.5.1 Energy Surfaces Give an Overview

An energy diagram is a plot of energy vs. the reaction coordinate. The reaction coordinate is a measure of the degree of a molecule's progress toward complete reaction. The energy diagram is actually a slice along the lowest-energy path on an energy surface. Although the actual energy surface for all but the simplest of reactions is very complex, we can learn much from a simple three-dimensional surface. The vertical axis is energy, and the two horizontal axes will each be the distance between atoms undergoing bond breaking or bond making.

For example, in a simple reaction, $Nu:^{-} + Y{-}L \rightarrow Nu{-}Y + L:^{-}$, a nucleophile collides with the Y end of Y–L and knocks off the leaving group L. One horizontal axis will be the distance between Nu and Y; the other horizontal axis will be the distance between Y and L. As the reaction progresses, the distance between Nu and Y decreases while that between Y and L increases. The reaction would go through a point of highest energy, a *transition state*, symbolized by ‡, in which Y is partially bonded to both Nu and L. When the reaction is over, the leaving group (or nucleofuge), L, has departed with its bonding pair of electrons.

For simplicity, we start with two dimensions by looking at the surface from above. The reaction's simple energy surface as viewed from the top is shown in Figure 2.8. The bold red line from reactants, R, to products, P, is the path of lowest energy that the reacting species might follow. As reactants proceed along this diagonal lowest-energy path, the Y–L bond begins to break (movement on the surface to the right stretches the Y–L bond), and the Nu–Y bond begins to form (downward movement on the surface brings Nu and Y closer together). The midpoint of the diagonal corresponds to a point where the Y–L bond is half-broken and the Nu–Y bond is half-formed.

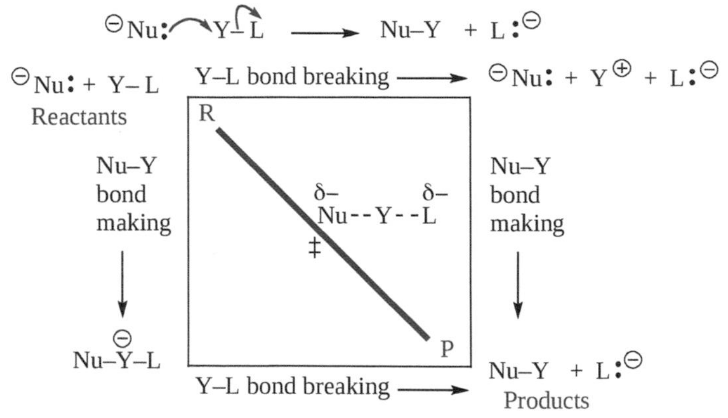

Figure 2.8 A simplified energy surface viewed from the top.

The third dimension, energy, becomes apparent in Figure 2.9 as we view the surface from off to the side. The lowest-energy path, again shown by the red line, starts back at R and goes up through the "mountain pass," the transition state, and then down to point P in the front corner of the surface (the path is dashed when it is behind the surface). It is downhill to the products from the transition state.

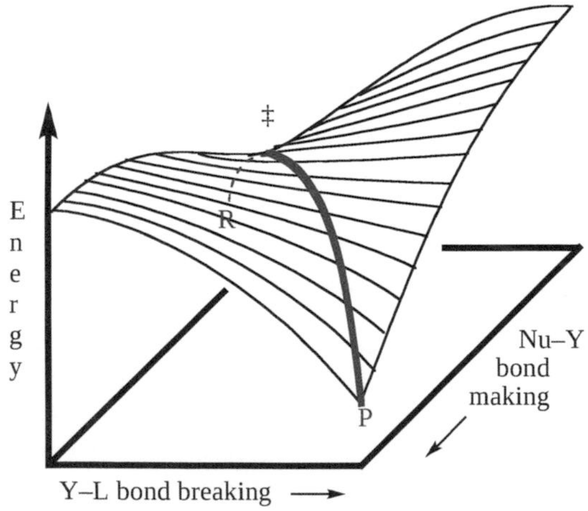

Figure 2.9 A simplified energy surface viewed from the side.

The slice through the surface from point R to P along the lowest-energy path is the energy diagram (Figure 2.10). For this particular simple surface, the reaction coordinate was the diagonal in the top view of the surface. Energy diagrams are easier to visualize than the lowest-energy path displayed on a more complex surface; there the path may weave around, fall into, and climb out of the energy minima of several intermediate compounds along the way to product.

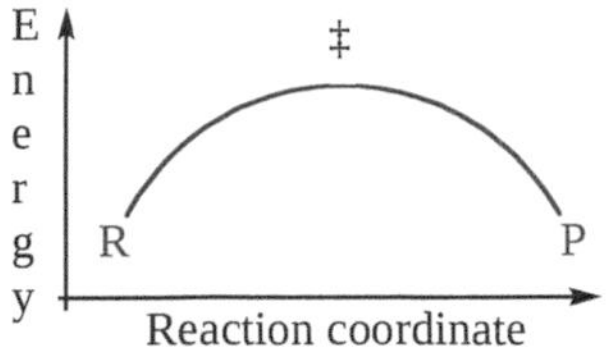

Figure 2.10 The lowest-energy slice of the simplified energy surface in Figure 2.9.

2.5.2 The Equilibrium Constant is Determined by ΔG°

The example in Figure 2.10 with the slight modification that P is more stable than R would produce an energy diagram as shown in Figure 2.11. Energy is expressed as *G*, free energy, the energy free to do work. The standard free-energy difference between the reactants and the products is $\Delta G°$ and is related to the equilibrium constant, K_{eq}, by the formulas given below where $R = 1.987 \times 10^{-3}$ kcal/mol-K (8.314×10^{-3} kJ/mol-K) and T = temperature in K.

$$\boldsymbol{\Delta G° = -RT \ln K_{eq}} = -2.303\ RT \log K_{eq}$$

or $$\log K_{eq} = -\Delta G°/2.303RT$$

or $$K_{eq} = 10^{-\Delta G°/2.303RT}$$

At a room temperature of 25°C, $T = 298$ K, so $2.303RT = 1.364$ kcal/mol (5.706 kJ/mol). If $\Delta G°$ is in kcal/mol, $K_{eq} = 10^{-\Delta G°/1.364}$ or, conversely, $\Delta G° = -1.364\log K_{eq}$. **At room temperature, every 1.364 kcal/mol (5.706 kJ/mol) change in $\Delta G°$ changes the equilibrium constant K_{eq} by a factor of 10.** Thus, we have a quick way to interconvert $\Delta G°$ in kcal/mol to K_{eq} at room temperature: Divide $\Delta G°$ by −1.364 to obtain the exponent of K_{eq}; conversely, multiply the exponent of K_{eq} by −1.364 to get $\Delta G°$ in kcal/mol (use −5.706 For kJ/mol).

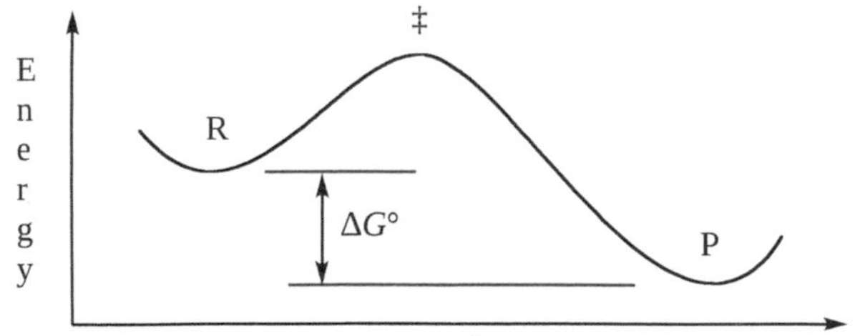

Figure 2.11 An energy diagram in which the products are more stable than the reactants.

As the standard free-energy difference, $\Delta G°$, gets more negative, K_{eq} gets larger, and the more the lower-energy compound P will predominate in the equilibrium mixture. If $\Delta G°$ is a positive value, then the product P is uphill in energy from the reactants. Table 2.1 gives representative values for 25.0°C, with P becoming more stable than R.

Example problem

The K_{eq} of a proton transfer reaction at 25.0°C is 10,000. What is $\Delta G°$ in kcal/mol?

Answer: We can use the formula $\Delta G° = -1.364\log K_{eq}$. Since a K_{eq} of 10,000 = 10^{+4}; then $\log K_{eq}$ is 4. This gives $\Delta G° = -1.364(4) = -5.45$ kcal/mol (-22.8 kJ/mol). A K_{eq} greater than 1 favors products, and so does a negative $\Delta G°$; therefore the negative sign cross-checks.

Table 2.1 $\Delta G°$ and K_{eq} Values for 25.0°C (Room Temperature)

$\Delta G°$ kcal/mol	K_{eq}	Reactant	Product	$\Delta G°$ (kJ/mol)
+5.45	10^{-4}	99.99	0.01	+22.8
+4.09	10^{-3}	99.9	0.1	+17.1
+2.72	10^{-2}	99	1	+11.4
+1.37	0.10	91	9	+5.73
+1.03	0.18	85	15	+4.30
+0.50	0.43	70	30	+2.10
0.00	1	50	50	0.00
−0.50	2.3	30	70	−2.10
−1.03	5.7	15	85	−4.30
−1.37	10	9	91	−5.73
−2.72	10^{2}	1	99	−11.4
−4.09	10^{3}	0.1	99.9	−17.1
−5.45	10^{4}	0.01	99.99	−22.8
−9.55	10^{7}	Essentially complete		−39.9

Exercise: Cover the value of K_{eq} in Table 2.1 and give the approximate K_{eq} from the $\Delta G°$; then cover the $\Delta G°$ and give the approximate $\Delta G°$ from the K_{eq}.

A reaction can have a positive $\Delta G°$ and still proceed. For example, if the reaction had an unfavorable equilibrium constant with $\Delta G°$ as +2.72 kcal/mol (+11.4 kJ/mol), the equilibrium mixture would contain a 99:1 R to P mixture. If we started the reaction with 100 mol of pure R, the reaction would be spontaneous and produce 1 mol of P to give the equilibrium mixture. By Le Châtelier's principle, a system at equilibrium will respond to any change by shifting to restore the equilibrium mixture. If a subsequent reaction removed the small amount of P that was formed, then more P would be produced to restore the equilibrium mixture. That following reaction would act as a *driving force* to overcome an unfavorable R to P equilibrium.

In order to understand what contributes to the standard free energy change of a reaction, it can be separated into its components:

$$\Delta G° = \Delta H° - T\Delta S°$$

The standard *enthalpy* change**, $\Delta H°$, is a measure of the heat absorbed or evolved in a reaction** and bears a negative sign if heat is given off (exothermic reactions). All exothermic reactions break weak bonds and make strong ones. Conversely, if heat is absorbed, the reaction is endothermic and $\Delta H°$ is positive. A bond-strength table is included in the Appendix. The heat of reaction can be calculated from the difference in the heats of formation of the reactants and products, but often those heats of formation are not known. A method to get an approximate heat of reaction is to consider just the bonds being formed and broken. Since bond breaking requires heat and bond making releases it, the heat of reaction can be approximated as:

$$\Delta H° = \Delta H_{\text{(bonds broken)}} - \Delta H_{\text{(bonds made)}}$$

Another use for the bond-strength table is to predict which of two products has most likely a lower heat of formation and is the preferred product of the reaction. The product that overall has the strongest bonds is preferred. Only the bonds that differ in each structure need be considered. We can tell which of two possible products is more stable by just finding out which way is exothermic for product A equilibrating with product B.

The standard *entropy* change, **$\Delta S°$, is a measure of the change in the randomness within the system.** If there is more disorder after the reaction, for example, one molecule breaking in two (with no complicating factors such as solvation), $\Delta S°$ is positive. The entropy term arises from a combination of translational, rotational, and vibrational motions. The entropy loss of two molecules combining into one at 25°C is estimated to be -30 entropy units (eu) or –30 cal/mol-K (–126 J/mol-K). Note the smaller unit, cal (or J) not kcal (or kJ). In many organic reactions the $-T\Delta S$ term is small and does not dominate over the $\Delta H°$ term, so if the reaction is exothermic ($\Delta H°$ is negative) then it is probably exergonic ($\Delta G°$ is negative) also. Thus, the heat of reaction serves as a guide for predicting the energetics of the process.

Example problem

Calculate $\Delta H°$ for $H_2C{=}CH_2$ + HBr → $H_3C{-}CH_2Br$

Answer: Bonds broken = C=C 146 kcal/mol (611 kJ/mol) + H–Br 88 kcal/mol (368 kJ/mol) = 234 kcal/mol (979 kJ/mol). Bonds made = C–C 83 kcal/mol (347 kJ/mol) + C–H 99 kcal/mol (414 kJ/mol) + C–Br 68 kcal/mol (285 kJ/mol) = 250 kcal/mol (1046 kJ/mol). Since $\Delta H°$ = bonds broken − bonds made = 234 − 250 = −16 kcal/mol (−67 kJ/mol). Because the sign of $\Delta H°$ is negative, the reaction is exothermic; it gives off heat. And even though it has joined two molecules into one, costing disorder, the overall reaction is still energetically favorable. It is important to notice that we broke the C=C double bond, at 146 kcal/mol (611 kJ/mol), and replaced it by making a C–C single bond, at 83 kcal/mol (347 kJ/mol).

2.6 KINETICS, RATE OF REACTION

We now know how to tell whether a reaction is energetically favorable, but a reaction may be downhill in energy and still not proceed at a reasonable rate.

2.6.1 Barrier Height, ΔG‡, Determines Reaction Rate

The rate of reaction is controlled by the barrier height to reaction, $\Delta G^‡$, the free energy of activation. No reaction occurs if the reactants do not have enough energy to overcome the reaction barrier. If sufficient energy is added, then the reaction proceeds. Paper can sit in air without burning for quite a long time at room temperature, but if the temperature is raised high enough it will burst into flames. In Figure 2.12, note that the $\Delta G^‡$ values for the forward and reverse reactions are different. Since B starts out at a lower energy it must climb more of an energy hill than A to get to the same transition state, denoted by the symbol ‡.

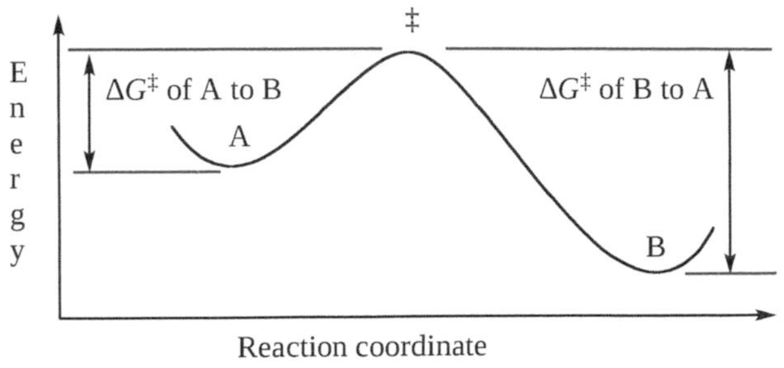

Figure 2.12 The free energy of activation, $\Delta G^‡$ for the forward and reverse reactions.

If the barrier for A forward to B is small, and for B reversing to A is very large, then the reaction is an irreversible *unimolecular* process, A → B, and the rate of B formation is denoted by k_f[A] where k_f is the forward rate constant times [A] the molar concentration of compound A. The left side of Figure 2.13 shows this irreversible unimolecular reaction and illustrates the reaction half-life. The half-life of a reaction is the amount of time it will take for half of the reactant to react. After two half-lives just 1/4 of the reactant is left, and after three half-lives just 1/8 of it is left. A reaction is 97% complete after five half-lives.

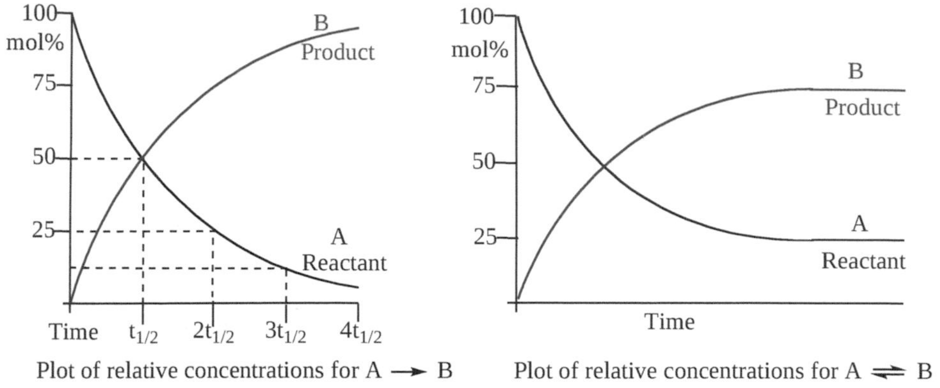

Figure 2.13 The relative concentrations of A and B vs. time for an irreversible reaction (left) and a reversible reaction (right) which reaches equilibrium when the forward rate equals the reverse rate.

An example reversible reaction is shown on the right side of Figure 2.20. If the reaction is reversible, the overall rate is the reverse rate, k_r[B], subtracted from the forward rate, k_f[A], or just $k_f[A] - k_r[B]$. The initial rate (when [B] = 0) is just k_f[A], but as the concentration of B grows, so does the extent of reverse reaction. At equilibrium, the A and B concentrations do not change, and the overall rate is zero, making $k_f[A] = k_r[B]$, which rearranges to $[B]/[A] = k_f/k_r = K_{eq}$. In Figure 2.20, the K_{eq} is 0.75/0.25 = 3.

The rate of an irreversible *bimolecular* (two molecule) reaction in which A collides with B to produce C is expressed as a rate constant, *k*, times the two reactant concentrations.

$$A + B \rightarrow C \qquad \text{Rate} = k[A][B]$$

Concentrations for reactions vary, but the average organic reaction is run at 1 M or less. Reactive intermediates often have large rate constants, which can more than make up for their low concentrations. The largest value that the bimolecular rate constant can have is 10^{10} L/mol-s, which corresponds to a reaction upon every collision. This limit, the rate at which the two molecules collide, is called the *diffusion-controlled limit.*

Using the formula in the supplementary section that follows, we can calculate the room temperature unimolecular rate constant *k* for various values of $\Delta G^‡$, shown in Table 2.2. A unimolecular reaction with a $\Delta G^‡$ of 15 kcal/mol (63 kJ/mol) would be 97% complete in 0.05 seconds. If the barrier were 25 kcal/mol (105 kJ/mol), then 97% completion would take almost 2 weeks. The number of molecules in the reaction mix at room temperature that have enough energy to traverse a 20 kcal/mol (84 kJ/mol) barrier is small (as in Figure 2.21); the average energy of a molecule at room temperature is only 1 kcal/mol (4.2 kJ/mol). However, if the reaction barrier is lower, many more molecules will have sufficient energy to react.

At room temperature, dropping the $\Delta G^‡$ by 1.364 kcal/mol (5.706 kJ/mol) increases the rate tenfold. A common landmark is that **the $\Delta G^‡$ of a reaction that proceeds at a reasonable rate at room temperature is about 20 kcal/mol (84 kJ/mol).** Our landmark value for a reasonable reaction rate at room temperature is highly dependent on $\Delta G^‡$, the molecularity of the reaction, the reactant concentrations, and on how long the experimenter considers a reasonable time to wait. This value is still useful as a reference point. An approximate general rule for the temperature dependence of a reaction is that **the rate doubles for every 10°C increase in temperature**.

Table 2.2 $\Delta G^{\ddagger}$, Rate Constants, Half-Lives, and Completion Times of a Unimolecular Reaction, A → B, at Room Temperature, 298 K (25°C)

$\Delta G^{\ddagger}$ in kcal/mol	k_{298} in s^{-1}	Half-Life	97% Complete
15 (63 kJ/mol)	64.1	0.01 sec	0.05 sec
16 (67 kJ/mol)	11.9	0.06 sec	0.3 sec
17 (71 kJ/mol)	2.2	0.3 sec	1.6 sec
18 (75 kJ/mol)	4.1×10^{-1}	1.7 sec	8.5 sec
19 (79 kJ/mol)	7.5×10^{-2}	9.2 sec	45.9 sec
20 (84 kJ/mol)	1.4×10^{-2}	49.6 sec	4.1 min
21 (88 kJ/mol)	2.6×10^{-3}	4.5 min	22.3 min
22 (92 kJ/mol)	4.8×10^{-4}	24 min	2 hr
23 (96 kJ/mol)	8.9×10^{-5}	130 min	10.8 hr
24 (100 kJ/mol)	1.6×10^{-5}	11.7 hr	2.4 days
25 (105 kJ/mol)	3.0×10^{-6}	63.2 hr	13.2 days

If the reaction goes by a stepwise mechanism, it may have an energy diagram similar to that shown in Figure 2.14, in which intermediates lie in minima, energy valleys, and transition states are always at maxima, the top of the energy hills. The highest energy hill, $\ddagger_2$, is a major factor in determining the rate of reaction along with the concentration of intermediate C, which may not be known. The *lifetime* of intermediate C is defined as the reciprocal of its rate constant for unimolecular decomposition. As the height of the lowest barrier to decomposition of C drops, the lifetime of C gets shorter. When the lifetime of the intermediate gets less than 10^{-12} sec. (only long enough for several molecular vibrations), then A + B → D is considered concerted, occurring in one step. Still, the various bond-forming and bond-breaking processes need not be exactly synchronous. If the first step in a multistep process is the slowest, this "bottleneck" step will be the *rate-determining step* and not any of the faster steps that follow in the overall process. For example, if the first transition state, $\ddagger_1$, in Figure 2.14 were the highest peak then converting A to C would be the rate determining step, not the decomposition of C to D.

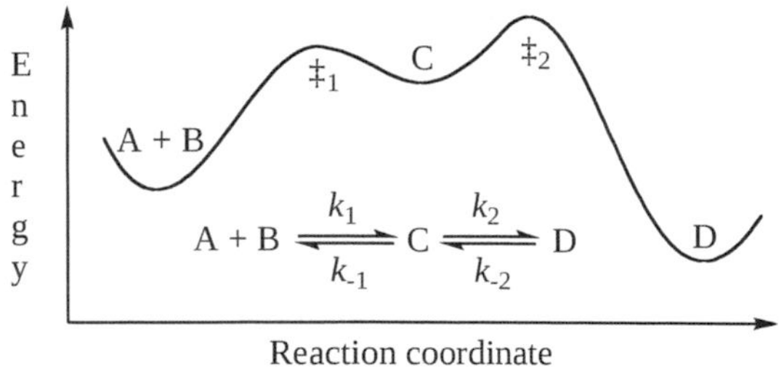

Figure 2.14 The energy diagram for a reaction that proceeds through an intermediate, C. Over the equilibrium arrows are the rate constants for each forward and reverse step.

A transition state contains partially broken and formed bonds, but the degree of this varies and depends on how close to the starting materials or products the point of transition is. The *Hammond postulate*, illustrated by Figure 2.15, states that the transition states of exothermic reactions resemble starting materials in energy and geometry whereas transition states of endothermic reactions resemble products. Transition states close to starting materials, A → B, have rather little bond breaking or making. Transition states close to products, C → D, have almost completed bond breaking and making.

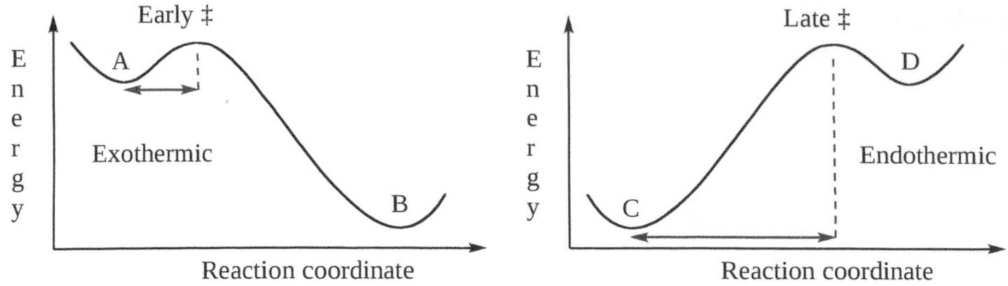

Figure 2.15 The transition state shifts position as the overall energy of the process changes.

Exercise

Hold a strip of paper in your hand so that both ends are at the same level and bow it up as in Figure 2.10. Notice that the high point, our transition state, is in the exact center. Now lower the end of the strip held in your right hand and notice that the transition point moves toward your left hand. Now raise your right hand higher than your left hand and notice how the transition state shifts and compares with that shown in Figure 2.15.

As shown in Figure 2.16, the stability of the product has a great influence on the energy of the transition states of endothermic reactions (D → E and F) but has very little influence on the energy of the transition states of exothermic reactions (A → B and C).

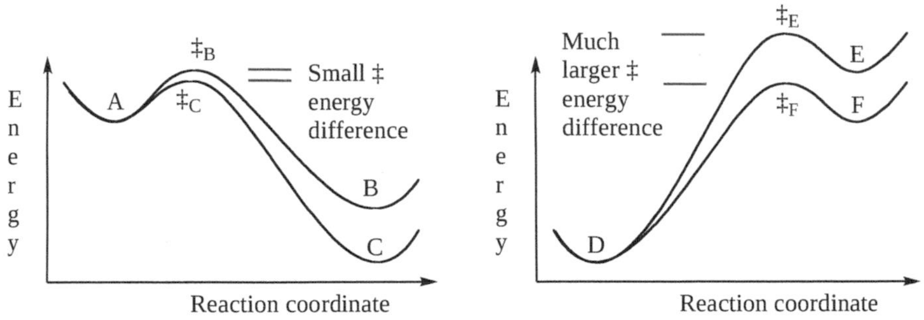

Figure 2.16 The degree to which the product stability influences the transition state.

Therefore, with few exceptions, the more exothermic a reaction is, the less the product stability affects the energy of the transition state and the less selective the reaction is. **The more reactive a species is, the less selective it is.** A general rule of reactivity is, **the more stable a compound is, the less reactive it is**. The converse is also true: The less stable a compound is, the more reactive it is. Basically, a highly reactive, unstable compound might react with almost anything, whereas a less reactive, more stable compound is more limited in what it can react with. It would react at the best site for reaction in preference to other, less favorable sites.

Some reactions can produce two different products, and it is necessary to determine whether the reaction is reversible to be able to predict the product. In Figure 2.17, both reactions have the same $\Delta\Delta G^{\ddagger}$ and $\Delta\Delta G°$, but the reaction on the right is more exothermic and therefore not reversible. If the reaction is completely reversible (Figure 2.17 left), equilibrium is established in which the major species in the reaction pot is G, the most stable, lowest-energy species. Such a reversible reaction that produces the more stable product is under *thermodynamic control*. The equilibrium product ratio is determined by the difference in free energy of the two products, $\Delta\Delta G°$. If the reaction is not reversible (Figure 2.17 right), then the product that is formed faster, F, will be the major product. When the relative rates of reaction (barrier heights), $\Delta\Delta G^‡$, determine the product ratio, the reaction is under *kinetic control*. As the temperature drops and the molecules have less energy, the reaction takes the lower-barrier pathway.

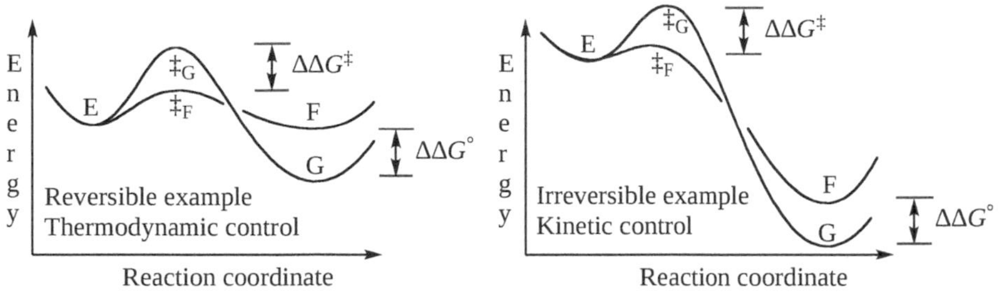

Figure 2.17 The energy diagrams for two competing reactions with the same $\Delta\Delta G^‡$ and $\Delta\Delta G°$ but with different exothermicity. In both cases, reaction E to F has the lower barrier to reaction, whereas reaction E to G produces the more stable product. As the reaction becomes more exothermic on the right, it becomes less reversible as the barriers for products to return to reactants increase.

When the relative rates of reaction control the product ratio, how much does a small difference in $\Delta G^‡$ make? At room temperature, a 1.364 kcal/mol (5.706 kJ/mol) difference in barrier heights, $\Delta\Delta G^‡$, means that F is formed about ten times as fast as G. This would mean that the product mixture would contain ten times as much F, the product of traversing the lower barrier, as it would G. In Figure 2.17, if the barrier height to form G were less than that for F, then product G would be both the kinetic **and** thermodynamic product.

2.6.2 The Basics of Barriers

What causes a reaction barrier? Just like $\Delta G°$, the $\Delta G^‡$ can be broken into two terms:

$$\Delta G^‡ = \Delta H^‡ - T\Delta S^‡$$

The $\Delta H^‡$ term arises because bond breaking must usually precede bond making to a slight degree. The old bond must be partially broken before the new one can begin to form. Since it costs energy to break a bond and we have not yet gained any energy from bond formation, extra energy in the form of $\Delta H^‡$ is required to start the reaction. Any reaction that requires the complete breakage of a bond before any bond formation can begin will have a larger $\Delta H^‡$ than a reaction whose bond breaking and making occur at almost the same time. The larger the $\Delta H^‡$ is, the larger the barrier is.

The $\Delta S^‡$ term is a measure of the degree of disorder in the transition state. If the transition state for the reaction requires that the reacting molecules be aligned relative to each other in a specific organization, then disorder is sacrificed and the value of $\Delta S^‡$ is a negative number. A negative $\Delta S^‡$ makes the $-T\Delta S^‡$ term a positive number, which adds to the $\Delta H^‡$ raising the barrier by increasing the value of $\Delta G^‡$. The reaction would proceed slowly because few collisions have the proper alignment. As the temperature increases and molecular motion gets more violent, the alignment problem increases.

There are many factors that will increase the barrier height. **As an intermediate increases in energy, the transition states leading up to it must also increase in energy.** Therefore, the stability of any intermediates will directly affect the barrier height. As shown in Figure 2.18, anything that tends to stabilize the intermediate, I, relative to the reactant, R, lowers the barrier, and conversely, anything that stabilizes the reactant relative to the intermediate raises the barrier. As an example, if the reactant has additional bonding that is lost in going to the transition state, the barrier is raised. If the transition state has additional bonding that is not present in the reactant, the barrier is lowered. We will spend the next few chapters discussing the stability of intermediates.

Several generalizations on barriers are useful: Formation of charge from neutral reactants tends to raise the barrier, whereas neutralization of charge lowers the barrier. If reactant groups have trouble achieving the proper alignment for reaction, the barrier is raised. If the site for reaction is not very accessible, the barrier is raised. If the intermediate and the transition state leading up to it are better stabilized by solvation than the reactants, the barrier is lowered by this solvent stabilization (see Section 2.7).

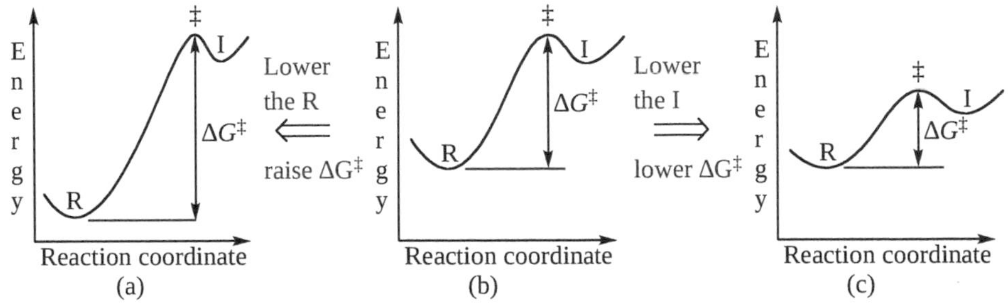

Figure 2.18 Use diagram *b* for reference. In *a* compared to *b*, the reactant R is stabilized relative to I; therefore the barrier increases. In *c* compared to *b*, I is stabilized relative to R; the barrier decreases.

2.6.3 Mathematical Perspective on Kinetics

(A Supplementary, More Advanced Explanation, Able To Be Skipped)

The *unimolecular* process, A → B, has the rate of B formation denoted by k[A] where k is the irreversible rate constant times [A] the molar concentration of compound A. The rate constant, k, is related to the barrier height, $\Delta G^\ddagger$, by the Eyring equation:

$$k = (\kappa T/h)e^{-\Delta G^\ddagger/RT}$$

where

κ = Boltzmann's constant = 1.38×10^{-23} J/K

T = temperature in K

h = Planck's constant = 6.63×10^{-34} J-s

e = 2.71828

R = gas constant = 1.99×10^{-3} kcal/mol-K (8.33×10^{-3} kJ/mol-K)

Using the above equation, we can calculate the rate constant k for various values of $\Delta G^\ddagger$, shown in Table 2.2. The half-life of a unimolecular reaction is equal to (ln 2)/k or 0.693/k, and a reaction is 97% complete after five half-lives. Table 2.2 uses the formula $\ln(c°/c) = kt$ to relate the rate constant, k, and the original concentration, $c°$, to the concentration, c, at time t seconds for a unimolecular reaction. Reactant concentrations are very important in a bimolecular reaction. If reactant B in a bimolecular reaction is in large excess, then its concentration will not change significantly over the course of the reaction and can be considered a constant, making the reaction *pseudo-first order*. We can then use the above formula to calculate the concentration of A at time t by substituting the pseudo-first-order rate constant, $k' = k$[B].

We can estimate the half-life and 97% completion times for a bimolecular reaction in which both reactants are at the same concentration by using the k from Table 2.2 and the formula $(1/c) - (1/c°) = kt$. For example, a reaction with a 20 kcal/mol (84 kJ/mol) barrier and both reactants at 1 M has a 97% completion time of 38.5 min at 25°C (298 K).

Example problem

If the concentration of both reactants of a bimolecular reaction were both at 0.1 M, how long would it take to be 97% complete at 25°C if $\Delta G^{\ddagger}$ is 20 kcal/mol (84 kJ/mol)?

Answer: 97% complete means 3% starting material remains, so c is 0.003 M if $c°$ was 0.1 M. Substitution of these values and a k of 0.014 from Table 2.2 into the formula $(1/c) - (1/c°) = kt$ gives the time as 23,095 sec or 6.4 hr.

Derivation of a rate expression using two assumptions

Any proposed reaction mechanism must fit the experimentally derived rate expression. The simplified rate expression for the reaction below, from Figure 2.14, is obtained if two assumptions are made: The second step is irreversible (k_{-2} is negligible), and the second step is slow compared to the first (k_2 is much smaller than k_1 or k_{-1}).

$$\mathrm{A} + \mathrm{B} \underset{k_{-1}}{\overset{k_1}{\rightleftharpoons}} \mathrm{C} \underset{k_{-2}}{\overset{k_2}{\rightleftharpoons}} \mathrm{D}$$

The reaction rate for the formation of D is k_2[C], where [C] is determined by the equilibrium constant, k_1/k_{-1} = [C]/[A][B]. Solving for [C] gives [C] = k_1[A][B]/k_{-1}. Substitution into the rate expression gives

$$\text{rate} = \frac{d[\mathrm{D}]}{dt} = k_2[\mathrm{C}] = \frac{k_2 k_1[\mathrm{A}][\mathrm{B}]}{k_{-1}}$$

Alternate derivation with only one approximation

The rate expression without any restrictions on the rate constants can be obtained by assuming that intermediate C does not build up, the *steady-state approximation*. The rate of formation of C, k_1[A][B] + k_{-2}[D], equals its rate of destruction, k_{-1}[C] + k_2[C]. Solving for [C] gives

$$[\mathrm{C}] = \frac{k_1[\mathrm{A}][\mathrm{B}] + k_{-2}[\mathrm{D}]}{k_{-1} + k_2}$$

Substitution for [C] into the reversible rate expression d[D]/dt = k_2[C] − k_{-2}[D] gives

$$\frac{d[\mathrm{D}]}{dt} = k_2[\mathrm{C}] - k_{-2}[\mathrm{D}] = \frac{k_2(k_1[\mathrm{A}][\mathrm{B}] + k_{-2}[\mathrm{D}])}{k_{-1} + k_2} - k_{-2}[\mathrm{D}]$$

This simplifies to a rate expression with a forward and a reverse term just like our simple equilibrium example.

$$\text{rate} = \frac{d[\mathrm{D}]}{dt} = \frac{k_2 k_1[\mathrm{A}][\mathrm{B}]}{k_{-1} + k_2} - \frac{k_{-1}k_{-2}[\mathrm{D}]}{k_{-1} + k_2}$$

If C → D is irreversible, k_{-2} = 0, the second term drops out, giving

$$\text{rate} = \frac{d[\mathrm{D}]}{dt} = \frac{k_2 k_1[\mathrm{A}][\mathrm{B}]}{k_{-1} + k_2}$$

If the second step is slow compared to the first, $k_{-1} >> k_2$, the expression becomes the same as our example with the two assumptions.

2.7 SOLVENT STABILIZATION OF IONS

Rates of reactions that produce a charged intermediate from a neutral reactant can be accelerated, as in Figure 2.18c, by running the reaction in a polar solvent. Since many nucleophiles, electrophiles, and reactive intermediates are charged, it is important to understand how charged species are stabilized. Nature abhors an isolated, localized charge.

There are two ways, intramolecular and intermolecular, that charges are spread out and thereby stabilized. Intramolecular stabilization is discussed in the next chapter (resonance and inductive effects). Intermolecular stabilization of charge is the essence of solvation.

Charge can be partially stabilized by polar solvent molecules. The clustering of the solvent dipoles of the opposite charge around the charged molecule becomes, in effect, an intermolecular sharing of the charge (Figure 2.19). Water can solvate cations and anions well; it has lone pairs with which to stabilize cations and polarized O–H bonds that can hydrogen bond to anions. A proton in water is commonly represented in organic mechanisms as the hydronium ion, H_3O^+, or sometimes as a proton bonded to four other water molecules as $H_9O_4^+$, but the aggregate is certainly much larger.

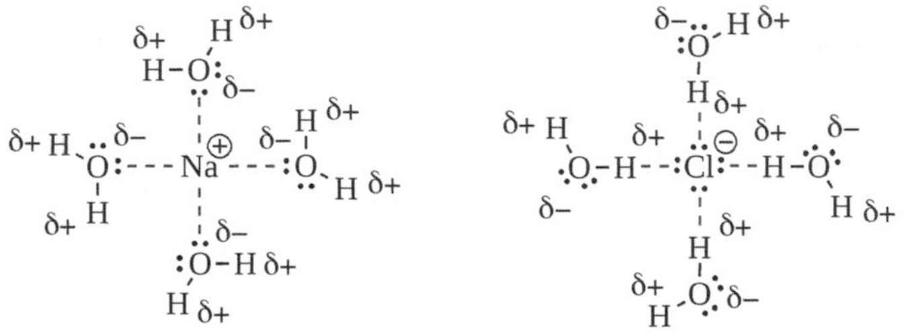

Figure 2.19 One example of solvation of the salt, sodium chloride, in water.

The arrangement of common solvents in the following list is from most to least polar. A polar solvent usually has a significant dipole moment (Section 1.8.2) from polarized bonds and lone pair dipoles. Solvents are also classified as either *protic* (solvents with acidic hydrogens) or *aprotic* (no acidic hydrogens). The most polar protic solvent, water, has two polarized O–H bonds and two lone pairs. Polar protic solvents can stabilize both cations and anions. Hexane, a nonpolar aprotic solvent, has no polarized bonds (C–H bonds are not polarized) and no lone pairs. **The more alkyl "grease" a solvent has, the less polar it is**. Remember that "**like dissolves like**." Ionic compounds such as salts dissolve in polar solvents and not in nonpolar ones. Nonpolar compounds dissolve in nonpolar solvents and not in polar ones. There are "highly polar aprotic" solvents, such as dimethyl sulfoxide (DMSO) and hexamethylphosphoramide (HMPA) that can stabilize cations much better than anions. The negative dipole end of these solvents is easily accessible and therefore can nestle in close to a cation, solvating it well. However, their positive dipole end is highly hindered and can't get very close to an anion, therefore solvating it poorly. Anions are less stabilized by these solvents, and thus are more reactive. Anions are much more basic in these solvents.

DMSO 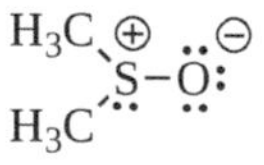 HMPA 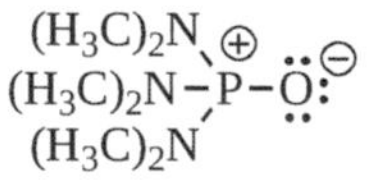

Polar protic
water, acetic acid (carboxylic acids), methanol, ethanol, isopropanol (alcohols)
Polar aprotic
dimethyl sulfoxide (DMSO), dimethylformamide (DMF), acetonitrile, ethyl acetate, acetone (ketones), hexamethylphosphoramide (HMPA), pyridine (amines)
Slightly polar to Nonpolar
dichloromethane, chloroform, tetrahydrofuran (THF), diethyl ether, benzene, hexane

Solvation is a hard–hard interaction, and therefore, the smaller and harder the ion, the more energy is released when it is solvated. The more energy released, the more stabilized and tightly solvated the ion is. Tightly solvated ions will be less useful in reactions because to react they must break out of this stabilizing solvation. The tightness of solvation will have a major effect in determining reactivity. The importance of solvation is often underestimated in chemical reactions.

Solvent polarity can restrict the possibilities for reaction paths. Whenever charged species are present in a reaction, the reaction barrier is highly dependent on the polarity of the solvent (Figure 2.20). Solvent stabilization of the reactants more than the products decreases reactivity. Solvent stabilization of the products more than the reactants increases reactivity. In addition to stabilizing ions, polar protic solvents can also allow proton transfer and equilibration between the various ionic species in solution.

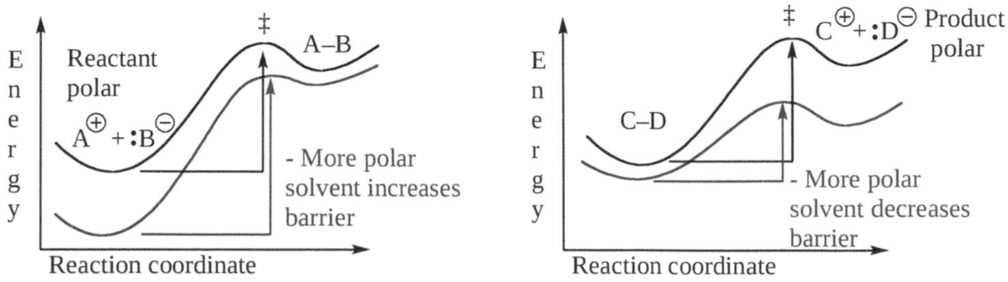

Figure 2.20 The effect of a more polar solvent on the barrier to reaction depends on which the polar solvent stabilizes more: the reactants (left) or the products (right).

Example problem

Which of the following solvents, ethanol, hexane, diethyl ether, would dissolve the most KOH? The least?

Answer: KOH is a salt and therefore very polar. It will dissolve the most in a polar solvent, the least in a nonpolar. The most polar solvent of the three is ethanol. Diethyl ether is slightly polar. Hexane is nonpolar and will dissolve the least KOH.

2.8 ENZYMATIC CATALYSIS—LESSONS FROM BIOCHEMISTRY

A *catalyst* is a substance that speeds up a reaction and can be recovered unchanged. The most common catalysts for organic reactions are acids or bases. An acid can protonate a reactant, giving it a full positive charge and making it a better electron sink. A base can remove a proton, making the reactant anionic and a better electron source. Even though another step has been added (protonation or deprotonation), with a better electron source or sink the overall reaction proceeds faster. The reaction barrier, $\Delta G^{\ddagger}$, is lowered by catalysis, and therefore more molecules now have enough energy to react.

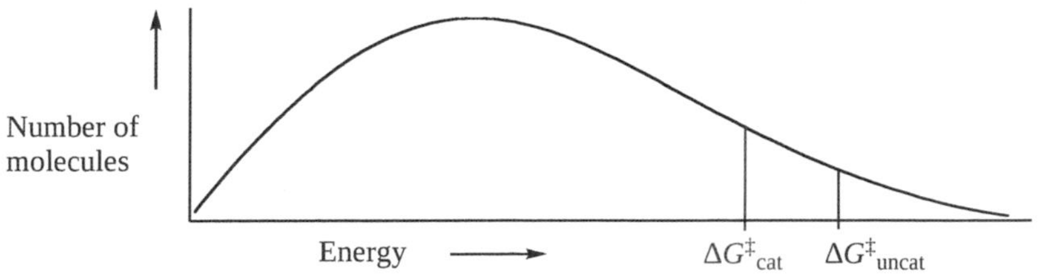

Figure 2.21 Boltzmann energy distribution and the $\Delta G^{\ddagger}$ for catalyzed and uncatalyzed processes. More molecules have the energy to cross the lower $\Delta G^{\ddagger}_{cat}$ barrier than the higher $\Delta G^{\ddagger}_{uncat}$ barrier.

Figure 2.21 shows the distribution of energies near room temperature. The area to the right of the barrier value is the proportion of molecules with sufficient energy to react. As the

$\Delta G^‡$ is lowered by a catalyst, many more molecules have sufficient energy to cross the lowered barrier. Recall that the $\Delta G°$, on the other hand, is determined by the identity of the reactants and products and so is not affected by a catalyst. Only the rate of reaction is changed by a catalyst, not the position of equilibrium.

To really understand the basics of catalysis, we can try to understand how some of the best catalysts work. Enzymes are biochemical catalysts with exceptional abilities to speed up reactions. There are enzymes in your body that are highly selective and work in water and at mild temperatures, and some are so fast that they can produce their products as fast as the reactants can diffuse to them.

2.8.1 Collision Frequency

Molecules have to collide to react. Most reactions are run in solution, so as a solution gets more concentrated, collisions increase and so does the rate. But there is an upper limit, often determined by solubility, on how concentrated the solution can get. A concentration landmark is that pure water is 55.6 mol/L; pure benzene, C_6H_6, is 11.1 mol/L. Often chemists must run reactions at 1 mol/L or less. In addition, not all collisions are effective because usually just certain parts of the reactant molecules must hit to react; otherwise they just bounce off each other. This orientation requirement is part of the $\Delta S^‡$ portion of the barrier, $\Delta G^‡ = \Delta H^‡ - T\Delta S^‡$. If the reactive partners are part of the same molecule (intramolecular reactions), they collide much more frequently as the molecule twists and turns than if they were separate. In fact, intramolecular reactions can get effective concentrations of well over a million mol/liter!

2.8.2 Enzyme Active Sites

It is believed that proximity and orientation are critical to the way enzymes accelerate reactions. The key is to increase the number of effective collisions between reactive partners. Not only are the reactive partners undergoing a huge number of collisions within the active site of the enzyme, but the active site is also orienting them correctly. The slow step for some enzymes is just getting the reactants into the active site, the diffusion-controlled limit.

An enzyme active site often has a unique shape; only molecules that have the right shape will fit. This allows the enzyme to select for the correct reactants among all those swimming in the cell soup. Even better, the correct reactant usually has to fit in the active site in the proper orientation to react. This first step takes care of the $\Delta S^‡$ component of the $\Delta G^‡$ barrier by binding the reactant in the proper orientation for reaction. What is left of the barrier is the $\Delta H^‡$ component, which can be lowered by stabilization of the transition state(s) or intermediate(s).

The enzyme active site can also twist the reactant into a more reactive conformation, and can provide the perfect microenvironment for reaction: proper solvation, nearby acid or base catalysts, needed electrophiles and nucleophiles.

2.8.3 Enzymes Bind the Transition State Best

Outside of the enzyme we are limited to two molecule collisions, because the odds of a simultaneous collision of three independent molecules in the proper orientation for reaction are very slim. Three molecule collisions are considered a mechanistic possibility only when two of the three reacting molecules are loosely associated via hydrogen bonding or pi-adduct formation. However, within the enzyme's active site, there can be several groups snuggled in around our reactant to contribute as needed to lower the reaction barrier by stabilizing the transition state.

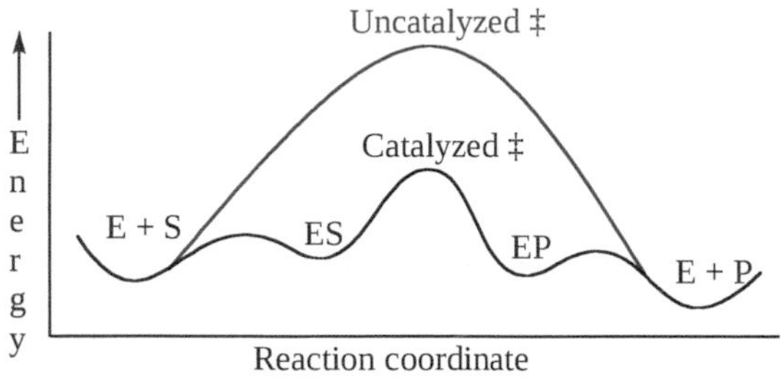

Figure 2.22 The enzyme E binds the substrate S in the active site to form an enzyme-substrate complex ES. The chemical step occurs to give the enzyme-product complex EP that dissociates into product P and free enzyme E, ready for another cycle.

Figure 2.22 illustrates a simple energy diagram for enzymatic catalysis. For the enzyme-catalyzed chemical step, the transition state fits best in the active site, and so is stabilized the most. The active site can contain the needed acid and/or base catalysts in just the right position. Groups inside the active site often stabilize any reaction intermediates by electrostatics or hydrogen bonding, thus bringing down the barriers leading to those intermediates. Amazingly, the reaction barriers in an enzyme-catalyzed process can be brought so low that their reaction rates are incredibly fast, approaching 10^{20} times faster than the uncatalyzed reaction in water. Later sections will discuss lessons we can learn from the chemical mechanisms of enzymatic catalysis.

2.9 SUMMARY

• Strong bonds require good overlap of the bonding orbitals. The formation of sigma bonds requires approximate colinearity of the reacting orbitals. The formation of pi bonds requires approximate coplanarity of the reacting orbitals.

• Principle of microscopic reversibility: The forward and reverse reactions follow the same lowest-energy route but in opposite directions. There is only one energetically best pathway. The transition state of lowest energy maximizes the extent of bonding.

• The HSAB principle: Hard bases favor binding with hard acids; soft bases favor binding with soft acids. Softness is characterized by high polarizability and often little charge; hardness is the reverse. The soft–soft interaction is a covalent bond, which is favored by good overlap of orbitals that are close in energy. The hard–hard interaction is an ionic bond, which is favored by highly charged species with small radii so that they can get close together. When molecules collide, two attractive forces lead to reaction: the hard–hard attraction (opposite charges attracting each other) and the soft–soft attraction (the interaction of filled orbitals with empty orbitals). Solvation is a hard–hard interaction, so the smaller and harder the ion, the more energy is released when it is solvated.

• At room temperature, a drop of 1.364 kcal/mol (5.706 kJ/mol) in $\Delta G°$ increases the equilibrium constant by a factor of 10. Likewise, dropping the $\Delta G^‡$ by 1.364 kcal/mol (5.706 kJ/mol) increases the reaction rate tenfold. The $\Delta G^‡$ of a reaction that goes at a reasonable rate at room temperature is about 20 kcal/mol. The rate of a reaction usually doubles for every 10°C increase in temperature.

• Free energy change is related to enthalpy and entropy by the equation $\Delta G° = \Delta H° - T\Delta S°$. The standard *enthalpy* change, $\Delta H°$, is a measure of the heat absorbed or evolved in a reaction and is approximated as $\Delta H° = \Delta H_{(\text{bonds broken})} - \Delta H_{(\text{bonds made})}$, whereas $\Delta S°$, the standard entropy change, is a measure of the change in the randomness within the system.

• A *catalyst* is a substance that speeds up a reaction and can be recovered unchanged. An enzyme catalyst increases reaction rates by the dramatic amplification of the number of effective collisions between the reactants within the enzyme's active site. Enzymes increase the rate of the chemical step by using binding to stabilize the transition state more than the reactants or products.

• The more reactive a species is, the less selective it is.

• The more stable a compound is, the less reactive it is.

• As an intermediate increases in energy, the transition states leading up to it must also increase in energy. Anything that stabilizes the intermediate drops the barrier to it.

CHAPTER 2 EXERCISES

2.1 There are two ways that a compound can be stable: one due to $\Delta G°$ and the other due to $\Delta G^‡$. Draw energy diagrams to explain.

2.2 Use HSAB theory to predict the direction of the following reaction:

$$2\ CH_3Li + CuCl \rightleftharpoons (CH_3)_2Cu^- Li^+ + LiCl$$

2.3 Using this energy diagram:

(a) What is the slow step?

(b) What is the $\Delta G°$ of C → D?

(c) What is the $\Delta G^‡$ of A → C?

(d) What is the $\Delta G^‡$ of D → C?

(e) Can C be isolated at 25°C?

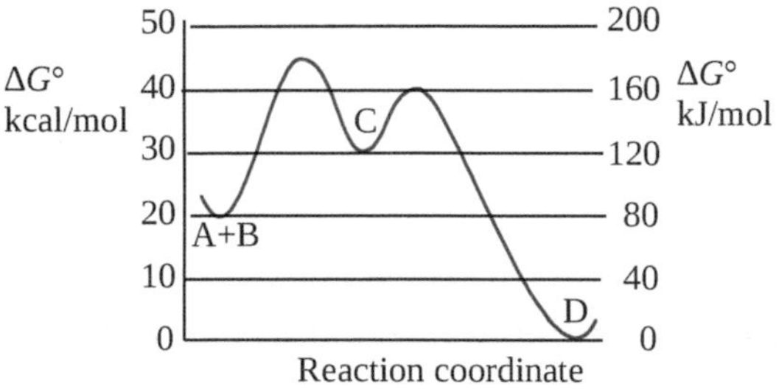

2.4 In problem 2.3, if the reactants A and B were charged and the intermediate C was not, would the barrier to reaction increase or decrease in a more polar solvent?

2.5 Calculate the $\Delta H°$ and decide whether this reaction is exothermic or endothermic.

$$CH_4 + F_2 \rightleftharpoons CH_3F + HF$$

2.6 Rank all of the following:

(a) Use #1 to indicate the softest Lewis base
F^- CH_3^- OH^- NH_2^-

(b) Use #1 to indicate the most polar solvent
$CH_3CH_2CH_2CH_2CH_3$ EtOEt H_2O CH_3OH $CH_3CH_2CH_2OH$

2.7 Rank all of the following:

(a) Use #1 to indicate most soluble in water
CH_3CH_2OH $CH_3CH_2OCH_2CH_3$ $CH_3CH_2CH_2OH$ $CH_3CH_2CH_2CH_2CH_3$

(b) Use #1 to indicate the weakest H–X bond
HBr HCl HI HF

2.8 Calculate the $\Delta H°$ and decide whether this reaction is exothermic or endothermic.

$$H_2C{=}CH_2 + Cl_2 \xrightleftharpoons{?} ClCH_2CH_2Cl$$

2.9 Using this energy diagram, what is:

(a) the kinetic product?
(b) the thermodynamic product?
(c) the $\Delta G^{\ddagger}$ for A → B?
(d) the $\Delta G^{\ddagger}$ for A → C?
(e) the $\Delta G°$ for A → C?
(f) the most likely product at 0°C?

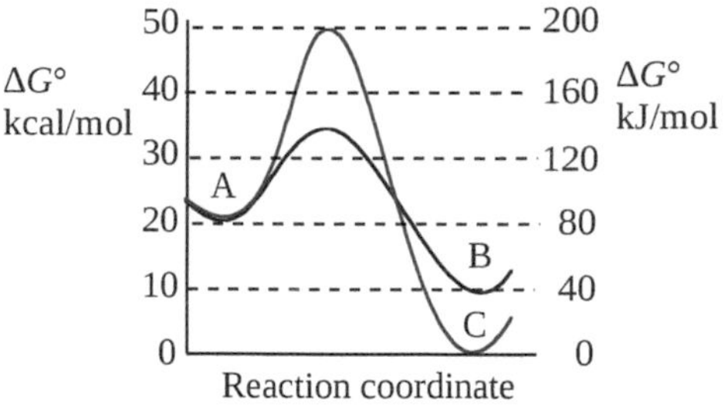

2.10 At room temperature, a reversible reaction equilibrates A into B with an 85:15 ratio respectively. What is the free-energy difference between A and B?

2.11 Explain why possibility B and not possibility A occurs even though they are technically the same process. Molecular models will help.

Possibility A

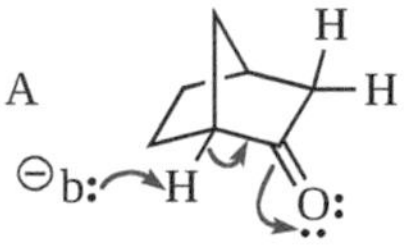

Possibility B

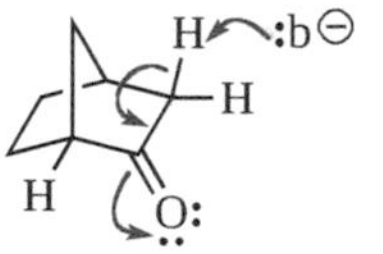

2.12 Using the bond-strength table in the Appendix, calculate the $\Delta H°$ of reaction for the following process. Would an equilibrium favor reactants or products?

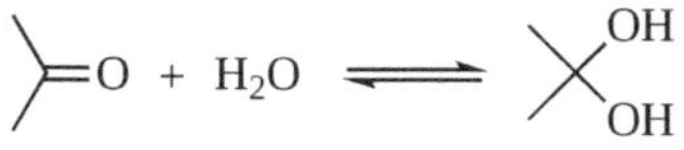

2.13 Use a $\Delta H°$ calculation to tell which of the two possible products is the more stable.

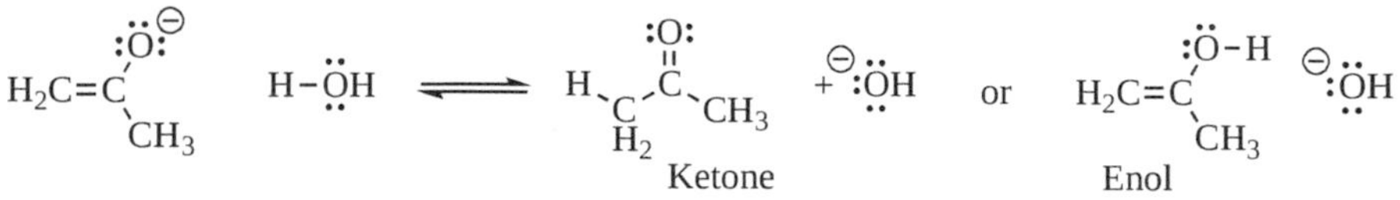

2.14 What are the relationships between stability, reactivity, and selectivity?

2.15 Below is an energy diagram for the rate-determining step of a reaction. How would a more polar solvent affect the rate? Draw a curve for the more polar solvent on the diagram.

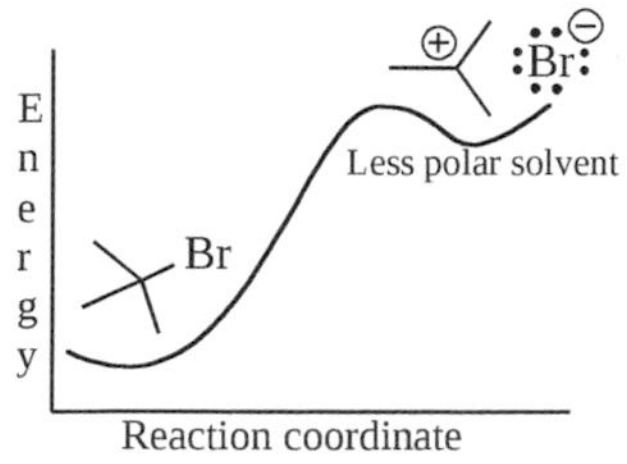

2.16 Draw an energy diagram for the exothermic reaction below (assume $\Delta H°$ dominates $\Delta G°$ as it does often). Compound I is a reactive intermediate.

$$A + B \xrightarrow{\text{Slow}} I \xrightarrow{\text{Fast}} C$$

2.17 The relative rate of two related unimolecular reactions is 10,000:1 under identical conditions at 25°C. What is the difference in barrier heights, $\Delta\Delta G^{\ddagger}$?

2.18 For the reaction below, the equilibrium at 25°C favors the reactant but at 160°C favors the products. In this case, $\Delta G°$ does vary with significantly with temperature, but assume $\Delta H°$ and $\Delta S°$ do not. Explain how the favored species can vary with temperature.

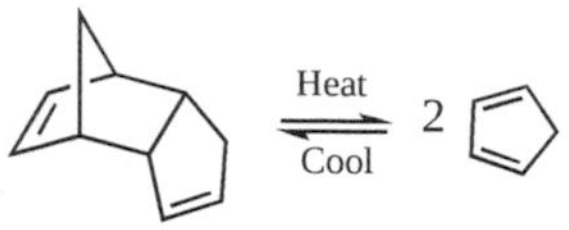

2.19 Using the bond-strength table in the Appendix, calculate the $\Delta H°$ of reaction for the following enolization process. Would an equilibrium favor reactants or products?

2.20 Using the energy diagram below for two similar reactions (one trace in red), decide what determines the preferred product E or F from A + B. Note that intermediate C on route to product F is more stable than intermediate D on route to product E. The barriers from each intermediate to the products differ also. Assume neither overall route is reversible.

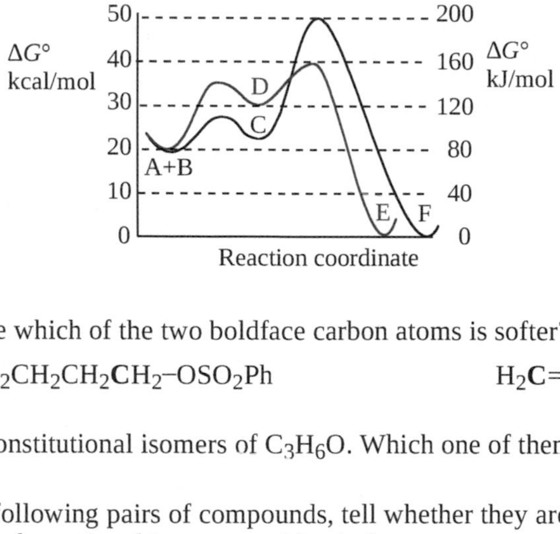

2.21 In each molecule which of the two boldface carbon atoms is softer?

I–**C**H$_2$CH$_2$CH$_2$**C**H$_2$–OSO$_2$Ph H$_2$**C**=CH–**C**HO

2.22 Draw the nine constitutional isomers of C_3H_6O. Which one of them has a stereoisomer?

2.23 For each of the following pairs of compounds, tell whether they are enantiomers, diastereomers, conformational isomers, or identical.

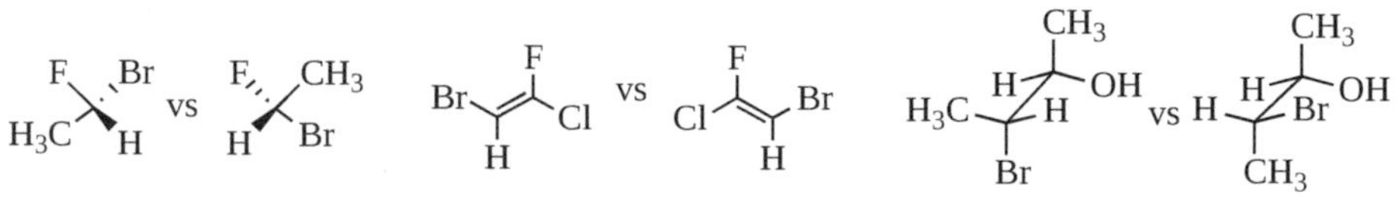

2.24 Which of the following structures are chiral?

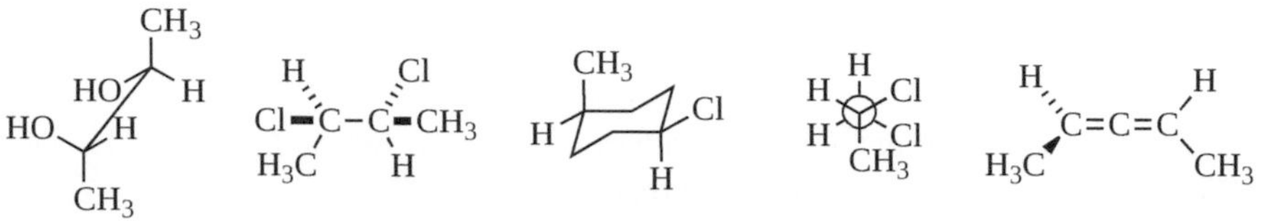

3

PROTON TRANSFER AND THE PRINCIPLES OF STABILITY

Electron Flow in Organic Chemistry: A Decision-Based Guide to Organic Mechanisms,
Third Edition. Paul H. Scudder.

Always Decide If The Reaction Is Run In Acid Or Base And Stay Within Those Bounds. Acidic Hydrogens Are Found On Heteroatoms And Adjacent To Polarized Multiple Bonds. As A First Step In Acidic Media, Protonate The Reactant, And In Basic Media, Deprotonate It. Web of Concepts for Factors Increasing Acidity.

3.1 INTRODUCTION TO PROTON TRANSFER

Most of the principles used to explain the acidity of compounds are also used to explain the reactivity and stability of organic compounds in general. Also, proton-transfer reactions are often the first step in most organic reactions. For this reason, if we thoroughly understand proton transfer, then we have a solid basis for understanding most organic reactions.

3.1.1 Acid and Base Generic Groups and the Proton-Transfer Path

A Brønsted–Lowry acid is a proton donor. A hydrogen attached to an electronegative atom is very often quite acidic because the electronegative atom can stabilize the lone pair created when the proton is removed. A Brønsted–Lowry base accepts the proton to form the conjugate acid. Most anions and lone pairs can serve as Brønsted–Lowry bases. The conjugate base and the conjugate acid are the base and acid for the reverse reaction. Initially, the base forms a hydrogen bond to the acid; the proton is transferred, and then the hydrogen-bonded conjugate acid/base pair dissociates. This text will symbolize a Brønsted–Lowry acid as HA and base as b in a proton-transfer (p.t.) reaction and use the **p.t. electron flow path. The tail of the first arrow starts from the electron pair on the base, and the arrowhead is near the proton removed. The second arrow starts from the H–A bond, breaks it, and deposits the bonding electron pair on the conjugate base:**

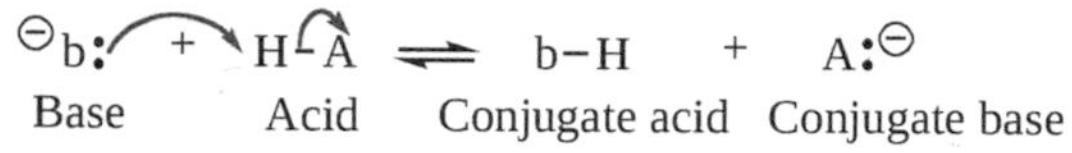

Figure 3.1 gives examples of four common charge types of proton-transfer reactions. The base can be neutral or negative, and the acid can be neutral or positive. It is very unusual for the base to be positive or the acid to be negative, because the proton-transfer reaction would produce either a dication or dianion, respectively, and those tend to be less stable.

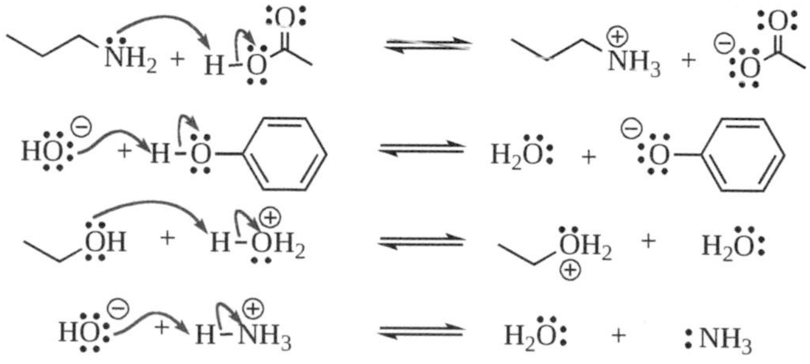

Figure 3.1 Four common charge types of proton-transfer reactions (from top): neutral base + neutral acid, anionic base + neutral acid, neutral base + cationic acid, anionic base + cationic acid.

In general, **strong acids have weak, stable conjugate bases**. The more stable the conjugate base is, the stronger the acid. Similarly, the more stable the conjugate acid is, the stronger the base. **Proton-transfer reactions favor neutralization, so they usually head toward the weaker acid and weaker base.** Some common acids and bases used in organic reactions are listed in the next section. We will first see how the strength of acids and bases are ranked, and then use that data to understand trends in acidity and basicity.

Example problem

Draw the acid, conjugate acid, and electron flow arrows for the reaction below.

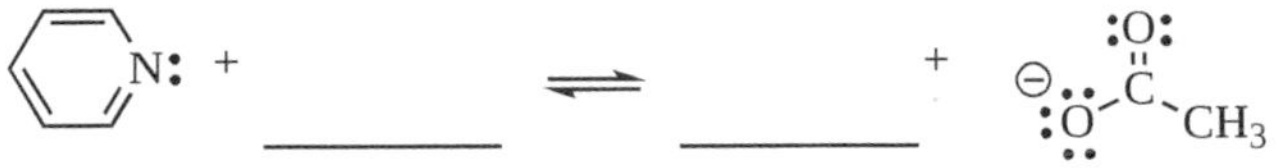

Answer: The missing acid is the conjugate base with a proton added on its most basic atom. The missing conjugate acid is the base with a proton added on its most basic atom. Note that charge has to balance, so the conjugate acid had to be positively charged.

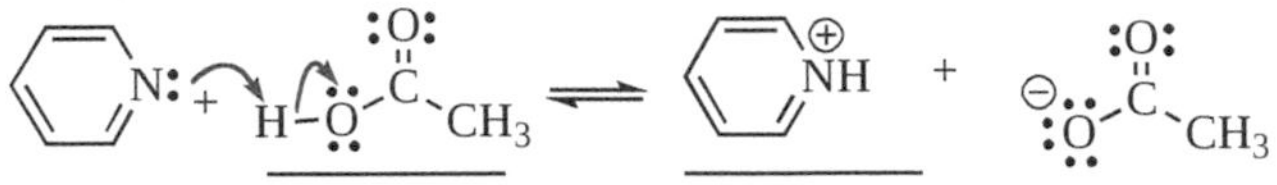

3.2 RANKING OF ACIDS AND BASES, THE pK_a CHART

3.2.1 The pK_a Chart and pH Relationships

$$\text{H-A} \rightleftharpoons \text{H}^{\oplus} + \text{A:}^{\ominus} \qquad K_a = \frac{[\text{H}^+][\text{A}^-]}{[\text{HA}]}$$

$$\text{pH} = -\log[\text{H}^+] \qquad [\text{H}^+] = 10^{-\text{pH}} \qquad \text{p}K_a = -\log K_a \qquad K_a = 10^{-\text{p}K_a}$$

The dissociation constant, K_a, of the acid HA is shown above. The larger the value of the K_a, the more ionized the acid. Since the range of values is so huge, we make use of a log scale similar to that of pH. The pK_a of an acid is the negative log of its K_a. A low pK_a corresponds to a large value for K_a, the dissociation constant. The pK_a chart (see Appendix) serves as a good ranking of acids. **Stronger acids have lower pK_a values. An acid with a negative pK_a is a strong acid**; a pK_a of 0 gives a K_a of 1.

The pK_a chart is a composite of many experimental observations under widely different conditions. For example, the solvent has a major effect on the stabilization of the conjugate base, and therefore on the strength of the acid. Several different solvents were needed to extend the range of the chart, so extrapolations between solvents were required, using a few acidities determined in both solvents. Most important was to get the order of acidities correct. The chart has the greatest extrapolation errors at its extremes. It is most accurate when the pK_a values were determined in water as the solvent, approximately between the pK_a values of −2 and 16. While we will write for simplicity HA ionizing to give $H^+ + A^-$, it is always true that the proton is transferred to a base. For our composite pK_a chart (in the Appendix) this base is assumed to be water. Please see the bibliography for references on pK_a resources.

When [HA] = [A^-], these terms cancel out of the K_a expression, leaving $K_a = [H^+]$; therefore, the pK_a of the acid matches the solution pH when [HA] = [A^-]. At this pH, the acid HA is half ionized. As the pH becomes more basic, [A^-] concentration increases and [HA] decreases, as Figure 3.2 illustrates. For each pH unit away from the pK_a the ratio changes by a factor of 10. The following buffer equation gives the relationship between pH, pK_a, and the ratio of [A^-]/[HA]. We can use this buffer equation or Figure 3.2 to see that if we wanted a buffer of pH 4 from an acid with a pK_a of 5, we would need a 1:10 ratio of [A^-]/[HA] because log(1/10) = −1. A buffer is designed to be near the pK_a of the weak acid because roughly equal concentrations of the acid and conjugate base allow the buffer solution to keep the pH relatively constant if small amounts of an acid or base are added. The solution can "buffer" or moderate the changing conditions. Biochemical systems are almost always buffered.

$$pH = pK_a + \log \frac{[A^-]}{[HA]}$$

Figure 3.2 has important biochemical implications. At physiological pH of 7, which is two pH units more basic than its pK_a, the carboxylic acid symbolized by H–A in the figure would be 99% the conjugate base. Both of the amino acids aspartate and glutamate have similar carboxylic acids on their side chain. The overall charge of a protein will depend on pH and pK_as of the groups attached to it.

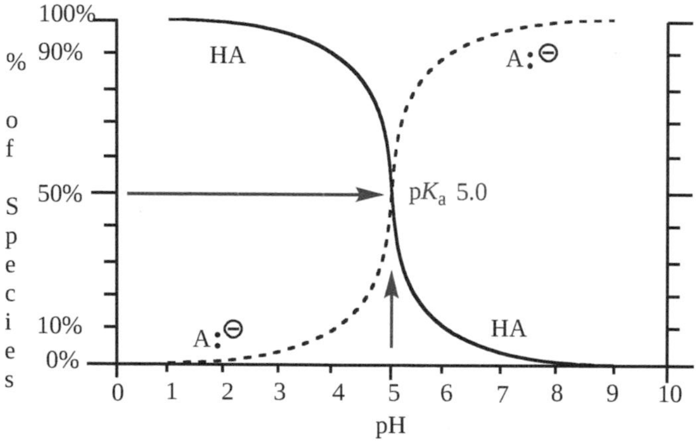

Figure 3.2 Species concentration as a function of pH for an acid of pK_a = 5. The acid and conjugate base are equal concentrations at the pH = pK_a. At one pH unit away from the pK_a the species are in a 10 to 1 ratio, so 91:9. Two pH units away from the pK_a will give a ratio of 99:1.

Acids can have more than one acidic proton, and thus have more than one pK_a. The first (lowest number) pK_a is for the most acidic H. Phosphoric acid, H_3PO_4, has three acidic hydrogens, and therefore has three pK_a values. The species concentration vs. pH graph for phosphoric acid is shown in Figure 3.3. Note how each acid and conjugate base concentration varies with pH and that the concentrations of each coupled acid/conjugate base pair cross (become equal) at their related pK_a. For a species vs. pH calculation from a drawn structure see: www.chemicalize.com.

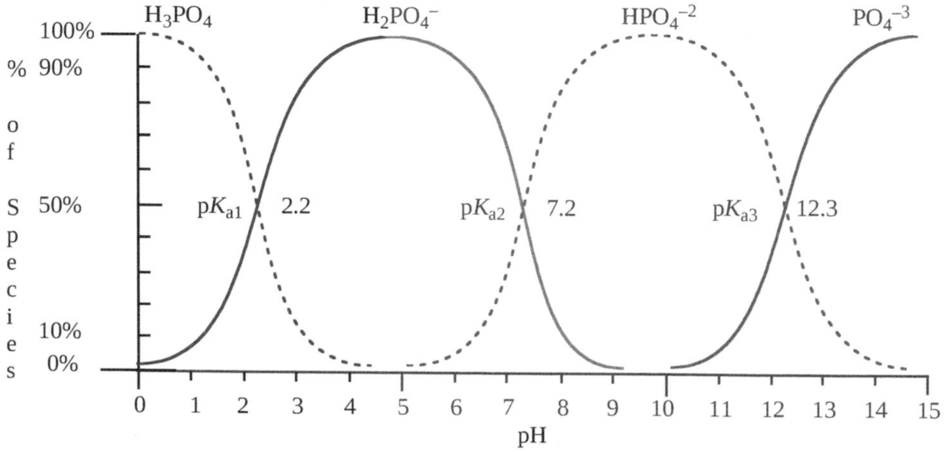

Figure 3.3 The species concentration as a function of pH for H_3PO_4, with pK_{a1} = 2.2, pK_{a2} = 7.2, and pK_{a3} = 12.3.

Example problem

Using Figure 3.3, which compounds and in what relative concentration would you have to use to make a phosphate buffer with a pH of 7.2? How about with a pH of 8.2?

Answer: pH 7.2 is at pK_{a2}; equal concentrations of $H_2PO_4^-$ and HPO_4^{2-} are needed. At pH of 8.2, one unit more basic, 10 times more conjugate base HPO_4^{2-} than conjugate acid $H_2PO_4^-$ is needed to get the correct pH. **For any mixture of an acid and its conjugate base, the pH will be within a few pH units of the pK_a of the acid used.**

3.2.2 Media pH Cross-check

The old phrase "a snowball's chance in hell" illustrates the fact that the environment can play a critical part in the probability of existence of any species. Since $K_w = [H^+][OH^-] = 10^{-14}$, it is easy to calculate one concentration given the other. For example, in pH 1 water, the $[H^+]$ is 10^{-1} M; therefore the concentration of hydroxide ion is 10^{-13} M. The probability that a reactive species will encounter hydroxide in a medium this acidic is very, very low. Likewise, it would be highly unlikely to find hydronium in very basic solutions. These principles generalize into several important control knowledge rules:

Acidic media contain powerful electrophiles and rather weak nucleophiles, whereas basic media contain excellent nucleophiles and weak electrophiles. Similarly, no medium can possibly be both strongly acidic and strongly basic.

You could prepare two different solutions, one with an excellent electrophile and the other with an excellent nucleophile, and pour them together. However, as in tossing a snowball into a blast furnace, you must be prepared for the process to be rapid and probably violent.

Forgetting the media restrictions is one of the most common mistakes of students learning to do mechanisms. Keeping Figures 3.2 and 3.3 in mind, you can easily decide whether the medium is acidic or basic by using the pK_a of the reagents. For example, if the solution contains EtOH/EtONa, and knowing that EtONa is ethanol's conjugate base, means that the approximate pH of the solution will be close to the ethanol's pK_a of 16, or very basic. We don't need to be exact, just in the ballpark. Often, all we need to know for the mechanism is whether the solution is acidic or basic.

3.2.3 Common Acids and Their pK_a Values

The next list gives the common acids used in organic reactions in rank order by pK_a.

	Strong acid	
pK_a =	−15	CF_3SO_3H, trifluoromethanesulfonic (triflic) acid
	≈ −9	H_2SO_4, sulfuric acid pK_{a1}
	−9	HBr, hydrobromic acid
	−6	HCl, hydrochloric acid
	−6.5	$HOSO_2C_7H_7$, toluenesulfonic acid
	−1.4	HNO_3, nitric acid
	0 (−1.7)	H_3O^+, hydronium ion
	2.2	H_3PO_4, phosphoric acid pK_{a1}
	4.8	CH_3COOH, acetic acid
	9.2	NH_4Cl, ammonium chloride
	Weak acid	

Another value for the pK_a of hydronium –1.7, is still debated and used. The strongest acid that can exist in a solvent is the protonated solvent (the leveling effect). When mixed in water, all acids above hydronium on the chart will just protonate water to form hydronium. For example, nitric acid added to water completely ionizes, forming hydronium and nitrate.

Example problem

Which of the following, HNO_3, HF, HCl, is the strongest acid? The weakest acid?

Answer: The acid with the lowest pK_a is the strongest. The strongest acid is HCl with a pK_a at 6; the next is HNO_3 at −1.4; the weakest is HF at 3.2.

3.2.4 Common Bases and Their pK_{abH} Values

The pK_a chart also serves as a good ranking of bases. For the reaction $b^- + H^+ \rightleftharpoons$ b–H, the proton acceptor is the base b^- and b–H is the conjugate acid. **The basicity of the base increases as the pK_{abH} increases** (pK_{abH} is the pK_a of the conjugate acid). **Strong bases have weak conjugate acids.** Poor solvation of an anionic base will increase its basicity (Section 3.3.8). The following list gives in rank order by the common bases used in organic reactions. Another value for the pK_{abH} of hydroxide, 15.7, is still debated and used. Please see the bibliography for references on pK_a resources. Similar to what was seen in Section 3.2.3, the strongest possible base in a solvent is the deprotonated solvent (the leveling effect).

	Strong base	
pK_{abH} =	≈ 50	CH_3Li, *n*-BuLi
	≈ 36	Et_2NLi, $NaNH_2$, NaH, KH
	≈ 26	$[(CH_3)_3Si]_2NLi$
	≈ 19	*t*-BuOK
	≈ 16	EtONa, CH_3ONa
	≈ 14 (15.7)	NaOH
	≈ 10	Na_2CO_3, Et_3N
	≈ 5	CH_3COONa, pyridine
	Weak base	

Example problem

Which of EtO^-, NH_3, NH_2^- is the strongest base? Which is the weakest?

Answer: The strongest base has the highest pK_{abH}. This means that NH_2^- at pK_{abH} 35 is the strongest base. The next strongest is EtO^-, at pK_{abH} 16. The weakest base is NH_3, with a pK_{abH} of 9.2. Remember to look up the conjugate acid: Add a proton to the base, then find it on the pK_a chart. For NH_3 as a base, you must look up $^+NH_4$ at 9.2 and not NH_3.

3.3 STRUCTURAL FACTORS THAT INFLUENCE ACID STRENGTH

The many factors contributing to acidity can be divided into intrinsic (related to just the compound itself) or extrinsic (related to the environment into which the compound is placed). For the following discussion, we will use the example of a neutral acid that produces an anionic conjugate base upon deprotonation. Figure 3.4 shows how stabilization or destabilization of the acid or conjugate base affects the overall energy diagrams to give an unfavorable or a favorable proton transfer. Section 3.3.8 will discuss that solvation has a major effect in stabilizing the conjugate base, thus contributing to effects 2 and 4 in Figure 3.4. Acids are much weaker in solvents like DMSO that poorly stabilize anions (effect 4).

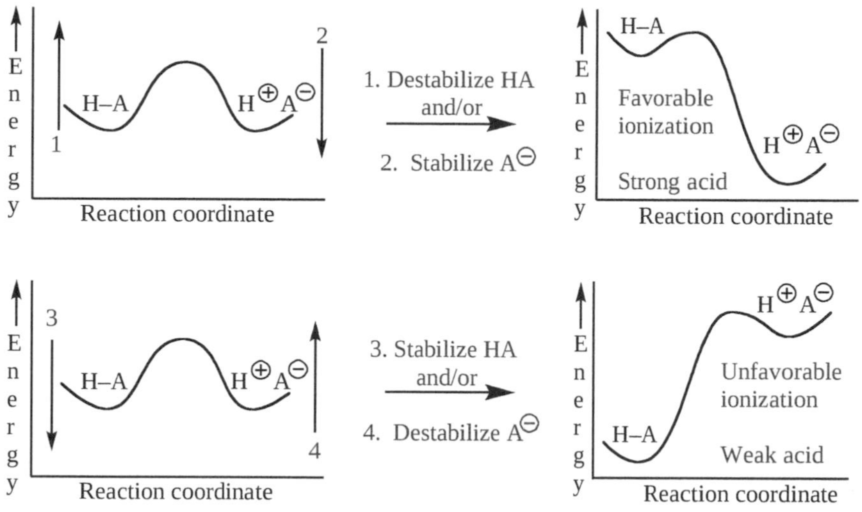

Figure 3.4 Energy diagrams for favorable and unfavorable ionizations of an acid.

3.3.1 Electronegativity

A hydrogen atom attached to a more electronegative atom will be more acidic. Electronegativity increases as we go to the right on the periodic table because the nuclear charge is increasing while the electron shell remains nearly the same. The anionic conjugate base is more stable if the lone pair formed from deprotonation is on a more electronegative atom. The increasing nuclear charge going to the right within a row of the periodic table means that the lone pair on the conjugate base resides in a more stable, lower-energy orbital.

H–CH_3 pK_a = 48 H–NH_2 pK_a = 35 H–OH pK_a = 14.0 (15.7) H–F pK_a = 3.2
< least acidic, least electronegative most acidic, most electronegative >

3.3.2 Strength of the Bond to H

If the H–A bond is stronger, it will be more difficult to break, and the acid will be weaker. The overlap to H decreases going down a column in the periodic table, so the H–A bond gets weaker. The acidity increases going down the column from fluorine to iodine: HF < HCl < HBr < HI. The overlap with the hydrogen's 1*s* orbital is worst with the 5*p* orbital of iodine, gets a little better with the 4*p* of bromine, better still with the 3*p* of chlorine, and is best with the 2*p* of fluorine.

3.3.3 Resonance Effects

Resonance effects allow us to delocalize the nonbonding electron pair formed from loss of a proton. An example of increasing acidity as more resonance stabilization is added is CH_3CN pK_a = 25, $CH_2(CN)_2$ pK_a = 11.2, $CH(CN)_3$ pK_a = −5. Resonance effects are dramatic and dominant. Tricyanomethane at pK_a = −5 is more acidic than nitric acid pK_a = −1.4, and is one of the strongest carbon acids. Since the cyano group delocalizes the anionic pair of electrons into its pi system, it is called a pi acceptor. The best electron-withdrawing groups (Section 3.5.1) are pi acceptors that can place the anion on an electronegative atom, like in the resonance form shown below. The stabilization by the cyano group also includes inductive and field effects, discussed next.

With neutral oxygen acids, the more the conjugate base can distribute the negative charge, the stronger the acid. For example, HOCl at pK_a = 7.54, is a much weaker acid than $HClO_4$ at pK_a = −15, which delocalizes the resulting anion over four oxygen atoms.

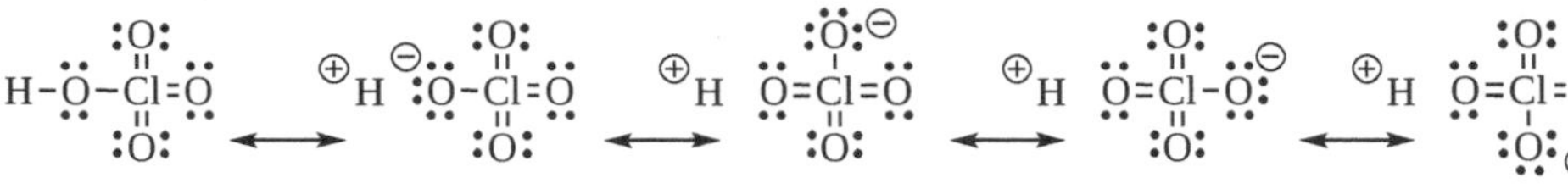

3.3.4 Inductive/Field Effects

Inductive effects are electron withdrawal or donation through the sigma bond framework of the molecule. Field effects are through-space electrostatic attractions and repulsions. Since it is hard to separate the through-bond from through-space effects, inductive and field effects are usually grouped together. A good relative ranking of these inductive/field effects for various groups, G, is the acidity of a series of carboxylic acids, G–CH_2–COOH. Most groups, being more electronegative than carbon, tend to increase acidity. For example, if G is the very electronegative trimethylammonium group, then $(CH_3)_3N^+$–CH_2–COOH, is more acidic than acetic acid (G = H) by three pK_a units, a factor of 1000. Inductive/field effects fall off rapidly with distance; the effect of the group in G–CH_2–CH_2–COOH is only 40% of that for G–CH_2–COOH. Inductive/field effects are approximately additive. For example, the acidity of Cl_2CH–COOH is about halfway between that of ClCH_2–COOH and Cl_3C–COOH.

The resonance and inductive/field effects can conflict, and resonance effects usually dominate. For example, an oxygen atom has lone pair of electrons it can donate into a pi system by resonance (pi donor). At the same time, the oxygen atom is electronegative, so is withdrawing by an inductive/field effect. Section 4.2.5 will cover the ranking of electron-releasing groups (ERG). In the three phenols shown below, stabilization of the anionic conjugate base by a good electron-withdrawing group will increase the acidity of phenol. The lowest pK_a (most acidic) is 4-nitrophenol, at 7.2. The unsubstituted phenol at pK_a 10.0 is next. Least acidic at pK_a 10.2 is 4-methoxyphenol. The inductive effect of the electronegative methoxy oxygen has fallen off with distance, whereas the resonance donation from the methoxy lone pair destabilizes the conjugate base and decreases the acidity of the phenol.

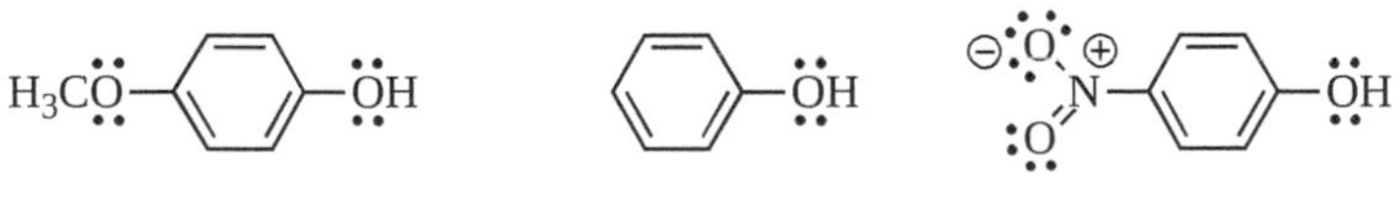

3.3.5 Hydrogen Bonding

Internal hydrogen bonding to the conjugate base will definitely increase acidity. Fumaric acid below has two acidic hydrogens, pK_{a1} = 3.10 and pK_{a2} = 4.60. The second carboxylic acid can inductively withdraw and lower the first pK_a but is held too far away to hydrogen bond. On the other hand, maleic acid's two acidic hydrogens differ much more in acidity, pK_{a1} = 1.912 and pK_{a2} = 6.33. For the first deprotonation, the other carboxylic acid in maleic acid can easily hydrogen bond to the conjugate base and stabilize the anion. The second deprotonation of maleic acid is more difficult than that of fumaric acid because of the destabilizing field/inductive effect of two like charges being held closer together.

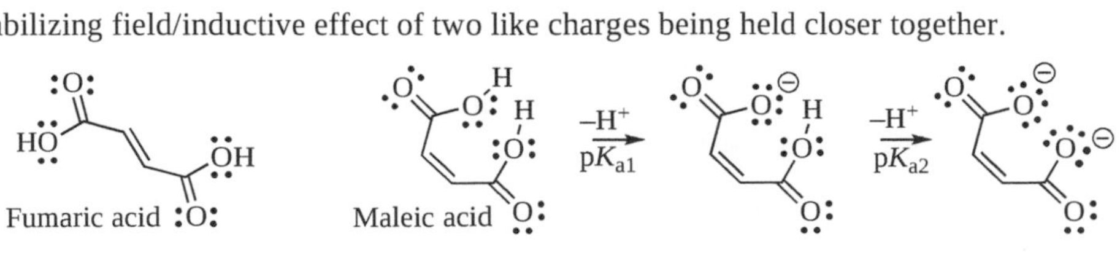

3.3.6 Aromaticity

Cyclopentadiene is unusually acidic for a carbon acid, with a pK_a of 16. The conjugate base has six electrons in an unbroken loop of *p* orbitals and so is stabilized by aromaticity (Section 1.9.3). Aromaticity stabilizes this ion by more than 25 pK_a units.

3.3.7 Charge

Positively charged acids are more acidic than their corresponding neutrals, as illustrated by the lower pK_a of the cations of the following compounds.

Cation	pK_a	Neutral	pK_a
NH_4^+	9.2	NH_3	35
H_3O^+	0.0 (–1.7)	H_2O	14.0 (15.7)
More acidic		Less acidic	

3.3.8 Extrinsic Factors—Solvation and Ion Pairing

Besides resonance and inductive/field effects, solvation can affect acidity. Protic solvents like water or alcohols stabilize the conjugate base by hydrogen bonding. In fact, our common pK_a chart is referenced to water and would have very different numbers if done in a polar aprotic solvent like dimethyl sulfoxide (DMSO) that can't hydrogen bond or solvate anions well. Anions are much more basic in DMSO. For example, hydroxide, pK_{abH} 14.0 (15.7) has a pK_{abH} in DMSO of 31.4. Bulky alkyl groups can sterically hinder solvation and will decrease acidity; for example, acidity decreases with CH_3OH > $(CH_3)_2CHOH$ > $(CH_3)_3COH$. Polar solvents also favor dissociation by stabilizing charged species by clustering the oppositely charged ends of the solvent dipoles around the charged species as shown in Figure 2.19. An acid will dissociate less in a nonpolar solvent.

Enzyme active sites can be very different from bulk water (often quite nonpolar), so the pK_as of groups within can differ by several orders of magnitude. Electrostatic stabilization or destabilization of charge by neighboring charged groups also alters the pK_as of acidic and basic groups.

Ion pairing (two close opposite charges) is a very significant source of stabilization via ionic bonding. Biochemical systems often make use of 2^+ ions for catalysis. A water molecule that is complexed to a cation like Zn^{2+} is significantly more acidic than bulk water, because the conjugate base, hydroxide, is stabilized by ion pairing with the 2^+ ion. The enzyme carbonic anhydrase does exactly this to create a hydroxide nucleophile at neutral pH to rapidly convert carbon dioxide to bicarbonate (Section 6.2.3). Electrostatic catalysis is the acceleration of a reaction by ion pairing with the electric field of an opposite charge.

3.3.9 Super Acid Systems

Although fluorosulfonic acid, FSO_3H, (pK_a of about −10) and trifluoromethanesulfonic acid, CF_3SO_3H, (estimated pK_a of −15) are extremely strong acids, even stronger acid systems exist. The combination of a strong acid and a Lewis acid to complex and stabilize the conjugate base can produce an acid with an estimated pK_a less than −20. For example, the combination of FSO_3H with the Lewis acid SbF_5 is a super acid (sometimes called magic acid) that is a strong enough acid system to protonate an alkane. The strongest super acid system to date is HF with SbF_5, which has the very nonnucleophilic conjugate base SbF_6^- and has been used to study carbocations (themselves strong acids) under stable ion conditions.

3.4 STRUCTURAL FACTORS THAT INFLUENCE BASE STRENGTH

Since strong bases have weak conjugate acids, the information in the previous section also applies to base strength, only in reverse. Anything that stabilizes the conjugate acid lowers its acidity, and thus makes the base stronger. The bottom of Figure 3.4 shows that stabilization of HA or destabilization of A^- makes the acid HA weaker, thus making the base A^- stronger. As mentioned in Section 3.3.8, bases are much stronger in polar aprotic solvents like DMSO that have poor solvation of any conjugate base anions formed. Therefore, as was shown in Section 3.2.4:

Stronger bases have a higher pK_{abH}.

3.4.1 Anions are More Basic Than the Corresponding Neutral Species

Anions are more electron rich and so are more basic as illustrated by the higher pK_{abH} of the anions compared to their neutrals in the following compounds.

Anion	pK_{abH}	Neutral	pK_{abH}
NH_2^-	35	NH_3	9.2
$PhNH_2^-$	28	$PhNH_3$	4.6
OH^-	14.0 (15.7)	H_2O	0.0 (–1.7)
EtO^-	16	EtOH	−2.4
More basic		Less basic	

3.4.2 Lone Pairs with Less *s* Character are More Basic

Keeping within the same atom, mixing more *s* orbital into the lone pair hybridization stabilizes the lone pair because a 2*s* orbital is lower in energy than a 2*p* orbital. The more stable lone pair is less basic and harder to protonate illustrated by the lower pK_{abH} of the following compounds. Sections 3.4.4–3.4.6 are exceptions to this and to the next trend.

sp^3 Lone pair	pK_{abH}	sp^2 Lone pair	pK_{abH}	*sp* Lone pair	pK_{abH}
	10.6	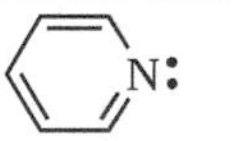	5.2	$H_3C{-}C{\equiv}N$:	−10
Most basic				Least basic	

3.4.3 Lone Pairs on Less Electronegative Atoms are More Basic

Lone pairs on electronegative atoms are held tighter in lower-energy orbitals because of the increased nuclear charge. These stabilized lone pairs are less available for protonation, as illustrated by the lower pK_{abH} of the following compounds.

$^-CH_3$ pK_a = 48	$^-NH_2$ pK_a = 35	^-OH pK_a = 14.0 (15.7)	^-F pK_a = 3.2
Most basic			Least basic
Least electronegative			Most electronegative

3.4.4 Conjugated Systems Preserve Conjugation If At All Possible

Lone pairs conjugated with pi systems are less basic than lone pairs not involved with conjugation. Conjugation provides extra bonding, and that stabilizes the electrons in the conjugated system, making them less basic. The following examples show the stabilization coming from conjugation dominating. Amides protonate on the oxygen lone pairs not involved in the allylic conjugation of the amide even though oxygen is more electronegative than nitrogen.

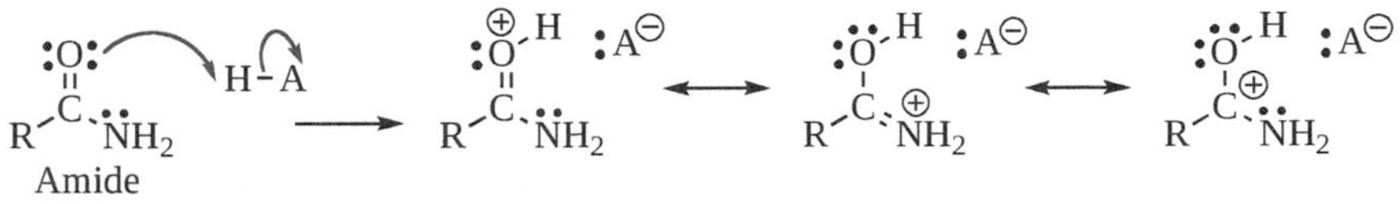

The base DMAP below protonates on the sp^2 nitrogen that is perpendicular to the pi system and not the lone pair involved in conjugation with the pyridine ring. Likewise, imidazole protonates on the lone pair not involved in the aromaticity of the ring system.

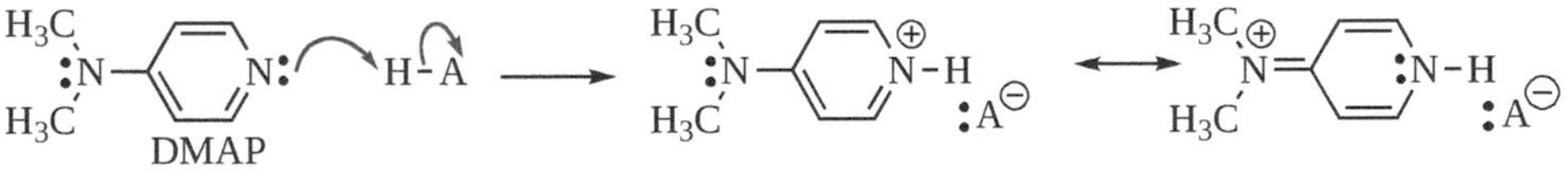

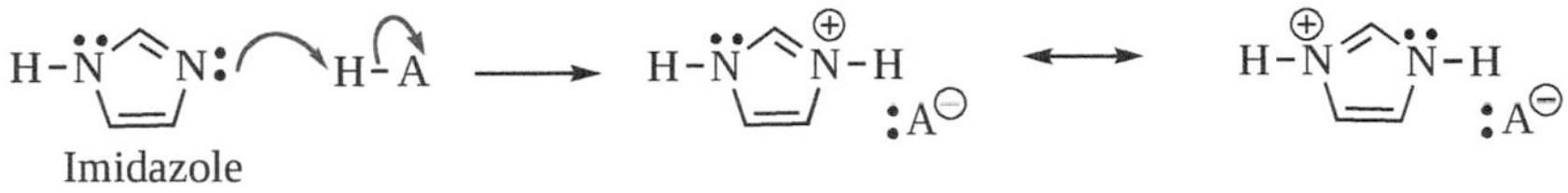

3.4.5 Inductive/Field Effects Affect the Stability of Conjugate Acids

Protonation of lone pairs adjacent to the partial positive charge of a carbonyl is less favored than protonation of the carbonyl itself. Protonation of the ester carbonyl produces a delocalized cation, whereas protonation of the ester's ether oxygen is destabilized by having a positive charge next to the partial positive carbonyl carbon.

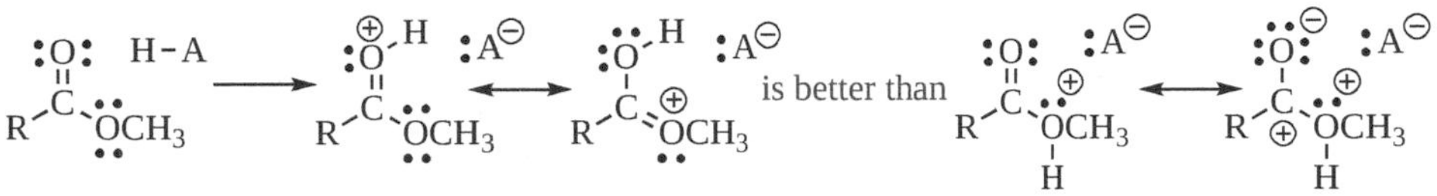

3.4.6 Resonance Effects Affect the Stability of Conjugate Acids

Arginine is an amino acid containing a guanidinium group that is positively charged at physiological pH. The resonance-stabilized cationic conjugate acid of this side chain group has a pK_{abH} of 13.6. At that very basic pH, it would be half protonated, half neutral. Arginine in a protein's active site can be used as a positive charge at physiological pH to attract a negatively charged substrate.

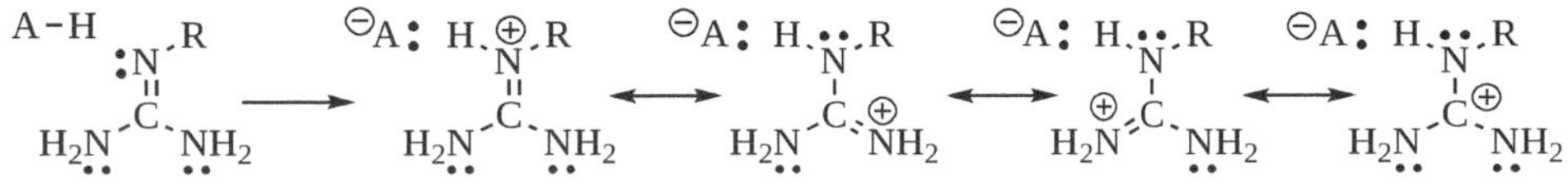

3.5 CARBON ACIDS AND RANKING OF ELECTRON-WITHDRAWING GROUPS

Carbon acids can serve as good examples for the principles of acidity. Often organic reactions in basic media start with deprotonation of an acidic C–H, so we will need to understand carbon acids in more detail. The major factor governing the acidity of the carbon–hydrogen bond is the stability of the resultant carbanion (conjugate base). A useful generalization is: **the more acidic the carbon acid (lower pK_a), the more stable the carbanion.** A pK_a chart then becomes a handy reference for carbanion stability.

3.5.1 Ranking of Electron-Withdrawing Groups

The major factor in the acidity of carbon acids is resonance delocalization of the conjugate base by electron-withdrawing groups (EWG). The pK_a chart becomes a guide (good enough for our purposes) for the approximate ranking of the common electron-withdrawing groups. The acidity of CH_3–EWG increases as groups better at withdrawing electrons more effectively stabilize the anionic conjugate base, $^-CH_2$–EWG. If the pK_as of CH_3–EWG's are compared, **the more acidic (lower pK_a) the CH_3–EWG is, the better the EWG group is at withdrawing electrons.** The following is a list of common electron-withdrawing groups ranked by the pK_a of CH_3–EWG.

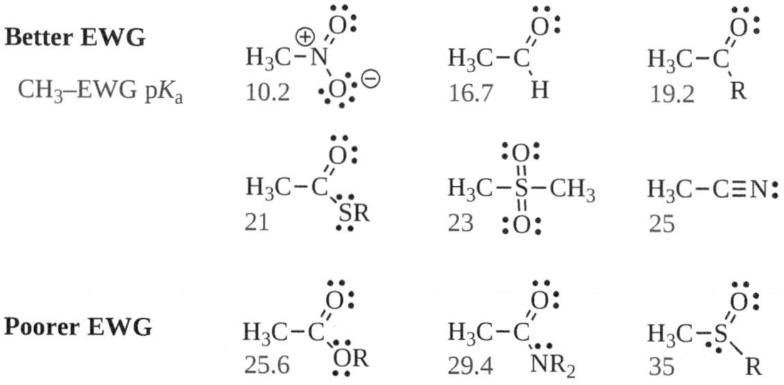

The best electron-withdrawing groups can stabilize the anion by resonance; for this reason they are called pi acceptors. They delocalize the anion and distribute the charge onto more electronegative atoms. The two most common electron-withdrawing groups are the nitrile and the carbonyl. The ability of the carbonyl to withdraw electron density depends on the nature of the group bound on the other side of it. If the group attached to the other side of the carbonyl also has a lone pair donating in electron density, the carbonyl becomes a poorer electron-withdrawing group. For this reason, ketones are better electron-withdrawing groups than esters, which are better than amides.

Example problem

Which of the following, CN, Ph, NO_2, is the best electron-withdrawing group? Which is the poorest?

Answer: The best electron-withdrawing group has the lowest pK_a for CH_3–EWG. Therefore CH_3NO_2, with a pK_a of 10.2, makes the nitro group the best. The next is the nitrile, CH_3CN, with a pK_a of 25. The worst electron-withdrawing group is phenyl because $PhCH_3$ has a pK_a of 40. The phenyl group is not really an electron-withdrawing group; its field/inductive effect is mildly withdrawing. It stabilizes the charge simply by delocalization, something it does almost equally well for either anions or cations.

Groups that are not pi acceptors stabilize an adjacent anion less effectively, and their CH_3–EWG pK_as are not yet available. Very electronegative groups, such as trimethylammonium cation, $^+N(CH_3)_3$, and trifluoromethyl, CF_3, can withdraw electron density through field/inductive effects. The halogens are resonance electron donors, but their inductive effect is withdrawing. The field/inductive effect of an electron-withdrawing group (resonance effects blocked) is occasionally needed and can be ranked by the effect on the pK_a of the carboxylic acid group, EWG–CH_2COOH (Section 3.3.4).

3.5.2 Carbanion Stability Trends

Hybridization has a dramatic effect on carbanion stability. Table 3.1 illustrates the strength, pK_a, electronegativity, and hybridization trends for some simple C–H bonds. Since the 2*s* orbital is closer to the nucleus and lower in energy than the 2*p* orbital, an anion residing in a hybrid orbital with greater %*s* (therefore a more electronegative orbital) is more stable. For carbon acids, the electronegativity of the carbon is more important than the C–H bond strength.

Resonance likewise has a major effect. Conjugation with double bonds or aromatic rings stabilizes anions by distributing the charge over more atoms. An electron-withdrawing group (EWG) directly bonded to the carbon bearing the negative charge can best stabilize the anion. The better the electron withdrawal is, the greater the stabilization. The stabilization is greatest by resonance (through pi bonds) to a pi acceptor. The more electron-withdrawing groups on the carbanion, the more stable it will be. However, additional groups do not have as much effect as the first group. The attachment of one carbonyl drops the pK_a of methane from 48 to 19.2; the attachment of a second carbonyl group drops the pK_a less, from 19.2 to 9.0.

Table 3.1 Hybridization of the C–H Bond and Acidity

Decreasing stability of carbanion formed →	HC≡CH	$H_2C{=}CH_2$	$H_3C{-}CH_3$
Hybridization of carbon	*sp*	sp^2	sp^3
%*s* character of carbon	50	33	25
pK_a of C–H bond	25	44	50
Electronegativity of C	3.29	2.75	2.48
Strength of C–H bond (kcal/mol)	131.9	110.7	101.1

Other factors contribute, but often to a much lesser degree. Inductive and field effects from electronegative atoms can also stabilize the carbanion somewhat. Hydrogen bonding is very rare for carbanions. Aromaticity greatly stabilizes carbanions, but the only common example is cyclopentadienyl anion $C_5H_5^-$ mentioned in Section 3.3.6.

Solvation of carbanions is complicated by their basicity. Less basic carbanions (p$K_{abH} \leq$ 29) can be made in protic media like alcohols that can hydrogen bond. But more basic carbanions will just deprotonate the protic solvent, and be quenched. Therefore, nonprotic solvents such as diethyl ether are commonly used as solvents for these very basic carbanions.

Example problem

Which of the following, $^-$C≡CR, $^-CH_2Ph$, $^-$CH(C≡N)$_2$, is the most stable carbanion? Which is the least?

Answer: The most stable carbanion will be the weakest base and have the lowest pK_{abH}. This means $^-$CH(C≡N)$_2$, with a pK_{abH} of 11.2, is the most stable. The conjugate base anion is stabilized by resonance with two nitrile electron-withdrawing groups. The next is $^-$C≡CR, with a pK_{abH} of 25. The acetylide anion is on an *sp* hybrid carbon, with an electronegativity of 3.29 (see Table 3.1 above). The least stable carbanion is $^-CH_2Ph$, with a pK_{abH} of 40.

3.5.3 Estimation of the pK_a for Related Compounds

The pK_a for compounds not on the pK_a chart must be estimated using closely related compounds. Since the hydrogens of a simple alkyl chain are not acidic, we need to focus only on the hydrogens on electronegative heteroatoms or hydrogens immediately adjacent to functional groups. We are looking for anything that can intrinsically stabilize the conjugate base. Hydrogens on electronegative atoms are reasonably acidic (Section 3.3.1). Hydrogens on atoms adjacent to the δ+ end of a polarized multiple bond (C=Y or C≡Y) are acidic because the conjugate base can delocalize the electron pair formed (Section 3.3.3).

Identify the functional groups in the compound, and then find the pK_a for hydrogens on or adjacent to those functional groups. In this way, the pK_a for $(CH_3)_2C{=}O$ at 19.2 on the chart must do for all simple ketones. Likewise, the pK_a for ethanol at 16 will have to do for 1-butanol. This is because any attached hydrocarbon pieces do not provide resonance or inductive stabilization, so they will not affect the pK_a more than a few units. For example, *tert*-butanol is just a few pK_a units less acidic than ethanol, and much of that may be due to steric hindrance of solvation of the conjugate base. To compare, 1-butanol is less than half a unit less acidic than ethanol. Don't get fooled by changes in the hydrocarbon part of the molecule since the "grease" contributes very little to the stabilization of the conjugate base, therefore does not significantly affect acidity. For compounds of similar structure, the error from this estimation process is usually just a few pK_a units. Try to get as close as you can to the structural piece that will stabilize the conjugate base. Look for anything that can stabilize the conjugate base by resonance delocalization or electronegativity. The resonance delocalization of the conjugate base with an electron-withdrawing group is a major factor, so make sure to look for the same functional group. Do not confuse the stabilization by a ketone with that of an ester or amide, simply because all three have a carbonyl group.

Example problem

Provide an approximate pK_a for each set of hydrogen atoms on the compound below.

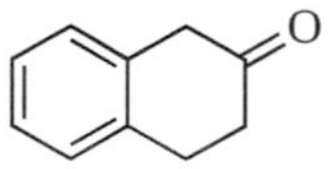

Answer: The trick is to pick a relevant model compound for each set of hydrogen atoms that is on the pK_a chart. For example, the conjugate base of set "a" has resonance with the ketone carbonyl and with the aromatic ring. The best model on the pK_a chart is shown for each and has the same stabilization of the conjugate base as the part of the compound we are interested in.

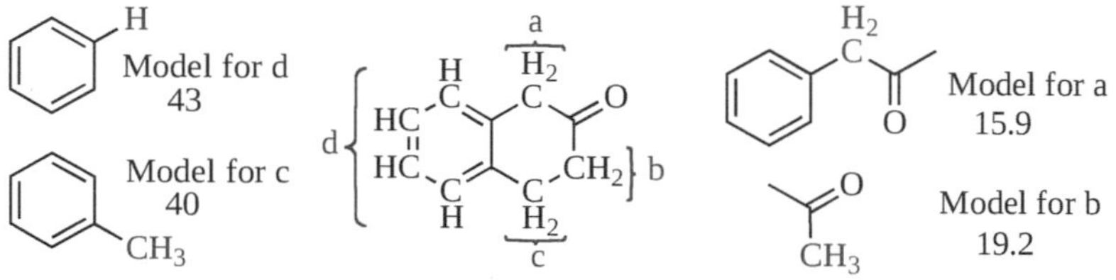

The previous examples illustrated the estimation of the pK_a for compounds that were structurally close to ones on the pK_a chart. It is necessary to recognize general structural features that affect acidity so that crude estimations of acidity can be made for compounds that do not have close relatives on the pK_a chart. Structural features that affect acidity are many, and examples can be found on the pK_a chart.

3.5.4 Finding the Most Acidic Hydrogen

Usually it is the most acidic hydrogen that is deprotonated in a proton-transfer reaction. However, finding the most acidic hydrogen of a compound is not always easy, especially if the compound has more than one functional group. A hydrogen on a carbon situated between two polarized multiple bonds, for example, N≡C–CH_2–C≡N, would be even more acidic because the conjugate base is very delocalized. In conclusion, check all hydrogens on electronegative atoms and those adjacent to functional groups; watch for any H whose loss would produce a highly delocalized system. Compare representative pK_as for each to find the most acidic H.

Example problem

What is the pK_a of $CH_3CH_2CH_2CH_2CH_2C{\equiv}N$?

Answer: The functional group is a nitrile, R–C≡N. The only nitrile on the pK_a chart is $CH_3C{\equiv}N$ at a pK_a of 25. Therefore in $CH_3CH_2CH_2CH_2CH_2C{\equiv}N$ the CH_2 next to the nitrile will have a pK_a close to 25. The other CH_2s are not acidic (pK_a of 50 or higher) because the conjugate base would not have any resonance stabilization.

3.6 CALCULATION OF K_{eq} FOR PROTON TRANSFER

Calculation of the equilibrium constant for the proton-transfer reaction is easy. **Proton transfers go toward the formation of the weaker base.** The reaction of hydroxide with hydrochloric acid is an example. A hydroxide pK_{abH} of 14.0 and a hydronium pK_a of 0.0 will be used for calculations in this text. Please see the bibliography for references on pK_a resources. The reaction goes from the stronger base, hydroxide, to the weaker base, chloride.

Base	Acid	Conjugate acid	Conjugate base
pK_{abH} = 14.0	pK_{aHA} = –6	pK_{aHA} = 14.0	pK_{abH} = –6

Because the reaction favors products, we expect the K_{eq} to be greater than 1. We can estimate the K_{eq} by knowing that each pK_a unit gives a factor of 10 in K_{eq}. So if the pK_{abH} in a proton transfer drops 2 pK_a units then the K_{eq} is 10^{+2}. In the above reaction, the pK_{abH} dropped 20 pK_a units [14.0 – (–6)] so the $K_{eq} = 10^{+20}$.

$$\log K_{eq} = pK_{abH} - pK_{aHA} \quad \text{or} \quad K_{eq} = 10^{(pK_{abH} - pK_{aHA})}$$

Subtract the pK_a of the acid from the pK_{abH} of the base to get the exponent of K_{eq}.

How unfavorable can a proton transfer be and still be useful in a reaction?

b^- +	H–A	b–H +	A^-
Base	Acid	Conjugate acid	Conjugate base

In this example, since the pK_a of bH is 16 and that of HA is almost 26, the K_{eq} would be very small, $10^{(16-26)}$ or 10^{-10}, indicating a strong preference for b^- and HA. If b^- and HA were initially equimolar, [b^-] = [HA], then it follows from the stoichiometry of the reaction that [bH] = [A^-]. Substitution into the K_{eq} expression produces:

$$K_{eq} = \frac{[A^-][bH]}{[HA][b^-]} = \frac{[A^-]^2}{[b^-]^2} \quad \text{or} \quad \sqrt{K_{eq}} = \frac{[A^-]}{[b^-]} = \sqrt{10^{-10}} = 10^{-5} \quad \text{or} \quad \frac{1}{100{,}000} = \frac{[A^-]}{[b^-]}$$

Therefore, the concentration of b^- at equilibrium is 100,000 times greater than A^-. On occasion a K_{eq} as unfavorable as 10^{-10} can still be useful. Products arising from A^- will be significant only if b^- produces a less favored product that easily reverts back to reactants, or if A^- is much more reactive than b^-. Consumption of A^- drives the equilibrium to replenish it.

If the proton transfer K_{eq} is greater than 10^{+10}, it can for all practical purposes be considered irreversible. An irreversible proton transfer is often the last step in a reaction and the driving force for the entire process. Each pK_a unit converts to a 1.364 kcal/mol (5.706 kJ/mol) contribution to the $\Delta G°$ of the reaction at room temperature.

Proton transfer is the first step in many important reactions. The ability to predict which protons will be pulled off in basic media or which sites will be protonated in acidic media is very important. The reaction solvent often behaves as the actual acid or base for proton transfer because the strongest acid that can occur in any medium is the protonated solvent, and the strongest base is the deprotonated solvent. Thus, the strongest acid in water is H_3O^+ and the strongest base is OH^- (the leveling effect).

The speed of proton transfer is usually very fast, often at the diffusion-controlled limit. Exceptions occur when the acidic H is very hindered, and when a carbanion deprotonates a C–H bond, which can be so slow that other reactions can easily compete.

Proton-Transfer K_{eq} Cross-Check and the ΔpK_a Rule Cross-Check

An unfavorable K_{eq} smaller than 10^{-10} is most likely too small to be useful. All favorable K_{eq} values are useful, and thus have no limit. The check of the proton transfer K_{eq} will be very important in mechanisms to decide if a proton transfer is reasonable. This lower limit principle generalizes into a very important piece of control knowledge: **the ΔpK_a rule cross-check: Avoid intermediates that are more than 10 pK_a units *uphill* from the reactants (either 10 pK_a units more basic or 10 pK_a units more acidic).** Reactions usually head toward neutralization, forming weaker acids and weaker bases, not stronger.

Example problem

Calculate the K_{eq} for proton transfer for the following reaction. Does it favor reactants or products?

$$EtO^- + PhOH \rightleftharpoons EtOH + PhO^-$$

Answer: The base in the forward reaction is EtO^- at a pK_{abH} = 16; the acid is PhOH at pK_a = 10. The K_{eq} equals $10^{(pK_{abH} - pK_{aHA})}$ or $10^{(16-10)} = 10^{+6}$, thus favoring products.

3.7 PROTON-TRANSFER MECHANISMS

Mechanisms are our best guess of the steps occurring in a reaction. Mechanisms can never be proven but may be disproved by experimental data. Mechanisms are like a grammatically correct sentence; to make sense, we need to use known words with good grammar. Our words are the electron flow paths and the grammar consists of the cross-checks for the paths. So far, we have one path, proton transfer (p.t.) and three cross-checks (media pH, most acidic H, and proton transfer K_{eq}). Before things get complicated with more paths and cross-checks, let's explore mechanisms.

3.7.1 Problem Space for Proton-Transfer Mechanisms

The set of all possible steps that could occur in a problem is its problem space. It includes disproven steps, untested possibilities, and routes consistent with existing data. Our search for an answer occurs within this problem space. Often instructors present just one reasonable route, leaving you to wonder whether your different route is OK or not. At this stage, we are concerned with a reasonable mechanism, not necessarily the absolute best. The problem space is often in the shape of a tree, with branches occurring at each decision point. We need an efficient way to search the problem space tree.

Consider the problem of driving from Los Angeles to San Francisco. The problem space is all the roads in the US. The first task in narrowing down the problem space is to decide in what direction the goal is. Since San Francisco is north of L.A., we can discard generally south routes, and the problem space reduces by half. The media pH cross-check is a similar mechanistic check: is the reaction run in acid or base? We need to stay on the roads (our electron flow paths). Some routes branching north from L.A. may be interstate highways and others dirt roads. Other cross-checks would now come in to aid us in picking a good route. We would like a good route that does not require white-knuckle mountain roads and is reasonably direct (going via Chicago is out). Working backward from our goal can help a lot. We may find in looking for alternatives that a certain process, although reasonable, does not go on to product; it is a dead end. Before the road gets too rough, backtrack to the last fork and try another route. Avoid the temptation to go off-road at the dead end, rather than going back to the main road.

3.7.2 Make Conscious Decisions

Being oblivious to forks in the road or deciding on impulse rather than using a map is an inefficient way to travel. Picking the first alternative that comes to mind is impulse, not thinking. Critical thinking is making a deliberate choice based on considering the worth of all the alternatives. Draw a map of alternatives you have already explored so you don't repeat the same mistakes. Be willing to reconsider choices already made, and if you are really stuck, consider asking for directions. Show your instructor your scratch sheet for the problem, so you can see where you took a wrong turn. Learn from your mistakes.

3.7.3 What Happened?—Mapping Changes

A great help in decreasing the size of the problem space is to map what changes have occurred in the reaction. With mechanisms, and especially proton-transfer reactions, draw out all the hydrogen atoms near the areas that change (line structure can hide changes because the Hs are not drawn). It is much easier to cross-check changes and your arrows if you always draw Lewis dot structures for any part of the molecule that comes near an arrow. You may have to number the carbon atoms in the reactant and in the product to identify what has changed. Ockham's razor (seek simplicity) guides us to make the fewest changes necessary when numbering our product.

3.7.4 Pruning the Tree—Cross-checks

Our cross-checks are tests of reasonableness. Our first check was north or south? Is the route going in the correct direction? Are we still on a good road? Are we climbing a ridiculously steep hill? For proton-transfer mechanisms the equivalent questions are the following. Is the reaction in basic or acidic media, and is our mechanism consistent with that? Did we pick a reasonable position on the molecule to protonate or deprotonate? Did we get the p.t. path arrows correct? Is the K_{eq} of the process reasonable?

3.7.5 Example Proton-Transfer Mechanism

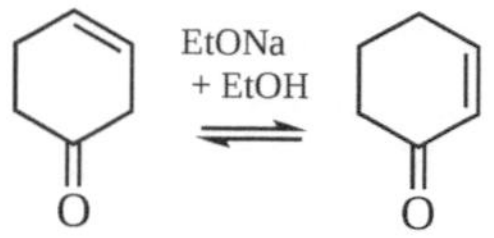

First, we need to figure out what has changed during the reaction. You need to see all the changes, and some may be hidden by the line structure. After drawing out the hydrogen atoms, we see that there are two ways we could number carbon atoms of the product, keeping the C=O as carbon number 1. The left one changes only carbons 2 and 4, whereas the right changes carbons 3, 4, 5, and 6. We choose the left, the simplest with the fewest changes. If we just picked the first numbering that popped into our head, there is a 50% chance of making the problem nearly impossible.

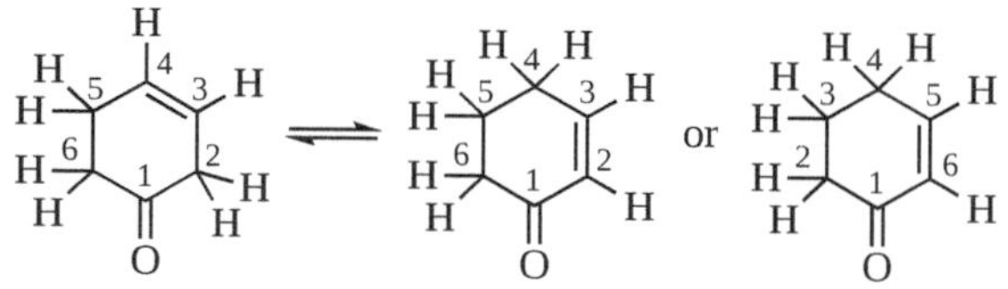

With the correct numbering we can see that carbon 2 lost an H and carbon 4 gained an H. On more complex problems, it is helpful to make a "to do" list of changes, but remember that list must be done in an order that makes chemical sense. The medium is basic, so acidic routes are out. Instead of protonating the molecule with an acid, we will deprotonate with a base. Both carbons 2 and 6 have acidic hydrogen atoms, since they are next to the polarized multiple bond, C=O. Although the H on carbon 2 is more acidic because it forms a more delocalized conjugate base, we should explore both possibilities. However good the deprotonation of carbon 6 may seem, it does not get us toward our goal. It is a very reasonable process that is a dead end. Since deprotonation of carbon 2 heads in the right direction, we need to know if the K_{eq} is within the range of our base. The pK_a of our reactant is known to be 15.2, and the conjugate acid of our ethoxide base is pK_a 16, which gives a K_{eq} of $10^{(16-15.2)} = 10^{+0.8}$, passing the cross-check, and actually downhill. The proton-transfer electron flow path forms the Lewis structure of the product of the first step. The major resonance form is the middle one (the minus on electronegative oxygen), but the right one gives us a hint as to why carbon 4 is basic. Therefore, if you are stuck, draw resonance forms.

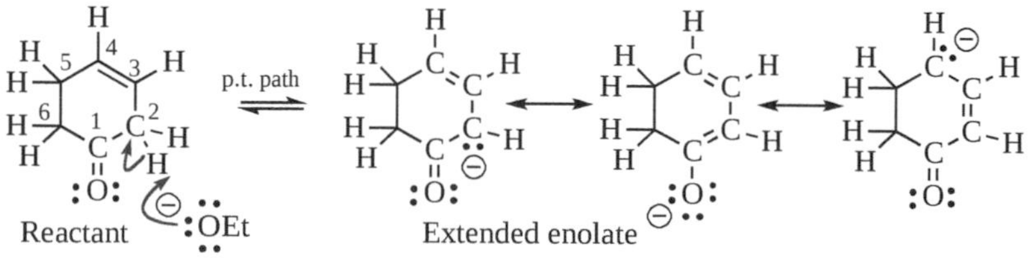

Our best acid in the solution is the solvent, ethanol. The proton-transfer path again gives us the Lewis structure of the product. The pK_a of the product is known to be 17.6, and the pK_a of ethanol (our acid) is at 16, so that gives a K_{eq} of $10^{(17.6-16)} = 10^{+1.6}$, passing the cross-check, and again downhill. Both proton-transfer steps in this case were downhill because ethanol's pK_a was higher than that of the reactant and lower than that of the product. The product is downhill from the reactant because it is conjugated and the reactant is not. Both the reactant (pK_a of 15.2) and the product (pK_a of 17.6) deprotonate to give the same extended enolate anion, but since the product is more stable, it has more of a climb to get to the extended enolate, so its acidity is less, indicated by a higher pK_a.

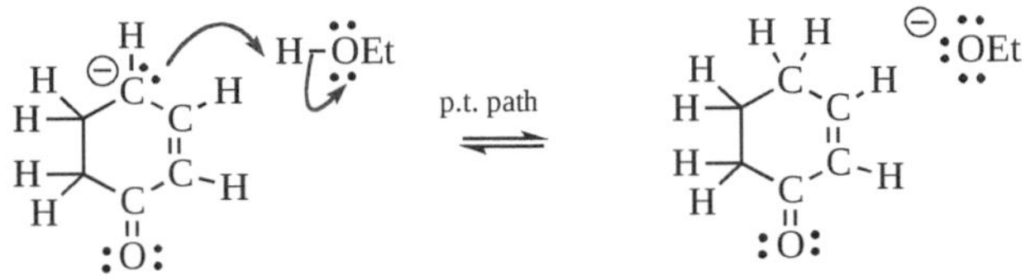

Now let's look at the problem space that we just navigated in Figure 3.5 to see all the branches and where else we could have gone under basic conditions. Two new enol structures from the protonation of the oxygen anion are very reasonable dead ends for this problem. A $\Delta H°$ calculation will show that the enol is uphill from the reactant.

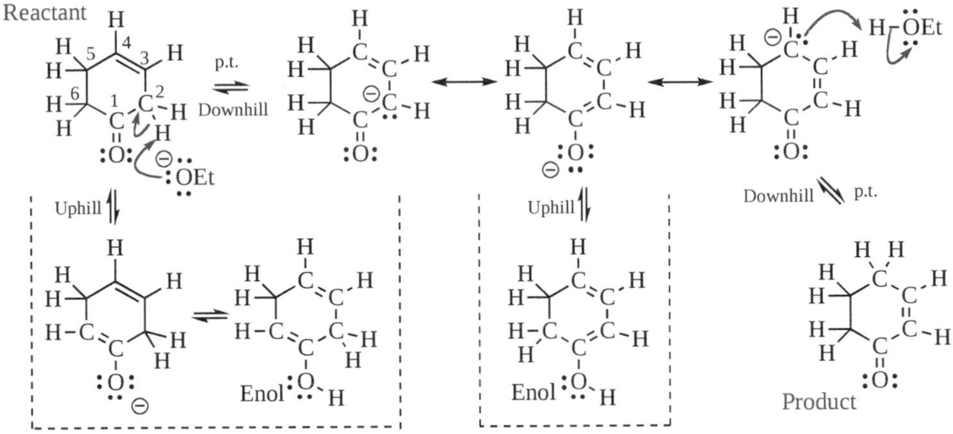

Figure 3.5 Reasonable basic problem space for the example proton-transfer mechanism. Acid routes have been dropped. Only proton-transfer routes are shown. Routes deprotonating carbons 3, 4, and 5 have been dropped because the proton-transfer K_{eq} is less than 10^{-10}, and thus not useful.

The decisions and operations that we went through to navigate this example proton-transfer mechanism problem space are flowcharted in Figure 3.6. One common mistake is to forget that line structure hides changes in hydrogen atoms. Another common mistake is to forget to check the acidity or basicity of medium, and draw a mechanism for the wrong pH.

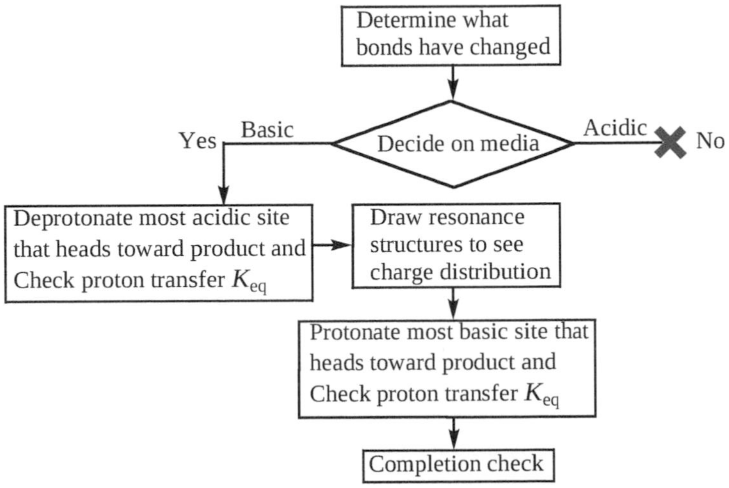

Figure 3.6 Flowchart for Section 3.7.5 basic media example proton-transfer mechanism.

3.8 COMMON ERRORS

Where could we have gone astray here? Most often, it is forgetting a cross-check.

3.8.1 Failure to Stay on the Path

We could have done the previous mechanism in one step with four arrows, but that would assume a three-molecule collision, something that is very rare. The arrows would not involve the C=O, which is essential for the reaction to proceed. This reaction goes through a resonance-stabilized intermediate, the stability of which we can check. Made-up combinations of arrows are hard to check and may just shift the lines and dots of our Lewis structure without saying anything chemically understandable. For our mechanistic sentences to make sense we must use known words, the electron flow paths.

Biochemical Note: An enzyme's active site can have weak acids and bases close to our reactant. It is often possible to have proton transfer occurring in the same step as another electron flow path. These are discussed separately in Section 7.4.3.

3.8.2 Failure to Check K_{eq}

Often beginners will forget to check the K_{eq} for proton transfer and use a base to remove a proton that is not acidic. It should be assumed that the hydrogen is not acidic if you cannot find the appropriate or a related compound on the pK_a chart. As an illustration, you will not find diethyl ether, $CH_3CH_2OCH_2CH_3$, on the pK_a chart; none of its hydrogens are acidic. The carbonyl hydrogen of an aldehyde, RC**H**O, also is not acidic. If you cannot find the compound, or one with a closely related functional group, on the pK_a chart, then suspect that its hydrogens may not be acidic. Therefore, it is very important that you know how to use the pK_a chart.

3.8.3 Media pH Errors

In our mechanism example, we could have started off on an impulse to do an acidic mechanism in basic conditions. What would have caught us was our first K_{eq} cross-check. Our best acid in the solution is ethanol. Using ethanol as an acid (pK_a = 16), the most favorable K_{eq} we could have gotten would have been protonating the carbonyl lone pair at 10^{-23}, which is not usable. We would be saved by our cross-check. Generally a good rule is: **In acid protonate the reactant, and in base deprotonate it.**

3.8.4 pK_a Span Errors

Beginners will sometimes try to do both acid-catalyzed and base-catalyzed processes at the same time, forgetting that a strong acid and base present in the same vessel would neutralize each other. **No medium can possibly be both strongly acidic and strongly basic.** The relative hydronium ion and hydroxide ion concentrations are defined by $K_w = [H_3O^+][OH^-] = 10^{-14}$. As one increases, the other must decrease. So when working in acidic media, avoid creating or using strong bases. Likewise, in basic media, avoid creating or using strong acids.

3.8.5 Drop-off H^+ Only Shorthand Notation

A shorthand notation you may see in some texts is to drop a proton off by itself or show a naked proton being picked up by a base. Under the conditions found in almost all reactions, the proton is transferred from acid to base and does not float off on its own. There is nothing wrong with using a shorthand notation, as long as you remember it is not reflecting reality.

3.8.6 Wrong pK_a Errors

Be careful when looking up a pK_a to determine a K_{eq}, because the wrong pK_a values will lead to an incorrect decision based on poor data. It is an unfortunate reality that many compounds can serve as either an acid or a base. A very common example is water. Water can act as a base to form hydronium or can act as an acid to form hydroxide. Since this book's pK_a chart lists both the acid and its conjugate base, some compounds appear twice. Ethanol is on the pK_a chart twice, once as an acid at 16, and once as a conjugate base at −2.4.

3.8.7 Strained Intramolecular Proton-Transfer Errors

Sometimes a proton can be transferred internally if the transition state is not too strained. Three- and four-membered transition states are energetically unfavorable. The optimum ring size for intramolecular proton transfer is a five- or six-membered transition state.

3.9 PROTON-TRANSFER PRODUCT PREDICTIONS

The problem seems simple, "Draw a reasonable proton-transfer reaction given the following conditions." However, several skills and correct decisions are required to get a reasonable answer. If any one of these is done incorrectly, the final answer will be wrong, so let's pick this problem apart. A proton transfer is simply passing a proton from one site to another, from acid to base (Section 3.1). We need to decide whether the conditions are basic or acidic, since we tend to protonate substrates in acid and deprotonate substrates in base (Sections 3.2.1 and 3.2.2). We need to identify the acid (Section 3.2.3) and decide where the most acidic hydrogen is located (Section 3.5). We also need to identify the base and its most basic site (Sections 3.2.4 and 3.4). For this, we need to understand the acidity and basicity trends (Sections 3.3 and 3.4), and use a pK_a chart if one is available (Appendix). Once we have found the most acidic H and most basic site, then comes the test of thermodynamic reasonableness (Section 2.5): Is the K_{eq} greater than 10^{-10} (Section 3.6)? If that test is passed all we have to do is draw out the proton transfer correctly, using good Lewis structures (Section 1.3) and arrows (Section 1.4), and avoiding common mistakes such as charge balance errors. As you can see, this one "simple" problem makes use of much of what you've learned and shows the interrelatedness of organic chemistry. So let's try one!

Example problem

Draw a reasonable proton-transfer reaction given the following conditions.

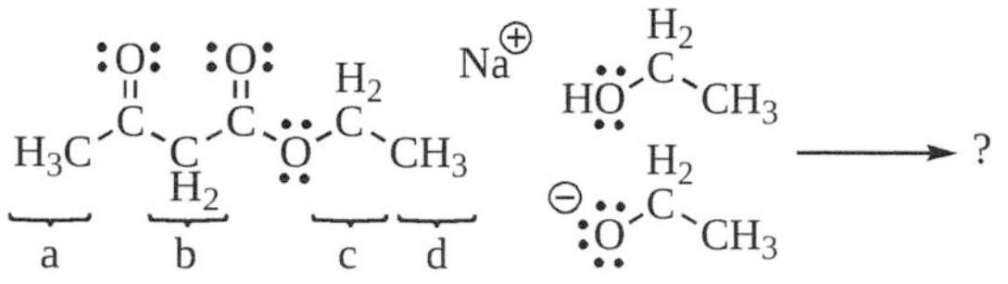

Answer: First decide whether the medium is acidic or basic. With both ethanol and ethoxide present, the media is very basic (visualize Figure 3.2 but with the pK_a of 16 of ethanol). Since ethoxide is much more basic than ethanol, ethoxide is our base. Now we must decide which is the most acidic H on the substrate: a, b, c, or d. Using acidity trends, we decide that the protons on carbon labeled "b" between the two electron-withdrawing groups will have the most delocalized conjugate base, and so are most acidic. We can check this with a pK_a chart: The H on "a" is adjacent to a ketone for a pK_a of about 19.2 ; the H on "b" is between a ketone and an ester for a pK_a of about 10.7; the Hs on "c" and "d" have no delocalization of the conjugate base and so are expected to be close to ethane at pK_a 50. Clearly, the protons on "b" are most acidic because they have the lowest pK_a. The check for thermodynamic reasonableness is next: $K_{eq} = 10^{(pK_{abH} - pK_{aHA})}$ or $10^{(16 - 10.7)} = 10^{+5.3}$ which is clearly a downhill process. After the check of K_{eq}, we can finally say that the following is a correctly drawn and reasonable proton-transfer reaction. The arrows could be drawn to form any resonance form of the conjugate base. The conjugate base is a delocalized anion, and just the major resonance form is shown below.

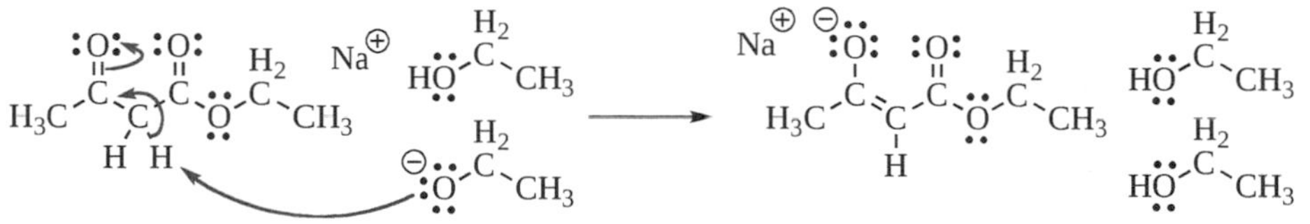

The K_{eq} is greater than 1, indicating the proton-transfer reaction favors products. Charge is balanced. Figure 3.7 shows each of the decisions in the problem space just navigated.

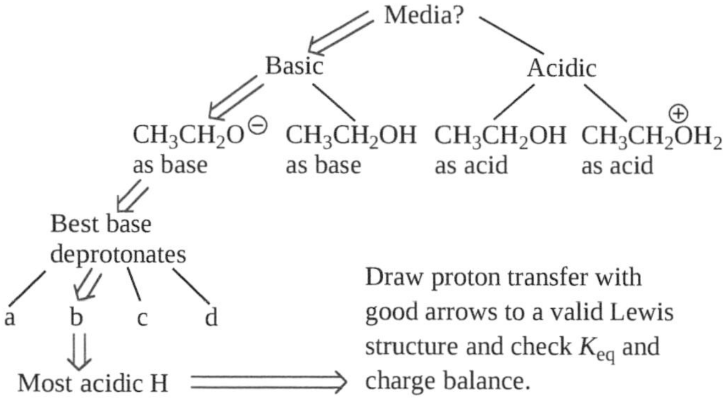

Figure 3.7 Problem space for predicting a reasonable proton transfer in basic media.

3.10 PROTON-TRANSFER SUMMARY

The take-home message for this chapter is that structure determines reactivity. Figure 3.8 relates the major factors that influence the acidity of a compound.

- Strong acids have a low pK_a and have weak conjugate bases.
- Strong bases have a high pK_{abH} and have weak conjugate acids.
- Proton transfer $K_{eq} = 10^{(pK_{abH} - pK_{aHA})}$
- A useful lower limit for proton transfer is: K_{eq} equal to or greater than 10^{-10}.
- Always decide if the reaction is run in acid or base and stay within those bounds.
- Acidic Hs are found on heteroatoms and adjacent to polarized multiple bonds.
- In acid protonate the reactant, and in base deprotonate it.

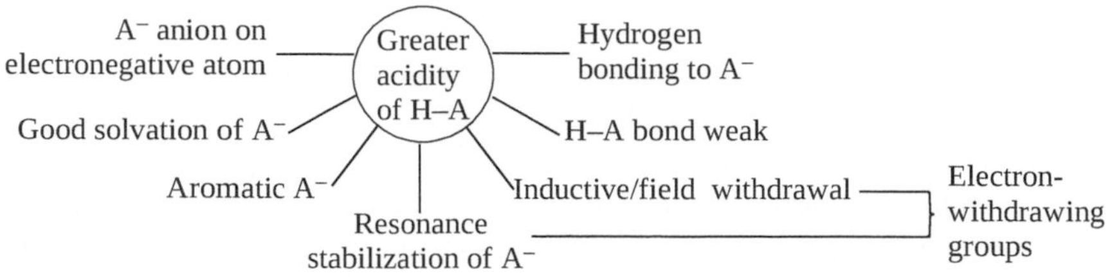

Figure 3.8 Web of concepts for factors increasing acidity.

CHAPTER 3 EXERCISES

3.1 Using the pK_a chart in the Appendix, calculate the numerical value of K_{eq} for these reactions. Predict whether it favors reactants or products and estimate $\Delta G°$ at 25°C.

(a) PhONa + PhSH $\rightleftharpoons$ PhSNa + PhOH

(b) 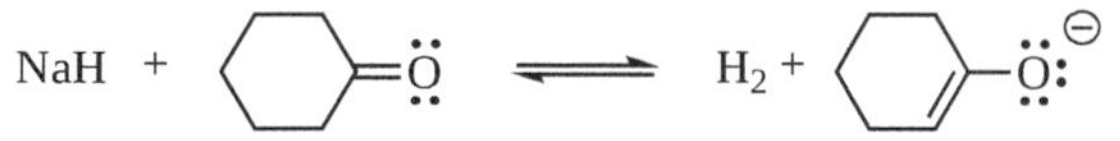

3.2 Using the pK_a chart in the Appendix, calculate the numerical value of K_{eq} for these reactions. Predict whether it favors reactants or products and estimate $\Delta G°$ at 25°C.

(a) $(CH_3)_3COK$ + HC≡CH $\rightleftharpoons$ $(CH_3)_3COH$ + HC≡CK

(b) CH_3ONa + CH_3COOH $\rightleftharpoons$ CH_3OH + CH_3COONa

3.3 Trends: With the help of the pK_a chart when needed, rank all species, beginning with the numeral 1 to designate:

(a) The most stable carbanion

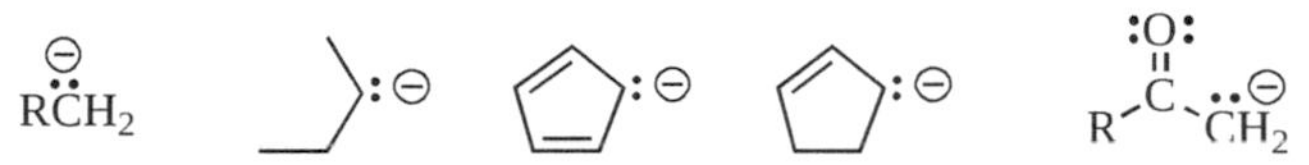

(b) The best electron-withdrawing group

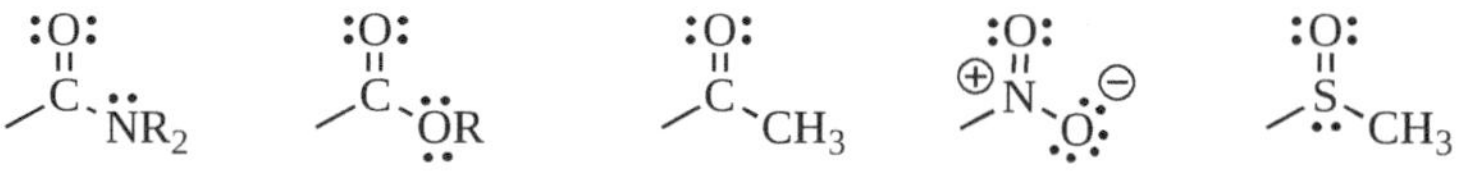

3.4 Trends: With the help of the pK_a chart when needed, rank all species, beginning with the numeral 1 to designate:

(a) The strongest base

^{-}CN $PhNH_2$ $^{-}NH_2$ HO^{-} $(CH_3)_3CO^{-}$

(b) The strongest acid

H_3O^{+} HF CH_3OH HCN NH_3

3.5 For the diprotic malonic acid $CH_2(COOH)_2$ (pK_{a1} = 2.8 and pK_{a2} = 5.7), using crossing and break points, draw a species vs. pH graph for malonic acid, like Figures 3.2 and 3.3. What is the major species in solution at pH 4.3, at pH 7.0?

3.6 Use your knowledge of anion stability and resonance to decide which of the following two alternatives is the lower-energy process.

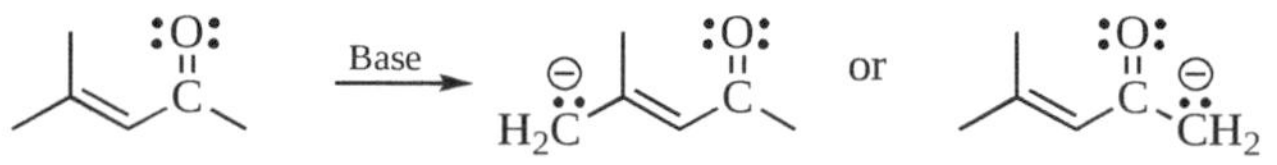

3.7 With the help of a pK_a chart, find the most acidic H in each of the following compounds.

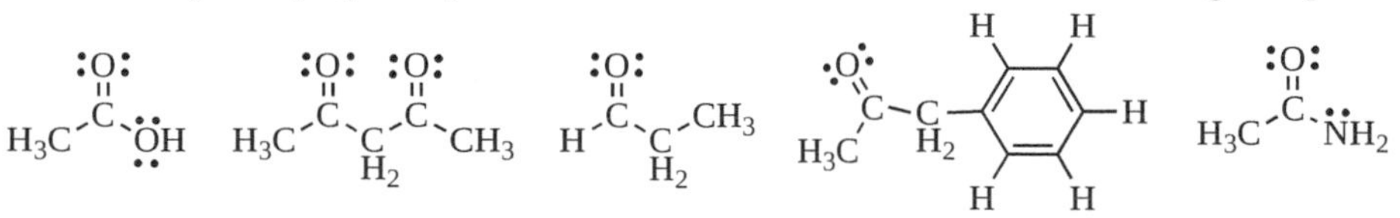

3.8 Use your knowledge of proton-transfer equilibria to decide which of the following two alternatives is the lower-energy process.

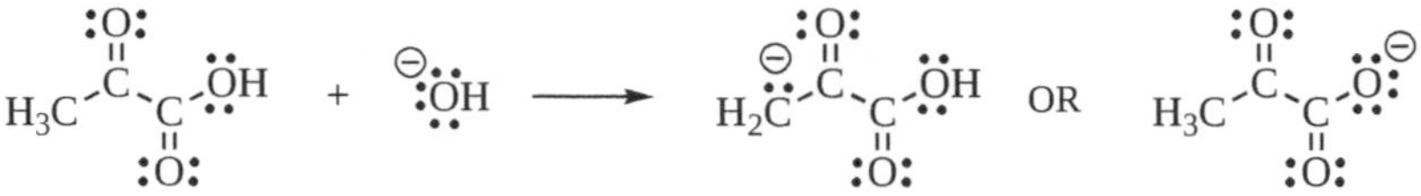

3.9 Ascorbic acid, vitamin C, has one very acidic H ($pK_{a1} = 4.2$), and all the others are not very acidic ($pK_{a2} = 11.6$). Use your knowledge of anion stability and resonance to find the acidic H and explain why it is so acidic. Draw all the resonance forms of the conjugate base.

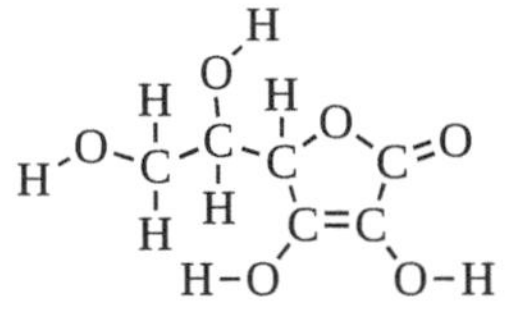

3.10 Decide whether these conditions are acidic or basic and give an approximate pH.
(a) A water solution of 2 M NH_4Cl and 2 M NH_3
(b) A water solution of 1.0 M sodium acetate and 1.0 M acetic acid
(c) A water solution of 1.0 M sodium acetate and 0.1 M acetic acid

3.11 Decide whether these conditions are acidic or basic and give an approximate pH.
(a) A water solution of 0.1 M PhONa and 1.0 M PhOH
(b) A water solution of 1 M KF and 1 M HF
(c) A water solution of 1.0 M KF and 0.1 M HF

3.12 Draw an energy diagram for the proton-transfer reaction of $CH_3COOH + NH_3$. Calculate the proton transfer K_{eq} to get an approximate $\Delta G°$ for the reaction.

3.13 What is the most acidic hydrogen of the following compound?

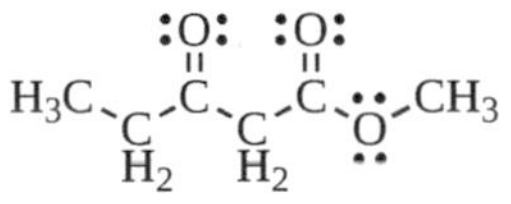

3.14 Fill in the blank for the missing partners in these proton-transfer reactions. Give K_{eq}, $\Delta G°$, and the equilibrium direction of each problem.

(a) 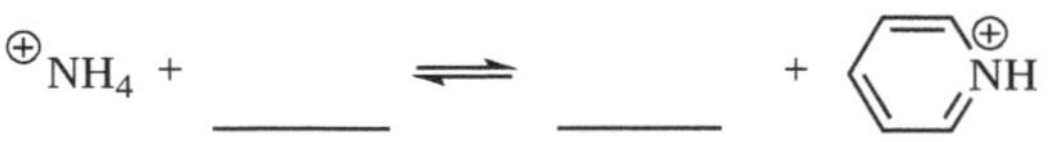

(b) 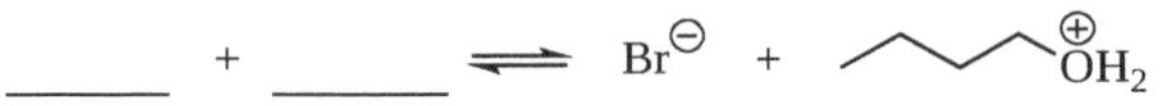

3.15 Fill in the blank for the missing partners in these proton-transfer reactions. Give K_{eq}, $\Delta G°$, and the equilibrium direction of each problem.

(a) 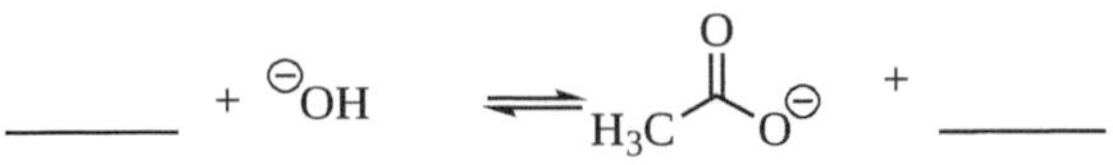

(b) 

3.16 Predict the pK_as of all the Hs in the following compounds.

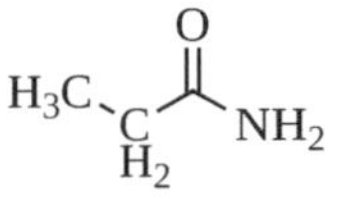

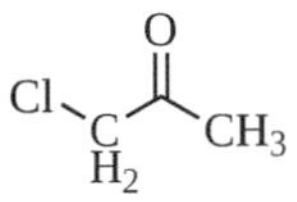

$H_3C-C\equiv CH$

3.17 Trends: With the help of the pK_a chart when needed, rank all species, beginning with the numeral 1 to designate:

(a) The weakest base

^-CN $\quad$ $PhNH_2$ $\quad$ $^-NH_2$ $\quad$ HO^- $\quad$ $(CH_3)_3CO^-$

(b) The weakest acid

H_3O^+ $\quad$ HF $\quad$ MeOH $\quad$ HCN $\quad$ NH_3

3.18 Trends: With the help of the pK_a chart when needed, rank all species, beginning with the numeral 1 to designate:

(a) The least stable carbanion

$H_3C-\ddot{C}H_2^{\ominus}$ $\quad$ $NC-\ddot{C}H_2^{\ominus}$ $\quad$ $Ph-\ddot{C}H_2^{\ominus}$ $\quad$ $H_2C=\ddot{C}H^{\ominus}$ $\quad$ $O_2N-\ddot{C}H_2^{\ominus}$

(b) The worst electron-withdrawing group

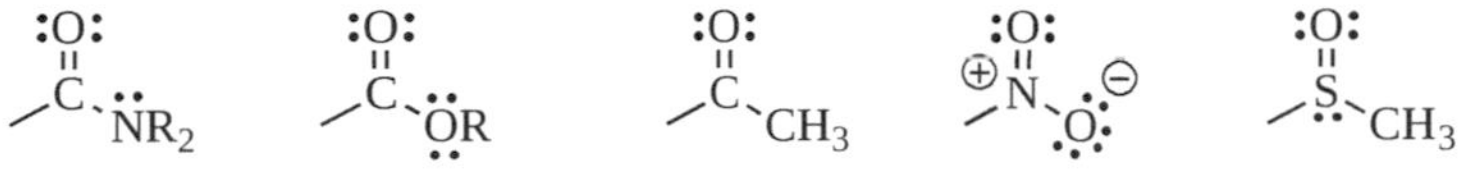

3.19 Draw a reasonable proton transfer for the reaction below, and calculate the value of K_{eq}. A cyanoketone pK_a is not listed; it is about halfway between the dicyano and the diketone.

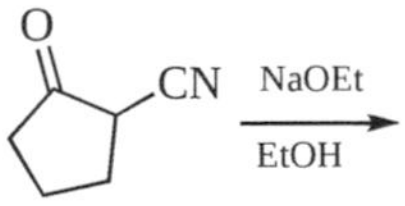

3.20 Is this a reasonable proton-transfer reaction? KOH + $PhCH_3$ → H_2O + $PhCH_2K$.

3.21 Draw the most reasonable proton-transfer product given the following conditions.

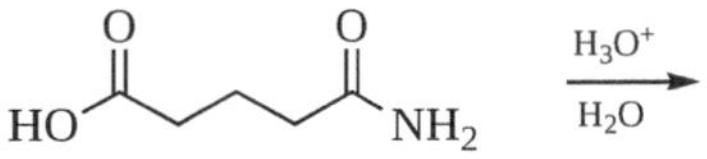

3.22 Draw the proton-transfer products for the following reaction and using the pK_a chart in the Appendix, calculate the value of K_{eq} and estimate $\Delta G°$ at room temperature. Draw a qualitative energy diagram.

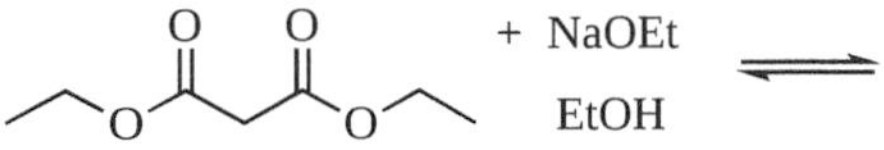

3.23 Draw the proton-transfer products for the following reaction and using the pK_a chart in the Appendix, calculate the value of K_{eq} and estimate $\Delta G°$ at room temperature. Draw a qualitative energy diagram.

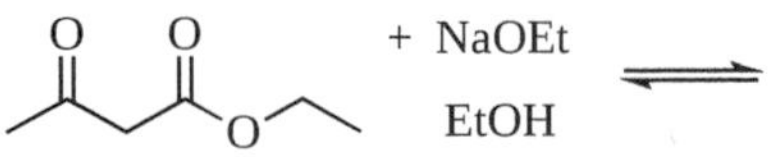

3.24 Provide a proton-transfer mechanism for this isomerization in basic media. Which is favored, reactant or product?

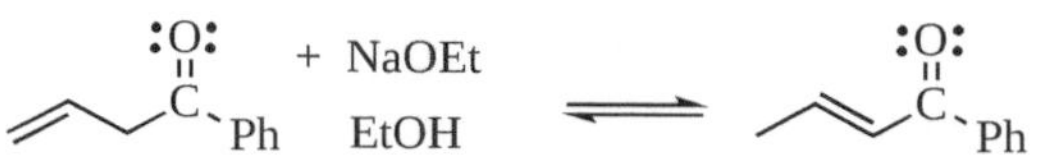

3.25 Draw the proton-transfer products and calculate the value of K_{eq} and $\Delta G°$ at room temperature. The cyanoester is not on the chart, but is easily estimated as halfway between the dicyano and the diester.

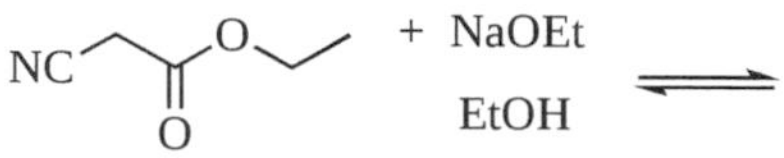

3.26 Decide which of the two anions below is more stable and explain why.

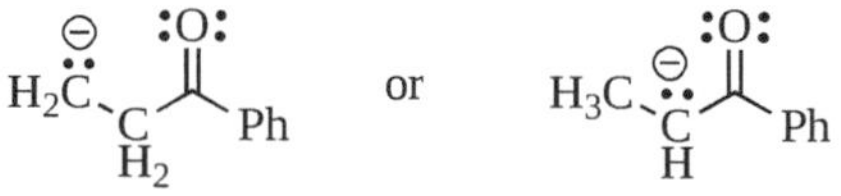

3.27 In cells, enolpyruvic acid is in equilibrium with pyruvic acid. In the lab in acidic media, the first step most likely is a protonation of enolpyruvic acid to give a resonance-delocalized cation intermediate, followed by a deprotonation to give pyruvic acid. With those hints, propose a reasonable mechanism. Draw both resonance forms of the intermediate cation and decide which form is major. Use H_3O^+ to protonate and H_2O to deprotonate.

Enolpyruvic acid Pyruvic acid

3.28 Given that the K_{eq} for the reaction below converting the keto form into the enol is about 10^{-8}, and that the pK_a for the keto form is 18.3, what is the pK_a of the enol? Both the enol and keto form have the same conjugate base, the enolate form.

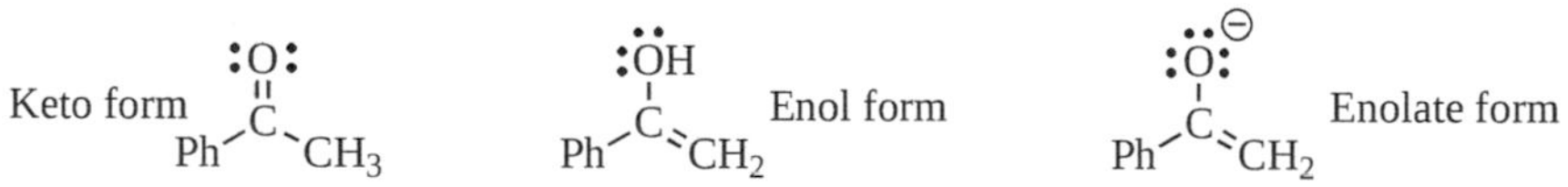

3.29 For the simplest amino acid, glycine, NH_2CH_2COOH (pK_{a1} = 2.31 and pK_{a2} = 9.24), using crossing and break points, draw a species vs. pH graph for glycine, like Figures 3.2 and 3.3. In acid, the nitrogen is protonated. What is the major species in solution at pH 6.0?

3.30 In each molecule below, circle the most acidic H.

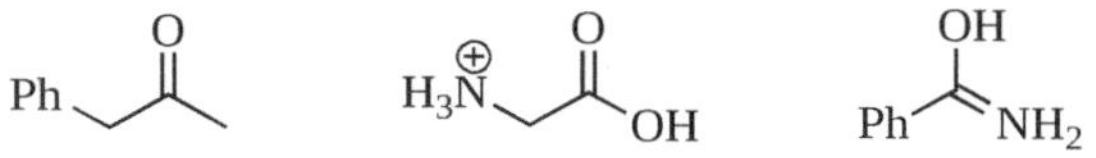

3.31 Sugars isomerize between aldehyde and ketone forms via an enediolate HO–C=C–O⁻ or enediol intermediate. Give a reasonable proton-transfer mechanism in basic water for the reaction below. Since the alcohol and the H adjacent to the C=O have almost the same acidity, there are two possible first steps. Explore both so you recognize the dead end.

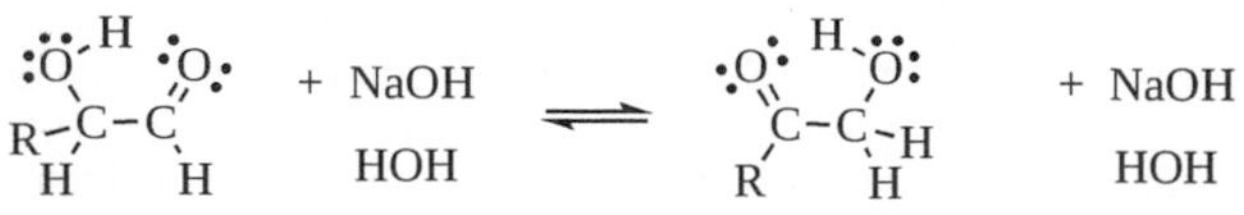

4

IMPORTANT REACTION ARCHETYPES

4.1 INTRODUCTION TO REACTION ARCHETYPES

There are really only four organic reaction types (Figure 4.1). The first three can occur intramolecularly (within the same molecule) or intermolecularly (via a bimolecular collision),

Electron Flow in Organic Chemistry: A Decision-Based Guide to Organic Mechanisms,
Third Edition. Paul H. Scudder.

whereas rearrangement is only an intramolecular process. The archetypes are:

Substitution—replace a leaving group with a nucleophile

Elimination—lose a leaving group and usually a proton to form a pi bond

Addition—add a nucleophile and an electrophile to a pi bond

Rearrangement—produce a constitutional isomer

If we understand idealized versions or archetypes of each of these four reaction types, we will have come a long way toward mastering the basics of most organic reactions. Figure 4.1 gives generic examples of each reaction type. We will find that substitutions at a tetrahedral center are significantly different from substitutions at a trigonal planar center. Since trigonal planar center substitutions usually go via a two-step process of addition then elimination, it is easier to discuss them after those processes.

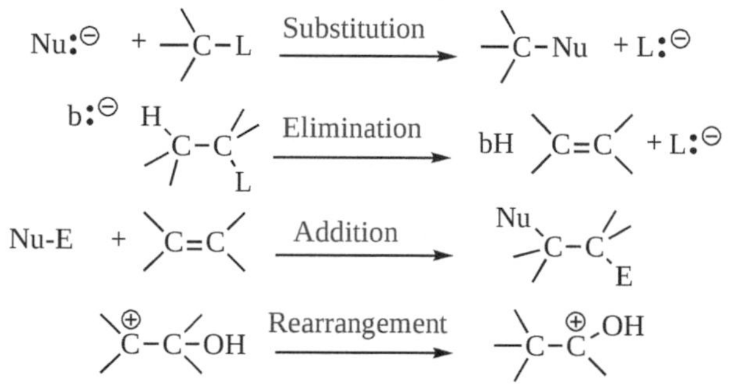

Figure 4.1 Examples of the four different reaction archetypes.

4.2 NUCLEOPHILIC SUBSTITUTION AT A TETRAHEDRAL CENTER

4.2.1 Substitution Electron Flow Paths

The energy surface in Figure 4.2 gives an overview of three possible routes to substitute a leaving group bound to a tetrahedral center. Our two new generic groups (a lone pair nucleophile and a leaving group on a tetrahedral carbon) are found as the reactants in the upper right corner. We will look at the reactivity of nucleophiles and leaving groups and what characterizes these two groups later in this section. The leaving group–carbon bond must break (the vertical axis) and the carbon–nucleophile bond must form (the horizontal axis) to reach the substitution products in the lower left.

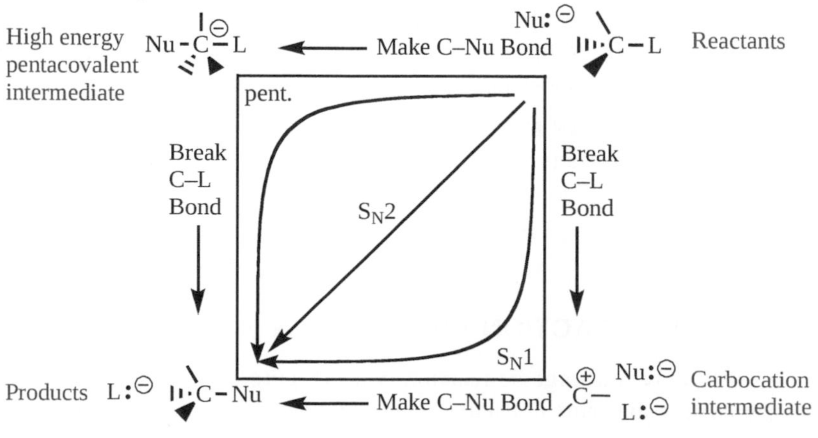

Figure 4.2 Top view of a simplified energy surface for substitution at a tetrahedral center.

Three idealized routes can occur. The pentacovalent intermediate route is not available to second-row elements like carbon but is well established for third-row elements like phosphorus, capable of *d* orbital bonding. There are two routes for carbon that differ only in the timing of when the nucleophile attacks and when the leaving group falls off: **S_N1** (substitution, nucleophilic, unimolecular): the leaving group departs first (rate determining) by path D_N, producing a reasonably stable carbocation; the carbocation is then trapped by a nucleophile, path A_N. **S_N2** (substitution, nucleophilic, bimolecular): the simultaneous attack of nucleophile and loss of the leaving group, path S_N2.

These three new electron flow paths are shown in Figure 4.3. Path D_N (dissociation, nucleofugic) is the loss of a leaving group; the arrow starts from the C–L bond, breaking it, forming a carbocation and a new lone pair on the leaving group. Path A_N (association, nucleophilic) is the microscopic reverse, a lone pair nucleophile adds to the carbocation, to form a C–Nu bond. This two-step combination of paths D_N + A_N forms the S_N1 route. The two paths, combined into one step of A_ND_N, form the S_N2 route. But before we can discuss these routes, we must understand what makes a good nucleophile, a good leaving group, and a good carbocation.

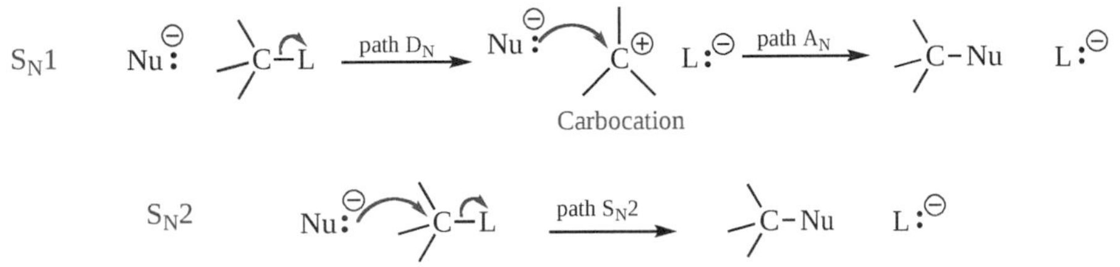

Figure 4.3 The electron flow paths for substitution at a tetrahedral carbon.

4.2.2 Ranking of Nucleophiles

Nonbonding electron pairs are good electron sources, especially lone pairs of anions, which are electron rich. The following list gives the relative nucleophilicity of many lone pair nucleophiles in an S_N2 reaction on methyl iodide. An anion is always a better nucleophile than its neutral counterpart; for example, methoxide anion, CH_3O^-, is almost 2 million times more nucleophilic than neutral methanol, CH_3OH.

We might suspect that the more basic species would generally be more nucleophilic, but nucleophilicity does not parallel basicity exactly because **softness must be considered first.** The electron pair on the nucleophile has to distort towards the electrophile for the bond making to happen. The valence electrons on softer atoms, by definition, are more polarizable and easier to distort. Solvation also has a major effect. The harder ions are more tightly solvated in polar solvents and cannot break free of the solvent "cage" easily to serve as a nucleophile as well as a poorly solvated softer ion. To verify this, we need only to look at the following list and see that iodide at pK_{abH} −10 is over a thousand times more nucleophilic than the much more basic (and harder) acetate, pK_{abH} 4.8. In comparing atoms with similar softness, then basicity plays the major role. **For lone pairs of the *same element,* the more basic the lone pair is, the more nucleophilic it is.** To rank nucleophiles in this way, remember to look up the conjugate acid on the pK_a chart.

Steric hindrance decreases nucleophilicity. As the nucleophile becomes larger, its bulk tends to get in the way of its acting as a nucleophile, especially when the electrophile is also large. For this reason, the larger *tert*-butoxide, $(CH_3)_3CO^-$, is a much poorer nucleophile than the smaller methoxide, CH_3O^-.

Relative nucleophilicity toward CH_3I (in CH_3OH as solvent)	
CH_3OH	1
F^-	500
CH_3COO^-	20,000
Cl^-	23,000
Et_2S	220,000
NH_3	320,000
PhO^-	560,000
Br^-	620,000
CH_3O^-	1,900,000
Et_3N	4,600,000
CN^-	5,000,000
I^-	26,000,000
Et_3P	520,000,000
PhS^-	8,300,000,000
$PhSe^-$	50,000,000,000

The following is a ranked list of the pK_{abH} of only the oxygen nucleophiles from the previous list. The one exception to this trend is that nucleophiles with adjacent lone pairs like HOOH, NH_2NH_2, and NH_2OH are more nucleophilic than expected; this α-effect is most likely due to poorer solvent stabilization.

Most nucleophilic and most basic O	$\mathbf{pK_{abH}}$
CH_3O^-	15.2
PhO^-	10
CH_3COO^-	4.8
CH_3OH	–2.2
Least nucleophilic and least basic O	

Electron availability decreases as the orbital the electrons occupy is made more stable. As the hybridization of the nitrogen lone pair goes from sp^3 to sp^2 to sp, the lone pair gets less basic and less nucleophilic as indicated by the following trend.

Most nucleophilic and most basic N	
Et_3N:	pK_{abH} = 10.7; lone pair hybridization = sp^3
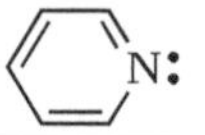	pK_{abH} = 5.2; lone pair hybridization = sp^2
$CH_3C{\equiv}N$:	pK_{abH} = –10; lone pair hybridization = sp
Least nucleophilic and least basic N	

An exception to this hybridization trend is that an ester's carbonyl oxygen sp^2 lone pairs are much more basic and more nucleophilic than its ether oxygen sp^3 lone pairs. When an ester's carbonyl lone pair is protonated, a delocalized cation is formed, as shown below. When an ester's ether oxygen lone pair is protonated, the cation is less stable because it is not delocalized and the electron withdrawal by the adjacent carbonyl destabilizes it.

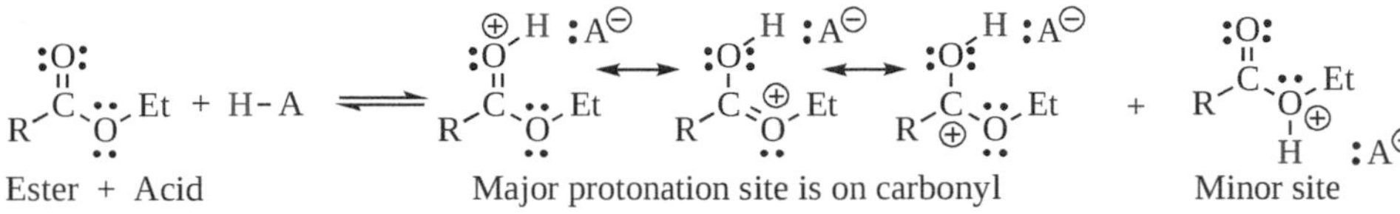

Solvation is a hard–hard interaction, and therefore **solvent effects drastically alter nucleophilicity**. Small, hard, highly shielded ions must "break out" of the solvent cage to be available as nucleophiles, and thus nucleophilicity will vary greatly with solvation. Solvation, in fact, appears to be mostly responsible for the following halide nucleophilicity trend in methanol. The ion most strongly solvated is fluoride, the weakest iodide. In polar aprotic solvents like dimethyl sulfoxide, the halide anion is poorly solvated, and the nucleophilicity order reverses from that in methanol, with fluoride being the best nucleophile. The trend is dominated by the strength of the bond formed (the C–F bond is strongest and C–I is weakest).

Best halide nucleophile in methanol $I^- > Br^- > Cl^- > F^-$ **Worst**

A pK_a chart can be used as a reference for nucleophilicity only if the difference in softness is considered. A partially plus carbon atom is a much softer electrophile than a proton. Soft ions are more nucleophilic in protic solvents because tighter solvation greatly decreases the nucleophilicity of the hard ions.

We can restate the HSAB principle. Hard nucleophiles favor binding with hard electrophiles; soft nucleophiles favor binding with soft electrophiles. Most nucleophilicity charts show relative rates of nucleophilic attack with methyl iodide as the electrophile. A carbon–iodine bond is very soft because the electronegativity difference is nearly zero (Section 1.2). Therefore softer (less electronegative, more polarizable) atoms have lone pairs that are better electron sources toward soft electrophiles.

In summary, to rank the nucleophilicity of nonbonding electron pairs reacting in protic solvents with soft electrophiles such as R–X, rank first by softness, then by basicity (within the same attacking atom). However, for nonbonding electron pairs reacting with harder electrophiles such as a proton or a carbonyl, rank by basicity. Very reactive electrophiles like carbocations are not selective and react with the most abundant nucleophile (commonly the solvent).

Example problem

Which of the following, PhO^-, PhS^-, CH_3COO^-, is the best nucleophile to react with CH_3I in methanol? Which is the worst?

Answer: Since the electrophile is soft and the solvent is protic, the softness of the nucleophile must be considered. Because sulfur is softer than oxygen, the softest and most nucleophilic is PhS^-, thiophenoxide. The remaining oxygen anions are then ranked by basicity: Phenoxide (PhO^-), $pK_{abH} = 10$, is more basic than acetate (CH_3COO^-), at $pK_{abH} = 4.8$. Therefore, the next best nucleophile is phenoxide, and the least is acetate.

4.2.3 Ranking of Leaving Groups

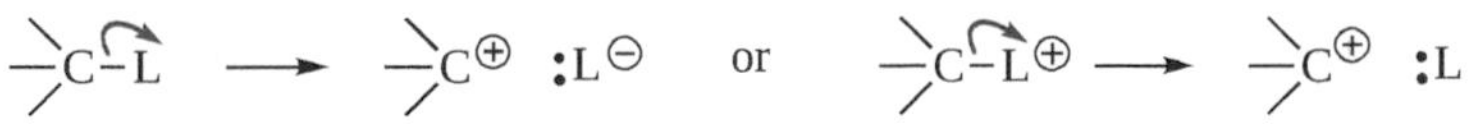

A group that separates from a compound and takes with it its bonding pair of electrons is called a *leaving group* (L) or nucleofuge (nucleus fleeing), and can depart as an anion or as a neutral species. As the leaving group gets better, so does the substitution reaction to replace it. If a substrate has two different tetrahedrally bound leaving groups, then the better leaving group is usually the first to react. It is common to rank leaving groups by their conjugate acid pK_a, the pK_{aHL}. **A good leaving group, L, has a zero or negative pK_{aHL}.** The pK_{aHL}s of poor leaving groups range from 13 to 30. The loss of a group like NH_2^- whose pK_{aHL} is greater than 30 is exceedingly rare.

Some groups do not fit this highly simplistic scheme for ranking leaving groups. If the L is highly electronegative, or if the carbon-leaving group bond is weaker or more polarizable, the bond is more easily broken. In ranking leaving groups by their pK_{aHL}, we are drawing a parallel between two reactions that are close but not identical:

Only three types of leaving groups fall significantly out of order when ranked by pK_{aHL}. Fluoride and cyanide are much *poorer* leaving groups than predicted (due to a strong C–L bond). The charge delocalized alkyl or aryl sulfonates, RSO_3^- or $ArSO_3^-$, are much *better* leaving groups than predicted by their pK_{aHL}s (they are better than iodide); common representatives of these follow. Also, molecular nitrogen, N_2, is an excellent L.

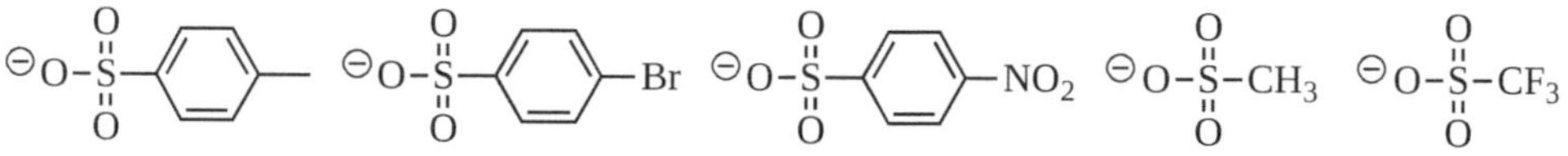

Tosylate, OTs Brosylate, OBs Nosylate, ONos Mesylate, OMs Triflate, OTf

When using a pK_a chart to rank leaving groups, always look up the conjugate acid of the leaving group on the chart. If EtO^- departs, look up EtOH, pK_a +16; if EtOH departs, look up $EtOH_2^+$, pK_a −2.4. Protonation of a group before it falls off makes it cationic, more electronegative, and a much better leaving group. The following is a ranked list of common leaving groups and their conjugate acid pK_a. **Note that H^-, NH_2^-, and CH_3^-, each with a pK_{aHL} above 30 are not leaving groups.**

Excellent	**pK_{aHL}**
N_2	<−10
$CF_3SO_3^-$	−15
$ArSO_3^-$	−6.5
$CH_3SO_3^-$	−6

Good	**pK_{aHL}**
I^-	−10
Br^-	−9
Cl^-	−6
EtOH	−2.4
H_2O	0.0
CF_3COO^-	+0.5

Fair	**pK_{aHL}**
$O_2NC_6H_4COO^-$	+3.4
$RCOO^-$	+4.8
$O_2NC_6H_4O^-$	+7.2
NH_3	+9.2
RS^-	+10.6
NR_3	+10.7

Poor	**pK_{aHL}**
HO^-	+14.0
EtO^-	+16
$RCOCH_2^-$	+19
$ROCOCH_2^-$	+26

Example problem

Which of these, R–Cl, R–Br, R–OH, has the best leaving group, and which has the worst?

Answer: The best leaving group is bromide, pK_{aHL} of −9. The next-best leaving group is chloride, pK_{aHL} of −6. The worst leaving group is hydroxide, pK_{aHL} = 14.0.

4.2.4 Carbocation Stability Trends

Figure 4.4 gives the geometries of the alkyl and vinyl carbocations. There are two common hybridizations for carbocations, sp^2 for alkyl carbocations and *sp* for vinyl carbocations. The carbon bearing the positive charge hybridizes to place the valence electrons in the most stable arrangement. Since hybrid orbitals with a higher percentage of *s* character (%*s*) are more stable, these are filled first, and an empty *p* orbital is left over. Alkyl carbocations can be relatively easy to form and are sp^2 and trigonal planar at the cationic center. Vinyl carbocations, however, are linear and more difficult to form because the cation is on a more electronegative *sp*-hybridized carbon.

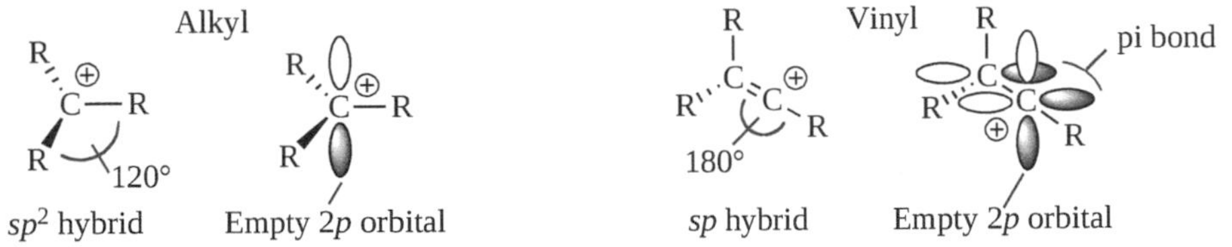

Figure 4.4 The geometries of an alkyl carbocation (left) and a vinyl carbocation (right).

Carbocations are usually formed either by loss of a leaving group or by addition of an electrophile to a pi bond. Carbocation formation is very rare in nonpolar solvents that cannot stabilize the cation. Polar solvent molecules can cluster their lone pairs around the cationic center, forming an intermolecular donation of electron density that stabilizes the carbocation. Some carbocations are stable enough to be on the pK_a chart; the more stable ones are the weaker acids. Recall that when the pH equals the pK_a, the conjugate acid and base concentrations are equal. A protonated amide with a pK_a of −0.5 requires very acidic water before it is half-cation and half-neutral. A protonated ester with a pK_a of −6.5 is much less stable. However, since most carbocations are not on the pK_a chart, we need another method to rank carbocation stability.

A lone pair on an atom directly bonded to the cationic center serves as a pi-donor group and stabilizes the carbocation by donating some of the lone pair electron density. A pi bond is formed between the pi-donor lone pair and the empty 2*p* orbital of the carbocation (Figure 4.5 top). The less electronegative the atom bearing the lone pair, the greater will be its capacity to donate. A lone pair in a nonbonding 2*p* orbital will be able to stabilize a carbocation better than a lone pair in a 3*p* or 4*p* orbital because of better overlap with the empty 2*p* carbon orbital. **The better the electron-releasing group, or the more of them attached to the cationic center, the more stable the carbocation is.**

Carbon–carbon double bonds can stabilize the carbocation by donating some electron density from the pi bond by resonance. The electrons in a pi bond are less available because they are usually in a lower-energy bonding molecular orbital compared to a nonbonding orbital of a lone pair. Resonance with a pi bond forms a conjugated system, distributing the charge over more atoms (Figure 4.5 middle). Stabilization of a carbocation by one double bond gives the allylic cation.

Alkyl groups stabilize the carbocation to a much lesser extent by donating some electron density from the sigma framework via *hyperconjugation* (Figure 4.5 bottom). The more substituted a carbocationic center is, the more stable it is (tertiary, R_3C^+ > secondary, R_2HC^+ > primary, RH_2C^+ > methyl, H_3C^+).

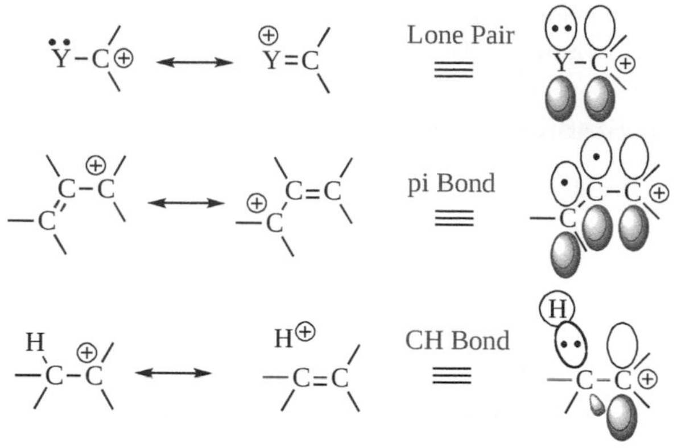

Figure 4.5 Carbocation resonance forms and the pi-type orbital overlap that those resonance forms represent. *Top*: Resonance with a lone pair is best. *Middle*: Resonance with a pi bond is next. *Bottom*: sigma bond hyperconjugation is poorest; the H^+, pi bond resonance form is minor.

Table 4.1 provides a ranking of carbocations by stability. We can state a general trend for carbocation stability: **Carbocations stabilized by resonance with a lone pair are more stable than those stabilized only by resonance with a carbon–carbon double bond, which in turn are more stable than those stabilized only by alkyl group substitution.** There are overlaps where three of a lesser type of stabilization are at least as good as one of the better type. For example, three R groups (tertiary alkyl) stabilize better than one pi bond (primary allylic). Figure 4.6 shows this simplified stability trend.

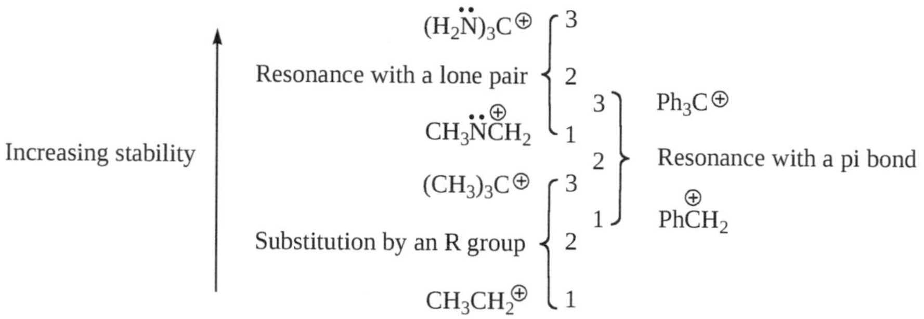

Figure 4.6 Simplified carbocation stability trend.

Vinyl carbocations are less stable than the corresponding alkyl carbocation of similar substitution (secondary vinyl cation is less stable than the secondary cation but more stable than the primary cation). Aromaticity greatly stabilizes carbocations; an example is tropylium cation, $C_7H_7^+$ (Section 1.9).

Example problem

Which of the following cations, $(CH_3)_2HC^+$, $Ph(CH_3)_2C^+$, $(CH_3)_3C^+$, is the most stable? The least stable?

Answer: The delocalized tertiary benzylic carbocation, $Ph(CH_3)_2C^+$, is the most stable. The *tert*-butyl carbocation, $(CH_3)_3C^+$, stabilized by three methyl groups, is of intermediate stability, whereas the isopropyl carbocation, $(CH_3)_2HC^+$, stabilized by only two methyl groups, is the least stable.

Table 4.1 The Carbocation Stability Ranking

Carbocation Name	Structure	Comments on Stabilization
Stable carbocations		
Guanidinium cation	$(H_2\ddot{N})_3C^{\oplus}$	Three N lone pairs delocalize plus
Tropylium cation	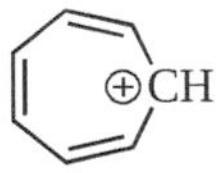	Stable by aromaticity
Protonated amide	$H_3C{-}C^{\oplus}(\ddot{O}H)(\ddot{N}H_2)$	Delocalized by N and O lone pairs
Moderately stable		
Protonated carboxylic acid	$H_3C{-}C^{\oplus}(\ddot{O}H)(\ddot{O}H)$	Delocalized by two O lone pairs
Triphenylmethyl cation	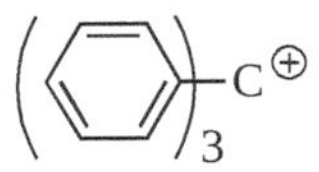	Delocalized by three phenyl groups
Protonated ketone	$H_3C{-}C^{\oplus}(\ddot{O}H)CH_3$	Delocalized by one O lone pair
Diphenylmethyl cation	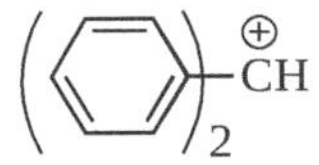	Delocalized by two phenyl groups
Average stability		
Tertiary alkyl cation (3°)	$(CH_3)_3C^{\oplus}$	Stabilized by three alkyl groups
Benzyl cation	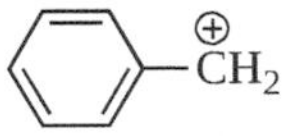	Delocalized by one phenyl group
Primary: allyl cation	$H_2C{=}HC{-}\overset{\oplus}{C}H_2$	Delocalized by one pi bond
Acylium cation	$H_3C{-}\overset{\oplus}{C}{=}\ddot{\underset{\cdot\cdot}{O}}$	Lone-pair-delocalized vinyl cation
Moderately unstable		**No resonance delocalization**
Secondary: alkyl cation (2°)	$(CH_3)_2\overset{\oplus}{C}H$	Stabilized by two alkyl groups
Secondary: vinyl cation	$H_3C{-}\overset{\oplus}{C}{=}CH_2$	Plus on more electronegative *sp* C
Unstable		**(Rarely formed)**
Primary alkyl cation (1°)	$CH_3\overset{\oplus}{C}H_2$	Stabilized by one alkyl group
Primary vinyl cation	$H{-}\overset{\oplus}{C}{=}CH_2$	Plus on more electronegative *sp* C
Phenyl cation	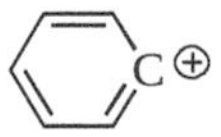	Bent vinyl cation (not delocalized)
Methyl cation	$^{\oplus}CH_3$	No stabilization

Nonclassical Carbocations

Certain primary or secondary carbocations can enter into a type of bonding with a pi bond or strained sigma bond that involves three orbitals overlapping to give a bonding molecular orbital that is occupied by two electrons, similar to a pi-adduct. These bridged species can be transition states, intermediates, or sometimes the lowest-energy structure of the carbocation. The 2-norbornyl cation (Figure 4.7) is a nonclassical ion, but much experimentation was required to determine that the bridged ion was the lowest-energy species, not a transition state between two interconverting secondary carbocations.

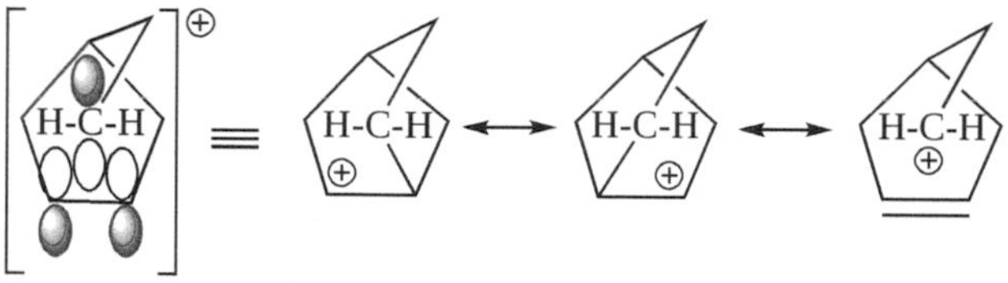

Figure 4.7 The 2-norbornyl nonclassical ion. The resonance forms on the right describe this unusual three-center, two-electron bonding molecular orbital.

To generalize, Figure 4.8 shows three ways that the sigma overlap of a lone pair, a pi bond, or a sigma bond can stabilize a carbocation. The most effective sigma-overlap stabilization comes from overlap of the empty orbital of the carbocation with a lone pair orbital forming an onium ion (Figure 4.8 top); for example, if Y is Br, then it is called a bromonium ion. For comparison, Figure 4.5 gave the more common pi overlap stabilization of a carbocation by a lone pair, a pi bond, and a CH bond.

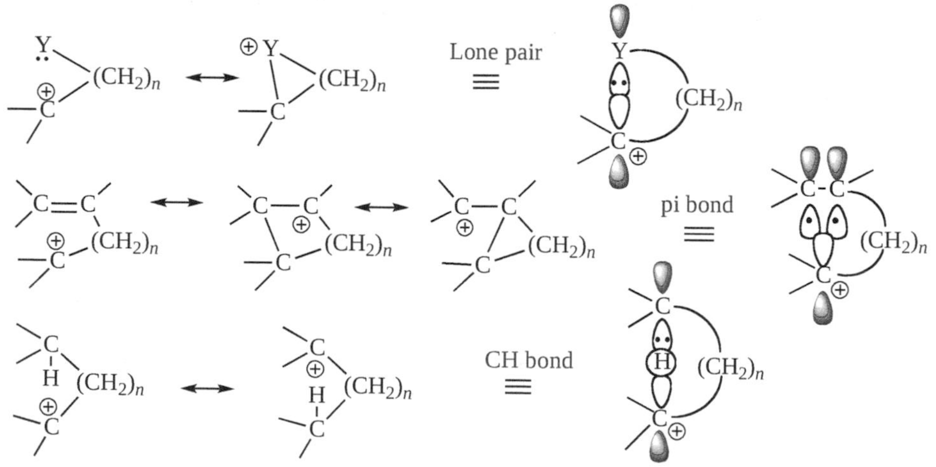

Figure 4.8 Summary of ways a filled orbital can stabilize a carbocation using sigma-type overlap. The arc symbolizes one or more CH_2 groups. Most common is the onium ion shown on the top.

4.2.5 Ranking of Electron-Releasing Groups

The ranking of electron-releasing groups (ERG), also known as electron-donor groups (EDG), is very important. The electron richness of electron sources will depend on this trend. The ranking is based on how available the electrons of the donor are; the more stabilized they are, the less available they are. Electrons in nonbonding orbitals are the most available. The best donors are anions that can donate an electron pair by resonance. The better donors all contain lone pairs that can form a strong pi bond to the empty *p* orbital of the cation, and are pi donors for that reason. Table 4.2 is a ranked list of electron-releasing groups.

Table 4.2 The Ranking of Electron-Releasing Groups, ERG

Excellent donors	**Features of group**
$-CH_2^{\ominus}$	Anionic and least electronegative
$-NH^{\ominus}$	Anionic and more electronegative than C
$-O^{\ominus}$	Anionic and more electronegative than N
Good donors	
$-N(CH_3)_2$	Neutral and less electronegative than O
$-NH_2$	Neutral and less electronegative than O
$-OH$	Neutral and more electronegative than N
$-OCH_3$	Neutral and more electronegative than N
$-NH-C(=O)-CH_3$	Amide carbonyl EWG decreases N lone pair availability
$-SCH_3$	Sulfur 3*p* orbital has poor overlap with a 2*p* carbon orbital
Poor donors	
—Ph	Delocalization through resonance with phenyl
—R	Hyperconjugation only, no lone pair or pi resonance
—H	No substituent at all
Very poor donors	
$-O-C(=O)-CH_3$	Electronegative oxygen, ester carbonyl EWG decreases O lone pair availability
$-Cl$	Electronegative and poor orbital overlap (a 3*p* with a 2*p*)

For second-row elements, as the nuclear charge increases, any electrons in the 2*p* shell are pulled closer to the nucleus. These tighter held electrons are the reason for the electronegativity effect. As the atom containing the 2*p* lone pair becomes more electronegative, it is less able to donate electron density. For this reason, carbon anions are better electron releasing groups than nitrogen anions, which are in turn better donors than oxygen anions. Anions are the best electron releasing groups because they donate both by resonance and by inductive/field effects. Neutral heteroatoms are more electronegative than carbon is; therefore, inductive electron withdrawal competes with resonance electron donation, making them poorer lone pair donors than anions.

An electron-withdrawing group attached to a lone pair donor decreases the electron availability of the lone pair. For example, an amide nitrogen lone pair is a much poorer electron-releasing group than an amine nitrogen lone pair. The electron-withdrawing carbonyl group of the amide diminishes the electron donation ability of the nitrogen lone pair. Delocalization of the nitrogen lone pair into the carbonyl stabilizes the lone pair and makes the nitrogen partially positive and more electronegative.

The formation of a strong pi bond is responsible for the stabilization afforded by the pi donor. However, pi bonds between orbitals from different shells necessarily have poor overlap and are weak. A sulfur atom is a poor pi donor because there is poor overlap of the sulfur 3*p* orbital with the carbon 2*p* orbital. The two worst lone pair donors are poorer donors than alkyl groups because their high electronegativity is not compensated by their weak pi donation ability. For example, a chlorine atom is not only very electronegative but also is a poor resonance pi donor because of poor 2*p*-3*p* overlap.

Resonance effects will usually dominate over inductive/field effects. When the resonance effect is blocked, the only groups that will still be donors are anions and alkyl groups. The electronegativity of most other groups will cause them to be inductive electron-withdrawing groups. Electron-withdrawing groups destabilize carbocations. Similarly, electron-releasing groups destabilize anions.

Example problem

Which of these, OCH_3, CH_3, NH_2, is the best electron-releasing group? The worst?

Answer: Both the ether and the amine are lone pair donors and so are better pi donors than a methyl group, which is the weakest. The amine is the best electron-releasing group because the amine nitrogen is less electronegative than the ether oxygen.

4.2.6 Stereochemistry of Substitution Reactions

Figure 4.9 illustrates the three possible stereochemical outcomes of a substitution reaction. In retention, the nucleophile replaces the leaving group without changing the stereochemistry, but the transition state is very crowded, so it is a very rare process. Two inversions will have the same effect as a retention process. The S_N2 substitution (Section 4.2.7) gives inversion because the nucleophile displaces the leaving group from the backside via a trigonal pyramidal transition state. The S_N1 substitution (Section 4.2.8) often gives racemization because the reaction goes through a trigonal planar carbocation intermediate, and the nucleophile can attack the carbocation from top or bottom equally.

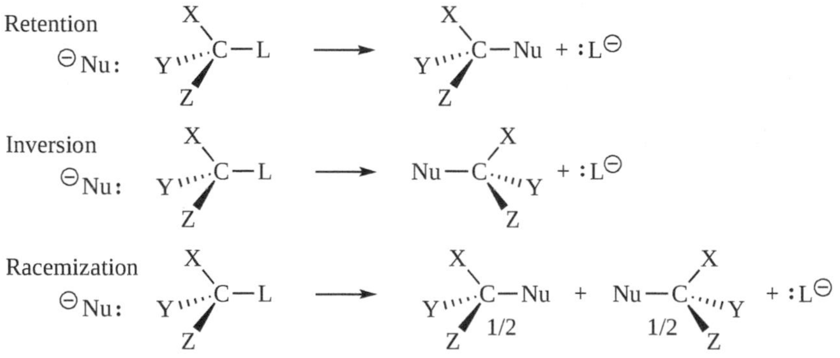

Figure 4.9 The three possible fates of a chiral center undergoing a substitution reaction: retention, inversion, or racemization (producing a 1:1 mixture of enantiomers, usually shown by ±).

4.2.7 S_N2 Electron Flow Pathway

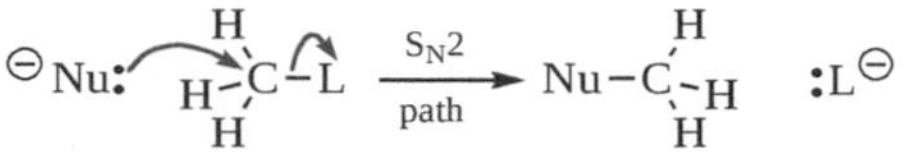

When the carbocation is unstable, the lowest energy path on Figure 4.2 is the S_N2 in which the nucleophile pushes out the leaving group. The first electron flow arrow starts from the nucleophile lone pair and ends at the carbon attacked. A second arrow breaks the carbon-leaving group bond and deposits the bonding pair of electrons as a lone pair on the leaving group. The nucleophile inverts the stereochemistry of the carbon attacked.

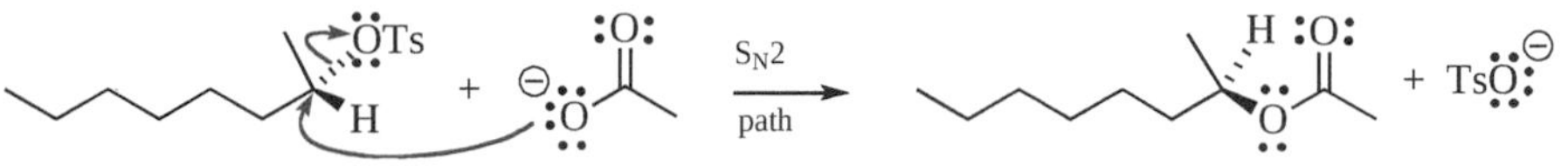

S_N2 Path Limitations

There are three requirements for the S_N2 reaction. In order to be pushed out, the leaving group must be at least fair, usually good. Section 4.2.3 gave a ranking of leaving groups; the pK_{aHL} cutoff for the S_N2 is usually no greater than 10. An exception to this is epoxides; strain relief of opening the three-membered ring makes up for the poorer alkoxide leaving group. The nucleophile also must be reactive enough to push out the leaving group. A list of nucleophile rankings for the S_N2 on methyl iodide in methanol solvent was given in Section 4.2.2. A poorer nucleophile can be compensated by a better leaving group and vice versa. Usually simple pi bond sources do not react by S_N2. Allylic sources like enolates, enols, and enamines do, since the attached electron-releasing group makes them more electron rich. Finally, the backside of the tetrahedral carbon attacked must be accessible to the nucleophile and not blocked by other groups.

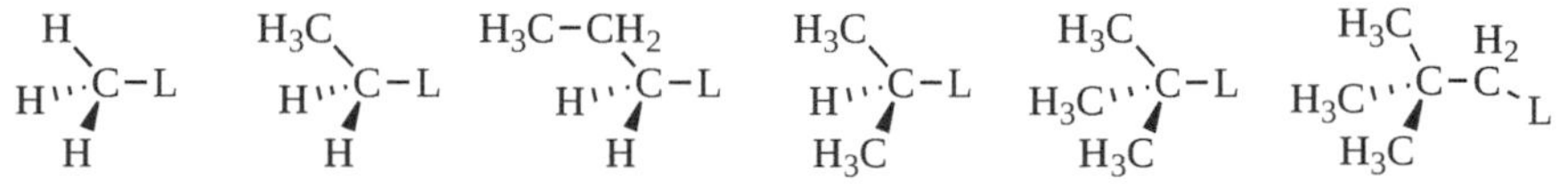

Figure 4.10 Relative rates of S_N2 reactions as a function of steric hindrance of the site of attack.

Figure 4.10 gives the relative rates of S_N2 reactions as the groups around the carbon attacked get bigger. Methyl and primary are fastest; secondary is slow. Tertiary is so slow to S_N2 that other processes like S_N1 easily win out. Neopentyl is the slowest, as a methyl group of the *tert*-butyl group always blocks the back side nucleophilic attack.

Example problem

Which of the following, CH_3Cl, CH_3I, and CH_3Br, will be the fastest S_N2 in methanol using hydroxide as a nucleophile? Which is the slowest?

Answer: The S_N2 on a better leaving group will be faster using the same nucleophile. The best leaving group is iodide, pK_{aHL} of −10, so is the fastest S_N2. The next-best leaving group is bromide, pK_{aHL} of −9. The worst leaving group and slowest S_N2 is chloride, pK_{aHL} = −6.

S_N2 Energetics and the ΔpK_a Rule Cross-check

If the reactants and the products of a reaction are not charged, we can calculate the $\Delta H°$ (Section 2.5) and thereby know whether the reaction is uphill or downhill in energy. Unfortunately, this calculation is of limited use for mechanisms, since most reaction steps involve charged species as nucleophiles, electrophiles, or intermediates. We need a way to tell whether a step involving a charged species is energetically uphill or down.

For example, the S_N2 displacement of a leaving group by a nucleophile is microscopically reversible. Therefore. a general rule is needed to determine whether the reaction will proceed to the left or to the right. We can use the relative basicity of the Nu and the L as a guide, because **usually the reaction will proceed to form the weaker base**, as was seen in the proton-transfer process. Sometimes a slightly stronger base can be formed if the bond broken is weaker than the one formed; the formation of the stronger bond would counterbalance the formation of a slightly stronger base. Even a 10-unit climb in pK_{abH}, corresponding to a 13.6 kcal/mol (57.0 kJ/mol) climb in energy, could occasionally be compensated for by strong bond formation. Similar to the proton-transfer K_{eq} cross-check in Section 3.6, we will state a general rule that will become one of our most important energetics cross-checks for polar reactions. **The substitution ΔpK_a rule cross-check: Do not create a product or intermediate more than 10 pK_a units more basic than the incoming nucleophile.** The ΔpK_a rule cross-check can be used to predict feasibility and the reversibility of an S_N2 reaction. If the reverse reaction climbs more than 10 pK_a units, the reaction probably does not reverse. If the forward reaction climbs more than 10 pK_a units, the reaction probably does not go at all.

Example problem

Is the following S_N2 reasonable, and is it reversible?

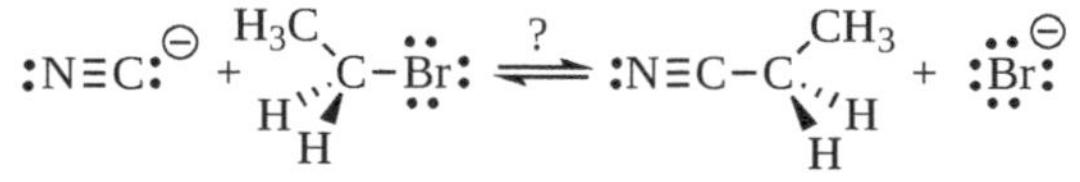

Answer: The cyanide nucleophile has a pK_{abH} = 9.2 and the bromide leaving group has a pK_{aHL} of −9. The forward direction drops 18.2 pK_a units, so is very favorable. The reverse reaction would have to climb the same 18.2 units, and so is ruled out by our ΔpK_a rule cross-check. The reaction goes forward, but is not reversible.

4.2.8 S_N1 Reaction: Electron Flow Pathways D_N + A_N

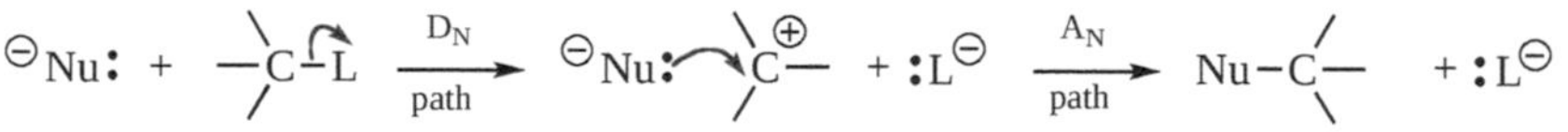

The S_N1 substitution reaction is a two-step process, a slow D_N step to break the carbon-leaving group bond forming a carbocation, followed by a fast A_N trapping of the carbocation to form the new bond. The D_N and A_N paths are just the reverse of each other. Carbocations have just three fates: They can be trapped by a nucleophile as discussed in this section; they can lose a proton to form the alkene (Section 4.3), or they can rearrange to another carbocation of equal or greater stability (Section 4.7).

Stereochemistry of S_N1 reactions

Figure 4.9 illustrated the three possible stereochemical outcomes of a substitution reaction. The S_N1 reaction often gives racemization because the trigonal planar carbocation can be trapped from either face equally. Sometimes S_N1 gives partial inversion (more inversion than retention) because the leaving group is ion paired with the carbocation, and the back side is more open to nucleophilic attack because the leaving group is still blocking the front. Polar solvents aid the breaking up of ion pairs, so they are expected to favor complete racemization in the S_N1 reaction.

Example problem

For the compound below, draw an S_N1 substitution mechanism and show both enantiomers of the product that is formed.

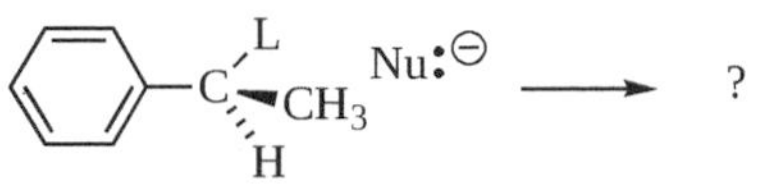

Answer: The S_N1 route first loses the leaving group through path D_N to form a trigonal planar carbocation that is trapped from either side by path A_N.

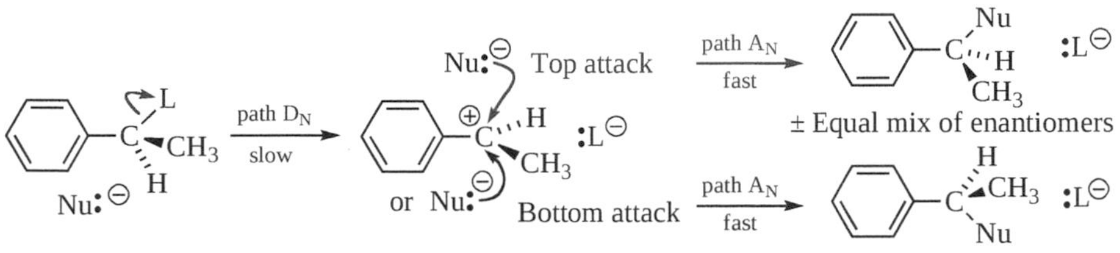

4.2.9 Ionization of Leaving Groups

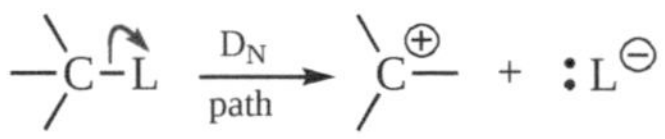

The slow step in the S_N1 is the loss of the leaving group. The ionization of a leaving group depends primarily on three factors, the ability of the **solvent** to stabilize the charges formed, the stability of the **carbocation**, and the quality of the **leaving group**.

Table 4.3 gives conditions for ionization of leaving groups. **To summarize, definitely expect ionization when these three conditions are satisfied: a good leaving group (pK_{aHL} ≤ 0), a polar solvent, and a carbocation of stability better than secondary.** Do not expect leaving group ionization to occur in a nonpolar solvent. If you do not have all three conditions satisfied, the S_N1 does not occur at a significant rate.

Table 4.3 The Ionization of Leaving Groups

Carbocation	pK_{aHL}	Solvent	Ionize?
1° Primary RCH_2^+	Any	Any	Almost never
2° Secondary R_2CH^+	Above 0	Any	Rarely
	Below 0	Polar	Slow
3° Tertiary R_3C^+	Any	Nonpolar	Rarely
	higher than 7	Polar	Rarely
	0.5 to about 6	Polar	Possible
	≤ 0	Polar	Fast

Example problem

Which of the following, $(CH_3)_3CBr$, Ph_2CHBr, and $(CH_3)_2CHBr$, will be the fastest S_N1 in methanol? Which is the slowest?

Answer: The solvents and leaving groups are the same. The S_N1 reaction will be faster via the more stable carbocation. The most stable carbocation and fastest S_N1 reaction is via the diphenylmethyl carbocation; next is via the *tert*-butyl carbocation. The least stable isopropyl cation is the slowest, not expected to form at a significant rate.

4.2.10 S_N1/S_N2 Substitution Spectrum

Figure 4.11 shows an S_N1/S_N2 spectrum exists formed by competition between two rates: the rate of ionization of the leaving group and the rate of nucleophilic attack on the substrate or partially ionized substrate. Table 4.4 outlines the extremes of the spectrum.

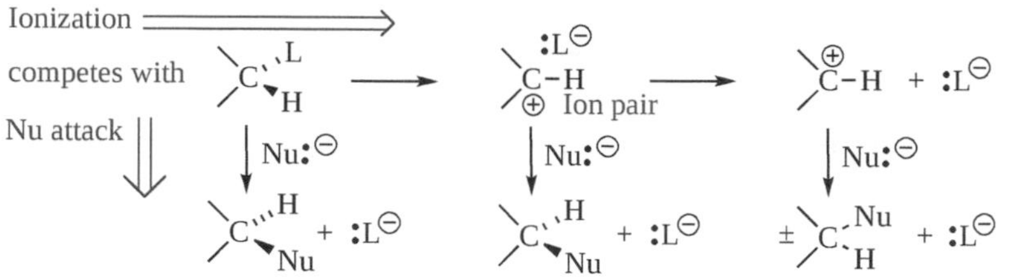

Figure 4.11 Ionization of the leaving group competes with nucleophilic attack in a substitution.

TABLE 4.4 The S_N1/S_N2 Spectrum Extremes

Variable	S_N1 (Path D_N then A_N)	S_N2 (Concerted A_ND_N)
Carbocation	Must be good (usually better than secondary) because it is an intermediate	Not formed, therefore carbocation stability is not a concern
Nucleophile	Weaker Nu (often neutral) tolerated since the sink is good	Good Nu (commonly anionic) required to push out the leaving group
Site of attack	Often hindered	Must be open
Stereochemistry	Racemization (or inversion on ion pair)	Inversion
Media	Often acidic	Often basic
Leaving group	Good to Excellent ($pK_{aHL} \leq 0$)	Usually ($pK_{aHL} \leq 10$)

Although the actual energy surface for the simplest substitution reaction is much more complex than the simplified surfaces drawn in the following figures, we can use these surfaces to understand how the changes in reaction conditions would affect the reaction path. The vertical axis of energy is added to Figure 4.2 to give Figure 4.12, where the front horizontal axis is still the degree of C–Nu bond making and the axis going back to front is the degree of C–L bond breaking. Reactants are in the back right corner at point R. The back left corner corresponds to only C–Nu bond making without C–L bond breaking, thus forming a high-energy pentacovalent intermediate since carbon can't expand its valence shell. The front

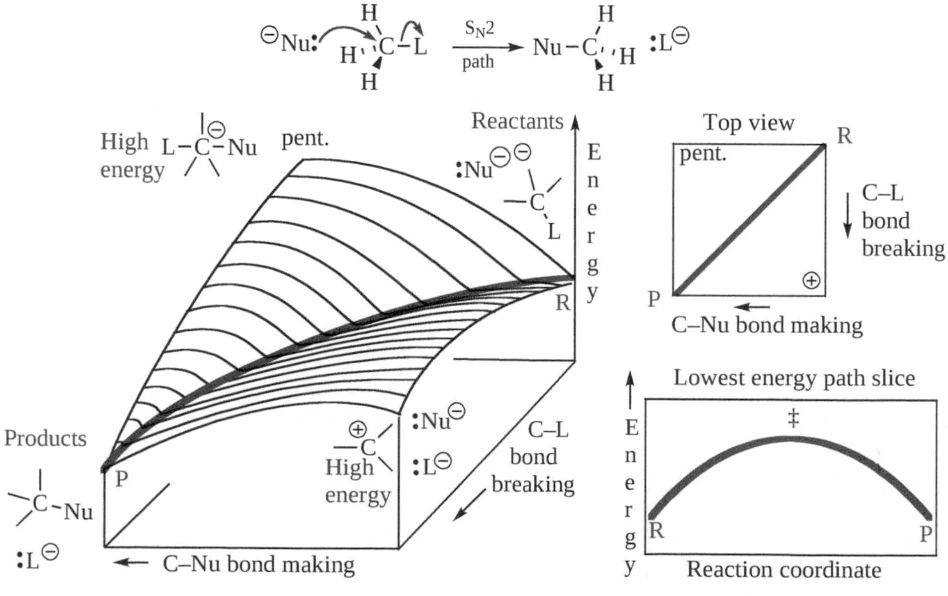

Figure 4.12 The S_N2 energy surface; C–L bond breaking and C–Nu bond making are concerted.

right corner corresponds to only C–L bond breaking and therefore is the carbocation; the front left corner is the substitution product P. The lowest-energy path on the surface is the red bold line from R to P.

It is possible to understand the trends by watching how the substitution surface folds and/or tilts in response to a change. In Figure 4.12, the carbocation is poor; its corner is high in energy, and since the pentacovalent corner is also high in energy, the surface folds down the middle. The reaction will go by the S_N2 process in which C–Nu bond making and C–L bond breaking occur to an equal extent, and the lowest-energy path will follow the *diagonal* through the "pass" between the two "mountains" created by the high-energy carbocation and high-energy pentacovalent intermediate.

A better nucleophile makes C–Nu bond making easier, and therefore lowers the entire left edge, since all points along that edge have the C–Nu bond made. Therefore, the effect of a better nucleophile is to lower the pentacovalent corner somewhat, favoring S_N2, and to lower the product, making the overall reaction more favorable.

On the other hand, if the carbocation is rather stable, the carbocation corner on the diagram will be lowered in energy, and now the lowest-energy path will go via the carbocation: the S_N1 substitution (Figure 4.13). The highest barrier is the loss of the leaving group.

Steric hindrance raises the pentacovalent intermediate to an even higher energy, since it is the most crowded, with five groups around the central atom. The trigonal planar carbocation with only three groups around the central atom is least affected by steric hindrance. Therefore, increasing steric hindrance raises the pentacovalent intermediate greatly and the carbocation little, resulting in a tilting of the surface toward carbocation formation.

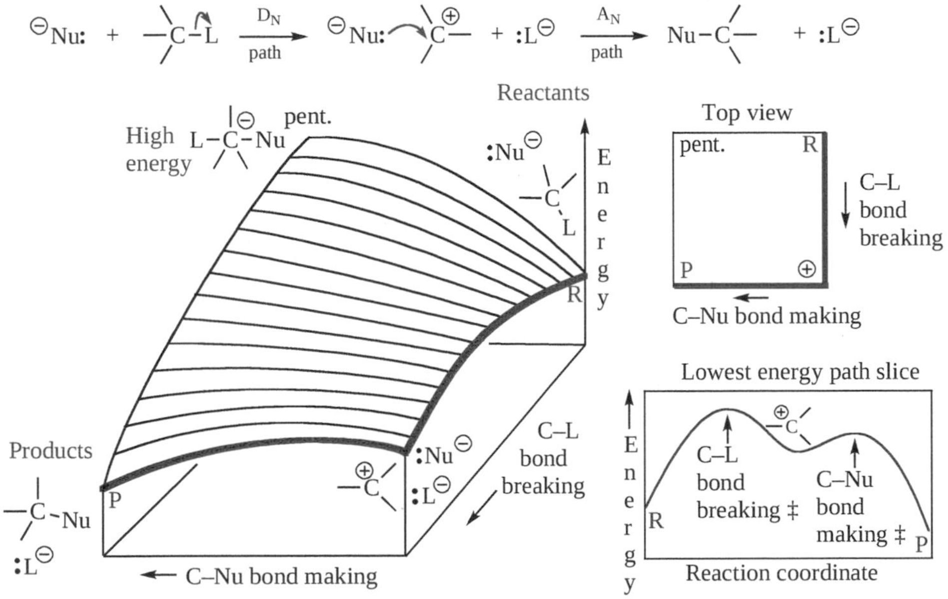

Figure 4.13 The S_N1 energy surface. The C–L bond breaking occurs first to form a carbocation, and then the C–Nu bond is made.

If the leaving group is better, C–L bond breaking is easier, and therefore, the entire front edge is lowered because all points along that edge have the C–L bond broken. The effect of a better leaving group lowers the carbocation corner, favoring S_N1, and lowers the product. This makes the overall reaction more favorable.

4.2.11 Substitution via a Pentacovalent Intermediate

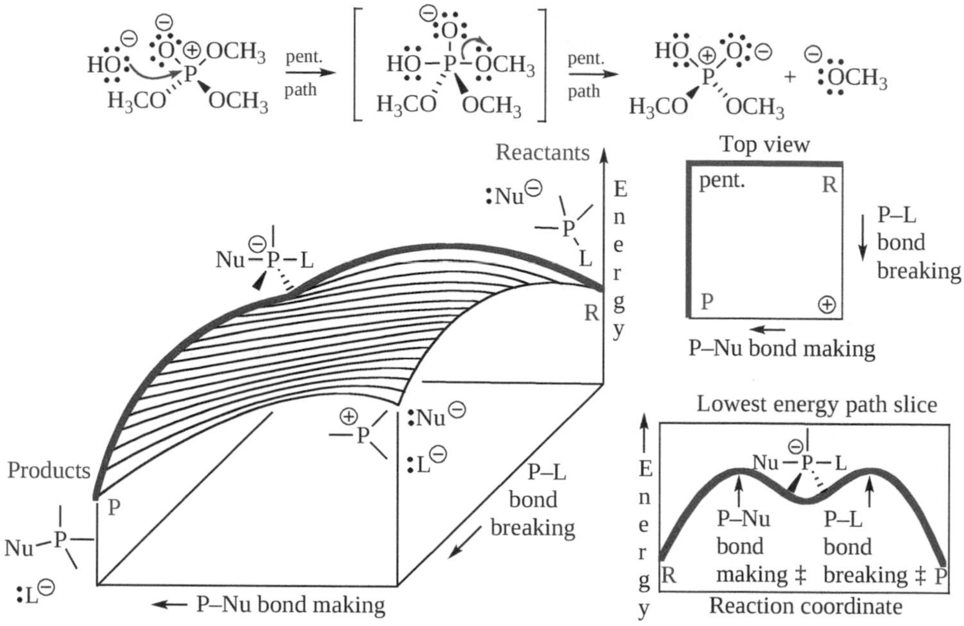

Figure 4.14 The energy surface for substitution via a pentacovalent intermediate.

Third-row and higher elements can stabilize a pentacoordinate intermediate by *d*-orbital bonding. Their larger size also allows for a five-bonded intermediate to be less crowded. These factors lower the pentacovalent intermediate corner far enough so that the reaction can proceed through that intermediate (Figure 4.14). Substitution at phosphorus commonly proceeds through a pentacovalent intermediate, via the pent. path, as shown in the above example. The pentacovalent intermediate shown in brackets is a trigonal bipyramid with the nucleophile in the apical position. By microscopic reversibility, the leaving group is also in the apical position. The better leaving groups lower the front edge of the diagram, and also lower the barrier for the second step, decreasing the lifetime of the intermediate. At the good leaving group extreme, the intermediate can last less than a vibration, and the route is difficult to distinguish from an S_N2 substitution.

4.2.12 Approaches to Substitution Mechanisms

With three substitution routes each with new cross-checks, our problem space has gotten more complex. The first task is always to map changes on a balanced reaction: What group was replaced, and what group replaced it? Is the medium acidic or basic? In acid, we can protonate leaving groups to improve them. In base, we can often deprotonate the nucleophile to improve it. Remember, the medium can't be both acidic and basic. In essence, we check for possible proton transfer first. Next we decide the substitution route. Visualize that you are standing on the reactant "street corner" of the energy surface of Figure 4.2 and need to look at the "traffic lights" (energetics) to decide which way to cross. If the leaving group in the problem is on a carbon, then the pentacovalent route is out because it is too high in energy. Next check the three restrictions for the D_N step: a carbocation stability usually better than secondary, a polar solvent, a good leaving group with a $pK_{aHL} \leq 0$. If it fails on any of the three, the S_N1 route is also too high in energy. It is time to check S_N2 restrictions: good access, a decent nucleophile, and a decent leaving group. We are checking each branch of the problem space tree to see which is the best route before we commit to a direction to proceed. Once you decide the best route, make sure you draw the arrows correctly and keep track of charge balance. Use the known electron flow paths, and don't combine them to just save drawing an extra structure. Skipped or combined steps are difficult to cross-check. With practice, the consideration of alternatives and cross-checks become automatic. You will develop a good intuition for these reactions.

Common Substitution Errors

The most common errors are the result of forgotten cross-checks:
Mixed media errors (Don't do acidic routes in basic media or vice versa).
Proton transfer energetics errors (make sure K_{eq} is greater than 10^{-10}).
S_N2 does not occur on leaving groups attached to a trigonal planar center.
S_N2 access restrictions (a tertiary or neopentyl center does not react by S_N2)
S_N2 energetics error (always check the ΔpK_a rule cross-check)
S_N1 or S_N2 leaving group restriction (don't drop off poor leaving groups)
S_N1 carbocation restriction (don't make an unstable carbocation)
Pentacovalent path does not occur for second-row elements.

Example Substitution Mechanism

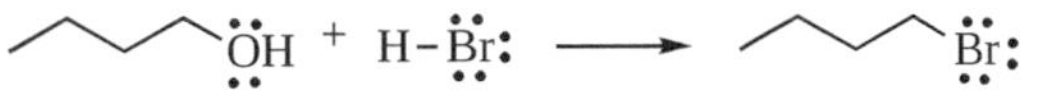

Often organic chemists don't bother to fully balance reactions, so that is your first task in order to be able to map the changes. With HBr present, the reaction is certainly in acidic medium. A substitution definitely has occurred; the OH has been replaced by Br. With a balanced reaction, we can see the leaving group was water.

We first check for favorable proton transfers. A proton-transfer reaction not only improves the leaving group by protonation, but also produces a better nucleophile, the conjugate base of HBr, the bromide anion. The proton transfer passes a K_{eq} check: $\log K_{eq} = (pK_{abH} - pK_{aHA}) = (-2.4 - (-9)) = +6.6$, so $K_{eq} = 10^{+6.6}$ is very favorable.

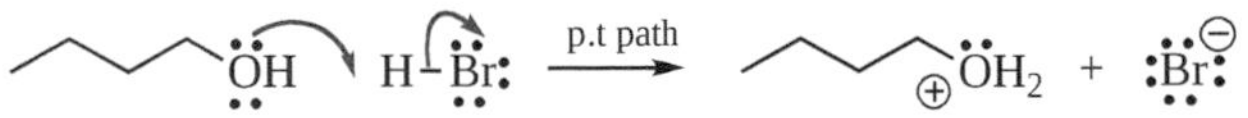

It is time to decide on a substitution route. As the leaving group is on a second-row element, the pentacovalent path is out. A check of D_N restrictions shows the leaving group is good, and the solvent polar, but the carbocation would be primary, and therefore very unstable, so S_N1 is also out. A check of S_N2 access passes since the carbon bearing the leaving group is unhindered. Finally, the ΔpK_a rule cross-check has an acceptable climb of $pK_{aHL} - pK_{aHNu} = (0 - (-9)) = +9$ pK_a units.

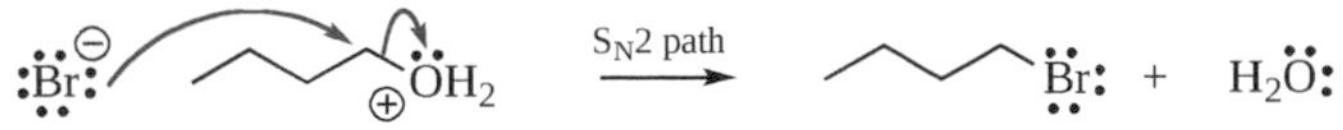

By using the cross-checks we have produced a reasonable mechanism and avoided the common errors that going with our first idea may have fallen prey to. Often it may seem that diving into a problem without considering all these restrictions is quicker, but much time is then lost exploring useless dead ends. A logical approach avoids those, and overall is faster and will get even faster with practice.

Nucleophilic Catalysis

Nucleophilic catalysis, often present in biochemical systems, is illustrated in the following example. The uncatalyzed reaction is slow because even though chloride is a decent leaving group, neutral water is a poor nucleophile. We expect an S_N2 reaction, not S_N1 because the carbocation would be primary, and the pentacovalent route is out because the leaving group is on a second-row element. This reaction is similar to the previous worked reaction mechanism, just with the conditions changed (excess water) to favor the reverse reaction.

Adding a small amount of KI impressively speeds up this reaction. Potassium iodide is not an acid or a base, and KCl does not speed the reaction at all. What is happening here? It must be due to the iodide anion, acting as a nucleophilic catalyst.

Iodide, being very soft and negatively charged, is a much better nucleophile than neutral water. Iodide can do an S_N2 substitution to replace chloride, giving the butyl iodide in a fast step. Since the C–I bond is much weaker than the C–Cl bond because the overlap is poorer, iodide is a much better leaving group than chloride. Water can now react with the butyl iodide in a fast S_N2 to give the butanol product, kicking out the iodide anion, which is now free to continue catalysis. Even though we have added an extra step, all the barriers are lower, and therefore the overall reaction is much faster.

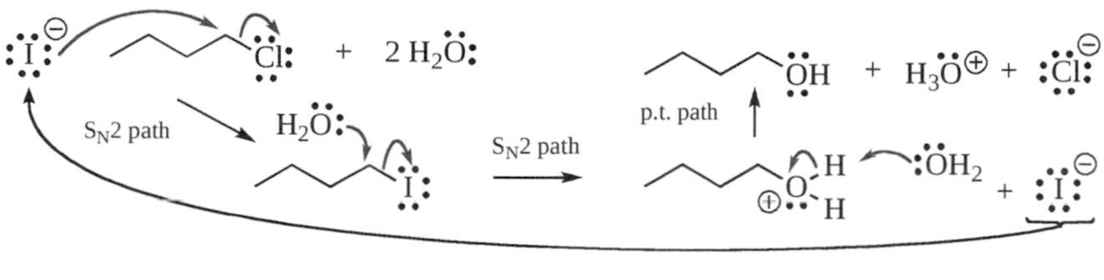

4.2.13 Substitution Summary

- For nucleophiles, rank first by softest as best then within the same atom, by most basic.
- For leaving groups, rank by the lowest pK_{aHL} with exceptions of sulfonates better, and fluoride and cyanide worse than expected.
- Carbocations are stabilized most by resonance with a lone pair, next by pi resonance, least by alkyl substitution. Three of the lesser stabilizations are better than one of the better.
- Electron-releasing groups are best anionic, next the less electronegative pi donors, alkyl substitution, and least, very electronegative pi donors with poor $2p$-$3p$ overlap.
- Substitution version of the ΔpK_a rule cross-check: Never kick out a leaving group more than 10 pK_a units more basic than the incoming nucleophile.
- S_N2 does not occur on leaving groups attached to a trigonal planar center, or on a highly hindered site like a tertiary or neopentyl center.
- S_N2 must pass three tests: good access (primary or secondary), a leaving group with a pK_{aHL} of 10 or less, and a decent nucleophile.

• S_N1 must pass three tests: forms a carbocation that is usually more stable than simple secondary, a polar solvent, and a leaving group with a pK_{aHL} of zero or less.

• The pentacovalent path is for third-row or higher elements and must pass a ΔpK_a rule cross-check.

• S_N2 goes via inversion at the site of nucleophilic attack, and S_N1 usually goes by racemization.

• Any route that goes via a carbocation like S_N1 may have rearrangement as a complication (Section 4.7)

The decision process flow chart for working a substitution problem is shown in Figure 4.15. Since many organic reactions are catalyzed by proton transfer as a first step, we explore that first. Using media pH as a guide and the substitution path restrictions, the next decision is the type of substitution. More substitution examples can be found in Sections 8.3.1 and 8.4.1.

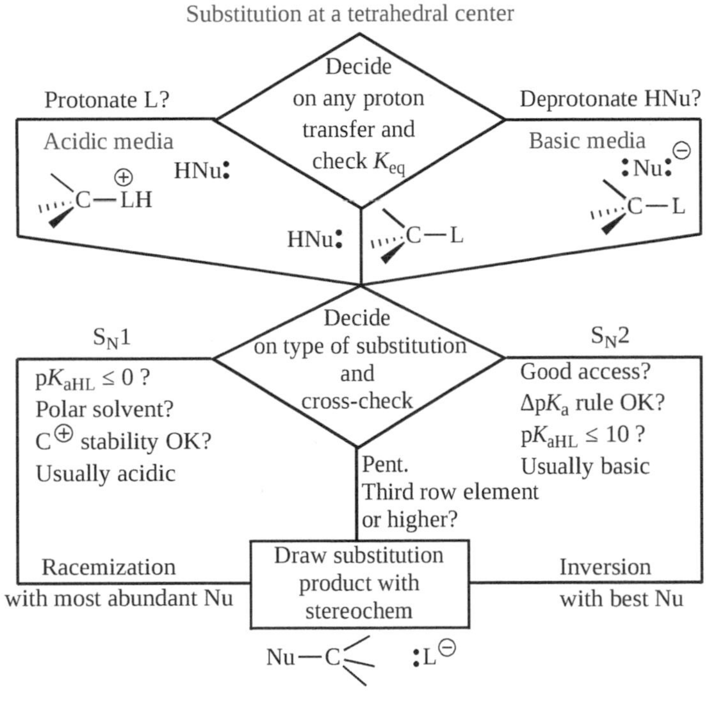

Figure 4.15 The flowchart for substitution at a tetrahedral center.

4.3 ELIMINATION REACTIONS CREATE PI BONDS

4.3.1 Elimination Electron Flow Paths

Elimination reactions often compete with substitutions, but we will worry about that later. Substitution vs. elimination is considered in depth in Section 9.5. There are no new generic classes of reactants to consider, but there are three new routes. The energy surface in Figure 4.16 gives an overview of three possible routes to eliminate a leaving group to form a pi bond. The reactants are found in the upper left corner. The leaving group–carbon bond must break (the vertical axis) and the carbon–hydrogen bond must break (the horizontal axis) to reach the elimination products in the lower right. Three idealized routes can occur.

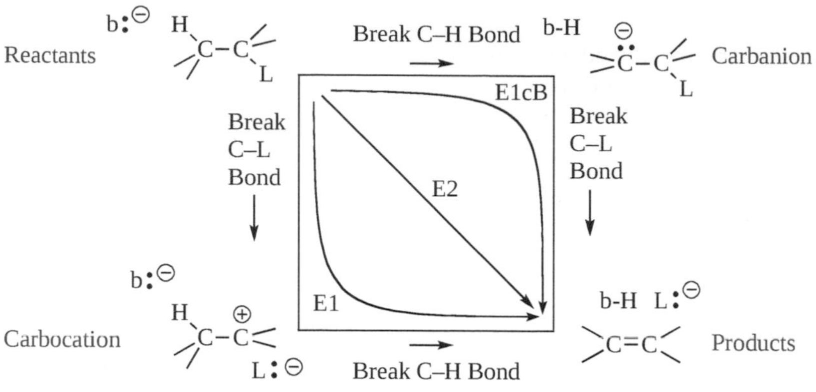

Figure 4.16 Top view of the simplified energy surface for elimination.

The three pathways differ only in the timing of when the proton is removed and when the leaving group departs. **Generic E1** (elimination, unimolecular)—the leaving group departs first (rate-determining) by path D_N, producing a reasonably stable cation; a proton is then lost by path D_E, forming an alkene. **Generic E2** (elimination, bimolecular)—occurs by a concerted loss of a proton and the leaving group. **Generic E1cB** (elimination, unimolecular, conjugate base)—proton transfer forms an anion (the conjugate base) usually stabilized by an electronegative atom or group, and then the leaving group departs (rate-determining) by path E_β. Our two new electron flow paths are D_E (Dissociation, Electrofugic) and E_β (beta elimination from a lone pair). An *electrofuge* (electron fleeing) is a group that leaves **without** its bonding pair of electrons. The electron flow arrows for the three elimination routes are shown in Figure 4.17. Since we already know how to rank bases (Section 3.2), carbanions (Section 3.5.2), leaving groups (Section 4.2.3), and carbocations (Section 4.2.4), we can proceed directly to discuss each elimination route in detail.

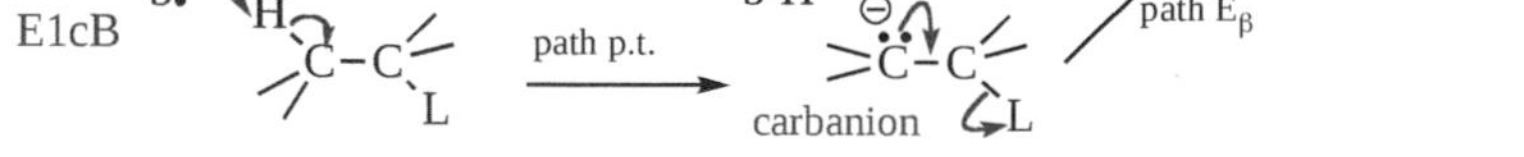

Figure 4.17 The electron flow paths for elimination.

4.3.2 The E1 Reaction: Electron Flow Pathways D_N + D_E

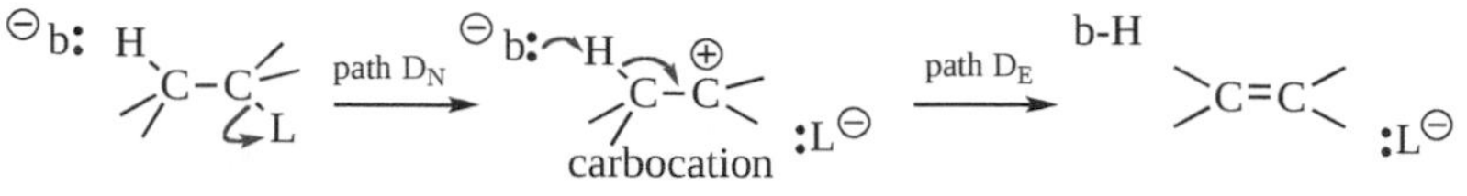

The **E1 elimination** has the same rate-determining first step, D_N, as the S_N1 substitution and is subject to the exact same **limitations: a carbocation stability usually better than secondary, a polar solvent, a good leaving group with a pK_{aHL} ≤ 0.** If it fails on any of the three, the E1 route is too high in energy. The second step, D_E, (Dissociation, Electrofugic) is a new electron flow path, and has the following form.

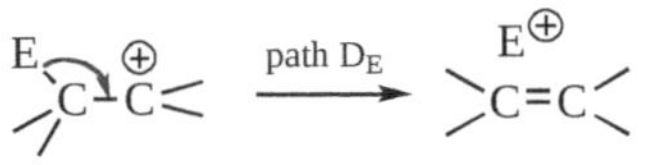

The arrow starts from the C–E bond, breaking it and forming a new pi bond to the cationic center. The C–E bond and the carbocation empty orbital must be able to align in the same plane to form a pi bond, but because the C–C single bond between them can rotate, this is usually not a problem. If the electrofuge is not stabilized, then it will not fall off by itself, and must be removed by bonding to another species. This is the case in the E1 example shown previously, as a proton is never spat out into a vacuum, but is picked up by a basic species. This path is energetically downhill unless the electrofuge is less stable than the carbocation. We have now seen two of three paths available to carbocations: A_N to trap a nucleophile, D_E to lose an electrofuge, and yet to come, rearrange to another carbocation.

The Alkene Stability Trend

The following is a ranking of alkene stability. The sp^2–sp^3 C–C bond is slightly stronger and more stable than an sp^3– sp^3 C–C bond, so the most substituted alkene, having four of the stronger bonds is the most stable. The *cis*-disubstituted double bond is more sterically hindered and less stable than the *trans*-disubstituted double bond. The *gem*-disubstituted double bond is about the same stability as the *cis*-disubstituted double bond. The monosubstituted 1-hexene is the least stable of the hexene isomers listed below, having just one sp^2–sp^3 C–C bond.

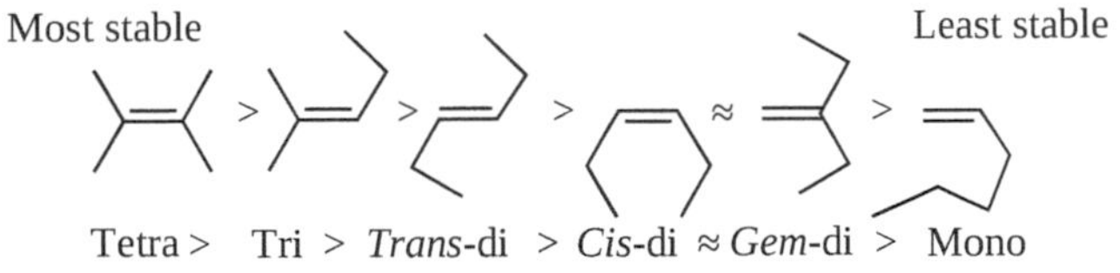

E1 Regiochemistry and Stereochemistry

Any H adjacent to the carbocation can be lost, often giving a different alkene. This loss is often reversible (acidic media), so the E1 often produces an equilibrium mixture of alkenes, the most stable alkene predominating (Zaitsev's rule). When a reaction produces a mixture of different positional constitutional isomers (regioisomers) like 1- vs. 2- hexene, the term *regiochemistry* is used to refer to those products. Also, when the E1 produces a disubstituted alkene, the trans stereoisomer predominates because it is less hindered, so more stable.

Example problem

For the compound below, draw an E1 elimination mechanism and show both major and minor products.

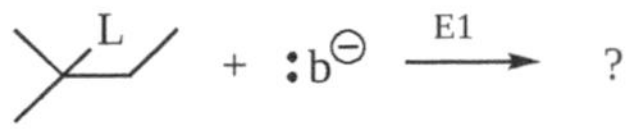

Answer: The first step of the E1 is the loss of the leaving group by path D_N to give a carbocation, which is deprotonated by a base to form a pi bond by path D_E. Either the methyls or methylene adjacent to the carbocation can lose the H, but the more stable, more substituted product is major.

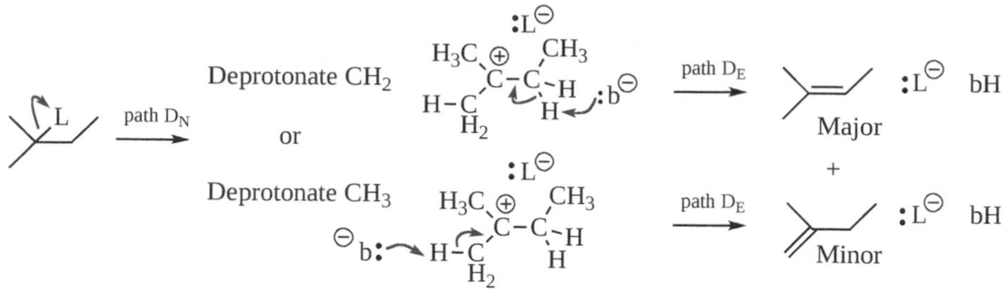

4.3.3 The E2 Electron Flow Pathway

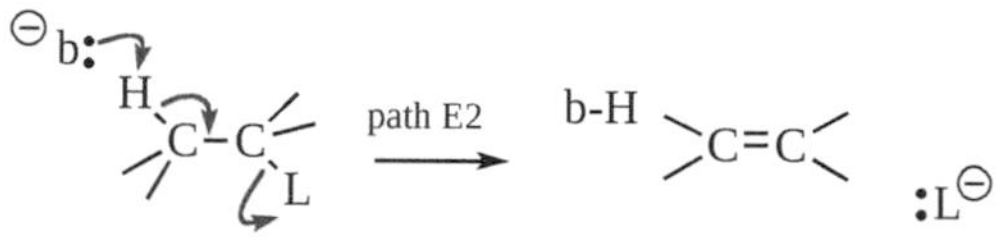

The E2 is a new electron flow path. The electron flow arrow starts from the base lone pair forming a bond to H. A second arrow breaks the C–H bond and forms a pi bond. The last arrow breaks the C–L bond and forms a lone pair on the leaving group. Since the *p* orbitals in the pi bond must be coplanar, and we are forming them in one step, then **the C–H and C–L bonds broken must be coplanar in the E2 transition state**. Those bonds can be in the same plane on the same side or on opposite sides. The staggered antiperiplanar arrangement shown, with the H and L coplanar but on opposite sides, is favored because the orbital overlap is better, and the base and leaving group have more space. Figure 4.18 illustrates several views of the E2 antiperiplanar transition state.

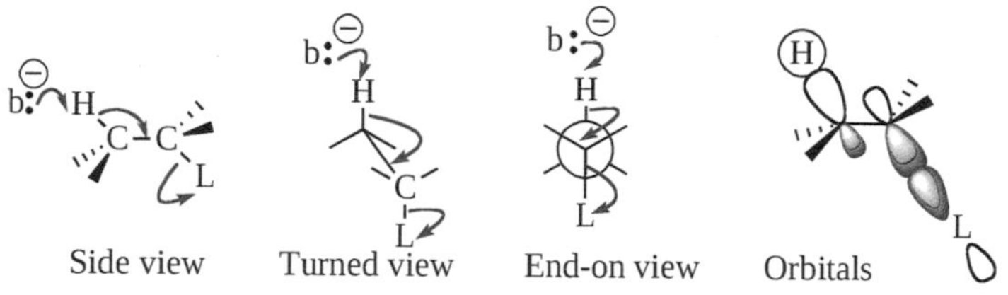

Figure 4.18 Views of the antiperiplanar transition state for the E2 elimination.

E2 Regiochemistry and Stereochemistry

Since the C–C single bond is free to rotate in open chain systems, the sterically less hindered E2 transition state will be the lowest energy. Because the E2 is rarely reversible, the lowest-energy transition state will determine the product. Both product stability and steric effects influence the energy of the transition state. In the E2 transition state the C–H bond and the C–L bond must lie in the same plane, preferably *anti* to each other for steric reasons. The protons on either carbon adjacent to the leaving group can be in proper alignment for reaction; thus different products are produced, as shown in Figure 4.19.

The following reactions show that the regiochemistry of the E2 elimination varies with the size of the base. The small base, methoxide, gives the *trans*-2-alkene as the major product because it is the most stable product with the least hindered transition state (a preference for one of several stereoisomers is *stereoselective*) If the base size is huge, like *tert*-butoxide, steric hindrance raises the energy of both of the 2-alkene transition states, so the 1-alkene is preferred because the hydrogens on the end are more accessible to the large base (Hofmann product). As the leaving group gets huge, sterically crowding the internal hydrogens, the 1-alkene is likewise favored (a preference for one of several regioisomers is *regioselective*).

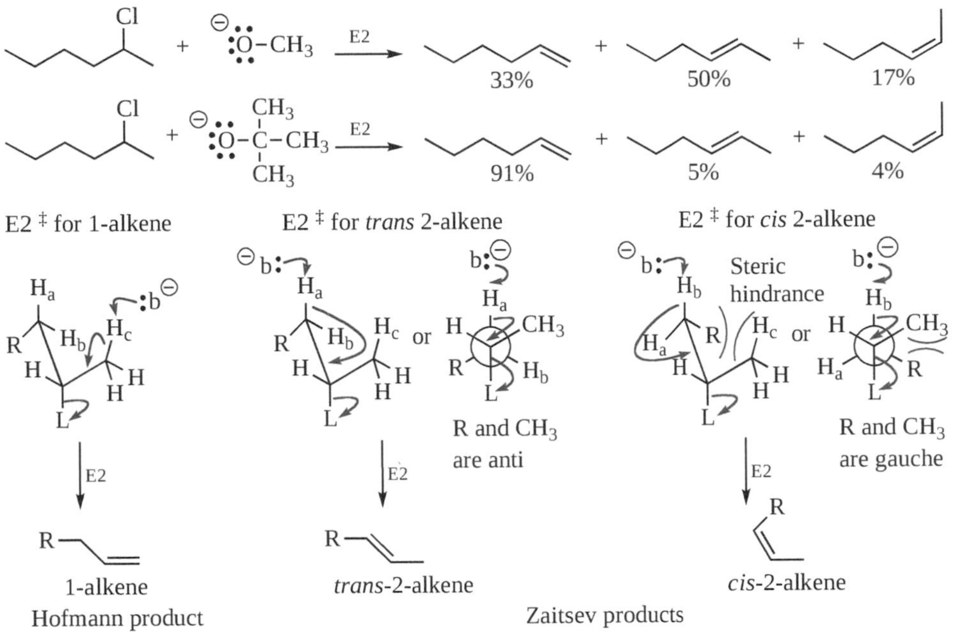

Figure 4.19 E2 elimination regioisomers and stereoisomers. The *cis*-2-alkene has the most crowded transition state, and the 1-alkene has the least crowded transition state.

4.3.4 The E1cB Reaction: Electron Flow Pathways p.t. + E_β

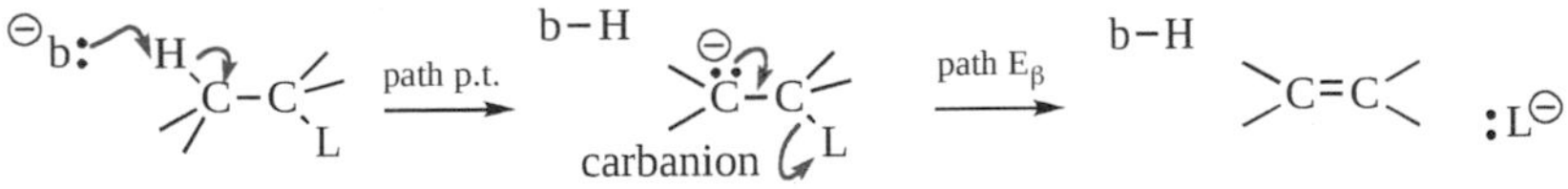

The E1cB elimination starts with a proton transfer as its first step. It is the **conjugate base** (cB) of the reactant that kicks out the leaving group. The second step, E_β (beta elimination from a lone pair), is a new electron flow path, and has the following general form. The arrow starts from the lone pair, forming a new pi bond. A second arrow starts from the C–L bond, breaking it and placing the electrons as a new lone pair on the leaving group. Whenever we make a new bond to an atom with a complete octet, we have to break one of its existing bonds, so as not to exceed its octet.

The K_{eq} of the proton transfer is the limiting factor for the E1cB. The E1cB is most often seen when an electron-withdrawing group increases the acidity of the C–H bond. **A useful K_{eq} of the proton transfer for E1cB is greater than 10^{-10}.**

E1cB Regiochemistry and Stereochemistry

The E1cB regiochemistry is solely determined by the position of the acidic hydrogen. Since the C–C single bond is free to rotate in open chain systems, the sterically less hindered product stereochemistry is usually formed. In the following example, the carbonyl acts as an electron-withdrawing group to make the adjacent hydrogen acidic enough for the E1cB to proceed.

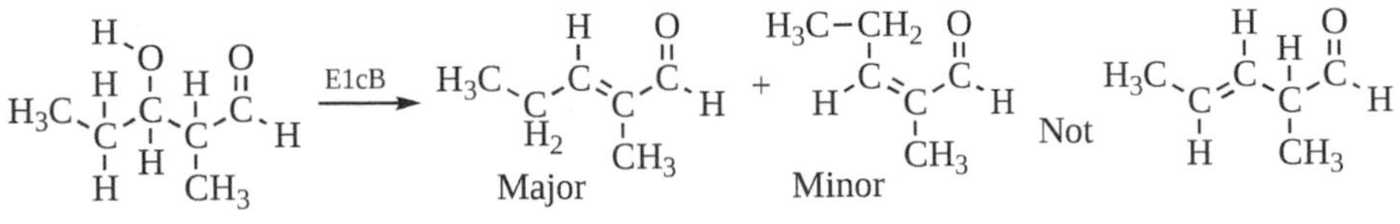

Example problem

For the reaction above, run in basic water, draw an E1cB elimination mechanism, and explain the regiochemistry and stereochemistry found.

Answer: The first step of the E1cB is the deprotonation of the acidic hydrogen next to the electron-withdrawing group by the p.t. path. This deprotonation determines the regiochemistry. The carbanion formed can then push out a poorer leaving group to form the alkene. The bond between the leaving group carbon and the carbanion can easily rotate. The major stereoisomer comes from the least crowded transition state (because the oxygen anion is surrounded by solvent) and is the more stable of the two possibilities.

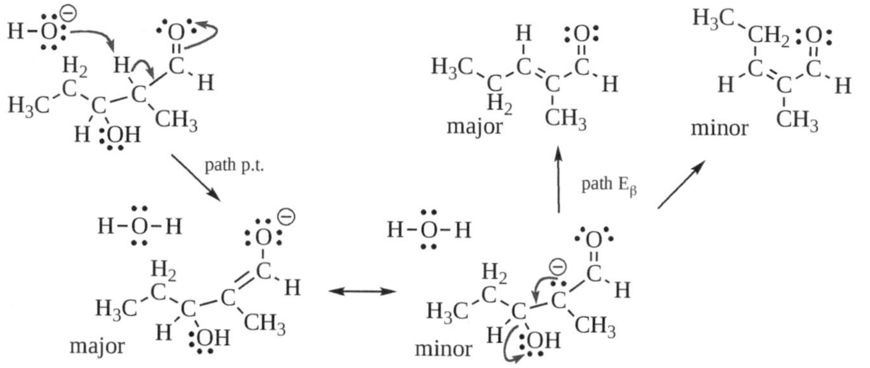

4.3.5 The E1/E2/E1cB Elimination Spectrum

The factors that decide which elimination path will occur are listed in Figure 4.20. For the E1 and the E1cB, solvation of the ionic intermediate is very important. The E1 and E1cB should be considered the two extremes of a spectrum of elimination mechanisms in which the E2 is in the center.

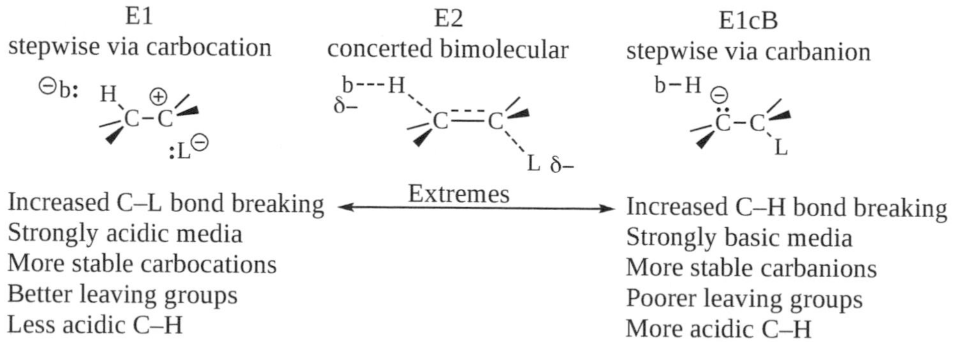

Figure 4.20 The E1/E2/E1cB spectrum.

Mechanisms occur that do not cleanly fit the label of E1, E2, or E1cB. If the reaction is partway between E2 and E1, it is called E1-like, for the C–L bond is broken to a greater extent than the C–H bond, and a partial plus builds up on the leaving group carbon; this adjacent partial plus makes the C–H bond more acidic and therefore more easily broken. If the reaction is partway between E2 and E1cB, it is called E1cB-like, for the C–H bond is broken to a greater extent than the C–L bond, and a partial minus builds up on the carbon adjacent to the leaving group; this partial minus aids in pushing off the leaving group.

The elimination spectrum can be visualized as an energy surface (Figure 4.21), where the front horizontal axis is the degree of C–H bond breaking, the axis going from back to front is the degree of C–L bond breaking, and the vertical axis is energy. Reactants are in the back left corner at point R; the back right corner corresponds to only C–H bond breaking and therefore is the carbanion; the front left corner corresponds to only C–L bond breaking and therefore is the carbocation; the front right corner, P, is the product alkene. The lowest-energy path is the bold red line from R to P (dashed when behind the surface).

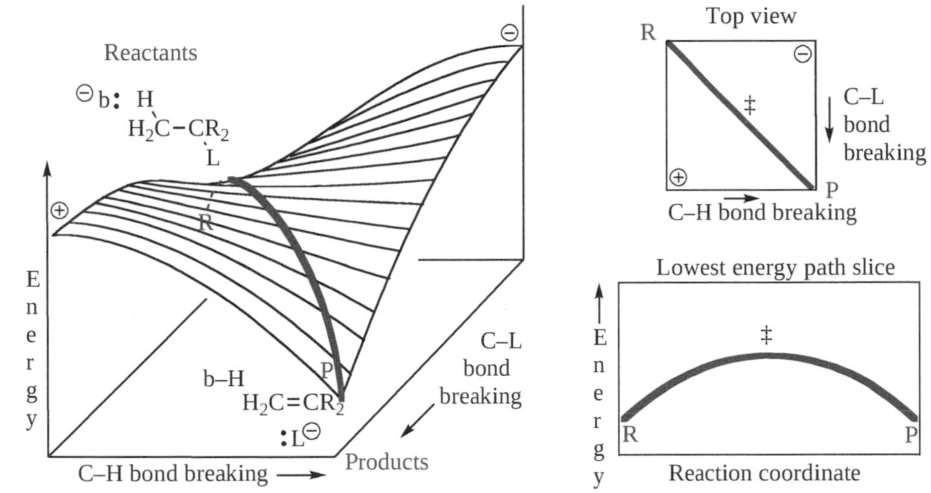

Figure 4.21 The E2 energy surface.

If both carbocation and carbanion are poor, the two corners corresponding to them will be high in energy; the surface folds down the middle. The reaction will go by the E2 process in which C–H bond breaking and C–L bond breaking occur to an equal extent, and the lowest-energy path will follow the diagonal through the saddle resulting from the high-energy carbocation and carbanion. If the carbocation is rather stable, the corner on the diagram corresponding to the carbocation will be lowered in energy, and now the lowest-energy path will go via the carbocation, the E1 elimination (Figure 4.22).

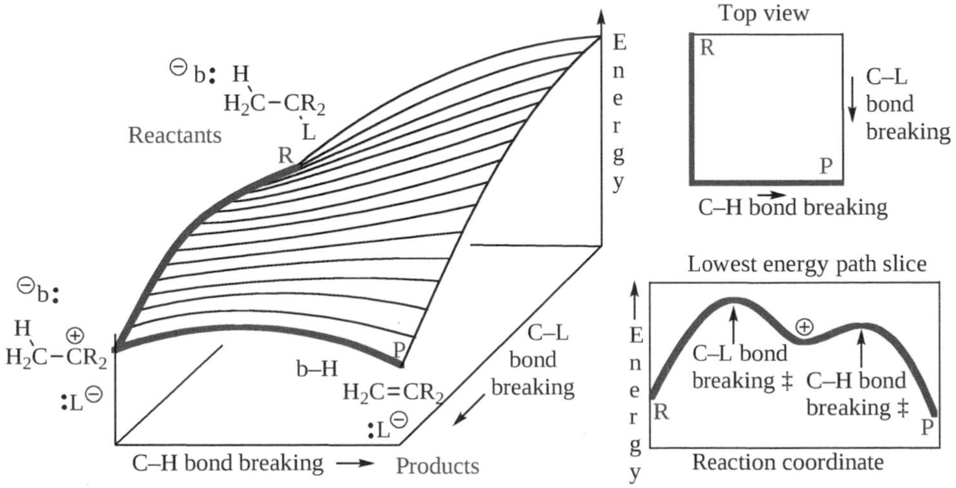

Figure 4.22 The E1 energy surface.

If, on the other hand, the carbanion is reasonably stable and the carbocation is not, the corner on the diagram corresponding to the carbanion will be lowered in energy instead, and now the lowest-energy path will go via the carbanion, thus the E1cB elimination (Fig, 4.23).

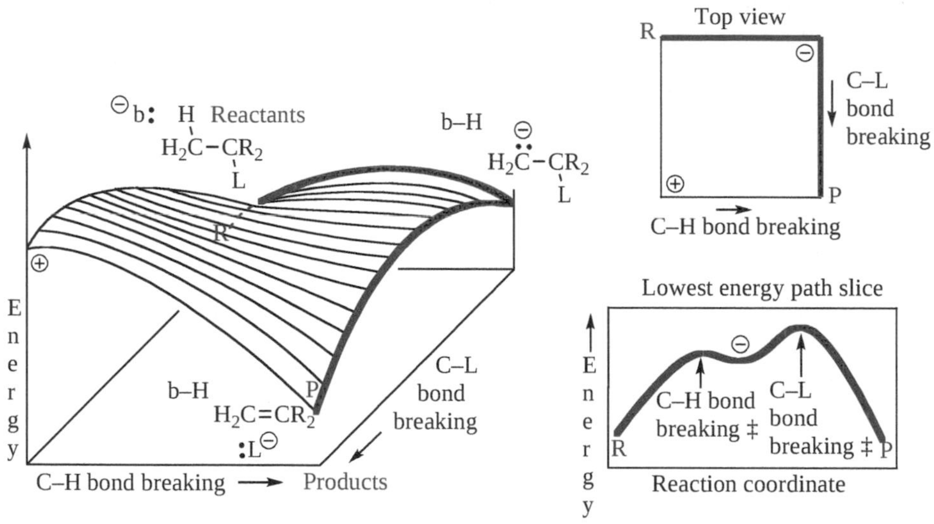

Figure 4.23 The E1cB energy surface.

Finally, if the carbocation is more stable than the carbanion but neither is very good, the two "mountains" on the diagram may be of different heights and will have a "pass" that is off to one side, not exactly on the diagonal (Figure 4.24). In this case, the lowest-energy path to product takes a curve through that pass. More C–L bond breaking than C–H bond breaking occurs in going to the transition state "pass," a partial plus builds up on the leaving group carbon at the transition state, and the reaction is called E1-like. Had the carbocation been slightly more stable, the lowest-energy path would have fallen into the carbocation energy well, and the process would be E1, not E1-like.

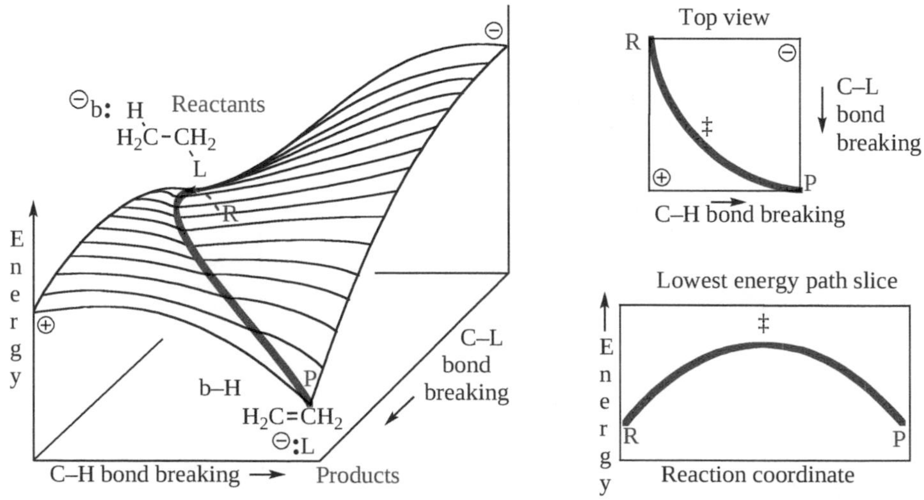

Figure 4.24 The E1-like energy surface.

In a similar manner, if the carbanion is more stable than the carbocation, the "pass" will be on the carbanion side and more C–H bond breaking than C–L bond breaking occurs in going to the transition state. A partial minus charge builds up on the carbon adjacent to the leaving group carbon at the transition state, helping to push out the leaving group, and the reaction is E1cB-like.

Also for the E1/E2/E1cB spectrum, similar to what was seen in the S_N1/S_N2 spectrum, a better leaving group will lower the front edge, making the overall process more favorable and tilting the surface toward the E1 path. If the H is more acidic or the base stronger (making deprotonation easier), the right-hand edge will be lowered, again making the elimination more favorable but this time tilting the surface toward the E1cB.

4.3.6 Eliminations in Cyclic Systems

Cyclohexane favors a chair conformation in which all the bonds are staggered. All the C–H bonds roughly in the plane of the ring are called equatorial; all those perpendicular are called axial. Cyclohexane can easily flip between two chair conformations, as shown in Figure 4.25; the barrier is only 10.8 kcal/mol (45 kJ/mol). All the axial positions become equatorial and vice versa upon ring flip. With substituted rings, the equilibrium favors the larger groups in the less crowded equatorial position.

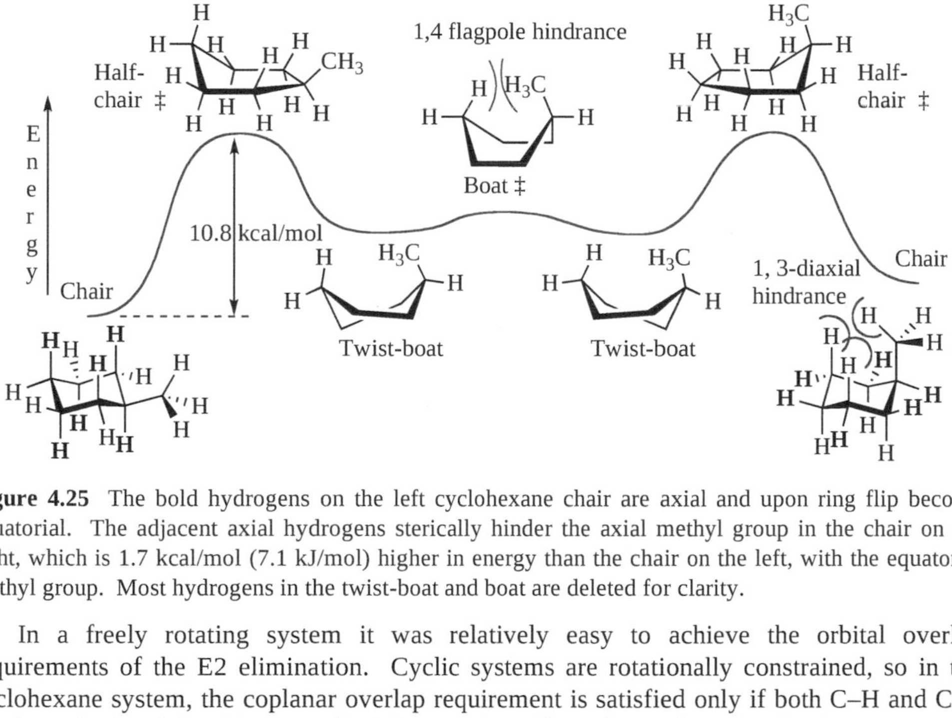

Figure 4.25 The bold hydrogens on the left cyclohexane chair are axial and upon ring flip become equatorial. The adjacent axial hydrogens sterically hinder the axial methyl group in the chair on the right, which is 1.7 kcal/mol (7.1 kJ/mol) higher in energy than the chair on the left, with the equatorial methyl group. Most hydrogens in the twist-boat and boat are deleted for clarity.

In a freely rotating system it was relatively easy to achieve the orbital overlap requirements of the E2 elimination. Cyclic systems are rotationally constrained, so in the cyclohexane system, the coplanar overlap requirement is satisfied only if both C–H and C–L bonds are both axial. This *trans*-diaxial restriction affects the product distribution.

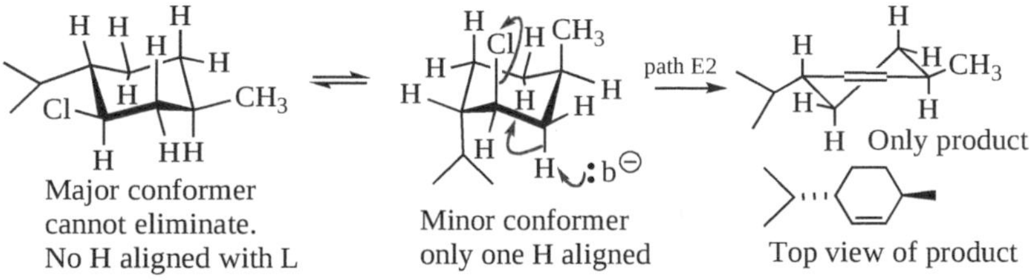

Another aspect of the coplanar overlap requirement is that twisted alkenes are high in energy and hard to make. Only if the ring is large enough (8 carbons or more) is a bridgehead alkene stable enough to be isolated. The following compound undergoes a *syn* coplanar E2 elimination and does not form the highly twisted bridgehead alkene.

4.3.7 Eliminations That Yield Carbonyls

The carbonyl group, C=O, is perhaps the most important functional group in organic chemistry. If we replace one of the carbons in our elimination surface with an oxygen atom, we get a related surface, shown in Figure 4.26.

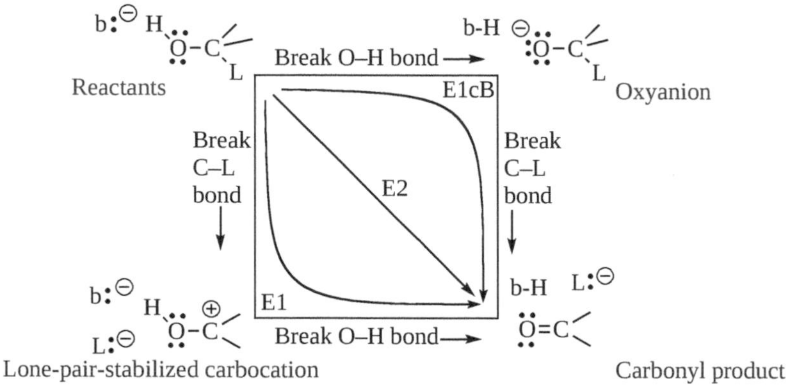

Figure 4.26 Top view of the carbonyl-forming elimination surface.

All elimination routes are improved by this swap of oxygen for carbon (Figure 4.27). The E1cB proton-transfer step now creates a very stable oxyanion, an anion on a very electronegative oxygen atom. The slow step of the E1, loss of the leaving group, is also sped up because the oxygen lone pair can stabilize the carbocation formed. All routes have formed a very strong C=O bond. The primary determinant at this point, since all routes are reasonable, is the pH of the reaction medium. The E1cB is found in base because the oxyanion is basic (pK_{abH} is about 12–16). The E1 is found in acid because the lone pair stabilized carbocation is just a protonated carbonyl (pK_a is near −7). The E2 is found in more neutral media.

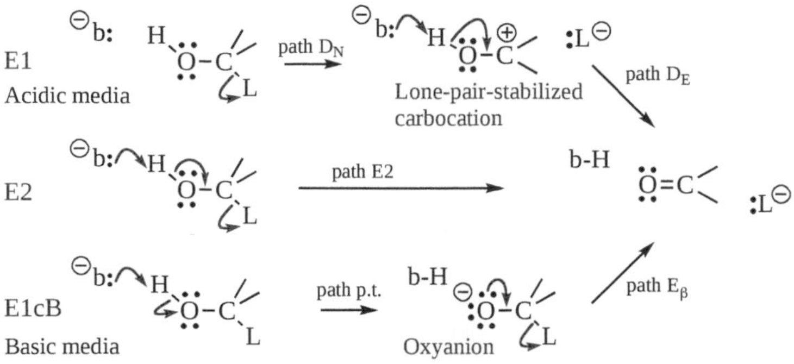

Figure 4.27 Three good carbonyl-forming elimination routes with the leaving group on carbon.

What would happen if we replaced the other carbon, the one bearing the leaving group, with oxygen? The three routes, shown in Figure 4.28, would have drastically different energetics. Loss of the leaving group in the first step of the E1 would produce a highly unstable cation with an incomplete octet on a very electronegative oxygen atom. The E1 is not possible here. The deprotonation step in the E1cB creates an anion next to oxygen, which is destabilized by the adjacent lone pairs on oxygen. This leaves the E2 as the only reasonable route in this system. Placing a leaving group on oxygen and then proceeding with this E2 is the most common method to oxidize an alcohol to a ketone.

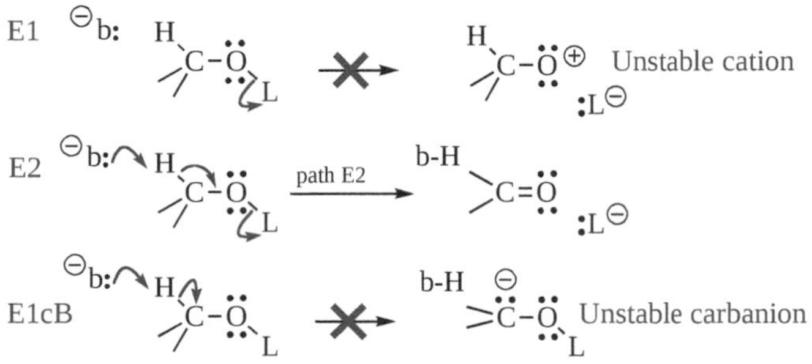

Figure 4.28 Only one carbonyl-forming elimination route exists with the leaving group on oxygen.

4.3.8 Approaches to Elimination Mechanisms

With the addition of the three elimination routes with new cross-checks, our problem space has again gotten more complex. The first task is always to map changes on a balanced reaction. Is the medium acidic or basic? In acid, we can again protonate leaving groups to improve them. Like our approach to substitution mechanisms, we will first check for a favorable proton transfer before deciding which elimination route to take. Again visualize that you are standing on the reactant "street corner" of the elimination surface and need to look at the "traffic lights" (energetics) to decide which way to cross. Is there an appropriately positioned electron-withdrawing group making the C–H acidic? If not, our number of alternatives drops down to two, E1 or E2. Check the three restrictions for the D_N step: a carbocation stability usually better than secondary, a polar solvent, a good leaving group with a $pK_{aHL} \leq 0$. If it fails on any of the three, the E1 route is too high in energy. It is time to check E2 restrictions: a decent leaving group and the ability to get the C–H and C–L bonds in the same plane. Again, we check each branch of the problem space tree to find the best route before committing to a direction to proceed. Once you decide the best route, make sure you draw the arrows correctly, keep track of charge balance, and use known electron flow paths.

Common Elimination Errors

The most common errors are the result of forgotten cross-checks. The following list shows the most common student errors on elimination reactions.

Mixed media errors (Don't do acidic routes in basic media or vice versa).

Proton-transfer energetics errors (make sure K_{eq} is greater than 10^{-10}).

E2 energetics error (always check the ΔpK_a rule cross-check)

E1 carbocation restriction (don't make an unstable carbocation)

E1 leaving group restriction (the pK_{aHL} should be zero or below)

E1cB lack of EWG error (something is needed to make the H acidic)

Charge Balance Errors Combined with Line Structure

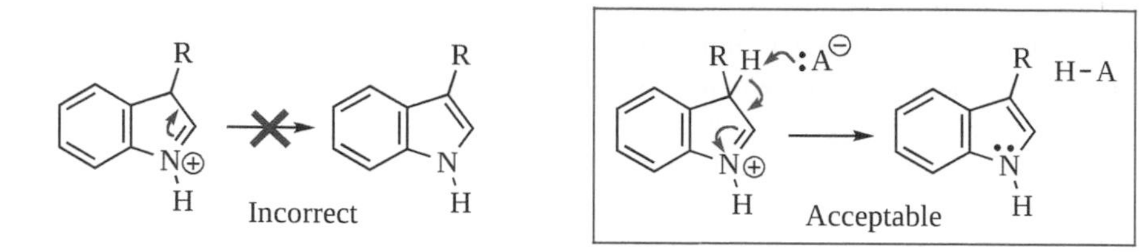

Line structure can be deceptive in eliminations because the important C–H bond is not explicitly drawn out, in this example, when there is a lone-pair-stabilized carbocation intermediate in the E1. The loss of a formal charge may be difficult to notice in reactions that drop off a proton because the proton is commonly not drawn as one of the products. Since the proton lost was not drawn in the line structure on the left, it was simply forgotten. The arrow made a double bond in the right place but flowed the electrons the wrong way: away from the positive center and not toward it. Besides the loss of the formal charge, the lone pair on nitrogen was lost; it was assumed in the product, but the incorrect arrow did not form it. The right side shows an acceptable electron flow path D_E, deprotonation of a carbocation to form a pi bond.

Internal Elimination Shortcut Errors

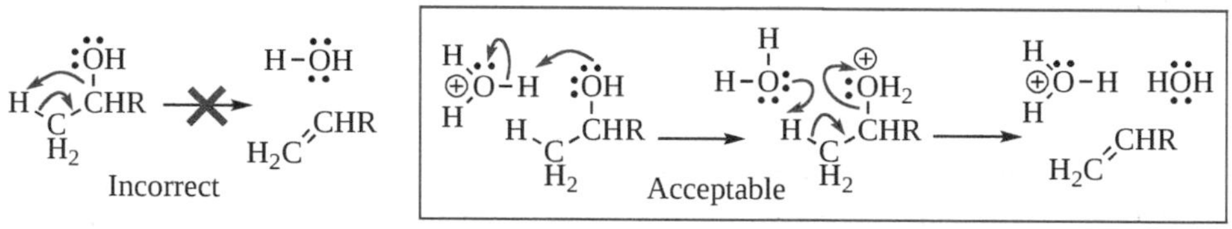

Alcohols do not do uncatalyzed eliminations of water at reasonable temperatures. The carbon–oxygen bond is not a base, and the carbon–hydrogen bond is not an acid. The process drawn on the left, with very few exceptions, does not occur thermally (see Section 13.9, for a concerted, four-center, four-electron process). The most common route for water elimination is by acid catalysis, as shown on the right: path p.t., protonation of a lone pair, followed by the E2 elimination. If the carbocation is reasonably stable, the reaction may proceed via E1.

Orbital Alignment Errors

How much angular "slop" there is in a transition state is currently a debated topic. There is agreement that perpendicular orbitals do not interact. However, most calculations of the actual shape of orbitals show them to be much fatter than they tend to be drawn in beginning textbooks. It now appears that a 10° deviation from the optimum has a negligible effect.

The most important place to watch orbital overlap is in the formation of pi bonds. Pi bonds that have about a 30° twist can be made but are reactive. Since substitution competes with elimination, it does not take very much of a twist to tilt the balance away from elimination. There are many examples to show that changing the leaving group from approximately coplanar with the hydrogen to 60° out of alignment shuts off the E2 elimination process entirely. Be especially careful with eliminations in rigid systems.

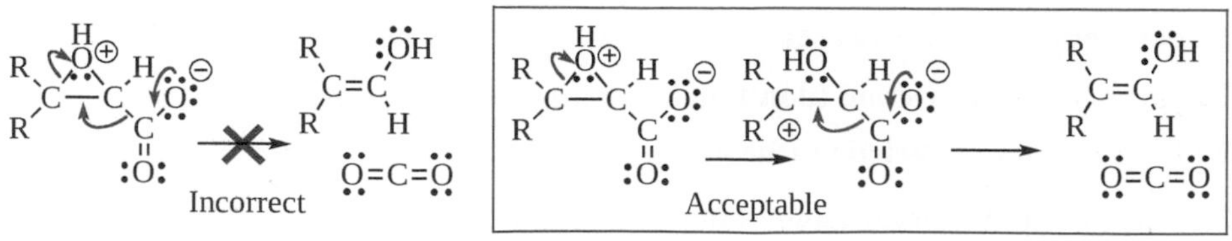

The loss of carbon dioxide (a decarboxylation reaction) on the left, written as a variant of the E2, contains a serious overlap error. The error is that the source and sink are about 60° out of being coplanar, clearly beyond the distortion limits of the E2 elimination (Figure 4.29). We can propose a reasonable alternative elimination, the E1 process, that produces the same product. If the leaving group departs first, the carbocation formed can rotate about the single bond until the carbon–carbon bond of the source is coplanar with the empty *p* orbital. Then the loss of carbon dioxide can occur easily to form the pi bond of the enol product.

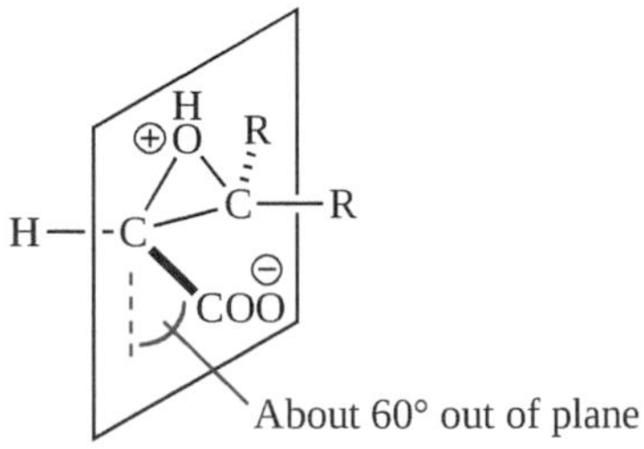

Figure 4.29 An illustration of improper orbital alignment. The C, C, and O of the three-membered ring define the plane. The carboxylate sticks out about 60° from the plane of the ring.

Chemical structures drawn flat on a page may not accurately reflect the three-dimensional nature of the compound drawn. If you have any suspicion that a reaction step has poor orbital alignment, or is physically impossible (inaccessibility of a reactive site, for example), build the molecular model.

Example Elimination Mechanism—Dehydration of an Alcohol

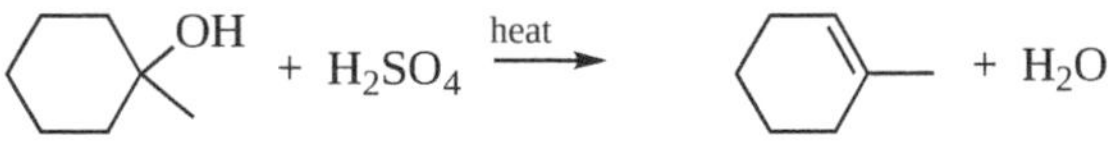

The first task in order to be able to map the changes is to check to see whether the reaction is balanced; it is, and we can see that the leaving group was water. With sulfuric acid present, the reaction is definitely in an acidic medium. An elimination reaction has occurred since we have a new pi bond. We first check for favorable proton transfers. A proton-transfer reaction will improve the leaving group significantly (hydroxide has a pK_{aHL} of 14.0, whereas water has a pK_{aHL} of 0.0). Although the listed pK_a of sulfuric acid is only approximate, the proton transfer easily passes a K_{eq} check: $\log K_{eq} = (pK_{abH} - pK_{aHA}) = (-2.4 - (-9)) = +6.6$, so $K_{eq} = 10^{+6.6}$, which is very favorable.

It is time to decide on an elimination route. To aid in this decision, recall the elimination energy surface (Figure 4.16). There is no electron-withdrawing group making the C–H adjacent to the leaving group carbon acidic, so the E1cB is out. The E1cB is rare in acidic media anyway. For the E1, a check of D_N restrictions shows the leaving group is good, the solvent polar, and the carbocation would be tertiary, and therefore sufficiently stable. Since the E1 passes all cross-checks, we do not have to worry about the E2 since the elimination surface is tilted toward E1. We consider E2 when both the carbocation and the carbanion corners are high in energy, causing the surface to fold down the middle, making the E2 our possible lowest-energy route.

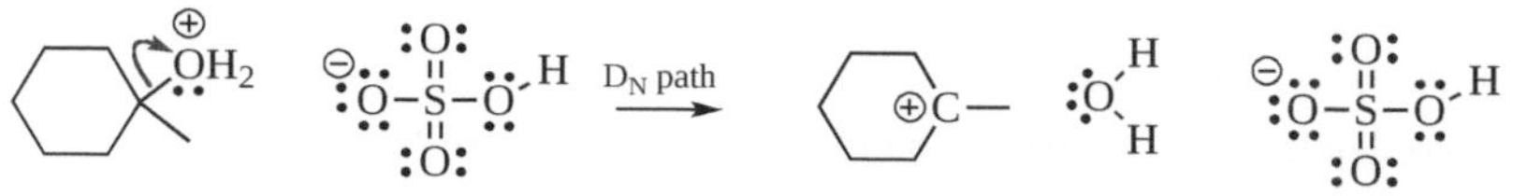

Now we finish the E1 route by path D_E to produce the most stable trisubstituted alkene. The most basic group present is water, so it removes the proton from the carbocation. The protonated water would be expected to protonate the next molecule of alcohol, continuing the acid catalysis and freeing up water as a product.

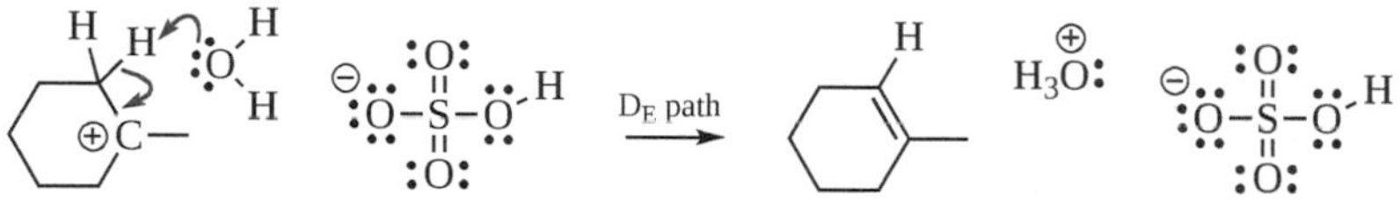

4.3.9 Elimination Summary

• Elimination version of the ΔpK_a rule cross-check: Never kick out a leaving group that is more than 10 pK_a units more basic than the incoming base.

• E1 must pass three tests: forms a carbocation stability usually better than simple secondary, a polar solvent, and a leaving group with a pK_{aHL} of zero or less. Any route that goes via a carbocation like E1 may have rearrangement as a complication (Section 4.7).

• E2 occurs when the carbanion and carbocation are not very stable: The energy surface folds down the middle. The C–H and C–L eliminated must be close to coplanar.

• E1cB requires an acidic hydrogen and a good base. Check that the proton-transfer K_{eq} is at least 10^{-10}. E1cB tolerates poorer leaving groups, but check the ΔpK_a rule cross-check.

More elimination examples can be found in Section 8.4.2 The decision process for working a elimination problem is shown in Figure 4.30.

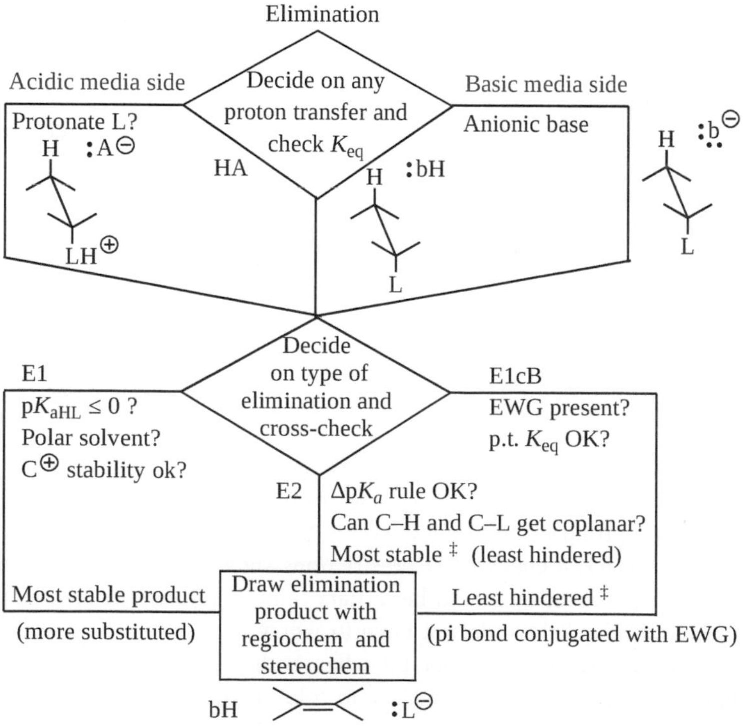

Figure 4.30 The flowchart for elimination.

4.4 ADDITION REACTIONS TO POLARIZED MULTIPLE BONDS

4.4.1 Introduction to Addition Reactions

An addition reaction is formally the reverse of an elimination reaction. We will see both electrophilic and nucleophilic additions in this section. There are few new generic classes of reactants to consider, and there are three new routes, but these are just the reverse of the elimination routes just covered. The new generic classes are discussed in much more detail in Chapter 5 and 6, but need to be introduced here. A carbon–carbon double bond can range from electrophilic to nucleophilic depending on what is attached to it (Figure 4.31). Another way to make a double bond electrophilic is to replace one of its carbon atoms with an electronegative heteroatom like oxygen, C=O, a carbonyl.

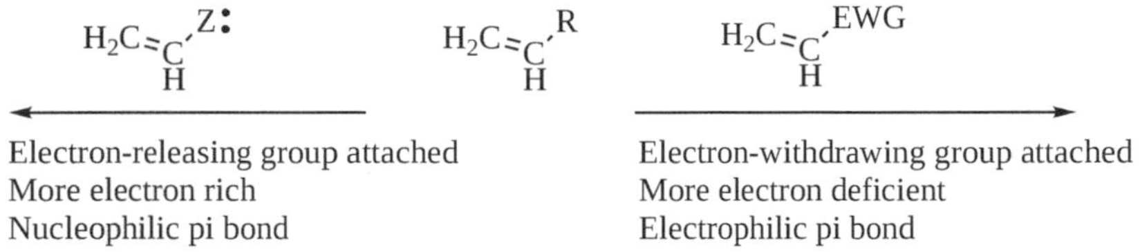

Figure 4.31 Electrophilic and nucleophilic pi bonds.

4.4.2 Addition Electron Flow Paths

The energy surface in Figure 4.32 gives an overview of three possible routes to add an electrophile and a nucleophile to a pi bond. The reactants are found in the upper left corner. In this example surface, the electrophile is a proton, delivered by an acid. The carbon–nucleophile bond must be made (the vertical axis) and the carbon–hydrogen bond must form (the horizontal axis) to reach the addition product in the lower right. Three idealized routes can occur.

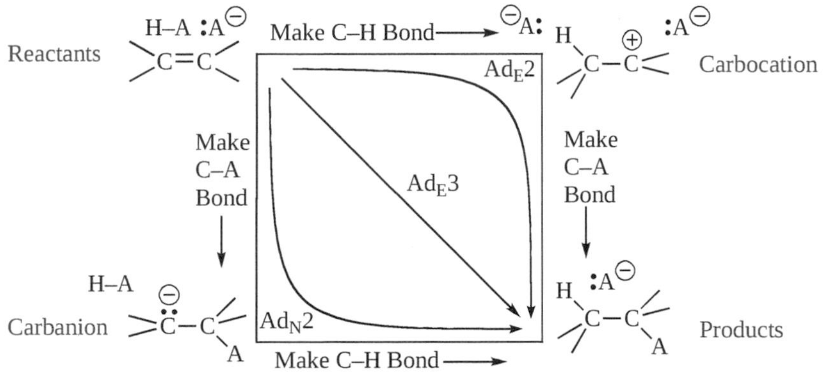

Figure 4.32 Top view of the simplified energy surface for addition to a carbon–carbon pi bond.

The three pathways differ only in the timing of when the electrophile is added and when the nucleophile is added. **Generic Ad_E2** (addition, electrophilic, bimolecular)—The electrophile adds first (rate-determining) by path A_E, producing a reasonably stable carbocation; a nucleophile traps the carbocation by path A_N, forming the addition product. **Generic Ad_E3** (addition, electrophilic, trimolecular)—The reaction occurs by a concerted addition of an electrophile and a nucleophile, path Ad_E3. **Generic Ad_N2** (addition, nucleophilic, bimolecular)—Nucleophilic addition (rate-determining) forms an anion usually stabilized by an electronegative atom or group, and then is protonated by path p.t. to give the addition product. Our three new electron flow paths are A_E (Association, Electrophilic), Ad_E3 (Addition, Electrophilic, Trimolecular) and Ad_N (Nucleophilic addition to a polarized multiple bond). The electron flow arrows for the three addition routes are shown in Figure 4.33. Since we already know how to rank all the intermediates, we can proceed directly to discuss each addition route in detail.

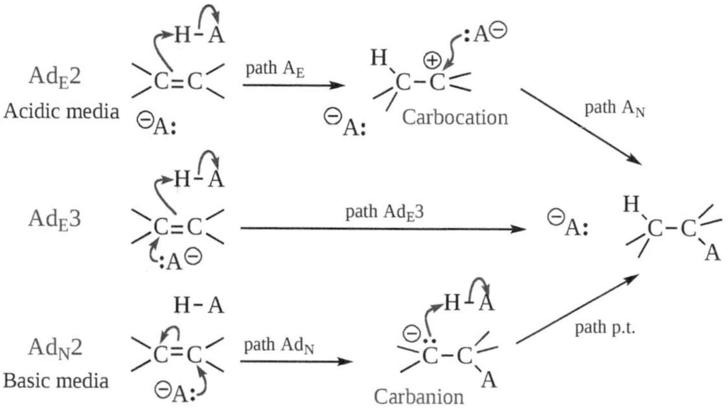

Figure 4.33 The electron flow paths for addition.

4.4.3 The Ad_E2 Reaction: Electron Flow Pathways A_E + A_N

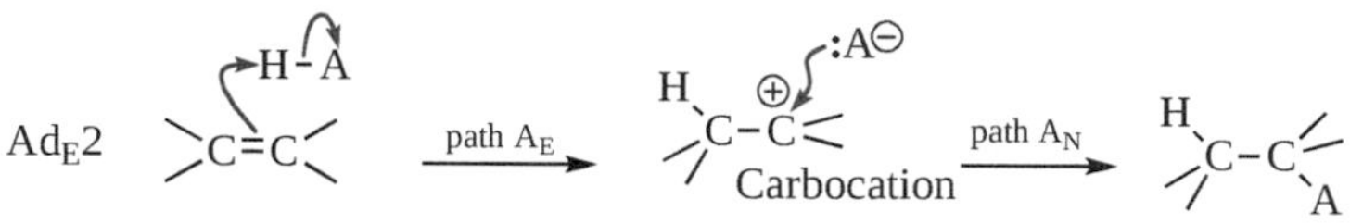

The Ad_E2 is the conceptual reverse of the E1 elimination. The first step, A_E, (Association, Electrophilic) is a new electron flow path and has the following general form. The arrow starts from the pi bond, breaking it and forming a new sigma bond to the electrophile, creating a carbocation next to the carbon–electrophile bond.

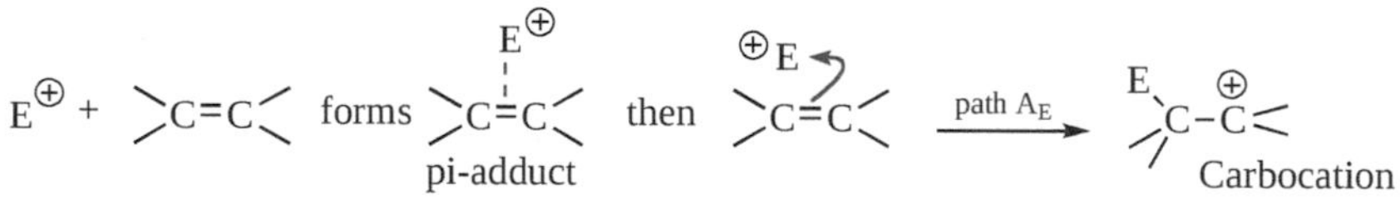

It is thought that the electrophile complexes to the electron cloud of the pi bond first, similar to the pi-adduct (or pi-complex) in Figure 1.17, and then the sigma bond is formed. An energy diagram for a typical Ad_E2 is given in Figure 4.34. The stability of the carbocation (Table 4.1) dominates the energy diagram and is the major contributor to the reaction barrier.

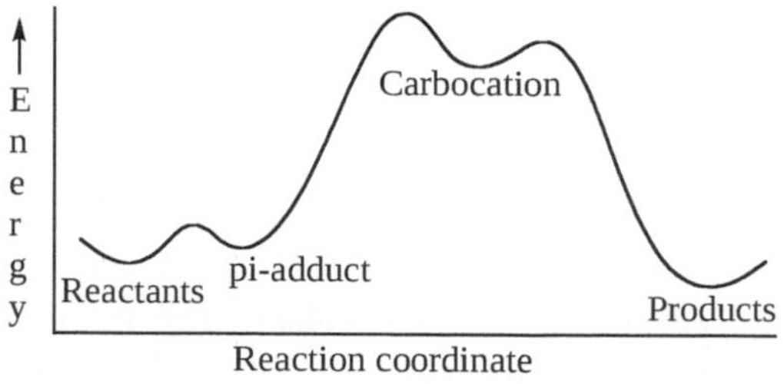

Figure 4.34 The energy diagram for a typical Ad_E2 addition.

Ad_E2 Regiochemistry and Stereochemistry

There are two regiochemical possibilities since the electrophile could have added to either carbon of the pi bond. Since the addition reaction barrier is dominated by the stability of the carbocation formed, the reaction follows the lowest-energy route, namely, to create the more stable of the two possible carbocations. The modern statement of Markovnikov's rule is that the electrophile will add to the pi bond to form the more stable carbocation. The nucleophile adds equally to the top or bottom face of the carbocation.

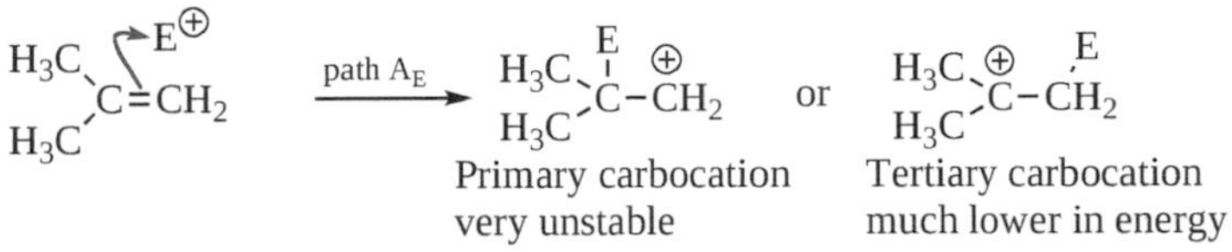

In the following example, H_3O^+ is used to represent acidic water (which is almost all water, not H_3O^+). The pi bond is protonated to form a carbocation, resonance stabilized by the phenyl. Water adds equally to either carbocation face to give a 1:1 mix of enantiomers of the protonated alcohol, which water then deprotonates to give the product.

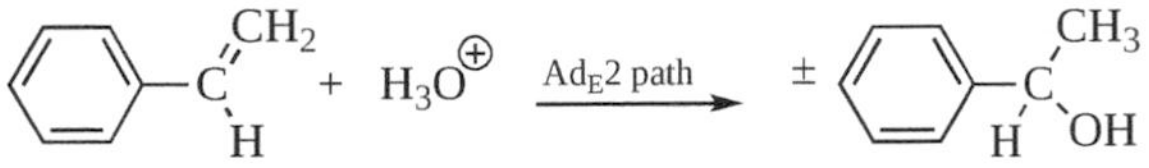

Example problem

For the reaction above, using the explanation in the previous paragraph, draw an Ad_E2 addition mechanism and show both enantiomeric protonated alcohol products.

Answer: The first step of the Ad_E2 is the addition of the electrophile by path A_E to give a carbocation, which is trapped by a nucleophile to form the product by path A_N. Deprotonation of the enantiomers shown below by water yields the neutral alcohol above.

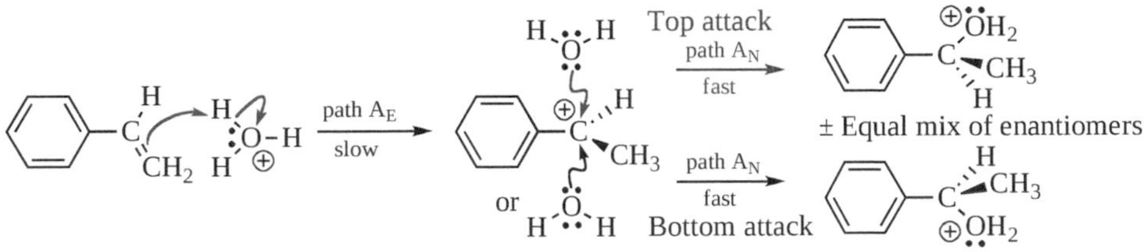

4.4.4 The Ad_E3 Electron Flow Pathway

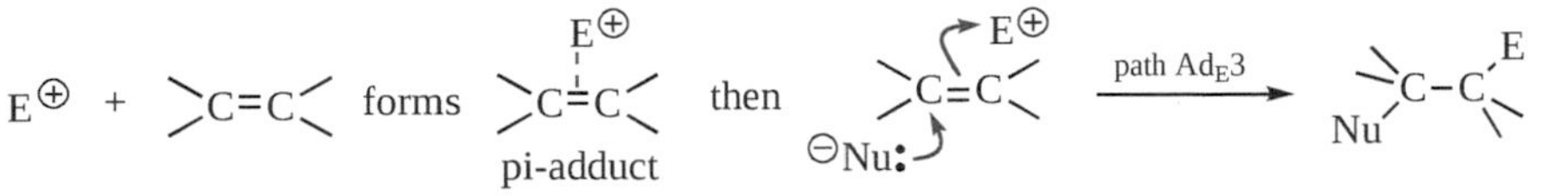

The Ad_E3 is conceptually the reverse of the E2 elimination, and a new electron flow path. Since three-molecule collisions are exceedingly rare, this reaction is thought to proceed by the nucleophile colliding with an electrophile/alkene pi-adduct (or pi-complex), making it just a two-body collision. The electrophile's pi complexation polarizes the pi bond, making it susceptible to nucleophilic attack. The Ad_E3 avoids formation of an unstable carbocation. In the most general form, the electron flow arrow starts from the nucleophile lone pair and attacks the partial-positive end of the pi bond. The next arrow breaks the pi bond and forms a new sigma bond to the electrophile. In the case shown below, a third arrow is needed to break the H–A bond.

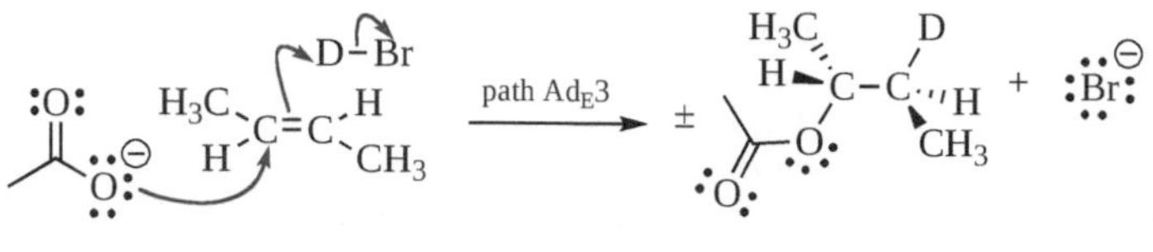

Ad_E3 Regiochemistry and Stereochemistry

We saw in the Ad_E2 that the electrophile first formed a pi-adduct and then added to the pi bond to form the most stable carbocation. In the Ad_E3, the carbocation does not form, but the polarization of the pi bond by the pi-adduct is similar. The nucleophile is attracted to the greatest partial plus of the pi-adduct, resulting in overall Markovnikov addition. Since the electrophile blocks one face of the pi bond, the nucleophile attacks from the opposite face, giving overall *anti* addition. In the following example, the chloride nucleophile for the Ad_E3 comes from ionization of HCl. The anti addition leaves the H trans to the Cl.

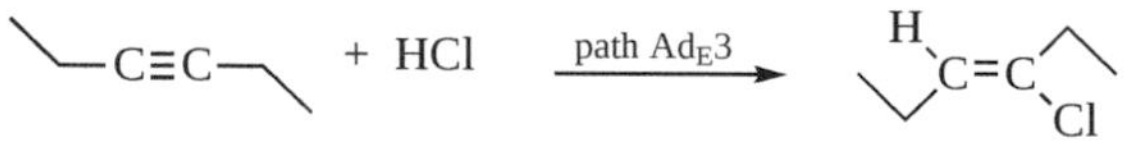

4.4.5 The Ad_N2 Reaction: Electron Flow Pathways Ad_N + p.t.

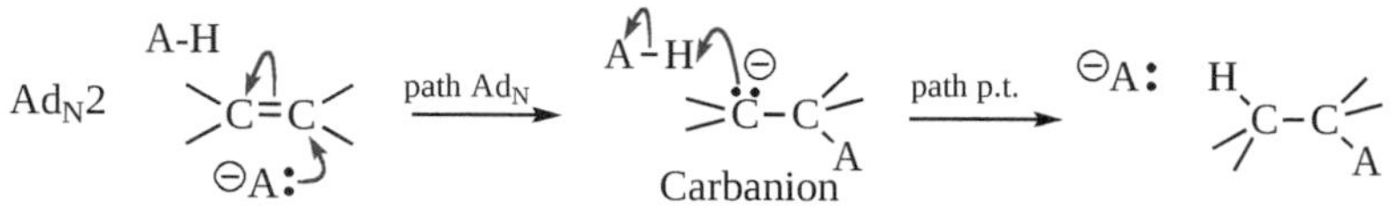

The Ad_N2 is the conceptual reverse of the E1cB elimination and has a new electron flow path, Ad_N, as its first step. The first arrow of the Ad_N starts from the nucleophile lone pair and makes a bond to the partial positive carbon of the pi bond. The second arrow starts from the pi bond, breaking it, and creates a carbanion next to the carbon–nucleophile bond. For the energetics of the Ad_N path to be reasonable, the intermediate anion formed needs to be on an electronegative atom (Section 4.4.7) or be stabilized by an adjacent electron-withdrawing group (below). **The cross-check for the Ad_N2 is the ΔpK_a cross-check (the anion formed should not be any more than 10 pK_a units more basic than the incoming nucleophile).**

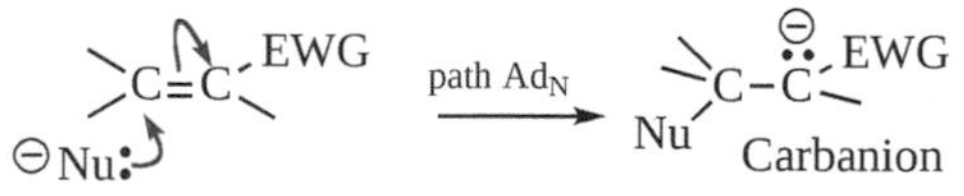

Ad_N2 Regiochemistry and Stereochemistry

The Ad_N2 regiochemistry is solely determined by the position of the electron-withdrawing group. The nucleophile attacks the double bond carbon not bearing the electron-withdrawing group. The stereochem is often a mixture because the carbanion can be protonated from either face. If the nucleophile is very basic and not tolerant of proton sources, then this protonation step occurs later in an acidic workup.

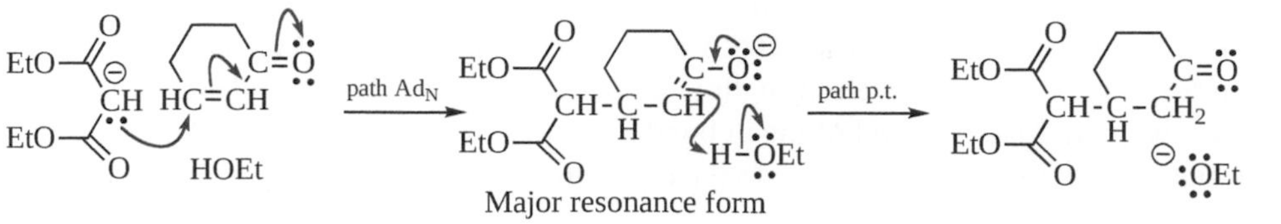

4.4.6 The $Ad_E2/Ad_E3/Ad_N2$ Addition Spectrum

The same factors that folded and tilted the elimination surface have a similar effect when the surface is tilted to favor addition. If the cation and anion are relatively unstable, the surface folds and the lowest-energy path is the Ad_E3 process. If an electron-withdrawing group stabilizes the carbanion, the Ad_N2 process is the lowest-energy route. If an electron-releasing group stabilizes the carbocation, the Ad_E2 is energetically best.

The Interrelationship of the Addition and Elimination Processes

Addition and elimination reactions, being complementary processes, occur on essentially the same energy surface. Conditions that lower the alkene corner favor elimination, whereas conditions that raise it or lower the addition product corner favor addition. A top view of the addition/elimination energy surface is shown in Figure 4.35.

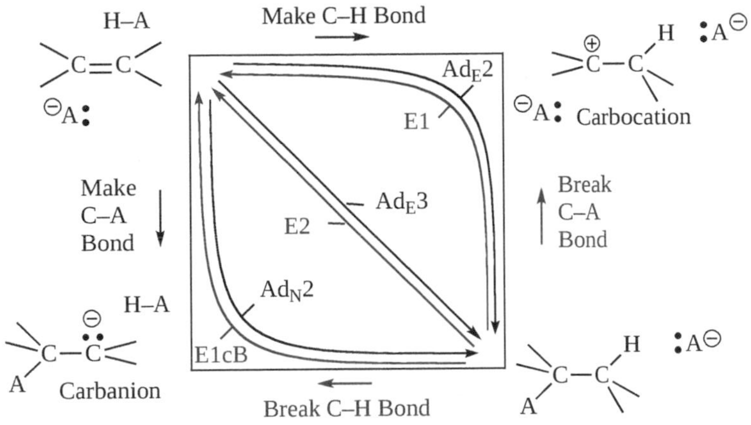

Figure 4.35 The top view of the addition/elimination surface showing how the different addition and elimination reactions interrelate.

4.4.7 Additions to Carbonyls

Additions to carbonyl compounds are perhaps the most important reactions in organic chemistry. If we replace one of the carbons in our addition surface with an oxygen atom, we get a related surface, shown in Figure 4.36.

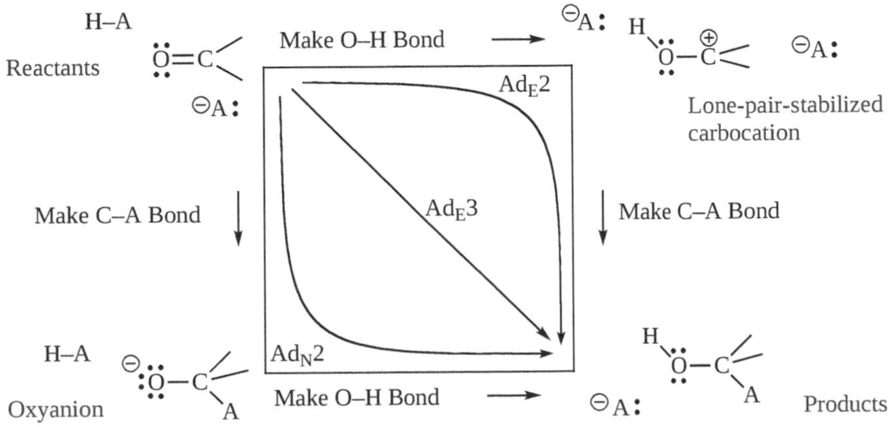

Figure 4.36 Top view of the carbonyl addition surface.

The three addition routes in Figure 4.37 are similar to those for a carbon–carbon pi bond with the exception that the carbonyl lone pair is protonated instead of the carbonyl pi bond. This is because the nonbonding electron pair on oxygen is much more energetically accessible than a stabilized bonding pair of electrons. The major resonance form of the lone-pair-stabilized carbocation is consistent with bond length data showing that the pi bond is still intact. Likewise in the Ad_E3, the hydrogen bond is to the oxygen lone pair. The hydrogen bond "partially protonates" the carbonyl activating it to nucleophilic attack.

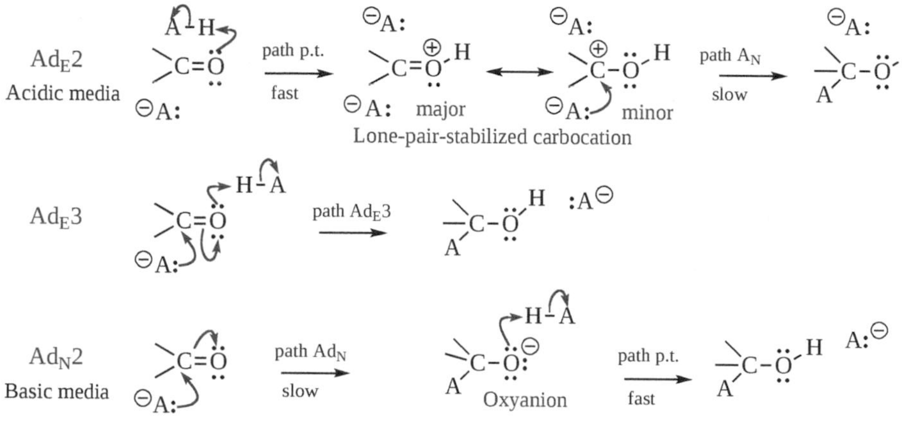

Figure 4.37 Electron flow paths for addition to a carbonyl.

Reversibility of Carbonyl Additions

The acidic media additions to carbonyls are often reversible. For example, addition of water to a carbonyl group usually favors the carbonyl. A $\Delta H°$ calculation gives the hydrate as slightly uphill of the carbonyl as two C–O bonds at 86 kcal/mol each, totaling to 172 kcal/mol (719 kJ/mol) is weaker than one C=O bond at 177 kcal/mol (741 kJ/mol).

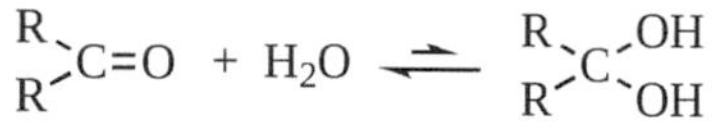

The Ad_N2 reaction on a carbonyl is often carried out in two separate and sequential reaction conditions if the nucleophile is a strong base (and would react with proton sources). The addition of the nucleophile occurs in basic aprotic media, followed by addition of a weak acid in the workup for the proton-transfer step. The reversibility of the Ad_N2 is usually predicted by the ΔpK_a rule cross-check. If the nucleophilic attack occurs in a protic solvent, a following irreversible proton-transfer step may make the overall reaction favorable. This very favorable p.t. step is the driving force for the cyanohydrin formation reaction below.

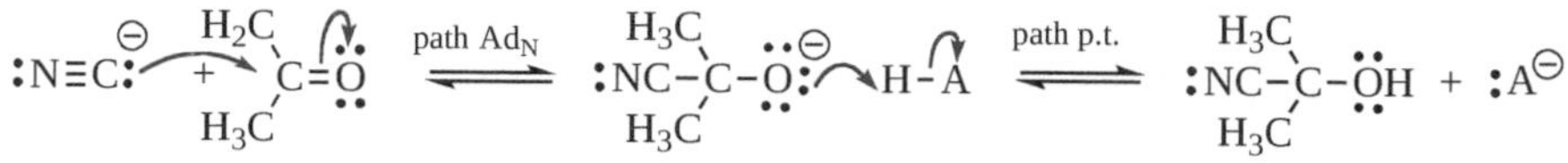

A look at two energy diagrams for this reaction that differ only in the proton-transfer step is instructive. The overall $\Delta G°$ from experiment is −2 kcal/mol (−8.4 kJ/mol). The pK_{abH} of the intermediate is calculated to be about 12.8, and with the acid as HCN (pK_a 9.2), the proton transfer is downhill 3.6 pK_a units, making the proton-transfer K_{eq} = = $10^{+3.6}$. Multiplying 3.6 by −1.364, gives an approximate $\Delta G°$ value of −5 kcal/mol (−21 kJ/mol) for the following proton transfer. By subtraction, we get the $\Delta G°$ for the formation of the intermediate to be about +3 kcal/mol (+12.6 kJ/mol) and can construct an energy diagram, shown in Figure 4.38.

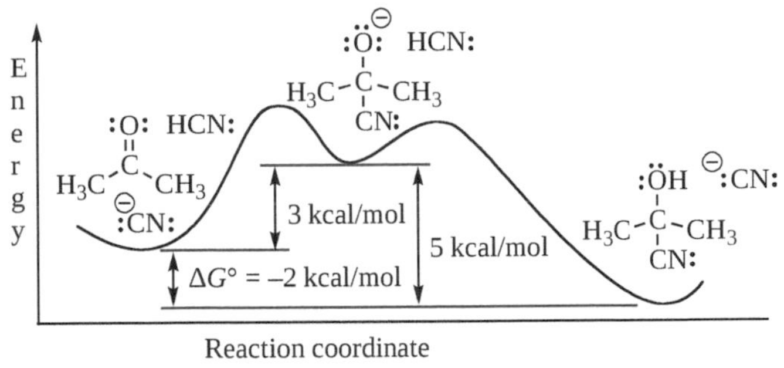

Figure 4.38 The energy diagram for cyanohydrin formation.

If we change the acid from HCN to water, the first part of the diagram remains the same since the acid is just a spectator in the first step. With a weaker acid, the proton-transfer step is now uphill (pK_{abH} 12.8 up to 14.0), so the equilibrium now favors reactants (Figure 4.39).

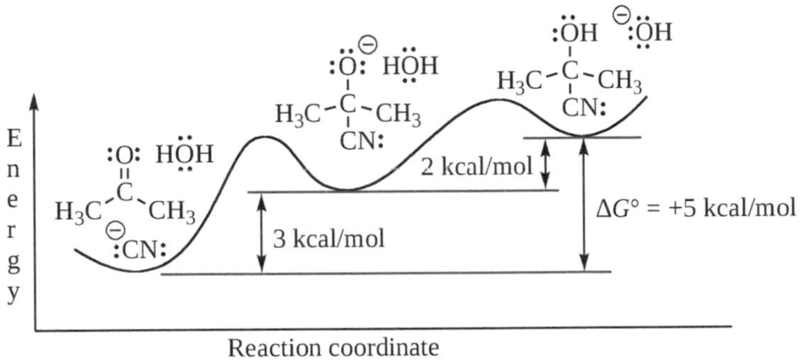

Figure 4.39 The energy diagram for cyanohydrin cleavage.

4.4.8 Approaches to Addition Mechanisms

Although we have three new addition routes each with cross-checks, the similarity to the elimination routes should help. The first task is always to map changes on a balanced reaction. Is the medium acidic or basic? As always, we will first check for a favorable proton transfer before deciding which addition route to take. Again visualize the reactant "street corner" on the addition surface to decide which addition route to use. Is there an appropriately positioned electronegative atom or electron-withdrawing group making Ad_N2 viable? If not, our number of alternatives drops down to two, Ad_E2 or Ad_E3. For the Ad_E2, check the two possible carbocations for stability (better than secondary). If neither carbocation is reasonably stable, the Ad_E2 route is too high in energy, and the reaction will go Ad_E3. On either Ad_E2 or Ad_E3 remember to follow Markovnikov's rule when adding the electrophile. Check each branch of the problem space tree to find the best route before committing to a direction to proceed, then make sure you draw the arrows correctly, keep track of charge balance, and use the known electron flow paths.

Common Addition Errors

Besides the usual proton transfer errors, media pH errors, and media pH span errors, the most common addition error is forgetting to check the carbocation stabilities. A carbocation stability assessment is needed to decide between Ad_E2 and Ad_E3, and also to use Markovnikov's rule. For Ad_N2, a common error is to forget a ΔpK_a rule cross-check.

Example Addition Mechanism

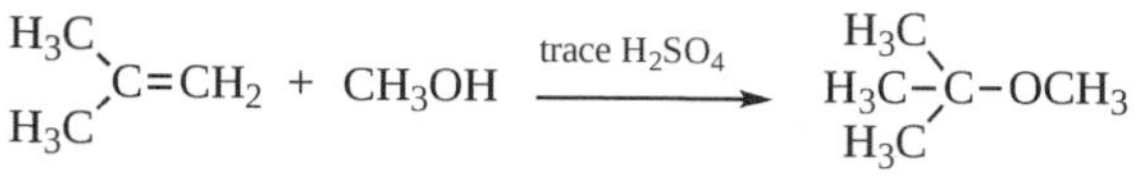

We are asked for the mechanism of formation for a controversial gasoline additive methyl *tert*-butyl ether or MTBE. With sulfuric acid present, the medium is certainly acidic. Mapping changes gives an H added to the pi bond methylene (CH_2) and the OCH_3 added to the pi bond quaternary carbon as expected from Markovnikov addition. The pi bond carries no electron-withdrawing group, so Ad_N2 is out. We need to decide between Ad_E2 and Ad_E3. Markovnikov addition of a proton forms a tertiary carbocation; the Ad_E2 route passes the carbocation stability check. We do not have to consider Ad_E3 if the surface is tilted toward Ad_E2. The sulfuric acid protonates the methanol solvent (leveling effect), so the acidic catalyst in this addition is protonated methanol.

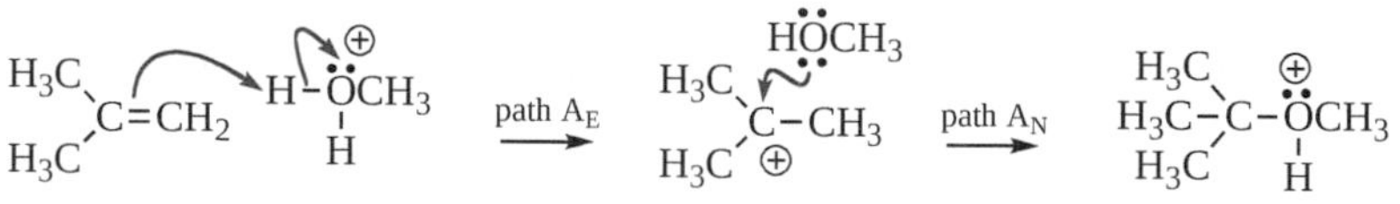

The Ad_E2 reaction gives the protonated product; deprotonation by methanol yields the neutral product and protonated methanol.

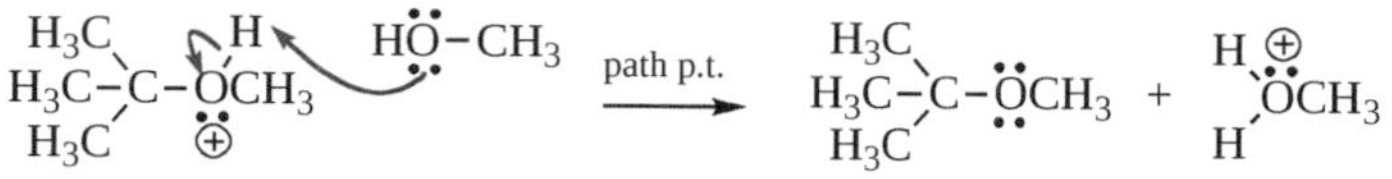

The decision process for working an addition problem is shown in Figure 4.40.

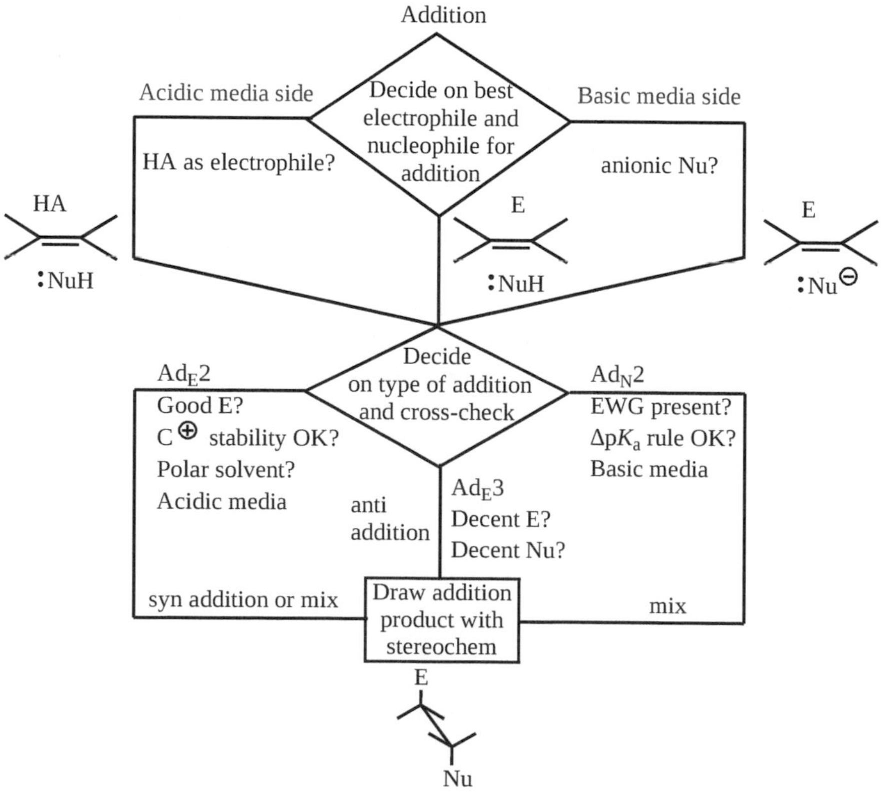

Figure 4.40 The general flowchart for addition. A proton-transfer step may follow.

4.4.9 Addition Summary

• Ad_E2 must form a carbocation of secondary or better, and tends to produce a mix of *syn* and *anti* addition. Markovnikov's rule is followed: The electrophile adds to form the most stable carbocation.

• Ad_E3 occurs when the carbanion and carbocation are not very stable: The energy surface folds down the middle. The electrophile and nucleophile end up on opposite faces of the pi bond (*anti* addition). Markovnikov's rule is followed.

• Ad_N2 follows the ΔpK_a rule cross-check: Never form an anion more than 10 pK_a units more basic than the incoming nucleophile. For this to happen, the anion formed must usually be on an electronegative atom or stabilized by an electron-withdrawing group.

• Species such as Y–L can serve first as the addition electrophile and the leaving group can serve later as the nucleophile. More addition examples can be found in Chapter 8.

4.5 NUCLEOPHILIC SUBSTITUTION AT A TRIGONAL PLANAR CENTER

4.5.1 Addition–Elimination Electron Flow Paths

A leaving group on a trigonal planar center is almost exclusively replaced by an addition–elimination mechanism. The S_N2 has rarely been seen on a trigonal planar center, and the S_N1 is seen with this system in only very acidic media with great leaving groups. The group C=C–L is usually unreactive to nucleophiles, because the anion that results from nucleophilic addition is not stabilized. This ΔpK_a rule cross-check requirement for the nucleophilic attack means that addition-elimination reactions are commonly seen with EWG–C=C–L or Y=C–L, where Y is an electronegative heteroatom such as oxygen. The reactivity trends of carboxylic acid derivatives, O=C–L, are discussed in detail in Section 6.5. Carboxylic acid derivatives are useful and typical reactants for our next reaction archetype, addition–elimination.

We can modify our familiar addition and elimination surfaces to give us a combined simplified addition–elimination energy surface (Figure 4.41). Although this system is further complicated by additional proton-transfer reactions, we can get an overview of the problem space with this simplified surface as a map. The reactants are in the upper left corner.

In acidic media, the carbonyl can be protonated to give a more reactive electrophile, the lone-pair-stabilized carbocation, shown on the top right of Figure 4.41. In acidic media, the weaker nucleophile can add to that more reactive electrophile to give the neutral tetrahedral intermediate. Proton transfer interconverts the different charge types of the tetrahedral intermediate. The neutral tetrahedral intermediate undergoes elimination and loses the leaving group to head toward products (or the nucleophile to return to reactants). In acid, protonation to improve the leaving group on the tetrahedral intermediate is common before elimination. Therefore, the most common acidic route is around the right side, the Ad_E2 addition followed by E1 elimination.

In basic media, there are no acids strong enough to protonate the carbonyl, but basic media usually contain much better nucleophiles, so protonation of the carbonyl sink is not necessary. In basic media, the nucleophile adds to the carbonyl to give the anionic tetrahedral intermediate, which then kicks out the leaving group to form the product. In summary, the most common basic route is straight down the left side of Figure 4.41, Ad_N addition, followed by possible leaving group protonation, before an E_β elimination.

Reactions in neutral media or in enzymes usually avoid the strongly acidic (or basic) intermediates such as the protonated carbonyl (lone-pair-stabilized carbocation corners) by proceeding via the diagonals on Figure 4.41, Ad_E3 addition followed by E2 elimination. Enzymatic reactions can also proceed via general acid or base routes covered in Section 7.4.3.

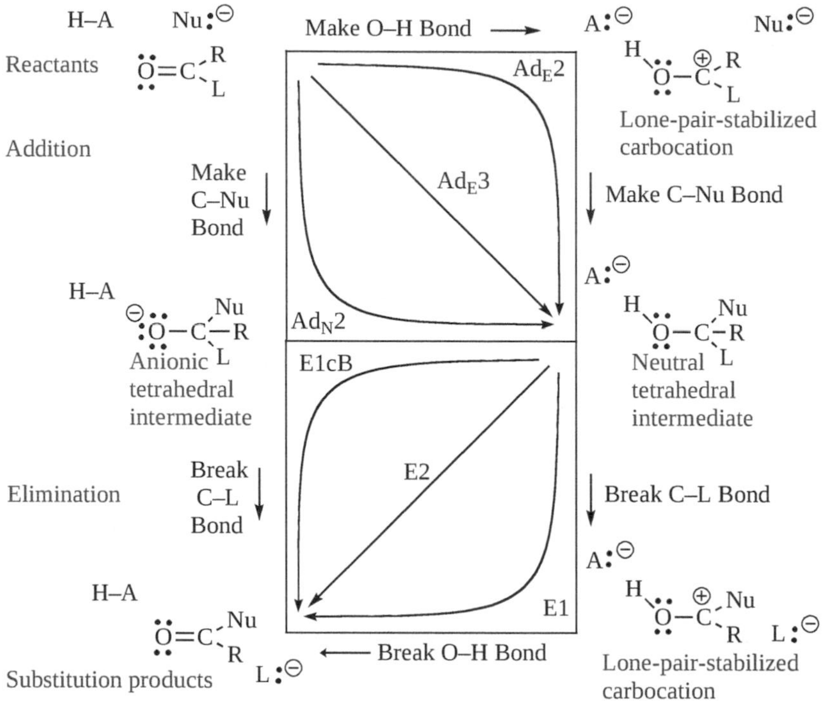

Figure 4.41 Top view of the simplified combined energy surface for addition–elimination.

The arrow pushing for the basic addition–elimination route for carboxyl derivatives follows. The ΔpK_a rule is a very important cross-check for this often-reversible route. The reaction will tend to form the weaker base. If the leaving group is a lot less basic than the nucleophile, then the elimination reaction may be irreversible (use the ΔpK_a rule cross-check).

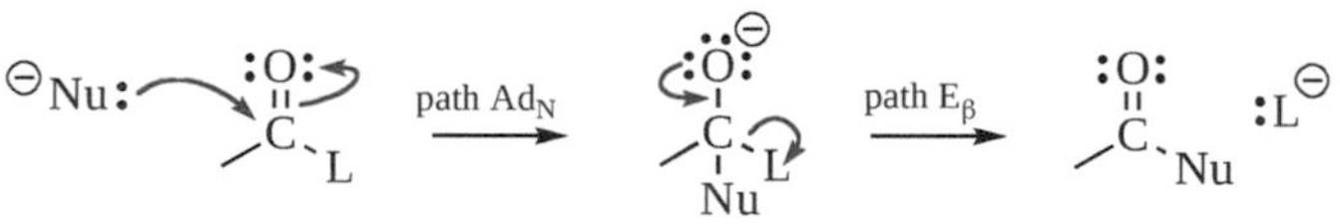

In protic media, the basic route also may go by the Ad_N2 addition to the neutral tetrahedral intermediate followed by the E1cB elimination to the product.

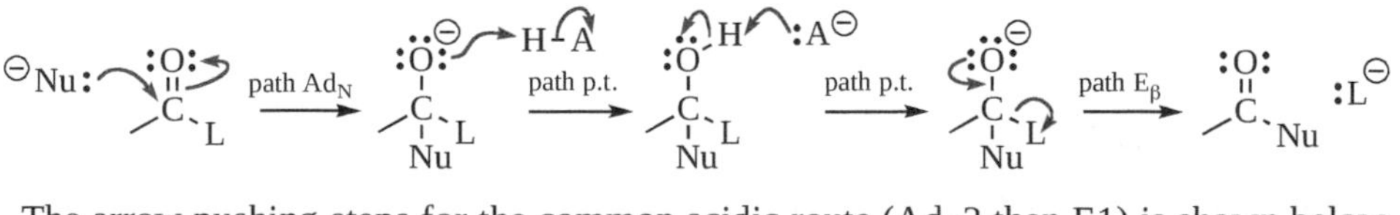

The arrow pushing steps for the common acidic route (Ad_E2 then E1) is shown below. It is often reversible. A reactant excess drives the equilibrium toward products. In acidic media an additional proton-transfer step to improve the leaving group before elimination is common.

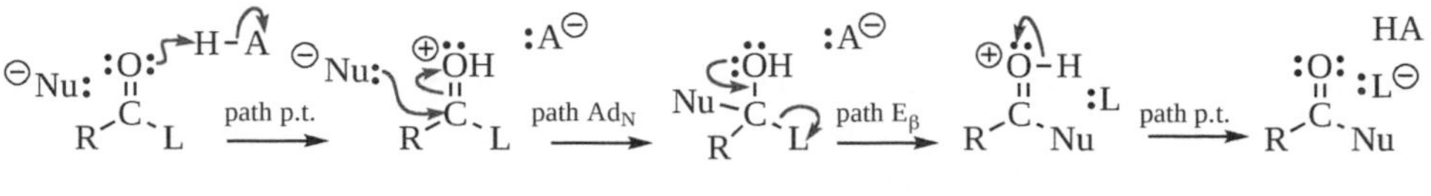

4.5.2 Drawing Energy Diagrams

The ΔpK_a rule cross-check is exceedingly helpful in predicting the energetics of a reaction. We need some rule, no matter how approximate, to tell us whether a particular step or alternative is uphill or downhill in energy. The mechanism for basic hydrolysis of an ester can be used to illustrate the ΔpK_a rule cross-check and show how to draw an energy diagram.

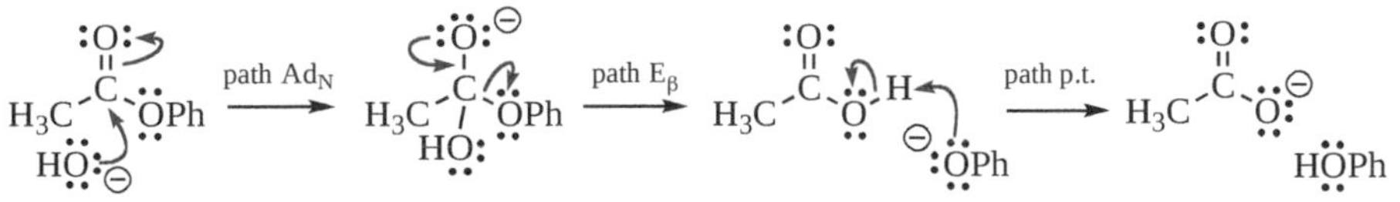

The first step in drawing an energy diagram is to write out the mechanism, so you know how many intermediates there are. The reactants, each intermediate, and the products each go in an energy valley, or well. This energy diagram must have four wells. The first and third structures are both carboxylic acid derivatives, differing only in the nature of the anion, not in the number or type of bonds. The phenoxide (PhO^-) anion has a more distributed charge than hydroxide, and so is less basic than hydroxide. We can place the phenoxide and carboxylic acid well lower than the hydroxide and ester (reactants) well. The tetrahedral intermediate is placed at the highest energy because two C–O bonds are not as strong as one C=O bond, and we have lost entropy combining two molecules into one. The proton-transfer K_{eq} between phenoxide and acetic acid is calculated (Section 3.6) to be $10^{(10-4.8)} = 10^{+5.2}$ corresponding to a $\Delta G°$ at room temperature of approximately (5.2)(−1.36) = −7.1 kcal/mol (–29.6 kJ/mol), almost irreversibly downhill (Section 2.1). This last proton-transfer step forming the resonance stabilized acetate ion is the driving force for the reaction, so we place the products' well lowest. Figure 4.42 shows the relative (and very qualitative) energy of the wells.

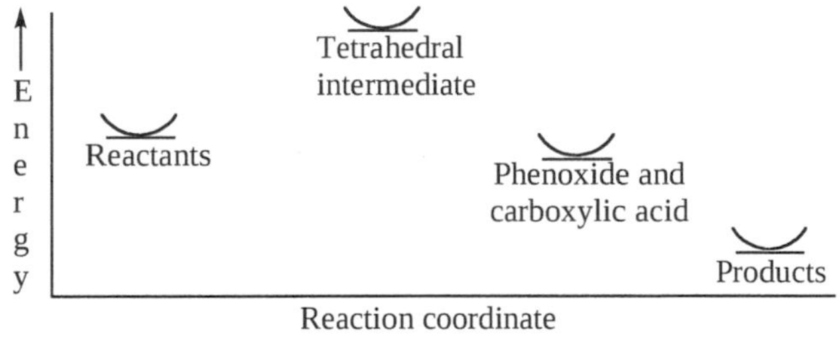

Figure 4.42 Energy well positions for the basic hydrolysis of phenyl acetate.

The only thing left to do is decide the rate-determining step, which will usually be one side or the other of the highest well, the tetrahedral intermediate in this case. If we imagine sitting in the well of the tetrahedral intermediate, which way, forward or back, would be easiest? The three possible groups that could be lost from the tetrahedral intermediate are PhO^- (pK_{aHL} 10), HO^- (pK_{aHL} 14.0), or CH_3^- (pK_{aHL} 48), and clearly the first is the best. Methyl anion is not a leaving group at all. Since the best leaving group from the tetrahedral intermediate is PhO^-, the expulsion of the better leaving group should have a lower barrier (viewed from the well of the intermediate). The rate-determining step is the attack of hydroxide ion on the ester. Figure 4.43 gives a reasonable energy diagram for the reaction.

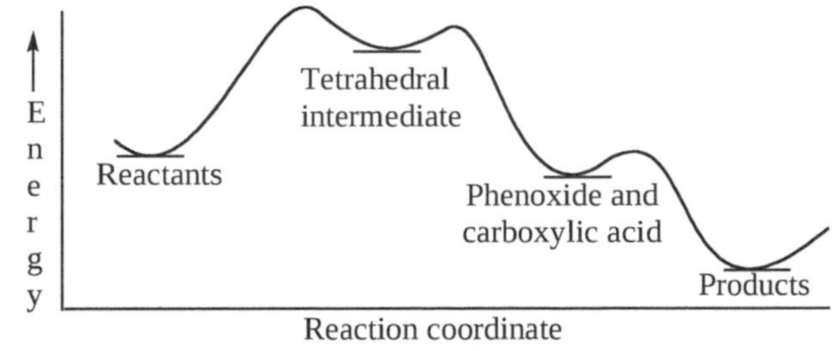

Figure 4.43 Energy diagram for the basic hydrolysis of phenyl acetate.

The plot of pK_{abH}, the pK_a of conjugate acid of the most basic partner in each step vs. the reaction coordinate, is shown in Figure 4.44 and has a profile similar to the energy diagram (Figure 4.41). Note that over the course of a reaction the pK_{abH} of the intermediates may increase slightly, but they will either hover about the initial pK_{abH} or very likely fall to yield a weaker base at the end of the reaction. The breaking of a weak bond and formation of a stronger bond can compensate for the formation of an intermediate with a higher pK_{abH}. **No reaction has huge jumps upward in the pK_{abH} of its intermediates.** There can be great downward jumps as strongly basic intermediates are neutralized.

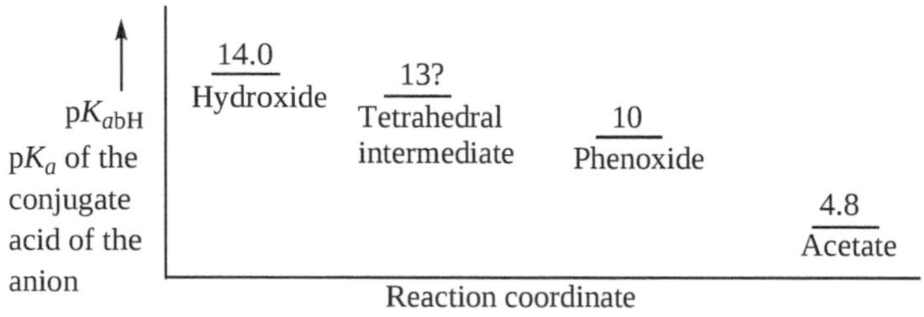

Figure 4.44 The pK_{abH} of the most basic species in each step of the basic hydrolysis of phenyl acetate vs. reaction coordinate. The pK_{abH} must be estimated for the tetrahedral intermediate.

Obviously, the match is not perfect. In the energy diagram (Figure 4.43) the energy of all species in each step is taken into account, whereas in the pK_{abH} diagram only the most basic partner was considered. The C=O bond replaced by two C–O bonds and the entropy lost as two reactants join to form the tetrahedral intermediate cause a significant error. However, overall we can draw a rough association between a reaction proceeding energetically downhill and the fact that the pK_{abH} of the products is almost always less than that of the reactants. Consequently, the following is an extremely useful, although very crude, approximation. **The energy goes downhill if the pK_{abH} drops significantly.** This gross generalization will allow us to choose which of two alternative reaction routes is probably of lower energy.

4.5.3 The Amazingly Useful pK_a Chart

The pK_a chart can be used for:

Ranking the strength of acids by lowest pK_{aHA} is strongest.

Ranking the strength of bases by highest pK_{abH} is strongest.

Calculating a proton-transfer K_{eq} by $\log K_{eq} = pK_{abH} - pK_{aHA}$.

Ranking the stability of carbanions by lowest pK_{abH} is more stable.

Approximate ranking of leaving groups by lowest pK_{aHL} is better.

Ranking of electron-withdrawing groups by lowest pK_{abH} of CH_3–EWG is better.

Estimating the energetics of reaction by the ΔpK_a rule cross-check (≤ 10 pK_a units uphill).

Checking the reversibility of a reaction (ΔpK_a rule cross-check on the reverse reaction).

Helping to rank nucleophiles (in conjunction with softness) (Section 4.2.2).

Example problem

Does the following reaction favor products or reactants, and is it reversible?

Answer: Since the pK_{abH} of methoxide is 15.2 and that of chloride is −6, chloride is obviously the weaker base. The reaction proceeds to the right to form the weaker base and is not reversible because the reverse reaction would climb 21.2 pK_a units.

4.5.4 Approaches to Addition–Elimination Mechanisms

This time we are just applying known addition and elimination routes to a new reactant. The first task is always to map changes on a balanced reaction. Is the medium acidic or basic? The pH of the medium is very important in deciding which addition–elimination route to take, basic or acidic. Again visualize the reactant "street corner" on the addition surface to decide which way to go, then make sure you draw the arrows correctly, keep track of charge balance, and use the known electron flow paths.

Common Addition–Elimination Errors

Besides the usual proton transfer errors, the most common addition–elimination errors are switching media pH in mid-mechanism, media pH errors, and media pH span errors. A common error in basic media is to forget the important ΔpK_a rule cross-check.

Errors from Shortcuts to Save Arrows or Compress Steps

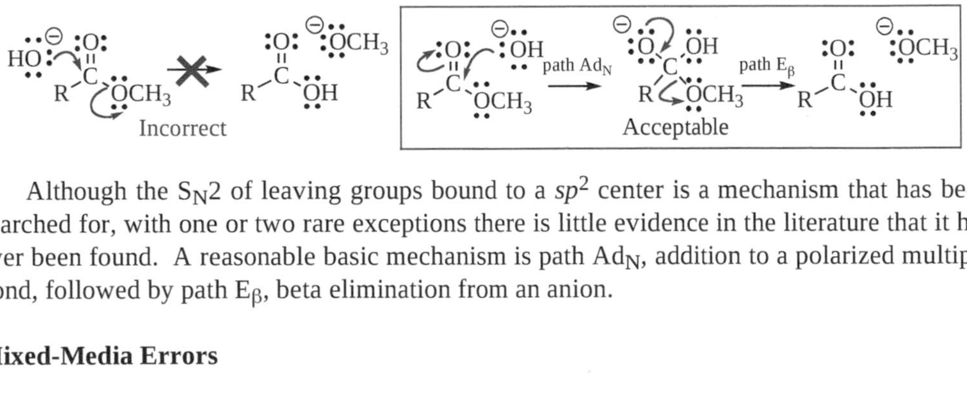

Although the S_N2 of leaving groups bound to a sp^2 center is a mechanism that has been searched for, with one or two rare exceptions there is little evidence in the literature that it has ever been found. A reasonable basic mechanism is path Ad_N, addition to a polarized multiple bond, followed by path E_β, beta elimination from an anion.

Mixed-Media Errors

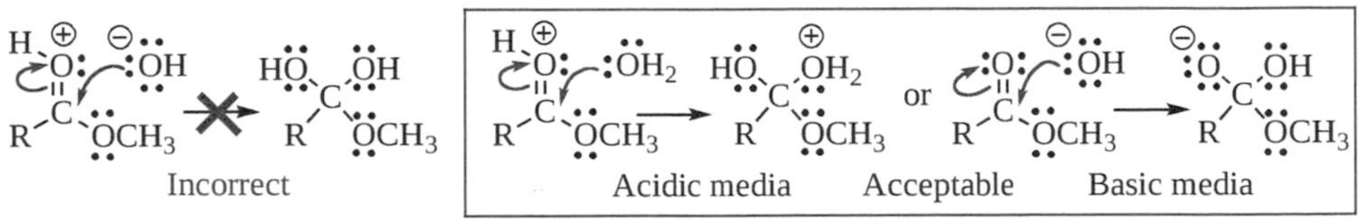

Sometimes students will optimize the source and the sink and forget that those two species cannot exist in the same medium. No medium can have both a strong base and a strong acid present at the same time. Since the protonated ester pK_a is −6.5 and the pK_{abH} of hydroxide is 14.0, they span over 20 pK_a units. Therefore, any medium that would be acidic enough to have some concentration of the protonated ester would have essentially no concentration of hydroxide ion. Shown to the right are two reasonable alternatives: an acidic medium with the protonated ester and neutral water as the nucleophile, or a basic medium with the neutral ester and hydroxide as the nucleophile.

Loss of a Poor Leaving Group Error

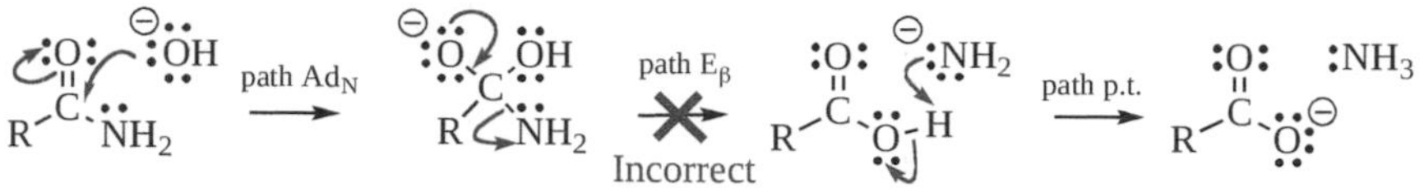

This is basically a ΔpK_a rule cross-check violation. Groups with a pK_{aHL} of over 30 are not leaving groups from tetrahedral intermediates, which don't get more basic than about a pK_{abH} of 16. Unfortunately, these errors occur even in textbooks; for example, the basic amide hydrolysis mechanism shown above kicks out an NH_2^- anion of pK_{abH} 35 from a tetrahedral intermediate of about pK_{abH} 14, a climb of over 20 pK_a units. For a reasonable alternative, see the worked basic amide hydrolysis example in Section 10.7.4.

Example of an Addition–Elimination Mechanism

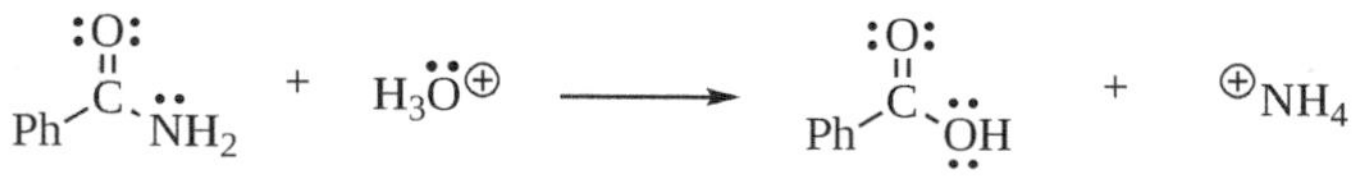

We are asked for the mechanism of acidic amide hydrolysis. The digestion of proteins is an amide hydrolysis. With acidic water present, we will choose the acid catalyzed addition–elimination route, Ad_E2 followed by elimination. We anticipate needing to add a proton-transfer step to improve the leaving group, since it is currently very poor. The first step is to activate the carbonyl for attack by the mediocre nucleophile water. An alternative proton-transfer to nitrogen is possible but unfavorable because this puts a plus charge next to the partial plus carbon of the carbonyl. Protonating the carbonyl instead makes a reactive resonance-stabilized electrophile. A cross-check gives K_{eq} to be $10^{(-1.5\ -\ 0.0)} = 10^{-1.5}$, favoring reactants, but easily achievable.

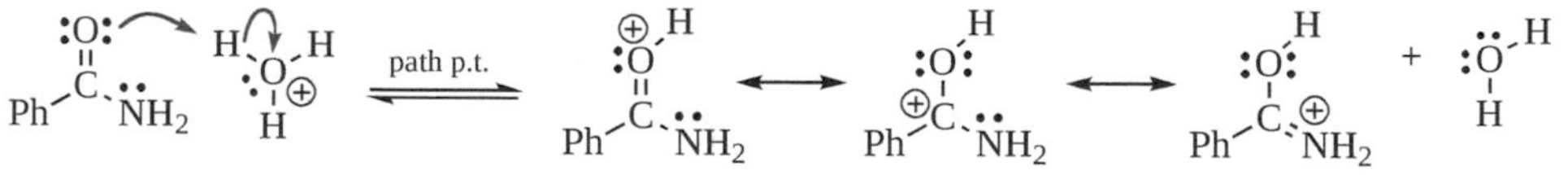

Water acts as a nucleophile on the protonated carbonyl completing the Ad_E2, and then loses a proton to give the neutral tetrahedral intermediate.

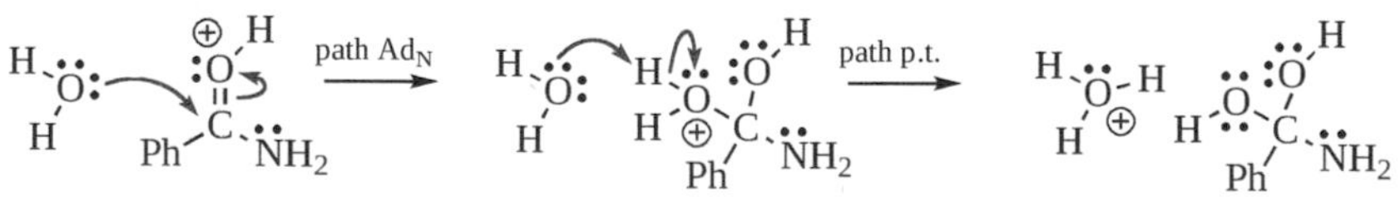

We need to do an elimination to restore the carbonyl, but the NH_2 anion is not a leaving group at $pK_{aHL} = 35$. Protonation will improve it significantly, giving neutral NH_3 as a leaving group, $pK_{aHL} = 9.2$. The calculated pK_{abH} of the intermediate is 8.3. A cross-check of the proton transfer below gives K_{eq} to be $10^{(8.3\ -\ 0.0)} = 10^{+8.3}$, well downhill.

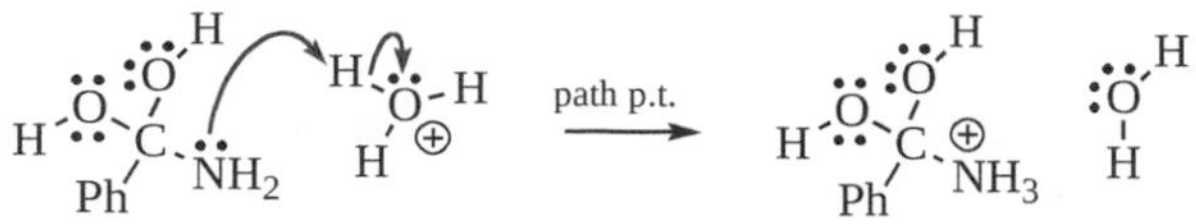

Now it is time to decide an elimination route. The E1 has a problem because although the carbocation would be stable and the water solvent is very polar, the leaving group is just not good enough. E1cB rarely occurs in acidic media because it needs a decent base for the initial deprotonation step. E2 remains a reasonable alternative, climbing 9.2 units overall.

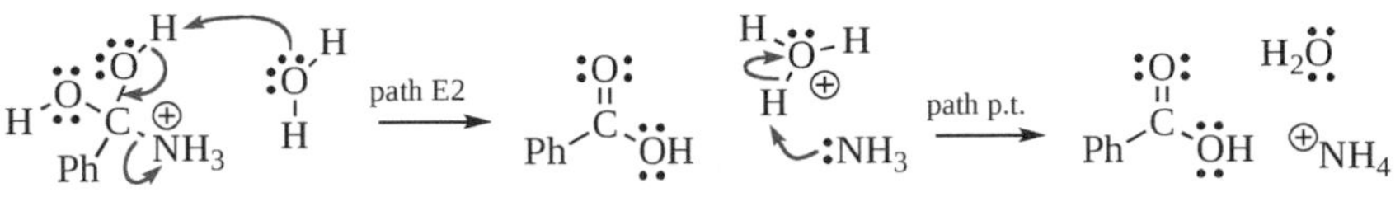

All we need to complete the mechanism is a very favorable proton transfer with a $K_{eq} = 10^{(9.2-0.0)} = 10^{+9.2}$, which makes the overall reaction irreversible.

4.5.5 Addition–Elimination Summary

The previously discussed limitations for both additions and eliminations apply. More addition–elimination examples can be found in Section 8.8. The decision process for working an addition–elimination problem is shown in Figure 4.45. Like before, proton transfer is the first decision to be made.

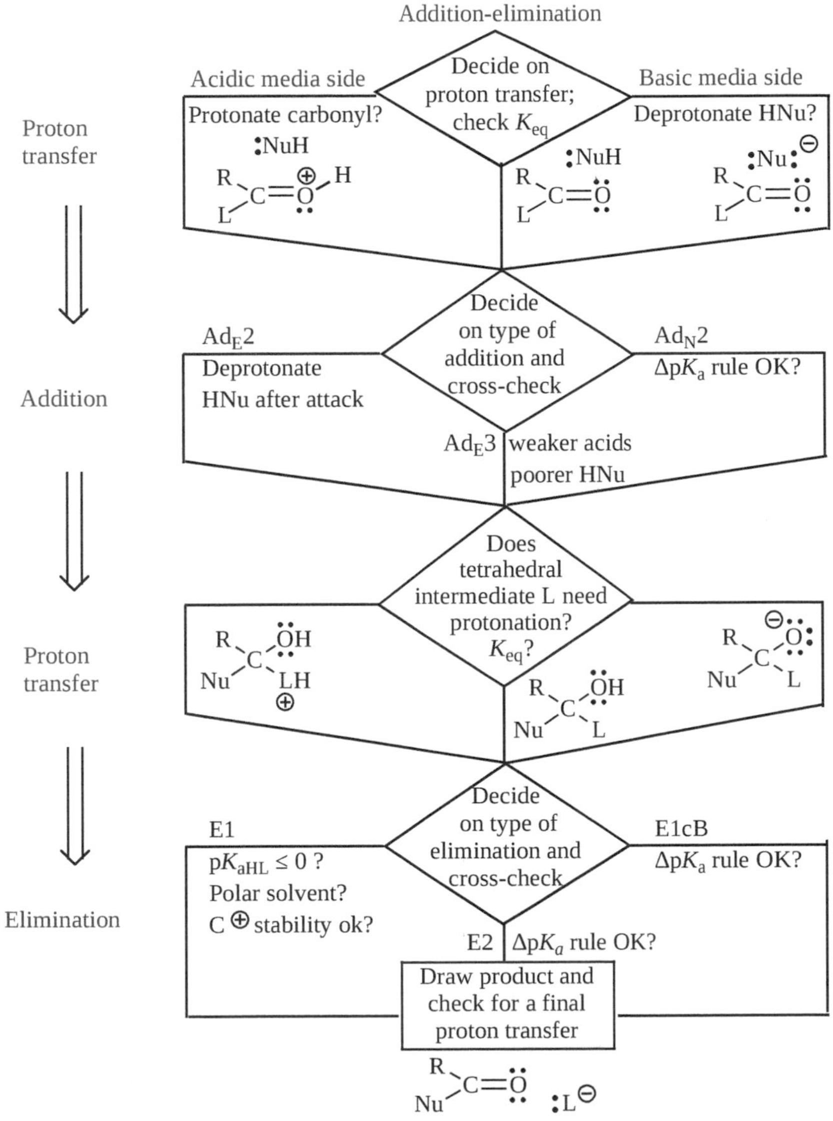

Figure 4.45 The flowchart for addition–elimination.

• Ad_E2 usually occurs in acid because it forms a very acidic protonated carbonyl.

• Ad_E3 occurs in buffered media, where the medium is not acidic enough to protonate the carbonyl and not basic enough to support strong nucleophiles.

• Ad_N2 occurs in base with good nucleophiles and follows the ΔpK_a rule cross-check: Don't form an anion more than 10 pK_a units more basic than the incoming nucleophile.

• E1 tends to occur in acid because the leaving groups are often improved by being protonated and must pass three tests: a reasonably stable carbocation (usually better than secondary), a polar solvent, and a leaving group with a pK_{aHL} of zero or less.

• E2 often occurs in buffered media when the energy surface folds down the middle.

• E1cB usually requires basic media and an acidic hydrogen. Check that the proton transfer K_{eq} is more than 10^{-10}. Because of the anionic tetrahedral intermediate, E1cB can kick out poorer leaving groups. The E1cB follows the ΔpK_a rule cross-check: Never kick out a leaving group more than 10 pK_a units more basic than the tetrahedral intermediate.

4.6 ELECTROPHILIC SUBSTITUTION AT A TRIGONAL PLANAR CENTER

The most common example of electrophilic substitution at a trigonal planar center is electrophilic aromatic substitution, which will serve as our archetype. Aromaticity was covered in Section 1.9, and the reactivity trends of aromatics will be covered in detail in Chapter 5. Chapter 8 has additional reaction examples. Aromatic rings are usually poor nucleophiles; therefore excellent electrophiles are needed for the reaction to proceed.

4.6.1 Electrophilic Aromatic Substitution

There is only one route by which aromatic rings react with electrophiles. The electrophile is believed to form a pi-adduct with the pi cloud of the aromatic ring; then the electrophile attacks an electron-rich carbon of the ring by path A_E, creating a delocalized carbocation called the *sigma complex*. The sigma complex loses an electrophile, commonly a proton by path D_E to restore the aromaticity to the ring. This combination of the A_E then D_E paths has the mechanistic abbreviation S_E2Ar (substitution, electrophilic, bimolecular, aromatic). Figure 4.46 shows a typical energy diagram for an electrophilic aromatic substitution. Usually the attacking electrophile is more reactive than a proton, and therefore the rate-determining step is electrophile attack. Proton loss from the sigma complex will be an easier process than loss of a reactive electrophile.

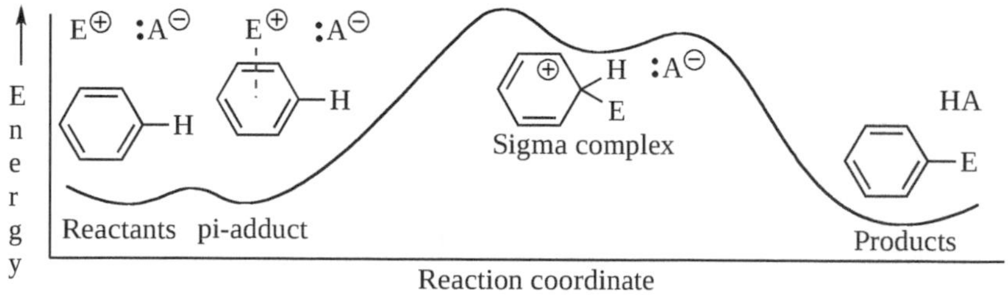

Figure 4.46 A common energy diagram for electrophilic aromatic substitution.

The arrow pushing for the electrophilic aromatic substitution is shown below. An arrow starts from a pi bond of the aromatic ring and creates the carbon electrophile bond in an A_E step. The first arrow of the D_E step starts from any slightly basic species (often the counterion of the electrophile) and a final arrow breaks the C–H bond *at the site of electrophile attack* and reforms the pi bond of the aromatic ring.

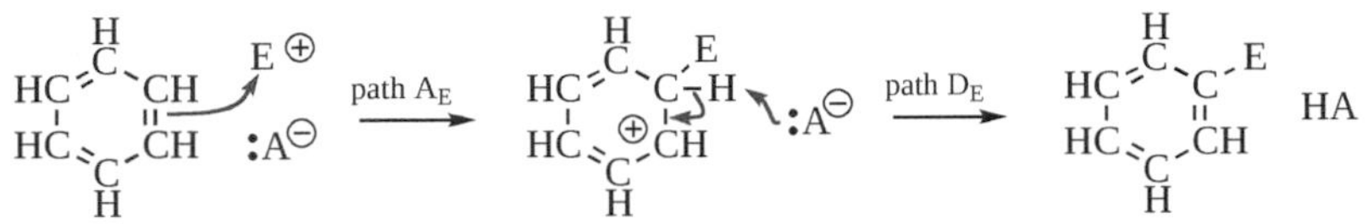

Electrophilic Aromatic Substitution Regiochemistry

Groups present on the ring before the electrophile attacks dictate not only the reactivity of the ring but also the position of attack. Any group that has a resonance interaction with an aromatic ring has the greatest effect at the *ortho* and *para* positions. This is illustrated by the following resonance forms.

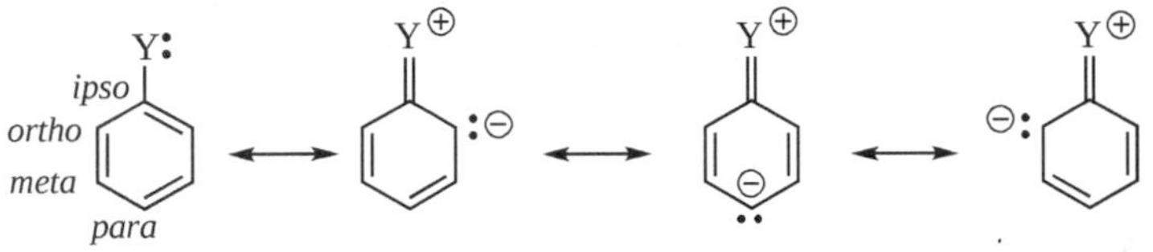

Electron-releasing groups (ERG) attached to the ring can stabilize the sigma complex if the electrophile attacks either *ortho* or *para* to the group. In one resonance form of either sigma complex, the positive charge is adjacent to the electron-releasing group. By stabilizing the sigma complex, the electron donors both increase the reactivity of the ring (by increasing the electron availability) and direct the incoming electrophile *ortho* or *para*. The *ortho* position is more hindered, so larger electrophiles usually go *para*.

Ortho attack:

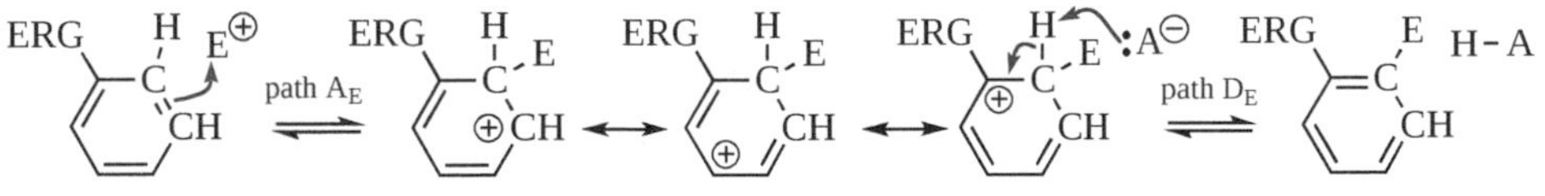

Para attack:

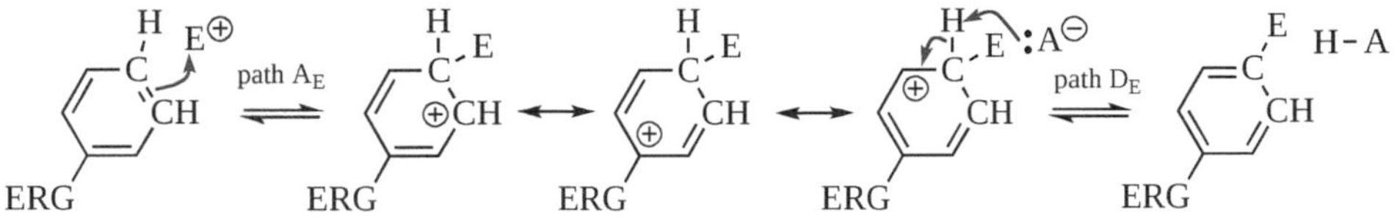

Electron-withdrawing groups (EWG) make the ring less electron rich, less reactive to electrophiles, and direct the incoming electrophile *meta*. The reactivity of the aromatic ring drops far enough so reactions with carbon electrophiles fail. The incoming electrophile is directed *meta*, going via the least-destabilized sigma complex, which has no resonance form with the positive charge adjacent to the electron-withdrawing group.

Meta attack:

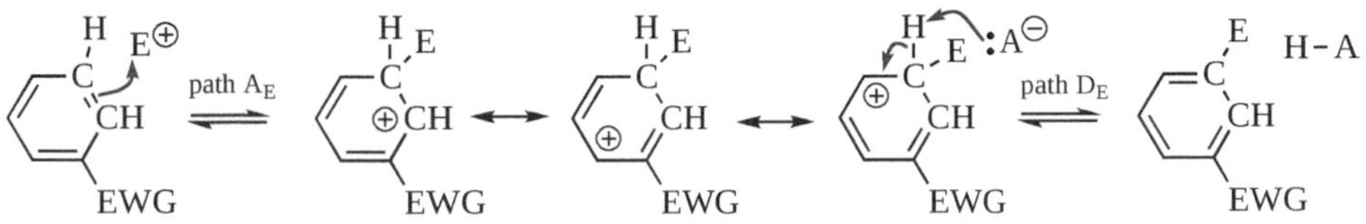

Occasionally the electrophile will attack the ring carbon that bears the group itself (*ipso* attack), but this site is definitely more hindered. If the group can form a cation equal to or more stable than the incoming electrophile, the group can depart from the sigma complex as a cation (path D_E) and thus be replaced by the electrophile. Groups less electronegative than carbon, like silanes or metals, prefer the *ipso* substitution route.

Ipso attack:

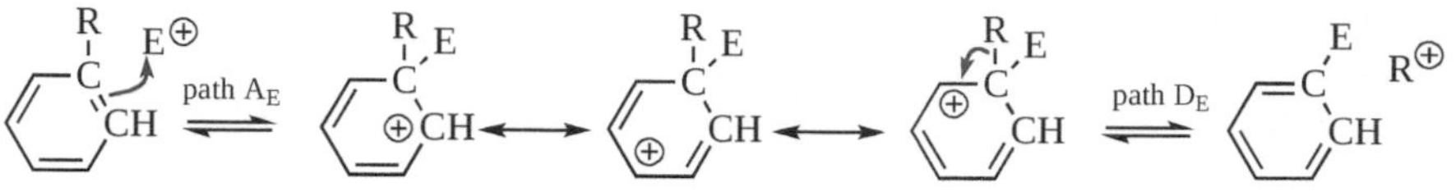

If more than one group is on the ring, the one that can stabilize the sigma complex the most will be the one that directs electrophile attack. The reaction proceeds through the most stable of the possible sigma complexes. For example, if two electron-releasing groups are on the ring, the electrophile attacks *ortho* or *para* to the best one.

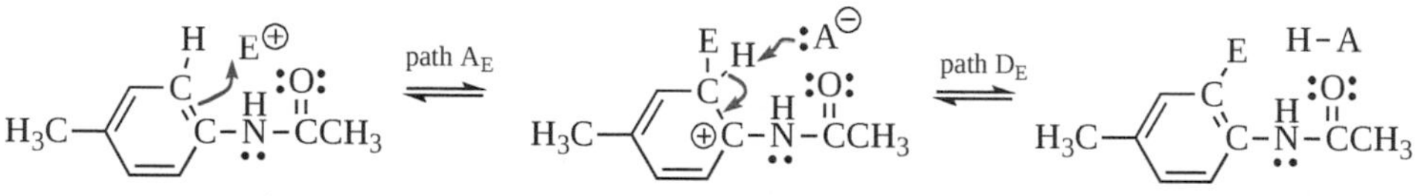

When two electron-releasing groups are *meta* to each other, attack at the position between them is extremely difficult because of steric hindrance.

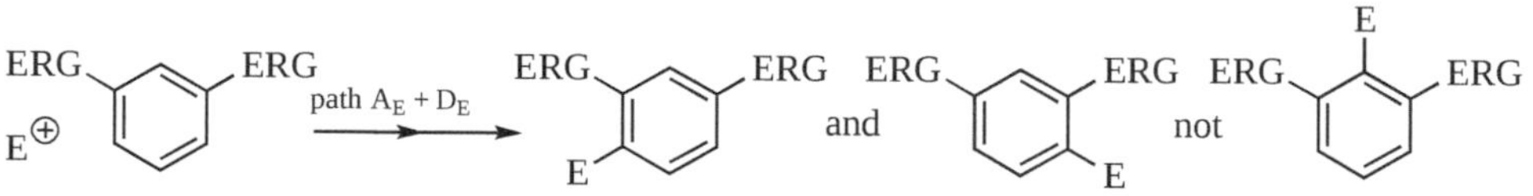

4.6.2 Approaches to Electrophilic Aromatic Substitution Mechanisms

Often the most difficult part of electrophilic aromatic substitution mechanisms is working out the generation of the reactive electrophile. The first task is always to map changes on a balanced reaction. The medium is almost always acidic because reactive electrophiles are present. The electrophilic addition to the aromatic ring is just a two-step process, A_E then D_E. Draw the H at the electrophile attack site because that is the H lost. Make sure you draw the arrows correctly, keep track of charge balance, and use the known electron flow paths.

Example Electrophilic Aromatic Substitution Mechanism

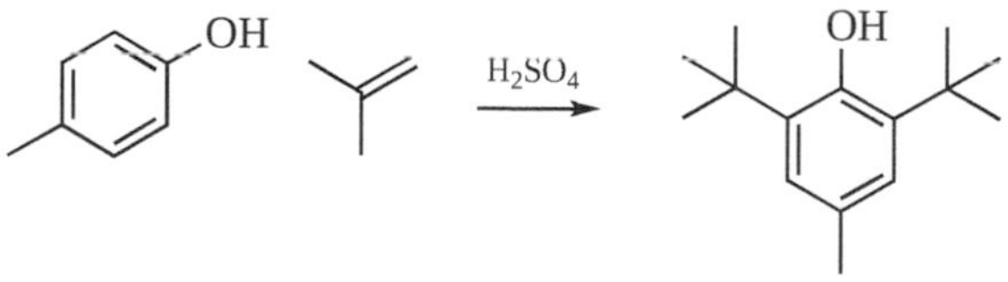

This is the synthesis of a somewhat controversial food preservative called BHT. It is an antioxidant that has been put in packaging to keep foods from becoming rancid from air oxidation. A balanced reaction would require two molecules of the alkene for each of the starting aromatic. The aromatic ring is relatively electron rich with two donors attached to it. With sulfuric acid present, the medium is definitely acidic. The first step is to generate the excellent electrophile needed for electrophilic aromatic substitution. For simplicity, let's symbolize sulfuric acid as H–A. The Markovnikov addition of a proton to isobutylene gives the *tert*-butyl carbocation, an excellent electrophile.

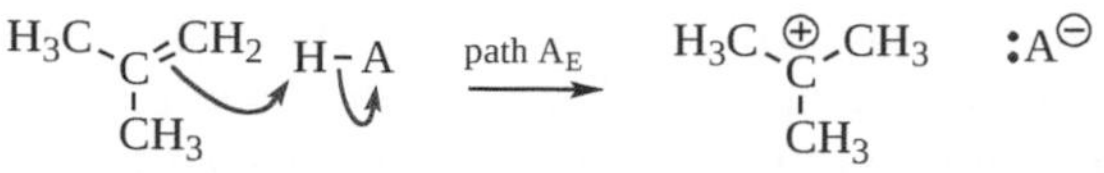

Now we have an excellent electrophile for the electrophilic aromatic substitution reaction. The hydroxyl lone pair is a much better electron-releasing group than the methyl, so the electrophile adds *ortho* to the best electron-releasing group, hydroxyl (the *para* position is already occupied by methyl). The D_E step removes the proton from the carbon that the electrophile attacked to rearomatize the ring.

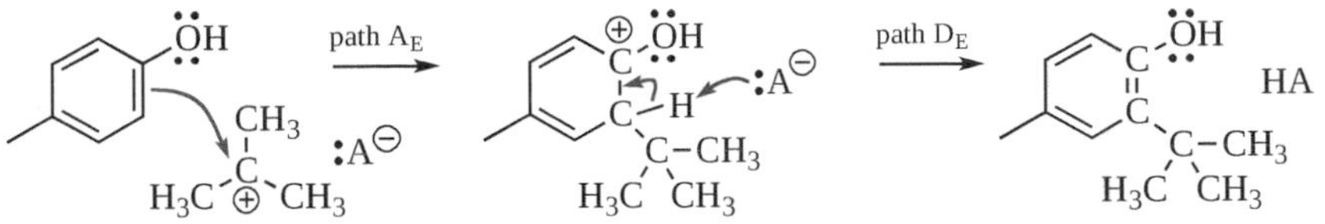

The electrophilic aromatic substitution reaction is repeated again on the other *ortho* position with the same electrophile to give BHT.

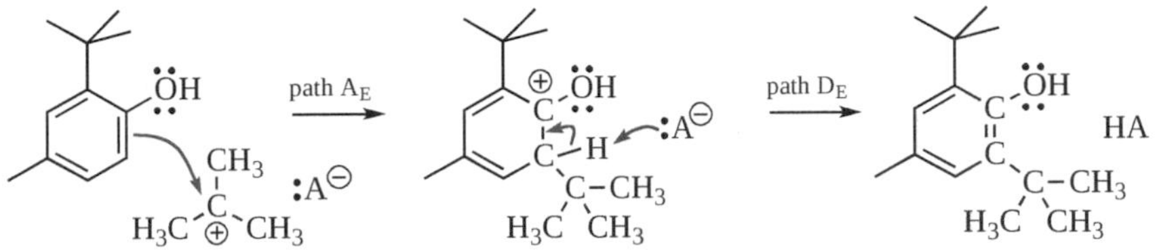

4.6.3 Electrophilic Aromatic Substitution Summary

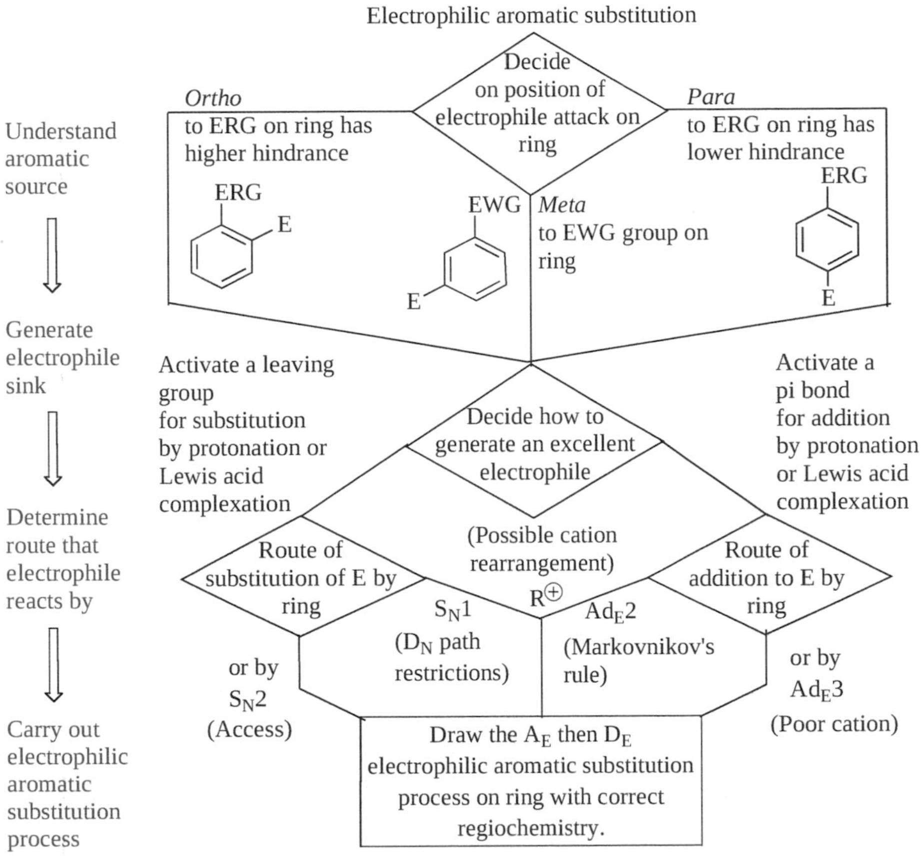

Figure 4.47 A general flowchart for electrophilic aromatic substitution.

The decision process for working an electrophilic aromatic substitution problem is shown in Figure 4.47. First we have to decide the best position on the ring to add the electrophile, and next we need to figure out how to generate the electrophile. Then we have to decide the process by which our electrophile reacts before adding it to the aromatic ring. Carbocation formation can have the possibility of rearrangement, our last archetype; Section 9.8 discusses this decision. More examples of electrophilic aromatic substitution reactions are in Chapter 8.

Common Electrophilic Aromatic Substitution Errors

Some texts will draw a circle in a six-membered ring to symbolize an aromatic. A semicircle with a plus may be also used to symbolize a sigma complex. Don't use the circle notation when doing mechanisms, because it is so easy to lose track of electrons.

Media pH errors and media pH span errors are common. **Since electrophilic aromatic substitution is almost exclusively an acidic media process, do not make any strong bases during the mechanism.** The proton on the aromatic ring becomes very acidic after the electrophile attaches and forms the carbocationic sigma complex. However, before the electrophile attacks, the aromatic H is not acidic at all, $pK_a = 43$, so do not get your steps out of order and try to pull the H off first.

Line structure can set you up for a common beginner's error. You can forget about the H that needs to be pulled off in a D_E step because line structure does not show the hydrogen atoms. **It is the proton at the site of electrophile attack that is lost, not any other.**

4.7 REARRANGEMENTS TO AN ELECTROPHILIC CARBON

4.7.1 Introduction to Rearrangement Reactions

Rearrangements encompass so many different structural features and mechanisms that it is not possible to write a generic rearrangement example that would fit all. Instead, for our archetype, we will pick two competing rearrangements to an electrophilic carbon that complicate our other acidic routes, primarily the S_N1, E1, and S_E2Ar. We have to be suspicious of the possibility of rearrangement occurring any time a compound is placed in strongly acidic medium.

4.7.2 Rearrangement Electron Flow Paths

The arrow pushing for the two rearrangement paths is shown below. When the carbocation is of sufficient stability, the leaving group will depart first (path D_N), and then rearrangement occurs in a path resembling the S_N1 process (1,2 rearrangement, path 1,2R). The arrow starts from the migrating bond and points to the carbocation, indicating that the migrating group is just sliding over while maintaining orbital overlap between the atom it leaves and the destination atom. Rearrangement creates an equal or more stable carbocation.

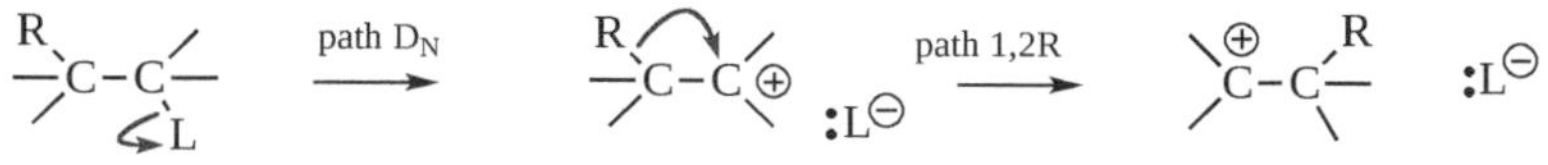

When the carbocation that would have formed in the D_N step is not very stable, rearrangement can still occur by a process similar to the S_N2. The migrating group displaces the leaving group from the back side. For this rearrangement with leaving group loss, the 1,2RL path, we start the first arrow from the bond of the migrating group and slide it over to the carbon bearing the leaving group. A second arrow starts from the C–L bond, breaking it and making a new lone pair on the leaving group.

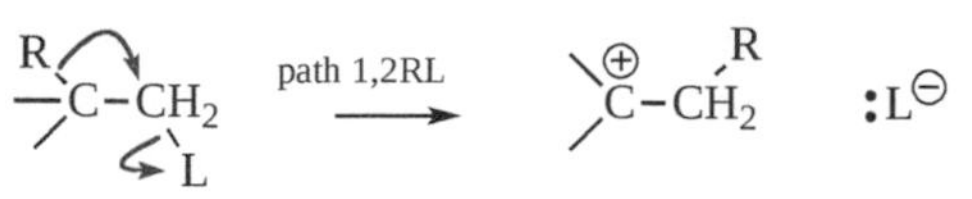

Rearrangement Path Restrictions and Regiochemistry

The main path restriction on rearrangements is that the migrating group must maintain orbital overlap with both the atom it migrates *from* and the atom it migrates *to*. At the transition state, the migrating group is partially bonded to both atoms. If the leaving group departs first, the usual D_N restrictions apply (carbocation stability usually better than secondary, good leaving group, polar solvent). A specific example is useful. Isobutyl alcohol reacts with concentrated hydrobromic acid to give *tert*-butyl bromide.

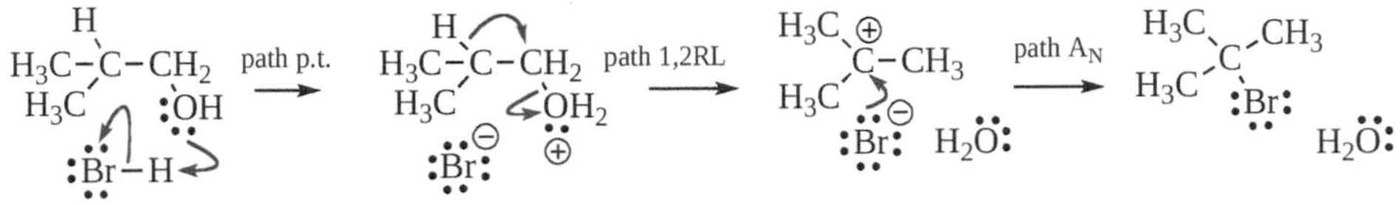

Our start is to check for any decent proton-transfer reactions. Protonating the alcohol with HBr has a favorable $K_{eq} = 10^{(-2.4 - (-9))} = 10^{+6.6}$, which greatly improves the leaving group. A check of the D_N restrictions finds a good leaving group, a polar solvent, but it would form an unstable primary alkyl carbocation, so we do not have all three criteria satisfied. To repeat, a highly unstable primary carbocation is not a reasonable alternative for a mechanism. The rearrangement can still go by the 1,2RL path. However, either a methyl or a hydride could migrate. If the methyl migrated, a secondary carbocation would be formed. The hydride migration produces a more stable tertiary carbocation. Since free rotation will line up either one, the lowest-energy route will form a more stable carbocation. The tertiary carbocation formed is trapped by path A_N, giving the product.

4.7.3 Approaches to Rearrangement Mechanisms

Predicting when rearrangement will occur is a bit of a challenge, covered more in Section 9.8; important factors are highly acidic media or strong Lewis acids, and a group that if it migrated would produce a compound of significantly greater stability. In mechanism problems, we have the distinct advantage of being given the product. If the product has a different bonding order from the reactant, then rearrangement has happened. We just have to find a reasonable route.

Common Rearrangement Errors

The most common error is to go with the first rearrangement route that you may think of and not consider important cross-checks like carbocation stability. Secondary alkyl carbocations are on the decision boundary. If your mechanism is going through a carbocation less stable than secondary alkyl, it is most likely incorrect. Orbital alignment needs to be watched in cyclic systems, where free rotation does not occur. Make sure the migrating group can maintain overlap in the migration; migrating groups usually slide over to the adjacent atom, and do not hop over anything.

Example Rearrangement Mechanism (Pinacol Rearrangement)

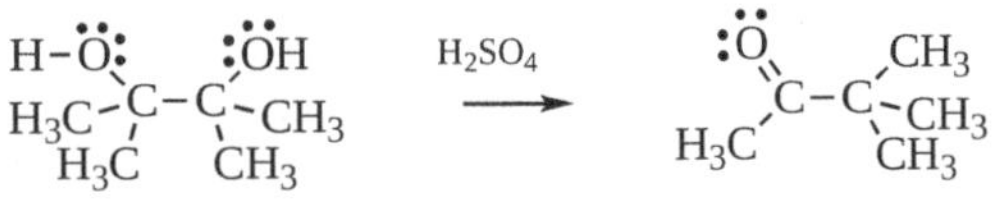

This is an example of a common type of acid-catalyzed rearrangement. A balanced reaction gives water as the other product. Initially, we explore proton transfer to improve the leaving group. Since the molecule is symmetrical, it matters not which hydroxyl is protonated. The proton transfer to the alcohol is downhill. If the conjugate acid is below (higher pK_a) the acid on the chart, then the proton transfer is favorable and there is no need to calculate the K_{eq}. Using H–A to represent sulfuric acid, we begin by protonating the hydroxyl group.

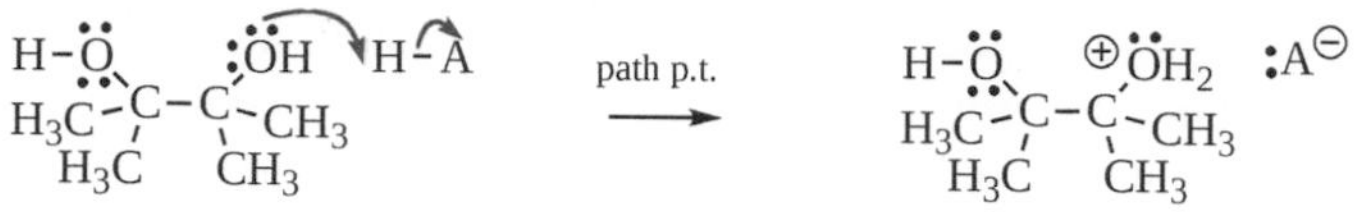

Now is the time to choose which of our two rearrangements to electrophilic carbon route to use. The D_N cross-check passes, the pK_{aHL} of the leaving group is zero or negative, the carbocation is tertiary, and the solvent is polar. We can proceed with the D_N step.

From the product, we see that a methyl has migrated; migrating the hydroxyl would just give the same connectivity as the reactant. The carbocation produced from the methyl migration is more stable than the one we started with because it is now resonance stabilized by the lone pair on oxygen.

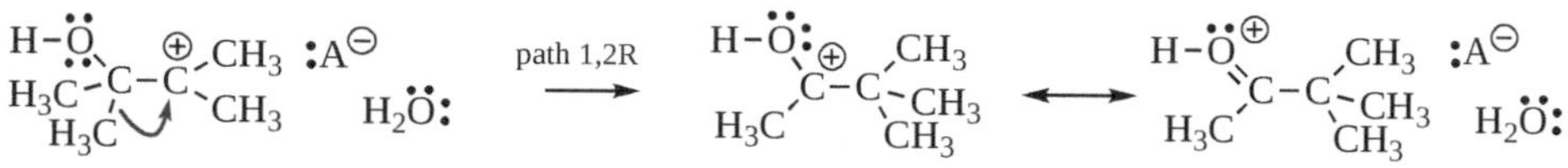

The rearrangement produces the protonated product, which is deprotonated by the strongest base in solution: neutral water.

4.7.4 Rearrangement Summary

Rearrangement competes with substitutions, eliminations, and additions in acidic media and rarely in basic. If steric hindrance slows down the other routes, rearrangement can compete. Strongly acidic media or powerful Lewis acids encourage rearrangement. The following are some systems that commonly rearrange in acid. More rearrangement examples are in Section 8.11, and the decision whether to migrate or not is considered in Section 9.8.

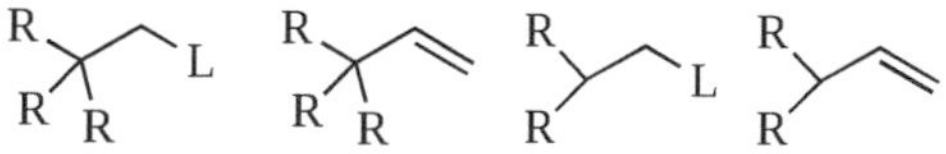

4.8 REACTION ARCHETYPE SUMMARY

Proton transfer (Chapter 3) and the four reaction archetypes in this chapter, **substitution, elimination, addition,** and **rearrangement,** make up the core of an undergraduate organic course. More complex reactions will combine these archetypes or play one off against another. The archetypes and their mechanisms are so central to all reactions that we will rarely have to bring in a new electron flow path to explain a reaction. In the next two chapters, we discuss a way to categorize our reactants into generic electron sources and sinks so that the sheer mass of material does not become overwhelming. Then in Chapter 7, all of the electron flow pathways will be discussed in detail. We will be able to understand almost all organic reactions as consisting of selected sequences of the dozen most common electron flow pathways. Chapter 8 then maps the interaction of the electron sources with electron sinks, using these pathways to help you understand the vast majority of organic reactions. Chapter 9 deals with the major decisions like substitution vs. elimination. Chapter 10 shows the analytical process for writing reasonable organic reaction mechanisms and for predicting the products of reaction. Chapter 11 reviews all the important cross-checks.

CHAPTER 4 EXERCISES

4.1 Trends: Rank all species, using the numeral 1 to designate:

(a) the most stable carbocation

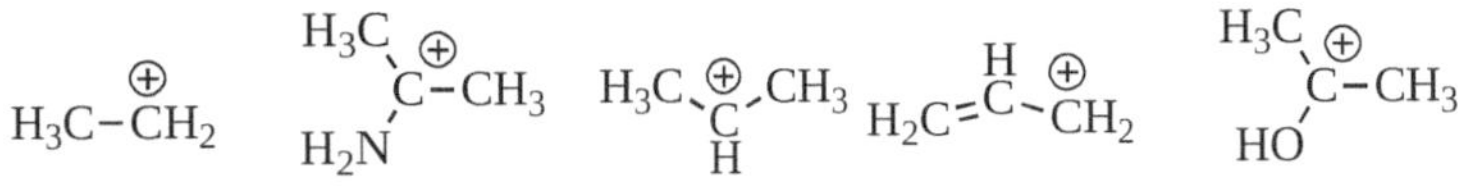

(b) the most stable carbanion

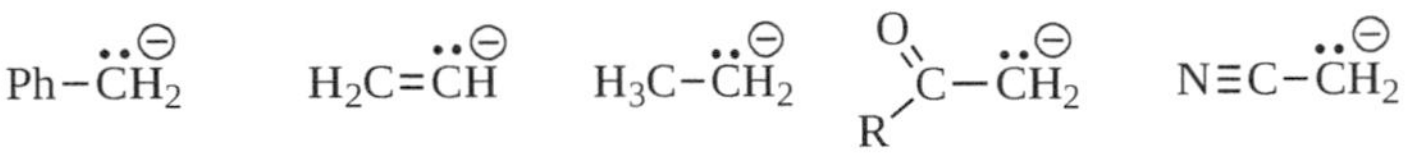

(c) the best electron-withdrawing group

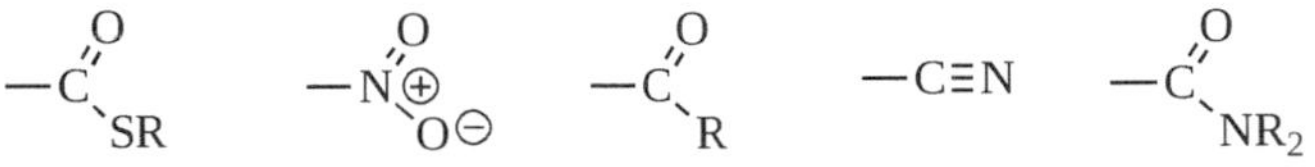

4.2 Trends: rank all species, using the numeral 1 to designate:

(a) the most reactive nucleophile toward CH_3I

$PhS^{\ominus}$ $PhO^{\ominus}$ $CH_3O^{\ominus}$ $PhSe^{\ominus}$ CH_3OH

(b) the best leaving group

—OCH_3 —Cl —OPh —OSO_2CH_3 —NH_2

(c) the best electron-releasing group

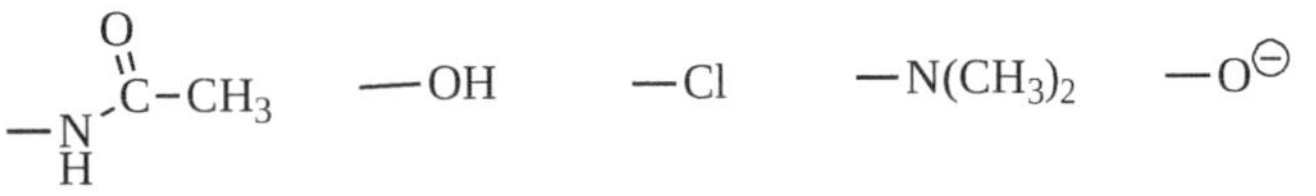

4.3 In the reaction below, find the leaving group then identify the nucleophile. With that information, draw the **substitution** product and decide which path is more likely, S_N1 or S_N2.

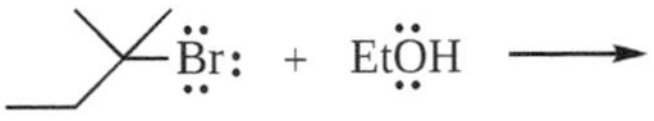

4.4 In the reaction below, find the leaving group then identify the nucleophile. With that information, draw the **substitution** product and decide which path is more likely, S_N1 or S_N2.

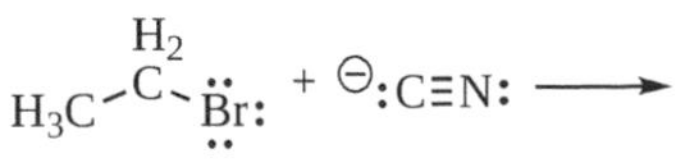

4.5 In the reaction below, find the leaving group. Draw Lewis structures for any carbon(s) next to the leaving group to see which H atoms could be lost. With that information, draw the elimination product and decide which path is more likely, E1, E2, or E1cB.

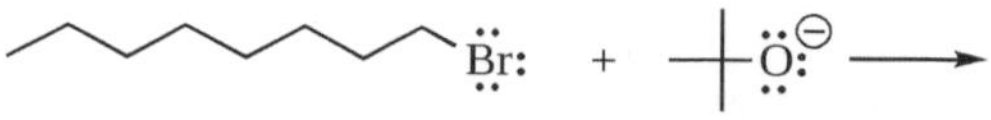

4.6 In the reaction below, find the leaving group. Draw Lewis structures for any carbon(s) next to the leaving group to see which H atoms could be lost. With that information, draw the elimination product and decide which path is more likely, E1, E2, or E1cB.

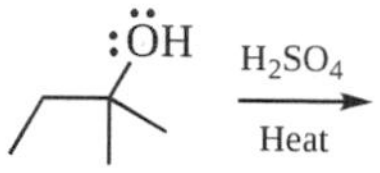

4.7 Draw the **addition** product. Decide which path is more likely, Ad_E2, Ad_E3, or Ad_N2.

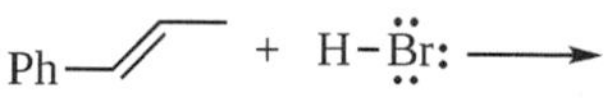

4.8 Draw the **addition** product. Decide which path is more likely, Ad_E2, Ad_E3, or Ad_N2.

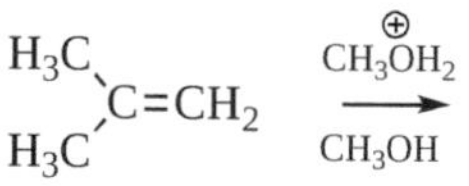

4.9 Draw the product of **addition–elimination**. What is the nucleophile? Decide whether the route is basic or acidic.

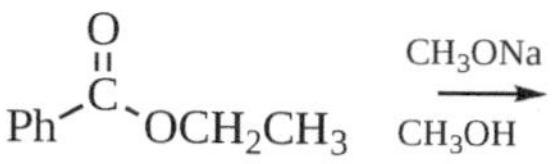

4.10 Draw the problem space for the following reactant prediction problem. Select the best of two possible substitution alternatives. *Hint*: look at the proton transfer reaction.

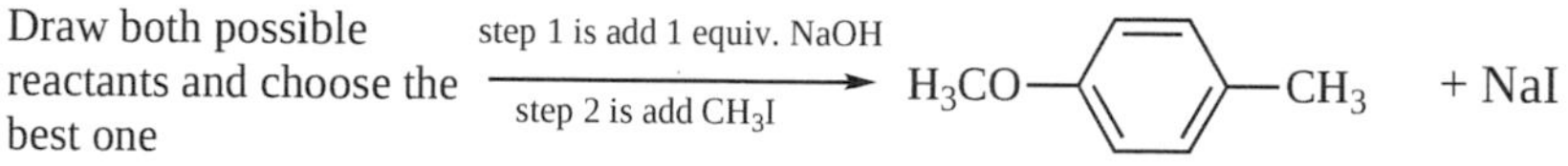

4.11 Draw the product of **electrophilic aromatic substitution**. The E adds just once.

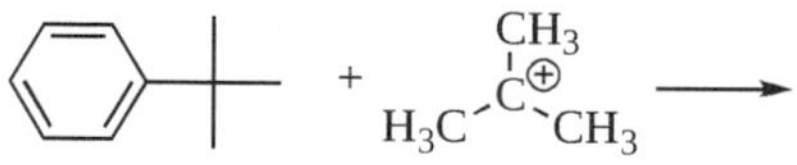

4.12 Draw the product of **electrophilic aromatic substitution**. The E adds just once.

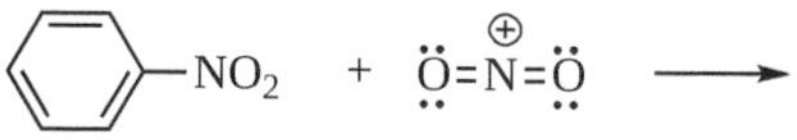

4.13 Use the appropriate electron flow pathway(s) to rationalize the product of these reactions. First, decide which of the four reaction archetypes has occurred. These mechanisms have only one electron source and one sink present in the reaction mixture.

(a)

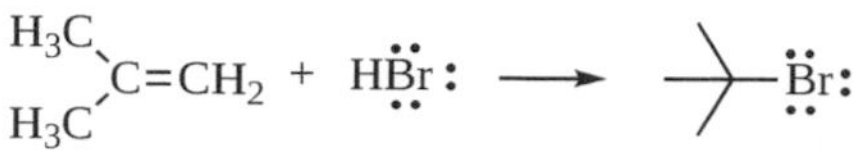

(b)

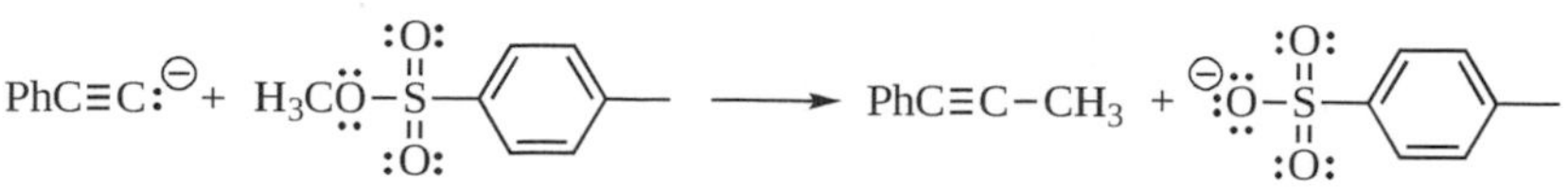

4.14 Use the appropriate electron flow pathway(s) to rationalize the product of these reactions. First, decide which of the four reaction archetypes has occurred. These mechanisms have only one electron source and one sink present in the reaction mixture.

(a) 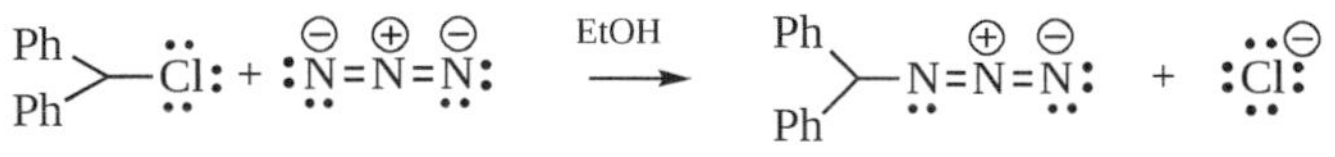

(b) 

4.15 First decide which of the four reaction archetypes has occurred. Use the appropriate electron flow pathways to rationalize the product of these **two-step** reactions. The first step is a favorable proton transfer that generates the electron source for the reaction; there is only one electron sink.

(a) 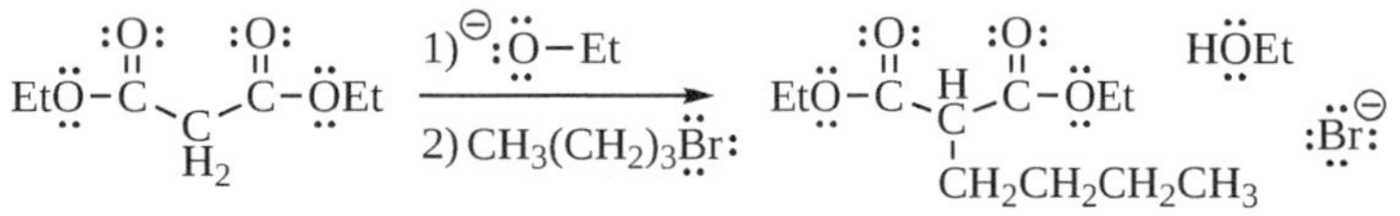

(b) 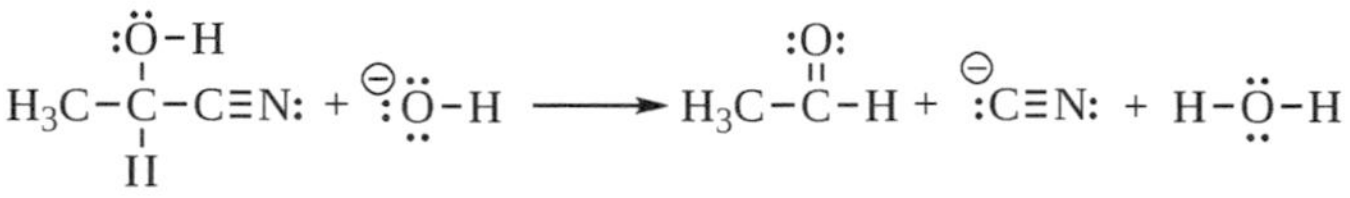

4.16 First decide which of the four reaction archetypes has occurred. Use the appropriate electron flow pathways to rationalize the product of these **two-step** reactions. The first step is a favorable proton transfer that generates the electron source for the reaction; there is only one electron sink.

(a) 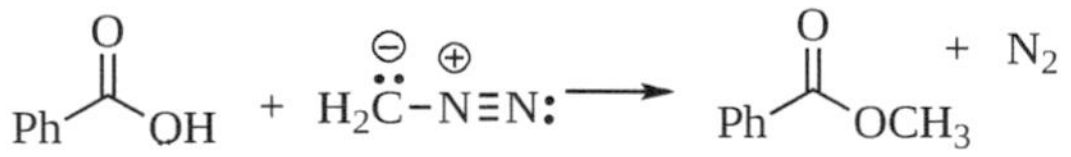

(b) $CH_3(CH_2)_{10}CH_2OH + HBr \rightleftharpoons CH_3(CH_2)_{10}CH_2Br + H_2O$

4.17 Use the ΔpK_a rule to predict if each of these reactions favors reactants or products.

(a) 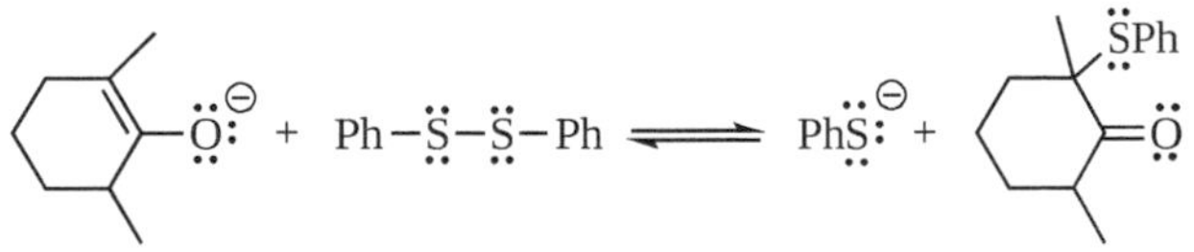

(b) $H_3C{-}SH + I^{\ominus} \rightleftharpoons HS^{\ominus} + H_3C{-}I$

4.18 Use the ΔpK_a rule to predict if each of these reactions favors reactants or products.

(a) 

(b) $PhH + LiOH \rightleftharpoons PhLi + H_2O$

4.19 Use the ΔpK_a rule to determine which of the two following alternatives is the lower-energy process.

$CH_3COOEt + HO^-$ reacting to form $H_2O + {}^-CH_2COOEt$

or

$CH_3COOEt + HO^-$ reacting to form $EtOH + CH_3COO^-$

4.20 Use the leaving group trend to decide which of the following two alternatives is the lower-energy process.

$ClCH_2CH_2CH_2Br + CN^-$ displacing bromide to form $ClCH_2CH_2CH_2CN$

or

$ClCH_2CH_2CH_2Br + CN^-$ displacing chloride to form $BrCH_2CH_2CH_2CN$

4.21 Use the carbocation stability trend to decide which of the following two alternatives is the lower-energy process.

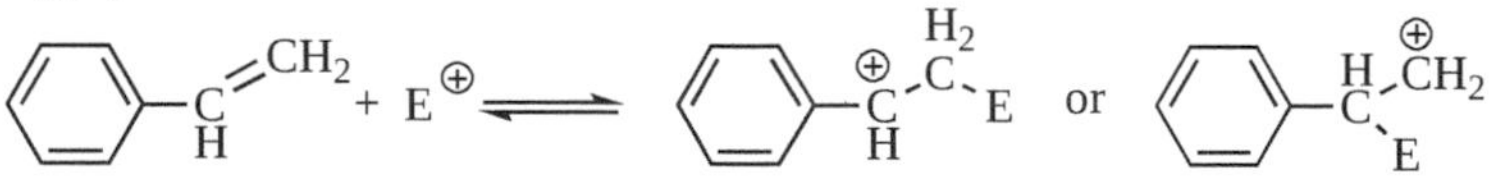

4.22 Explain why sodium sulfite below protonates on oxygen but alkylates with methyl iodide on sulfur. Start by drawing good Lewis structures for reactants and products

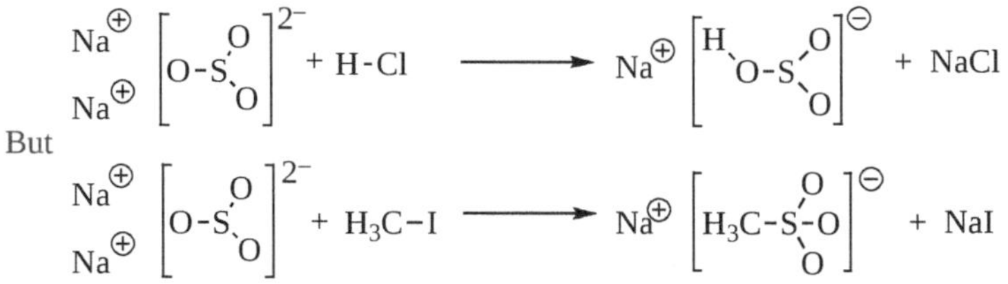

4.23 Decide which of the following two alternatives is the lower-energy process.

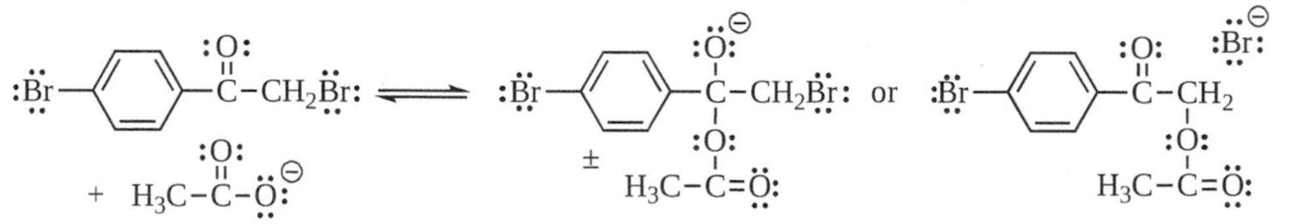

4.24 Account for the different rates in the ionization of the leaving group X. *Hint:* What factors influence the ease of ionization of a leaving group?

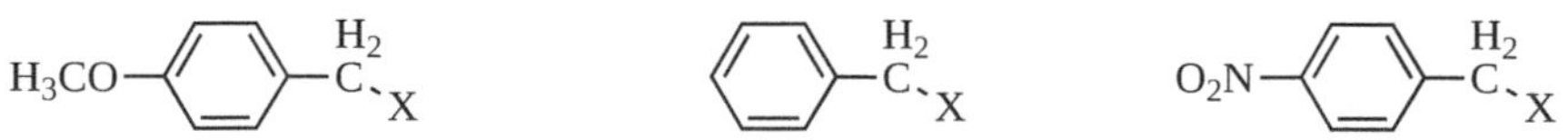

Fast Average Slow

4.25 Explain the fact that thiocyanates react at sulfur whereas cyanates react at nitrogen. Draw Lewis structure for both products.

$$H_3C{-}I + SCN^{\ominus} \longrightarrow H_3C{-}SCN + I^{\ominus}$$

$$H_3C{-}I + OCN^{\ominus} \longrightarrow H_3C{-}NCO + I^{\ominus}$$

4.26 Trends: rank all species, using the numeral 1 to designate:

(a) the most reactive for S_N1

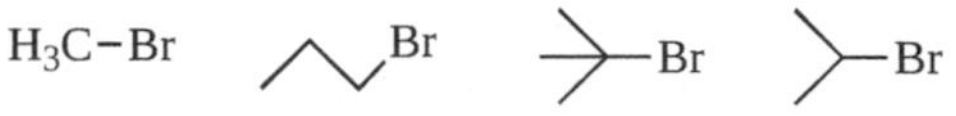

(b) the most reactive for S_N2

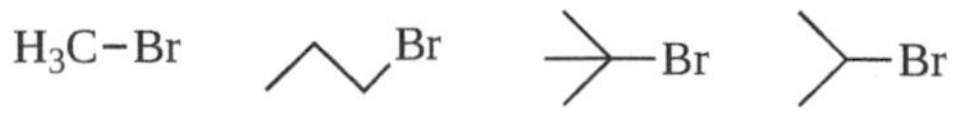

5

CLASSIFICATION OF ELECTRON SOURCES

5.1 GENERALIZED RANKING OF ELECTRON SOURCES

In this chapter we shall explore the classification and ranking of electron sources into the 12 common "generic" classes of electron sources. These generic classifications are based on chemical reactivity, and just like you do not expect a pig to fly, these classifications allow you to narrow down the chemical behavior of members of the class. A big problem in organic chemistry is the sheer magnitude of information, so efficient ways of organizing the material are very valuable. To repeat for emphasis, **there are only 12 generic classes from which a set of arrows, an electron flow, can start.** Some classes are quite rare, and others have several dozen examples. It is very important that you learn to classify according to generic class. **It is much easier to handle multiple examples of 12 familiar classes than hundreds of different special cases**. The first thing that you should do in learning a new reaction is to classify the electron source into one of these 12 classes. Three questions must be answered about each class: What characterizes the members of the class? What is the most electron-rich site? What determines reactivity within each class?

Electron Flow in Organic Chemistry: A Decision-Based Guide to Organic Mechanisms,
Third Edition. Paul H. Scudder.

Before discussing the individual classes, we need to get an overview of all of them. Since most reactions have a hard–hard and a soft–soft component, it is important to notice both how much negative charge a source bears and how soft the source is. **Electrons that reside in more stable, lower-energy bonding orbitals are less available for use as electron sources in reactions.** We can divide good electron sources into three groups: nonbonding electrons, electron-rich sigma bonds, and electron-rich pi bonds.

The best electron sources are usually nonbonding electron pairs. They are electron rich, and no bonds need be broken to use them as electron sources. Other excellent electron sources are highly ionic sigma bonds and also pi bonds highly polarized by excellent electron-releasing groups.

The most stabilized electrons, such as those in strong sigma bonds or aromatic pi bonds, make poor electron sources. For example, carbon–carbon single bonds are generally poor sources. There are exceptions, of course; an aromatic bearing an excellent electron-releasing group is an improved but still mediocre source.

Although an electron flow path is shown for each source in this chapter, it is included only to point out the electron pair from which the electron flow must start. The detailed discussion of the electron flow paths is presented in Chapter 7.

5.2 NONBONDING ELECTRONS

5.2.1 Lone Pairs as Brønsted–Lowry Bases

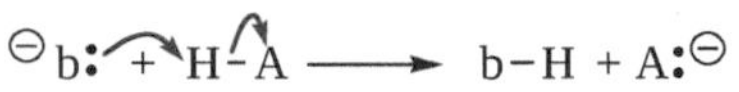

A nonbonding electron pair can serve as a Lewis base and attack an electron-deficient carbon or serve as a Brønsted–Lowry base (symbolized as a lower case “b”) and pull off a proton. The decision of whether a lone pair serves as a base or a nucleophile will not concern us in this section, only how to recognize the generic class. As a Brønsted–Lowry base, the electron flow starts from the lone pair of the base. Section 3.2.4 showed how to rank Brønsted–Lowry bases and gave common examples. A synthetically useful proton transfer has a K_{eq} greater than 10^{-10}. See Section 3.6 for the calculation of the proton transfer K_{eq}.

5.2.2 Heteroatom Lone Pairs as Nucleophiles

The electron flow starts from the lone pair. Nonbonding electron pairs are good electron sources, especially the lone pairs of anions. Nucleophilicity is a gauge of the rate of a Lewis base’s attack on an electron-deficient carbon. A proton is much harder than an electron-deficient carbon, so the relative hardness of the nonbonding electron pair is important in any discussion of basicity vs. nucleophilicity (see Section 9.5). To rank the nucleophilicity of a nonbonding electron pair, we must consider the major factors of charge, size, softness, and solvation and the minor factors of electronegativity and strength of the bond that is formed. A detailed discussion of nucleophilicity is in Section 4.2.2, so just the major points are reviewed here.

• A charged anion is a better nucleophile than its neutral counterpart. Because softness is important, nucleophilicity does not parallel basicity exactly. For lone pairs of the *same element*, the more basic the lone pair is, the more nucleophilic it is.

• Steric hindrance decreases nucleophilicity. As the nucleophile becomes larger, its bulk tends to get in the way of its acting as a nucleophile, especially when the electrophile is also bulky.

• Electron availability decreases as the lone pair electrons become held more tightly (hardness); therefore anions on more electronegative atoms are usually less nucleophilic. Softness (less electronegative, more polarizable) is important because the more polarizable the attacking lone pair is, the more easily it can be distorted to form the new bond with the electrophile.

• Solvation matters a lot. Softness also affects how tightly solvated the nucleophile is, since solvation is a hard–hard interaction. Small, hard, highly shielded ions must "break out" of the solvent cage to be available as nucleophiles. Soft ions are more nucleophilic in protic solvents because tight solvation greatly decreases a hard ion's nucleophilicity.

• Nucleophilicity also depends on the electrophile. Hard nucleophiles favor binding with hard electrophiles; soft nucleophiles favor binding with soft electrophiles (HSAB principle, Section 2.4). Most nucleophilicity charts show relative rates of nucleophilic attack with methyl iodide, which is a very soft electrophile because the carbon–iodine electronegativity difference is nearly zero. A pK_a chart can be used as a reference for nucleophilicity only if the difference in softness is considered. A partially plus carbon atom is a much softer electrophile than a proton.

In summary, to rank the nucleophilicity of nonbonding electron pairs reacting in protic solvents with soft electrophiles such as R–L, first rank by softness, then by basicity (within the same attacking atom). However, for nonbonding electron pairs reacting with harder electrophiles such as a proton or a carbonyl, rank by basicity. Very reactive electrophiles like carbocations are not selective and react with the most abundant nucleophile (commonly the solvent).

5.2.3 Easily Oxidizable Metals

$$\mathrm{M\!:} + \mathrm{R{-}Br} \longrightarrow \mathrm{M^{2+}} + \mathrm{R\!:^{\ominus}} + \mathrm{:\!\ddot{B}r\!:^{\ominus}}$$

The valence electrons of a reactive metal such as lithium, sodium, magnesium, or zinc can be donated to another species to form an ionic bond. Reactions of oxidizable metals occur by *electron transfer* from the metal surface to the electron sink. This process oxidizes the metal, producing the metal ion, and reduces the electron sink. For example, sodium metal will lose one electron to achieve the noble gas configuration of neon in the sodium cation. In the case of magnesium and zinc that lose two electrons, the electron transfer very likely occurs one electron at a time (Section 12.8). In the *halogen metal exchange reaction* shown, the R–Br bond is broken in the reduction process, producing an ionic or partially ionic organometallic salt.

Metals differ greatly in how easily they are oxidized. The most reactive metals are those with a very negative standard electrode potential. Of the commonly used metals in organic reactions, the alkali metals lithium, sodium, and potassium are highly reactive; the next most reactive is magnesium, then zinc.

5.3 ELECTRON-RICH SIGMA BONDS

We can consider a partially ionic sigma bond as an already partially "broken" bond. An organometallic bond, like the ones formed in Section 5.2.3 for example, has electrons that are very available. Organometallics, group 1A metal hydrides, and complex metal hydrides can be some of the most reactive electron sources. Conversely, a carbon–carbon covalent bond is completely covalent; its electrons reside in a very stable sigma bonding orbital and are not available. The main exception to this is the carbon–carbon bonds of cyclopropane, which are destabilized by ring strain and poor overlap, and thus are more reactive to electrophiles.

5.3.1 Organometallics

The electron flow starts from the polarized sigma bond. Organometallics vary in their ability to act as electron sources. Very ionic organometallics are excellent electron sources, whereas the most covalent organometallics are poor sources. Therefore, checking the electronegativity difference between the metal and carbon will allow us to rank the reactivity of the organometallic when the R group is the same. **A good RM source will have a large electronegativity difference.** The following is a list of several organometallics ranked by their electronegativity differences (R is the same for all).

Electronegativity difference	**Most reactive**
1.62	R–Na
1.57	R–Li
1.24	R–MgX
0.90	R_2Zn
0.86	R_2Cd
0.65	$R_2Cu^-Li^+$
0.55	R_2Hg
0.22	R_4Pb
	Least reactive

To determine the reactivity of an organometallic on the basis of R when the metal is the same, compare the carbanion stability (Section 3.5.2). As the number of electron-donating alkyl groups on the carbanion increases, the carbanion gets less stable and more reactive. The alkyllithium general reactivity trend is tertiary > secondary > primary > methyl. The less stable the carbanion, the more reactive it is as an electron source. For carbanions that are resonance stabilized by electron-withdrawing groups, the identity of the metal is of less importance than the stabilization of the carbanion.

Simple alkyl carbanions, like CH_3Li, are sp^3 hybridized at the carbanionic center and commonly exist as the tetramer, $(CH_3Li)_4$. This tetramer exists as a tetrahedron of lithium ions with a carbanion snuggled in between the three lithium atoms of each face.

Example problem

Which of the following, CH_3Li, CH_3MgBr, and CH_3CdCl, is the most reactive organometallic? Which is the least?

Answer: The most ionic RM is the most reactive. The least electronegative metal in this group is lithium, so CH_3Li is the most reactive. The next most reactive is CH_3MgBr; the least reactive and the most covalent organometallic is CH_3CdCl.

5.3.2 Group 1A Metal Hydrides

The electron flow starts from the polarized sigma bond. Metal hydrides can serve as good electron sources, but their behavior varies with the electronegativity of the metal. Alkali metal hydrides, NaH and KH, are ionic and function primarily as bases, reacting with acidic H's to form hydrogen gas; little reduction occurs.

5.3.3 Complex Metal Hydrides

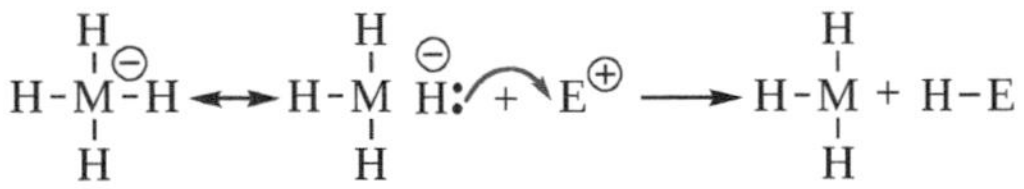

The softer complex metal hydrides, AlH_4^- and BH_4^-, donate hydride to reduce the substrate and are better nucleophiles than bases. However, they will rapidly deprotonate acidic H's on heteroatoms, generating flammable hydrogen gas, so they must be used in dry aprotic solvents. To summarize, the harder alkali metal hydrides are good bases and poor nucleophiles, whereas the softer complex metal hydrides are good nucleophiles and weaker bases. As with the organometallics in the previous section, the electronegativity difference governs the reactivity of the metal hydrides. The metal hydride will become somewhat less reactive if the metal is made more electronegative by attaching an inductively withdrawing group. As the hydride source becomes less reactive, it also becomes more selective. The complex metal hydride reactivity trend is **AlH_4^- > $HAl(OR)_3^-$ > BH_4^- > H_3BCN^-**.

5.4 ELECTRON-RICH PI BONDS

5.4.1 Allylic Sources

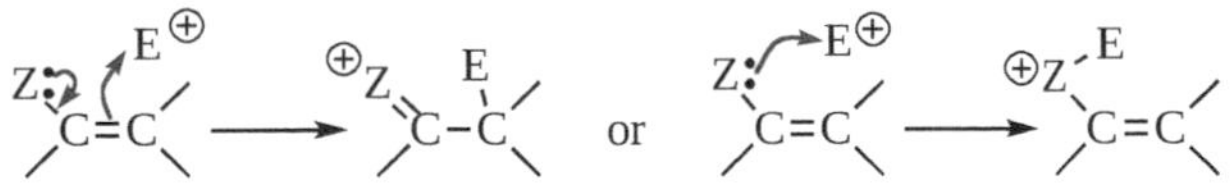

Pi bonds tend to be weaker than sigma bonds and are therefore usually easier to break. We can make a pi bond more reactive by attaching a good lone pair donor. The electron flow can come from either end of the allylic system. When pi bonds are used as electron sources, an attached lone pair electron-releasing group can greatly increase the availability of the pi bond's electrons and stabilize the cationic product usually formed upon an electrophilic attack on the double bond. We have in essence "extended" via delocalization with a double bond the properties of the lone pair donor group (vinylogy). These sources contain three adjacent *p* orbitals, and almost all are much more reactive than a simple alkene.

Resonance structures show us that either the double bond or the heteroatom can serve as an electron source. Because allylic sources can "bite" at two different sites they are called *ambident nucleophiles*. Usually the double bond is the attacking nucleophile; the decision about which end of the allylic source attacks the electrophile will be discussed in Section 9.4.

We will see later that some electron sinks are in equilibrium with a species that can serve as an allylic electron source, Z=C–C–H $\rightleftharpoons$ H–Z–C=C; this equilibrium is called *tautomerization* and is discussed in Chapter 7.3.8 under path combinations.

The rankings of allylic electron sources reflect the electron-releasing group trends already covered in Table 4.2 and thus do not present new trends for you to learn. **The better the electron-releasing group is, the better the allylic electron source is.** The following are several allylic electron sources, ordered by their rank as sources:

Best 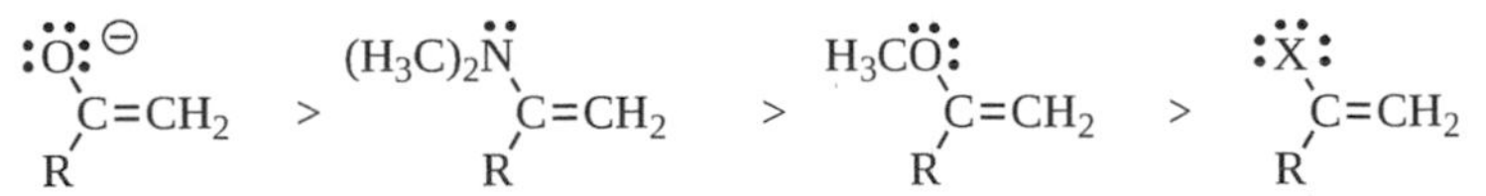 Worst

Enolate > Enamine > Enol ether > Vinyl halide

The last one in the series, the vinyl halide, is actually a poorer electron source than a simple alkene because halogens are poorer donors than alkyl groups.

Example problem

Which of the following, CH_2=CH–O^-, CH_2=CH–NR_2, and CH_2=CH–Cl, is the most reactive allylic source? Which is the least?

Answer: The best electron-releasing group is the oxygen anion, so the best allylic source of the three is the enolate, CH_2=CH–O^-; the next best is the enamine, CH_2=CH–NR_2. The least is the vinyl chloride, CH_2=CH–Cl, because chlorine is a very poor donor.

Vinylogous allylic sources, :Z–C=C–C=C or :Z–C=C–C=O, commonly react at the center atom and therefore will be treated as a subset of this source. A few examples of these vinylogous allylic electron sources are shown below: extended enolates, and enolates of acetoacetates and malonates.

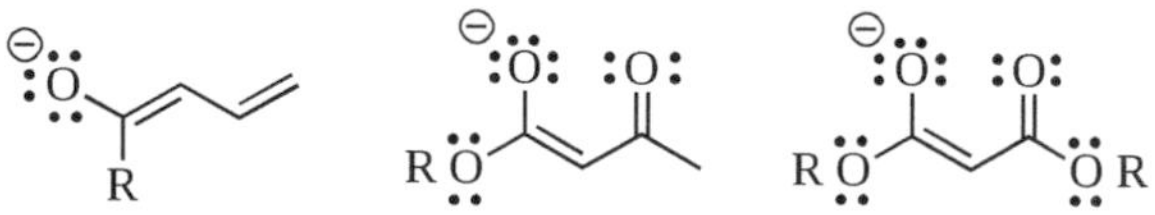

5.4.2 Allylic Alkyne Sources

:Z–C≡C–

If we add a second pi bond perpendicular to an allylic system we create this rare electron source that should just be treated as a subset of the allylic sources.

5.5 SIMPLE PI BONDS

5.5.1 Alkene Sources

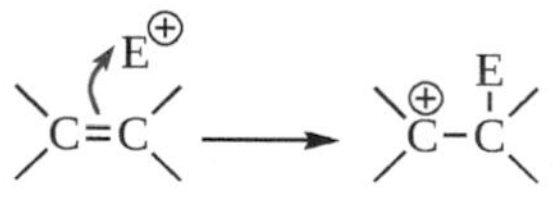

The pi bond is the electron source. For allylic sources, the electron pair donor stabilizes the resultant cation. The simple alkene reactivity trend reflects the stabilization of the resultant carbocation by alkyl substitution and delocalization with other double bonds. The ranking of simple double bonds as electron sources basically is: the more substituted, the more reactive.

Best 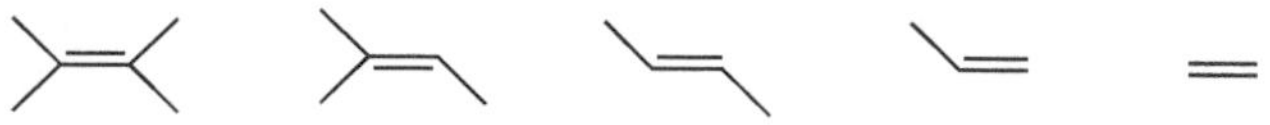 Worst

When a substituted double bond is attacked by an electrophile, the two possible carbocations formed can differ in stability. **The formation of the more stable carbocation will be the lower-energy process and will determine the site of electrophilic attack** (a modern version of Markovnikov's rule). Figure 5.1 shows that if the group on the double bond acts as an electron-releasing group, the carbocation adjacent to and stabilized by the electron-releasing group will be formed. Electron-releasing groups stabilize cations, and electron-withdrawing groups destabilize cations. Avoid forming cations adjacent to electron-withdrawing groups.

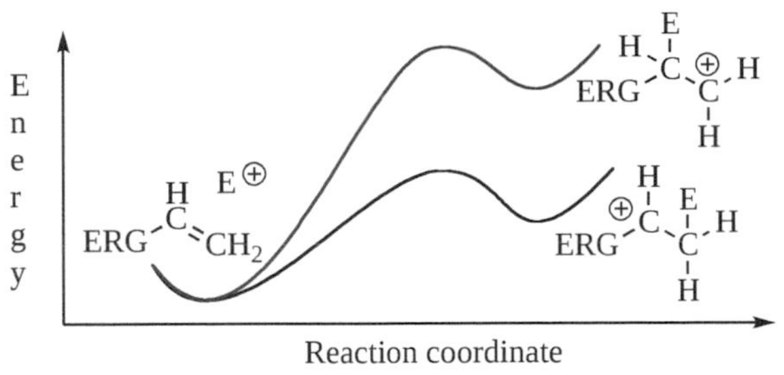

Figure 5.1 Energy diagrams for electrophilic addition to substituted alkenes. The carbocation adjacent to the electron-releasing group is stabilized and favored.

5.5.2 Diene Sources

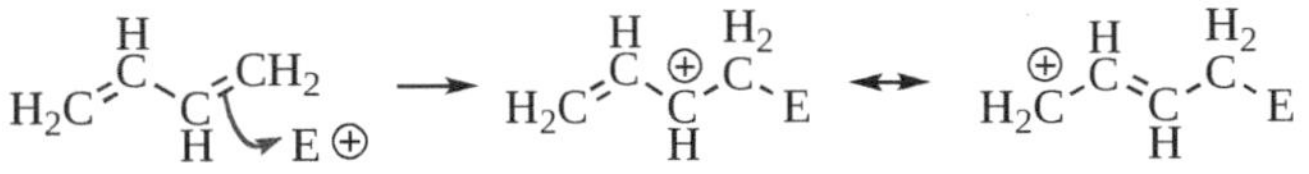

Dienes are conjugated systems of two pi bonds. The simplest diene, 1,3-butadiene shown above, adds the electrophile to the end to produce a resonance stabilized allylic carbocation. As with alkenes, the more substituted the diene is, the more reactive. To make the choice on where to add an electrophile to a diene pi system, draw all possible carbocations and then pick the more stable one according to the trends you learned in Section 4.2.4. We will often use the trends to make decisions about the route a reaction will take.

Example problem

Predict where an electrophile would attack the following compound.

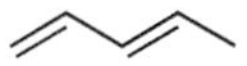

Answer: Draw out all four possible carbocations that would be formed upon electrophilic attack and rank their stability.

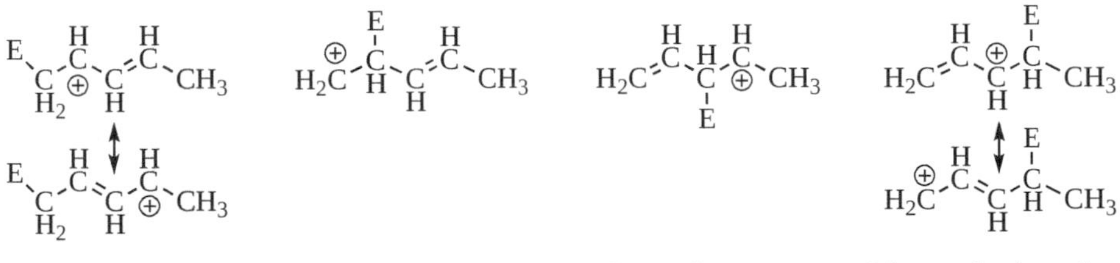

Disubstituted allylic carbocation | Primary carbocation | Secondary carbocation | Monosubstituted allylic carbocation

The ranking of these carbocations from most to least stable is disubstituted allylic > monosubstituted allylic > secondary > primary. Since the electrophile will add to form the most stable cation, the electrophile addition will form the disubstituted allylic carbocation.

5.5.3 Alkyne Sources

The two perpendicular pi bonds of a triple bond can be considered separately, and thus this source becomes merely a subset of the double-bond sources. Alkynes are slightly poorer

electron sources than alkenes because the vinyl cation formed upon electrophile attack is relatively unstable. Acetylide anions, RC≡C:$^-$ M$^+$, which react on the anion and not the triple bond, are treated as organometallic reagents (Section 5.3.1) and not as alkyne sources.

Example problem

Which of the following, $CH_3C{\equiv}CCH_3$, $RC{\equiv}C{-}NR_2$, and HC≡CH, is the most reactive triple-bond electron source? Which is the worst?

Answer: The best electron-releasing group attached to the triple bond is the NR_2, so the best triple-bond electron source of the three is $RC{\equiv}C{-}NR_2$; the next best is $CH_3C{\equiv}CCH_3$ because methyl is a weak donor. The worst is HC≡CH, because the cation formed upon electrophilic attack lacks any form of stabilization.

5.5.4 Allene (Cumulene) Sources

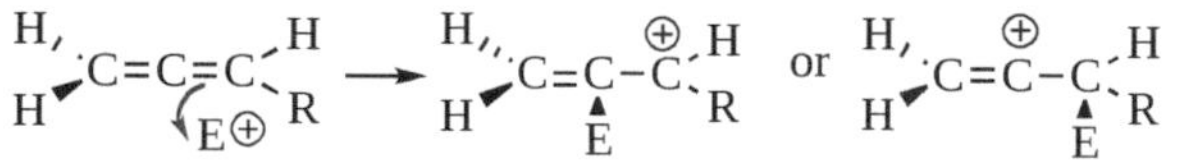

The least common of the simple pi bond sources are allenes, $R_2C{=}C{=}CR_2$, whose two perpendicular double bonds must again be considered separately. Allene reactions have a bit of a strange twist to them, literally. There are two possible sites for the electrophile to attack on the allene above, Figure 5.2. It would seem to be no contest between a resonance-stabilized allylic carbocation and the less stable vinyl carbocation. But a look at the orbital arrangement below for the reaction reveals that all is not right with the allylic system; in fact, it is twisted 90° out of alignment, so no resonance stabilization can occur at the transition state for addition.

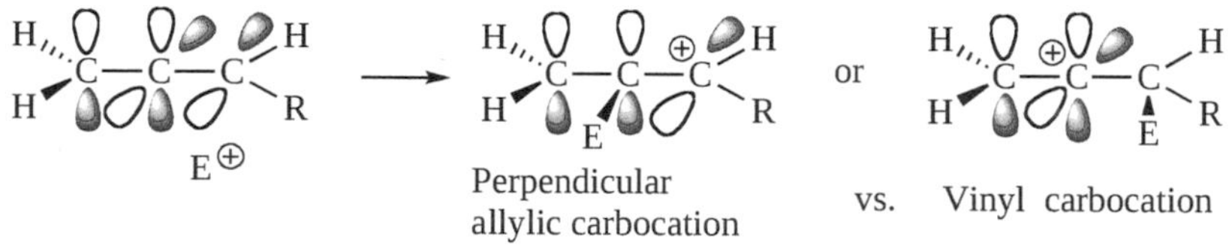

Figure 5.2 Electrophilic attack on a cumulene source has two possible outcomes that vary by R.

The resonance stabilization of an allylic carbocation is about 15 kcal/mol (74 kJ/mol). The loss of this resonance stabilization means that the vinyl cation is of comparable stability. When the R group is hydrogen or alkyl, the electrophile adds to the end as shown below.

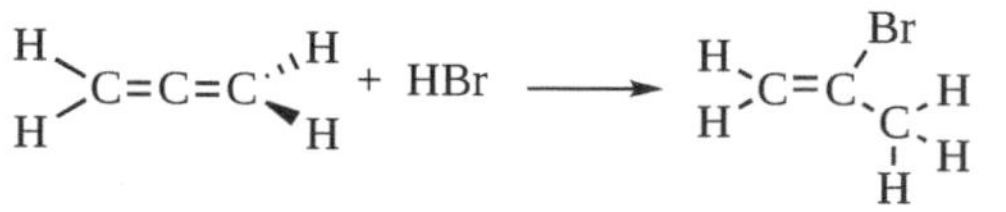

However, if the R group on the allene is capable of additional resonance stabilization of the carbocation, then electrophile attack shifts to the middle carbon. For example, phenylallene adds a proton to the center carbon to form a carbocation that is instead stabilized by resonance with the phenyl. Then bond rotation allows the allyl carbocation to form. The chloride then reversibly adds to form the more stable conjugated product.

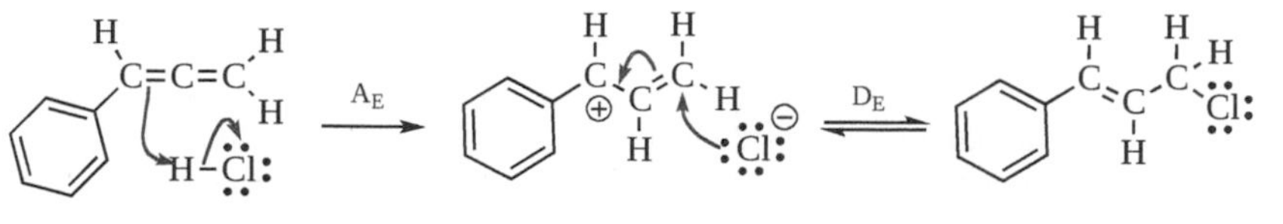

5.6 AROMATIC RINGS

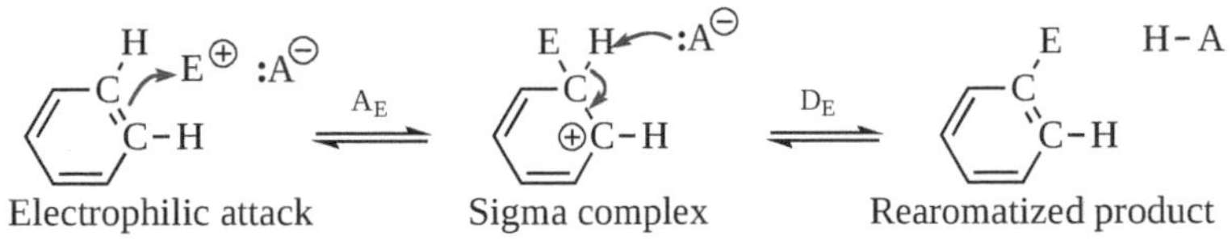

The pi cloud of the aromatic ring reacts with an electrophile to produce a cationic sigma complex that then loses an electrophile, commonly H^+, to restore the aromatic stabilization of the ring. Almost all aromatic rings are poor electron sources because the electrons in an aromatic ring are very stabilized (Section 1.9.3) and are therefore not very available. However, an aromatic ring can be made more reactive by the attachment of one or more electron-releasing groups.

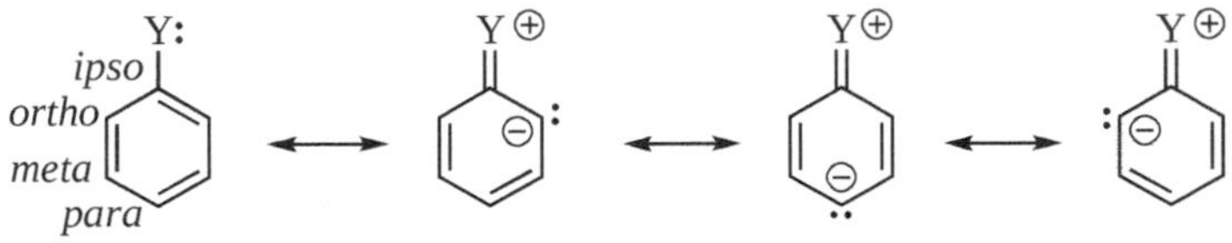

To summarize Section 4.6, any group that has a resonance interaction with an aromatic ring has the greatest effect at the *ortho* and *para* positions. An electron-releasing group on the ring increases the electron availability at the *ortho* and *para* positions, increases reactivity, and stabilizes the carbocation formed upon attack at those positions. The less crowded *para* position is favored when the electron-releasing group is large or when the electrophile is large.

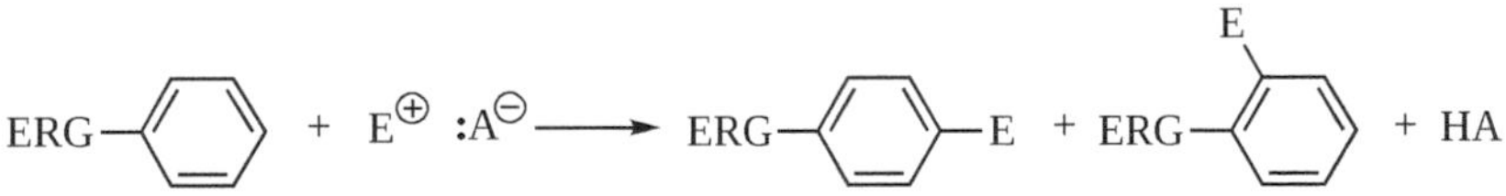

An electron-withdrawing group attached to the ring decreases the electron availability at the *ortho* and *para* positions (making the *meta* position the least electron poor), decreases reactivity, and destabilizes the carbocation formed upon attack at the *ortho* and *para* positions, so *meta* attack is preferred.

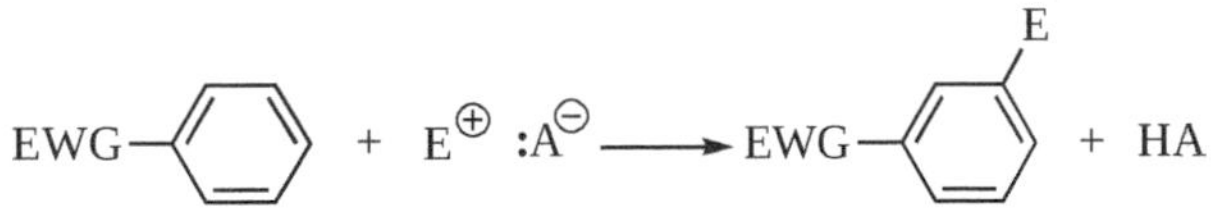

Therefore to rank the reactivity of a benzene ring to electrophilic attack, the better electron-releasing groups (Section 4.2.5) make the ring more reactive. The better electron-withdrawing groups (Section 3.5.1) make the ring less reactive. Combining these two trends back to back will give an overall reactivity trend for an aromatic ring substituted with a group, G, shown in Figure 5.3 below.

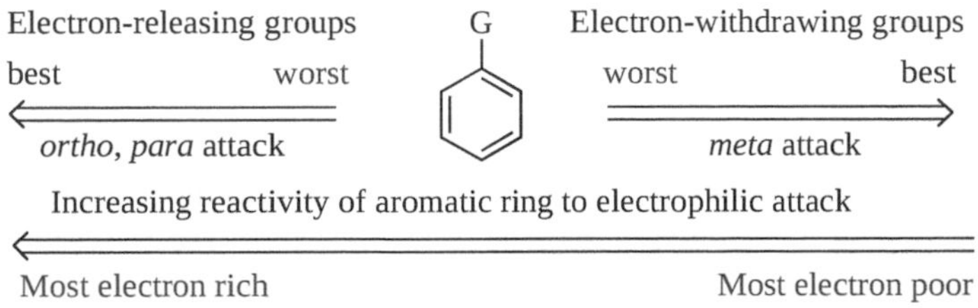

Figure 5.3 Reactivity of aromatic electron sources as a function of attached group.

Other Aromatics as Electron Sources

The stability of the sigma complex is the guiding principle when determining the site of electrophilic attack on any aromatic ring. The more approximately equal energy resonance structures the carbocation has, generally the more stable it is. Don't count unstable resonance forms, like ones that place the positive charge on an atom bonded to a destabilizing EWG. Pyridine undergoes electrophilic attack *meta* to the nitrogen.

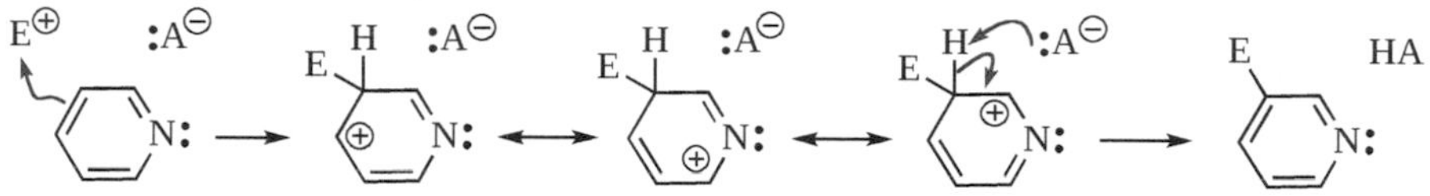

Resonance forms that have an electronegative heteroatom with an incomplete octet are especially bad. For example, no attack occurs on the *ortho* or *para* positions of pyridine.

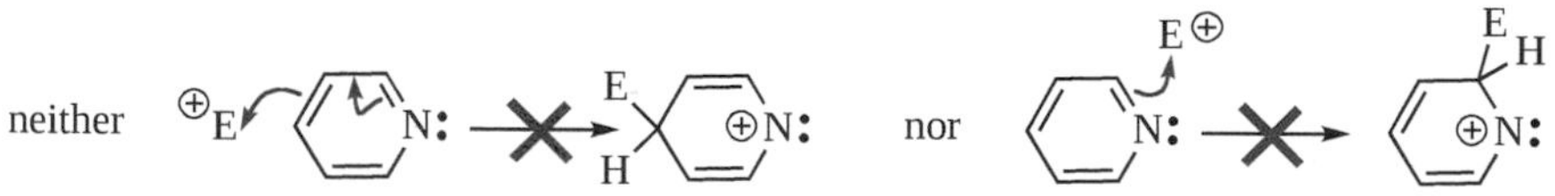

However, the attack of the electrophile on the nitrogen lone pair of pyridine is a competing reaction. Note that the nitrogen lone pair of pyridine is perpendicular to the pi system of the aromatic ring and therefore cannot be used to stabilize any sigma complex.

Five-membered ring heteroaromatics undergo electrophilic aromatic substitution at the 2-position (adjacent to the heteroatom). Furan example:

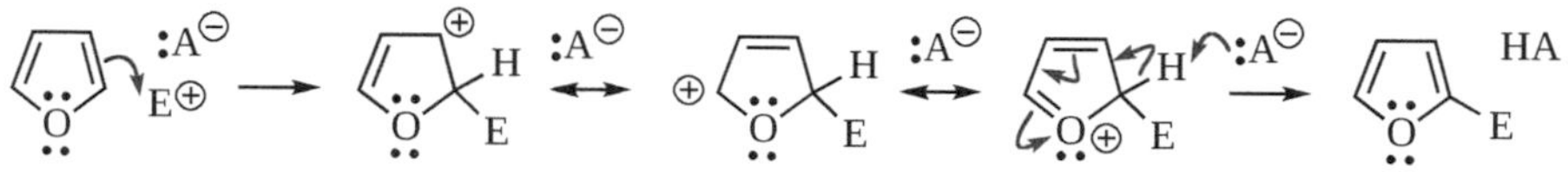

Attack at the further away 3-position gives a sigma complex that has one fewer resonance form, so it is less stable. The product from the most stabilized sigma complex dominates.

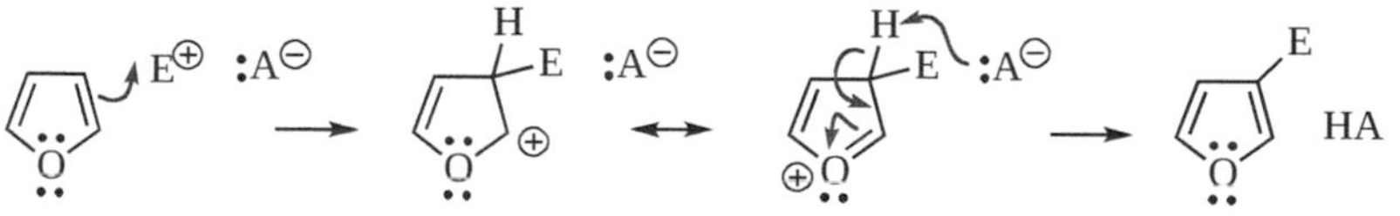

Naphthalene undergoes electrophilic aromatic substitution at the position next to the second ring. The sigma complex leading to the obtained product has more resonance structures, indicating a more delocalized charge than that of the alternative.

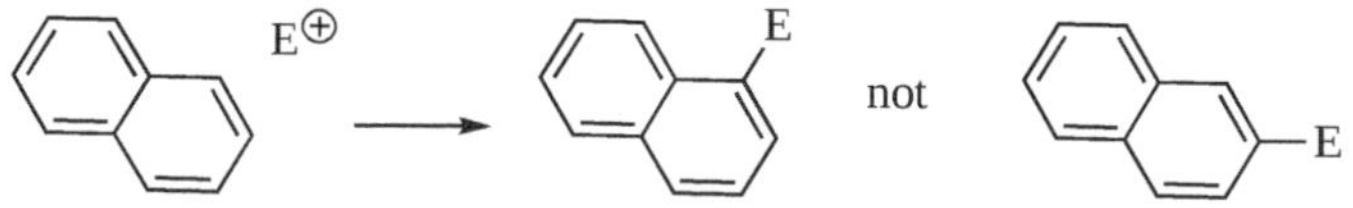

Example problem

Which of the following, $PhCH_3$, $PhNR_2$, and PhCOR, is the most reactive aromatic ring toward electrophilic attack? Which is the least?

Answer: The best electron-releasing group is the NR_2, so the most reactive ring of the three is $PhNR_2$; the next best donor is methyl, so the next most reactive ring is $PhCH_3$. The least reactive ring is PhCOR, because ketones are electron-withdrawing groups and deactivate the ring.

5.7 SUMMARY OF GENERIC ELECTRON SOURCES

5.7.1 Generic Electron Source Table

Table 5.1 lists 12 generic electron sources that start the electron flow in a reaction. Conspicuously absent are covalent sigma bonds. **Regular sigma bonds are just too stabilized to act as electron sources** (only when destabilized in a highly strained three-membered ring may a carbon–carbon single bond behave as a source). Usually, the only single bonds that serve as sources are highly ionic ones, single bonds to metals.

Table 5.1 The Twelve Generic Electron Sources

Generic Class	Symbol	Examples
Nonbonding electrons (5.2)		
Lone pairs on heteroatoms	Z:	I^-, HO^-, H_2O, H_3N, *t*-BuO^-
(Nu / base dual behavior)		CH_3COO^-
Easily oxidizable metals	M:	Na, Li, Mg, Zn metals
Electron-rich sigma bonds (5.3)		
Organometallics	R–M	RMgBr, RLi, $R_2Cu^-Li^+$
Metal hydrides	MH_4^- (Nu)	$LiAlH_4$, $NaBH_4$,
	MH (Bases)	NaH, KH
Electron-rich pi bonds (5.4)		
Allylic sources	C=C–Z:	Enols: C=C–OH
		Enolates: C=C–O^-
		Enamines: C=C–NR_2
Allylic alkyne sources	C≡C–Z:	Et–C≡C–NEt_2
Simple pi bonds (5.5)		
Alkenes	C=C	$H_2C=CH_2$
Dienes	C=C–C=C	$H_2C=CH\text{-}HC=CH_2$
Alkynes	C≡C	HC≡CH
Allenes	C=C=C	$H_2C=C=CH_2$
Aromatic rings (5.6)		
Aromatics	ArH	

5.7.2 Generic Electron Source Classification Flowchart

As noted in the beginning of this chapter, the reason for generic classification is so that you know how a compound is going to behave. Generic classification will allow you to predict what will happen when you encounter a compound you have not seen before. This powerful classification tool is useful only if it is actually employed to organize all new incoming organic reactions, and not relegated to just another thing to know.

The remaining task is to come up with a method for classification of a given structure into its most appropriate generic class. Often several functional groups are present in a reactant, and several reactants may be present in a reaction. One functional group may possibly fall into more than one class, so we need to pick the most applicable. We need to recognize the generic sources in a way such that we find the most reactive first. A flowchart of the classification process for electron sources is shown in Figure 5.4.

1. Oxidizable metals, organometallics, and metal hydrides are usually very reactive.
2. Next look for lone pair donors and the electron-rich pi bonds.
3. Then search for the simple pi bonds.
4. The least reactive, aromatics, are sorted out last.

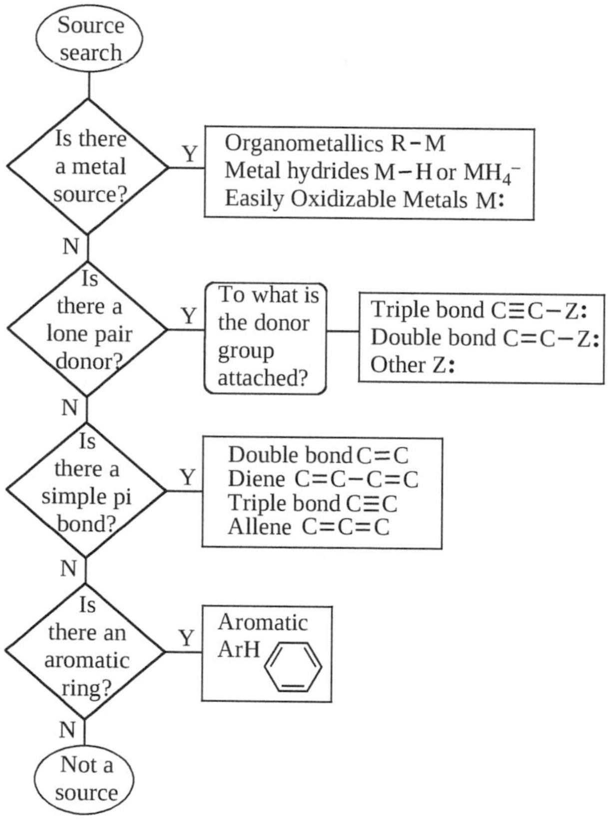

Figure 5.4 Classification flowchart for electron sources.

Example problem

To which generic classes do these belong: Ph–NH_2, CH_2=CH–Li, CH_2=CH–NH_2?

Answer: We notice that in Ph–NH_2 there is a nitrogen lone pair and an aromatic ring. Following our classification scheme, we drop it into the lone pair class, Z, rather than into the less reactive aromatic class. Likewise for CH_2=CH–Li , we classify it as an organometallic, R–M, rather than the much less reactive alkene. For CH_2=CH–NH_2, the nitrogen lone pair is conjugated with the double bond, giving an allylic source, C=C–Z. Since it is an ambident nucleophile, a decision is necessary as to which end of the allylic system to use as a source (Section 9.4). On each of these examples, a generic classification gives us insight into how each compound will react, similar to the way we've developed expectations for how different animals will behave. **Understanding structure allows us to understand reactivity.**

5.7.3 Common Sources Reactivity Summary

A much smaller set of sources is responsible for most reactions and is summarized in Table 5.2, which gathers examples, reactivity trends, common reactions, necessary decisions, and the related generic groups. Some have a decision discussed in Chapter 9.

Table 5.2 Summary Reactivity Table for Common Electron Sources

Symbol and Name	Common Examples in Decreasing Reactivity Order	Reactivity Trend Principle	Common Reactions, Decision?	Related Generic Groups
Z: Heteroatom lone pairs as nucleophiles	RS^- I^- CN^- RO^-	Softer anions are more nucleophilic. If same atom, more basic, more Nu.	Substitutions and Additions. Nucleophile vs. base decision	See allylic sources
Z: Heteroatom lone pairs as bases	R_2N^- t-BuO^- R_3N CH_3COO^-	The higher pK_{abH} is a stronger base.	Proton-transfer $K_{eq} \geq 10^{-10}$	NaH and KH as MH bases
R–M Organo-metallics	CH_3Li CH_3MgI $(CH_3)_2Cu^-\ Li^+$	The more ionic RM bond is more reactive.	Substitutions and Additions. Deprotonate acidic H's	See enolates (allylic sources)
M–H Group 1A metal hydrides	NaH KH	Both are very reactive strong bases but poor nucleophiles.	Deprotonates acidic H's	See complex metal hydrides
MH_4^- Complex metal hydrides	$LiAlH_4$ $LiAlH(OR)_3$ $NaBH_4$ $NaBH_3CN$	The more ionic MH bond is more reactive.	Substitutions and Additions. Deprotonates acidic H's on heteroatoms	NaH and KH as MH bases
C=C–Z: Allylic sources	Enolates $C{=}C{-}O^-$ Enamines $C{=}C{-}NR_2$ Enol ethers C=C–OR	The better electron releasing group on the pi bond makes the allylic source more reactive.	Substitutions and Additions. Carbon vs. heteroatom decision	Extended enolates. Allylic alkyne sources
C=C Simple pi bonds	$R_2C{=}CR_2$ RHC=CHR $H_2C{=}CH_2$	The more stable the carbocation formed, the more reactive the pi system.	Electrophilic Additions. Markovnikov's rule used for regiochemistry	Other pi systems: C=C–C=C C≡C C=C=C
ArH Aromatic rings	PhOR PhH PhCl PhCOR	An ERG on the ring makes it more reactive; an EWG makes ring less reactive.	Electrophilic aromatic substitution. Regiochemistry of addition	Hetero-aromatics, condensed aromatics

CHAPTER 5 EXERCISES

5.1 Give the generic class of each of the following electron sources.

$LiAlH_4$ CH_3Li NH_3 $CH_3OCH{=}CH_2$ Et_2NLi

5.2 Give the generic class of each of the following electron sources.

CH_3O^- CH_3COO^- EtMgBr $(CH_3)_2NCH{=}CH_2$ H_2O

5.3 Draw Lewis structures for the species in problem 5.1 and indicate on the structures which atom of the electron source would be attacked by the electrophile in a reaction.

5.4 Draw Lewis structures for the species in problem 5.2 and indicate on the structures which atom of the electron source would be attacked by the electrophile in a reaction.

5.5 Give the generic class of each of the following electron sources.

NaH $CH_3C{\equiv}CH$ $(CH_3)_2C{=}CH_2$ $t\text{-}BuO^-$ $PhCH_3$

5.6 Give the generic class of each of the following electron sources.

PhBr $^-C{\equiv}N$ PhLi $^-CH_2COOCH_3$ CH_3OH

5.7 Draw Lewis structures for the species in problem 5.5 and indicate on the structures which atom of the electron source would be attacked by the electrophile in a reaction.

5.8 Draw Lewis structures for the species in problem 5.6 and indicate on the structures which atom of the electron source would be attacked by the electrophile in a reaction.

5.9 Give the generic class of each of the following electron sources.

$NaBH_4$ $^-O{-}CH{=}CH_2$ HO^- $PhC(CH_3)_3$ NaOEt

5.10 Give the generic class of each of the following electron sources.

$^-SC{\equiv}N$ $H_2C{=}C{=}CH_2$ PhSH $H_2C{=}CHCH{=}CH_2$ KH

5.11 Draw Lewis structures for the species in problem 5.9 and indicate on the structures which atom of the electron source would be attacked by the electrophile in a reaction.

5.12 Draw Lewis structures for the species in problem 5.10 and indicate on the structures which atom of the electron source would be attacked by the electrophile in a reaction.

5.13 Circle the one atom on each molecule that is best for electrophilic attack.

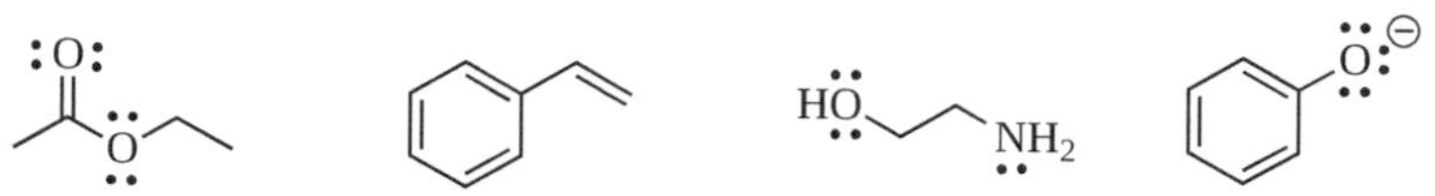

5.14 Rank all species, beginning with the numeral 1 to designate:

(a) The best nucleophile to react with CH_3I in CH_3OH

$PhSe^-$ PhO^- PhS^- CH_3COO^- CH_3OH

(b) The most reactive organometallic

$CH_3{-}K$ $CH_3{-}MgBr$ $(CH_3)_2Hg$ $(CH_3)_2Cd$ CH_3Li

5.15 Rank all species, beginning with the numeral 1 to designate:

(a) The best allylic electron source

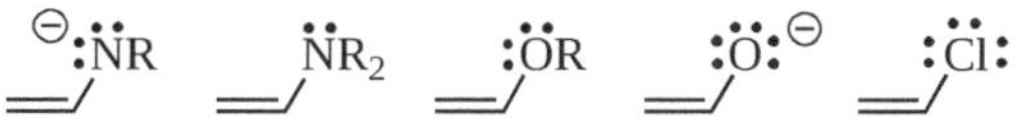

(b) The most reactive aromatic ring toward electrophilic attack

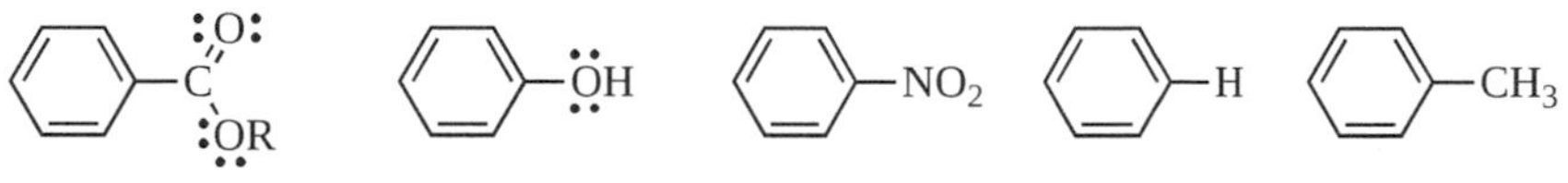

5.16 Rank which of the following pi bonds are most reactive to electrophilic attack.

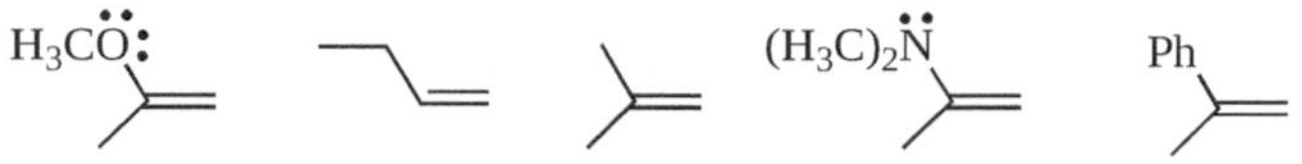

5.17 Use Markovnikov's rule to predict the site of electrophilic attack on each of the pi bonds in the compounds in problem 5.16.

5.18 At which position(s) on the aromatic ring would you expect electrophilic attack to occur for the following compounds?

$PhCF_3$ PhOH $PhNO_2$ PhCOOH PhCl

5.19 Use the nucleophilicity trend to decide which of the following two alternatives is the faster process in methanol solvent.

$CH_3CH_2CH_2Br + CH_3O^-$ displacing bromide to form $CH_3CH_2CH_2OCH_3$ or

$CH_3CH_2CH_2Br + CH_3S^-$ displacing bromide to form $CH_3CH_2CH_2SCH_3$

5.20 When an acyl chloride is added to aqueous ammonia two products are possible. Use the nucleophilicity trend to decide which of the following two alternatives is the faster process.

$CH_3COCl + H_2O$ displacing chloride to form CH_3COOH or

$CH_3COCl + NH_3$ displacing chloride to form CH_3CONH_2

5.21 Rank all species, beginning with the numeral 1 to designate:

(a) The strongest base

NH_2^- PhO^- CH_3COO^- CH_3O^-

(b) The most reactive complex metal hydride

$NaBH_3CN$ $LiAlH_4$ $NaBH_4$

5.22 Circle the one atom on each molecule that is best for electrophilic attack.

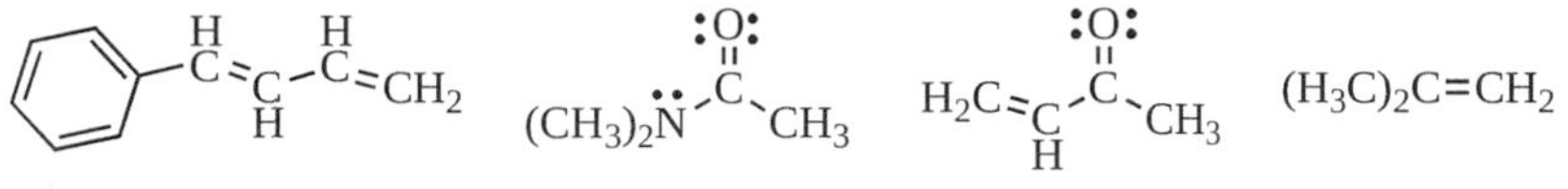

5.23 Decide whether the following aromatic rings would be attacked *ortho/para* or *meta*.

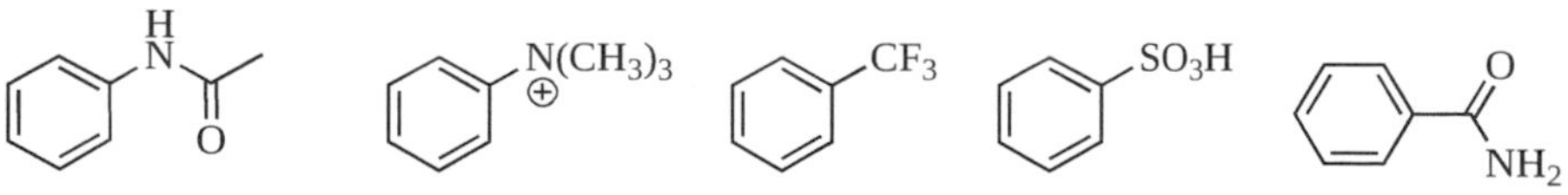

6

CLASSIFICATION OF ELECTRON SINKS

6.1 GENERALIZED RANKING OF ELECTRON SINKS

In this chapter, we will study the 18 generic classes of electron sinks in which an electron flow ends. Because most reactions have a hard–hard and a soft–soft component, again it is important to notice both how much positive charge the sink bears and how soft it is. **The larger the partial plus is, the better the electron sink can attract a negatively charged nucleophile.**

The three general classes of electron sinks are an empty orbital, a weak single bond to a leaving group, and a polarized multiple bond. Excellent electron sinks are electron-deficient cationic species having an empty orbital, such as protons and carbocations that can easily form a new bond with an electron source. Similarly, strong Lewis acids like $AlCl_3$ or BF_3 are also good electron sinks. Medium electron sinks are groups that accept the electron flow from the electron source by breaking a weak bond and forming a stable species. The bond that is broken is either a pi bond of a polarized multiple bond or a sigma bond to a leaving group. All of these electron sinks have a full, partial, or inducible plus charge on the atom that gets attacked by the electron source, the nucleophile. The poorer electron sinks would be any of the above species with poorer leaving groups, less electronegative atoms, or poorer electron-withdrawing groups. Here are a few examples listed with their bond polarization:

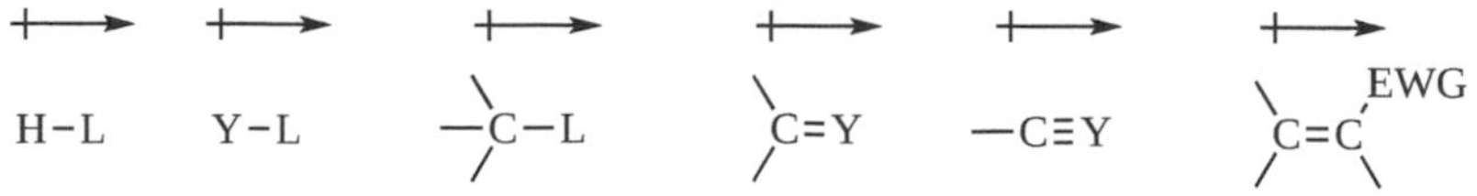

Electron Flow in Organic Chemistry: A Decision-Based Guide to Organic Mechanisms,
Third Edition, Paul H. Scudder.

Although an electron flow path is shown for each sink in this chapter, it is included only to point out the atom that would be attacked by the electron source. A detailed discussion of the electron flow paths is provided in Chapter 7.

6.2 ELECTRON-DEFICIENT SPECIES

6.2.1 Carbocations

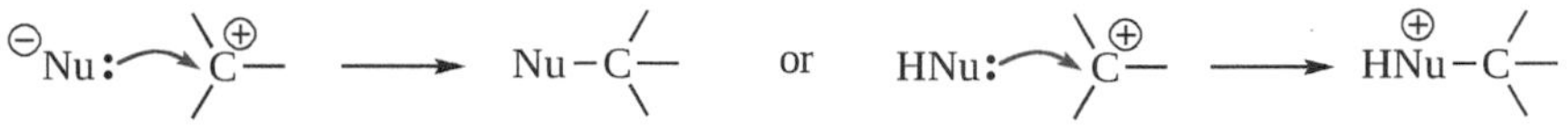

The best electron sinks, occurring almost exclusively in acidic media, are reactive cations such as carbocations (stability ranking in Section 4.2.4). Most are such good electron sinks that they can react with even very poor electron sources like aromatic rings. The most stable carbocations are the least reactive. Highly stabilized carbocations like guanidinium, $^+C(NH_2)_3$, are so stable that they can exist in basic media and make very poor electrophiles. The amino acid arginine has this cationic group on its side chain and serves in some enzyme active sites as a positive charge at physiological pH. The following are some of the more common reactive carbocations.

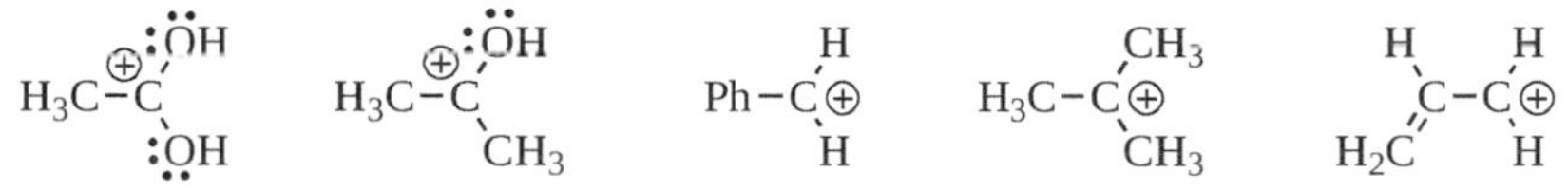

Protonated carboxyl Protonated ketone Benzyl cation *t*-Butyl cation Allylic cation

6.2.2 Inorganic Lewis Acids

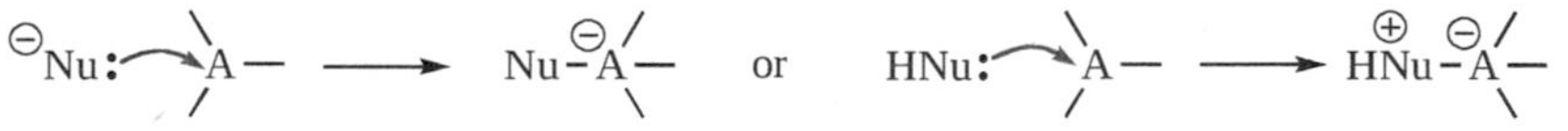

Electron-deficient Lewis acids like $AlCl_3$ or BF_3 behave much like carbocations. Like the $^+CR_3$ carbocation, BR_3 has an empty 2*p* orbital, exactly the same orbital configuration (*isoelectronic*), and will tend to react in a very similar way. The strength of a Lewis acid varies with substituents; attached pi resonance donors reduce the Lewis acidity by resonance donation into the empty *p* orbital of the Lewis acid. A crude ranking of Lewis acid strengths is $BX_3 > AlX_3 > FeX_3 > SbX_5 > SnX_4 > ZnX_2 > HgX_2$. Another strong Lewis acid is $TiCl_4$.

6.2.3 Metal Ions M^{2+}

Although not really electron deficient, metal ions may bear a high positive charge and can act as electrophilic catalysts because they can complex with many electron sinks and make them more positive, and thus more prone to nucleophilic attack. The extent of catalysis by the cationic counterion of an anionic nucleophile is sometimes underestimated. It is much more favorable to form an ion pair in most organic solvents than an isolated charge. Some metal ions that show electrophilic catalysis are Ag^+, Hg^{2+}, Li^+, and Mg^{2+} ions.

Many biochemical reactions are catalyzed by complexing divalent metal ions such as Mg^{2+} or Zn^{2+} to activate the substrate. Carbonic anhydrase, which converts carbon dioxide into bicarbonate in red blood cells, has a Zn^{2+} ion in the active site that complexes to water making it easy to remove a proton and form hydroxide at neutral pH. It is estimated that the pK_a of the zinc-bound water is about 7, so is easily deprotonated by an active site histidine. Carbonic anhydrase is one of the fastest of all enzymes, reacting with carbon dioxide as fast as it can diffuse in, producing bicarbonate. The bicarbonate is then transported to the lungs, where this equilibrium reaction reverses, releasing the carbon dioxide to be exhaled.

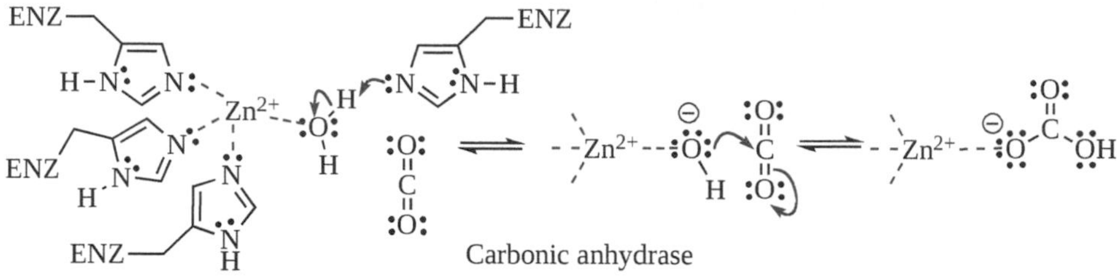

Carbonic anhydrase

6.2.4 Metal Ions as Oxidants Mn^{7+} and Cr^{6+}

High oxidation state transition metal ions such as MnO_4^- and CrO_4^{2-} can serve as electron acceptors in oxidation reactions, for they can have many available lower oxidation states. Metal ions as electron sinks will be covered in more detail in Section 8.10.

6.3 WEAK SINGLE BONDS

6.3.1 Acids

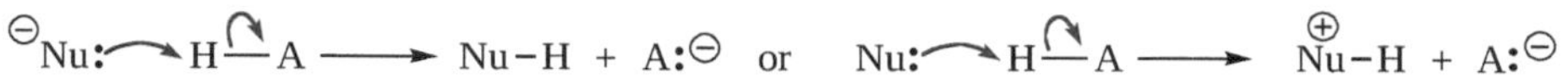

All Brønsted–Lowry acids belong to this class. Two of four common reaction charge types are shown above (see Figure 3.1). The more stable the conjugate base is, the stronger the acid is (see Section 3.2). Strong acids often will protonate the solvent; therefore the strongest acid that can exist in a solvent is the protonated solvent. See Section 3.6 for the calculation of the proton transfer K_{eq}, which should be greater than 10^{-10} to be useful.

6.3.2 Leaving Groups on Heteroatoms

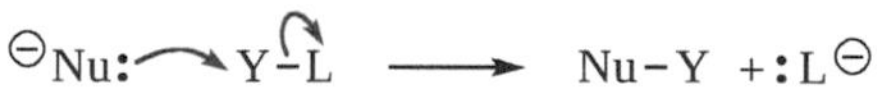

Bonds between heteroatoms and good leaving groups are often weak. Lone pair–lone pair repulsion causes weak bonds in RO–OR and RS–SR. These weak bonds are easily broken by a nucleophile that attacks Y and displaces the leaving group. The leaving group rarely departs first since an incomplete octet on an electronegative heteroatom is unstable. **Do not make any cationic species with an incomplete octet on an oxygen or nitrogen atom.** These reactions are often exothermic since the new bond formed is usually stronger than the weak Y–L bond broken. In unsymmetrical compounds like I–Cl, the δ+ end of the molecule gets attacked by the nucleophile (in this case iodine because Cl is more electronegative). A subset of the Y–L class that will be discussed in Sections 7.2.4 and 8.12 is the Nu–L class where Y bears a nucleophilic lone pair. Examples are halogens like Br_2 and Cl_2.

6.3.3 Leaving Group Bound to an *sp*³-Hybridized Carbon

The *sp*³-bound leaving groups can be directly displaced by the nucleophile as shown, or the leaving group can ionize off first to form a carbocation. As the leaving group gets better, the following equilibrium begins to produce more of the carbocation, an excellent electron sink. The weaker electron sources must wait until the carbocation is formed before the reaction can occur. As we saw in Chapter 4, the extent of this equilibrium depends heavily on the stability of the carbocation produced, the quality of the leaving group, in addition to solvent polarity. Remember that vinylic, C=C–L, sp^2-bound leaving groups behave much differently from sp^3-bound leaving groups.

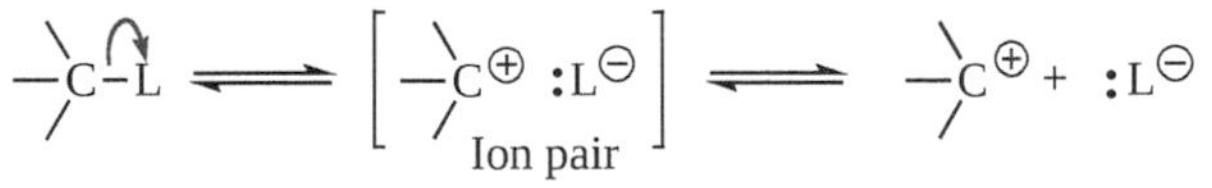

Any electron sink that contains a leaving group will become a much better sink as the leaving group improves. One common way to improve the quality of the leaving group (Section 4.2.3) is with acid catalysis. Acid catalysis is the complexation of a proton or strong Lewis acid (for example, BF_3 or $AlCl_3$) with an electron sink, which increases the electron sink's ability to accept electrons.

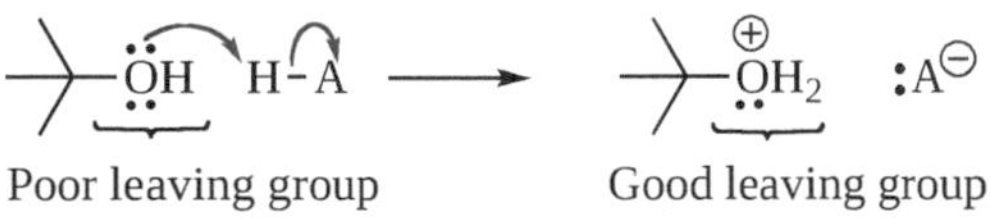

In the previous acid catalysis example, the pK_{aHL} of the leaving group improves by 14 pK_a units (hydroxide loss at a pK_{aHL} of 14.0 compared to water loss at a pK_{aHL} of 0.0). In fact, no reaction at all would occur without catalysis because hydroxide is a poor leaving group. **With few exceptions, no reaction occurs when the *sp*³-bound L sink bears a poor leaving group (p*K*$_{aHL}$ greater than 12).** Because diethyl ether, $CH_3CH_2OCH_2CH_3$, is so unreactive, it is a common solvent for reactions involving very reactive nucleophiles. The strained ring ethers (three- and four-membered) are exceptions. Ring strain raises the energy of the reactant relative to a product where the ring is broken by a Nu and the strain is relieved. The five-membered ring ether (tetrahydrofuran, THF) is not strained and is unreactive in base.

The best sp^3-bound leaving group sinks are allylic leaving groups, CH_2=CH–CH_2–L, and benzylic leaving groups, Ph–CH_2–L, because both are especially activated toward direct displacement (see Section 7.1.6) and ionization (see Section 7.1.2). Some additional examples of sp^3-bound leaving group sinks are:

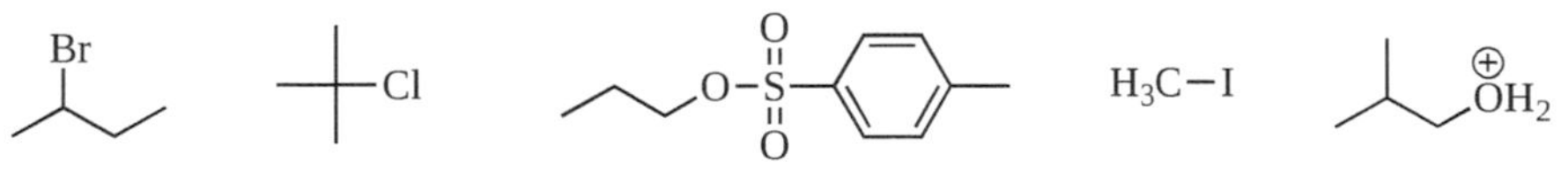

Compounds with more than one leaving group are just a subset of this class. Ketals, $R_2C(OR)_2$, and orthoesters, $RC(OR)_3$, are common examples that are not reactive in base but react in acidic media when the oxygen atom is protonated to improve it as a leaving group.

Example problem

Which of the following, CH_3OCH_3, CH_3I, or CH_3Cl, is the most reactive toward nucleophilic attack? Which is the least?

Answer: The better the leaving group is, the more reactive this sp^3-bound leaving group sink is. The most reactive is CH_3I, with a pK_{aHL} of −10. CH_3Cl is next, with a pK_{aHL} of −6. The least reactive is the ether, CH_3OCH_3, with a pK_{aHL} of 15.2.

6.4 POLARIZED MULTIPLE BONDS WITHOUT LEAVING GROUPS

6.4.1 Heteroatom–Carbon Double Bonds

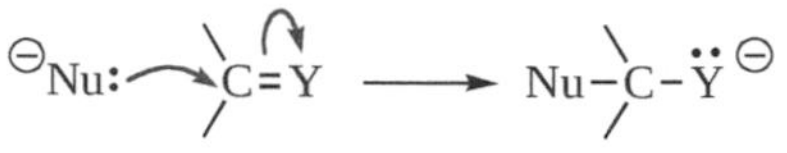

The electronegativity of Y polarizes the double bond to put a partial plus charge on carbon, making it susceptible to nucleophilic attack. The electronegativity of Y stabilizes the anionic product of the nucleophilic attack. The atom Y is most commonly oxygen, occasionally nitrogen. Alkyl substitution decreases reactivity by weak electron donation into the carbonyl reducing the partial plus and by making the carbonyl less accessible; thus aldehydes are more reactive than ketones. Conjugation stabilizes the carbonyl and thus makes it less reactive; for example, $PhCOCH_3$ is less reactive than CH_3COCH_3.

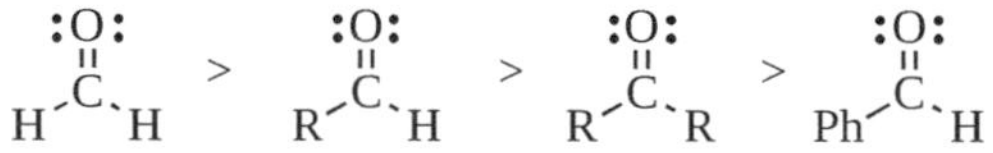

Acid catalysis, often by proton transfer or hydrogen bonding to Y, enhances the ability of this electron sink to accept electrons. Since the lone pair on an sp^2 nitrogen is easier to protonate than a lone pair on an sp^2 oxygen (see pK_a chart, Appendix), acidic catalysis occurs much more readily with nitrogen. The carbocation thus formed is a much better electron sink. Additional examples of this sink are:

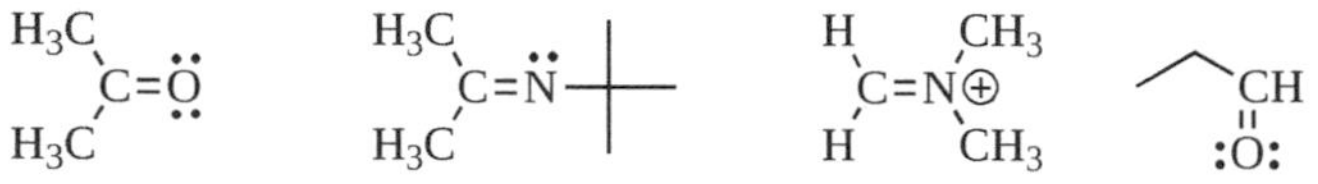

Example problem

Which of the following is the most reactive toward nucleophilic attack? Which is the next most reactive? Which is the least reactive?

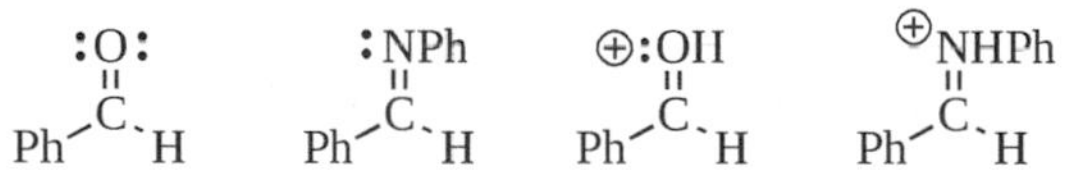

Answer: The most reactive is the protonated aldehyde. The protonated imine is next, for it is slightly more stable. The least reactive is the neutral imine because the anion formed upon nucleophilic attack is more basic than the anion formed upon attack of the neutral aldehyde.

6.4.2 Heteroatom–Carbon Triple Bonds

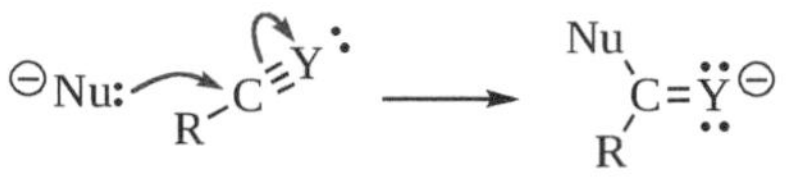

The most common representatives of this electron sink are the nitriles with Y as N. Nitriles are much less reactive than ketones. The electronegativity of Y polarizes the triple bond so that carbon bears a partial plus charge. The two perpendicular double bonds can be treated separately; thus attack by a nucleophile is identical to the doubly bonded electron sink, C=Y. Some examples are Ph–C≡N, and acylium, CH_3–C≡O^+.

6.4.3 Heterocumulenes

The heterocumulene derivatives, C=C=Y or Z=C=Y, are a minor variation of the polarized multiple-bond sink and undergo an identical nucleophilic attack at the C=Y bond. The only difference is that the intermediate is a resonance hybrid that protonates to form the most stable product. The product with the stronger bonds predominates.

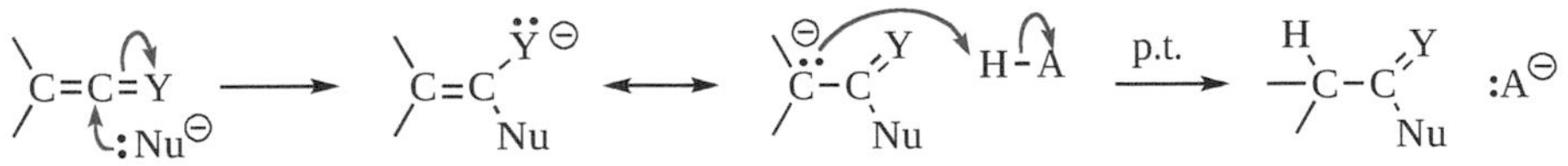

Some examples of heterocumulene sinks are:

6.4.4 Conjugate Acceptors

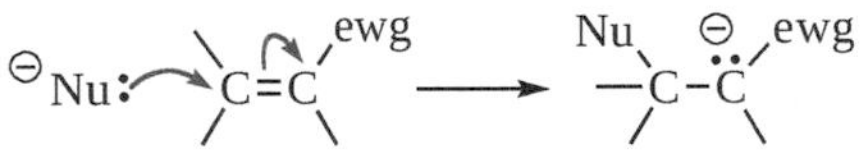

The electron-withdrawing group polarizes the carbon–carbon double bond in a way similar to the C=Y electron sink and makes this sink a subset of that group (with Y equaling C–ewg). The electron-withdrawing group must be present for the double bond to accept electron density from a nucleophile. The electron-withdrawing group makes the C–ewg carbon more electronegative. **Trading C–ewg for Y is simply exchanging an electronegative carbon atom for an electronegative heteroatom. The better the electron-withdrawing group is, the better the electron sink will be.** More than one good electron-withdrawing group on a double bond, C=C(ewg)$_2$, gives a very reactive electron sink because the resulting anion is stabilized to a much greater extent. Some conjugate acceptors are so reactive that the conjugate addition product adds to another molecule of conjugate acceptor and polymerizes (cyanoacrylates or super glue).

A complication arises in that the nucleophile now has a choice; it can add in a conjugate fashion, or it can add to the electron-withdrawing group directly. Section 9.6, Ambident Electrophiles, discusses this problem in depth. Examples of these sinks are:

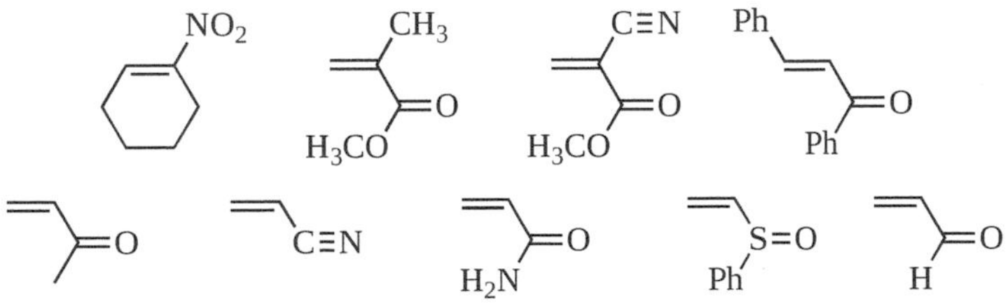

6.4.5 Triply Bonded and Allenic Conjugate Acceptors

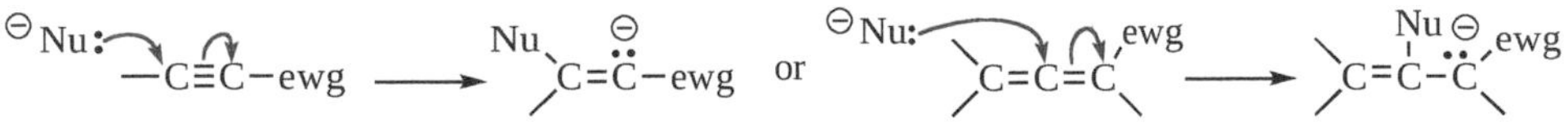

These two electron sinks are the least common of the six that make up the set of multiple-bonded electron sinks with no attached leaving group. They behave very similarly to the conjugate acceptors previously discussed. The electronegative C–ewg group replaces the electronegative Y in accepting the electron flow from the nucleophile.

6.5 POLARIZED MULTIPLE BONDS WITH LEAVING GROUPS

6.5.1 Carboxyl Derivatives, L–C=Y

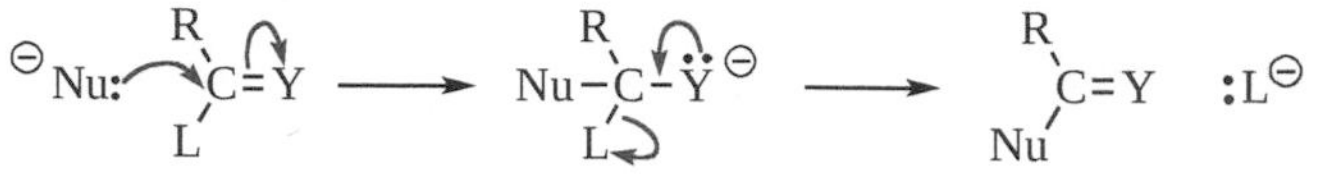

The most common representatives of this group are the carboxylic acid derivatives with Y as oxygen. As we saw in Section 4.5, the replacement of the leaving group is a two-step process: First nucleophile addition occurs, then the leaving group is eliminated. The rate-determining step is almost always the nucleophilic attack; thus the rate depends on the ability of this electron sink to attract and accept electron density from a nucleophile.

If the leaving group is a good pi-donor it will delocalize into and diminish the partial plus on the carbon of the carbonyl, decreasing its hardness and its ability to attract a negatively charged nucleophile. Conjugation of a leaving group lone pair tends to make these systems more stable and less reactive (Figure 6.1). The barrier to reaction increases since this delocalization in the reactant is not possible in the tetrahedral intermediate. In an acyl chloride, the most reactive carboxyl derivative, this delocalization of the leaving group lone pair into the carbonyl is very minor because of poor 3*p*–2*p* overlap between the chlorine and the carbonyl carbon. **As the L becomes a better pi-donor, carboxyl derivatives become less reactive.** The electron-releasing group ranking was covered in Section 4.2.5.

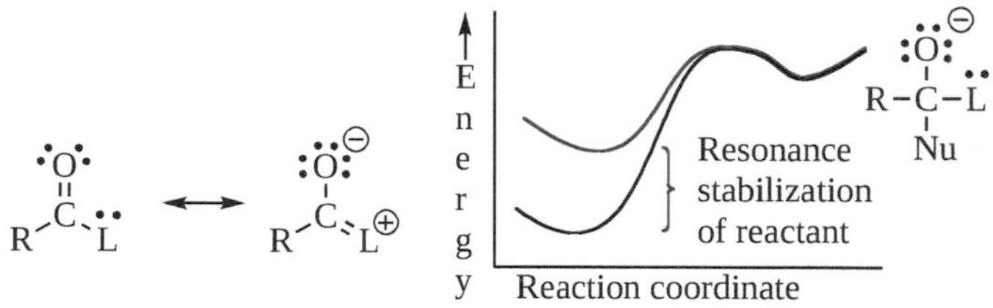

Figure 6.1 Reactant stabilization relative to the intermediate increases the reaction barrier.

Carboxylic acids are doubly poor sinks, for they can protonate the nucleophile (thus making it useless) and form the very unreactive carboxylate anion. Anions rarely attack anions because of the repulsion of like charges. **Only R–Li and $LiAlH_4$ are reactive enough to attack the carboxylate anion** (which is probably paired with a metal cation to cancel the negative charge repulsion of the anionic nucleophile). The less reactive nucleophiles will react only with the most reactive carboxyl derivatives. Using the donor trend, we can place ketones as more reactive than esters. Acyl halides, with the halide being both a poor donor and a good leaving group, are the most reactive carboxyl derivatives. A carboxyl derivative with a good leaving group is very reactive because the loss of the leaving group from the intermediate is often the driving force for the overall process. Acyl chlorides are more reactive than anhydrides for this reason. In general, the better the leaving group on the carboxyl derivative, the more reactive it will be. The better donor group deactivation and the better leaving group activation of carboxyl derivatives do not often conflict, so either trend can be used to judge the reactivity of this electron sink.

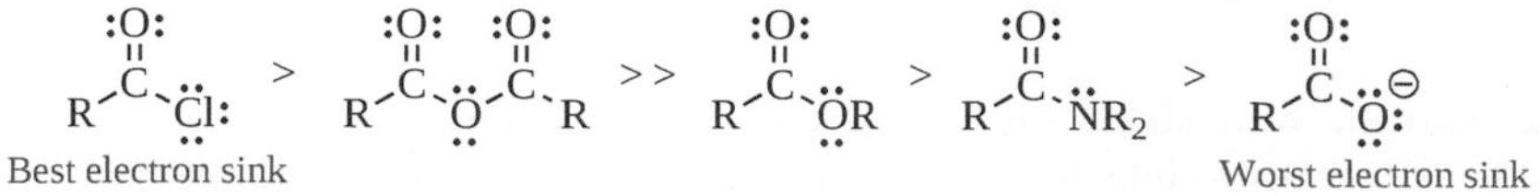

One minor variation of this group is the carbonate derivatives, which have two leaving groups attached to the carbonyl instead of just one. These systems can be quite unreactive since the partial plus on the carbonyl is often diminished by two electron-releasing groups. Both leaving groups can be replaced by a nucleophile. If the leaving groups are different enough (for example, Cl and OEt), the monosubstitution product can be isolated.

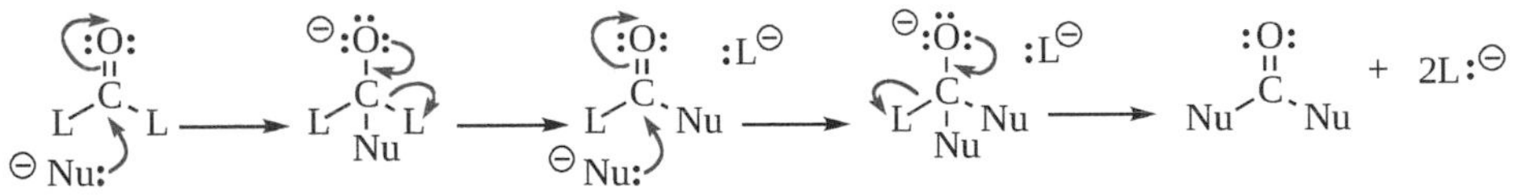

Example problem

Which of the following is the most reactive toward nucleophilic attack? Which is the least reactive?

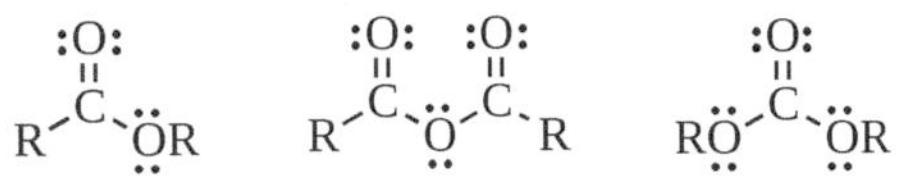

Answer: The most reactive bears the weakest electron-releasing group. See Table 4.2 for the ranking of electron-releasing groups. The anhydride is the most reactive. The ester is the next most reactive; least reactive is the carbonate.

6.5.2 Vinyl Leaving Groups

This less common group L–C=C–ewg bears the same relationship to the carboxyl derivatives L–C=Y discussed in Section 6.5.1 as the conjugate acceptors C=C–ewg bear to the C=Y series. Again we have merely substituted C–ewg for the electronegative Y atom in the group. To show how similar the reactivities are, if the ewg is C=Y, then the L–C=C–C=Y group is merely a vinylogous L–C=Y group. Aromatic rings that bear electron-withdrawing groups and a leaving group can be included in this classification. Again, more than one electron-withdrawing group can be attached, thereby increasing the reactivity.

The electron-withdrawing group is absolutely necessary for the acceptance of electron density from a nucleophile. Simple olefinic halides, RCH=CH–X, are very unreactive because there is no good place to put the electrons from the nucleophile.

6.5.3 Leaving Groups on Polarized Triple Bonds

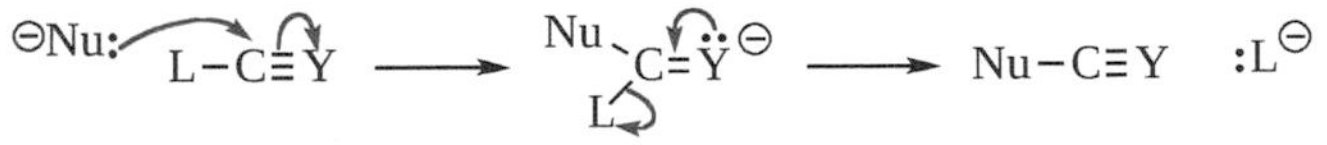

Two very rare electron sinks, **L–C≡Y** and **L–C≡C–ewg**, the first represented by the acid bromide of HOCN, Br–C≡N, react as expected for a multiply bonded electron sink with an attached leaving group. The C–ewg again replaces Y in Br–C≡C–COOEt.

6.6 SUMMARY OF GENERIC ELECTRON SINKS

6.6.1 Reactivity Comparison Between Various Classes

All the electron sinks are summarized in Table 6.1; however, the general reactivity trends are not as clear-cut as they were for electron sources. Look first for electron-deficient species, then for the others. The reactivity trend within each class is more important than the class to which a species belongs. For example, **acyl chlorides,** the top end of the reactivity spectrum for polarized multiple bonds with leaving groups, **are more reactive than most alkyl halides**; however, **esters are less reactive than most alkyl halides. Acyl chlorides are more reactive than aldehydes. Nitriles tend to be about as reactive as amides.**

6.6.2 Generic Electron Sink Table

Table 6.1 The 18 Generic Electron Sinks

Generic class	Symbol	Examples
Electron-deficient species (6.2)		
Carbocations	$-\overset{\oplus}{C}<$	$^{+}C(CH_3)_3$
Lewis acids	$-A<$	BF_3, $AlCl_3$, $FeBr_3$
Metal ions	M^{2+}	$Hg(OAc)_2$
Weak single bonds (6.3)		
Acids	H–L (or H–A)	HCl, $ArSO_3H$, CH_3COOH
Leaving groups on a heteroatoms	Y–L	RS–SR, HO–OH, Br–Br, HO–Cl
Leaving groups bound to sp^3 carbon	$-\!\!\!>$C–L	CH_3CH_2–Br, $(CH_3)_3C$–Br, $(CH_3)_2CH$–Cl, CH_3–I Ketals $R_2C(OCH_3)_2$ (2 poor Ls)
Polarized multiple bonds without leaving groups (6.4)		
Heteroatom-carbon multiple bonds	>C=Y	Aldehydes: RHC=O Ketones: R_2C=O Imines: R_2C=NR
	–C≡Y	Nitriles: RC≡N
Conjugate acceptors	C=C–ewg	Enones: C=C–C=O, Conjugated esters: C=C–COOR
	C≡C–ewg	RC≡C–COOR
Heterocumulenes	C=C=Y	Ketenes: R_2C=C=O
	Z=C=Y	Ph-N=C=O, RN=C=NR, O=C=O
	C=C=C–ewg	$(CH_3)_2C$=C=$CHCOOCH_3$
	ewg–C=C=C–ewg	PhCOCH=C=CHCOPh
Polarized multiple bonds with leaving groups (6.5)		
Carboxyl derivatives	L–C(=Y)–	Acyl chlorides: RCOCl Anhydrides: RCOOCOR Esters: RCOOEt Carbonate derivatives: ClCOCl
Vinyl leaving groups	L–C=C–ewg	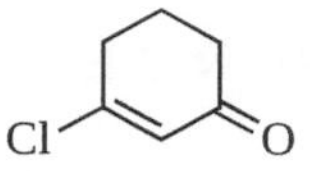
Leaving groups on polarized triple bonds	L–C≡Y L–C≡C–ewg	Br–C≡N Br–C≡C–COOR

6.6.3 Generic Electron Sink Classification Flowchart

Classification of a given structure into its most appropriate generic class relies on your ability to recognize electron-deficient species, leaving groups, and polarized multiple bonds. Figure 6.2 shows a flowchart for the classification of electron sinks.

1. Carbocations and Lewis acids are usually highly reactive, and thus are found first.
2. Next look for any leaving groups. Is the leaving group bound to carbon and, if so, what is the hybridization of that carbon? Then sort into the two general classes: weak single bonds and polarized multiple bonds with leaving groups.
3. Search for any polarized multiple bonds without leaving groups.
4. Finally, metal ions as sinks are sorted last since they tend to be the least reactive.

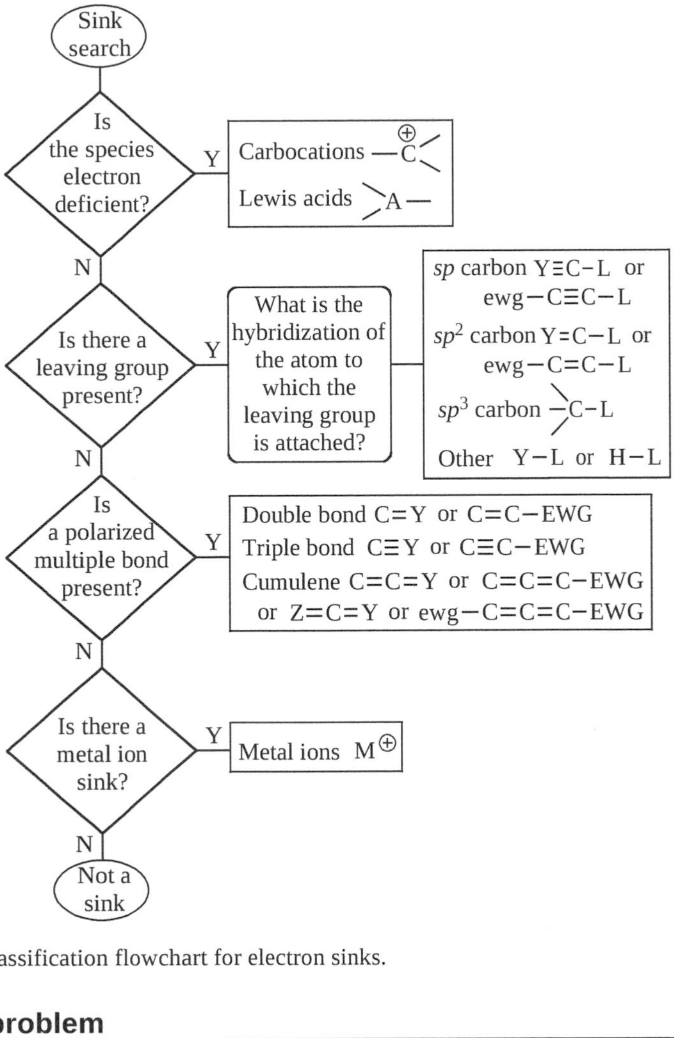

Figure 6.2 Classification flowchart for electron sinks.

Example problem

To which generic classes do these belong, PhCOOH, CH_2CHCHO?

Answer: We have two problems: translating the typed formula into a recognizable structure and doing a correct classification. The first compound is a carboxylic acid, which should tip us off to put it in the acids, the H–L or H–A class. The next compound is an aldehyde conjugated to a pi bond. Since an aldehyde is a polarized multiple bond and also an ewg, we would put this compound in the conjugate acceptor class, C=C–EWG.

6.6.4 Common Sinks Reactivity Summary

A much smaller set of sinks is responsible for most reactions and is summarized in Table 6.2, which gathers reactivity trends, common reactions, and related generic groups.

Table 6.2 Summary Reactivity Table for Common Electron Sinks

Symbol and Name	Common Examples in Order of Decreasing Reactivity	Reactivity Trend Principle	Common Reactions. Decision?	Related Generic Groups
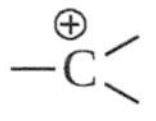 Carbocation	$(CH_3)_2CH^+$ $(CH_3)_3C^+$ Ph_3C^+ $(H_2N)_3C^+$	More stabilized, less reactive. Donors stabilize Lone pairs > pi bonds > R > H	C^+ trap Nu. Rearrange to equal or more stable C^+? Drop off H^+?	Inorganic Lewis Acids 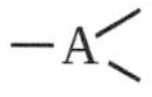
H–L or H–A Acids	H_2SO_4 HBr HCl	The lower pK_a is stronger	Proton-transfer $K_{eq} \geq 10^{-10}$	Carbon acids H_3C–EWG
Y–L Leaving group on a heteroatom	Chlorine Cl_2 Bromine Br_2 Iodine I_2 Peroxides ROOR	The most reactive break a weak bond and form a much stronger bond	Often Substitution Nu attacks δ+ end of Y–L if different	Leaving groups on phosphorus or silicon
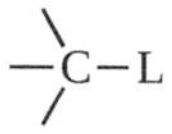 Leaving group on sp^3 carbon	Sulfonates R–OSO_2R Alkyl iodides R–I Protonated alcohols ROH_2^+ Alkyl bromides R–Br Alkyl chlorides R–Cl	The better L is more reactive	Substitution vs. Elimination Decision (Section 9.5)	2 Ls on C R_2CL_2 3 Ls on C RCL_3
 Polarized multiple bond without L	Protonated carbonyls $R_2C{=}OH^+$ Iminium ions $R_2C{=}NR_2^+$ Aldehydes RHC=O Ketones $R_2C{=}O$ Imines $R_2C{=}NR$	The larger partial plus carbon is more reactive	Addition. Following Substitution or Elimination? (Section 8.5)	Polarized triple bonds R–C≡Y without leaving groups
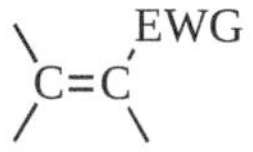 Conjugate acceptor	Conj. nitros C=C–NO_2 Enones C=C–COR Conj. esters C=C–COOR Conj. nitriles C=C–CN	The better the EWG, the more reactive	1,2 vs. 1,4? Direct vs. conjugate addition (Section 9.6)	Triply bonded conjugate acceptors C≡C–EWG
 Polarized multiple bond with L	Acyl halides RCOCl Anhydrides RCOOCOR Thioesters RCOSR Esters RCOOR Amides $RCONR_2$	The poorer pi donor L is, the more reactive	Addition-Elimination. Add a second Nu after addition-elimination? (Section 9.2)	Carbonate derivatives $L_2C{=}Y$ L on triple bonds L–C≡Y

CHAPTER 6 EXERCISES

6.1 Give the generic class of each of the following sinks.

PhNCO	BF_3	HOBr	$(CH_3)_2CHBr$	$(CH_3)_2CO$
$(CH_3CO)_2O$	CH_3Br	$AlCl_3$	$CH_2{=}CHCOCH_3$	CO_2

6.2 Give the generic class of each of the following sinks.

CH_3CN	$[(CH_3)_2NCH_2]^+$	CH_3CH_2I	CH_3COCl	$(CH_3)_3C^+$
H_2SO_4	$PhSO_3CH_3$	CH_3SO_2Cl	$SOCl_2$	$CH_2{=}CHCN$

6.3 Draw Lewis structures for all the species in problem 6.1 and designate which atom(s) on each would be attacked by the electron source in a reaction.

6.4 Draw Lewis structures for all the species in problem 6.2 and designate which atom(s) on each would be attacked by the electron source in a reaction.

6.5 Give the generic class of each of the following sinks.

$COCl_2$	CH_3COOH	H_2O_2	CH_3COOEt	$CH_3OSO_3CH_3$
PhCHO	EtOCOCl	Cl_2	Cl-C≡N	$H_2C{=}C{=}O$

6.6 Circle the best atom on each molecule for nucleophilic attack.

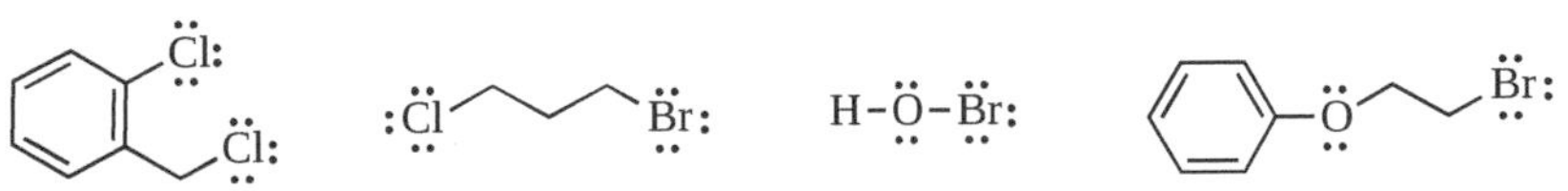

6.7 Draw Lewis structures for all the species in problem 6.5 and designate which atom(s) on each would be attacked by the electron source in a reaction.

6.8 Trends: rank all species; use the numeral 1 to designate:

(a) The most reactive toward nucleophilic attack

$CH_3{-}OCH_3$ $CH_3{-}Br$ $CH_3{-}N(CH_3)_2$ $CH_3{-}I$ $CH_3{-}Cl$

(b) The most reactive toward nucleophilic attack

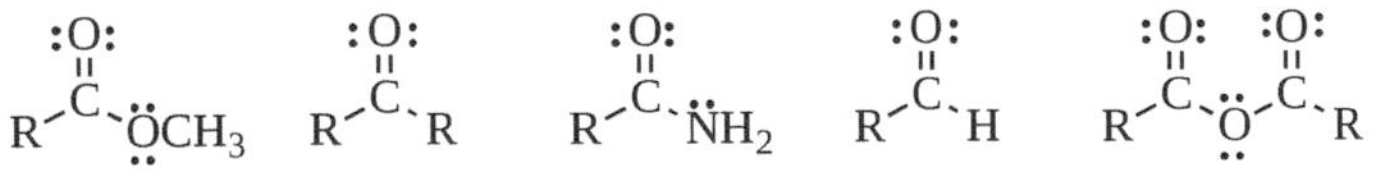

6.9 Trends: rank all species; use the numeral 1 to designate:

(a) The most reactive toward nucleophilic attack

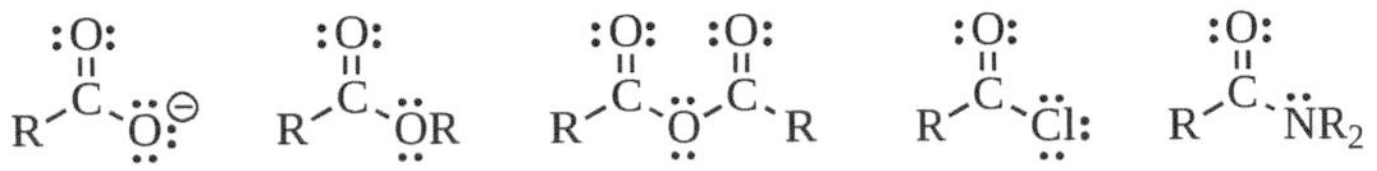

(b) The most reactive conjugate acceptor

6.10 Explain in general terms why ketals, $R_2C(OR)_2$, fail to react (act as electron sinks) in basic media but react rapidly in acidic.

6.11 Sodium borohydride reduces ketones quickly but reacts very slowly to reduce esters. Comparatively, how quickly would you expect an aldehyde to react?

6.12 Thioesters, common in living systems, are more reactive than esters. Explain.

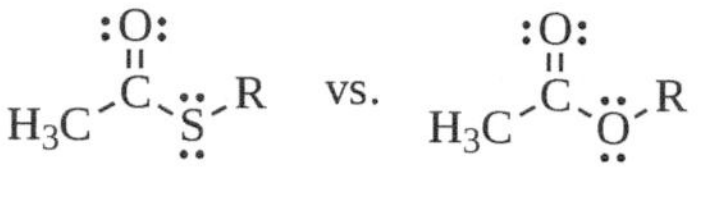

6.13 Circle the best atom on each molecule for nucleophilic attack.

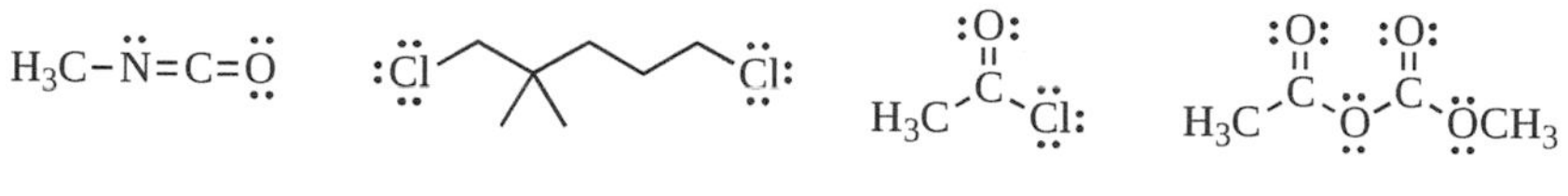

6.14 Trends: rank all species; use the numeral 1 to designate:

(a) The most reactive toward nucleophilic attack

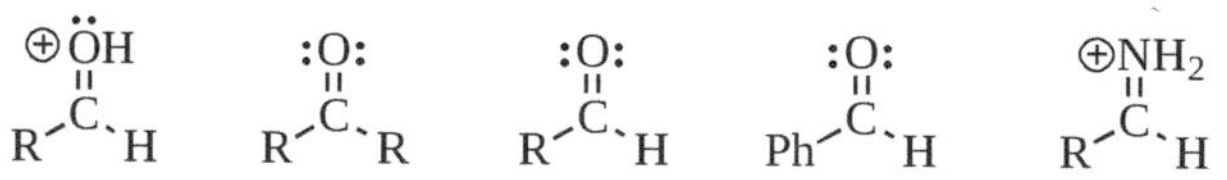

(b) The most reactive carbocation

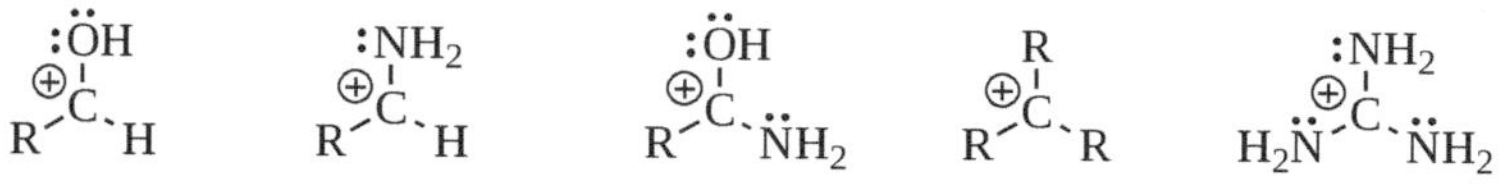

6.15 Trends: rank all species; use the numeral 1 to designate:

(a) The most reactive toward nucleophilic attack

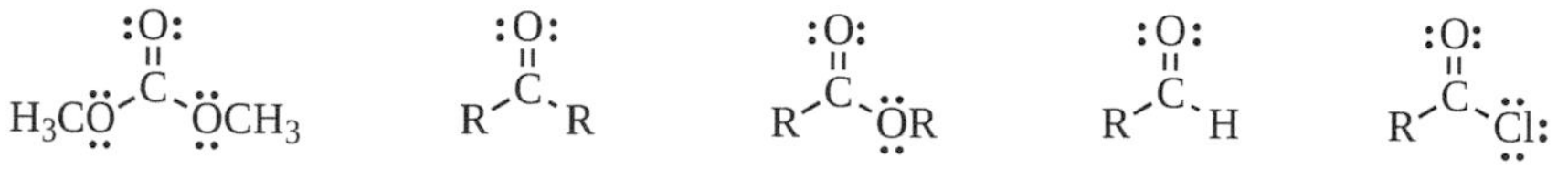

(b) The strongest acid

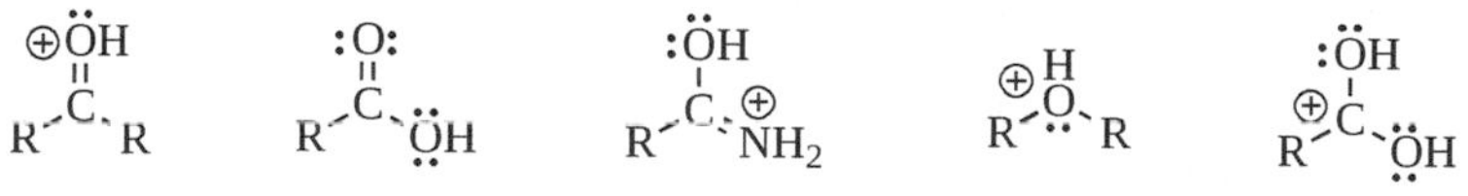

6.16 Carbamic acids are unstable and fall apart easily. The salt is stable. Explain.

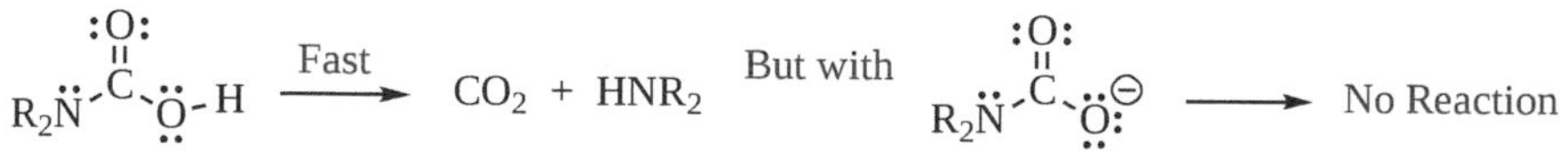

6.17 Some carbocations are stable enough to appear on the pK_a chart. How does their reactivity correspond to their acidity? Rank the following, use the numeral 1 to designate the most reactive, and explain the trend. The pK_a is listed below the structures.

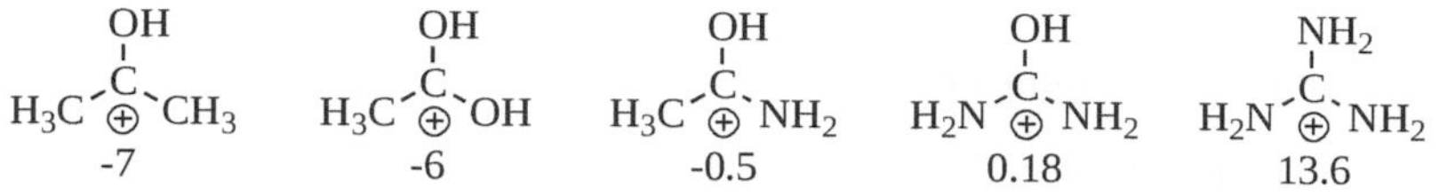

6.18 Circle the best atom on each molecule for nucleophilic attack.

7

THE ELECTRON FLOW PATHWAYS

Electron Flow in Organic Chemistry: A Decision-Based Guide to Organic Mechanisms,
Third Edition. Paul H. Scudder.

Chapter 3 covered the proton transfer electron flow path and reviewed the factors that contribute to acidity. Chapter 4 introduced all the rest of the major electron flow paths along with the four reaction archetypes: substitution, elimination, addition, and rearrangement. This chapter gathers together all the major electron flow paths, introduces a few minor paths, and reviews common path combinations. Section 7.4, Variations on a Theme, shows how the 18 electron flow paths might be reasonably extended and modified.

7.1 THE DOZEN MOST COMMON PATHWAYS

Although there are thousands of different organic reactions, they can be explained by mechanisms that are a combination of relatively few pathways for electron flow. The following are twelve of the most common generic electron flow pathways. These pathways should become an important part of your mechanistic "vocabulary." You will need to have an excellent command of these dozen pathways to be able to combine them to postulate a reasonable mechanism for almost any reaction.

These electron flow pathways have been shown to be reasonable by many detailed mechanistic studies. Therefore, if we confine our use of arrows to stepwise combinations of these pathways, taking into account each pathway's restrictions, we are pretty much assured that whatever we write will also be judged reasonable, if not correct.

Beware of shortcuts that combine steps to avoid redrawing a structure. Any sequence of arrows on one structure implies that that particular electron flow occurs in one step, which in turn demands that all the interacting orbitals be properly lined up. Any sequence of more than three arrows may have major $\Delta S^{\ddagger}$ problems. There are examples in published works of absolute nonsense written with arrows, so always be suspicious of any set of arrows that does not fit a known pathway.

All of the electron sources and sinks previously discussed can react by at least one of these pathways or a simple sequence of them. We must learn what sort of functionality is required for each pathway and also notice the limitations on each pathway. Eventually we will be trying to decide among alternate routes, and the limitations are important so that we may narrow down the possibilities. Each of these pathways has four charge types since the source and/or sink can be charged or neutral: Nu with E, Nu^- with E, Nu with E^+, and Nu^- with E^+. Although any combination of E with Nu can be viewed as either an electrophilic attack or a nucleophilic attack, for consistency's sake we will call it nucleophilic attack if the organic reactant is the sink and electrophilic attack if an inorganic reactant is the sink.

Because of microscopic reversibility, many of these pathways are merely the reverse of others. The transition states for each related pair of paths are similar (they need not be identical, for the reaction conditions are often slightly different). A figure that illustrates the **approximate** orbital alignment and the transformation of the orbitals of the reactants into those of the product is given for each pair of paths.

7.1.1 Path p.t., Proton Transfer to a Lone Pair

Proton transfer can occur from any acid to any basic group or to and from the solvent. A common shortcut in writing mechanisms is to draw just the proton rather than the acid or the protonated solvent, but you should remember that a naked proton will never be floating free in solution. Always check the proton transfer K_{eq} (Section 3.6).

Proton transfer is either a deprotonation of the reactant, or the microscopic reverse, protonation of the reactant. The base is usually a lone pair on a heteroatom, symbolized by A: or more generally by b: (not to be confused with the element boron). Likewise, an acid can be b–H or H–A. The four charge types of proton transfers (as in Figure 3.1), result from the combinations of a neutral or negatively charged the base with a neutral or positively charged acid. Protons on electronegative heteroatoms (H–Y) are usually rather acidic because the electronegativity of the heteroatom can stabilize the conjugate base. An electronegative carbon atom, C–EWG, can replace an electronegative heteroatom, Y. Below are some common examples of two common charge types. Deprotonation:

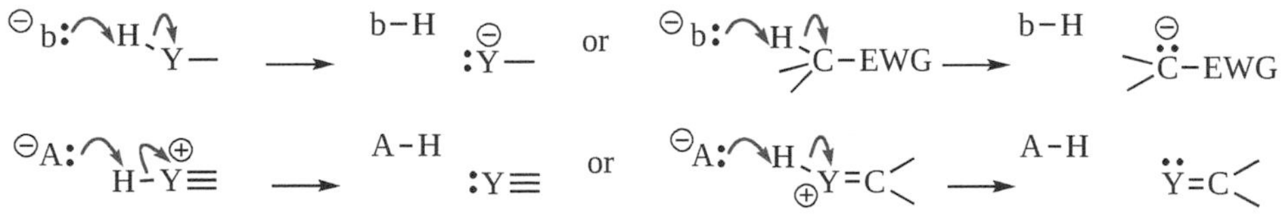

Or the reverse of the above, protonation:

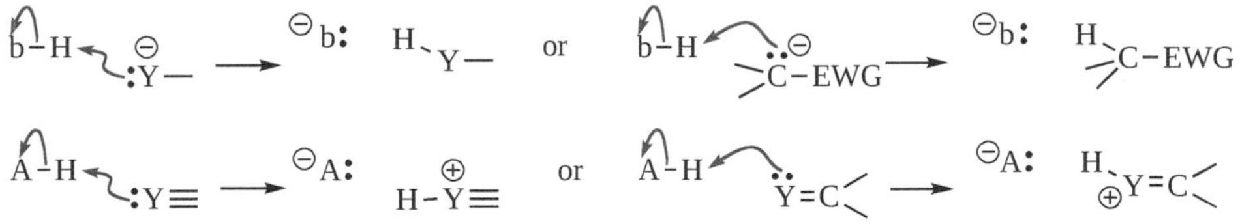

The nonbonding lone pair electrons of the carbonyl group, C=O, are more available than the electrons of the pi bond; therefore, proton transfer occurs to and from the nonbonding lone pairs rather than to and from the more stabilized carbonyl pi bonding electrons.

Proton transfer can occur in any medium, depending on the charge type. **A solution that contains equal concentrations of a base and its conjugate acid has a pH that is equal to the pK_{abH}** (Section 3.2.1). The pH of the medium will be close to the pK_{abH} of the anion needed for reaction because the anion must have a high enough concentration to react at a reasonable rate. As an illustration, if a ketone enolate, pK_{abH} = 19.2, is needed for reaction, the medium will definitely be very basic.

Path Cross-checks (see Chapter 3 for more discussion)

Energetics: The proton transfer K_{eq} should be greater than about 10^{-10}; although more uphill proton transfers occur, there are usually more probable competing processes.

Path Details

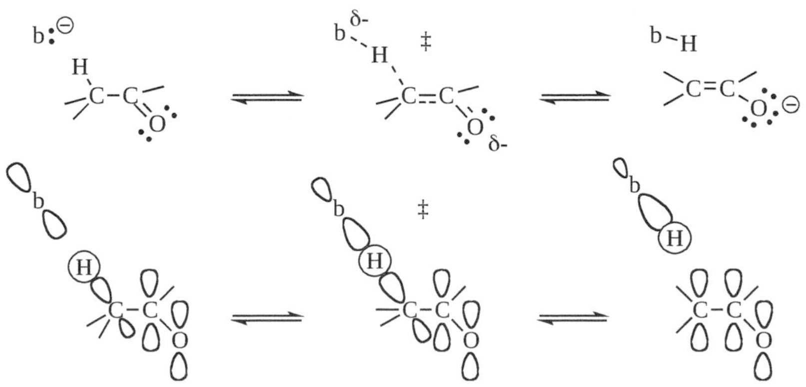

Figure 7.1 Loss of a proton adjacent to a carbonyl with the transition state shown in the center.

Overlap: (viewed as a reactant deprotonation) With H–C–EWG as the acid, the C–H bond and the adjacent *p* orbital of the pi system of the EWG must be roughly coplanar (Figure

7.1). Little angular dependence is expected for H−Y as the acid, since the base is overlapping with a spherical hydrogen 1*s* orbital and deprotonation forms a lone pair on the heteroatom.

Media: Acidic, basic, or neutral. The pH is near the pK_{abH} of the conjugate base formed.

Solvent: Often protic or polar to stabilize anion formed. The strongest acid possible in a solvent is the protonated solvent. The strongest base is the deprotonated solvent.

Steric: Proton must be accessible to base.

Kinetics: Proton transfer catalyzes many reactions. Proton transfer between heteroatom lone pairs is very fast, often at the diffusion-controlled limit. Under reversible (equilibrium) conditions, the most acidic proton is removed preferentially. However, if the deprotonation is done under irreversible conditions, the proton removed is determined by kinetics, not thermodynamics (Section 9.3). Anion basicity always competes with nucleophilicity. Proton transfer is slow enough between organometallics and protons adjacent to carbonyls (carbon bases with carbon acids) that addition of the organometallic to the carbonyl is the dominant process, path Ad_N.

7.1.2 Path D_N, Ionization of a Leaving Group
(Dissociation, Nucleofugic)

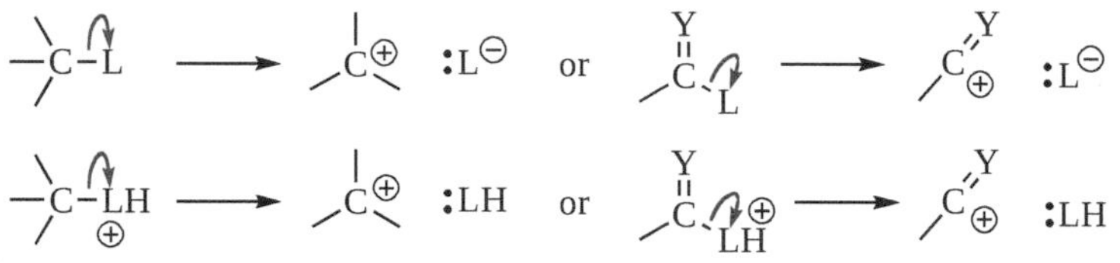

Path Cross-checks

Energetics: Often uphill since a bond is broken and usually charged pieces form. Both pieces should be reasonably stable ($pK_{aHL} \leq 0$, and a carbocation usually secondary or better), and a solvent should be polar enough to stabilize charges (Section 4.2.9). If a lone pair pushes out the leaving group, the cross-checks are different (see path E_β, Section 7.1.10).

Path Details

Overlap: Leaving group departs along the axis of the bond (Figure 7.2). Cleavage is a simple extension of a bond stretching vibration.

Cation stability: Usually better than secondary, R_2CH^+. Primary, RCH_2^+, is very rare.

Leaving group: For ⋟C−L, the L must be good, usually with a $pK_{aHL} \leq 0$. However, for Y=C−L or even more so for C=C−L, the L must be excellent (improved by Lewis acid complexation or protonation) to compensate for forming a less stable carbocation.

Media: Commonly acidic, occasionally neutral, rarely basic. The formation of unstable cations like the vinyl cation occurs only in strong acid.

Solvent: Polar to stabilize cation.

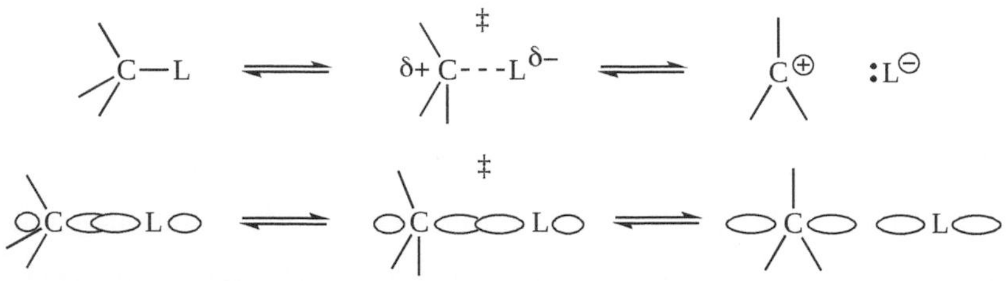

Figure 7.2 Ionization of a leaving group with the transition state in the center.

Steric: Sterically crowded leaving groups tend to depart a little faster because of the strain released in going to the planar carbocation.

7.1.3 Path A_N, Trapping of an Electron-Deficient Species

(Association, Nucleophilic, and the reverse of the above D_N, Ionization)

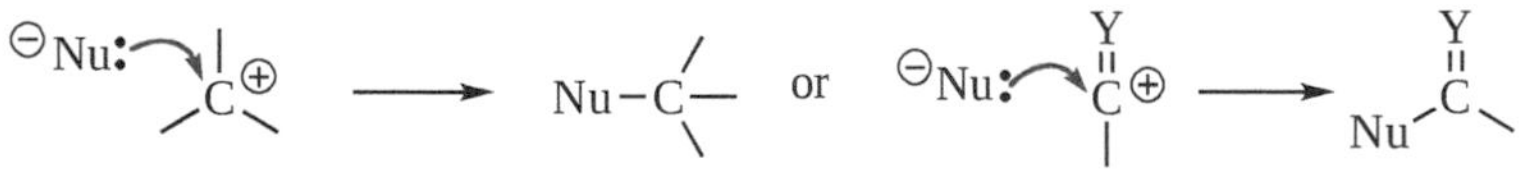

Included in this group are any of the isoelectronic Lewis salt-forming reactions.

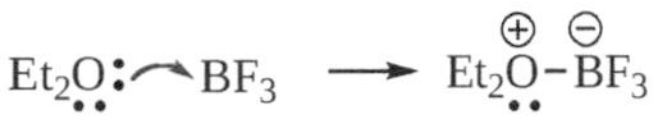

Path Cross-checks

Energetics: Usually downhill since charge is neutralized, but may be reversible by path D_N (Section 7.1.2) if D_N path cross-checks are met.

Path Details

Overlap: Same as path D_N; Nu must approach the empty *p* orbital along its axis so that good overlap occurs (Figure 7.2, but view the figure from right to left).

Media: Commonly acidic, occasionally neutral, rarely basic.

Solvent: Polar to stabilize cation.

Steric: Relatively little problem because access to the cationic center is usually very good because the cation is flat.

7.1.4 Path A_E, Electrophile Addition to a Multiple Bond

(Association, Electrophilic)

An alkene is an average electron source, and an aromatic compound is usually worse; therefore to get electrophilic addition to alkenes and aromatic compounds to occur, one needs a good electron sink. Often, a loose association of an electrophile with the pi electron cloud (called a pi–adduct or pi–complex) occurs before the actual sigma bond formation step. The best electrophiles, carbocations, add easily. For an overview of electrophilic additions to alkenes, see Section 4.4.

Path Cross-checks

Energetics: Usually uphill and dominated by the stability of the carbocation. **The electrophile adds to give the most stable of the possible carbocation intermediates** (Markovnikov's rule). See Table 4.1 for the ranking of carbocation stability. If the electrophile is the most stable of the possibilities, one expects the reverse reaction to be rapid and the equilibrium to favor reactants. Usually the carbocation formed should be more stable than secondary, or the Ad_E3 addition path will be favored.

Path Details

Overlap: Electrophiles must approach along the axis of the *p* orbital attacked, not from the side (Figure 7.3).

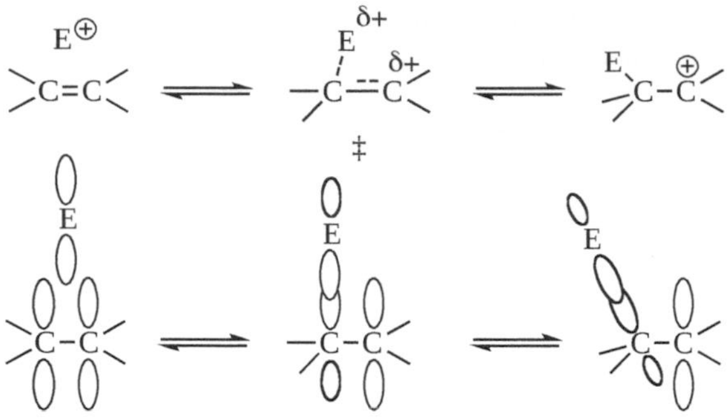

Figure 7.3 Electrophile addition to a double bond with the transition state shown in the center.

Media: Most commonly acidic since good electrophiles are necessary for reaction. Occasionally neutral, rarely basic.

Solvent: Often polar to stabilize carbocation.

Steric: Will direct electrophile attack only if there is no difference in the two possible carbocations formed.

Variations: The electrophile is commonly a proton. The reaction is usually uphill in energy unless the acid is very strong or the cation is very stable, and can easily reverse since most carbocations are very strong acids. As the carbocation becomes less stable, the equilibrium shifts toward reactants. This equilibrium can be shifted to form more of the carbocation by making the acid stronger. For protonation of a pi bond, see Figure 7.4 (but view the figure from right to left).

7.1.5 Path D_E, Electrofuge Loss from a Cation To Form a Pi Bond

(Dissociation, Electrofugic, the Reverse of the Previous Reaction)

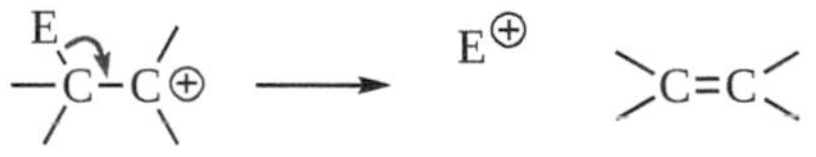

Path Cross-checks

Energetics: Downhill unless the electrophile is less stable than the carbocation. *Overlap*: Since a pi bond is formed, the C–E bond broken must be roughly coplanar with the carbocation empty *p* orbital so a good pi bond is formed (Figure 7.3, right to left).

Path Details

Media: Acidic or neutral.

Electrophile lost (electrofuge): Must be reasonably stable compared to carbocation.

Solvent: Polar to stabilize cation.

Steric: Little problem.

Variations: The electrophile lost is commonly a proton. This reaction is downhill in energy unless the A–H is a very strong acid. Figure 7.4 shows the orbital overlap.

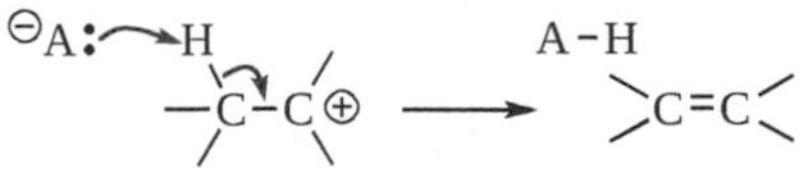

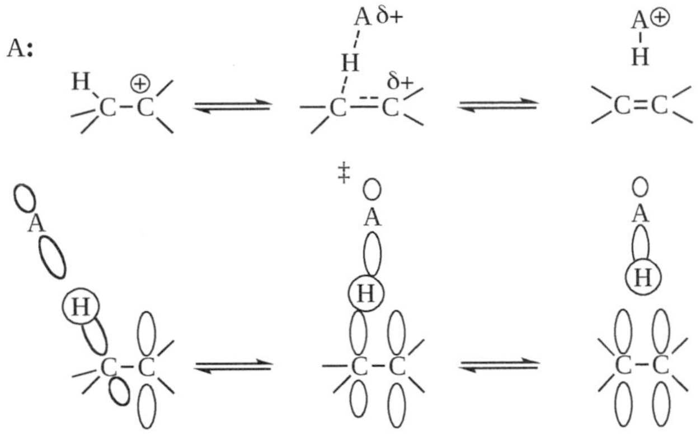

Figure 7.4 Loss of a proton from a carbocation with the transition state shown in the center.

7.1.6 Path S_N2, The S_N2 Substitution

(Substitution, Nucleophilic, Bimolecular)

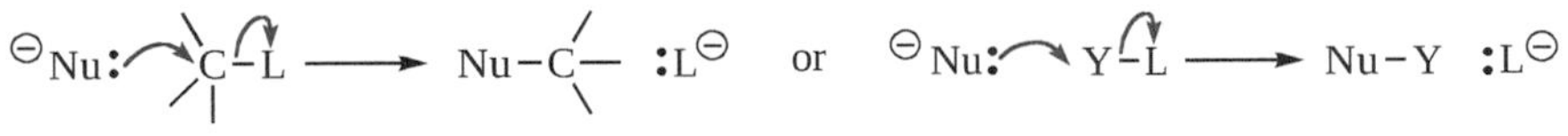

The leaving group is forced out by the nucleophile, as introduced in Section 4.2.7. The transition state is a five-coordinate carbon. **The tetrahedral configuration at the carbon atom is inverted.** The transition state has a carbon *p* orbital partially bonded to the nucleophile on one side and the leaving group on the other (back side attack, see Figure 7.5). The "2" in the name S_N2 indicates that there are two reactant molecules involved in the rate-determining step (the nucleophile and the molecule attacked).

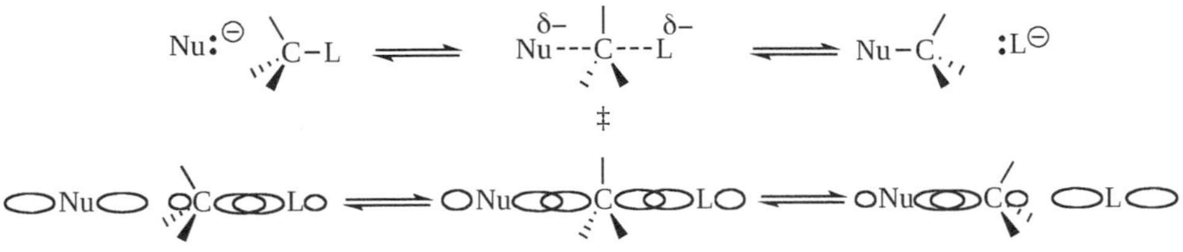

Figure 7.5 The S_N2 substitution with the trigonal bipyramidal transition state shown in the center.

Path Cross-checks

Energetics: Do not kick out a leaving group that is more than 10 pK_a units more basic than the incoming nucleophile. This substitution version of the ΔpK_a rule cross-check can also be used to decide whether the forward or reverse reaction is favored, since the reaction tends to form the weaker base.

Leaving group: Good or better, occasionally fair ($pK_{aHL} \leq 12$). An exception is epoxides, where the strained three-membered ring is destabilized relative to ring opening.

Steric: It is critical that the nucleophile have good access to the carbon bearing the leaving group. Methyl, CH_3–L, and primary, CH_3CH_2–L, are best; unhindered secondary, $(CH_3)_2CH$–L, is possible. **Tertiary**, $(CH_3)_3C$–L, **does not react by this pathway**. Neopentyl, $(CH_3)_3C$–CH_2–L, also fails to react by this path because a methyl group always blocks the nucleophile's line of approach. **Leaving groups on a trigonal planar center like aromatic, Ar–L, and vinylic, C=C–L, leaving groups do not react via the S_N2 process.**

Path Details (see Section 4.2.7 for more discussion)

Overlap: Nu, carbon attacked, and L are usually close to collinear; distortions are tolerated, for three-membered rings form easily.

Media: Basic or neutral, occasionally acidic.

Solvent: Polar or medium polarity.

Special substrates: Allylic, H_2C=CH–CH_2–L, benzylic, PhCH_2–L, and :Z–CH_2–L react even more quickly than methyl. The overlap of the *p* orbital (of the double bond, aromatic, or lone pair) with the adjacent *p* orbital of the S_N2 transition state forms a pi-type bond that lowers the energy of the transition state, speeding up the reaction.

7.1.7 Path E2, The E2 Elimination

(Elimination, Bimolecular, Concerted Loss of a Proton and Leaving Group)

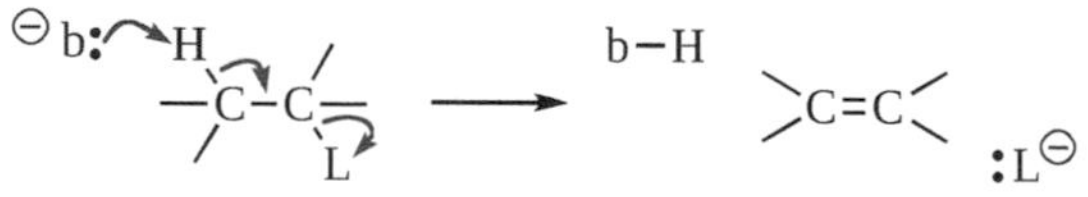

Path Cross-checks (see Section 4.3 for an overview of elimination)

Energetics: Use elimination version of the ΔpK_a rule cross-check that the leaving group should not be more than 10 pK_a units more basic than the incoming base.

Overlap: Both the C–H bond and the C–L bond must be roughly coplanar (usually achieved by free rotation), shown in Figure 7.6, to form an untwisted pi bond. Forming an unstable, twisted pi bond takes more energy, so other processes such as substitution can successfully compete.

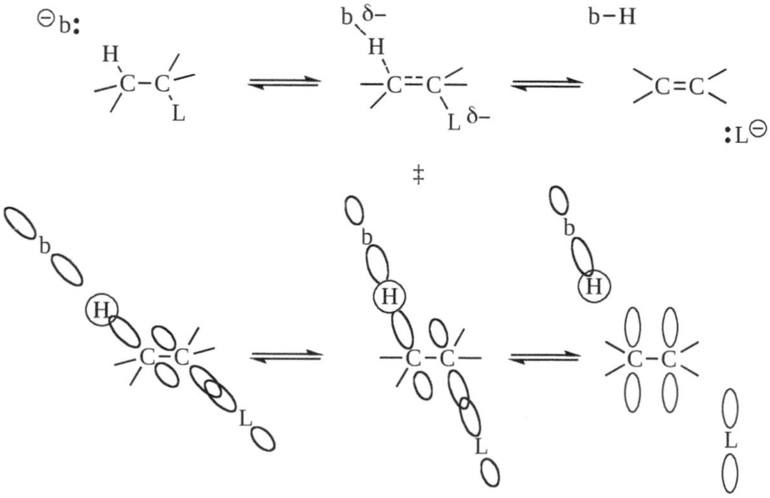

Figure 7.6 The E2 elimination with the transition state shown in the center.

Path Details

Leaving group: Fair or better.

Media: Normally basic, occasionally neutral, rarely acidic.

Solvent: Medium polarity.

Steric: Base must be able to approach the C–H bond easily. For this reason, the *anti* elimination, H and L on opposite sides, is preferred over *syn*-elimination, H and L on the same side. The leaving group crowds the incoming base in the *syn*.

Variations: A heteroatom can replace either carbon so that the elimination forms a carbon–heteroatom pi bond (Section 4.3.7). For other variations, see Section 7.4.1.

7.1.8 Path Ad_E3, The Ad_E3 Addition

(Addition, Electrophilic, Trimolecular, and reverse of the above E2 elimination)

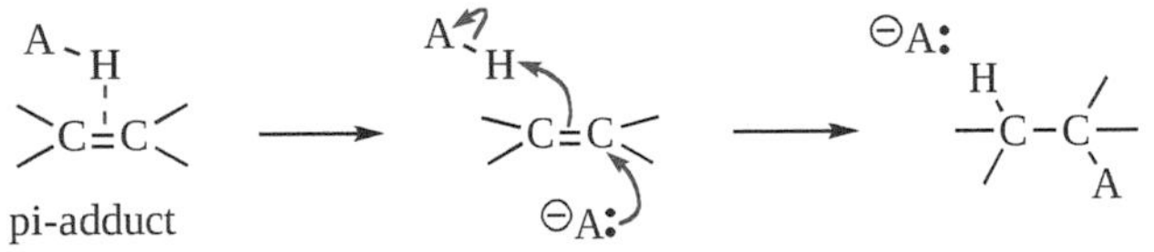

Electrophile complexation to the pi bond or hydrogen bonding to the pi bond serves further to polarize and activate the double bond toward an otherwise slow nucleophilic attack. (A dotted line indicates only complexation; it does not change the electron count, and should not be confused with a solid line describing a bonding pair of electrons.)

Path Cross-checks

Energetics: Favorable in acidic media because the $\Delta H°$ is usually exothermic. The H–A is usually pi-complexed or hydrogen bonded to the pi bond to activate it for nucleophilic attack. A three-molecule collision of HA, pi bond, and Nu would be rare. Expect Ad_E3 when other addition routes like the Ad_E2 are disfavored (Section 4.4.2).

Path Details

Overlap: The H added, the pi bond, and the A^- must be roughly coplanar. Figure 7.6, viewed from right to left, with the substitution of A for both b and L, approximates the overlap requirements. The A^- attacks the largest partial plus of the pi-adduct. Markovnikov's rule is followed even though there is no full positive charge formed.
Media: normally acidic, occasionally neutral.

Steric: The nucleophile is often the conjugate base of the acid, A^-, and must be able to approach the pi bond easily. For this reason, *anti* addition is preferred over *syn*. The H–A complexation crowds the incoming nucleophile in the latter.

7.1.9 Path Ad_N, Nucleophilic Addition to a Polarized Multiple Bond

(Addition, Nucleophilic)

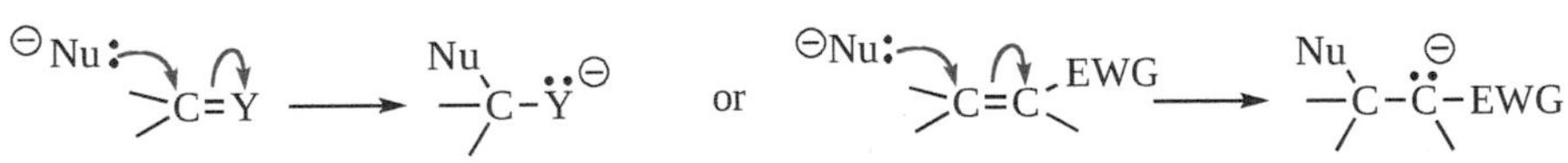

In this pathway, electrons flow from the source to the multiply bonded carbon, break the pi bond, and produce a stable anion. An electronegative carbon atom, C–EWG, can replace an electronegative heteroatom, Y, and the electron flow does not change. When the electron sink is a carbon–carbon multiple bond with an electron-withdrawing group, the reaction is called *conjugate addition*. The electron-withdrawing group is essential, for without a means to stabilize the carbanion formed, electron flow cannot occur.

Path Cross-checks (see Section 4.4.5 for more discussion)

Energetics: Utilize the ΔpK_a rule cross-check. The anion formed must usually be more stable than the Nu or the reaction will reverse by path E_β, the microscopic reverse of this path. However, in a protic solvent the initially formed anion may be rapidly and irreversibly protonated before the reverse reaction can occur. In acidic media, protonation of Y can also occur prior to or simultaneous with nucleophilic attack (see Section 7.4.3, Extent of Proton Transfer Variations).

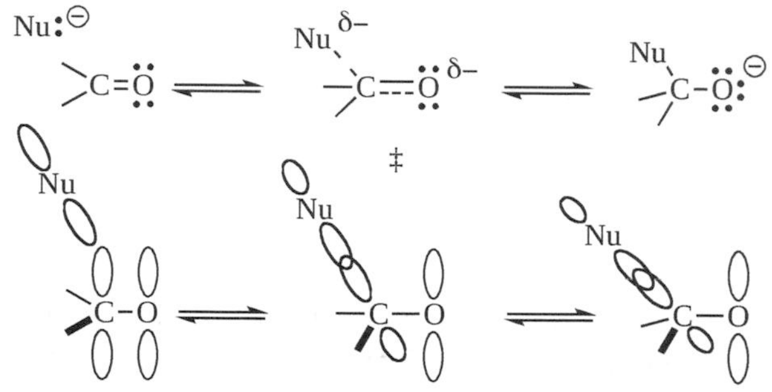

Figure 7.7 Ad_N addition to polarized multiple bond with the transition state shown in the center.

Path Details

Overlap: Nu must approach the δ+ carbon attacked from behind and in the plane of the *p* orbitals of the pi bond (Figure 7.7).

Media: No limitation; however, if the Nu gets protonated it's no longer a good Nu.

Solvent: Often polar to stabilize the anion formed.

Steric: The approach of the Nu to the trigonal planar sink is not usually blocked; the Nu attacks the carbonyl from either top or bottom. However, rate differences are common. Example: Aldehydes are much more reactive than ketones toward Nu attack.

Variations: Because for R–C≡Y, the two pi systems of the triple bond are perpendicular to each other, they behave independently. The pathway is the same as for double bonds with few path differences:

Overlap: Not a problem because of the cylindrical shape of the triple bond.

Resultant anion: Slightly more stable being sp^2 hybridized (for the same Y).

Steric: Usually not a problem.

7.1.10 Path E_β, Beta Elimination from an Anion or Lone Pair

(The E_β is the reverse of the previous Ad_N addition)

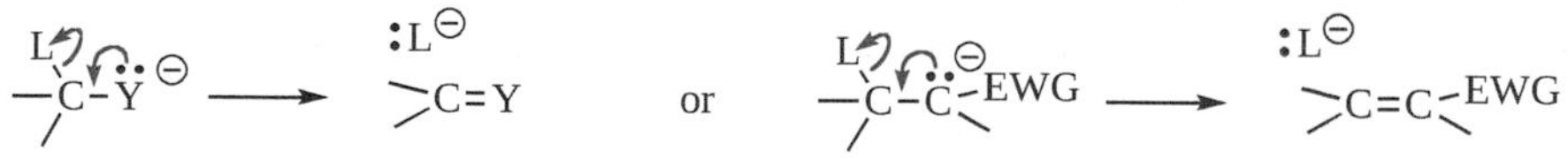

Path Cross-checks

Energetics: Use ΔpK_a rule cross-check (Sections 3.6 and 4.5.2). The leaving group should not be more than 10 pK_a units more basic than the anion pushing it out.

Overlap: Anion *p* orbital and C–L bond must be roughly coplanar to form an untwisted pi bond (usually not a problem if free rotation is possible). View Figure 7.7 from right to left.

Path Details

Leaving group: Can be fair to poor if the pushing anion is less stable. This means a powerful source is able to force out a poorer leaving group. Use the ΔpK_a rule cross-check.

Solvent: Polar to stabilize anion.

Media: No limitation. However, if the anion or lone pair becomes protonated, the electron pair needed to push out the leaving group is lost.

7.1.11 Path 1,2R, 1,2 Rearrangement of a Carbocation

Rearrangement will usually occur whenever a full positive charge is formed on a carbon that has an adjacent group capable of shifting over to it to form an equal or more stable cation. If the carbon bears only a partial plus, the tendency to rearrange is less. Rearrangement of alkyl groups to an anionic or radical center does not occur.

Path Cross-checks

Energetics: An energetically favorable rearrangement produces a more stable cation (further discussion in Section 4.7). Carbocations of similar stability can also be formed by migration, but rarely is a significantly less stable carbocation formed this way.

Overlap: The most important path restriction for rearrangements is that **the migrating group must maintain orbital overlap with both the atom it migrates from and the atom it migrates to**. At the transition state, the migrating group orbital will overlap with orbitals on **both** atoms, as shown in Figure 7.8. Usually a group migrates to the neighboring atom from the one it starts out on—a 1,2 shift.

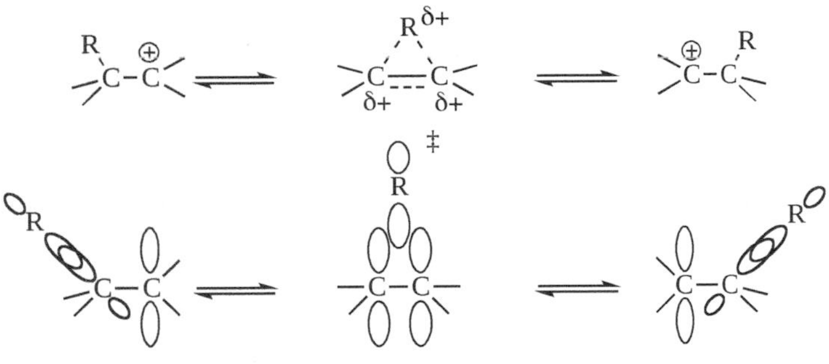

Figure 7.8 The 1,2 rearrangement of a carbocation. The transition state is in the center.

Path Details

Media: Commonly acidic, occasionally neutral, and rarely basic.

Solvent: Polar to stabilize cation.

Migratory aptitudes: At the transition state the migrating group bears a partial positive charge. In freely rotating systems, the groups that can best tolerate this partial plus migrate the best; the order is $-H > -Ph > -CR_3 > -CHR_2 > -CH_2R > -CH_3$. However, in rigid or conformationally restricted systems, the migrating group is the one that is best able to achieve proper alignment with the *p* orbital of the carbocation, regardless of the previous trend.

7.1.12 Path 1,2RL, 1,2 Rearrangement with Loss of Leaving Group

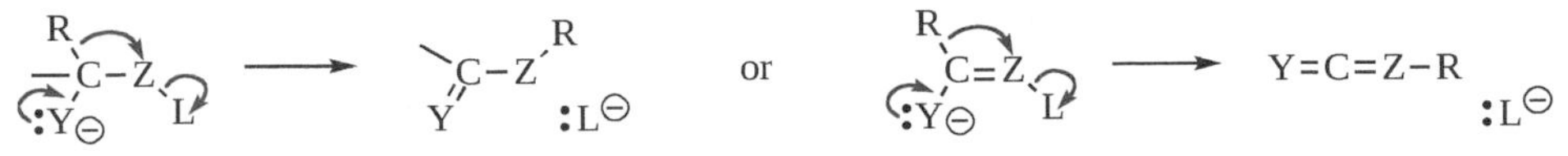

Often rearrangement occurs as shown with a path reminiscent of the generic S_N2, a backside displacement of the leaving group by the migrating group (further discussion in Section 4.7). There are some rearrangements in which it appears that the leaving group departs first, as in the S_N1 rather than the S_N2, so that the electron sink the group migrates to is the carbocation (we would then call it path D_N followed by 1,2R).

Path Cross-checks

Energetics: This path avoids forming an unstable cation and therefore is a useful alternative to path D_N followed by path 1,2R. Expect to use this path if a D_N path fails its cross-checks because of cation instability; for instance, if it would have formed a primary carbocation, or formed an electronegative heteroatom with an incomplete octet.

Overlap: As shown by Figure 7.9, the migrating bond and the leaving group bond must lie in the same plane for good overlap. Again, the migrating group slides from the starting orbital to the neighboring orbital, maintaining overlap as it goes.

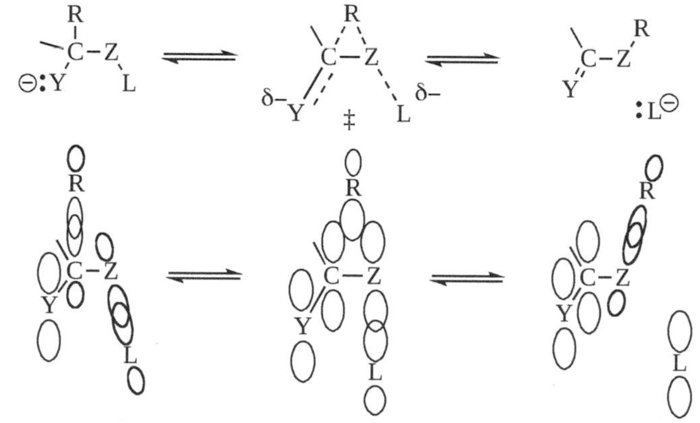

Figure 7.9 Rearrangement with loss of a leaving group showing the transition state in the center.

Path Details

Leaving group: Excellent to fair. With excellent leaving groups, the pushing Y^- group is not needed and can be replaced by H or R. Heteroatom Z can be replaced by a carbon. Cross-checks if no pushing Y: good L, form a cation usually secondary or better, polar solvent.

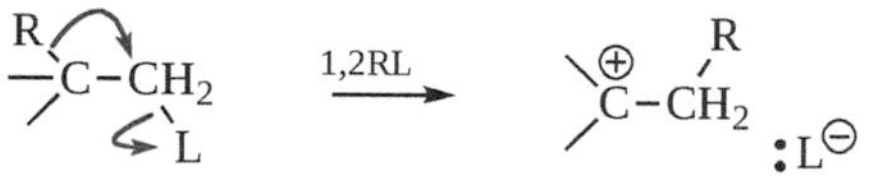

Variations: Sometimes the leaving group can be replaced by a polarized multiple bond as the electron sink, as shown in the following reaction. In essence, rearrangement can proceed with any of our three general classes of electron sinks: an empty *p* orbital, a weak single bond to a leaving group, or a polarized multiple bond. See Section 4.7 for more examples.

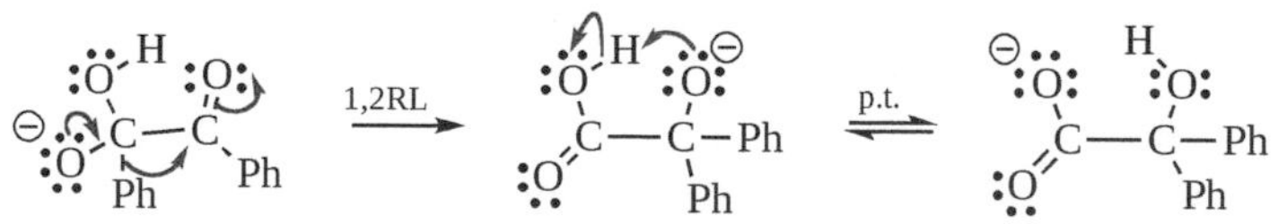

7.2 SIX MINOR PATHWAYS

7.2.1 Path pent., Substitution via a Pentacovalent Intermediate

(For Atoms Capable of Pentacoordinate Bonding Only, Not for Carbon)

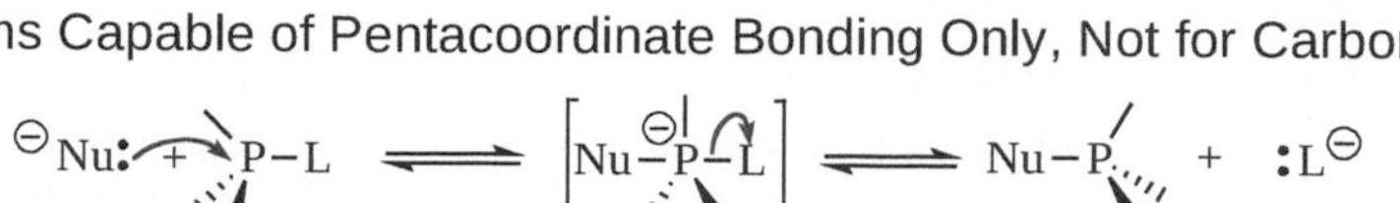

A nucleophile adds, forming a trigonal bipyramidal pentacovalent intermediate that then ejects the leaving group in the microscopic reverse of the nucleophilic addition reaction. The nucleophile, central atom, and leaving group are collinear and are on axial (or apical) bonds. The pentacovalent intermediate is often short-lived with good leaving groups and is enclosed in brackets. If the intermediate lifetime gets very short, then this path becomes indistinguishable from an S_N2. Three common elements that can react by this pathway are silicon, phosphorus, and sulfur. Hydrolysis of phosphate esters like RNA occurs by this path.

Path Cross-checks

Energetics: Pentacoordinate intermediates are possible for third-row or higher elements, like Si, P, and S, that have available *d* orbitals. This path is not for second-row elements like carbon. The leaving group can't be more than 10 pK_a units more basic than the nucleophile.

Path Details

Overlap: Attacking Nu forms an axial bond; therefore, by microscopic reversibility, the departing L must leave from an axial position. The orbital overlap is similar to that shown for the S_N2 (Figure 7.5), but the structure in the center of the figure is now an intermediate rather than a transition state.

Leaving group: Good to fair. If the leaving group is very good, the intermediate would be expected to have a short lifetime and may become just a transition state (S_N2).

Media: Acidic, basic, or neutral.

Solvent: Commonly polar or medium polarity.

Steric: Access usually not a problem since bond lengths are longer.

7.2.2 Path 6e, Concerted Six-Electron Pericyclic Reactions

The largest class of thermal concerted reactions involves six-electron cyclic transition states. Reactions include cycloadditions (two pieces forming a ring) and rearrangements. Although the following reactions may all look different, they all involve six electrons (three arrows) going around in a circle (pericyclic). These pericyclic reactions are discussed in more detail with their cross-checks in Chapter 13.

Thermal rearrangements:

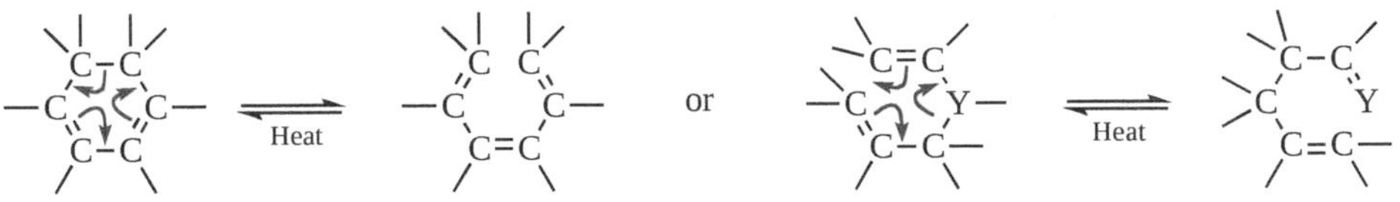

Thermal cycloadditions or cycloreversions (the reverse reaction):

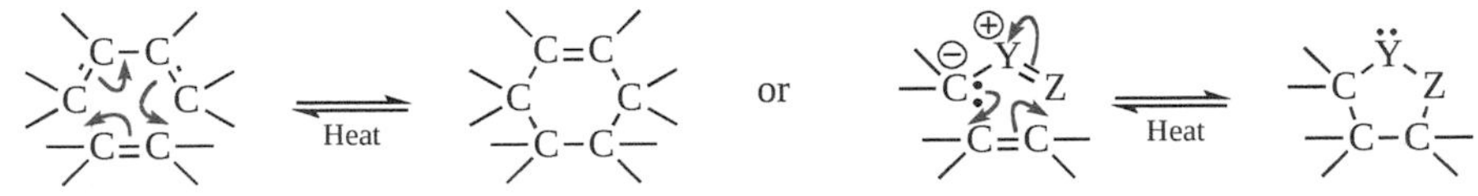

Metal-chelate-catalyzed additions:

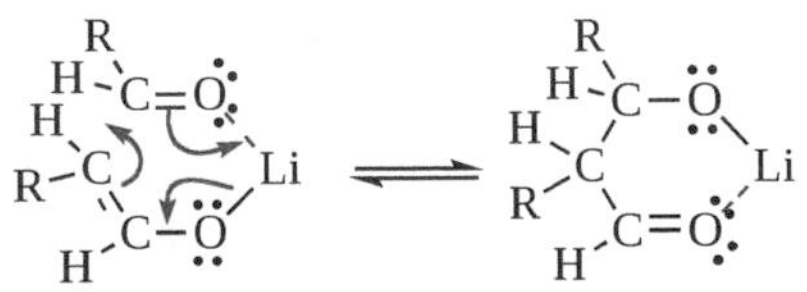

The transition state for these reactions has been described as aromatic since it has six electrons in orbitals that form a loop. In the neutral cases, the direction that the arrows are drawn does not matter. The reverse reactions of all of these examples also occur. The major path limitation is to achieve the proper alignment for the cyclic transition state. There is a large diversity of compounds that react by this pathway; in many of the reactions, there are examples of reactions where a carbon atom has been replaced by a heteroatom or vice versa.

There are many reactions in which a metal ion is present and serves to hold the reacting partners together by complexation. Some of the metals for which this process is common are lithium, magnesium, aluminum, and the metalloid boron. In the following hydride transfer example, in addition to serving as a counterion for $R_2HC{-}O^-$, aluminum acts as a Lewis acid catalyst, complexing the carbonyl and making it a better electron sink. This reaction equilibrium is driven by what solvent the reaction is run in. Run the reaction in isopropanol; it will reduce added ketones. Run the reaction in acetone; it will oxidize added alcohols.

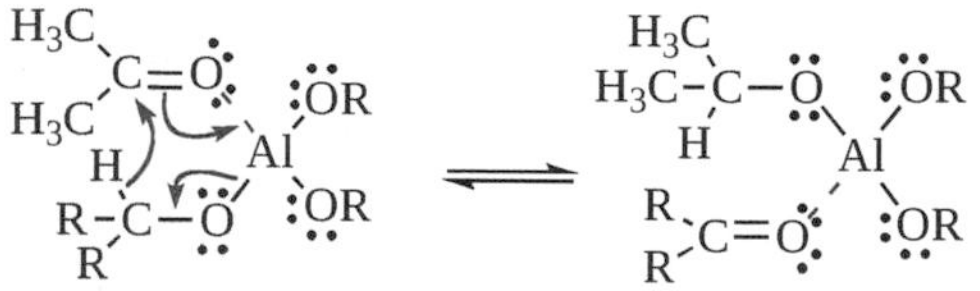

Path Cross-checks and Details

Overlap: These six-electron pericyclic reactions have very ordered transition states, so achieving that arrangement is necessary for the success of the reaction.

Energetics: Thermal cycloaddition reactions often are exothermic (negative $\Delta H°$ term), but since the product usually has fewer degrees of freedom than the reactant, also have a negative $\Delta S°$ (making the $-T\Delta S°$ term positive). Since $\Delta G° = \Delta H° - T\Delta S°$, at higher temperatures the negative enthalpy term can be overwhelmed by the positive entropy term. The reaction can reverse direction as $\Delta G°$ goes from negative (downhill) to positive (uphill).

7.2.3 Path Ei, Thermal Internal *Syn* Elimination

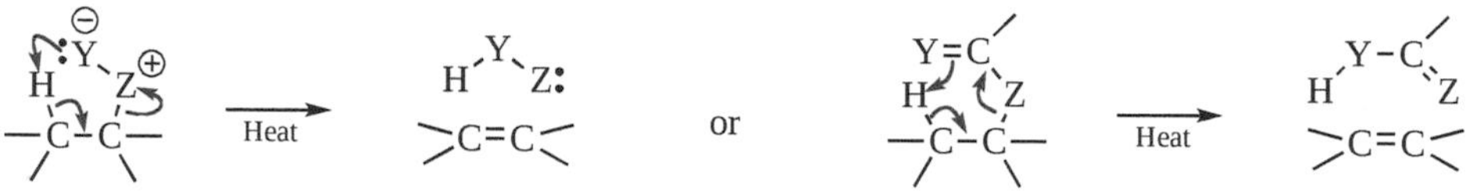

This reaction is just a thermal, internal *syn* E2 elimination and a relative of path 6e. There are many possible reactants for this internal *syn* elimination path. Five-membered transition state examples include Y equaling oxygen and Z being $^+NR_2$ (amine oxides), SPh (sulfoxides), or SePh (selenoxides). Six-membered transition state examples include both Y and Z being oxygen (esters) or Y is sulfur and Z being oxygen (xanthates).

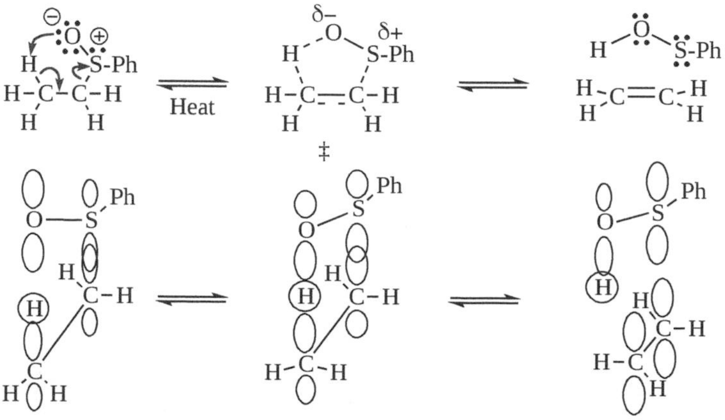

Figure 7.10 Thermal internal *syn* elimination with the transition state shown in the center.

Path Cross-checks

Overlap: Like the E2, the C–H bond and the C–Z bond involved in the elimination must be roughly coplanar for strong, untwisted pi bond formation (Figure 7.10).

Energetics: Higher temperatures are needed if the leaving group is poorer. Selenoxides eliminate easiest, sulfoxides are next, and amine oxides require the highest temperatures.

Path Details

Anion: Average.

Leaving group: The weaker the C–Z bond strength, the better the leaving group.

Media and solvent: No limitation; can occur in the gas phase.

Variations: An additional six-membered transition state example is cyclic decarboxylation. The overlap limitation for the new carbon–carbon pi bond requires the breaking C–C bond to align coplanar with the carbon *p* orbital of the carbonyl.

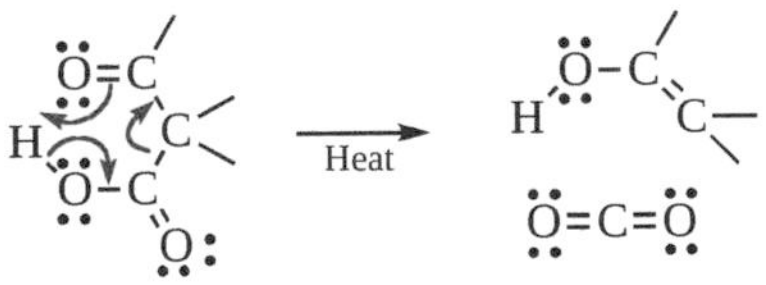

7.2.4 Path NuL, Nu–L Additions (Three-membered ring formation)

This pathway shows some variation in the timing of when the leaving group falls off. Occasionally the addition may be concerted as shown (with bromine, for example), but more often the leaving group loss and the nucleophilic attack on the pi bond are separate steps. The leaving group may fall off before attack (a carbene if Nu is carbon).

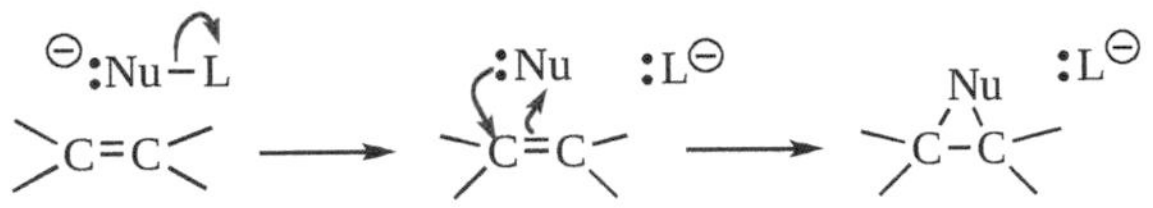

If the multiple bond is capable of stabilizing a carbocation, as in erg–C=C, then the leaving group may depart with nucleophilic attack.

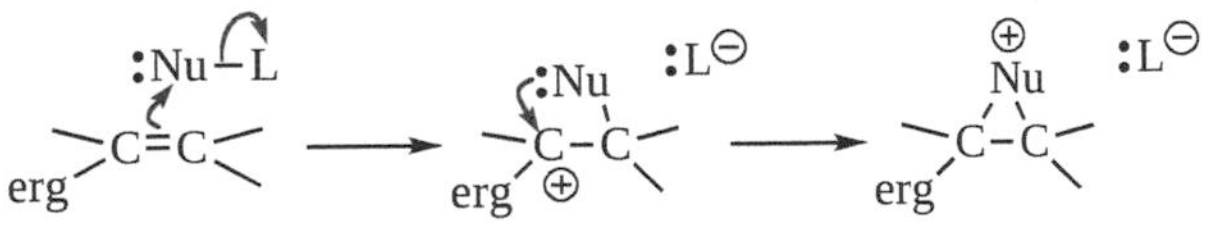

If the multiple bond is capable of stabilizing an anion, as C=Y is, then the leaving group may fall off after nucleophilic attack (for a thorough discussion, see Section 8.12).

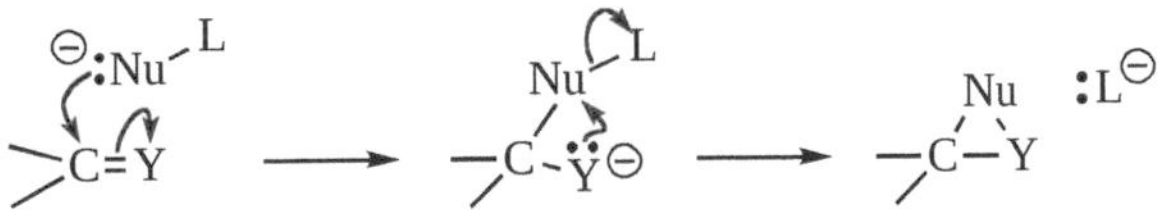

Path Details (Concerted)

Overlap: Nu–L must approach the pi bond from above or below and not from the side. The Nu, the L, and the double-bond carbons should be close to coplanar, as shown in Figure

7.11. The nucleophile and leaving group orbitals are close to being perpendicular, so they react independently. A concerted example is shown, but most reactions are stepwise.

Leaving group: Good to fair, may leave before addition occurs.

Steric: Usually no problem as the alkene is flat.

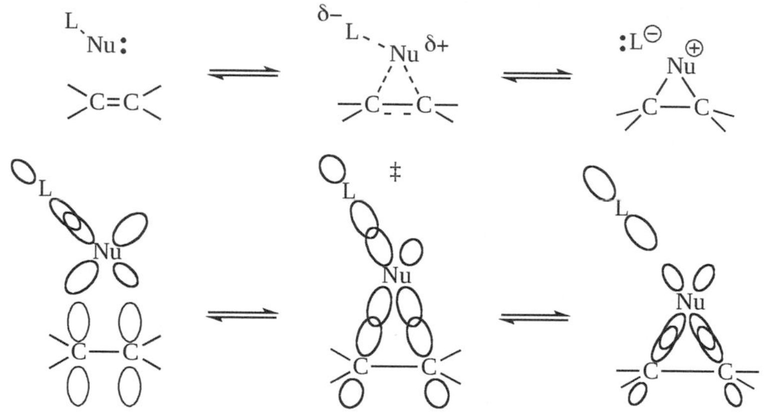

Figure 7.11 The addition of a Nu–L species to form a three-membered ring with the transition state shown in the center.

7.2.5 Path 4e, Four-Center, Four-Electron

Four-center, four-electron processes do not occur thermally. The transition states for these unfavorable reactions have been described as having antiaromatic destabilization because they have four electrons in a normal closed loop. There are **three exceptions** that follow this path; all have some unusual orbital arrangement to allow them to bypass the problem of antiaromatic destabilization of their transition states. The first example is the addition of borane to pi bonds (Figure 7.12). The less common but similar trivalent aluminum hydrides, such as diisobutylaluminum hydride, $[(CH_3)_2CHCH_2]_2AlH$, add by this pathway. The overlap path limitation is that all four atoms must be roughly coplanar.

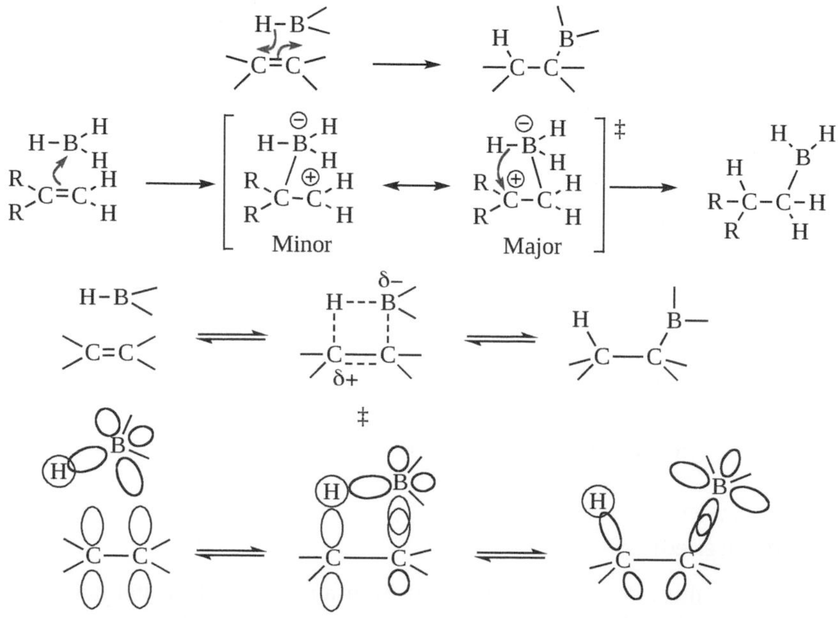

Figure 7.12 Addition of a borane to a double bond with the transition state shown in the center.

A pi-adduct of the empty *p* orbital on boron with the double bond forms first, then collapses by this four-electron process to the product. Boron is the electrophile, and hydrogen is the nucleophile (boron is less electronegative than hydrogen). Because the nucleophile and electrophile orbitals are perpendicular (and therefore do not interact), the loop is not closed. The direction (regiochemistry) of boron's electrophilic addition is determined by the formation of the most stable partial plus.

Alternatively, the two bond formation steps may not be synchronous but occur so close together in time that the intermediate has an insignificant lifetime. Electron density from the double bond flows into the empty orbital on boron, creating an electron-deficient center on a carbon to which the now electron-rich boron can donate a hydride. The addition is *syn*: both boron and hydrogen add to the same face of the double bond. More details about this hydroboration reaction are in Sections 8.9.1 and 8.12.1.

A second exception is the elimination of phosphine oxides from oxaphosphetanes. The pentacoordinate phosphorus serves as the electron source in an internal elimination of the oxygen bound to phosphorus. This phosphorus-oxygen bond is very strong, which makes this reaction energetically downhill. More details about this reaction are in Section 8.12.5.

The last exception has an unusual twist, literally. The transition state is a strange loop with a half-twist, a Möbius loop. In contrast to a normal loop, Möbius loops are predicted to be stable with 4*n* electrons in them. Given enough heat, the cyclobutene sigma bond twists open to form a diene. This and other pericyclic reactions are discussed in Chapter 13.

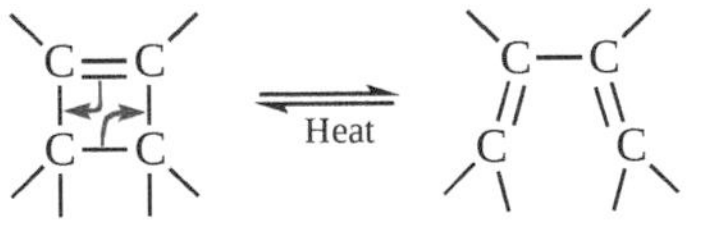

7.2.6 Path H$^-$ t., Hydride Transfer to a Cationic Center

Hydride has the electronic configuration of helium, but is still relatively reactive. Salts like sodium hydride are very basic because hydrogen, the conjugate acid, has a pK_a of 35. With that high a pK_{abH}, hydride makes a very poor leaving group. However, a carbocation can abstract a hydride to form an equal or more stable carbocation. It can be viewed as the hydrogen atom with its two electrons being passed between the two sites, maintaining bonding overlap the whole time, and never actually being set free in solution.

Path Cross-checks

Overlap: Only that the carbocation must approach within bonding distance to the hydride transferred. This pathway is related to path 1,2R, rearrangement of a carbocation, in which a group migrates with its pair of bonding electrons to an adjacent carbocation. In this minor path, a hydrogen with its bonding electron pair is passed to a nonadjacent carbocation; the empty *p* orbital of the carbocation must get close enough to allow the hydride to be partially bonded to both atoms at the transition state, as in Figure 7.13.

Path Details

Carbocation formed: of equal or greater stability than starting carbocation.

Media: Acidic or occasionally neutral.
Solvent: Commonly polar or medium polarity.
Steric: Not a problem because the carbocation is flat.
Energetics: An energetically favorable transfer forms a more stable carbocation. Similar stability carbocation formation is also facile, but not formation of a less stable carbocation.
Related path: If the carbocation is replaced by a partial plus center, then a pushing Y^- is required on the hydride donor, and the reaction tends to go by path 6e (see last example in Section 7.2.2).

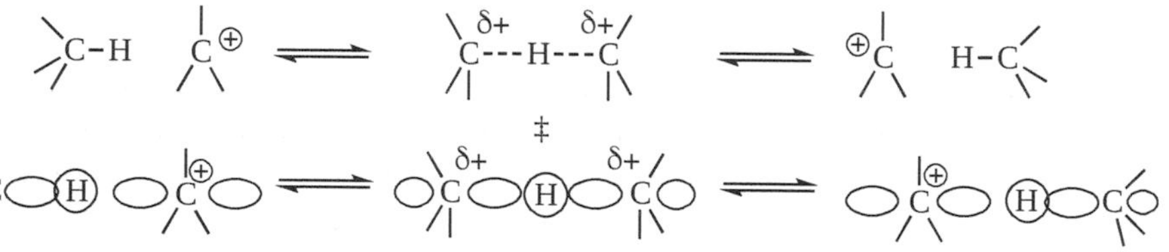

Figure 7.13 Hydride abstraction by a cation with the transition state in the center.

Biochemical Example: NADH Reduction of Acetaldehyde

Nicotinamide adenine dinucleotide (NADH) is a biochemical source of hydride. In the following example NADH reduces acetaldehyde to ethanol via minor pathway H^- t., hydride transfer to a cationic center. A Zn^{2+} ion acts as a Lewis acid to polarize the acetaldehyde carbonyl (similar to protonating the carbonyl). The Lewis acid makes the carbonyl a better electron sink by increasing the partial positive charge on carbon. In fact, the electrophilic catalysis by 2+ and 3+ metal ions can accelerate additions to carbonyls by over a million times. The formation of the aromatic pyridinium ring in the NAD^+ product helps balance the energetics of this easily reversible reaction.

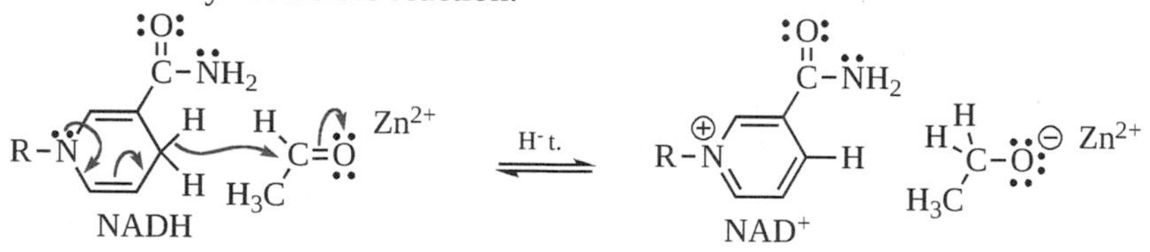

Cross-checks for Suspected Additional Minor Paths

Any suspected new electron flow path should be well tested before it is accepted as a new pathway. It may be just a combination of (Section 7.3) or a variation on (Section 7.4) already known paths. Watch for $\Delta S°$ problems if too many things are happening at once. Look for steric and strain problems and check the orbital alignment with molecular models—orbitals that will become double bonds in the product must be able to get close to coplanar in the starting material. Check any intermediates for stability, especially if charged. Check that the electronics fit HOMO–LUMO (Section 13.2.3) and HSAB theory (Section 2.4). Check the energetics with a $\Delta H°$ calculation or ΔpK_a rule cross-check. Be skeptical anyway, it's healthy.

7.3 COMMON PATH COMBINATIONS

Several pathway combinations occur so frequently that they have been given a name. Most were introduced and discussed in Chapter 4 and are grouped here as a summary. **These "phrases" in your mechanistic vocabulary are the next step toward your eventual construction of a grammatically correct mechanistic "sentence."**

7.3.1 S_N1 (Substitution, Nucleophilic, Unimolecular), $D_N + A_N$

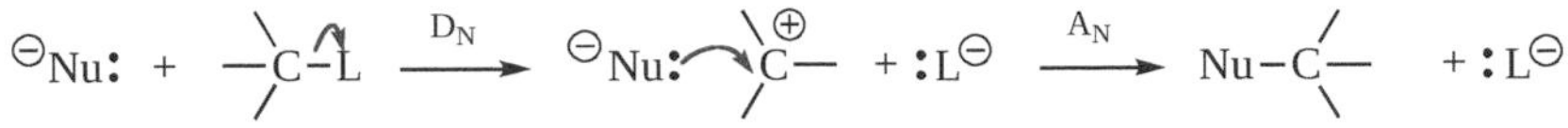

The rate-determining ionization of a good leaving group, path D_N, produces a carbocation-leaving group ion pair that may or may not fully dissociate before the carbocation is trapped by a nucleophile, path A_N. Substrates that would make a poor cation, CH_3–L, and Ar–L, do not react via the S_N1 process. The nucleophile attacks the carbocation or ion pair and gives the product in a fast second step. If the leaving group were on a chiral center, a racemic mixture can result because the nucleophile can attack the free carbocation from either the top or bottom face equally. Often, however, nucleophilic attack on the ion pair results in inversion, since the leaving group partially blocks one face. If the nucleophile is the solvent, this process is called solvolysis (see Section 4.2 for more on the various substitution types).

Lone-Pair-Assisted S_N1, $E_\beta + A_N$

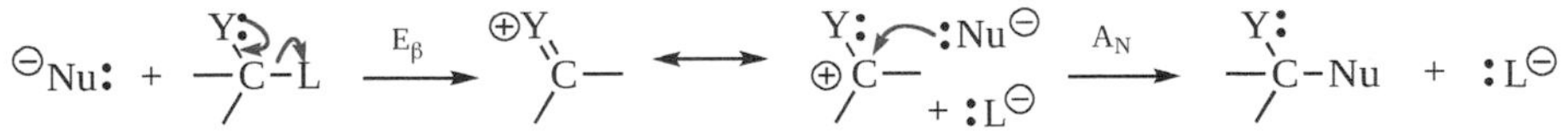

The lone-pair-assisted S_N1 has a properly aligned lone pair to expel the leaving group, path E_β. The resultant delocalized cation is then trapped by a Nu, path A_N. The E_β tolerates poorer leaving groups because it has more "push" from the Y lone pair. Use the ΔpK_a cross-check.

7.3.2 Ad_E2 (Addition, Electrophilic, Bimolecular), $A_E + A_N$

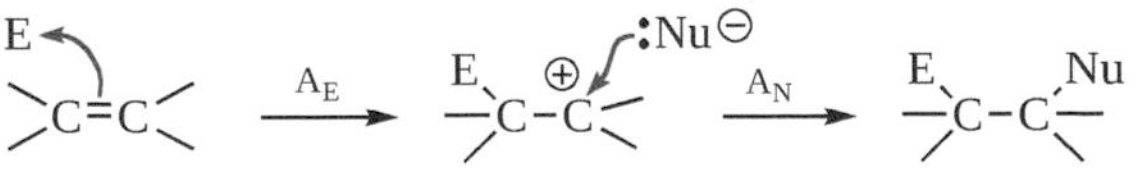

The Ad_E2 is the major addition pathway of alkenes and dienes. Electrophile addition to a pi bond or protonation of a pi bond, path A_E, produces the most stable cation, which is then trapped by a nucleophile, path A_N. (Section 4.4.2 discusses the various addition types). If the electrophile is a proton, this reaction is the reverse of the E1 reaction. The reaction commonly produces a mixture of *syn* and *anti* addition.

Hetero-Ad_E2, p.t. + Ad_N

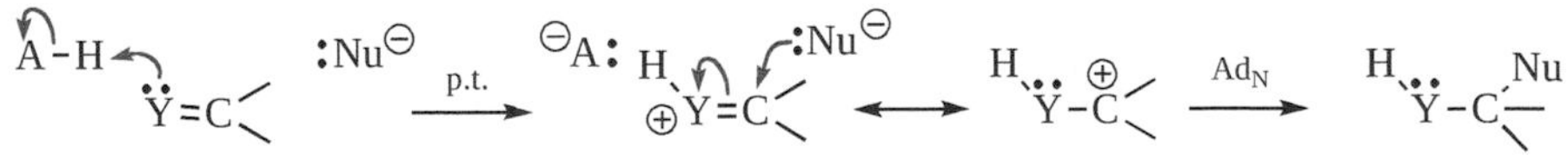

This reaction is a minor variant of the Ad_E2 in which the electrophile, usually a proton, attacks the lone pair rather than the less available heteroatom-carbon pi bond. The lone pair is protonated, path p.t., to give a highly polarized multiple bond that can be viewed as a stabilized carbocation. The nucleophilic attack on this carbocation could also be viewed as path A_N, trapping of a cation by a nucleophile, instead of path Ad_N.

7.3.3 E1 (Elimination, Unimolecular), $D_N + D_E$

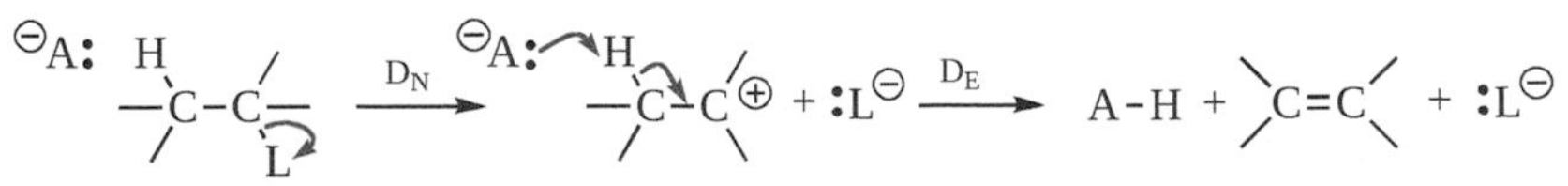

After rate-determining ionization of a good leaving group, path D_N, the cation is deprotonated by path D_E to produce a pi bond. This elimination is the reverse of the Ad_E2 addition. The E1 process competes with the S_N1 process. **If A^- acts as a base, E1 occurs; if it acts as a nucleophile, S_N1 occurs.** Section 9.5 will discuss such decisions in detail.

Lone-Pair-Assisted E1, E_β + p.t.

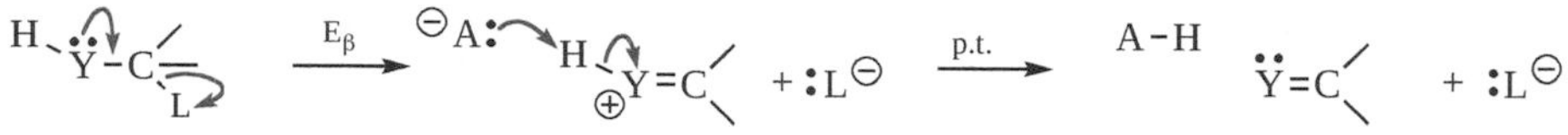

This reaction is the reverse of the hetero-Ad_E2 reaction. The lone-pair-assisted E1 uses a properly aligned lone pair to expel the leaving group, path E_β. The resultant cation is then deprotonated, path p.t. The lone-pair-assisted E1 has more "push" from the Y lone pair, so this elimination can tolerate poorer leaving groups. Use the ΔpK_a cross-check.

7.3.4 S_E2Ar Electrophilic Aromatic Substitution, A_E + D_E

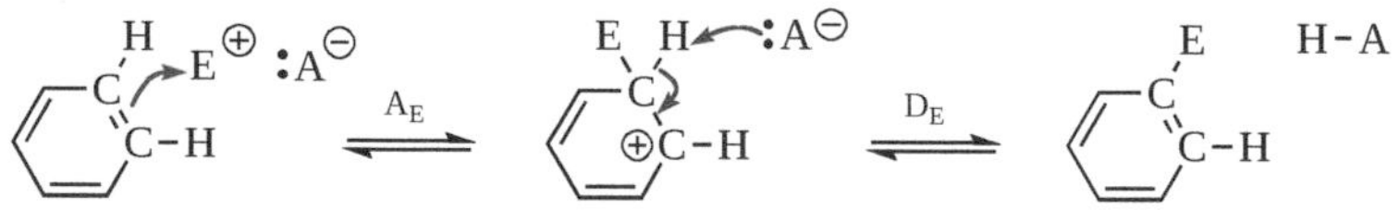

This reaction was introduced in Section 4.6, Electrophilic Substitution at a Trigonal Planar Center. The electrophile adds to the pi bond of the aromatic ring, path A_E, followed by deprotonation of the cation formed, path D_E, restoring aromatic stabilization. See Section 5.6 Aromatic Rings, for a discussion of electrophilic aromatic substitution on heteroaromatics like pyridine and on condensed aromatics like naphthalene.

7.3.5 E1cB (Elimination, Unimolecular, Conjugate Base), p.t. + E_β

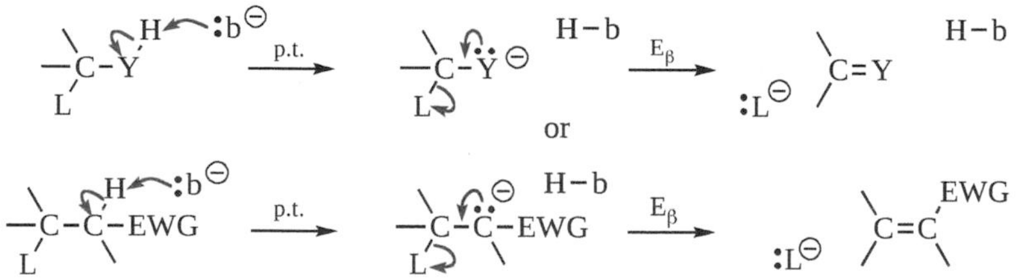

A stabilized anion forms first via path p.t., and then the conjugate base kicks out a fair leaving group in the slow step via the beta elimination from an anion path E_β. See Section 4.3 for more discussion of the various elimination types. Use the ΔpK_a cross-check.

7.3.6 Ad_N2 (Addition, Nucleophilic, Bimolecular), Ad_N + p.t.

This reaction is the reverse of the E1cB elimination above. A rate-determining addition of a nucleophile to the polarized multiple bond occurs by path Ad_N, producing a stabilized anion that is then protonated by path p.t. See Section 4.4 for more discussion of the various addition types. Use the ΔpK_a cross-check.

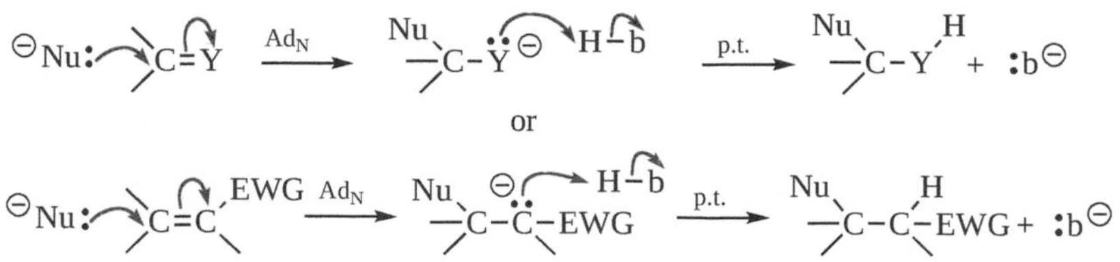

7.3.7 Addition–Elimination, $Ad_N + E_\beta$

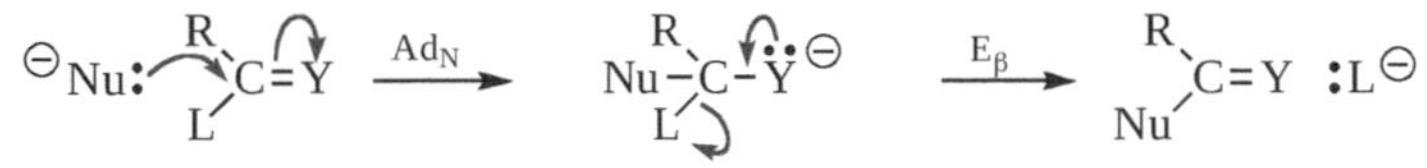

This is the almost exclusive route in basic media for replacement of a leaving group bound to a double bond. This reaction was introduced in Section 4.5, Nucleophilic Substitution at a Trigonal Planar Center. The most common substrates for this reaction are the carboxyl derivatives. A nucleophile adds first via path Ad_N, and then the leaving group departs via the beta elimination from an anion path E_β. Use the ΔpK_a cross-check.

7.3.8 Tautomerization, taut.

$$\text{H–C–C=Z} \rightleftharpoons \text{C=C–Z–H}$$

Tautomerization is the shift of an H from a carbon adjacent to a carbon–heteroatom double bond to the heteroatom itself (and the reverse). It is an acid- or base-catalyzed equilibrium. Two examples are the keto/enol pair (Z = oxygen) and the imine/enamine pair (Z = nitrogen). Base catalysis goes via the enolate anion. Note: either resonance form of the intermediate is fine to use, but the arrows will be slightly differ between the two forms.

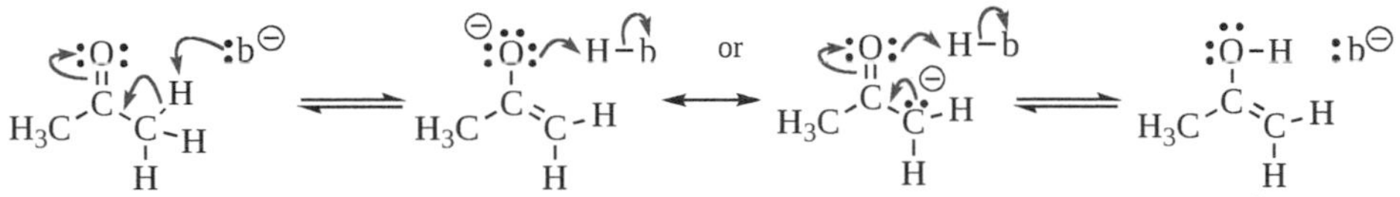

Keto — Major Enolate Minor — Enol

Acid catalysis goes via the lone-pair-stabilized carbocation.

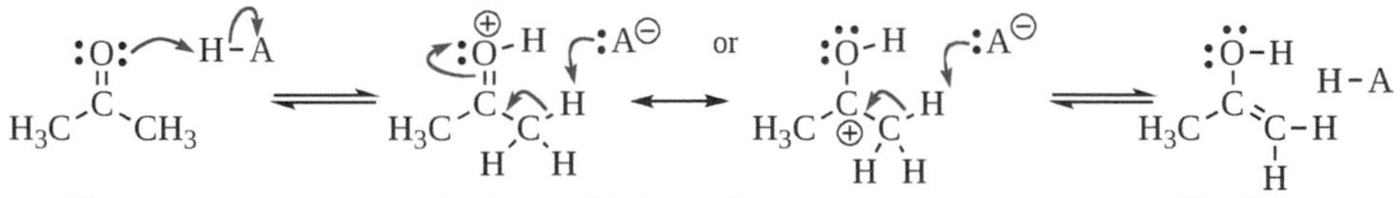

Keto — Major Carbocation Minor — Enol

Common Error

This shortcut certainly saves writing some structures and arrows, but there is good experimental evidence that this tautomerization definitely needs to be catalyzed by acid or base and will not proceed on its own. A four-center, four-electron process with very few exceptions (minor path 4e) does not occur thermally (see Section 7.2.5). The base- or acid-catalyzed process shown above, depending on pH, is much more reasonable. An energy surface of tautomerization, including a less common push-pull route, is in shown Figure 7.18.

Exercise

With the above structures as a guide, draw the arrows for the tautomerization of the enol back to the keto form for both acidic and basic media. What pathways did you use?

Answer: The acidic route from the enol to the keto form is path A_E then p.t. Note that this backward route uses the reverse of each path.

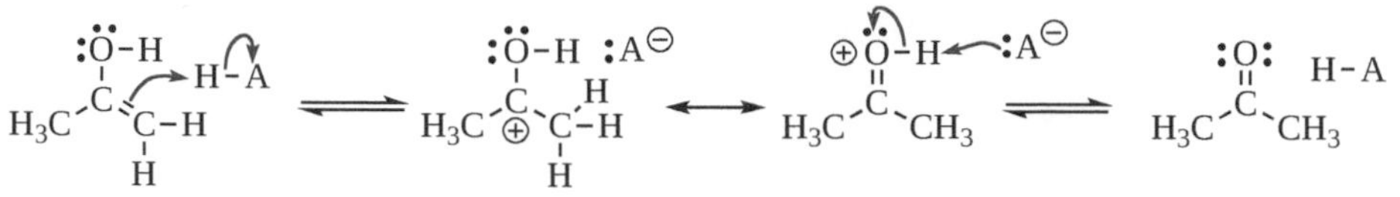

The basic route from the enol to the keto form is via proton transfer, path p.t. twice.

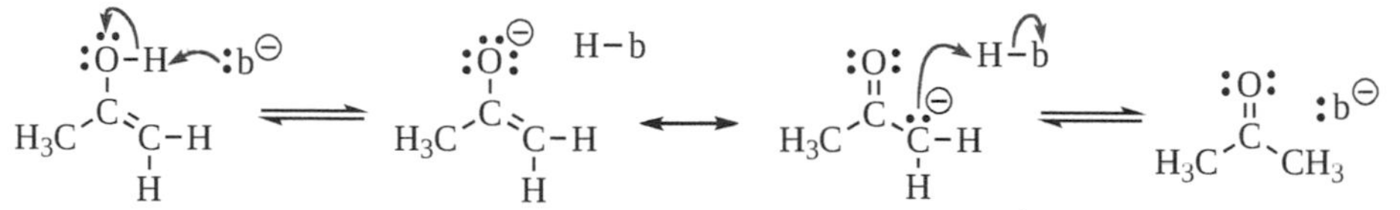

7.4 VARIATIONS ON A THEME

The variety of the acids, bases, electrophiles, nucleophiles, and leaving groups may sometimes make an otherwise familiar reaction look unrecognizable. It is very important that you be able to recognize and classify groups into their respective generic classes.

7.4.1 Atom Variations

The pathways just discussed should be considered generic; the route for electron flow is important, but the identity of the atoms in each path can easily change. For example, all of the electron flow paths were written generally, with Nu standing for any nucleophile, b for any base, L for any leaving group, and Y for any electronegative atom like oxygen, nitrogen, sulfur, or even C–EWG, an electronegative carbon atom. This section gives examples of how the atoms can vary while the process remains the same.

E2 Variants

Besides having several charge types, the E2 pathway has many variants, a few of which will be discussed in this section as illustrations of how the identity of the atoms can change while the electron flow stays the same. As was shown in Section 4.3.7, a heteroatom can be exchanged for carbon and not affect the process at all:

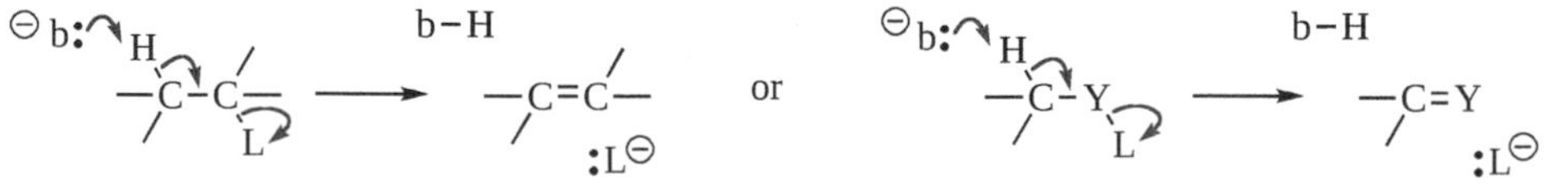

The dehalogenation reaction is a variant of the E2 where the H and L are now halogens, commonly bromine, and the base has been replaced by iodide. The orbital alignment restrictions are the same as for the E2 reaction:

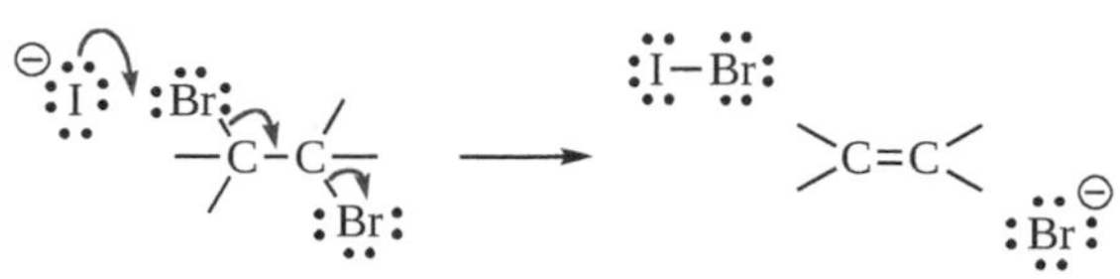

A fragmentation reaction is a reaction that breaks a molecule apart, usually into at least two pieces, by breaking a carbon–carbon bond. A fragmentation can be considered to be a type of elimination, similar to the E1cB or the E2, in which a weakened carbon–carbon single bond is broken. A fragmentation has the same path restrictions as the elimination it resembles. The fragmentation reaction below is an E2 elimination with another electron source replacing the base acting on a C–H bond. All orbitals that form double bonds must be coplanar and usually must be arranged as shown. The source in the right example can also be a carboxylate anion, in which each Y is oxygen, and then the fragmentation would be called a decarboxylation and produce CO_2.

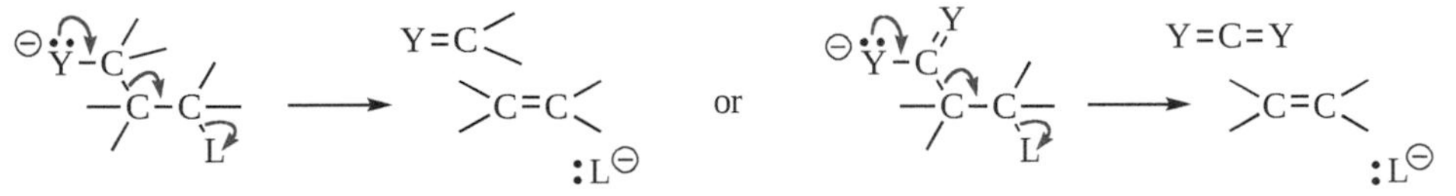

Enolates as Leaving Groups

The reaction shown on the right in the following structures (if Y is oxygen) also produces CO_2. It can be considered a variant of deprotonation adjacent to a polarized multiple bond. Again, we have changed the identity of the atoms, but the electron flow remains the same. The overlap limitation for deprotonation, shown previously in Figure 7.1, similarly requires that the breaking C–C bond must align with the *p* orbital of the C=Z pi bond. Occasionally, an electron flow can be viewed in more than one way; this decarboxylation reaction could also be considered an elimination in which the leaving group is an enolate.

The reverse of the aldol reaction (Section 8.5.5) is an E1cB type of fragmentation in which the leaving group is an enolate:

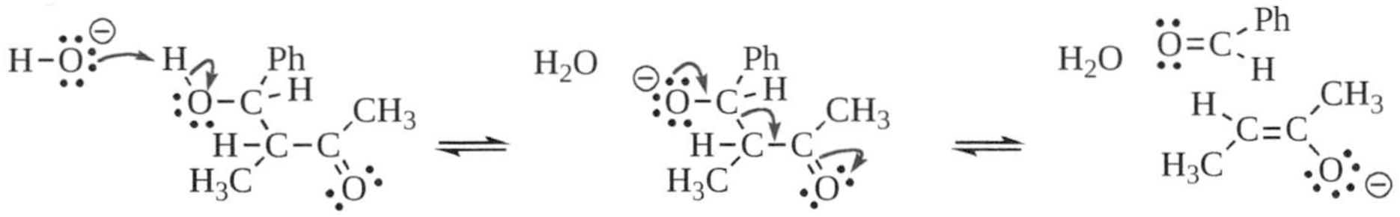

7.4.2 Vinylogous Variations

Additional conjugation may not change the system significantly (the principle of vinylogy, Section 1.9.2). Vinylogous paths are often denoted with a "prime" mark.

Vinylogous Substitutions at a Tetrahedral Center

The S_N2' reaction is just a vinylogous S_N2 reaction, where the nucleophile attacks a pi bond adjacent to the leaving group:

However, the overlap restrictions now include two more atoms and therefore may be more difficult to achieve. For the S_N2' the Nu, *p* orbital of carbon attacked, and C–L bond should be coplanar (often *syn*). This variant is found mostly in rigid systems where the C=C–C–L is locked into the proper alignment. Freely rotating systems go by the normal S_N2 reaction unless the S_N2 site is sterically blocked, and the S_N2' site is open. Triply bonded systems C≡C–C–L (propargyl leaving groups) do not have as critical an alignment problem; one of the two pi bonds will be close to the needed orientation.

The S_N1 also has a vinylogous version. In the following reaction, loss of the leaving group produces a resonance-delocalized allylic carbocation. Trapping of the carbocation on the other end of the allylic system gives an S_N1' reaction, a vinylogous S_N1. Trapping of the carbocation on the carbon that held the leaving group competes (the normal S_N1).

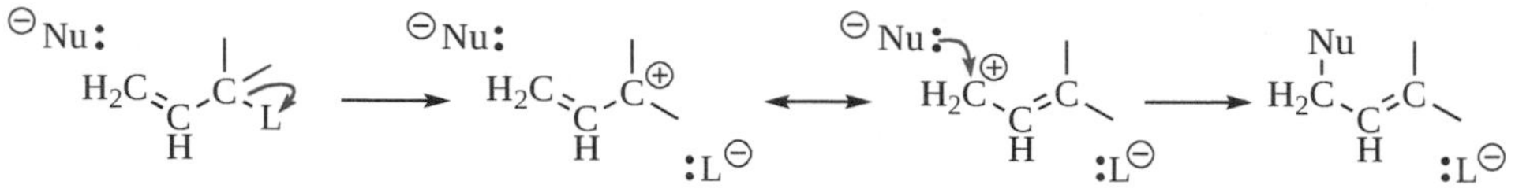

Vinylogous Substitutions at a Trigonal Planar Center

These electron sinks are much less common. If the EWG is a carbonyl and the leaving group is chloride, this electron sink is a vinylogous acid chloride (Cl–C=C–C=O). Just like all L–C=Y these groups substitute by addition-elimination.

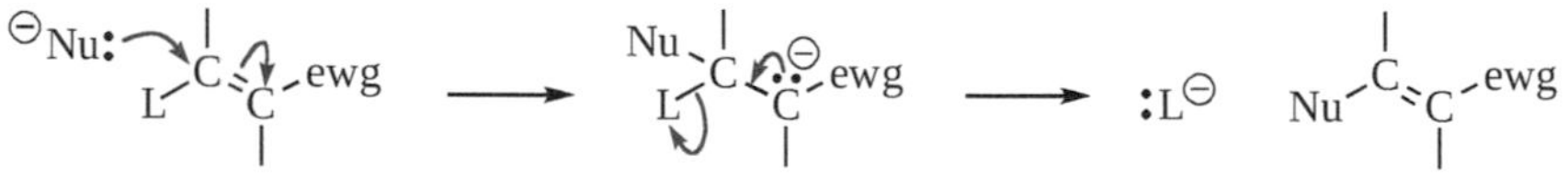

Vinylogous Eliminations

These vinylogous eliminations with one double bond between the C–H and C–L reactions are sometimes called 1, 4-eliminations; a vinylogous E2 example follows. As in the E2, orbital alignment is important: both C–H and C–L must be coplanar with their adjacent *p* orbital on the pi bond. Freely rotating systems have little problem with this.

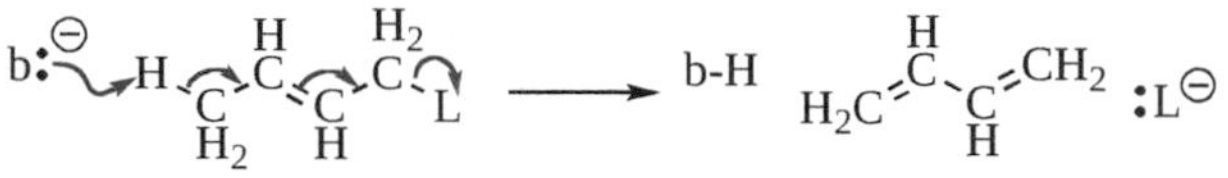

Vinylogous Additions

These vinylogous additions are the reverse of the vinylogous eliminations and are sometimes called 1, 4-additions. The electrophile adds to form the allylic carbocation that can add the nucleophile at either end. The overall process shown below, a vinylogous Ad_E2, often produces a mixture of 1, 2 and 1, 4 addition products (see Section 8.3.6 for details).

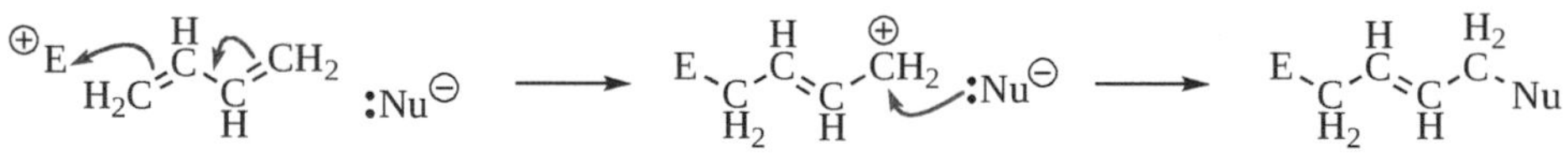

Vinylogous nucleophilic additions are also common with the sink C=C–EWG. The vinylogous Ad_N2 on this sink is called a 1, 4-addition or *conjugate addition* and is covered in Section 8.7, and the competition between the 1, 2 and 1, 4 addition is discussed in Section 9.6.

7.4.3 Extent of Proton Transfer Variations

Another topic that is appropriate under the variations-on-a-theme rubric is the extent of proton transfer between groups in protic solvents. Hydrogen bonding is very important, common, and should always be considered for mechanisms of reactions that are run in protic solvents, especially enzymatic reactions. A hydrogen bond to a carbonyl or to a leaving group lone pair might possibly be considered a partial protonation leading to greater reactivity. After a brief review of the hetero-Ad_E3, this section discusses two protic-media electron flow paths that utilize hydrogen bonding in the transition state, the general base-catalyzed addition to a polarized multiple bond and the reverse reaction, the general acid-catalyzed beta elimination. Finally, a hydrogen-bonded, push–pull path for enolization is discussed.

If the rate of reaction catalyzed by an acid or a base depends on the pH of the medium and not the identity of the acid or base, then the reaction is called *specific acid* or *specific base catalyzed*. In the case of water, the specific acid is the protonated solvent, hydronium, and the specific base is the deprotonated solvent, hydroxide. Specific acid catalyzed reactions do not show a rate change if the identity or concentration of the acid is changed as long as the pH remains the same. *Specific acid catalysis* indicates that the proton-transfer equilibria are fast and not involved in the rate-determining step. For similar base-catalyzed reactions, the term *specific base catalysis* is used.

However, if the proton transfer is part of the rate-controlling step, then the concentration and pK_a of the acid matter, and the reaction is called *general acid catalyzed*. *General acid catalysis* indicates a reaction mechanism that has proton transfer occurring in the rate-determining step (for example, attack on a species hydrogen bonded to HA). For similar base-catalyzed reactions, the term *general base catalysis* is used.

We have already seen one general acid-catalyzed electron flow path, the hetero Ad_E3, in Chapter 4. Here it is in more detail, along with other general acid and general base catalyzed processes. The transition states of all these reactions involve a collision of a hydrogen-bonded pair with one other reactant and therefore avoid a highly unlikely entropically disfavored three-molecule collision.

General Acid-Catalyzed Addition to a Polarized Multiple Bond

The hetero-Ad_E3 (Figure 7.14) involves nucleophilic attack on a hydrogen-bonded complex between the heteroatom lone pair and a weak acid and is called a general acid catalyzed addition. Since this is a one-step reaction the weak HA bond is broken in the rate-determining step. The identity and concentration of the weak acid will affect the rate of reaction. Note that the stereoelectronic requirements are different since the hydrogen-bonded lone pair lies in the pi nodal plane and the nucleophilic attack occurs perpendicular to this.

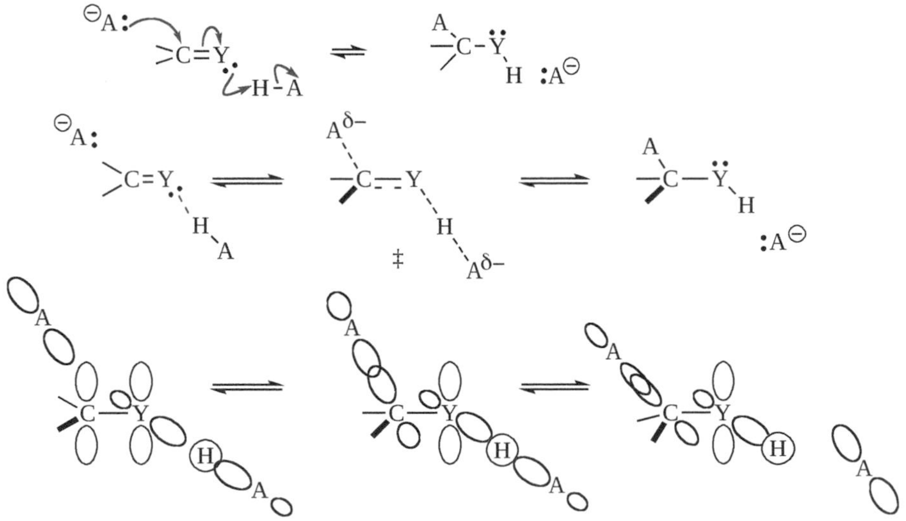

Figure 7.14 The hetero-Ad_E3 addition with the transition state in the center, a general acid catalysis.

Common Errors

The following set of arrows is a commonly drawn shortcut, and although it gets all the lines and dots in the right places, it ignores the fact that the hydrogen bond is usually to the lone pair and not to the less available pi bond.

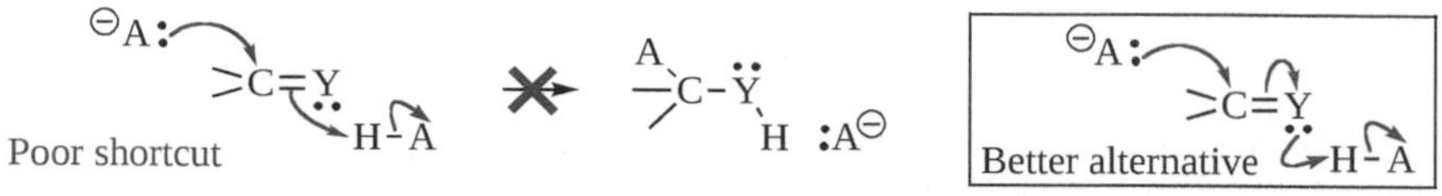

General Base-Catalyzed Addition to a Polarized Multiple Bond

This pathway (Figure 7.15) is very similar to path Ad_N, with the difference that the nucleophile is poorer and is hydrogen bonded to a base when this pair collides with the polarized multiple bond. In this pathway, the electron flow comes from the base to break the Nu–H bond, which in turn enhances the nucleophilicity of the nucleophile's lone pair. This lone pair attacks the multiply bonded carbon, breaks the pi bond, and produces a stable anion similar to path Ad_N.

Two processes are occurring to a varying extent: The hydrogen is being pulled off by the base, and the nucleophile lone pair is attacking the polarized multiple bond. If the nucleophile is especially poor, proton removal would need to be completed before a good enough nucleophile would be generated for the attack to proceed; this would then be path p.t. followed by path Ad_N. If the nucleophile is very good, proton removal would be unnecessary, and the reaction would go by path Ad_N. This path is used when there is a weak base present like acetate, $pK_{abH} = 4.8$.

An electronegative heteroatom, Y, can also be replaced by an electronegative carbon atom, C–EWG. The electron-withdrawing group is extremely necessary, for without a means to stabilize the carbanion formed, the electron flow cannot occur. The enhancement of nucleophilicity by hydrogen bonding may occur in a protic solvent whenever the nucleophile bears a reasonably acidic hydrogen.

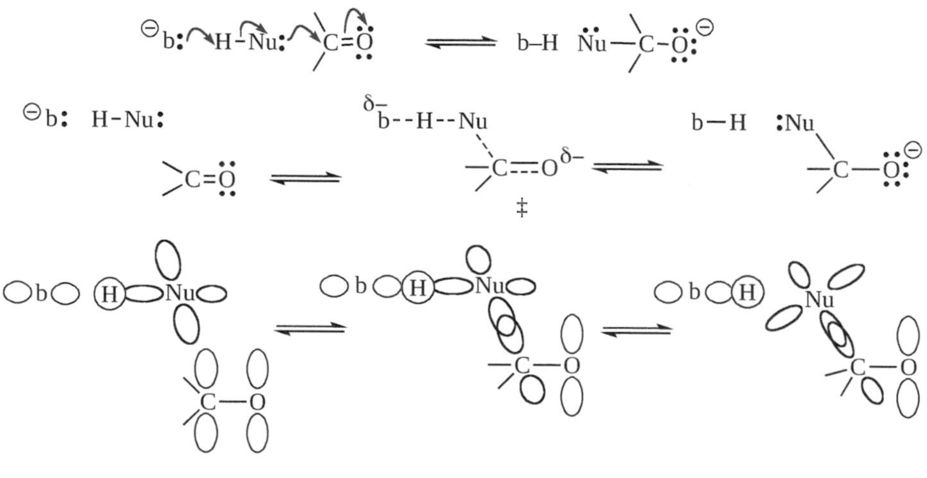

Figure 7.15 The general base-catalyzed addition to a polarized multiple bond with the transition state shown in the center.

Path Details

Overlap: Nu must approach the δ+ carbon from behind and close to the pi plane.
Media: A weak base is needed; the medium is usually weakly basic to neutral.
Solvent: Protic and polar to stabilize the anion formed.
Steric: The approach of the Nu to the sink is usually not a problem.
Energetics: Anion formed should be equal to or more stable than the base or the reaction can reverse. The ultimate test is the ΔpK_a rule cross-check (Sections 3.6 and 4.5.2).

Common Errors

The following set of arrows is the frequently drawn shortcut; it ignores the role of the lone pair and incorrectly implies that the sigma bond is acting as a nucleophile. The four-arrow description is more rigorous and agrees better with known enzymatic catalysis.

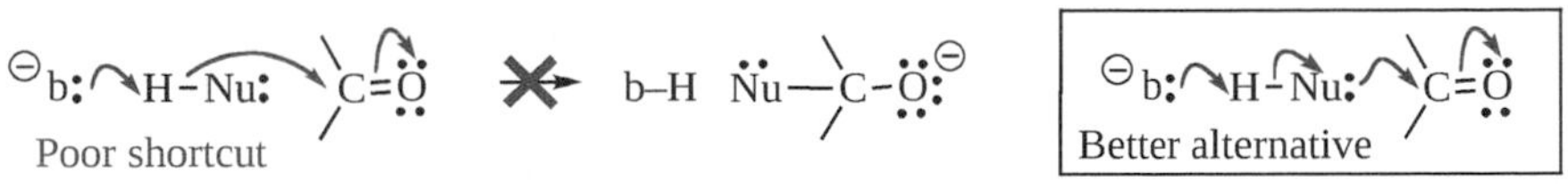

Occasionally, an arrow in the middle of the flow is forgotten, quite often the breaking of a C–H or O–H bond. In the left example, the origin of the middle arrow is unclear; it appears to come from a lone pair position, but the lone pairs have been omitted. This general base-catalyzed addition to a polarized multiple bond should be best drawn with four arrows.

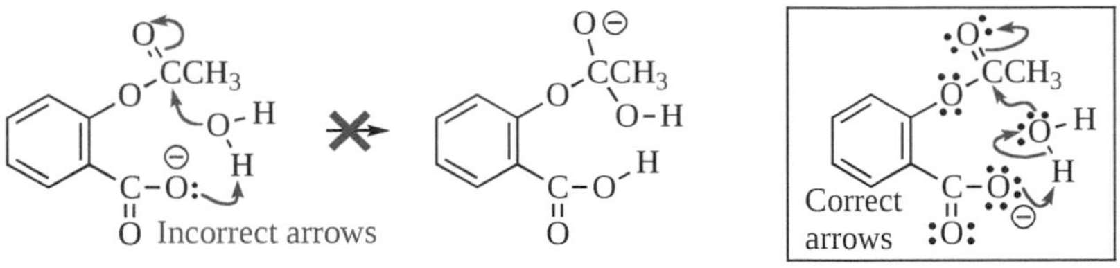

General Acid-Catalyzed Beta Elimination

(The Reverse of the Previous Reaction)

The difference between this path and path E_β is that the leaving group is poorer and must hydrogen bond to a weak acid (like RCOOH) to be good enough to depart. The hydrogen bond enhances the quality of the leaving group. This is the middle of three paths that involve different degrees of proton transfer. Catalysis by hydrogen-bonded species may occur whenever a reasonably basic leaving group is lost in a protic solvent. If the leaving group were very poor, it would need to be fully protonated before it would be good enough to depart; this would be path p.t. followed by path E_β. If the leaving group is good, then proton transfer is unnecessary and the reaction would proceed by path E_β.

Path Details

Overlap: Anion *p* orbital and C–L bond close to coplanar (Figure 7.15, right to left).
Leaving group: Average, but needs a lone pair to hydrogen bond to the weak acid.
Media: A weak acid is needed; the medium is usually weakly acidic to neutral.
Solvent: Protic and polar to stabilize anion.
Energetics: Use the ΔpK_a rule cross-check.

Common Errors

To save drawing an extra arrow, chemists may draw the following set of arrows, a shortcut that does get all the lines and dots in the right places but has some problems that are more obvious in this direction than they were in the microscopic reverse.

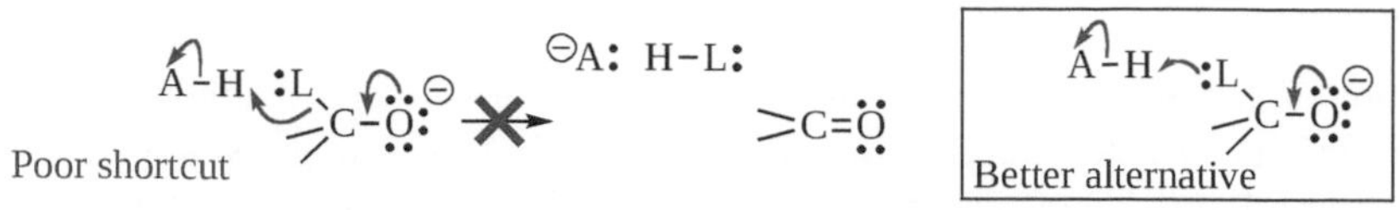

The problems with the three-arrow shortcut are that it ignores the necessary role of hydrogen bonding to the lone pair of the leaving group, that it protonates the unavailable sigma bond electron pair, and that it does not quite agree with intramolecular and enzymatic studies. This process is more rigorously drawn with four arrows to emphasize the role of the lone pair.

Summary of Polarized Multiple Bond Reactions

Figure 7.16 shows the top view of an energy surface that relates all three **basic media** mechanisms for addition of a nucleophile to a polarized multiple bond. The reactants are in the upper left corner, and the products are in the lower right corner. If the base is weak and the nucleophile is good, the nucleophile adds (path Ad_N), going to the top right corner, and later gets deprotonated (path p.t.), going down to product. If the nucleophile is poor or the base very strong, deprotonation occurs first (path p.t.), going down, followed by addition to the polarized multiple bond (path Ad_N), going across to product. This is called specific base catalysis. If the nucleophile is mediocre and a weak base is present, then a general base-catalyzed process (the diagonal route) occurs in which partial proton removal from the nucleophile enhances its ability to attack the C=Y.

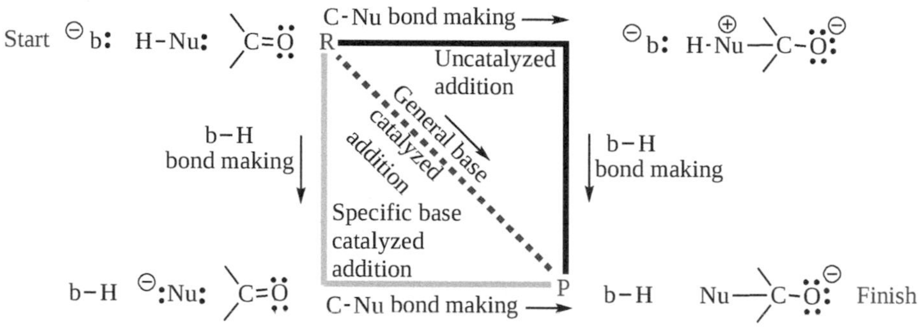

Figure 7.16 Polarized multiple bond addition/elimination mechanisms in **basic** media. Reactants are in the upper left. The black route is an uncatalyzed addition. The gray route is the specific base catalyzed, and the dashed red diagonal is the general base-catalyzed process.

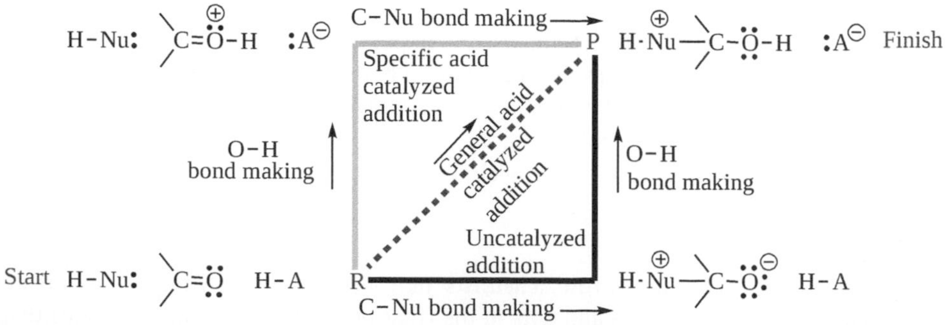

Figure 7.17 Polarized multiple bond addition/elimination mechanisms in **acidic** media. Reactants are on the lower left. The black route is an uncatalyzed addition. The gray route is the specific acid catalyzed, and the dashed red diagonal is the general acid-catalyzed process.

Figure 7.17 is the top view of an energy surface composed of three **acidic media** mechanisms that depend on the acidity of the media. Specific acid catalysis is protonation followed by nucleophilic attack (path p.t. then Ad_N), and can occur in strong acids with weak nucleophiles. General acid catalysis (diagonal) is protonation along with nucleophilic attack (hetero Ad_E3 path), and occurs in weak acids. Nucleophilic attack followed by protonation (path Ad_N then p.t.), would occur with good nucleophiles in weakly acidic media.

Push–Pull Catalysis of Enolization

As mentioned in Section 2.8, enzymes can have several catalytic groups in the cavity of their active site. When the reactant fits into that active site, these catalytic groups may function together in a push–pull catalysis. For example, the top view of the energy surface for enolization is shown in Figure 7.18. The reactants are in the lower left corner, and the products are in the upper right. The specific acid-catalyzed process forms the O–H bond before the C–H bond breaks (up, then right), whereas the specific base-catalyzed process breaks the C–H bond before the O–H bond forms (right, then up). The push–pull catalysis is the diagonal route in which the C–H bond breaks as the O–H bond forms.

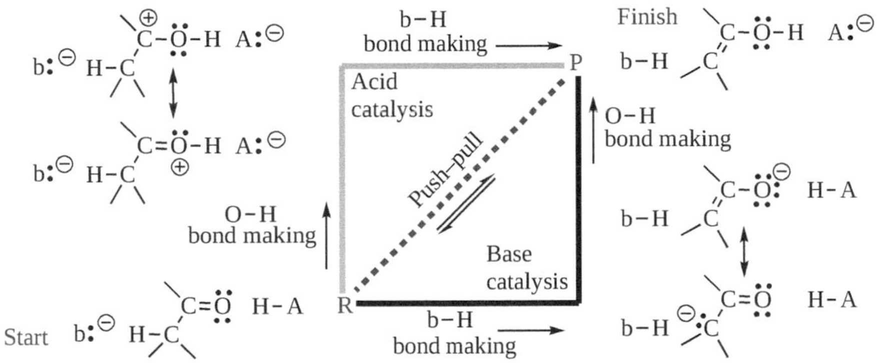

Figure 7.18 The energy surface for enolization. Reactants are on the lower left. The gray path is acid catalyzed, the black path is base catalyzed, and the dashed red diagonal path is the push–pull path, which combines the arrows for the base and acid catalyzed enolization processes on the same structure.

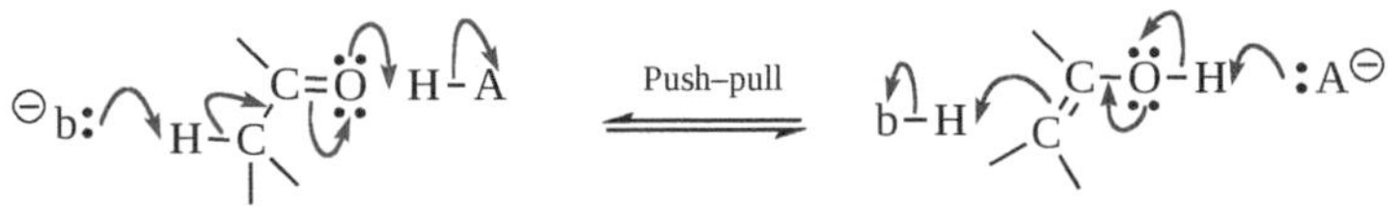

We have now seen general acid- and general base-catalyzed processes for polarized multiple bond additions, eliminations, and enolizations. **There is good reason to expect that all reaction types, additions, eliminations, substitutions, and rearrangements have general acid- or general base-catalyzed routes under appropriate polar–protic conditions.** We can also expect that enzyme active sites with appropriately positioned acidic and basic groups can easily do general acid, general base, and push–pull catalysis.

Highly hydrogen-bonding solvents like water should have routes available in which hydrogen-bonded water serves as a proton donor. Converting a hydrogen bond into an actual bond during the bond-forming step is similar to a general-acid catalyzed process. A mechanism sometimes written for aqueous basic amide hydrolysis has a hydrogen-bonded assisted loss of the leaving group step written to avoid the loss of a very poor leaving group.

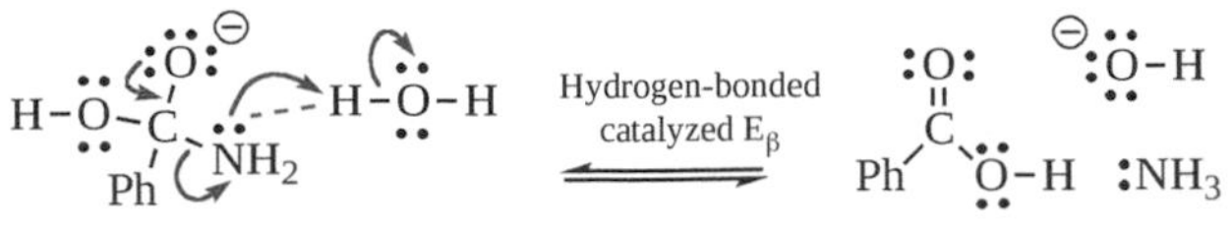

Charge Type Varieties

Almost all of the pathways have several charge types. Nucleophiles can be anionic or neutral; electrophiles can be cationic or neutral; and leaving groups can depart as anions or neutrals. Reactions catalyzed by a Brønsted–Lowry acid can often be catalyzed by a Lewis acid or by a positively charged metal ion. Don't get fooled by an unfamiliar charge type.

7.5 TWELVE MAJOR PATHS SUMMARY AND CROSS-CHECKS

p.t., Proton Transfer to and from an Anion or Lone Pair (Section 7.1.1)

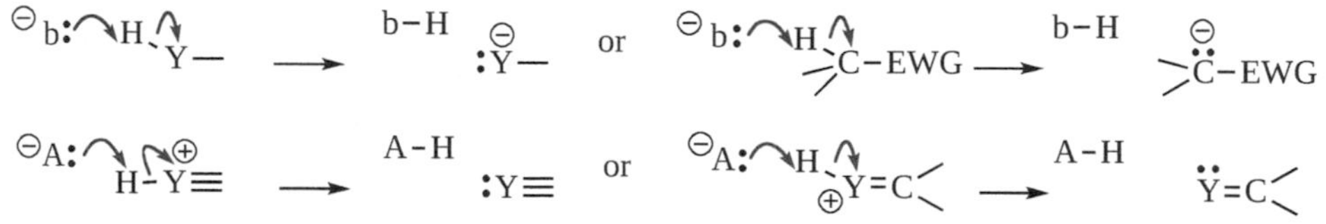

Include both deprotonations (above) and protonations (below). Cross-check: K_{eq} or ΔpK_a rule.

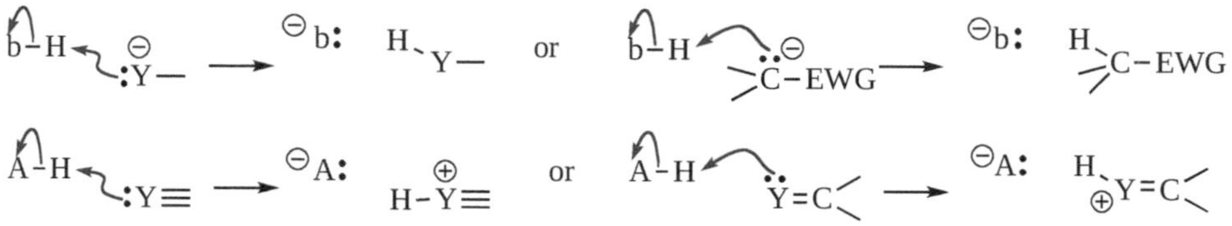

D_N, Ionization of a Leaving Group (Section 7.1.2)

Cross-check: $\Delta pK_{aHL} \leq 0$, cation usually more stable than 2°, polar solvent.

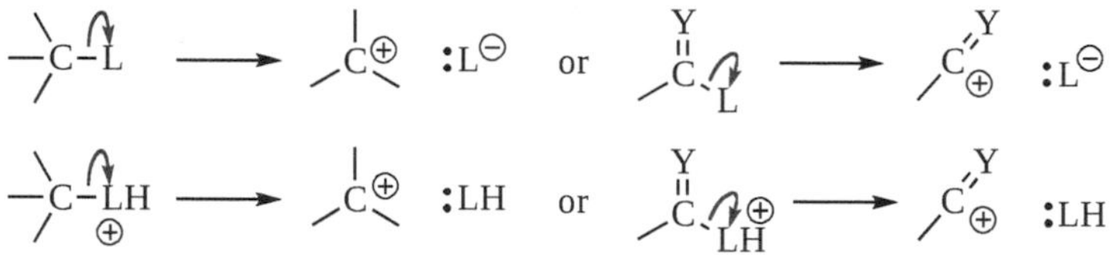

A_N, Trapping of an Electron-Deficient Species (Section 7.1.3)

Cross-check: None, usually OK.

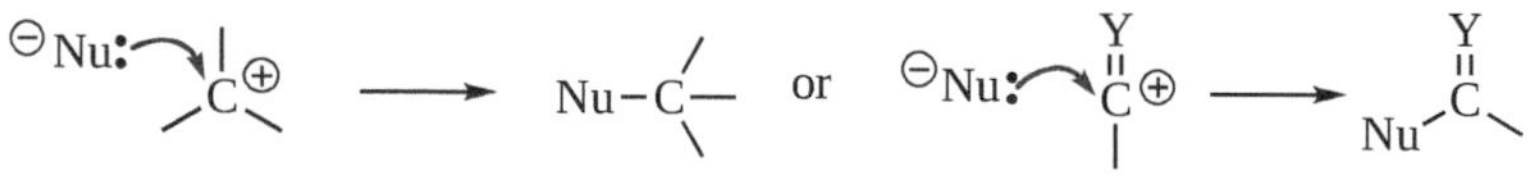

A_E, Electrophile Addition to a Multiple Bond (Section 7.1.4)

Cross-check: Markovnikov's rule — make the most stable carbocation.

D_E, Electrofuge Loss from a Cation To Form a Pi Bond (Section 7.1.5)

Cross-check: Electrofuge lost should be reasonably stable.

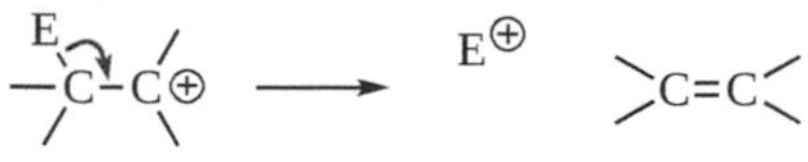

S_N2, The S_N2 Substitution (Section 7.1.6)

Cross-check: ΔpK_a rule on Nu and L, $\Delta pK_{aHL} \leq 12$, good access to C attacked.

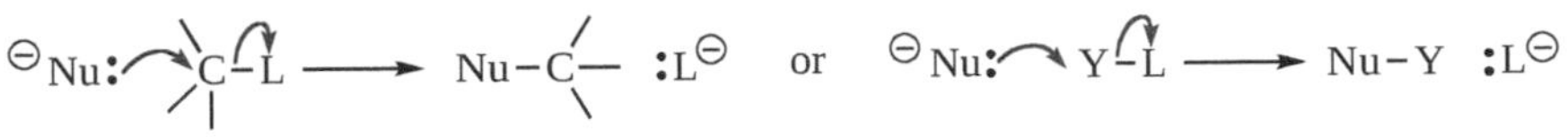

E2, The E2 Elimination (Section 7.1.7)

Cross-checks: ΔpK_a rule on incoming base and outgoing L, C–H bond and C–L need to be nearly coplanar.

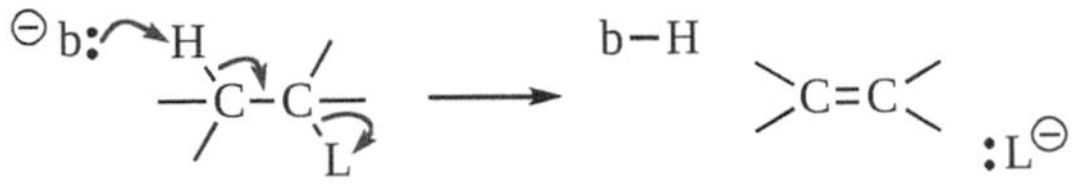

Ad_E3, The Ad_E3 Addition (Section 7.1.8)

Cross-check: Markovnikov's is rule followed because of polarization of pi-adduct.

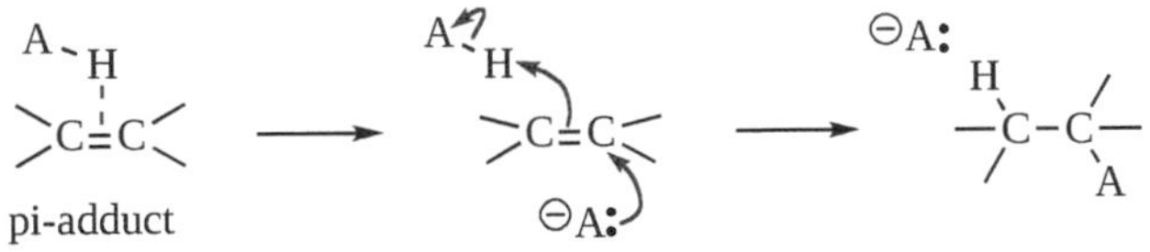

Hetero Ad_E3 (See 7.4.3) Cross-check: ΔpK_a rule on incoming Nu and anion formed.

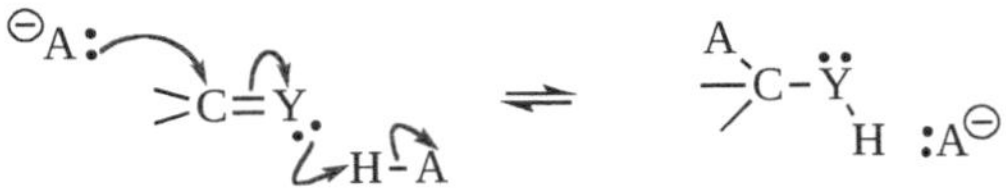

Ad_N, Nucleophilic Addition to a Polarized Multiple Bond (Section 7.1.9)

Cross-check: ΔpK_a rule on incoming Nu and outgoing anion formed.

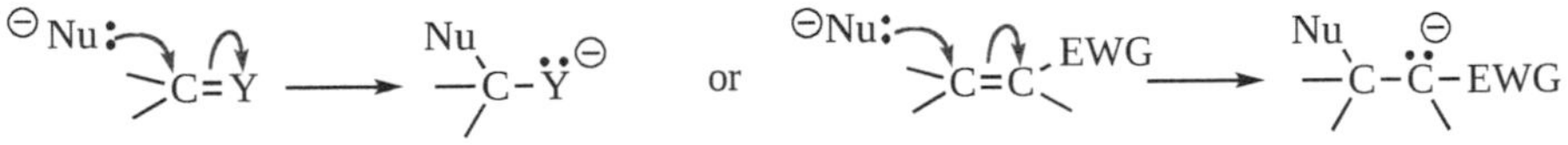

E_β, Beta Elimination from an Anion or Lone Pair (Section 7.1.10)

Cross-check: ΔpK_a rule on L and anion reactant.

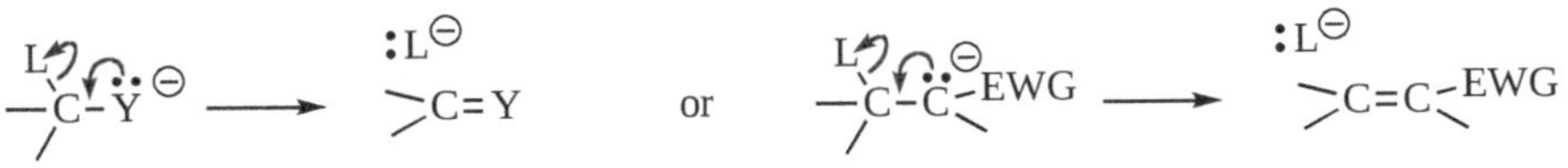

1,2R, Rearrangement of a Carbocation (Section 7.1.11)

Cross-check: Usually forms carbocation of equal or greater stability.

Path 1,2RL, Rearrangement with Loss of Leaving Group (Section 7.1.12)

Cross-check: (if no pushing Y): Good L, cation usually secondary or better, polar solvent.

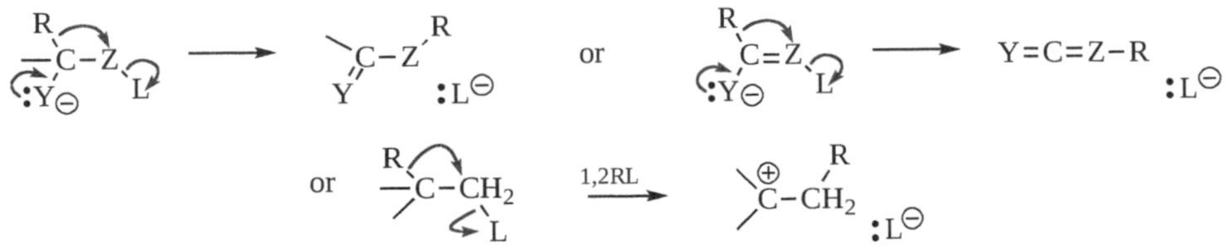

7.6 SIX MINOR PATHS SUMMARY

Path pent., Substitution via a Pentacovalent Intermediate (Section 7.2.1)

(Not for carbon, only for third-row and higher elements)

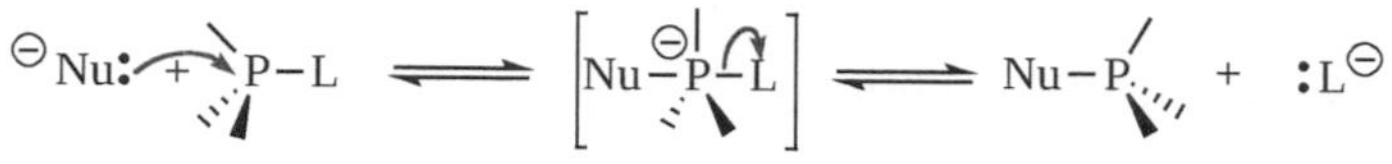

Path 6e, Concerted Six-Electron Pericyclic Reactions (Section 7.2.2)

Thermal rearrangements:

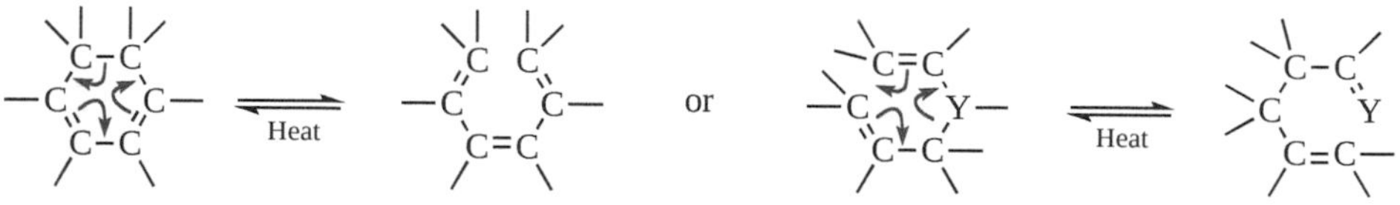

Thermal cycloadditions or cycloreversions (the reverse reaction):

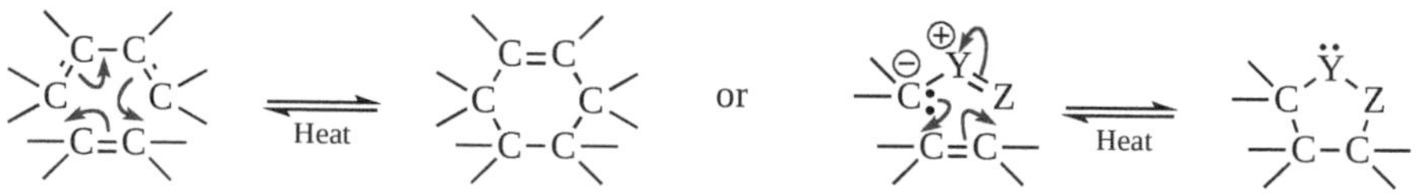

Metal-chelate-catalyzed additions:

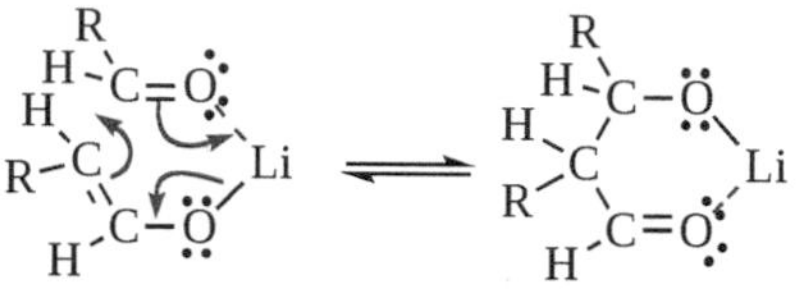

Path Ei, Thermal Internal *Syn* Elimination (Section 7.2.3)

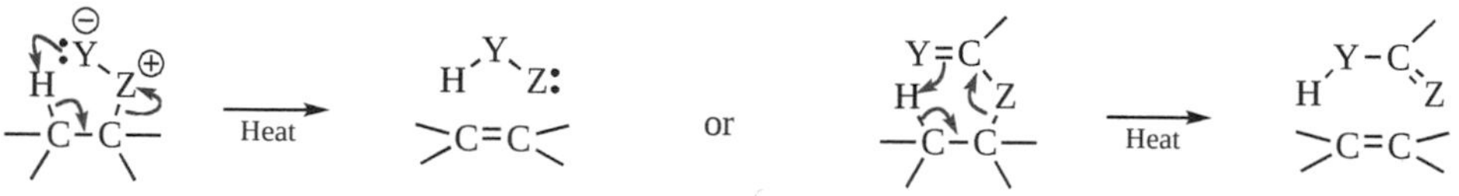

Path NuL, Nu–L Additions (Section 7.2.4) (three-membered ring formation)

Path 4e, Four-Center, Four-Electron (Section 7.2.5) (three cases only)

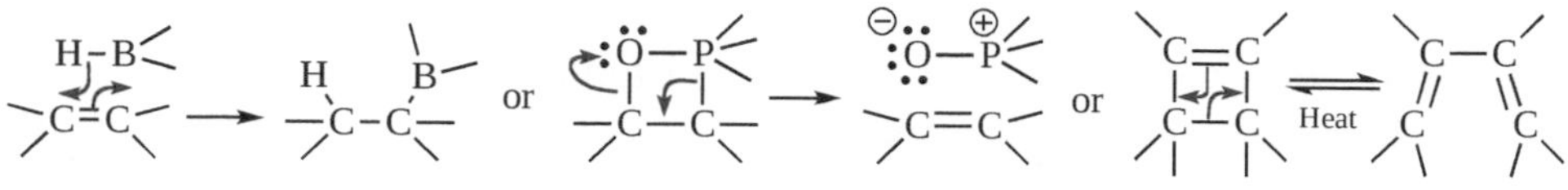

Path H⁻ t., Hydride Transfer to a Cationic Center (Section 7.2.6)

(forms a carbocation of equal or greater stability than the starting carbocation)

7.7 COMMON PATH COMBINATIONS SUMMARY

S_N1 (Substitution, Nucleophilic, Unimolecular), $D_N + A_N$ (Section 7.3.1)

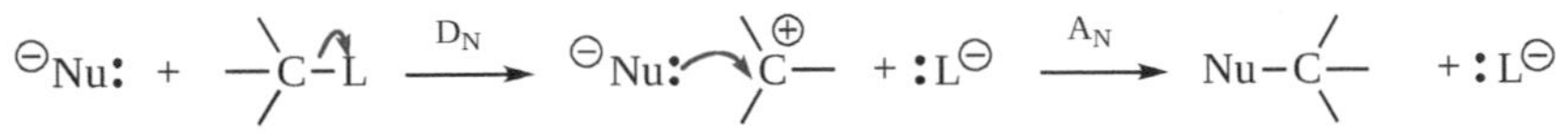

Lone-Pair-Assisted S_N1, $E_\beta + A_N$

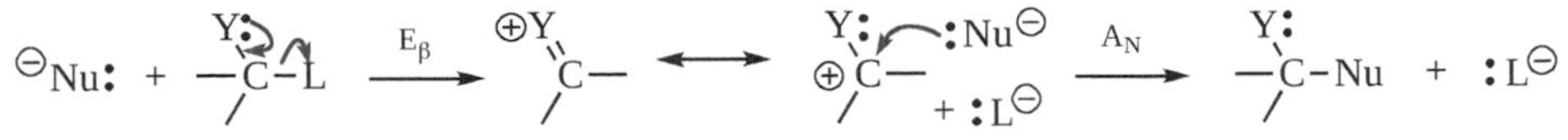

Ad_E2 (Addition, Electrophilic, Bimolecular), $A_E + A_N$ (Section 7.3.2)

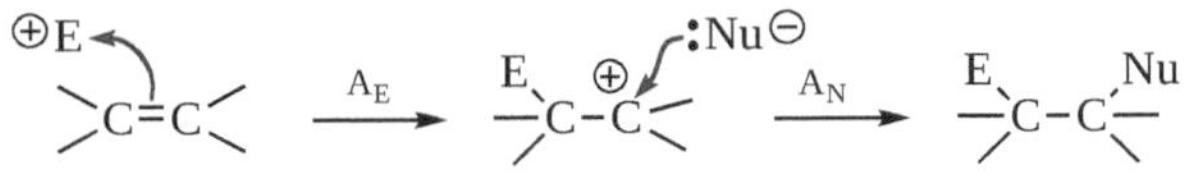

Hetero-Ad_E2, p.t. + Ad_N

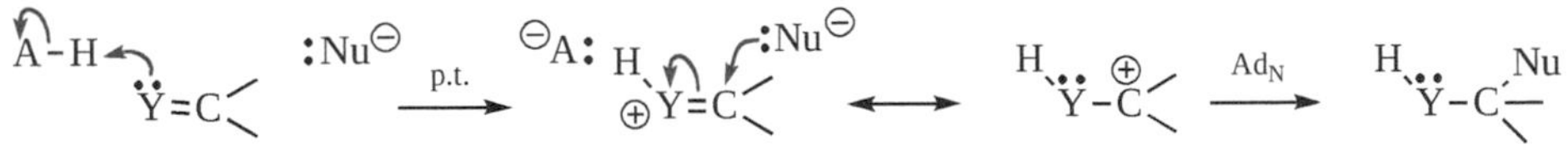

E1 (Elimination, Unimolecular), $D_N + D_E$ (Section 7.3.3)

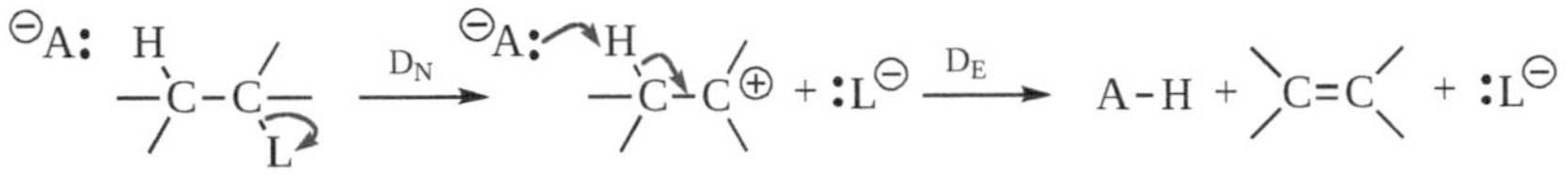

Lone-Pair-Assisted E1, E_β + p.t.

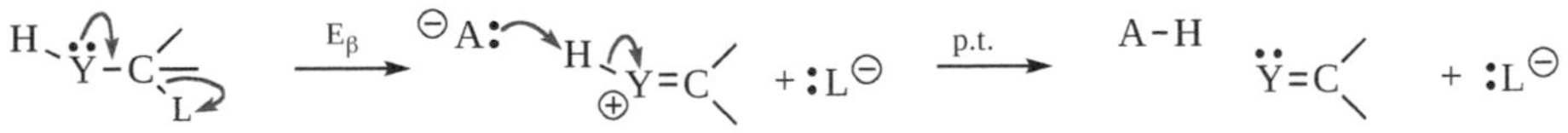

S_E2Ar Electrophilic Aromatic Substitution, $A_E + D_E$ (Section 7.3.4)

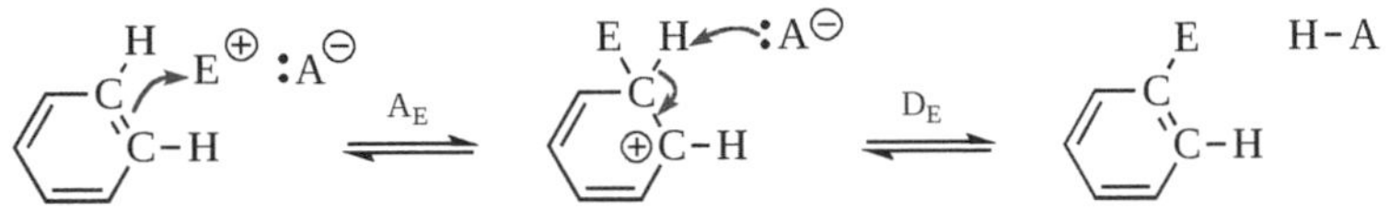

E1cB (Elimination, Unimolecular, Conjugate Base), p.t. + E_β (Section 7.3.5)

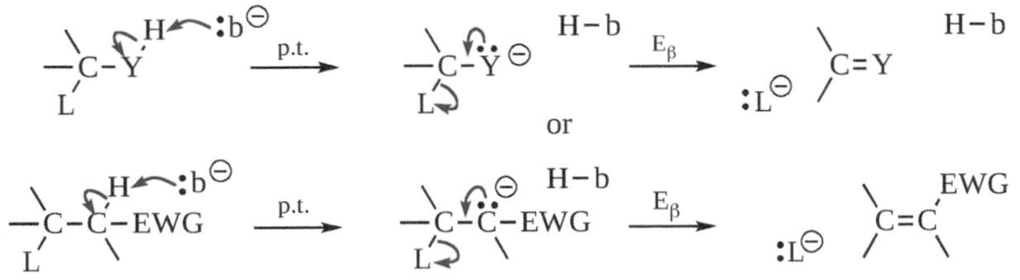

Ad_N2 (Addition, Nucleophilic, Bimolecular), Ad_N + p.t. (Section 7.3.6)

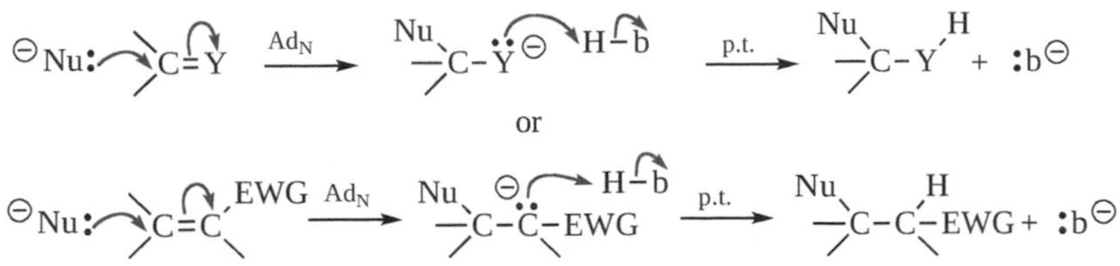

Addition–Elimination, $Ad_N + E_\beta$ (Section 7.3.7)

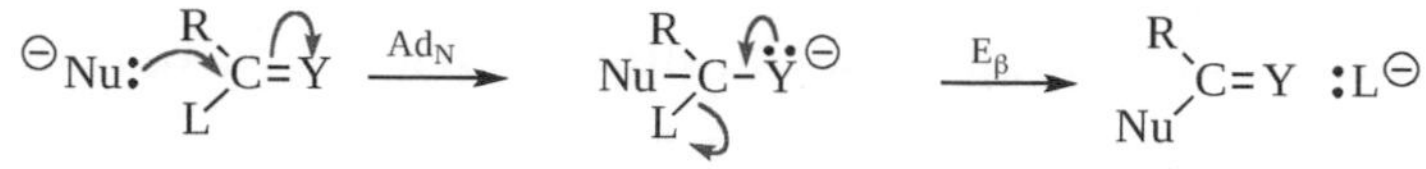

Tautomerization, taut. H–C–C=Z ⇌ C=C–Z–H (Section 7.3.8)

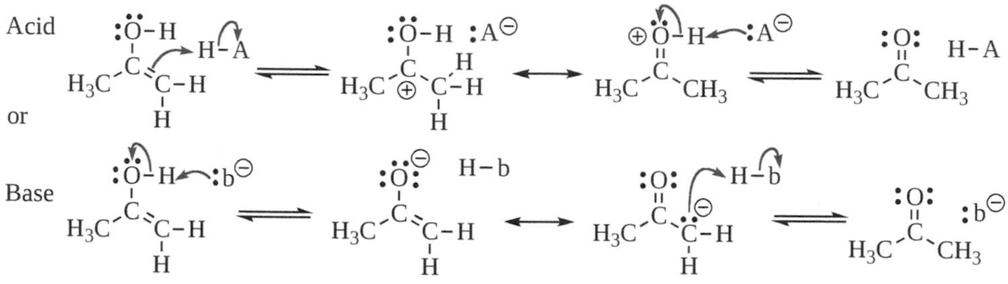

Variations on a Theme are Common

Atom variations (Section 7.4.1) swap out one atom for another, but the electron flow can remain the same. **Vinylogous variations** (Section 7.4.2) add an extra pi bond and arrow but do not significantly alter the electron flow. **Proton transfer variations** (Section 7.4.3) range from hydrogen bonding (partial protonation) to full protonation or deprotonation. General acid and general base catalyzed processes are common in enzymes.

CHAPTER 7 EXERCISES

7.1 Go back to Chapter 1, problem 1.5, and place the name of the path over each transformation arrow.

7.2 Go back to Chapter 1, problem 1.6, and place the name of the path over each transformation arrow.

7.3 Go back to Chapter 1, problem 1.7, and place the name of the path over each transformation arrow.

7.4 Go back to Chapter 1, problem 1.8, and place the name of the path over each transformation arrow.

7.5 Draw a general example of each of these most common pathways.
(a) path p.t., proton transfer to a lone pair
(b) path D_N, ionization of a leaving group
(c) path A_N, trapping of a cation by a nucleophile.

7.6 Draw a general example of each of these most common pathways.
(d) path A_E, electrophile addition to a multiple bond
(e) path D_E, electrofuge loss from a cation to form a pi bond
(f) path S_N2, substitution.

7.7 Draw a general example of each of these most common pathways.
(a) path E2 elimination
(b) path Ad_E3 addition
(c) path Ad_N, nucleophilic addition to a polarized multiple bond.

7.8 Draw a general example of each of these most common pathways.
(d) path E_β, beta elimination from an anion or lone pair
(e) path 1,2R, rearrangement of a carbocation
(f) path 1,2RL, rearrangement with loss of a leaving group.

7.9 Draw a general example of each of these more common minor pathways.
(a) path 6e, concerted six-electron pericyclic
(b) path pent., substitution via a pentacovalent intermediate
(d) path E_i, thermal internal *syn* elimination.

7.10 Draw a general example of each of these common path combinations.
(a) S_N1 substitution
(b) Ad_E2 addition
(c) E1 elimination
(d) Electrophilic aromatic substitution.

7.11 Draw a general example of each of these common path combinations.
(a) E1cB elimination
(b) Ad_N2 addition
(c) Addition–Elimination route of carboxyl derivatives
(d) Tautomerization.

7.12 Sort the path combinations in problems 7.10 and 7.11 into those that prefer predominantly acidic media, basic media, or may appear in both.

7.13 A few major electron flow pathways have critical requirements. What are the path requirements that, if not met, the following paths will probably not happen?
(a) E2 elimination
(b) 1,2R, Rearrangement of a carbocation
(c) Ad_N addition.

7.14 A few major electron flow pathways have critical requirements. What are the path requirements that, if not met, the following paths will probably not happen?
(a) p.t., Proton transfer to a lone pair
(b) D_N, Ionization of a leaving group
(c) S_N2 substitution.

7.15 Many reactions produce product mixtures. Give the product of an S_N2 and also the product of an S_N2' on the following compound. Use Nu to symbolize the nucleophile.

$$R{-}C{\equiv}C{-}\underset{L}{\overset{H}{C}}{-}R$$

7.16 From biochemical studies, a general acid catalyst is most effective if its pK_a is close to the pH of the medium. Explain this by examining the extremes. What would happen if the acid catalyst were very acidic? Then consider if the acid catalyst were not very acidic?

7.17 Classify the reaction below into one of the four archetypes and list the possible paths. Using the path restrictions, pick the only path or common path combination that fits for this reaction. Notice how this analysis process turns what appears to be an open-ended problem into a multiple-choice problem. Write out all the steps of the mechanism.

$$(CH_3(CH_2)_3)_3P{:} + H_3CH_2C{-}\ddot{\underset{..}{I}}{:} \longrightarrow (CH_3(CH_2)_3)_3\overset{\oplus}{P}{-}CH_2CH_3 + {:}\ddot{\underset{..}{I}}{:}^{\ominus}$$

7.18 Classify the reaction below into one of the four archetypes and list the possible paths. Using the path restrictions, pick the only path or common path combination that fits for this reaction. Notice how this analysis process turns what appears to be an open-ended problem into a multiple-choice problem. Write out all the steps of the mechanism.

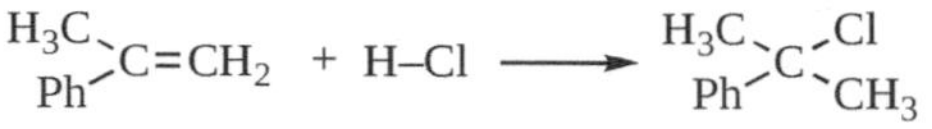

7.19 Classify the **acid-catalyzed** reaction below into one of the four archetypes and list the possible paths. Using the path restrictions, pick the only path or common path combination that fits for this reaction. Notice how this analysis process turns what appears to be an open-ended problem into a multiple-choice problem. Write out all the steps of the mechanism.

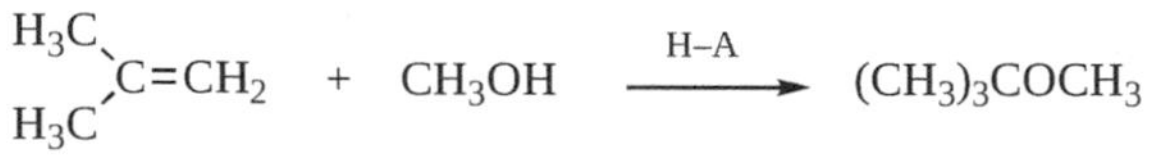

7.20 Classify the **base-catalyzed** reaction below into one of the four archetypes and list the possible paths. Using the path restrictions, pick the only path or common path combination that fits for this reaction. Notice how this analysis process turns what appears to be an open-ended problem into a multiple-choice problem. Write out all the steps of the mechanism.

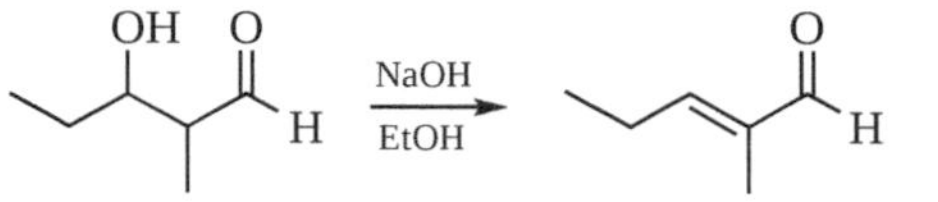

7.21 A reasonable mechanism for the base-catalyzed (sodium ethoxide) Claisen condensation reaction is as follows: proton transfer, addition, elimination, proton transfer. Write out that mechanism with all its steps. Remember to choose between the three types of additions, Ad_E2, Ad_N2, or Ad_E3, then between the three types of eliminations, E1, E2, or E1cB.

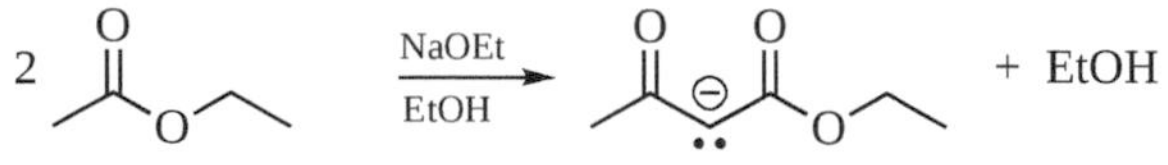

7.22 A reasonable mechanism for the acid-catalyzed (sulfuric acid) Fischer esterification reaction is as follows: addition, proton transfer, proton transfer, and elimination. Write out that mechanism with all its steps. Remember to choose between the three types of additions, Ad_E2, Ad_N2, or Ad_E3, then choose between the three types of eliminations, E1, E2, or E1cB.

7.23 A reasonable mechanism for this **acid-catalyzed** (sulfuric acid) elimination reaction is as follows: proton transfer, then elimination. Write out that mechanism with all its steps. Remember to choose between the three types of eliminations, E1, E2, or E1cB.

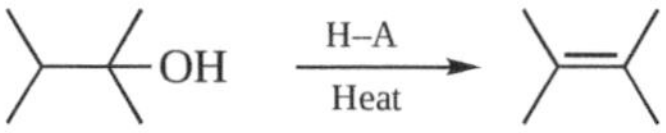

7.24 A reasonable mechanism for this base-catalyzed (sodium ethoxide) reaction in ethanol is: p.t., substitution, p.t., and then substitution. Write out that mechanism with all its steps. Remember to choose between the two types of substitutions for carbon, S_N1 or S_N2.

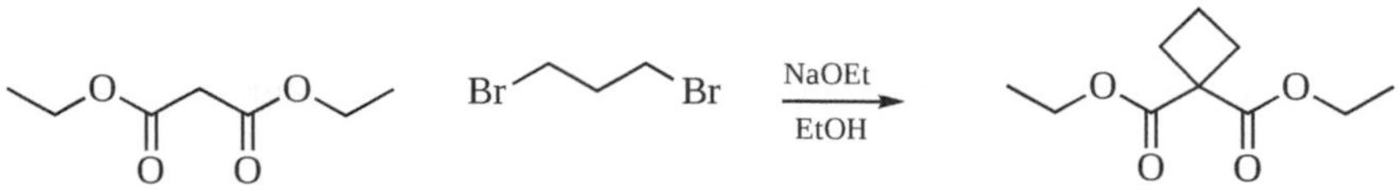

8

INTERACTION OF ELECTRON SOURCES AND SINKS

The flow of electron density, symbolized by a set of arrows, comes from the generic electron sources discussed in Chapter 5, via the pathways of Chapter 7, and ends in one of the generic electron sinks covered in Chapter 6. This chapter links each combination of the common sources and sinks with appropriate pathways, and it discusses specific examples of each process. These reactions are more examples of those reaction archetypes, proton transfer, substitution, elimination, addition, and rearrangement, discussed in Chapters 3 and 4.

Electron Flow in Organic Chemistry: A Decision-Based Guide to Organic Mechanisms,
Third Edition. Paul H. Scudder.

8.1 SOURCE AND SINK CORRELATION MATRIX

A correlation matrix is the best way to display all the simple interactions between two groups so that no interactions are accidentally left out. In Chapters 5 and 6, we identified the electron sources and sinks that will be reaction partners. Those interactions of the generic sources with sinks are displayed with a matrix, each cell of the matrix corresponding to a specific combination of source and sink. Almost all of the reactions in organic chemistry can be described as simple bimolecular collisions between one of the 12 generic electron sources (Section 5.7) and one of the 18 generic electron sinks (Section 6.6). Many reaction mechanisms are multistep processes in which the product of the initial collision of source and sink then collides with another reactant and continues on. A giant matrix containing all the electron sources and sinks can be drawn, but it would be 12 × 18 and thereby contain 216 individual cells. This large matrix contains many cells in which no reaction occurs and many cells that have been little researched.

For simplification, a much smaller 7 × 7 matrix (Table 8.1) will be used that contains representatives of the more common sources and sinks. Within each cell of the matrix is the subsection in Sections 8.2–8.8 that covers the appropriate pathways for electron flow for that particular combination of source and sink. Any important combinations not included in the simplified matrix are covered in Sections 8.9–8.12.

Table 8.1 Correlation Matrix for the More Common Electron Sources and Sinks: An Index to Sections 8.2–8.8

	Common Electron Sources						
Electron Sinks	Lone pair Z:	Base and Adjacent CH	Complex Metal Hydride MH_4^-	Organo-metallic R–M	Allylic :Z–C=C	Simple Pi Bonds C=C C≡C	Aromatics 
H–A	**8.2.1**	**8.2.2**	**8.2.3**	**8.2.4**	**8.2.5**	**8.2.6**	**8.2.7**
Y–L	**8.3.1**	**8.3.2**	**8.3.3**	**8.3.4**	**8.3.5**	**8.3.6**	**8.3.7**
sp^3 C–L	**8.4.1**	**8.4.2**	**8.4.3**	**8.4.4**	**8.4.5**	**8.4.6**	**8.4.7**
C=Y	**8.5.1**	**8.5.2**	**8.5.3**	**8.5.4**	**8.5.5**	**8.5.6**	**8.5.7**
C≡Y	**8.6.1**	**8.6.2**	**8.6.3**	**8.6.4**	**8.6.5**	**8.6.6**	**8.6.7**
C=C–EWG	**8.7.1**	**8.7.2**	**8.7.3**	**8.7.4**	**8.7.5**	**8.7.6**	**8.7.7**
L–C=Y	**8.8.1**	**8.8.2**	**8.8.3**	**8.8.4**	**8.8.5**	**8.8.6**	**8.8.7**

8.2 H–A SINKS REACTING WITH COMMON SOURCES

8.2.1 Lone Pair Sources Reacting with Acids

Acids react by simple protonation of the electron source via proton transfer (see Section 3.6 for the calculation of the K_{eq}).

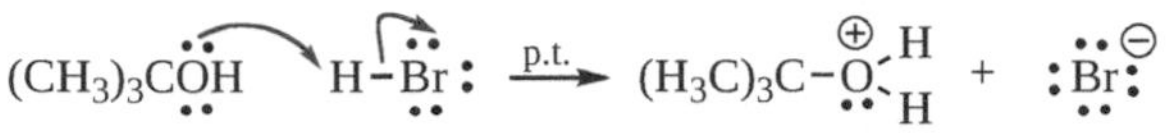

8.2.2 Bases Reacting with Acids Having an Adjacent CH

This simple neutralization reaction always favors the weaker base. Thus, the most acidic H is deprotonated, even if the acid has an adjacent CH. Because of the higher electronegativity of heteroatoms, protons on heteroatoms are usually lost before protons on carbon. The base *n*-butyllithium deprotonates the nitrogen of diisopropyl amine. The adjacent CH is about 14 orders of magnitude less acidic. Always check the K_{eq} to see if the proton transfer is reasonable. The K_{eq} for the reaction below is highly favorable, $10^{(50-36)} = 10^{+14}$.

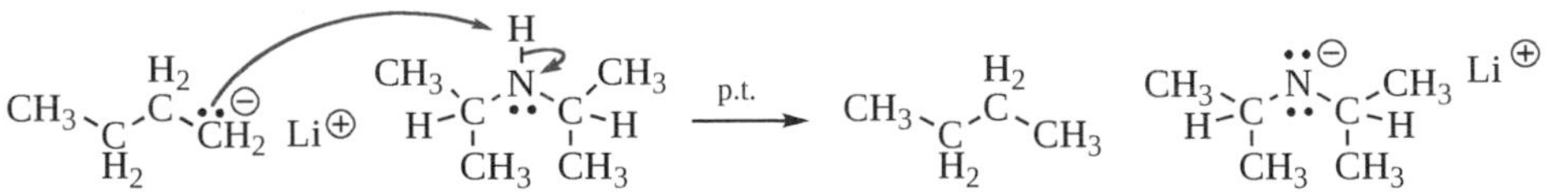

8.2.3 Complex Metal Hydrides Reacting with Acids

Aluminum hydrides react violently with protic solvents and acids (path p.t.) to produce hydrogen gas and can be used only in dry *aprotic* media, like diethyl ether. The borohydrides, BH_4^- are weaker sources and react much more slowly with protic solvents than do the aluminum hydrides. Thus, borohydrides can be used in alcohol solvents. After the first step, the trivalent aluminum or boron species is now electron deficient and can react with a lone pair donor to form a tetravalent Lewis acid–base salt (path A_N) that can donate hydride again. This repeats until all hydrogen–metal bonds have been used.

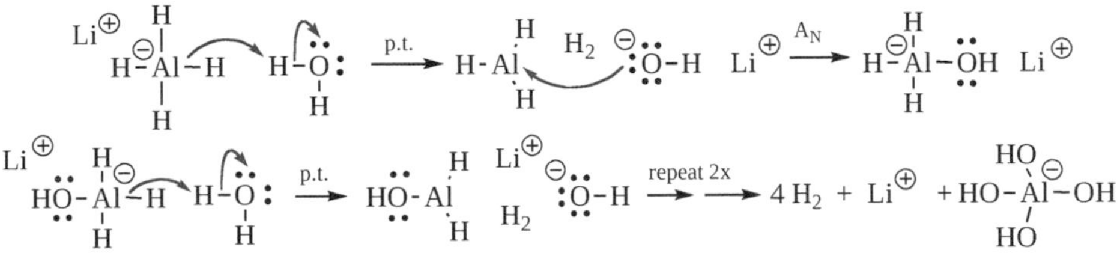

8.2.4 Organometallics Reacting with Acids

Organometallics can act as bases and get protonated by any acidic hydrogen (path p.t.). If the acidic hydrogen is a C–H bond, a new and more stable organometallic is slowly produced. Reaction of RM with any acidic O–H bond is fast, and often undesired, so organometallics are usually formed in anhydrous aprotic solvents like diethyl ether.

8.2.5 Allylic Sources Reacting with Acids

Allylic sources are ambident nucleophiles (Sections 5.4 and 9.4) and can therefore attack an electrophile at either of two sites. Since proton transfer is commonly reversible and rapid, an equilibrium mixture is quickly achieved. When the source is anionic, the protonation occurs most rapidly on the heteroatom (hard–hard, path p.t.) but can also occur on carbon. Equilibration to the more stable product occurs by proton transfer (Section 7.3.8, tautomerization). A $\Delta H°$ calculation (Section 2.5) will verify that the carbon-protonated species is the more stable product.

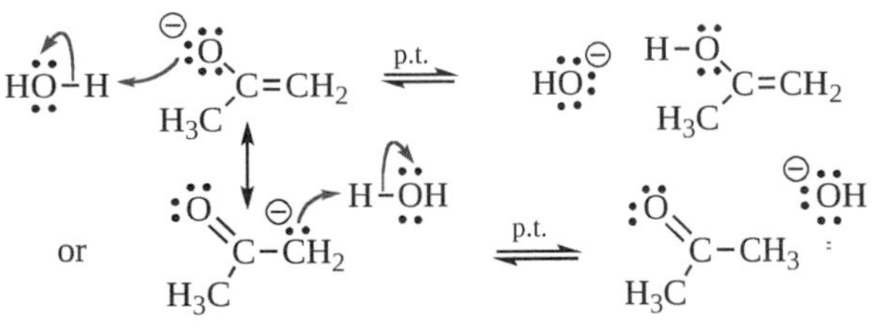

When the source is neutral, protonation can occur to produce the heteroatom-protonated species or the delocalized heteroatom-substituted carbocation. Protonation on the more basic heteroatom lone pair is the lower-energy process, but the heteroatom-protonated species usually just returns to reactants.

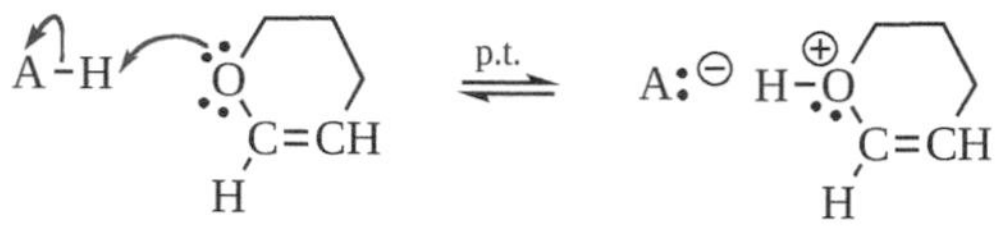

However, protonation on carbon creates a delocalized heteroatom-substituted carbocation that can easily be captured (path A_N) by solvent or another nucleophile.
Ad_E2 example (with follow-up proton transfer to solvent):

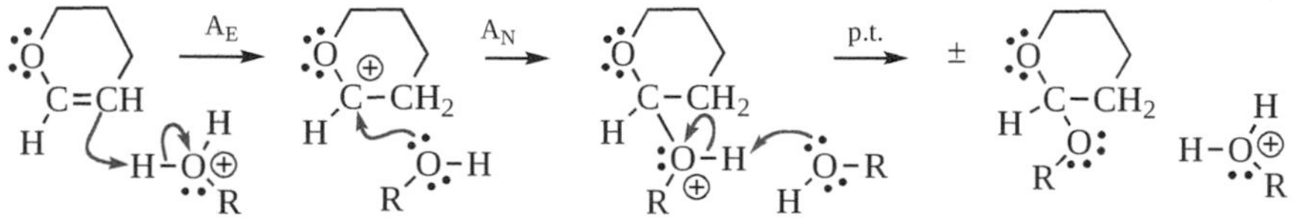

8.2.6 Simple Pi Bonds Reacting with Acids

Electrophilic Addition of H–A to Alkenes

As was discussed in detail in Sections 4.4.1–4.4.4, here are two mechanisms by which acids add to alkenes, the Ad_E2 and Ad_E3 processes. The reaction is carried out in the dark to minimize free radical side reactions (Chapter 12). In the Ad_E2, the proton adds by pathway A_E to produce the more stable carbocation intermediate (Markovnikov's rule); in a second step the cation is trapped by a nucleophile, path A_N. The Ad_E2 often gives a mixture of *syn* and *anti* addition since the nucleophile can approach the carbocation from either top or bottom face (Section 7.3.2). Ad_E2 example:

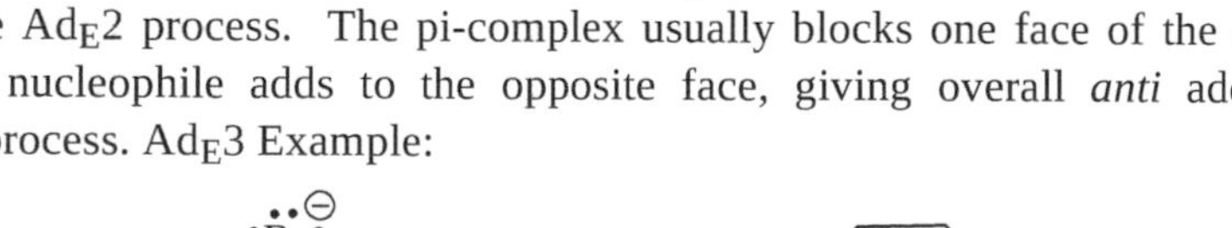

The Ad_E3 occurs when the cation is not as stable (Section 7.1.8). Instead of proton transfer, a pi-adduct forms, which polarizes and activates the double bond. The nucleophile attacks the largest partial plus of the complex and gives the same Markovnikov orientation of product as the Ad_E2 process. The pi-complex usually blocks one face of the double bond, therefore the nucleophile adds to the opposite face, giving overall *anti* addition as the predominant process. Ad_E3 Example:

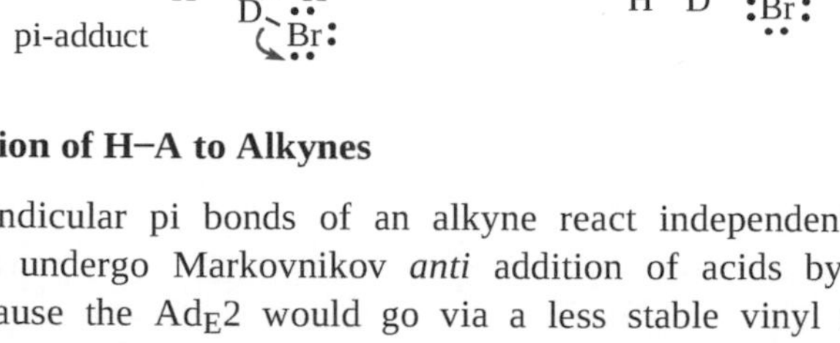

Electrophilic Addition of H–A to Alkynes

The two perpendicular pi bonds of an alkyne react independently of one another. Alkynes commonly undergo Markovnikov *anti* addition of acids by the Ad_E3 process (Section 7.1.8) because the Ad_E2 would go via a less stable vinyl cation intermediate. Slightly more vigorous conditions will favor the double-addition product.

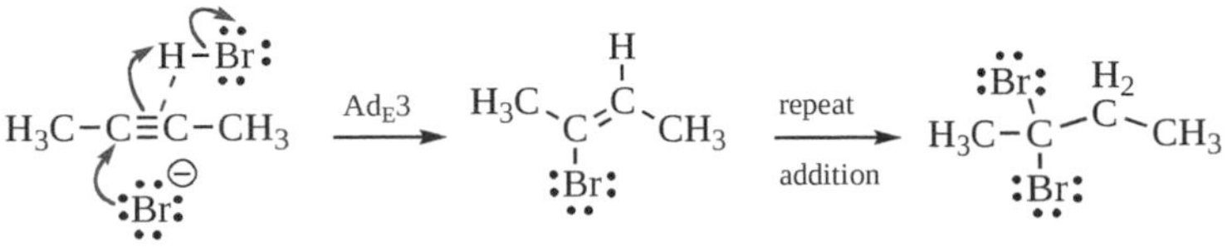

Electrophilic addition can be followed by tautomerization (Section 7.3.8).

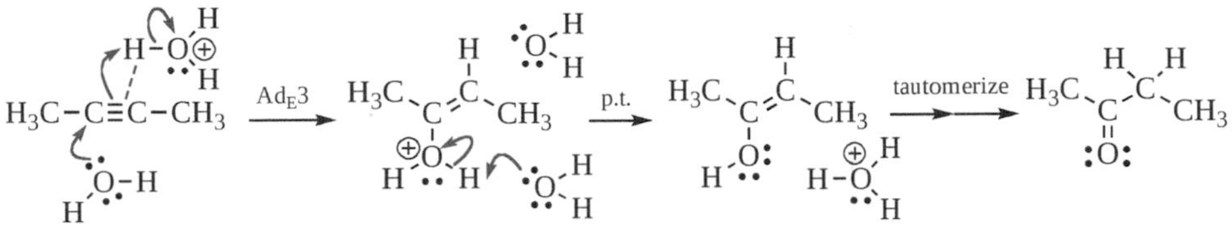

Electrophilic Addition of H−A to Dienes C=C−C=C

An electrophile will attack a diene to produce the delocalized allylic cation. The partial charge on each end of the allylic system will be different if the substitution at each end of the allylic unit is different. Since the attraction of unlike charges contributes greatly to the bringing of nucleophile and electrophile together (hard–hard), **the larger the partial plus is on an atom, the greater a negatively charged nucleophile is attracted to it.**

For the simplest diene, 1,3-butadiene, electrophilic attack creates an allylic cation where the greatest partial plus is on the end bearing the methyl electron releasing group.

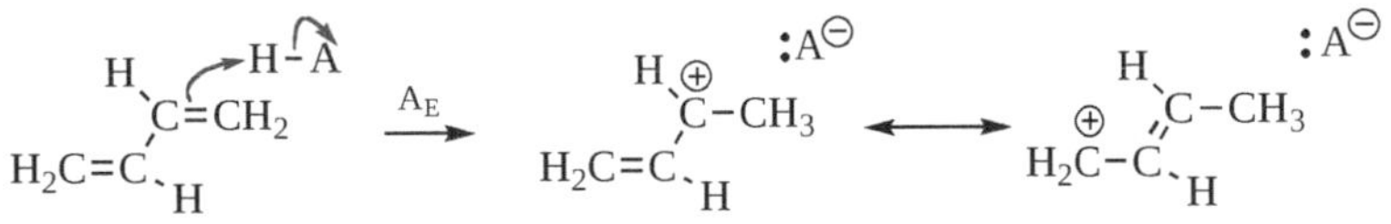

Major Minor

If the rate of nucleophile attack governs the product distribution (the nucleophilic attack is not reversible at that temperature), the major product will be the 1,2-addition product resulting from Nu attack at the greatest partial plus (kinetic control, Section 2.6.1). However, if the addition of the nucleophile is reversible (higher temperature or the Nu is a good L), then the 1,4-addition product will be formed because the more substituted double bond is the more stable product (thermodynamic control, Section 2.6.1). Suspect kinetic control if the reaction is run significantly below 0°C. Diene example (Ad_E2):

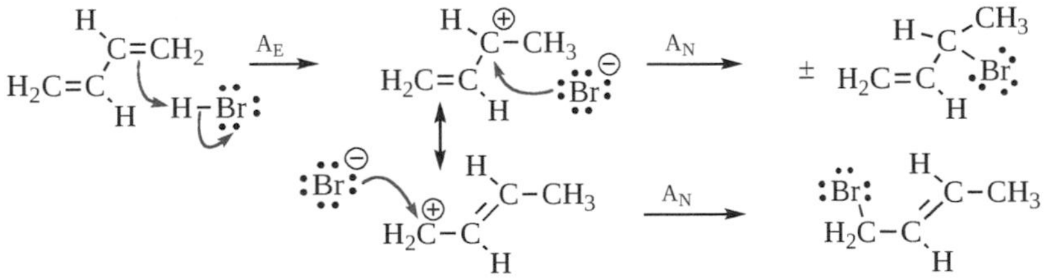

1, 2

1, 4

8.2.7 Aromatics Reacting with Acids

Aromatics Reacting with H−A

A strong acid can protonate an aromatic ring to give a cationic sigma complex that can rapidly protonate any conjugate base to rearomatize the ring. This reaction is noticeable only if *ipso* attack triggers rearrangement, or loss of a group on the ring, or if a deuterated acid is used. One possible mechanism for desulfonation of an aromatic ring is shown below.

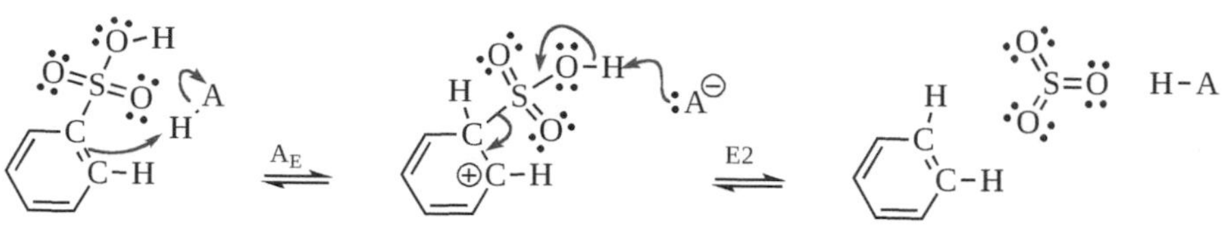

8.3 Y–L SINKS REACTING WITH COMMON SOURCES

8.3.1 Lone Pair Sources Reacting with Y–L

Since Y often represents an electronegative atom, Y^+ is usually a poor cation, so S_N1 is ruled out, commonly leaving S_N2. The ΔpK_a rule cross-check is helpful in determining the position of equilibrium in the displacement of a leaving group by a nucleophile.
Example (with follow-up proton transfer):

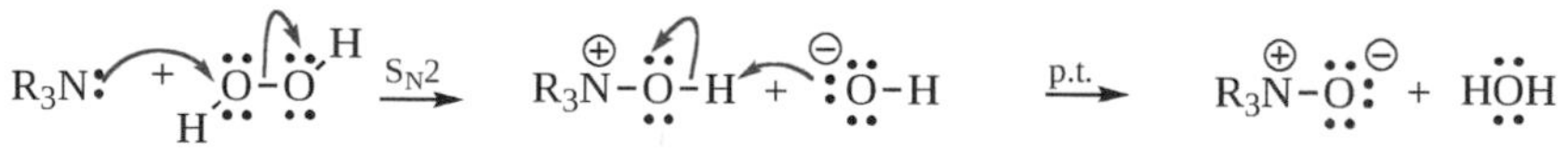

Sulfur example: form a tosylate from tosyl chloride, an alcohol, with an amine base:

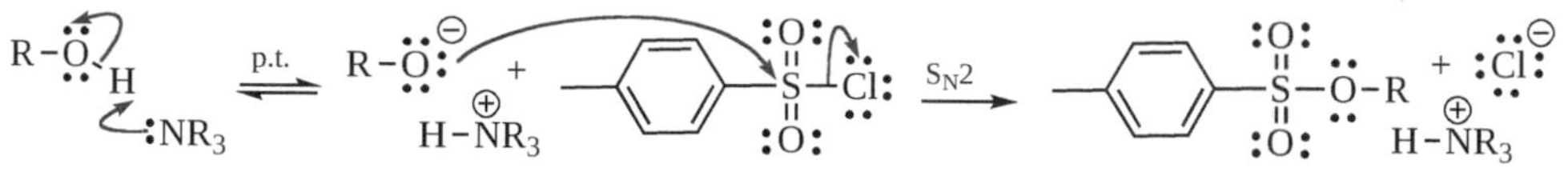

The amine may deprotonate the alcohol before (shown) or after the attack on sulfur. Sulfur probably goes by the S_N2, but an addition–elimination or pentacovalent route is possible. Most likely the substitution at phosphorus goes via the pentacovalent intermediate (path pent.). Second, the phosphorus–oxygen bond is so strong that follow-up reaction often occurs in which O–P serves as a leaving group. Overall, an OH group is converted to a better leaving group, then is displaced by a nucleophile, commonly the leaving group on the Y–L. Phosphorus example: with follow-up S_N2 substitution, covered in Section 8.4.1:

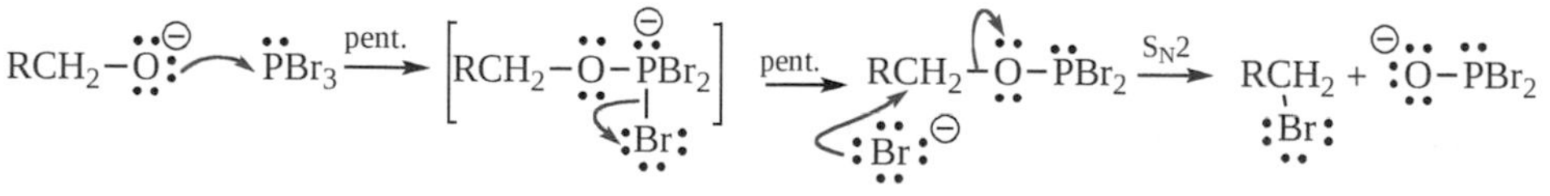

Biochemical Example

Adenosine triphosphate, ATP^{4-}, is the body's energy currency and a Y–L sink. It activates compounds by phosphorylation. The mechanism of the enzyme glycerol kinase (Figure 8.1) is postulated to go by a pentacovalent phosphorus intermediate. A base in the active site makes the oxygen lone pair more nucleophilic by deprotonating it as it attacks (general base catalysis, Section 7.4.3). The magnesium ion acts as an electrophilic catalyst (Section 6.2.3), improving the leaving group, ADP^{3-}.

8.3.2 Bases Reacting with Y–L Having an Adjacent CH, Oxidations

Since Y often represents an electronegative atom, Y^+ is usually a poor cation. Therefore, the E1 path is out. The bases used in this reaction are usually weak, leaving the E2 or the Ei as the common route (Section 4.3.7). For examples of this reaction see Section 8.12.3, Nu–L Reacting with sp^3 C–L, and Section 8.10.2, Metal Ions as Oxidants.

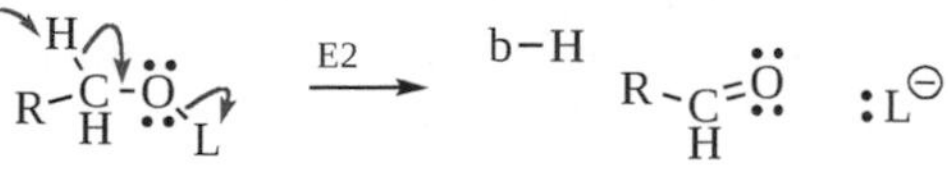

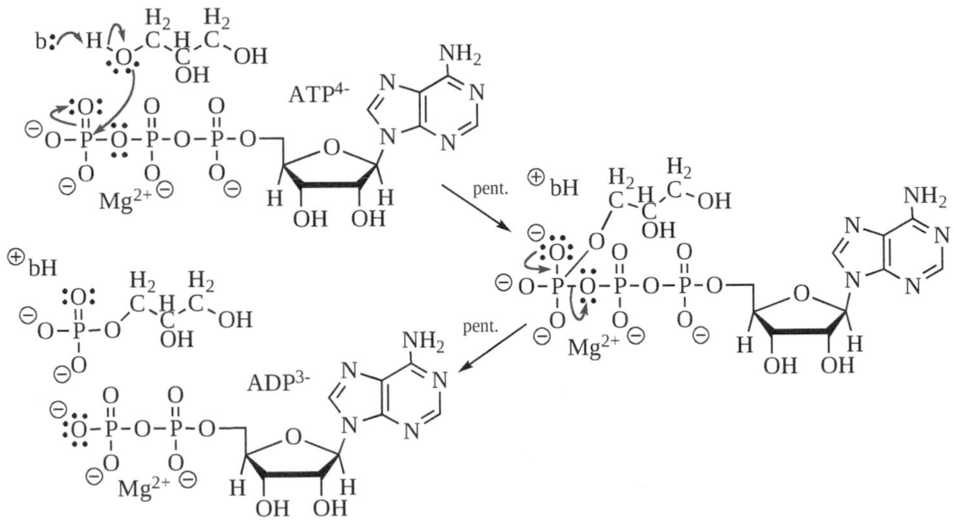

Figure 8.1 Phosphorylation of glycerol by ATP^{4-}.

8.3.3 Complex Metal Hydrides Reacting with Y–L

The substitution reaction occurs in neutral or basic media via the S_N2 pathway to replace the leaving group with hydrogen. Often the metal cation can form a complex with the leaving group to aid in its leaving.

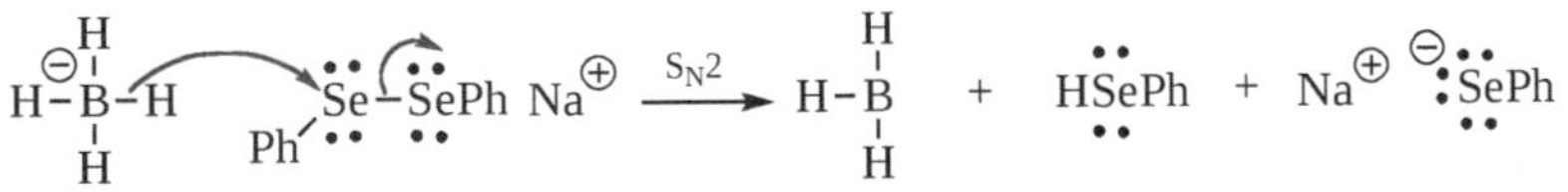

8.3.4 Organometallics Reacting with Y–L

Since organometallics react with acids (Section 8.2.4), they can be used only in basic or neutral media and in aprotic solvents. The most common reaction is S_N2 substitution.

8.3.5 Allylic Sources Reacting with Y–L

An allylic electron source can be considered to be a vinylogous lone pair. Sometimes the lone pair itself is the nucleophile and can undergo reactions similar to the lone pair sources in the first column of the matrix. However, most often the nucleophile is the end carbon atom of the pi bond. The heteroatom lone pair is much harder than the pi bond, and the decision of which to use as the nucleophile is based on the hardness of the electrophile (HSAB principle) and will be discussed at greater length in Section 9.4. The Y–L reaction is almost exclusively by the S_N2 pathway because Y^+ is usually a poor cation. However, the substitution at phosphorus and silicon usually goes via the pentacovalent intermediate (path pent.). The following examples show the dual reactivity of the same allylic source, the enolate ion.
Z as Nu (hard E, Si–Cl bond polarized because of a large electronegativity difference):

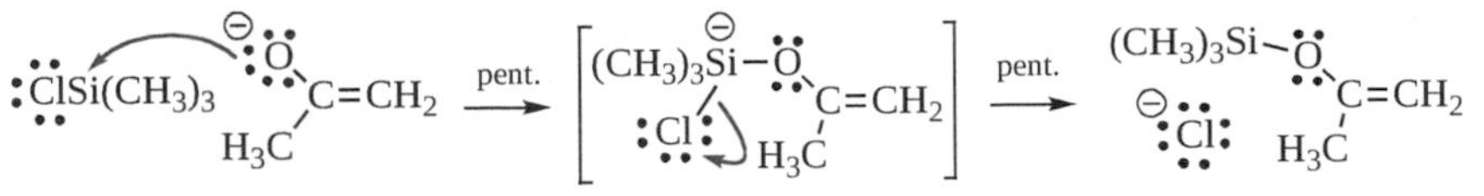

C as Nu (soft E, Br–Br bond not polarized at all):

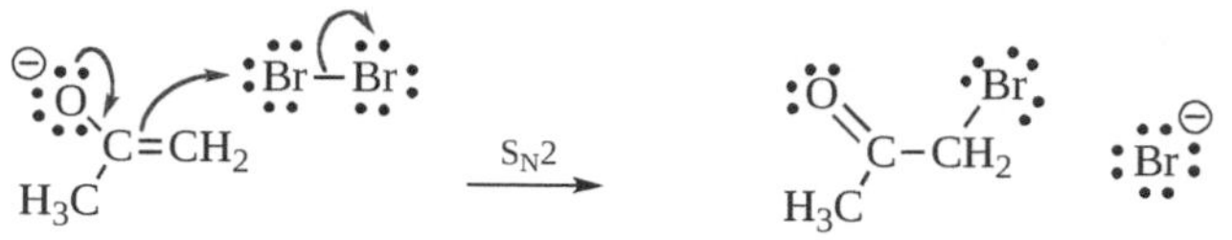

8.3.6 Simple Pi Sources Reacting with Y–L

Electrophilic Addition of Y–L to Alkenes

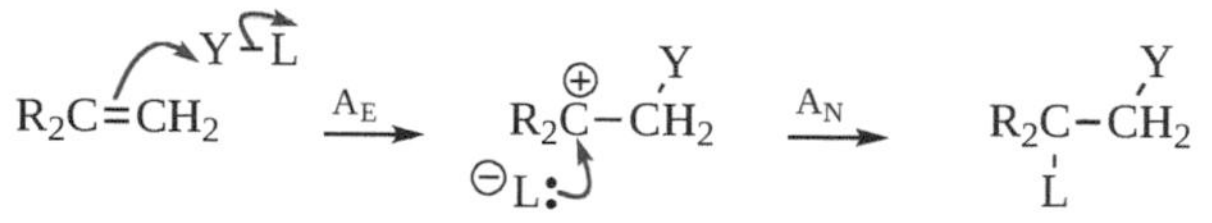

This reaction and the electrophilic addition of acids compose almost all of the ionic reactions of simple alkenes as electron sources. When Y bears a lone pair, the intermediate carbocation can be further stabilized by sharing the lone pair electrons with the cationic center in a bridge to form a three-membered ring, resonance form B in the following example. Although the Ad_E2 process (A_E, followed by A_N) was drawn above, it would often be proper to use the Nu–L path (Section 7.2.4) to get to the bridged onium ion directly, followed by path S_N2 to open the bridge. An example is bromination, in which the bromide leaving group traps the cation when run in nonnucleophilic solvent.

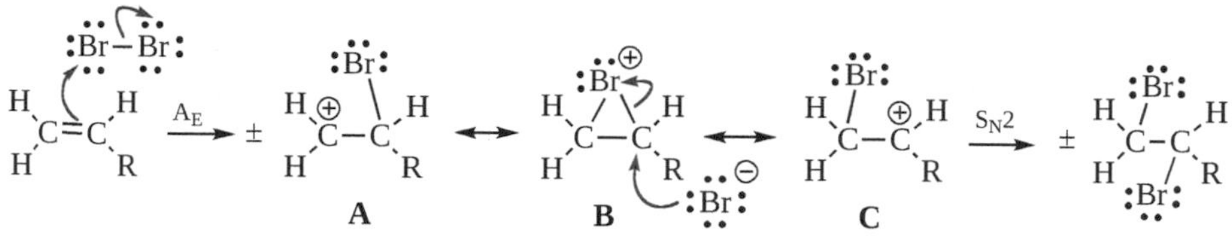

Bridging gives all atoms an octet, and some energy is released in forming a new bond, but this is at the expense of creating ring strain and putting a partial positive charge on the heteroatom. A highly electronegative element like fluorine is relatively poor at bridging. Less electronegative elements have a greater tendency to share their lone pairs and bridge. Stabilization of the carbocation by electron-donor groups favors the open form. Bridging is favored if the cation is poorly stabilized.

The second step in the electrophilic addition of Y–L to alkenes is the attack of a nucleophile on the cationic intermediate by the original leaving group or the solvent. The bridging group is displaced by an S_N2 path, so the nucleophile comes in from the opposite face, giving overall *anti* addition. The site of attack will depend on which resonance form above, A or C, contributes more to the overall hybrid. The nucleophile will usually attack the carbon with the greater partial plus. Since the bromine could have attacked the pi bond from either the top or bottom face, the product is racemic. Solvent trap example (same reaction run in water solvent):

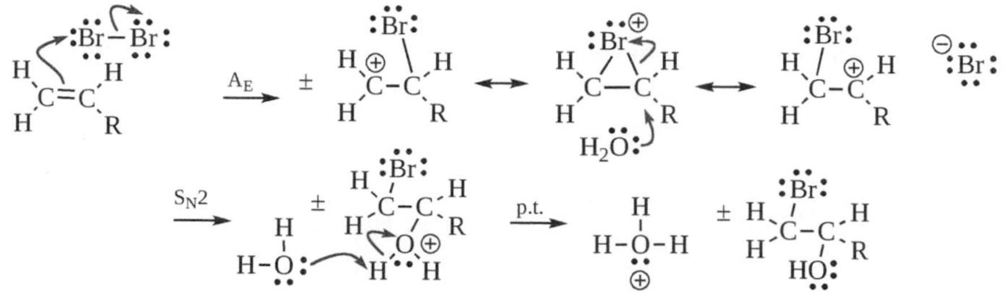

Electrophilic Addition of Y–L to Alkynes

Alkynes undergo the same set of reactions as alkenes but are slightly less reactive because the intermediates involved are less stable. For the Ad_E2 process, the vinyl cation intermediate formed is less stable than the alkyl carbocations formed when electrophiles attack alkenes. Addition to the vinyl cation produces a mixture of *syn* and *anti* addition. Stabilizing the vinyl cation by bridging is less favorable since the bridged ion is more strained and may have some antiaromatic character.

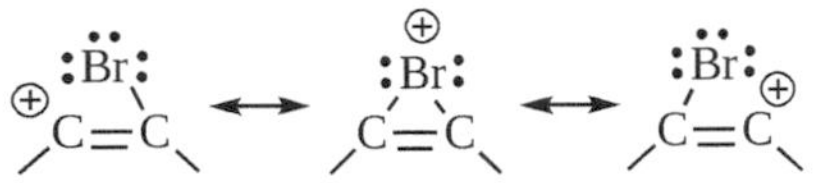

The Ad_E3 process competes with the Ad_E2 and can make the overall addition *anti*. The Ad_E2 process is seen when there is some additional stabilization of the vinyl cation. Double additions are frequent. Often the product of an electrophilic addition is less reactive than the original alkyne, and thus the reaction can be stopped before a second addition occurs (if run with an equimolar amount of electrophile and alkyne). Ad_E3 example (note *anti* addition):

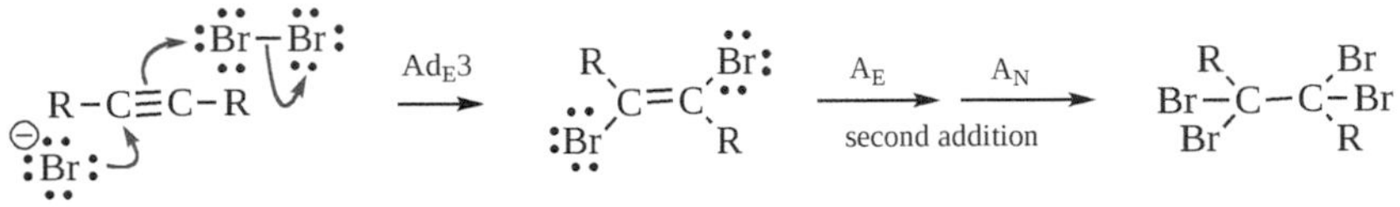

Electrophilic Addition of Y–L to Dienes

Since the allylic cation formed upon addition of an electrophile to an alkene is relatively stable, dienes add Y–L by the Ad_E2 process. Both 1,2 and 1,4 addition can occur. At higher temperatures, the more stable 1,4 product predominates because the products equilibrate.

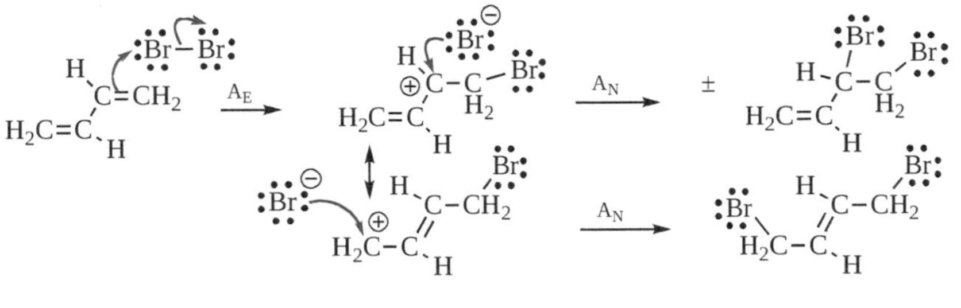

8.3.7 Aromatic Sources Reacting with Y–L

Because of the lower reactivity of aromatic sources, many of the electrophiles are cationic, and their generation and use are often similar to an S_N1 process, as in the following nitration reaction. Protonation on nitric acid's more basic oxygen atoms is reversible and does not produce the electrophile. Only when nitric acid's OH is protonated can water be lost to give the resonance-delocalized NO_2^+ electrophile needed for the reaction.

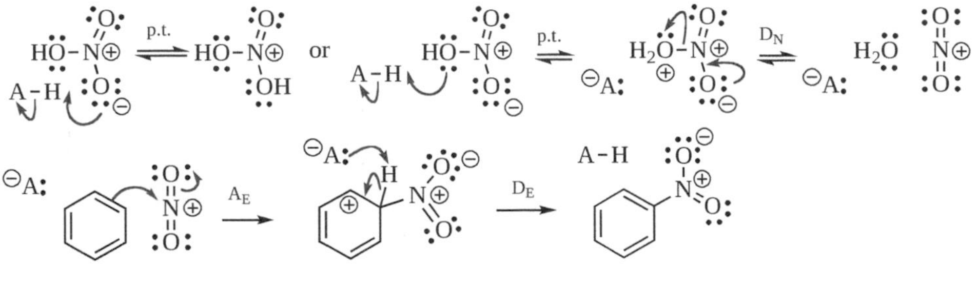

Other reactions merely require that a Lewis acid catalyze the departure of the leaving group similar to an S_N2 process on the electrophile, as in the following chlorination reaction, catalyzed by aluminum chloride.

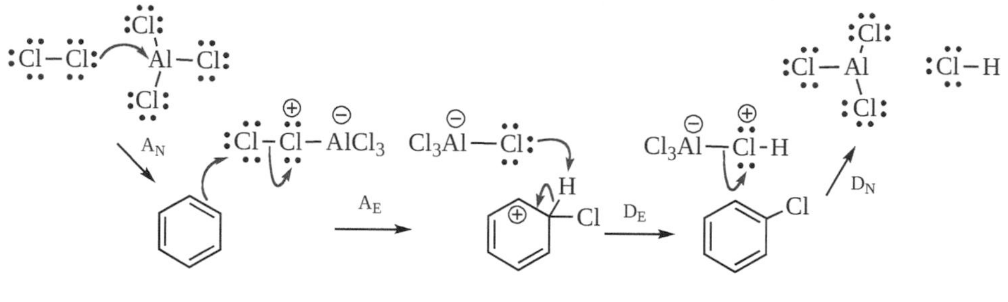

8.4 *sp*3 C–L SINKS REACTING WITH COMMON SOURCES

8.4.1 Lone Pair Sources Reacting with *sp*3 C–L

The substitution vs. elimination decision is considered in depth in Section 9.5.

Alkylation

As was covered in Section 4.2.1, there are two routes that differ only in the timing of when the nucleophile attacks and when the leaving group falls off: **Generic S_N1**—A good leaving group departs first by path D_N, producing a reasonably stable cation; the cation is then trapped by the nucleophile, path A_N. **Generic S_N2**—Simultaneous attack of nucleophile and loss of the leaving group. It is best to consider a S_N1/S_N2 spectrum where there is competition between the rate of ionization of the leaving group and the rate of nucleophilic attack on the substrate or partially ionized substrate (Figure 4.11).

S_N2 example (poor cation, good access):

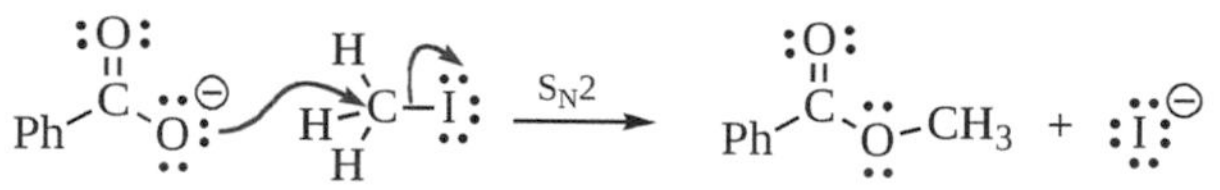

S_N1 example (good cation, hindered site) with follow-up proton transfer:

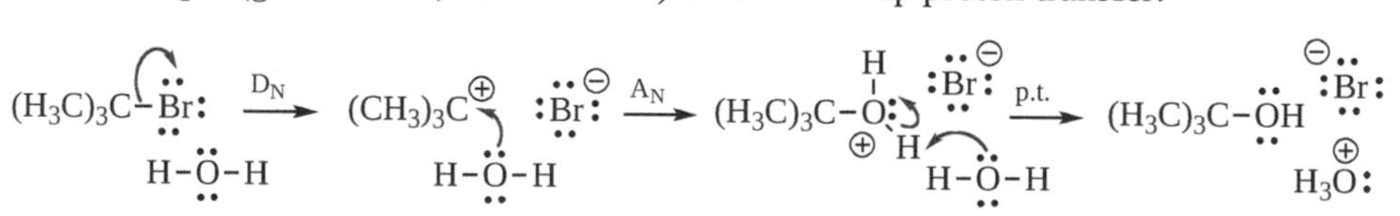

S_N2 Biochemical Example

S-adenosylmethionine is the primary biochemical methyl group donor and provides an excellent example of a biological S_N2 reaction (Figure 8.2).

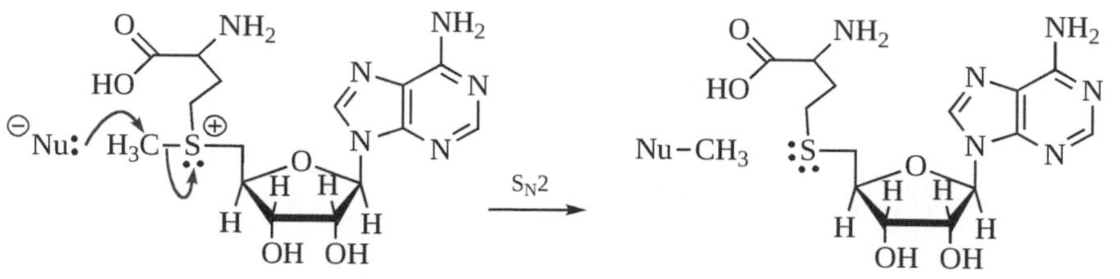

Figure 8.2 Methylation of a nucleophile by *S*-adenosylmethionine.

S$_N$1 Biochemical Example

Glycogen phosphorylase mobilizes the body's stored glycogen to glucose-1-phosphate for use in glycolysis, the metabolic sequence in producing energy from glucose. The first step is a proton transfer to improve the leaving group, another sugar (Figure 8.3). An E$_β$ step starts the lone-pair-assisted S$_N$1 to give an oxygen lone-pair-stabilized carbocation that is trapped by phosphate in an A$_N$ step to give glucose-1-phosphate.

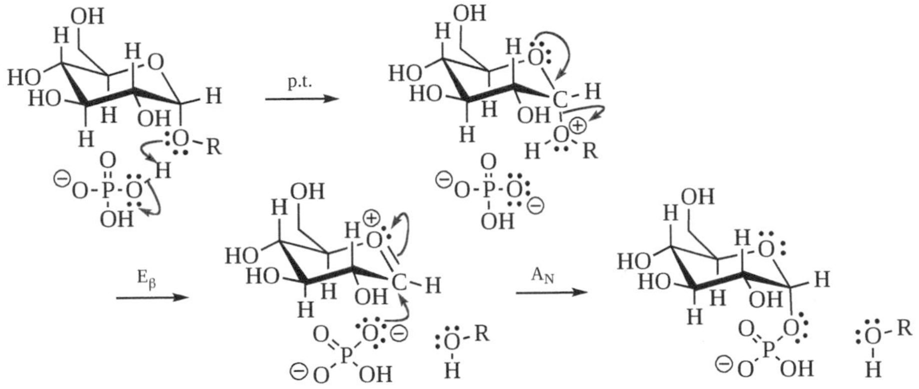

Figure 8.3 Phosphorylation of glycogen.

8.4.2 Bases Reacting with *sp*3 C–L Having an Adjacent CH, Eliminations

As covered in Section 4.3.1, there are three pathways that differ only in the timing of when the proton is pulled off and when the leaving group falls off: **Generic E1**—A good leaving group departs first by path D$_N$, producing a reasonably stable cation; the proton is then lost by path D$_E$, forming the alkene. **Generic E2**—Simultaneous loss of proton and leaving group. **Generic E1cB**—Proton transfer forms an anion stabilized by an electronegative atom or group, and then anion pushes out the leaving group by path E$_β$. Path selection between substitution and elimination is considered in depth in Section 9.5.

Alkene Formation via the E1/E2/E1cB Spectrum

The different elimination paths often produce different alkene constitutional isomers as products (*regiochemistry*). Figure 4.18 showed in the E2 the C–H bond and the C–L bond must lie in the same plane, preferably *anti* to each other for steric reasons. The protons, shown in bold in the scheme below, on either carbon adjacent to the leaving group can be rotated into proper alignment for the E2 reaction; thus three different products can be produced. By rotating the RCH$_2$ to align the other bold H of the RCH$_2$ to be antiperiplanar with the C–L bond, the *cis* product can be formed by E2 also.

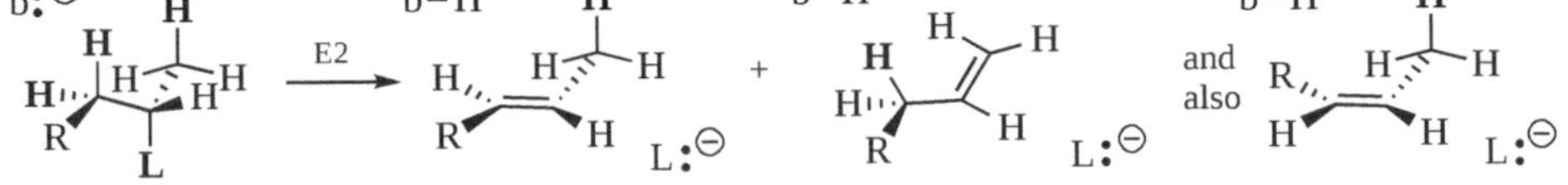

The regiochemistry of the elimination depends on the type of elimination process that occurs. **The E1 process favors the formation of the more substituted alkene** because reversible protonation of the double bond occurs and creates an equilibrium mixture that favors the more stable product. **The E2 regiochemistry is controlled by the need to minimize steric interactions in the transition state;** the size of the base is important because one proton may be more accessible than another, as in Figure 4.19. **The E1cB regiochemistry is determined by the loss of the most acidic proton.** Elimination reactions can produce different stereoisomers, for example, *cis* and *trans* alkenes. Since the *trans* isomer is usually of lower energy because of steric reasons, it usually predominates over the *cis* isomer in the product mixture.

E1 example—good L and cation, forms most substituted alkene (Zaitsev's rule):

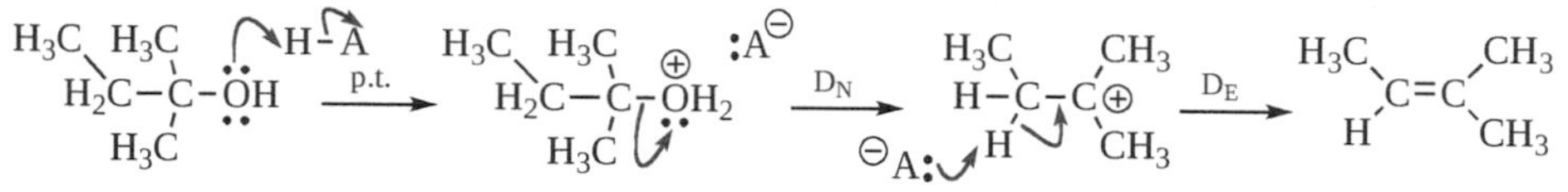

E2 example—concerted:

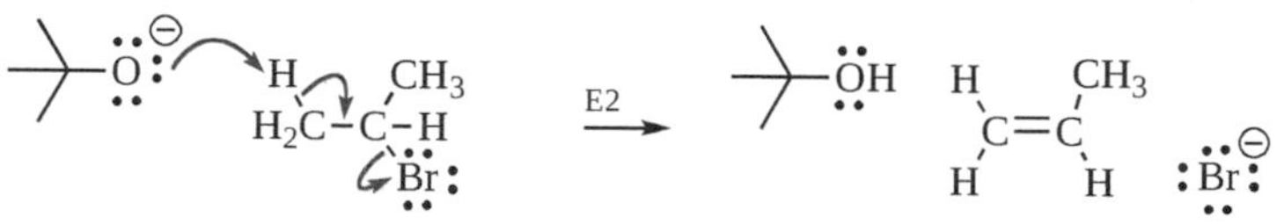

E1cB example—EWG makes C–H acidic; anion needed to kick out poor L:

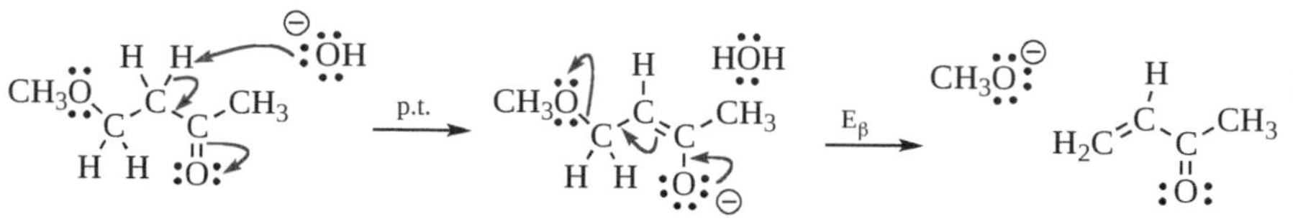

Heteroatom Variations (sp^3 C–L Having an Adjacent YH)

Anionic eliminations of the following type will also be considered as belonging to the E1cB class of eliminations. Since protons on heteroatoms are usually rather acidic because of the electronegativity of the heteroatom, the first step, deprotonation, can occur with a rather weak base. The second step is beta elimination from an anion, E_β, the reverse of nucleophile addition to a polarized multiple bond.

Hetero E1cB example (see Figures 4.38 and 4.39 for an energy diagram):

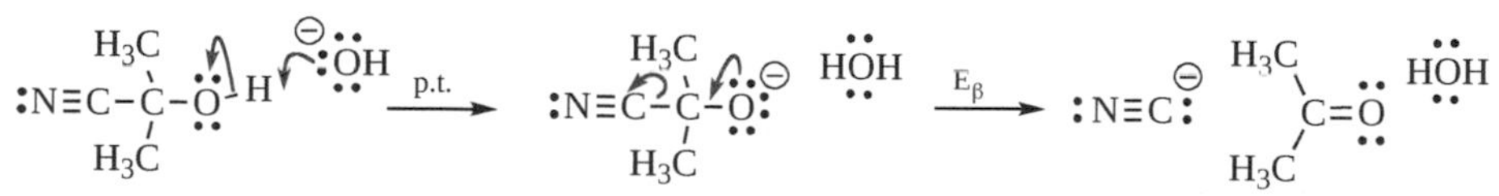

Likewise, a lone pair on a heteroatom can aid the loss of the leaving group in the E1 process. The following example shows first the protonation of the leaving group followed by the lone-pair-assisted E1, path E_β followed by proton transfer.

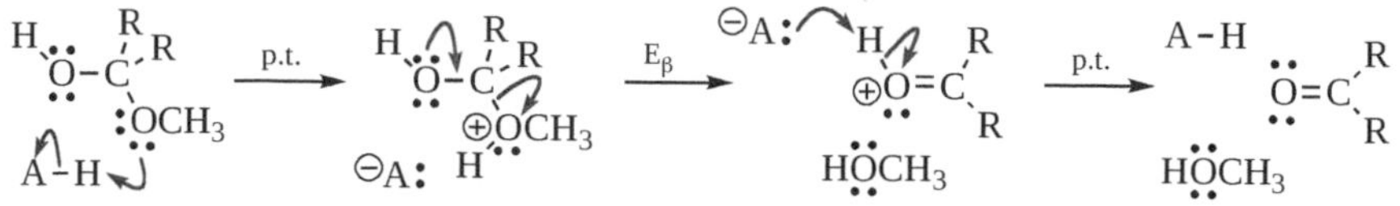

A hetero E2 example was discussed in Section 7.4, Variations on a Theme.

E1cB Biochemical Example

The synthesis of fats is an important way an organism stores the energy from foods. The dehydration step in fatty acid biosynthesis is an example of an E1cB sequence in which the conjugate acid of the initial base serves to protonate the leaving group on departure. This second step is a general acid-catalyzed beta elimination (Section 7.4.3).

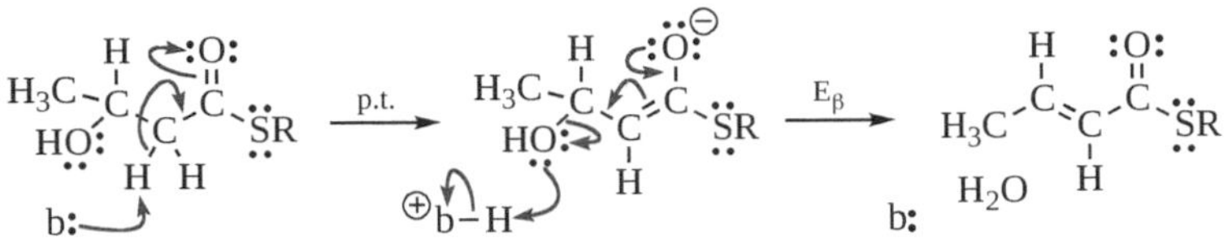

8.4.3 Complex Metal Hydrides Reacting with sp^3 C–L

The substitution reaction occurs in neutral or basic media and in aprotic solvents via the S_N2 pathway to replace the leaving group with hydrogen. Often, the metal cation can form a complex with the leaving group to aid in its leaving.

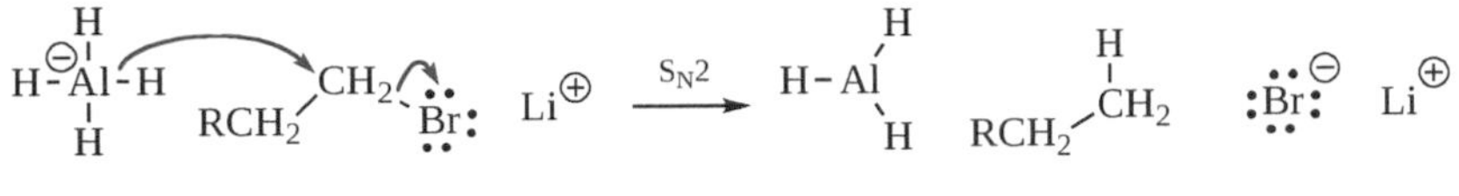

8.4.4 Organometallics Reacting with sp^3 C–L

Since organometallics react with acids (Section 8.2.4), they can be used only in basic or neutral media and in aprotic solvents. The most common reaction is substitution via S_N2. Substitution is favored by sulfonate leaving groups.

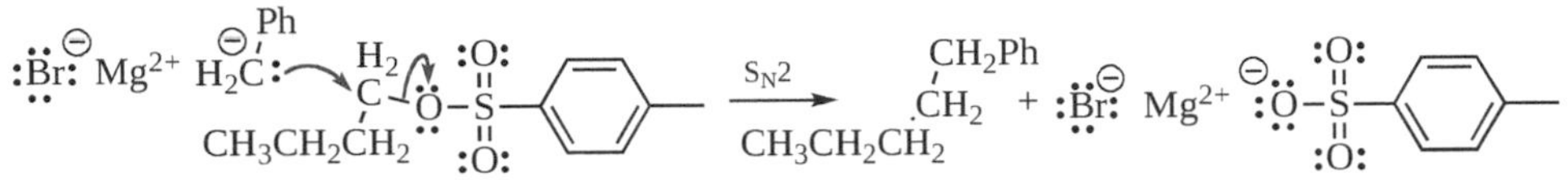

8.4.5 Allylic Sources Reacting with sp^3 C–L

An allylic electron source can be considered to be a vinylogous lone pair. Sometimes, the lone pair itself is the nucleophile and can undergo reactions similar to the lone pair sources in the first column of the matrix. However, most often the nucleophile is the end carbon atom of the pi bond. The heteroatom lone pair is much harder than the pi bond, and the decision of which to use as the nucleophile is based on the hardness of the electrophile (HSAB principle) and will be discussed at greater length in Section 9.4.

Enamines, N–C=C, usually alkylate on carbon with this sink, and produce an iminium ion, $^+$N=C–C–R, which is normally hydrolyzed to the ketone in an acidic water workup (the reverse of imine formation, Section 8.5.1). Enamine alkylation example:

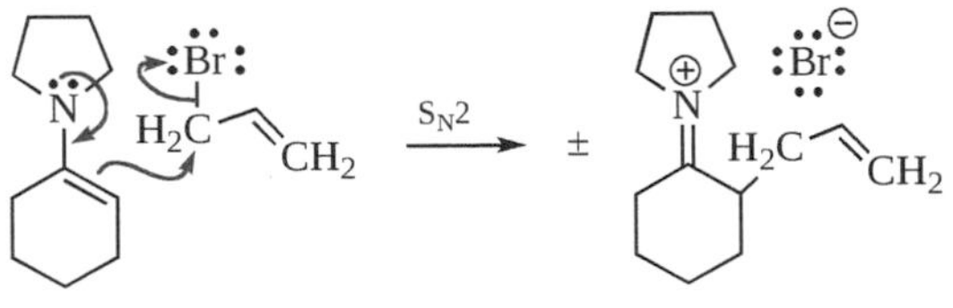

However, with simple enolates, $^-$O–C=C, alkylation can be a messy reaction. The alkylation product and the enolate can usually undergo proton transfer, producing new electron sources and leading to undesired products. The best way to run this reaction is to add the enolate source into an excess of sink and hope that the alkylation proceeds faster than proton transfer.

Allylic Electron Sources with Additional Delocalization

One can speed up alkylation relative to elimination by stabilizing an enolate anion with an additional EWG, making the anion softer and less basic. For example, the acidic CH_2 of malonates, $CH_2(COOEt)_2$, or acetoacetates, CH_3COCH_2COOEt, deprotonates easily and makes an excellent nucleophile. The ester can later be hydrolyzed to the acid (Section 8.8.1) and removed by decarboxylation via path Ei (Section 7.2.3). This ester is a detachable EWG making the enolate less basic and softer, favoring substitution over elimination (Section 9.5).

Acetoacetate example:

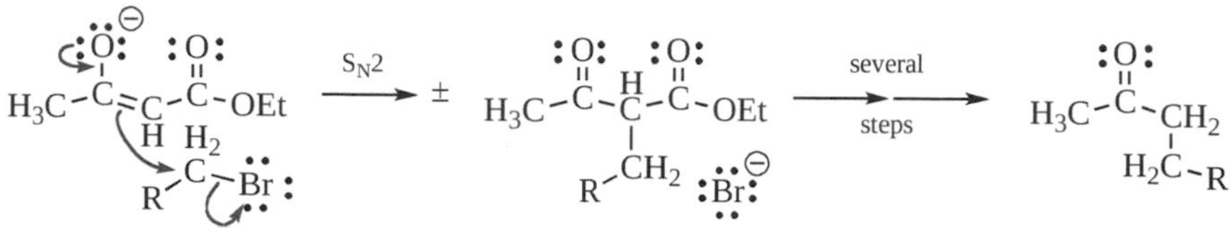

8.4.6 Simple Pi Sources Reacting with *sp*3 C–L

Simple pi bonds are usually not good enough nucleophiles to react with alkyl halides. If a Brønsted–Lowry acid or a Lewis acid is added to improve the electron sink, then addition can happen, but it is often complicated with polymerization.

Biochemical Example

The synthesis of terpene natural products begins with the following coupling reaction, Figure 8.4. The magnesium ion acts as a Lewis acid and complexes to the pyrophosphate (diphosphate) leaving group (OPP) to activate its departure. The delocalized allylic carbocation adds to the pi bond of isopentenyl pyrophosphate, which then loses a proton to form the coupling product geranyl pyrophosphate. The coupling can be repeated with another isopentenyl pyrophosphate, adding another five-carbon piece.

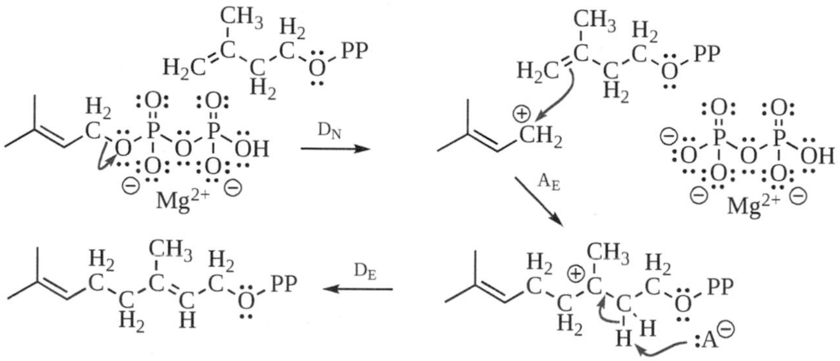

Figure 8.4 Biosynthesis of geranyl pyrophosphate from dimethylallyl pyrophosphate.

8.4.7 Aromatic Sources Reacting with *sp*3 C–L

Friedel–Crafts Alkylation

Aromatic rings are usually not good enough nucleophiles to react with alkyl halides. If a Brønsted–Lowry or Lewis acid is added to improve the electron sink, then electrophilic aromatic substitution occurs. If the carbocation formed would be more stable than secondary, then the loss of the leaving group starts the process (an S_N1 viewed from the electrophile). With methyl, primary, or secondary alkyl halides, electrophilic aromatic substitution usually occurs via direct displacement of the leaving group (an S_N2 viewed from the electrophile).

This reaction has several limitations and problems. First, the reaction fails for deactivated rings. Second, if it can, the carbocation almost always rearranges to a more stable carbocation. Third, once the alkyl group is attached, it activates the ring for further attack, so sequential electrophile additions to give multiple substituted aromatics are possible. The electrophiles are generated in a variety of ways:

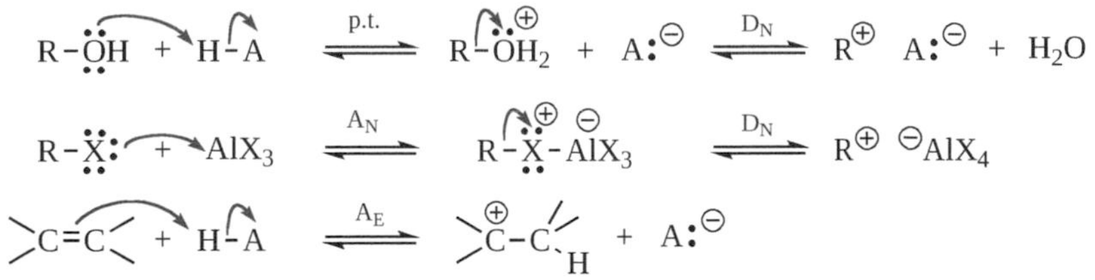

Alkylation example (an S_N2 viewed from the electrophile):

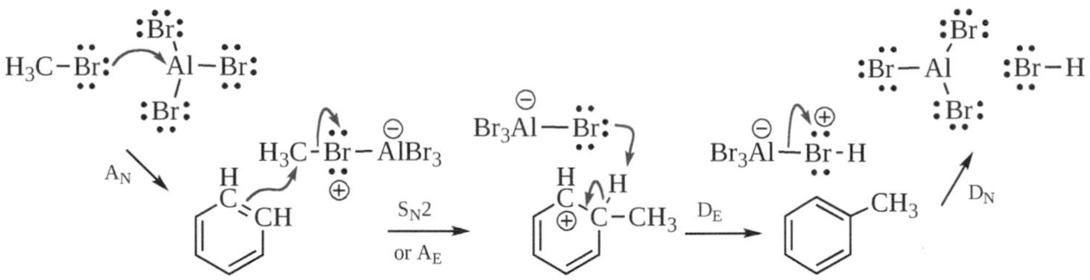

Alkylation example (an S_N1 viewed from the electrophile):

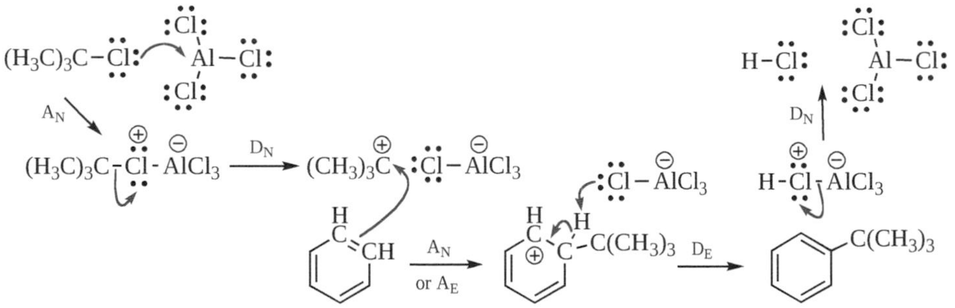

8.5 C=Y SINKS REACTING WITH COMMON SOURCES

Section 4.47 introduced the theory of additions to carbonyls, C=O, the most common example of C=Y sinks. All three routes, Ad_N2, Ad_E2, and Ad_E3 were found to occur.

8.5.1 Lone Pair Sources Reacting with C=Y

These reactions are very common (see Figures 7.16 and 7.17 for the interrelationship of the available paths in acidic and in basic media). **In basic media, the attack of the nucleophile on a polarized multiple bond forms an anion that in the workup of the reaction is usually protonated (path combination Ad_N2).** In protic solvents, if the nucleophilic attack forms a stronger base, a following irreversible proton-transfer step may make the overall reaction favorable (see Figures 4.38 and 4.39). Alternatively, the intermediate anion can be avoided by combining the two steps shown below into the hetero Ad_E3 addition, discussed in Section 7.4.3 Extent of Proton-Transfer Variations.

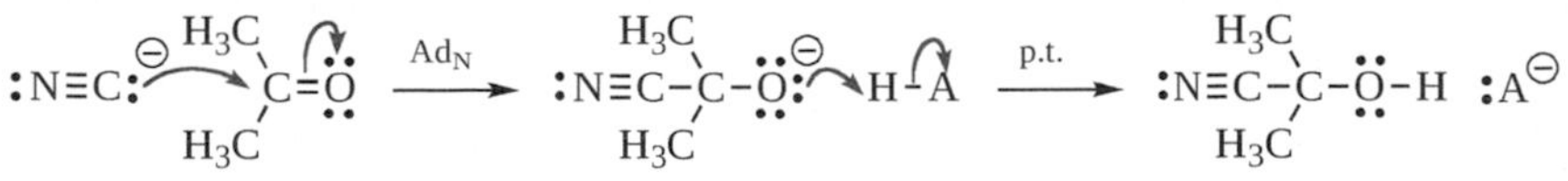

In acidic media, polarized multiple bonds often undergo acid catalyzed addition, and **a common mode of addition is the Ad_E2.** Deprotonation of the nucleophile by solvent gives the neutral compound. Common examples of this easily reversible Ad_E2 reaction are the formation of *hydrates* (NuH is H_2O) and, if NuH is ROH, *hemiacetals* (from aldehydes) and *hemiketals* (from ketones). Usually this reaction favors reactants.

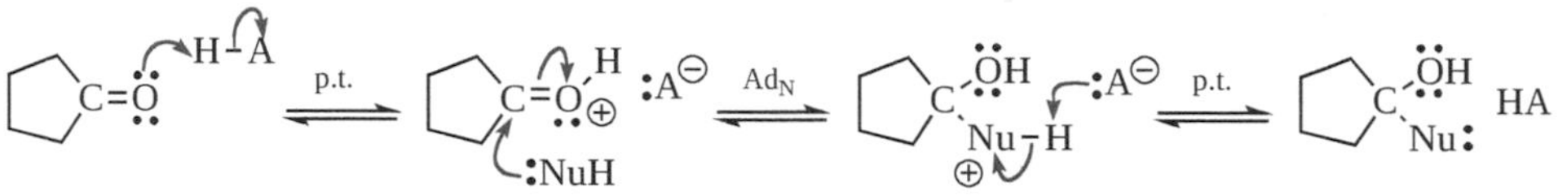

A follow-up reaction often occurs depending on the nucleophile and conditions. Six new functional groups can be easily made from the aldehyde or ketone (Table 8.2).

Table 8.2 New functional groups formed from aldehydes and ketones reacting with water, alcohols, and amines

Nucleophile	Mechanistic Process	Product Name	Structure
H_2O	Ad_E2	Hydrate	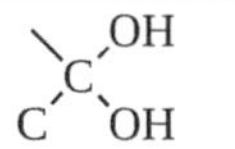
ROH	Ad_E2	Hemiacetal from aldehyde Hemiketal from ketone	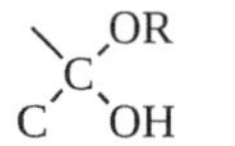
ROH	Ad_E2 then S_N1	Acetal from aldehyde Ketal from ketone	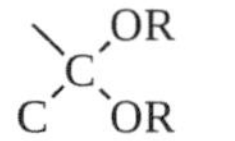
ROH	Ad_E2 then E1	Enol ether	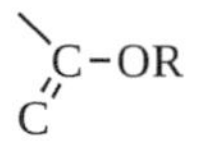
RNH_2	Ad_E2 then E1	Imine	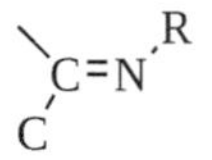
R_2NH	Ad_E2 then E1	Enamine	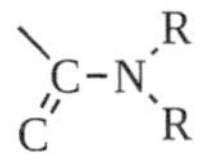

Biochemical Example

Five- and six-membered ring hemiacetals and hemiketals are common configurations of many sugars. Ribose, which forms a part of the structure of RNA, closes to form a five-membered ring preferentially.

Ribose open form   Ribose hemiacetal

If the following equilibria are **driven by the removal of water**, a follow-up reaction can occur in acid in which the OH group gets protonated (path p.t.), then the water is pushed out by the lone pair (path E_β), and the carbocation formed either is trapped (S_N1) by a nucleophile or loses a proton to a base (E1). The nucleophile determines which route the reaction takes.

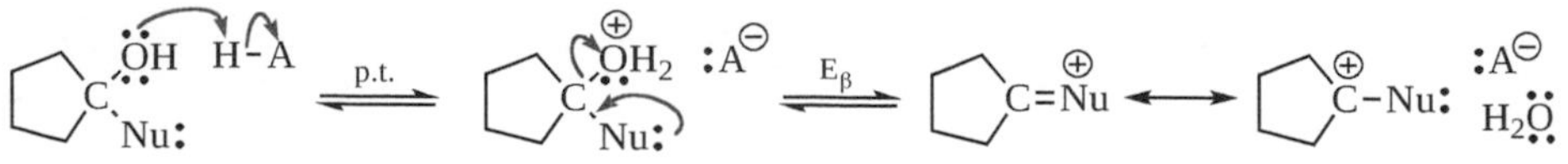

The carbocation can add a second nucleophile (path A_N followed by path p.t.).
This is common for alcohol sources (NuH is ROH), in acidic media, with removal of water to form *ketals* and *acetals*.

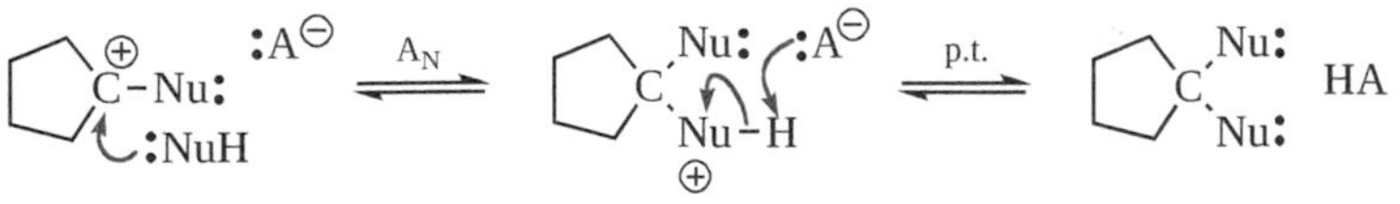

In compounds with multiple functional groups, ketals can be used to protect a ketone from basic nucleophiles and can later be removed by acidic water.
Cyclic ketal example:

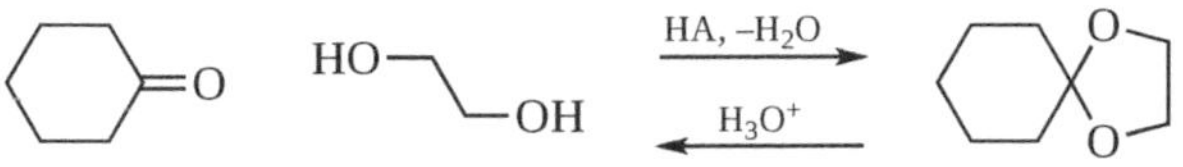

Alternatively, the C–H bond next to the carbocation can be deprotonated (path D_E). This is common for secondary amine sources (NuH is R_2NH), in acidic media, with removal of water to form *enamines*.

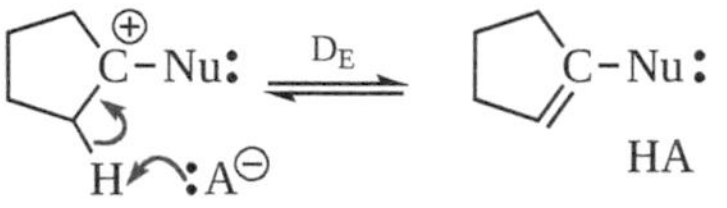

Common example (cyclic secondary amine):

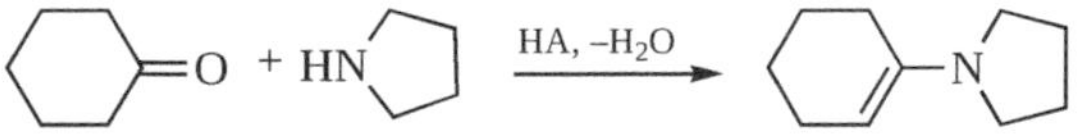

However, if there were two Hs on the original nucleophile (for example RNH_2), we could now remove that second H on the nucleophile to yield a stronger heteroatom–carbon double bond (path D_E). This is common for primary amine sources (NuH_2 is RNH_2) in acidic media, with removal of water to form *imines*. For a full mechanistic analysis see Section 10.5.2.

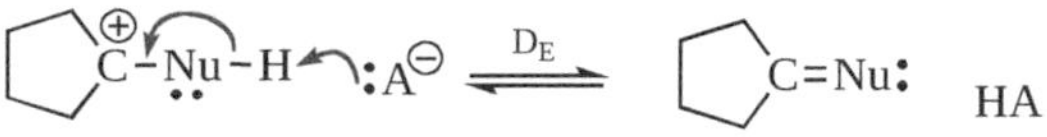

Finally, if the NuH_2 is $HONH_2$ the reaction forms an oxime (N is a better Nu than O).

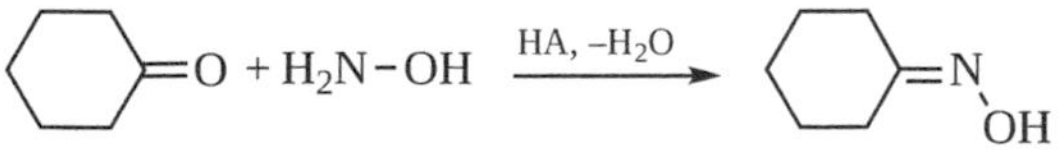

Biochemical Example

Pyridoxal (vitamin B_6) forms imines (Schiff bases) with amino acids as the first step in catalyzing their transformation (for one example, see Section 2.2.1). After pyridoxal has performed its particular catalysis on the attached amino acid, the imine can readily hydrolyze by the reverse of the formation reaction (shown in the worked mechanism example 10.5.2) to release the altered amino acid.

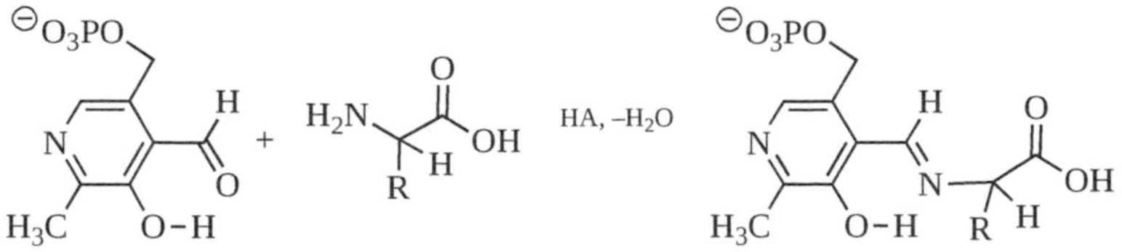

8.5.2 Bases Reacting with C=Y Having an Adjacent CH, Enolates

The polarized multiple bond makes adjacent C–H bonds acidic; deprotonation generates the enolate, an allylic electron source. Often there is more than one possible regiochemistry or stereochemistry of the enolate, which can often be selected for by the reaction conditions (see Section 9.3).

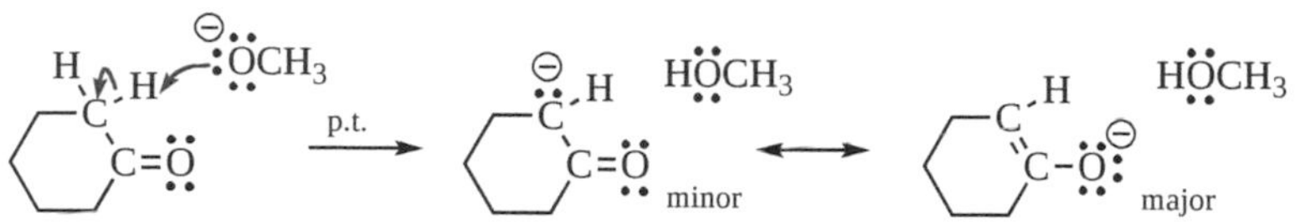

8.5.3 Complex Metal Hydrides Reacting with C=Y

Hydride is added to the multiple bond (path Ad_N) as expected. Usually the metal cation complexes with the Y heteroatom; nucleophilic attack on the polarized multiple bond will then yield a product that is stabilized by ion pairing.

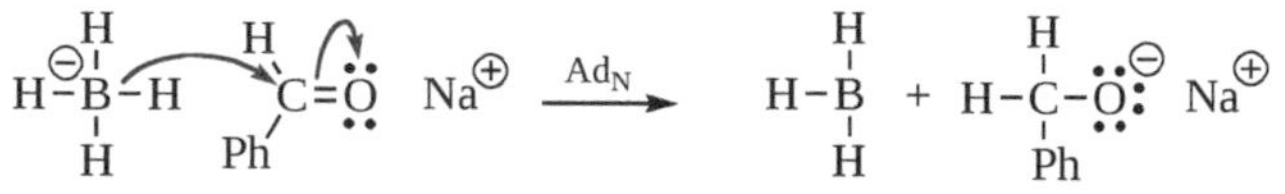

Chiral Lewis acids combined with hydride sources like borohydride can reduce carbonyls to give one enantiomer selectively. The chiral Lewis acid complex with the carbonyl makes one face of the carbonyl more accessible to the hydride nucleophile. The chirality of the Lewis acid determines the chirality of the product.

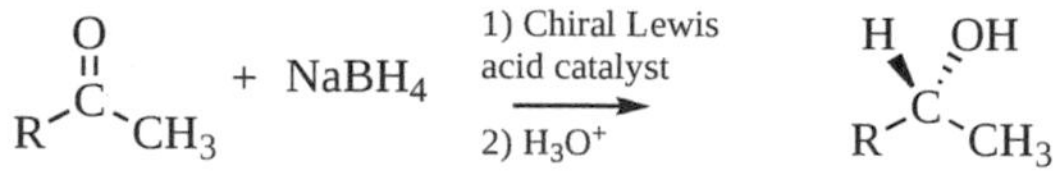

8.5.4 Organometallics Reacting with C=Y

Organometallics react with this sink by addition to the multiple bond (path Ad_N). The more covalent, less reactive organometallics, like R_2Cd, react very slowly with almost all of these sinks, whereas organomagnesiums, RMgX, and organolithiums react quickly. Complexation of the metal ion to the Y heteroatom catalyzes this reaction. Organometallics react much faster as nucleophiles with polarized multiple bonds than as bases with the adjacent C–H bonds. (carbon–acid, carbon–base proton transfer is slow).

C=Y example:

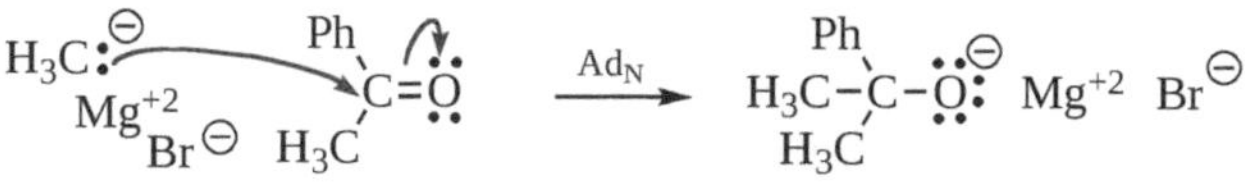

Chiral Lewis acids combined with less reactive organometallics like organotins (organostannanes) can add to carbonyls to give one enantiomer selectively. The chiral Lewis acid complex with the carbonyl makes one face of the carbonyl more accessible to the organometallic nucleophile. The chiral catalyst determines the product chirality. Chiral synthesis is very important for pharmaceuticals where the other enantiomer of the drug may be ineffective or worse.

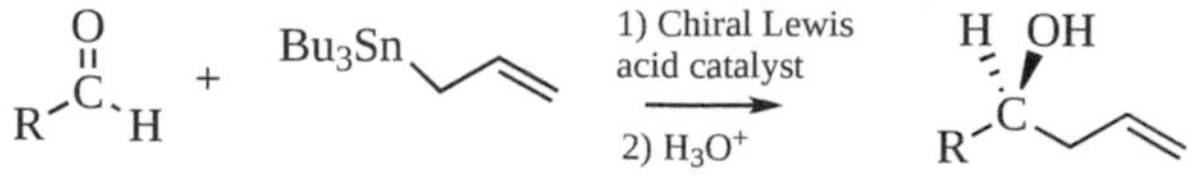

8.5.5 Allylic Sources Reacting with C=Y

C=Y as Sink, Aldol Reaction

Under equilibrium conditions (thermodynamic control), the allylic source adds to the polarized multiple bond (path Ad_N), but problems exist. One issue is that the allylic source can also serve as a base and may deprotonate the sink, creating a mixture of sources and sinks and thus a messy statistical mixture of products. Clean products result if the source is just the deprotonated sink or if the sink has no acidic protons. With ketones, the equilibrium of the attack step favors the starting materials, and therefore, the reaction goes to completion only if driven by a following elimination. In the next Ad_N2 example, the source is the deprotonated sink. The product is an aldehyde–alcohol, or aldol, a name now used for the general process of an enol (in acidic media) or enolate (in basic media) reacting with an aldehyde or ketone.

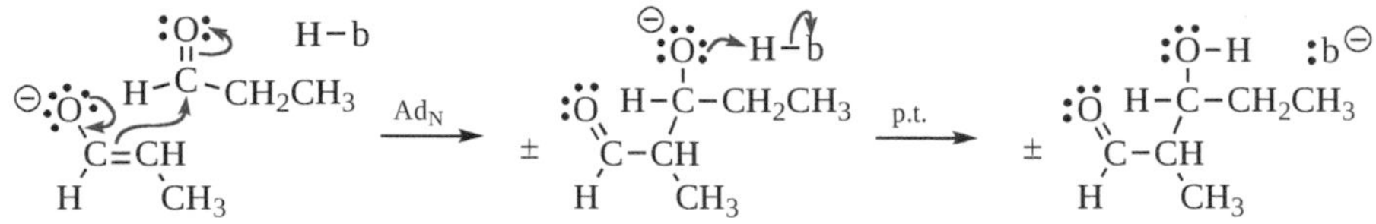

If the reaction is heated, elimination usually follows via the E1cB path combination, because the carbonyl is an electron-withdrawing group.

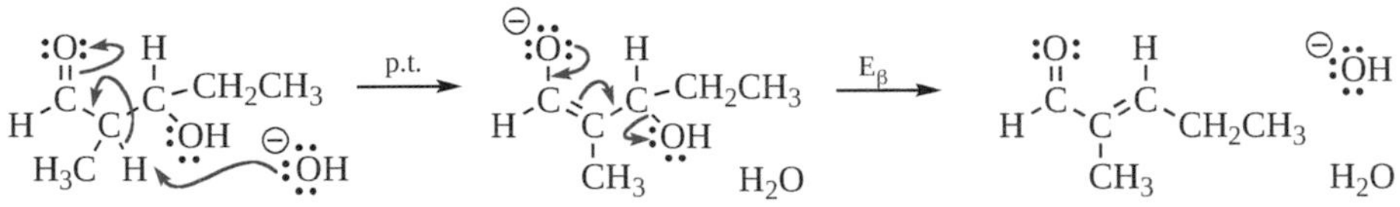

Under kinetic control, the aldol reaction is very stereospecific (reactant stereochemistry determines product stereochemistry). The lithium enolate is generated in an aprotic solvent, and then the carbonyl compound is added. The reaction proceeds via the metal-chelated minor path 6e, Figure 8.5. The minimization of steric effects in the chair transition state and the stereochemistry of the enolate (Section 9.3) determine the stereochemistry of the product.

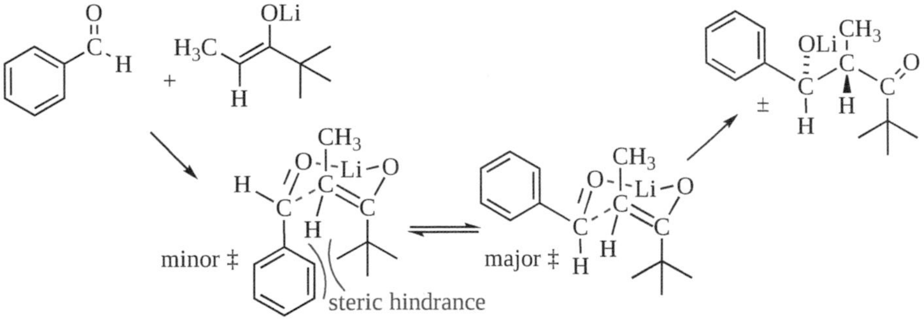

Figure 8.5 The stereospecific aldol reaction goes via the least hindered chair transition state.

The following is an example of an aldol reaction with a nonenolizable sink. The additional conjugation with the aromatic ring makes the follow-up elimination favorable.

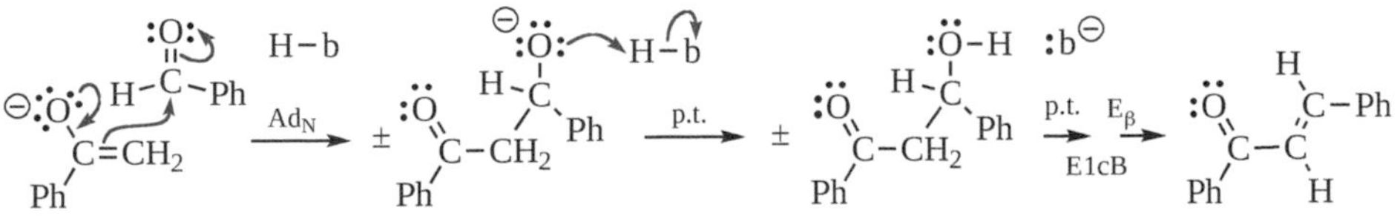

Biochemical Example

Fats and carbohydrates are metabolized down to carbon dioxide via an acetyl unit, $CH_3C{=}O$, which is attached to a coenzyme, HSCoA, as a thioester called acetyl CoA. Acetyl CoA enters the citric acid cycle and eventually is converted to two molecules of carbon dioxide. The first step in the citric acid cycle is the aldol of acetyl CoA with oxaloacetate (Figure 8.6). The elegance of this biochemical aldol mechanism is that the acidic and basic groups within the enzyme's active site provide a route that avoids any strongly acidic or basic intermediates. The enzyme accomplishes an aldol reaction at neutral pH, without an acidic protonated carbonyl or basic enolate intermediate via push–pull catalysis (Section 7.4.3).

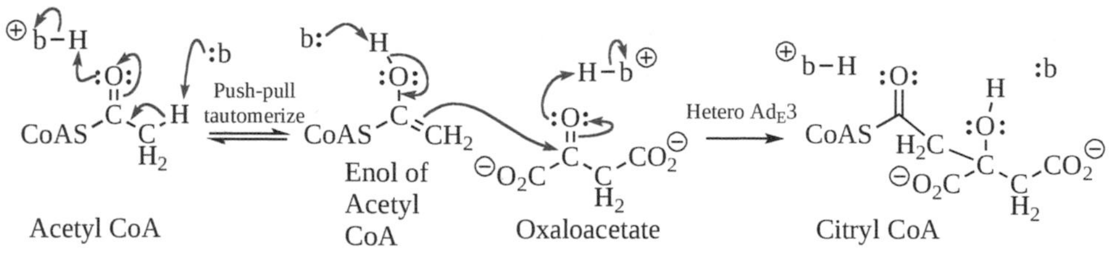

Figure 8.6 The biosynthesis of citryl CoA.

8.5.6 Simple Pi Bond Sources Reacting with C=Y

Simple pi bonds are usually not good enough nucleophiles to react with polarized multiple bonds. If a Brønsted–Lowry acid or a Lewis acid is added to improve the electron sink, then addition can occur via the lone-pair-stabilized carbocation as the sink. Figure 8.7 shows a mechanistic example from a short synthesis of the human hormone estrone.

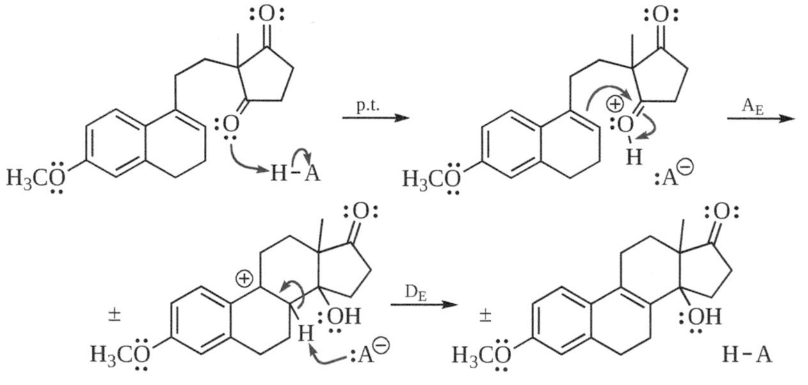

Figure 8.7 The electrophilic ring closure step in a synthesis toward estrone.

8.5.7 Aromatic Sources Reacting with C=Y

Aromatic rings are usually not good enough nucleophiles to react with polarized multiple bonds. If a Brønsted–Lowry acid or a Lewis acid is added to improve the electron sink, then electrophilic aromatic substitution occurs, usually via the lone-pair-stabilized carbocation as an electrophile.

Friedel–Crafts—Aromatic Rings with Lone-Pair-Stabilized Carbocations as Sinks

The electrophile is often a carbonyl that is protonated or complexed to a Lewis acid. The product can enter into a second Friedel–Crafts alkylation, so that often the product is a result of two attacks. The synthesis of the insecticide DDT (Figure 8.8) is an example.

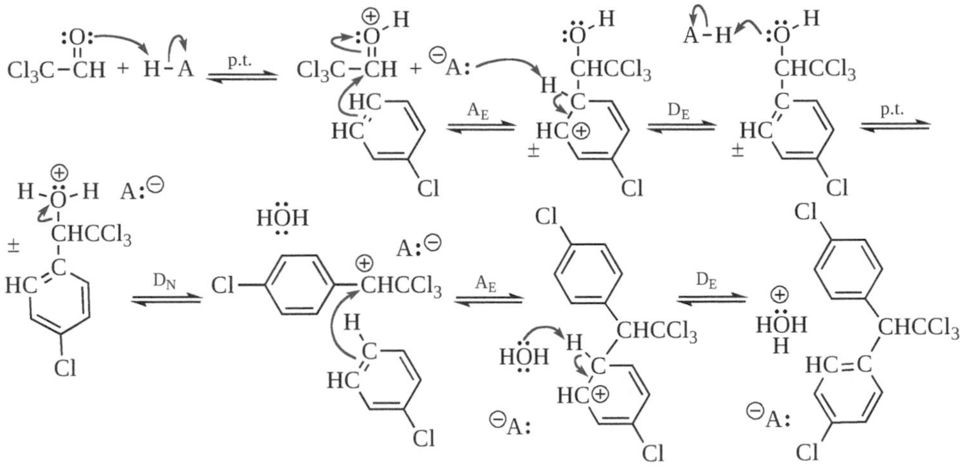

Figure 8.8 The synthesis of DDT catalyzed by sulfuric acid.

8.6 R−C≡Y SINKS REACTING WITH COMMON SOURCES

8.6.1 Lone Pair Sources Reacting with R−C≡N

The two pi bonds in a triple bond are perpendicular to each other, so they act independently. In the next example, the attack of the Nu is aided by protonation from the protic solvent, the hetero Ad_E3 path. Tautomerization then gives the more stable amide.

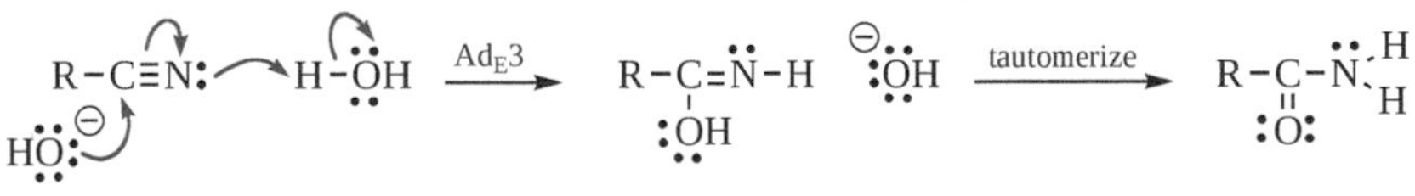

8.6.2 Bases Reacting with R−C≡N Having an Adjacent CH, Nitrile Enolates

The polarized multiple bond of the nitrile makes the adjacent CH bonds acidic (pK_a = 25); deprotonation by a hindered strong base gives the nitrile enolate, an allylic source.

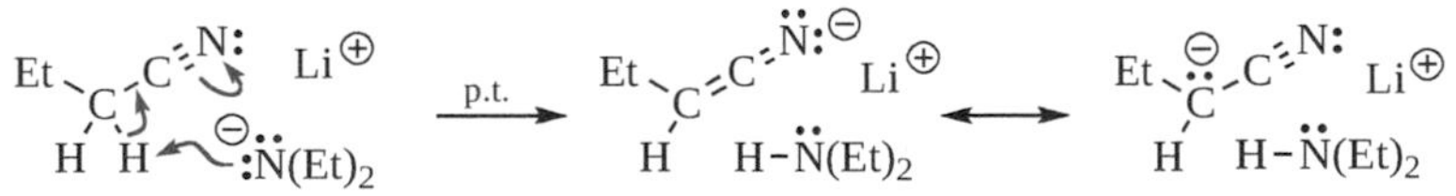

8.6.3 Complex Metal Hydrides Reacting with R−C≡N

With $LiAlH_4$ a second attack on the nitrile occurs. The product of the first addition can form a Lewis salt, which (most likely after acting as a hydride source) catalyzes the second addition. The product is the amine, after the acidic water in the workup protonates the nitrogen and dissociates it from the metal. This double addition can be prevented by using one equivalent of a less reactive aluminum hydride like $LiHAl(OEt)_3$ or by the use of one equivalent of $[(CH_3)_2CHCH_2]_2AlH$, diisobutyl aluminum hydride (see Section 9.2.2 for the multiple addition decision of nucleophiles to nitriles).

Double addition example (path Ad_N, then A_N, followed by the Lewis salt serving as a hydride source, then Ad_N again):

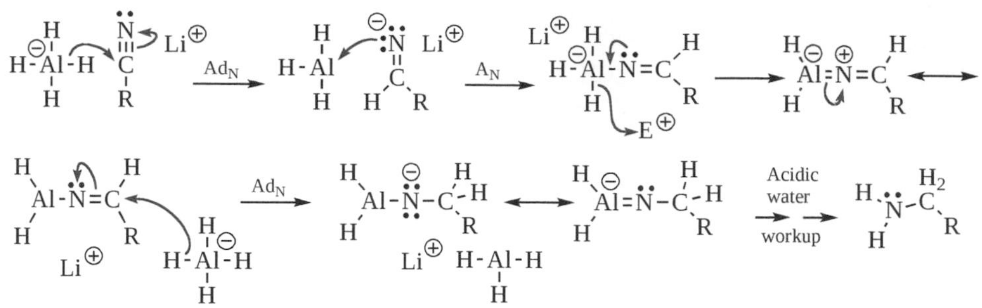

8.6.4 Organometallics Reacting with R−C≡N

Organometallics react with this sink by addition to the multiple bond (path Ad_N). The more covalent, less reactive organometallics, like R_2Cd, react very slowly with almost all of these sinks, whereas organomagnesiums, RMgX, and organolithiums react quickly. Complexation of the metal ion to the Y heteroatom catalyzes this reaction. Organometallics react much faster as nucleophiles than bases (carbon–acid, carbon–base proton transfer is slow). The resulting nitrogen anion is protonated in the acidic workup to give the imine, which then is hydrolyzed to the ketone. The mechanism of this acidic water workup is the reverse of imine formation, discussed in Section 10.5.2.

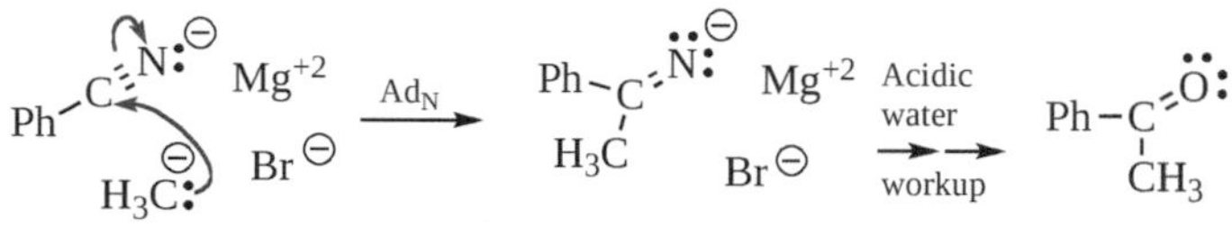

8.6.5 Allylic Sources Reacting with R−C≡N

The following example is an intramolecular reaction similar to the aldol reaction. A deprotonated nitrile (nitrile enolate) acts as a nucleophile and adds via an Ad_N path to another nitrile. The acidic water workup is the reverse of imine formation, Section 10.5.2.

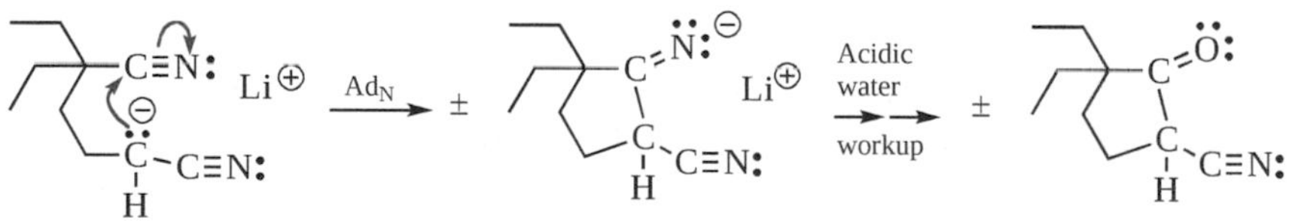

8.6.6 Simple Pi Bond Sources Reacting with R−C≡N

Simple pi bonds are not good enough nucleophiles to react with nitriles. It is basically a case of little push and little pull, so no electron flow occurs. If the pi bond is made more nucleophilic by addition of a pi donor (making it an allylic source) then the reaction is similar to the one above. Alternatively, a Brønsted–Lowry acid or a Lewis acid could be added to improve the electron sink, but few examples of this reaction are known.

8.6.7 Aromatic Sources Reacting with R−C≡N

Aromatic rings are usually not good enough nucleophiles to react with nitriles. If a Brønsted–Lowry acid or a Lewis acid is added to improve the electron sink, then electrophilic aromatic substitution can occur. The acidic water workup first protonates the aluminum salt of the addition product to form the imine, which is then hydrolyzed to the ketone by the reverse of imine formation discussed in Sections 8.5.1 and 10.5.2.

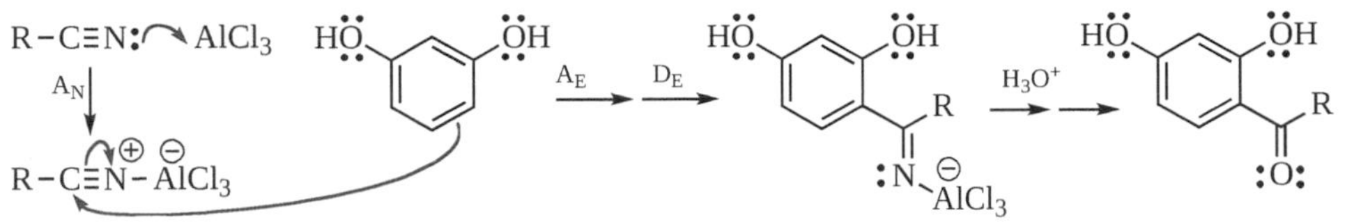

8.7 C=C−EWG SINKS REACTING WITH COMMON SOURCES

8.7.1 Conjugate Addition by Lone Pair Sources to C=C−EWG

If an electronegative carbon atom replaces the electronegative Y atom, the nucleophilic addition to the C=C is called a *conjugate addition* (a vinylogous nucleophilic addition). Competition between direct addition to the EWG and conjugate addition is in Section 9.6. The reaction below goes by either a vinylogous Ad_N2 with an irreversible proton transfer from the protic solvent as the driving force, or all at once by a vinylogous hetero Ad_E3 (add another pi bond to Figure 7.14) where protonation of the C=O occurs along with the nucleophilic attack on the C=C. The enol from the addition tautomerizes to the product's more stable keto form.

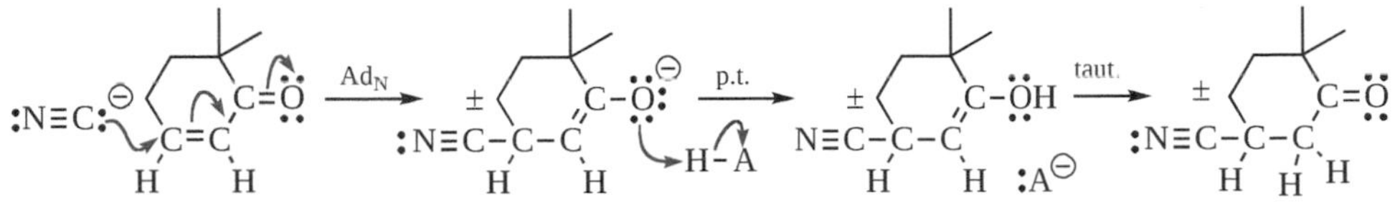

8.7.2 Bases Reacting with C=C−EWG Having an Adjacent CH, Extended Enolates

These deprotonations generate extended enolates, which are vinylogous allylic sources. Deprotonation can also occur on the methyl next to the carbonyl in the following example. However, the extended enolate is more delocalized. Section 9.3 covers the regiochemistry and stereochemistry of enolate formation in more detail.

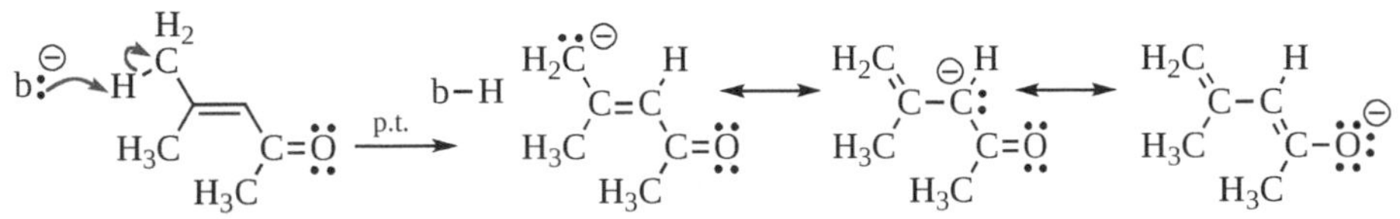

8.7.3 Conjugate Addition by Complex Metal Hydrides to C=C−EWG

Aluminum hydride, AlH_4^-, is relatively hard; therefore conjugate addition is usually only a minor side reaction, as most of the hydride addition occurs to the harder electron-withdrawing group and not the double bond. Borohydride, BH_4^-, since it is softer, has a much greater tendency toward conjugate addition but is often difficult to predict (see Section 9.6) and may form product mixtures. Reagents and conditions have been found to give the conjugate addition. K-Selectride conjugate addition example (R is sec-butyl):

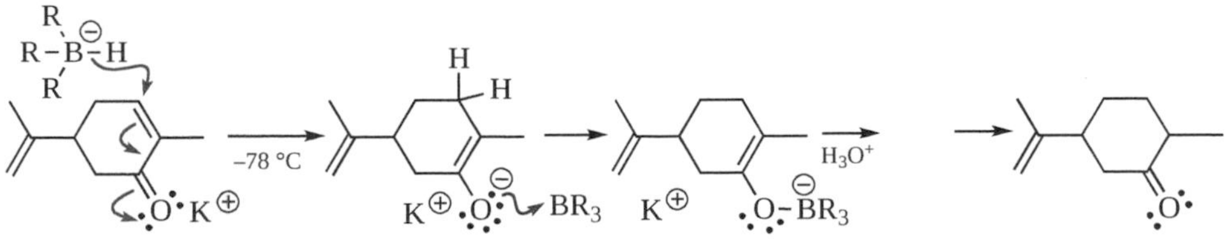

8.7.4 Conjugate Addition by Organometallics to C=C−EWG

Since the δ+ at the conjugate position is less, the carbon atom is softer, and softer organometallics tend to favor conjugate attack (see Section 9.6).

Example (the organocuprate is a soft organometallic):

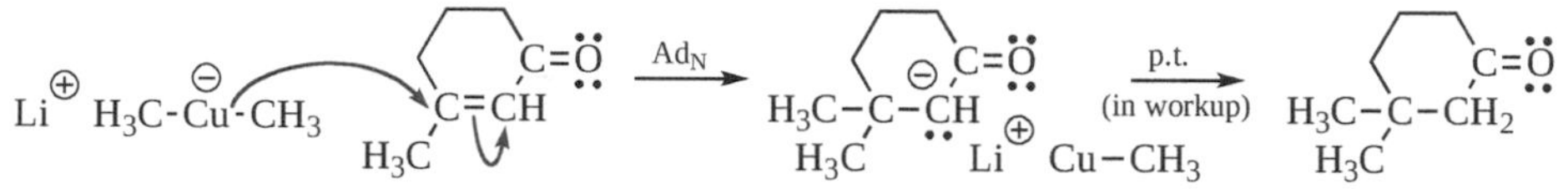

8.7.5 Conjugate Addition by Allylic Sources to C=C−EWG

The reaction is Ad_N2 addition to the polarized multiple bond (Ad_N, then p.t.). This reaction is easy to predict because the most stable product is formed under these equilibrium conditions. HSAB theory predicts that the combination of the two softest sites is favored. This conjugate addition is often called a Michael addition and forms the first step in the Robinson annulation diagramed in Figure 8.9. The later steps are an aldol addition (Ad_N2) to form the new ring followed by elimination by E1cB to give the bicyclic product. The aldol condensation mechanism is shown in detail in Section 10.5.3.

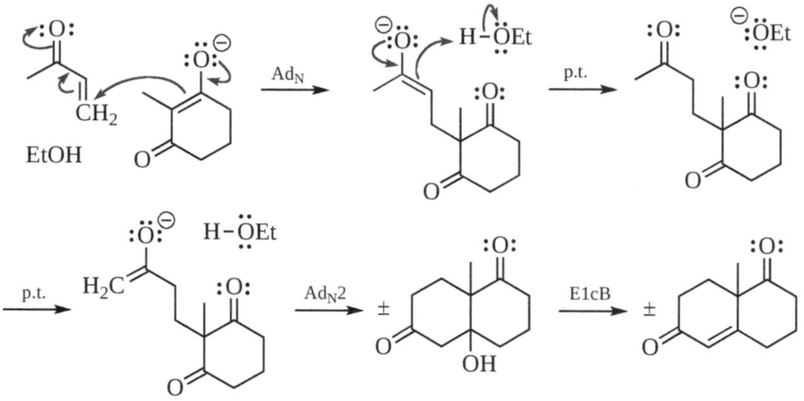

Figure 8.9 The Robinson annulation, conjugate addition (Ad_N2) followed by an aldol (Ad_N2) and an elimination (E1cB).

8.7.6 Conjugate Addition by Simple Pi Sources to C=C−EWG

Simple pi bonds are not good enough nucleophiles to react with conjugate acceptors. It is basically a case of little push and little pull, so no electron flow occurs. If the pi bond is made more nucleophilic by adding a pi donor (making it an allylic source), then the reaction is similar to the one above. Alternatively, a Brønsted–Lowry acid or a Lewis acid could be added to improve the electron sink; few examples of this reaction are known.

8.7.7 Conjugate Addition by Aromatic Sources to C=C−EWG

Aromatic rings are usually not good enough nucleophiles to react with conjugate acceptors. If a Brønsted–Lowry acid or a Lewis acid is added to improve the electron sink, then electrophilic aromatic substitution can occur.

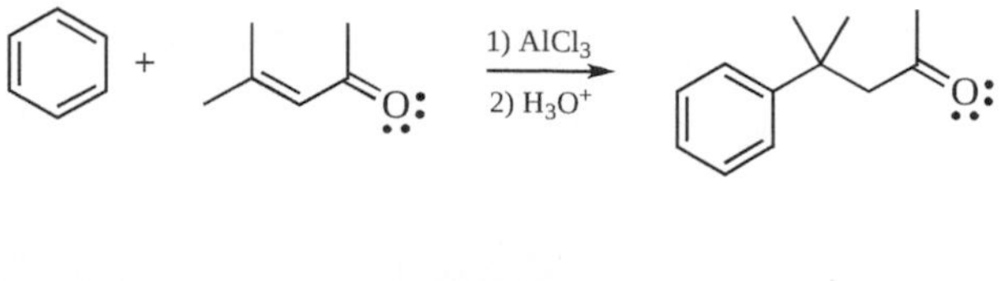

8.8 L–C=Y SINKS REACTING WITH COMMON SOURCES

8.8.1 Lone Pair Sources Reacting with L–C=Y

The most common representatives of the L–C=Y class of electron sinks are the carboxyl derivatives with Y equal to oxygen. **In basic media, there is only one pathway: the addition–elimination path, path Ad_N + E_β** (see Section 4.5.1). The leaving group should be a more stable anion than the nucleophile, or the reaction will reverse at the tetrahedral intermediate. A follow-up reaction of a second addition to the polarized multiple bond occasionally occurs. With lone pair sources a second addition is rare because the nucleophile is usually a relatively stable species; the second tetrahedral intermediate tends to kick it back out (see Section 9.2).

Basic media addition–elimination example (path Ad_N + E_β):

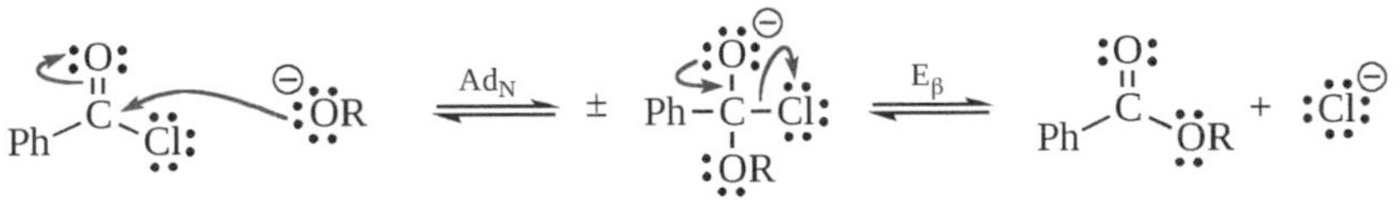

In acidic media, the carbonyl lone pair can get protonated to produce a much better electron sink. This better electron sink can be attacked by weaker nucleophiles. Then proton transfer to the leaving group makes it a better leaving group. The reaction is reversible; the position of equilibrium is determined by mass balance.

Acidic media esterification example (Ad_E2, p.t. from Nu, p.t. to L, E1):

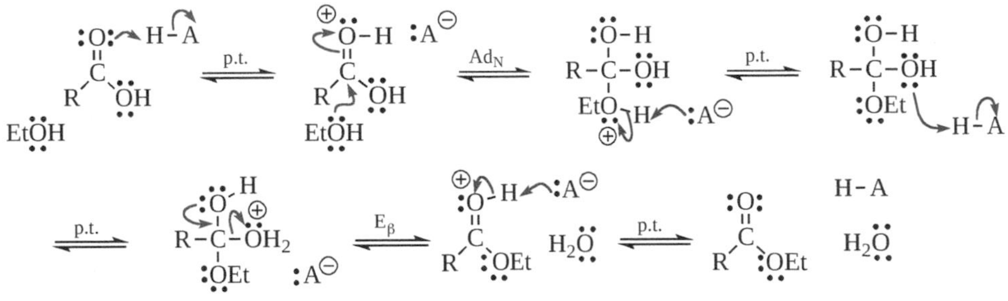

8.8.2 Bases Reacting with L–C=Y Sinks Having an Adjacent CH, Enolization

These enolates are usually generated by deprotonation below 0°C. At higher temperatures, the enolate can eject the leaving group via path E_β to form a reactive heterocumulene (usually an undesirable side reaction). Enolization example:

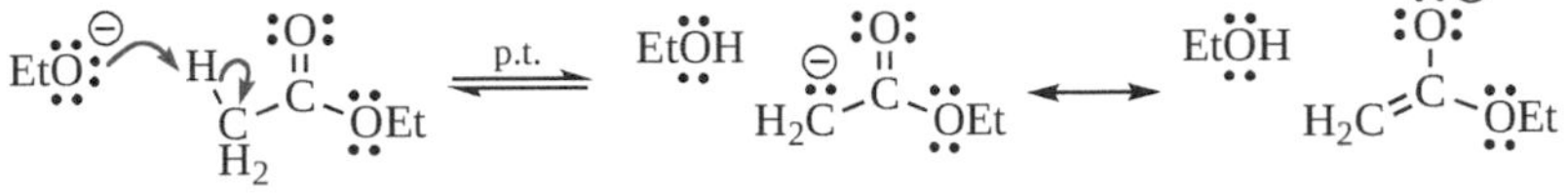

8.8.3 Metal Hydrides Reacting with L–C=Y

The product of the initial reduction is most often more reactive than the starting material; therefore a second addition is very common (Ad_N, then E_β, then Ad_N, covered in Section 9.2). An Ad_N then E_β product can be obtained with acyl halides and one equivalent of a less reactive metal hydride at low temperature. Borohydrides selectively react with aldehydes and ketones in the presence of less reactive esters and amides.

Second-addition example (path Ad_N then E_β, Lewis salt formation with the aluminum species to create the nucleophilic aluminate, followed by a second Ad_N):

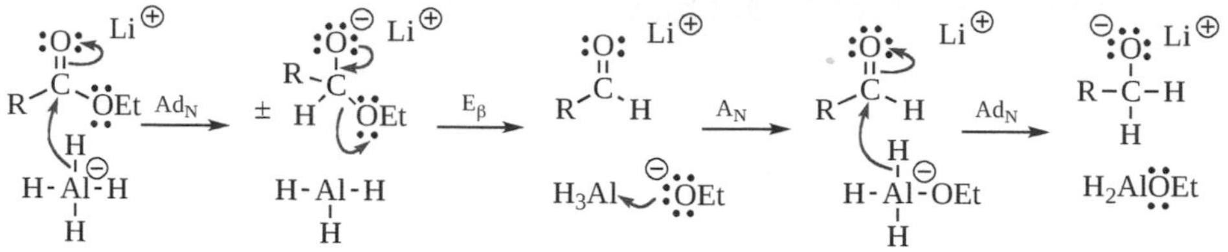

If the L is poor as in amides, Lewis salt formation allows a stable aluminum oxide to be kicked out by the nitrogen electron releasing group instead of kicking out the very poor amine leaving group pK_{aHL} = 36 (path Ad_N, then A_N, followed by the Lewis salt serving as a hydride source, then E_β, and then Ad_N).

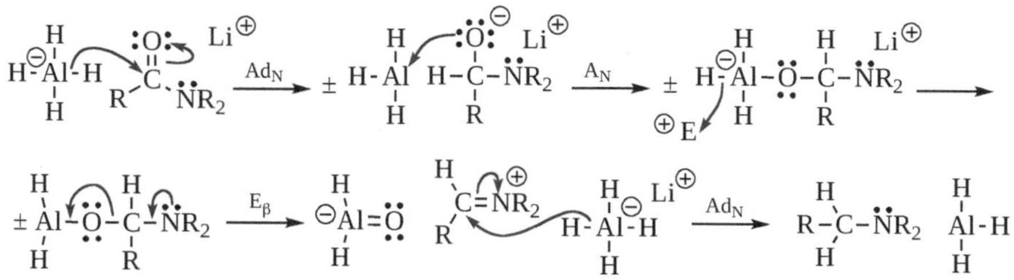

8.8.4 Organometallic Sources Reacting with L−C=Y Sinks

Organometallics follow the addition–elimination pathway (Ad_N + E_β) characteristic of this sink. However, the products of the addition–elimination path often undergo a second attack by the organometallic because the Ad_N + E_β product is often more reactive than the original substrate (see Section 9.2). The reactivity ranking of the electron sinks and sources is important in this decision.

Second-addition example (path Ad_N, then E_β, then Ad_N again):

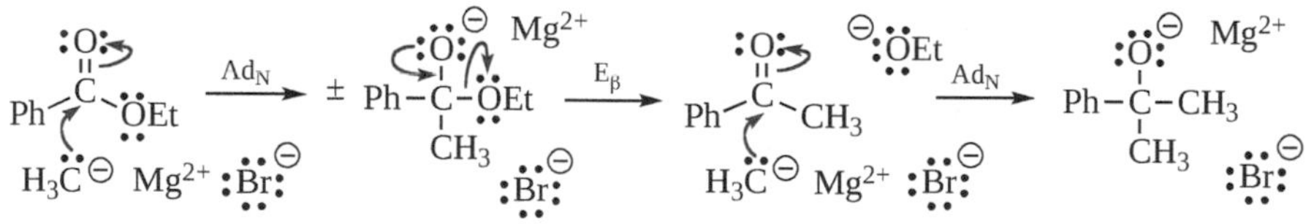

Single-addition example (path Ad_N, then E_β):

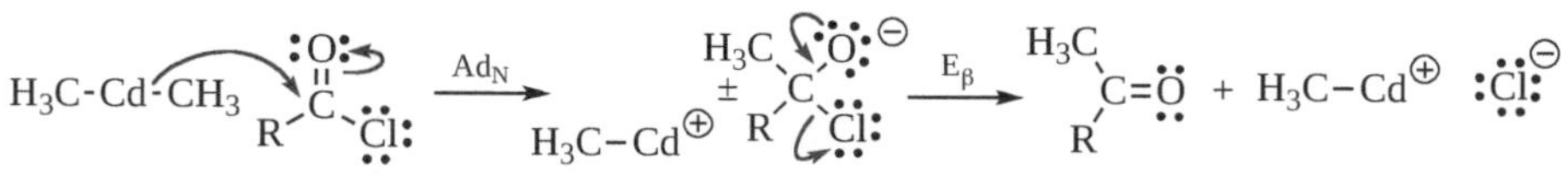

8.8.5 Allylic Sources Reacting with L−C=Y Sinks, Acylation

Allylic sources attack carboxyl derivatives such as esters almost exclusively on carbon by the path Ad_N, then E_β. As in Section 8.5.5, the enolate source should be the deprotonated sink or the sink should not be enolizable. Otherwise, this condensation reaction would then produce a statistical mix of all four possible products. A final proton transfer to give a stable resonance delocalized anion is usually the driving force for the reaction. A mild acidic workup then gives the neutral product.

Claisen condensation example (the source is a deprotonated sink), details in Section 10.7.2:

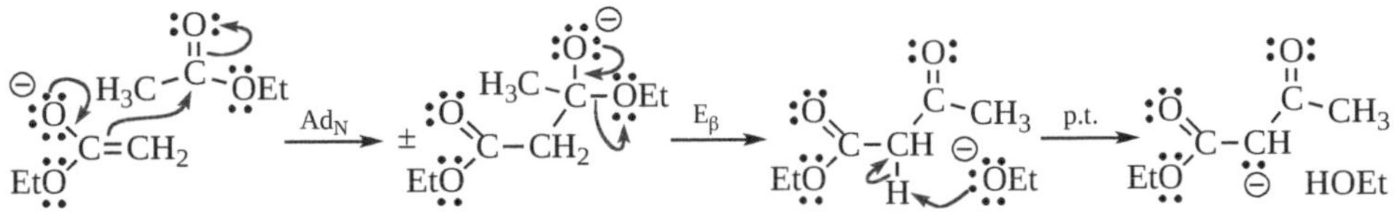

Biochemical Examples

Fats are metabolized down to a thioester called acetyl CoA. Acetyl CoA enters the citric acid cycle and eventually is converted to two molecules of carbon dioxide. The cleavage step in metabolizing long-chain fatty acids is the reverse of the Claisen reaction of the previous section. A thiol group attached to the enzyme is the nucleophile for general base-catalyzed addition, starting the reverse Claisen condensation.

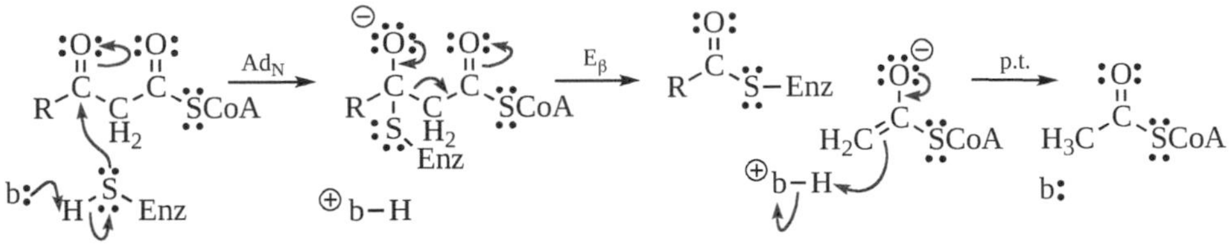

The release of the shortened fatty acid from the enzyme is an example of a lone pair nucleophile attacking an L–C=O sink as in Section 8.8.1. Again, a general base catalyzes the addition to the thioester.

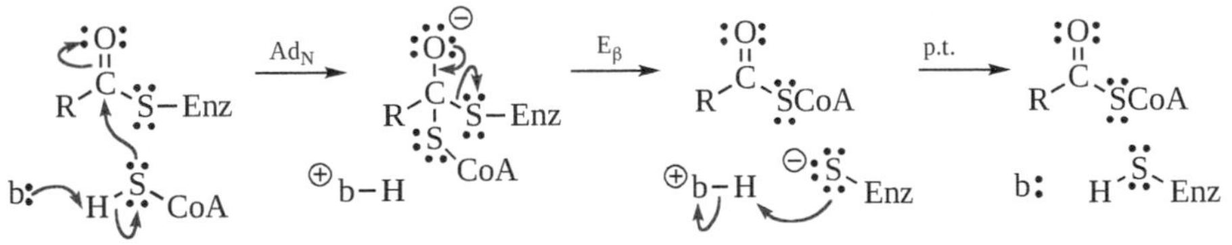

8.8.6 Simple Pi Bond Sources Reacting with L–C=Y Sinks

Simple pi bonds are usually not good enough nucleophiles to react with this sink. If a Brønsted–Lowry acid or a Lewis acid is added to improve the electron sink, then addition can happen but it is often complicated with other reactions. We can improve the electrophilicity of an acyl chloride by complexing it with a Lewis acid like $AlCl_3$. The $AlCl_3$ can complex with either oxygen or chlorine lone pairs to form a reactive complex. Also, the leaving group can depart to give the reactive acylium ion.

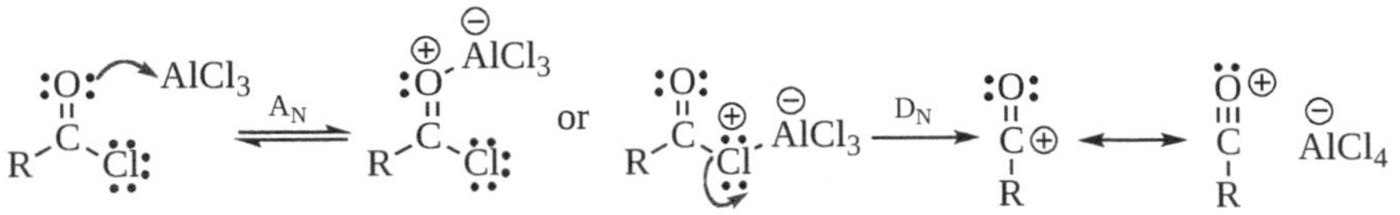

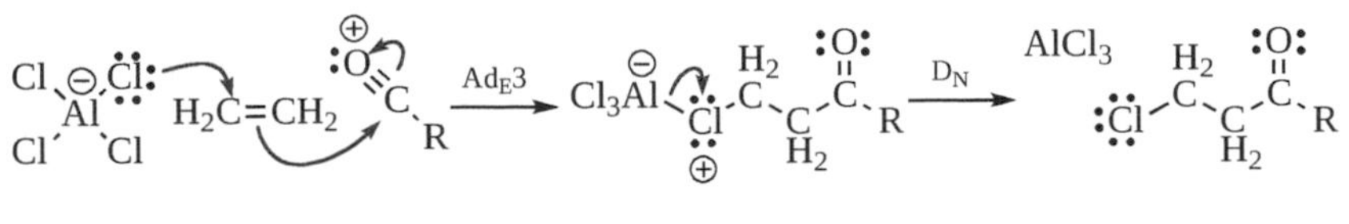

8.8.7 Aromatic Sources Reacting with L–C=Y Sinks

Aromatic rings are usually not good enough nucleophiles to react with this sink. If a Brønsted–Lowry acid or a Lewis acid is added to improve the electron sink, then electrophilic aromatic substitution occurs, usually via the acylium cation as an electrophile.

Friedel–Crafts Acylation

The electrophile is the acylium ion, R–C≡O⁺, generated by Lewis acid-catalyzed ionization of a leaving group (path D_N) from acyl halides or acid anhydrides (shown in the previous section). The proton that is lost comes from the same carbon that the electrophile attacked. The reaction fails for deactivated rings (Ar–EWG, *meta* directors). After the electrophile adds, it deactivates the ring toward further attack. No rearrangement of the electrophile occurs.
Acylation example:

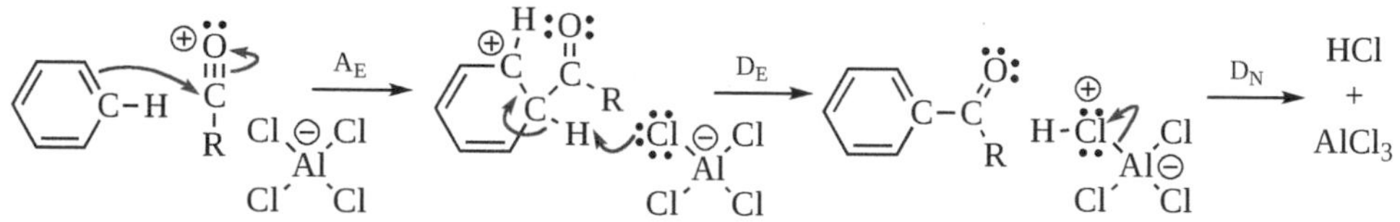

8.9 MISCELLANEOUS REACTIONS

8.9.1 Electron-Deficient Species as Electron Sinks

Lone Pairs as Sources

The trapping of a cation by a nucleophile, A_N, can be considered the electrophilic addition of a cation to a lone pair.

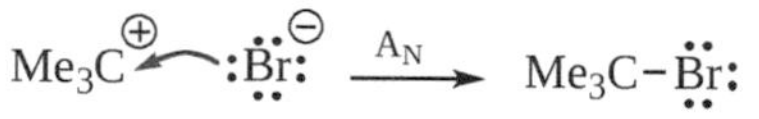

The electrophilic addition of a Lewis acid to a lone pair to form a Lewis salt is an isoelectronic reaction. The first steps of borate ester hydrolysis show a good example of this:

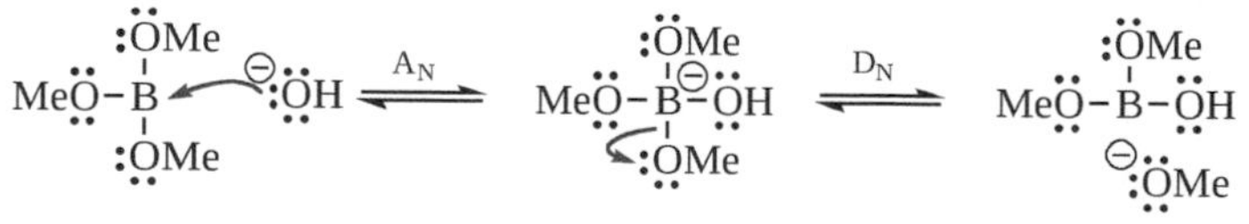

Pi Bonds as Sources

Unwanted product mixtures can result since the product of the carbocation addition to an alkene via path A_E is also a carbocation that can rearrange and/or attack another alkene molecule (polymerization).

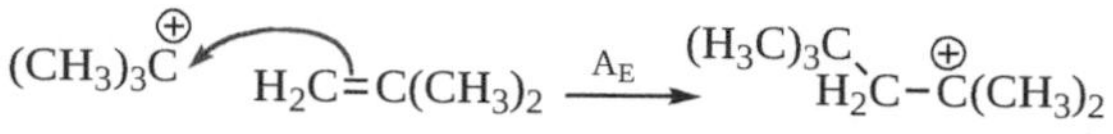

Borane, $2BH_3 \rightleftharpoons B_2H_6$, isoelectronic with carbocations, adds exclusively *syn* to multiple bonds. Initially a pi-adduct is believed to form, followed by rapid hydride donation to the largest ∂^+ atom (4e path, Figure 7.12). The reaction repeats to produce dialkyl and occasionally trialkylboranes, R_3B.

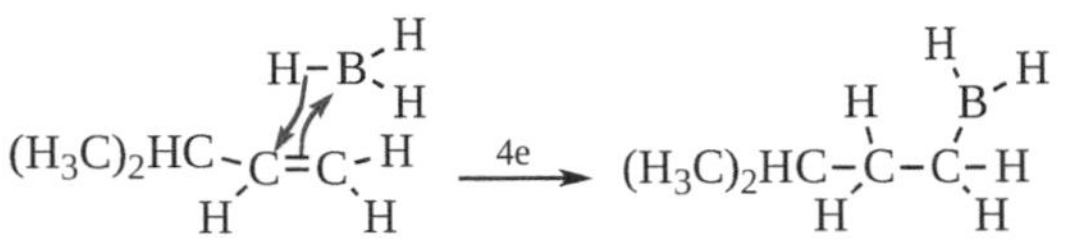

One equivalent of R_2AlH (also isoelectronic) at low temperature adds once to nitriles or many carboxylic acid derivatives to give aldehydes upon acidic aqueous workup (Section 9.2.2).

8.9.2 Carbonate Derivatives as Electron Sinks

These are represented by the following structures and can undergo (depending on the quality of the leaving group) single, double, and triple additions by nucleophiles via the addition–elimination path. The tendency toward a single addition–elimination is highest with haloformates (center of the following structures) because one leaving group is so much better than the other. This is multiple addition decision is discussed in Section 9.2.

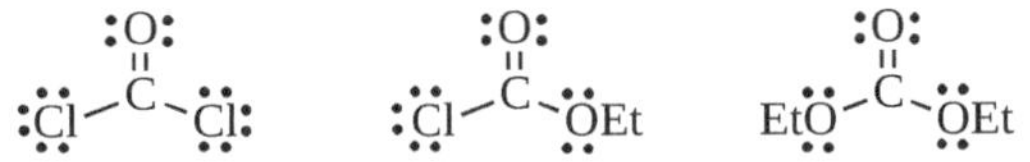

Carbonate derivatives with both a good leaving group and an OH are not stable with respect to an E1cB or E2 elimination reaction to produce carbon dioxide.

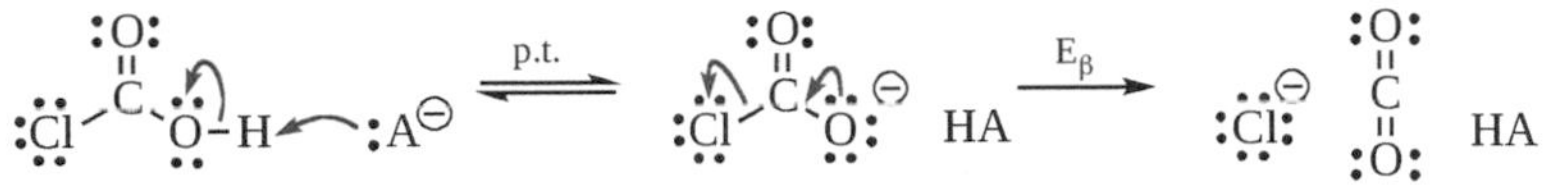

8.9.3 Heterocumulenes as Electron Sinks

Heterocumulenes are attacked at the central carbon. In acidic media, proton transfer may occur first. This is a common way to activate an amino acid COOH for laboratory peptide synthesis (R = cyclohexyl). Lone pair source Ad_E2 example:

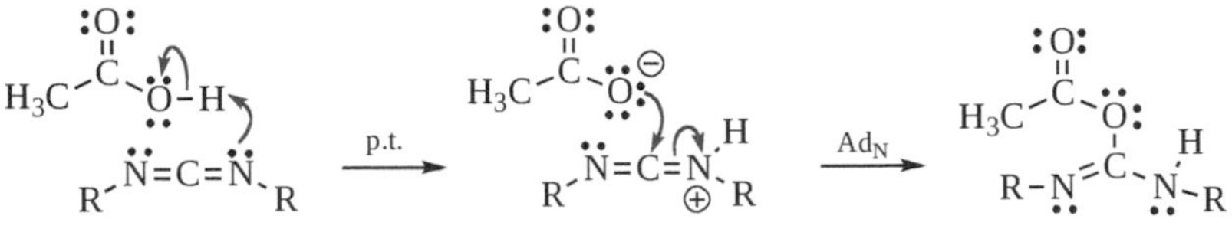

In basic media, the resultant anion from the Nu attack is then protonated in the acidic workup. Organometallic Ad_N2 example (acidic water workup protonates the carboxylate anion):

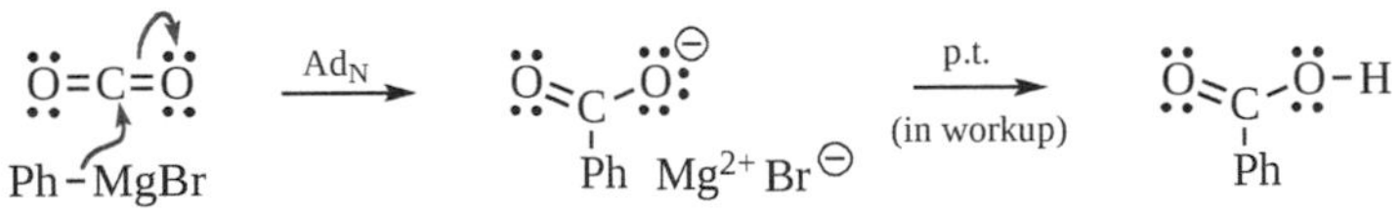

8.9.4 Nucleophilic Aromatic Substitution

There are three routes available for this very poor electron sink, Ar–L. If the leaving group is good—for example, chloride—but there are no electron-withdrawing groups on the ring, temperatures over 300°C will be required to get reaction. Nitro groups and other good electron-withdrawing groups on the ring lower the temperature for reaction by stabilizing the intermediate carbanion. $Ad_N + E_\beta$ example:

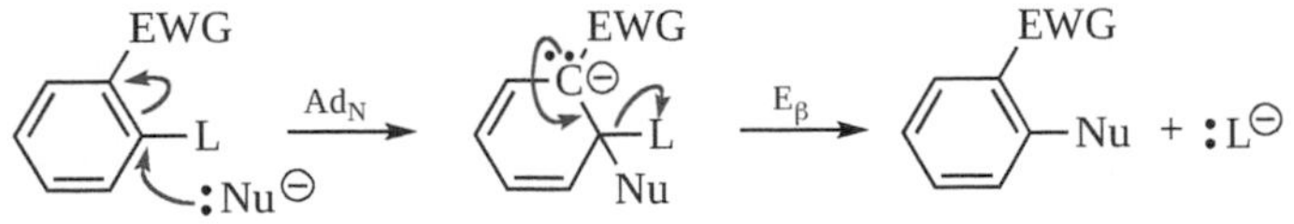

An unusual reaction occurs with aromatic halides and very strong bases; the substitution reaction can go via elimination to the highly reactive benzyne followed by addition, path E2 then Ad_N2.

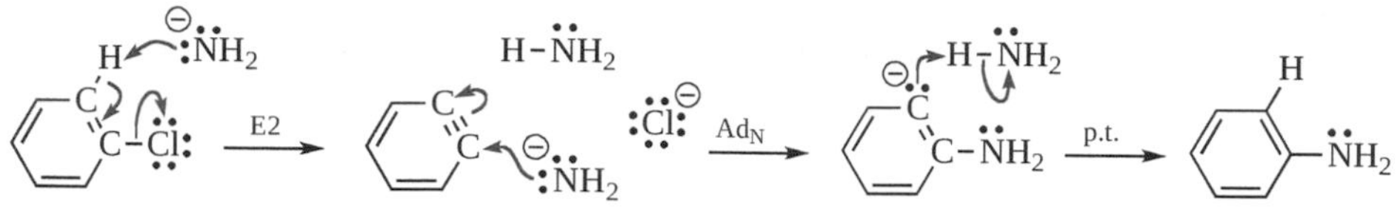

If the leaving group is excellent, like N≡N in diazonium ions, an S_N1-like path has been proposed. However, many mechanisms for substitution on these diazonium ions are complex; some involve electron transfer, radical intermediates, or copper catalysis (see Section 12.10).

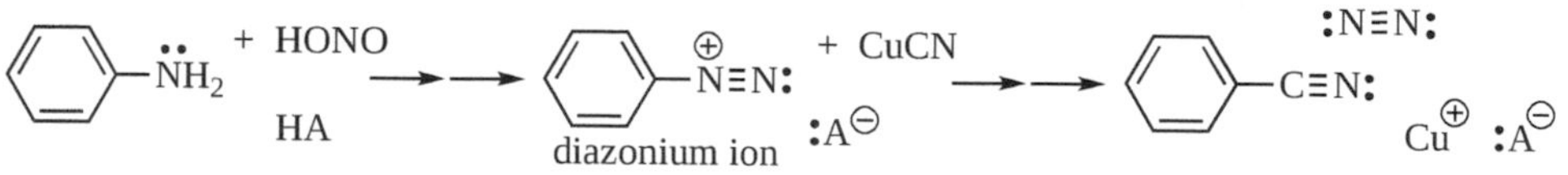

8.10 METAL IONS AS ELECTRON SINKS

8.10.1 Metal Ions as Electrophiles and Electrophilic Catalysts

The role of metal ions as electrophilic catalysts is often underestimated. The formation of good ion-pair stabilization in the transition state in an anionic nucleophilic attack lowers the energy of the transition state. Even a powerful hydride nucleophile such as $LiAlH_4$ will not add to a carbonyl if the lithium ion is sequestered and kept from complexing with the carbonyl.

A good example of a metal ion as an electrophile is the oxymercuration reaction. The nucleophile is often solvent, usually water or an alcohol. The charge is distributed, so little rearrangement occurs. The very toxic mercury is removed in a subsequent reduction step.

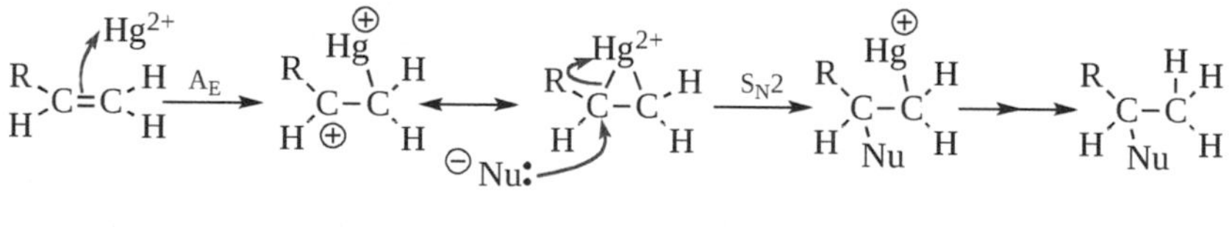

8.10.2 Metal Ions as Oxidants

Chromium trioxide, CrO_3, and chromic acid, H_2CrO_4, are commonly used to oxidize alcohols. The alcohol adds to the chromium species to give a chromate ester, which then undergoes an E2 elimination, which oxidizes the alcohol and reduces the metal ion.

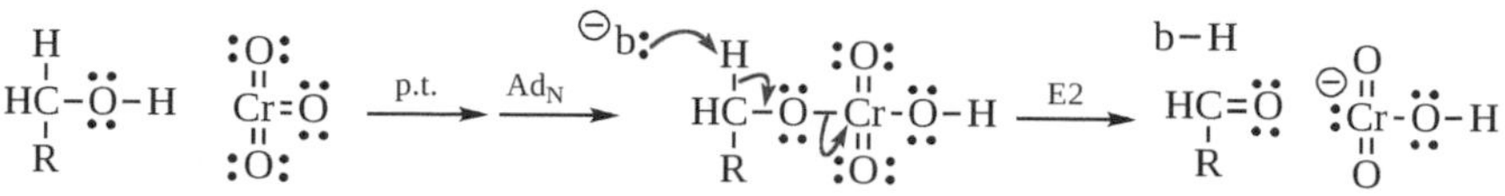

If water and a little acid are present, the initially formed aldehyde adds water to form a hydrate, which is further oxidized to the carboxylic acid.

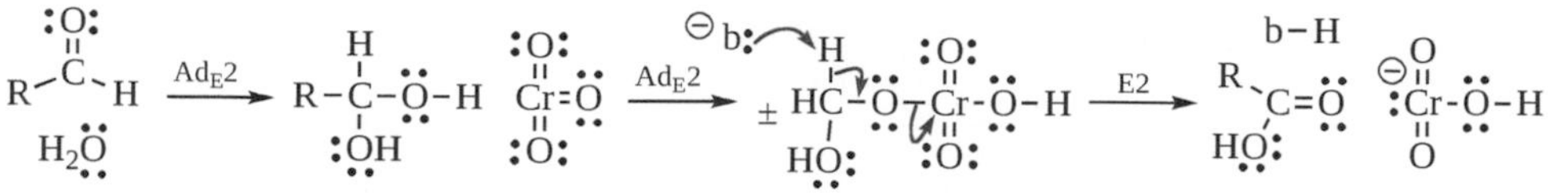

8.10.3 Hydrogenation via Transition Metal Catalysts

The exploration of transition metals as redox catalysts is a large and very active area of research. Transition metal mechanisms are quite different from those discussed so far. Catalytic hydrogenation of pi bonds is an important reaction for organic synthesis, and will serve as a good example. The mechanism proposed for catalytic hydrogenation by platinum, palladium, nickel, rhodium, iridium, and ruthenium is shown in Figure 8.10.

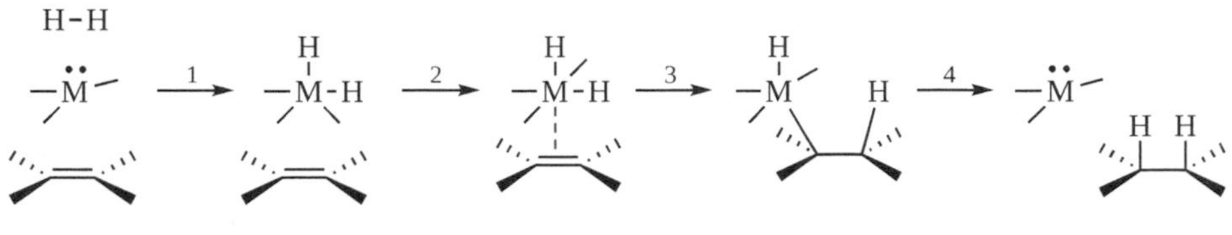

Figure 8.10 The proposed mechanism for catalytic hydrogenation.

Step 1 is the *oxidative addition* of hydrogen; the metal loses two electrons to form a bond to each of the two hydrogen atoms. Step 2 is the *complexation* of the alkene to the metal, activating the pi bond. Complexation usually occurs from the least hindered face of the pi bond. Step 3 is the *insertion* of the metal hydride bond into the pi bond, forming a metal–carbon bond. Step 4 is the *reductive elimination* (the reverse of oxidative addition), which returns the metal to its original oxidation state and releases the alkane. The reductive elimination is usually very fast. Both hydrogens are added to the same face of the pi bond, a *syn* addition. The appendix section, A Bridge to Transition Metal Organometallics, provides an approach to understanding these reactions.

8.11 REARRANGEMENTS TO AN ELECTROPHILIC CENTER

Rearrangements to an electrophilic carbon were briefly introduced in Section 4.7, and the electron flow paths were covered in Sections 7.1.11 and 7.1.12. To recap, the most important path restriction on rearrangements is that orbital overlap must be maintained by the migrating group with both the atom it migrates from and the atom it migrates to. In the transition state, the migrating orbital overlaps the orbitals of both. Rearrangement can proceed with any of our three general classes of electron sinks: an empty *p* orbital, a weak single bond to a leaving group, or a polarized multiple bond. The migration to an empty *p* orbital on carbon was the worked example in Section 4.7.3. The example below shows a migration to polarized multiple bond (benzil to benzilic acid rearrangement):

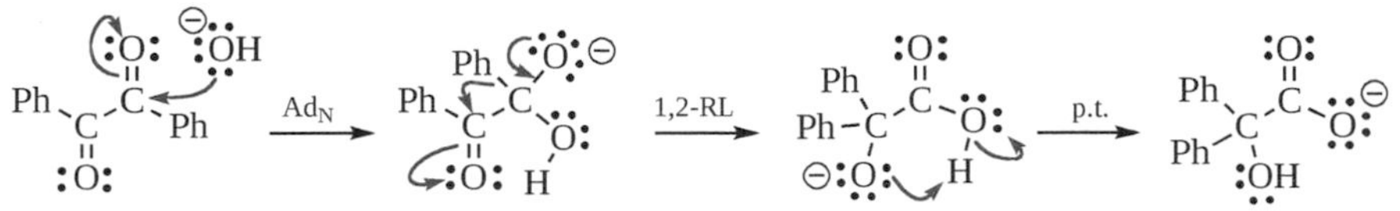

With a weak single bond to a leaving group, rearrangement can occur by a path (1,2RL) reminiscent of the generic S_N2. Good orbital alignment is crucial; the group that is best lined up to displace the leaving group from the back side is the one that preferentially migrates. Occasionally the leaving group will depart first (path D_N), and then rearrangement occurs (path 1,2R) in a path resembling the S_N1 process. Examples of migration to oxygen and migration from boron will be given in Section 8.12, Nu–L Pathways. The migration can occur to a vinylic center as the following examples show.

Migration to nitrogen in basic media: (in the example below, the migration is followed by Nu attack on the heterocumulene, Section 8.9.3, and then decarboxylation, Section 8.9.2):

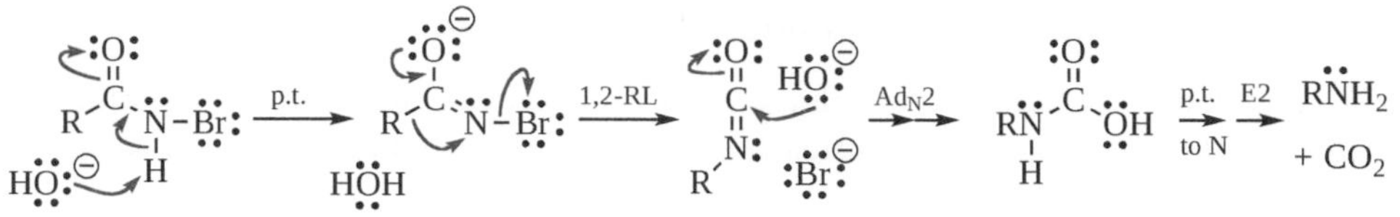

Migration to nitrogen, acidic media example: in this example, the leaving group returns as a nucleophile to trap the hetero-substituted (lone pair resonance stabilized) vinyl cation:

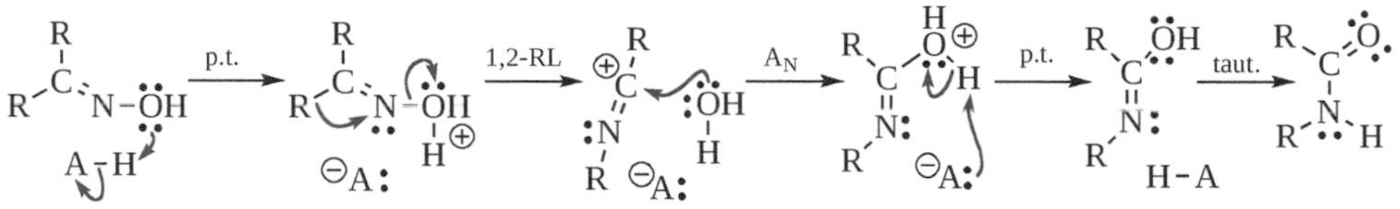

More examples of rearrangements to an electrophilic center are shown in the next section and the decision to rearrange or not is discussed in Section 9.8.

8.12 Nu–L REACTIONS

The unique Nu–L class of reagents can be recognized by a nucleophilic center to which a leaving group is directly bonded. **This unusual arrangement in which the same atom can serve both as a sink and a source is characteristic of the Nu–L class**, and thus their reactions are reasonably similar, most often three-membered ring formation or addition followed by rearrangement. Below are several examples of the Nu–L class.

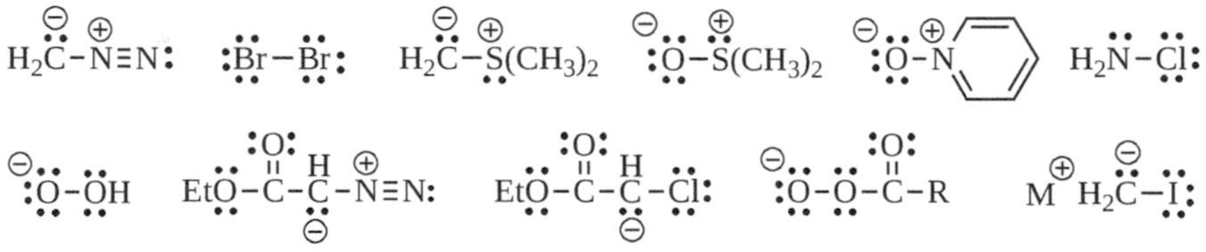

8.12.1 Nu–L Reacting with Trialkylboranes

Trialkylboranes, R_3B, react with Nu–L reagents to form Lewis acid–base salts that rearrange by migration of the alkyl group to the Nu with loss of the leaving group (path 1,2RL). The process usually repeats until all the alkyl groups have reacted. The $B(NuR)_3$ product that is formed is then hydrolyzed in the workup to give the free RNuH and a water soluble borate (see Section 8.9.1). Because of the many Nu–L possibilities, this is a very general and useful set of reactions.

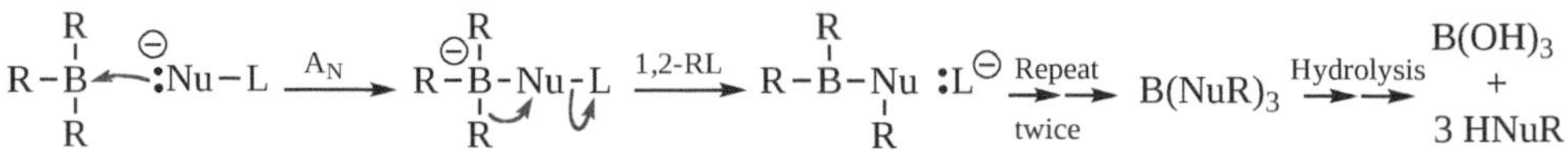

8.12.2 Nu–L Reacting with Acids

The Nu portion of the Nu–L can be protonated with acids, producing a reactive species with which the conjugate base of the acid may react.

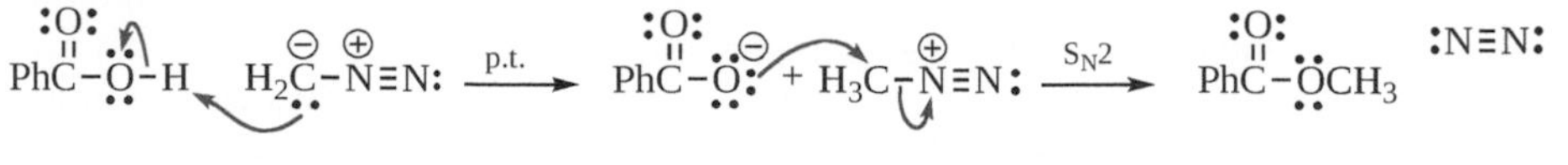

8.12.3 Nu–L Reacting with *sp*3 C–L

The Nu–L can substitute the leaving group on carbon and can be followed by an elimination reaction.

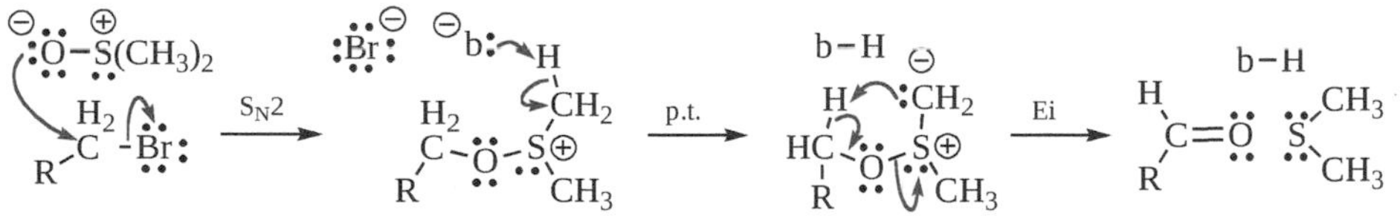

8.12.4 Nu–L Reacting with C=C

Leaving Group Drops Off First: Carbenes

Carbenes bear a resemblance to carbocations in that there is an empty *p* orbital that can behave as an electron sink. However, a full orbital that can serve as an electron source is on the same atom. Trichloromethyl anion loses chloride, forming the reactive dichlorocarbene, a neutral, electron-deficient, electrophilic intermediate. Stabilization in dichlorocarbene results from the pi bonding-like interaction of the full lone pair orbitals of chlorine with the empty *p* orbital of the carbene (Figure 8.11). If the donors on the carbene are good enough, the carbene becomes nucleophilic. With few exceptions, carbenes react stereospecifically with double bonds to produce three-membered rings.

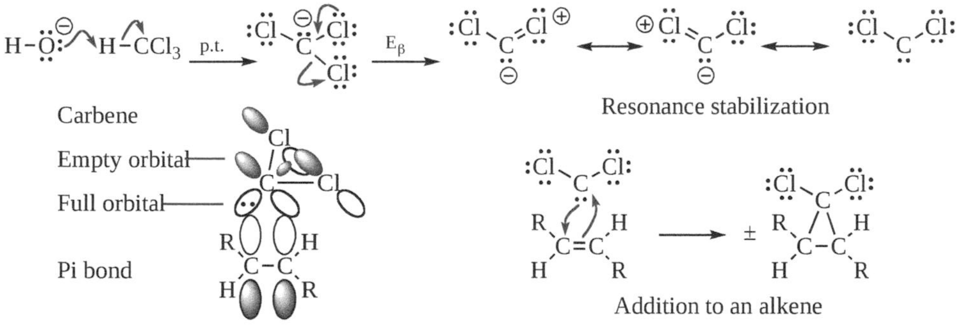

Figure 8.11 Dichlorocarbene addition.

Leaving Group Loss Concerted with Attack (path NuL)

These reactions are characteristically stereospecific. The formation of a bromonium ion from attack of bromine on a pi bond was the first example of this path (Section 8.3.6).

The organometallics of this subset are called carbenoids because of their similar reactivity to carbenes. Carbenoids are generated by the reaction of a geminal organic dihalide such as CH_2I_2 with a metal by halogen–metal exchange. The presumed sequence is shown in the following example. The structure of the organometallic is currently under debate.

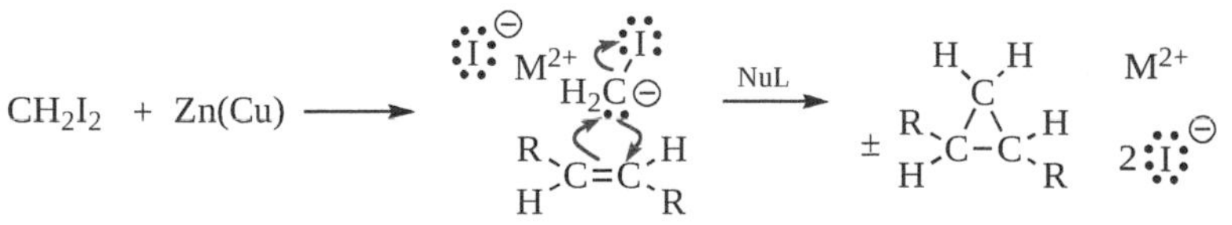

Although they are not considered carbenoids, peracids are mechanistically similar in their additions to alkenes. The intramolecular hydrogen bond in the peracid (shown by the two resonance forms) partially breaks the O–H bond, making it a better nucleophile.

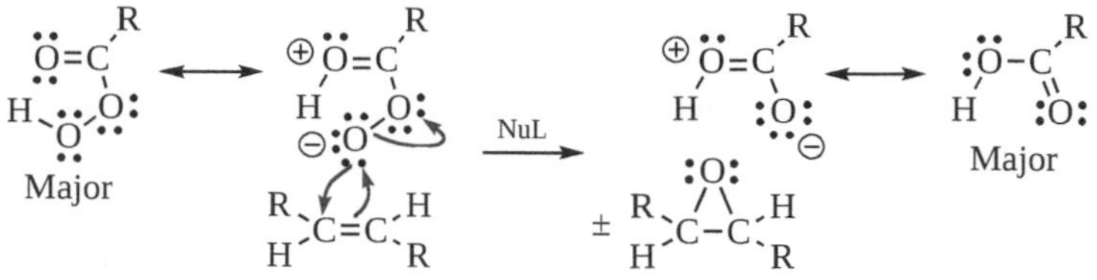

8.12.5 Nu–L Reacting with C=Y

Leaving Group Drops Off After Attack

This mechanism implies a new intermediate, one that stabilizes the anion formed upon nucleophilic attack. The Nu–L species acts first as an electron source, then in a second step functions as an electron sink. Common members of this subset, the ylides, have adjacent plus and minus charges, $^{-}$Nu–L^{+}. This mechanism is a combination of the addition to a polarized multiple bond (path Ad$_N$) followed by an internal S$_N$2.

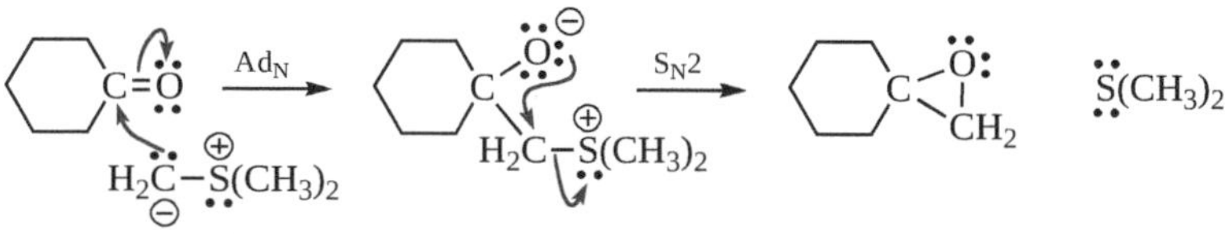

Phosphorus ylides are almost the only representative of a mechanistic variant. Rather than being displaced in an internal S$_N$2 reaction, the phosphorus bonds to oxygen and then undergoes an elimination (path 4e) to the very stable phosphine oxide.

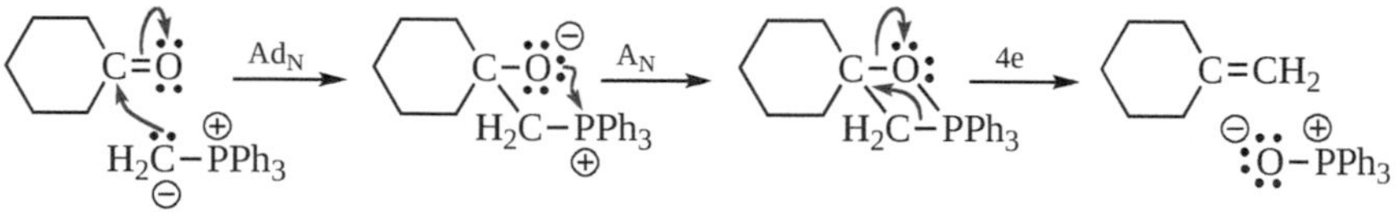

The Nu–L species can attack the polarized multiple bond (path Ad$_N$), then rearrangement with loss of the leaving group occurs to break the weak O–O bond (path 1,2RL), similar to reactions discussed in Section 8.11.

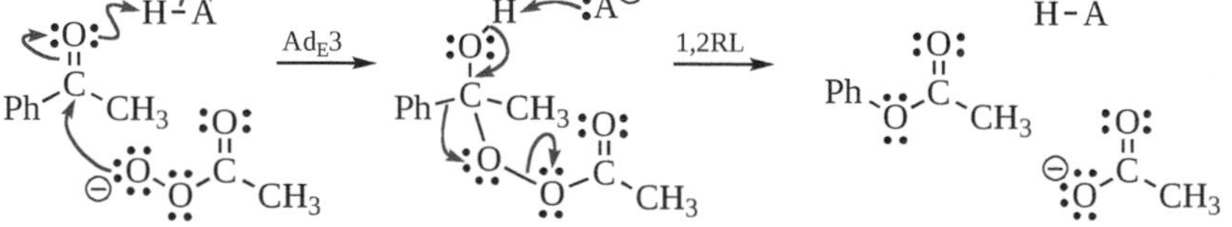

8.12.6 Nu–L Reacting with C=C–EWG

The allylic source, formed by the conjugate addition of peroxide ion to the polarized pi bond, then acts as a nucleophile in an internal S$_N$2 to make an epoxide.

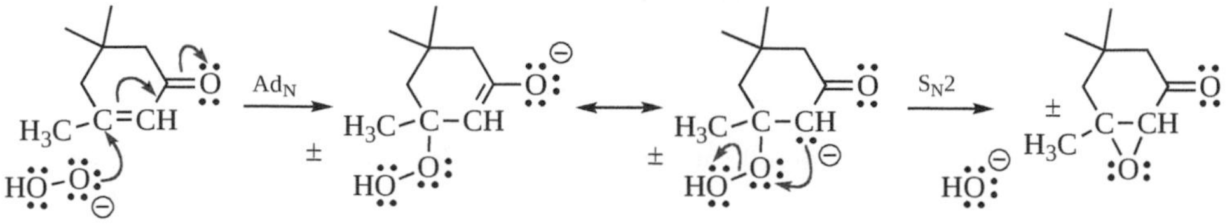

8.12.7 Nu–L Reacting with L–C=O

The expected addition–elimination path Ad_N + E_β occurs. The diazomethane example below gives a diazoketone, which after deprotonation is also a Nu–L species. These can add to alkenes like carbenes or rearrange like the azide example shown just below this example.

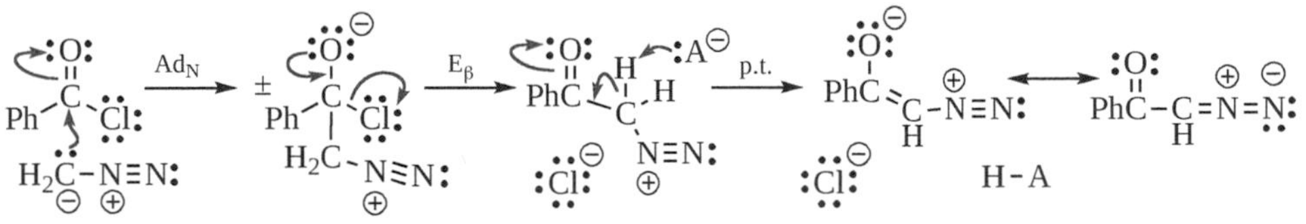

Azide ion addition–elimination is followed by a rearrangement with loss of nitrogen (path 1,2RL) in the following example.

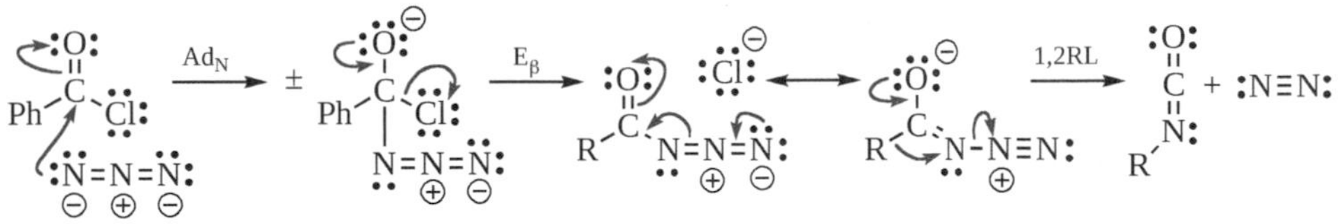

8.12.8 Thiamine-Catalyzed Decarboxylation of Pyruvate

Biochemical Example

Thiamine is an amazing coenzyme unique in biochemistry. Thiamine is an elegantly optimized catalyst that serves as an attachable/detachable electron-withdrawing group to allow electron flow when it would otherwise be impossible. One reaction thiamine catalyzes is the decarboxylation of pyruvate, CH_3COCOO^-, to acetaldehyde, CH_3CHO (Figure 8.12). Initially, a deprotonation forms the nucleophilic thiamine carbene. This thiamine carbene adds to the more electrophilic ketone carbonyl of pyruvate. A fragmentation step via an E_β path (in which the leaving group is an enamine) can now occur because the thiamine accepts the electron flow from the carboxylate anion (discussed in Section 7.4, Variations on a Theme). The electrons flow back out of the thiamine to pick up a proton (or possibly another electrophile). Then thiamine detaches as a leaving group by E2 to start the process all over again. Without the thiamine coenzyme, the decarboxylation of pyruvate does not occur at all, for the reaction would have to produce the highly unstable acyl anion, $[CH_3\text{–}C\text{=}O]^-$.

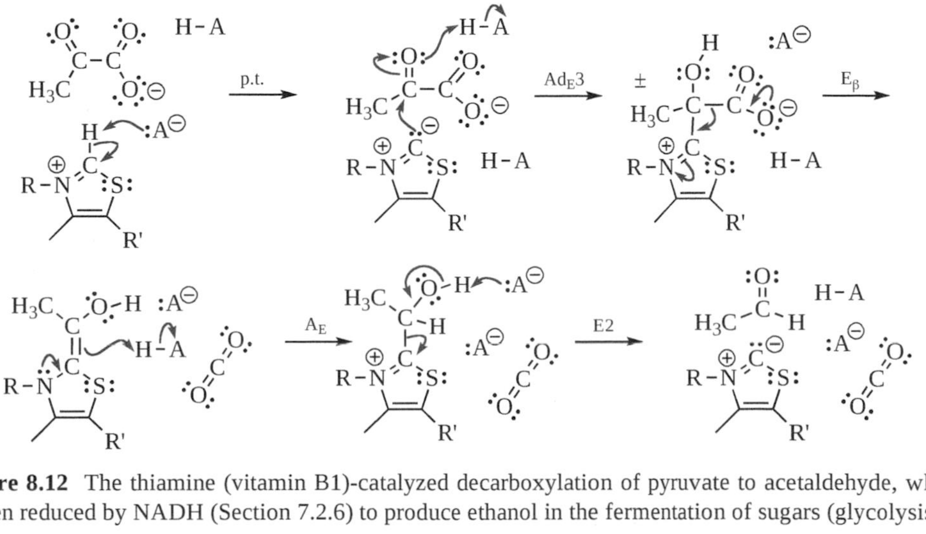

Figure 8.12 The thiamine (vitamin B1)-catalyzed decarboxylation of pyruvate to acetaldehyde, which is then reduced by NADH (Section 7.2.6) to produce ethanol in the fermentation of sugars (glycolysis).

8.13 PRODUCT MATRIX SUMMARY

Table 8.3 Product matrix

Source⇒ Sink⇓	Lone pair Nu $Z{:}^{\ominus}$	Base b:$^{\ominus}$ and adjacent CH	Metal hydride $MH_4^{\ominus}$	Organometal. R–M	Allylic [a] C=C–Z:$^{\ominus}$	Simple pi bonds C=C	Aromatic ArH
H–A	p.t. Z–H	p.t. b–H	p.t. H–H	p.t. R–H	p.t.[a] HC–C=Z	add. H–C–C–A	H exchange Ar–H
Y–L	subst. Z–Y	elim. >C=Y	subst. H–Y	subst. R–Y	subst.[a] Y–C–C=Z	add. Y–C–C–L	Ar subst.[c] Ar–Y
–C–L	subst. –C–Z	elim. >C=C<	subst. –C–H	subst. –C–R	subst.[a] –C–C–C=Z	usually no reaction[d]	Ar subst.[c] –C–Ar
>C=Y	add. Z–C–Y$^{\ominus}$	p.t. $^{\ominus}$C–C=Y	add. H–C–Y$^{\ominus}$	add. R–C–Y$^{\ominus}$	$^{\ominus}$Y add.[a] –C–C–C=Z	usually no reaction[d]	Ar subst.[c] Ar–C–YH
—C≡Y	add. Z–C=Y$^{\ominus}$	p.t. $^{\ominus}$C–C≡Y	add.[b] H–C=Y$^{\ominus}$	add. R–C=Y$^{\ominus}$	add.[a] C=Y$^{\ominus}$, C–C=Z	usually no reaction[d]	Ar subst.[c] Ar–C=YH
>C=C< [e] EWG	1,4 add.[e] Z–C–C$^{\ominus}$–EWG	p.t. C=C–EWG, C$^{\ominus}$	1,4 add.[e] H–C–C$^{\ominus}$–EWG	1,4 add.[e] R–C–C$^{\ominus}$–EWG	1,4 add.[a,e] C–C$^{\ominus}$–EWG, C–C=Z	usually no reaction[d]	Ar subst.[c,e] Ar–C–C–EWG
L–C=Y	add.-elim. Z–C=Y	p.t. L–C=Y, $^{\ominus}$C	add.-elim.[b] H–C=Y	add.-elim.[b] R–C=Y	add.-elim.[a] C=Y, C–C=Z	usually no reaction[d]	Ar subst.[c] Ar–C=Y

[a] Ambident source, Section 9.4. [b] Multiple addition possible, Section 9.2. [c] With Lewis or Brønsted–Lowry acid catalyst. [d] Reacts with strong acid catalyst but may give polymers or mixtures. [e] Ambident sink, Section 9.6.

The product matrix (Table 8.3) is a summary of most of Chapter 8. It is the source and sink matrix (Table 8.1) that was an index to Sections 8.2 through 8.8 filled in with the most common reaction product in generic form. It cannot be stressed too much that **everything depends on your ability to recognize the generic class of the sources and sinks** in the reaction mixture. Ambident sources and sinks can each react in more than one spot, but Table 8.3 had room for only one product. Sometimes a species has a dual reactivity and therefore may fit into more than one generic class. A common example of this is that most anions can behave as a nucleophile or as a base. Chapter 9 discusses all these common major decisions.

CHAPTER 8 EXERCISES

Ever notice how the book problems within a chapter appear easier than the ones at the end of the chapter? And perhaps the ones on an exam may appear harder still? It might just be a matter of "mental priming" of your analysis process. Within the chapter, you know that the problem has something to do with what you just read; you do not have to search for the principles to apply to solve the problem. At the chapter's end, you have a broader area to search, but usually the problems relate somehow to the chapter just covered. On an exam that covers several chapters, you may have difficulty trying to figure out which part of what you have learned applies to the problem. Therefore a crucial part of problem analysis is determining what information is relevant. We have learned how to classify sources and sinks and to use that to predict reactivity. The real reward of classifying sources and sinks is in priming your analysis of a problem you have never seen before. As was done in this chapter, focus on the sink first, then find a source.

In all of these problems, make sure the missing structure balances the reaction equation.

(Refer to the Hints to Selected Problems from Chapters 8, 9, 10, and 11 if you need help).

8.1 In the following problem, identify the electron sink and look at the product to identify the electron source and predict the pathway(s) to the product.

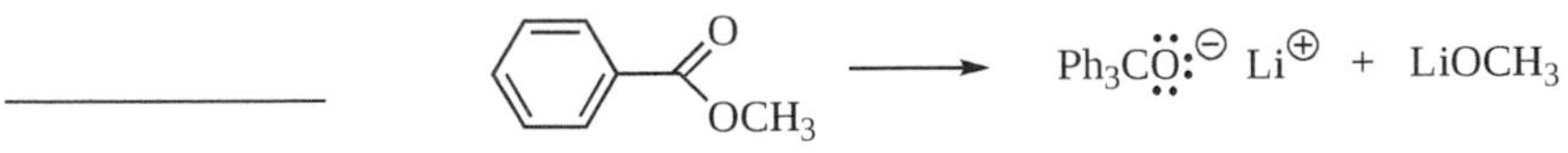

8.2 In the following problem, identify the electron sink and electron source and predict the pathway(s) to the product.

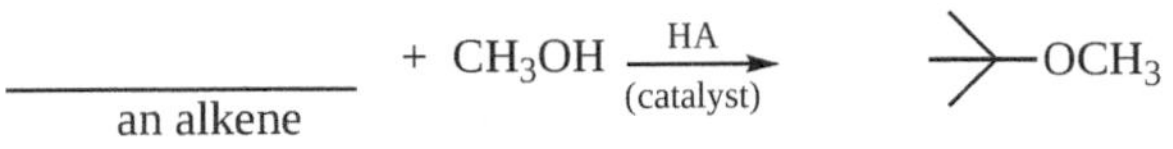

8.3 In the following problem, identify the electron sink and electron source and predict the pathway(s) to the neutral product.

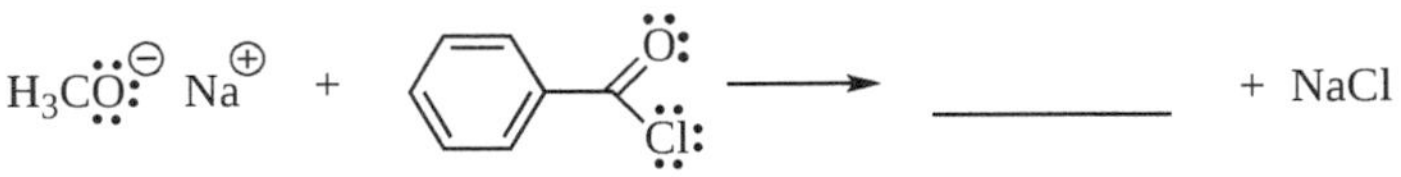

8.4 In the following problem, identify the electron source and look at the product to identify the electron sink and predict the pathway(s) to the product.

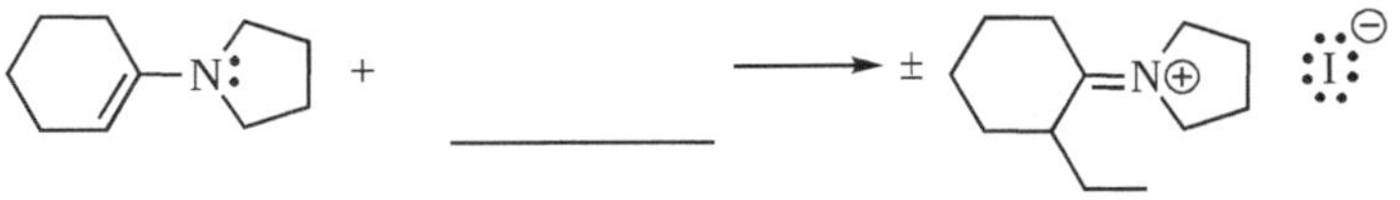

8.5 In the following problem, identify the electron source and look at the product to identify the electron sink and predict the pathway(s) to the product.

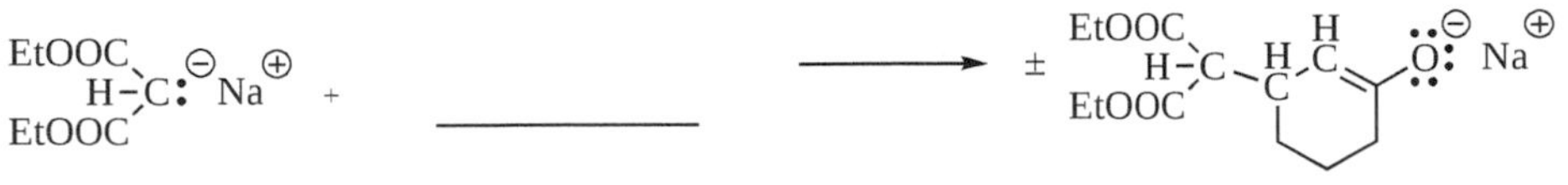

8.6 In the following problem, identify the electron sink and electron source and predict the pathway(s) to the product.

8.7 In the following problem, identify the electron sink and look at the product to identify the electron source and predict the pathway(s) to the product.

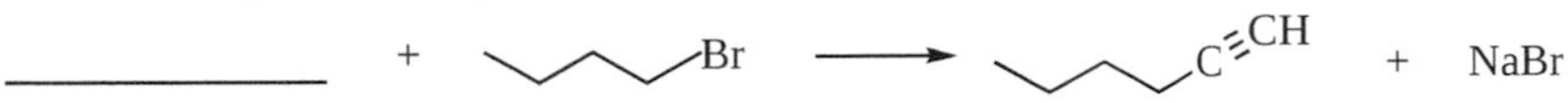

8.8 In the following problem, identify the electron source and look at the product to identify the electron sink and predict the pathway(s) to the product.

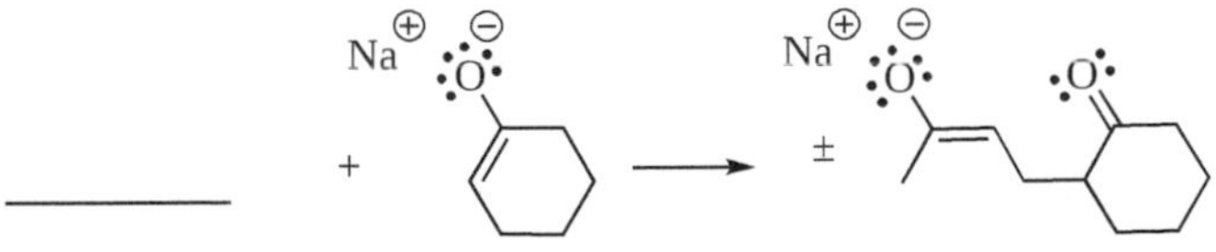

8.9 In the following problem, identify the generic electron sink and electron source and predict the proton transfer product.

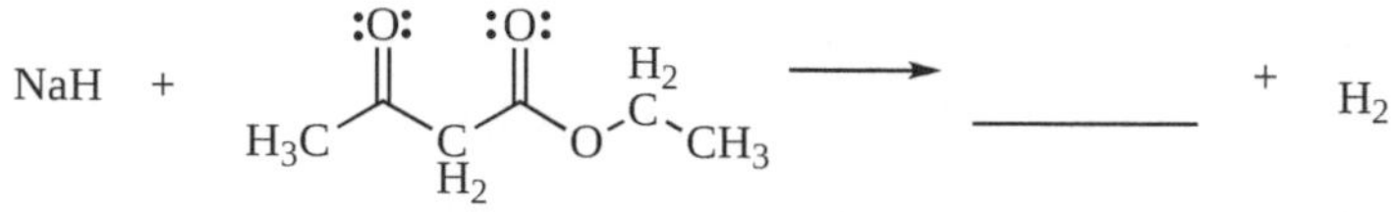

8.10 In the following problem, identify the electron source and look at the product to identify the electron sink and predict the pathway(s) to the product.

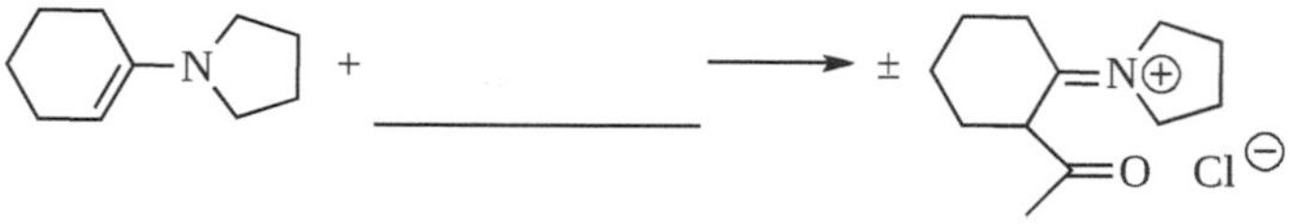

8.11 In the following problem, identify the electron sink and electron source and predict the product and the pathway(s) to the product.

8.12 In the following problem, identify the electron sink and electron source and predict the product and the pathway(s) to the product.

$$Ph{-}Li + I_2 \longrightarrow \_\_\_\_\_\_\_\_ + LiI$$

8.13 In the following problem, identify the electron source and look at the product to identify the electron sink and predict the pathway(s) to the product.

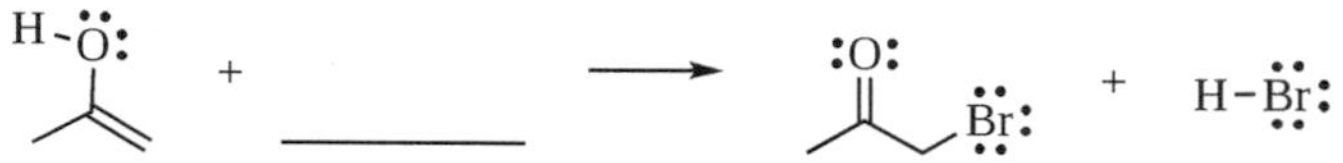

8.14 In the following problem, identify the electron sink and electron source and predict two possible products. A Chapter 9 decision is required to decide which one is favored.

8.15 In the following problem, identify the electron sink and electron source and predict the pathway(s) to the product.

8.16 In the following problem, identify the best electron sink and electron source and predict the product and the pathway(s) to the pesticide product.

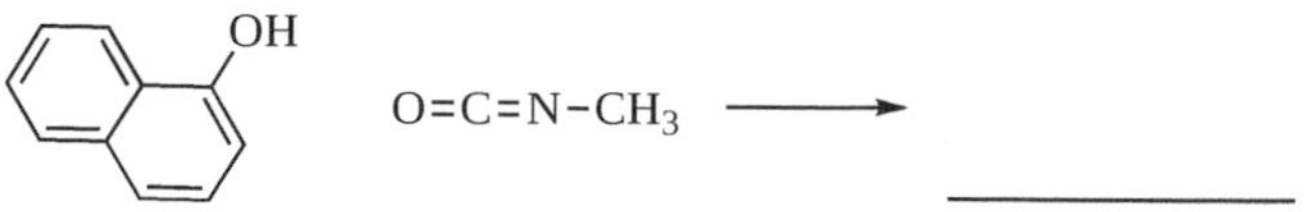

8.17 In the following problem, identify the electron sink and electron source and predict the pathway(s) to the bleach product.

$$2\ NaOH + Cl_2 \longrightarrow \_\_\_\_\_\_\_\_ + NaCl + H_2O$$

8.18 Identify the electron sink and electron source and predict the substitution product and the pathway(s) to the product.

9

DECISIONS, DECISIONS

9.1 DECISION POINT RECOGNITION

In Chapter 4, the D_N decision to ionize or not to ionize the leaving group was part of your ability to predict whether the first step in an E1 or S_N1 would proceed. Several other similar decisions are collected in this chapter. The most difficult choice is between a species serving as a nucleophile or as a base (Section 9.5). Other decisions are essentially a choice of regiochemistry (Sections 9.3, 9.4, and 9.6). One (Section 9.2) is a question of extent of

Electron Flow in Organic Chemistry: A Decision-Based Guide to Organic Mechanisms,
Third Edition. Paul H. Scudder.

reaction, to stop or keep going. Another (Section 9.7) is the competition between internal reactions and external ones. The last (Section 9.8) is the competition between nucleophilic trapping and cation rearrangement.

The most important question that you must continually ask yourself when examining any step in a process that has known and unknown branch points is, What else can happen? This may seem to add complexity, but it is the essence of critical thinking. Obvious routes often do not work and therefore lead to frustration and the question, "Where did I go wrong?" The answer may be, "forgetting to watch for alternative routes and proceeding with the first idea that occurred to you."

Draw out every step of the process and look for alternative routes from each step; try to find at least one alternative for each step. For hints, check resonance forms, proton transfer, alternative sources and sinks, less likely combinations, and higher-energy species. Specifically look for the more obvious decision points covered in this chapter.

If you find that you habitually miss a decision point or two, then a good idea would be to make up a checklist of reminders until new habits form. Refer to this checklist whenever you are working a problem. Continue to personalize this checklist by adding reminders to watch for common errors that you find yourself repeating.

9.2 MULTIPLE ADDITIONS

9.2.1 Nucleophilic Addition to L–C=O

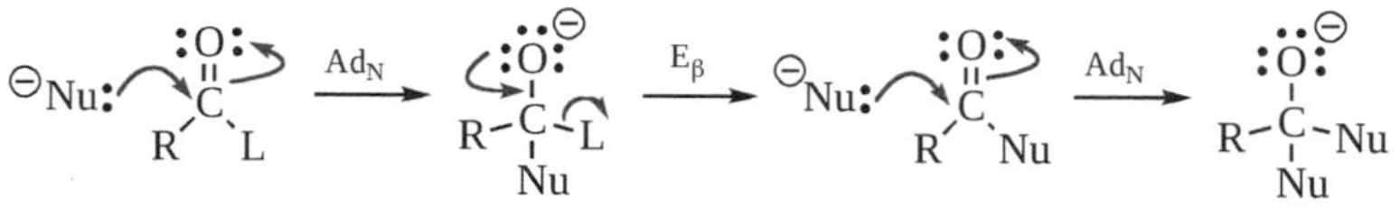

The products of an addition–elimination path very often undergo a second attack by the nucleophile (Ad_N + E_β then a second Ad_N) because the product of the first addition–elimination is also a reactive electron sink. If the addition–elimination product is more reactive than the original starting material, a second addition almost certainly will occur. A ranking of the reactivity of electron sinks and sources is very important for path selection. For each reaction, the useful limit must be found for each reactivity spectrum. How far down the reactivity trend for the electron source can we go and still have the reaction with a particular electron sink occur at a useful rate? There are four ways that the reaction may stop after the initial Ad_N + E_β:

1. **Reversible Unfavorable Equilibrium:** The Nu is such a weak base that the second addition easily reverses and the equilibrium favors the more stable carbonyl. Heteroatom lone pair sources almost never add twice.

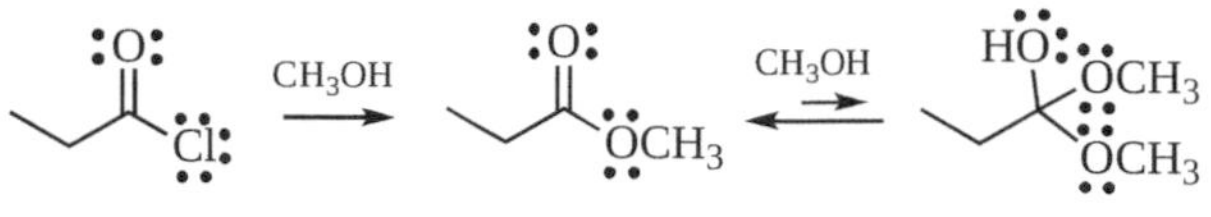

2. **Selective Nucleophile:** The initial electron sink must be more reactive than the product of the addition, and the electron source, although reactive enough to add once, is not reactive enough to make a second addition to the less reactive compound. Organometallics in the reactivity range of organocadmiums, organozincs, and organocoppers (metal electronegativities are 1.69, 1.65, and 1.90, respectively) add quickly to the more reactive acyl chlorides but slowly to the ketone products.

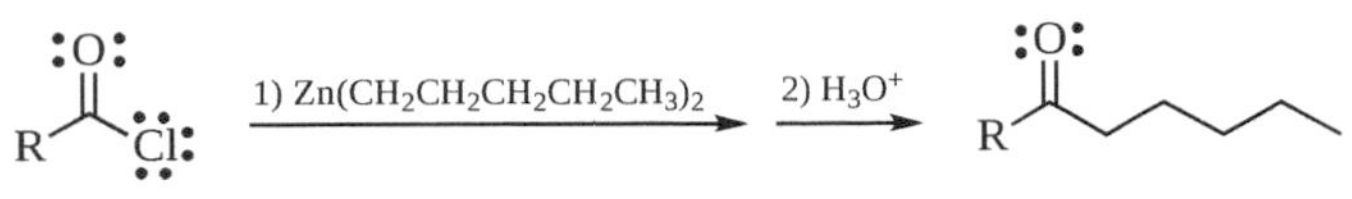

3. **Poor Leaving Group:** The leaving group is so poor (in amides, pK_{aHL} = 36, for example) that it cannot be ejected easily from the tetrahedral intermediate. Since the carbonyl does not reform, there is no site for nucleophilic attack. The carbonyl forms in the acidic workup (no Nu is left by then) by protonating the nitrogen, making it into a better L.

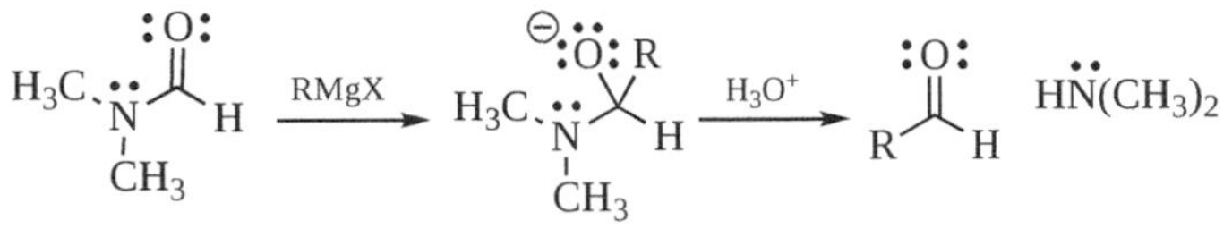

4. **Low Temperature:** One equivalent or less of the electron source was added to the sink at a low enough temperature (usually −78°C, cooled with dry ice). Since the addition step usually has a lower barrier than the elimination step, the loss of the leaving group from the tetrahedral intermediate by elimination path E_β does not occur to a significant extent before the electron source is used up. Tetrahydrofuran is the preferred solvent.

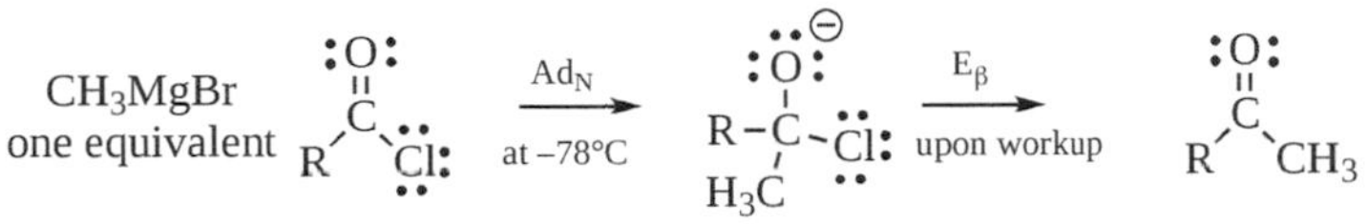

Example problem

At 25°C, which of these nucleophiles add once to an ester: CH_3Li, $LiAlH_4$, CH_3O^-?

Answer: At 25°C, since the first addition–elimination product of CH_3Li adding to an ester is a ketone, which is more reactive than the ester, the nucleophile would add a second time. At 25°C, since the first addition–elimination product of $LiAlH_4$ adding to an ester is an aldehyde, which is much more reactive than the ester, the nucleophile would add a second time. The addition–elimination of methoxide to the ester creates another ester, not something more reactive. Since the pK_{abH} of the incoming Nu and the outgoing L are almost the same, the reaction is reversible at 25°C. Also a second addition is uphill because two C–O bonds are weaker than a C=O bond. Therefore the second addition of CH_3O^- is unfavorable and reversible, meeting our criteria for stopping.

9.2.2 Additions of Nucleophiles to Nitriles

With few exceptions, organometallics add only once to nitriles (Section 8.6.4). Only if there is a very strong Lewis acid present to activate the anionic intermediate of the first nucleophilic attack will a second addition take place. An example of this second addition is the reduction of nitriles with $LiAlH_4$, which proceeds all the way to the amine (Section 8.6.3). The AlH_3 formed from the initial reduction acts as a Lewis acid to catalyze the second addition. This second addition can be prevented by using one equivalent of a less reactive aluminum hydride like $LiAlH(OEt)_3$ or diisobutylaluminum hydride, $[(CH_3)_2CHCH_2]_2AlH$. The reactions occurring in the acidic water workup are the reverse of imine formation (Section 10.5.2). One equivalent of $LiAlH(OEt)_3$ example:

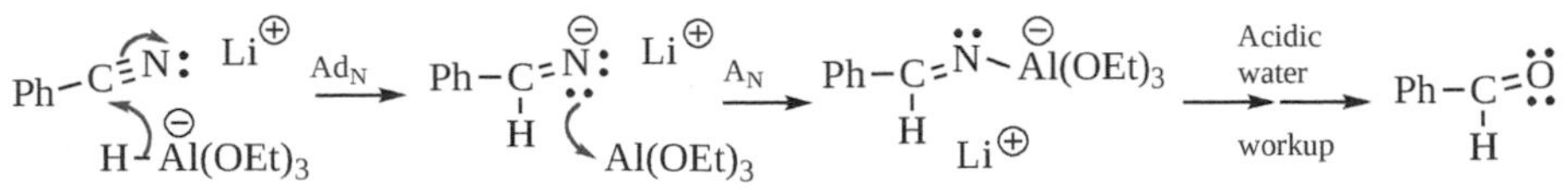

One equivalent of $[(CH_3)_2CHCH_2]_2AlH$ example:

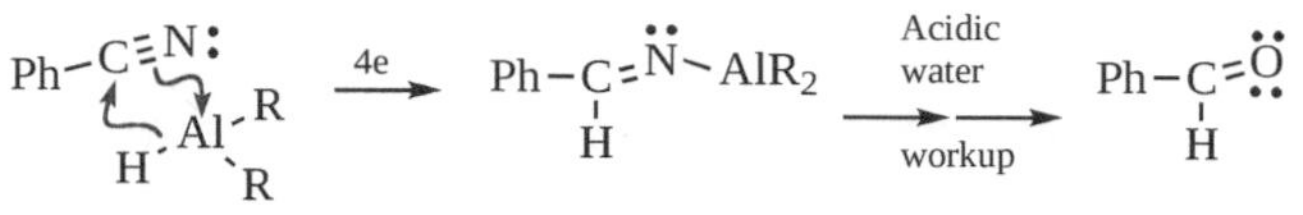

9.3 REGIOCHEMISTRY AND STEREOCHEMISTRY OF ENOLATE FORMATION

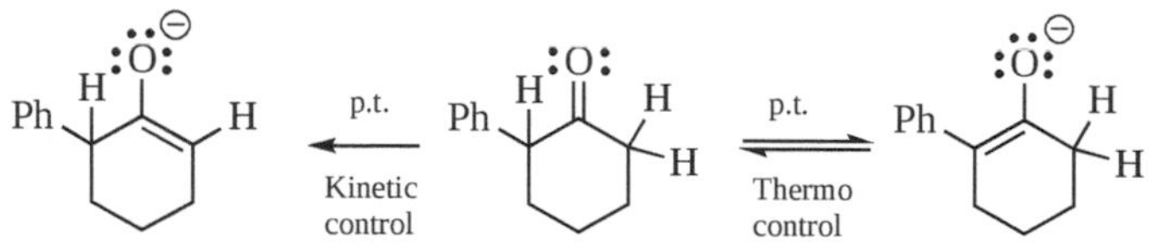

An unsymmetrical ketone can deprotonate to form two possible regioisomeric enolates, and the choice of which enolate forms depends on whether the deprotonation was carried out under kinetic or thermodynamic control. The thermodynamic enolate is the most stable of the possible enolates. In general, the more substituted enolate is the more stable (reflecting the double-bond stabilities); however, exceptions are common. The kinetic enolate is usually the result of deprotonating the most accessible acidic hydrogen. The two hydrogens shown on the methylene outnumber and are more accessible than the single, more substituted hydrogen.

Deprotonation of the ketone must be fast, complete, and irreversible for kinetic control of enolate formation. No equilibration of the enolates can be allowed to occur. Optimum conditions for kinetic control of deprotonation are: Add the ketone slowly to an excess of very strong base (usually *i*-Pr_2NLi, the anion of diisopropyl amine, pK_{abH} = 36) in an aprotic solvent (such as dry tetrahydrofuran or dimethoxyethane). Since the K_{eq} for deprotonation of a ketone with this base is $10^{(36-19.2)} = 10^{+16.8}$, the reaction is complete and irreversible.

Any equilibrium will produce the thermodynamically most stable enolate. The most stable enolate will have the greatest charge delocalization. In the above example, the thermodynamically favored enolate is conjugated to the phenyl; the kinetically favored enolate is not. Common conditions for thermodynamic control are to use average bases (like sodium ethoxide or potassium *tert*-butoxide, pK_{abH}s 16–19) in alcohol solvents. Proton-transfer equilibria rapidly occur among base, solvent, ketone, and enolate. Sodium hydride or potassium hydride in an ether solvent are also thermodynamic reaction conditions that allow equilibration between the reactant ketone and the enolate. Enones have two possible enolates; weaker bases give the thermodynamically more stable extended enolate, whereas kinetic conditions produce the cross-conjugated enolate.

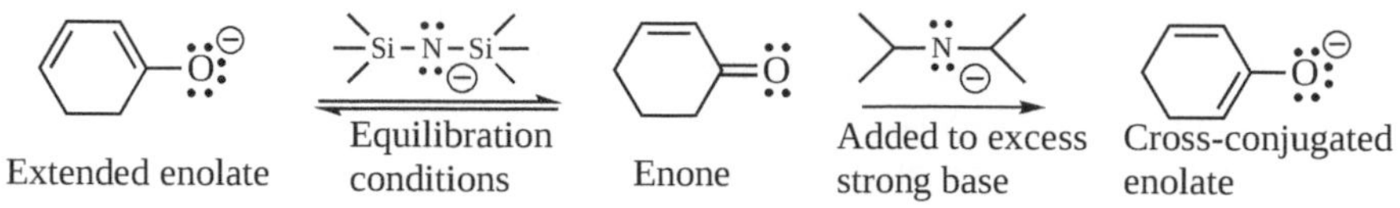

In addition to regiochemistry, acyclic carbonyl compounds can produce two possible stereoisomeric enolates, *cis* or *trans,* as shown below. Steric interactions determine the favored enolate stereochemistry. Under reversible conditions, especially if R is large, *cis* enolates are more stable than *trans* as they minimize steric interactions. Under irreversible conditions in ether solvents, the *trans* enolate is often favored because the steric size of the base/cation aggregate around the oxygen dominates, especially if R is smaller, as happens with esters.

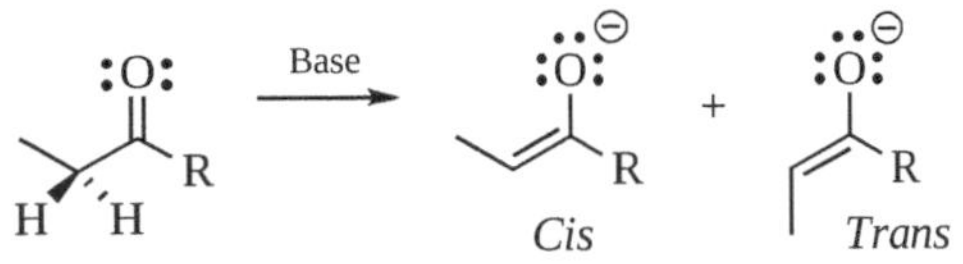

9.4 AMBIDENT NUCLEOPHILES C=C–Z:

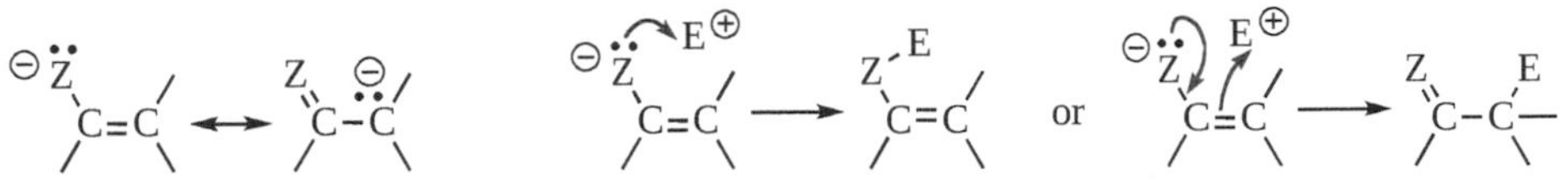

All allylic sources have the capability to act as a nucleophile on either δ– end of the source. Since the two ends of the allylic source commonly differ in electronegativity, the charge on each end differs. The two ends commonly differ in polarizability also. Therefore, the Z end is usually much harder than the softer C end. For soft electrophiles, the soft–soft component is most important; therefore the softest electrophile will favor the C end. For hard electrophiles, the hard–hard component is most important, therefore the atom with the largest partial minus (usually the Z end) will attract a hard electrophile (charge control). Solvation is also a hard–hard interaction, and the tighter the solvation around the Z end, the more hindered and poorer the Z nucleophile is. A $\Delta H°$ calculation usually predicts the C-reacted compound to be thermodynamically more stable than the Z-reacted compound (mainly because of the greater C=Z bond strength in the C-reacted product compared to the C=C in the Z-reacted). It is important to determine which is the dominant effect, product formation based upon product stability or upon kinetic direction from HSAB theory. To do this we need to determine whether the reaction is under kinetic or thermodynamic control. Figure 9.1 gives a flowchart for the decision for a common ambident nucleophile, an enolate anion (Z equals oxygen).

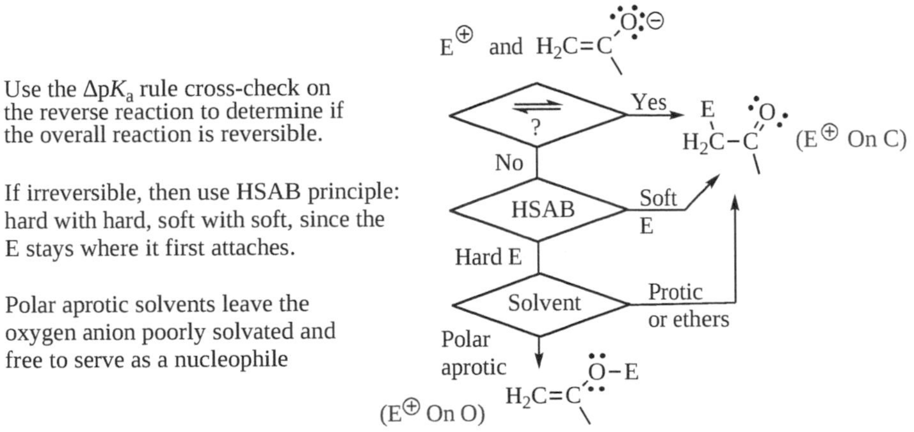

Figure 9.1 Flowchart for the ambident nucleophile decision of an enolate.

9.4.1 Alkylation of Enols, Enolates, and Enamines with sp^3 C–L

Enolate general example:

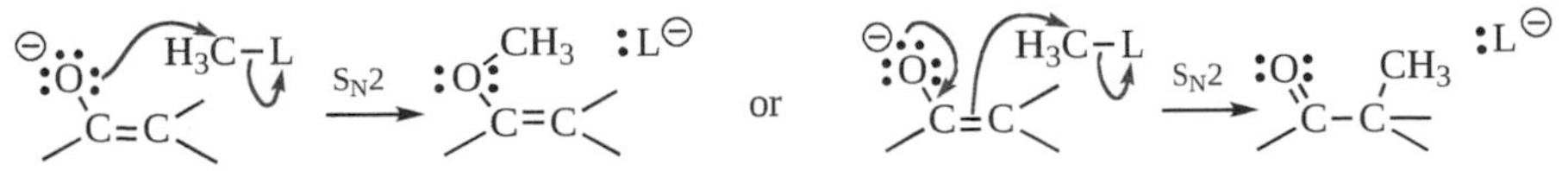

Alkylation on oxygen Alkylation on carbon

Most of these alkylations are not reversible, so whatever process is faster dominates (kinetic control). The oxygen atom is much harder than the carbon atom of the enolate and bears more of the charge because of its greater electronegativity. The decision whether to use the heteroatom or the carbon atom as the nucleophile is based on the hardness of the electrophile and the solvent. Soft electrophiles like R–Cl, R–Br, and R–I will alkylate on the softer carbon of the enolate. Hard electrophiles will attack the heteroatom (hard with hard) in polar aprotic solvents that poorly solvate the heteroatom. Since carbocations are harder electrophiles (greater charge) than *sp*^{3} C–L, the tendency to alkylate on the heteroatom will increase as the reaction mechanism shifts from S_N2 toward S_N1. With *sp*^{3} C–L, as the leaving group gets more electronegative, the partial plus on carbon increases, and the electrophile gets harder; therefore the amount of alkylation on the heteroatom lone pair increases—for example, oxonium ions, $(CH_3)_3O^+$, sulfate esters, $CH_3OSO_2OCH_3$, and sulfonate esters, $ArSO_2OCH_3$, tend to alkylate on the heteroatom in DMF, DMSO, and HMPA. A special case and not really considered an enolate, phenoxide usually alkylates on oxygen because alkylation on carbon would interrupt the aromaticity of the ring:

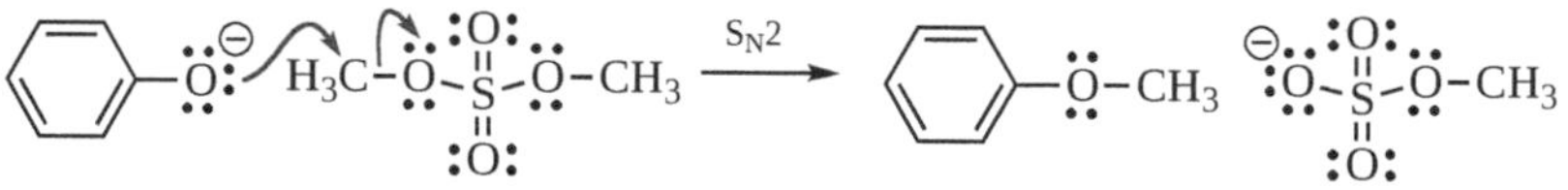

With anionic allylic sources, highly polar aprotic solvents increase the amount of lone pair alkylation because poor solvent stabilization of the anion leaves the heteroatom end less hindered by solvent. Conversely, groups bound to the heteroatom increase the steric hindrance about it, and therefore decrease the tendency toward heteroatom alkylation. For example, enamines (R_2N–C=C) tend to alkylate on carbon rather than nitrogen (as shown in Section 8.4.5).

Since alkylation is usually not reversible, the products are the result of kinetic control. However, iminium ions, $CH_2=N(CH_3)_2^+$, are stabilized electrophiles and reversibly add to enols to give the thermodynamically more stable C-alkylated product.

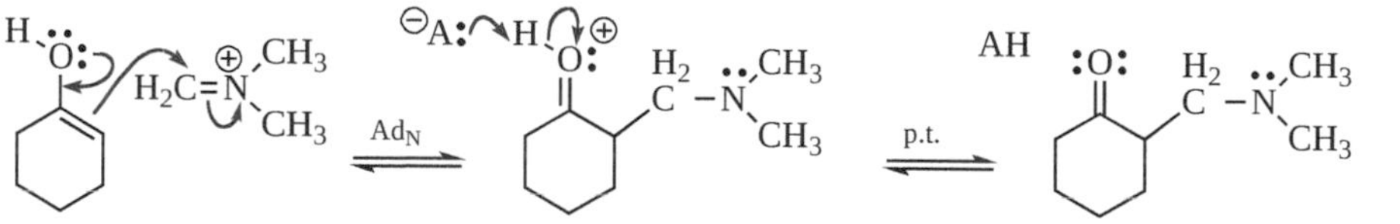

In the following list are the extremes for the alkylation reaction of allylic sources; however, product mixtures occur rather often.

Alkylation on Z (less common)	Alkylation on C (much more common)
Hard electrophiles	Soft electrophiles
Highly polar aprotic solvent	Ether or alcohol solvent
Heteroatom accessible	Carbon atom more accessible
Irreversible reactions	Reversible reactions

9.4.2 Reaction of Enols and Enolates with Y–L

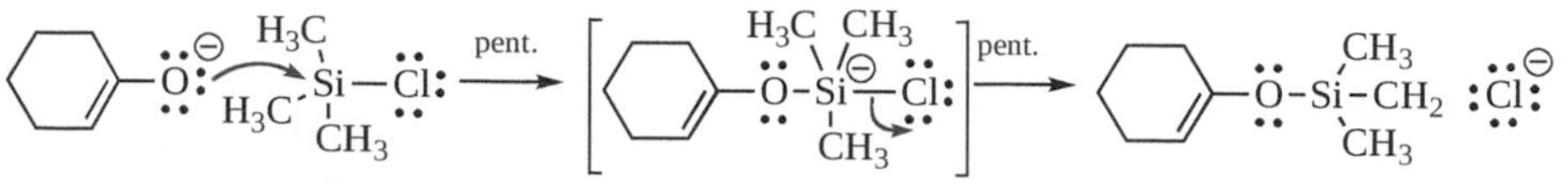

In the above section, the attacking electrophile was a carbon atom of varying hardness, and since the C–C and C–O bonds were about the same strength, the C=C vs. C=O bond strength dominated the thermodynamic choice. With the Y–L sink, the difference in the Y–C vs. Y–O bond strength can be controlling. With enolates for example, the Si–O bond is 36 kcal/mol (150 kJ/mol) stronger than the Si–C bond, so the O-silated product is both thermodynamic and kinetic favored. Trimethylsilyl chloride, $(CH_3)_3Si–Cl$, having a highly polarized silicon–chlorine bond, reacts exclusively and irreversibly on the oxygen of enolates. Likewise, the P–O bond is over 70 kcal/mol (293 kJ/mol) stronger than the P–C bond, so all P–L species go on the oxygen of enolates also.

On the other hand, the soft Br_2 brominates enolates on carbon. Since the incoming Nu pK_{abH} is much higher than the outgoing L pK_{abH}, the reaction is expected to be irreversible. Also, the Br–C bond is 20 kcal/mol (84 kJ/mol) stronger than the Br–O bond, making the carbon brominated product both the thermodynamic and kinetic product.

9.4.3 Acylation of Enols, Enolates, and Enamines with O=C–L

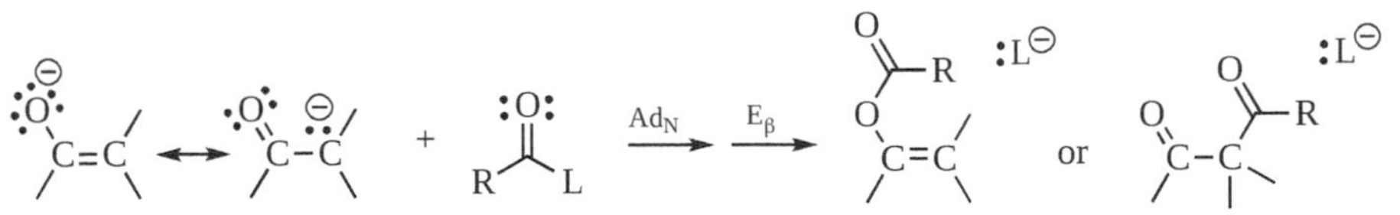

Acylation on oxygen or Acylation on carbon

For the most common conditions, **basic media**, acylation almost always goes on carbon. Acylation on the heteroatom produces a product that can be attacked by another molecule of the allylic source to produce the C-acylated compound. The product of any equilibrium is the thermodynamic C-acylated product.

The basic media exceptions can be easily understood if we invoke HSAB theory and realize that the kinetic and thermodynamic products are different. As L becomes a poorer donor, the partial plus on the acyl carbon increases, making it harder. Acylation on the heteroatom of the allylic source is fast for acyl halides and anhydrides where the acyl carbon is harder (greater partial plus) than the acyl carbon of esters. If the reaction is under kinetic control (allylic source added to an excess of acyl halide or anhydride), the Z-acylated product is formed; however, if equilibration occurs (excess of allylic source), the product will be the C-acylated, thermodynamic product. Reaction conditions matter.

In **acidic media**, the electron sink is most often the carbocation produced from protonating the acylating agent, and therefore the sink is very hard. Attack by the Z end (harder end) of the allylic source is very fast. For enols, the Z-acylated kinetic product can be isolated. Since the Z-acylated enol is itself an allylic source (but weaker), it can be forced by more vigorous conditions to equilibrate to the more stable C-acylated product. For enamines, the Z-acylated enamine is a good acylating agent; any excess of enamine will attack it and equilibrate it to the more stable C-acylated product.

In summary, acylation on Z is achieved with hard electrophiles (acyl halides) under kinetic control; acylation on C is achieved with any good O=C–L electrophile under thermodynamic control. As with alkylation of allylic sources, C acylation of allylic sources is more common.

9.4.4 Amides and Amidates

In amides the oxygen is the better nucleophile because it is partially minus and the nitrogen is partially plus. The partial plus on the nitrogen repels any like charged electrophile. However, in the amidate anion both nitrogen and oxygen share the negative charge, and therefore the softer nitrogen is commonly the better nucleophile.

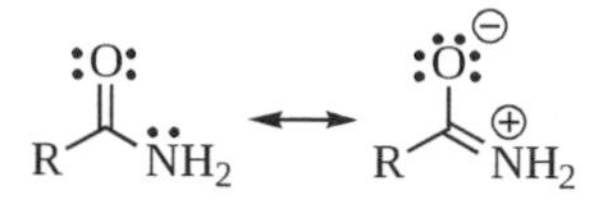

Amide resonance forms

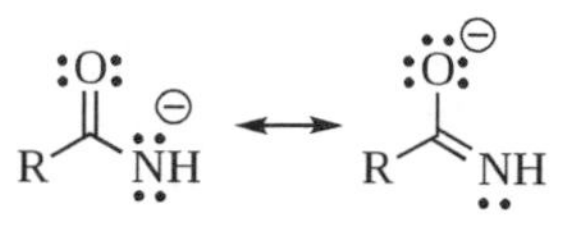

Amidate resonance forms

9.5 SUBSTITUTION VS. ELIMINATION

9.5.1 Substitution vs. Elimination Energy Surface and Variables

Nucleophilicity vs. basicity is perhaps the most difficult decision to make because almost any anion can serve as either a base (elimination) or a nucleophile (substitution):

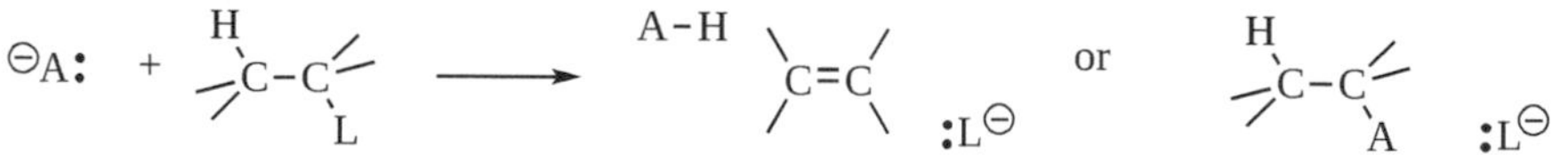

Elimination

Substitution

Again HSAB theory is useful. In general, the C–H bond is significantly harder than the C–L bond. Therefore the softer anions will tend toward substitution, and the harder anions will tend toward elimination. However, the situation is not that simple, for the accessibility of the substitution site and the size of the anion play a major role also. If the anion is hindered, it is a poor nucleophile. If the site is so hindered that the nucleophile cannot attack it, the balance tilts toward elimination. **The substitution/elimination decision becomes a function of three major variables: nucleophilicity, basicity, and steric hindrance.**

The substitution energy surface can be placed adjacent to the elimination surface since they share the C–L bond-breaking axis (Figure 9.2). The factors that tilted each of the surfaces can be used to understand the substitution vs. elimination competition.

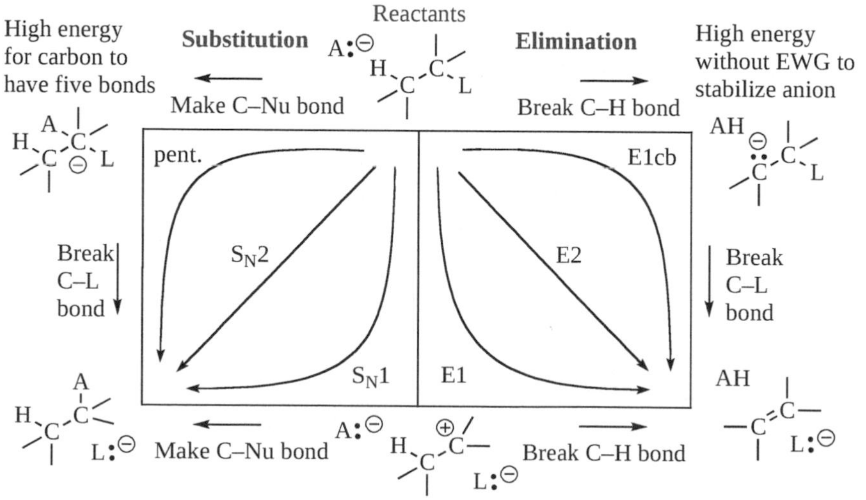

Figure 9.2 Top view of a simplified substitution/elimination surface. Reactants are in top center.

9.5.2 S_N1 vs. E1 Competition

With a reasonably stable carbocation and energy as a vertical axis, we get Figure 9.3, the S_N1 vs. E1 surface. In acidic media, the S_N1 competes with the E1 process; the first step of both processes is to lose the leaving group to form the carbocation. Carbocations are excellent electron sinks and tend to react quickly with low selectivity. Equilibrium thermodynamics is therefore the best way to bias this surface. The substitution product corner is raised if the Nu is poor and also is a good leaving group so that it falls off again, returning to the carbocation. The alkene product corner is now the lowest point on the surface and is favored by any equilibrium. An example is elimination of alcohols by concentrated phosphoric acid.

Attempts to bias this surface toward substitution present a slight problem. The substitution product corner cannot be lowered very much by making the nucleophile better because the acidic media that the D_N ionization prefers will protonate any good nucleophiles. The best way to increase substitution is to use a low-basicity nucleophile and to shift the equilibrium by mass balance: Make the nucleophile the solvent if possible (solvolysis).

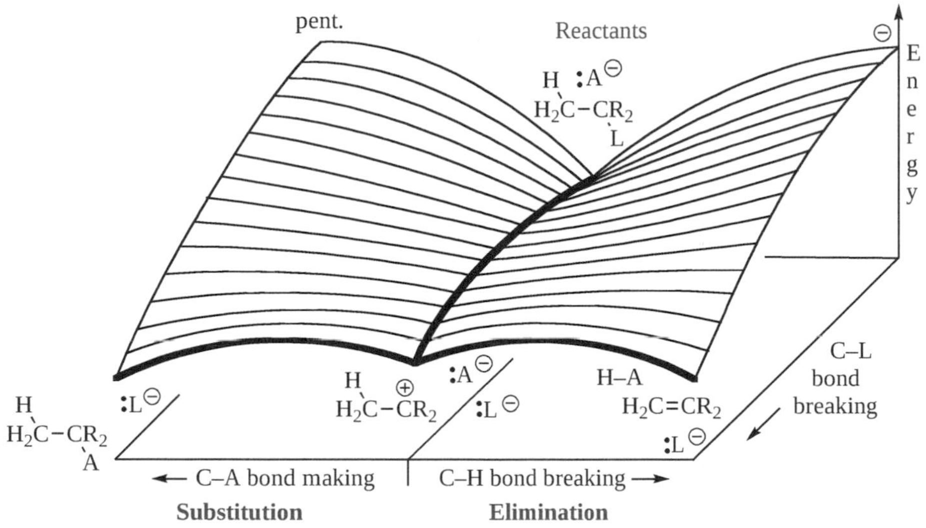

Figure 9.3 The simplified S_N1/E1 energy surface. Reactants are in top center.

9.5.3 S_N2 vs. E2 Competition

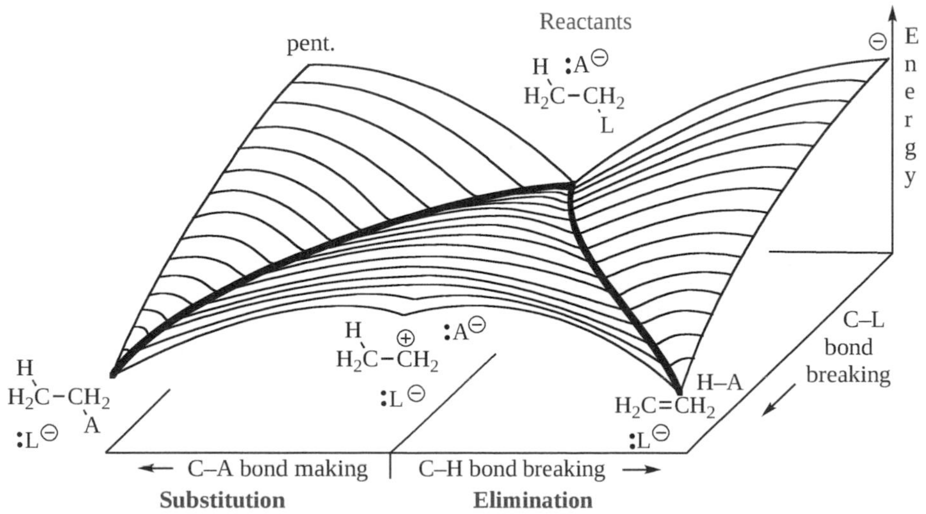

Figure 9.4 The simplified S_N2/E2 energy surface. Reactants are in top center.

There normally is competition in basic media between the S_N2 and E2. In contrast to the previous surface, Figure 9.4 shows that the split in paths occurs much earlier because the carbocation is much higher in energy and not an intermediate. The choice of reaction path depends heavily on kinetics because the system is usually not reversible.

To bias this surface toward elimination, one makes the S_N2 a higher-energy route by using hard, highly hindered strong bases, which are poor nucleophiles. Strong bases make the deprotonation energetically more favorable and thus lower the energy of the right edge. Poor nucleophiles make the substitution reaction more difficult and thus raise the energy of the left edge of the diagram. The entire surface then tilts toward elimination. Highly hindered strong bases with low nucleophilicities, $(CH_3)_3CO^-K^+$ or $(i\text{-}Pr)_2N^-Li^+$, are used to optimize elimination.

Conversely, good, unhindered, soft nucleophiles of low basicity bias the surface toward substitution by tilting it exactly the opposite way. Poor bases raise the energy of the right edge; good nucleophiles lower the energy of the left edge; the entire surface tilts toward substitution. As mentioned above, the groups around the C–L site make a large difference in whether substitution or elimination occurs. As the C–L site gets more hindered, the left side is raised in energy and the surface then tilts toward elimination.

To summarize, substitution vs. elimination is a multivariable decision that breaks down into major and minor variables. The three major variables are C–L site hindrance, nucleophilicity, and basicity; the minor variables are reaction temperature (higher temperature favors elimination) and the electronegativity of the leaving group (more electronegative leaving groups make the C–L carbon harder). A less common variable is that electron-withdrawing groups can make the C–H so acidic that elimination dominates.

9.5.4 3D Correlation Matrix for Substitution vs. Elimination

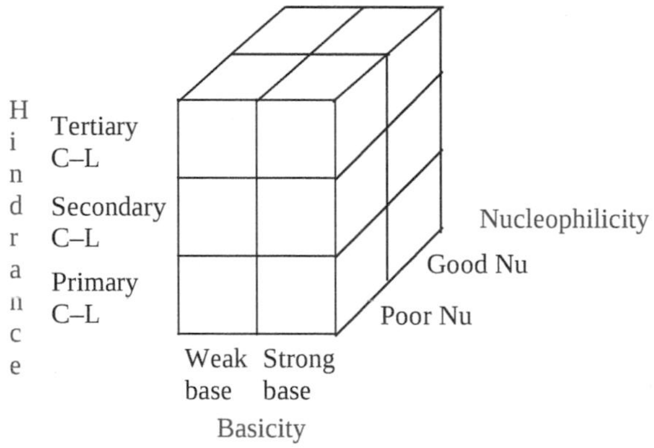

Figure 9.5 A three-dimensional correlation matrix for substitution vs. elimination.

We need to consider each different type of C–L site as a function of the other two major variables, nucleophilicity and basicity. A three-dimensional correlation matrix is the best way to do this (Figure 9.5).

We divide each axis into convenient portions corresponding to the most common cases (although we actually have a continuum in all three dimensions). Hindrance can vary with adjacent substitution as noted below. A rough decision boundary for basicity is about $pK_{abH} = 10$, but this varies with the system. For nucleophilicity, the poor nucleophiles are often highly hindered anions or are neutral solvents with poor polarizability (Section 4.2.2). To make the individual cells in this matrix easier to view, we will consider each vertical layer separately, and show them as individual two-dimensional matrices in Figure 9.6.

The unhindered primary layer has the strong base, poor nucleophile box as its mixture quadrant with all others as substitution. Notice how the mixture quadrant in

the primary matrix moves around to the strong base, good nucleophile quadrant in the secondary matrix, then again to the weak base, good nucleophile quadrant in the tertiary matrix, each time leaving behind an elimination quadrant.

These guidelines hold for polar protic media like alcohols. As nucleophilicity increases in polar aprotic media (see Section 4.2.2), expect all the mixture quadrants to become mostly substitutions. For each mixture quadrant more information is required to be able to make a decision on substitution vs. elimination. The major variables have balanced out, and therefore the minor variables can tilt the balance. In the primary mixture quadrant, if the leaving group is a halide, elimination is the major product. If, however, the leaving group is a sulfonate, its electronegativity increases the partial plus on carbon (making it harder); substitution is then the major product. In the secondary mixture quadrant, elimination is common for any reaction that is heated. Elimination has a slightly higher energy barrier than substitution and is therefore favored by higher temperatures. In the tertiary mixture quadrant, a fine-tuning of basicity is important, for even mild bases like acetate cause elimination to occur.

Substitution vs. elimination decision

Example reagents

	Weak base	Strong base
Good nucleophile	$I^{\ominus}$ $RS^{\ominus}$	$HO^{\ominus}$ $EtO^{\ominus}$
Poor nucleophile	H_2O EtOH	$(CH_3)_3CO^{\ominus}$ $R_2N^{\ominus}$

Unhindered primary

	Weak base	Strong base
Good nucleophile	Subst. by S_N2	Subst. by fast S_N2
Poor nucleophile	No rxn. or very slow S_N2	Mixture (depends on L)

Secondary

	Weak base	Strong base
Good nucleophile	Subst. by S_N2	Mixture (depends on temp.)
Poor nucleophile	No rxn. or slow S_N2 or S_N1	Elimin. by fast E2

Tertiary

	Weak base	Strong base
Good nucleophile	Mixture (depends on basic.)	Elimin. by fast E2 or E1
Poor nucleophile	Subst. by S_N1 (polar solvent)	Elimin. by fast E2 or E1

Figure 9.6 Substitution/elimination correlation matrix layers from Figure 9.5 sorted by hindrance. The matrix in the upper left gives some common examples of each type of reagents.

For the weak base, poor nucleophile quadrant, substitution occurs by the S_N2 path for primary and secondary substrates, but because the electron source is poor by definition, the reaction can be very slow if the leaving group is not very good. Tertiary substrates in this quadrant substitute by the S_N1 mechanism. Even if a group is somewhat removed from the site, it can still hinder the approach of a nucleophile. Although $R{-}CH_2{-}L$ is a general description of a primary site, the size of R is very important because it can change which hindrance matrix best applies. The matrix for hindered primary leaving groups, like isobutyl, $(CH_3)_2CH{-}CH_2{-}L$, is similar to the matrix for secondary leaving groups. The matrix for hindered secondary leaving groups is similar to that of tertiary leaving groups. Vinyl leaving groups, $H_2C{=}CH{-}L$, do not substitute under normal conditions and with a strong base eliminate to the alkyne. For $CH_3{-}L$ systems, substitution occurs in all quadrants because elimination is not a structural possibility.

We treat as an exception our less common major variable, the presence of an electron-withdrawing group. Leaving groups beta to electron-withdrawing groups, EWG–CH_2CH_2–L, commonly eliminate because the increased acidity of the H makes the C–H easier to break, thereby favoring elimination by E1cB.

All multivariable decisions are messy, and if this presentation appears too clean to hold universally, you are correct in your suspicion. We are predicting the predominant product in a product mixture; rarely is the predicted product the only product. There are examples where a major variable joined with the other minor variables (and an experimental technique or two) can outweigh the other two major variables. Alcohol dehydration is such a case. Alcohols are heated in concentrated sulfuric or phosphoric acid, and the alkene is distilled out of the reaction mixture into another flask. The elimination is forced: The medium is acidic and dehydrating; sulfate and phosphate are poor nucleophiles and good leaving groups; heating favors elimination and shifts any equilibrium by removing lower boiling elimination products.

9.6 AMBIDENT ELECTROPHILES C=C–EWG

9.6.1 Conjugated Ketone Systems—Enones

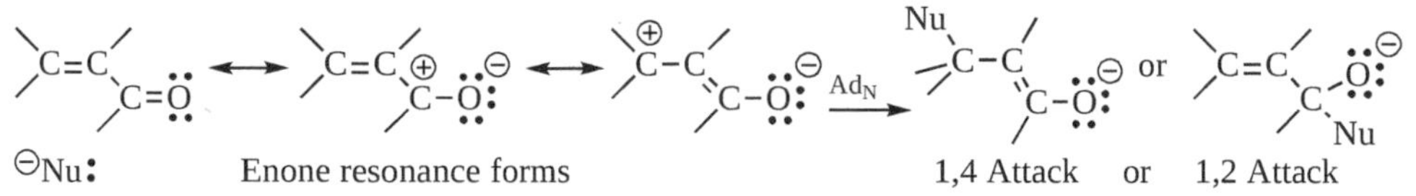

A conjugate acceptor system with two δ+ sites can be attacked at either site. The thermodynamic product will be the most stable of the possible products. A $\Delta H°$ calculation shows that the thermodynamic product is the **1,4 or conjugate addition product**, primarily because of the greater bond strength of the C=O bond than the C=C bond. Any equilibrium would produce the conjugate addition (vinylogous addition) product.

Because the effect of an electronegative group diminishes with distance, the carbonyl carbon in this system will have the greatest partial plus and therefore will be harder and attract a hard, negatively charged nucleophile best; the **1,2 product or direct addition product** will be the kinetic product for hard nucleophiles. Soft nucleophiles therefore prefer conjugate attack. Finally, if one site is very sterically hindered, attack at the more open site will dominate. Several nucleophiles and their main site of attack on an enone are presented in Table 9.1 and diagramed in Figure 9.7.

Table 9.1 The Mode of Attack of Various Nucleophiles on Enones

Nucleophile	Favored attack	Rationale
R–Li	1,2	Very hard organometallic
R–Mg–X	1,2 (usually)	Hard organometallic
R_2CuLi	1,4	Soft organometallic
R–Mg–X/CuX	1,4	Makes soft organometallic by transmetallation
Enolates	1,4	Soft Nu, reversible addition
Lone pair Nu	1,4	Stable Nu, reversible addition
$LiAlH_4$	1,2 (mainly)	Hard hydride source
$NaBH_4$ or $KBHR_3$	1,4 (mainly)	Softer hydride source

In summary, highly reactive, harder organometallics add irreversibly to an enone and produce the 1,2 product; the softer organocuprates, R_2CuLi, add to enones giving the 1,4 product. The more stable delocalized anions add reversibly and therefore produce the 1,4 product. Organomagnesium reagents give predominately 1,2 additions except when the carbonyl of the enone is hindered (by a phenyl or larger group).

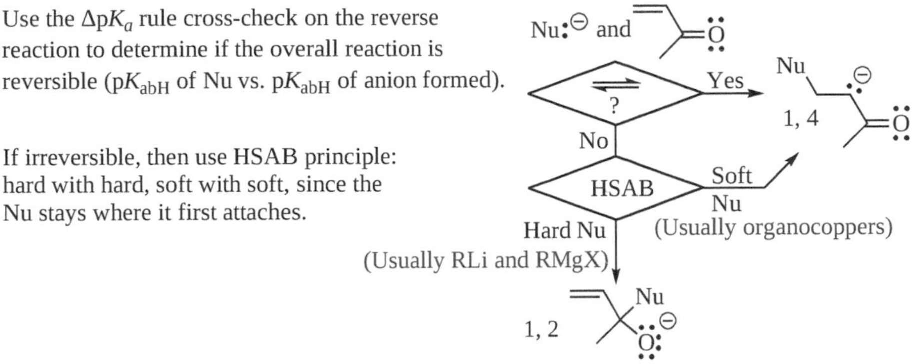

Figure 9.7 Flowchart for the ambident electrophile decision of an enone.

9.6.2 Miscellaneous Ambident Electrophiles

As the carbonyl becomes more reactive, 1,2 attack will become more predominant; in contrast to what was found for enones, borohydride attacks 1,2 on conjugated aldehydes. All organomagnesium reagents add 1,2 on conjugated aldehydes. With lone pair nucleophiles (often very hard), conjugated acyl halides undergo 1,2 attack (hard carbonyl, often irreversible 1,2 addition), whereas conjugated esters undergo 1,4 attack (softer carbonyl, reversible 1,2 addition). Only 1,2 attack occurs on phenyl-substituted carbonyls like $PhCOCH_3$, because conjugate attack would disrupt the aromaticity of the aromatic ring.

9.7 INTERMOLECULAR VS. INTRAMOLECULAR

One of the determining factors in reaction rates is how frequently the reaction partners collide. If collisions are more frequent, the reaction rate will be faster. Increasing the concentration of a reactant invariably increases the collision frequency and therefore the reaction rate. If the nucleophile and electrophile are part of the same molecule, they may collide much more often than is possible in even the most concentrated solutions. Therefore intramolecular reaction rates can easily exceed intermolecular rates. The energy of the cyclic intramolecular transition state depends on the rigidity and size of the loop of atoms in the cyclic transition state and on the orbital alignment restrictions of the process involved.

For the S_N2 reaction shown in the following example, the determining factor for ring closure in a freely rotating chain is its size. The formation of very small rings requires the bending of sigma bonds, creating significant ring strain. In the formation of large rings (rings greater than eight atoms), the ends that must be brought together to form the ring may collide so infrequently that bimolecular reactions can easily compete. The optimum ring size for closure is a five-membered ring; six is not as good; seven starts to get too large. A three-membered ring forms easily, for although there is much ring strain, the number of effective collisions is great since the ends are so close. Four-membered ring formation is poor because the ring strain has not dropped off appreciably, but the number of effective collisions has.

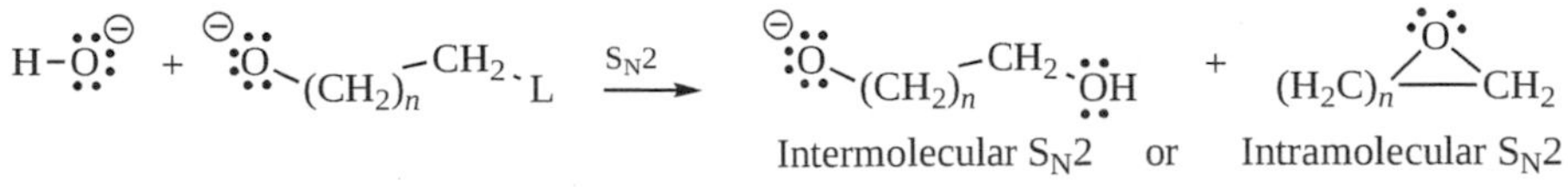

Although there is some variation among systems, the general order is 5 > 6 > 3 > 7 > 4 > 8 to 10-membered transition states. In more concentrated solutions intermolecular reactions may be competitive with four-membered ring formation. In summary, **any intramolecular process with a 5-, 6-, 3-, or 7-membered transition state is almost always faster than the corresponding intermolecular process.**

In forming a ring on a substrate that has some conformational rigidity, you must be sure that the ends can reach each other. In fact, any intramolecular process should be checked with molecular models to see whether the reacting partners can interact as needed for the reaction.

9.8 TO MIGRATE OR NOT TO AN ELECTROPHILIC CENTER

Predicting when migration to an electrophilic center will occur is not an easy task. The main three reactions of carbocations are to trap a nucleophile (path A_N), to lose an electrofuge like a proton and form a pi bond (path D_E), or to rearrange to a carbocation of equal or greater stability (path 1,2R) (Section 4.7). Often it is a matter of kinetics. Can the carbocation survive long enough to rearrange, or does A_N or D_E happen first? Carbocation rearrangements are favored by strongly acidic media precisely because those conditions have very poor nucleophiles and weak bases, so that carbocation has time to rearrange (a good base or nucleophile would get protonated in strong acid). Less stable carbocations are rarely intermediates. With few exceptions, tertiary carbocations will not rearrange to secondary; secondary carbocations will not rearrange to primary. Those are just too far uphill in energy.

To rearrange, the migrating group orbital must align with the empty *p* orbital of the carbocation. Another way to view it is that the migrating group is participating in hyperconjugation with the carbocation before migration. As Figure 9.8 shows, the migrating group maintains bridging-type bonding in the transition state between the starting and ending atoms in the migration (see the 1,2R path, Section 7.1.11 and the 1,2RL path, Section 7.1.12).

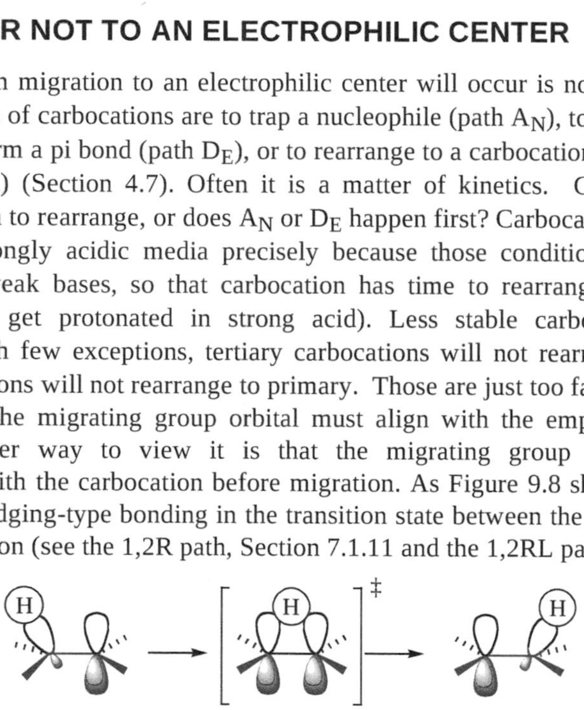

Figure 9.8 The migration of a hydride to a cationic center maintains overlap throughout the move.

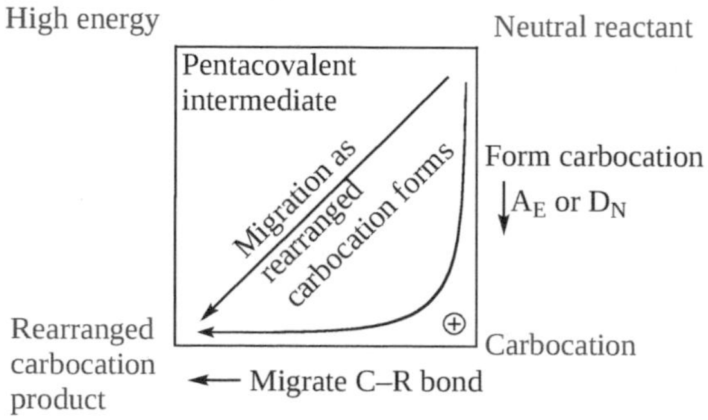

Figure 9.9 Top view of a simplified energy surface for carbocation rearrangements.

Rearrangement commonly is found with paths D_N and A_E, which make up path combinations, S_N1, E1, Ad_E2. Figure 9.9 shows an energy surface map of the spectrum between the diagonal one-step and two-step rearrangement processes.

With tertiary or more stable carbocations, we would have one extreme, with the carbocation corner low enough in energy to be the two-step intermediate. Secondary carbocations seem to be the decision boundary. The following arrows are for the corresponding two-step processes via an intermediate carbocation.

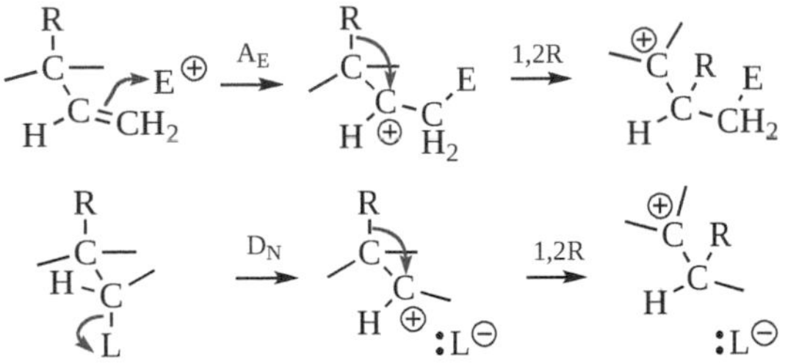

At the other extreme, the one-step rearrangement (the diagonal on Figure 9.9 that bypasses the initial carbocation) would predominate when that carbocation would be less stable. As in the spectrum for S_N1 and S_N2, as the carbocation that would be formed by the D_N or A_E becomes less stable, the corresponding carbocation corner of the rearrangement surface begins to rise in energy. If the carbocation is very poor, like primary, the surface folds down the middle and the lowest-energy route is the diagonal. The following arrows are for the corresponding one-step diagonal processes.

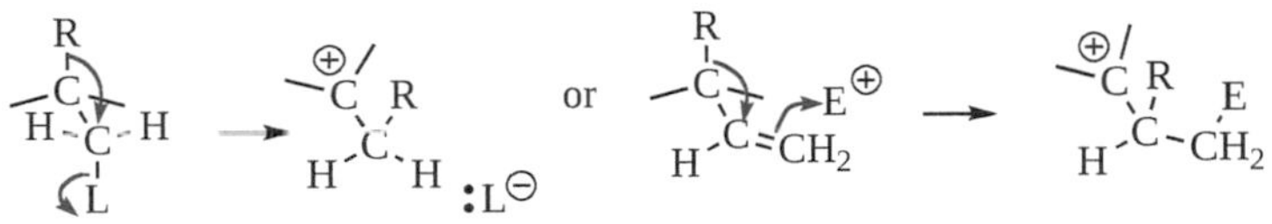

In between, with a secondary carbocation, we would expect that the lowest-energy path would begin to break the C–L bond, creating a partial plus, but before the C–L bond is fully broken, the R would migrate. The lowest-energy route would curve on the surface between the two extremes. The faster carbocation rearrangements tend to make a more stable carbocation. Certain systems that are more prone to fast rearrangement are shown below.

Neopentyl Isobutyl Strained ring Pinacol

Example problem

To which carbocations would the above compounds rearrange when the leaving group is lost?

Answer: The neopentyl system would migrate a methyl to give a tertiary carbocation. Isobutyl systems migrate an H to give the *tert*-butyl carbocation. The strained ring compound produces the 1-methylcyclopentyl carbocation. (Relief of severe ring strain can often be enough to create a less stable carbocation, like secondary from tertiary.) Migration of a methyl in the pinacol system can occur by D_N then 1,2R if all the loss of leaving group cross-checks are met, otherwise the 1,2RL process would produce the identical product.

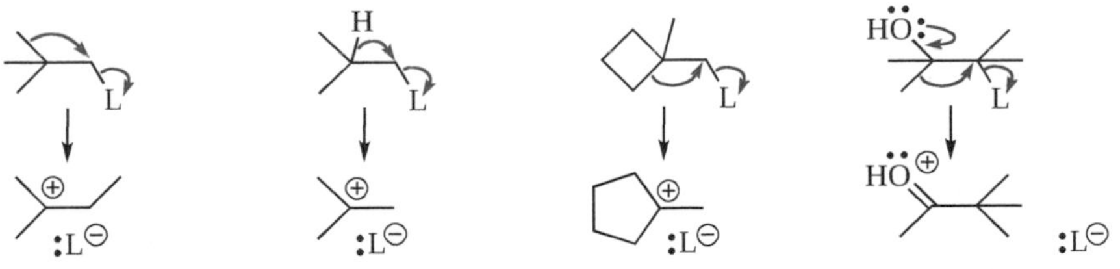

One last common system that rearranges by a different mechanism is shown below. Unlike the previous rearrangements that go through a planar carbocation that could be trapped on either side, this process replaces the leaving group with retention because it is two sequential S_N2 reactions. A double inversion gives overall retention. The phenonium ion does have two different CH_2 carbons that could be attacked by the nucleophile to open the strained three-membered ring, and that must be considered if those two carbons are different.

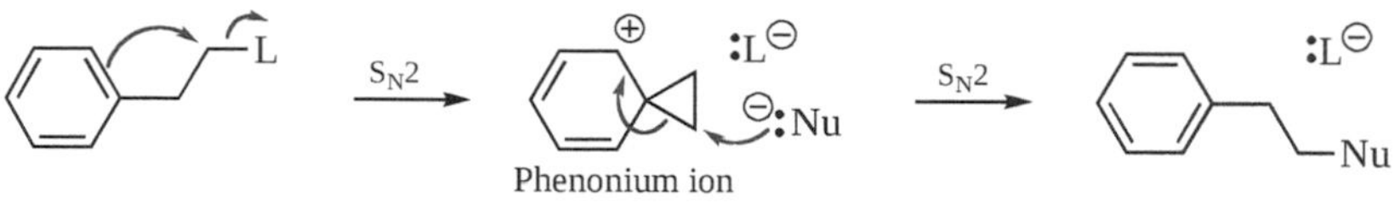

9.9 SUMMARY

In summary, a caution: In complex multivariable decisions like those covered in this chapter, the best we can hope for is to be able to predict correctly in straightforward cases. Many of these reactions produce mixtures of products, and we are often just trying to predict the major product.

Finally, a common error for many of these major decisions is to build a rationale based on kinetic control when the system is under thermodynamic control (or vice versa). Always check for reversibility with the ΔpK_a rule crosscheck.

CHAPTER 9 EXERCISES

(Refer to the Hints to Selected Problems from Chapters 8, 9, 10, and 11 if you need help).

9.1 For the following reactions decide whether substitution or elimination dominates.

	Substrate	Reagent	Temperature
(a)	$CH_3CH_2CH_2CH_2OH$	HBr	(assume 25°C if not given)
(b)	$CH_3CH_2CH_2CH_2OTs$	*t*-BuO$^-$	
(c)	$CH_3CH_2CH_2CH_2Br$	*t*-BuO$^-$	
(d)	$(CH_3)_2CHCH_2Br$	I^-	

9.2 For the following reactions decide whether substitution or elimination dominates.

	Substrate	Reagent	Temperature
(a)	$(CH_3)_2CHBr$	EtOH	
(b)	$(CH_3)_2CHBr$	EtO$^-$	55°C
(c)	$(CH_3)_2CHCH_2CH_2Cl$	$[(CH_3)_2CH]_2N^-$	
(d)	$(CH_3)_2CHBr$	CH_3COO^-	

9.3 For the following reactions decide whether substitution or elimination dominates.

	Substrate	Reagent	Temperature
(a)	$(CH_3)_2CHBr$	EtS$^-$	
(b)	$(CH_3)_2CHCHBrCH_3$	CH_3COO^-	55°C
(c)	$CH_3CH_2CBr(CH_3)_2$	CH_3OH	
(d)	$(CH_3CH_2)_3CCl$	CH_3O^-	

9.4 For each of the following reactions a decision discussed in this chapter is required. Choose which product would be preferred.

(a) 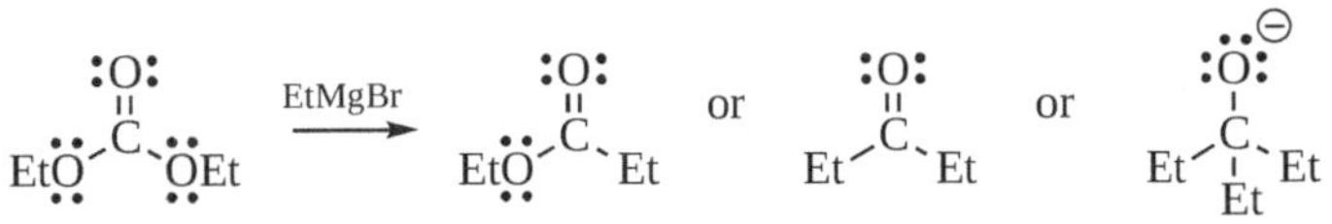

(b) 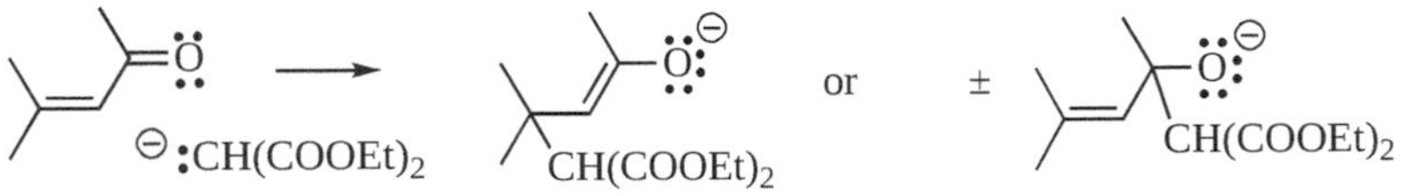

(c) 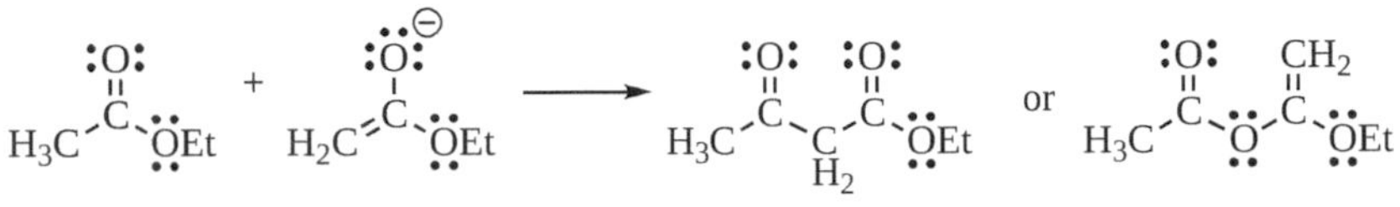

(d) 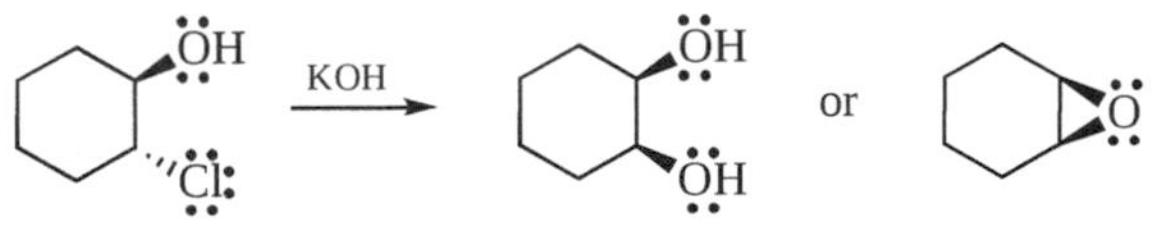

9.5 For each of the following reactions a decision discussed in this chapter is required. Choose which product would be preferred.

(a) 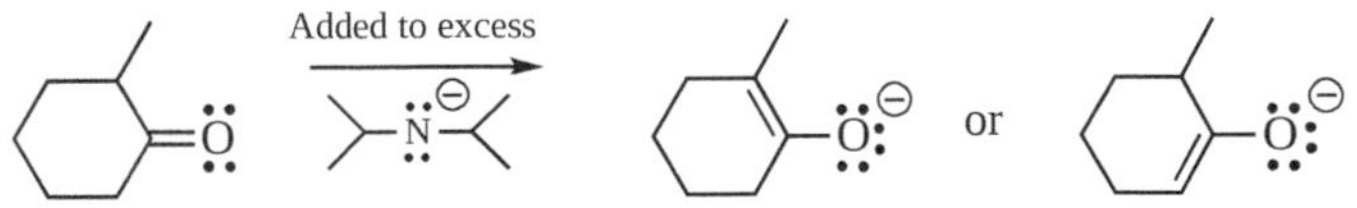

(b) 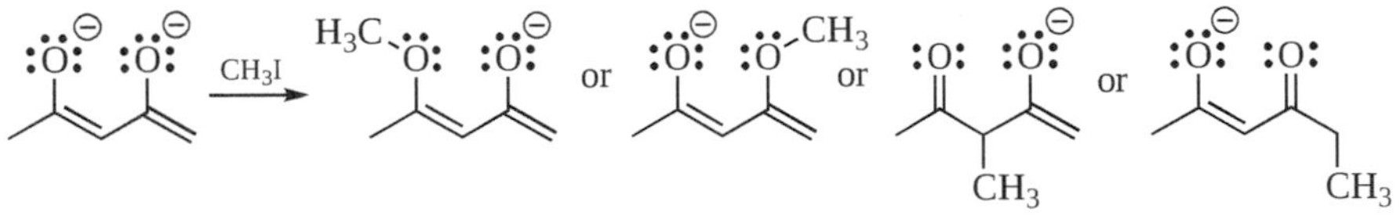

(c) 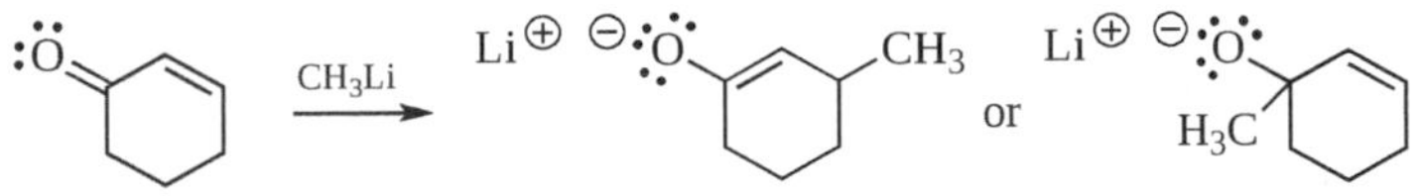

(d) 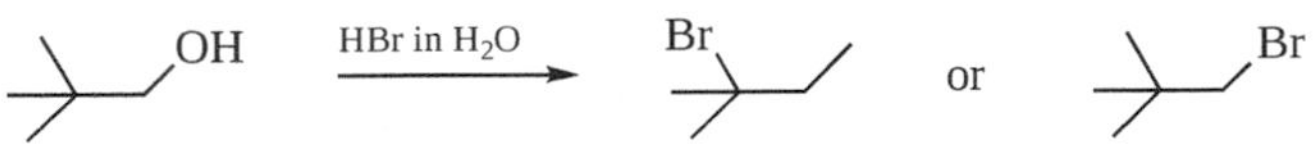

9.6 Provide a mechanism for the formation of the preferred products in problem 9.4. Combine the source and sink with the appropriate pathway(s).

9.7 Provide a mechanism for the formation of the preferred products in problem 9.5. Combine the source and sink with the appropriate pathway(s).

9.8 Each of the following reactions has only one electron source and sink. Give the product of the following reactions; a decision discussed in this chapter is required. Assume a mildly acidic workup for the organometallic reactions that just protonates any anionic species but goes no further.

(a) 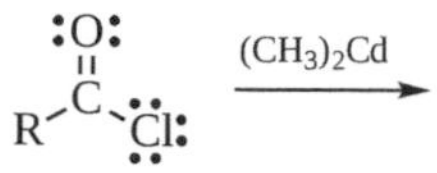

(b) 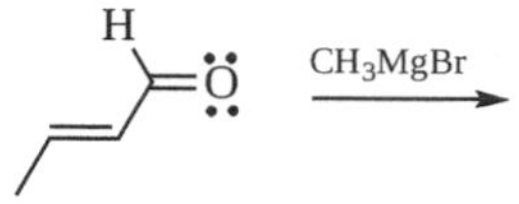

(c) 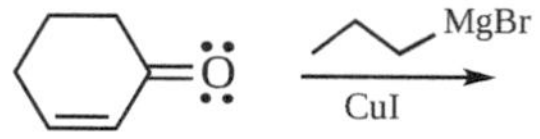

9.9 Each of the following reactions has only one electron source and sink. Give the product of the following reactions; a decision discussed in this chapter is required. Assume a mildly acidic workup for the organometallic reactions that just protonates any anionic species but goes no further.

(a) 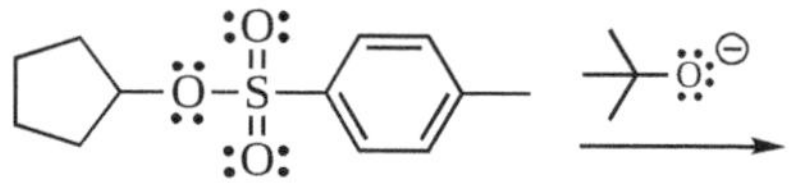

(b) 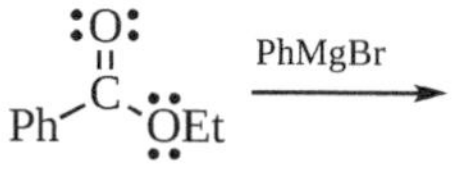

(c) 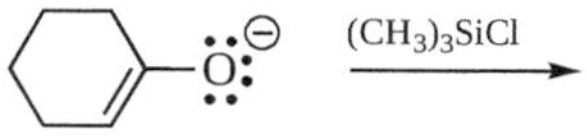

9.10 Provide a mechanism for the formation of the preferred products in problem 9.8. Combine the source and sink with the appropriate pathway(s).

9.11 Provide a mechanism for the formation of the preferred products in problem 9.9. Combine the source and sink with the appropriate pathway(s).

9.12 The following compounds rearrange under the reaction conditions; give the product and a mechanism.

(a) 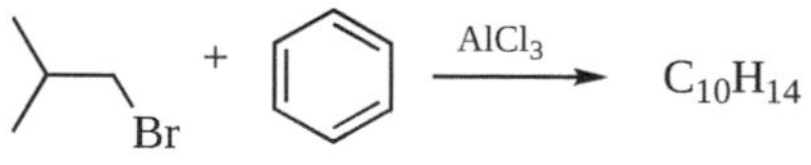

(b) 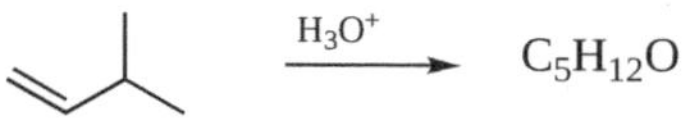

9.13 The following compounds rearrange under the reaction conditions; give the product and a mechanism.

(a) 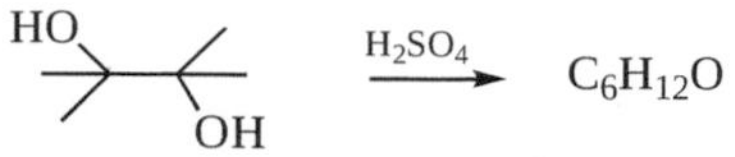

(b) 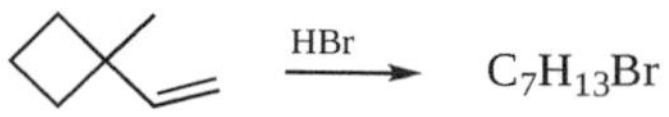

9.14 In the following problem, identify the electron sink and electron source and predict two possible products. When one equivalent of the hydride source is added at low temperature, which product would be preferred?

9.15 The following compound rearranges in acid. Give a reasonable mechanism.

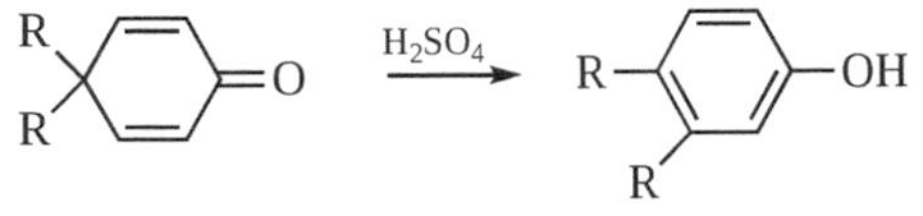

9.16 For each of the following reactions a decision discussed in this chapter is required. Choose which product would be preferred.

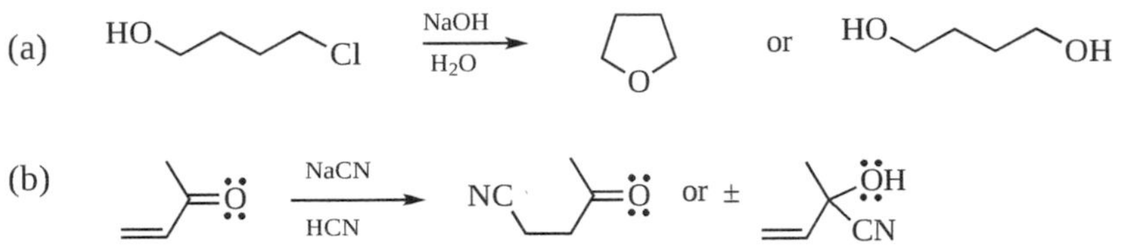

9.17 For each of the following reactions a decision discussed in this chapter is required. Choose which product would be preferred.

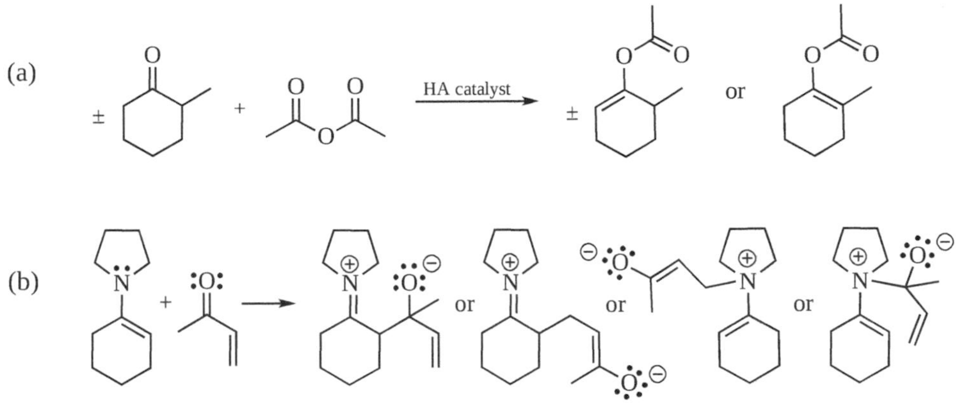

9.18 Provide a reasonable mechanism for the formation of the preferred products in problem 9.16. Combine the source and sink with the appropriate pathway(s).

9.19 Provide a reasonable mechanism for the formation of the preferred products in problem 9.17. Combine the source and sink with the appropriate pathway(s).

9.20 A decision discussed in this chapter is required to choose which product would be preferred below. Provide a reasonable mechanism for the formation of the preferred product.

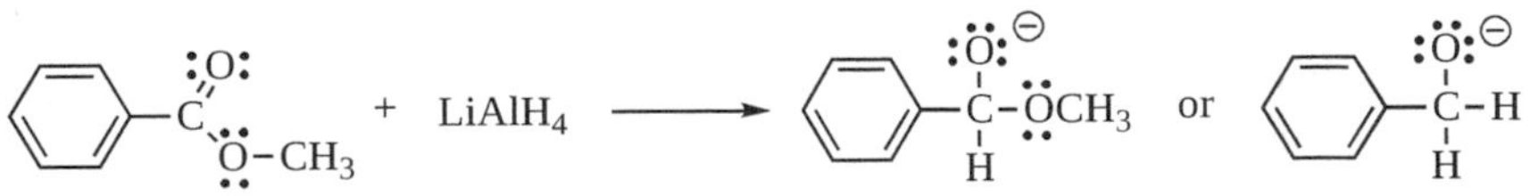

10

CHOOSING THE MOST PROBABLE PATH

Electron Flow in Organic Chemistry: A Decision-Based Guide to Organic Mechanisms,
Third Edition, Paul H. Scudder.

10.1 PROBLEM-SOLVING IN GENERAL

The following hints are intended to help you avoid the most common mental “traps” or conceptual blocks encountered in solving a problem in organic chemistry or actually in any field. Specific decisions were covered in Chapter 9 and procedures for various problem types are covered later in this chapter. All of these procedures are extra work when they are compared to “just having the answer pop into your head,” but they make the problem-solving process more systematic and more reliable; they make it easier to retrace your thinking to check the answer and simpler to “debug” when things go wrong.

The more complex a problem is, the less likely you will be able to see the answer from the start. Therefore you must begin work on the problem without any clear idea of how to proceed. If so far you have always seen the answer from the start, working on a complex problem is, in a way, like flying on instruments through a cloud bank when you are used to visual navigation. Often students will just stare at the problem and wait for the answer to hit them. With difficult problems this can be a very long wait. These tips allow you to pursue the answer actively rather than wait for it to come to you.

These analytical methods and habits are not unique to organic chemistry but are applicable to many scientific fields and to medicine. Proposing a reasonable organic mechanism and giving a good medical diagnosis employ similar critical thinking skills.

10.1.1 Study The Material Before Attempting The Problems

Study with a pencil to underline important ideas; make margin notes about things that do not make sense; write down questions to ask the instructor; make a written summary of what you have read. Understand the trends, vocabulary, and principles that are needed to solve the problem. Target your study to concentrate on weak areas as indicated by exams and homework. Patch your conceptual holes! Because organic chemistry is cumulative, like a foreign language, any important concepts you miss in the beginning will come back to bite you in the end.

10.1.2 Establish An Informational Hierarchy While You Study

You tend to remember the things that you use the most and those that are most recent. The only way you can raise the mental awareness of the important concepts over the more recent insignificant fact is to **review the most important material more often**. As any course of study progresses, the incoming facts and important concepts must be put in their place in the hierarchy. You manage an information overload by allowing the little fact to be forgotten after it has served its purpose of being an example for a more important principle. The older, important concepts must be continually raised above the newer facts by additional review (even though you feel you know the concepts very well; you are just reinforcing their importance by keeping them active). Continuously review the classification of the sources and sinks and the electron flow pathways.

10.1.3 Gather A Collection Of Commonly Needed Problem-Solving tools

As a start toward this, the Appendix gathers together important tools: a pK_a chart, a bond-strength table, a list of sources and sinks, the electron flow pathways, trends, general rules, and other useful items. Continue to customize your toolbox with any valuable tools that you need. Chapter 11 collects all the cross-check tools for seeing if your hypothesis is reasonable.

10.1.4 Read The Problem Carefully—Understand The Problem

Be sure to know what is asked. Can you rephrase the question in your own words? Make a model or draw a figure. Can you break the problem into separate, smaller, more easily solved units? Can you give a rough estimate of the overall process? Can you see the first few steps? What might the last steps be? Maybe work from both ends toward the middle. Be willing to start on the problem before you can see the entire solution in your head.

10.1.5 Map Changes—Understand Bonds Made And Broken

For molecules drawn with skeletal structures, draw out all the atoms near any changes in the structure. It is easy to forget about hydrogen atoms that are not drawn. Number the atoms of the skeleton if it is not obvious. Pick the simplest numbering scheme, the one with the fewest changes. What bonds were broken? What bonds have to be made? Is the system set up at the beginning to do any of those on your list? Would a proton transfer get you started? Keep an open mind over which process occurs first.

10.1.6 Do Not Ignore What You Do Not Understand

If you do not know what a reagent in a reaction is doing, don't ignore it. It may be critical for the solution of the problem. Solvents can often be ignored, so know the common abbreviations for them.

10.1.7 Gather All Applicable Information About The Problem First

Look to see that there is enough information to solve the problem. Search for relationships between the data and what is wanted. Write down any chemical equations that relate to the problem. What are the assumptions and limitations that come with the problem? What principle(s) was the problem designed to illustrate?

10.1.8 Classify Into Generic Groupings

Back off from the problem and get an overview of it. Have you seen a problem of the same generic class before? Often changes in the hydrocarbon "grease" around a reactive site will disguise the similarities to reactions that you have seen before: don't slip on the grease! What are the *real* similarities and differences? Decide what parts of that related problem are useful. Have you seen a problem with a similar start or finish point?

10.1.9 Recognize Possible Intermediate Goals

For a mechanism question, number the atoms in the starting material and try to find those atoms in the product. Find out what bonds need to be broken or made. On what atoms do you need to generate nucleophiles or electrophiles? What are all the possible starting points? Are there any possible last steps? Middle steps?

10.1.10 Always Write Down Any Possibility That You Consider

Work on paper, not "in your head." The biggest problem-solving trap is to shoot down a possibility "in the air" without writing it down first. **Writing down something that you are considering forces you to take more time to think about it** and produces a written record that it has been checked out.

10.1.11 Have A Systematic Method To Your Answer Search

Random trial and error does not work on complex problems. A systematic search for the answer may seem like a long and tedious process, but it is much quicker and more reliable than random shots in the dark.

10.1.12 Look For Alternatives—Generate All Paths, Then Select The Best

Pick the best path and try it out, but always keep in mind the other alternatives in case the first one does not work out. In writing down the alternatives, you may get other ideas. Don't waste a lot of time at an apparent dead end before going back to check out other possibilities.

10.1.13 Recognize The Generic Form Of Each Step

Classify each step or group of steps as a proton transfer, substitution, elimination, addition, or rearrangement. Then check for other alternative forms of the same process that may fit the reaction conditions better. For example, your first impulse in a particular problem might be to write some sort of elimination because it seems to get you closer to the answer. If you recognize that what you need is a type of elimination, you would know that eliminations could go by E1, E2, E1cB, or Ei mechanisms. You can then pick the most appropriate one, rather than remaining with the first one that occurred to you.

10.1.14 Work Carefully And Cross-check The Work As You Go

Simple errors can lead to some rather absurd predictions. Arrow pushing is a form of electron bookkeeping; therefore the arrows and Lewis structures must correlate and be accurately drawn for the results to make sense. The total charge on both sides of the transformation arrow should be the same. Pay attention to detail.

10.1.15 Don't Skip Steps—Look For Any Hidden Decision Points

One common trap is to skip a step rather than take the time to write out the intermediate. If there is another route from the unwritten intermediate, the decision point will be missed and difficult to locate on a cross-check. Be continually on the lookout for the major decisions discussed in Chapter 9. The most common and also the most commonly missed decision is the nucleophile vs. base decision.

10.1.16 Make A Scratch Sheet Into An Idea Map

Keep your scratch sheets neat. Be organized on the scratch paper so you can trace your thoughts and be able to go back to other ideas and check your thinking **without getting lost**.

Remember, you are exploring unknown territory; draw yourself a map of the problem space as you go. You don't want to forget that you already went down a particular road and why you discarded it.

10.1.17 Beware Of Memorization

Memorization is never a satisfactory or reliable substitute for understanding what you are doing. Beware of mindlessly “turning the crank” on a problem; it might not be connected to anything at all.

10.1.18 Watch For Bad Habits

Do not stop at the first answer you find but continue to look for alternative better solutions. Make up a list of your own problem-solving bad habits and refer to it when you work a problem. For example, if you usually forget to check for nucleophilicity vs. basicity, write yourself a reminder. Know the common “potholes” on the route to your answer.

10.1.19 Stay On The Pathways

Although it is tempting to get the problem over with by using a burst of arrows and as few steps as possible, our objective is to write a **reasonable** hypothesis, not necessarily the shortest one. **Construct your mechanistic “sentences” with known, tested “words,” the electron flow pathways.** Mechanistic steps that you invent as a beginner may not be reasonable. Selecting from known pathways turns an apparent open-ended question into a simpler multiple-choice question.

10.1.20 Don't Force The Answer!

Be patient. The correct answer will seem to have a natural flow to it; nothing will have to be forced. Some dead ends can seem frustratingly close to the final answer, and there is a tendency, born out of this frustration, to “hammer home” the “last step” even though you know it is not right. Put the hammer away when working a jigsaw puzzle.

10.1.21 If Stuck, Don't Just Stare At The Page, Draw Something On It!

Reread the problem, draw Lewis structures to see what atoms have changed, then draw resonance forms to see where the charges are, recall relevant energy surfaces, explore any other possible equilibria and less probable possible paths, build a molecular model, do anything to search actively for that crucial hint that is needed. **Blank paper gives no hints.**

10.1.22 Recognize The Limits Placed By The Reaction Conditions

Is the reaction in acidic or basic media, in protic or aprotic solvents? Classify the electron sink, and then sort out those pathways that do not use that sink to help you reduce the number of paths you need consider. Is the reaction site accessible or do large groups block it? Are there constraints on orbital alignment that need to be met?

10.1.23 Beware Of Limits That You Place On The Problem

Sometimes we have a tendency to make an initial guess as to the region in which the solution will be found, and later when the solution is not found within those limits to forget that those limits were not imposed by the problem. Without being aware of it, we can artificially confine our search to an area that does not contain the answer.

10.1.24 If Stuck, Examine The Other Possibilities At Each Decision Point

Force yourself to consider other alternatives. One tends to be reluctant to cross-check a hard decision, especially if it was a hassle to make. When cross-checking decisions, make sure that you **go back all the way to the first decision** at the start. Check all decisions, especially those that seem obvious.

10.1.25 When Done, Always Go Back And Check The Answer

Can you see the answer now at a glance? Is there a second way to get the same result? Does it make sense? Use the law of microscopic reversibility to check alternatives on reversible steps. Is the assumed causality correct?

10.1.26 Practice, Practice, Practice

Any skill requires practice. Only by working on many problems and different types of problems can get you good at problem-solving. Just reading the material is not sufficient; you need to be able to **use** what you've learned. You need to work actively with the material to build a network of interrelated concepts. Working problems allows you to find your weak areas before an exam reveals them.

10.1.27 What Can Be Learned From The Methods You Used To Solve The Problem That Would Be Applicable To Other Problems?

Learn not only from your mistakes but also from your successes!

10.2 GENERAL MECHANISTIC CROSS-CHECKS

For a mechanism to be considered reasonable it must fit into the body of knowledge that has accumulated over the years. The following are some good cross-checks for any mechanistic proposal in order for you to judge whether or not the mechanism is reasonable.

10.2.1 Electron Flow Pathway Check

The most important check is to make sure that each mechanistic step is a valid electron flow path. The mechanistic "sentence" you write must use known "words" to be understandable. There is a temptation, arising out of frustration, to rearrange the lines and dots of reactant into the lines and dots of product with a mechanistically meaningless barrage of arrows just to get the problem over with. **It is important that you assemble your mechanistic hypothesis from known steps, the 18 electron flow pathways.** Check that the electron flow starts at a lone pair or bond that is a good electron source. Make sure that the flow continues without interruption and ends at a good electron sink. Did you consider alternate pathways that achieve the same overall transformation (the energy surface street corner analogy) and then pick the most appropriate?

10.2.2 Completeness Check

When you generate the possible paths make sure that you have not left out any important alternatives, then pick the most probable and check that decision. A common problem-solving error is to race off with the first possibility that looks reasonable, failing to check all the possibilities. **This completeness check is especially important for the first step where the initial direction that you set out to explore is determined.** If your first step is in error, you may be reluctant, several steps later, to go back to the start and reconsider your beginning steps. Rather than retreat to the start, the typical student is more likely, as frustration builds, to force an answer just to be done with the problem.

10.2.3 Media Check

An important cross-check is the media restriction. Paths that form reactive cations almost exclusively occur in acidic media. Likewise paths that form reactive anions are the domain of basic media. No medium can be both a strong acid and a strong base; it would neutralize itself. **The reactive species in equilibrium-controlled reactions have a limited range of acidities.** For example, in neutral water the hydronium ion concentration and the hydroxide ion concentration are both 10^{-7} mol/L. Their relative concentrations are defined by $K_w = [H^+][OH^-] = 10^{-14}$. Their p$K_a$ values span 14 pK_a units. Would a reaction mechanism proposal be reasonable if it required both hydronium and hydroxide ions? At what point does the span of pK_a values become unreasonable because the needed species have too low of a concentration to react at a usable rate?

To attempt to answer these questions, let's return to the example given in Section 3.6, the deprotonation of an ester (pK_a 25.6) by ethoxide ion (pK_{abH} 16); the pK_a span is 9.6. This small concentration of the ester enolate can then enter into a reaction with the original ester (see the product prediction in Section 10.7.2 for a discussion of the entire surface). If we try to run this reaction catalyzed by phenoxide anion (pK_{abH} 10) instead of ethoxide, the reaction fails; no product is isolated, only unreacted starting materials. There are two ways that we could interpret this result: first that the base was too weak, and no appreciable concentration of ester enolate was formed. Another interpretation is that the driving force for the reaction was not sufficient, since the anion of product is now a slightly stronger base than phenoxide. The reaction is no longer forming a weaker base, but this difference of 0.7 pK_a units should have given an equilibrium mix. The pK_a span with phenoxide ion was 15.6 pK_a units.

Therefore as a first approximation, **a pK_a span of 10 units is acceptable, but a span of 15 pK_a units is probably not.** In your mechanisms, make sure that the span of pK_a's of any species that are in equilibrium do not stray too far from the acceptable range. If several steps occur in the same medium, check the pK_a span of all species within those steps. In a way, we are just restating the ΔpK_a rule cross-check (which set a limit of 10 for a single step) and applying it to all species in equilibrium throughout the reaction.

Certainly the pK_a span at which the reaction rate becomes too slow to be useful is open to debate. Exceptions are expected to arise when the driving force for a reaction is an irreversible proton transfer at the diffusion-controlled limit in water or a similar protic solvent. With a maximum bimolecular rate constant of 10^{10} L/mol-sec and the usual reactant concentrations of one molar, a reactive intermediate could conceivably have a concentration as low as 10^{-15} mol/L, and the reaction still proceed at a usable rate.

10.2.4 Energetics and Stability Check

We have three ways to check the energetics of a process. For proton-transfer reactions we can calculate a K_{eq}. A K_{eq} greater than 1 indicates that products are favored and that the $\Delta G°$ is negative. If neutral compounds are transformed into neutrals, we can calculate the $\Delta H°$ for the reaction from the bond strengths. If the reactants or products are ionic or neutral, we can use a ΔpK_a rule cross-check. The K_{eq} is the most reliable of these, the next probably $\Delta H°$, and least the ΔpK_a rule. Not checking the proton transfer K_{eq} is a common error. Use the trends to gauge the stability of any intermediates formed in the reaction. Common errors include the failure to judge the stability of carbocations, carbanions, and leaving groups.

10.2.5 Charge and Typo Check

Make sure that the Lewis structures and their formal charges are correct. An error on a formal charge can be especially dangerous since the attraction of opposite charges is used to predict reactive sites. The charge must balance on both sides of the reaction arrow. Don't accumulate charges on your intermediates as the reaction progresses. Avoid generating any species with adjacent like charges. Look for line-structure errors, particularly for the "vaporization" of an H that was not drawn in the line structure.

10.3 THE PATH-SELECTION PROCESS

One can view the reaction process as a wandering over an energy surface via the lowest energy path toward the lowest accessible point on the surface. The reacting partners have a limited amount of energy; only certain transition states and intermediates can be achieved and the reaction will end up at the **lowest accessible** point on the surface, which may not necessarily be the lowest point on the surface.

The most useful mental models approximate some physical reality. Since we will be exploring an energy surface with peaks, passes, and valleys unknown to us, the best mental model would be a mountainous wilderness, but one luckily crossed with trails. Imagine being dropped off in the middle of this wilderness at a trail junction, assured that one of the many trails leads down and out. We know that the desired route will be the lowest energy route. We certainly do not want to panic and get off the trails into real unknown territory or to guess randomly at a direction and wander off. We must carefully and logically explore the nearby wilderness drawing a map as we go.

Armed with the tools that we have learned in the previous chapters, we can predict where the lowest energy path may go. The **electron flow pathways are the trails to guide us** in this wilderness of an unfamiliar energy surface. Some trails may dead end in high valleys that have no reasonable exit other than the one we came in. Others may lead us to a low point (side product) on the energy surface but not the lowest accessible point (major product). The reacting partners are also exploring these dead ends, and very few reactions produce only one product.

What do we need in order to succeed? We need the ability to draw a good map (if you don't want to repeat some paths or wander around in circles), the ability to recognize whether we are going uphill or down (energy awareness), the ability to recognize all the paths and when they branch (you can't stay on the trails if you can't recognize them), and finally the ability to cross-check what you have done (was the hike to a certain point a silly thing to do or is it on the right track?)

The exploration process is really quite simple. The hard part is to stick with it when things are not going well and you are feeling lost; it's easy to get discouraged and do something rash just to be done with the struggle. The other time that it is hard to stick with it is when things are going so well that the end is in sight, and it feels that it might be fine to abandon the trails and just make a straight run for the end. Both arise out of the natural tendency to act on your first impulse because the alternative, rigorous logical analysis, is hard work.

So what is the search process? First understand where you are. Observe your surroundings; take inventory; classify what you have to work with. Second, start drawing your map by marking down the start point and note all the paths that lead from it. Find a downhill path or at best the least uphill path. Follow that path to the next point and mark that on your map. Again observe your surroundings. Now cross-check that going to this new point is indeed a reasonable thing to have done (here is where you make use of all those trends). Cross-check that the energy of the route is reasonable (with the ΔpK_a rule cross-check, K_{eq} calculation, or $\Delta H°$ calculation—whichever is appropriate). Now look for all the new paths leading off this new point, and start the process again. Always keep in mind your starting point and do not climb too high before exploring other routes.

How do you go about learning this method of exploration? First learn the needed skills (Chapters 1–9), and then try them out on easy problems with the help of an experienced guide. On these guided tours, we will explore the entire surface, including all its dead ends, so that you feel you know the territory. We will gradually explore more difficult surfaces until you feel confident enough to venture out on your own. Let's look at each step of the exploration process (Figure 10.1) in more detail.

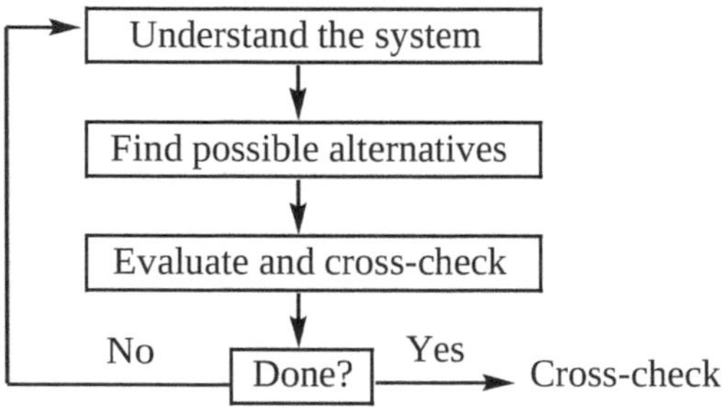

Figure 10.1 The overall generate and select analytical process used throughout this text.

10.3.1 Understand the System

Write down the Lewis structure of the reactants, complete with formal charges, and draw any major resonance forms. Look for leaving groups, polarized single and multiple bonds, acids and bases. Classify into generic sources and sinks and then rank them. **The reaction usually occurs between the best source and sink.** Above all, note if the media is acidic or basic. In basic media, find the best base, and then locate any acidic hydrogen within range (not much more than 10 pK_a units above the pK_{abH} of the base). In acidic media, identify the best sites for protonation. Likewise, do not create a species that is much more than 10 pK_a units more acidic than your acid. Understand what bonds have been made or broken, but do not lock into an arbitrary order as to which occurred first.

10.3.2 Find Possible Alternatives

Generate all the possible paths. The trails are few since only those pathways that have the same sources and sinks as the reacting partners need to be considered. The media will also restrict paths. Do not use acid media paths when the medium is definitely basic or vice versa. Draw the Lewis structure of the product. Never discard a route without first taking a small amount of time to write down the result (this will force you to consider it more carefully). Again, use scratch paper to draw a map of this new problem space that you are exploring so that you can keep track of where you have been. Note branches in the path that you have passed but not explored (and may want to come back to), and write down all the possibilities branching from the spot that you are currently considering. Remember the "street corner" analogy that we used in Chapter 4 when exploring the common reaction archetypes.

Dealing with complex problems one elementary electron flow path at a time can make the problem feel much more complex. As was shown in Section 7.3 there are many common path combinations like the S_N1, E1, E1cB, Ad_E2, and Ad_N2 that allow us to think in greater strides, instead of single steps. The energy surfaces introduced in Chapter 4 allow you to think in terms of the reaction archetypes, which gather all the relevant reactions into one image. This use of surfaces allows you to see the alternatives within the archetypical process and assemble a mechanism from even bigger conceptual pieces. Use these surfaces to get a "big picture" analysis of the process.

For example, if a reaction connected a nucleophile to a ketone, then the addition surface is our big building block needed to create a reasonable mechanism. There might be steps necessary to prepare for the addition, and maybe a step or two to finish up to get the product, but a big chunk of the process is already in place. Often these starting and finishing setup reactions are just proton transfer reactions. Using a surface reminds us that there is a decision to be made between the various addition routes, Figure 10.2.

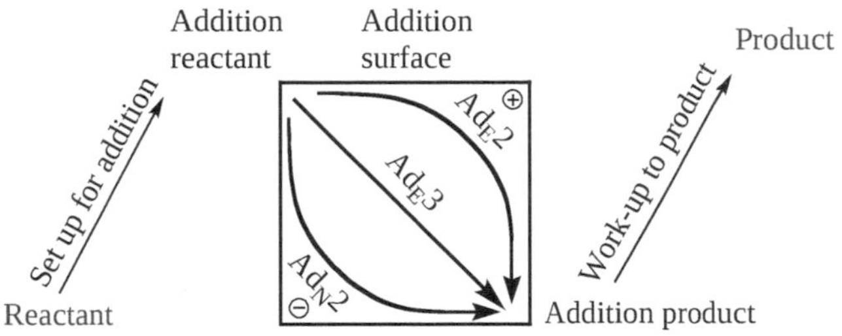

Figure 10.2 The use of the addition surface to achieve an overview of an addition to a ketone.

This use of surfaces becomes more important as the complexity of the reaction mechanism increases. For example, a common reaction of the Y=C–L sink is an addition of a nucleophile followed by elimination of the leaving group. We can take our generic addition surface and append an elimination surface to it to get an overview of what decisions will be required in the more complex mechanism, Figure 10.3.

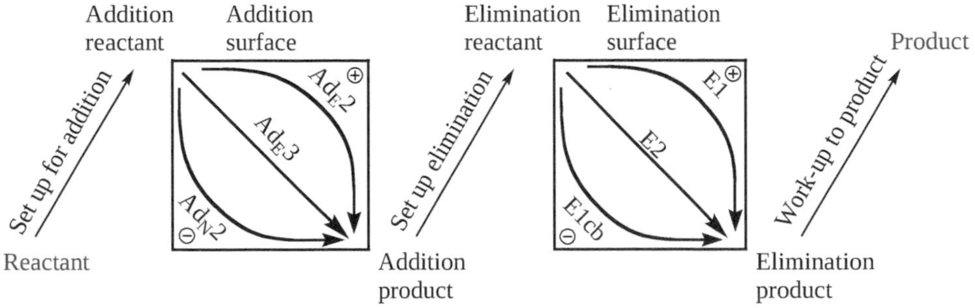

Figure 10.3 The joining of two surfaces surface to give an overview of the addition and elimination.

Assembling reaction mechanisms using these bigger blocks is similar to a chess player strategizing by groups of moves, standard openings for example. The fewer blocks one needs to assemble to get a reasonable mechanism, the easier the entire process is to visualize. In summary, decide what reaction archetypes have occurred then build an initial graphical roadmap of the mechanism's problem space using the correct surfaces. Finally, at each decision point generate the choices that each surface presents, for example E1, E2, or E1cB, and then select the best path.

Drawing such a map may seem like it would slow down the process of reaching the goal, but actually it speeds it up because much less time is spent lost, wandering aimlessly, taking random shots at the answer. Don't work from just the beginning; what could have been the last step? If you can see and map the first and last steps, then the problem just got shorter and easier to solve. If you come to a major decision, look at the product. Did the anion add as a nucleophile or did it act as a base? If it is not attached to the product, it functioned as a base.

10.3.3 Evaluate and Cross-check

The trends allow us to gauge the relative height of the possible passes and high valleys so that we explore only the most reasonable routes. Especially cross-check the media restriction and the stability of intermediates. We can often use either the ΔpK_a rule cross-check on our alternatives or the calculation of the $\Delta H°$ or the K_{eq} to make sure the energetics of the process are reasonable. **Generate and select** the best path. As with any problem for which you cannot see the route to the answer at the start, you must be very careful at the beginning to select the best route, not just the first one you see. Double-check any major decision. Chapter 11, Cross-checks and Decision Boundaries, gathers and discusses all these cross-checks in one place.

10.3.4 Repeat The Process

With mechanism problems, the final product is given, so the goal is unambiguous. The problem-solving process is a deliberate search from the starting point, which eventually triggers an insight coming from the goal. A novice can learn the methods of deliberate searching; insight comes by being familiar with the territory and will get better with practice.

With product-prediction problems, there will always be some doubt as to whether you are at the lowest accessible point on the surface or are in the energy well of a side product. Hopefully, the end should be obvious because you will definitely be downhill from your starting point, and it will seem that all alternatives for several steps around are definitely uphill. Again, knowing when you are done also improves with practice.

10.4 REACTION MECHANISM STRATEGIES

Mechanism problems are easier than product-prediction problems because helpful information can be derived from the product. Given the starting and ending points you need only supply the lowest energy path between them. You may not be able to see the answer to a long mechanism problem at the start; just let a logical analysis carry you to the answer. Don't panic and try to force a quick (and incorrect) answer.

Commonly, organic equations are not balanced, so make sure the chemical equation balances. The balanced equation may identify another product; for example, water may have been lost. If there is a small piece missing, try to find out where it came from.

Next identify the carbons that belong to the starting materials within the product (number the carbons if you have trouble) so that you can tell what bonds have been made or broken. If more than one numbering scheme is possible, go with the scheme that requires the fewest changes. If the starting material structure just will not fit into the product structure (considering any small pieces lost), rearrangement may have occurred.

Look at each new bond and then at the polarization of the corresponding sites in the starting materials so that you can decide what the original nucleophile–electrophile pairs might have been for those bonds. If more than one new bond has been made, consider all possible sequences of bond formation, not just the first one that occurs to you. Look at adjacent functionality to get a hint as to whether it would stabilize an electrophile's or nucleophile's partial or full charge. Ask how proton transfer might have generated the needed nucleophiles and electrophiles. Consider any reasonable proton transfer and always check the K_{eq} value (usually $\geq 10^{-10}$).

Use the general trends to guide the decision process. At every decision point, generate all the possible paths, and then select the best one on the basis of trends. For example, if there is a choice of making two possible cations, rank their stability and pick the route that goes through the most stable cation. If nothing looks reasonable, return to any decision point where the trends indicated that an alternative was also feasible. Use "hindsight" (knowing the final product) as a guide through any difficult decisions.

Finally, do not combine steps to save drawing out the structure again. The structure you skip might be a critical branch point in the decision tree, which you may then miss. Keep to the pathways to avoid the trap of merely rearranging the lines and dots of the structures on the page. Be as rigorous in your own practice as you would be on an exam; you are trying to develop good intellectual habits. The prime directive is this:

If it does not make sense, it's wrong.

10.5 WORKED MECHANISM EXAMPLES

The problems at the end of Chapters 8 and 9 were short and simple mechanism and product-prediction problems. Chapter 8 problems had only one source and sink and no major decisions to be made, and those in Chapter 9 were the same except one major decision was required. We are now going to cover problems of gradually increasing complexity: more than one source and sink, longer problems, more decisions, and more alternate routes. For each example, cover each step and try to provide the expected information; compare yours with what is given, then move on to the next portion.

10.5.1 Alkyne Hydrobromination

Give a reasonable mechanism for the following reaction.

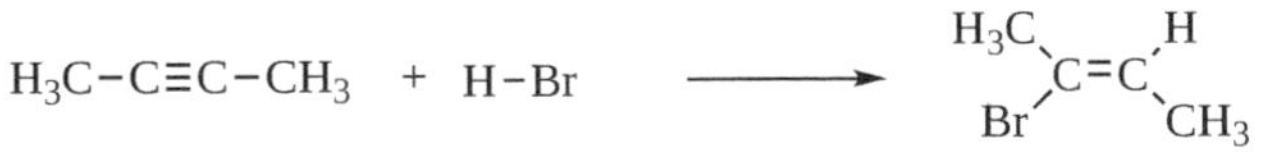

Understand the system

Balanced? Yes. Generic process? An addition has occurred. Medium? Strongly acidic, HBr pK_a is –9. Sources? The triple bond and the bromide ion. Sinks? The H−Br bond is easily broken because hydrogen bromide is a strong acid. Bromide anion is also expected to be present as a conjugate base from any proton transfers. Stereochemistry of addition is *anti* since the H and the Br end up on opposite sides of the double bond.

Find possible routes

We are at a well-known trail junction, the starting point of the addition surface from Section 4.4.2. Figure 10.4 is a simplified addition surface serving as a problem space map to guide our decisions and to remind us of alternative possibilities, because the first answer we think of may not be the best. From the Figure 10.4 surface, there are three possible routes that we must decide between.

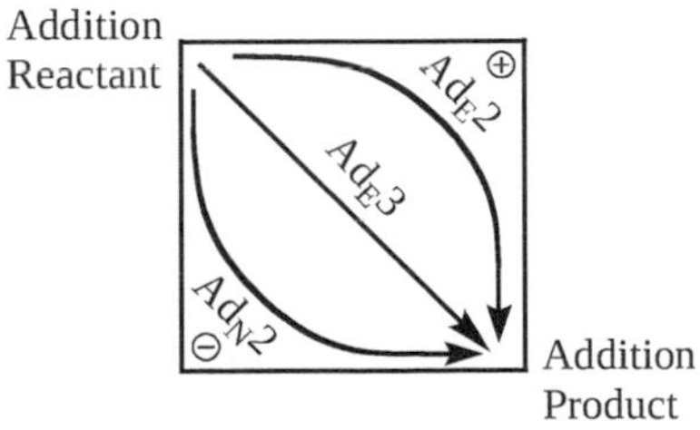

Figure 10.4 Anticipated addition surface serves as a problem space map for example 10.5.1.

The choices are the Ad_N2 addition (path Ad_N followed by p.t.),

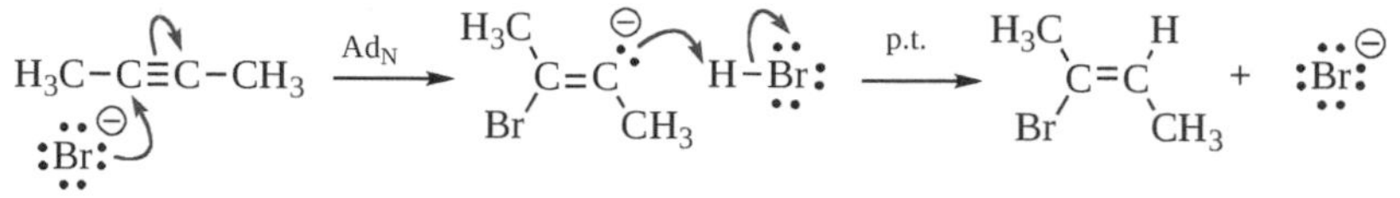

the Ad_E2 addition (path A_E followed by A_N),

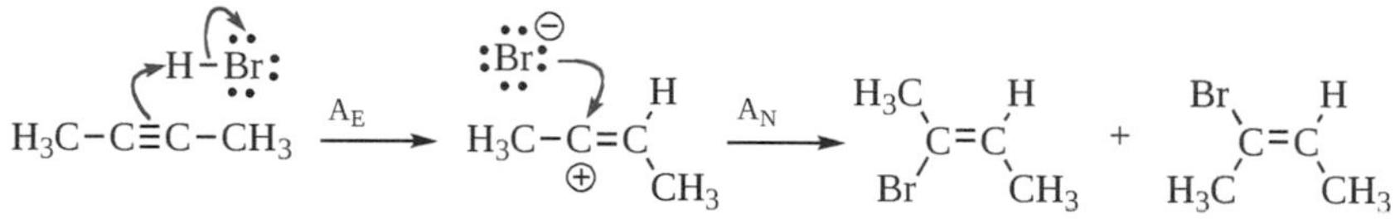

and the Ad_E3 addition.

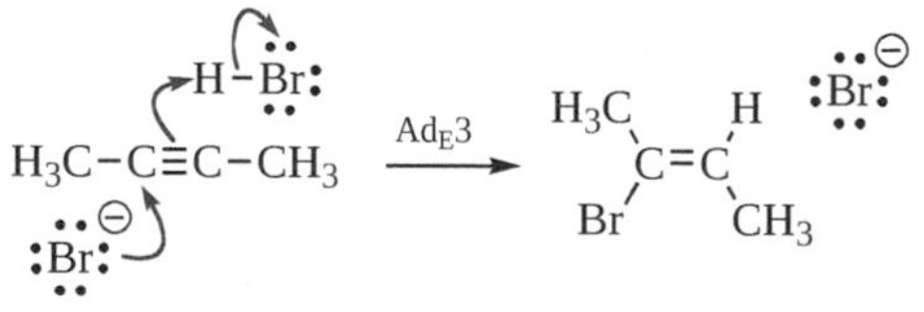

Evaluate and cross-check

We must judge the possible routes as to how reasonable they are to the sources, sinks, and the media. The Ad_N2 is not appropriate for acidic media because it produces a highly basic anion, pK_{abH} 44. The ΔpK_a rule cross-check would likewise throw out the Ad_N2 since the incoming bromide nucleophile at pK_{abH} −9 would never be expected to form a product anion of pK_{abH} 44. This route is so uphill that it would never occur. In this example to illustrate the sorting process, we showed all the possibilities, but in later mechanism examples we will consider only those routes that are contenders for the lowest energy route.

The Ad_E2 and Ad_E3 processes are appropriate for the acidic medium. The Ad_E2 goes via an unstable secondary vinyl carbocation. A second problem is that the vinyl carbocation is expected to be linear and could therefore trap bromide from either the top or the bottom face to produce a mixture of stereoisomers.

The Ad_E3 avoids the unstable carbocation and rationalizes the stereochemistry of the product. The attack is *anti*; the bromide attacks the bottom face since the pi complex of the alkyne with HBr blocks the top attack. In this case, the addition surface has folded so that the Ad_E3 is the lowest energy route proceeding up through a pass between the two high corners of an unstable anion and an unstable cation. For a 3D example of this surface, see the E2 surface (the reverse reaction) in Figure 4.21. The Ad_E3 route explains the stereochemistry and is reasonable for the alkyne substrate and strongly acidic media.

Surface overview: We can look back at the problem space that we just navigated to get an overview. The red arrow shows the most reasonable route on Figure 10.5 below.

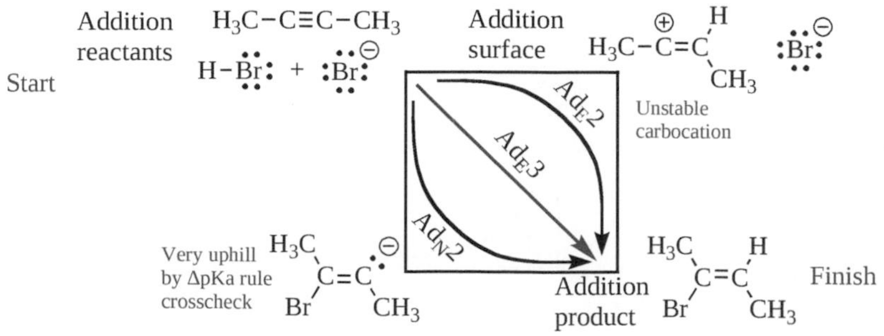

Figure 10.5 Final addition surface serving as a problem space map for example 10.5.1.

10.5.2 Imine Formation

Give a reasonable mechanism for the following biochemically important reaction.

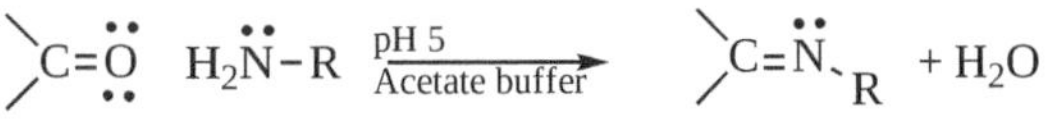

Understand the system

Balanced? Yes. Medium? Mildly acidic. A common acetate buffer of pH 5 is created from almost equal concentrations of acetic acid and sodium acetate. Sources? The amine lone pair and the carbonyl lone pair. Sinks? The ketone is a C=Y. Acidic H's? None within range of pH 5. Basic sites? An oxygen and/or nitrogen lone pair could become protonated. Leaving groups? None. Resonance forms? The carbonyl charge separated form, $^{+}$C—O^{-}, shows the carbonyl carbon is a good sink on carbon.

Bond changes? We must connect the pieces before forming the pi bond. The amine nitrogen bonds to the carbonyl, and then a pi bond to nitrogen must form by losing oxygen as water. Proton transfers? Nitrogen lost two H's and oxygen gained two H's to make water.

Find possible routes

It helps to understand the overall transformation before diving into the individual steps. We can get a good overview of the problem space by generalizing what has to happen: first an addition has occurred to connect the two reactants, then probably some proton transfers to set up the next step, an elimination of water to form a C=N bond. This allows us to draw a simplified problem space to guide our route decisions, Figure 10.6.

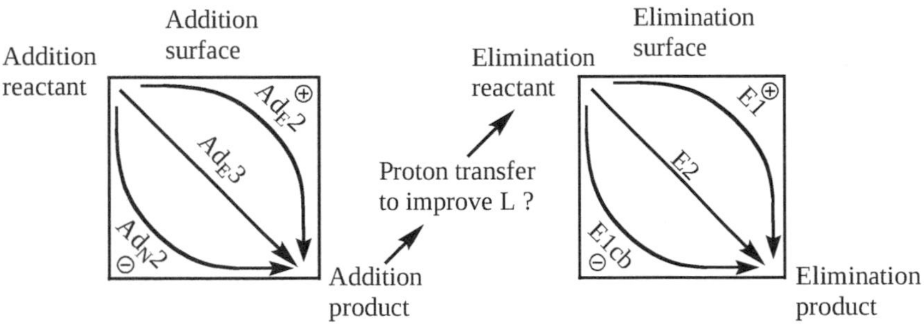

Figure 10.6 Addition and elimination surfaces as a simple problem space map for example 10.5.2.

We can see from our simplified problem space that there are three possible addition routes that should be considered:
the hetero Ad_E2 addition (path p.t. followed by Ad_N),

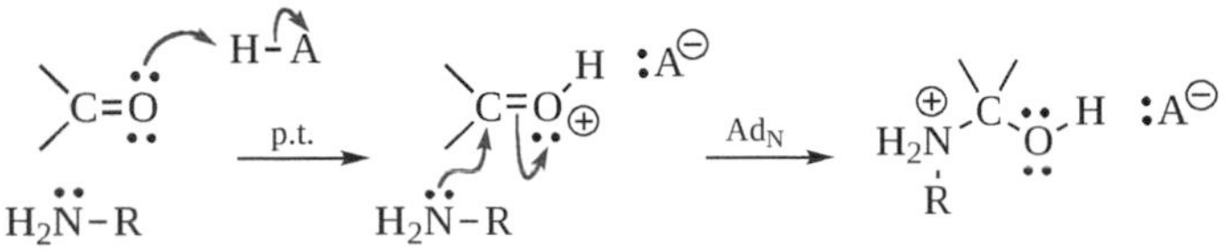

the hetero Ad_E3 addition,

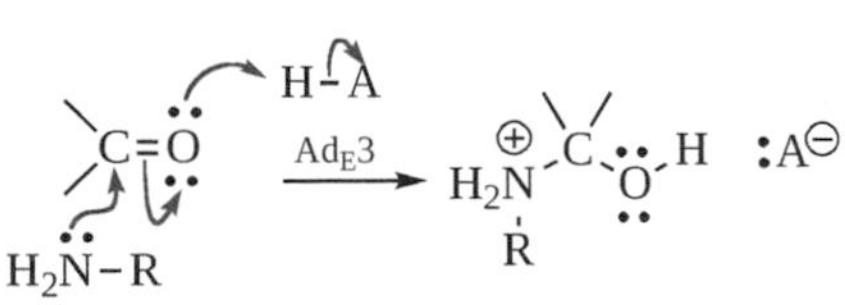

and the Ad_N2 addition (path Ad_N followed by p.t.),

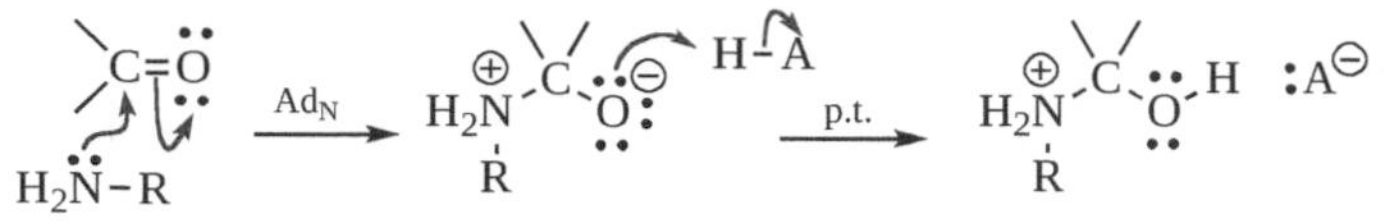

Evaluate and cross-check

The first step of the Ad_E2 is a proton transfer that creates a strong acid, $pK_a = -7$. Our best acid in acetate buffer is acetic acid, $pK_a = 4.8$, and so this proton transfer K_{eq} would be $10^{-11.8}$, which is slightly beyond our useful range.

The Ad_E3 has an incoming nucleophile whose conjugate base pK_a is about 10.6 and the addition is removing the proton from acetic acid, $pK_a = 4.8$, thereby creating a weaker base, and passing our cross-check.

The Ad_N2 first step produces an ion pair that would be expected to have a pK_{abH} near 12. Our nearest model, choline $pK_a = 13.9$, is expected to be less acidic because it has an extra methylene $(CH_3)_3N^+CH_2CH_2OH$. The incoming nucleophile (conjugate base $pK_a = 10.6$) is only slightly less basic than the ion pair produced by the addition. The Ad_N2 second step also removes the proton from acetic acid, $pK_a = 4.8$, thereby creating a weaker base, so both steps pass our cross-checks.

We conclude either the Ad_E3 or Ad_N2 would be reasonable alternatives.

Understand the system

We need to set up the elimination. The nitrogen still has two protons on it and the oxygen has only one. Hydroxide at $pK_{aHL} = 14.0$ is not a good leaving group but could be improved by protonation. The protons on the cationic nitrogen have a pK_a of 11, within range of acetate, $pK_{abH} = 4.8$. Even though oxygen's electronegativity makes its lone pair less basic than nitrogen's, at $pK_{abH} = -2.4$, the OH is still in range of being protonated by acetic acid at $pK_a = 4.8$.

Find possible routes

We need to deprotonate the nitrogen and protonate the oxygen, but which order is best? Draw both out and compare.
Oxygen first, then nitrogen:

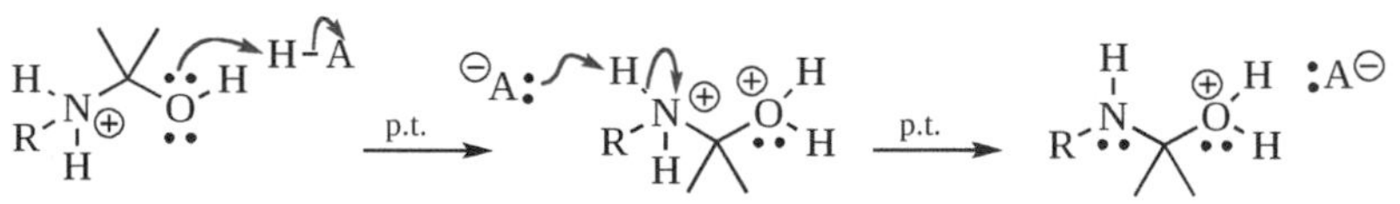

or nitrogen first, then oxygen.

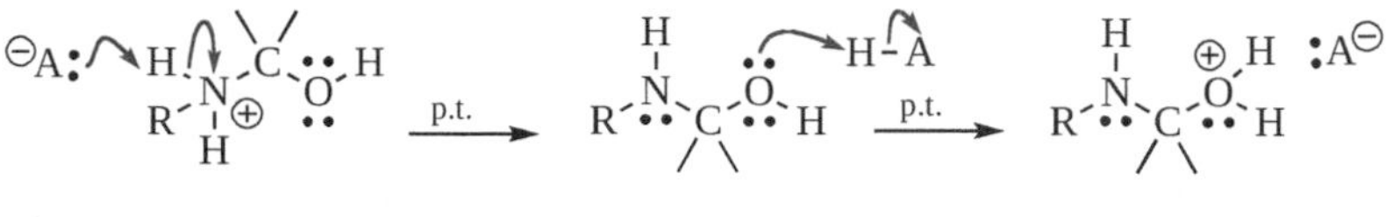

Evaluate and cross-check

If we protonated the oxygen first, we'll have the unfavorable electrostatics of having two like charges close to each other. Since we would like to avoid the electrostatic repulsion of like charges, it would be best to deprotonate the nitrogen before protonating the oxygen.

Understand the system

We are now ready to consider an elimination to remove the proton on nitrogen and kick out the water leaving group to get the final product. The proton on nitrogen is not very acidic and our media pH is only 5. Water, pK_{aHL} = 0.0, is a good leaving group.

Find possible routes

We can see from our example problem space, Figure 10.6, that there are three possible elimination routes that should be considered.

Lone-pair-assisted E1

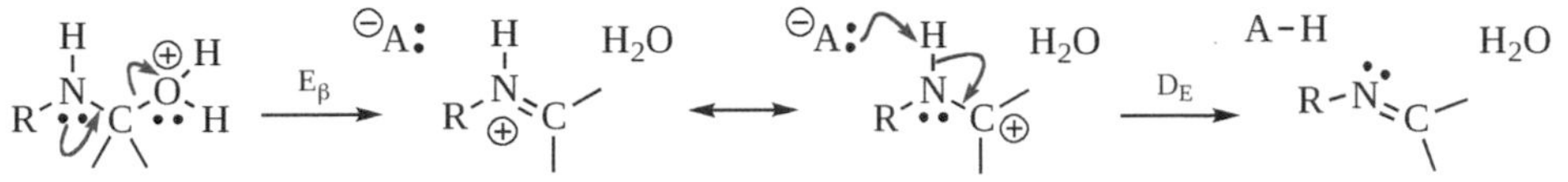

E2

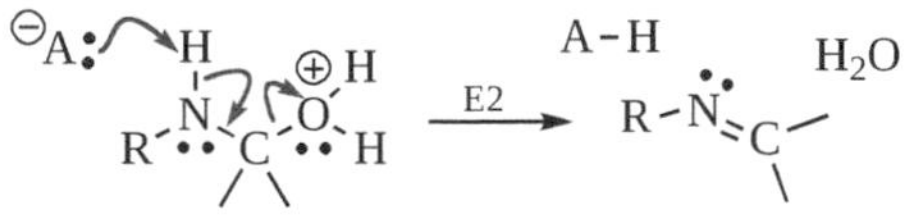

E1cB

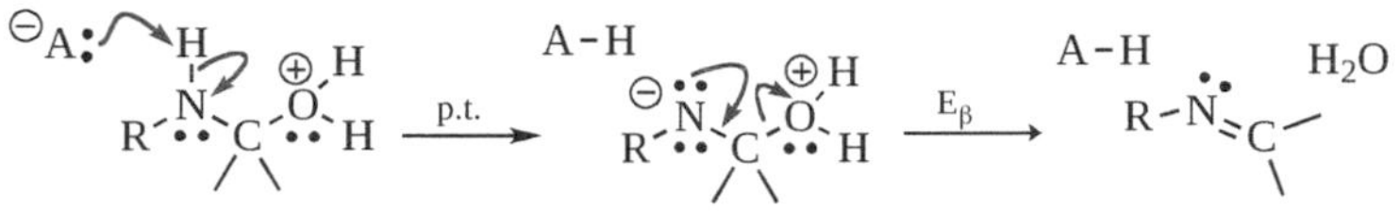

Evaluate and cross-check

The lone-pair-assisted E1 usually requires three conditions to be fulfilled: a polar solvent, good leaving group, and a carbocation of secondary or better stability. The buffered water solvent is very polar. The water leaving group is pK_{aHL} = 0.0, and the carbocation formed has lone pair stabilization by resonance. Therefore the lone-pair-assisted E1 route, is very reasonable. A variation on the E1 theme is general acid catalyzed beta elimination (Section 7.4.3). The carbinolamine oxygen picks up a proton from an acid as the leaving group departs. This can be considered a partial L protonation by hydrogen bonding also.

The E2 route needs to pass a ΔpK_a test and an alignment test. The E2 passes the ΔpK_a test, for with the base as acetate, pK_{abH} 4.8, and the pK_{aHL} of the leaving group is 0.0, the E2 reaction has created a much weaker base. Since the C–N single bond is free to rotate, and then the N–H and the C–L can get the coplanar alignment that an E2 requires. Therefore, the E2 route is also reasonable.

The proton-transfer first step of the E1cB has problems. The nitrogen anion conjugate base has no resonance, and only the ion pairing with the positive oxygen to stabilize it. Without the ion pairing, the conjugate base pK_{abH} would be 36. Field and inductive effects are less influential than resonance. For example, resonance stabilization of a nitrogen anion by a phenyl ($PhNH_2$) brings the pK_{abH} down to 28, still a long way from our strongest base in this reaction, acetate at pK_{abH} = 4.8. We conclude that the proton transfer, thus also the E1cB, fails to be a reasonable hypothesis.

Overview

We are done, but let's take an overview of the process. Two addition and two elimination routes were reasonable. One acceptable reaction path is shown vertically in the center of Figure 10.7, and the alternate routes are shown to the side.

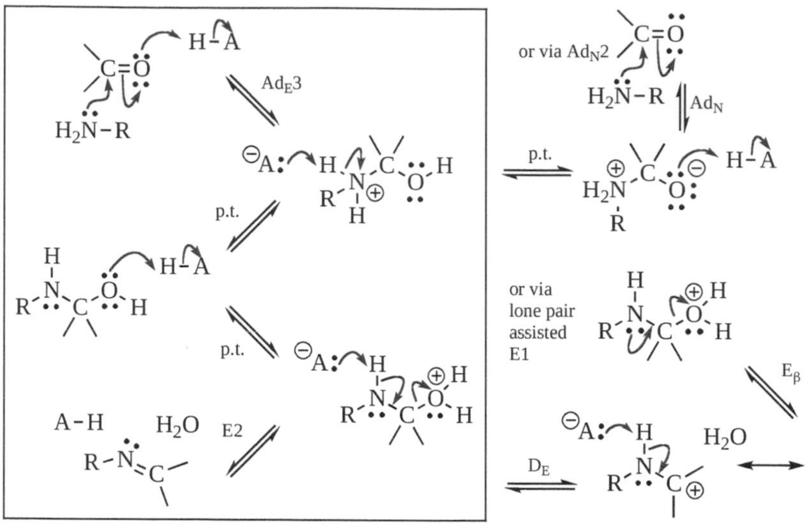

Figure 10.7 Imine formation overview.

Surface overview: We can look back at the problem space that we just navigated to get an overview. The red arrows show the most reasonable route(s) on Figure 10.8 below.

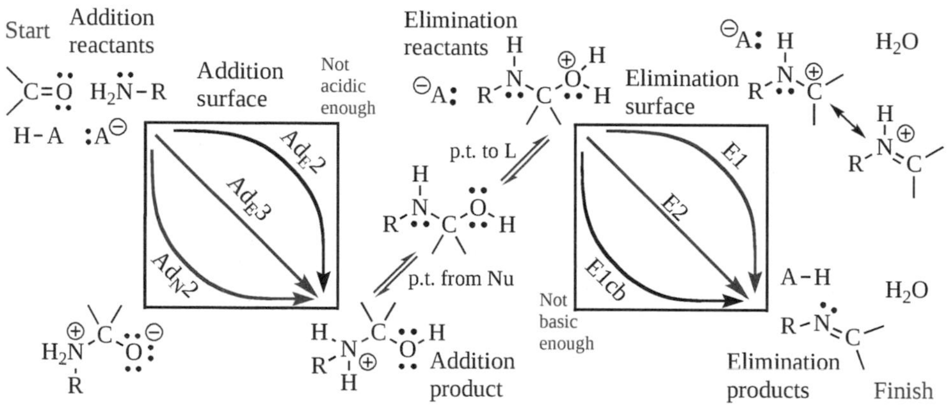

Figure 10.8 Final addition and elimination surfaces as a problem space map for example 10.5.2.

Biochemical example problem

Imine formation is a very important biochemical process. It has an interesting behavior that shows a maximum in the pH rate profile, Figure 10.9. Using the previous mechanism postulated for imine formation, examine extremes in pH to understand the figure. What would you expect to happen in strong acid to slow the reaction down? How about in strong base?

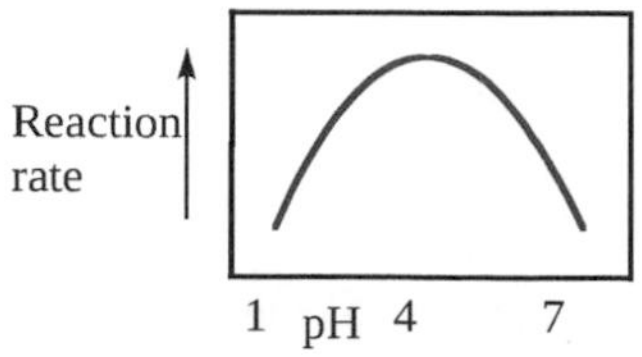

Figure 10.9 A pH rate profile for imine formation.

Answer: In strong acid, the amine nucleophile in the first step would be protonated thus no longer nucleophilic. Without a good concentration of nucleophile the reaction would slow down greatly. In strong base, there would be no acid good enough to protonate the leaving group for the elimination step. With a poorer leaving group, the reaction would also slow greatly. At pH 4, there is enough free amine to serve as a nucleophile, and enough weak acid to protonate the leaving group, so the reaction proceeds nicely.

10.5.3 Aldol Condensation

Give a reasonable mechanism for the following aldol condensation.

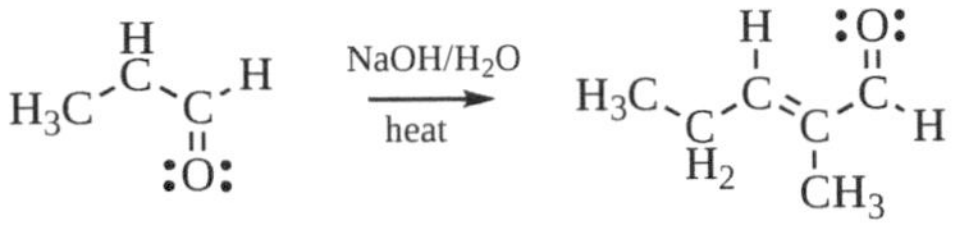

Understand the system

Balanced? No, the left-hand side requires 2 mols of aldehyde, and the right-hand side needs a water molecule to balance. Find the pieces of the two reactants within the product. Note that the reaction has occurred between the aldehyde carbonyl (carbon 1) of one reactant and the carbon next to the aldehyde (carbon 2) on the other.

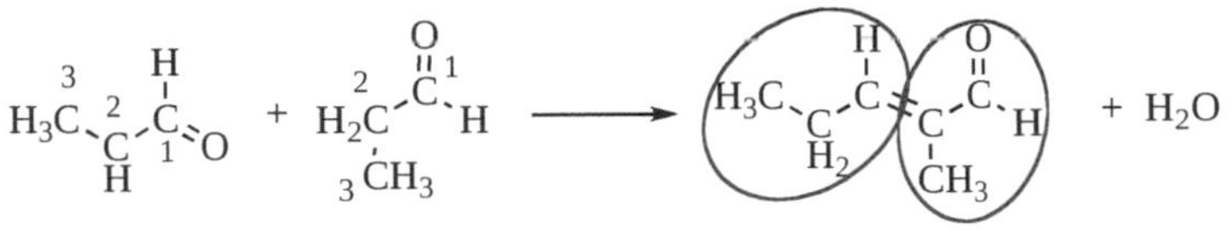

Generic process? An addition and elimination have occurred. Medium? Definitely basic, predominant anion is hydroxide, pK_{abH} 14.0, whose pK_a would give a useful proton-transfer K_{eq} up to about pK_a 24. Sources? The carbonyl lone pair, water lone pair, and hydroxide anion. Best source? Hydroxide anion, a lone pair source can behave as a nucleophile or as a base. Sinks? Polarized multiple bond, the aldehyde carbonyl. Acidic H's? Water and the CH_2 next to the aldehyde, pK_a 16.7, are within range of hydroxide. Leaving groups? None.

Find possible routes

We need to connect the two partners before we can make the double bond, so the addition must have occurred before the elimination to form a pi bond. Since bonds are made by combination of nucleophile and electrophile, we can see the addition is not ready to proceed since although the aldehyde carbonyl is an electrophilic electron sink, carbon 2 of the other partner is a mere methylene and not a nucleophile. We need to generate a nucleophile on carbon 2 first. The problem space for the overall process might look like Figure 10.10.

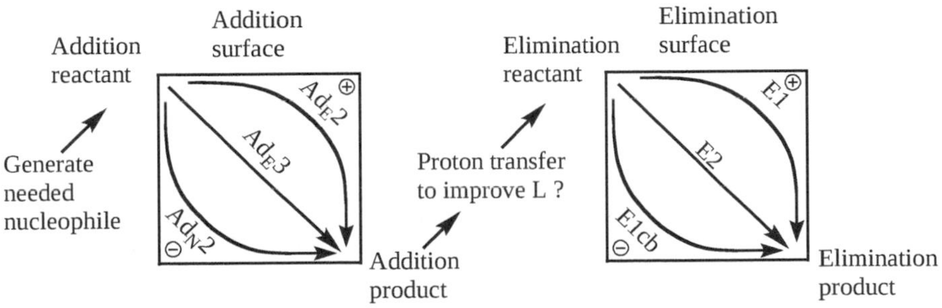

Figure 10.10 Anticipated addition and elimination surfaces as a problem space map for example 10.5.3.

As a first step, there are just two paths that fit the basic medium, sources and sinks: deprotonation to form an anion (path p.t.), which conveniently makes carbon 2 nucleophilic, and is therefore a step closer to the product.

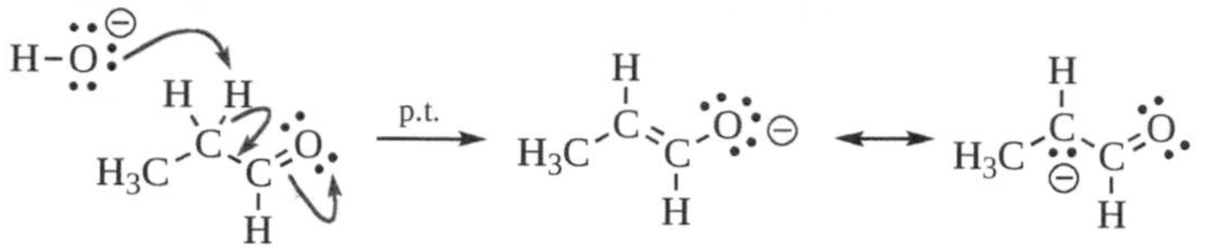

or addition to a polarized multiple bond (path Ad_N).

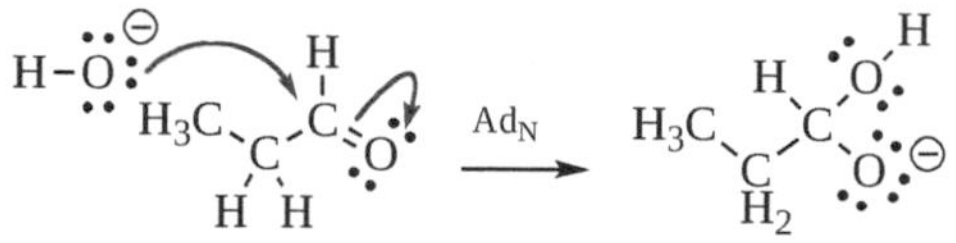

Evaluate and cross-check

Since the pK_a for both products can be estimated, and they both come out to be around pK_{abH} 16–17 vs. 14–15, they are too close in energy to distinguish with our crude ΔpK_a rule cross-check. We should explore both routes. A use of "product hindsight" shows that no additional hydroxyl groups are attached, thus making this nucleophilic attack path less probable. If this were a product prediction problem and not a mechanism problem we would need to explore both routes.

Understand the system

We need to record only that which has changed from our original observations. We now have a new source, the delocalized anion just formed. This enolate anion is nucleophilic on carbon 2, just what is needed for our addition process. Resonance forms indicate the ambident nature of the enolate, an allylic source:

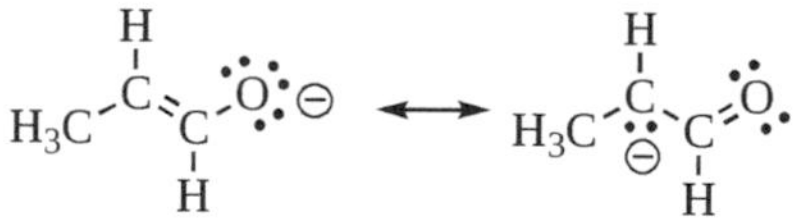

Find possible routes

Now we have a good nucleophile and a good electrophile for our addition process to occur. The three addition routes are:
the hetero Ad_E2 addition (path p.t. followed by Ad_N),

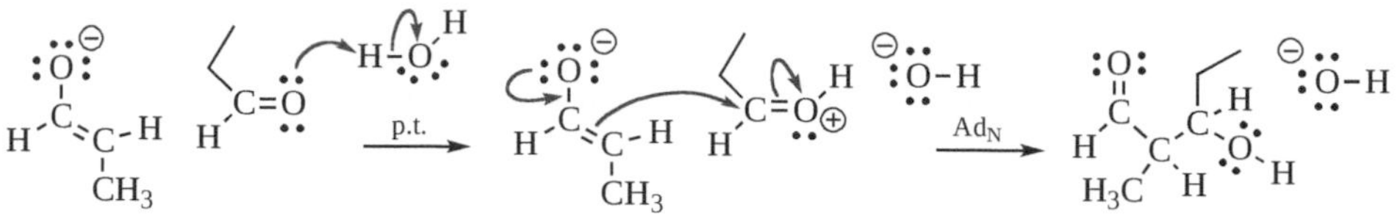

the hetero Ad_E3 addition,

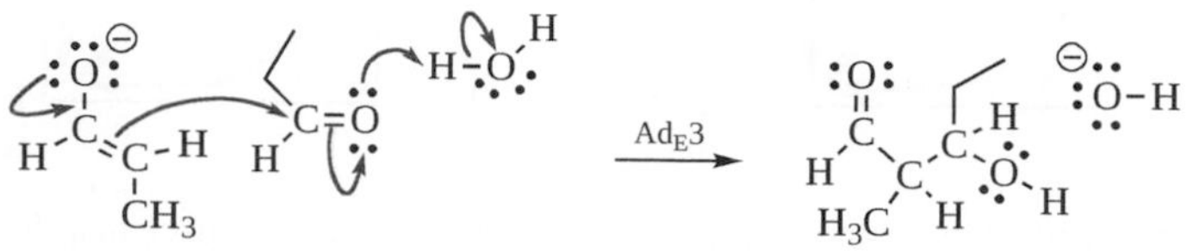

and the Ad_N2 addition (path Ad_N followed by p.t.),

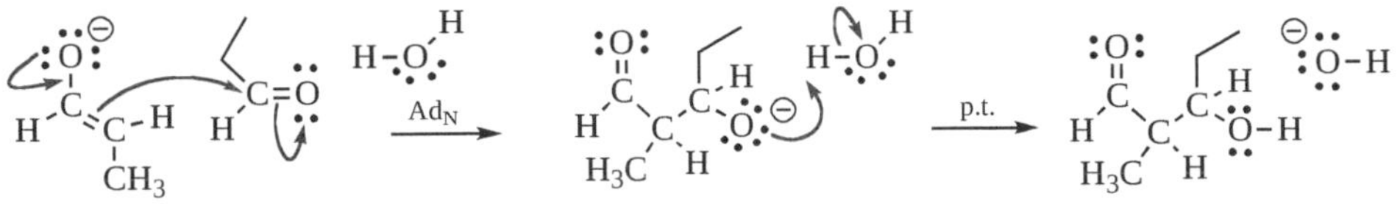

Evaluate and cross-check

For the Ad_E2, the best acid in basic water is water itself with a pK_a of 14.0, which is very weak. The protonated aldehyde product has a pK_a of –8, giving the proton transfer K_{eq} a value of 10^{-22}, very much beyond our minimum useful threshold of 10^{-10}. Another tip that something is wrong is that we have both a strong base and a strong acid present in an intermediate step, violation our pK_a span limit of no more that 14 pK_a units. The Ad_E3 addition is reasonable by the ΔpK_a rule cross-check, incoming aldehyde enolate nucleophile is pK_{abH} 16.7 and the product of the nucleophilic attack is a hydroxide, pK_{abH} 14.0. The Ad_N2 addition is also reasonable by the ΔpK_a rule cross-check, incoming nucleophile is pK_{abH} 16.7 and the product of the nucleophilic attack is a secondary alkoxide, having a pK_{abH} of about 17. Either Ad_E3 or Ad_N2 addition is energetically reasonable by the ΔpK_a rule cross-check.

Other things that can happen

There are other paths that fit the medium, sources and sinks: protonation of an anion, p.t., and addition to a polarized multiple bond by the oxygen of the enolate by Ad_E3 or Ad_N2. There are two sites that can serve as a base or as a nucleophile on this ambident allylic source (see Section 9.4). Even though we have a carbon as a nucleophile path above that leads toward the product, for completeness we should explore the other alternatives. We have three other choices to evaluate:
proton transfer to oxygen,

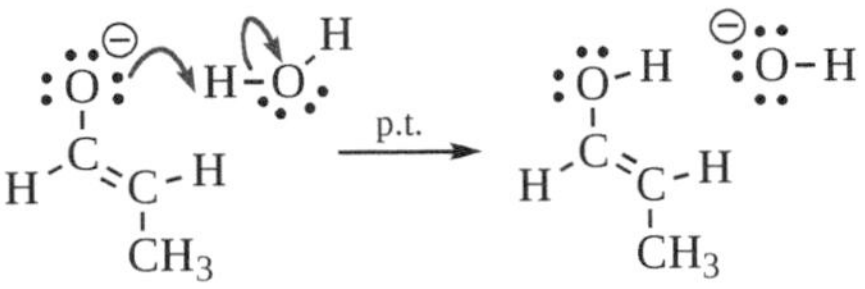

or to carbon,

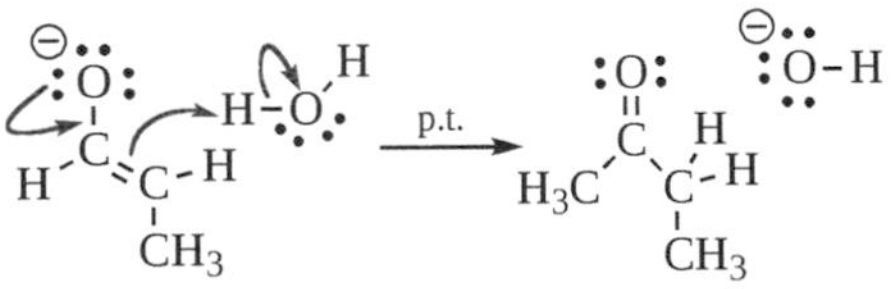

or addition to a polarized multiple bond by oxygen by either Ad_E3 or Ad_N2.

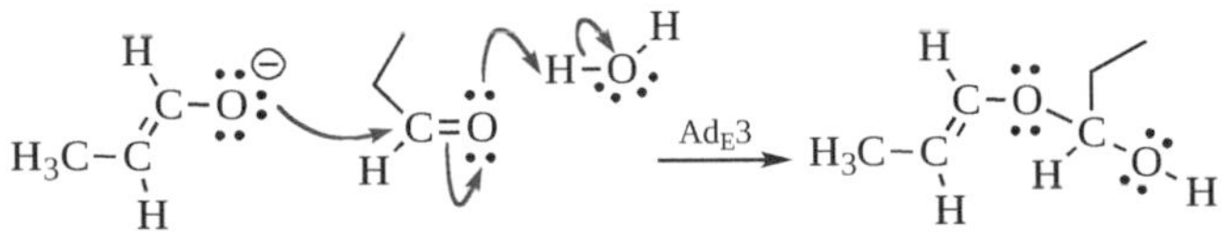

Evaluate and cross-check

Protonation of the anion takes us to either the enol or back to starting material, both reasonable alternatives. A $\Delta H°$ calculation gives the enol as uphill in energy from the carbonyl. We know from our product that carbon not oxygen has been the nucleophile (as expected, Section 9.4). The oxygen attack on the carbonyl is reversible by the ΔpK_a rule cross-check, and a $\Delta H°$ calculation confirms the hemiacetal product will be uphill in energy.

Understand the system

An elimination of water gets us to the product. We now are close enough to product to restrict paths to elimination pathways. From our simplified problem space in Figure 10.10, we need to determine if the media can improve our leaving group and then move to the reactant corner of the elimination surface, Section 4.3.1.

Resonance forms:

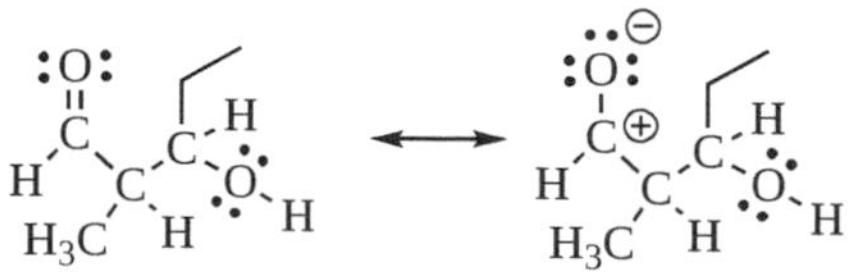

Find possible routes

The basic media does not contain an acid strong enough (water is our best acid) to improve the leaving group by protonation; creating a protonated alcohol of pK_a –2.4 would have a K_{eq} of $10^{-16.4}$ with water as the acid. Therefore, because the hydroxide leaving group is poor, pK_{aHL} 14.0, and the E1 elimination path is not viable. There then are just two elimination paths that fit the medium, sources and sinks: the E2 elimination,

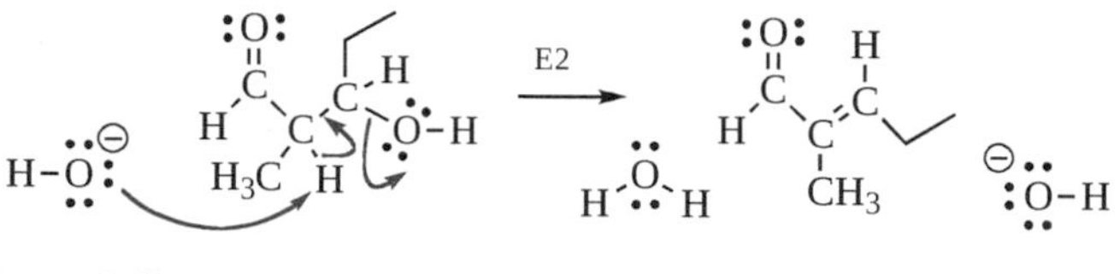

or the E1cB (path p.t. followed by path E_β).

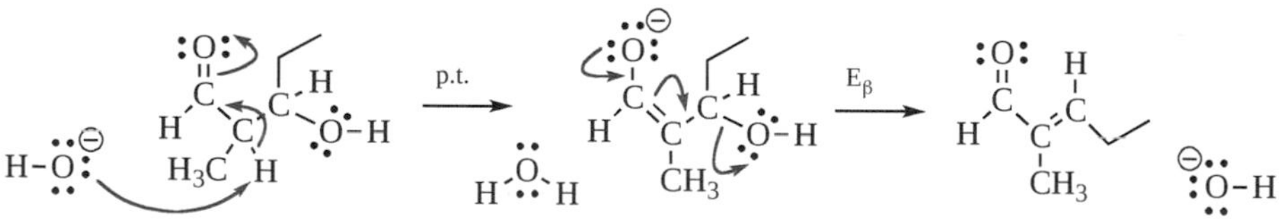

Evaluate and cross-check

Both elimination processes are reasonable, but the E1cB path is more probable because the H is acidic by virtue of being next to an EWG, and the leaving group is poor, pK_{aHL} 14.0, so an adjacent anion can help it leave. An anion of pK_{abH} 16.7 kicking out hydroxide, pK_{abH} 14.0, is downhill by the ΔpK_a cross-check. The pK_a span on all steps has been within reason.

Overview

We are done, but let's take an overview of the reasonable routes (Figure 10.11). The reaction path is shown vertically in the center and the alternate routes are shown to the side. The path abbreviation for the forward reaction only is shown. To summarize the route: proton transfer generates the nucleophile, which reacts by Ad_N2, and then the E1cB process gives us the product. Notice how this five-path process seems to simplify by using path combinations, p.t., then Ad_N2, and then the E1cB. Start thinking in bigger blocks.

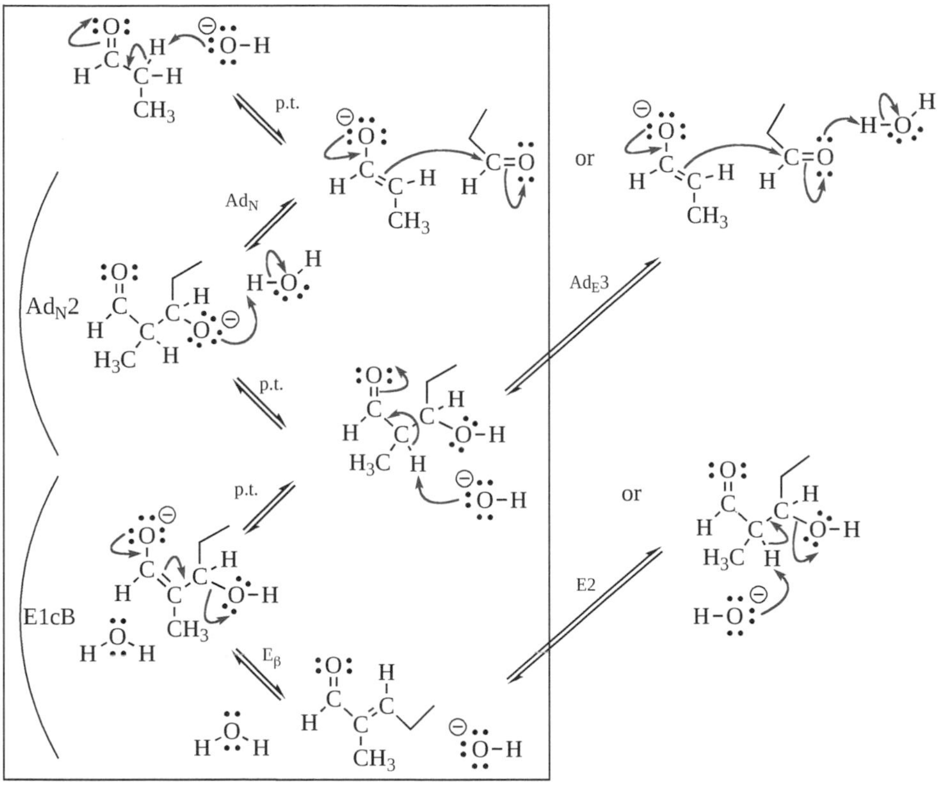

Figure 10.11 Aldol condensation overview. The most likely reaction path is shown boxed.

Surface overview: We can look back at the problem space that we just navigated to get an overview. The red arrows show the most reasonable route(s) on Figure 10.12 below. The addition and elimination energy surfaces can be used like big building blocks for creating this mechanism.

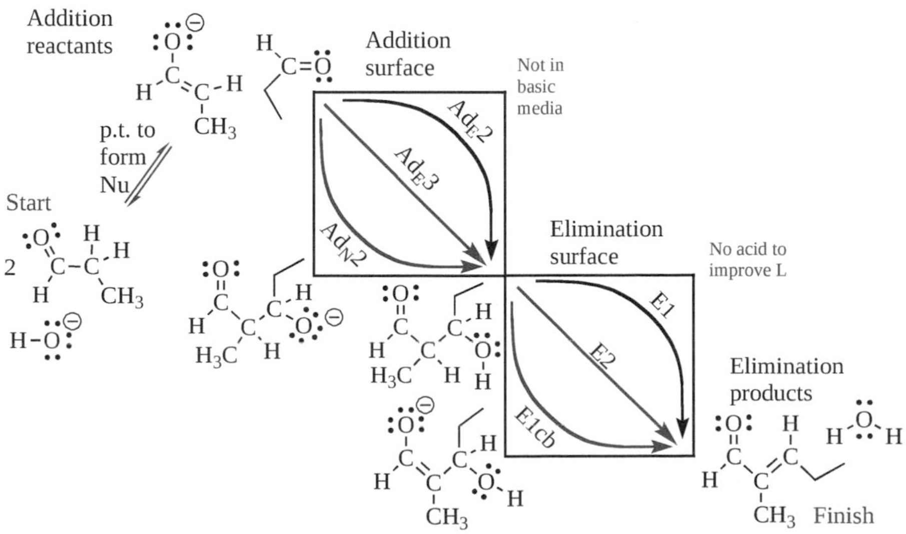

Figure 10.12 Final addition and elimination surfaces as a problem space map for example 10.5.3.

10.5.4 Electrophilic Aromatic Substitution

Give a reasonable mechanism for the following reaction.

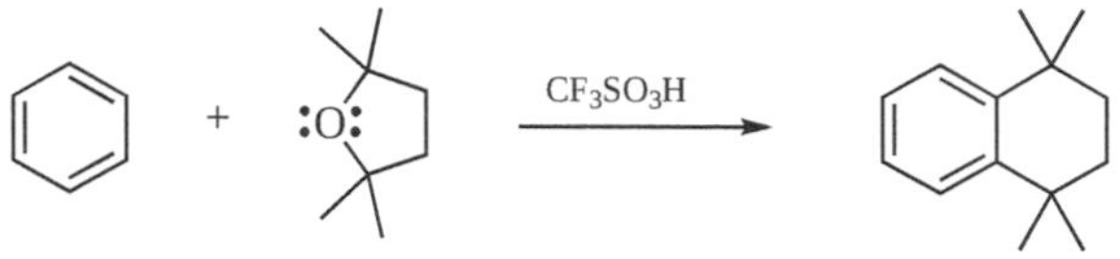

Understand the system

Balanced? No, water must be added to the right-hand side to balance the equation. Generic process? Benzene has substituted for the ether oxygen at both tertiary carbons. Two benzene CH bonds have been broken and have ended up on the water formed.
Medium? Highly acidic. Sources? The ether lone pairs are best with the aromatic ring a poor second. There are two sinks, HA and an sp^3 bound L. Acidic H's? Trifluoromethanesulfonic acid (triflic acid) is a very strong acid with an estimated pK_a of –15. Leaving groups? The ether oxygen, but only if protonated. Resonance forms? Only a benzene pi bond shift.

Every bond can be considered to be a combination of a nucleophile and an electrophile. Our best source is the ether lone pairs and our best sink is the triflic acid. Since the reaction is in very acidic media try protonation as a first step. In base, try deprotonation as a first step.

To the ether electrophile it is a substitution process, but to the benzene nucleophile, it is an electrophilic aromatic substitution that has two steps, electrophile addition then rearomatization. The problem space for the overall process must incorporate both the substitution and rearomatization and also the fact that it happens twice, Figure 10.13.

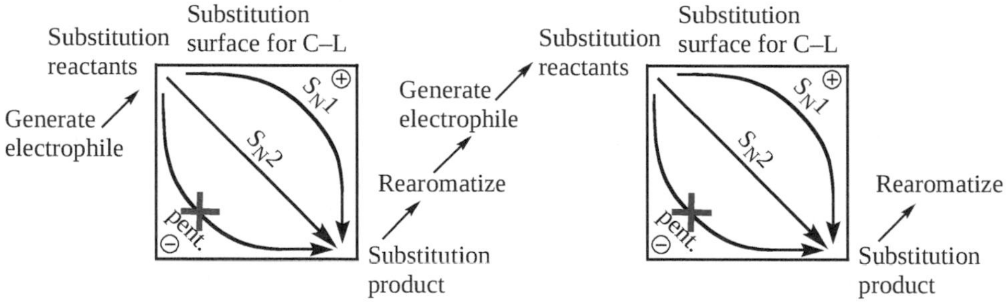

Figure 10.13 Anticipated problem space for electrophilic aromatic substitution example 10.5.4.

Find possible routes

There is one reasonable path that heads toward product, proton transfer to the lone pair.

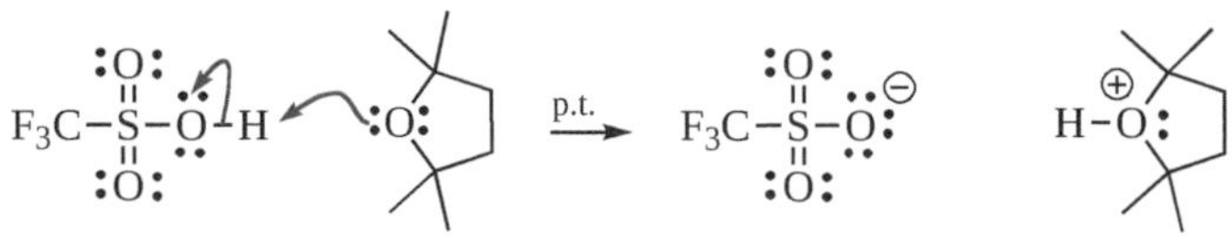

Evaluate and cross-check

The reaction is downhill in energy if the acid catalyst has a pK_a of less than –3.5, the pK_a of the protonated ether. With the triflic acid pK_a of –15, this proton transfer is downhill by about 11.5 pK_a units. Another reaction that can happen is the reversible protonation of benzene, but all that will do is exchange the protons on the ring on return.

Understand the system

The protonated ether is now a good leaving group. A beginner might be tempted to over optimize, to make a phenyl anion (pK_{abH} = 43) as a nucleophile and attack the electrophile, but anions that basic can't be formed in acidic media. Remember the pK_a span limit of 14 pK_a units. Powerful electrophiles and weak nucleophiles occur in acidic media, and the reverse is true in basic media. Our weak nucleophile is the aromatic ring.

Find possible routes

Of the three common routes on the substitution energy surface, Figure 4.2, only two substitution routes are energetically possible because the site of attack is a carbon atom (the pentacoordinate path is out).
The two alternatives are the S_N2,

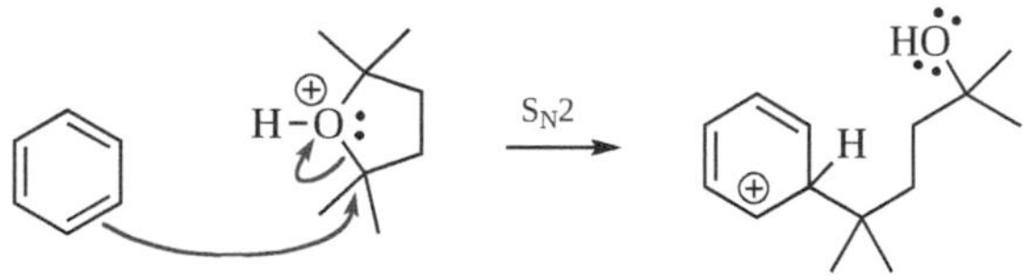

or the S_N1, path D_N followed by A_N (or from the point of the benzene, A_E).

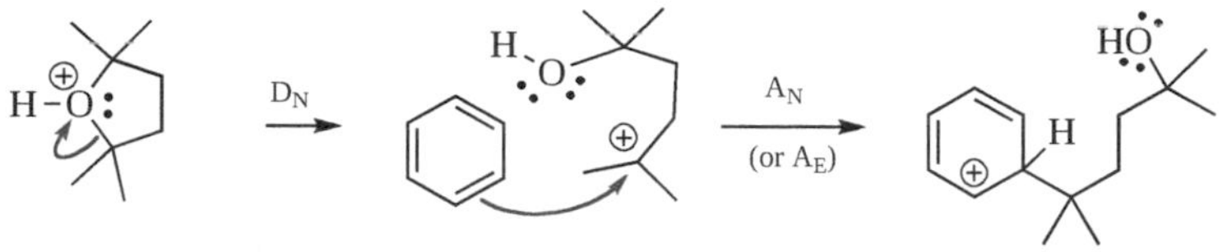

Evaluate and cross-check

The S_N2 has access problems, for the site of attack is very sterically hindered. The cutoff for a reasonable S_N2 is a secondary site, as discussed in Section 7.1.6, so this tertiary site is just too hindered to proceed by S_N2.

The S_N1 must pass three criteria for the D_N step to proceed. The solvent needs to be polar, which this acidic media is. The leaving group has to be very good, with a p$K_{aHL} \leq 0$; the protonated ether with a pK_{aHL} of –2.4 definitely qualifies. Finally, the carbocation must be reasonably stabilized, usually better than secondary, and this D_N would produce a tertiary carbocation, passing the last criteria. The S_N1 is very reasonable.

Understand the system

The aromatic ring has become a delocalized carbocation, but this sigma complex is no longer aromatic because of the newly generated tetrahedral center at the point of attack. Usually the next step in an electrophilic aromatic substitution is to restore the stability of the aromatic system. A common beginner mistake at this point is to remove the wrong H. This is especially true if line structure is used, which does not show the hydrogens. For this reason, always draw out the Lewis structure for any part of the molecule that interacts with an arrow. It is too easy to forget about hydrogens that are not shown.

Find possible routes

Restoring the aromaticity of the benzene ring requires a base to remove the proton at the site of electrophilic attack. The alcohol lone pair needs to get protonated to become a good leaving group if a second substitution is to occur. We can transfer this proton from the sigma complex to the alcohol either intermolecularly or intramolecularly.

The intermolecular route is:

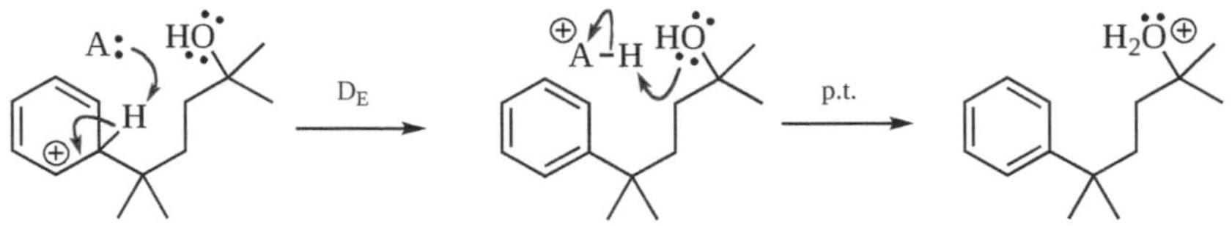

or the intramolecular route is:

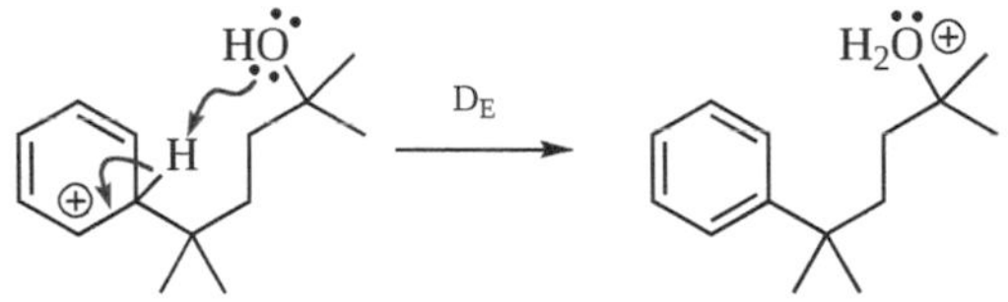

Evaluate and cross-check

Since the sigma complex is also a strong acid, any neutral species in solution is basic enough to deprotonate it. Although both routes are reasonable, the alcohol lone pair is within reach (having a seven-membered transition state) and reasonably basic for this media, with a pK_{abH} of about –,2.4. Protonation of the alcohol also serves to improve the leaving group for the next step.

Find possible routes

Of the three common routes on the substitution energy surface, Figures 4.2 or 10.13, only two substitution routes are energetically possible because the site of attack is a carbon atom (again, the pentacoordinate path is out). Although we suspect that the S_N2 is also out from the previous steps analysis, it will be considered here for completeness.
The two alternatives are the S_N2,

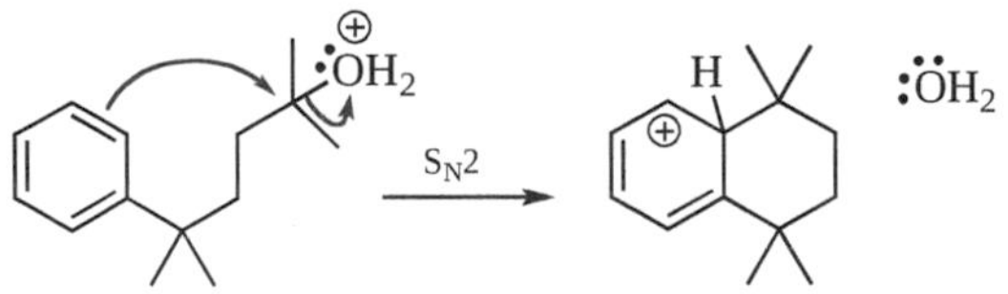

and the S_N1 (path D_N followed by A_N).

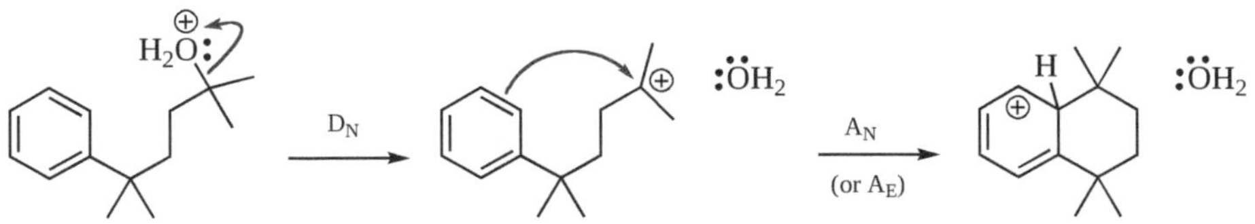

Evaluate and cross-check

Again, the S_N2 has access problems, for the site of attack is very sterically hindered, so this tertiary site is just too hindered to proceed by S_N2.

The S_N1 must pass three criteria for the D_N step to proceed. The solvent needs to be polar, which this acidic media is. The leaving group has to be very good, with a p$K_{aHL} \leq 0$; a water leaving group with a pK_{aHL} of 0.0 qualifies. Finally, the carbocation must be reasonably stabilized, usually better that secondary, and this D_N would produce a tertiary carbocation, passing the last criteria. The S_N1 is very reasonable.

Find possible routes

Restoring the aromaticity of the benzene ring gives the final product. The water we just created is basic enough, pK_{abH} of 0.0, to remove the very acidic proton at the site of electrophilic attack, thus finishing the last step of electrophilic aromatic substitution.

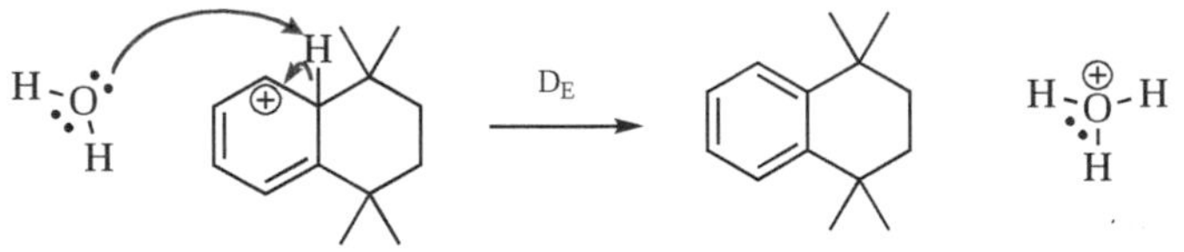

Overview

Now let's back off from the problem far enough to see the entire process as a whole (Figure 10.14). The mechanism for this transformation or "mechanistic sentence" can be made from our "mechanistic phrases" or path combinations.

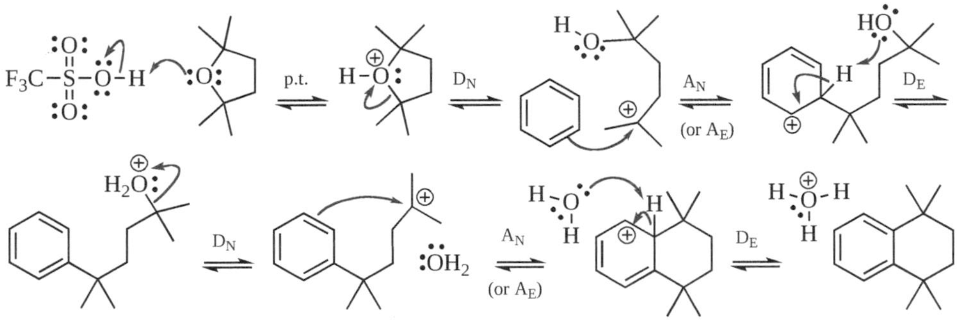

Figure 10.14 Electrophilic aromatic substitution example 10.5.4 overview.

A proton transfer from the acid catalyst improves the leaving group, then an S_N1 substitution followed by a D_E rearomatization step that sets the alcohol up for a second S_N1 substitution, ending with a second D_E rearomatization step. The only alternative route is to use solvent to shuttle the proton from the sigma complex to the alcohol.

Surface overview: We can look back at the problem space that we just navigated to get an overview. The red arrows show the most reasonable route(s) on Figure 10.15.

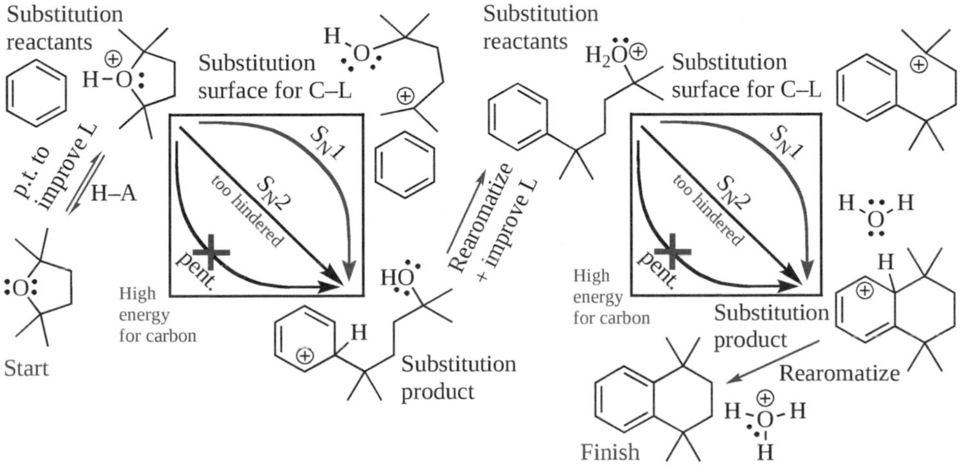

Figure 10.15 Final substitution surfaces serving as a problem space map for example 10.5.4.

10.5.5 Glucose to Fructose

Give a reasonable mechanism for the following biochemically important reaction. Glucose is your blood sugar, and the hormone insulin carefully controls its concentration in the blood. When glucose is brought into your cells, it must be converted to fructose to be metabolized. The entire metabolic process turns glucose into carbon dioxide and water and in the process produces ATP, adenosine triphosphate, which is your cell's energy currency. The glucose to fructose isomerization reaction can be done in the lab with dilute aqueous base.

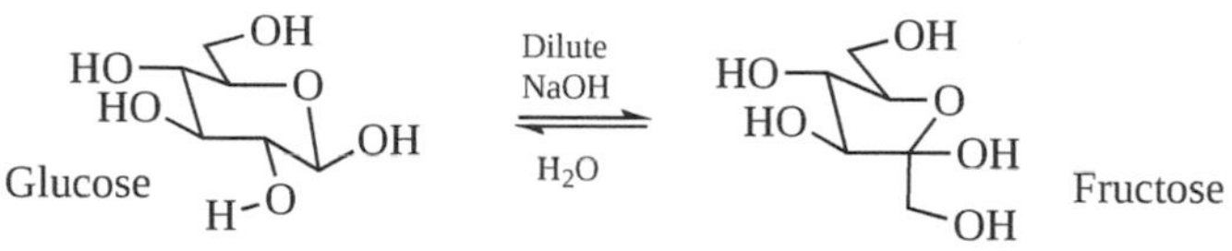

Understand the system

Because the reactant and the product are structural isomers of one another, all pieces are accounted for in this balanced reaction. No overall addition of a nucleophile has occurred (there is nothing new attached), so we assume that hydroxide serves as a base. The carbons are now numbered to aid in the reaction analysis.

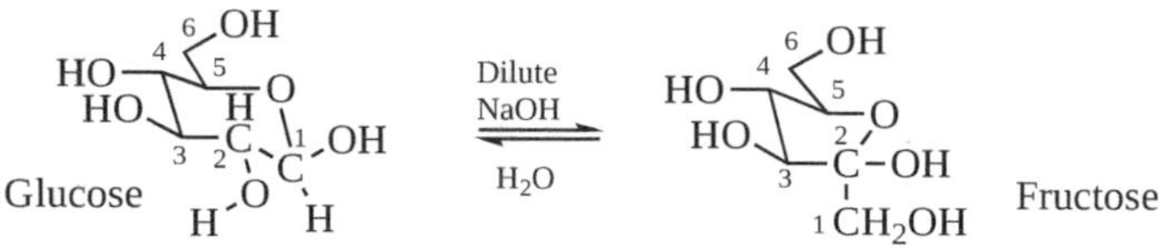

It appears nothing has happened at carbons 3, 4, 5, and 6, so we leave those alone. The C–O bond to carbon 1 has been broken and a new C–O bond made to carbon 2. However there is more; the line structure shorthand partially disguised that a C–H bond on carbon 2 has been broken and a new C–H bond made on carbon 1. The functional group at carbon 1 of the starting glucose is a hemiacetal, and the functional group at carbon 2 of the product fructose is a hemiketal.

The novice might be tempted to get this mechanism over with quickly by swapping C–O and C–H bonds between carbons 1 and 2. Even though it can be drawn with arrows, such a concerted 1, 2 swap has never been observed. It is not one of the established electron flow pathways. We may not be able to visualize what a detailed problem space might look like, so we will have to trust where the known electron flow pathways might lead us.

Overall, we start with basic elimination reactions that will break our starting hemiacetal. We know at the end of the mechanism will be an addition reaction that forms the product hemiketal. What happens in between is uncertain at this point, but we have a rough idea of where to start and where to end on our problem space, Figure 10.16.

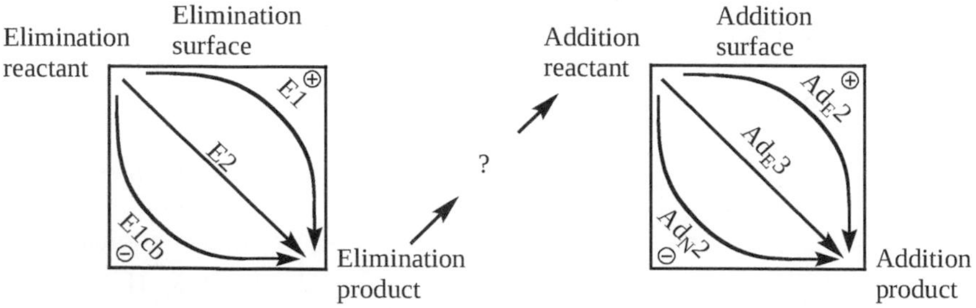

Figure 10.16 Anticipated addition and elimination surfaces serving as a glucose to fructose partial problem space map.

Find possible routes

First try proton transfer. All the hydroxyls are within a few pK_a units of the hydroxide base, so all are available for proton transfer. Only carbon one bears both a hydroxyl and anything that can serve as a leaving group. All the other hydroxyls can only deprotonate then reprotonate. Our two basic media elimination routes are the E1cB,

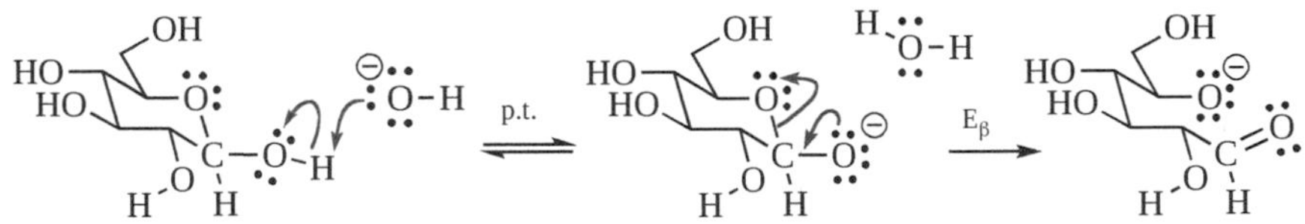

and the E2.

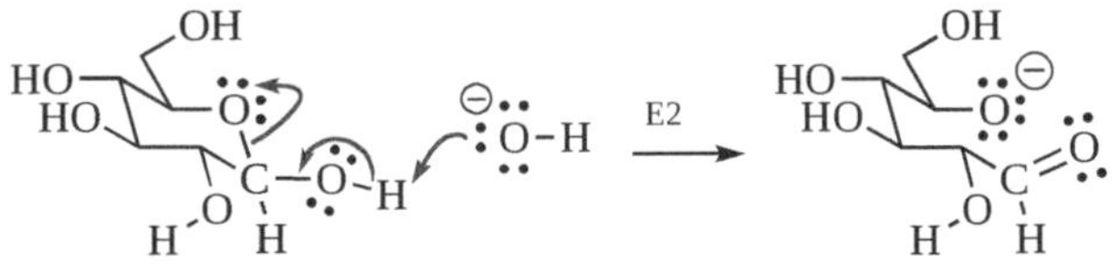

Evaluate and cross-check

We ruled out the E1 because the leaving group is not good enough at a pK_{aHL} of about 14 (the pK_a of $HOCH_2CH_2OH$ is 14.2), and the basic media has no acids strong enough to improve the leaving group by protonation. The E1cB path combination (path p.t. followed by path E$_\beta$) accomplishes one goal of the transformation; it breaks the C–O bond to carbon 1. The product alkoxide is in equilibrium with the alcohol form in this medium. The literature value for the pK_a of glucose is 12.3. The alkoxide at this pK_{abH} has kicked out a leaving group at about pK_{aHL} 14, which checks with the ΔpK_a rule cross-check.

The E2 passes the same ΔpK_a rule cross-check since the hydroxide base at pK_{abH} of 14.0 has kicked out a leaving group of pK_{abH} of 14, a similar base. Since the OH bond can freely rotate, the E2 coplanar alignment restriction is not a problem. Both the E1cB and E2 eliminations are reasonable.

Understand the system

The newly formed aldehyde is an electron-withdrawing group that makes the proton on carbon 2 acidic; deprotonation of the hydrogen on carbon 2 accomplishes the second goal of the transformation.

Find possible routes

Although hydrate formation by attack of hydroxide on the aldehyde is fast and reversible, it is also a dead end; a Δ$H°$ calculation shows us that the hydrate is uphill from the aldehyde. The alkoxide just kicked out could attack the aldehyde to return to starting material. The alkoxide is also a base and can remove the proton next to the aldehyde.

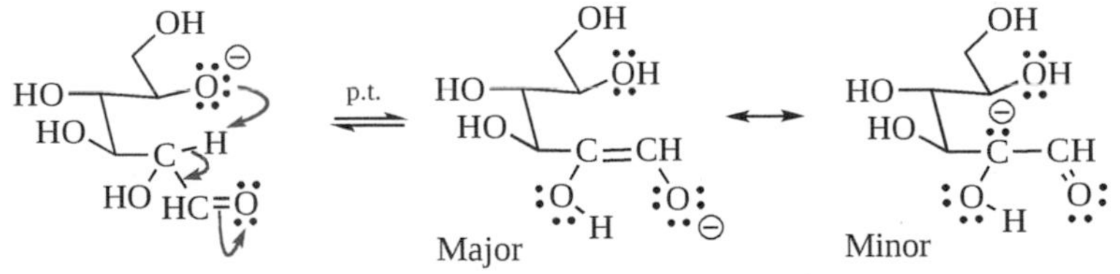

Evaluate and cross-check

The proton-transfer step is the only unexplored new route, and it removes the CH on carbon 2, which is one of our goals. It is intramolecular and therefore expected to be fast. The alkoxide, whose pK_{abH} is 14, has removed a proton adjacent to an aldehyde at the same pK_a or perhaps slightly higher, forming an enediolate. We could also have transferred this proton from carbon 2 to the alkoxide via the solvent, but this intramolecular proton transfer is expected to be faster because it has an unhindered six-membered transition state.

Understand the system

The newly formed enediolate intermediate has two resonance forms. It is internally hydrogen bonded. The enediolate is halfway along; two out of four goals have been achieved. Now another H is needed on carbon 1 and a C–O bond needs to be made to carbon 2. Because there is no electrophilic site at carbon 2, the C–O bond cannot yet be made.

Find possible routes

The enediolate just formed can easily undergo reversible proton transfer with the solvent to form the neutral enediol, but that does not get us closer to the product. Even though this is a good process, it is a dead end, so we should explore other routes that carry the process toward product.

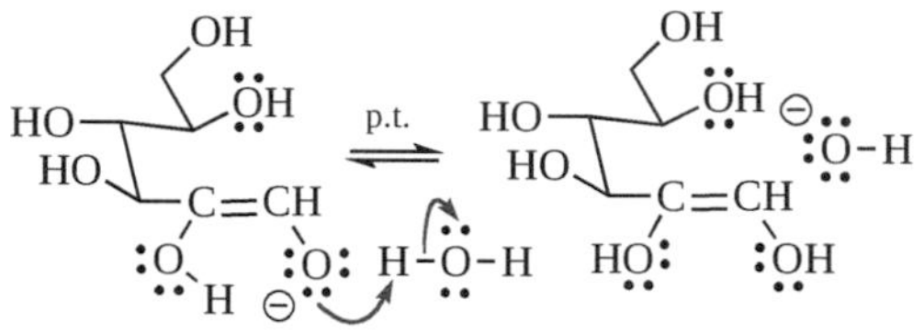

The internally hydrogen-bonded enediolate intermediate undergoes rapid intramolecular proton transfer. This proton transfer could also be mediated by water.

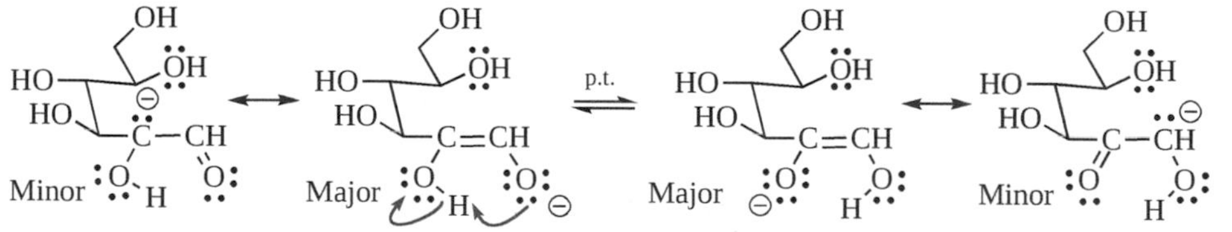

Evaluate and cross-check

Since the transition state is a five-membered ring, this intramolecular process is expected to be exceedingly fast. This proton-transfer step is between two groups with the about same pK_a for a proton-transfer K_{eq} close to 1.

Understand the system

The minor resonance form should provide a hint toward product. Proton transfer to this anion not only puts the H where it is needed on carbon 1 but also generates a site for nucleophilic attack on carbon 2.

Find possible routes

The proton transfer can occur via solvent, but an intramolecular proton transfer can occur that is probably faster and sets the molecule up for the ring-closure addition reaction.

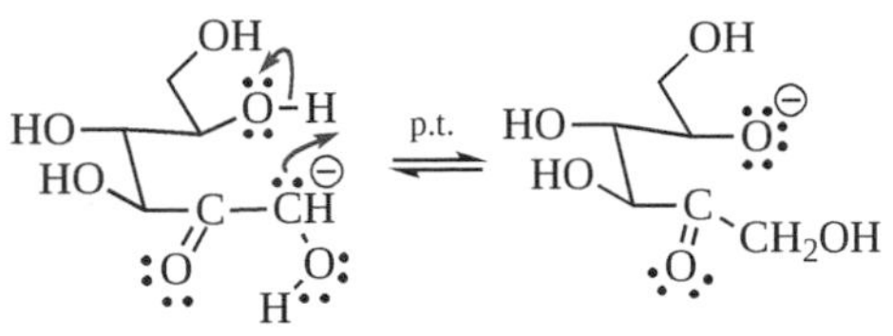

Evaluate and cross-check

Since the transition state is a seven-membered ring, this intramolecular process is expected to be reasonably fast. This proton-transfer step is between two groups with about the same pK_a for a proton-transfer K_{eq} of approximately 1. The water solvent can also mediate this proton-transfer step. The water that is hydrogen bonded to the anion can be deprotonated, then in turn deprotonate the alcohol.

Understand the system

The carbonyl is a new site for nucleophilic attack; the proton transfer has generated a good nucleophile.

Find possible routes

The C–O bond can now be made because we have a good nucleophile, a good sink, and known routes for the addition reaction,
the Ad_N2,

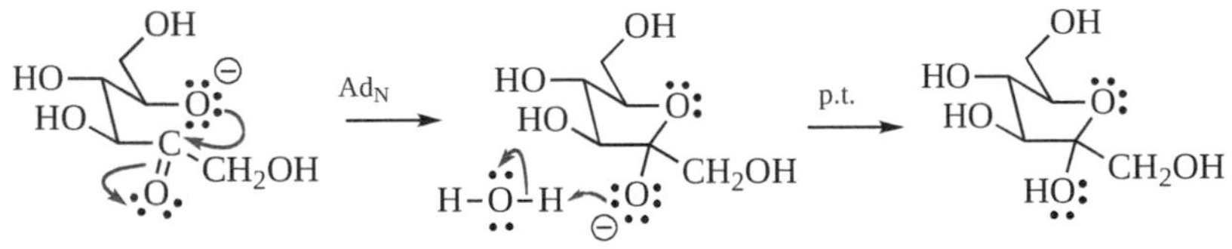

or the hetero AdE_3.

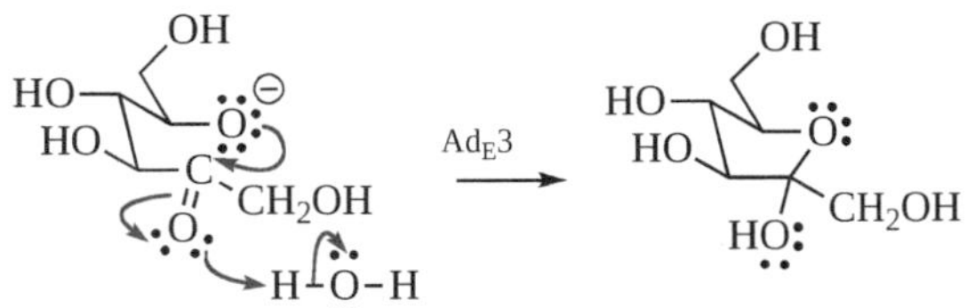

Evaluate and cross-check

The Ad_N2 addition of the alkoxide at pK_{abH} 14 has created a hemiketal alkoxide with a pK_{abH} at about the same value, which passes the ΔpK_a rule cross-check. At this point, we can see that protonation of the product yields fructose, with a reasonable K_{eq}. The alternate addition, the Ad_E3 path, passes the same ΔpK_a rule cross-check, so is also acceptable. The Ad_E2 is unlikely because there is not a strong enough acid present to protonate the carbonyl, pK_{abH} at -7.

Overview

Overall, the mechanism for this transformation was an E1cB (or E2) elimination, three proton transfers that move the carbonyl from carbon 1 to carbon 2 via the enediolate, and an Ad_N2 (or Ad_E3) addition (Figure 10.17).

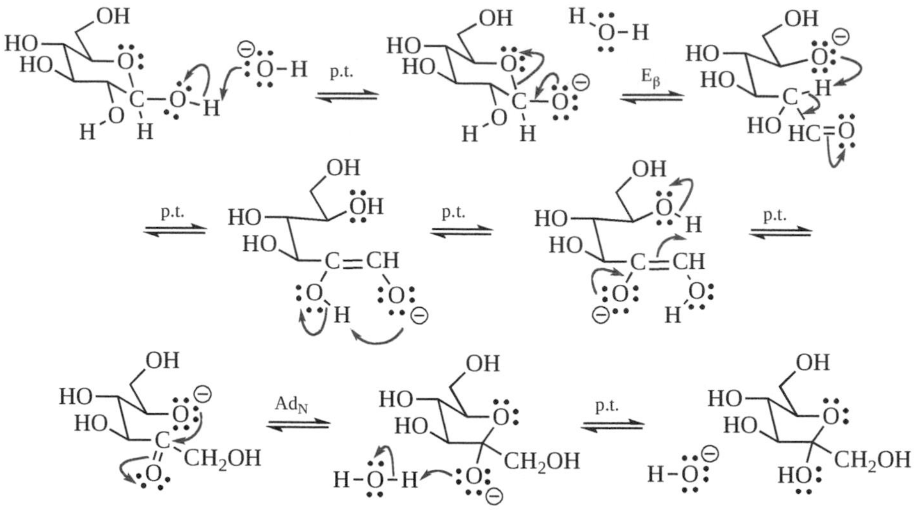

Figure 10.17 Glucose to fructose overview: E1cB top line, several p.t. steps, and Ad_N2 bottom line.

Surface overview: We can look back at the problem space that we just navigated to get an overview. The red arrows show the most reasonable route(s) on Figure 10.18. This map may seem complex compared to Figure 10.17 because of the many alternatives shown, but it allows you to see possible viable alternative routes.

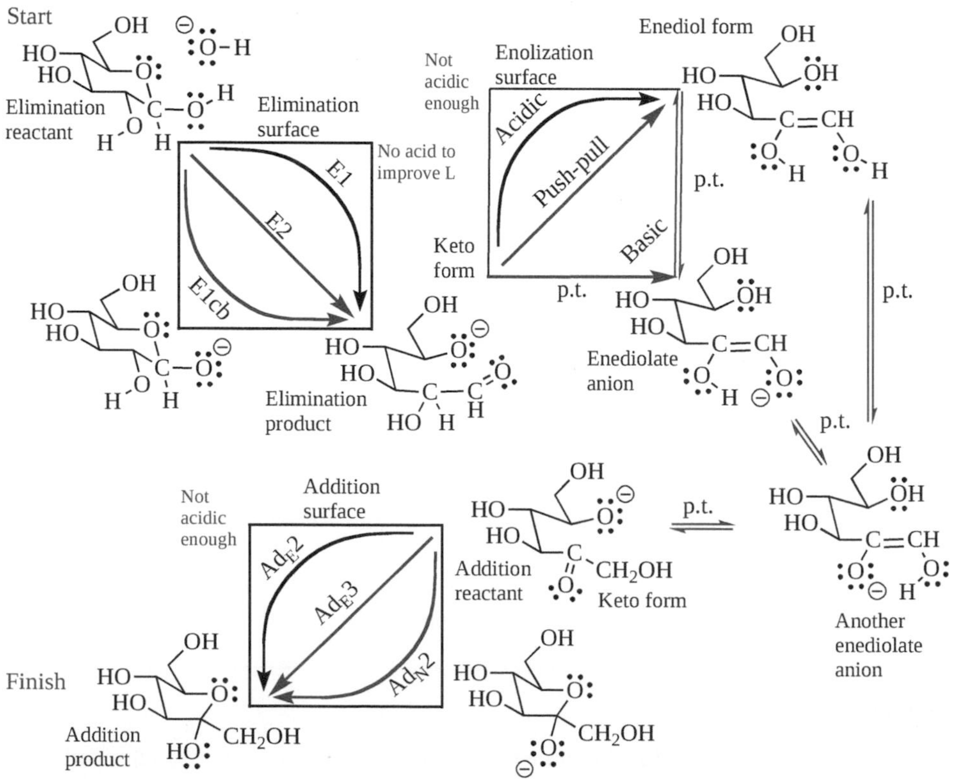

Figure 10.18 Final elimination, addition, and enolization surfaces serving as a problem space map for example 10.5.5.

10.6 PRODUCT PREDICTION STRATEGIES

10.6.1 Understand the System

Since structure determines reactivity, how are compounds in the mixture expected to behave? To understand this, we need to classify the reactants into generic sources and sinks, and then find the most reactive source and sink (Chapters 5 and 6). Being able to go from the specific to the general is an essential skill of a scientist. Ask yourself, "Have I seen anything like this?" Generally, how does a compound like this usually react? Often a first step is the reaction of an acid or base, so look for any reasonable proton transfer. Remember to consider both the original species and the product of any proton transfer when you go to classify species into the generic classes of sources and sinks. Look for leaving groups, polarized single bonds, and polarized multiple bonds. Once we have the best source and sink, we can go looking for an electron flow pathway to combine them.

10.6.2 Find Possible Routes

Next using the electron sink as a guide, find the few electron flow pathways that fit those sources and sinks (Chapter 7). Look at restrictions to limit the choice of pathways, such as acidic vs. basic media. Narrow down the choice of pathways further by using steric, electronic, and solvent limitations if available. Combine the best source with the best sink through an appropriate pathway to yield a preliminary product (Chapter 8). Only about half of all the possible sources and sinks are common, and those common combinations were outlined in the correlation matrix of Table 8.1 and discussed in the sections that followed. Be aware of alternatives like competing paths (Chapter 4) and decisions like substitution vs. elimination (Chapter 9).

10.6.3 Evaluate and Cross-Check

The difficult part is to decide whether or not the preliminary product is really the final product or just an intermediate. The ΔpK_a rule cross-check can be helpful to determine whether the process has climbed in energy or not. If you have formed an unstable anion or cation, of course it is just an intermediate. If there remains an easily lost leaving group or another reactive electron sink, look for a good source to react with it. You are following your energetics intuition downhill to a reasonable product. Basically look for any good process that can occur, if none are found, stop before you do anything rash.

Lastly, have faith in the predictive process and allow it to carry you to the answer. Do not try to shorten the process to get to the answer by combining or skipping steps. Don't panic and force an answer because you can't see the answer from the beginning.

10.7 WORKED PRODUCT PREDICTION EXAMPLES

In these examples, if several routes go to the same compound, for simplicity only the more reasonable will be shown. Try to work each step, before looking at the answer.

10.7.1 Predict the Product of Reaction of Ethyl Acetoacetate, Ethoxide, and 1-Bromobutane

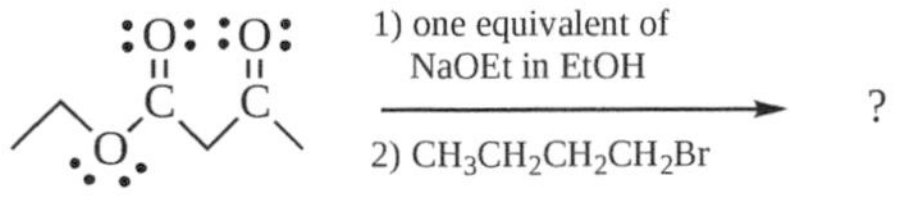

Understand the system

The first step is in basic media; ethoxide has a pK_{abH} of 16. Sources? The ethoxide anion. Leaving groups? Only ethoxide on the ester. Sinks? The ketone is a polarized multiple bond, and the ester is a polarized multiple bond with a leaving group. Acidic H's? The CH_2 between the two carbonyls is the most acidic, pK_a 10.7; the methyl on the carbonyl is the next most acidic, pK_a 19. Resonance forms help to understand the polarization of the reactants.

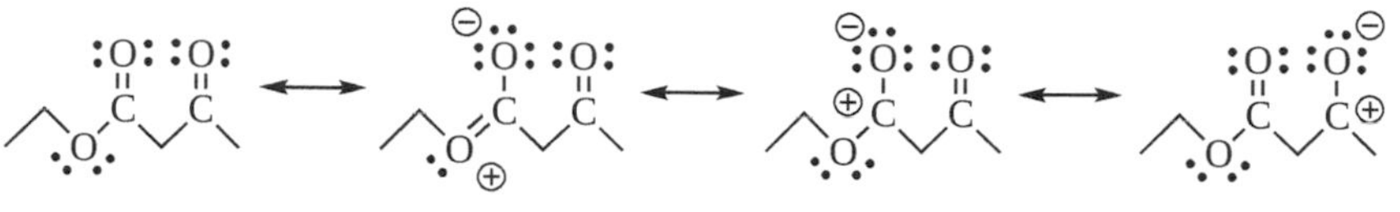

Find possible routes

There are two possible electron source alternatives, ethoxide can serve as a nucleophile or as a base, and there are two electron sink possibilities with each, giving four routes to consider.

Ethoxide as Nu on the ketone, path Ad_N:

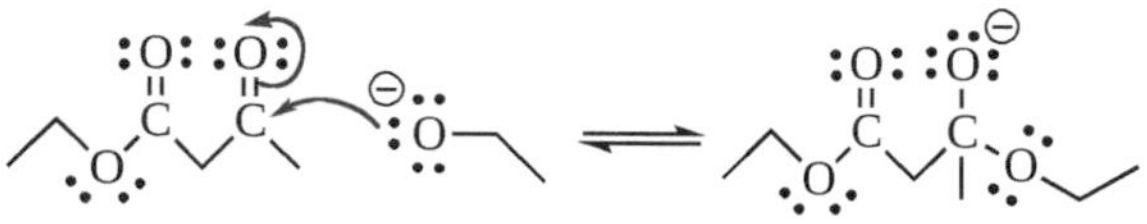

Ethoxide as Nu on the ester, path Ad_N:

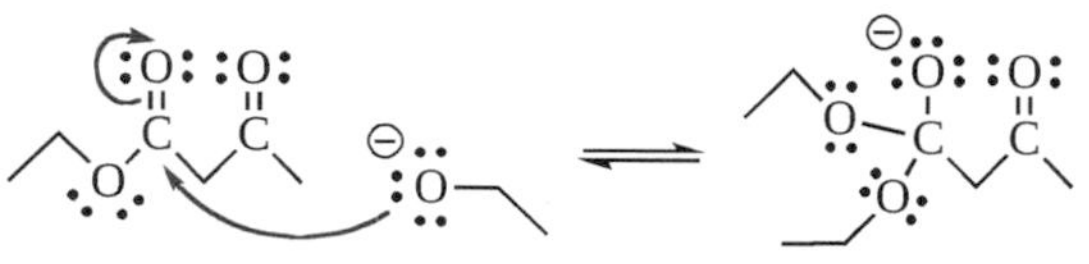

Ethoxide as base on the methyl, path p.t.:

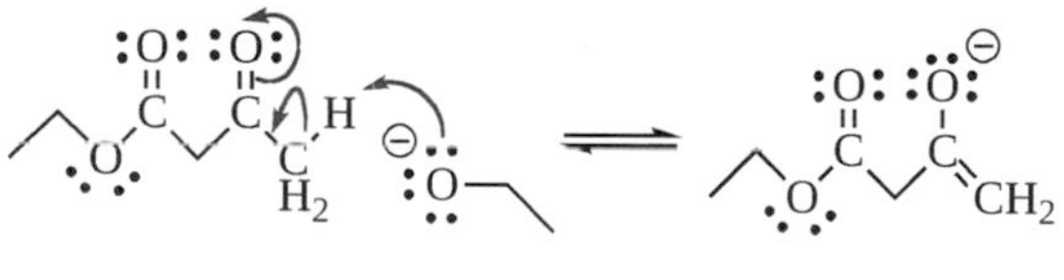

Ethoxide as base on the methylene between the carbonyls, path p.t.:

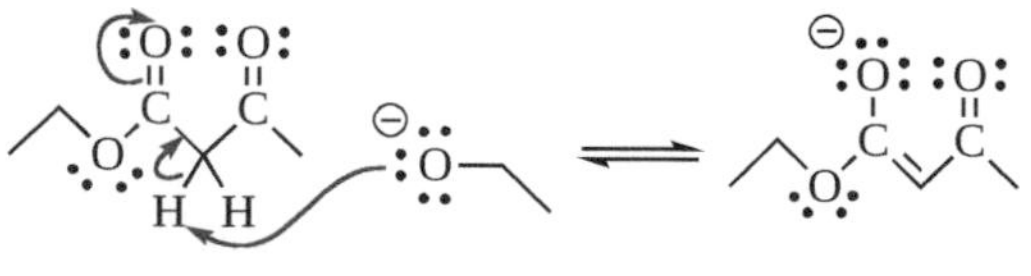

Evaluate and cross-check

The reversible addition of the ethoxide to the ketone produces a hemiketal anion that is uphill in energy because a carbonyl is a stronger bond than two C–O single bonds. The similar reversible addition of the ethoxide to the ester produces a tetrahedral intermediate anion that although uphill in energy can kick out ethoxide leaving group. If the ester were not an ethyl ester, this process would result in a noticeable exchange (transesterification). As the leaving group is the same as the nucleophile, the process proceeds without change.

Ethoxide as a base on the methyl reversibly produces a product anion of pK_{abH} = 19.2, from a base of pK_{abH} = 16, slightly uphill but within range. Ethoxide as a base on the methylene between the carbonyls produces a product anion of pK_{abH} = 10.7, from a base of pK_{abH} = 16, definitely energetically favorable (more favorable than deprotonation of the methyl by 8.5 pK_a units). Since this favorable proton transfer making the ethyl acetoacetate enolate anion appears to be the best process, we carry this allylic source on as our first choice into step two.

Understand the system

Medium? Still basic. Sources? The ethyl acetoacetate enolate anion has a pK_{abH} of 10.7 and is an allylic source. Leaving groups? Bromide is a reactive sp^3 bound leaving group. Resonance forms help us understand the polarization of our reactants.

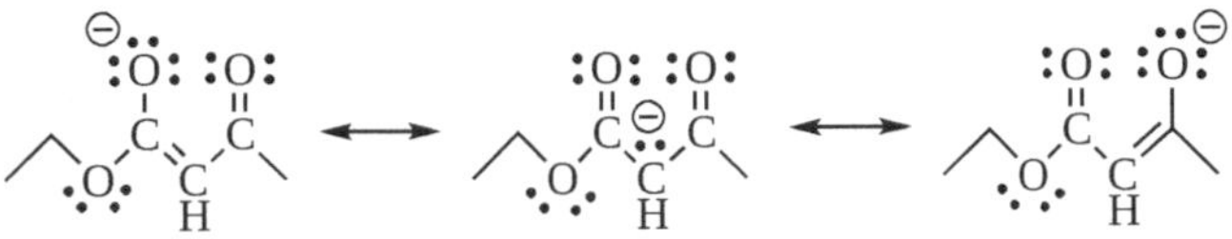

Find possible routes

In the interest of trimming the search tree, we can use what is known about the sources and sinks to limit the alternatives to the more reasonable choices.

Any of the three possible anionic atoms in the resonance forms above could serve as our source, but usually allylic sources attack on carbon because it is a softer, better nucleophile, and the product preserves the strong carbonyl double bond.

Our electron sink also presents two alternative routes, substitution or elimination. With a pK_{abH} of 10.7 and a soft carbon anion, this is a good nucleophile, and a moderate base. Our 1-bromobutane presents a primary unhindered site and the ethyl acetoacetate enolate anion is a good nucleophile and not a very strong base, so the decision is clearly in the primary substitution quadrant, Section 9.5. There are two possible substitutions, S_N1:

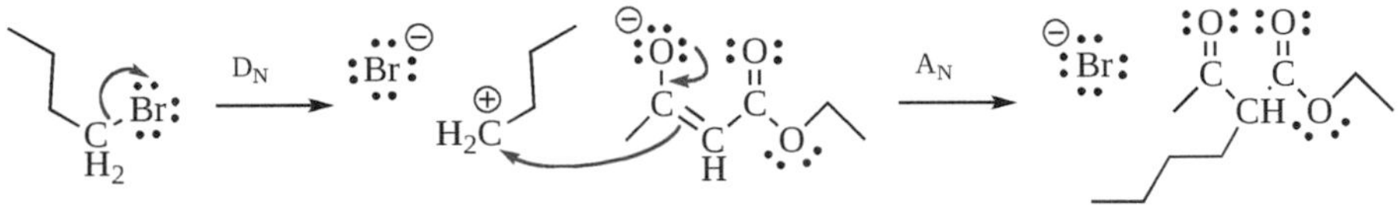

and S_N2:

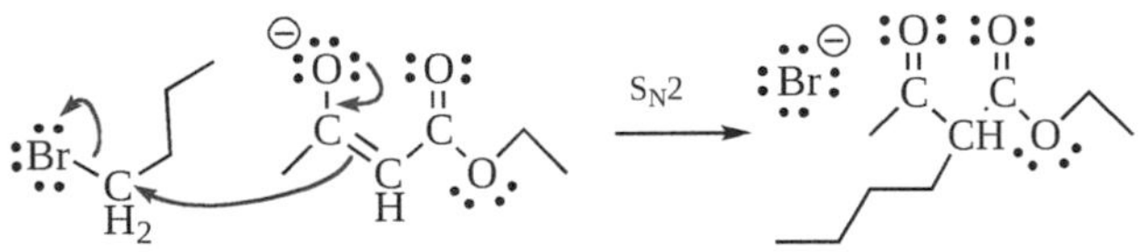

Evaluate and cross-check

Of the two substitutions, the S_N2 is predicted to be the lowest energy path because the primary site has good access, the leaving group is good, and the enolate is a good nucleophile. A possible S_N1 is ruled out because the resultant primary carbocation would not be stable at all. Try not to invoke primary carbocations in any of your mechanisms.

Overview

We started with an ethoxide anion of a pK_{abH} of 16 and now have bromide anion of pK_{abH} −9; the reaction is definitely downhill in energy and irreversible by the ΔpK_a rule cross-check. We have used up the single equivalent of ethoxide base and turned it into a poorly basic bromide anion. All the paths from this point are all uphill; it is a good time to stop. Figure 10.19 gives an overview of the problem space with all the alternative routes on the right. The reaction path is shown vertically on the left, and all the side routes that were considered are shown on the right. One side route not discussed above is reversible protonation to the enol below.

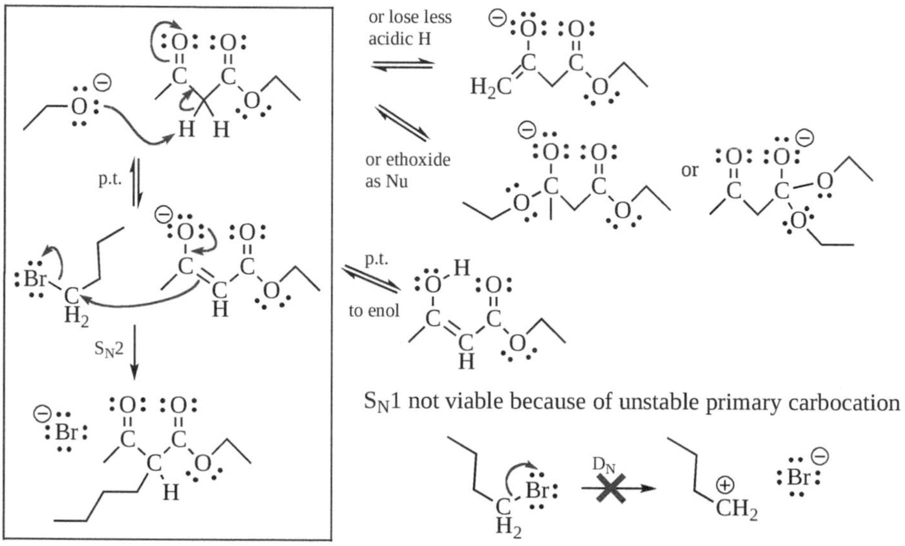

Figure 10.19 Ethyl acetoacetate, ethoxide, and 1-bromobutane overview for example 10.7.1. The most likely reaction path is boxed.

Surface overview: We can look back at the problem space that we just navigated to get an overview. The red arrows show the most reasonable route(s) on Figure 10.20 below.

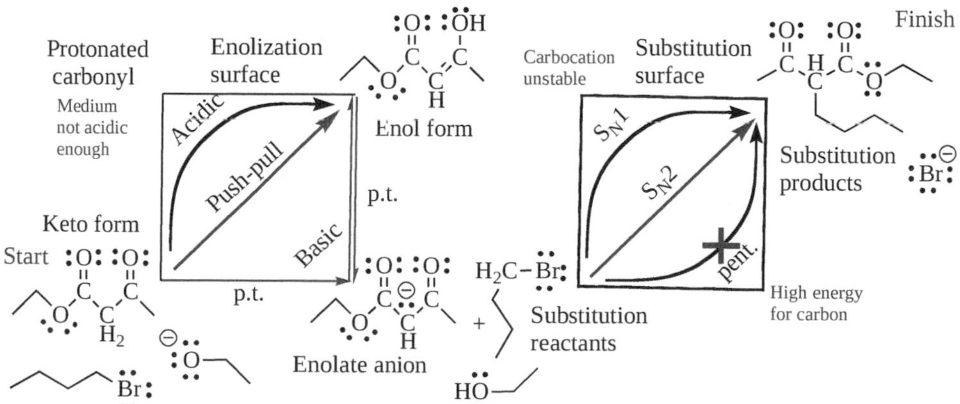

Figure 10.20 Final enolization and substitution surfaces serving as a problem space map for example 10.7.1. Reversible formation of the enol, a poorer Nu than the enolate anion, also occurs as shown in the ethanol solvent.

10.7.2 Predict the Product of Reaction of an Ester and Ethoxide

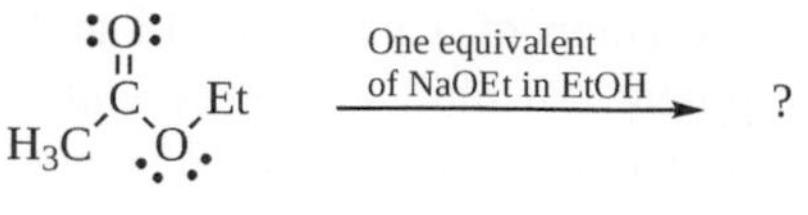

Understand the system

This reaction and its reverse are similar to important biochemical steps in the synthesis and metabolism of fatty acids. Medium? Basic, ethoxide's pK_{abH} is 16. Sources? Ethoxide. Sinks? The ester is a polarized multiple bond with a leaving group. Acidic H's? The methyl next to the ester carbonyl has a pK_a 25.6. Leaving groups? Ethoxide from the ester (poor leaving group). Resonance forms?

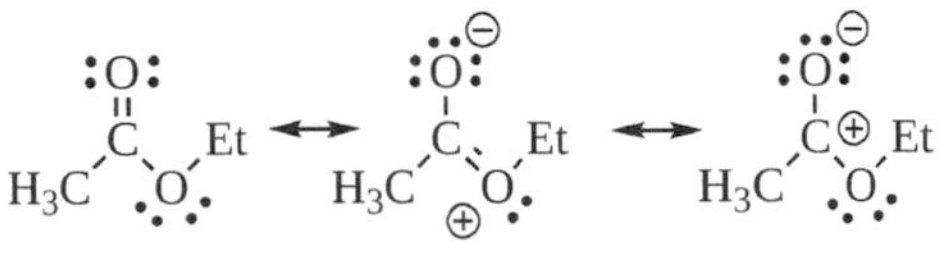

Find possible routes

Three routes are possible: Ad_N2 addition to a polarized multiple bond (Ad_N + p.t.),

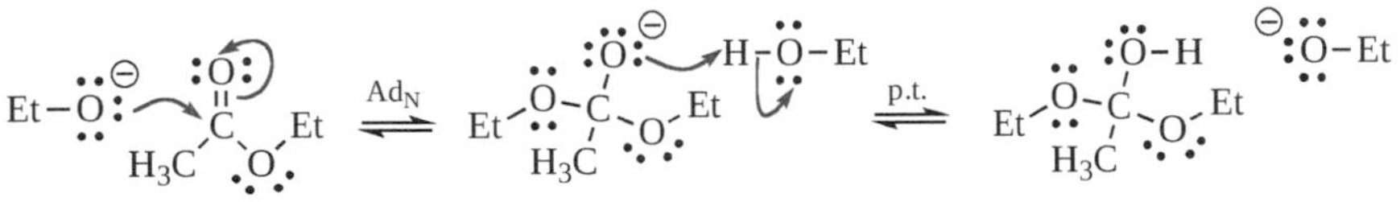

addition–elimination (Ad_N + E_β)

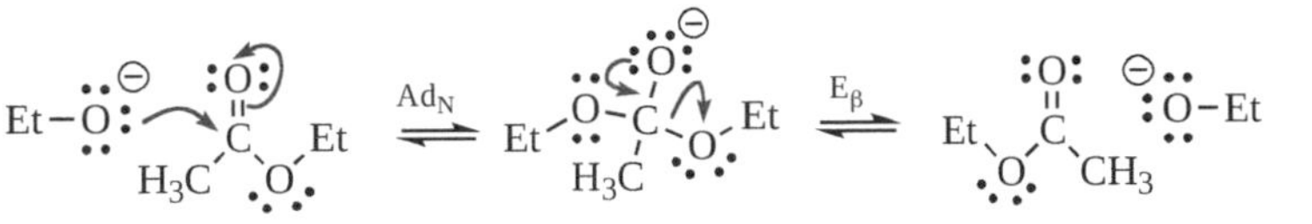

or deprotonation to form an anion.

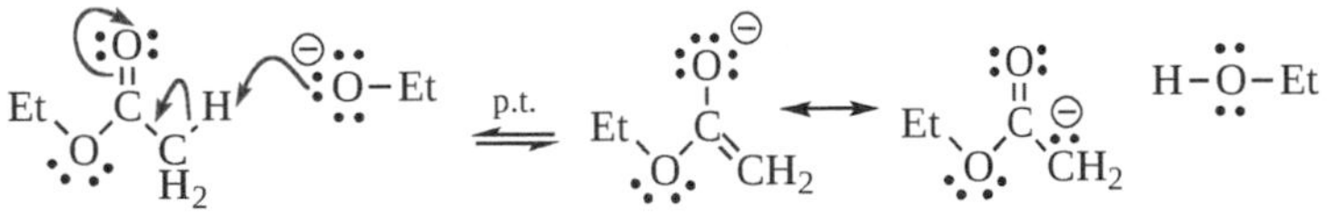

Evaluate and cross-check

Since the K_{eq} for proton transfer between the ester (pK_a = 25.6) and ethoxide (pK_{abH} = 16) is $10^{(16\ -\ 25.6)} = 10^{-9.6}$ and therefore almost out of the useful range, it is reasonable to explore ethoxide as a nucleophile first.

The Ad_N2 addition creates a hemiorthoester that is just an ester version of a hemiketal. The tetrahedral intermediate is too hindered to serve as a nucleophile, and if it is protonated it yields a species that is uphill in energy from the starting ester. A $\Delta H°$ calculation shows that a stronger carbonyl double bond (177 kcal/mol) has been replaced with two C–O single bonds (2 x 86 = 172 kcal/mol), so this reaction is expected to be uphill by about 5 kcal/mol. Since this equilibrium will favor reactants, we should explore further.

Lone pair nucleophiles usually react with carboxyl derivatives by addition via path Ad_N, followed by beta elimination of the leaving group from the tetrahedral intermediate anion formed, path E_β. Because the leaving group is ethoxide, the same as the nucleophile, loss of either ethoxide returns us to the starting materials. This process is noticeable only if the nucleophile and leaving group are different (transesterification). The basic medium nucleophilic paths all return to the starting ester with this compound.

After exploring the more probable processes and ending up back at the start, we should now explore the energetically less favorable alternative, ethoxide as base even though the K_{eq} is so unfavorable.

Understand the system

The ester enolate is an allylic source that can serve as a base or a nucleophile. The original ester has acidic hydrogens within range of the enolate. The ester carbonyl is an electrophilic sink (a polarized multiple bond with attached leaving group).

Find possible routes

Proton transfer easily occurs to the ester enolate on either carbon or oxygen. Protonation on carbon just exchanges hydrogens and returns us to reactants. The two new choices are that the enolate can act as a base to form the enol of the ester,

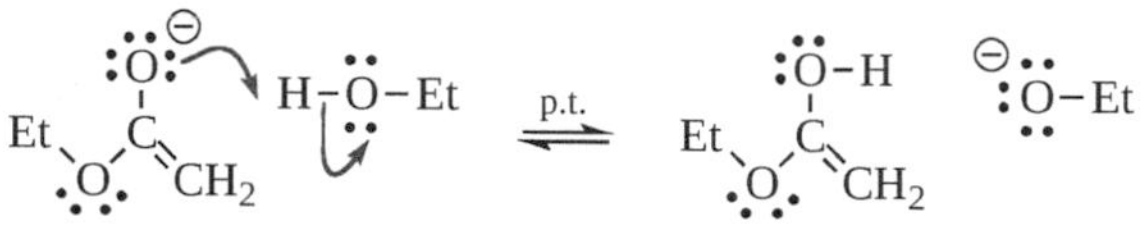

or the enolate can act as a nucleophile to form a new product by path Ad_N.

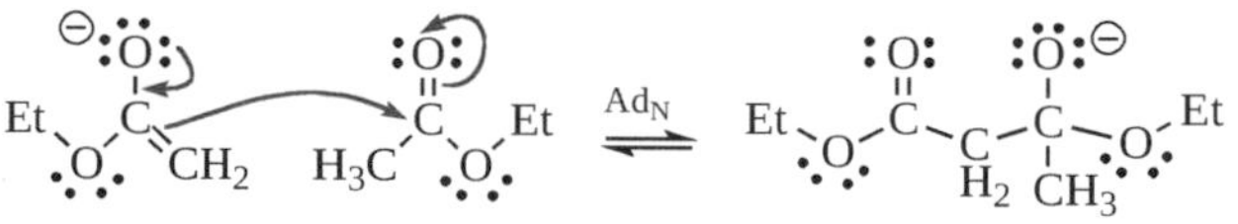

Evaluate and cross-check

The enol of the ester is uphill in energy from the ester since the C=C is much weaker than the C=O bond, and is less basic and less nucleophilic than the enolate. The enolate Ad_N path did something new and is downhill because it forms a weaker base; the ester enolate pK_{abH} is 25.6; that of the tetrahedral intermediate is about 14.

Understand the system

The tetrahedral intermediate is basic (pK_{abH} = 14) and has two possible leaving groups, the enolate (pK_{aHL} = 25.6) or ethoxide (pK_{aHL} = 16), attached to the carbon adjacent to the oxyanion. Methyl (pK_{aHL} = 48) is not a leaving group.

Find possible routes

The tetrahedral intermediate can lose the enolate and return to reactants. The tetrahedral intermediate is too hindered to serve as a good nucleophile. Two unexplored paths are available. The tetrahedral intermediate can serve as a base (path p.t.),

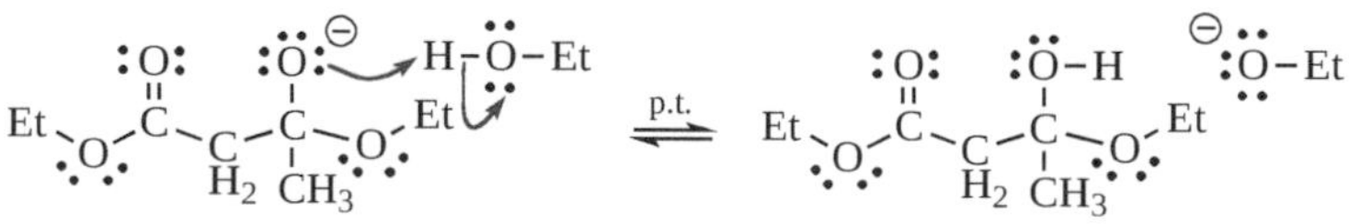

or it can lose the best leaving group it has, ethoxide (path E_β).

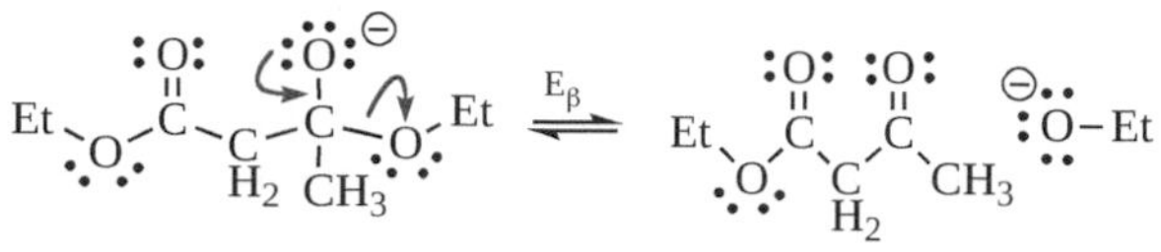

Evaluate and cross-check

Protonation gives the hemiketal, which by a $\Delta H°$ calculation is energetically uphill. The hemiketal will simply return to reactants. The loss of the ester enolate would take us back to the previous structure. The leaving group trends tell us that ethoxide is the best leaving group and would be lost instead of the ester enolate. The loss of ethoxide has produced a slightly stronger base by about 2 units, well within the ΔpK_a rule cross-check.

A $\Delta H°$ calculation is possible at this point since we now have a neutral species; two esters have formed the ketoester and ethanol. We have broken a C–H and a C–O and in return formed a C–C and an O–H for (99 + 86) – (83 + 111) = –9 kcal/mol (–38 kJ/mol). This tells us that the reaction has probably gone downhill in energy, but we need to explore further to see if there is anything farther downhill.

Understand the system

There are two sites for nucleophilic attack, the ketone and the ester carbonyl; the ketone carbonyl is the more reactive. The methylene between the two carbonyls is acidic (pK_a = 10.7). The ester enolate and ethoxide are both basic and nucleophilic.

Find possible routes

The ketone carbonyl can be attacked by ethoxide to give the previous tetrahedral intermediate. If the ketone carbonyl is attacked by the ester enolate, the intermediate formed is expected to reverse (Multiple additions, Section 9.2). The ketoester can be deprotonated by either ethoxide or the ester enolate.

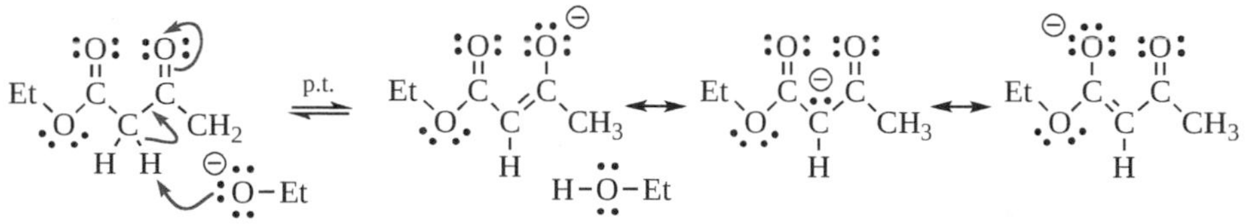

Evaluate and cross-check

This deprotonation has a very favorable K_{eq} of $10^{+5.3}$ because it produces a stable anion (pK_{abH} is 10.7) stabilized by two electron-withdrawing groups. Because it is so stabilized, this anion is a weak base and is expected to be energetically downhill from the original starting reagents. If this highly delocalized anion acted as a nucleophile and attacked another ester, it would just preferentially fall off again because it is a better leaving group than ethoxide. Therefore an exploration of the surface has produced a stable anion downhill from the reagents that appears not to be able to do anything more under the reaction conditions. This stable anion is the product of the reaction. Mechanistic studies have shown that the deprotonation step to form this highly delocalized anion is the driving force for the reaction.

Overview

The overall reaction has been a deprotonation to produce the nucleophile for an addition–elimination reaction that is followed by a deprotonation to yield a stabilized anion (Figure 10.21). The reaction path is shown vertically on the left, and the side routes are shown to the right. This reaction is known as the Claisen condensation.

Surface overview: We can look back at the problem space that we just navigated to get an overview. The red arrows show the most reasonable route(s) on Figure 10.22. Note that the addition surface is connected by common intermediates to the elimination surface (as was done in Figure 4.41).

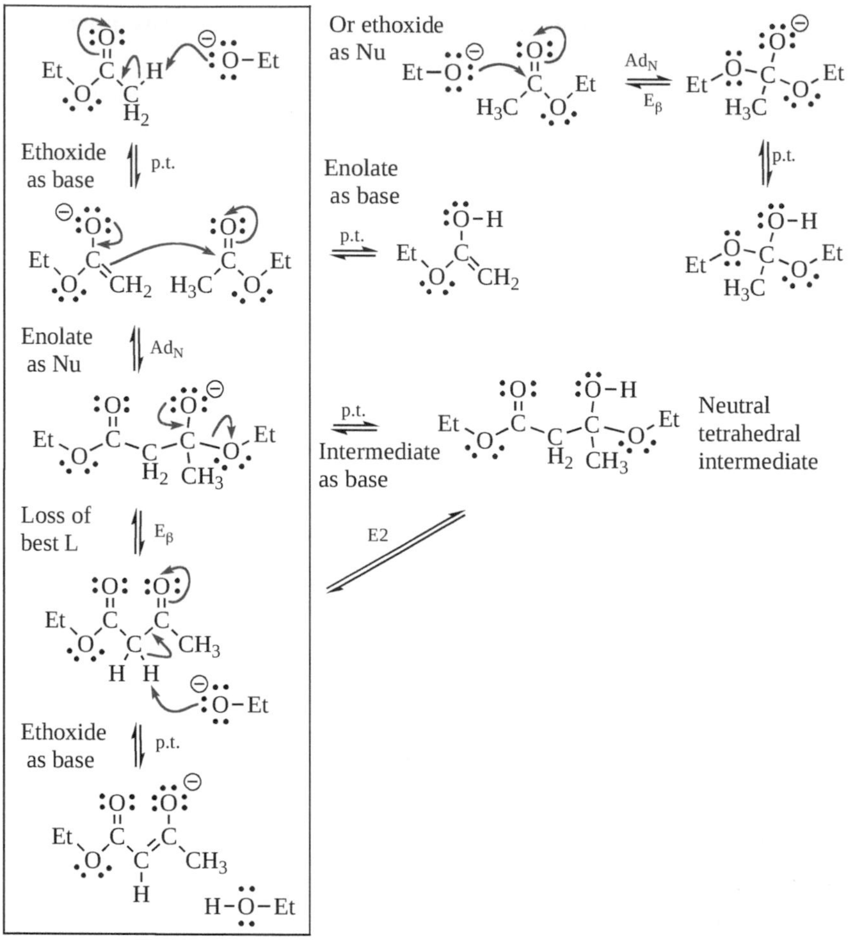

Figure 10.21 Ester and ethoxide overview for example 10.7.2. The most likely reaction path is boxed.

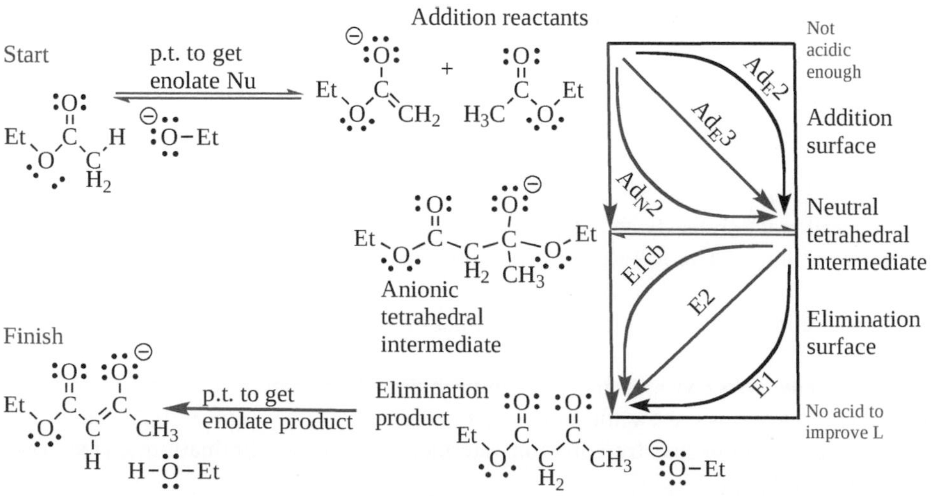

Figure 10.22 Final enolization and substitution surfaces serving as a problem space map for example 10.7.2. Reversible formation of the neutral tetrahedral intermediate also occurs in ethanol as shown.

10.7.3 Predict the Product of Reaction of a Ketal and Acidic Water

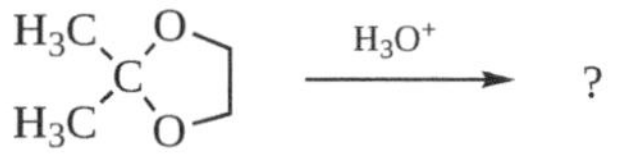

Understand the system

In this example, a ketal is mixed with an excess of acidic water. Medium? Acidic water, pK_a = 0.0. Sources? The lone pairs on the oxygens. Sinks? The acidic H's of the acid. Leaving groups? No decent ones without protonation.

Find possible routes

The only path available at the start is proton transfer to the lone pair of the ketal.

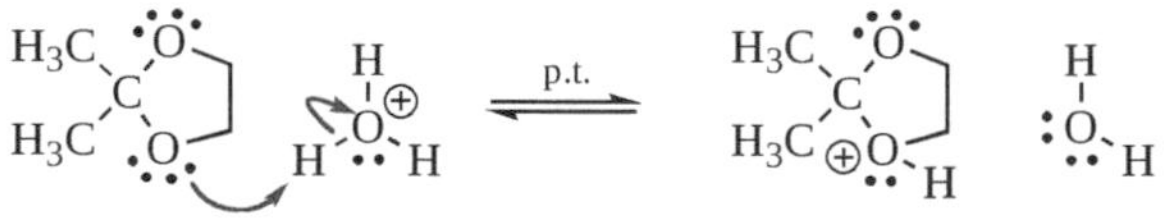

Evaluate and cross-check

There is only one source and one sink and one path. The proton transfer above has a K_{eq} of $10^{(-3.5 - 0.0)} = 10^{-3.5}$, certainly uphill, although still a good process.

Understand the system

Protonation of the ketal oxygen makes it a good leaving group with a pK_{aHL} = –2.4. Since the reaction is in an acidic medium, the leaving group is good, and the delocalized tertiary carbocation formed upon loss of the leaving group would also be relatively stable, the reacting species can now enter the substitution/elimination surface, Figure 9.2.

Find possible routes

The leaving group can ionize (path E_β).

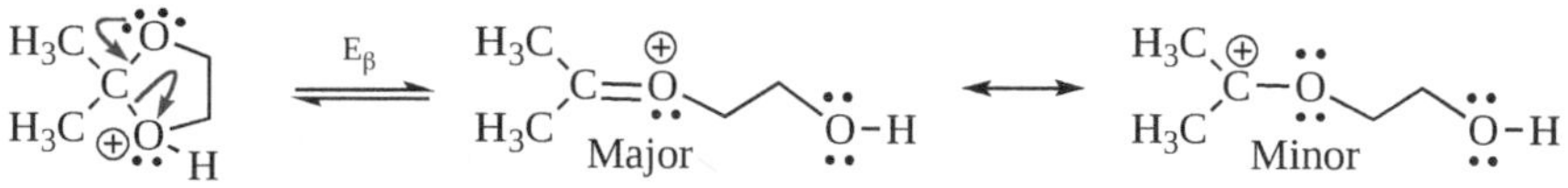

Either C–O bond to the protonated oxygen conceivably could break.

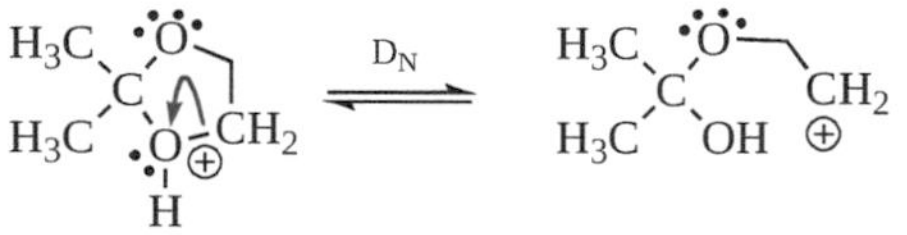

The same bonds could break by an S_N2 path with water as the nucleophile.

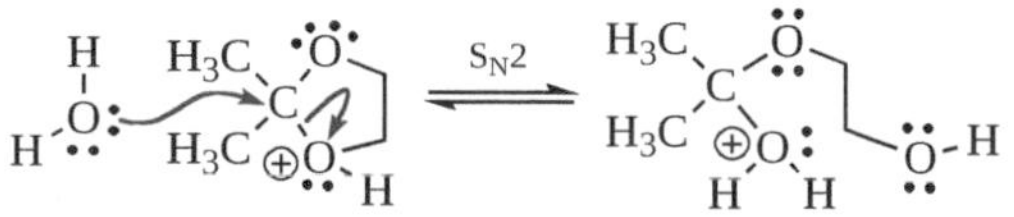

or

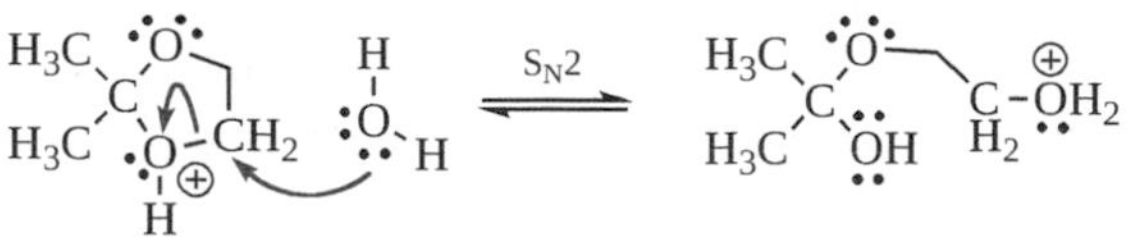

Evaluate and cross-check

The stability of both product carbocations must be checked: the lone-pair-stabilized tertiary cation is much more stable and therefore is favored over the unstabilized primary cation. The lone-pair-assisted ionization of the leaving group creates a somewhat less stable cation than the protonated ketal, so ionization of the leaving group is uphill in energy.

The S_N2 path alternatives need to be explored next. An S_N2 attack on a tertiary center is too hindered to proceed and can be ruled out. The S_N2 attack on the CH_2 is reasonable and must be considered. In fact, it took a mechanistic study with ^{18}O labeled water to settle this S_N1 vs. S_N2 question. The product with ^{18}O attached to the CH_2 group was not seen, ruling out the S_N2 mechanism. Therefore the best route is the E_β step to give the lone-pair-stabilized tertiary carbocation.

Understand the system

The lone-pair-stabilized tertiary cation is an excellent electron sink. The newly formed alcohol group and also the water solvent are both mildly basic and nucleophilic. However, nucleophilic attack on the carbocation by the alcohol group simply returns us to the previous intermediate, so let's go forward.

Find possible routes

This cation can lose an adjacent proton from the methyl (path D_E) to form the enol ether completing an E1 process,

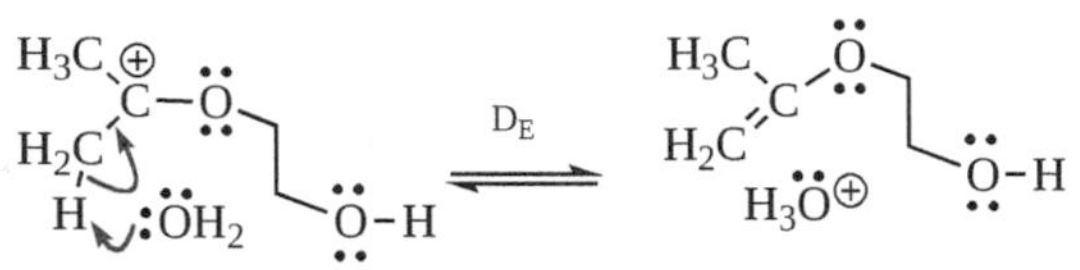

or the most abundant nucleophile, water, can trap the carbocation by path A_N, which completes an S_N1 process,

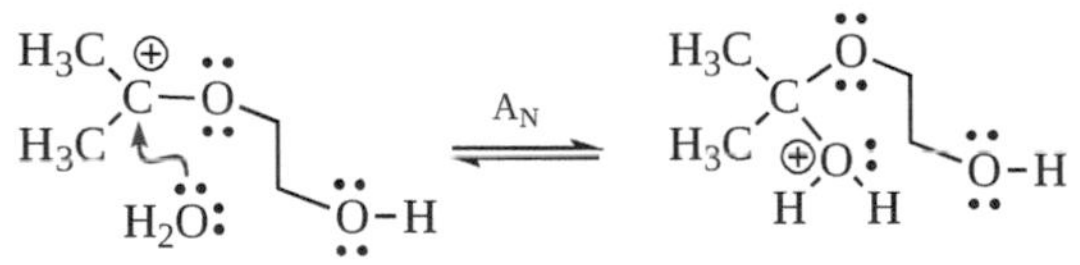

followed by proton loss to produce the hemiketal.

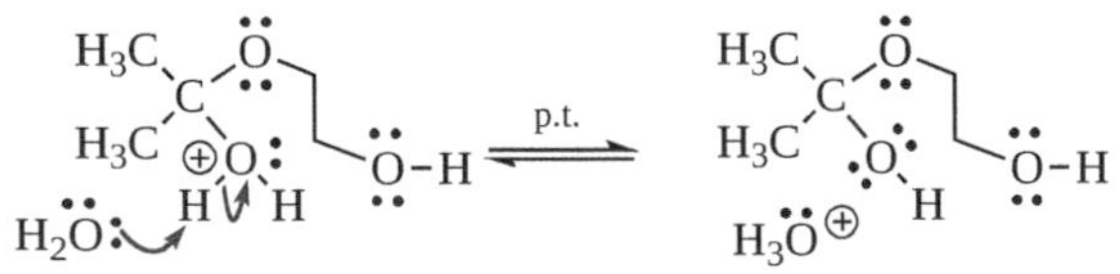

Evaluate and cross-check

Since both processes are reversible, the enol ether is uphill from the starting ketal. A $\Delta H°$ calculation gives +11 kcal/mol (46 kJ/mol), and there is an excess of nucleophile, S_N1 is favored. The proton transfer from the protonated alcohol to water is reasonable since it is downhill by about 2.4 pK_a units.

Now we need to consider whether the hemiketal is downhill in energy from the ketal. A quick calculation of the $\Delta H°$ of reaction yields zero, which means they are about the same in energy. It is best to keep exploring the energy surface to see if there is a lower-energy product.

Understand the system

The lone pairs on oxygen are basic, and there are no decent leaving groups without protonation. The media is still acidic.

Find possible routes

Protonation can occur on any lone pair of the three oxygen atoms. Protonation on the tertiary alcohol oxygen is fine, but it returns us to a previous structure. Protonation on the primary alcohol oxygen is also fine, but we know from the ^{18}O study that its methylene is not a site for attack. What is left is to protonate one of the ether oxygen's lone pairs.

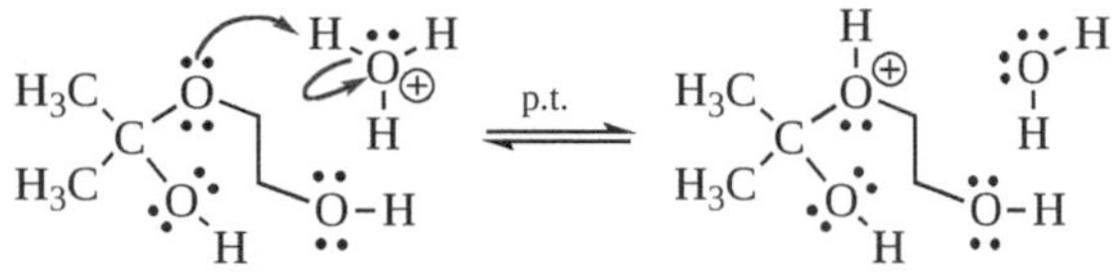

Evaluate and cross-check

Protonation on the slightly less basic ether lone pair converts it into a good leaving group. Similar to the first proton-transfer step with the ketal, K_{eq} = $10^{-3.5}$, the hemiketal is easily protonated in this acidic medium.

Understand the system

Protonation of the hemiketal oxygen makes it a good leaving group (pK_{aHL} = –2.4).

Find possible routes

Again, the reacting species enter the Figure 9.2 S_N1/E1 surface (the S_N2 was previously ruled out by the ^{18}O study). The protonated hemiketal can lose the leaving group in a process identical with that of the ketal (path E_β) to give the lone-pair-stabilized tertiary cation.

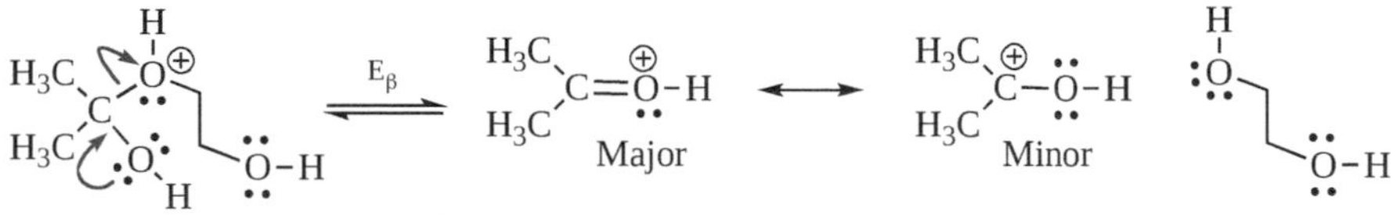

The lone-pair-stabilized cation can lose a proton from carbon to form the enol (path D_E),

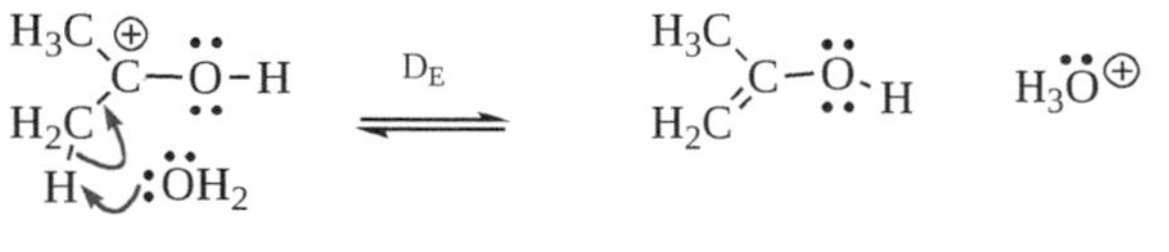

or can be trapped by water (path A_N) to yield, after deprotonation, the hydrate.

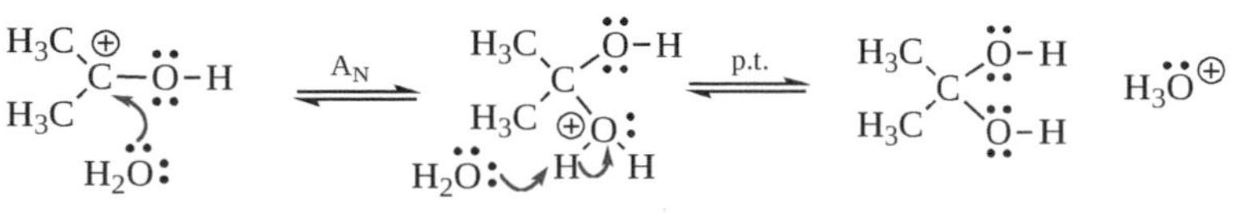

But additionally the lone-pair-stabilized tertiary cation can lose a proton from the oxygen to form the ketone (path D_E), and thereby completing a lone-pair-assisted E1 elimination.

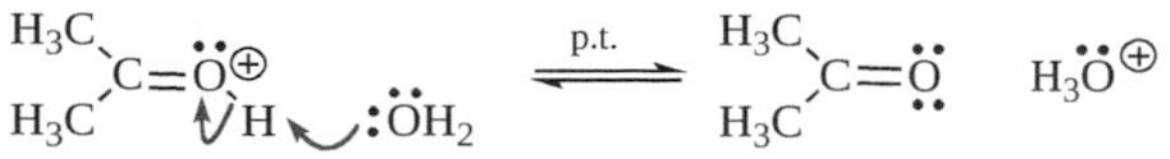

Evaluate and cross-check

Since an equilibrium process favors the most stable product, we need to again calculate $\Delta H°$ to find out which of the three products are preferred. The hydrate is at about the same energy as the ketal, and the enol is calculated to be 11 kcal/mol (46 kJ/mol) higher, the ketone is 5 kcal/mol (21 kJ/mol) lower in energy than the ketal and is therefore the product.

Overview

Occasionally a reaction will be driven to a product uphill in energy by displacement of the equilibrium, commonly an excess of one reagent combined with removal of one product. Ketals are formed from ketones in this manner: an excess of alcohol is used and the water formed is removed by distillation. The equilibria are the same as those discussed above, driven backward by mass balance effects. Watch reaction conditions for driven equilibria.

We have wandered over a rather complex energy surface (Figure 10.23), exploring every possible side reaction until there was nothing more that was reasonable to do. Looking at the stability of all the possibilities on the surface, we arrived at the correct, most favored product. The reaction path is shown vertically on the left and the side routes are shown to the right side. The path name for the forward reaction only is shown. Although the elimination surface tilts toward the E1 because of the stability of the carbocation, an E2 is also reasonable.

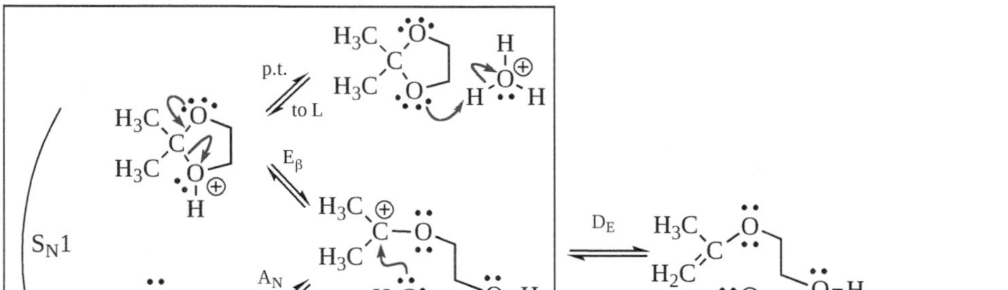

Figure 10.23 Ketal hydrolysis overview, example 10.7.3. The most likely reaction path is boxed.

Surface overview: We can look back at the problem space that we just navigated to get an overview. The red arrows show the most reasonable route(s) on Figure 10.24.

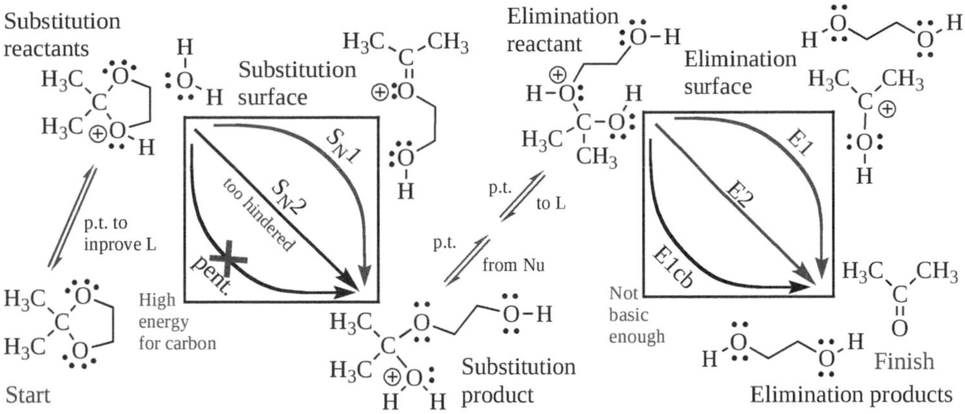

Figure 10.24 Final substitution and elimination surfaces serving as a problem space map for example 10.7.3. Solvent likely mediates the proton transfer from nucleophile to leaving group.

10.7.4 Predict the Product of Reaction of an Amide and Basic Water

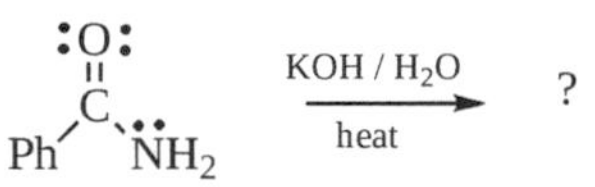

Understand the system

Biochemically similar reactions are responsible for the cleavage of the amide links of proteins into their constituent amino acids. Medium? Basic water. Sources? Hydroxide anion is the best source. Leaving groups? None, without protonation to improve the N. Sinks? The amide is a polarized multiple bond with leaving group. Although the NH_2 places it in that class, it is a very poor leaving group (NH_2^- pK_{aHL} is 35). Acidic H's? The NH_2 of the amide is pK_a 17. Bases? Hydroxide, pK_{abH} = 14.0. Resonance forms?

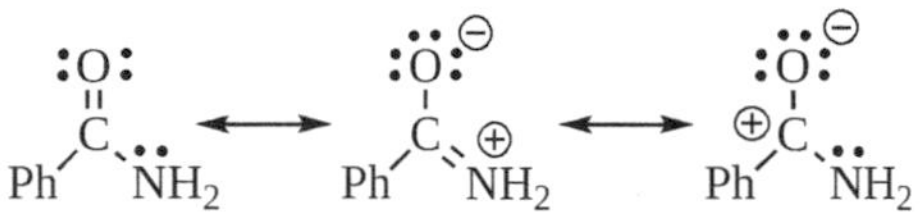

Find possible routes

There are two possible: We can use hydroxide as a base and remove an amide NH to form the amidate anion (path p.t.),

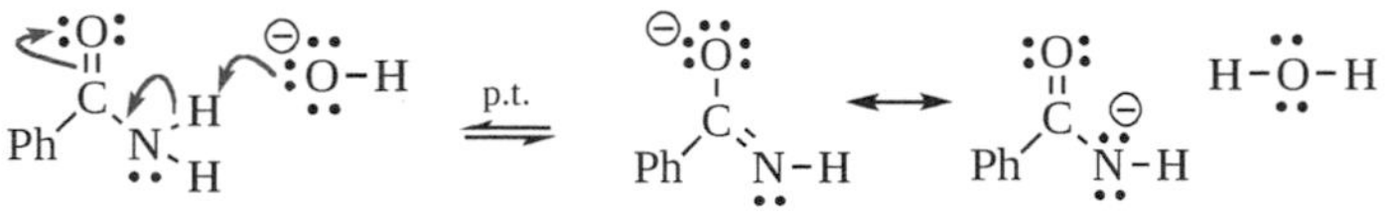

or use hydroxide as a nucleophile and attack the amide carbonyl (path Ad_N).

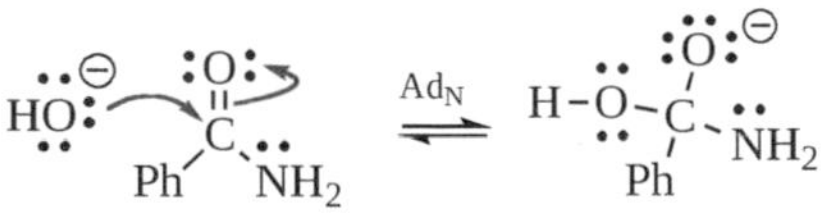

Evaluate and cross-check

The K_{eq} for proton transfer is 10^{-3}, which favors reactants. If the amidate anion were to act as a nucleophile, the only site for attack would be another amide, and then it would just be kicked right back out again since it would be the best leaving group. If the amidate acts as a base, it returns to the reactants. This short exploration down the amidate route reveals it as a dead end. Hydroxide as a nucleophile might be more promising.

The ΔpK_a rule cross-check is hard to use because we do not have a good value for the pK_a of the product of the addition of hydroxide. Steric hindrance from R raises the pK_{abH} of an alkoxide anion (poorer intermolecular solvation), whereas the inductive/field effect from the nitrogen should lower the pK_{abH}. If these effects balance out, the tetrahedral intermediate product of nucleophilic attack would have a pK_{abH} of about 13 ($HOCH_2OH$ is pK_a 13.3).

Understand the system

There is no leaving group other than the original nucleophile (kicking out NH_2^- with a pK_{aHL} = 35 from a tetrahedral intermediate 22 pK_a units less basic is not an option), so we should examine the possibilities for proton transfer. Hydroxide is a good base. In basic water, the most common proton source is neutral water.

Find possible routes

Proton transfer can occur from water to the alkoxide anion to make the alcohol. Water can also protonate the nitrogen lone pair. There are four common charge types of the tetrahedral intermediate, shown below. Also if the medium were basic enough, the other OH can be deprotonated to give a dianion.

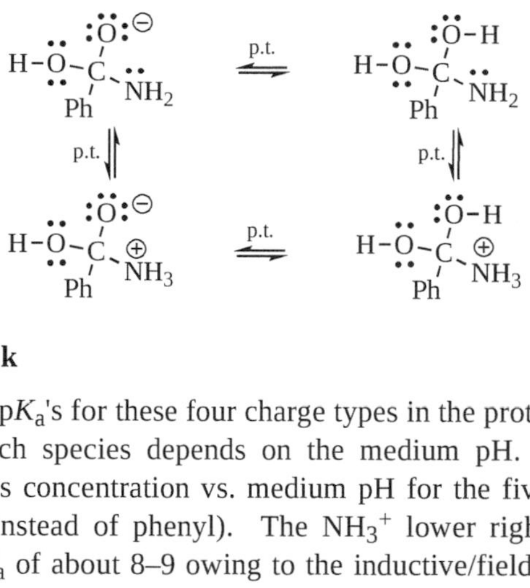

Evaluate and cross-check

We need to estimate pK_a's for these four charge types in the proton-transfer square above. The concentration of each species depends on the medium pH. Figure 10.25 shows the calculated relative species concentration vs. medium pH for the five charge types of a very similar system (methyl instead of phenyl). The NH_3^+ lower right structure in the above square should have a pK_a of about 8–9 owing to the inductive/field effect of two hydroxyls. This pK_{a1} is calculated at 8.3. Its conjugate base is the upper right structure in the above square. The OH in the upper right structure is estimated at a pK_a of about 11–12 because the electronegative nitrogen should lower the pK_a of $HOCH_2OH$ at 13.3. This pK_{a2} is calculated at 11.3. Its conjugate base is the upper left structure in the above square. The next OH deprotonation creates a dianion. It is difficult to gauge the reasonableness of the dianion since its pK_{abH} is difficult to estimate because of charge repulsion. This pK_{a3} is calculated at 13.4, which may still be a bit low. The calculation for Figure 10.25 shows that around pH 9.8 the neutral tetrahedral intermediate predominates at about 90% of the species, and that these proton-transfer reactions in the above proton transfer square are all reasonable. The top two species in the proton-transfer square above can easily revert to starting materials, so in the absence of an obvious lower energy route, we should explore paths forward from all species. Note how each species concentration changes with pH, and that there is small but usable concentration of the zwitterionic species (bottom left in the above proton-transfer square).

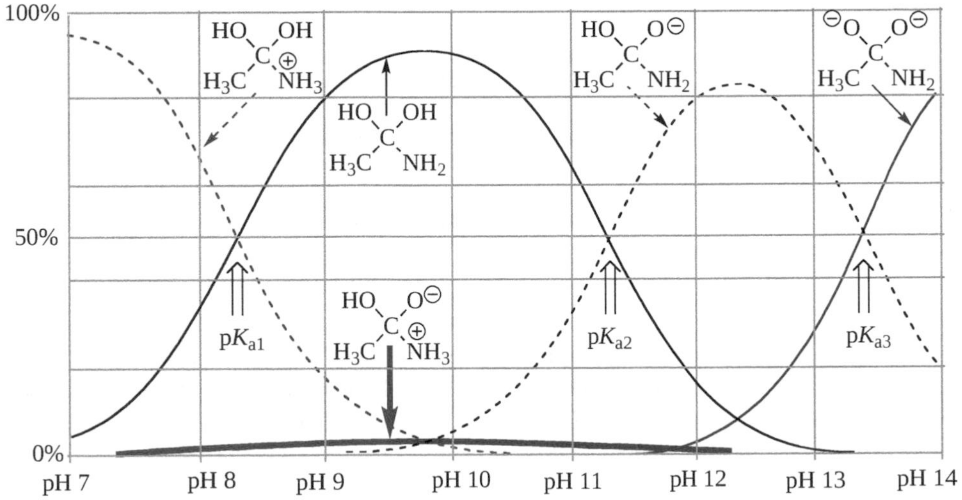

Figure 10.25 Five charge types of the tetrahedral intermediate as a function of pH. The pK_a's are calculated to be pK_{a1} = 8.3, pK_{a2} = 11.3, pK_{a3} = 13.4 (plot modified from chemicalize.com).[1]

Understand the system

The NH_3^+ can serve as a fair leaving group (pK_{aHL} 9.2). **Note well: The unprotonated amine, pK_{aHL} 35, is not a leaving group at all.** Since the reaction is run in water, which is great at hydrogen bonding, a hydrogen-bonded leaving group could remove a proton from water as it departs and therefore avoid forming a highly basic NH_2^- anion. This was discussed in Section 7.4.3, Variations on a Theme, as the hydrogen-bonded catalyzed beta elimination.

Find possible routes

Four charge types make elimination paths possible:
Beta elimination from an anion (path E_β),

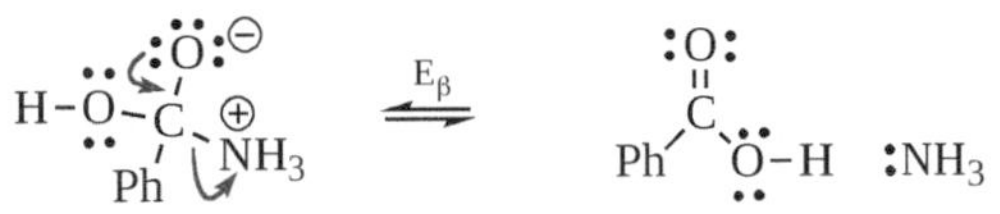

E2 elimination,

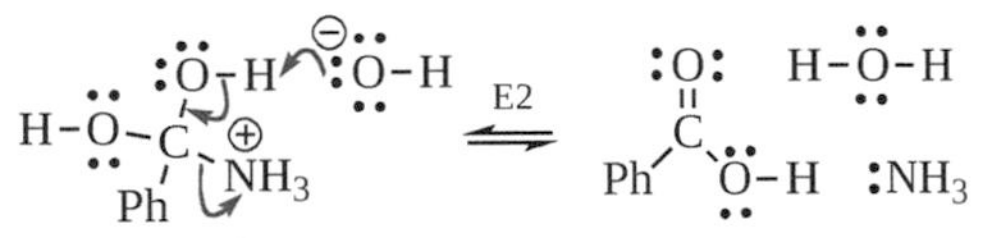

The hydrogen-bond catalyzed beta elimination.

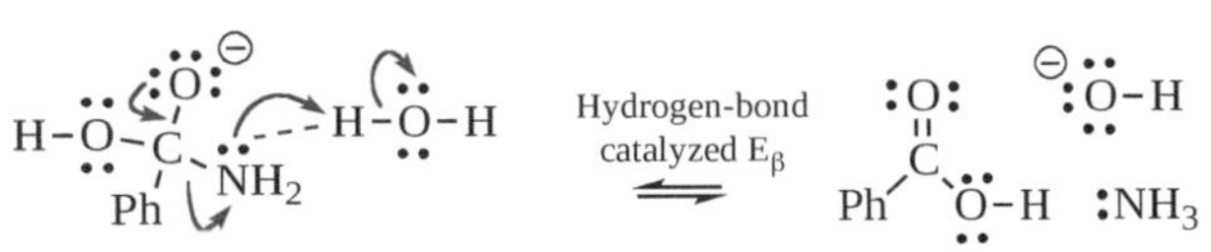

Or the hydrogen-bond catalyzed E2 elimination.

---

[1] Chemicalize was used for prediction of the pK_a vs. pH properties on 9/2019, https://chemicalize.com/ developed by ChemAxon (http://www.chemaxon.com).

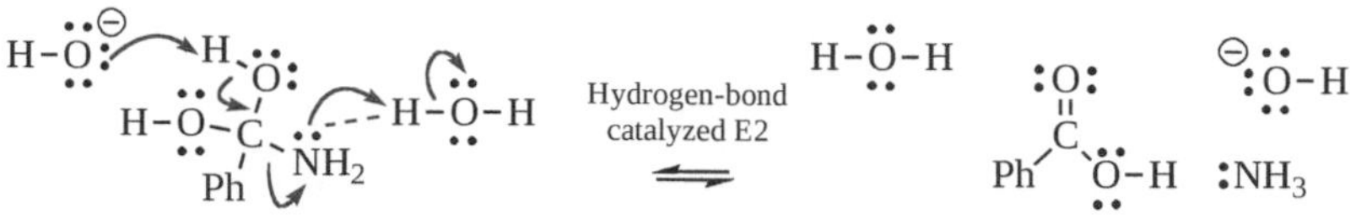

Evaluate and cross-check

All are viable routes, passing the ΔpK_a cross-check by avoiding making a highly basic anion from less basic reactants. The E2 forms a weaker base of pK_{abH} 9.2 from a starting anion of pK_{abH} 14.0. The beta elimination from the zwitterion, a slightly weaker base, is still downhill. The hydrogen-bond catalyzed beta elimination is slightly uphill because hydroxide at pK_{abH} 14.0 is more basic than the reactant anion, pK_{abH} about 13. The hydrogen-bond catalyzed E2 elimination is balanced with incoming and outgoing pK_{abH} values the same. To determine the most probable, look at Figure 10.5 to see which are the major species in the solution near a basic pH of 12. The anionic tetrahedral intermediate is dominant at this pH.

Find possible routes

We can spot an irreversible proton-transfer step, log $K_{eq} = pK_{abH} - pK_{aHA} = 14 - 4.2 = 9.8$, thus $K_{eq} = 10^{+9.8}$, that creates an unreactive carboxylate anion. With this proton transfer we are done.

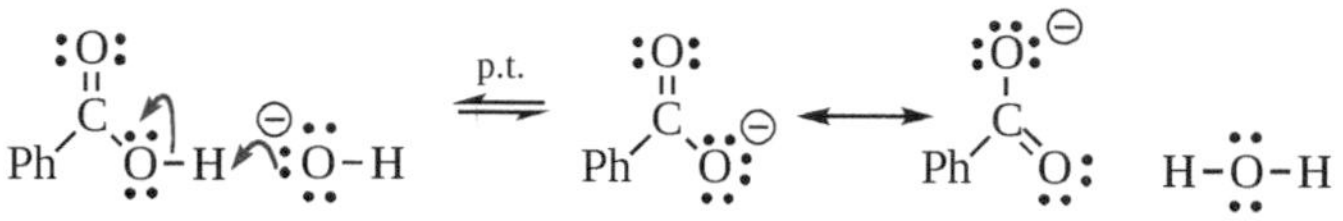

Overview

Figure 10.26 shows the most probable overall reaction path vertically, and the alternate paths are to the side.

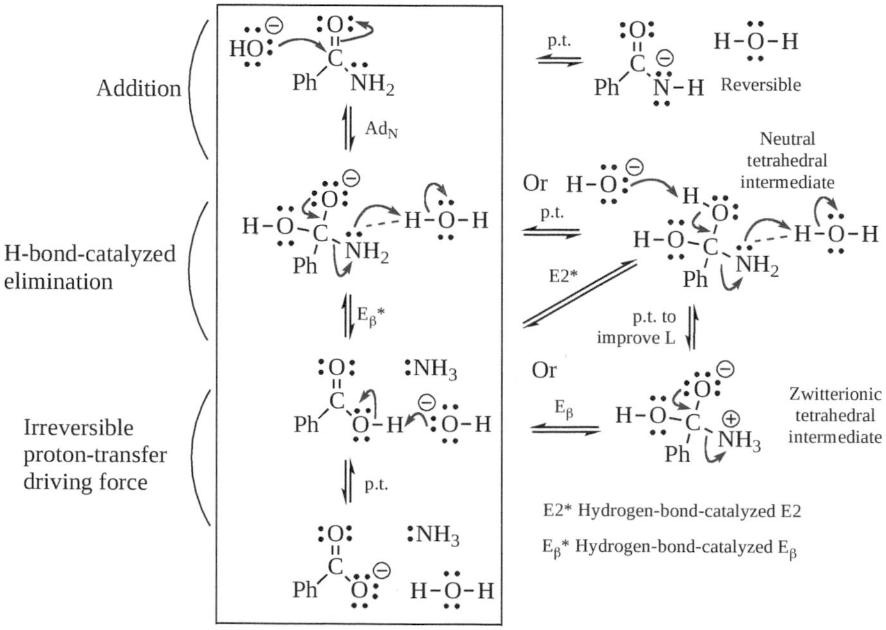

Figure 10.26 Amide hydrolysis overview for example 10.7.4. The most likely reaction path is boxed.

Surface overview: We can look back at the problem space that we just navigated to get an overview. The red arrows show the most reasonable route(s) on Figure 10.27 below.

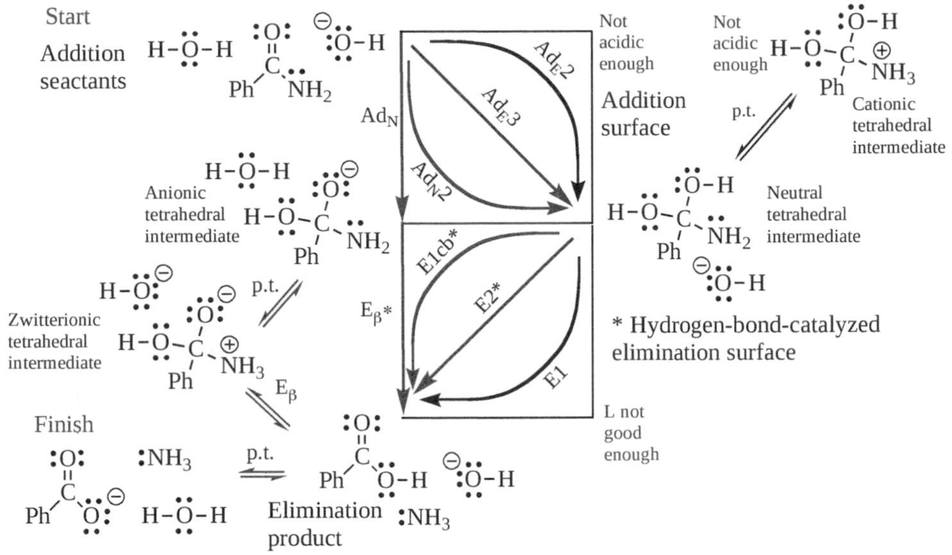

Figure 10.27 Final addition and elimination surfaces as a problem space map for example 10.7.4. The elimination surface is all hydrogen-bond catalyzed since the leaving group is very poor and needs partial or full protonation to be kicked out from the tetrahedral intermediate. See Section 7.4.3 for discussion.

10.7.5 Predict the Product of Reaction of a Carboxylic Acid and $SOCl_2$

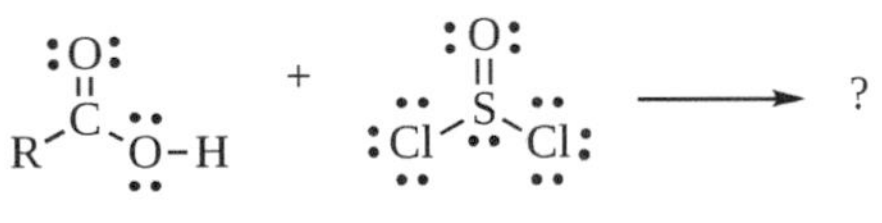

Understand the system

Medium? Acidic, because $SOCl_2$ produces HCl and SO_2 with trace water. Sources? The carbonyl lone pairs are the best source (much better than the lone pairs of the OH). Leaving groups? Chloride. Sinks? The best, thionyl chloride is a Y–L. The carboxylic acid is a carboxyl derivative, but the OH is a not a good leaving group. Acidic H's? The carboxylic acid's OH. Bases? None. Resonance forms? By VSEPR $SOCl_2$ is tetrahedral, often drawn with an expanded octet resonance form containing a *d–p* pi bond.

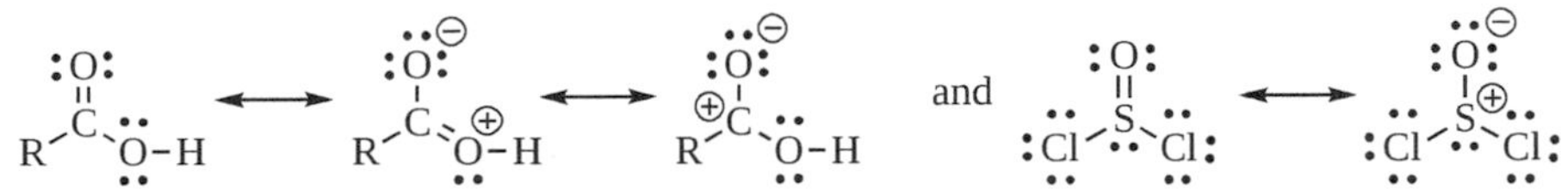

Find possible routes

We combine the best source (carbonyl lone pair) and the best sink (S–Cl weak bond) with the best path. The most probable path for a Y–L is the S_N2 substitution. It is currently debated whether sulfur goes through a pentacoordinate intermediate. The S_N1 route might also work since the cationic intermediate is isoelectronic with SO_2. The S_N2 substitution is:

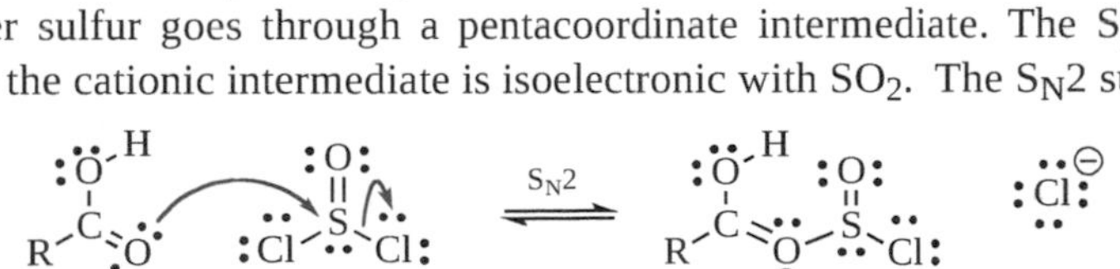

Evaluate and cross-check

If we had used the carboxylic acid's OH lone pair, the reaction would have produced a less stable cation, a positive oxygen between two electron-withdrawing groups, the carbonyl and the sulfoxide. The nucleophile's pK_{abH} is –6 and the leaving group's is also –6. The attack has broken a weak S–Cl bond and formed an S–O bond with better overlap. The reaction is most likely downhill in energy.

Understand the system

The carbocation formed is tertiary and substituted by two lone pair donors.

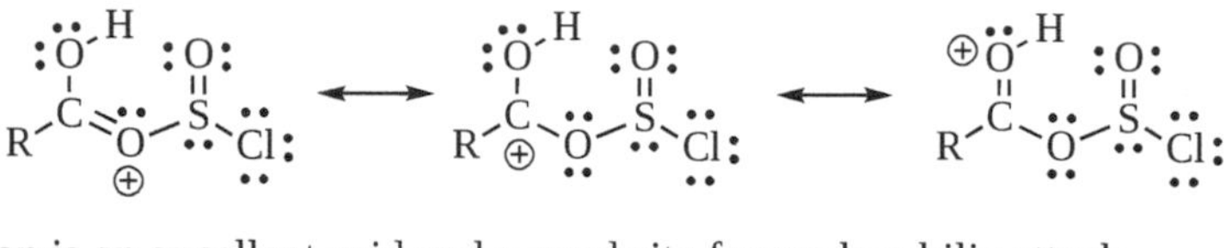

The carbocation is an excellent acid and a good site for nucleophilic attack.

Find possible routes

The two alternatives are deprotonation (path p.t.),

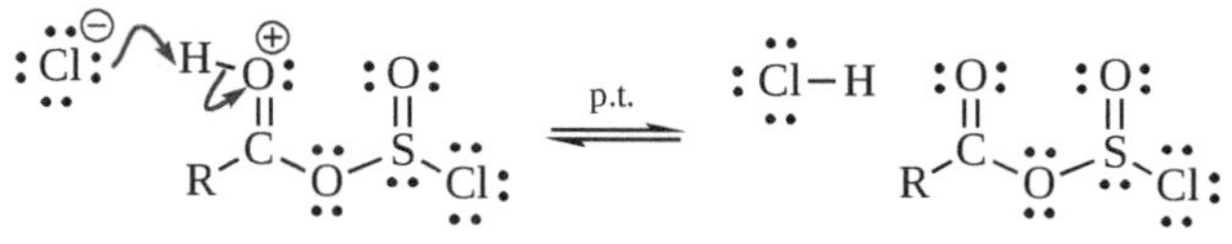

or nucleophilic attack (path Ad_N). The best nucleophile is now chloride, which can attack the protonated carbonyl to give a tetrahedral intermediate.

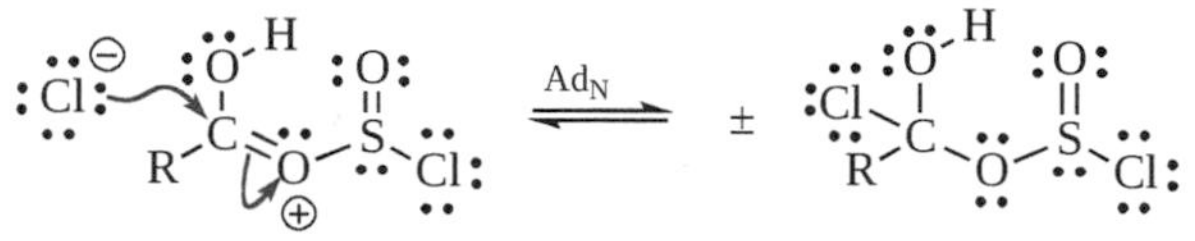

Evaluate and cross-check

Both processes should be favorable. The proton transfer occurs between two groups that have about the same pK_a to yield the acyl chlorosulfite. The nucleophilic attack to give the tetrahedral intermediate has neutralized charge also should be favorable. Both processes are consistent with a very acidic medium. We should explore both routes.

Understand the system

Let's explore the product of the proton-transfer route first. The reaction is generating HCl, pK_a –6, and the media is becoming strongly acidic. The resonance forms for the acyl chlorosulfite are similar to those of the starting carboxylic acid.

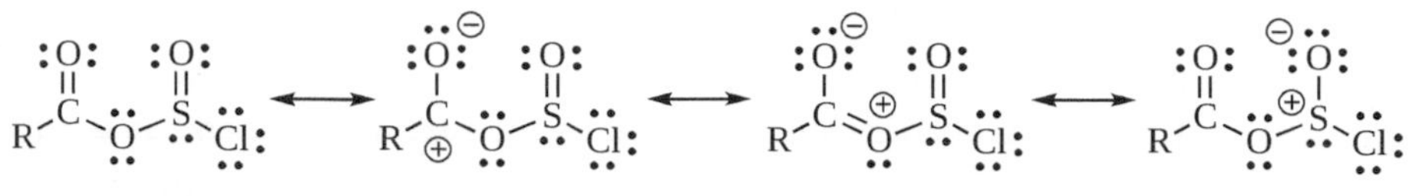

The acyl chlorosulfite carbonyl is a good site for nucleophilic attack. The best nucleophile is still chloride. Chloride attack on sulfur just returns to the starting point.

Find possible routes

The carbonyl addition surface reminds us that there are three route alternatives available.

the Ad_N2 addition (path Ad_N then path p.t.),

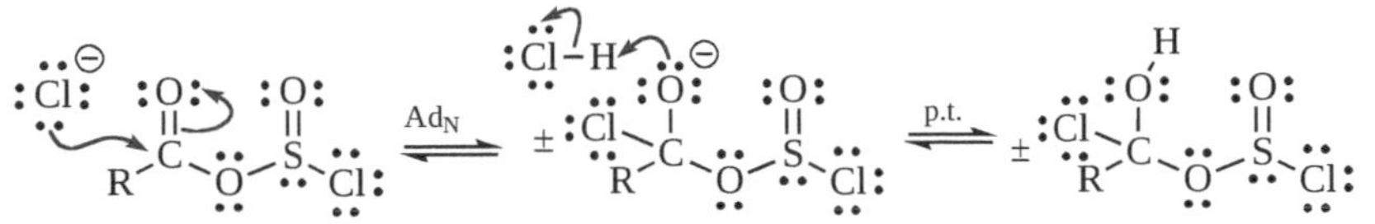

or the hetero Ad_E3 addition,

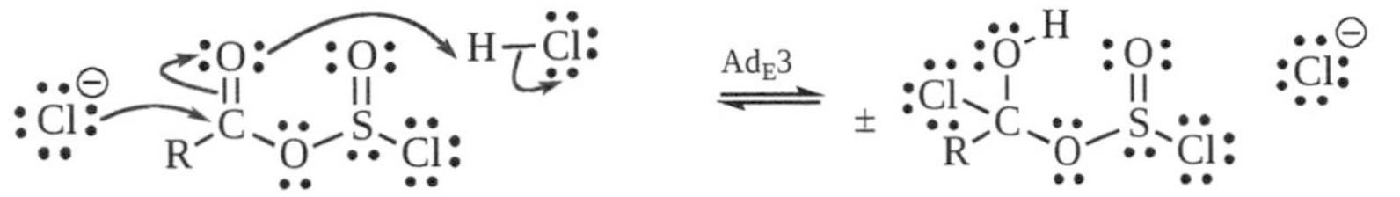

or the hetero Ad_E2 (path p.t. followed by path Ad_N).

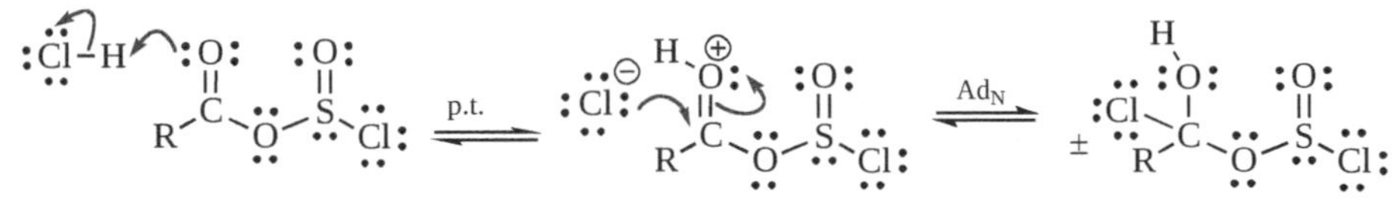

Evaluate and cross-check

The Ad_N2 is not a favored route because the attack of chloride, pK_{abH} of –6, to give an anionic tetrahedral intermediate, pK_{abH} about 10, goes against the ΔpK_a rule cross-check (a climb of 16 pK_a units) and forms a basic anion in very acidic media. The Ad_E3 and the hetero Ad_E2 are both reasonable by the ΔpK_a rule cross-check and consistent with very acidic media.

Understand the system

We have come full circle back to the original tetrahedral intermediate. The tetrahedral intermediate has two attached leaving groups, the chloride and chlorosulfite. Loss of chloride reverts to reactants but loss of chlorosulfite yields new compounds.

Find possible routes

There are four possible elimination routes that produce the unstable chlorosulfite anion, which immediately loses the good leaving group chloride to give sulfur dioxide gas. Chlorosulfite can be lost in an E2 elimination,

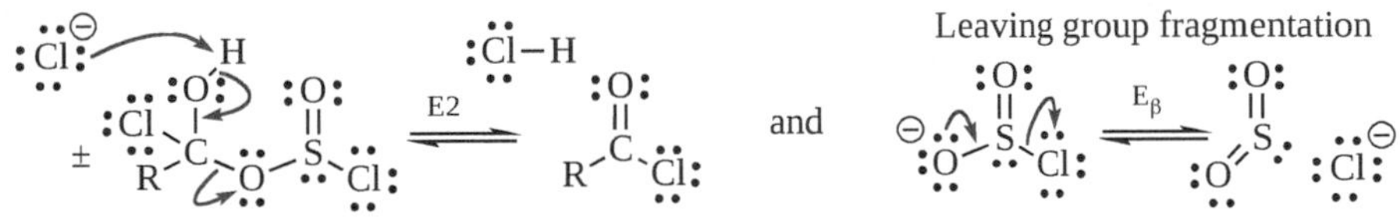

or in a lone-pair-assisted E1 (path E_β followed by path p.t.),

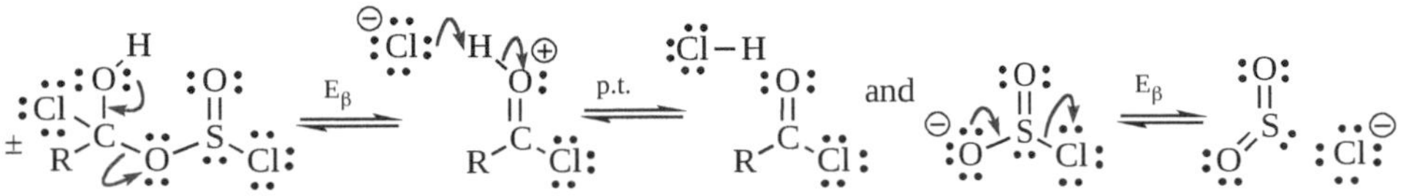

or by an E1cB (path p.t. followed by path E_β).

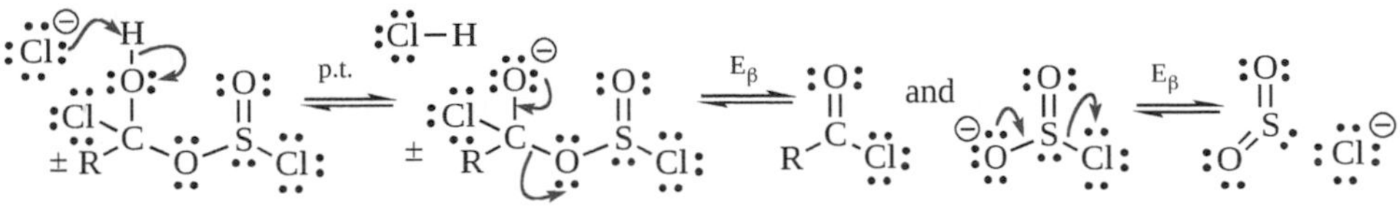

or by a *syn* internal elimination (path Ei),

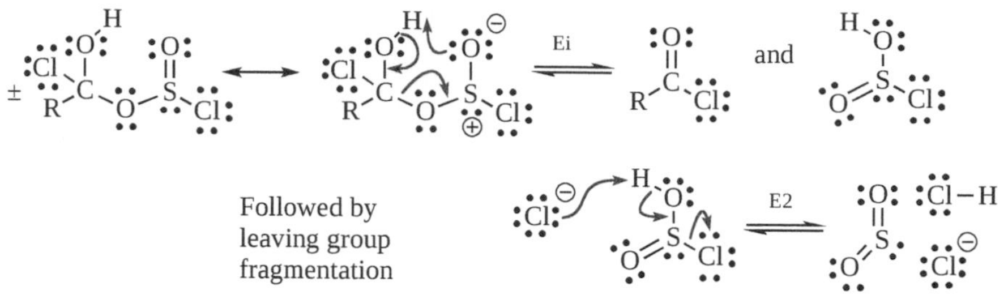

Evaluate and cross-check

The E1cB is probably not a good route because the deprotonation by chloride, pK_{abH} −6, to give the anionic tetrahedral intermediate, pK_{abH} about 10, goes against the ΔpK_a rule cross-check (a climb of 16 pK_a units) and forms a basic anion in a strongly acidic medium.

The other three eliminations are compatible with a strongly acidic medium. Because chlorosulfite is so unstable, no pK_a is available for its conjugate acid, and it is difficult to rank as a leaving group. The closest pK_a we can get to the unstable chlorosulfite is benzenesulfinic acid, $PhSO_2H$, whose pK_a is 1.5. Substitution of an electronegative chlorine for phenyl will make the acid much more acidic, dropping the pK_a at least several units; thus chlorosulfite should be a good leaving group. It is difficult to rank which elimination route is lowest in energy. The Ei elimination has an intramolecular advantage and thus may be the fastest.

Overviews

Now two gases, HCl and SO_2, have been evolved and should escape the reaction mixture driving the transformation to completion. We can look back at the problem space that we just navigated to get an overview. The red arrows show the most reasonable route(s) on the surfaces in Figure 10.28 below.

Figure 10.29 is a more conventional summary with the most probable route drawn in the center with alternative routes drawn to either side. The path name for the forward reaction only is shown. The low-probability routes have been omitted for clarity.

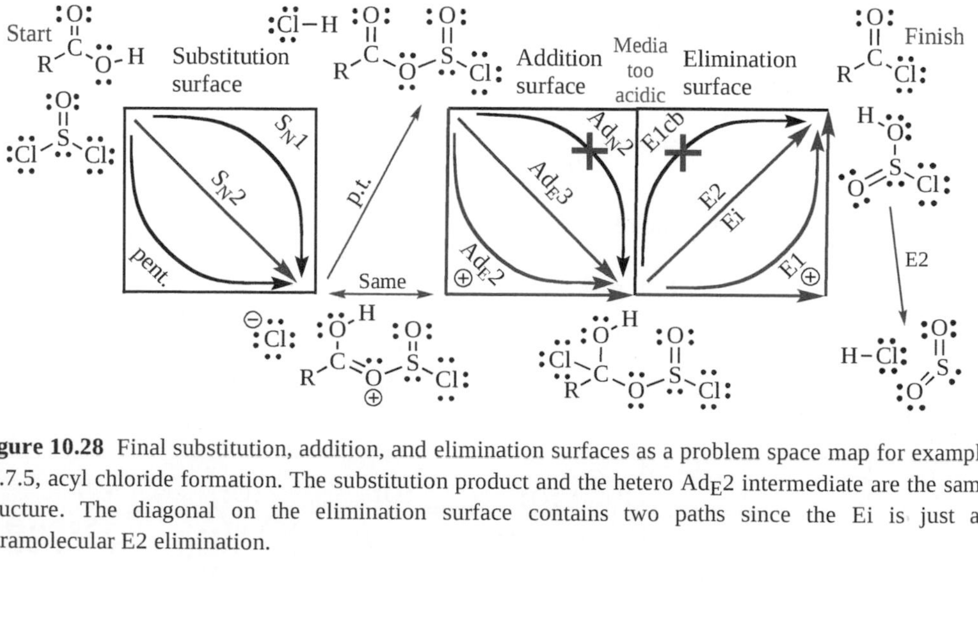

Figure 10.28 Final substitution, addition, and elimination surfaces as a problem space map for example 10.7.5, acyl chloride formation. The substitution product and the hetero Ad_E2 intermediate are the same structure. The diagonal on the elimination surface contains two paths since the Ei is just an intramolecular E2 elimination.

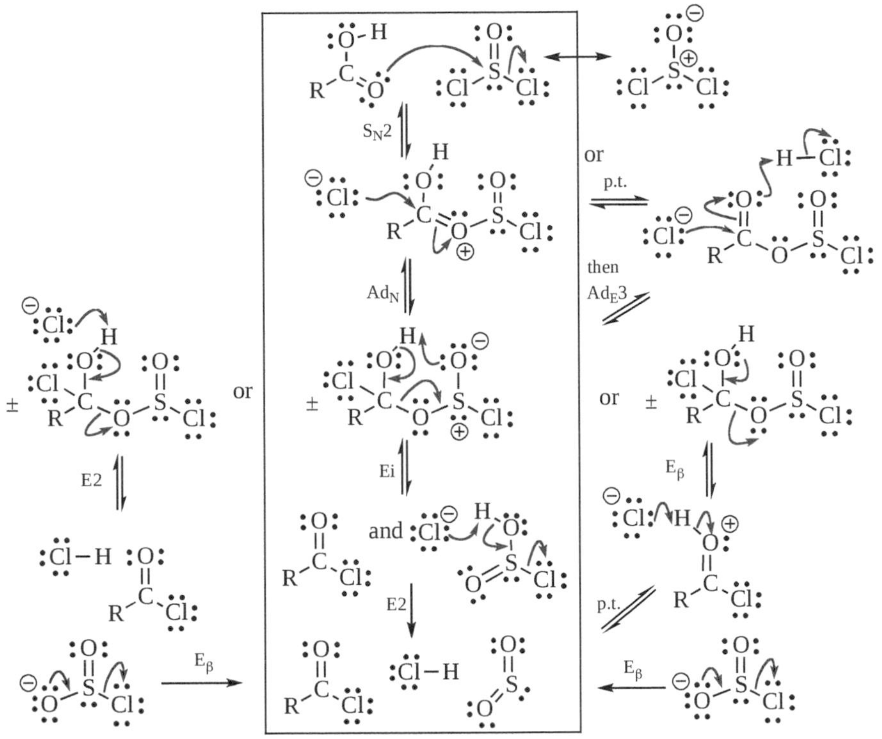

Figure 10.29 Overview for example 10.7.5, acyl chloride formation. The most likely reaction path is boxed.

10.8 METHODS FOR TESTING MECHANISMS

So far we have seen how to generate mechanistic hypotheses. The next step in the scientific method is testing them to rule out ones incompatible with experimental observations. We use the tools presented in this section to **disprove** these hypotheses, not to try to prove one correct. A reaction mechanism always remains a hypothesis and therefore can never be proven true. Experimental design is important; the experiment should be set up such that it will clearly exclude one of the mechanistic alternatives. The experiment must be carefully carried out so as to get a clean result. We repeat the testing process until only one mechanistic hypothesis remains, and within the limits of our abilities to test, fails to be disproven. If the original set of hypotheses is not complete, then the single hypothesis we narrow down to may still be incorrect because we missed a more reasonable alternative.

10.8.1 Initial Studies

Identify the players. The first thing that needs to be done in studying a reaction is to know the exact stoichiometry of the reactants and major products. The stereochemistry of the reactants and products needs to be determined. Any side products should be identified.

10.8.2 Catalysis, Inhibition

What gets the reaction going? Sometimes nothing happens until a catalyst is added. Does the reaction require acidic or basic catalysis? Is a weak acid or base capable of catalyzing the reaction? Is a proton source, like trace water, required? Do Lewis acids catalyze the reaction? Do free radical inhibitors or initiators or light have an effect (See Chapter 12)? Any postulated mechanism must explain the need for a required catalyst.

10.8.3 Isolation, Detection, or Trapping of Intermediates

"Who's on first?" There are several spectroscopic methods that are capable of distinguishing reaction intermediates of a reaction in progress (Chapter 14). For nuclear magnetic resonance (NMR) , infrared (IR) , and ultraviolet–visible (UV–VIS) spectra we can run the reaction in the instrument. Electron paramagnetic resonance (EPR) can detect free radicals (Chapter 12). There are even instruments that can record optical spectra at picosecond or less intervals. Also, it is occasionally possible to isolate intermediates from especially mild reaction conditions or low temperatures. Reactive intermediates of unimolecular reactions can be characterized spectroscopically by running the reaction in a glass such as frozen argon.

Reagents can be added to the reaction to trap the supposed reactive intermediate specifically and preferentially. For example, an enzyme was suspected of reversibly forming an iminium ion with the substrate before cleaving it. Sodium borohydride was added, reducing the iminium ion to an amine, stopping the forward reaction, and preventing the reversal. The enzyme–product mix was broken down by hydrolysis and the isolated substrate was found attached to an amino acid of the enzyme. Not only did this confirm an iminium intermediate, but also it helped identify the amino acid responsible for the catalysis.

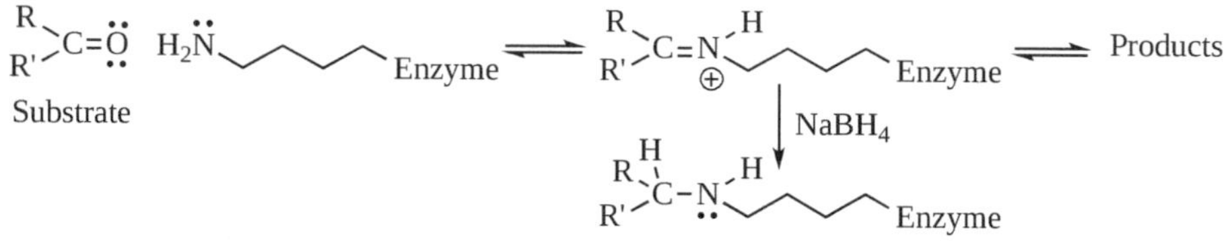

10.8.4 Isotopic Labels

Find out where the atoms went. Labels are a convenient probe of a reaction mechanism, but care must be taken that whatever is chosen as a label does not alter the system under study. The label must be stable to the reaction conditions and not accidentally fall off into solution or scramble its position. Isotopic labels are a useful way of following an atom through a reaction. Common labels are deuterium ^{2}H (D), ^{3}H, ^{13}C, ^{14}C, ^{15}N, ^{18}O, and ^{32}P. The radioactive labels, ^{3}H, ^{14}C, and ^{32}P, can be detected in very small quantities, so are useful in tracing products in biochemistry. However extensive degradation schemes are usually required to determine the position of the radiolabel on the molecule. Mass Spectra (MS) can be used to clarify the position of a label such as deuterium, ^{15}N, or ^{18}O from the fragmentation pattern of the molecule. With carbon-13 NMR spectroscopy, finding the location of a ^{13}C label is especially easy to carry out. Labels can be very useful in determining the origins of a particular group in a rearrangement, a cleavage, or a biosynthesis.

Doubly labeled compounds provide an interesting test for an intermolecular vs. intramolecular transfer of a group. The group and another part of the same molecule are both labeled, and the reaction is run with an equal amount of unlabeled material. The product of an intramolecular transfer would stay doubly labeled or unlabeled. The intermolecular process would give a statistical mixture of products (with half the product being singly labeled).

Example problem

How would you use a label to disprove that a methyl ester underwent basic hydrolysis in KOH/water by an S_N2 mechanism?

Answer: Label the oxygen in the KOH/water with ^{18}O. An S_N2 on the methyl with labeled hydroxide will give the ^{18}O label in the methanol, whereas other mechanisms such as the addition–elimination path combination would not.

10.8.5 Regiochemistry, Stereochemistry, and Chirality

Probe the orientation of groups in the product. Does the reaction have a preferred regiochemistry, like following Markovnikov's rule? Is stereochemistry lost or retained in the reaction? Is the process stereospecific? Does addition or elimination occur *anti* or *syn*? Is the chirality at the reacting center preserved, inverted, or lost?

For example, addition of bromine to *cis*-2-butene gives equal concentrations of (2R,3R)-2,3-dibromobutane and (2S,3S)-2,3-dibromobutane (a racemic mix), whereas addition of bromine to *trans*-2-butene gives the *meso* compound (2S,3R)-2,3-dibromobutane. Since the stereochemistry of the reactant is determining product stereochemistry, this reaction is stereospecific, and whatever mechanism is proposed must account for that.

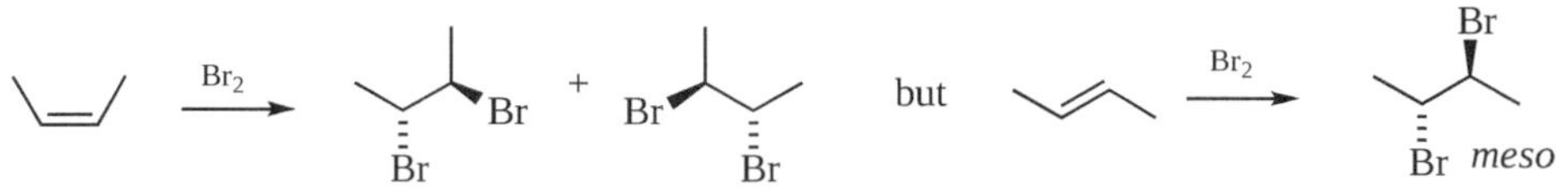

Example problem

The addition of DBr to cyclohexene gives the *anti*-addition product. Is this result more consistent with an Ad_E2 or an Ad_E3 mechanism?

Answer: The Ad_E3 is consistent with *anti*-addition. The Ad_E2 would be expected to give a *syn/anti* mixture with *syn*-addition preferred. The Ad_E2 intermediate ion pair collapses preferentially *syn*; the symmetrically solvated carbocation has no preference.

10.8.6 Kinetics

Who are rate-determining-step players? Kinetics can yield information about the rate-determining step of a reaction. At a constant temperature, the concentration of a particular reactant is changed and the effect on the rate of reaction is noted. For example, a reaction that is second order in compound A would quadruple its rate if the concentration of A were doubled. If the reaction were first order in A, the rate would double if the A concentration were doubled. In this way, the kinetic order with respect to each reactant is determined. These kinetic orders can then be compared to that expected for the slow step of a postulated mechanism. Often, however the kinetic order can easily fit several possibilities. Some examples of kinetic order have been presented already and are shown below. The magnitude of rate constant, *k*, is of course unique to each reaction and differing substrates.

Order	Example	Rate expression
First	E1	Rate = k[R–L]
Second	S_N2	Rate = k[R–L][Nu]
Third	Ad_E3	Rate = k[alkene][HA][A^-]

Example problem

Propose a mechanism that is consistent with the fact that the rate of the acidic bromination of acetone is independent of the concentration of bromine.

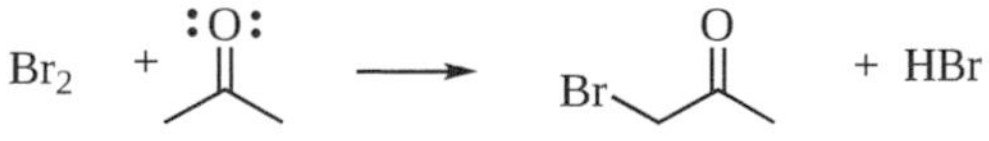

Answer: The independence of the rate on the bromine concentration means that bromine reacts after the slow step occurs. The rate is already determined before the bromine enters into the reaction. Acetone is a polarized multiple bond and the methyl group is not nucleophilic. Under acidic conditions, the oxygen lone pair can be protonated, and this heteroatom-to-heteroatom proton transfer is usually fast. The protonated carbonyl is a great EWG, making the methyl hydrogens acidic. Loss of a methyl hydrogen produces the reactive enol, an allylic source that is nucleophilic on carbon that can attack bromine. The kinetic data tell us that the formation of the enol has to be slower than the enol attacking bromine. This is consistent with the enol tautomer being less stable than the keto form, so enol formation is uphill in energy. The final proton transfer is again expected to be fast.

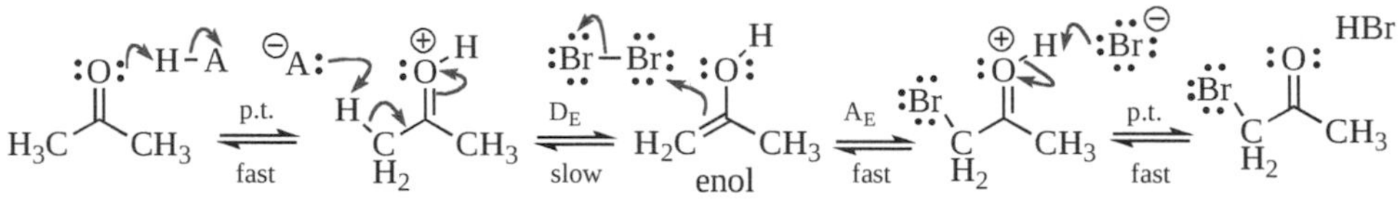

10.8.7 Solvent Polarity Effects

Solvent effects probe charge concentration or dispersal. Solvent polarity can stabilize charged species, and can affect the rate of a reaction, Section 2.7. If the reaction forms a charged intermediate or product from neutral reactants, increased solvent polarity will speed up the reaction by lowering the energy of the charged species. However, if charge is neutralized or dispersed during the reaction, a more polar solvent will slow down the rate because reactant stabilization increases the reaction barrier.

S_N2 charge type	Charge at ‡	Products	Solvent polarity effect
Nu: and R–L	increasing	$^+$Nu–R and L:$^-$	Polar increases rate
Nu:$^-$ and R–L$^+$	decreasing	Nu–R and L:	Polar decreases rate
Nu:$^-$ and R–L	dispersing	Nu–R and L:$^-$	Polar decreases rate
Nu: and R–L$^+$	dispersing	$^+$Nu–R and L:	Polar decreases rate

Example problem

The substitution of an alkyl chloride by an anionic nucleophile is slowed as the reaction solvent becomes less polar, is this effect more compatible with an S_N1 or S_N2 mechanism?

Answer: The S_N1 rate-determining step is the ionization of the leaving group to form a carbocation and a chloride ion. The solvent must stabilize these charged species, or the reaction will be slowed or stopped. The S_N2 reaction rate would be expected to increase slightly since the charge is dispersed in the transition state. The observed solvent effect is inconsistent with an S_N2 mechanism and is consistent with the S_N1 (see Section 2.7).

10.8.8 Substituent Effects

Substituent effects probe the electronic demand of the slow step. Electron-withdrawing or releasing groups can greatly alter the rate of a reaction and sometimes can even cause a change in mechanism. A common way to study substituent effects is to connect a benzene ring that has a meta or para substituent group to the site undergoing reaction.

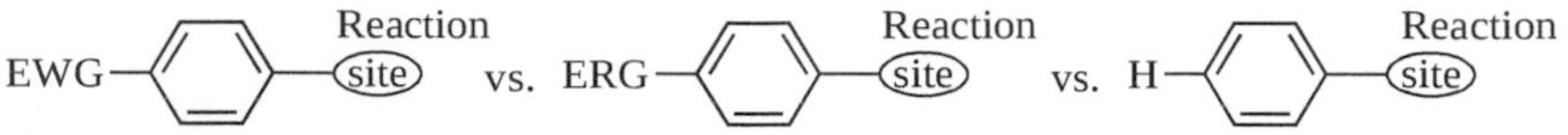

This substituent group is then varied, and the rate of the reaction is compared to the rate when the group is simply H. If electron-releasing groups accelerate the rate of reaction, the conclusion is drawn that either a positive charge is increased or a negative charge is decreased in the slow step. If electron-withdrawing groups accelerate the rate, either a negative charge is increased or a positive charge is decreased in the slow step.

Example problem

The rate of the substitution reaction of $Ar(CH_3)_2C{-}Cl$ with water is greatly favored by electron-releasing groups on the aryl (Ar) group. Is this substituent effect explained better by an S_N1 or S_N2 mechanism?

Answer: Since electron-releasing groups help, then positive charge is increased or a negative charge is decreased in the rate-determining step. The S_N1 rate-determining step is the formation of a carbocation that donors on the aromatic ring would stabilize. The charge-dispersed S_N2 transition state would not be expected to have a significant substituent effect.

10.8.9 Primary Deuterium Isotope Effects

Is C–H bond breaking in the slow step? The fact that a C–H bond is just slightly easier to break than a C–D bond can be used to probe reaction mechanisms that have the C–H bond breaking in their rate-determining step. In this case, the rate of the deuterium-substituted reactant will be two to seven times slower.

Example problem

The rate of the elimination reaction of fully deuterated 2-bromopropane with ethoxide is 6.7 times slower than the rate of the undeuterated. Does this observation fit with an E1 or E2?

Answer: Since the E1 rate-determining step is leaving group loss and not C–H bond breakage, it would not show a primary deuterium isotope effect. However, an E2 reaction would, because C–H (D) breaking occurs in the transition state of the concerted E2.

10.8.10 Barrier Data

Both $\Delta H^‡$ and $\Delta S^‡$ and therefore $\Delta G^‡$ can be determined ($\Delta G^‡ = \Delta H^‡ - T\Delta S^‡$) from how the rate constant changes as a function of temperature. Figure 10.30 shows an Eyring plot in which the natural log of k/T (the rate constant divided by temperature in Kelvin) is plotted against $1/T$. We get $\Delta H^‡$ from the slope and $\Delta S^‡$ from the intercept.

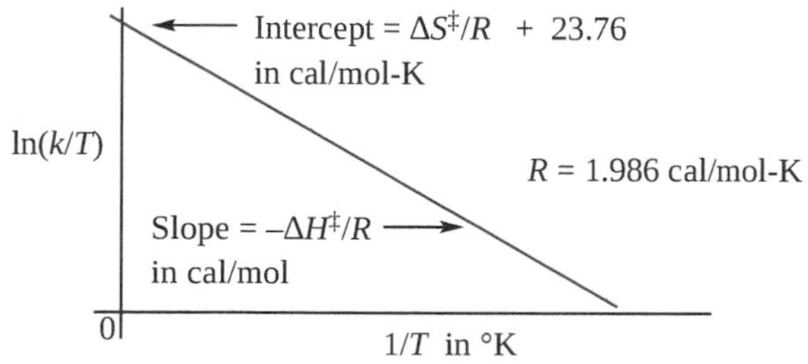

Figure 10.30 Example Eyring plot to determine $\Delta H^‡$ and $\Delta S^‡$ from plotting $\ln(k/T)$ vs. $1/T$.

Since bond breaking in the transition state increases the $\Delta H^‡$, and bond formation diminishes it, processes that have simultaneous bond breakage and formation are expected to have a low $\Delta H^‡$. Transition states that require a specific spatial arrangement of reactants will tend to have a negative $\Delta S^‡$ because the degree of disorder decreases in going to the transition state. A low $\Delta H^‡$ and negative $\Delta S^‡$ for a gas-phase reaction is indicative of a concerted process with an ordered transition state. However, reactions in solution can have a negative $\Delta S^{\ddagger}$ because the degree of solvent disorder decreases in going to a highly solvated transition state, even as the actual reaction disorder increases.

10.9 LESSONS FROM BIOCHEMICAL MECHANISMS

Section 2.8 discussed two major contributors to enzymatic catalysis; lowering the $\Delta G^{\ddagger}$ increases the number of molecules with sufficient energy to react, and confining the reacting partners within an enzyme's active site greatly increases the number of effective collisions. Binding in the substrate in the proper orientation for reaction significantly reduces the $T\Delta S^{\ddagger}$ contribution to the $\Delta G^{\ddagger}$, thus lowering the reaction barrier. The lowering of the $\Delta H^{\ddagger}$ component was addressed briefly with the statement that enzymes bind the transition state best. Section 2.8 ended with the promise that we would return for a general look at the chemical step of enzymatic catalysis, and how the $\Delta H^{\ddagger}$ for an enzyme catalyzed reaction can be significantly lowered.

Solution reactions have several disadvantages, but the greatest difficulty is the inherent disorder of a liquid. Since the probability of a three-molecule collision is rare, almost all collisions involve just two molecules, and paths like the Ad_E3 addition result from a reactant collision with weakly bound pi complex or hydrogen-bonded pair. On the other hand, an enzyme active site can prearrange all the catalytic groups so they are in just the correct position and orientation. The solvation within the active site is also pre-organized to bind the transition state or reactive intermediate best; in this way solvation energy lowers the barrier. Since the amino acid side chains in the active site can vary from very polar to very nonpolar, the solvation provided to the substrate can vary in a similar way. The binding of the substrate in the active site can also destabilize the reactant relative to the transition state, thereby effectively lowering the reaction barrier.

Not only do enzyme cavities create their own solvent conditions, but in so doing they can adjust the pK_as of catalytic groups. A carboxylate anion placed in a nonpolar cavity is more basic than a free carboxylate. A cationic acidic group increases in acidity if placed in a nonpolar cavity, because poor solvation raises its energy. A good general acid catalyst will have a pK_a close to the physiological pH of 7. If the acid is too acidic, it will be mostly deprotonated at physiological pH, and therefore very few will still have the acidic H to serve as an acid. Likewise, a good general base catalyst will have its pK_{abH} close to 7. Two groups that fit these criteria are the imidazole group of histidine (pK_{abH} about 6) and the carboxylic acid group of glutamic and aspartic acids (pK_a about 4–5), shown below.

Common enzyme active site acids | Common enzyme active site bases

The breaking down of sugars (glycolysis) to create energy is an important part of all cellular metabolism. Section 10.5.5 gave the conversion of glucose to fructose. The enzyme fructose-1,6-bisphosphate aldolase breaks that sugar in half and is an example of an enzyme using electrophilic covalent catalysis, Figure 10.31. The amine of an active site lysine forms an iminium ion (Section 10.5.2) with the carbonyl of the fructose. This more reactive iminium ion sink allows a reverse aldol E2 fragmentation to occur via the less basic enamine rather than through a more basic enolate ion. The resulting enamine is tautomerized (Section 7.3.8) to the iminium ion that is then hydrolyzed by the reverse of its formation. This releases dihydroxyacetone phosphate (DHAP) and R-glyceraldehyde-3-phosphate (G3P), and returns the enzyme active site lysine to the free anime, ready for the next cycle.

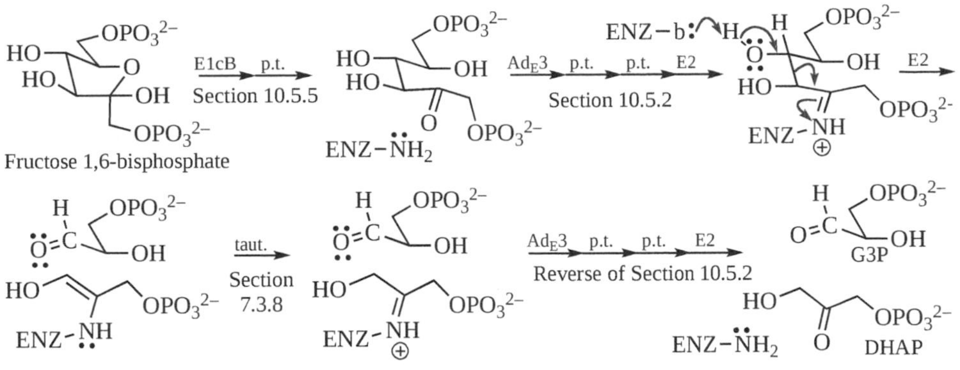

Figure 10.31 The enzyme fructose-1,6-bisphosphate aldolase breaks fructose into G3P and DHAP.

Enzyme mechanisms usually avoid high energy, unstable cationic or anionic intermediates that increase the reaction barrier. Triosephosphate isomerase, Figure 10.32, catalyzes the tautomerization of the achiral DHAP made in the reaction above to the chiral G3P by the mechanism shown below. Not only does the push–pull mechanism avoid forming a highly basic enolate, but also the binding mode of DHAP determines which face of the enediol will be protonated in the second step, and therefore which enantiomer of G3P forms. Triosephosphate isomerase has reached catalytic perfection as it can produce just a single enantiomer of G3P as fast as DHAP can diffuse in. It is estimated this enzyme can lower the barrier to reaction by about 12 kcal/mol. This push–pull catalytic route is another type of general acid and general base catalysis discussed in Section 7.4.3, where the important bond making and breaking is occurring in the same step as proton transfer.

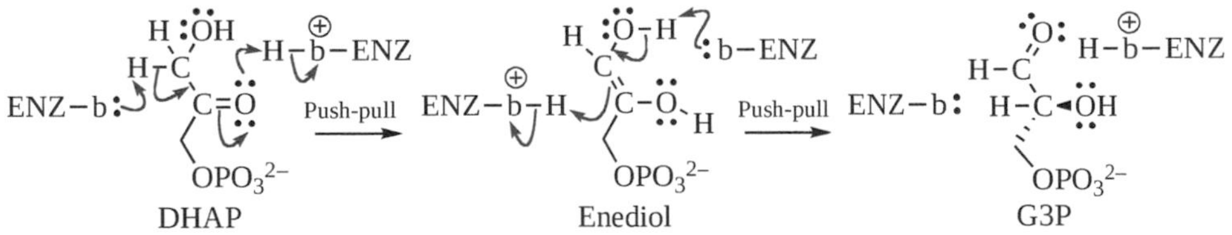

Figure 10.32 The enzyme triosephosphate isomerase converts DHAP into G3P.

Figure 10.33 shows how in the glycolysis of blood sugar, glucose metabolites can produce the cell's energy currency, ATP^{4-}, using the coenzyme NAD^+ (Section 7.2.6) to capture the energy from the oxidation (Section 8.3.2) of the G3P formed in Figure 10.32. The enzyme's cysteine SH group adds to the G3P aldehyde via a general base catalyzed addition. Hydride is then transferred to the cationic NAD^+ forming a very reactive thioester. Inorganic phosphate does an addition–elimination on the thioester to form an acyl phosphate, 1,3-BPG. In 1,3-BPG we now have a leaving group on the phosphorus activating it to be harvested by ADP^{3-} via an S_N2 or pentacovalent intermediate substitution process. The NADH thus formed is then recycled back into NAD^+ during oxidative phosphorylation by taking the oxygen you inhale and reducing it to water and in the process generating even more ATP^{4-}.

Enzymes may use covalent catalysis, forming a covalent bond to the substrate that is later broken before product release. This catalysis can be nucleophilic or electrophilic. In Section 4.2.12, we saw a nucleophilic catalysis by iodide of a substitution reaction; iodide was a better nucleophile than the original nucleophile, water, and a better leaving group than the original leaving group, chloride. Enzymes often use specialized reagents for this covalent catalysis, coenzymes like thiamine (vitamin B_1), which in Figure 8.12 was used to decarboxylate pyruvate to acetaldehyde. Most of the vitamins are critically needed coenzymes.

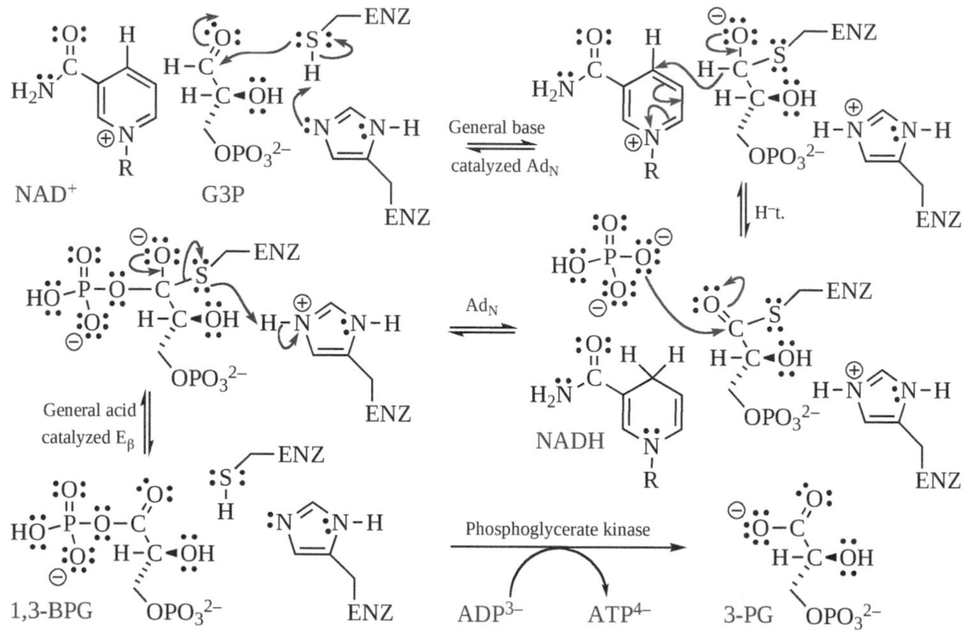

Figure 10.33 Enzymes glyceraldehyde-3-phosphate dehydrogenase, GAPDH, and phosphoglycerate kinase make ATP in glycolysis. GAPDH puts a leaving group on phosphate for ADP^{3-} to displace.

Enzymes, in addition to providing hydrogen bonding to anions formed, use electrostatic catalysis to polarize the reactant and compensate for a charge formed on the substrate. We can't protonate a carbonyl (pK_{abH} = –7) at physiological pH, so an enzyme uses Lewis acids such as Mg^{2+} to polarize the carbonyl instead. The Mg^{2+} can be viewed as a super proton at neutral pH for electrostatic/electrophilic catalysis (Section 6.2.3). Similarly, the acetaldehyde carbonyl's complexation to the Zn^{2+} ion in alcohol dehydrogenase (Section 7.2.6) makes the acetaldehyde carbonyl much more electrophilic, capable of reduction by NADH.

A most remarkable enzymatic use of divalent metal ion catalysis is in the glycolysis enzyme enolase, which catalyzes an elimination of water from 2-phosphoglycerate (2-PG) to give phosphoenolpyruvate (PEP) (Figure 10.34). Isotopic labels show the H exchanges faster than elimination, suggesting an E1cB mechanism, but how does the enzyme avoid forming a very basic anion? The enzyme uses two Mg^{2+} cofactors to increase the acidity of that H by stabilizing the conjugate base. Then pyruvate kinase, similar to phosphoglycerate kinase above, transfers an activated phosphate from PEP to ADP^{3-} to give pyruvate and more ATP^{4-}.

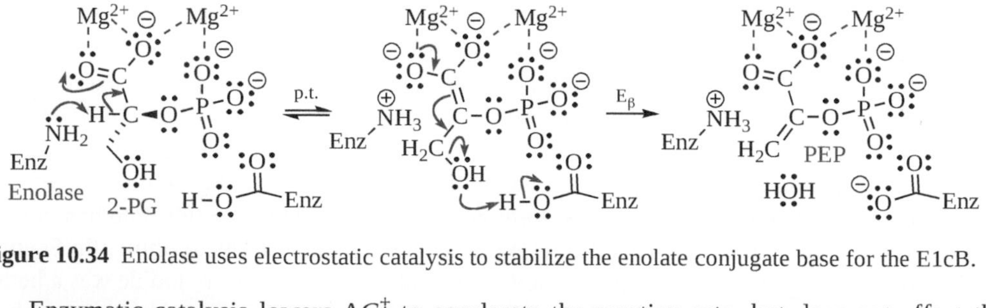

Figure 10.34 Enolase uses electrostatic catalysis to stabilize the enolate conjugate base for the E1cB.

Enzymatic catalysis lowers $\Delta G^{\ddagger}$ to accelerate the reaction rate, but does not affect the $\Delta G°$ that controls the position of equilibrium. However, cells also have an elegant ability to overcome an unfavorable equilibrium by coupling an endothermic reaction with exothermic ATP hydrolysis. For example, the reaction of acetate with coenzyme A (CoASH) is uphill by about 7 kcal/mol. The hydrolysis of ATP to ADP and phosphate is downhill by about the same amount. By coupling these two reactions as shown below, the cell achieves a total $\Delta G°$ of almost zero for a K_{eq} of about one. The phosphorylation of acetate by ATP improves the leaving group for the second reaction with CoASH. There are many enzymes whose catalytic process remains a mystery; enzymes still have much to teach us about reaction mechanisms.

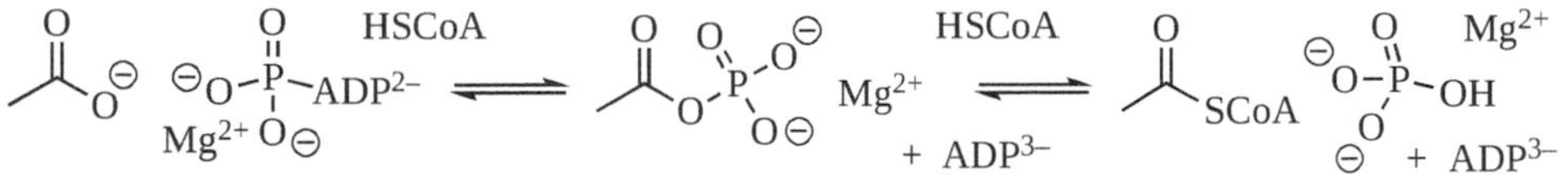

10.10 SUMMARY

The entire emphasis of this chapter is to use the trends to guide a deliberate decision process that generates reasonable alternatives then selects the best one. This decision process has a limited number of inputs (Figure 10.35). Mechanism problems have the additional input of the structural information from the product. We can see what bonds were made and broken to guide our decisions. Although this "generate and select" process might seem slower than what you may be used to, it is the essence of the scientific method. As the problems you attempt get more difficult, a deliberate and organized search process for solutions not only is more efficient, but also becomes very necessary. If you can internalize this "best-first search" process it will serve you well in many fields beyond organic chemistry.

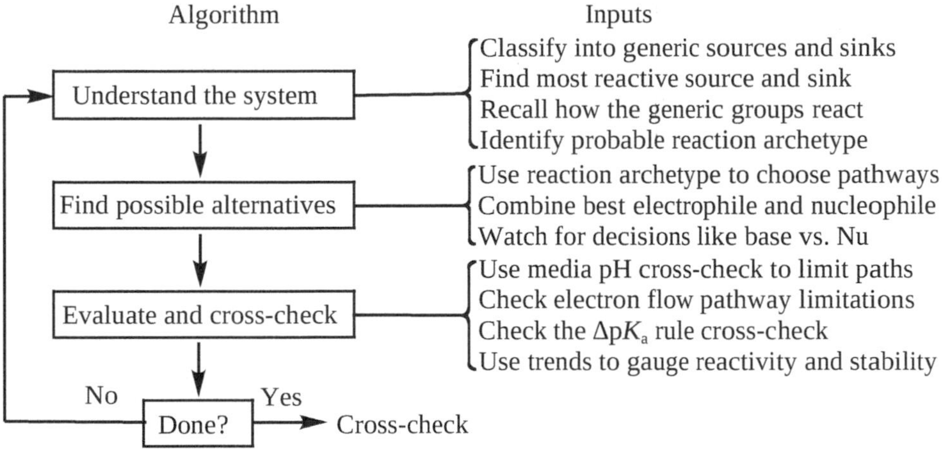

Figure 10.35 The organic problem solving algorithm and its common inputs.

CHAPTER 10 EXERCISES

(Refer to the Hints to Selected Problems from Chapters 8, 9, 10, and 11 if you need help).

10.1 Provide a mechanism for the following transformations.

(a)

(b)

(c)

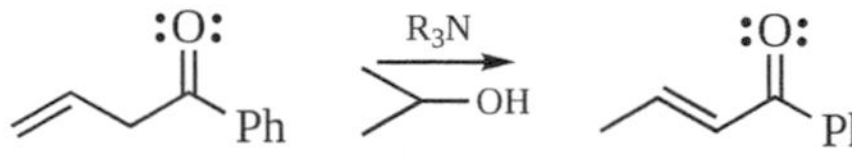

10.2 Provide a mechanism for the following transformations.

(a) 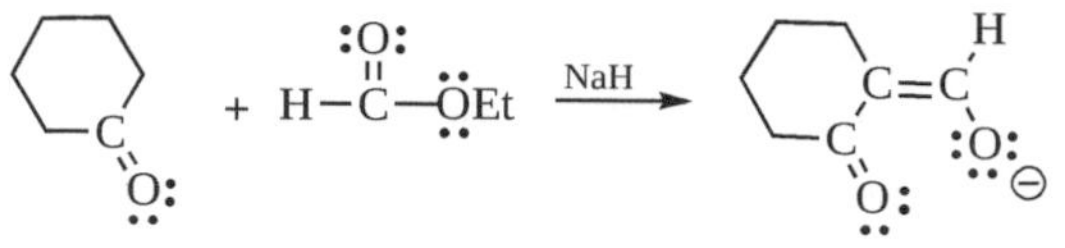

(b) 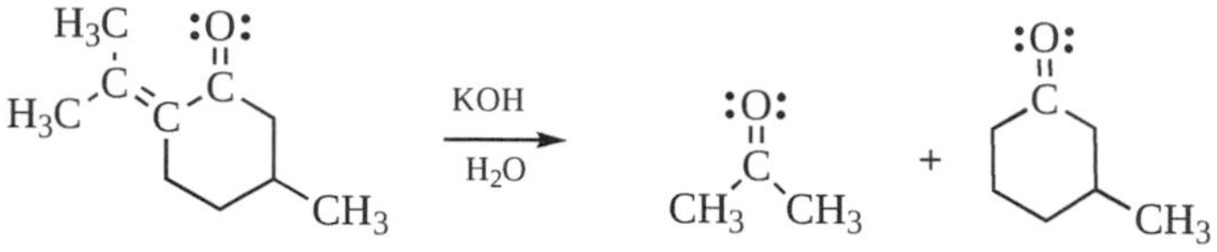

(c) 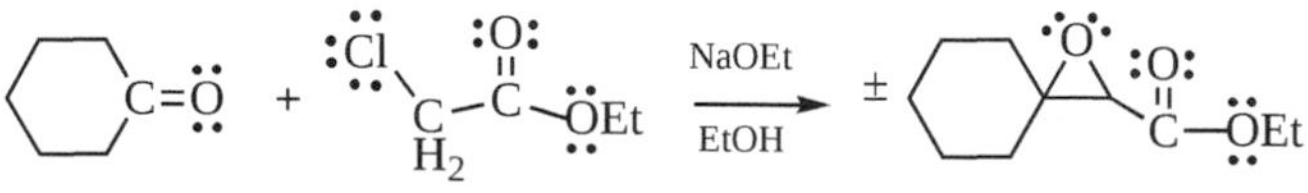

10.3 Provide a mechanism for the following transformations (medium difficulty).

(a) 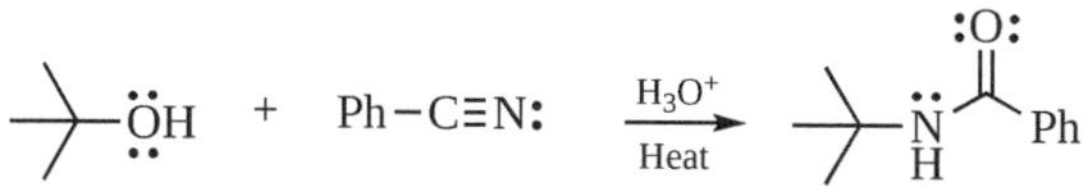

(b) 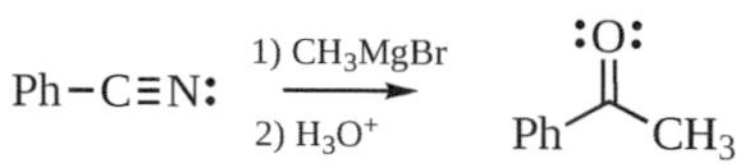

(c) 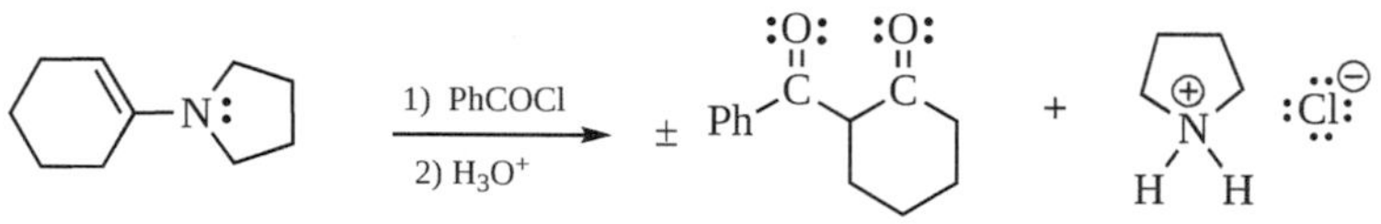

(d) 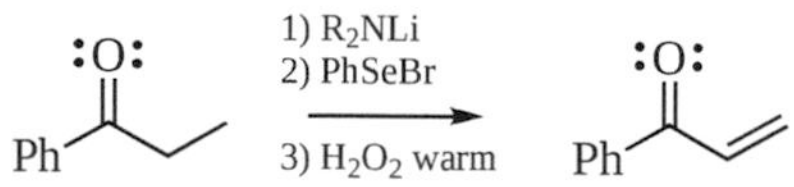

10.4 Provide a mechanism for the following transformations (medium difficulty).

(a) 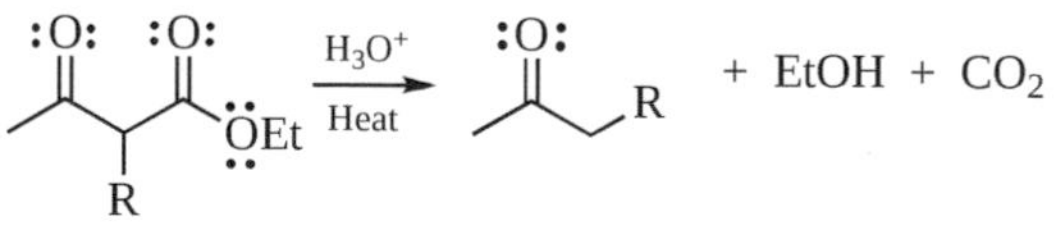

(b) 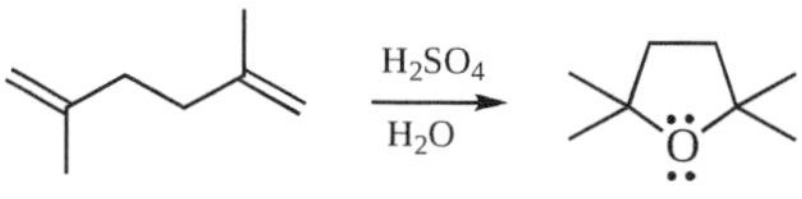

(c) 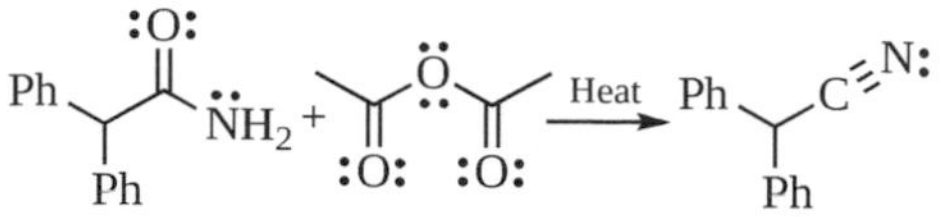

(d) 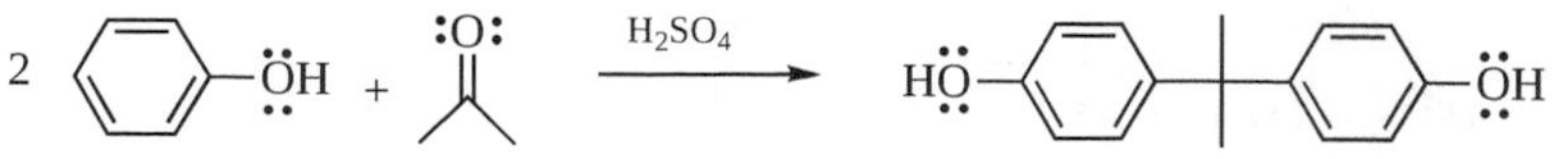

BPA

10.5 Give the product of the following reactions.

(a) 

(b) 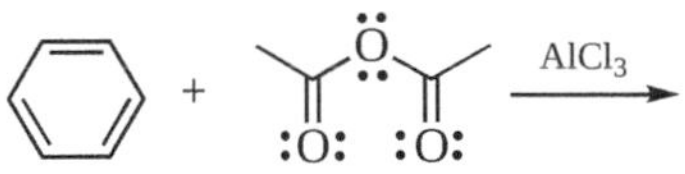

(c) $PhMgBr + D_2O \longrightarrow$

(d) 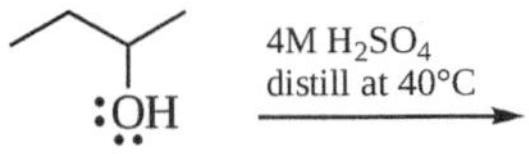

10.6 Give the product of the following reactions.

(a) 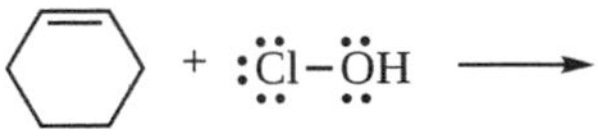

(b) 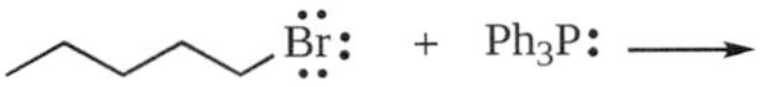

(c) 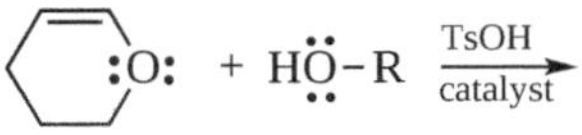

(d) 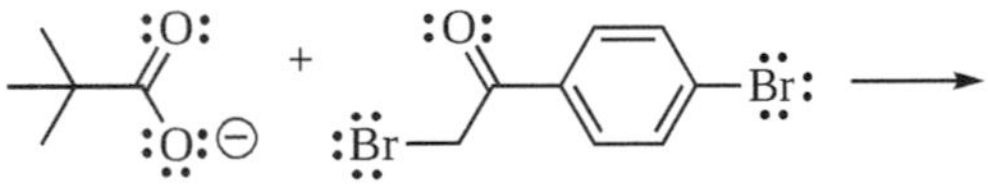

10.7 Give the product of the following reactions (more difficult).

(a) 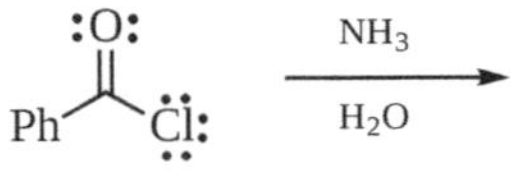

(b) $Ph{-}N{=}C{=}O + EtOH \xrightarrow{\text{Trace } R_3N}$

(c) 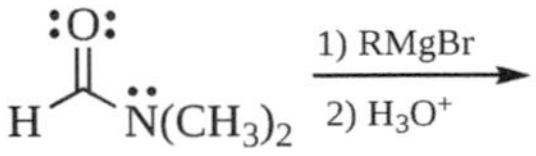

10.8 Give the product of the following reactions (more difficult).

(a) 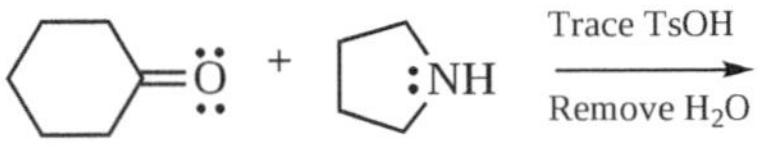

(b) 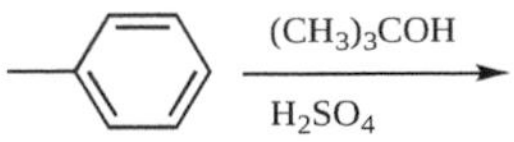

(c) 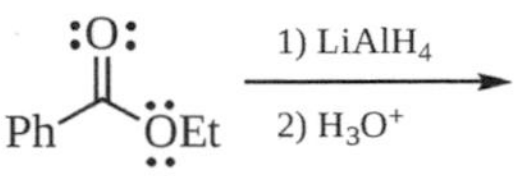

10.9 Provide a mechanism for the following transformations that is consistent with the experimental results provided.

(a) Similar systems run in ^{18}O water show no ^{18}O in the alcohol product.

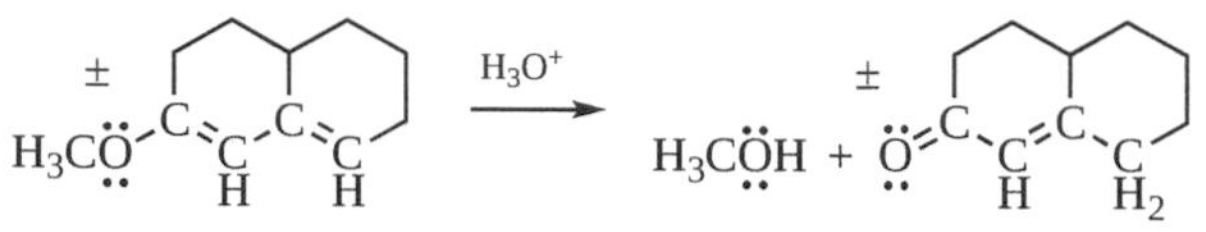

(b) This reaction requires cyanide ion as a catalyst and fails to go if it is absent.

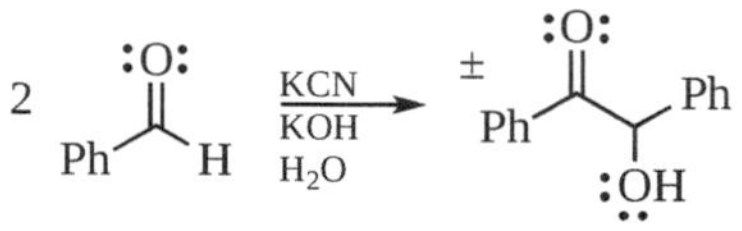

(c) Only this stereoisomer of product is formed.

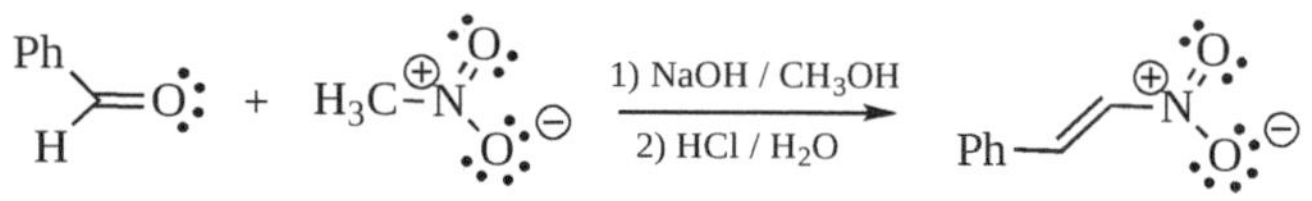

10.10 Provide a mechanism for the following transformations that is consistent with the experimental results provided.

(a) The following reaction is unusual in that it goes with complete <u>retention</u> of configuration. Propose a mechanism that explains this.

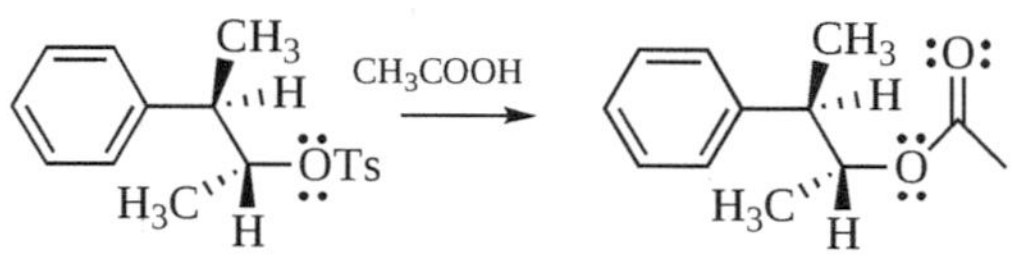

(b) Only this stereoisomer of product is formed at –72°C.

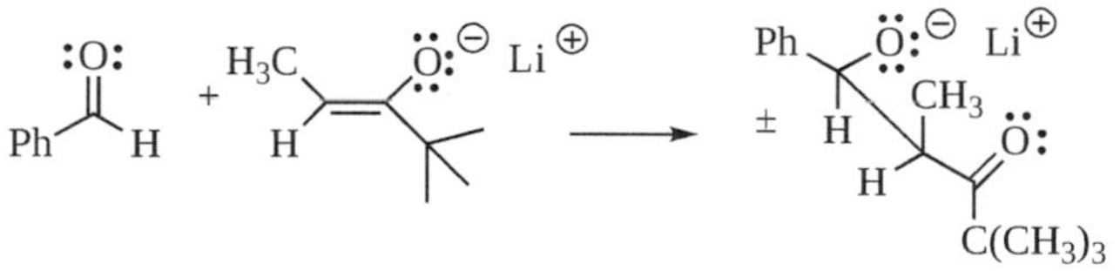

(c) Chiral optically active reactant gives a <u>racemic</u> product with overall <u>retention</u> of stereochemistry.

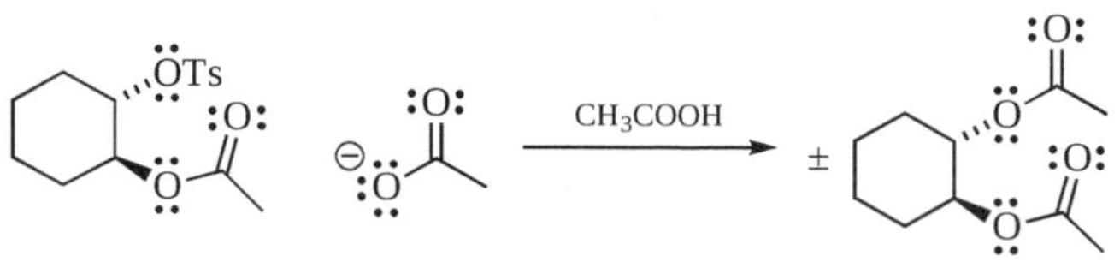

10.11 Provide a mechanism for the following transformations (more difficult).

(a)

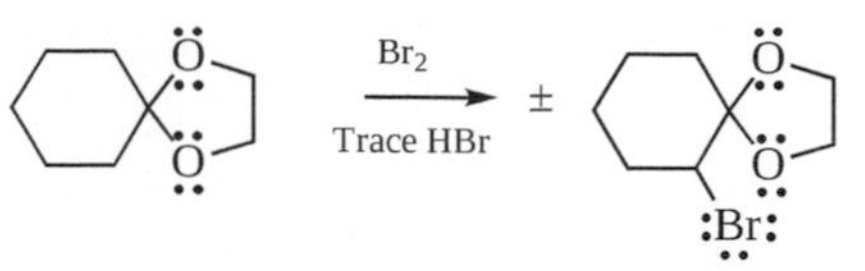

(b)

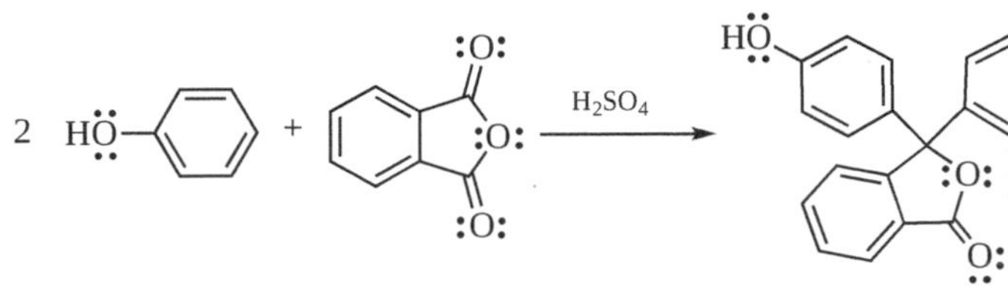

Phenolphthalein, the pH indicator

(c)

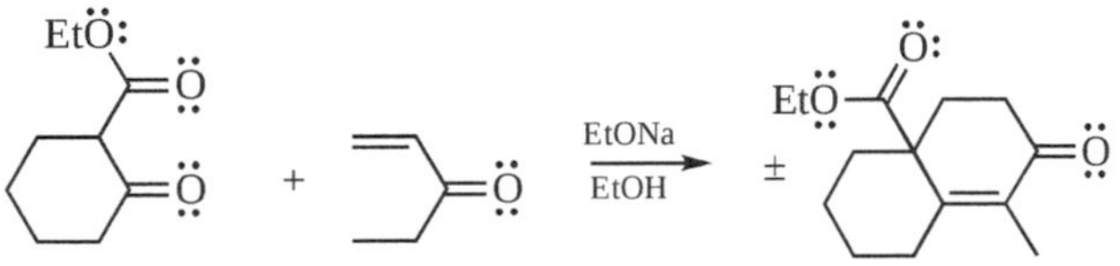

10.12 Provide a mechanism for the following transformations (more difficult).

(a)

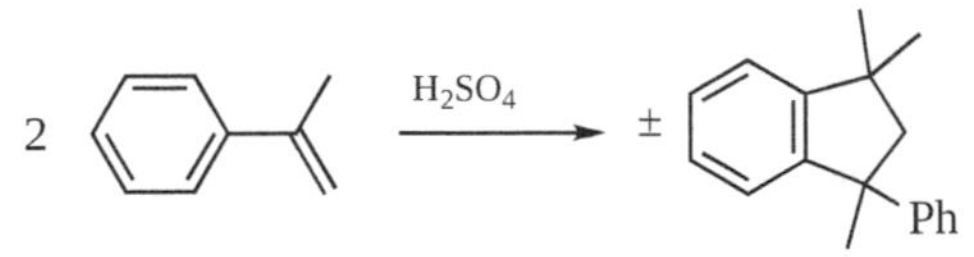

(b)

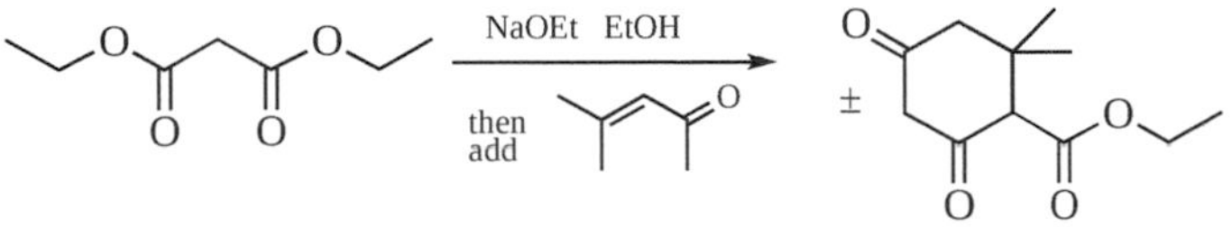

(c)

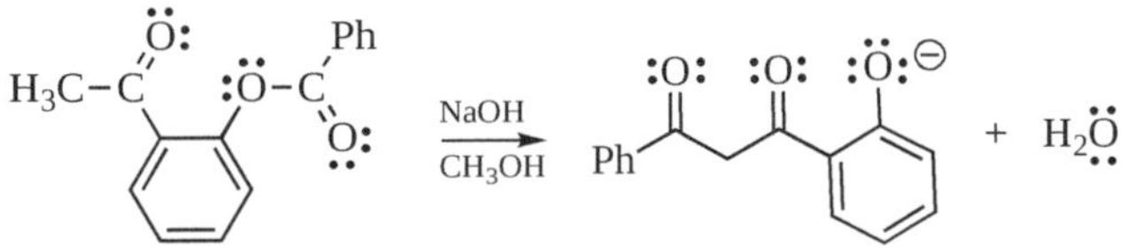

(d)

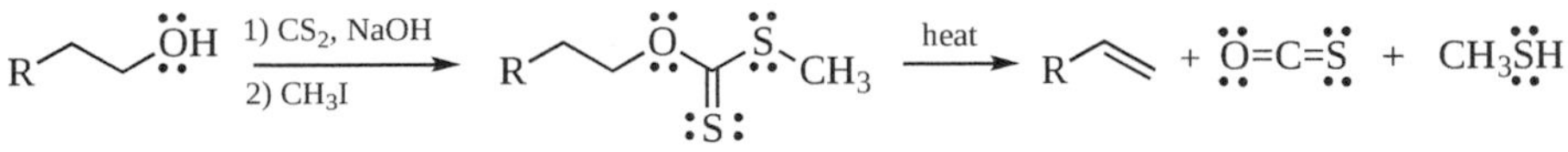

10.13 Part of the dark reactions of photosynthesis is interconversion of sugars with an enzyme called transketolase using thiamine pyrophosphate, TPP, as a catalyst (Section 8.12.8). Provide a reasonable mechanism for this enzymatic reaction. In addition to water, there are weak general acids and general bases present in the active site at pH 7.

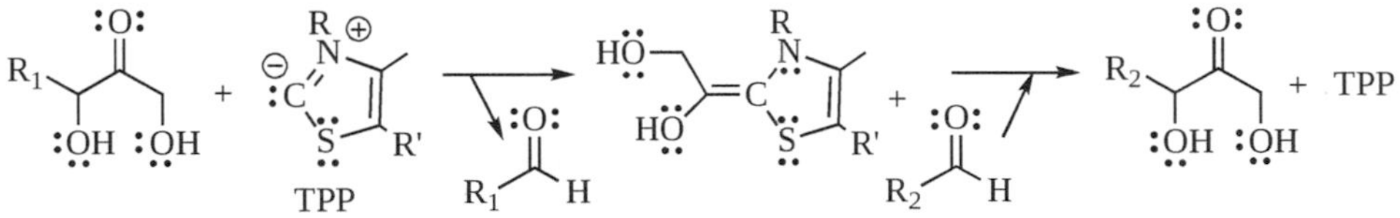

10.14 The biosynthesis of limonene (odor of orange peel) is shown. A monoterpene cyclase enzyme converts geranyl diphosphate by the steps shown. Provide a reasonable mechanism for this enzymatic reaction. In addition to water, there are weak general acids and general bases present in the active site at pH 7. Diphosphate, OPP, is a leaving group.

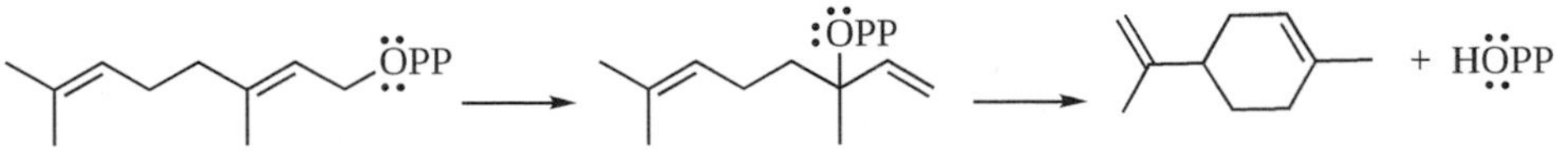

Geranyl diphosphate

Limonene

11

CROSS-CHECKS AND DECISION BOUNDARIES

Electron Flow in Organic Chemistry: A Decision-Based Guide to Organic Mechanisms,
Third Edition, Paul H. Scudder.

11.1 CROSS-CHECKS ARE CRITICAL TO A REASONABLE HYPOTHESIS

Formulating a reasonable hypothesis is a critical part of the scientific method, but how do we determine what is "reasonable?" Those critical cross-checks for reasonability are covered in this chapter. They need to become automatic if you wish to have your mechanisms judged as possible explanations for how a reaction works. Reaction mechanisms are never proven. They always remain hypotheses, capable of being disproved by a new piece of experimental data. Just because a mechanism rearranges the lines and dots of the Lewis structures of the reactants into the Lewis structures of the products does not mean it is chemically possible. If your mechanism is to be deemed reasonable; it must be consistent with already existing experimental data. This chapter gathers and explores some common-sense guidelines that all mechanisms need to follow. It provides evidence as to why we think that the guidelines are valid. These guidelines have boundaries that are fuzzy because the balance can depend on reaction conditions. In light of that, this chapter recommends approximate decision boundaries to allow you to determine what is most likely reasonable and what is probably not.

A hypothesis must also be energetically reasonable; therefore, it is very important to be aware of the energetics of a mechanistic proposal. The lowest-energy reaction path is the fastest. If our mechanistic hypothesis requires too much energy, it will not be the most reasonable. Most of the following cross-checks test for reasonable energetics.

A beginning student is often tempted to run with the first idea that comes to mind when working a complex problem, what is called in artificial intelligence a depth-first search. In contrast, experienced scientists would ask, "what are my options and which is the best alternative?" In artificial intelligence, this is called a best-first search. Using these cross-checks must become an automatic habit, like looking both ways before crossing the street. For example, failure to check the stability of an intermediate can be fatal to the proposed mechanism. Using these cross-checks might even allow you to find errors in your textbook, as several examples in this chapter came from published texts. Just because a mechanism is published does not mean it is correct; maybe the reviewers missed the error. Mechanisms are just our best guess at the time. We have learned many important mechanistic principles in the last several decades. Those older published mechanisms that appear to ignore a basic cross-check may need reexamining.

Ernest Hemingway once commented, "In order to be a great writer, a person must have a built-in, shockproof crap detector." To reason through reaction mechanisms that you have never seen, you must be able to decide what is chemically reasonable and what is not. You need a functioning, shockproof, chemical crap detector. Having the entire Internet in your pocket is both a blessing and a curse if you can't tell truth from nonsense.

11.2 FUNDAMENTALS CROSS-CHECKS

Typo and Transcription Error Cross-check — Have you drawn what you intended? If you miscopy the problem, then it usually becomes very difficult to solve. The first step in solving a mechanism is to decide what has changed, so draw a Lewis structure for any part of the reactant or product that has changed. This will allow you to notice all changes. Line structure can hide hydrogen atoms that may be critical to the mechanism.

Lewis Structure Cross-check — has the octet been exceeded for second row elements? Make sure that the Lewis structures and their formal charges are correct. A most common error is to exceed the octet for second row elements like C, N, and O. Drawing Lewis structures near electron flow arrows allows for a good electron and atom count. Always start your arrows from drawn electron pairs, either lone pairs or bonds. Make sure that the Lewis structures and their formal charges are correct. An error on a formal charge can be especially dangerous since the attraction of opposite charges is used to predict reactive sites. Make sure you split ionic compounds correctly. For the example below, the sodium cation has the stable noble gas configuration of neon, and the hydride anion has the stable noble gas configuration of helium.

NaH becomes $Na^{\oplus}$ $:H^{\ominus}$ NOT $^{\ominus}Na:$ ✕ $H^{\oplus}$

Make sure any resonance forms agree with the electronegativity trends. Let the more electronegative atom win in the minor resonance form. The carbonyl resonance forms below serve as examples.

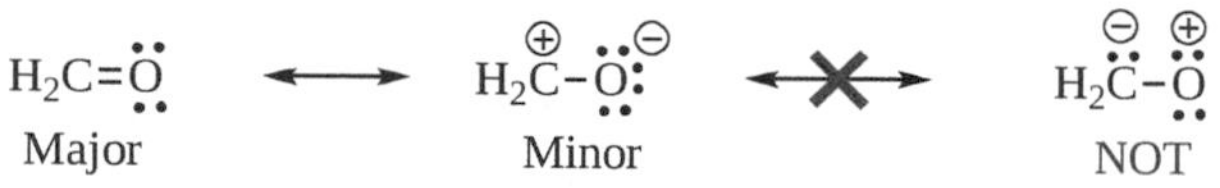

Line Structure Error Cross-check — Have the hydrogen atoms not drawn in the line structure been forgotten about and "vaporized"? Hydrogen atoms that are not drawn are easy to forget. Using Lewis structures near any electron flow arrows can avoid this error. An example student error is losing the wrong H in electrophilic aromatic substitution, as shown below.

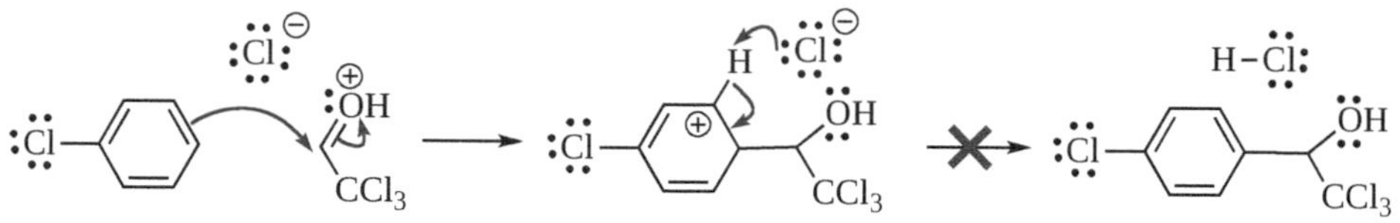

To avoid this line structure-induced error, we would first draw out Lewis structures near any changes. It then becomes obvious which H needs to be lost to restore the aromaticity of the ring. It is the H at the site of electrophilic attack that needs to go, as follows.

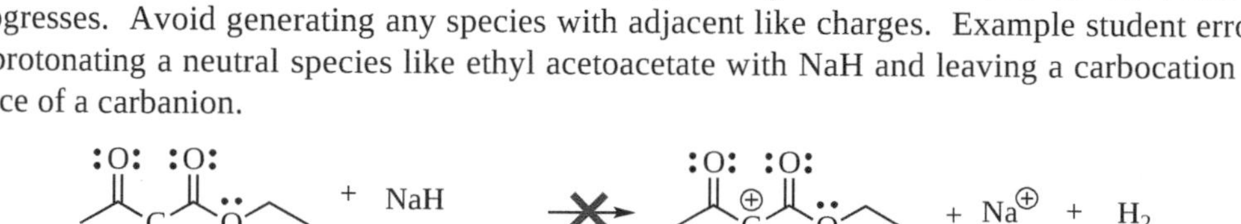

Charge Balance Cross-check — The charges on both sides of the transformation arrow must balance. Do not accumulate charges on your intermediates as the reaction progresses. Avoid generating any species with adjacent like charges. Example student error: deprotonating a neutral species like ethyl acetoacetate with NaH and leaving a carbocation in place of a carbanion.

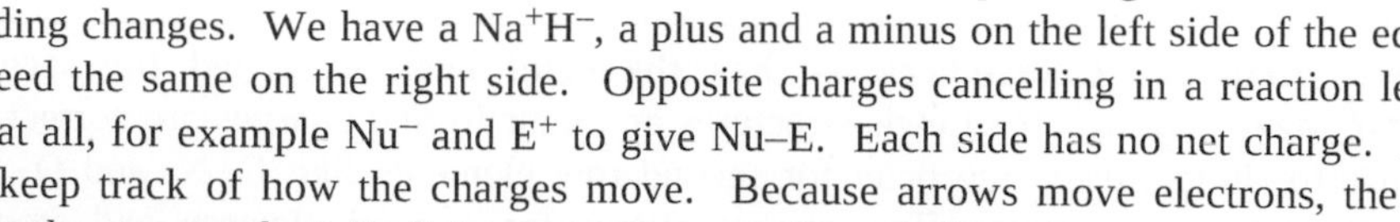

To avoid this error, we can again keep track of atoms and charges using Lewis structures near any bonding changes. We have a Na^+H^-, a plus and a minus on the left side of the equation, so we need the same on the right side. Opposite charges cancelling in a reaction leave no charges at all, for example Nu^- and E^+ to give Nu–E. Each side has no net charge. Arrows help us keep track of how the charges move. Because arrows move electrons, the formal charge on the atom at the tail of the arrow increases by one (becomes more positive), and the formal charge on the atom at the head of the arrow decreases by one (becomes more negative). The correct charge-balanced reaction is as follows:

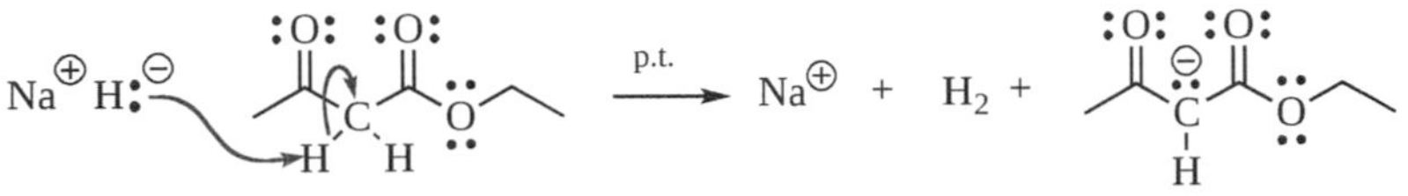

Atom Correspondence Cross-check — Do you have all the pieces and where do they go? Occasionally, organic equations are not balanced. Sometimes critical reactants are drawn above the arrow. Make sure you distinguish reactants from solvents and other pieces that do not become part of the product.

11.3 PROTON TRANSFER K_{eq} CROSS-CHECK

We can check the energetics of proton transfer reactions by calculating a K_{eq} (Section 3.6). A K_{eq} greater than 1 indicates that products are favored and that the $\Delta G°$ is negative. Proton transfers tend to form the weaker base. As a rough decision boundary, a K_{eq} much smaller than 10^{-10} is probably not useful or reasonable. Figure 11.1 shows that we can go downhill all we want, but we have limited energy available to climb a big energy hill.

Often, proton transfer is the first step in an organic mechanism. A common error is to forget to check the proton transfer K_{eq} and head off in an unreasonable direction at the very start. Always cross-check the first step.

The proton transfer K_{eq} decision boundary is just a subset of the ΔpK_a rule cross-check, Section 11.4, which uses the same decision boundary of 10 pK_a units uphill. The values in the pK_a chart in the appendix are scaled for water as a solvent. Solvation plays a critical role in basicity. Compounds in polar aprotic solvents have vastly different pK_a values. For example, ethanol has a pK_a of 16 in water, but 29.8 in DMSO. DMSO poorly solvates anions, thus does not stabilize the conjugate base. If the reaction is run in such a polar aprotic solvent, then you must use pK_a values determined in DMSO (https://organicchemistrydata.org/).

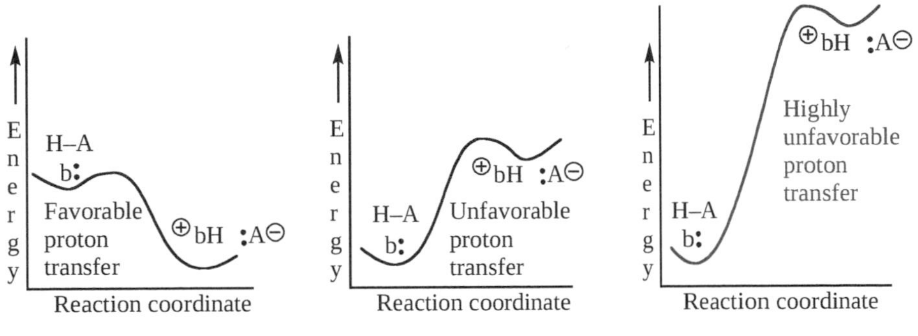

Figure 11.1 Energetics diagrams of various proton transfer reactions.

The first step in the electrophilic aromatic substitution of benzene with t-butanol in sulfuric acid (shown below) is to protonate the alcohol to form a good leaving group. Students who have tried to use sulfuric acid as a base to deprotonate benzene instead are obviously not using this cross-check, for the proton transfer K_{eq} is an impossible 10^{-53}. Sulfuric acid is a great acid but a really rotten base. To avoid major errors like this one, be sure you protonate your substrate in acidic media and deprotonate your substrate in basic media. Always check the proton transfer K_{eq} to see if it is reasonable.

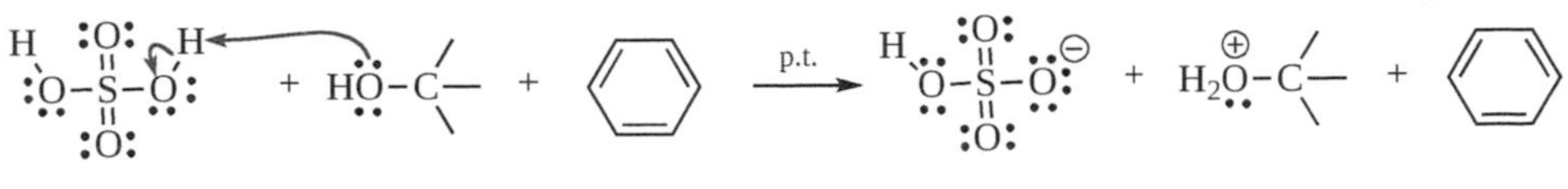

11.4 ΔpK_a RULE CROSS-CHECK

ΔpK_a Rule Cross-check — Avoid intermediates that are more than 10 pK_a units *uphill* from the reactants (either 10 pK_a units more basic or 10 pK_a units more acidic).

Reactions tend toward neutralization. A reaction has a limited amount of energy to climb uphill toward intermediates. On the other hand, the reaction can go very far downhill and release a lot of energy (as in an explosion). We need a cross-check to tell us when the energy climb to an intermediate is too great for the energy available. Using this ΔpK_a rule cross-check requires an extensive pK_a chart (see Appendix), but also a habit to make good use of it.

The ΔpK_a rule cross-check was designed into Prof. William Jorgensen's CAMEO expert system for predicting organic reactions (see bibliography). This cross-check was used to eliminate much of what he called "naive chemistry," such as this example below. This reaction has an incoming nucleophile pK_{abH} of 4.8 and an outgoing leaving group pK_{abH} of 16) for a climb of pK_a 11.2 units. This reaction fails completely.

This 10 pK_a unit decision boundary should be considered a fuzzy limit, not an absolute. There are possible exceptions for addition reactions that are run in hydroxylic solvents, with anion trapping by proton transfer. An example is the thiophenoxide ion conjugate addition to enones in water, a climb of 12.7 pK_a units, which is just beyond our 10 pK_a unit decision boundary. It might be possible, but a better alternative is suggested by the conditions.

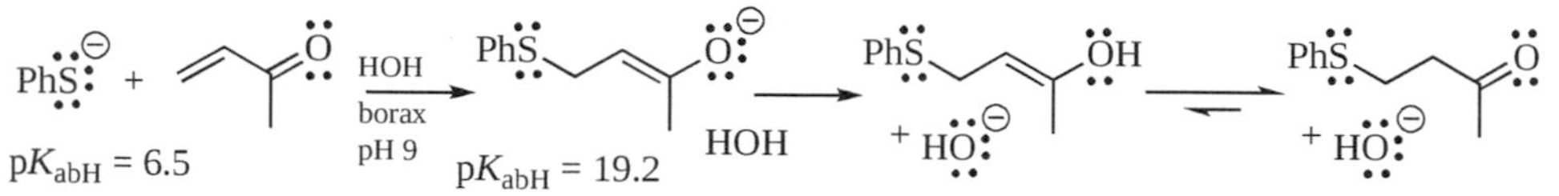

The reaction is run in a borax/water buffer that is a complex mixture that includes boric acid (pK_a 9.24), which can serve as an acid, H–A. Therefore, this reaction can occur by a hydrogen-bonded Ad_E3 addition, bypassing the basic enolate anion altogether. The Ad_E3 conjugate addition shown below would have a climb of 2.74 pK_a units using boric acid as HA, well below our decision boundary. Even water as HA works, with a climb of 7.5 pK_a units.

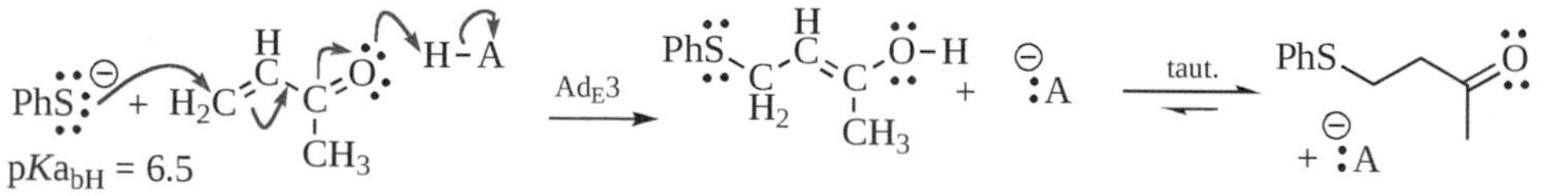

11.5 LEAVING GROUP CROSS-CHECK

Leaving Group Cross-check — This is also just a subset of the ΔpK_a rule cross-check. The pK_{aHL} should be zero or negative for S_N1, about 12 or below for S_N2, and can be as high as 25 or so for E1cB. In eliminations, the ΔpK_a rule cross-check is used to decide if there is enough push for kicking out a poorer leaving group.

The mechanism for basic aqueous amide hydrolysis is sometimes written forming a highly basic R_2N^- anion, pK_{abH} 36 from a not so basic tetrahedral intermediate of about pK_{abH} 11. An example textbook error is shown below. This is a climb in basicity from compound 2 to compound 3 of over 25 pK_a units and clearly disregards the pK_a cross-check.

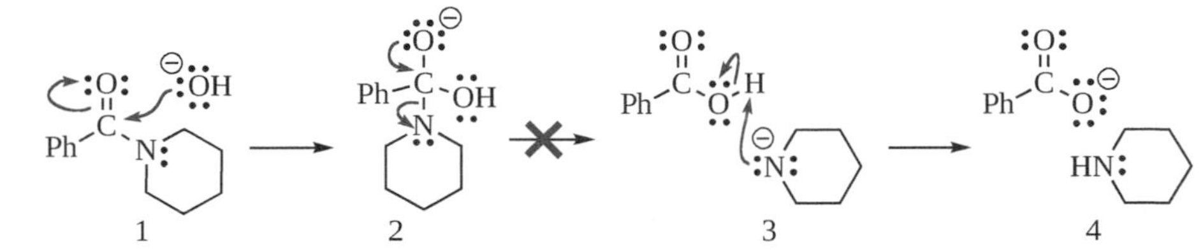

By the ΔpK_a rule cross-check, the only option available to intermediate 2 is to kick back out the hydroxide nucleophile (uphill 4.7 pK_a units), returning to compound 1, and indeed this happens. The hydroxide addition in basic amide hydrolysis has been shown to be reversible.

What evidence do we have that, *in the absence of a proton source*, intermediates like compound 2 are stable and that this elimination of the R_2N^- anion does not occur? Organometallics add to amides like *N,N*-dimethylformamide only once. In the absence of a proton source, the tetrahedral intermediate does not eliminate $(CH)_2N^-$ anion to form the aldehyde that would get hit again by the organometallic. Upon acidic workup, the nitrogen in the tetrahedral intermediate (analogous to compound 2 above) is protonated and becomes a much better leaving group, pK_{aHL} = 10.7. The protonated amine can then be lost from the zwitterionic tetrahedral intermediate to form the aldehyde and the neutral amine. This reaction is known as the Bouveault aldehyde synthesis.

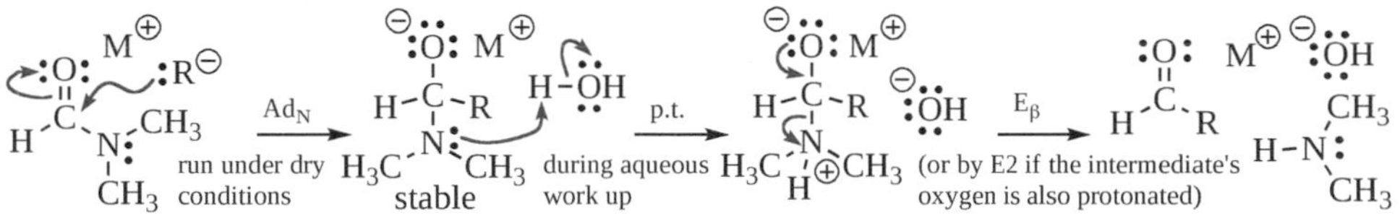

How can the textbook error example be fixed? We can easily protonate something in basic media, but just watch the proton transfer K_{eq}. A possible alternative conforming to the ΔpK_a rule cross-check recognizes that the nitrogen in the tetrahedral intermediate is relatively basic and can be protonated by water. The proton transfer K_{eq} is uphill by about five pK_a units, but within bounds. The zwitterion is internally stabilized by opposite-charge attraction. The protonated amine is no longer a very poor leaving group. The leaving group is then R_2NH, pK_{aHL} = 10.7, about the same basicity as the zwitterionic tetrahedral intermediate. The basic aqueous amide hydrolysis mechanism is covered in detail in Section 10.7.4, including two additional reasonable hydrogen-bonded catalyzed elimination alternatives.

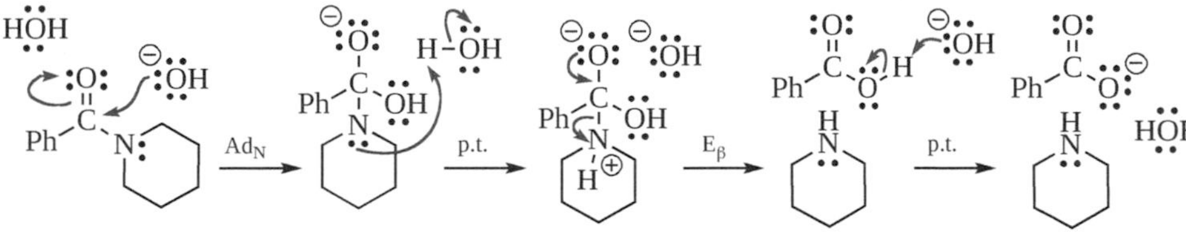

11.6 MEDIA pH SPAN CROSS-CHECK

Media pH Span Cross-check — No medium can possibly be both strongly acidic and strongly basic; it would neutralize itself. Therefore, the reactive species in equilibrium-controlled reactions have a limited range of acidities or basicities.

Let's explore extremes; we can easily have a solution that is a mixture of a weak acid and its conjugate weak base; this is just a buffer. For example, a pH 4.8 buffer can be made from acetic acid and an equimolar amount of sodium acetate, the conjugate base. Acetic acid has a pK_a of 4.8, and its conjugate base, acetate anion, therefore has a pK_{abH} of 4.8 (Section 3.2.1).

With no difference in the acid pK_a and the base pK_{abH}, then log K_{eq} = pK_{abH} – pK_{aHA} = 0 or K_{eq} is 1 (Section 3.6). The relative equimolar concentrations of acetic acid and its conjugate base, sodium acetate, do not change when they are mixed.

On the other hand, if we mix equal molar amounts of the strong base NaOH with the strong acid HCl, the acid and base irreversibly neutralize each other to produce NaCl and water at neutral pH. If we look at it from the opposite direction, a solution of neutral salt water has no strong acids or strong bases left in it. The neutralization reaction was complete. With the strong acid HCl (whose pK_a is –6) and the strong base NaOH (whose pK_{abH} is 14.0) there is a difference of 20 pK_a units between the pK_a and pK_{abH} that makes the proton transfer $K_{eq} = 10^{+20}$, and that irreversibly favors the neutralization products.

If the proposed pK_a span between strongest acid pK_a and strongest base pK_{abH} is zero, as we have in the buffer first example, it is totally reasonable to expect both the acid and base to be simultaneously present. However, if the pK_a span is 20 units, as it is in the second example, then it is unreasonable for the strong acid and the strong base to be simultaneously present in the solution. So where is the decision boundary? Water has 10^{-7} molar hydronium ion and likewise 10^{-7} hydroxide ion present in it at neutral pH 7 (K_w is $[H^+][OH^-] = 10^{-14}$). The pK_a span in this case is 14 pK_a units. Would a reaction mechanism proposal be reasonable if it required both hydronium and hydroxide ions? At what point does the span of pK_a values become unreasonable because the needed species have too low a concentration to react at a usable rate? It is known that neutral water is very poor at acid or base catalysis because of these low concentrations. We usually need higher concentrations for our acid or base to be useful.

To attempt to answer these questions, let's return to the example given in Section 3.6, the deprotonation of an ester (pK_a 25.6) by ethoxide ion (pK_{abH} 16); the pK_a span is 9.6. This small concentration of the ester enolate can then enter into a reaction with the original ester (see the product prediction in Section 10.7.2 for a discussion of the entire surface). If we try to run this reaction catalyzed by phenoxide anion (pK_{abH} 10) instead of ethoxide, the reaction fails; no product is isolated, only unreacted starting materials. There are two ways that we could interpret this result: first, the base was too weak and no appreciable concentration of ester enolate was formed. Another interpretation is that the driving force for the reaction was not sufficient, since the anion of product is now a slightly stronger base than phenoxide. The reaction is no longer forming a weaker base, but this difference of just 0.7 pK_a units should have given an equilibrium mix. The pK_a span with phenoxide ion was 15.6 pK_a units.

Recognizing that a decision boundary is a fuzzy guideline and not a sharp edge, we can set a reasonable pK_a span between the most acidic and most basic species that can coexist in useful concentrations. Therefore, as a first approximation to **the pK_a span decision boundary, a pK_a span of 10 units is acceptable, but a span of 15 pK_a units is probably not.** In your mechanisms, make sure that the span of pK_as (from most basic to most acidic) in equilibrium do not stray too far from the acceptable range. If several steps occur in the same medium, check the pK_a span of all species within those steps, since they are all equilibrating.

Some reactions need both acidic and basic species to be present to proceed. If it is too acidic, then there are not enough of the basic or nucleophilic species, and the reaction slows to a crawl. The same is true if it is too basic, and there are not enough of the acidic species to catalyze a different step. A good example of this pH rate dependence is the formation of an imine from a carbonyl, which is shown in Figure 10.9 in Section 10.5.2.

Certainly the pK_a span at which the reaction rate becomes too slow is open to debate. Exceptions are expected to arise when the driving force for a reaction is an irreversible proton transfer at the diffusion-controlled limit in water or a similar protic solvent. With a maximum bimolecular rate constant of 10^{10} L/mol-s and the usual reactant concentrations of one molar, a reactive intermediate could conceivably have a concentration as low as 10^{-15} mol/L, and the reaction still might proceed at a usable rate.

Using this pK_a span cross-check, we avoid the common beginner mistake of over-optimizing the reaction by combining the hottest electrophile and hottest nucleophile possible, then forgetting that the solution could not possibly have each present at the same time. An example of this beginner over-optimization error is having a protonated carbonyl (pK_a –6.5) being attacked by hydroxide (pK_{abH} 14.0) for a span of over 20 pK_a units. This has about the same pK_a span as in the previous example of HCl and NaOH not existing simultaneously in the same solution.

Over-optimized with a pK_a span crosscheck error

Media pH Cross-check —The route proposed must be consistent with the pH of the medium. The way to fix this error is to have strong acids exist only in acidic media and strong bases exist only in basic media. You must always keep track of whether the reaction is run in acidic or basic media. Acidic media contain powerful electrophiles and rather weak nucleophiles, whereas basic media contain excellent nucleophiles and weak electrophiles. Paths that form reactive cations usually occur in acidic media. Likewise, paths that form reactive anions are usually the domain of basic media. This cross-check meshes well with the pK_a span cross-check. Reasonable basic or acidic media alternatives to the over-optimized example above (protonated carbonyl with hydroxide) are as follows: in base, hydroxide attacks the unprotonated carbonyl; in acid, water attacks the protonated carbonyl.

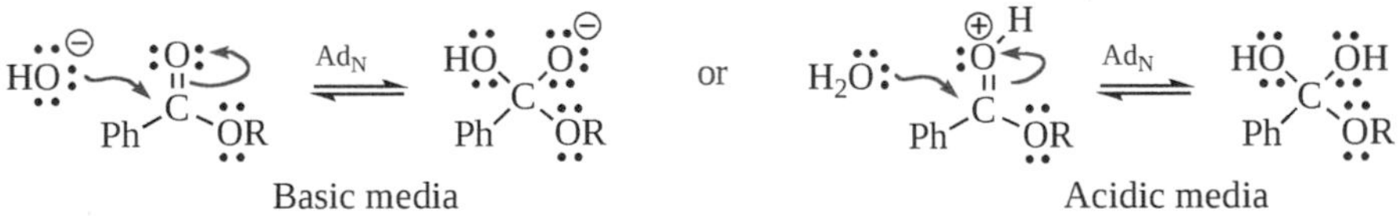

Step Order Cross-check — One final caution, reactions often have several steps, each in different medium. **Do the steps in the correct order.** You cannot go back to step one after beginning step two. Make sure you do not mix up the reaction conditions, for they are often drastically different. For example, organometallic reagents are usually run in dry aprotic solvents like ethers, but often have a dilute aqueous acidic work up as the second step. Any unreacted organometallic from the first step is destroyed in the acidic work up.

11.7 STABILITY OF INTERMEDIATES CROSS-CHECK

Has the reaction gone via the most stable intermediate? You need to know the relative stability of intermediates such as carbocations, carbanions, and radicals to be able to select between reasonable alternatives. The modern statement of Markovnikov's rule is that electrophilic additions to alkenes go via the most stable carbocation. Electrophilic aromatic substitutions go via the most stable sigma complex. Primary carbocations drawn in textbooks in the past are now thought too unstable to be intermediates in reactions; even secondary carbocations are probably rare.

A common error is to forget to check the stability of an intermediate. Proceeding through a very high energy intermediate is almost always energetically unreasonable; just as it is possible to get from Florida to the Pacific Ocean via the moon (as Apollo 8 did), such a high-energy-requiring process is exceedingly rare. Use the trends and cross-checks to gauge the stability of any intermediates formed in the reaction. If the stability trend indicates that the intermediate is highly unstable, then it is rarely formed, so find another route. For example, this intermediate stability error from a textbook fails both the ΔpK_a rule cross-check and the pK_a media span cross-check:

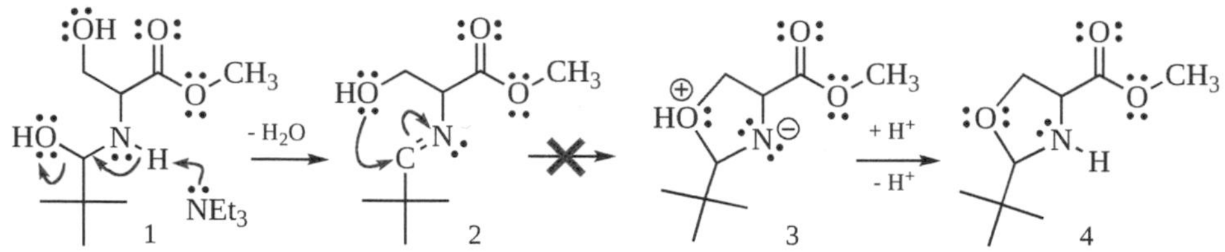

1 2 3 4

Intermediate 3, which has both a strong acid (protonated ether pK_a –3.5) and a strong base (deprotonated amine pK_{abH} 36), is being formed from a relatively stable neutral compound 2. The ΔpK_a rule cross-check indicates a climb of almost 40 pK_a units, and this number is also the impossible pK_a span of the intermediate 3.

To fix this textbook mistake, we must consider which decision was in error. The reaction from compound 2 to compound 3 is an addition to a polarized multiple bond, which has three alternatives that are shown in Figure 4.36 (which is for C=O but C=N would be similar): Ad$_N$2, Ad$_E$2, or Ad$_E$3. The author chose Ad$_N$2, which fails two cross-checks. The Ad$_E$2 is usually seen in acidic media, for the protonated polarized multiple bond is a good acid. But these conditions are mildly basic, using triethylamine (pK_{abH} 10.7). The Ad$_E$3 addition, shown below, is a good alternative passing all the cross-checks. The proton transfer is uphill by only 5.3 pK_a units (from pK_{abH} 10.7 to pK_{abH} 16), and the internal Ad$_E$3 attack on the hydrogen-bonded imine is expected to be fast, and is downhill by the same 5.3 pK_a units.

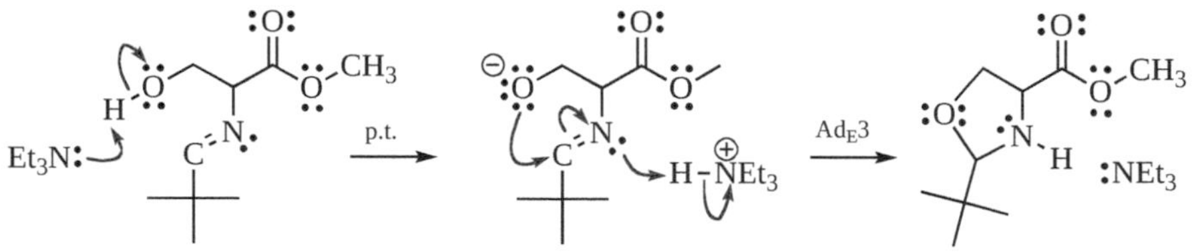

Carbocation Stability Cross-check — The ranking of carbocation stability is shown in Table 4.1. In section 4.4.3, the modern Markovnikov's rule was used to decide the Ad$_E$2 regiochemistry; the addition went via the most stable carbocation. In Section 4.2.9, the D$_N$ ionization of a leaving group usually requires a carbocation more stable than secondary. When we are trying to decide between Ad$_E$2 and Ad$_E$3, again **the carbocation stability decision boundary is the secondary carbocation.** Carbocations less stable than secondary usually require excellent leaving groups like triflates and optimum conditions to form. It is best to avoid highly unstable carbocations like primary alkyl, primary vinyl, phenyl, or methyl carbocations in your mechanisms. For example, the phenyl diazonium ion, Ph–N≡N$^+$, in spite of having the world's best leaving group would form one of the worst carbocations, and therefore is reasonably stable in solution, and some can (with great care) occasionally be isolated. Below is an example of an older textbook mechanism that goes via an unstable primary carbocation and the newer textbook version that bypasses that very unstable carbocation.

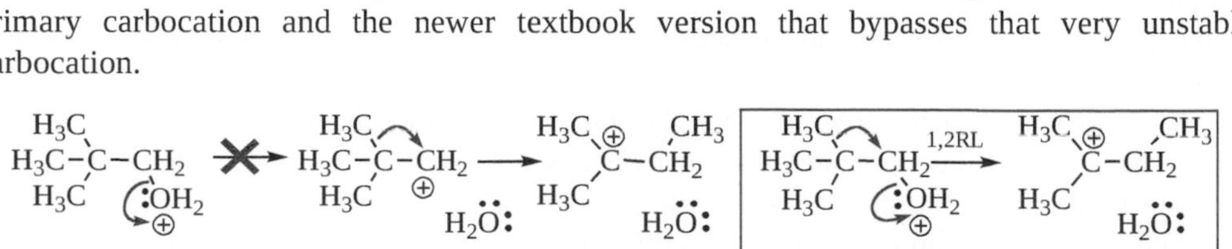

Older mechanism via unstable 1° carbocation

Newer mechanism avoids 1° carbocation

Cation Don'ts — Electron withdrawing groups, EWGs, stabilize anions but destabilize cations. **A cation next to an EWG is destabilized.** Because of the electronegativity of the EWG, that places a full plus next to a full or partial plus; adjacent like charges repel each other and create an unstable ion.

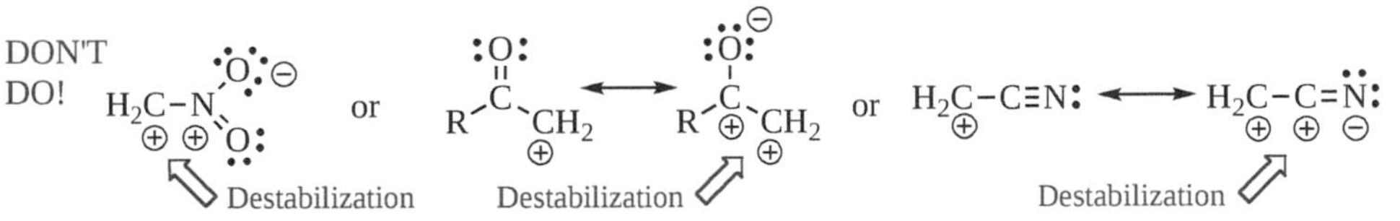

Avoid an Incomplete Octet on an Electronegative Atom. It is fine to have a positive charge on an electronegative atom as long as it keeps its octet.

Our route choice at the Cl_2-$AlCl_3$ complex shown below is just the normal substitution decision of S_N1 vs. S_N2, and we should pick S_N2 because of the instability of the Cl^+ ion. Just because one can fleetingly generate a Cl^+ in the high vacuum of a mass spectrometer with an ionization voltage of 70 electron volts (1610 kcal/mol), it does not mean that such an unstable ion can be created in a reaction solution at normal temperatures.

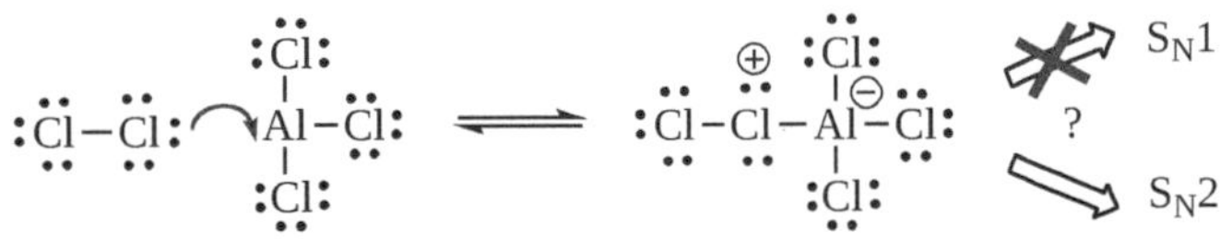

Carbanion Stability Cross-check —The most stable carbanion is the weakest base, having the lowest pK_{abH}. Since we often make reactive carbanions from even more reactive ones, we need to use the ΔpK_a rule cross-check to judge a decision boundary, because that will depend on the carbanion reactant. The following highly favorable reaction is 26 units downhill, but would not occur in the opposite direction, a climb of 26 units.

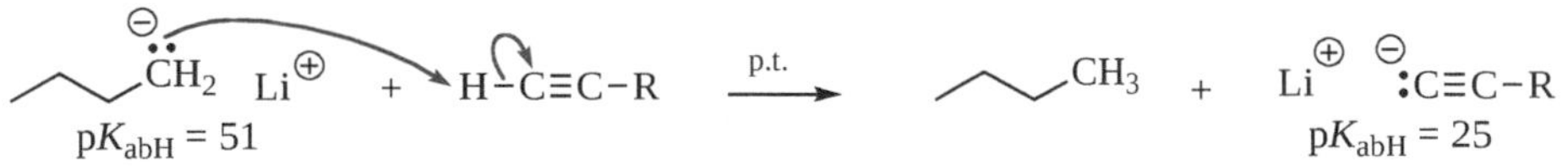

Anion Don'ts — Electron releasing groups, ERGs, stabilize cations but destabilize anions. **An anion next to an ERG is destabilized.** That places a full minus next to a full or partial minus; adjacent like charges repel each other and create an unstable ion.

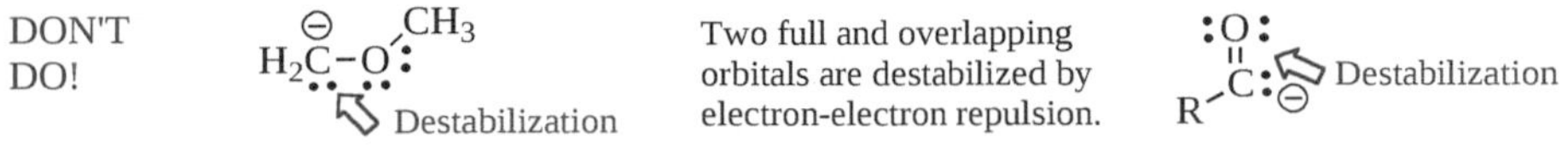

Don't make an anion by exceeding the octet of a second row element. Second row elements don't have the *d* orbitals or the size to exceed their octet. Please, no five-bonded carbon atoms or ten electron oxygen atoms.

Coulomb's Law Cross-check — Anions are electrostatically attracted to cations. Like charges repel each other. Because of like charge repulsion, cations do not attack cations nor do anions attack anions. Like charges that are adjacent are rarely formed. Because of this like charge destabilization, even nonadjacent organic dianions are uncommon, but are able to be made if strong enough bases are used. The resonance forms of ethyl acetoacetate dianion are shown below. Although the first pK_a is just 10.7, it takes a much stronger base than NaH (pK_{abH} 35) to remove a second proton to form the less stable dianion shown below.

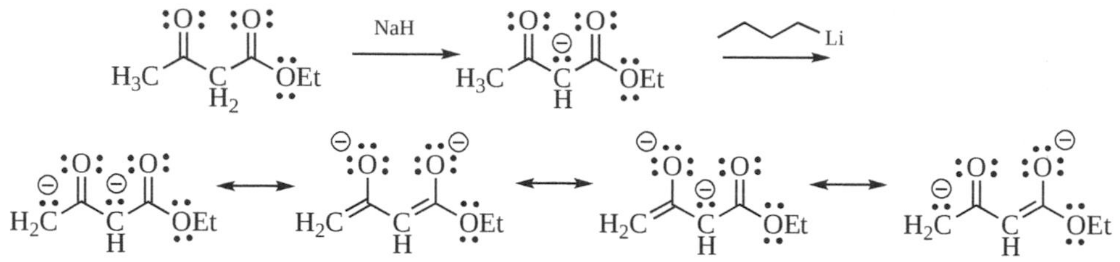

The Coulomb's law cross-check is complicated by the fact that solvents with a high dielectric constant can screen charges, thus reducing any repulsive forces. Dianions and dications are readily formed in polar solvents like water. For example bicarbonate ion, HCO_3^-, has a pK_a of 10.3 in water to form CO_3^{2-}, the carbonate dianion.

Another way to screen the electric field of a negative charge is to ion pair it with a compensating positive charge. In this way, an enzyme can cancel or even reverse the electrostatic repulsion of a negative species by pairing it with a 1^+ or 2^+ ion in its active site (enzymatic electrophilic catalysis, Section 10.9).

11.8 ELECTRON FLOW PATHWAY AND COMPLETENESS CROSS-CHECKS

An important check is to make sure that each mechanistic step is a valid electron flow path. The mechanistic "sentence" you write must use known "words" to be understandable. There is a temptation, arising out of frustration, to rearrange the lines and dots of reactants into the lines and dots of products with a mechanistically meaningless barrage of arrows just to get the problem over with. **It is important that you assemble your mechanistic hypothesis from known steps, the 18 electron flow pathways** (Chapter 7). Check that the electron flow starts at a lone pair or bond that is a good electron source (Chapter 5). Make sure that the flow continues without interruption and ends at a good electron sink (Chapter 6). Did you consider alternate pathways that achieve the same overall transformation (the energy surface street corner analogy of Chapter 4) and then pick the most appropriate? Use the appropriate energy surfaces to expand the mechanistic alternatives that you have to choose from (Chapters 4 and 10).

When you generate the possible paths, make sure to **include any important alternatives, then pick the most probable and check that decision. It is very useful to think of the energy surface for the process and remind yourself that there are usually three choices available.** For example, an addition can be Ad_N2, Ad_E2, or Ad_E3. A common problem-solving error is to race off with the first possibility that looks reasonable, failing to check all the possibilities. If your reactant has more than one reactive center, decide which is more reactive. Make sure you consider reactions at either site if they are close in reactivity. Essentially, what else can it do? Finally, check for possible differences in product stereochemistry or regiochemistry.

A path completeness check is especially important for the first step, where the initial direction that you set out to explore is determined. If your first step is in error, you may be reluctant, several steps later, to go back to the start and reconsider. Rather than retreat to the start, you may be more likely, as frustration builds, to force an answer just to be finished, ignoring whether the process is actually chemically reasonable. Keeping to known electron flow paths increases the probability that your mechanism would be considered reasonable, especially if it passes this chapter's cross-checks.

Example text error: Some older biochemistry texts have drawn the electron flow arrow in a proton transfer from H^+ to the site they want to protonate. A full-headed arrow moves two electrons from its tail to its head. An H^+ has no electrons at all to move. Have the electron flow arrows point toward electrophiles (or toward the tail of another arrow), and do not have arrows come from electrophiles. All electron flow arrows must start from drawn electron pairs, either a bond or a lone pair.

11.9 CROSS-CHECKS FOR SPECIFIC PATHS

Although most of these cross-checks can be found scattered throughout Chapter 7, it is useful to have them collected together in one section here. In addition to proton transfer reactions (Chapter 3 and Section 11.3), there are four types of organic reactions: substitution, elimination, addition, and rearrangement that have specific cross-checks. Use the energy surfaces from Chapter 4 to decide which route within an organic reaction type is most reasonable. For example, the elimination surface reminds you that eliminations can go by E1, E2, or E1cB. As was done in Chapter 10, consider linking the energy surfaces together to give you a general overview of what is possible for mechanistic routes in a reaction.

Substitution at an *sp*^{3} Center — The reactants starting point on the *sp*^{3} substitution surface (Figure 4.2) requires a choice between S_N2 and S_N1 for carbon bound leaving groups. The pentacovalent route on the surface is not an option for carbon but works for third-row elements like phosphorus. Make sure you check these alternatives and make a deliberate decision on which path to take.

S_N2 substitutions are often basic media routes, because a good nucleophile is often also a good base. The nucleophile must react with to the C–L carbon, so S_N2 reactions don't occur on highly hindered centers like tertiary butyl and neopentyl. The S_N2 should pass the ΔpK_a rule cross-check. The S_N2 does not occur on *sp*^{2} centers like C=C–L or Y=C–L.

S_N1 substitutions are often acidic media routes, because the leaving group is improved by protonation. The S_N1 has the loss of the leaving group as the slow step, and that requires three conditions to be met: a good leaving group ($pK_{aHL} \leq 0$), a reasonable carbocation (usually better than secondary), and a reasonably polar solvent to stabilize charges.

Eliminations — At the reactant starting point for the elimination surface (Figure 4.16), a three way decision, E1, E2, or E1cB, must be made, usually based on whether the media is acidic, basic, or close to neutral. Make sure you check these alternatives and make a deliberate decision on which path to take.

E1 eliminations are preferentially acidic media routes because the leaving group is improved by protonation. The E1 has the same three conditions that need to be met as the S_N1, for the E1 has the same slow step, the loss of the leaving group.

E1cB eliminations are mainly basic routes because the first step is a deprotonation of the reactant by a good base. This deprotonation step must pass the ΔpK_a rule cross-check. The E1cB can kick out a poorer leaving group, but use the ΔpK_a rule cross-check to be sure.

E2 eliminations are often invoked when the E1 and E1cB fail their cross-checks. The energy surface folds because both the cation and anion corners of the surface are raised in energy due to unstable intermediates. The lowest energy process is then down the middle, the E2 elimination. The E2 requires the carbon–hydrogen and carbon-leaving group bonds involved to be nearly coplanar in order to form a new pi bond. Twisted pi bonds are unstable and hard to make. The smallest easily isolatable twisted pi bond is trans-cyclooctene. Larger rings are fine. For eliminations in cyclic systems it is best to make a molecular model to check for proper alignment.

Substitution vs. Elimination Cross-check — On product predictions, remember to check if the reaction conditions favor substitution or elimination. This decision is based on three main variables: nucleophilicity, basicity, and hindrance (Section 9.5). Please learn these three trends. Whether an anion serves as a nucleophile or a base is probably the most common and often the most difficult decision you will have to make in an organic product prediction problem.

Additions — At the reactant starting point for the addition surface (Figure 4.32), a three way decision, Ad_N2, Ad_E2, or Ad_E3 must be made, usually based on whether the media is acidic, basic, or close to neutral. Make sure you check these alternatives and make a deliberate decision on which path to take.

Ad_N2 additions are preferentially basic media routes requiring good nucleophiles and should pass the ΔpK_a rule cross-check.

Ad_E2 additions are preferentially acidic media routes that often form an acidic intermediate as their first step. One cross-check is Markovnikov's rule, additions go via the most stable carbocation intermediate. The intermediate carbocation must be reasonably stable, preferably better than secondary.

Ad_E3 additions are often invoked when the Ad_N2 and Ad_E2 fail their cross-checks. The energy surface folds because both the cation and anion corners of the surface are raised in energy due to unstable intermediates. The lowest energy process is then down the middle, the Ad_E3 addition. The Ad_E3 addition follows Markovnikov's rule even though a discrete carbocation is not formed because the pi bond polarization by the electrophile is similar.

Three molecule collision cross-check — One valid complaint about the Ad_E3 addition is that the odds of three independent molecules bouncing around in a flask ending up in the exact same spot at the same time with sufficient energy and the correct orientation to react is exceptionally rare. Rationalizing the Ad_E3 occurring at all invokes the temporary stickiness of hydrogen bonding or pi complexation. If our nucleophile hits such a complex, then that requires just two things to collide: the complex and the nucleophile, not three independent molecules. The only other time we see three molecule collisions is when the molecules are confined to a very small box, the active site of an enzyme. A main reason enzymes accelerate reactions is due to the extremely high number of effective collisions within the confines of an enzyme's active site.

Rearrangements — One of the more common rearrangements is to an electrophilic center (Sections 8.11 and 9.8). The rearrangement reaction's electron sink is the electrophilic center that the group migrates to. For the rearrangement to be energetically favorable, the rearranged product should be more stable than the reactant. The Figure 9.9 energy surface shows that the rearrangement decision is similar to the S_N2 vs. S_N1 decision, where we must decide whether to form an intermediate carbocation before migration or to bypass that carbocation with a concerted migration process (the diagonal of the surface). If the carbocation is unstable, like primary or worse, then the energy surface folds, and diagonal concerted process is then of lower energy.

11.10 SUMMARY

Typo and Transcription Error Cross-check — Have you drawn what you intended?

Lewis Structure Cross-check — Don't exceed the octet of second-row elements.

Line Structure Error Cross-check — Have the hydrogen atoms not drawn in the line structure been forgotten?

Charge Balance Cross-check — The charges on both sides of the transformation arrow must balance.

Atom Correspondence Cross-check — Is the reaction balanced? Do you have all the pieces, and where do they go?

Proton Transfer K_{eq} Cross-check — Proton transfers tend to form the weaker base. **Unfavorable proton transfers with a K_{eq} less than 10^{-10} are most likely too unfavorable to be useful.**

Δp*K*a Rule Cross-check — A reaction has a limited amount of energy to climb uphill toward intermediates. **Avoid intermediates that are more than 10 pK_a units *uphill* from the reactants (either 10 pK_a units more basic or 10 pK_a units more acidic).**

Leaving Group Cross-check — This is also just a subset of the ΔpK_a rule cross-check. The pK_{aHL} should be zero or negative for S_N1, about 12 or below for S_N2, and can be as high as 25 or so for E1cB, where a ΔpK_a rule cross-check can decide if there is enough push for kicking out a poorer leaving group.

Media pH Span Cross-check — No medium can be both strongly acidic and strongly basic; it would neutralize itself. **A pK_a span of 10 units is acceptable between the most acidic and most basic species that coexist in solution, but a span of 15 pK_a units is probably not.**

Media pH Cross-check — The route proposed must be consistent with the pH of the medium. **Acidic media contain powerful electrophiles and weak nucleophiles, whereas basic media contain excellent nucleophiles and weak electrophiles.**

Carbocation Stability Cross-check — In the D_N ionization of a leaving group decision and that between Ad_E2 and Ad_E3, **the carbocation stability decision boundary is the secondary carbocation. A cation next to an EWG is destabilized. Don't make an incomplete octet cation on an electronegative atom.**

Carbanion Stability Cross-check — The most stable carbanion will be the weakest base, with the lowest pK_{abH}. Use the ΔpK_a rule cross-check. **An anion next to an ERG is destabilized. Don't make an anion by exceeding the octet of a second row element.**

Coulomb's Law Cross-check — Opposite charges attract; like charges repel each other. Don't jam a minus charge into another minus charge. Bases and nucleophiles are almost always neutral or anionic. Acids and electrophiles are almost always neutral or cationic.

Electron Flow Pathway and Completeness Cross-check — Speak with known words. **Assemble your mechanistic hypothesis from the known electron flow pathways.** When you generate the possible alternative routes, remember that the energy surfaces usually show three routes to consider. **Find any important alternatives, then pick the most reasonable one and check that decision.** This completeness check is especially important for the first step, where the initial direction that you set out to explore is determined.

Three Molecule Collision Cross-check — The odds of three independent molecules colliding at the same time with sufficient energy and in the correct orientation to react are extremely poor (except within the confines of an enzyme's active site). Two species in an Ad_E3 collision of a nucleophile with a pi-complexed or hydrogen-bonded species are fine.

Specific Path Cross-checks — Every path has limitations that must be considered. If the route passes one cross-check but fails another cross-check, look for another alternative.

A <u>reasonable</u> mechanism must pass all the required cross-checks, not just a few.

The whole purpose of this chapter has been to encourage you to develop the habit of cross-checking your proposed mechanisms so that they might be considered **reasonable** explanations of how a reaction works. This learned habit of cross-checking is transferable to many other fields since it is the foundation of excellent critical thinking and of a good medical diagnosis. This critical thinking habit may be the most valuable skill you learn in organic.

CHAPTER 11 EXERCISES

(Refer to the Hints to Selected Problems from Chapters 8, 9, 10, and 11 if you need help).

11.1 In each of the reactions below, a step is crossed out that failed a cross-check. What cross-check was missed? Use the energy surface for that process to find an alternative that would satisfy the cross-check. Draw a mechanism that passes all cross-checks.

(a)

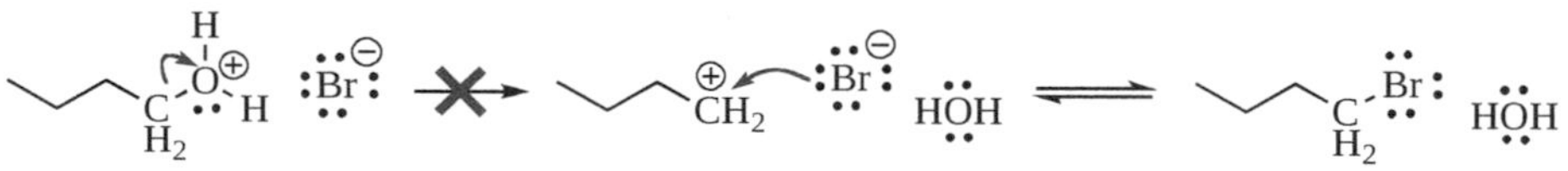

(b)

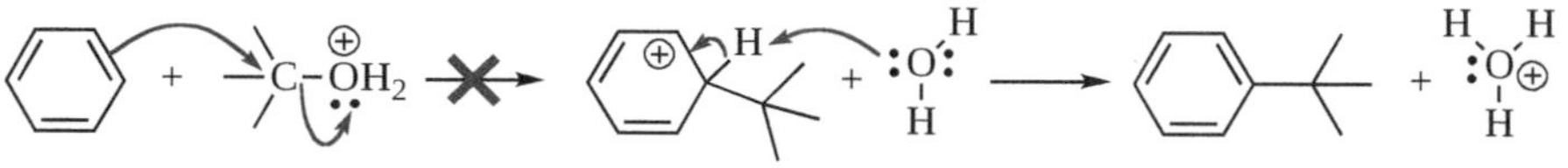

11.2 In each of the reactions below, a step is crossed out that failed a cross-check. What cross-check was missed? Use the energy surface for that process to find an alternative that would satisfy the cross-check. Draw a mechanism that passes all cross-checks.

(a)

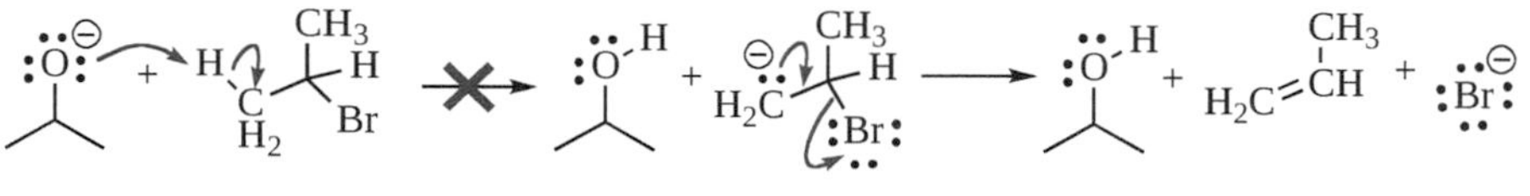

(b)

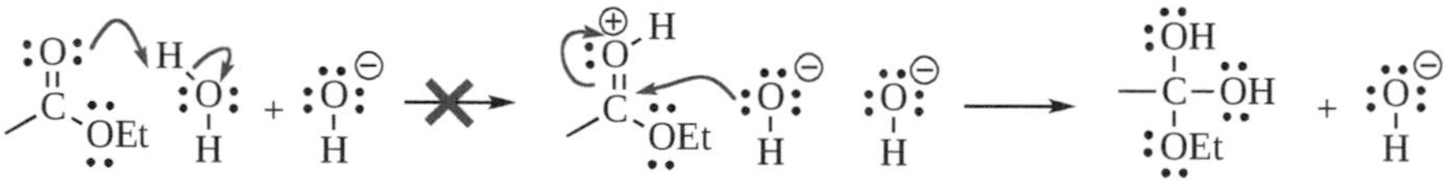

11.3 The reaction below fails a cross-check and the reaction does not happen. Why?

(a)

(b)

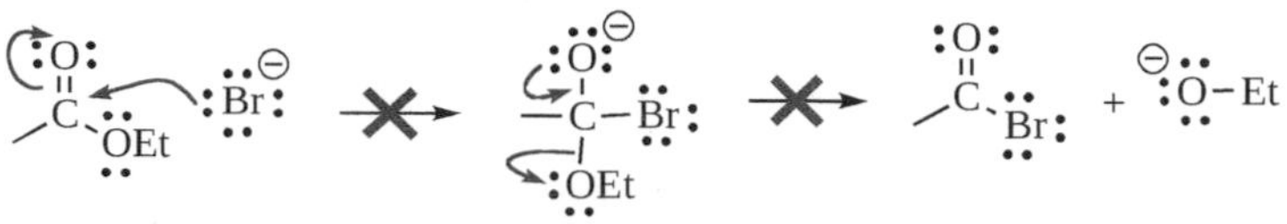

11.4 The reaction below fails a cross-check and the reaction does not happen. Why?

(a)

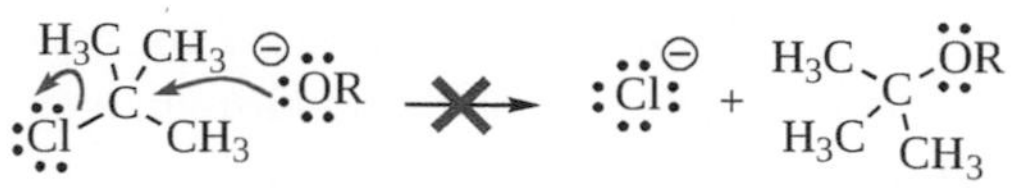

(b)

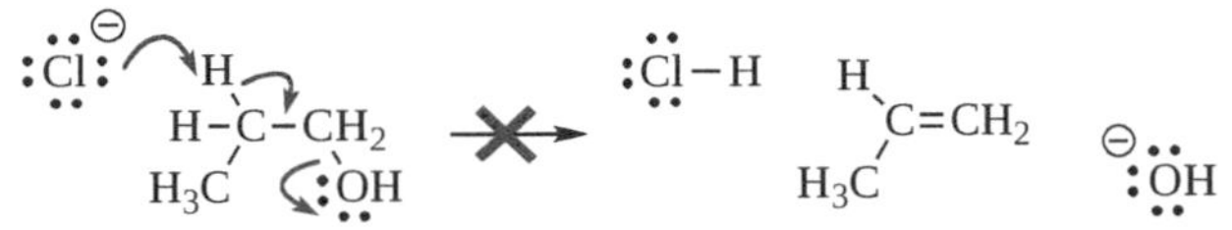

11.5 The reaction on the left is very poor, but a similar reaction on the right works. Why?

(a)

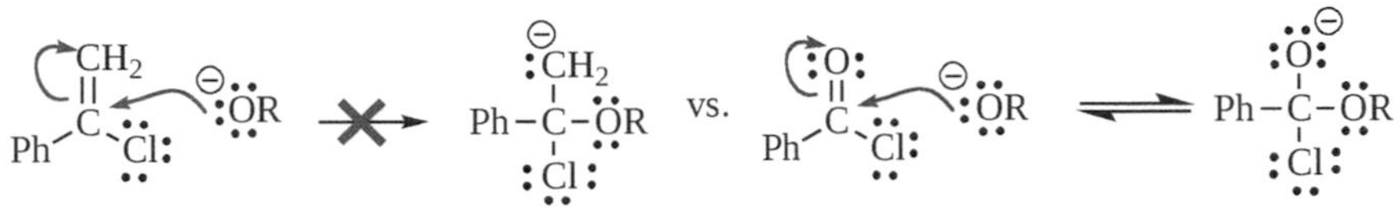

(b)

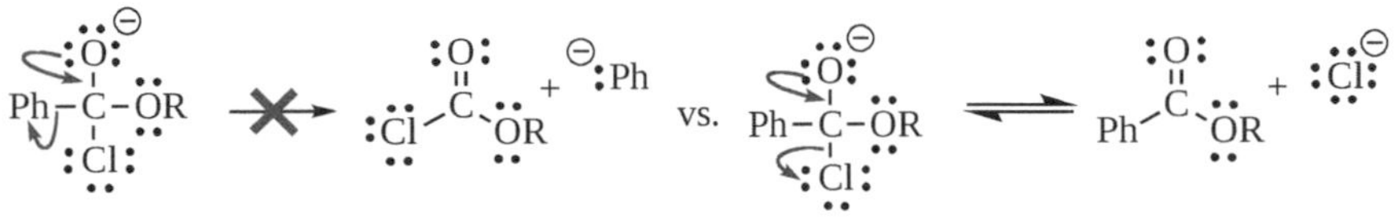

11.6 The reaction on the left is very poor, but a similar reaction on the right works. Why?

(a)

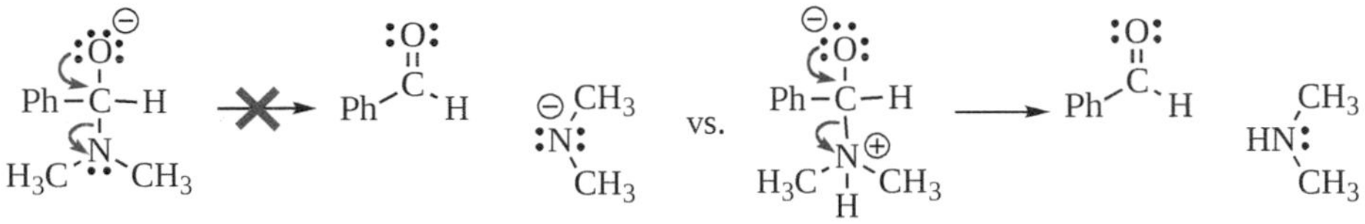

(b)

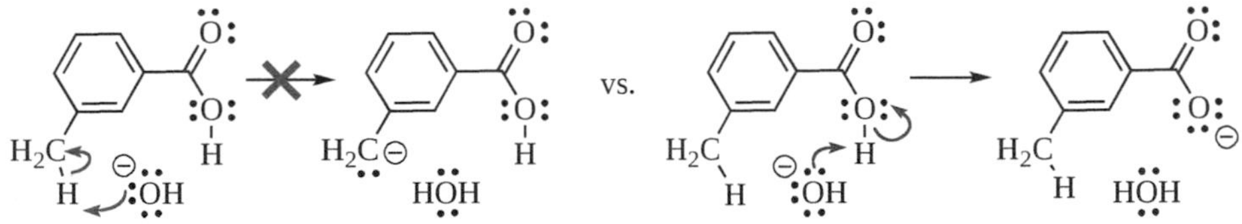

11.7 Which steps in the following commonly written mechanism are reasonable and which one is not? What cross-check did it fail? Draw a mechanism that passes all cross-checks.

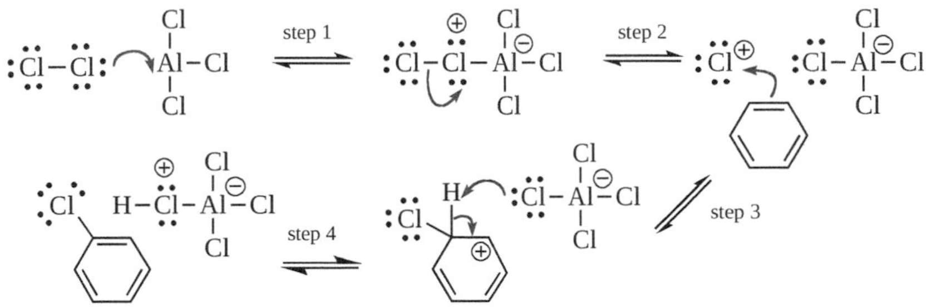

11.8 The student's mechanism written for the reaction below is incorrect.
The problem given was:

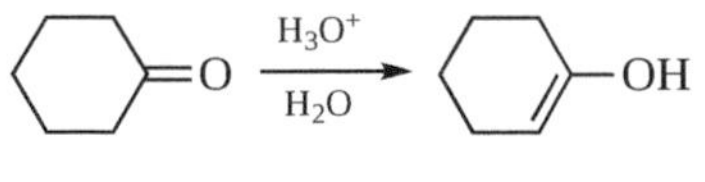

The incorrect mechanism was:

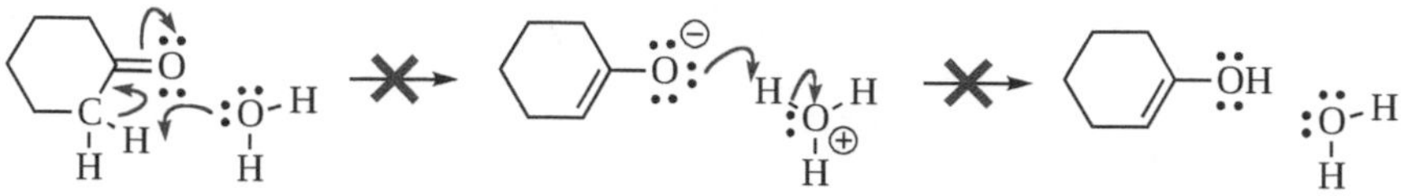

Draw a mechanism that passes all cross-checks.

11.9 The student's mechanism written for the reaction below is incorrect. The problem given was:

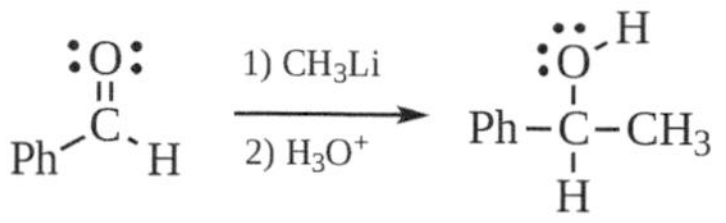

The incorrect mechanism was:

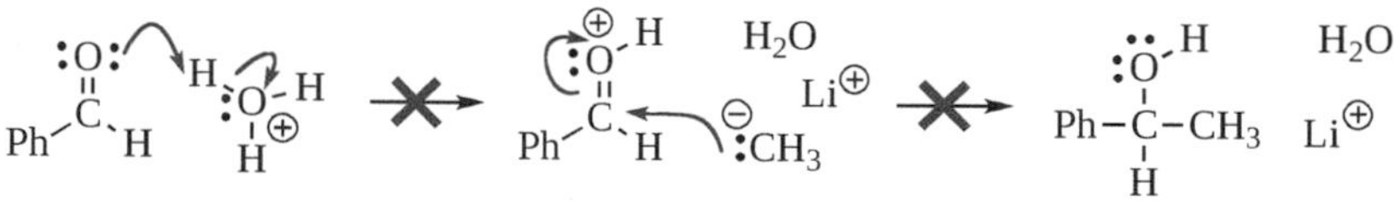

Draw a mechanism that passes all cross-checks.

11.10 Which steps in the following commonly written mechanism are reasonable, and which one is not? What cross-check did it fail? Draw a mechanism that passes all cross-checks.

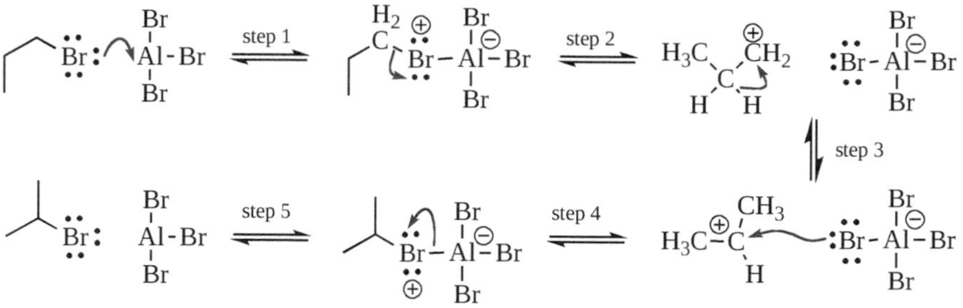

11.11 (More difficult) This possible mechanism shown below has failed a cross-check. What cross-check was missed? Draw a mechanism that passes all cross-checks.

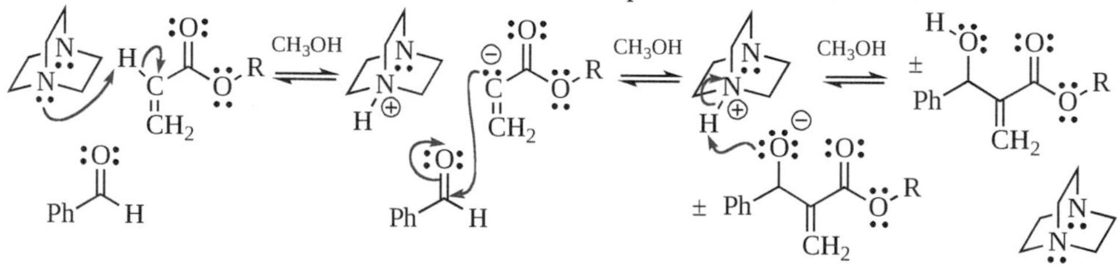

11.12 A student gave the following answers for two product prediction problems on an exam. One was marked correct and the other was not. Fix the incorrect answer and give reasonable mechanisms for both reactions. Explain what principle these two problems were intended to illustrate.

11.13 In the student-proposed mechanism on an exam shown below, steps are crossed out that failed a cross-check. Which cross-checks were missed? The test question:

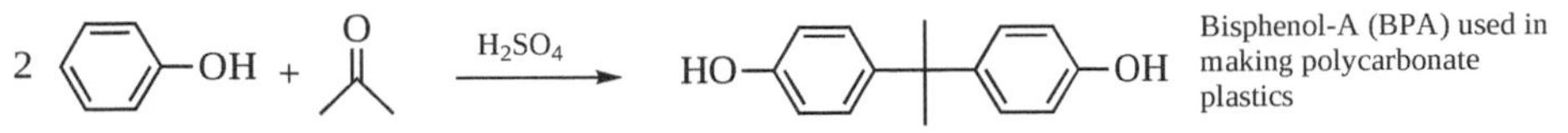

The student answer:

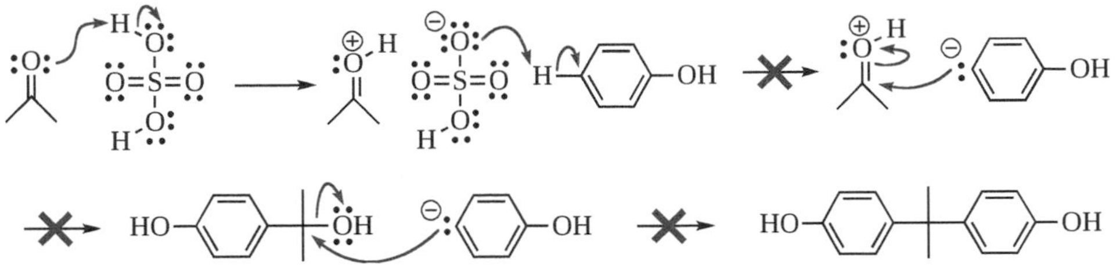

Draw a mechanism that passes all cross-checks.

11.14 In a mechanism on an exam, the student got stuck in a dead end shown below. What decision was missed?
The test question was to provide a mechanism for for this reaction:

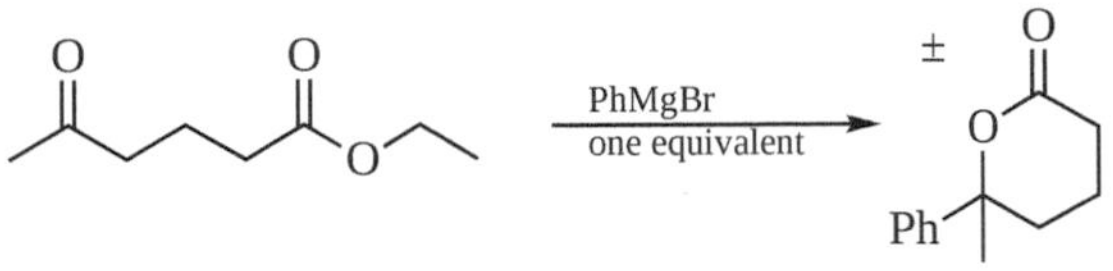

The student answer:

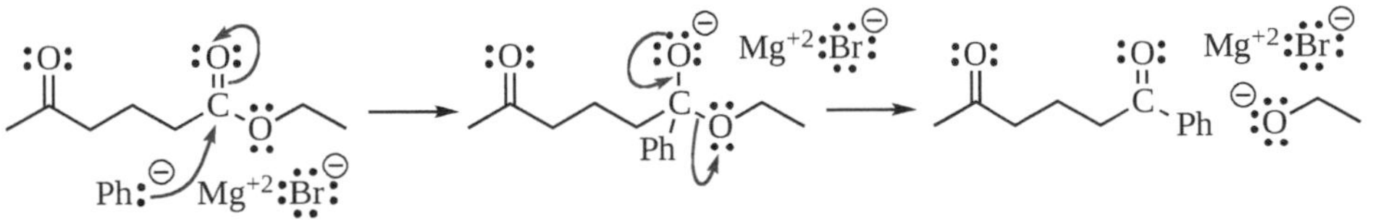

Draw a mechanism that passes all cross-checks.

11.15 In the mechanism on an exam, the student got stuck in the dead end shown below. What decision was missed? The test question:

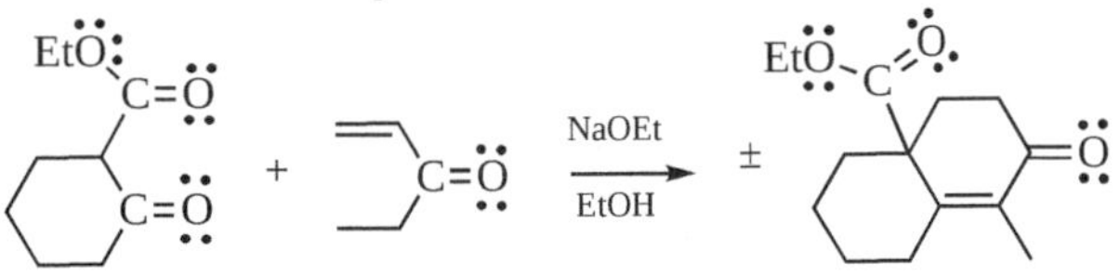

The student answer:

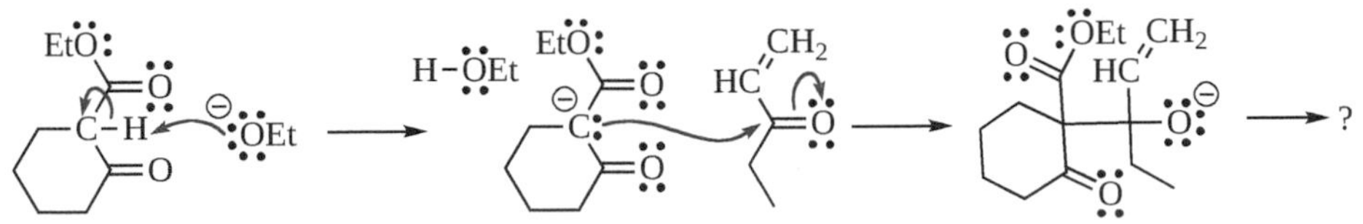

Draw a mechanism that passes all cross-checks.

11.16 In the mechanism on an exam, the student got stuck in the dead end shown below. What decision was missed? The test question:

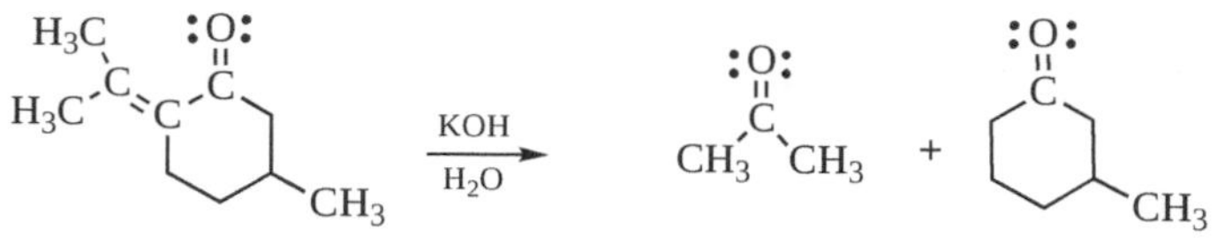

The student answer:

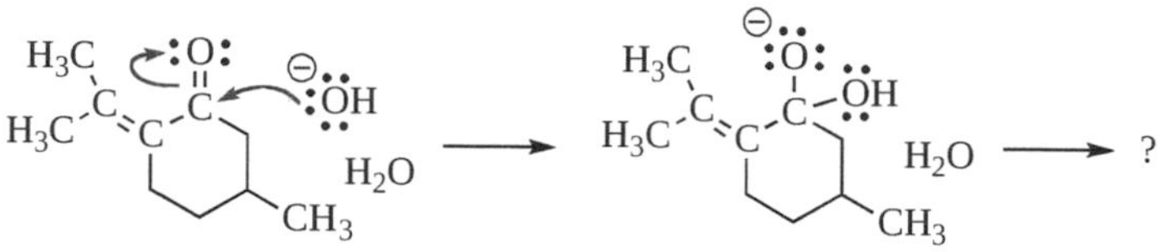

Draw a mechanism that passes all cross-checks.

11.17 In the mechanism on an exam, the student got stuck in the dead end shown below. What decision was missed? The test question:

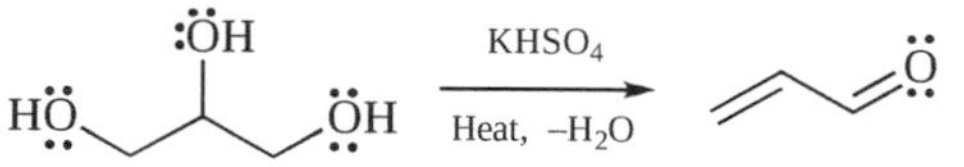

The student answer used HA for the acid catalyst:

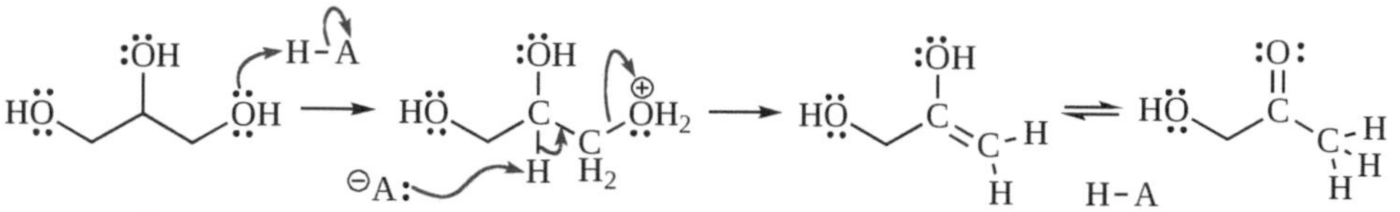

Draw a mechanism that passes all cross-checks.

11.18 In the student-proposed mechanism below, a step that failed a cross-check is crossed out. What cross-check was missed? Draw a mechanism that passes all cross-checks.

11.19 In the student-proposed mechanism below, a step that failed a cross-check is crossed out. What cross-check was missed? Use the energy surface for that process to find a reasonable alternative. Draw a mechanism that passes all cross-checks.

11.20 In the student-proposed mechanism below, a step that failed a cross-check is crossed out. What cross-check was missed? Use the energy surface for that process to find a reasonable alternative. Draw a mechanism that passes all cross-checks.

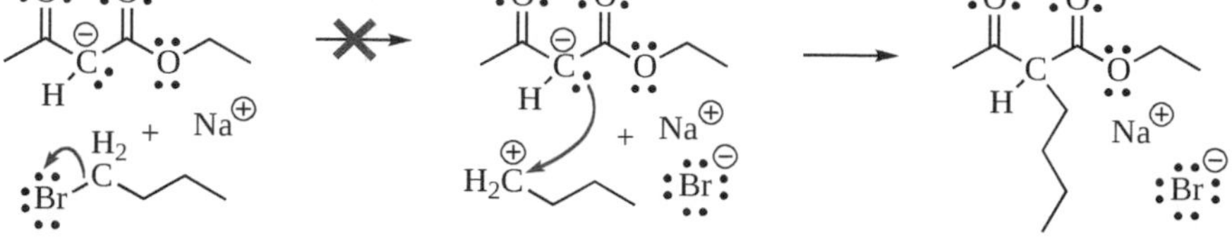

11.21 (More difficult) Of the two alternatives, which is the favored product and explain why?

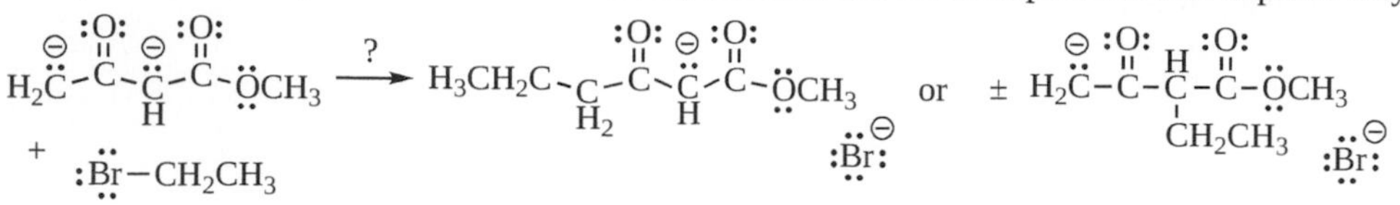

12

ONE-ELECTRON PROCESSES

12.1 RADICAL STRUCTURE AND STABILITY

One-electron processes belong to a conceptually different class of reactions and are best treated in this separate chapter. All one-electron processes require an initiation step, discussed in Section 12.2, which should serve as a flag to alert you to shift into using one-electron pathways. Because odd-electron species are usually very reactive, their concentration in a reaction will stay very low. Therefore, reactions of radicals with other radicals will be less common than reactions of radicals with the even-electron species that make up the bulk of the reaction mixture. **The most common radical reactions involve the radical reacting with an even-electron species.** The reaction usually gives an even-electron product and a new radical. These paths are discussed in Section 12.3.

Radical formation is a concern for all aerobic organisms. Oxygen is a ground-state diradical and therefore is very reactive and responsible for oxidative cell damage. Cells have radical-specific scavengers that minimize the damage that radicals can do.

Electron Flow in Organic Chemistry: A Decision-Based Guide to Organic Mechanisms,
Third Edition. Paul H. Scudder.

12.1.1 Radical Structure

Simple alkyl radicals like the methyl radical are believed to be nearly planar, shallow pyramids at the radical center, rapidly inverting with a barrier of <2 kcal/mol (8 kJ/mol). The vinyl radical is planar and rapidly inverts. Thus, all stereochemistry at the radical center is usually lost. In conjugated species, like the allyl radical, the radical is believed to be in a *p* orbital to maximize overlap with the rest of the pi system. Resonance forms allow us to predict the sites of radical reactivity in conjugated radicals.

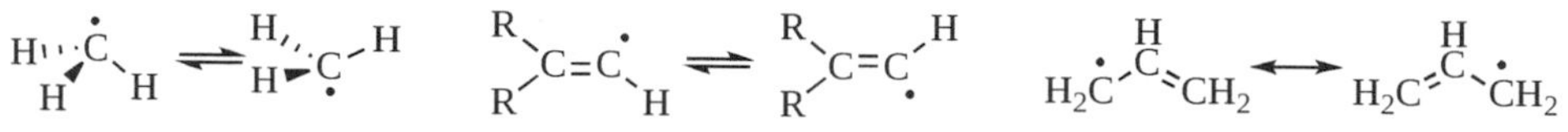

Methyl radical Vinyl radical Allyl radical

12.1.2 Bond Dissociation Energies

The energy required to break homolytically the R–H bond into R· and H· radicals is related to the stability of the R· radical. Lower values indicate greater radical stability. Common C–H and O–H bond dissociation energies in kcal/mol (kJ/mol) are given in Table 12.1 (adapted with permission from S. J. Blanksby, and G. B. Ellison, Acc. Chem. Res. 2003, 36, 255–263. Copyright 2003 by the American Chemical Society).

Table 12.1 Bond Dissociation Energies

Radical Name	Bond Broken	Dissociation Energy, kcal/mol (kJ/mol)
Break sp^3 C–H bond		
Methyl	CH_3–H	105.0 (439)
Ethyl	CH_3CH_2–H	101.1 (423)
Isopropyl	$(CH_3)_2CH$–H	98.6 (413)
tert-Butyl	$(CH_3)_3C$–H	96.5 (404)
Break sp^2 C–H bond		
Vinyl	CH_2=CH–H	110.7 (463)
Phenyl	C_6H_5–H	112.9 (472)
Acyl	CH_3CO–H	89.4 (374)
Delocalized radicals		
Acylmethyl	$HCOCH_2$–H	94 (393)
Phenoxyl	C_6H_5O–H	90 (377)
Benzyl	$C_6H_5CH_2$–H	89.8 (376)
Allyl	CH_2=$CHCH_2$–H	88.8 (372)
Misc. radicals		
Hydroxyl	HO–H	118.8 (497)
Hydroxymethyl	$HOCH_2$–H	96.1 (402)
Peroxy	HOO–H	87.8 (367)

12.1.3 Radical Stabilities

Although steric relief in going to the nearly planar radical contributes to the dissociation energy, it is generally believed that the stability of simple alkyl radicals follows the trend: tertiary > secondary > primary > methyl. However, as the top section of Table 12.1 shows, there is less than 9 kcal/mol (38 kJ/mol) difference between the two extremes of this trend. Both vinyl and phenyl radicals are destabilized relative to methyl. Radicals are stabilized much more by resonance delocalization than alkyl substitution. Whereas lone pair donors stabilize radicals somewhat, groups that increase conjugation have the greatest stabilizing effect. Since a radical is a neutral species, solvent polarity generally has little effect.

The *n*-propyl radical $CH_3CH_2CH_2\cdot$ has the same stability as the ethyl radical $CH_3CH_2\cdot$ because lengthening the chain has little effect on the stability of a primary radical. However the same length allyl radical $H_2C{=}CHCH_2\cdot$ is 12.3 kcal/mol (51.4 kJ/mol) more stable because of the resonance delocalization of the radical by the pi bond. Figure 12.1 shows the distribution of the odd electron over the molecule. The resonance forms are an attempt to describe this singly occupied orbital in the language of lines and dots.

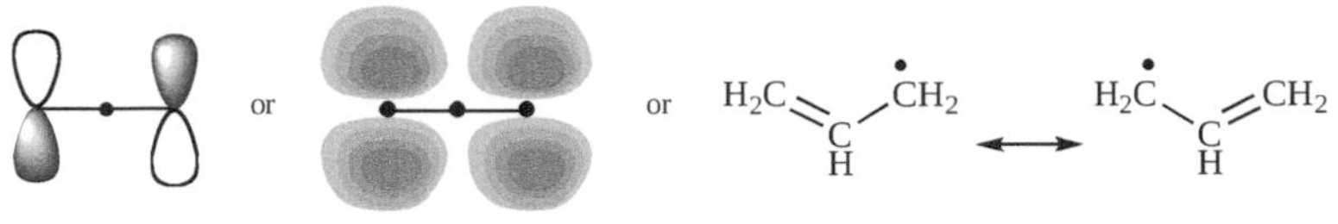

Figure 12.1 Alternate descriptions of the allyl radical.

Radicals can be stabilized by either a pi donor or pi acceptor because both delocalize the odd electron. Radicals that have both a pi donor and a pi acceptor are especially stable. The following resonance forms demonstrate the delocalization of a radical by a pi donor and by a pi acceptor. Section 13.2.5 gives a molecular orbital explanation of this radical stabilization.

pi donor stabilization

pi acceptor stabilization

Example problem

Which of the following, $(CH_3)_3C\cdot$, $CH_3CH_2\cdot$, $PhCH_2\cdot$, is the most stable radical; which is the least stable?

Answer: The delocalized radical, $PhCH_2\cdot$, is the most stable. The least stable is the primary radical, $CH_3CH_2\cdot$, because it is less substituted than the tertiary radical, $(CH_3)_3C\cdot$. Bond strengths from Table 12.1 are $PhCH_2$–H, 89.8 kcal/mol (376 kJ/mol); $(CH_3)_3C$–H, 96.5 kcal/mol (404 kJ/mol); CH_3CH_2–H, 101.1 kcal/mol (423 kJ/mol).

12.2 RADICAL PATH INITIATION

Homolytic Cleavage by Heat or Light

Since almost all stable species are even electron and spin paired (molecular oxygen is the obvious exception), radical processes must start either by single electron transfer or by homolytic cleavage. It is important to recognize these initiation steps in order to know when to use the one-electron paths from this chapter rather than the two-electron paths described previously. Half-headed arrows are used to symbolize the movement of one electron. Homolytic cleavage is the simple extension of a bond-stretching vibration (Figure 12.2). The process is **always** endothermic, for the barrier must be at least equal to the strength of the bond cleaved.

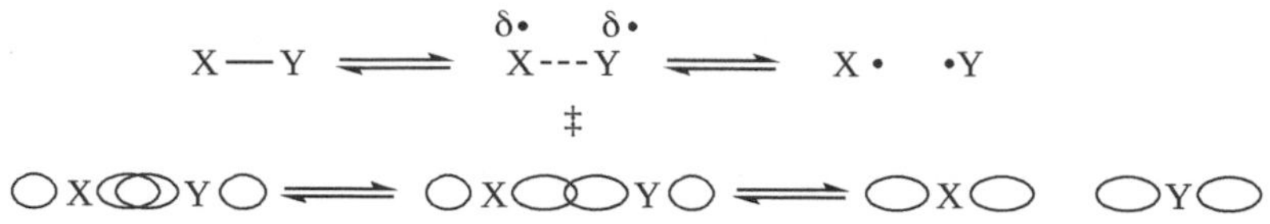

Figure 12.2 Homolytic cleavage of an X–Y bond with the transition state shown in the center.

The easiest bonds to break are weak sigma bonds between heteroatoms. Peroxide oxygen–oxygen single bonds, RO–OR, and the halogen–halogen single bonds, X–X, are easily broken by heat or light ($h\nu$). Azo compounds, R–N=N–R, are also cleaved by heat or light into two R radicals and molecular nitrogen. A common radical initiator is azobisisobutyronitrile (AIBN), $(CH_3)_2C(CN)$–N=N–$C(CN)(CH_3)_2$. The following are some common radical initiators and their cleavage into radicals.

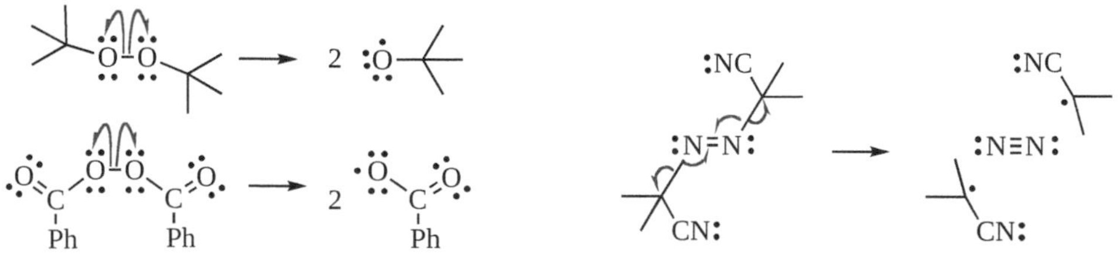

12.3 MAJOR PATHS FOR RADICALS REACTING WITH NEUTRALS

These two important reactions make up the core of most radical processes.

12.3.1 Abstraction

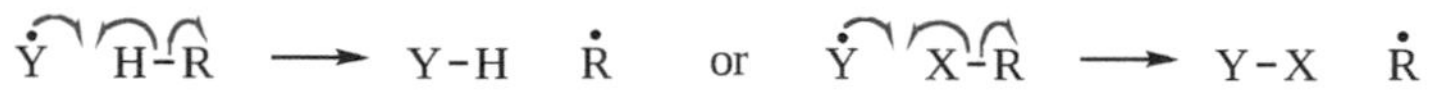

The initial radical abstracts an atom from a neutral molecule, giving a new radical (Figure 12.3). The atom abstracted is commonly a hydrogen or a halogen atom.

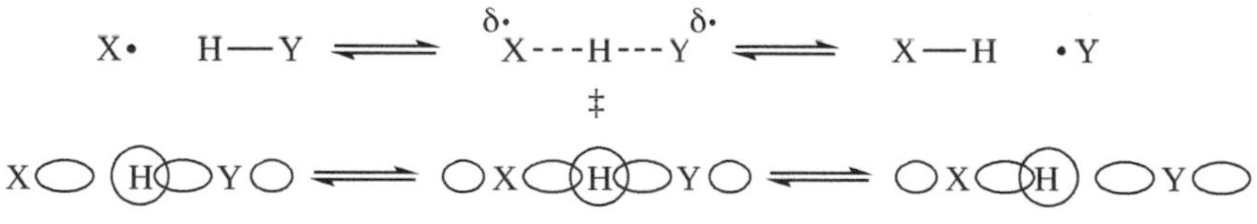

Figure 12.3 The abstraction of a hydrogen atom with the transition state shown in the center.

Overlap: No overlap restrictions other than the incoming radical center must collide with the atom to be abstracted.

Selectivity: There is a preference to abstract the hydrogen atom that would produce the most stable radical (breaking the weakest bond is energetically favored). Table 12.1 can usually be used to predict which C–H bond abstraction is preferred. Exceptions occur with abstractions by very electronegative radicals such as chlorine radical. With $CH_3CH_2CH_2$–EWG, chlorine radical preferentially abstracts the CH_2 away from the EWG since that CH_2 bears less of a partial plus.

Energetics: Bond strength tables (Table 12.1 and the bond strength table in the Appendix) allow the calculation of the $\Delta H°$ for the process. Highly exothermic abstractions tend to have poor selectivity. For example, C–H abstractions by a bromine radical (endothermic) are much more selective than C–H abstractions by a chlorine radical (exothermic).

Hydrogen abstraction can occur intramolecularly and is useful for functionalizing rather remote unactivated sites. With freely rotating alkyl systems, there is a great preference for a five- or six-membered cyclic transition state. With rigid or conformationally restricted systems, molecular models are necessary to predict the site of intramolecular hydrogen abstraction.

Variations: Halogens are also commonly abstracted by radicals, and the reaction favors the weakest C–X bond, I > Br > Cl > F. Sulfur and selenium are also capable of being abstracted by radicals, with a transition state reminiscent of an S_N2 substitution.

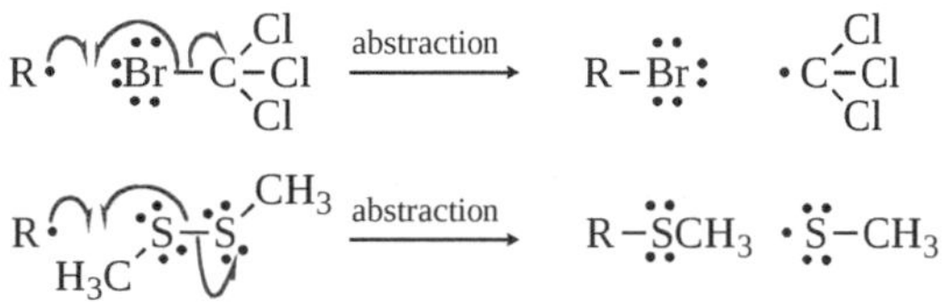

Example problem

Calculate the $\Delta H°$ for abstraction of an H from methane by fluorine radical.

Answer: A methane C–H bond is broken, 105 kcal/mol (439 kJ/mol), Table 12.1, and an H–F bond is formed, 136 kcal/mol (569 kJ/mol), from Appendix. The $\Delta H°$ is bonds broken − bonds made = 105 − 136 = −31 kcal/mol (–130 kJ/mol). This abstraction is very exothermic.

Addition

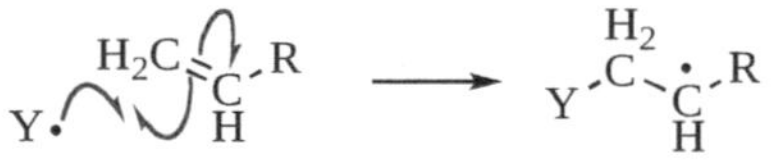

The initial radical adds to a pi bond to form a new radical (Figure 12.4).

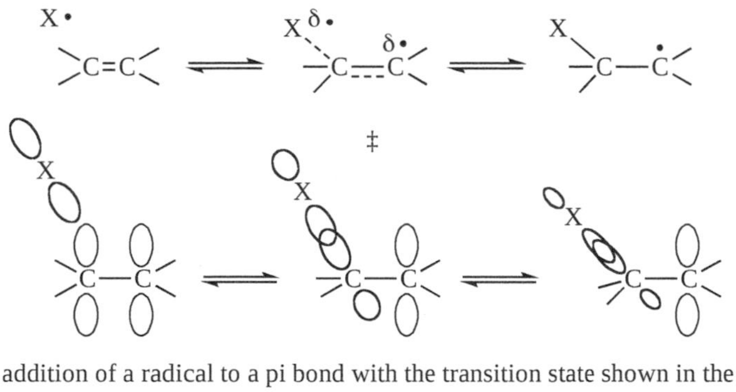

Figure 12.4 The addition of a radical to a pi bond with the transition state shown in the center.

Overlap: Calculations indicate that the preferred direction of attack of the radical on the end of the pi bond is at an angle of about 70° above the plane of the double bond (or approximately at 110° for the X–C–C angle in the figure).

Selectivity: Addition usually forms the more stable radical. Exceptions are relatively common since the radical stability difference can be less than a polar or steric effect.

Energetics: Since a pi bond is broken and usually a much stronger sigma bond is formed, this path is normally exothermic.

Variations: Cyclization is an intramolecular addition reaction. The preferred ring sizes for cyclization are usually 5 > 6 > 7 > 8; ring opening in these systems is rare. The closure to a four-membered ring has not been observed. Closure to a three-membered ring does occur, but the ring opening is preferred by a factor of 12,000.

Example problem

Calculate the $\Delta H°$ of an ethyl radical adding to ethene.

Answer: A C=C bond is broken, 146 kcal/mol (611 kJ/mol), and two C–C single bonds are formed, 83 kcal/mol (347 kJ/mol) for each C–C. The $\Delta H°$ for the reaction is 146 − 166 = –20 kcal/mol (–84 kJ/mol). The addition is exothermic. The values are from the Appendix.

12.4 UNIMOLECULAR RADICAL PATHS

12.4.1 Elimination or Fragmentation

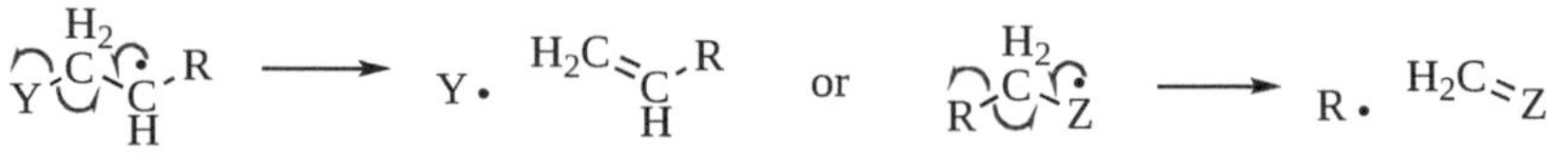

This path is the reverse of the addition path.

Overlap: By microscopic reversibility, the bond cleaved must be coplanar with the orbital bearing the odd electron (Figure 12.4, but view the figure from right to left).

Energetics: This path is normally endothermic but can be favored by the formation of a strong pi bond and higher temperatures. When more than one group could be lost, the loss of the more stable radical is favored.

12.4.2 Decarboxylation

Radical decarboxylation is a very fast and common fragmentation reaction because carbon dioxide is very stable:

12.4.3 Decarbonylation

The loss of carbon monoxide is an unusual reaction. If the radical formed upon decarbonylation is not stabilized, addition or abstraction can easily compete.

12.4.4 Rearrangement

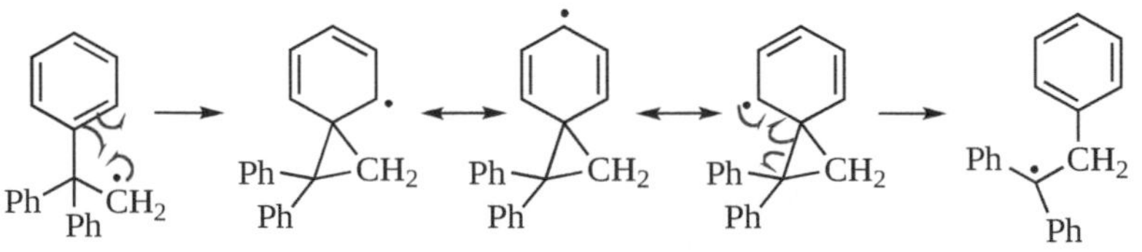

Rearrangements of radical species are rare. Migration occurs with unsaturated (aryl, vinyl, acyl) or halogen groups that can form a bridged intermediate. Rearrangement with unsaturated groups goes via intramolecular radical addition to form the three-membered ring followed by ring opening. The driving force for rearrangement is the formation of an equal or more stable radical or the relief of steric strain. Another way a radical can rearrange is by internal atom abstraction or fragmentation.

12.5 TERMINATION RADICAL PATHS

These two paths combine radicals to give nonradical products. These terminate the radical chain mechanisms discussed in Section 12.6, and so are often undesired but important reactions.

12.5.1 Radical–Radical Coupling

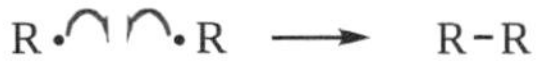

The coupling of two radicals to form a sigma bond is the reverse of the homolytic cleavage reaction and occurs without any activation barrier. Therefore, we might expect this reaction to be exceedingly fast. However, the rate of a reaction depends not only on the barrier height but also on the concentrations of the reacting species. Since the radical concentration in a usual reaction is very low, radical–radical collisions are infrequent.

An obvious exception occurs when the reaction mixture is so inert that there are few species for radicals to react with except each other. Another exception is when both radicals are formed close together (for example, azo compound cleavage) and can react before they diffuse apart. Often collision with a solvent molecule will send one newly formed radical into the other, causing reaction; this is called the solvent cage effect. A third exception is that radicals react with molecular oxygen (which behaves as a diradical) to form R–O–O·, hydroperoxide radicals. Generally radical reactions are run in the absence of oxygen to prevent this reaction from occurring.

12.5.2 Radical–Radical Disproportionation

Alternative to coupling, two radicals on collision can undergo disproportionation. Disproportionation occurs when one radical abstracts a hydrogen atom adjacent to the other radical center.

12.6 RADICAL PATH COMBINATIONS

Chain reactions are characteristic of radicals. The radical chain is started with an *initiation* step. The radical is formed in a low concentration, so it will almost always collide and react with the much higher concentration nonradical (even electron) compounds present. A chain *propagation* sequence occurs in which the radical species needed for the first propagation step is generated in the last step, and the sequence repeats many times. For a chain to be sustained, the overall process must be exothermic, and each individual step must have an easily achievable barrier. The chain undergoes *termination* whenever a coupling or disproportionation reaction removes the radical(s) needed to propagate the chain. If a propagation step is endothermic by more than 20 kcal/mol (84 kJ/mol), this propagation step will be slow enough that termination steps can easily compete, stopping the chain. The average number of times a chain reaction repeats before termination, commonly in the thousands or more, is called the *chain length*.

The way to control a radical chain reaction is to control the initiation and termination steps. Radical chain reactions can be favored by adding radical initiators. Likewise, chain reactions can be greatly diminished by adding compounds called *radical inhibitors* that react with radicals to increase chain termination. The sensitivity of radical reactions to radical initiators and inhibitors provides a convenient way to test for these mechanisms.

12.6.1 S_H2, Substitution Chain (Substitution, Homolytic, Bimolecular)

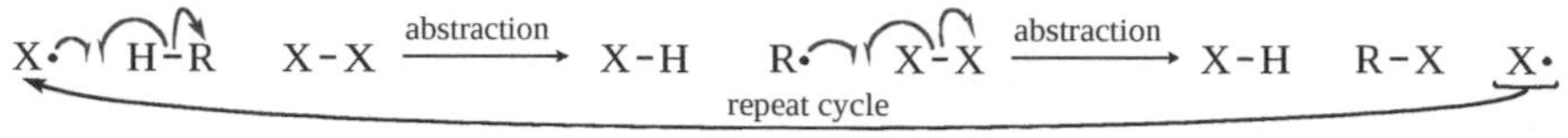

The two propagation steps in the substitution chain are both radical abstractions. Radical chain halogenations illustrate different types of chain energetics. Methane C–H is 105 kcal/mol (439 kJ/mol). The CH_3–X bond strengths are 115 (481), 83.7 (350), 72.1 (302), and 57.6 (241) for F, Cl, Br, and I, respectively, in kcal/mol (kJ/mol). The total reaction is the sum of the propagation steps. Radical species that appear on both sides of the transformation arrow cancel out. Initiation or termination steps are not included. Halogenation initiation is usually heat- or light-induced homolytic cleavage of the X–X bond into two X· radicals.

Fluorination radical chain (explosive)

Propagation step	CH_4 + F· → H–F + ·CH_3	$\Delta H°$ = −31 (−130)
Propagation step	·CH_3 + F_2 → F–CH_3 + F·	$\Delta H°$ = −78 (−326)
Total reaction	CH_4 + F_2 → F–CH_3 + H–F	$\Delta H°$ =−109 (−456)

Chlorination radical chain (fast)

Propagation step	CH_4 + Cl· → H–Cl + ·CH_3	$\Delta H°$ = +2.0 (+8)
Propagation step	·CH_3 + Cl_2 → Cl–CH_3 + Cl·	$\Delta H°$ = −25.7(−107)
Total reaction	CH_4 + Cl_2 → Cl–CH_3 + H–Cl	$\Delta H°$ = −23.7 (−99)

Bromination radical chain (slower)

Propagation step	CH_4 + Br· → H–Br + ·CH_3	$\Delta H°$ = +17 (+71)
Propagation step	·CH_3 + Br_2 → Br–CH_3 + Br·	$\Delta H°$ =−26.1(−109)
Total reaction	CH_4 + Br_2 → Br–CH_3 + H–Br	$\Delta H°$ =−9.1 (−38)

Iodination radical chain (does not occur)

Propagation step	CH_4 + I· → H–I + ·CH_3	$\Delta H°$ = +34 (+142)
Propagation step	·CH_3 + I_2 → I–CH_3 + I·	$\Delta H°$ =−21.6 (−90)
Total reaction	CH_4 + I_2 → I–CH_3 + H–I	$\Delta H°$ =+12.4 (+52)

The first propagation step of the bromination radical chain is significantly endothermic. A small change in product radical stability is reflected in the barrier for reaction and consequently in the rate of reaction. Thus the first propagation step (and the chain reaction) that goes the fastest forms the most stable product radical. Bromination will select for allylic and benzylic > tertiary > secondary > primary > vinyl and phenyl.

Allylic bromination is usually done with *N*–bromosuccinimide (NBS), which keeps the concentration of bromine low by reacting with the HBr formed in the first propagation step to produce the bromine needed for the second propagation step. This low bromine concentration suppresses the addition chain reaction, discussed next, by allowing time for the addition step to reverse before a bromine molecule is encountered.

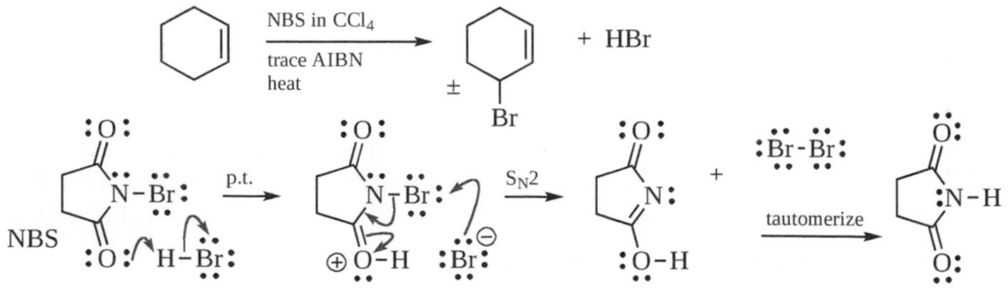

The radical chain dehalogenation with $Bu_3Sn–H$ is a very useful reaction. This radical chain dehalogenation reaction prefers the weakest C–X bond: C–I > C–Br > C–Cl > C–F. The initiator for this reaction is commonly AIBN. An example mechanism involving this reaction is shown in Section 12.7, Approaches to Radical Mechanisms.

Dechlorination with Bu_3SnH

Propagation step	$R–Cl + Bu_3Sn\cdot \rightarrow Bu_3Sn–Cl + R\cdot$
Propagation step	$R\cdot + Bu_3Sn–H \rightarrow R–H + Bu_3Sn\cdot$
Total reaction	$R–Cl + Bu_3Sn–H \rightarrow Bu_3Sn–Cl + R–H$

The oxidation radical chain (auto-oxidation) is very important in the spoiling of foodstuffs. The mechanism of auto-oxidation is complex, for hydroperoxide radicals easily add to multiple bonds, and those products also can cleave. The following is a much-simplified scheme. Free-radical chain inhibitors are often added to the packaging of foods to retard this reaction.

Auto-oxidation

Propagation step	$O_2 + R\cdot \rightarrow R–O–O\cdot$
Propagation step	$R–O–O\cdot + R–H \rightarrow R–O–O–H + R\cdot$
Total reaction	$O_2 + R–H \rightarrow R–O–O–H$

Example problem

Predict the major radical substitution chain products of the reaction drawn below.

$$(H_3C)_3CO–Cl + PhCH_2CH_2CH_3 \xrightarrow{\text{heat}} ?$$

Answer: The thermal homolysis of the weak oxygen–chlorine bond in *t*-butyl hypochlorite produces two radicals that can abstract an H atom from the benzylic methylene of 1-phenylpropane. This gives a resonance-delocalized secondary benzylic radical, the most stable of all alternatives. The propagation loop closes as the secondary benzylic radical abstracts the accessible chlorine atom from *t*-butyl hypochlorite creating a *t*-butoxy radical to start the process over again. The products are 1-chloro-1-phenylpropane and *t*-butanol.

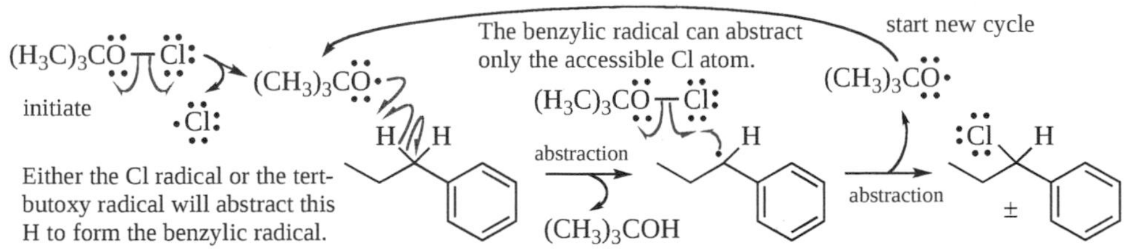

Ad_H2, Addition Chain (Addition, Homolytic, Bimolecular)

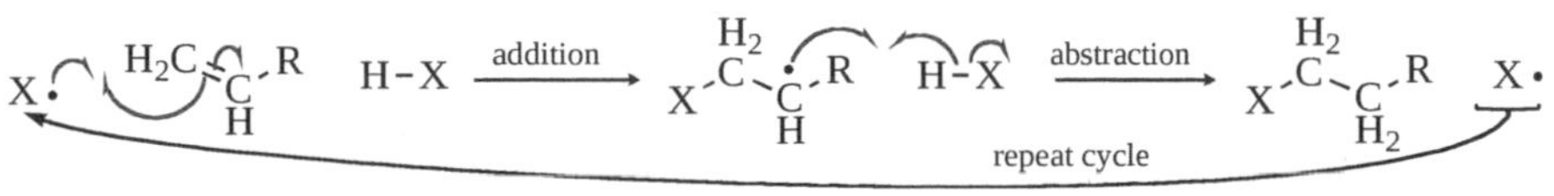

The first propagation step in the addition chain is a radical addition. The second propagation step in the addition chain is a radical abstraction. Radical addition of HBr is a typical reaction. Bromine radical adds to the multiple bond to form the most stable of the possible radicals. The radical addition of HBr gives the opposite regiochemistry of addition as the polar (Markovnikov) addition of HBr.

Radical addition of HBr

Propagation	$CH_2{=}CH_2 + Br\cdot \rightarrow \cdot CH_2{-}CH_2{-}Br$	$\Delta H° = -5\ (-21)$
Propagation	$\cdot CH_2{-}CH_2{-}Br + H{-}Br \rightarrow H{-}CH_2{-}CH_2{-}Br + Br\cdot$	$\Delta H° = -11\ (-46)$
Total	$CH_2{=}CH_2 + H{-}Br \rightarrow H{-}CH_2{-}CH_2{-}Br$	$\Delta H° = -16\ (-67)$

The radical addition of HCl is not as useful since polymerization tends to compete with the addition chain. The radical addition of HI or HF does not occur because in each case a propagation step is too endothermic.

Example problem

Check the $\Delta H°$ of the propagation steps for the radical chain addition of RSH to an alkene to see whether it is an energetically reasonable process.

Answer:

Step 1	$CH_2{=}CH_2 + RS\cdot \rightarrow \cdot CH_2{-}CH_2{-}SR$	$\Delta H° = -2\ (-8)$
Step 2	$\cdot CH_2{-}CH_2{-}SR + H{-}SR \rightarrow H{-}CH_2{-}CH_2{-}SR + RS\cdot$	$\Delta H° = -16\ (-67)$
Total	$CH_2{=}CH_2 + H{-}SR \rightarrow H{-}CH_2{-}CH_2{-}SR$	$\Delta H° = -18\ (-75)$

The radical chain addition of RSH to an alkene is an energetically reasonable process because there are no steps more than 20 kcal/mol (84 kJ/mol) endothermic, and the entire process is exothermic.

Polymerization Chain

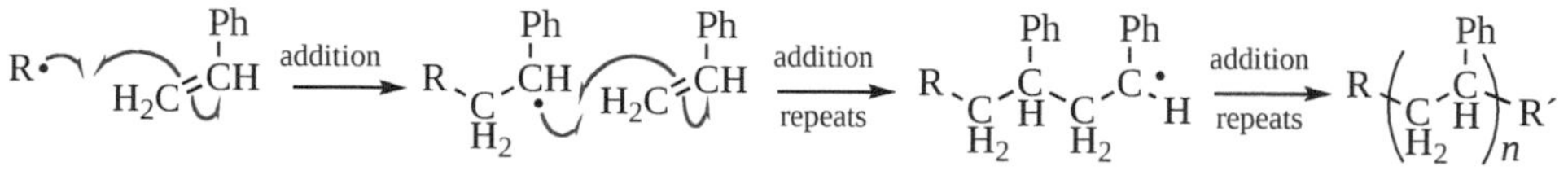

Polymerization is the repetition of an addition propagation step, like the polymerization of styrene to give polystyrene above. The initiator radical adds to a multiple bond to produce another radical that adds again and again until termination occurs. The identity of the end groups, R and R, may be uncertain, because they depend on the initiation and termination processes. Polymerization is favored by a high concentration of the multiple bond reactant.

Polymerization of ethylene to polyethylene

Initiation step	$CH_2{=}CH_2 + R\cdot \rightarrow \cdot CH_2{-}CH_2{-}R$
Repeated propagation step	$CH_2{=}CH_2 + \cdot CH_2{-}CH_2{-}R \rightarrow \cdot(CH_2{-}CH_2)_2{-}R$
Total reaction (after termination step)	$n\ CH_2{=}CH_2 + R\cdot \rightarrow H{-}(CH_2{-}CH_2)_n{-}R$ (and other terminations)

12.7 APPROACHES TO RADICAL MECHANISMS

12.7.1 Common Radical Mechanism Errors

Because radicals are in very dilute concentrations, the usual radical mechanisms involve a radical colliding with an even-electron molecule in a chain process. A common error is to have a termination step instead of creating a regenerating loop. The three common radical path combinations, S_H2, Ad_H2, and radical polymerization, all have propagation steps in which radicals collide with an even-electron species creating a new radical. Usually, the last propagation step creates the starting radical for the first propagation step, thus making a loop.

Initiators are added to the reaction in trace amounts, just to get the radical chain loop started. The loop often goes for many thousands of cycles before termination. Do not include this trace initiator in your propagation steps. Make your radical loop out of reactants present in good amounts.

12.7.2 Radical Mechanism Example

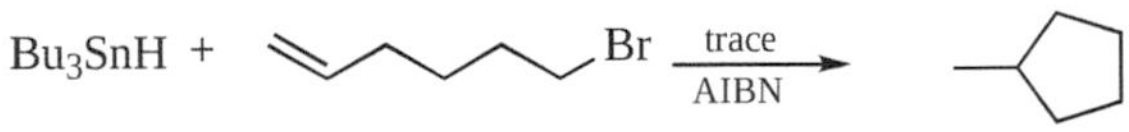

First, balancing the reaction and numbering the carbons of the reactant and product will give us more information about the reaction. Since carbons 1, 5, and 6 in the reactant were functionalized in the reactant, the most logical numbering of the product is:

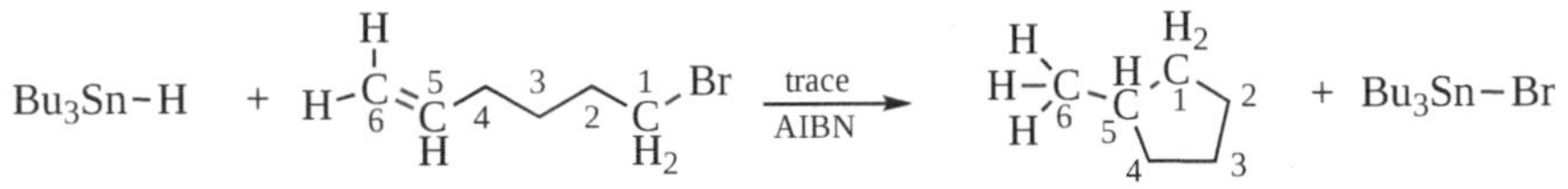

In figuring out what bonds have changed, it is helpful to draw in all the hydrogens on those atoms. Shown in red above, the changes are: carbon 6 has gone from a methylene to a methyl, so a new C–H bond was formed there; the bromine atom has been lost from carbon 1; A new single bond formed between carbons 1 and 5, and the carbon 5 to 6 bond is now single.

A radical initiator, AIBN, is present, and the product is suggestive of an intramolecular radical addition chain. To begin, homolytic cleavage of the AIBN initiator is used to create radicals that will begin the chain.

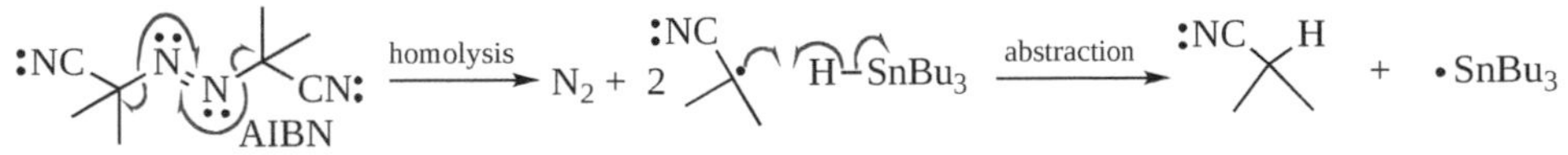

Now the propagation steps can start by having our tributyltin radical abstract a bromine atom from the alkyl bromide.

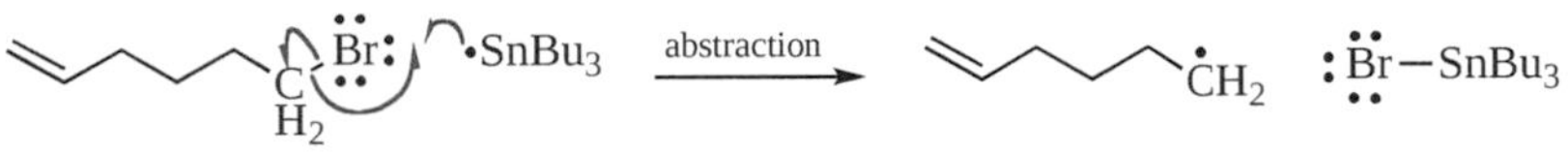

This gives us the tributyltin bromide product and a primary carbon radical with which to do an intramolecular radical addition to a pi bond. The five-membered ring formation is kinetically preferred even though it forms a slightly less stable radical (by 2.5 kcal/mol).

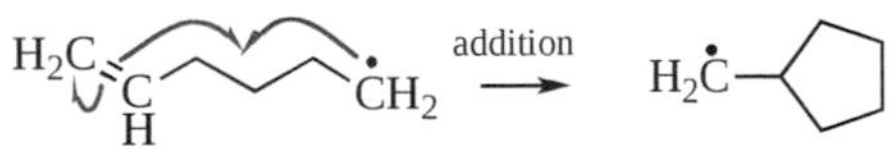

Now it is time to close the loop and create the radical from the first propagation step, so that the loop can continue.

Ideally, the propagation loop turns over many thousands of times before a termination step quenches the radical.

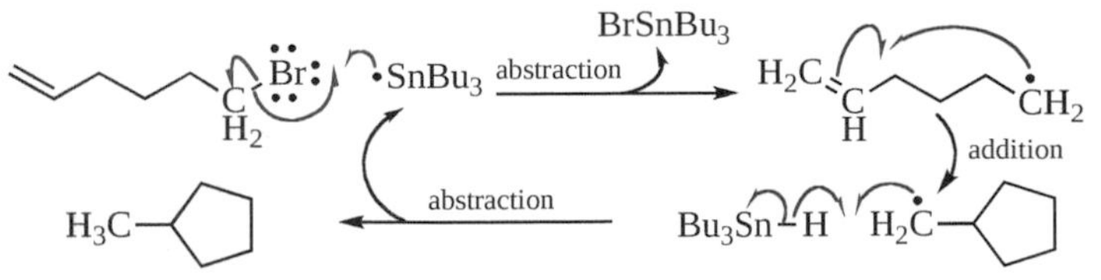

12.8 SINGLE-ELECTRON TRANSFER, S.E.T., AND CHARGED RADICALS

$$A + e^- \longrightarrow A^{\cdot -} \qquad \text{or} \qquad A\!: - e^- \longrightarrow A^{\cdot +}$$

Single-electron transfer is the one-electron oxidation or reduction of a species; no covalent bond formation occurs. Single-electron transfer to or from another chemical species or an electrode forms charged radicals. The chemical species must be picked using standard electrode potentials such that the electron transfer is energetically favorable. Single-electron transfers using an electrode are more versatile since the electrical potential of the electrode can be accurately set over a wide range.

Single-electron transfer from an alkali metal such as lithium, sodium, or potassium is a common method of producing a radical anion. A radical anion has its extra electron in an antibonding orbital, which weakens the corresponding bonding orbital and makes the radical anion more reactive than the neutral species.

Single-electron transfer to commercially available stable amine radical cation salts can produce a radical cation. Again, the radical ion is more reactive than the neutral species.

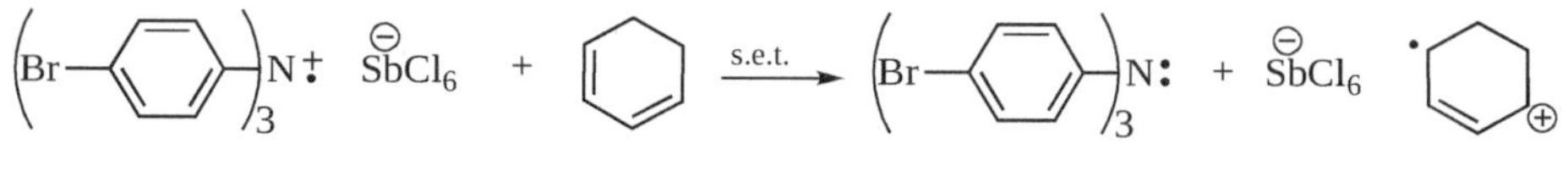

12.8.1 Organometallic Formation

The formation of organomagnesiums is believed to occur by single-electron transfer from magnesium that triggers the loss of the halide, followed by a second single-electron transfer.

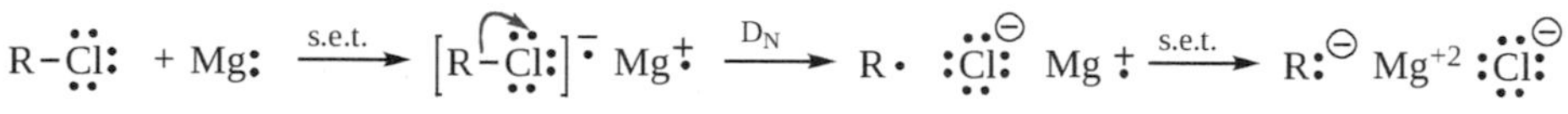

12.8.2 A Molecular Orbital Explanation of Single-Electron Transfer
(A supplementary, more advanced explanation, able to be skipped)

Single-electron transfer reduction adds one electron into the lowest unoccupied molecular orbital (LUMO) of the acceptor molecule. The lower in energy the acceptor LUMO is, the easier the molecule is to reduce; therefore, conjugated systems are easier to reduce than unconjugated systems. Single-electron transfer oxidation removes one electron from the highest occupied molecular orbital (HOMO) of the donor molecule. The higher in energy the donor HOMO is, the easier the molecule is to oxidize. For this reason, conjugated systems are also easier to oxidize than unconjugated systems. As electrons in the donor HOMO get higher in energy (as in nonbonded electron pairs on less electronegative elements), they can approach the energy of the LUMO of the acceptor. The single-electron transfer is then easily triggered by light or, if the gap is small enough, by heat. The LUMO of R–L is often the antibonding orbital of the C–L bond. When an electron is transferred into this LUMO, the result is two electrons in the bonding orbital and one in the antibonding for a net one-half bond order. This weakened C–L bond is now much easier to break. Chapter 13 has greater detail on molecular orbitals and how they relate to chemical reactivity.

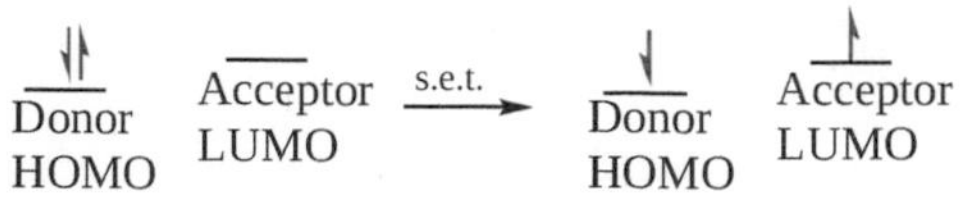

12.9 DISSOLVING METAL REDUCTIONS

Alkali metals added to liquid ammonia (boiling point –33°C) with a small amount of an alcohol in the presence of another reactant provides the opportunity for both single-electron transfer and two-electron processes to occur and are called dissolving metal reductions. Sodium metal dissolves in liquid ammonia to give a dark blue solution characteristic of a solvated electron.

As shown below, this highly reducing solution reacts with alkynes via single-electron transfer, forming a radical anion, protonation of this radical anion to give the neutral vinyl radical, and a second single-electron transfer to the radical, yielding a vinyl anion that then protonates. The second protonation gives the more stable *trans* alkene. In this manner, alkynes are reduced to *trans* alkenes. This dissolving metal reduction of an alkyne complements that of catalytic hydrogenation (Section 8.10.3), where certain special transition metal catalysts can reduce an alkyne to produce only *cis* alkenes.

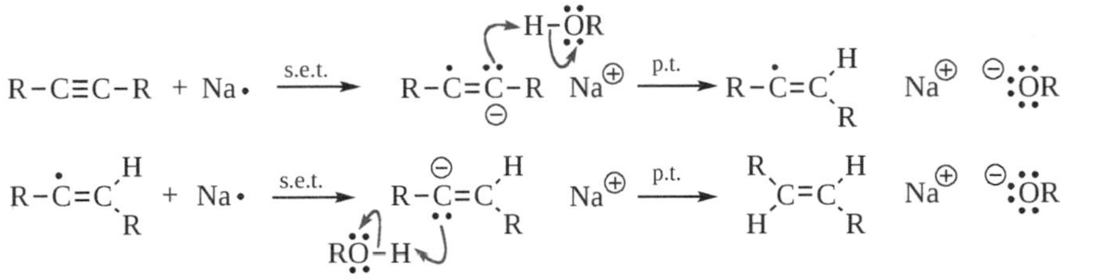

Aromatics and other conjugated systems can also be reduced by sodium in liquid ammonia with a trace of alcohol as a proton source.

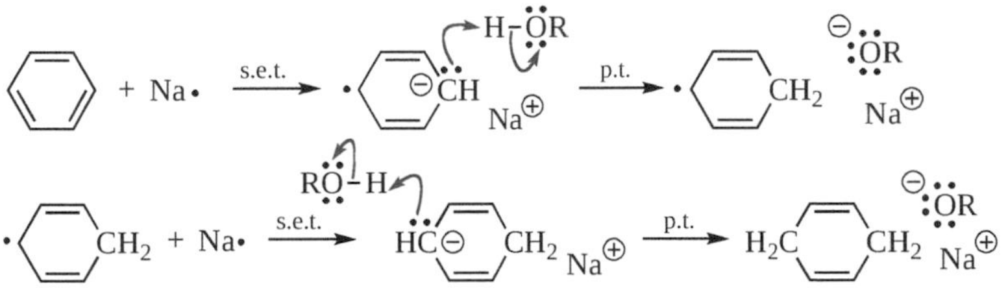

Electron-withdrawing and donating groups on the ring determine the regiochemistry of the reduction. An anion conjugated to an electron-releasing group is avoided, and an anion conjugated to an electron-withdrawing group is favored. Alkyl groups behave as mild electron donors.

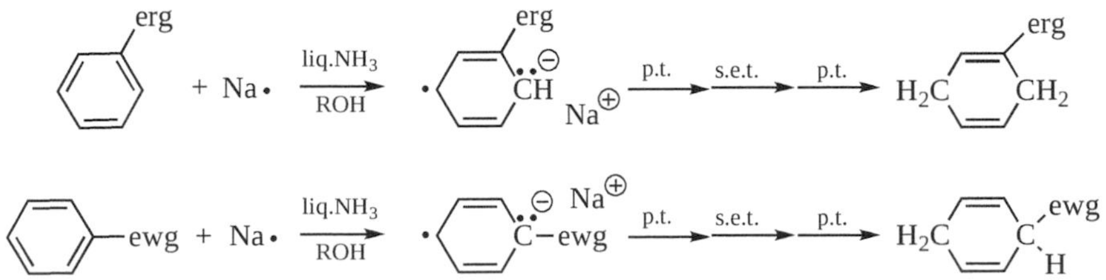

Another mixed process is a way to remove halogens or other leaving groups. Single-electron transfer forms a radical anion, which loses the leaving group anion to give the radical; a second single-electron transfer to the radical yields the anion, which protonates.

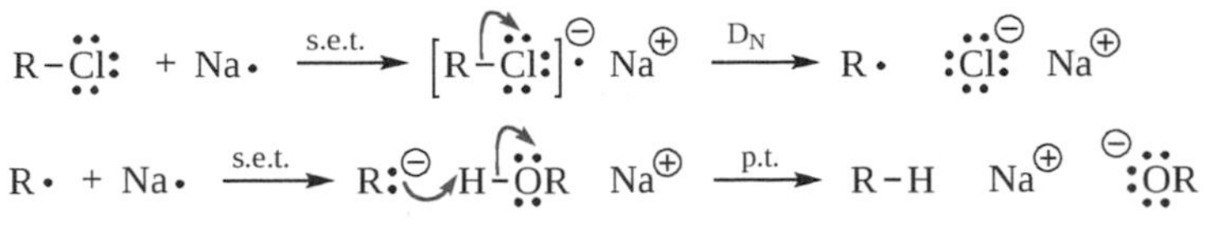

12.10 ELECTRON TRANSFER INITIATED PROCESSES

12.10.1 $S_{RN}1$Process (Substitution, Radical–Nucleophile, Unimolecular)

The $S_{RN}1$ process is a chain reaction that has both single-electron transfers and two-electron processes. A single-electron transfer forms a radical anion, which then loses the leaving group, forming the neutral radical that is attacked by a nucleophile to form a new radical anion. This new radical anion serves as the electron source for the initial single electron transfer. Aryl diazonium ions, $ArN{\equiv}N^+$, may react with nucleophiles this way.

Initiation step	$R{-}L + e^{\ominus} \rightarrow R{-}L^{\bullet\ominus}$	then
Propagation step	$R{-}L^{\bullet\ominus} \rightarrow R\bullet + :L^{\ominus}$	
Propagation step	$R\bullet + :Nu^{\ominus} \rightarrow R{-}Nu^{\bullet\ominus}$	
Propagation step	$R{-}Nu^{\bullet\ominus} + R{-}L \rightarrow R{-}L^{\bullet\ominus} + R{-}Nu$	
Total reaction	$R{-}L + :Nu^{\ominus} \rightarrow R{-}Nu + :L^{\ominus}$	

With R as aryl, initiation can be by single-electron transfer from an electrode or an alkali metal. Initiation by light, hν, is also possible. The $S_{RN}1$ process can be stopped with radical inhibitors. Systems that would not have reacted by an S_N1 or S_N2 process may react by an $S_{RN}1$ mechanism since the R–L bond is considerably weakened in the radical anion compared to the neutral.

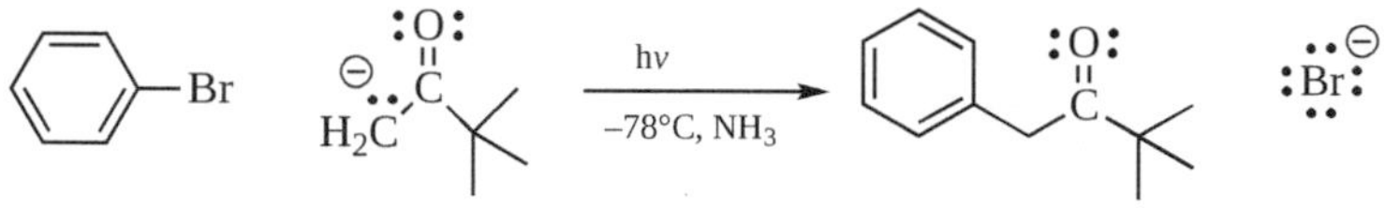

12.10.2 Other Electron Transfer Initiated Processes

The conjugate addition of some organometallics to carbonyl derivatives may occur by single-electron transfer. Enones are relatively easier to reduce than simple ketones, making the single-electron transfer more favorable. Organocopper reagents appear to use single-electron transfer followed by radical coupling when reacting with enones. The intermediates do not escape the solvent cage.

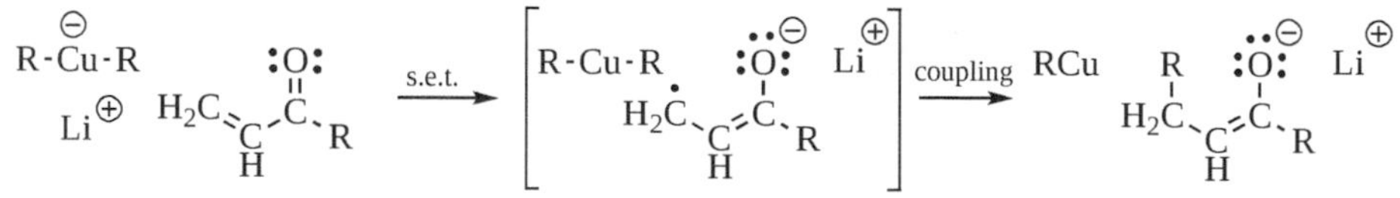

12.11 ONE-ELECTRON PATH SUMMARY

12.11.1 Initiation Is Required

Single-electron transfer (s.e.t.)

$$A + e^- \longrightarrow A^{\bullet -} \quad \text{or} \quad A: - e^- \longrightarrow A^{\bullet +}$$

Homolytic cleavage by heat or light

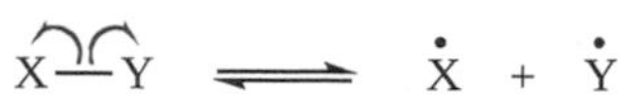

12.11.2 Radical Paths

Abstraction

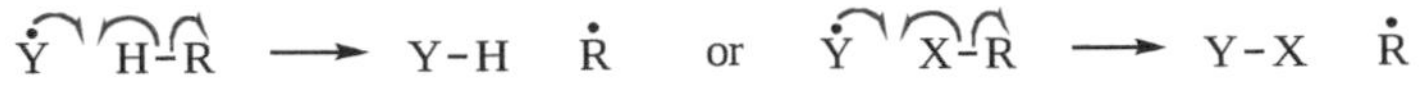

Addition

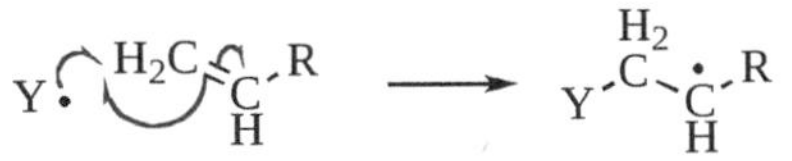

Elimination

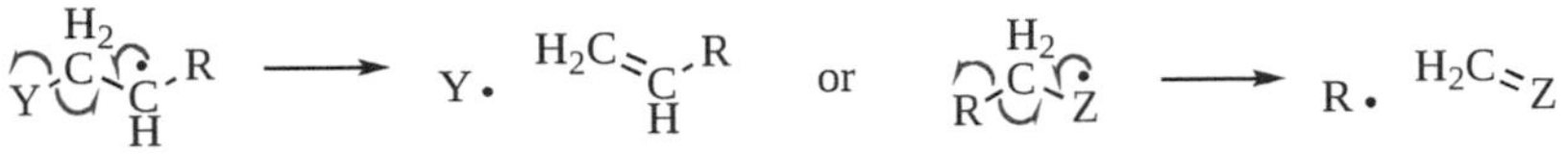

Decarboxylation

Decarbonylation

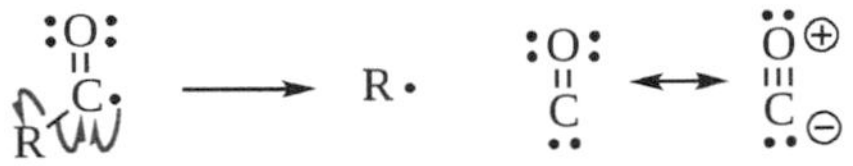

Rearrangement

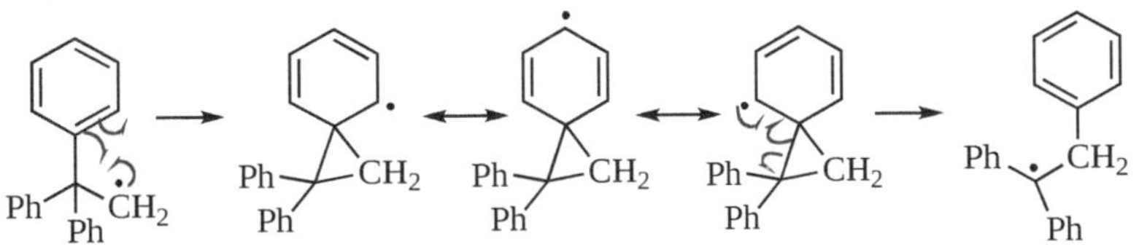

12.11.3 Radical Termination Paths

Coupling

Disproportionation

12.11.4 Common Radical Chain Reactions

Substitution Chain, S_H2

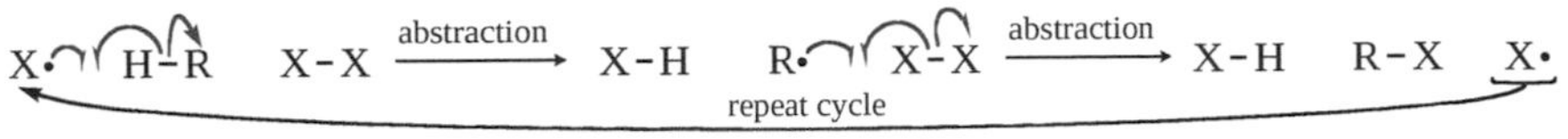

Addition Chain, Ad_H2

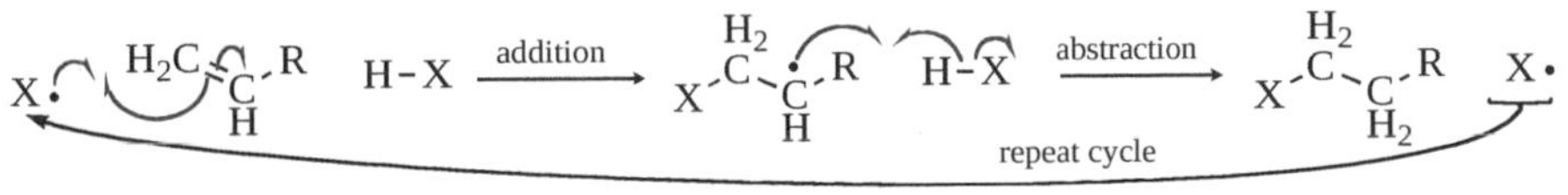

Polymerization Chain

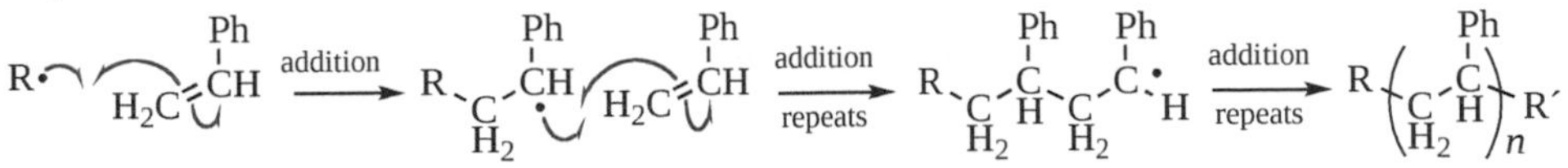

CHAPTER 12 EXERCISES

12.1 Rank all species, using the numeral 1 to designate the most stable radical.

$$\cdot CH{=}CH_2 \quad \cdot CH_2CH{=}CH_2 \quad \cdot CH_2CH_3 \quad \cdot CH(CH_3)_2 \quad \cdot CH_3$$

12.2 Use radical stabilities to predict the regiochemistry of the radical addition chain reaction of HBr to propene (initiated by AIBN).

12.3 Calculate the $\Delta H°$ for the abstraction of an H from ethane by a $\cdot$OOH radical.

12.4 Calculate the $\Delta H°$ of an iodine radical adding to a carbon–carbon double bond.

12.5 Check the $\Delta H°$ of the propagation steps for the radical chain addition of HF to ethene to see whether it is an energetically reasonable process.

12.6 Use radical stabilities to predict the selectivity of the following radical chain substitution reaction (initiated by light). Stop after one bromine is attached.

$$Br_2 \;+\; C_6H_5CH_2CH_2CH_3 \;\rightarrow$$

12.7 Predict which you would expect to be more selective: radical chain chlorination or bromination of an alkane. Explain.

12.8 BHT is a radical chain terminator. Explain why H abstraction from oxygen yields a relatively unreactive radical.

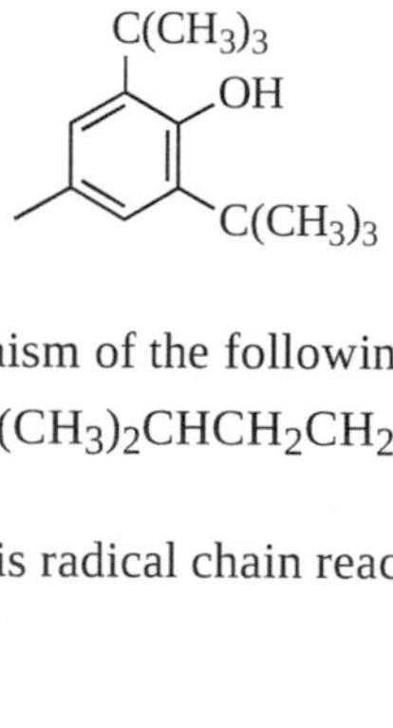

12.9 Give the product and a mechanism of the following radical reaction (initiated by light).

$$Br_2 \;+\; (CH_3)_2CHCH_2CH_2CH_3 \;\rightarrow$$

12.10 Give a likely mechanism for this radical chain reaction (initiated by light).

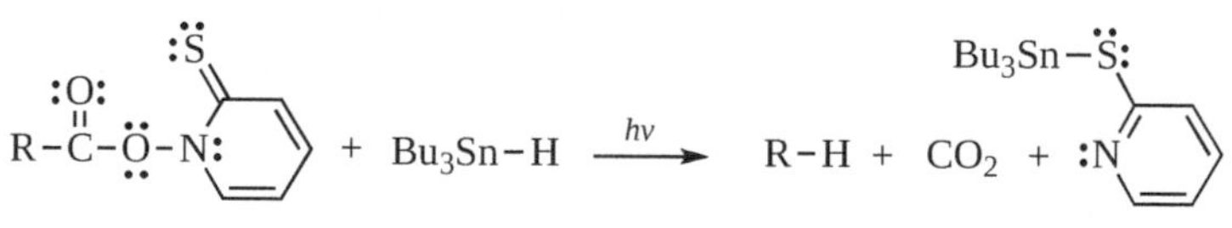

12.11 Write out an $S_{RN}1$ process for phenyl diazonium ion reacting with iodide anion to produce nitrogen gas and phenyl iodide. Use s.e.t. from iodide anion to start the process.

12.12 Give a likely mechanism for this radical chain reaction (initiated by heat or light).

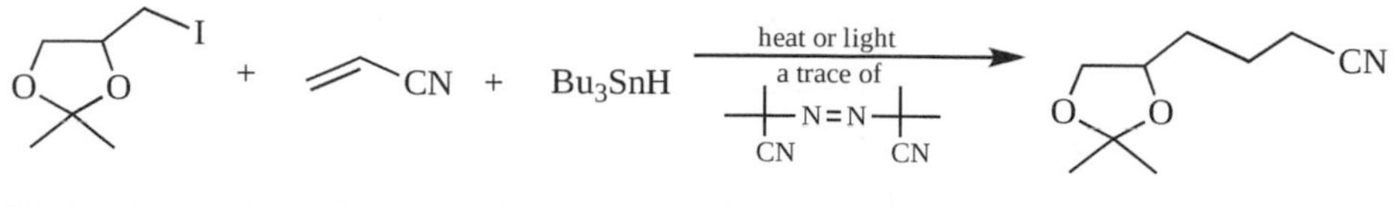

12.13 Give the product and a mechanism for this radical chain addition initiated by light.

$$BrCCl_3 \;+\; H_2C{=}CEt_2 \;\xrightarrow{h\nu}$$

13

QUALITATIVE MOLECULAR ORBITAL THEORY AND PERICYCLIC REACTIONS

Electron Flow in Organic Chemistry: A Decision-Based Guide to Organic Mechanisms, Third Edition. Paul H. Scudder.

This chapter introduces qualitative molecular orbital theory and pericyclic reactions, based on the concepts in Section 1.7. Pericyclic reactions have cyclic transition states and electron flow paths that seem to go around in a loop. The regiochemistry and stereochemistry of these reactions are predictable by molecular orbital interactions, so to understand them we need to understand molecular orbital (MO) theory, at least on a qualitative basis.

13.1 REVIEW OF ORBITALS AS STANDING WAVES

13.1.1 Electrons as Waves

At the turn of the century in 1900, the behaviors of electromagnetic waves like light vs. particles like an electron were clearly separate in classical physics. In 1905, Einstein explained the photoelectric effect by saying light sometimes behaved like a particle. In 1924, De Broglie predicted that a moving particle like an electron could behave like a wave. It was soon confirmed that a beam of electrons could be diffracted similar to light. Diffraction was considered a property of waves, and so the wave-particle duality was firmly established. The idea that an electron has predictable wavelike properties became the foundation of quantum mechanics and modern atomic theory. A negative electron in an atom is constrained in space to be near the positive nucleus by the attraction of opposite charges. Whenever we constrain a wave to a limited space, we get standing waves. Therefore, our next step is to understand standing waves. We will start with a wave confined in one-dimensional space, or as quantum mechanics would call it, a particle in a one-dimensional box. From one-dimensional standing waves, we will move to two- and then three-dimensional standing waves. These standing waves are described by mathematical equations called wave functions. Schrödinger's wave functions describe these three-dimensional standing waves, the specific quantized orbitals that an electron may occupy in an atom. We will take a qualitative approach to wave functions.

13.1.2 Standing Waves in One and Two Dimensions

Since the electron has a wave-like behavior, the electrons in a molecule will have properties similar to a wave confined to a limited space. The stable configurations will be standing waves. Atomic and molecular orbitals are like standing waves on a guitar string, because they too are waves confined to a limited space. Since the ends of the guitar string are fixed, only the waves that have zero amplitude at the ends will work. The lowest frequency, longest wavelength, that fits is called the fundamental (Figure 13.1), or first harmonic. The fundamental has one half-wavelength between the ends. The next one to fit is two half-wavelengths, the second harmonic. Three half-wavelengths gives the third harmonic, and four half-wavelengths gives the fourth harmonic.

As we fit more half-wavelengths into this limited space, the actual wavelength, λ, gets shorter, and the frequency, ν, gets higher. With light, the frequency is inversely proportional to the wavelength, $\nu = c/\lambda$ (where c is the speed of light), and frequency is also directly proportional to the energy, $E = h\nu$ (where h is Planck's constant). The fundamental is of lowest energy, and each higher harmonic gets higher in energy. The nodes, indicated by vertical dotted lines in Figure 13.1, are the regions where the wave goes through zero, the horizontal dotted line. The nodes divide the standing wave into half-wavelengths. Note that the nodes are symmetrical with the center of the system. When there is an odd number of half-wavelengths, as in the first and third harmonics, then the wave will have mirror symmetry about the center. These linear standing waves are very similar to MOs for linear conjugated pi systems in Section 13.3. As the number of p orbitals in the conjugated pi system increases, the standing wave is confined to a bigger space, so the wavelength of the fundamental gets longer and its energy drops. The converse is true; if we have a smaller box (shorter string), the fundamental wavelength is shorter, and thus is of higher energy and frequency. The pushing of a guitar string down at a fret does exactly this to raise the frequency.

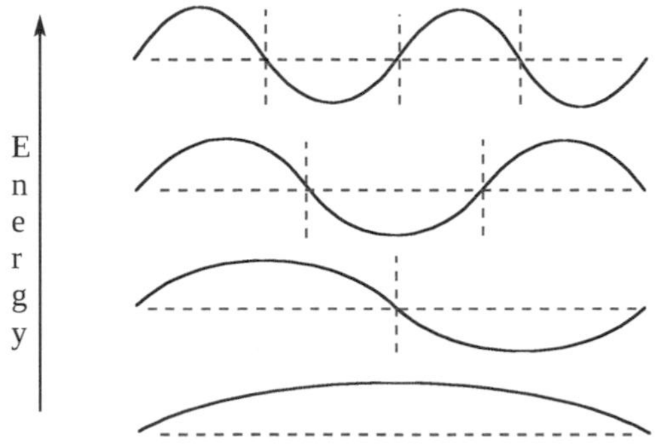

Figure 13.1 The standing waves of a vibrating string. The fundamental (first harmonic) has no nodes; the second harmonic has half the wavelength and twice the frequency of the fundamental and has one node, shown by a dotted vertical line. The gradually increasing number of nodes causes the frequency and energy to also increase.

The standing waves in Figure 13.1 are sine waves. The value of the sine wave has a positive mathematical sign when it is above the dashed horizontal zero line (peak), and it has a negative mathematical sign when it is below the zero line (trough). With orbitals, discussed in the next section, most texts just use shading to indicate mathematical sign, to avoid confusion between it and charge.

Electrons confined to an atom or molecule behave similarly; as the energy of the orbital increases, the number of nodes increases, like the higher harmonics on a vibrating string. Electrons are confined in three-dimensional space, so a quick look at standing waves in two dimensions will help the transition to standing waves in three dimensions. Instead of a guitar string, now visualize the head of a drum, Figure 13.2. The fundamental again has no nodes, so the entire drumhead moves up and down. A second harmonic has one side of the drumhead moving down as the other side moves up; the single node is a line crossing the exactly in the middle of the drum. A third harmonic has two nodal lines, perpendicular to each other. This means that third harmonic has the diagonal quadrants going up as the opposite ones are going down.

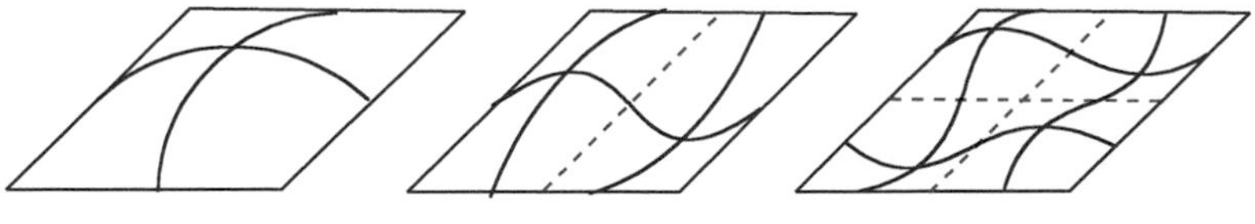

Figure 13.2 The standing waves of a drumhead. The fundamental (first harmonic) has no nodes; a second harmonic has a single nodal line crossing the drum in the middle; a third harmonic has two perpendicular nodal lines crossing the drum, intersecting in the exact middle.

13.1.3 Standing Waves in Three Dimensions: Atomic Orbitals

Atomic orbitals are the possible standing waves an electron can have in three dimensions. The lowest-energy orbital, 1*s*, has zero nodes. The orbitals of the second electron shell all have one node. We use shading to indicate where the wave function has a different mathematical sign, like the peak and trough of our standing wave. The 2*s* orbital's node is spherical, so the orbital looks like a ball inside a shell. Each 2*p* orbital has a single nodal plane through the nucleus. The orbitals of the third electron shell all have two nodes. The 3*s* has two nodal spheres, so that it looks like a ball in a shell inside another shell. The 3*p* has a nodal surface inside each lobe in addition to a plane going through the nucleus. Four of the 3*d* orbitals have two perpendicular nodal planes, each intersecting at the nucleus. The fifth, the $3dz^2$, has two nodal cones intersecting at the nucleus.

We try to make orbital drawings show two things: mathematical phase (with shading) and electron probability density (with shape). The wave function itself has mathematical signs that are important when we go to overlap it with another wave function. We need to be able to show if the two orbital wave functions overlap in phase (peak with peak, bonding) or out of phase (peak with trough, antibonding). The square of the wave function is the probability density of finding the electron (the odds of finding it in a particular region of space) and does not have negative areas. But the probability density is a diffuse cloud, making drawing such a fuzzball a problem. A compromise is to draw the surface that would contain about 90% of the electron density. So we take the shape of the probability density (usually stylized, like Figure 13.3) and shade it to show the sign of the original wavefunction, so that the drawing shows both phase and probability density, becoming the familiar orbitals from general chemistry.

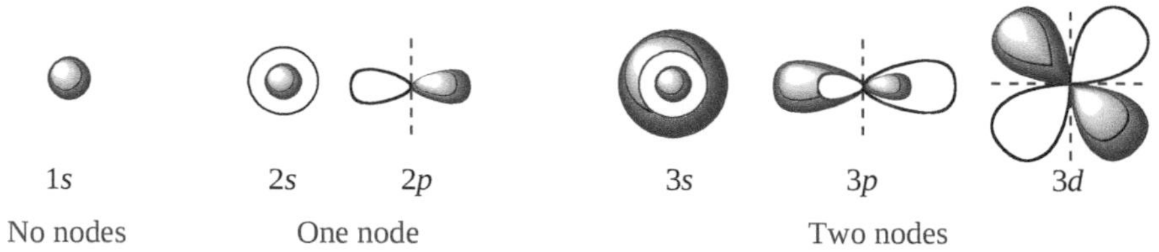

Figure 13.3 Representative stylized cross-sectional drawings of atomic orbitals, which are standing waves in three dimensions. Shading indicates mathematical sign.

13.1.4 Mixing Atomic Orbitals into Molecular Orbitals

The combination of atomic orbitals on different atoms produces molecular orbitals. Figure 13.4 shows how the 1*s* atomic orbital on each of two hydrogen atoms can be combined by addition and subtraction to give the molecular orbitals for the hydrogen molecule, H_2. To subtract the 1*s* orbital, we mix it in with its mathematical sign (shading) reversed. The addition combines the atomic orbital waves in phase (peak with peak), and the subtraction combines the waves out of phase (peak with trough). The number of orbitals is conserved. If we combine two atomic orbitals, AOs, we must get two molecular orbitals, MOs. A covalent bond is the sharing of an electron pair between two nuclei. The bonding molecular orbital has no nodes (the highest probability of finding the electrons is between the two nuclei) and is lower in energy. The bonding orbital is lower in energy simply because the "box" has gotten bigger, so the fundamental has a longer wavelength. The antibonding molecular orbital has a node between the two nuclei (the probability goes through zero there) and is the higher-energy molecular orbital. Orbitals are filled from lowest energy up. Since the hydrogen molecule has two electrons, one from each atom, the bonding molecular orbital is filled and the antibonding orbital is empty. For a mnemonic to remember the two hydrogen molecular orbitals, think of a black-eyed cow.

Since the antibonding molecular orbital is raised more than the bonding is lowered, if we try to fill both molecular orbitals (with four electrons), the overall energy is increased compared to the isolated atoms. This is why helium does not form a diatomic molecule. With both the bonding and antibonding orbitals filled with two electrons each, there is no net bond and the helium atoms drift apart. With three electrons, the antibonding orbital has just one electron in it, so the bond order of He_2^+ is 1/2, resulting in a very weak bond.

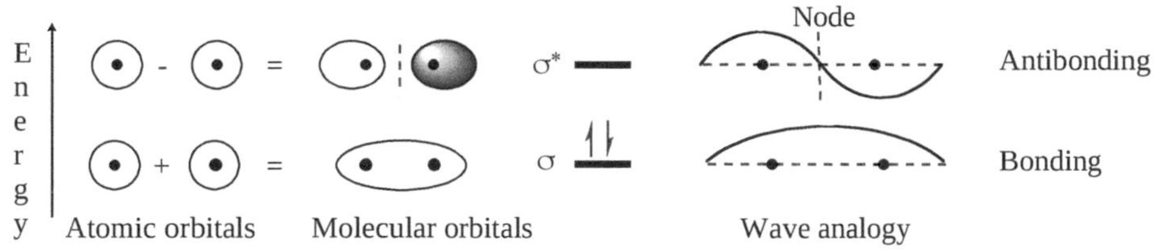

Figure 13.4 The bonding and antibonding molecular orbitals of hydrogen. The nuclei positions are approximated by dots. The standing wave extends beyond the nuclei because the orbitals do.

Figure 1.6 repeated below, showed how to use the three $2p$ orbitals on each N atom to form the three bonding (one sigma and two pi) molecular orbitals and the three antibonding molecular orbitals of N_2. The Lewis structure of N_2 also shows two lone pairs of electrons, which come from the in-phase and out-of-phase combinations of the $2s$ orbitals (not shown).

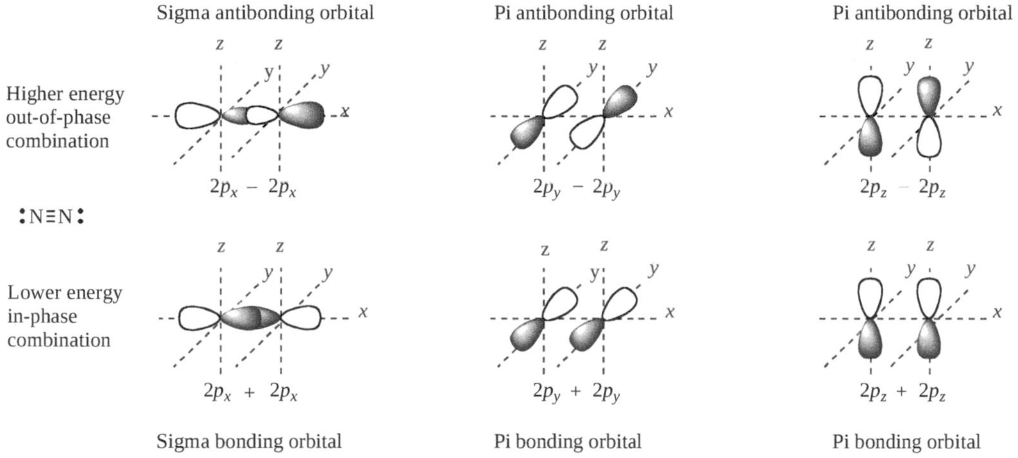

Figure 1.6 The in-phase and out-of-phase mixing of the $2p_x$, $2p_y$, and $2p_z$ atomic orbitals on two nitrogen atoms to give the bonding and antibonding molecular orbitals of molecular nitrogen.

Larger molecules will have many molecular orbitals. We are usually only concerned with MOs from valence atomic orbitals and not the core electrons because those are too low in energy to interact. MOs for strong sigma bonds are low in energy, and their corresponding antibonding MOs are high in energy. Pi bonding MOs are not as stabilized and are more reactive. The most available filled MO for reaction is the Highest Occupied Molecular Orbital (HOMO), and the most available empty MO for reaction is the Lowest Unoccupied Molecular Orbital (LUMO). Figure 13.5 shows the usual energy order of molecular orbitals of larger molecules. Notice the symmetry above and below the nonbonding level.

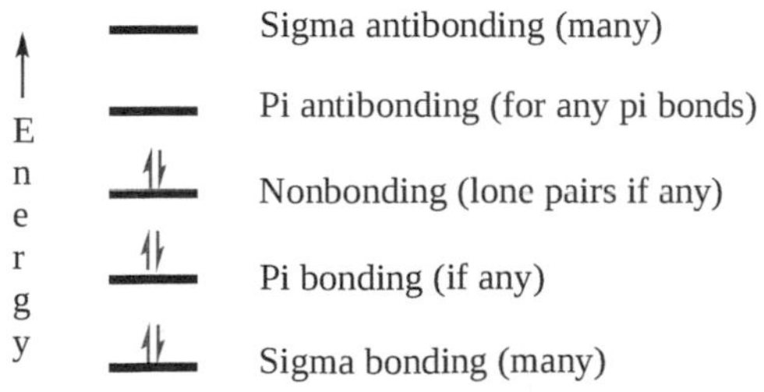

Figure 13.5 The usual energy order for molecular orbitals larger molecules.

13.2 ORBITAL INTERACTION DIAGRAMS

An understanding of interaction diagrams is not absolutely necessary for using the principle of electron flow to predict organic reaction products. However, it is useful for understanding reactivity trends and the stability of reactive intermediates. This section relies on the principles discussed in Section 1.7, An Orbital View of Bonding.

13.2.1 Atomic Orbital Interaction Diagram for Two Atoms

One way to describe the mixing of orbitals between two atoms shown in Figure 13.4 is to use an orbital interaction diagram. The orbitals to be interacted are drawn on each side of the diagram, and their interaction is shown in the center of the diagram. **When two orbitals interact, the result is two new orbitals.** Shown on the left side of Figure 13.6 is the interaction of two hydrogen 1*s* orbitals containing one electron each to form a bonding orbital into which both electrons are placed. The two electrons in the bonding orbital have a lower energy than they had before the atomic orbitals interacted to form the bonding and antibonding molecular orbitals. The electrons are stabilized; energy is released, and a sigma bond is formed between the two hydrogen atoms. Shown on the right side of Figure 13.6 is the interaction of two helium 1*s* orbitals containing two electrons each. Now both the bonding and antibonding orbitals must be filled. The two electrons in the antibonding orbital are destabilized more than the two electrons in the bonding orbital are stabilized, resulting in net destabilization and no bond between the helium atoms. The two helium atoms drift apart; there is no bond to hold them together.

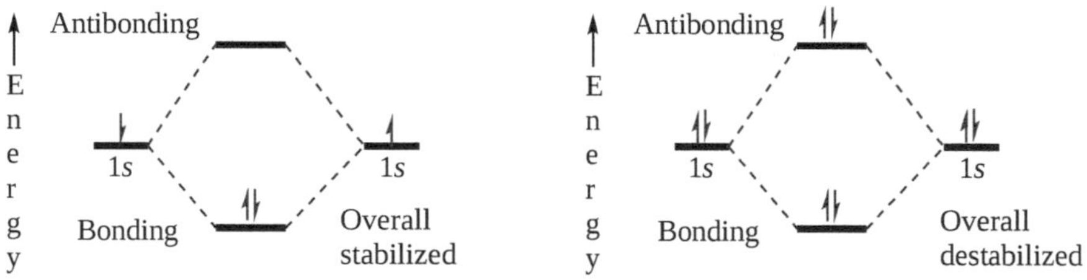

Figure 13.6 The left side shows the orbital interaction diagram for two hydrogen atoms and the right side shows the interaction of two helium atoms (see molecular orbitals drawn in Figure 13.4).

The better the overlap between two orbitals, the greater their interaction. If the orbitals are oriented such that their antibonding overlap exactly cancels their bonding overlap, then there is no interaction (Figure 13.7). Alternatively, if the orbitals are too far apart to overlap, there is no interaction.

Figure 13.7 Orbitals that don't interact because their bonding and antibonding overlap cancels.

The best overlap occurs between orbitals of about the same size. The better overlap of a 2*p*–2*p* pi bond is one reason that it is stronger than a 2*p*-3*p* pi bond, which has poorer overlap because of the size difference in the orbitals. Another example, the H–F bond made from overlap of a 1*s* orbital on H and a 2*p* orbital on F is very strong. At the other extreme, the H–I bond made from overlap of a 1*s* orbital on H and a much larger 5*p* orbital on I is much weaker. Both the bond strength trend and its effect on acidity can be seen in the following chart.

Best overlap →			Poorest overlap
H–F	H–Cl	H–Br	H–I
1*s*-2*p*	1*s*-3*p*	1*s*-4*p*	1*s*-5*p*
136 kcal/mol	103 kcal/mol	88 kcal/mol	71 kcal/mol
(569 kJ/mol)	(431 kJ/mol)	(368 kJ/mol)	(297 kJ/mol)
Strongest bond →			Weakest bond
pK_a = 3.2	pK_a = –6	pK_a = –9	pK_a = –10
Weakest acid →			Strongest acid

The closer in energy the two interacting orbitals are, the greater their interaction. A greater interaction results in both a larger rising of the antibonding orbital and a larger lowering of the bonding molecular orbitals formed. Figure 13.8 shows a good example of the decrease in stabilization of the bonding molecular orbital as the energy difference between the interacting orbitals increases.

The most stabilization is achieved by interacting a full orbital and an empty orbital. When a full orbital interacts with an empty orbital, a large stabilization of the bonding electron pair is possible, Figure 13.8. The unoccupied antibonding orbital destabilization does not affect the overall stability since we are concerned about stabilizing the electrons present. The greatest stabilizing interaction occurs when a filled MO overlaps well with and is close in energy to an empty MO. This is this is almost always a HOMO–LUMO interaction.

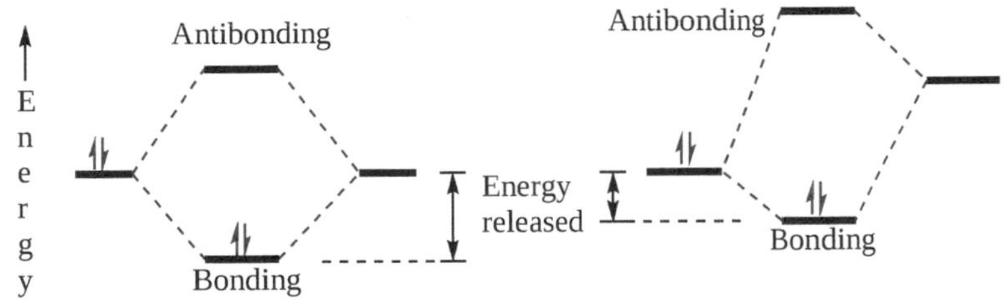

Figure 13.8 Energy is released by interacting a full with an empty orbital. Orbitals of the same energy have a large interaction (left). Orbitals of differing energies have a smaller interaction (right).

13.2.2 Electronegativity Effects

Electronegativity affects orbital energies and electron distribution. Because oxygen has a higher nuclear charge than carbon, its 2*p* orbital is of lower energy than the carbon 2*p* orbital. As illustrated in Figure 13.9, the lower-energy oxygen 2*p* orbital tends to pull down the energy of both the π bonding and π* antibonding molecular orbitals. The lowered π* orbital energy will have an important influence on reactivity.

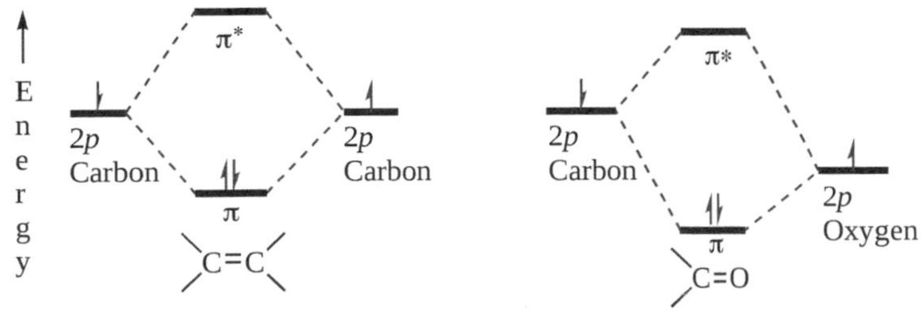

Figure 13.9 The increased electronegativity of oxygen lowers the energy of both π and π* of the C=O.

When two orbitals of differing energy interact, the molecular orbital contains a greater percentage of the atomic orbital closest to it in energy. This is easily demonstrated (Figure 13.10) by examining the two extremes. As we have seen for a carbon–

carbon pi bond, when the two orbitals are of equal energy each contributes equally to the molecular orbital. The other extreme is when the orbitals are so far apart in energy that they do not interact: The higher orbital would contain only the higher atomic orbital. An intermediate case in which the energies of the two orbitals differ but the orbitals still interact is shown in the center of Figure 13.10.

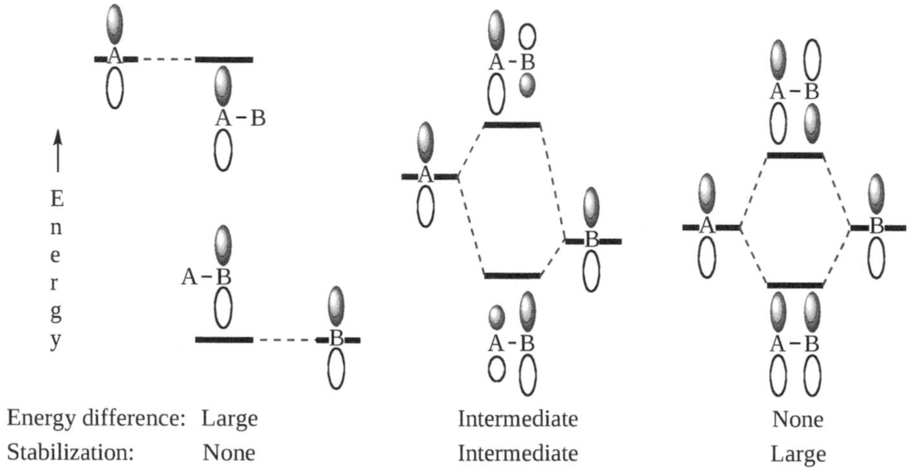

Figure 13.10 The interaction of orbitals of differing energy. The orbital with the larger contribution is drawn larger.

An example of the intermediate case of Figure 13.10 is found in the π molecular orbitals of a carbonyl, the carbon–oxygen double bond. The lower-energy oxygen 2*p* orbital contributes more to the π than to the π* molecular orbital, and the higher-energy carbon 2*p* orbital contributes more to the π* than to the π molecular orbital (Figure 13.11). Because of the distortion of the bonding molecular orbital, the electrons in the carbonyl pi bond have a greater probability of being found around the oxygen. The valence bond resonance forms are just another way of describing this electron distribution in the molecular orbital.

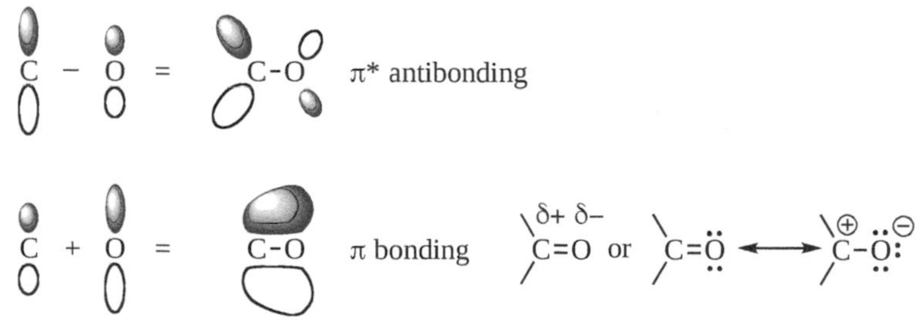

Figure 13.11 The pi molecular orbitals of a carbonyl group.

13.2.3 Carbanion Stabilization from a HOMO–LUMO Perspective

The best electron withdrawal occurs when the energy of the pi acceptor's LUMO is close to that of the full 2*p* orbital of the carbanion. For example, compare how two groups, the C=C and the C=O, stabilize an adjacent carbanion. The pK_a of methane is about 48; if we attach a C=C, which can stabilize the anion by delocalization, the pK_a drops to 43. If we instead attach a C=O, which can stabilize by delocalization and is a more electronegative group, the pK_a drops all the way to 19.2, indicating a much greater stabilization of the anion. As shown by Figure 13.12, the LUMO of the carbonyl is lower in energy than that of a simple carbon–carbon double bond and therefore interacts more with the HOMO of the carbanion.

The lower in energy the LUMO of the electron-withdrawing group (EWG) is, the better the group will be at withdrawing electrons. The lower-energy LUMO can interact better with filled orbitals. The better electron-withdrawing groups contain a low-lying π* orbital that can form a strong pi bond to the full *p* orbital of the anion.

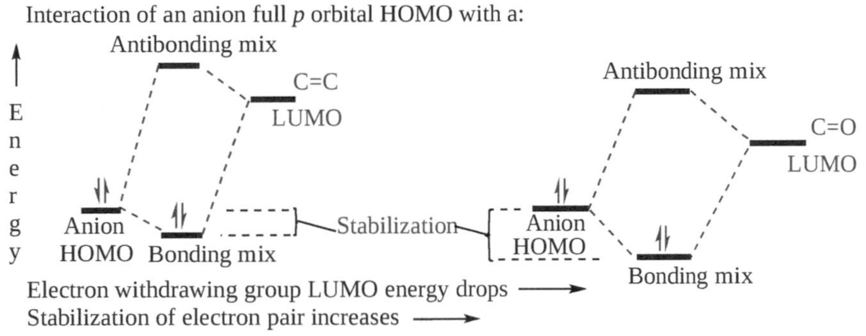

Figure 13.12 Interaction diagrams of an anion with empty orbitals from a C=C and from a C=O.

13.2.4 Carbocation Stabilization from a HOMO–LUMO Perspective

Another way to look at the stabilization of a carbocation is to consider interaction of the carbocation empty *p* orbital (LUMO) with the full electron-releasing group orbital (HOMO). The best donors to an empty 2*p* orbital will be full orbitals close to the same energy (the stabilizing interaction is greater). For lone pair donors, as the atom bearing the lone pair becomes more electronegative, the pi-donor effect decreases because the energy of the HOMO drops, taking it further away from the empty 2*p* orbital. Stronger bonds have lower-energy HOMOs, farther away in energy from the empty carbon 2*p* orbital. A C=C pi bond will be a better electron-releasing group than a carbon–hydrogen sigma bond (Figure 13.13).

The strained C–C sigma bonds of a three-membered ring have a HOMO much higher in energy than unstrained C–C sigma bonds so are the exception to sigma bonds being poor at stabilizing cations. A three-membered ring in a cyclopropylcarbinyl system, $C_3H_5{-}CH_2^+$ stabilizes a carbocation better than a phenyl ring in the benzyl carbocation, $C_6H_5{-}CH_2^+$.

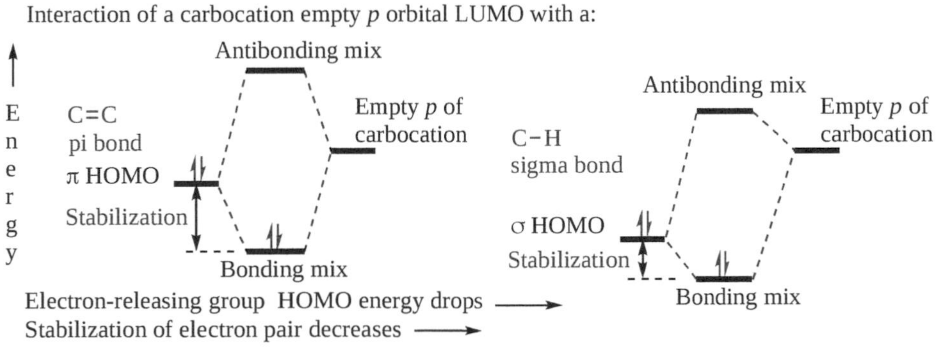

Figure 13.13 Interaction diagrams of a carbocation with a C=C pi bond and a C–H sigma bond.

13.2.5 Free Radical Stabilization from a HOMO–LUMO Perspective

Orbital interaction diagrams can easily explain how an adjacent radical center is stabilized by a pi acceptor (Figure 13.14a). The pi acceptor has a low-lying empty orbital, LUMO, close in energy to the singly occupied molecular orbital, SOMO. The interaction of the SOMO with the empty LUMO forms a new pair of orbitals, bonding and antibonding. The new bonding orbital is lower in energy than either the original SOMO or LUMO. The single electron now resides in this new bonding orbital and is stabilized relative to a system that does not have this SOMO–LUMO interaction.

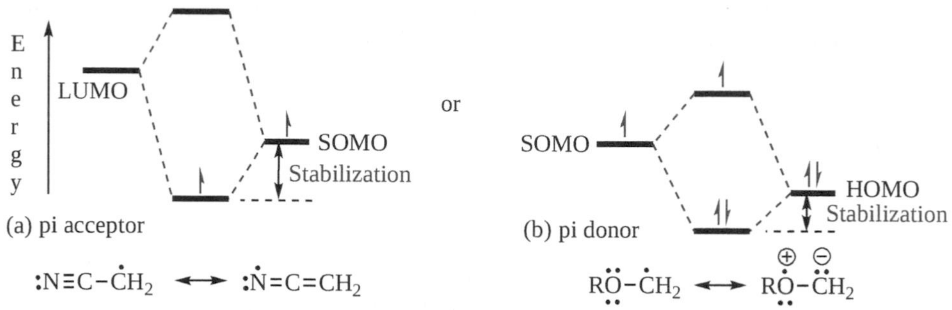

Figure 13.14 (*a*) The interaction of a singly occupied MO with the lowest unoccupied MO of a pi acceptor. (*b*) The interaction of a singly occupied MO with the highest occupied MO of a pi donor.

Similarly, orbital interaction diagrams can also explain the stabilization of a radical center by a pi donor (Figure 13.14b). The pi donor has an accessible full orbital, HOMO, close in energy to the orbital with the single electron, SOMO. The interaction of the SOMO with a full HOMO destabilizes one electron and stabilizes two for a net stabilization of one electron overall. We have created a pi bond with one electron in the antibonding orbital; the pi bond order is thus half (resonance structures are unable to show this partial pi bond).

13.2.6 Nucleophile–Electrophile Interactions

Nucleophile–electrophile interactions involve the overlap and interaction of the least stabilized filled orbital (highest occupied MO, HOMO) of the nucleophile with the most stabilized empty orbital (lowest unoccupied MO, LUMO) of the electrophile to yield the new bonding orbital and a new antibonding orbital. The closer in energy the two interacting orbitals are, the greater their interaction. The better the overlap is between the two orbitals, the greater the interaction. Just as we used interaction diagrams earlier to explain bonding stabilization, we can use them now to predict the bonding stabilization that occurs in a transition state when two different molecules react to form a new bond. The full orbital of the nucleophile interacts with the empty orbital of the electrophile to form a new bonding orbital and also a new antibonding orbital. The electrons of the nucleophile flow into the new bonding orbital and fill it. The nucleophile and electrophile approach each other so as to maximize the overlap of the HOMO with the LUMO, resulting in greater interaction and greater stabilization of the transition state. A generalized example is shown in Figure 13.15.

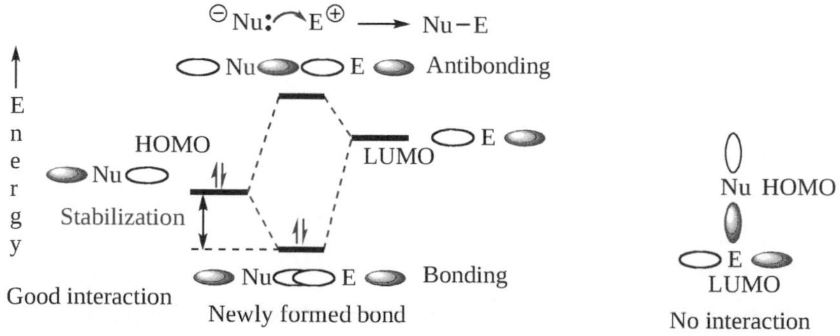

Figure 13.15 The interaction of the HOMO of the nucleophile and the LUMO of the electrophile to produce a sigma bond. The orbital arrangement on the right of the figure gives no interaction because the orbitals have equal positive and negative overlap (recall Figure 13.7).

Although resonance structures generally give enough information about the partial pluses and minuses to explain a molecule's reactivity (on the basis of opposite charges attracting each other), we occasionally need to refer to HOMO–LUMO interactions for additional help. Few explanations in this text rely just on HOMO–LUMO arguments.

S_N2 from a HOMO–LUMO perspective

When a nucleophile displaces a leaving group in an S_N2 substitution reaction, the nucleophile HOMO interacts with the LUMO of the leaving group bond as shown in Figure 13.16. Nucleophile attack produces the transition state wherein the nucleophile overlaps the largest lobe of the LUMO to get the greatest stabilization possible. The nucleophile displaces the leaving group from the backside, and therefore the configuration about the carbon attacked is inverted in this type of process. Front side displacement would involve much poorer interaction with the LUMO because of nearly equal positive and negative overlap.

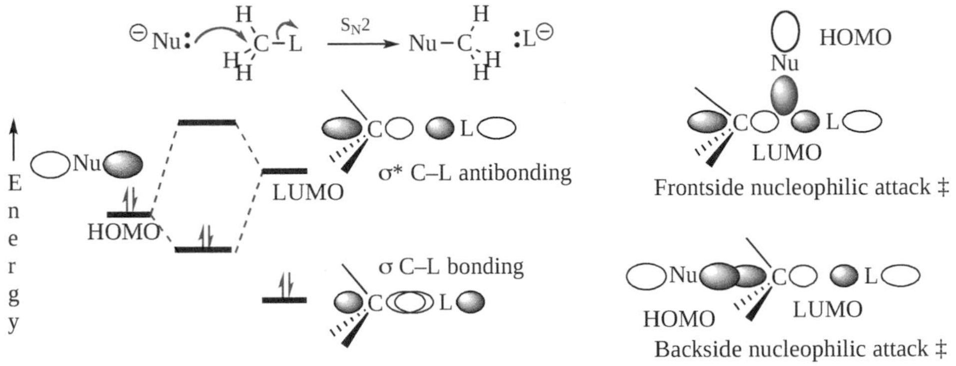

Figure 13.16 The interaction of the HOMO of the nucleophile and the LUMO of the leaving group bond to carbon to produce a new sigma bond.

Additions to alkenes from a HOMO–LUMO Perspective

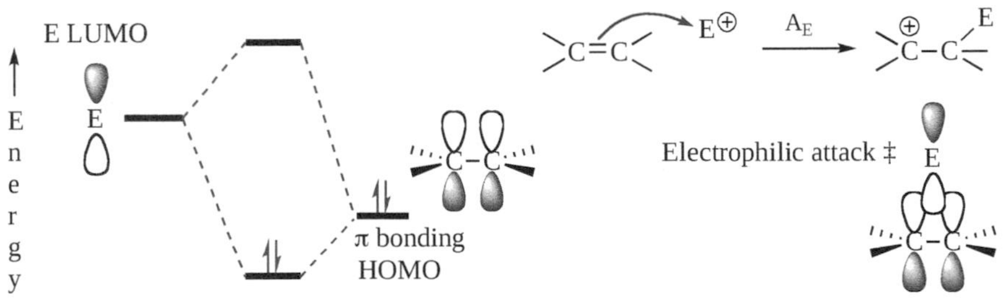

Figure 13.17 The interaction of the LUMO of the electrophile and the HOMO of the pi bond.

When an electrophile attacks an alkene (Figure 13.17), it will approach the pi bond in a way to achieve the best overlap of its LUMO with the alkene HOMO. The attack is vertical, along the axis of the *p* orbitals. We have interacted an empty electrophile orbital with a full pi bonding orbital to achieve stabilization of the transition state.

Additions to carbonyls from a HOMO–LUMO perspective

When a nucleophile attacks a carbonyl group (Figure 13.18), it will approach the carbonyl group so as to achieve the best overlap of its HOMO with the carbonyl's LUMO whose largest lobe is on carbon. The attack is not vertical, but close to the tetrahedral angle. We have interacted a full orbital with an empty one to achieve stabilization.

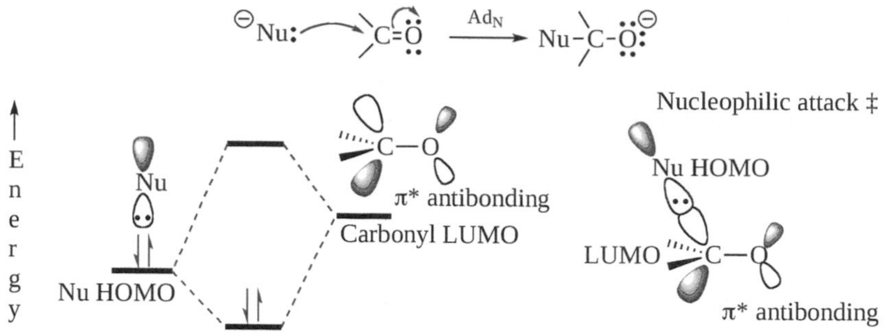

Figure 13.18 The interaction of the HOMO of the nucleophile and the LUMO of the carbonyl to produce a sigma bond.

Likewise, when an electrophile attacks a carbonyl group, it will approach the carbonyl group so as to achieve the best overlap of its LUMO with the carbonyl's HOMO, the lone pair on oxygen (Figure 13.19). The molecular orbitals are telling us what we already knew from the minor resonance form of the carbonyl group. Nucleophiles attack the δ+ end, and electrophiles attack the δ− end of the carbonyl group. The orientations of the orbitals tell us additional information on the position of attack.

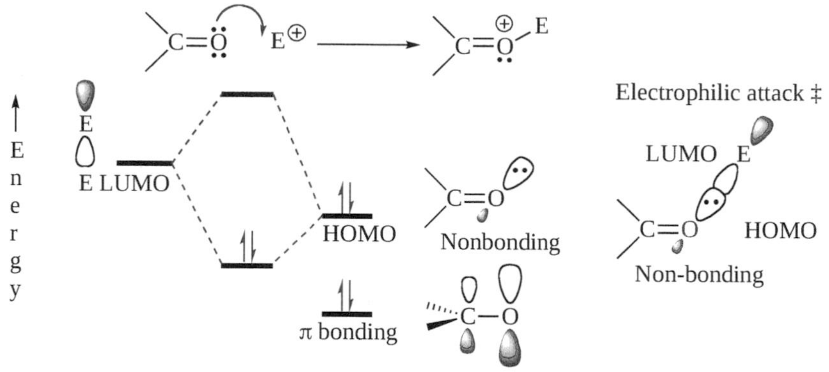

Figure 13.19 The interaction of the LUMO of the electrophile and the HOMO of the carbonyl to produce a sigma bond. The carbonyl HOMO is the nonbonding lone pair and not the pi bond.

13.3 MOLECULAR ORBITAL THEORY FOR LINEAR PI SYSTEMS

13.3.1 Two *p* Orbitals, a Simple Pi Bond

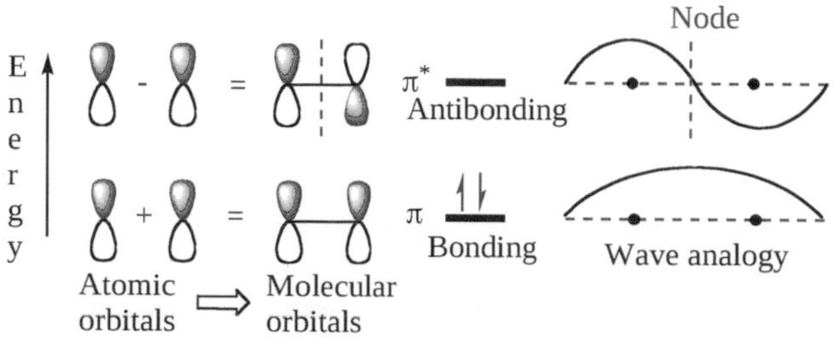

Figure 13.20 Pi molecular orbitals for ethene, $CH_2{=}CH_2$. The dots in the wave analogy are approximate nuclei position. The standing wave extends beyond the nuclei because the orbitals do.

Since the pi bonding portion of a molecule is perpendicular to the sigma bonded framework, it can generally be treated independently. In a double bond, **two *p* orbitals produce two molecular orbitals,** π (bonding) and π* (antibonding), shown in Figure 13.20. The node in the wave corresponds to the mathematical sign change. The lowest-energy molecular orbital always has no nodes between the nuclei (we ignore the node in the plane of the atoms common to all pi systems). The electron pair in the bonding MO is shared between the nuclei. The antibonding MO has a node between the nuclei.

13.3.2 Three *p* Orbitals, the Allyl Unit

Three *p* orbitals, all aligned parallel to one another, make up the allyl unit and **produce three molecular orbitals.** The MOs for the following systems differ from the completely carbon allylic system only in the previously discussed electronegativity effects (the bonding MOs will have a larger contribution from the heteroatom *p* orbital and the antibonding MOs will have a lesser contribution, Figures 13.10 and 13.11).

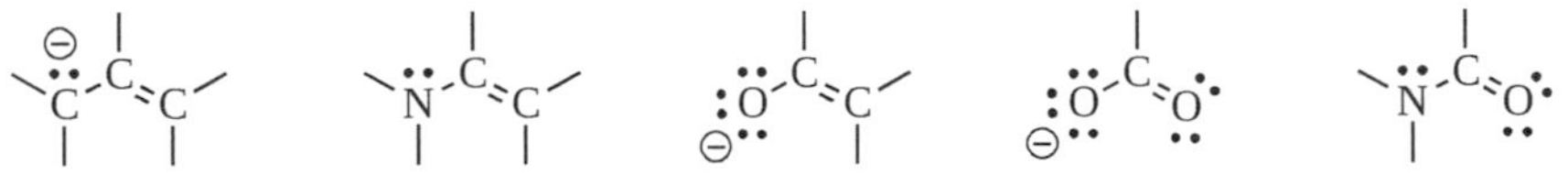

To determine how many electrons occupy any pi system, start by counting each double bond as contributing two electrons, but do not count any pi bonds that are perpendicular to the pi system of interest: For example, the second pi bond of a triple bond does not count. Because the maximum that a single atom can contribute is one filled *p* orbital containing two electrons, if a heteroatom has several lone pairs it can contribute only two electrons. Only one *p* orbital on an atom can align itself with the pi system. If an atom is doubly bonded and contains a lone pair, the lone pair cannot be counted since it must be in an orbital that is perpendicular to the double bond. All the examples above have four electrons in a three-*p* orbital pi system.

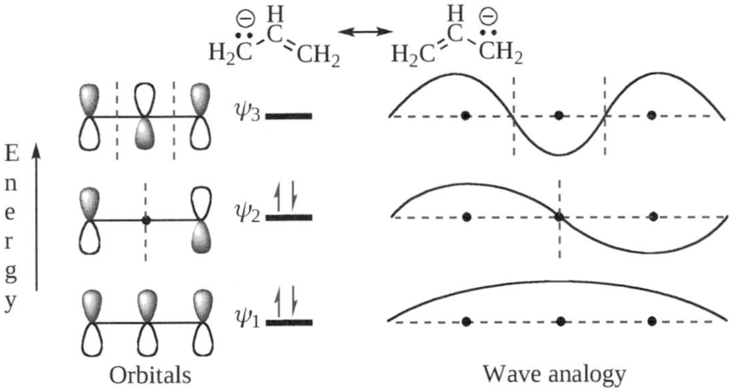

Figure 13.21 The pi molecular orbitals of the allyl anion. The dots are the approximate position of the nuclei. The bonding MO is ψ_1. The nonbonding MO ψ_2 has a node in the center thus no contribution from the center *p* orbital. The antibonding MO ψ_3 has nodes between every atom.

Shown in Figure 13.21 are several different ways to describe the MOs for C=C–C:$^-$, the allyl anion. The MOs are labeled from the lowest energy to the highest and are drawn with the lowest-energy MO at the bottom. The lowest-energy MO, ψ_1, has no nodes and will always be bonding. The highest-energy molecular orbital, ψ_3, has a node between each nucleus and will always be antibonding. In pi systems containing an odd number of *p* orbitals, there is a *nonbonding* MO, in this case ψ_2, in which the wave puts a node through the central atom. If an atom lies on a node in an MO, it does not contribute an atomic orbital to that MO. Nodes are always symmetrical about the center; in waves with an odd number of nodes, one of the nodes must go through the center. The allyl anion has four electrons in the pi system, filling both ψ_1 and ψ_2. The valence bond resonance forms, C=C–C:$^-$ ↔ $^-$:C–C=C, describe the electron distribution in ψ_2 of the anion. The two electrons of the anion can be found on the ends of the pi system and not in the middle. For C=C–C$^+$ ↔ $^+$C–C=C, the allyl cation, there would be only two electrons in the pi system, filling just ψ_1, leaving the others empty. The shape and energy of the MOs are independent of the occupancy. For the allyl radical, C=C–C$^•$ ↔ $^•$C–C=C, the stabilization of a radical by conjugation with a carbon–carbon pi bond can be understood by the interaction of a singly occupied *p* orbital with a carbon–carbon double bond. Although the singly occupied orbital is not changed in energy, the two electrons in the π orbital are stabilized by the interaction (Figure 13.22). The odd electron resides in ψ_2 of the allylic system and is called the singly occupied molecular orbital.

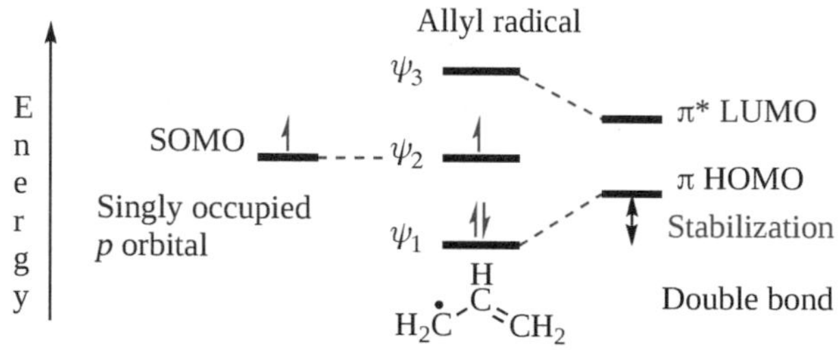

Figure 13.22 A singly occupied *p* orbital interacts with the π and π* orbitals of a pi bond to form the three molecular orbitals of an allylic radical.

The SOMO describes the distribution of the odd electron over the molecule. The odd electron appears at the ends of the allylic system and not in the middle because there is a node in the middle of ψ_2 (as shown in Figure 13.21). The resonance forms are an attempt to describe this SOMO in the language of lines and dots (Figure 13.23).

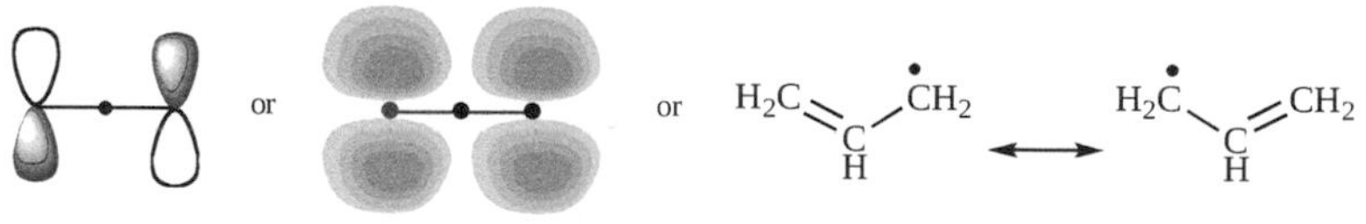

Figure 13.23 Alternate descriptions of ψ_2 of the allyl radical.

13.3.3 Four *p* Orbitals, the 1,3-Diene Unit

Figure 13.24 gives the MOs for 1,3-butadiene, $H_2C{=}CH{-}HC{=}CH_2$, which has **four *p* orbitals** in the pi system and thus **four MOs**. The four electrons are placed in the two bonding MOs, leaving the two antibonding MOs empty. Again note the similarity to a vibrating string. The *highest occupied molecular orbital* (HOMO) is ψ_2, and the *lowest unoccupied molecular orbital* (LUMO) is ψ_3.

An alternative way to order the MOs is to count the bonding and antibonding overlap within each MO. ψ_1 is bonding between all the *p* orbitals, so is the lowest in energy. ψ_2 is bonding between the first *p* orbital and the second, antibonding between the second and the third, and bonding between the third and the fourth, giving two bonding and one antibonding interaction. ψ_3 is antibonding between the first *p* orbital and the second, bonding between the second and the third, and antibonding between the third and the fourth. Finally, ψ_4 is antibonding between all atoms, so is the highest-energy MO of all.

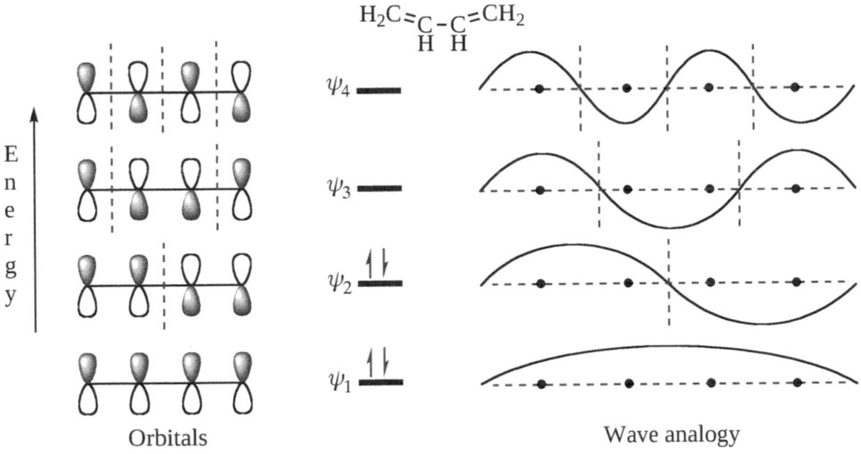

Figure 13.24 The pi molecular orbitals of the 1,3-butadiene, $H_2C{=}CH{-}HC{=}CH_2$. The highest occupied MO (HOMO) is ψ_2, and the lowest unoccupied MO (LUMO) is ψ_3.

As we saw in Section 13.2, if we combine two atomic orbitals, we must get two molecular orbitals. The final number of MOs equals the starting number of AOs. The lower-energy in-phase bonding combination is stabilized slightly less than the out-of-phase antibonding combination is destabilized. By looking at the wave analogy and orienting the phase of the *p* orbitals, we can create a similar set of harmonics. Figure 13.25 shows how the interaction diagram of two ethylene molecular orbitals can produce the same four molecular orbitals of 1,3-butadiene. Each π bonding orbital can be combined in phase or out of phase, and so can each antibonding π^* orbital.

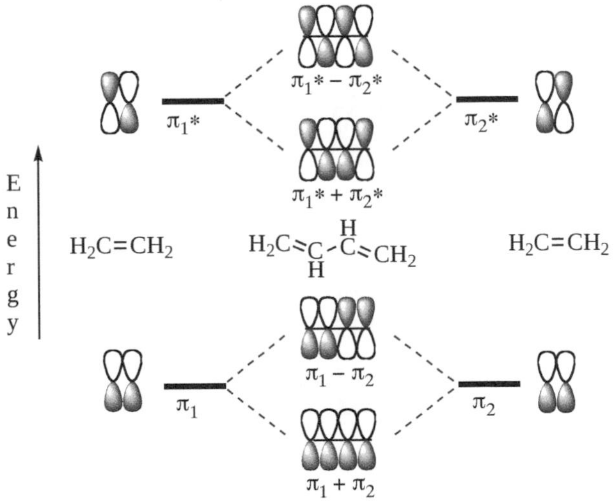

Figure 13.25 The interaction of two ethylenes to give the molecular orbitals for 1,3-butadiene.

13.3.4 The Progression of Energies and Orbitals for Linear-Conjugated Systems

Figure 13.26 shows how the relative energy of the pi orbitals changes as a second pi bond and then a third pi bond is added to form a longer conjugated system. **Additional conjugation raises the HOMO in energy and lowers the LUMO in energy**, as can be seen from comparing the energy of the MOs of ethene, 1,3-butadiene, and 1,3,5-hexatriene. Also, the energy gap between the HOMO and LUMO decreases with additional conjugation. The molecule can absorb light by promoting an electron from the HOMO to LUMO. Even 1,3,5-hexatriene absorbs in the high-energy ultraviolet, but with enough conjugation the HOMO–LUMO gap becomes small enough that the molecule can absorb longer-wavelength visible light and appear colored.

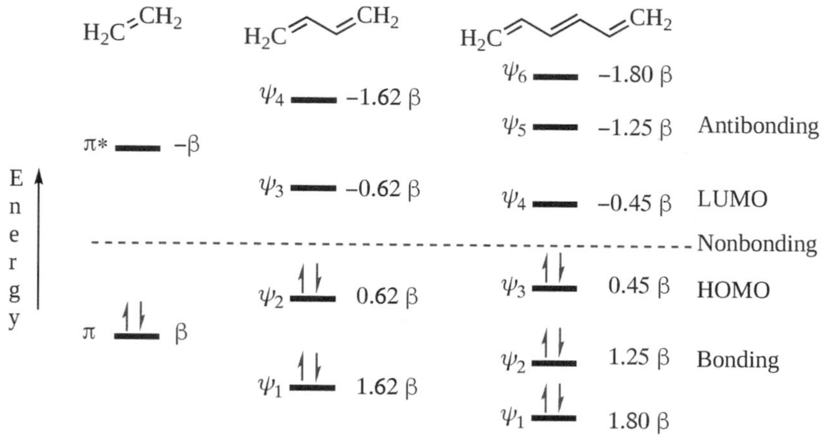

Figure 13.26 A comparison of the relative orbital energies of ethene (left), 1,3-butadiene (middle), and 1,3,5-hexatriene (right). The electron stabilization in the π of ethene is defined as β.

A summary of the progression of orbitals and energies for linear conjugated systems is shown in Figure 13.27. The orbitals are always evenly distributed above and below the nonbonding level. The nodes are always symmetric about the center of the pi system.

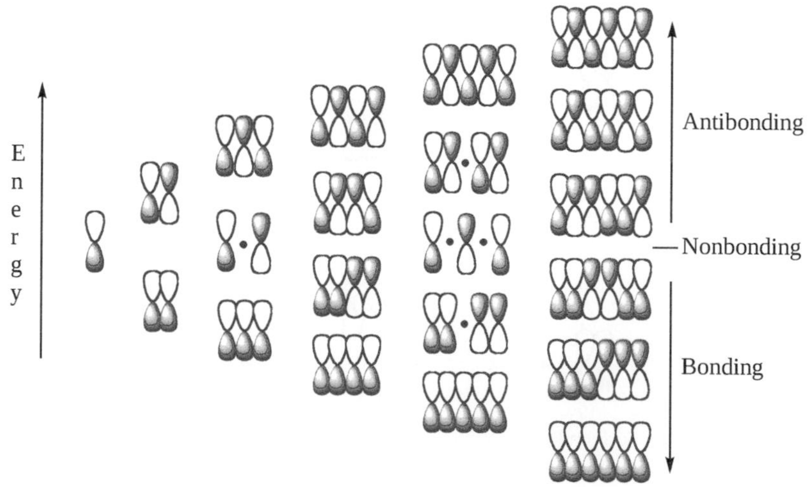

Figure 13.27 A summary of the molecular orbitals and relative energies for linear conjugated pi systems of one to six *p* orbitals. The phase of the ends of a linear pi system molecular orbital alternates as the energy of the MO increases. The fundamental (lowest MO) is always symmetric and all bonding.

It is the phase of the ends of the HOMO and LUMO that will be important in pericyclic reactions. The ends of the fundamental (first harmonic) MO in any system are always in phase (both end *p* orbitals have the same shading on top) because the fundamental has no nodes. The second harmonic (one node) has the ends out of phase because the phase switches across the node, by definition. The third harmonic with two nodes has two phase switches, so the ends are back in phase. To generalize, all the MOs with an even number of nodes also have their ends in phase, but all the MOs with an odd number of nodes have the ends of opposite phase. So since we gradually add nodes one at a time **as we go up in energy, the phasing of the ends of the pi system always alternates.** This alternation of the phase of the ends of the pi system will be used to predict the outcome of pericyclic reactions.

13.4 MO THEORY FOR CYCLIC-CONJUGATED PI SYSTEMS

The molecular orbitals for benzene are shown in Figure 13.28. The ring is planar, and every atom in the ring bears a *p* orbital; we have a closed loop of overlapping *p* orbitals. As with the linear MOs, the more nodes the MO has, the higher its energy. We have basically taken our linear system and brought the ends together so that now we have standing waves of a loop, similar to standing waves on a drumhead, like in Figure 13.2. Note that two pairs of MOs, called *degenerate pairs*, have the same energy, for they have the same number of nodes. When we place the six pi electrons in the MOs for benzene, a degenerate pair is completely filled; benzene has a filled shell.

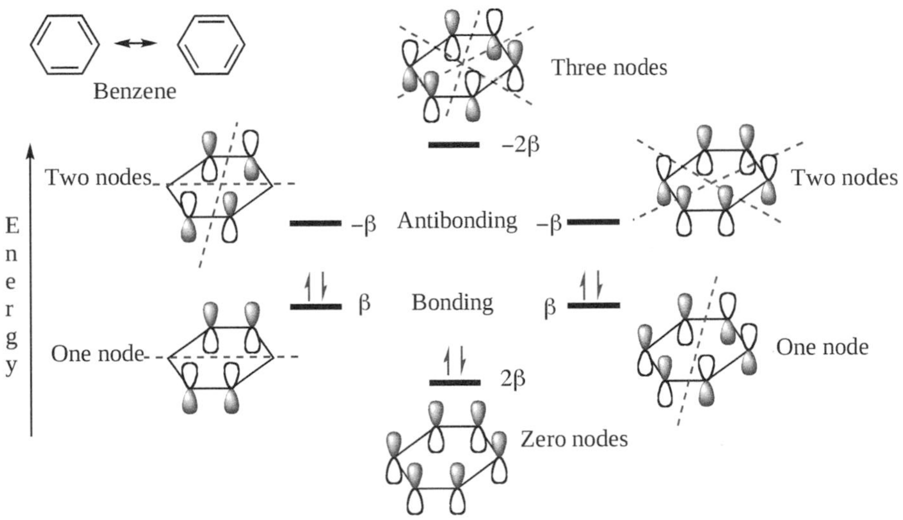

Figure 13.28 The pi molecular orbitals of benzene. The view is from the side. The nodes are planes that go through the center and cut both sides of the loop. There is just one zero node (fundamental) MO.

Cyclopentadienyl cation, on the other hand, is a very unstable molecule. Its MO energy levels are shown in Figure 13.29. When the four pi electrons are placed in the MOs for cyclopentadienyl cation, the degenerate pair is only half filled; cyclopentadienyl cation has an incomplete shell. The instability of cyclopentadienyl cation comes as a direct result of these unpaired electrons in the half-filled shell.

The MO energy levels of cyclopentadienyl radical are shown in Figure 13.30. Cyclopentadienyl radical has one more electron than the cation, thus five electrons total. It has one unpaired electron so is a reactive radical. Chapter 12 covered radical reactions.

The MO energy levels of cyclopentadienyl anion are shown in Figure 13.31. Cyclopentadienyl anion has two more electrons than the cation, thus six electrons total, and therefore has a filled degenerate pair and is predicted correctly to be aromatic and stable; it has enough electrons to fill the shell.

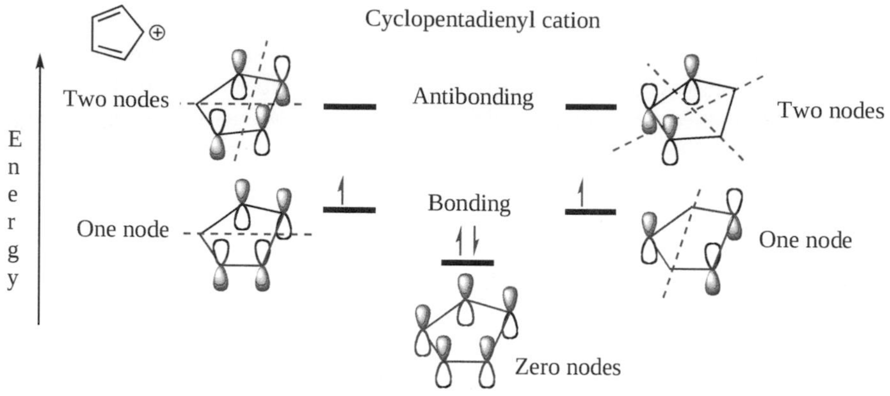

Figure 13.29 The pi molecular orbitals of cyclopentadienyl cation.

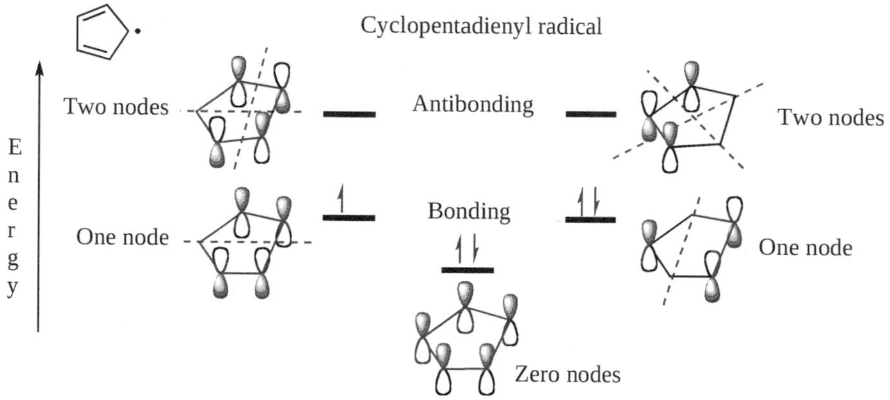

Figure 13.30 The pi molecular orbitals of cyclopentadienyl radical.

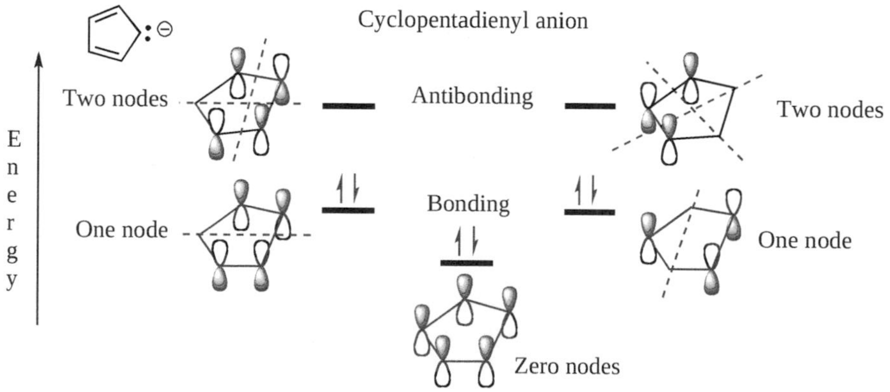

Figure 13.31 The pi molecular orbitals of cyclopentadienyl anion.

The stabilization of benzene can be placed on a semiquantitative scale. If we define the stabilization of one electron in the pi orbital of ethene as β, then the stabilization of two electrons in a double bond is just 2β. The stabilization of six pi electrons in three double bonds is 6β. Benzene's MO energy levels are at 2β, 1β, 1β, and −1β, −1β, −2β; the first three are filled with two electrons each for a stabilization of 8β total (2 × 2β + 4 × 1β). The six electrons of benzene pi loop are then 2β more stabilized than the six electrons in three isolated double bonds.

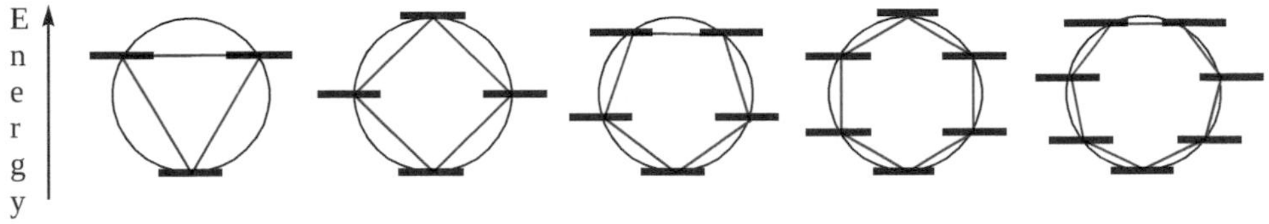

Figure 13.32 The pi molecular orbital energies of cyclic systems.

The MO energy levels for any simple cyclic conjugated polyene can be estimated by drawing the regular polygon of the ring point down (for the one unique fundamental MO) inscribed in a circle of 2β radius (Figure 13.32). The vertices of the polygon will correspond to the energies of the MOs on a vertical scale. The center of the circle corresponds to the nonbonding level (zero β) with the bonding levels below it and the antibonding above.

Since the degenerate pair must be completely filled or completely empty in order for the molecule to be aromatic, stable arrangements arise when 2, 6, 10, ... or $4n + 2$ pi electrons are in the cyclic conjugated pi system. Unstable half-filled degenerate shells will occur with 4, 8, 12, ... or $4n$ pi electrons. The $4n$ systems tend to distort from planarity to diminish pi overlap and this destabilizing antiaromatic conjugation.

13.5 PERTURBATION OF THE HOMO AND LUMO

The relative availability of the electrons in the HOMO determines the nucleophilicity of the pi system. The higher in energy the HOMO is, the more available that electron pair is for overlap with an electrophile's LUMO. Likewise, if the electrophile's LUMO is lower in energy, closer to the nucleophile's HOMO, the interaction of the two is greater, as in Figure 13.8. Recall that the closer in energy two interacting orbitals are, the greater their interaction. A favorable HOMO–LUMO interaction in a pericyclic reaction will predict a favorable reaction route. So it is important to understand what perturbs the energy of the HOMO and LUMO of a system.

One way to perturb a pi system is to swap out a carbon atom in the system for a more electronegative element such as oxygen, as shown for C=O in Figure 13.9. Replacement of carbon by **a more electronegative atom lowers both the HOMO and LUMO of the system**. In this way, C=O bonds are more electrophilic than carbon–carbon double bonds.

When two groups are conjugated with each other, the new conjugated system has properties that are different from the original two groups. Figure 13.26 demonstrated that **simple conjugation with another pi bond raised the HOMO and lowered the LUMO**.

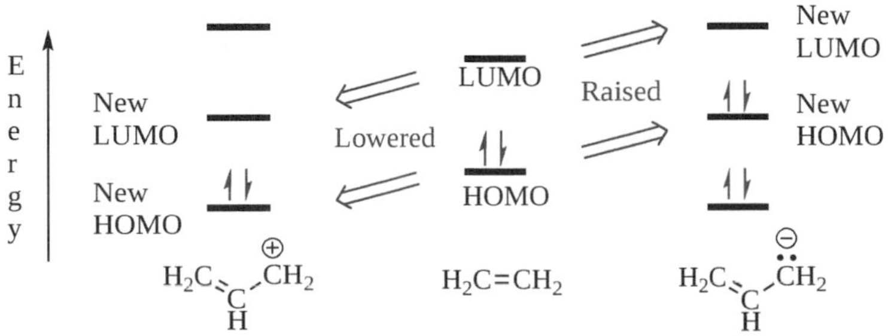

Figure 13.33 In the center is the HOMO and LUMO of an unperturbed pi bond. The left side shows that both the HOMO and LUMO drop in energy when a pi acceptor is attached. The right side shows that both the HOMO and LUMO are raised in energy when a pi donor is attached.

For example, we can look at the change that occurs in the molecular orbitals of a carbon–carbon pi bond as we attach a group that conjugates with it. One extreme would be to attach an atom with a fully occupied *p* orbital, an excellent pi donor. In general, any group that conjugates with a pi system and donates electron density to it is a pi donor. The other extreme would be to attach an atom with an empty *p* orbital, an excellent pi acceptor. Any group that conjugates with a pi system and withdraws electron density from it is a pi acceptor.

The right side of Figure 13.33 shows what happens if a good pi donor is attached to a double bond. The pi donor HOMO interacts with both the double-bond HOMO and LUMO to produce a new allylic system with a new HOMO and LUMO, which are raised from those of the double bond, displayed in the center of the figure. **Conjugation of a double bond with a pi donor raises the HOMO of the system, and makes the combination more nucleophilic.**

The left side of Figure 13.33 shows what happens if a good pi acceptor is attached to a double bond. The LUMO of the pi acceptor interacts with both the double bond HOMO and LUMO to produce a new allylic system with a new HOMO and LUMO, which are lowered from those of the double bond. **A pi acceptor bonded to a double bond makes the combination more electrophilic by lowering the LUMO of the system.**

Why does this matter? Hard and soft can be explained in terms of the HOMO and LUMO energies and their separation. The hard–hard interaction is the electrostatic attraction of opposite charges, an ionic bond in salt for example. Hard species have stable low-energy HOMOs and a large energy separation between the HOMO and LUMO. This large energy separation between the HOMO and LUMO means that there is little covalent interaction.

However almost all organic compounds are held together with covalent bonds that arise from soft–soft interactions. This soft–soft interaction is favored by a good interaction between the HOMO of the Lewis base and the LUMO of the Lewis acid. The closer in energy these two orbitals are, the larger the interaction, and therefore the covalent bond formed is stronger. If the electrophile's LUMO is lower in energy, then it is softer, and the stabilization from overlap with an average nucleophile's HOMO is greater. The left side of Figure 13.34 shows a good soft–soft interaction making a strong covalent bond.

The much poorer hard–soft interaction shown on the right side of Figure 13.34 has two problems: The more diffuse charge (if any) of the soft species makes a poor ionic bond. The large energy gap between the LUMO of the hard Lewis acid and the HOMO of the soft Lewis base significantly decreases their interaction, resulting in a weak covalent bond.

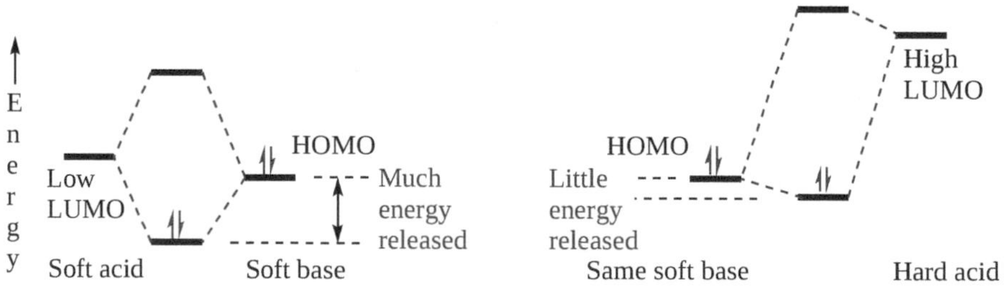

Figure 13.34 HSAB, the soft–soft vs. the soft–hard interaction.

An example of the principle in Figure 13.34 is that the LUMO of a C=O is significantly lowered by the electronegativity of oxygen compared to a carbon atom. Therefore, the carbonyl is very reactive to an organometallic nucleophile whereas a regular alkene is not.

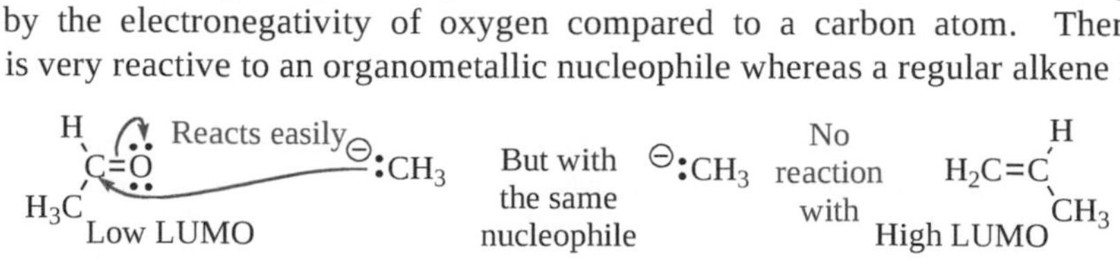

Optimally, the HOMO and LUMO are close in energy and overlap well, thus have a large interaction, releasing much energy, and forming a strong covalent bond. It is possible to reduce the energy gap between the nucleophile HOMO and electrophile LUMO, making both more reactive. Consider the case of two neutral reactants, A and B, which initially have no difference in their HOMO and LUMO levels, as shown on the left in Figure 13.35. If the HOMO and LUMO levels in each are far enough apart, there is no reaction. The soft–soft interaction is poor, and since the reactants are not charged there is no hard–hard interaction. For example, two molecules of ethylene do not normally react with each other. However, if a pi donor is added to reactant A, raising both HOMO and LUMO, and a pi acceptor is added to reactant B, lowering both HOMO and LUMO, then reaction occurs easily. The HOMO of A′ is now close in energy to the LUMO of B′, producing a good soft–soft interaction. Reactant A′ is now a good electron source, and reactant B′ is now a good electron sink. In essence, raising the source and lowering the sink has made bond formation, the flow of electrons from A′ to B′, a good process.

In summary, the best electron sources have an accessible high-energy HOMO. When the energy of the HOMO of the nucleophile is high, the nucleophile is softer, and the stabilization resulting from overlap with the electrophile's LUMO is much greater. The best electron sinks have an accessible low-energy LUMO.

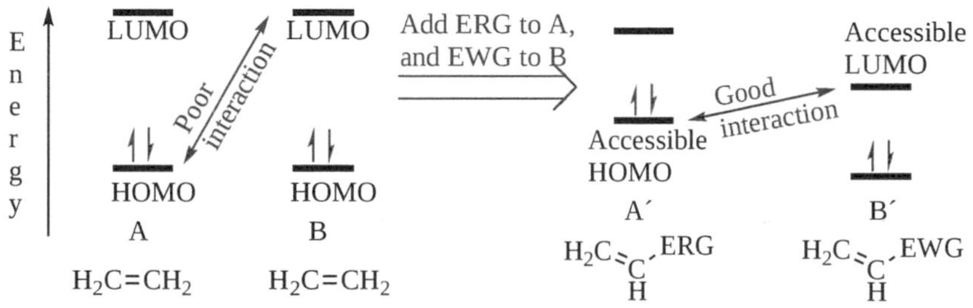

Figure 13.35 The perturbation of two reactants by adding a pi donor (ERG) to one and a pi acceptor (EWG) to the other, thus producing a more reactive pair.

13.6 DELOCALIZATION OF SIGMA ELECTRONS (MORE ADVANCED)

Advanced molecular orbital calculations for the nitrogen molecule (:N≡N:) give a simple set of molecular orbitals (Figure 13.36). The perpendicular pairs of π and π* molecular orbitals from the nitrogen–nitrogen triple bond were shown back in Figure 1.6. The sigma system from these calculations may be a bit of a surprise; the sigma bonding electrons are not constrained to individual bonds but are delocalized over the entire molecule. Instead of four localized orbitals, sigma bonding, sigma antibonding, and two lone pair orbitals, there are four sigma molecular orbitals created from atomic orbitals: the fundamental $2s + 2s$, the second harmonic $2s - 2s$, the third harmonic $2p_x + 2p_x$ and the fourth harmonic $2p_x - 2p_x$. These gradually increase in nodes: fundamental with no nodes, second harmonic with one, third harmonic with two, and fourth with three.

Sigma molecular orbitals pose a difficult representational problem, for they usually are as large and as complex as the framework of the molecule itself. Commonly we treat the sigma framework as if it were completely localized, chopping up the representational problem into intellectually convenient little pieces, each piece corresponding to a particular bond or electron pair. Toward this end, the atomic orbitals are mixed to give hybridized orbitals that are more convenient for the discussion of these localized bonding orbitals. These localized bonding orbitals are easy to use and correspond to the lines and dots of our Lewis formula.

There are very few penalties for using hybrid orbitals and treating the sigma framework as localized, when it actually is not.

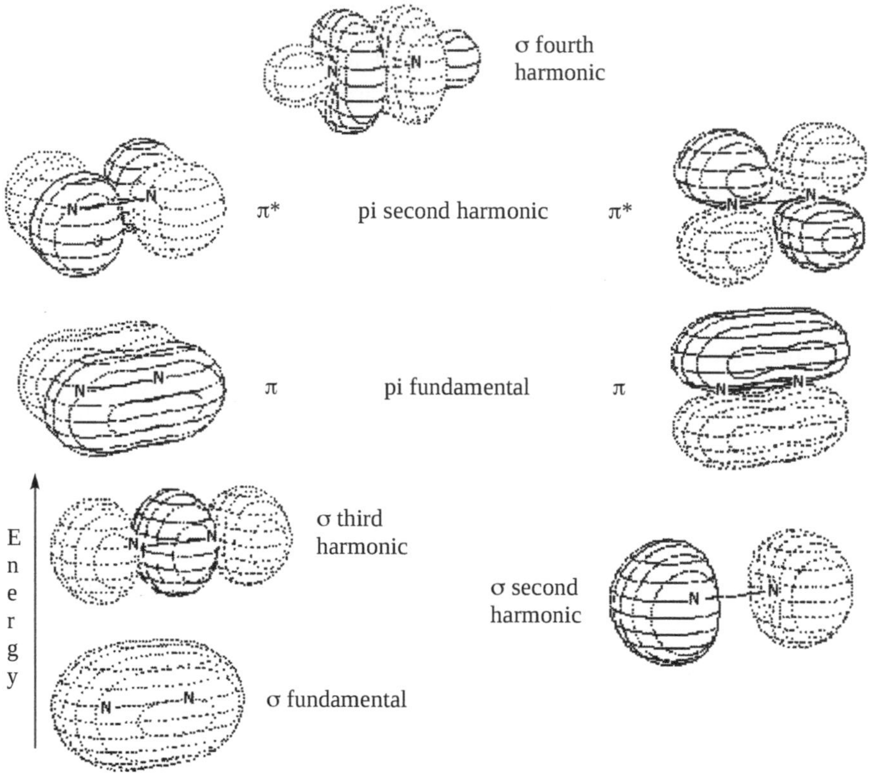

Figure 13.36 The valence molecular orbitals of nitrogen. The lowest five orbitals are filled.

13.7 CONCERTED PERICYCLIC CYCLOADDITION REACTIONS

13.7.1 Introduction to Pericyclic Reactions

Pericyclic reactions have cyclic transition states and electron flow paths that appear to go around in a loop. There are three main types of pericyclic reactions shown in Figure 13.37: cycloaddition reactions, electrocyclic reactions, and sigmatropic rearrangements. Cycloaddition reactions form a ring from two pi systems coming together. Electrocyclic reactions close the ends of a single pi system to form a ring or the reverse, open a ring to form a single pi system. Sigmatropic rearrangements move a sigma bond from one part of a pi system to another. The examples shown all have a transition state in which six electrons move around in a loop. The rules for prediction of the products will depend on many factors, including the number of electrons in the process.

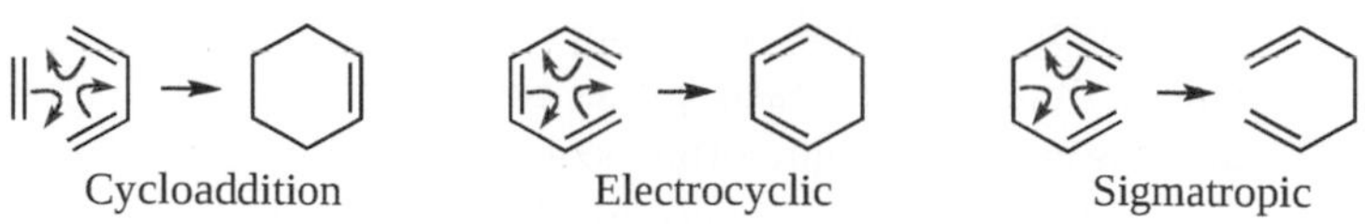

Figure 13.37 Examples of the three types of pericyclic reactions.

13.7.2 Cycloaddition Reactions

Four-Electron (4*n*) Cycloaddition Reactions

Concerted four electron suprafacial–suprafacial cycloaddition reactions are not "thermally allowed." All this really means is that the transition state for an allowed process is much lower in energy. Suprafacial just means on the same side or face of the molecule; antarafacial means on opposite faces. HOMO–LUMO theory explains why the concerted 2 + 2 cycloaddition of two ethylenes is not favored. Bonding overlap between a HOMO and a LUMO lowers the transition state energy of the allowed process, making it favorable. Figure 13.38 shows no such transition state stabilization happens because any bonding overlap is exactly canceled out by an equivalent antibonding overlap, yielding no net stabilization.

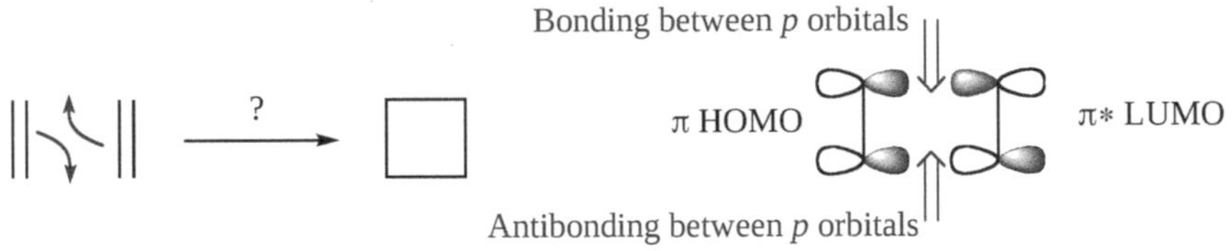

Figure 13.38 The thermal 2 + 2 concerted cycloaddition of two ethylenes is not favored because there is no net interaction possible between the HOMO of one partner and the LUMO of the other.

Six-Electron (4*n* + 2) Cycloaddition Reactions

The Diels–Alder reaction is an allowed thermal six-electron (4 + 2) suprafacial–suprafacial cycloaddition reaction between a diene (contributing four pi electrons to the transition state) and dienophile (contributing two pi electrons). It is an extremely useful cycloaddition because it goes in high yield and with predictable stereochemistry.

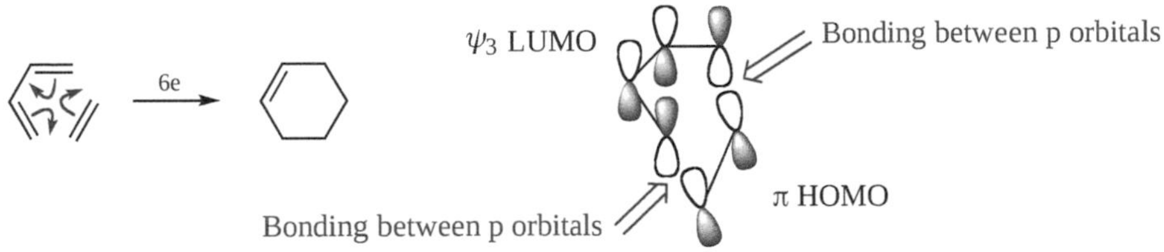

Figure 13.39 Both ends of the HOMO and LUMO have bonding overlap in a 4 + 2 cycloaddition.

The net reaction breaks the pi part of two pi bonds (63 kcal/mol each) and makes two stronger sigma bonds (83 kcal/mol each) and is therefore exothermic. The six-membered transition state for the reaction resembles a folded cyclohexane (a boat conformation). Figure 13.39 shows the reaction with line structures and with orbitals. Notice how the *p* orbitals of the pi bonds rehybridize into the sigma bonds of the product. The Diels–Alder reaction has much higher yields if an electron-withdrawing group is conjugated with the dienophile pi bond. Surprisingly, the product is the one with the electron-withdrawing group underneath, *endo*, rather than *exo*, sticking out away. Favorable orbital interactions between the electron-withdrawing group and the diene are used to rationalize this endo stereochemistry, see Figure 13.40. The stereochemistry of both the starting pieces is preserved in the product; for example, if two groups on the dienophile are *cis*, they remain *cis* in the product.

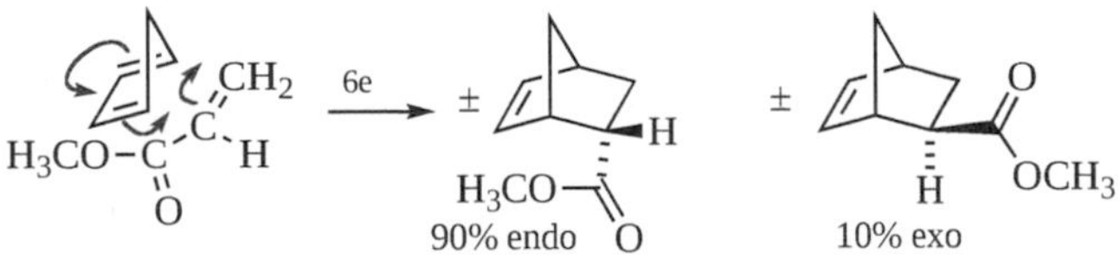

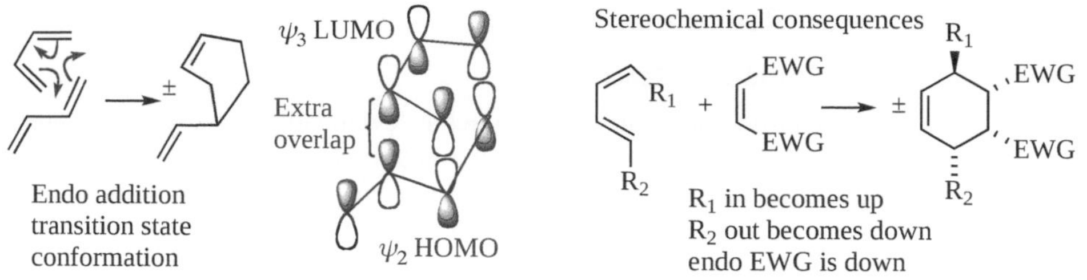

Figure 13.40 The Diels–Alder reaction of two dienes with the extra overlap of endo addition.

The five-membered transition state 4 + 2 cycloaddition is less common than the six. The four-pi electron piece is called a 1,3-dipole because it reacts as if it has a minus and a plus separated by one atom, $^+$A–B–C$^-$. The orbital arrangement for the 1,3-dipolar addition is similar to that for the Diels–Alder reaction:

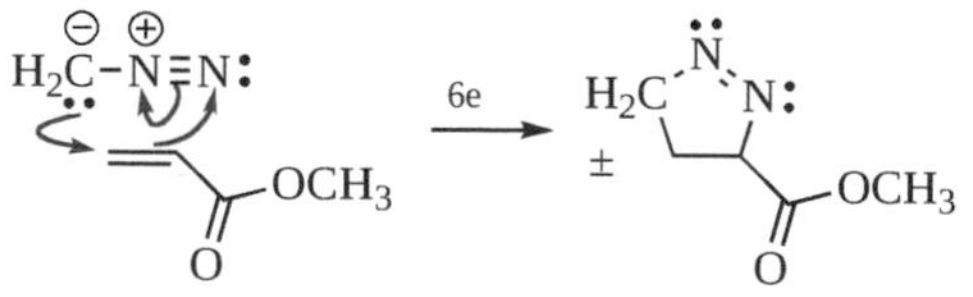

Cycloreversions are the reverse of cycloadditions and occur by the same mechanism. Because the cycloreversion breaks the molecule into two pieces, increasing degrees of freedom and therefore disorder, the $\Delta S°$ of reaction is positive. The $\Delta G°$ for the cycloreversion process becomes favorable (negative) with increased temperature because the $-T\Delta S°$ term dominates over the $\Delta H°$ term (recall $\Delta G° = \Delta H° - T\Delta S°$). The reverse Diels–Alder reaction of dicyclopentadiene into two cyclopentadienes is a common example.

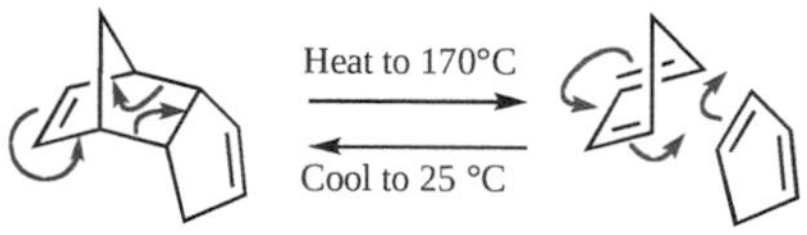

4 + 2 Cycloadditions from a HOMO–LUMO Perspective

The soft–soft interaction of filled with empty orbitals is the major interaction in the transition state because there is little hard–hard attraction. In an unsubstituted system, two soft–soft interactions stabilize the transition state: The filled ψ_2 molecular orbital of the diene interacts with the unfilled π* of the dienophile; also the empty ψ_3 of the diene interacts with the filled π of the dienophile. Figure 13.41 gives the interaction diagram.

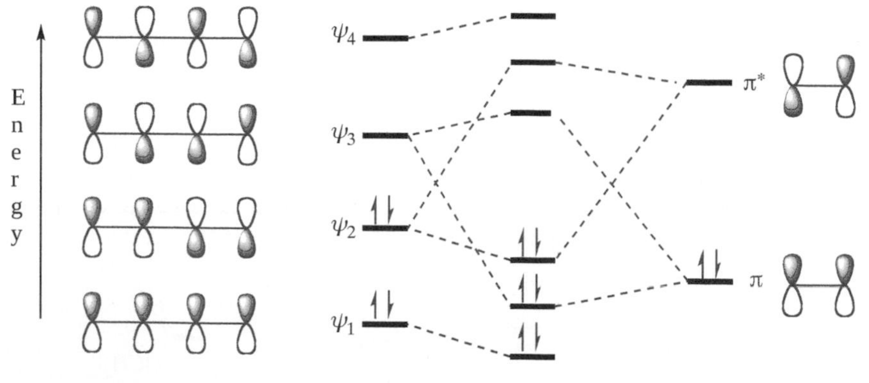

Figure 13.41 The Diels–Alder (4 + 2 cycloaddition) interaction diagram.

However, for the simplest system, 1,3-butadiene reacting with ethene, the HOMO–LUMO interactions are average, for the interacting orbitals are still reasonably far apart in energy. The interaction can be increased (Section 13.5), usually by placing an EWG on the dienophile (lowering the LUMO, π*), and sometimes by placing a donor on the diene (raising the HOMO, ψ_2). In addition, these substituents polarize the pi bond they are attached to and introduce a hard–hard factor into the reaction. With substituents on the diene and dienophile, different constitutional isomers are possible. The regiochemistry is determined by the best overlap of the HOMO–LUMO pair closest in energy. The following structures show the regiochemistry for two common combinations (the controlling interaction is the unfilled π* of the dienophile with the filled ψ_2 of the diene).

If the donor were on the dienophile and the EWG on the diene (inverse polarity), the controlling interaction would be the filled π of the dienophile with the empty ψ_3 of the diene. The regiochemistry for the inverse polarity Diels–Alder reaction is shown below.

If the dienophile has a substituent that extends its pi system, the best transition state for the reaction maximizes overlap between the two pi systems. To achieve this greater degree of interaction the dienophile must place the substituent underneath the diene, endo. The Diels–Alder reaction of two dienes has a transition state with an additional interaction between the unfilled diene ψ_3 MO and the filled diene ψ_2 MO (Figure 13.40).

We saw that the ends of the two pi systems that become the new sigma bond must both overlap in a bonding manner. With the 4*n* cycloaddition they did not (Figure 13.38), so the 2 + 2 reaction was not allowed. With the 4*n* + 2 cycloaddition (Figure 13.39), the ends did match, so the 4 + 2 reaction was allowed. Figure 13.27 showed us that the phasing of the ends of each MO alternated as we increased in energy. In adding two more electrons to the 4*n* electron system, we moved up an MO so the ends now matched. In summary, face-to-face cycloadditions with 4*n* electrons are not thermally allowed, but the **corresponding 4*n* + 2 electron cycloadditions are thermally allowed.**

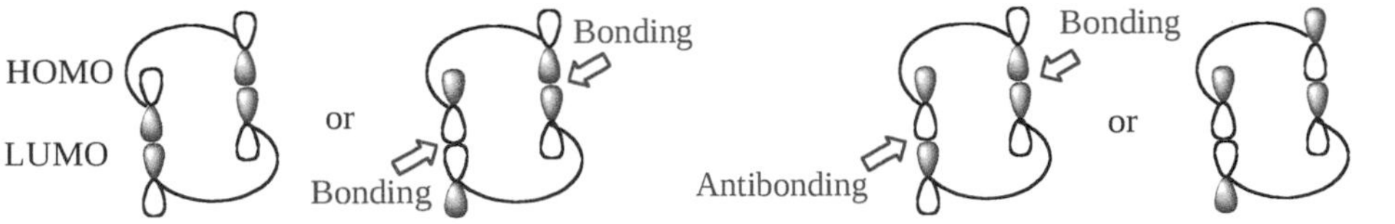

HOMO–LUMO interactions for allowed process create two new bonds on each end.

HOMO–LUMO interactions for disallowed process have both bonding and antibonding

Figure 13.42 The favorability of a cycloaddition between two pi systems is determined by how the phase of the ends of the HOMO and LUMO of two pi systems interact.

Figure 13.42 shows graphically the interaction of just the ends of two pi systems in the transition state for a cycloaddition for an allowed process (bonding between both ends) and a disallowed one (bonding on one end but not the other). If it appears that the interaction on

both ends is antibonding, reverse the shading on all the p orbitals of one MO, and this will give both ends bonding. For example, on the fundamental, it did not matter if we colored in the all the tops of the *p* orbitals or all the bottoms, as long as all were done in the same way.

13.8 CONCERTED PERICYCLIC ELECTROCYCLIC REACTIONS

13.8.1 Four-Electron (4*n*) Electrocyclic Reactions

In an electrocyclic reaction both ends must rotate 90° to convert a sigma bond into a pi bond or the reverse. The direction of rotation can be identical, for example, both clockwise or both counterclockwise, which is called *conrotatory* (con). They can rotate in opposite directions, either both in or both out, which is *disrotatory* (dis). Dashed black arrows will indicate orbital movement, not electron flow. The dots below represent the axis of rotation.

HOMO–LUMO principles allow us to predict whether the ends rotate conrotatory (con) or disrotatory (dis). If the process is favorable, as it is in Figure 13.43 when the rotation is in the same direction, then the σ-HOMO will overlap with the π-LUMO in a bonding manner to give a bonding molecular orbital (HOMO) of the diene product.

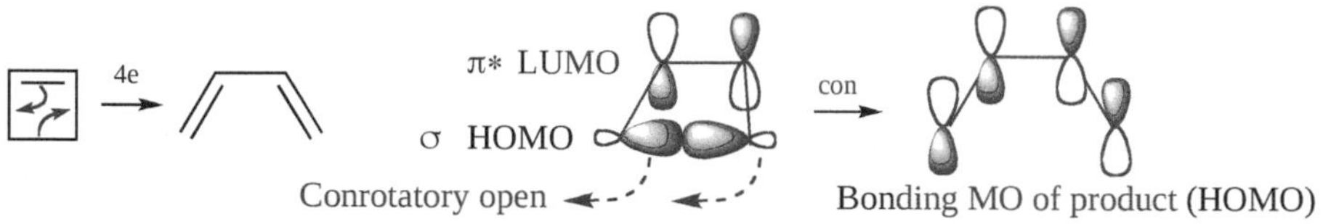

Figure 13.43 The thermal conrotatory ring opening of cyclobutene.

Stereochemical labels on the cyclobutene confirm that this 4-electron electrocyclic ring opening is conrotatory. Another way to view the reaction is from the closure of the diene. The ends of the HOMO of the diene must rotate con to create a bonding overlap for the new sigma bond. Because of microscopic reversibility, opening and closure follow the same route; if the opening is conrotatory, then the closure will likewise be con.

13.8,2 Six-Electron (4*n* + 2) Electrocyclic Reactions

As was seen for the four-electron electrocyclic ring opening (path 4e), the direction that the ends rotate when the sigma bond twists open is important. With two more electrons in the pi system, the six-electron electrocyclic ring opening (path 6e) of 1,3-cyclohexadiene prefers to open in a disrotatory manner as shown in Figure 13.44.

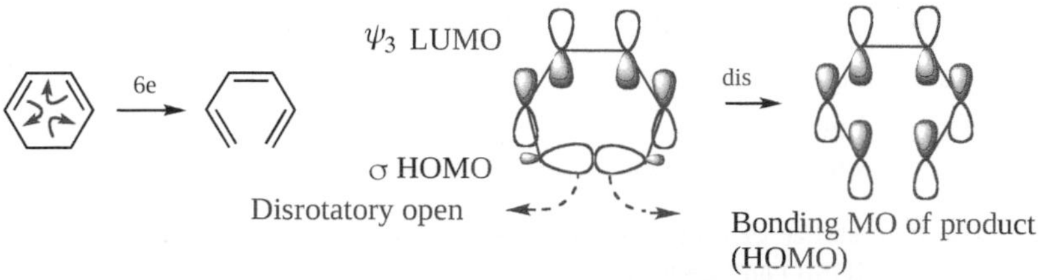

Figure 13.44 The thermal disrotatory ring opening of 1,3-cyclohexadiene.

The ring twists open so that the σ-HOMO overlaps with the diene ψ_3-LUMO in a bonding manner to produce a bonding molecular orbital (HOMO) of the product. Another way to view the reaction is from the closure of the triene. The ends of the HOMO of the triene would have to rotate dis to create a bonding overlap for the new sigma bond. Once we decide con or dis, there is still another decision to be made, since each rotational mode has two ways it can happen. With alkyl groups, steric effects dominate; the rotation that does not bump the alkyl groups is preferred.

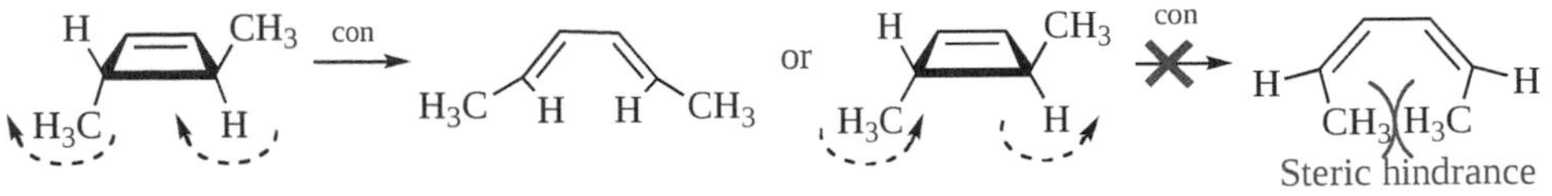

For other groups, secondary electronic effects can contribute to the rotational preference in the cyclobutene ring opening. Electron-releasing groups tend to rotate outward, and electron-withdrawing groups tend to rotate inward; the rationale is above this text's level.

We just saw the rules reverse for pericyclic reactions: con for four-electron systems to dis for six-electron systems (more generally from $4n$ electrons to $4n + 2$ electrons). Looking at the closures of the HOMOs (Figures 13.43 and 13.44), we can generalize the electrocyclic ring closure for any loop of *p* orbitals. If the ends of the HOMO are out of phase, then a con closure is required to get a sigma bonding orbital. If the ends of the HOMO are in phase, then a dis closure is now required (Figure 13.45). Figure 13.27 showed the phase of the ends alternate as the MOs increase in energy, starting with the in-phase fundamental. If we add two more electrons we get the next MO up as our HOMO. Thus the ends of the HOMO of all 4n species will be out of phase, and the ends of the HOMO of all $4n + 2$ species will be in phase. Check Figure 13.27 to verify this for yourself.

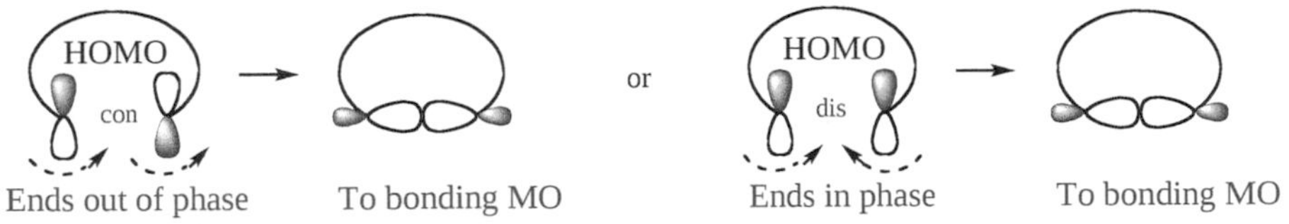

Figure 13.45 The thermal electrocyclic ring closure rotational mode of a loop of *p* orbitals is determined by the phase of the ends of the HOMO.

Let there be light! When an electron is promoted by light from the HOMO to the LUMO, the HOMO of the excited state is now the old LUMO. The HOMO has moved up one MO in energy, and the phase of the ends have reversed. We expect the rules to reverse, and they do. The $4n$ photochemical electrocyclic reactions go dis, and the $4n + 2$ photochemical electrocyclic reactions go con, as predicted. This all boils down to an easy-to-remember rule: **$4n$ electrocyclics go thermal con, with the rules reversing for $4n + 2$, and also reversing for photochemical reactions.** Recognize that two reversals bring you back to the con starting point, so photochemical $4n + 2$ reactions go con.

13.9 CONCERTED PERICYCLIC SIGMATROPIC REARRANGEMENTS

13.9.1 Four-Electron (4*n*) Cyclic Rearrangements

The migrating group in a sigmatropic reaction can stay on the same face of the pi system it started from (suprafacial) or migrate to the opposite face (antarafacial) (Figure 13.46). For consistency in this section, we will draw the sigma bond as the HOMO and the pi systems as the LUMO, even though it can be done reversed. We draw the sigma HOMO and pi LUMO initially in phase so that the newly formed pi bond will be bonding. Then we migrate the hydrogen to the same phase lobe of the other end of the pi system to form a new bonding sigma bond. The migrating group must maintain overlap with both ends at the transition state. This is easy for suprafacial migrations once the ends are close. Antarafacial migrations of hydrogen best occur if the pi system is long enough to spiral around on itself to pass the hydrogen between top and bottom faces. Dashed black arrows will show orbital movement, not electron flow.

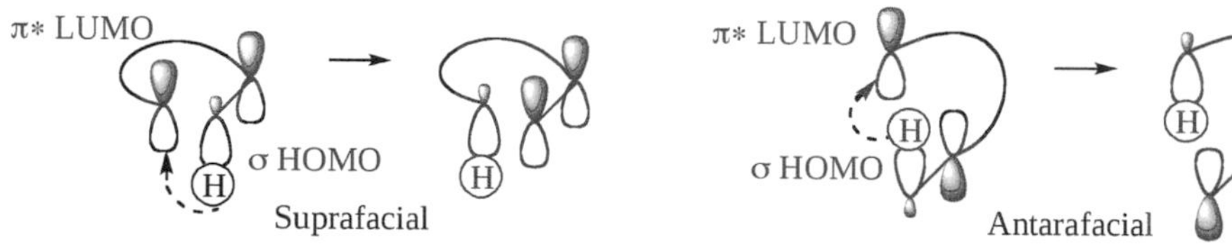

Figure 13.46 The suprafacial (same face) and antarafacial (spiral to migrate between opposite faces) sigmatropic rearrangement of a hydrogen atom across the ends of two different pi systems (pi LUMO ends in-phase on right, out-of-phase on left). The arc in the reactant symbolizes the rest of the pi system.

A common error in drawing a mechanism for tautomerization is passing the hydrogen internally, which is actually a concerted four-electron (two arrow) thermal 1,3 hydrogen sigmatropic rearrangement. The hydrogen in Figure 13.47 cannot migrate suprafacially because it would form an antibonding orbital. The ends of the π* LUMO are out of phase, so the allowed process is antarafacial. The hydrogen has to go from bottom to top, but the pi system is too short for the H to maintain continuous overlap on the allowed route. There is no low-energy allowed concerted process, so the reaction occurs stepwise.

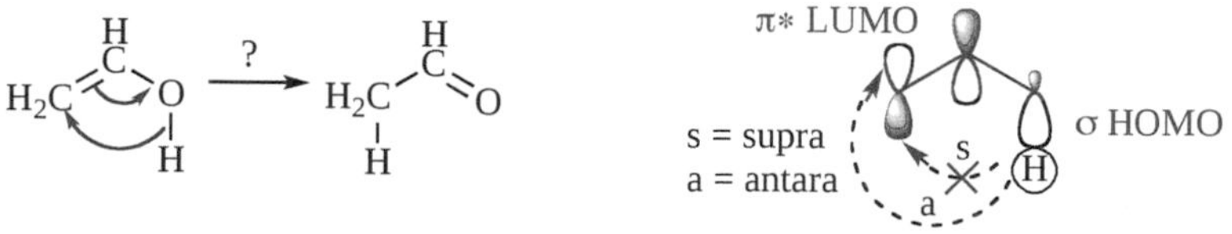

Figure 13.47 Concerted tautomerization—the thermal 1,3 hydrogen shift has a high barrier. HOMO–LUMO requires antarafacial, but the migrating H atom can't maintain overlap with both ends. Any OH twisting to achieve overlap would then make a high energy twisted pi bond instead.

13.9.2 Six-Electron (4*n* + 2) Cyclic Rearrangements

Let's look at the similar six-electron (three arrow) thermal 1,5 hydrogen sigmatropic rearrangement. We expect the ends of the pi system to change phase as we add two more electrons. This means the rules will reverse and the allowed process will be suprafacial. Figure 13.48 shows the HOMO–LUMO prediction of suprafacial.

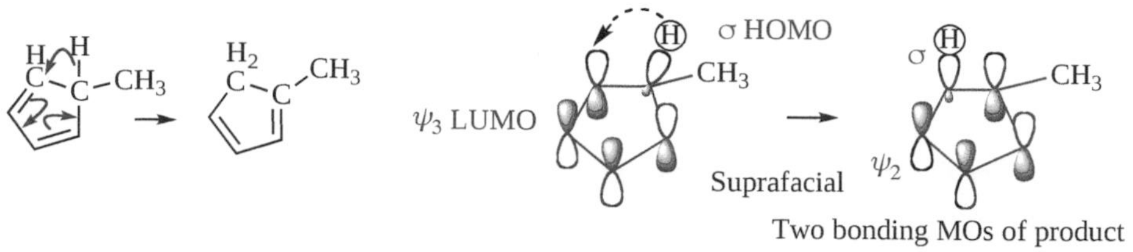

Figure 13.48 The 1,5 hydrogen shift occurs easily. Always set up the HOMO–LUMO analysis with the HOMO and LUMO with bonding overlap, then migrate to maintain bonding overlap. HOMO–LUMO analysis requires suprafacial migration in this case.

13.9.3 Sigmatropic Migration with Inversion

If the migrating group is carbon, not hydrogen, another variable is tossed in the mix. A carbon atom can migrate with retention (just like our H), or the carbon can undergo inversion (back side attack). The rules will reverse if the migrating center undergoes inversion (Figure 13.49). The *p* orbital on the migrating center gives the option of a phase change. The transition state in the center of the figure shows that overlap can be maintained on this suprafacial inversion route. Therefore, there are two types of four electron allowed paths: antara retention (Figure 13.46 right) and supra inversion (Figure 13.49 si). Likewise there are two types of six-electron allowed paths: supra retention (Figure 13.46 left) and antara inversion (not shown). Again, if the pi system is too short for the migrating group to maintain continuous overlap on the allowed antarafacial path, then the reaction will not have a low energy route.

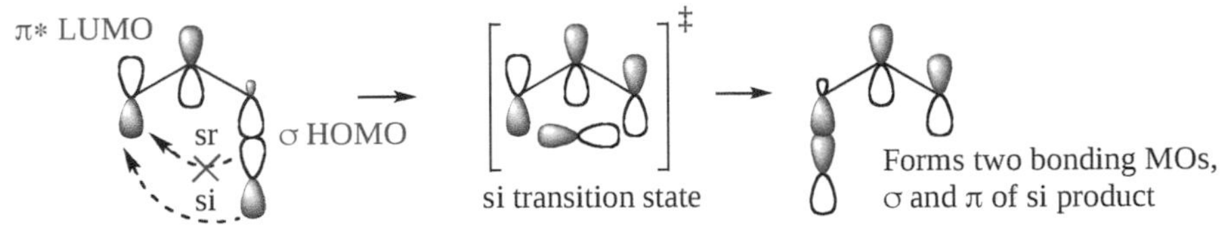

Figure 13.49 The 1,3 carbon shift. HOMO–LUMO requires suprafacial inversion (si) not supra retention (sr).

Again we remember (or derive) **one** sigmatropic allowed process, and then use the idea of phase reversals to generate all the other possibilities. **A 4*n* antara retention is thermally allowed, with the rules reversing for supra, reversing for 4*n* + 2, reversing for inversion, and also reversing for photochemical reactions.** Two reversals are equivalent to no change at all.

Example problem

Is the following concerted sigmatropic rearrangement thermally allowed?

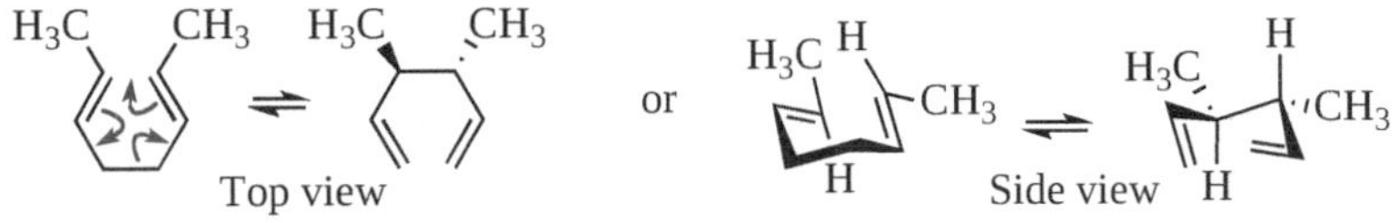

Answer: The process uses three arrows and so is a six-electron (4*n* + 2) process. The side view shows that the rearrangement occurs on the same face and with retention. We know that 4*n* antara retention is thermally allowed and that this process is 4*n* + 2 supra retention. There are two reversals: 4*n* to 4*n*+2 and antara to supra; the reversals cancel, so we predict that this reaction is also thermally allowed. Experiment shows this reaction easily proceeds thermally.

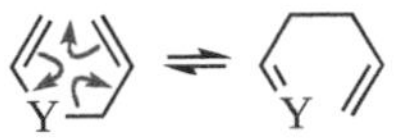

Two common six-membered rearrangements of 1,5-dienes have Y as either oxygen (Claisen rearrangement) or carbon (Cope rearrangement). Both are reversible and favor the more stable product. The preceding example problem demonstrated that they were thermally allowed 4*n* + 2 suprafacial retention processes. As in the Cope example problem above, the transition state for the reaction commonly resembles the chair conformation of cyclohexane. Figure 13.50 shows a biochemical example of the Claisen rearrangement.

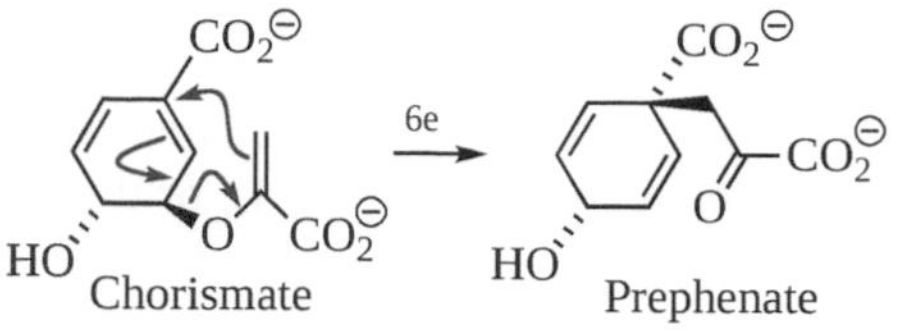

Figure 13.50 The concerted thermal rearrangement of chorismate to prephenate is a biochemical example of a Claisen rearrangement that is part of the bacterial biosynthesis of the amino acids phenylalanine and tyrosine.

13.10 PERICYCLIC REACTIONS SUMMARY

The thermal favorability of pericyclic reactions can be predicted by looking for a bonding interaction between the HOMO of one reacting partner and the LUMO of the other partner. Pericyclic reactions usually involve the ends of the pi system(s). Therefore, the relative phase of the ends of the pi systems dictate whether a bonding or antibonding interaction results. The ends of the pi system are in phase in the fundamental MO and out of phase in the next higher, second harmonic (one-node) MO. This phase alternation continues as the energy increases. Therefore, if we derive or remember one rule for a pericyclic reaction, then any time an MO phase change occurs, the rule will reverse. Two reversals cancel each other. Count arrows to tell whether the pericyclic reaction is $4n$ or $4n + 2$.

For example, $4n$ face to face (supra–supra) cycloadditions are not thermally allowed. If we add two electrons, we fill the next highest MO, which has a phase reversal. This means that $4n + 2$ cycloadditions are thermally favored. Additionally, cycloadditions may have an HSAB component determining regiochemistry.

Thermal electrocyclic reactions of $4n$ species go conrotatory, whereas thermal $4n + 2$ electrocyclic reactions go disrotatory. The rules reverse for photochemical reactions of each.

Thermal sigmatropic reactions of $4n$ species go supra-inversion or antara-retention. Phase reversals occur between retention/inversion at the migrating center, between antarafacial/suprafacial migration, with $4n$ vs. $4n + 2$ electrons, and between thermal and photochemically excited species.

To summarize for thermal $4n + 2$ pericyclic reactions:

Thermally allowed $4n + 2$ cycloadditions go supra–supra.

Thermally allowed $4n + 2$ electrocyclic ring openings go disrotatory.

Thermally allowed $4n + 2$ sigmatropic rearrangements go supra-retention.

The rules reverse for every MO phase reversal that occurs in going from $4n + 2$ to $4n$ electrons, retention to inversion, suprafacial to antarafacial, disrotatory to conrotatory, or thermal to photochemically excited systems. Two reversals are the same as none. A double reversal of the above is that $\underline{4n}$ $\underline{\text{con}}$rotatory electrocyclic ring openings are thermally allowed.

CHAPTER 13 EXERCISES

13.1 Draw the pi MOs of pentadienyl cation. What is the HOMO and LUMO?

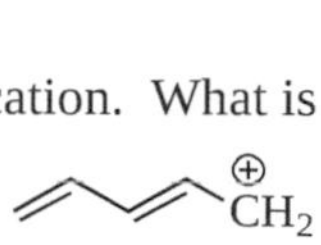

13.2 Which thermal electrocyclic ring closure mode, conrotatory or disrotatory, would the cation in problem 13.1 follow? Draw the product.

13.3 Would the thermal electrocyclic closure of 1,3,5-cycloheptatriene go con or dis? Draw the product.

13.4 Predict the regiochemistry of the Diels–Alder addition of these two reactants.

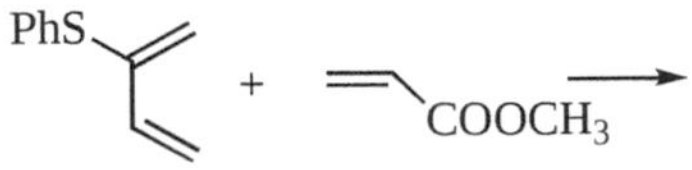

13.5 Is the following sigmatropic rearrangement thermally allowed?

13.6 Give an allowed thermal mechanism for the reaction below.

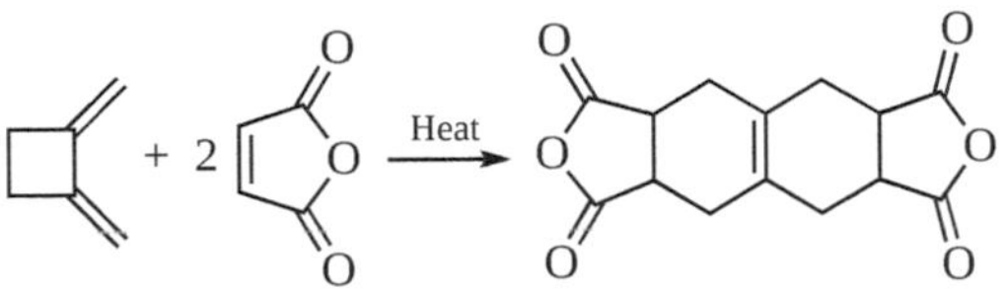

13.7 Give the product of this electrocyclic ring opening showing stereochemistry.

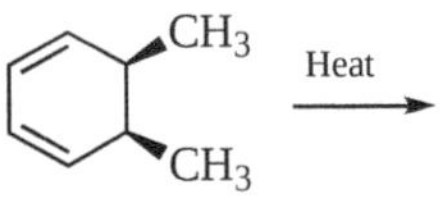

13.8 Draw the photochemical ring closure product of the compound below.

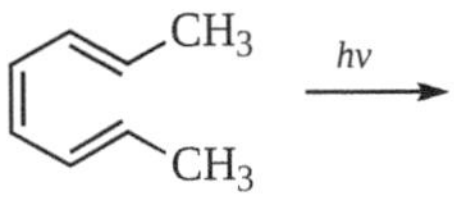

13.9 The compound below does not react with Diels–Alder dienophiles. Explain.

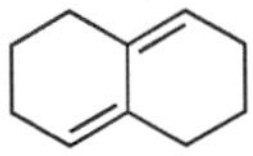

13.10 Give the product of this electrocyclic ring opening showing stereochemistry.

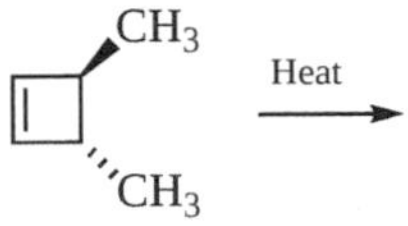

13.11 Draw the photochemical ring closure product of the compound below.

13.12 Give the regiochemistry of the following inverse electronic demand Diels–Alder.

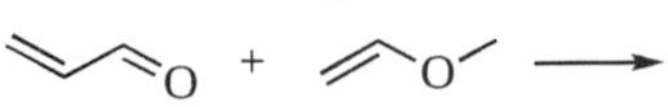

13.13 Ozone is a 1,3 dipole and adds to alkenes to give a molozonide by the 6e path, then breaks by the reverse process, and adds together again to form the ozonide by the 6e path. The ozonide is cleaved by dimethyl sulfide to give two carbonyls. Show the mechanism.

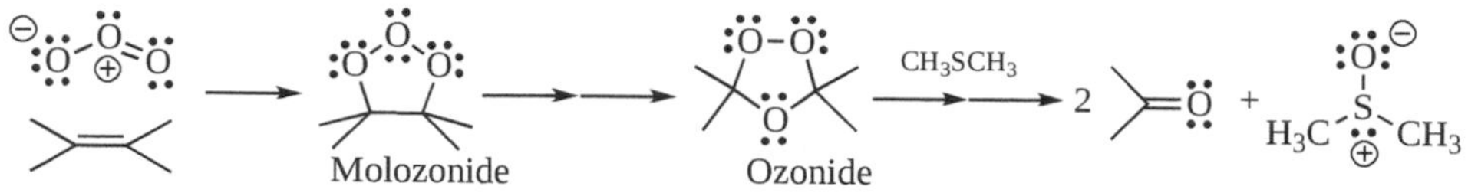

13.14 Give an allowed thermal mechanism for the reaction below.

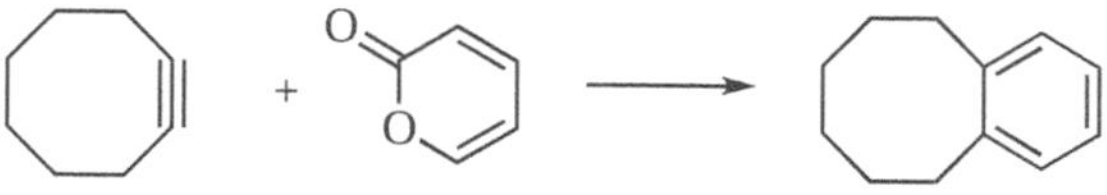

13.15 Predict the stereochemistry of the Diels–Alder addition of these two reactants.

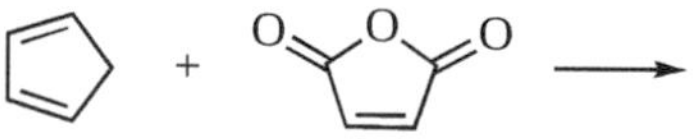

13.16 Predict the regiochemistry and stereochemistry of this Diels–Alder cycloaddition.

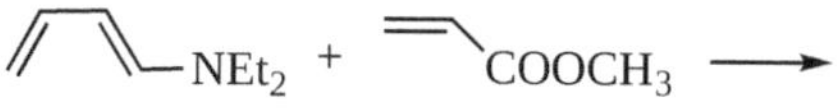

13.17 Give the product of this thermally allowed rearrangement.

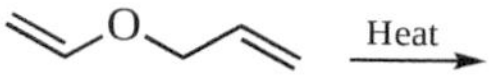

13.18 Provide a reasonable mechanism for the following reaction that has a pericyclic rearrangement as its last step.

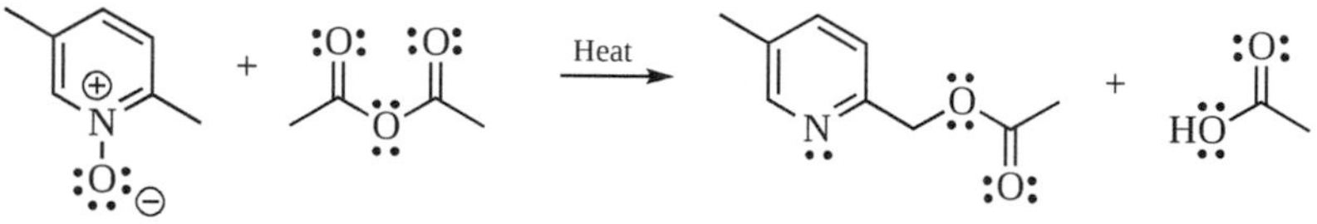

13.19 Vitamin D is synthesized in the skin by UVB light. The steps are a photochemical electrocyclic ring opening followed by a thermal sigmatropic hydrogen shift. Given those hints, propose a reasonable mechanism for this very important biochemical process.

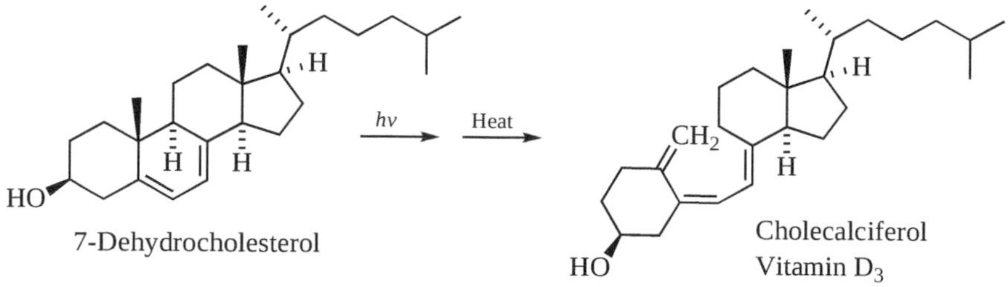

13.20 Provide a reasonable mechanism for the following thermally allowed reaction.

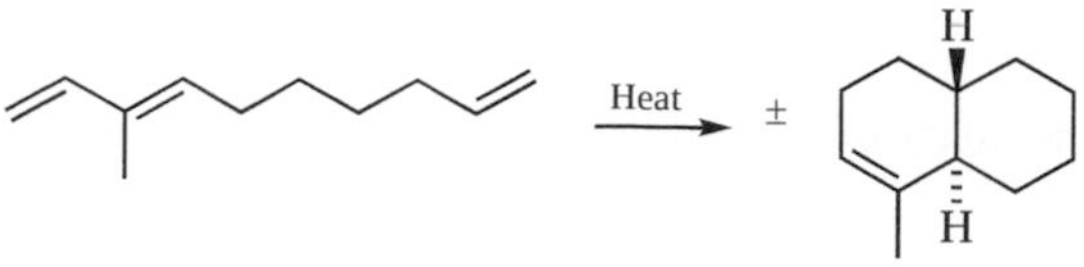

14

ORGANIC STRUCTURE ELUCIDATION STRATEGIES

14.1 A GENERAL APPROACH TO STRUCTURE ELUCIDATION

Many chemists build larger molecules from smaller ones, so a critical laboratory question is, "Did I get what I intended to make?" Another chemist may isolate a new compound from nature that has an important property, like being a new antibiotic, so the question still becomes, "What is its structure?" Therefore, most chemists need to be able to solve structure problems: What compound will fit the spectral data? A proper attitude toward this type of

Electron Flow in Organic Chemistry: A Decision-Based Guide to Organic Mechanisms,
Third Edition. Paul H. Scudder.

problem solving is needed, and nobody has said it better than Sherlock Holmes:[1] "It is a capital mistake to theorize before one has data. Insensibly one begins to twist facts to suit theories, instead of theories to suit facts." Strategies are definitely needed. Generate possibilities, and then select the best one that fits the data. If none fit, generate more possibilities and examine your assumptions. If you are still stuck, gather more data.

Spectra problems are much like jigsaw puzzles. The first step is to find all the pieces, spread them out, and look at them. The next step is usually to find all the pieces that make up the border of the puzzle. Then pieces adjacent to the border are filled in, moving toward the center until the puzzle is complete. Don't "hammer home" any pieces. The most common error is to try to put the puzzle together before identifying all the pieces. Make sure you understand all the parts of the puzzle before trying to put it together. Solving a difficult structural problem is a struggle, so approach the problem with patience and a strategy. You will have to maintain flexibility so that you don't "lock yourself out" of the correct answer. Structure elucidation also is like a medical diagnosis; if it does not fit all the facts and make sense, it's probably incorrect.

The first thing we often try to get is the molecular formula from the mass spectral formula weight and the percent composition from a combustion analysis if available. If we don't have a molecular formula, we can still get a minimum count on the number of carbon and hydrogen atoms from the NMR spectra, to be discussed in upcoming sections. If we have all the pieces, then their weights should sum up to the molecular weight. If not, just subtract the weights of the known pieces to get the weight of what is missing.

14.2 MASS SPECTRAL ANALYSIS—MS

The mass spectrometer functions like an atomic bathroom scale. We get an exact weight (mass) of our molecule and also the masses of some of the fragments when it breaks apart. The molecular ion is the most important because we can get the molecular formula from an exact mass of the compound. We know the exact masses of every isotope of the common elements to six or more decimal places. Carbon-12 is the standard set at exactly 12.000000 atomic mass units (amu). Hydrogen-1 is then 1.007825 amu, and nitrogen-14 is 14.003074 amu, and oxygen-16 is 15.994915 amu. Molecules that differ in elemental composition differ in exact mass. For example, both N_2 and CO have the same nominal mass of 28 amu, but the exact masses differ: 28.0061 and 27.9949, respectively. Therefore, if we have a high-resolution mass spectrometer capable of getting the molecular ion mass to three or more decimal places, we could easily narrow down our possibilities for a molecular formula.

Table 14.1 Significant M+1 and M+2 Isotopes of Common Elements

Element	M (%)	M+1 (%)	M+2 (%)	Average mass
Carbon	12 (100%)	13 (1.1%)		12.0112
Silicon	28 (100%)	29 (5.9%)	30 (3.35%)	28.0855
Sulfur	32 (100%)	33 (0.8%)	34 (4.4%)	32.0660
Chlorine	35 (100%)		37 (32.5%)	35.4527
Bromine	79 (100%)		81 (98%)	79.9094

Even with a lower-resolution mass spectrometer, we can get hints about the elemental composition. The mass of an element shown in the periodic table is just an average of all of that element's isotopes. In a mass spectrum, we see each isotope since each has a different

[1] Arthur Conan Doyle, "A Scandal in Bohemia," *The Adventures of Sherlock Holmes*, 1891

mass. Molecular ions (M) have higher weight peaks at M+1, and M+2 due to minor isotopes, which can tip you off to the presence of other elements. Common examples are shown in Table 14.1 (with the intensity of the taller peak set at 100%). Hydrogen, nitrogen, oxygen, fluorine, and iodine have no significant isotope peaks (<0.4%). The presence of chlorine or bromine is important to identify in the mass spectrum.

Whether the molecular weight is odd or even is an important clue. The molecular ion will have an even molecular weight unless it contains an odd number of nitrogen atoms. The presence of one nitrogen atom allows an additional hydrogen atom in the molecular formula, making the molecular weight odd.

To get a mass spectrum, a tiny bit of sample is introduced into the internal high vacuum of the mass spectrometer. Then a high-energy electron beam hits the sample in the source area and knocks an electron out of the compound forming a radical cation called the molecular ion. The radical-cation molecular ion usually fragments in multiple ways into cationic fragments and radical/neutral fragments, both of which can give good structural hints. Only the cationic fragments are separated by the mass analyzer and end up at the detector of the mass spectrometer. The weight of each corresponding radical/neutral fragment is found by subtracting the weight of the cationic fragment from the molecular ion. Don't try to put together two cationic fragments; they came from different fragmentations forming different radical/neutral partners. Assume all MS ions have just a single charge (m/z has an z of 1), since they usually do.

Small Molecule Example—Part 1, Mass Spectrum

An example MS fragmentation is shown in Figure 14.1 where a neutral piece of m/z 15 has broken off the molecular ion at m/z 58 to form a new cation at m/z 43. This fragmentation is so favorable that the m/z 43 peak is the most abundant peak in the spectrum, called the base peak. Sometimes the fragmentation of the molecular ion is so easy that the mass spectrum shows only fragments and the molecular ion is missing.

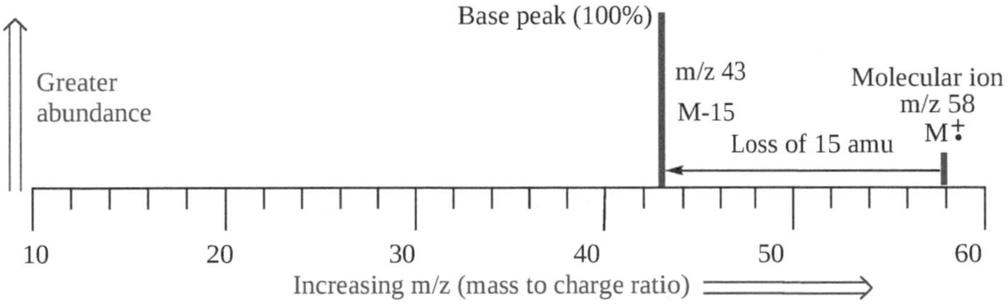

Figure 14.1 The layout of a mass spectrum and important peaks of an unknown small molecule.

Tables are available of common fragment ions and fragments lost. A glance at one of these tables shows that the break in Figure 14.1 is probably the loss of a methyl radical (H_3C• at 15 amu) to form a cation that is possibly $C_3H_7^+$ or CH_3CO^+ (each 43 amu). With an even molecular ion at 58 amu, we know the compound has zero or an even number of nitrogen atoms. We definitely need more information.

The mass spectral fragmentation pattern can be like a fingerprint for a particular compound. A unique complex fragmentation pattern can be matched to those in a mass spectral database and can serve to identify tiny amounts of an unknown compound. For this reason, mass spectrometers are extremely valuable in forensic work and have been put on spacecraft sent to Mars to look for organic compounds that could be signs of life.

14.3 ELEMENTAL ANALYSIS

If we burn a weighed compound in oxygen and measure the weight of CO_2 and H_2O formed, we can calculate the percentage of carbon and hydrogen in the compound (combustion analysis). Other methods allow for the calculation of the percentage of halogens, nitrogen, and other elements in the compound. The one element that cannot be found this way is oxygen, so it is assumed that the remaining percentage is oxygen. For example, if the combustion analysis is 62.0% carbon and 10.4% hydrogen, then the oxygen percentage is 100% − (62.0% + 10.4%) = 27.6%. Assume no other elements if the data are not given.

14.4 GETTING A MOLECULAR FORMULA

In Section 14.2 we saw how with a high-resolution mass spectrometer, we could easily narrow down our possibilities for a molecular formula using the exact masses of each element. With a lower-resolution mass spectrometer, we can get the molecular formula if we combine the molecular ion's weight with the elemental analysis. If we multiply the mass spectral molecular ion weight by an atom's percentage from the elemental analysis, we get the weight of that atom in the molecule. Dividing that weight by the atom's atomic weight, we get the number of those atoms in the molecular formula.

$$\text{Number of carbons} = \frac{\text{(molecular weight) (\% carbon)}}{\text{atomic weight of carbon}}$$

For example, ethane's MS formula weight is 30 and it is 80% carbon by combustion. The number of carbons in ethane is (30)(0.80)/(12) = 24/12 = 2. This is much easier than going via an empirical formula.

A molecular formula not only tells you how many of each type of atom is present but also gives the DU, the degree of unsaturation, which is the number of rings and/or pi bonds present. The number of hydrogen atoms actually present is subtracted from the number of hydrogen atoms expected if there were no rings or pi bonds; this gives the total number of hydrogen atoms missing. A compound has two less hydrogen atoms for every ring or pi bond. For propane below, every C is paired with two hydrogen atoms; each end is capped with a hydrogen atom, giving $2n + 2$ hydrogen atoms for n carbons for any molecule without rings or pi bonds. The next two structures show that to make a ring, we lose the 2 capping hydrogen atoms; pi bonds also lose 2 hydrogen atoms. Oxygen does not change the H count. For every nitrogen atom add one hydrogen. Count halogens as if they were hydrogen atoms.

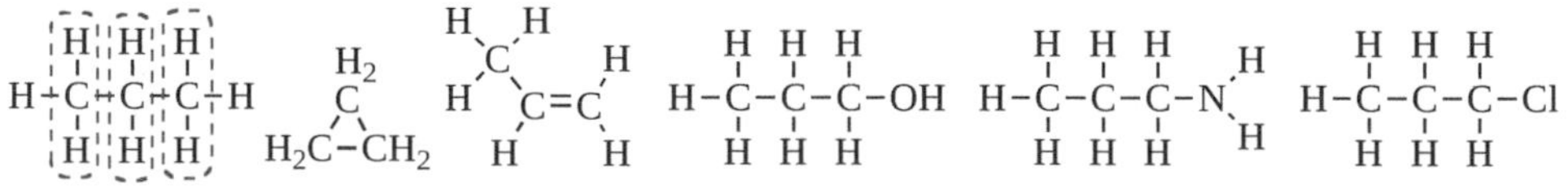

Small Molecule Example—Part 2, Combustion Analysis

The previous compound with an MS molecular ion weight of 58 amu has a combustion analysis is 62.0% carbon and 10.4% hydrogen. Calculate the molecular formula and the number of rings and/or pi bonds present.

Section 14.3 showed us that if there is 62.0% carbon and 10.4% hydrogen, then the oxygen percentage is 100% − (62.0% + 10.4%) = 27.6%. The Section 14.4 formula gives:

$$\text{Number of carbons} = \frac{(58)(0.62)}{12} = \frac{35.966}{12} = 3 \text{ carbons}$$

$$\text{Number of hydrogens} = \frac{(58)(0.104)}{1} = \frac{6.032}{1} = 6 \text{ hydrogens}$$

$$\text{Number of oxygens} = \frac{(58)(0.276)}{16} = \frac{16.008}{16} = 1 \text{ oxygen}$$

This gives a molecular formula of C_3H_6O. Since three carbons would have $2n + 2$ or 2*3+2 = 8 hydrogen atoms if there were no rings or pi bonds, and we have just 6, there are two hydrogen atoms missing. Thus, this molecule has one ring or pi bond, a DU of 1.

Even if we just had the molecular weight but did not have an elemental analysis, there are ways to approach the molecular formula using the other spectra discussed in the next sections. For example, if the IR (Section 14.5) shows a peak for a particular functional group, like a carbonyl, we can add that to our growing list of puzzle pieces. The C-NMR spectrum (Section 14.6) shows a peak for each unique type of carbon atom in the structure. This gives us a minimum count on the number of carbon atoms. Likewise, the H-NMR spectrum (Section 14.7) shows how many unique types of hydrogen atoms that are there. Combining that information with the integral of each of those H-NMR peaks gives us a rough count on the number of hydrogen atoms. As we discover pieces this way, we keep a running total of accumulated piece weight and use the difference from the molecular weight to see what the weight of missing pieces might be. If we find all the puzzle pieces this way, we may even solve the structure before getting the molecular formula. The take-home message is, although the problem is certainly easier to solve with a molecular formula, don't let the lack of one stop you from finding the puzzle pieces and solving the structure.

14.5 INFRARED SPECTRA—IR

Vibrational spectra like IR give information about functional groups, Figure 14.2. The vertical scale is percent transmittance with 100% T at the top; the horizontal scale is inverse centimeters, cm^{-1}, which is proportional to frequency (also energy). IR can be good evidence that a group is absent if an expected strong spectral band is missing, but since IR spectra often have many bands in a particular region, they give much weaker evidence that a group might be present. The intensity of an IR band is dependent on the vibration changing the dipole moment, so symmetrical stretches can have weak or missing IR bands. For example, 1-butyne has a weak triple bond stretch at 2100 cm^{-1}, but symmetrical 2-butyne has nothing in that region at all. Therefore, always keep an open mind when it comes to IR band assignments. The frequency of the band depends on two things, mass and bond strength. An oscillating atom connected to another by a bond is similar to a bouncing mass hanging on a spring. If the mass increases, the oscillation will be slower; the frequency decreases (O–H vs. O–C). If the spring is stronger, then the oscillation is faster; the frequency increases (C=O vs. C≡N).

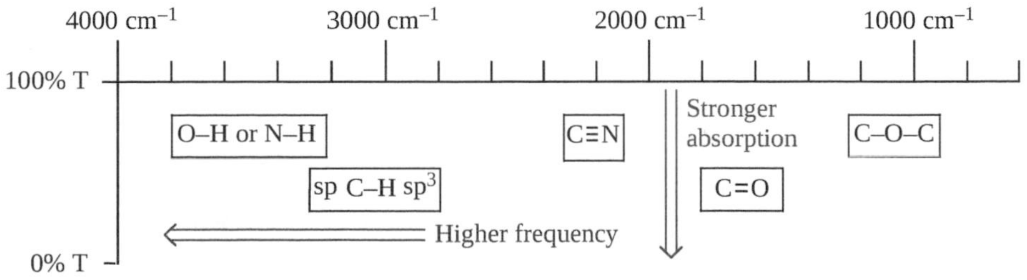

Figure 14.2 The layout of an infrared (IR) spectrum and position of major bands.

First, look at the molecular formula to find out what functional groups are possible. There is no sense in looking for nitrogen-containing functional groups if the compound has no nitrogen. There is also no sense in looking for functional groups containing a double or triple bond if the molecular formula indicates the compound has no rings or pi bonds. Using the many available IR band tables, try to rule out possible functional groups. Note that negative information is often more important than positive: We can rule out possibilities with negative information. If there is no band between 3600 and 3300 cm^{-1} then the structure does not have an OH or NH (if a weak band appears in that region it may just be a wet sample). Triple bond stretches of unsymmetrical alkynes and nitriles occur between 2250 and 2100 cm^{-1}. The C=O stretch is a strong peak near 1700 cm^{-1} whose exact position gives useful information. The different C=O stretch bands of ketones (1740–1710 cm^{-1}), esters (1750−1735 cm^{-1}), anhydrides (1820–1750 cm^{-1}), acyl chlorides (1800–1770 cm^{-1}), amides (1680−1640 cm^{-1}), and carboxylic acids (1760–1710 cm^{-1}) are useful in characterizing each compound. The weaker C=C stretch of unsymmetrical double bonds is near 1640 cm^{-1} and of aromatic rings near 1610 cm^{-1}. The C–O stretch of ethers and esters is around 1300–1000 cm^{-1}. Peaks in the fingerprint region below 1000 cm^{-1} are much harder to assign. Extract what functional group information you can from the IR, but do not try to assign more than a few major peaks.

Small Molecule Example—Part 3, Infrared Spectrum

Continuing with our previous example problem, we had a molecular formula of C_3H_6O, which has one ring or pi bond. For illustration in this example, our compound has such a small molecular weight that we are able to draw out all nine of the possibilities so you can see them be eliminated by the spectral data. Usually, there are too many isomers to list all of them. Our nine C_3H_6O possibilities (ignoring stereochemistry) are:

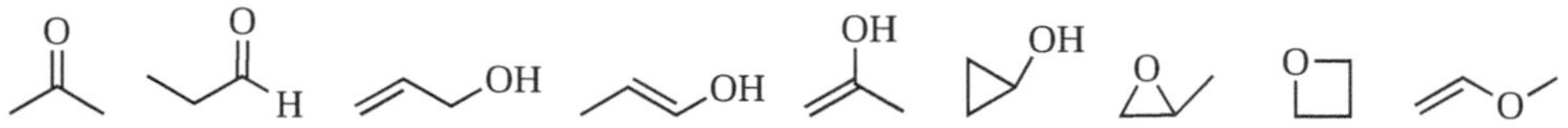

1 2 3 4 5 6 7 8 9

How would an IR spectrum help us narrow the field? Since the molecular formula has an oxygen and a DU = 1 the oxygen could be an alcohol, an ether, or a carbonyl. Each one of these three functional groups has a characteristic IR band. A strong peak near 1700 cm^{-1} is the C=O stretch, suggesting compounds 1 or 2. A band around 3600–3300 cm^{-1} is an O–H stretch, suggesting compounds 3, 4, 5, or 6. If the IR is lacking those bands, but has a C–O stretch around 1300–1000 cm^{-1}, that suggests compounds 7, 8, or 9. Our IR spectrum can narrow the choices considerably, but we usually need more data to solve the structure.

14.6 CARBON NMR SPECTRA—^{13}C NMR OR C-NMR

In a Nuclear Magnetic Resonance (NMR) spectrometer, our sample is placed in a strong magnetic field and hit with radio frequency power. The most commonly observed nuclei are ^{13}C and ^{1}H. Those nuclei have a magnetic moment, and when they are placed in a magnetic field, that magnetic moment interacts with the applied magnetic field and the nucleus ends up in one of two energy states. The lower energy state is when the magnetic moment aligns with the applied magnetic field, and the higher energy state is when the magnetic moment aligns against it. The difference in energy between the two states is very small, so the number of nuclei in each is almost equal, with more nuclei in the lower energy state. If the combination of the applied magnetic field and radio frequency is just right, we can observe a signal from the nuclei changing from one state to the other.

For a fixed magnetic field, each nucleus has its own "frequency address," so we can pick a nucleus to observe. Even within the correct frequency address for ^{13}C, there are slight differences because the electron cloud around each ^{13}C nucleus shields it from the applied magnetic field to a slightly different extent. This difference (chemical shift) is small, at most 250 parts per million (ppm), but the ability to distinguish slightly different carbon nuclei in a molecule is key to the immense value of C-NMR to structure elucidation.

Therefore, a C-NMR spectrum gives information on the different types of carbons present in the molecule, Figure 14.3. We run the C-NMR in a way that each carbon produces a narrow single line and not a complex pattern. The main information from a C-NMR spectrum is the position of the peak relative to the tetramethylsilane (TMS) internal standard, set at 0 ppm. Since the sharp C-NMR peaks are spread out over a relatively large range (0–250 ppm), small differences (less than 0.01 ppm) in the types of carbons can be distinguished. Usually, the more electron density around a carbon, the more shielded it is. The C-NMR chemical shift ranges are: simple alkyl 10–60 ppm, alkynes 65–90 ppm, alkyl C–O 50–90 ppm, alkene and aryl 100–170 ppm, carboxylic acid derivatives 150–185 ppm, and aldehydes and ketones 180–220 ppm.

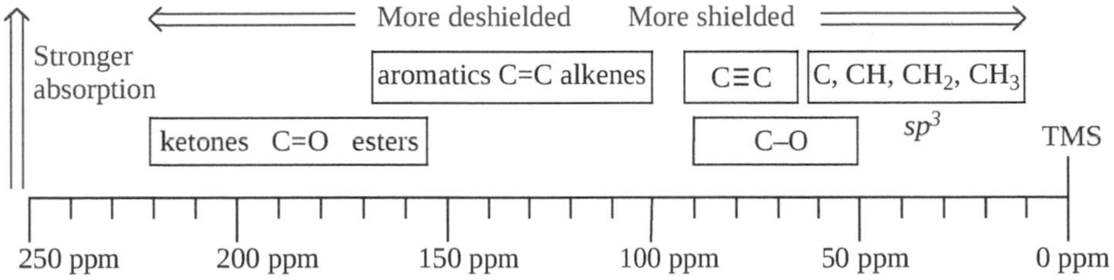

Figure 14.3 The layout of a C-NMR spectrum and relative position of common peaks.

Carbon environments in a molecule can be made identical by symmetry. Two carbon nuclei can have the same chemical shift if the molecule has an internal mirror plane or an axis of rotation making the carbons equivalent. For example, an internal mirror plane in $HOCH_2CH_2CH_2CH_2OH$ means that there are only two unique carbon environments. Check the number of carbons in the molecular formula against the number of C-NMR peaks to determine whether some symmetry is present. Even if you do not have a molecular formula, the structure must have at least as many carbons as there are C-NMR peaks. It is best to start with the C-NMR to get the initial pieces, then use the H-NMR discussed next to help assemble these C-NMR pieces. Shown below are some compounds with symmetry and their C-NMR peak shifts. In deciding symmetry, remember to consider resonance forms, like in benzene below, which has both mirror planes and rotation axes making all the carbons equivalent. Note that it doesn't take much to break the symmetry in the molecule on the right.

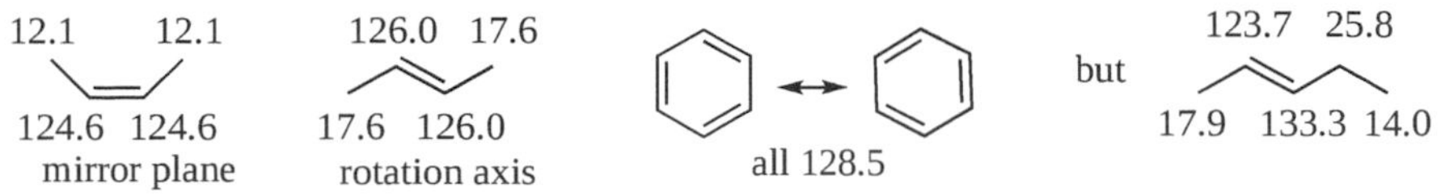

Example C-NMR symmetry problem

How many C-NMR peaks does each of these molecules have?

mono *ortho* *meta* *para*

Answer: Look for resonance forms and symmetry elements that make any carbon atoms equivalent. Aromatic resonance moves the ring pi bonds. As drawn, the mono (toluene) and *ortho*-xylene isomer each have a horizontal mirror plane; the *meta*-xylene isomer has a vertical mirror plane, and the *para*-xylene isomer has vertical and horizontal mirror planes.

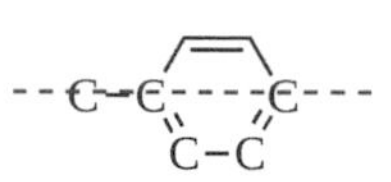
mono has 5 CMR peaks

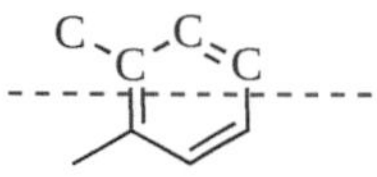
ortho has 4 CMR peaks

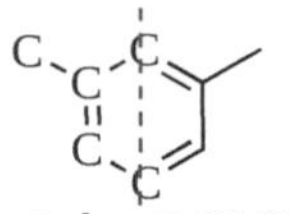
meta has 5 CMR peaks

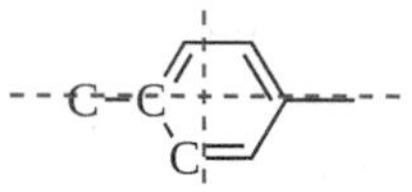
para has 3 CMR peaks

Small Molecule Example—Part 4, C-NMR Spectrum

Continuing with our C_3H_6O example IR problem with nine possibilities, we took a C-NMR and found only two peaks. Since we have fewer peaks than the three carbons in the formula, there must be symmetry. Two of the carbons must be in identical environments to have the same chemical shift. Only compounds 1, 6, and 8 have a mirror plane or rotational axis that would make two carbons identical, so only those would have just two carbon-13 NMR peaks. Now we finally have enough information with the IR spectrum to narrow down to one compound. Compound 1 has a carbonyl; compound 6 is an alcohol, and compound 8 is an ether, so each one would have a different IR peak.

Let's nail this one down. The two C-NMR peaks are at 30.6 and 206.7 ppm. The IR has a very strong peak at 1715 cm^{-1} consistent with a C=O stretch and the C-NMR has a peak for a carbonyl at 206.7 ppm. We have now used one carbon, one oxygen, and the only DU from our molecular formula of C_3H_6O. We now have a C=O middle piece of our structural puzzle. Subtracting the C=O from the molecular formula gives C_2H_6 leftover. We need two ends, and they must be identical because they share one C-NMR peak at 30.6 ppm. Two identical CH_3 end pieces fit the C_2H_6 remainder. The C-NMR shift at 30.6 is good for a CH_3 next to a carbonyl. The IR spectrum narrowed it to compounds 1 or 2, which both have a carbonyl, so the C-NMR confirmed it as 2-propanone (acetone), compound 1. Compound 2 would have three C-NMR peaks as it has no symmetry. One last confirming piece of data is the base peak in the MS was the loss of a CH_3 fragment (M – 15) to give CH_3CO^+ ion (43 amu).

A technique known as DEPT will distinguish between carbon peaks coming from CH_3, CII_2, CII, or C. The appearance of the DEPT 135° spectrum is similar to the carbon-13 spectrum with a few major differences. The quaternary carbon peaks are missing. The CH and the CH_3 peaks point up and the CH_2 peaks point down. A DEPT 90° just shows the CH peaks, allowing us to distinguish between CH and CH_3 peaks.

Example DEPT interpretation problem

Use the spectra below to tell if the carbon peaks come from CH_3, CH_2, CH, or C.

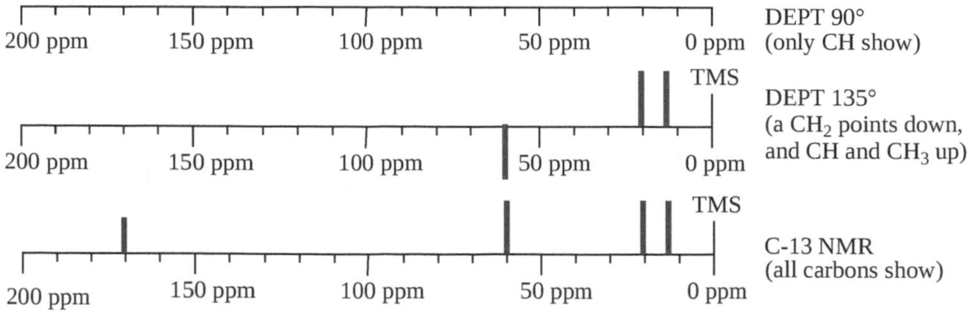

Answer: The peak at 171 ppm, which is missing from the DEPT 135° is a quaternary carbon. The peak at 60 ppm, which is pointing down the DEPT 135° is a CH_2 carbon. Since there are no peaks in the DEPT 90° there are no CH peaks in the spectrum. That means the peaks at 14 and 21 ppm, which are pointing up the DEPT 135° are both CH_3 carbons. The spectrum would be reported as: 14 ppm, (CH_3), 21 ppm (CH_3), 60 ppm (CH_2), and 171 ppm (C). We are starting to get significant structural puzzle pieces. The example problem in the next section gives the H-NMR spectrum of this compound.

Example C-NMR problem

Mass Spectra: molecular ion at m/z 72.1 gives the molecular weight. Since it is even, then zero or an even number of nitrogen atoms are present.

Combustion analysis: 66.63% carbon and 11.18% hydrogen do not sum to 100%; therefore the remaining 22.19% must be oxygen. Also, no other elements are present.

Molecular formula: (72.1)(0.6663)/12 = 4 carbons, (72.1)(0.1118)/1 = 8 hydrogens, and (72.1)(0.2219)/16 = 1 oxygen for a formula of C_4H_8O. Since no rings or pi bonds would be 10 Hydrogens but we have only 8 Hs, we are missing 2 Hs, so there is one ring or pi bond.

C-NMR (DEPT): 25.8 (CH_2) and 67.9 ppm (CH_2).

Answer: The first thing to notice is that we have half the number of C-NMR peaks as carbons in the molecular formula; there must be symmetry. There are no pi bonds (100–170 ppm) from the C-NMR shifts, so we must have a ring. There are only CH_2 peaks in the DEPT. Our jigsaw puzzle has no edges since CH_2s are middle pieces. Draw a line for the symmetry, then put all unique pieces on that line and identical pairs on both sides of the line. The 67.9 ppm C-NMR shift is good for a carbon deshielded by oxygen.

14.7 PROTON NMR SPECTRA—^{1}H NMR OR H-NMR

Proton NMR spectra can give three pieces of information about the hydrogen atoms in a molecule: The chemical shift (position) gives details about the magnetic environment that each set of protons is in, the integration (peak area) tells the relative number of protons in each environment, and the coupling pattern (peak complexity) of a proton resonance shows the number and type of neighbors it has.

Chemical shift reflects local electron density, Figure 14.4. Any attached electronegative groups remove the shielding electron density, and the peak for that proton is more deshielded (higher ppm). Resonance effects from erg or ewg groups can respectively add electron density (shield) or remove electron density (deshield). To tell if a resonance is deshielded by an electronegative group we need to know where it would come if undisturbed, necessitating a few numbers to be memorized: CH_3, CH_2, and CH come between 0.9 and 1.5 ppm; vinyl C=CH comes around 5.7 ppm; and aromatics near 7.2 ppm. The magnitude of the effect is also helpful to know: A carbonyl will pull a methyl group from 0.9 down to 2.1 ppm, while an ether oxygen will pull it all the way down to 3.3 ppm. Some groups have characteristic proton shifts; the carboxylic acid COOH is at 12 ppm, and the aldehyde CHO is at 9 ppm. The deshielding effects of attached groups are roughly additive, so two weakly deshielding groups may deshield an H to the same position as one more deshielding group. For this reason, keep your options open when looking at chemical shift tables.

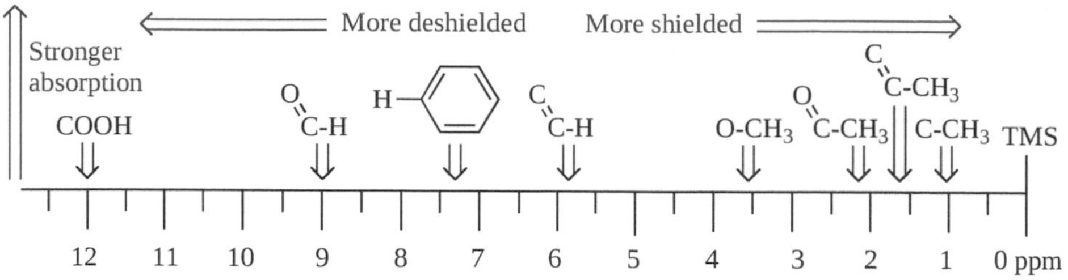

Figure 14.4 The layout of a H-NMR spectrum and midrange position of common peaks.

The **integration** often looks like a staircase and gives the relative areas of the peaks, thus the relative numbers of protons in each environment, Figure 14.5. Newer spectrometers often put a calculated peak area number below the peak; divide the smallest peak integral value into those for the other peaks to get a relative integration.

The integral must sum to the same number of hydrogens in the formula. Correlate the DEPT data with the H-NMR integration. Remember that the C-NMR pieces and H-NMR pieces must be consistent; don't assign a C-NMR methyl to a two-hydrogen H-NMR peak.

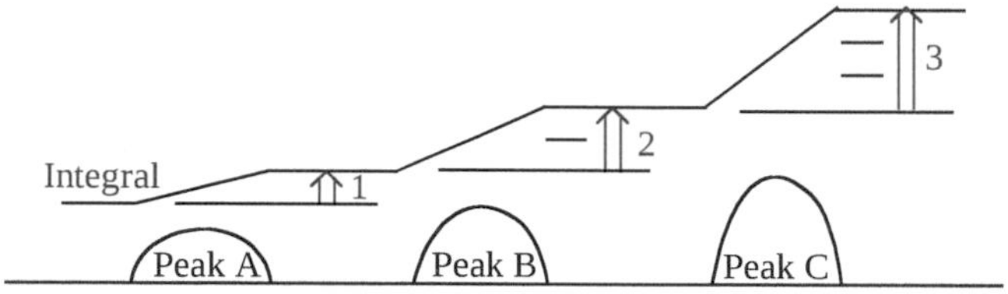

Figure 14.5 The step height of the integral line corresponds to the relative area of the peak below. Peak A has the smallest step. Peak B has twice that step height so has twice as many protons. Peak C has three times the step of peak A. The relative areas are 1:2:3, but this could easily be a multiple of this ratio. This integral could easily represent 2H for A, 4H for B, and 6H for C.

The **coupling** of nuclei to their immediate neighbors gives very valuable information on how to fit the pieces back together. The spin state of the neighbor affects the other nucleus. For example, if the neighbor's spin state is aligned with (adding to) the main field, then it is a tiny bit easier for the other nucleus to reach resonance. If the neighbor's spin state is aligned against the main field, then it is a tiny bit harder for the other nucleus to reach resonance. This coupling to one neighbor results in the other nucleus's resonance being split into two peaks corresponding to the neighbor's spin state. This is shown in the splitting diagram of the resonance of H_a into a doublet, on the left side of Figure 14.6. Shown in the middle of Figure 14.6, is the triplet pattern formed when there are two neighboring protons with equal coupling to H_a. With two neighbors with unequal influence, like the *cis* and *trans* protons shown on the right side of Figure 14.6, there are no overlaps, and coupling gives four peaks, a doublet of doublets for the resonance of H_a. These splitting diagrams are created by adding each coupling into the figure sequentially below the next, usually starting with the largest coupling.

The magnitude of the coupling constant depends on how well the two adjacent orbitals overlap each other. The coupling constant is at a maximum when the two adjacent orbitals are held in the same plane, and at a minimum when they are held perpendicular to each other. If the neighbors are connected by a freely rotating single bond, the coupling constant averages 7 Hz. The coupling constants for the adjacent hydrogen atoms on a double bond, H–C=C–H, can determine the double bond's stereochemistry. The cis coupling constant is about 10 Hz, and the trans is about 17 Hz. When the adjacent protons overlap differently with their partner, the coupling constants will be different, causing a more complex pattern as in Figure 14.6.

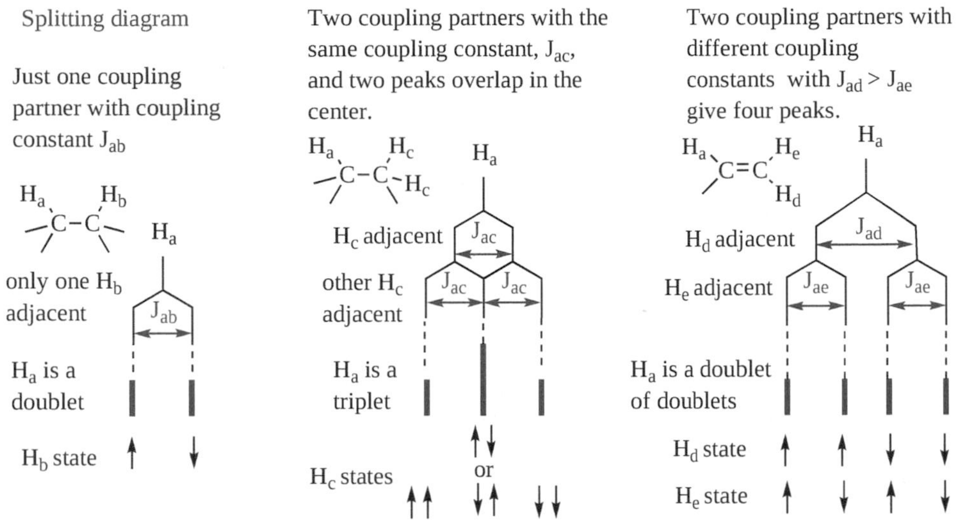

Figure 14.6 Splitting diagrams for one, two identical, and two different neighboring hydrogen atoms. With two identical H's the two center peaks overlap making a single peak twice as high.

For the simpler H-NMR spectra, *n* number of adjacent equivalent protons will split the resonance into *n*+1 lines. Common patterns and peak intensities are shown in Figure 14.7. A proton whose peak is a quartet has three adjacent protons; a triplet has two adjacent protons; a doublet has one adjacent proton; and a singlet none.

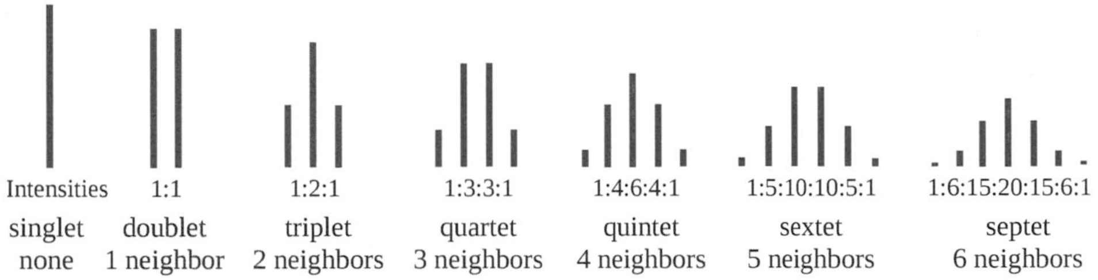

Figure 14.7 Common vicinal H–C–C–H coupling patterns and relative peak intensities.

This coupling of neighboring protons means that at the very minimum coupling comes in pairs, as each pair of neighbors mutually split each other. You cannot have just one triplet in a spectrum and no other peaks split; the neighbor must appear somewhere. The H-NMR of ethanol, CH_3CH_2OH, shown below in Figure 14.8 is a useful example. The methyl is coupled to the adjacent methylene that has two hydrogen atoms, and this splits the methyl into a triplet. The methyl, with three identical hydrogen atoms, splits the adjacent methylene into a quartet. Since they make up a coupled pair, both triplet and quartet have the same peak spacing or coupling constant, J. If the spacing is different, the peaks are not coupled to each other; look elsewhere in the spectrum for the partners.

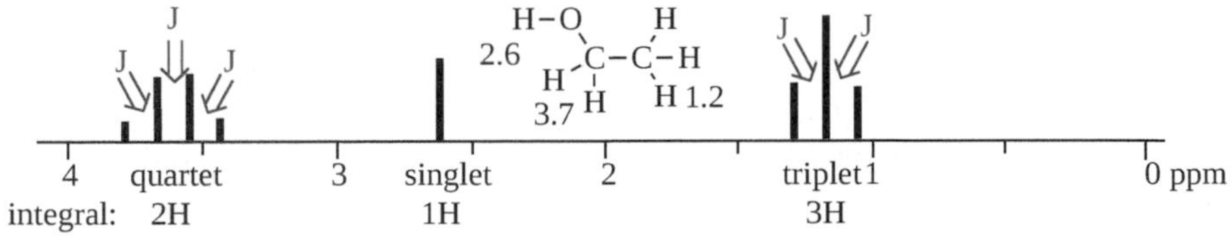

Figure 14.8 The H-NMR of ethanol. Coupled peaks have the same pattern spacing, J.

One further precaution: Acidic hydrogen atoms on heteroatoms like OH and NH groups tend to exchange protons faster than the NMR spectrometer can take a measurement; the NMR sees just an average of those hydrogen shifts and no coupling. Ethanol's OH is a singlet under normal conditions and does not couple to the adjacent methylene, Figure 14.8.

Figure 14.9 shows the most common pairs of coupled peaks and the NMR notation for each pattern. Since pattern recognition is a significant part of structure elucidation, remembering these common splitting patterns will help you recognize pieces quickly.

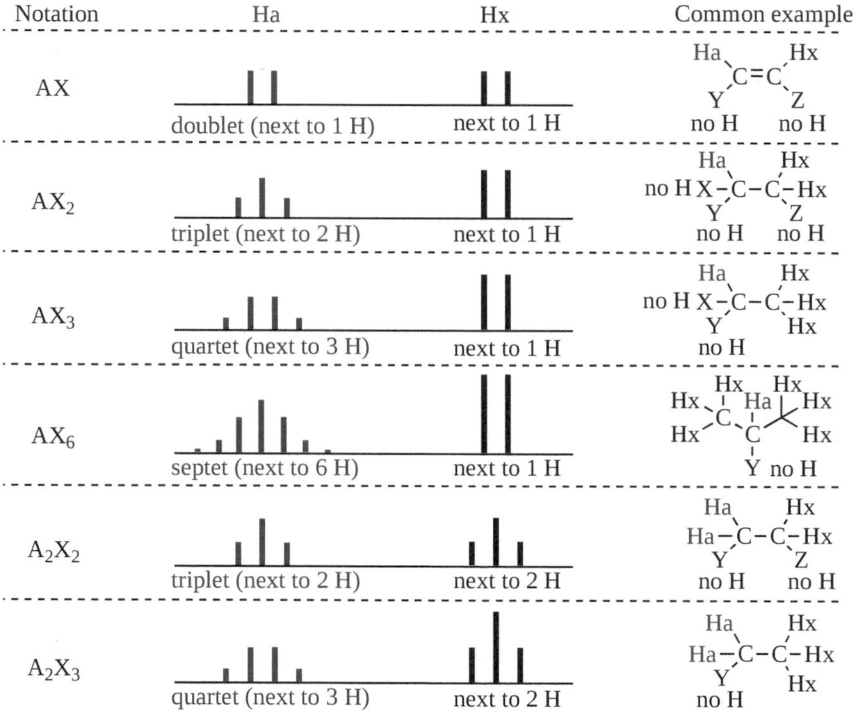

Figure 14.9 The most common pairs of splitting patterns (spin systems) and their NMR notations.

But keep an open mind; the smaller peaks in a larger pattern may get lost in the noise. As two proton resonances get closer together, the pattern distorts; the inner lines of each pattern get taller and outer lines get smaller, until finally, the two patterns merge into one. Therefore, protons at the same chemical shift do not show coupling. Thus, symmetry in the C-NMR can show up as a lack of coupling in the H-NMR. The H-NMR of 1,2-dichloroethane, $ClCH_2CH_2Cl$, is only one singlet, not two triplets, and it has just one C-NMR peak. Benzene is just a single line in both H-NMR and C-NMR.

Example H-NMR problem

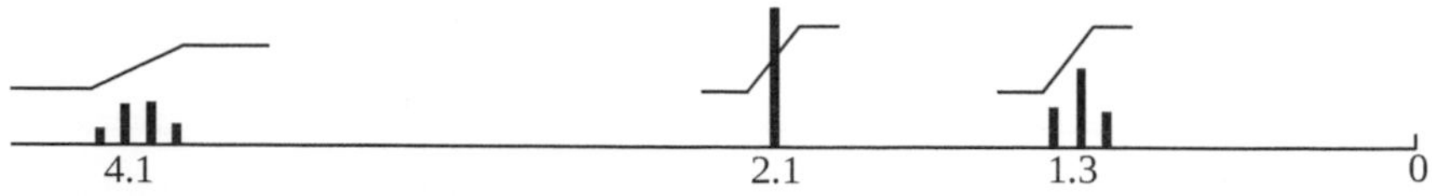

Mass spectra: molecular ion at m/z 88 gives the molecular weight.

IR spectra: no band 4000–3000 cm^{-1} rules out OH, and a strong band at 1740 cm^{-1} suggests a C=O.

Combustion analysis: 54.53% carbon and 9.15% hydrogen do not sum to 100%; therefore the remaining 36.32% must be oxygen. Also, no other atoms are present.

Molecular formula: (88)(0.5453)/12 = 4 carbons, (88)(0.0915)/1 = 8 hydrogens, and (88)(0.3632)/16 = 2 oxygens for a formula of $C_4H_8O_2$. Since no rings or pi bonds would be 10 Hydrogens but we have only 8 Hs, we are missing 2 Hs, so there is one ring or pi bond.

C-NMR: (DEPT): 14 (CH_3), 21 (CH_3), 60 (CH_2), and 171 ppm (C).

H-NMR: 1.3 ppm 3H triplet, 2.1 ppm 3H singlet, and 4.2 ppm 2H quartet.

Can you put this molecule together? Try it before going to the answer in the next section.

14.8 PIECING A SPECTRAL JIGSAW ALL TOGETHER

Working a spectral puzzle is done the same way as doing a jigsaw puzzle, from the edges in. First, we have to do the equivalent of turning all the pieces right side up. An NMR table is a useful way of gathering NMR data. Under "possibilities" the data in parenthesis is the number of adjacent protons, and the arrow signifies that the resonance is very deshielded. Write down all possibilities so that you do not artificially restrict the search for pieces. Check that your pieces are consistent with those from the C-NMR and DEPT spectra. Find those C-NMR pieces within the H-NMR!

NMR Table

Shift ppm	Integration # of Hs	Coupling	Adjacent H's	Possibilities (from C-NMR DEPT pieces)
1.3	3H	triplet	2	$-CH_3$ (2H)
2.1	3H	singlet	0	$-CH_3$
4.1	2H	quartet	3	$\leftarrow CH_2$ (3H)
	Sum 8H			C=O (from C-NMR)

Find all the pieces before trying to assemble any. The IR shows a C=O at 1740 cm^{-1}, therefore this must be our quaternary carbon NMR peak at 171 ppm. See if you have all the pieces. When we subtract those four carbon pieces from the molecular formula, one oxygen atom is left, our last piece. Our spectral jigsaw puzzle has these pieces:

Next, find all the edges of the molecule, the methyls, the phenyls, the halogens, and any group that makes just one bond to the central section. In this case, it is a pair of methyls, so we write them down with plenty of space in the middle for the rest of the pieces.

Like solving a jigsaw puzzle, we work inward from the edges, using the information on the number of adjacent hydrogens from the coupling data. The CH_2 adjacent to three hydrogens will fit next to the CH_3 adjacent to two Hs.

Now use the shift information (indicated by the arrow in the NMR chart) to place the most electronegative group next to the methylene piece that is very deshielded. We put the methylene on the oxygen atom since it is more electronegative than a carbonyl.

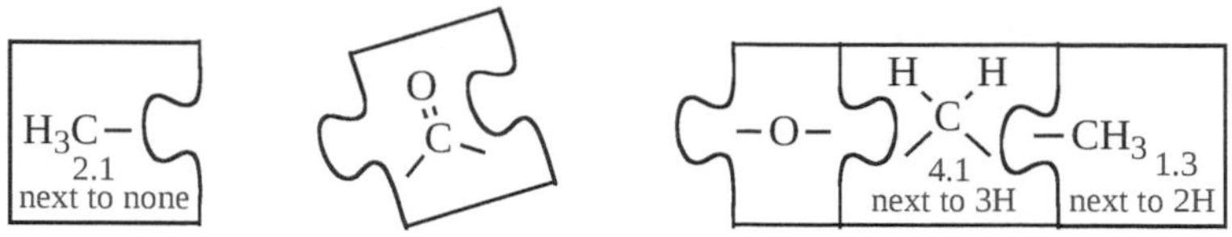

The deshielding of the methyl singlet can be explained by attaching it to the carbonyl.

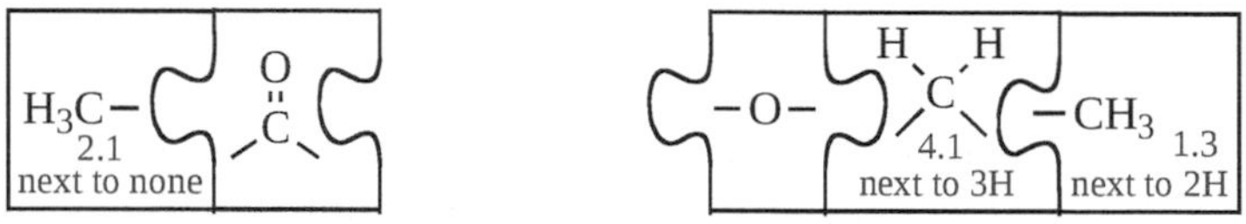

Finally, since we have used all the pieces, we can plug the two halves together. When you end up with more than two pieces, remember that the middle piece(s) might be joined to each other and to the ends in several ways. For example, two end pieces and two unsymmetrical middle pieces can be assembled in eight different ways. Draw all possibilities out then start eliminating them using the spectral data. Use chemical shifts to help decide alternatives if the coupling information is not there. Generate and select the best possible structure.

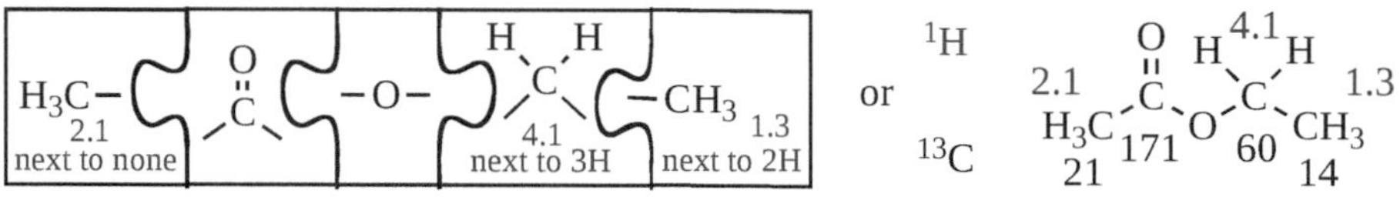

We have built inward until we have placed all the pieces. Now comes the hard part: not stopping and quitting, but going back and looking for alternative structures that also could fit the data. Could the pieces have gone together any other way? Were any isomers possible, or is there stereochemistry that needs to be accounted for? Do not "fall in love" with your first answer. The order in which you find answers may not correlate with their correctness. Check to see if you forced any pieces to fit before saying you are finished. In this case, there are no other ways to assemble the pieces, so we are done.

There was a way to go wrong though. You might have not noticed that the chemical shift of the quartet required that the oxygen be attached to the methylene and so might have had the central ester piece reversed. Taking the first answer that you get without checking other possibilities is not critical thinking. Generate all possibilities, and then select the one that best fits the data. Spectral puzzles are fun because when you finally hit upon the right answer, it will "click." You will see hidden clues in the spectra that are obvious now that you have the final structure. A forced answer is usually a wrong answer.

Example problem with symmetry using all spectra so far

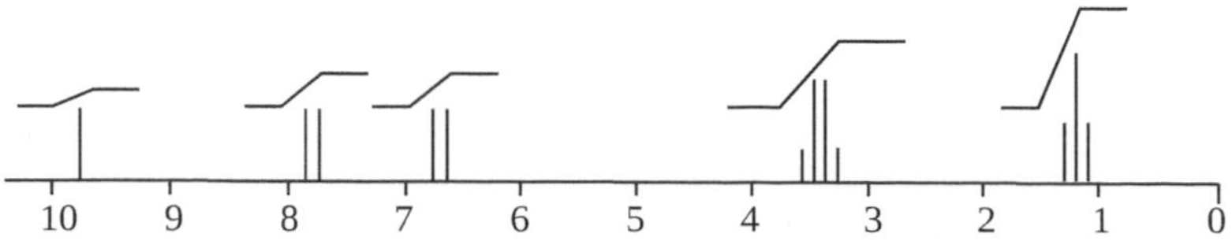

The MS has a molecular ion at m/z 177. The combustion analysis gives: C, 74.54%, H, 8.53%, and N, 7.90%, which totals to 90.97%. The H-NMR is above. The IR spectrum has a major band at 1714 cm^{-1} and nothing between 3800 and 3100 cm^{-1}. C-NMR (DEPT): 13 (CH_3), 45 (CH_2), 111 (CH), 125 (C), 132 (CH), 152 (C), 190 ppm (CH).

Take a moment to think about this problem before continuing on.

Answer: Since the combustion analysis does not sum to 100%, the missing 9.03% must be oxygen. Multiplying the 177 MS molecular ion weight by an atom's percentage from the elemental analysis, then dividing by the atom's atomic weight, gives the number of those atoms in the molecular formula, which is $C_{11}H_{15}NO$. A fully saturated formula has one more H due to the presence of one N, or $(2n + 2 + 1) = 25$ H. Ours has 15H, so is missing 10 H, and that gives a DU of 5 rings/pi bonds. IR spectra: nothing 3800–3100 cm^{-1} means no OH or NH, and the major band at 1714 cm^{-1} is a C=O stretch.

Eleven carbons give seven C-NMR peaks. If the C-NMR has fewer peaks than there are carbons in the molecular formula, then the molecule has some symmetry. The most common symmetry elements are a mirror plane or a rotation axis, and we can use the integration of the H-NMR to figure out which C-NMR peaks are doubled. In general, we can assume anything that deshields the H will also deshield the C it's bonded to. For example, the methyl at 13 ppm in the C-NMR best fits with the H-NMR shift at 1.2, but that integrates for six hydrogens, not three, so it must be doubled. Therefore, the molecule has two identical methyl groups that are next to two hydrogens on the neighboring atom. Likewise, the CH_2 at 45 in the C-NMR fits the H-NMR shift at 3.4 ppm integrating for four hydrogens, and must also be doubled. That CH_2 won't fit the H-NMR shift at 6.7 or 7.7 since those shifts are in the sp^2 proton region but the carbon shift is not. The deshielded H-NMR peaks at 6.7 and 7.7 both integrate for 2H but there are only two sp^2 CH peaks left in the C-NMR, at 111 and 132, so both of these must also be doubled. Finally, the C-NMR peak at 190 has the shift of a C=O but the DEPT says it is a CH, therefore it must be an aldehyde, and that checks with the H-NMR shift at 9.7 ppm. Below is an NMR table with a new symmetry column indicating how many times the C-NMR piece must be present to get the H-NMR integration.

NMR Table

Shift ppm	Integral # of Hs	Coupling pattern	Adjacent Hs	Possibilities DEPT pieces	Possible C-NMR peaks	Symmetry to achieve 11 C
1.2	6H	Triplet	2	$-CH_3$ (2H)	13 CH_3	Two identical
3.4	4H	Quartet	3	$\leftarrow CH_2$ (3H)	45 CH_2	Two identical
6.7	2H	Doublet	1	=CH (1H)	111 =CH	Two identical
7.7	2H	Doublet	1	=CH (1H)	132 =CH	Two identical
9.7	1H	Singlet	0	HC=O (no H)	190 HC=O	One
	sum 15H				125 =C	One
					152 =C	One

Always identify pieces before attempting to assemble the molecule. Two identical isolated HC=CH pieces satisfy the doublet coupling and the integration. The $-CH_3$ (2H) is an end jigsaw piece coupled to the $\leftarrow CH_2$ (3H) piece to give an identical pair of $\leftarrow CH_2CH_3$ (no H). One end of the molecule is the pair of identical $\leftarrow CH_2CH_3$ (no H). The aldehyde –CHO (no H) is the only other end piece we have. We need to assemble the molecule with symmetry. Again, draw a line for the symmetry and then put all unique pieces on that line and identical pairs on both sides of the line. The only electronegative group to deshield the two identical ethyl groups is the nitrogen atom, and the shift fits. The $(CH_3CH_2)_2N-$ gives us a new end piece, which with the aldehyde gives us two ends, each possibly connected to two unique quaternary carbons. The two identical HC=CH shifts of 6.7 and 7.7 are in the aromatic region, and the coupling could be explained with a 1,4-disubstituted benzene ring, and that with the aldehyde would satisfy the DU of 5. The molecule assembles in three steps. Now the most important step—Look for another possible solution and cross-check the one you have. This structure passes all checks, with no alternative structures found.

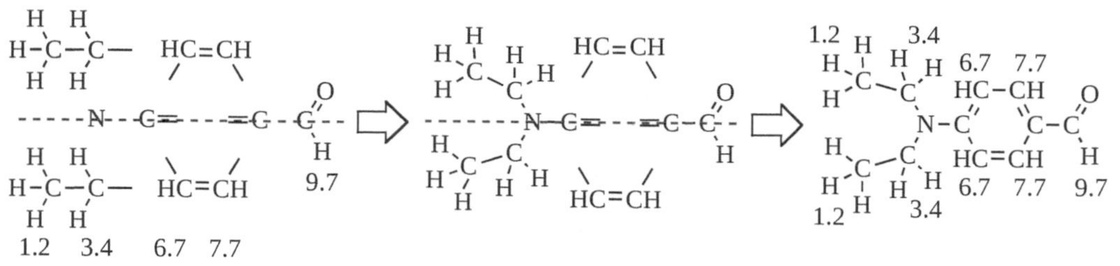

14.9 EXTRA HELP—A TASTE OF 2D NMR

Free online NMR prediction programs will create a calculated spectrum from a drawn structure. These programs are reasonably good for seeing if a proposed structure would produce the observed NMR, but the more complex the structure, the more errors can accumulate. Also, the spectra from different NMR prediction programs may not match since the programs use different algorithms, but they should be close. Complex problems may require 2D NMR to solve. To demonstrate how this is done, we will use simplified computed spectra from an online NMR prediction program, nmrdb.org.

Suppose we have an example compound C_4H_8ClI that we want to identify the structure of and fully assign both H-NMR and C-NMR peaks. We know that chlorine and iodine atoms are both end pieces since they usually form just one bond. The calculated H-NMR and C-NMR spectra for this compound are shown below (nmrdb.org).

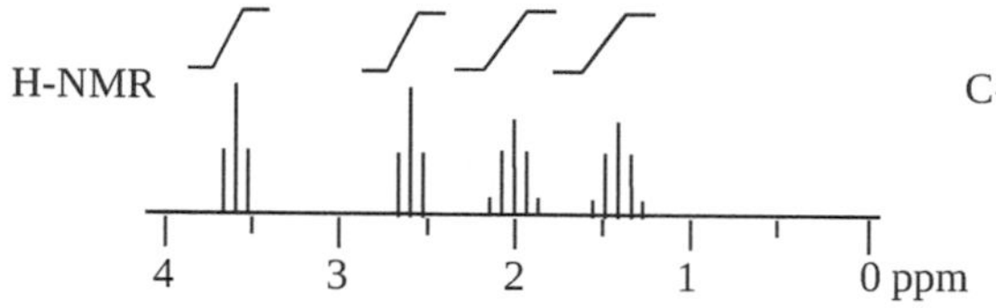

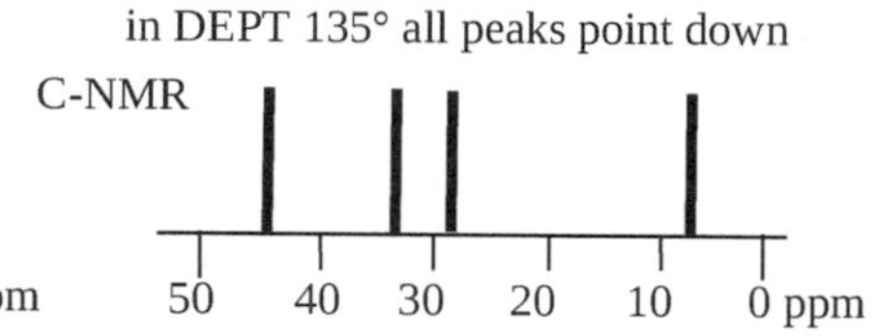

Right off we can see that there are four unique sets of protons and also four unique carbon atoms. The compound has no symmetry. The DEPT 135° has all four peaks pointing down, so they are all CH_2 groups. The H-NMR integration agrees that all peaks integrate for the same amount. The NMR table looks like the following.

NMR table

Shift ppm	Integration # of Hs	Coupling	Adjacent H's	Possibilities (use C-NMR DEPT pieces)
1.4	2H	pentet	4	—CH_2 (4H)
2.0	2H	pentet	4	—CH_2 (4H)
2.6	2H	triplet	2	← CH_2 (2H)
3.6	2H	triplet	2	← CH_2 (2H)
	Sum 8H			

The two halogens are end pieces since they usually bond to just one other thing. The pieces left are four CH_2 middle groups. The compound must be Cl–$CH_2CH_2CH_2CH_2$–I. Now all that is left is to assign the NMR peaks. Since chlorine is our most electronegative atom present, the most deshielded methylene at 3.6 ppm would be a CH_2Cl. The most deshielded carbon at 45 ppm is probably the same methylene. That leaves the other end CH_2I as the 2.6 ppm triplet, but which proton shifts go with the other two methylenes, and how about assigning the rest of the carbons? Instead of guessing, let's get more data.

14.9.1 H H Correlation Spectroscopy—COSY

COSY is a 2D NMR spectrum that correlates which hydrogen atoms are coupled to each other, shown in Figure 14.10. Displayed on both the horizontal and vertical axes is shown the H-NMR spectrum with a common 0 ppm corner. Within the 2D square are a series of peaks, which lie along the diagonal from the origin that corresponds to each peak on the top axis relating to the same peak on the vertical axis.

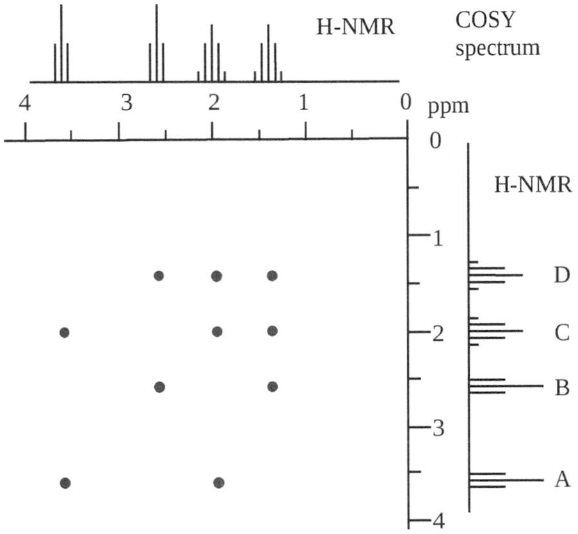

Figure 14.10 The calculated COSY spectrum for Cl–$CH_2CH_2CH_2CH_2$–I from nmrdb.org.

To interpret a COSY spectrum, we first label peaks on both H-NMR spectral axes and locate and label the diagonal, Figure 14.11. Notice that the off-diagonal pairs of peaks form boxes with two of the peaks on the diagonal. Draw in those boxes for guides for the next step. It is the off-diagonal peaks that convey important information. If there is a singlet in the spectrum, it will not have any off-diagonal peaks since it is not coupled to any other peaks.

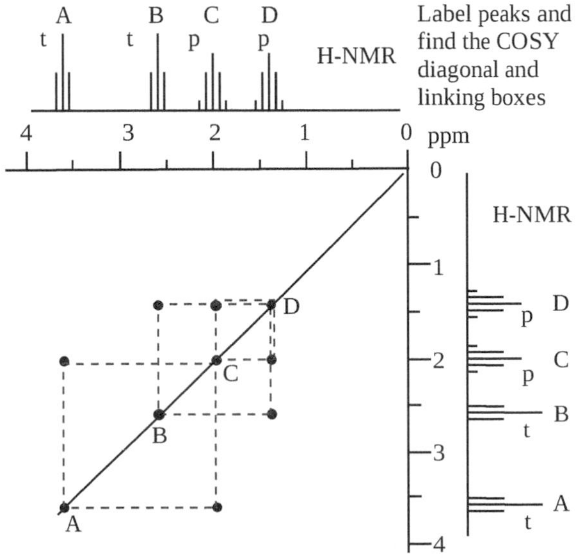

Figure 14.11 Label peaks on the spectrum, locate the diagonal and find each of the three linking boxes (shown with dashed red lines) for the calculated COSY spectrum of Cl–$CH_2CH_2CH_2CH_2$–I.

Our CH_2Cl peak in Cl–$CH_2CH_2CH_2CH_2$–I has been fully assigned earlier, so it will serve as a great anchor point (called peak A). We now "walk the COSY" spectrum by going from point A along a linking box edge to an off-diagonal peak, then follow the next edge of that linking box back to the diagonal, Figure 14.12. Depart that new diagonal peak on a different linking box to its off-diagonal peak then follow the next linking box edge back to the diagonal. Repeat this process until you have mapped all the linking boxes. Since the COSY is symmetrical across the diagonal, it does not matter to which side you walk those boxes.

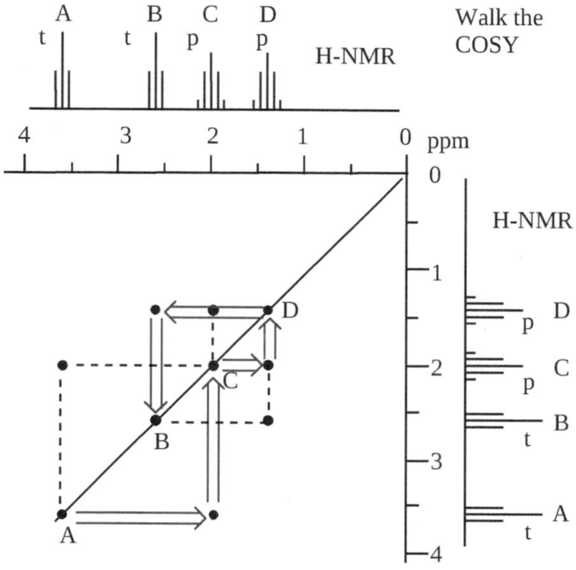

Figure 14.12 Starting from an known anchor point, our point A, walk the COSY spectrum by going along the linking box edge to the off-diagonal peak then follow a different edge of the same linking box back to a diagonal peak. That gets you to C. Repeat for each linking box to hit the next diagonal peak.

We have just walked up the carbon chain of Cl–$CH_2CH_2CH_2CH_2$–I from the CH_2Cl end to the CH_2I end. The H-NMR assignments mapped from this COSY walk are:

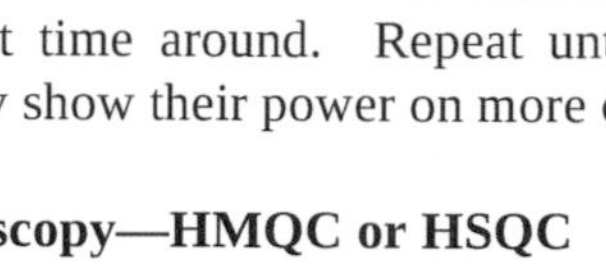

In more complex compounds, there might be two or more separate sets of spin systems in our compound. If walking the COSY misses some of the linking boxes, then find a second anchor point on the diagonal to one of those boxes to start from and repeat for those linking boxes that were missed the first time around. Repeat until all of the linking boxes are mapped. COSY 2D spectra really show their power on more complex spectra.

14.9.2 C H Correlation Spectroscopy—HMQC or HSQC

What about assigning the carbon NMR? We made a good guess at the CH_2Cl carbon being at 45 ppm but the rest are up for grabs. Again we need more data. An HMQC or HSQC spectrum is a 2D NMR spectrum that correlates which hydrogen atoms are coupled to which carbon atoms. Usually displayed on the horizontal axis is the H-NMR spectrum, and on the other axis is the C-NMR spectrum, with a common 0 ppm corner, shown in Figure 14.13. Within the 2D square are a series of peaks that line up with the peaks on the sides.

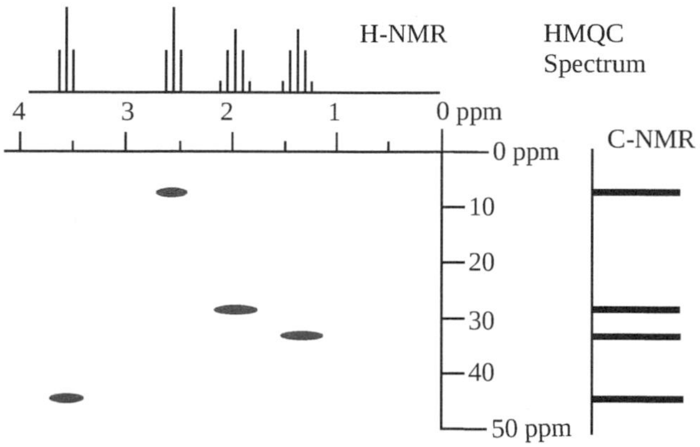

Figure 14.13 The calculated (nmrdb.org) HMQC 2D spectrum correlates H-NMR with C-NMR peaks.

If the carbon is quaternary, no peak appears in the center. Drawing lines down from the H-NMR peak and in from the sides from each C-NMR peak to those peaks inside the 2D square will aid in identifying the correlation of close peaks, as shown in Figure 14.14.

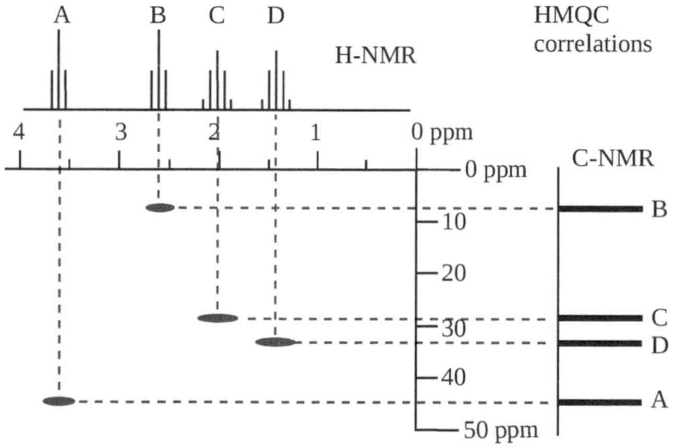

Figure 14.14 Drawing lines in from the sides of the HMQC spectrum to connect to the peaks within the box will identify which H-NMR peaks correlate with which C-NMR peaks.

We can now label the carbon peaks on the right of the figure with the appropriate letter. We have now fully assigned all the carbons in the structure. This HMQC 2D spectrum also really shows its power on more complex spectra. The full assignment is:

H-NMR

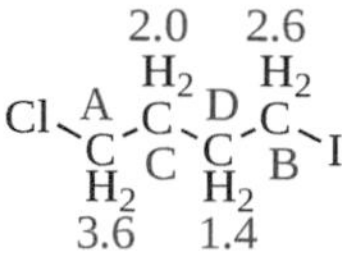

Label	H-NMR	C-NMR
A	3.6	45
B	2.6	7
C	2.0	29
D	1.4	33

C-NMR

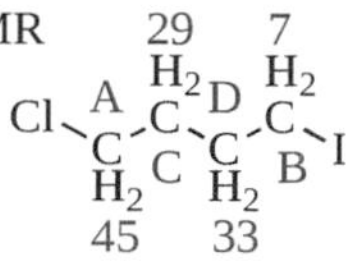

These calculated spectra illustrated the use of COSY and HMQC spectra to fully assign both the H-NMR and C-NMR spectra of 1-chloro-4-iodobutane. There are many types of 2D spectra, and they can get quite complex, but they are very valuable in solving difficult structure elucidation puzzles. In fact, half of the 2002 Nobel Prize in chemistry was awarded for the use of 2D NMR techniques in determining the three-dimensional structure of important biomolecules in solution. Some medically very important and complex compounds, isolated from nature, took everything we had to solve their structure. What a rewarding challenge!

14.10 STRUCTURE ELUCIDATION SUMMARY

Strategy: Find all the pieces first. Work in from the edges. Keep an open mind. Generate alternatives and select the best. Have patience and don't force the answer. Cross-check everything. Look for alternatives even when you think you are finished.

Mass spectra—MS: The molecular ion, M, gives the molecular weight, and an odd M can indicate the presence of nitrogen. The M + 2 peaks can indicate the presence of certain elements, such as Cl, Br, and S. Fragmentation patterns can give hints about structure. A fragmentation pattern library match can identify a compound.

Elemental Analysis: Combustion analysis gives the ratio of each element in a compound. A common elemental analysis gives percentages of C, H, and N, but not O.

Getting a molecular formula: Multiplying the molecular ion weight by an atom's percentage from the elemental analysis, dividing by the atom's atomic weight, gives the number of those atoms in the molecular formula. Find the DU, the number of rings/pi bonds.

Infrared spectra—IR: IR gives information about functional groups, and can rule one out if the expected IR band for it is missing. However, the intensity of an IR band is dependent on the vibration changing the dipole moment, so symmetrical stretches can have weak or missing IR bands. Try to rule out functional groups suggested by the molecular formula, and don't try to identify all IR peaks, just the largest. Below 1000 cm^{-1} is the fingerprint region where peaks are much harder to assign.

Carbon-13 NMR spectra: C-NMR spectra give information on the different types of carbons present in the molecule. Fewer C-NMR peaks than carbons in the molecular formula indicate molecular symmetry (often a mirror plane or rotation axis). A technique known as DEPT will distinguish between carbon peaks coming from CH_3, CH_2, CH, or C.

Hydrogen NMR spectra: H-NMR spectra give three pieces of information about the hydrogen atoms in a molecule: chemical shift (magnetic environment), integration (relative area of peaks), and coupling (number of adjacent Hs). In simple systems, *n* number of adjacent equivalent protons will split the resonance into n+1 lines. Protons at the same chemical shift do not show coupling. Exchange eliminates coupling.

Piecing a spectral jigsaw all together: Work like you would on a jigsaw puzzle. Find all the pieces and identify all the edge pieces. Work from the edges in toward the center of the molecule using the coupling patterns. Cross-check as you go. Don't hammer pieces into place. Look for alternatives at the end.

Extra help—a taste of 2D NMR: Complex problems often require more spectral data. H H correlation spectroscopy (COSY) is a 2D NMR spectrum that correlates which hydrogen atoms are coupled to each other. The off-diagonal peaks link coupled groups of hydrogens. C H correlation spectroscopy (HMQC or HSQC) is a 2D NMR spectrum that shows which hydrogens are on which carbons. The many types of 2D NMR spectra are very valuable in solving difficult structure elucidation puzzles.

CHAPTER 14 EXERCISES

14.1 Find the molecular formula and the DU given the MS molecular ion of 88.1 amu and combustion analysis data of 54.53 %C and 9.15 %H.

14.2 Find the molecular formula and the DU given the MS molecular ion of 74.1 amu and combustion analysis data of 64.82 %C and 13.60 %H.

14.3 The IR of $C_5H_{10}O$ has no peaks from 3600–3100 cm^{-1} but a strong peak at 1715 cm^{-1}. Find the DU and give the possible functional group(s) for this molecule.

14.4 The IR of $C_4H_{10}O$ has no peaks from 3600–3100 cm^{-1} but a strong peak at 1190 cm^{-1}. Find the DU and give the possible functional group(s) for this molecule.

14.5 How many C-NMR peaks would each of these compounds have?

14.6 How many C-NMR peaks would each of these compounds have?

14.7 List the C-NMR peaks with the DEPT assignment and propose a structure for C_4H_8O from the spectra:

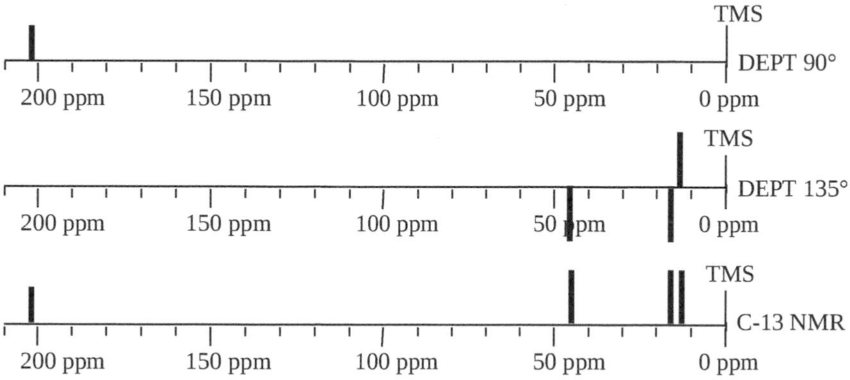

14.8 List the C-NMR peaks with the DEPT assignment and give a structure for C_5H_{12} from the spectra:

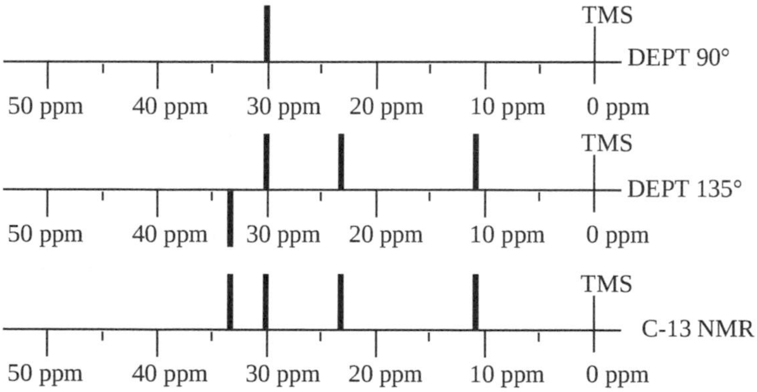

14.9 The C-NMR of $C_6H_{14}O$ is 23 and 68 ppm. Propose a consistent structure.

14.10 The C-NMR of $C_8H_{19}N$ is 14, 21, 33, and 50 ppm. Propose a consistent structure.

14.11 Make an NMR table, give a structure, and assign all the NMR peaks for the following compound. The MS molecular ion is at m/z 148. The combustion analysis is 81.04% C and 8.16% H. The H-NMR is 1.4 ppm 3H triplet (7Hz), 4.0 ppm 2H quartet (7Hz), 5.2 ppm 1H doublet (7Hz), 6.2 ppm 1H doublet (7Hz), and a 5H multiplet from 7.1 to 7.6 ppm.

14.12 Make an NMR table and give a structure for $C_7H_{12}O_3$ whose H-NMR is below. The IR spectrum has nothing 3800–3100 cm^{-1}, a major band at 1750 cm^{-1}, and 1159 cm^{-1}.

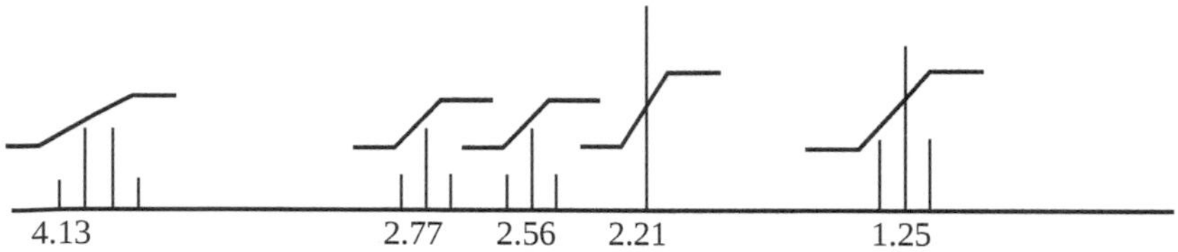

14.13 Make an NMR table, give a structure, and assign all the H-NMR peaks. The MS molecular ion is at m/z 134. The combustion analysis is 89.5% C and 10.5% H. The H-NMR is 0.8 ppm 3H triplet, 1.2 ppm 3H doublet, 1.6 ppm 2H quintet, 2.5 ppm 1H hexet, and a broad 5H peak at 7.1 ppm. The C-NMR (DEPT) is 12.3 (CH_3), 21.9 (CH_3), 31.3 (CH_2), 41.8 (CH), 125.8 (CH), 127.1 (CH), 128.3 (CH), and 147.7 ppm (C).

14.14 Fill in the NMR table, give a structure, and assign all the NMR peaks. The MS molecular ion is at m/z 120. The analysis is 90% C and 10% H. The H-NMR is a singlet at 2.3 ppm (integral = 3), a singlet at 6.8 ppm (integral =1). The C-NMR (DEPT) is 21 (CH_3), 127 (CH), and 137 ppm (C).

14.15 Fill in the NMR table, and give a structure for $C_{13}H_{16}O_2$. The IR spectrum has nothing 4000–3000 cm^{-1}, a strong peak at 1720 cm^{-1}, and one at 1180 cm^{-1}. The C-NMR (DEPT) is 19 (CH_3), 28 (CH), 71 (CH_2), 118 (CH), 127 (CH), 128 (CH), 131 (CH), 135 (C), 145 (CH), and 167 ppm (C). The H-NMR is tabulated below. The calculated (nmrdb.org) HMQC below allows the assignment of which H is on which C (but not in the 7.3 multiplet).

NMR Table

ppm	# of Hs	Coupling	Adjacent H's	Pieces
1.0	6H	doublet, 7Hz		
2.0	1H	nonet, 7Hz		
3.9	2H	doublet, 7Hz		
6.6	1H	doublet, 18Hz		
7.3	5H	multiplet		
7.6	1H	doublet, 18Hz		

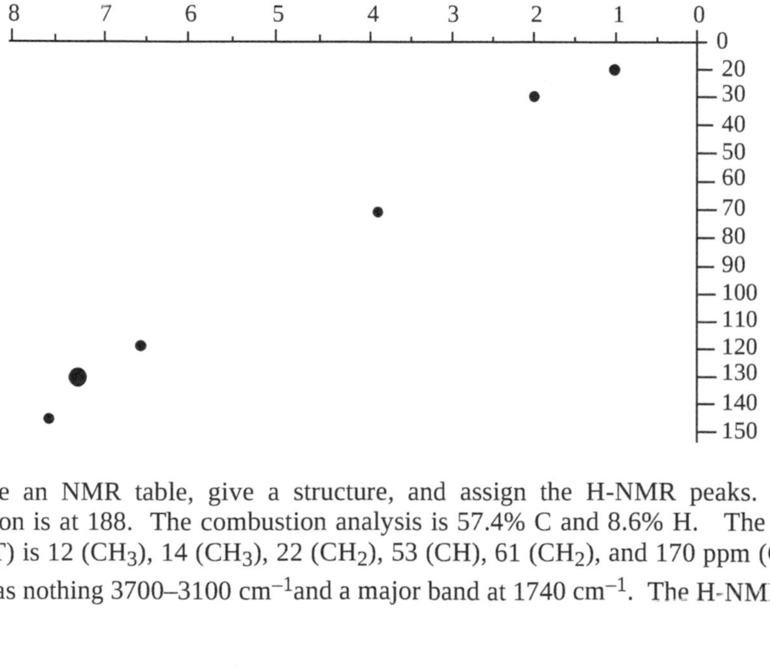

14.16 Make an NMR table, give a structure, and assign the H-NMR peaks. The MS molecular ion is at 188. The combustion analysis is 57.4% C and 8.6% H. The C-NMR in ppm (DEPT) is 12 (CH_3), 14 (CH_3), 22 (CH_2), 53 (CH), 61 (CH_2), and 170 ppm (C). The IR spectrum has nothing 3700–3100 $cm^{-1}$and a major band at 1740 cm^{-1}. The H-NMR follows.

14.17 Make an NMR table and give a structure for $C_8H_{14}O_3$, whose H-NMR is shown below. The C-NMR (DEPT) is 13 (CH_3), 14 (CH_3), 17 (CH_2), 45 (CH_2), 49 (CH_2), 61 (CH_2), 167 (C), and 203 ppm (C). The IR spectrum has nothing 4000–3000 cm^{-1}, two major bands at 1742 and 1717 cm^{-1}, and a band at 1200 cm^{-1}. The calculated (nmrdb.org) COSY below allows for a definite H assignment.

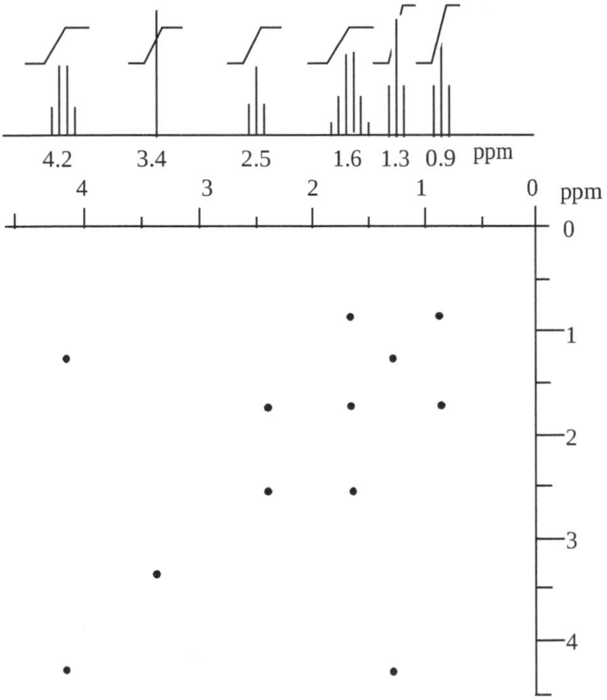

14.18 The H-NMR for $C_{10}H_{12}O$ is below. The IR spectrum has nothing at 3800–3100 cm^{-1} and a major band at 1685 cm^{-1}. Make an NMR table and give two possible structures.

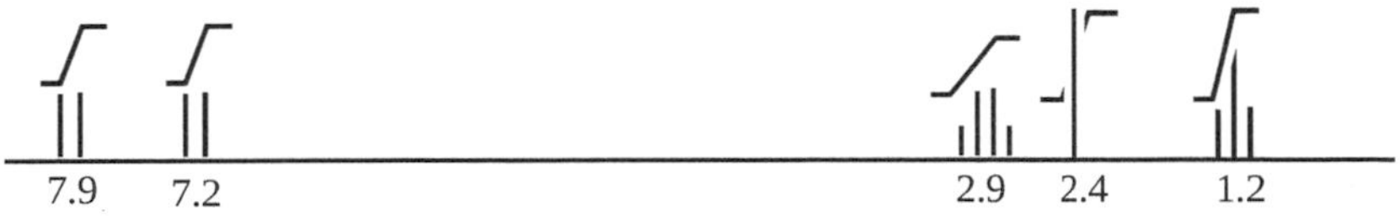

14.19 Give an NMR table and a structure for $C_{10}H_{12}O_2$, whose H-NMR is below. The C-NMR (DEPT) is 10 (CH_3), 22 (CH_2), 67 (CH_2), 128 (CH), 130 (CH), 132 (C), 134 (CH), and 167 ppm (C). The IR has nothing at 3800–3100 cm^{-1} and a major band at 1687 cm^{-1}.

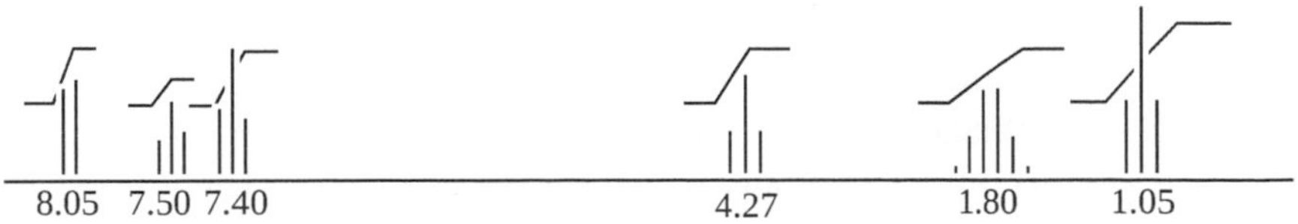

APPENDIX (A COLLECTION OF IMPORTANT TOOLS)

BIBLIOGRAPHY

Undergraduate Texts

Clayden, J.; Greeves, N.; Warren, S. *Organic Chemistry*, 2nd ed.; Oxford University Press, 2012.

Grossman, R. *The Art of Writing Reasonable Organic Reaction Mechanisms*, 3rd ed.; Springer, 2019.

Sykes, P. *A Guidebook to Mechanism in Organic Chemistry*, 6th ed.; Prentice-Hall, 1986.

Weeks, D. *Pushing Electrons*, 4th ed.; Brooks/Cole, 2014.

Graduate Texts

Anslyn E. & Dougherty D. *Modern Physical Organic Chemistry*, University Science Books, 2006.

Carey, F. & Sundberg, R. *Advanced Organic Chemistry*, 5th ed.; Springer, 2007.

Carroll, F. *Perspectives on Structure and Mechanism in Organic Chemistry*, 2nd ed.; Wiley, 2011.

Smith, M. *March's Advanced Organic Chemistry: Reactions, Mechanisms, and Structure*, 8th ed.; Wiley, 2020.

Molecular Orbital Theory

Fleming, I. *Molecular Orbitals and Organic Chemical Reactions, Reference Edition*; Wiley, 2010.

Jorgensen, W. & Salem, L. *The Organic Chemist's Book of Orbitals*; Academic, 1973.

Biochemical Mechanisms

Fersht, A., *Structure and Mechanism in Protein Science*; World Scientific, 2017.

McMurry J. & Begley T. *The Organic Chemistry of Biological Pathways*, 2nd ed.; Roberts, 2016.

Walsh, C. *Enzymatic Reaction Mechanisms*; Freeman: New York, 1979.

Inorganic Chemistry

Crabtree R., *The Organometallic Chemistry of the Transition Metals*, 7th ed.; Wiley, 2019.

Bond Strengths

Benson, S. W. *J. Chem. Educ.* **1965**, *42*, 502.

Blanksby, S. & Ellison, B. Accounts of Chemical Research, **2003,** *36*, 255.

Cottrell, T. *The Strengths of Chemical Bonds*, 2nd ed.; Butterworths, 1958.

Streitwieser, A., Heathcock, C., Kosower E., *Introduction to Organic Chemistry*, 4th ed.; Macmillan, 1992.

pK_a Values

Smith, M. *March's Advanced Organic Chemistry: Reactions, Mechanisms, and Structure*, 8th ed.; Wiley, 2020.

Speight, J., ed., *Lange's Handbook of Chemistry*, 17th ed.; McGraw-Hill, 2017.

Water pK_a value controversy: Walde, P., et al., *Helvetica Chimica Acta*, **2014,** *97*, 1–31.

For pK_a values in DMSO see: https://organicchemistrydata.org/hansreich/resources/pka/

For a species vs. pH calculation from a drawn structure see: www.chemicalize.com.

Electron Flow in Organic Chemistry: A Decision-Based Guide to Organic Mechanisms,
Third Edition, Paul H. Scudder.

Hard–Soft Acid–Base Theory

Pearson, R. G., *J. Chem. Educ.* **1968**, *45*, 581.

Nucleophilicity

Pearson, R. G., *J. Am. Chem. Soc.* **1968**, *90*, 319.

The ΔpK_a Rule and Computer Prediction of Organic Reactions

Salatin, T. D.; Jorgensen, W. L., *J. Org. Chem.* **1980**, *45*, 2043.

Calculation of NMR Spectra

To calculate an H-NMR, C-NMR, COSY, and/or HSQC/HMBC spectrum from a drawn structure see https://www.nmrdb.org/

ABBREVIATIONS USED IN THIS TEXT

Ac	Acetyl $CH_3C{=}O$
Ar	Any aryl (aromatic) group
b	Brønsted–Lowry base, proton acceptor
δ+	A partial positive charge
δ−	A partial negative charge
E	Electrophile, Lewis acid
Et	Ethyl CH_3CH_2
ERG	Electron-releasing group
EWG	Electron-withdrawing group
G	Unspecified group
HA	Brønsted–Lowry acid, proton donor
HOMO	Highest occupied molecular orbital
i-Pr	Isopropyl $(CH_3)_2CH$
L	Leaving group
LUMO	Lowest unoccupied molecular orbital
M	Metal atom
Me	Methyl CH_3
MO	Molecular orbital
n-Bu	Normal-butyl group $CH_3CH_2CH_2CH_2$
Nu	Nucleophile
Ph	Phenyl group, C_6H_5, a monosubstituted benzene
R	Any alkyl chain
t-Bu	Tertiary butyl $(CH_3)_3C$
Ts	*p*-Toluenesulfonyl, $CH_3C_6H_4SO_2$
X	Chlorine, bromine, or iodine
Y, Z	Heteroatoms, commonly oxygen, nitrogen, or sulfur
±	Racemic mixture
→→	Multistep process
‡	Transition state
······	Partially broken bond (or weak complexation)

For the path acronyms, please see the path summary in this Appendix.

Functional Group Glossary

Name	Functional Group	Example
Acetal	$>C(H)(OC)(OC)$	$H_3C-CH(OCH_3)_2$
Acid anhydride	$-C(=O)-O-C(=O)-$	$H_3C-C(=O)-O-C(=O)-CH_3$
Acyl halide	$-C(=O)-X$	$H_3C-C(=O)-Cl$
Acyloin	$-C(=O)-C(OH)<$	$H_3C-C(=O)-CH(OH)CH_3$
Alcohol	$\geq C-O-H$	H_3C-O-H
Aldehyde	$-C(=O)-H$	$H_3C-C(=O)-H$
Alkane	$\geq C-C\leq$	H_3C-CH_3
Alkene	$>C=C<$	$H_2C=CH_2$
Alkoxide	$\geq C-O^{\ominus}$	$H_3C-O^{\ominus}$
Alkyl halide	$\geq C-X$	H_3C-Br
Alkyne	$-C\equiv C-$	$HC\equiv CH$
Allene (a cumulene)	$>C=C=C<$	$(H_3C)(H)C=C=C(H)(H)$
Amide	$-C(=O)-N<$	$H_3C-C(=O)-NH_2$
Amidate	$-C(=O)-N^{\ominus}-$	$H_3C-C(=O)-NH^{\ominus}$
Amine	$\geq C-N<$	H_3C-NH_2
Amine oxide	$\geq N^{\oplus}-O^{\ominus}$	$(H_3CH_2C)_3N^{\oplus}-O^{\ominus}$
Aryl halide	C_6H_5-X	C_6H_5-Cl
Azo compound	$-N=N-$	$H_3C-N=N-CH_3$
Borane	$\geq B$	H_2B-H
Borate ester	$CO-B(OC)(OC)$	$H_3CO-B(OCH_3)(OCH_3)$
Carbamate (urethane)	$-O-C(=O)-N<$	$H_3C-O-C(=O)-N(CH_3)_2$
Carbene	$-\ddot{C}-$	$Cl-\ddot{C}-Cl$
Carbodiimide	$-N=C=N-$	$(H_3C)_2HC-N=C=N-CH(CH_3)_2$
Carbonate	$-O-C(=O)-O-$	$H_3C-O-C(=O)-O-CH_3$
Carboxylate	$-C(=O)-O^{\ominus}$	$H_3C-C(=O)-O^{\ominus}$

Carboxylic acid	
Diazonium	
Diene	
Disulfide	
Enamine	
Enediol	
Enol	
Enol ether	
Enolate	
Enone	
Epoxide (oxirane)	
Ester	
Ether	
Halohydrin	
Hemiaminal	
Hemiacetal	
Hemiketal	
Hydrate	
Hypochlorite	
Imine	
Isocyanate	
Ketal	
Ketene	

Ketone	$>C-C(=O)-C<$	$H_3C-C(=O)-CH_3$
Nitrile	$>C-C\equiv N$	$H_3C-C\equiv N$
Nitro compound	$C-\overset{\oplus}{N}(=O)-O^{\ominus}$	$H_3C-\overset{\oplus}{N}(=O)-O^{\ominus}$
Organometallic	R-M	H_3C-Li
Orthoester	$C-C(OC)_3$	$H_3C-C(OCH_3)_3$
Oxime	$>C=N-OH$	$(H_3C)_2C=N-OH$
Peroxide	$C-O-O-C$	$(H_3C)_3C-O-O-C(CH_3)_3$
Phosphate ester	$O=P(OC)_3$	$O=P(OCH_3)_3$
Phosphonate ester	$O=P(C)(OC)_2$	$O=P(CH_3)(OCH_3)_2$
Selenoxide	$C-Se(=O)-C$	$C_6H_5-Se(=O)-C_6H_5$
Sulfate ester	$CO-S(=O)_2-OC$	$H_3CO-S(=O)_2-OCH_3$
Sulfide	$C-S-C$	$H_3C-S-CH_3$
Sulfinic acid	$C-S(=O)-OH$	$C_6H_5-S(=O)-OH$
Sulfonate ester	$C-S(=O)_2-OC$	$H_3C-S(=O)_2-OCH_3$
Sulfone	$C-S(=O)_2-C$	$H_3C-S(=O)_2-CH_3$
Sulfonic acid	$C-S(=O)_2-OH$	$H_3C-S(=O)_2-OH$
Sulfonyl halide	$C-S(=O)_2-X$	$H_3C-S(=O)_2-Cl$
Sulfoxide	$C-S(=O)-C$	$H_3C-S(=O)-CH_3$
Thioester	$-C(=O)-S-C$	$H_3C-C(=O)-S-CH_3$
Thiol	$C-S-H$	H_3C-S-H
Urea	$>N-C(=O)-N<$	$H_2N-C(=O)-NH_2$
Vinyl halide	$>C=C<X$	$H_2C=CH-Cl$
Xanthate	$-O-C(=S)-S-$	$H_3C-O-C(=S)-SCH_3$

Composite pK_a Chart[1]

The acidic H is in boldface on the acid.
The basic lone pair is shown on the base.

Oxygen Acids

pK_a	Acid	Base
−15	F_3C-SO_2-OH	$F_3C-SO_2-O^{-}$
−15	O_3Cl-OH	O_3Cl-O^{-}
−12	$R-N^{+}(=O)-OH$	$R-N^{+}(=O)-O^{-}$
−10	$F-SO_2-OH$	$F-SO_2-O^{-}$
−10	$R-C(=O^{+}H)-H$	$R-C(=O)-H$
−9	$HO-SO_2-OH$	$HO-SO_2-O^{-}$
−7	$R-C(=O^{+}H)-R$	$R-C(=O)-R$
−6.5	$Ar-SO_2-OH$	$Ar-SO_2-O^{-}$
−6.5	$R-C(=O^{+}H)-OR$	$R-C(=O)-OR$
−6.4	$ArOH_2^{+}$	$ArOH$
−6	$R-C(=O^{+}H)-OH$	$R-C(=O)-OH$
−6	$Ar-O^{+}(H)-R$	$Ar-O-R$
−3.5	$R-O^{+}(H)-R$	$R-O-R$
−2.4	$CH_3CH_2OH_2^{+}$	CH_3CH_2OH
−1.5	$Ar-C(=O^{+}H)-NH_2$	$Ar-C(=O)-NH_2$
−1.5	$(CH_3)_2S{=}OH^{+}$	$(CH_3)_2S{=}O$
−1.4	$^{-}O-N^{+}(=O)-OH$	$^{-}O-N^{+}(=O)-O^{-}$
−0.5	$R-C(=O^{+}H)-NH_2$	$R-C(=O)-NH_2$
0 (−1.7)	H_3O^{+}	H_2O
0.18	$H_2N-C(=O^{+}H)-NH_2$	$H_2N-C(=O)-NH_2$

pK_a	Acid	Base
0.5	$F_3C-C(=O)-OH$	$F_3C-C(=O)-O^{-}$
0.7	$C_5H_5N^{+}-OH$	$C_5H_5N^{+}-O^{-}$
1.7	$O_2NCH_2-C(=O)-OH$	$O_2NCH_2-C(=O)-O^{-}$
1.8	$(H_3C)_3N^{+}CH_2-C(=O)-OH$	$(H_3C)_3N^{+}CH_2-C(=O)-O^{-}$
1.9	$HC\equiv C-C(=O)-OH$	$HC\equiv C-C(=O)-O^{-}$
2.0	$^{-}O-SO_2-OH$	$^{-}O-SO_2-O^{-}$
2.2	$(HO)_2P(=O)-OH$	$(HO)_2P(=O)-O^{-}$
2.3	$H_3N^{+}CH_2-C(=O)-OH$	$H_3N^{+}CH_2-C(=O)-O^{-}$
2.5	$N\equiv CCH_2-C(=O)-OH$	$N\equiv CCH_2-C(=O)-O^{-}$
2.5	$CH_3-C(=O)-C(=O)-OH$	$CH_3-C(=O)-C(=O)-O^{-}$
2.9	$ClCH_2-C(=O)-OH$	$ClCH_2-C(=O)-O^{-}$
3.1	$F_3CCH_2-C(=O)-OH$	$F_3CCH_2-C(=O)-O^{-}$
3.3	$O{=}N-OH$	$O{=}N-O^{-}$
3.4	$O_2N-C_6H_4-C(=O)OH$	$O_2N-C_6H_4-C(=O)O^{-}$
3.6	$H_3C-C(=O)-CH_2-C(=O)-OH$	$H_3C-C(=O)-CH_2-C(=O)-O^{-}$
3.6	$H_3CO-CH_2-C(=O)-OH$	$H_3CO-CH_2-C(=O)-O^{-}$
4.2	$C_6H_5-C(=O)OH$	$C_6H_5-C(=O)O^{-}$
4.5	$H_3CO-C_6H_4-C(=O)OH$	$H_3CO-C_6H_4-C(=O)O^{-}$
4.6	$(CH_3)_3N^{+}-OH$	$(CH_3)_3N^{+}-O^{-}$
4.8	$H_3C-C(=O)-OH$	$H_3C-C(=O)-O^{-}$
6.4	$HO-C(=O)-OH$	$HO-C(=O)-O^{-}$
7.2	$^{-}O-P(=O)(OH)-OH$	$^{-}O-P(=O)(OH)-O^{-}$
7.2	$NO_2-C_6H_4-OH$	$NO_2-C_6H_4-O^{-}$

[1] Main Source: Smith, M. B. *March's Advanced Organic Chemistry*, 8th ed.; Table 8.1, ©2020 Wiley. Other sources: Speight, J., *Lange's Handbook of Chemistry*, 17th ed., Table 2.60, © 2017 McGraw-Hill. ACS Division of Organic Chemistry, https://organicchemistrydata.org/hansreich/resources/pka/. Also see Bibliography.

Oxygen acids (continued)

pK_a	Acid	Base
7.5	Cl—OH	Cl—O:$^{\ominus}$
10.0	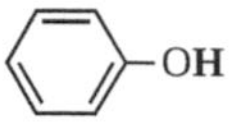	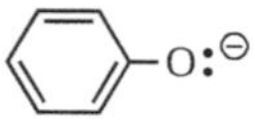
10.2	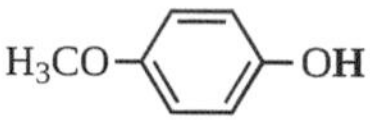	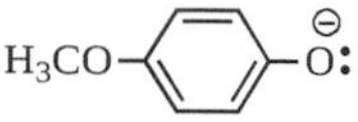
10.3	$^{\ominus}$O—C(=O)—OH	$^{\ominus}$O—C(=O)—O:$^{\ominus}$
11.6	HO—OH	HO—O:$^{\ominus}$
12.2	$(H_3C)_2C=N-OH$	$(H_3C)_2C=N-O:^{\ominus}$
12.3	$^{\ominus}$O(HO)P(=O)O$^{\ominus}$	$^{\ominus}$O($^{\ominus}$:O)P(=O)O$^{\ominus}$
12.5	CF_3CH_2OH	$CF_3CH_2O:^{\ominus}$
13.3	$HOCH_2OH$	$HOCH_2O:^{\ominus}$
13.9	$(CH_3)_3\overset{\oplus}{N}CH_2CH_2OH$	$(CH_3)_3\overset{\oplus}{N}CH_2CH_2O:^{\ominus}$
14 (15.7)	H_2O	HO:$^{\ominus}$
14.2	$HOCH_2CH_2OH$	$HOCH_2CH_2O:^{\ominus}$
15.2	CH_3OH	$CH_3O:^{\ominus}$
16	CH_3CH_2OH	$CH_3CH_2O:^{\ominus}$
17	$(CH_3)_2CHOH$	$(CH_3)_2CHO:^{\ominus}$
19	$(CH_3)_3COH$	$(CH_3)_3CO:^{\ominus}$

Nitrogen Acids

pK_a	Acid	Base
−10	R—C≡NH$^{\oplus}$	R—C≡N:
−5	$Ar_3NH^{\oplus}$	Ar_3N:
1	$Ar_2NH_2^{\oplus}$	$Ar_2\ddot{N}H$
1.0		
4.6	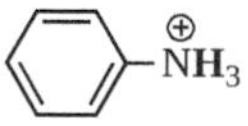	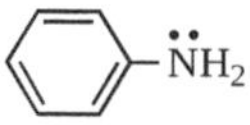
5.2	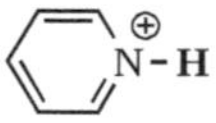	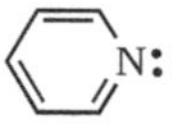
5.4	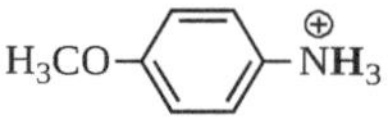	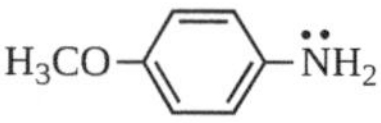
6.0	$HO-NH_3^{\oplus}$	$HO-\ddot{N}H_2$
7.0	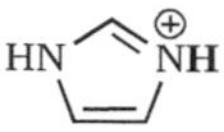	
7.9	$H_2N-NH_3^{\oplus}$	$H_2N-\ddot{N}H_2$
9.2	$NH_4^{\oplus}$	:NH_3
9.3		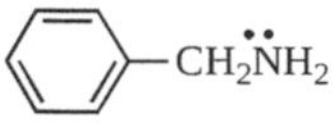
9.5	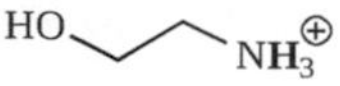	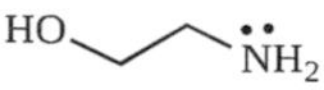

Nitrogen acids (continued)

pK_a	Acid	Base
9.6	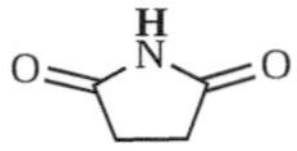	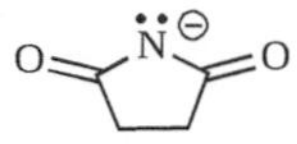
9.6	$^{\ominus}$O—C(=O)—$CH_2\overset{\oplus}{N}H_3$	$^{\ominus}$O—C(=O)—$CH_2\ddot{N}H_2$
10	PhS(=O)(=O)NH_2	PhS(=O)(=O)$\ddot{N}H^{\ominus}$
10.7	$Et_3NH^{\oplus}$	Et_3N:
13.6	$(H_2N)_2C=NH_2^{\oplus}$	$(H_2N)_2C=\ddot{N}H$
17	R—C(=O)—NH_2	R—C(=O)—$\ddot{N}H^{\ominus}$
25.8	$((CH_3)_3Si)_2NH$	$((CH_3)_3Si)_2N:^{\ominus}$
28	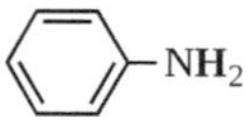	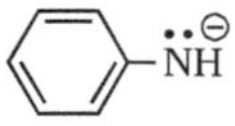
35	NH_3	$^{\ominus}\ddot{N}H_2$
36	R_2NH	$R_2\ddot{N}^{\ominus}$

Carbon acids

pK_a	Acid	Base
−5	$HC(CN)_3$	$^{\ominus}$:$C(CN)_3$
3.6	$^{\ominus}$O—$\overset{\oplus}{N}$(=O)—CH_2—$\overset{\oplus}{N}$(=O)—O$^{\ominus}$	$^{\ominus}$O—$\overset{\oplus}{N}$(=O)—$\ddot{C}H^{\ominus}$—$\overset{\oplus}{N}$(=O)—O$^{\ominus}$
5	H—C(=O)—CH_2—C(=O)—H	H—C(=O)—$\ddot{C}H^{\ominus}$—C(=O)—H
9	H_3C—C(=O)—CH_2—C(=O)—CH_3	H_3C—C(=O)—$\ddot{C}H^{\ominus}$—C(=O)—CH_3
9.2	HC≡N	$^{\ominus}$:C≡N
10.2	H_3C—$\overset{\oplus}{N}$(=O)—O$^{\ominus}$	$^{\ominus}$$H_2\ddot{C}$—$\overset{\oplus}{N}$(=O)—O$^{\ominus}$
10.7	H_3C—C(=O)—CH_2—C(=O)—OEt	H_3C—C(=O)—$\ddot{C}H^{\ominus}$—C(=O)—OEt
11.2	$CH_2(CN)_2$	$^{\ominus}$:$CH(CN)_2$
12.5	H_3C—S(=O)(=O)—CH_2—S(=O)(=O)—CH_3	H_3C—S(=O)(=O)—$\ddot{C}H^{\ominus}$—S(=O)(=O)—CH_3
13	EtO—C(=O)—CH_2—C(=O)—OEt	EtO—C(=O)—$\ddot{C}H^{\ominus}$—C(=O)—OEt
15.9	CH_3—C(=O)—CH_2Ph	CH_3—C(=O)—$\ddot{C}H^{\ominus}Ph$
16		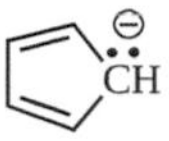
16	H_3C—C(=O)—CH_2Cl	H_3C—C(=O)—$\ddot{C}H^{\ominus}Cl$

Carbon acids (continued)

pK_a	Acid	Base
16.7	$H_3C{-}C(=O){-}H$	$^{\ominus}{:}H_2C{-}C(=O){-}H$
18	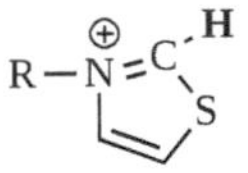	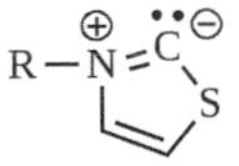
19.2	$H_3C{-}C(=O){-}CH_3$	$^{\ominus}{:}H_2C{-}C(=O){-}CH_3$
21	$CH_3{-}C(=O){-}SR$	$^{\ominus}{:}CH_2{-}C(=O){-}SR$
23	$H_3C{-}SO_2{-}CH_3$	$^{\ominus}{:}H_2C{-}SO_2{-}CH_3$
24	$HCCl_3$	$\ominus{:}CCl_3$
25	$H_3C{-}C{\equiv}N$	$^{\ominus}{:}H_2C{-}C{\equiv}N$
25	$HC{\equiv}CH$	$\ominus{:}C{\equiv}CH$
25.6	$H_3C{-}C(=O){-}OR$	$^{\ominus}{:}H_2C{-}C(=O){-}OR$
29	$H_3C{-}\overset{\oplus}{S}(CH_3)_2$	$^{\ominus}{:}H_2C{-}\overset{\oplus}{S}(CH_3)_2$
29.4	$H_3C{-}C(=O){-}NR_2$	$^{\ominus}{:}H_2C{-}C(=O){-}NR_2$
30	$H_3C{-}\overset{\oplus}{P}(CH_3)_3$	$^{\ominus}{:}H_2C{-}\overset{\oplus}{P}(CH_3)_3$
31	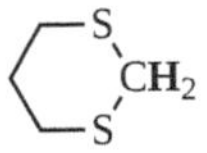	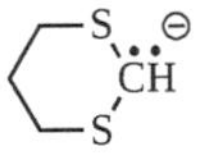
31.5	$HCPh_3$	$\ominus{:}CPh_3$
33.5	H_2CPh_2	$\ominus{:}CHPh_2$
35	$H_3C{-}S(=O){-}CH_3$	$^{\ominus}{:}H_2C{-}S(=O){-}CH_3$
40	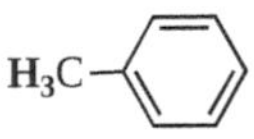	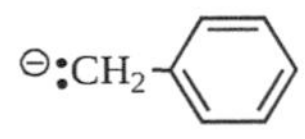
43	$H_3C{-}CH{=}CH_2$	$^{\ominus}{:}H_2C{-}CH{=}CH_2$
43	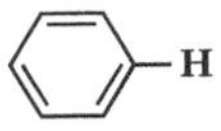	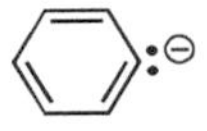
44	$H_2C{=}CH_2$	$^{\ominus}{:}HC{=}CH_2$
48	H_4C	$\ominus{:}CH_3$
50	$H_3C{-}CH_3$	$^{\ominus}{:}H_2C{-}CH_3$
51	$H_3C{-}CH_2{-}CH_3$	$H_3C{-}\overset{\ominus}{C}H{-}CH_3$
>52	$HC(CH_3)_3$	$\ominus{:}C(CH_3)_3$

Miscellaneous acids

pK_a	Acid	Base
−10	HI	$:I^{\ominus}$
−9	HBr	$:Br^{\ominus}$
−7	$RSH_2^{\oplus}$	$R\ddot{S}H$
−6	HCl	$:Cl^{\ominus}$
−5.4	$R_2SH^{\oplus}$	$R_2S:$
0.5	$F_3\overset{\ominus}{B}{-}\overset{\oplus}{F}{-}H$	$F_3\overset{\ominus}{B}{-}F:$
2.7	$Ph_3PH^{\oplus}$	$Ph_3P:$
3.2	HF	$:F^{\ominus}$
3.3	$H_3C{-}C(=O){-}SH$	$H_3C{-}C(=O){-}S:^{\ominus}$
3.9	H_2Se	$HSe:^{\ominus}$
6.5	PhSH	$PhS:^{\ominus}$
7.0	H_2S	$HS:^{\ominus}$
8.7	$Et_3PH^{\oplus}$	$Et_3P:$
10.6	EtSH	$EtS:^{\ominus}$
35	H_2	$H:^{\ominus}$

A strong acid has a low pK_a

A strong base has a high pK_{abH}

Proton-Transfer Equilibria

Proton transfers go toward the formation of the weaker base.

$$\log K_{eq} = pK_{abH} - pK_{aHA}$$

$$K_{eq} = 10^{(pK_{abH} - pK_{aHA})}$$

Subtract the pK_a of the acid from the pK_{abH} of the base to get the exponent of K_{eq}. If the proton-transfer K_{eq} is equal to or greater than 10^{-10}, the proton transfer is within the useful range.

The ΔpK_a Cross-check:

Avoid intermediates that are more than 10 pK_a units *uphill* from the reactants (either 10 pK_a units more basic or 10 pK_a units more acidic). Reactions tend toward neutralization. The energy drops if the pK_{abH} drops significantly.

Hard and Soft Acid–Base Principle:

Hard bases favor binding with hard acids; soft bases favor binding with soft acids.

BOND STRENGTH TABLE

Average Single Bond Energies in kcal/mol (kJ/mol)

	I	Br	Cl	S	P	Si	F	O	N	C	H
H	71 (297)	88 (368)	103 (431)	83 (347)	77 (322)	76 (318)	136 (569)	111 (464)	93 (389)	99 (414)	104 (435)
C	51 (213)	68 (285)	81 (339)	65 (272)	63 (264)	72 (301)	116 (485)	86 (360)	73 (305)	83 (347)	
N			46 (192)				65 (272)	53 (222)	39 (163)		
O	48 (201)	48 (201)	52 (218)		141 (590)	108 (452)	45 (188)	47 (197)			
F	58 (243)	60 (251)	61 (255)	68 (285)	117 (490)	135 (565)	37 (155)				
Si	56 (234)	74 (310)	91 (381)			53 (222)					
P	44 (184)	63 (264)	78 (326)		48 (201)						
S		52 (218)	61 (255)	60 (251)							
Cl	50 (209)	52 (218)	58 (243)								
Br	42 (176)	46 (192)									
I	36 (151)										

Average Multiple Bond Energies in kcal/mol (kJ/mol)

C=C	146 (611)	C≡C	200 (837)
C=N	147 (615)	C≡N	213 (891)
C=O	177 (741)		
C=S	128 (536)		
N=N	100 (418)	N≡N	226 (946)
N=O	145 (607)		
O=O	119 (498)		

Note: this table contains average values that should be considered very approximate (± several kcal/mol); only the values that correspond to bond strengths of simple diatomic molecules (like HCl) have little error.

Table data from:
Streitwieser, A., Heathcock, C., Kosower E., *Introduction to Organic Chemistry*, 4th ed., Macmillan, 1992. Blanksby, S. & Ellison, B. Accounts of Chemical Research, **2003,** *36,* 255. Cottrell, T., *The Strength of Chemical Bonds*, 2nd ed., Butterworths, London, 1958. Benson, S., *J. Chem. Educ.* **1965**, *42*, 502. The kJ/mol values were calculated from kcal/mol values or the reverse. One kcal/mol is 4.184 kJ/mol.

Generic Classification Guide

A big problem in organic chemistry is the sheer magnitude of information, so efficient ways of organizing the material are very valuable. **There are only 12 generic classes from which a set of arrows, an electron flow, can start. Likewise, there are only 18 generic classes of electron sinks in which an electron flow ends. It is much easier to handle multiple examples of familiar classes than hundreds of special cases.**

The best electron sources are electron rich, like anions and lone pairs, such that no bonds need be broken to use them as sources. Other excellent electron sources are highly ionic sigma bonds and pi bonds polarized by excellent electron-releasing groups (Table 5.2).

Table 5.2 Summary Reactivity Table for Common Electron Sources

Symbol and Name	Common Examples in Decreasing Reactivity Order	Reactivity Trend Principle	Common Reactions, Decision?	Related Generic Groups
Z: Heteroatom lone pairs as nucleophiles	RS^- I^- CN^- RO^-	Softer anions are more nucleophilic. If same atom, more basic, more Nu.	Substitutions and Additions. Nucleophile vs. base decision	See allylic sources
Z: Heteroatom lone pairs as bases	R_2N^- t-BuO^- R_3N CH_3COO^-	The higher pK_{abH} is a stronger base.	Proton-transfer $K_{eq} \geq 10^{-10}$	NaH and KH as MH bases
R–M Organo-metallics	CH_3Li CH_3MgI $(CH_3)_2Cu^-\ Li^+$	The more ionic RM bond is more reactive.	Substitutions and Additions. Deprotonates acidic H's	See enolates (allylic sources)
M–H Group 1A metal hydrides	NaH KH	Both are very reactive strong bases but poor nucleophiles.	Deprotonates acidic H's	See complex metal hydrides
MH_4^- Complex metal hydrides	$LiAlH_4$ $LiAlH(OR)_3$ $NaBH_4$ $NaBH_3CN$	The more ionic MH bond is more reactive.	Substitutions and Additions. Deprotonates acidic H's on heteroatoms	NaH and KH as MH bases
C=C–Z: Allylic sources	Enolates $C{=}C{-}O^-$ Enamines $C{=}C{-}NR_2$ Enol ethers C=C–OR	The better electron releasing group on the pi bond makes the allylic source more reactive.	Substitutions and Additions. Carbon vs. heteroatom decision	Extended enolates. Allylic alkyne sources
C=C Simple pi bonds	$R_2C{=}CR_2$ RHC=CHR $H_2C{=}CH_2$	The more stable the carbocation formed, the more reactive the pi system.	Electrophilic Additions. Markovnikov's rule used for regiochemistry	Other pi systems: C=C–C=C C≡C C=C=C
ArH Aromatic rings	PhOR PhH PhCl PhCOR	An ERG on the ring makes it more reactive; an EWG makes ring less reactive.	Electrophilic aromatic substitution. Regiochemistry of addition	Hetero-aromatics, condensed aromatics

Electron sinks have a full, partial, or inducible plus charge on the atom that gets attacked by the electron source. The larger the partial plus is, the better the electron sink can attract a negatively charged electron source. Table 6.2 repeated below gives the common electron sinks, and Figure A.1 shows a flowchart of the classification process.

Table 6.2 Summary Reactivity Table for Common Electron Sinks

Symbol and Name	Common Examples in Order of Decreasing Reactivity	Reactivity Trend Principle	Common Reactions. Decision?	Related Generic Groups
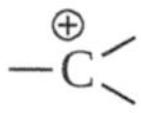 Carbocation	$(CH_3)_2CH^+$ $(CH_3)_3C^+$ Ph_3C^+ $(H_2N)_3C^+$	More stabilized, less reactive. Donors stabilize Lone pairs > pi bonds > R > H	C^+ trap Nu. Rearrange to equal or more stable C^+? Drop off H^+?	Inorganic Lewis Acids 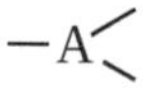
H–L or H–A Acids	H_2SO_4 HBr HCl	The lower pK_a is stronger	Proton-transfer $K_{eq} \geq 10^{-10}$	Carbon acids H_3C–EWG
Y–L Leaving group on a heteroatom	Chlorine Cl_2 Bromine Br_2 Iodine I_2 Peroxides ROOR	The most reactive break a weak bond and form a much stronger bond	Often Substitution Nu attacks δ+ end of Y–L if different	Leaving groups on phosphorus or silicon
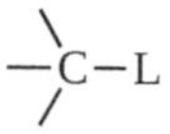 Leaving group on sp^3 carbon	Sulfonates R–OSO_2R Alkyl iodides R–I Protonated alcohols ROH_2^+ Alkyl bromides R–Br Alkyl chlorides R–Cl	The better L is more reactive	Substitution vs. Elimination Decision (Section 9.5)	2 Ls on C R_2CL_2 3 Ls on C RCL_3
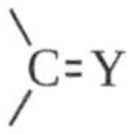 Polarized multiple bond without L	Protonated carbonyls $R_2C{=}OH^+$ Iminium ions $R_2C{=}NR_2^+$ Aldehydes RHC=O Ketones $R_2C{=}O$ Imines $R_2C{=}NR$	The larger partial plus carbon is more reactive	Addition. Following Substitution or Elimination? (Section 8.5)	Polarized triple bonds R–C≡Y without leaving groups
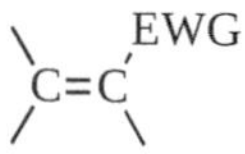 Conjugate acceptor	Conj. nitros C=C–NO_2 Enones C=C–COR Conj. esters C=C–COOR Conj. nitriles C=C–CN	The better the EWG, the more reactive	1,2 vs. 1,4? Direct vs. conjugate addition (Section 9.6)	Triply bonded conjugate acceptors C≡C–EWG
 Polarized multiple bond with L	Acyl halides RCOCl Anhydrides RCOOCOR Thioesters RCOSR Esters RCOOR Amides $RCONR_2$	The poorer pi donor L is, the more reactive	Addition–Elimination. Add a second Nu after addition–elimination? (Section 9.2)	Carbonate derivatives $L_2C{=}Y$ L on triple bonds L–C≡Y

Flowcharts For the Classification of Electron Sources and Sinks

For some individuals, a graphical presentation of a process in the form of a flowchart greatly helps them understand and visualize the overall process, Figures 5.4 and 6.2 below.

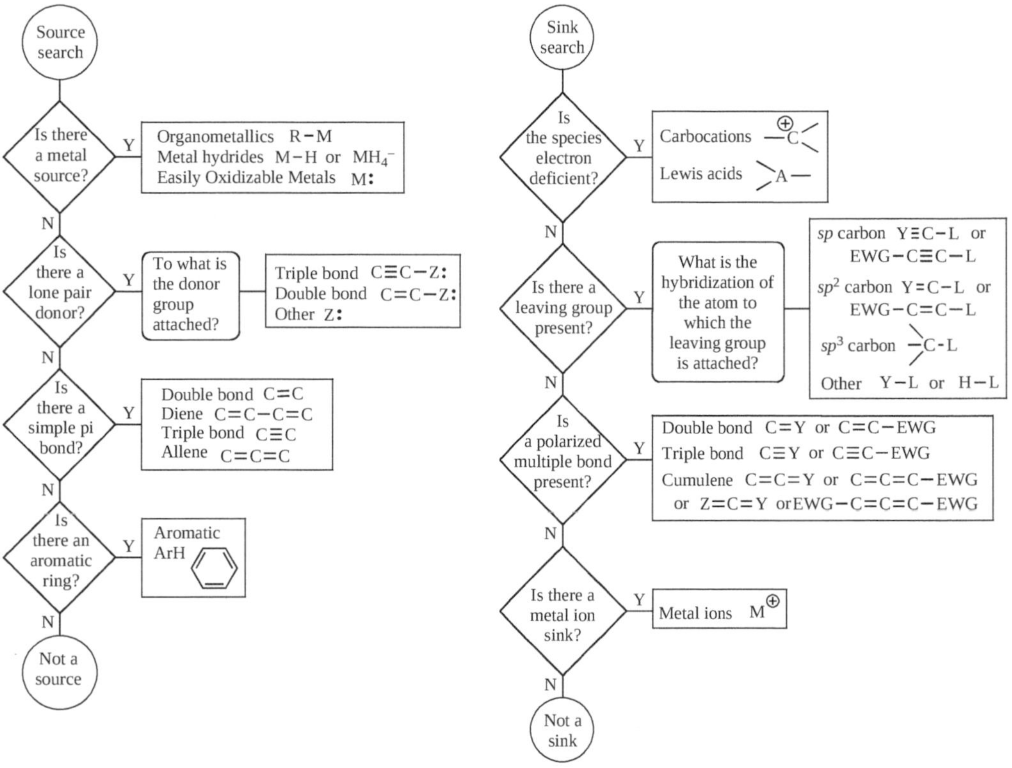

Figures 5.4 and 6.2 Flowcharts for the classification of electron sources (left) and sinks (right).

Pathway Summary and Cross-checks

Major Paths

p.t., Proton Transfer to and from an Anion or Lone Pair (Section 7.1.1)

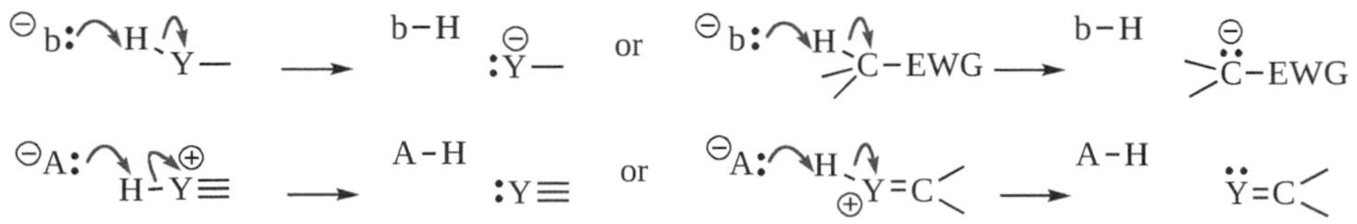

Include both deprotonations (above) and protonations (below). Cross-check: K_{eq} or ΔpK_a rule.

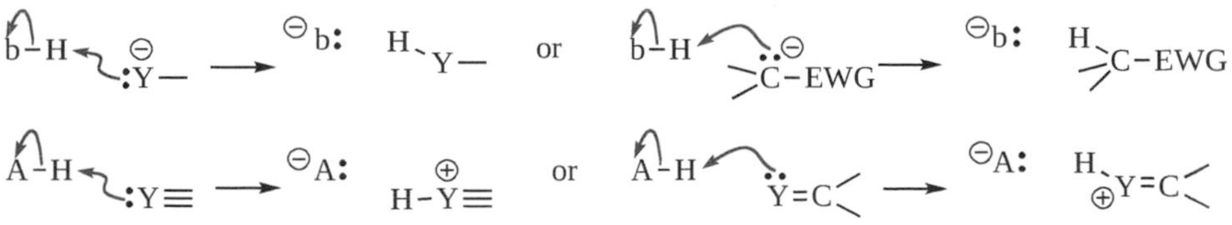

D_N, Ionization of a Leaving Group (Section 7.1.2)

Cross-check: $\Delta pK_{aHL} \leq 0$, cation usually more stable than 2°, polar solvent.

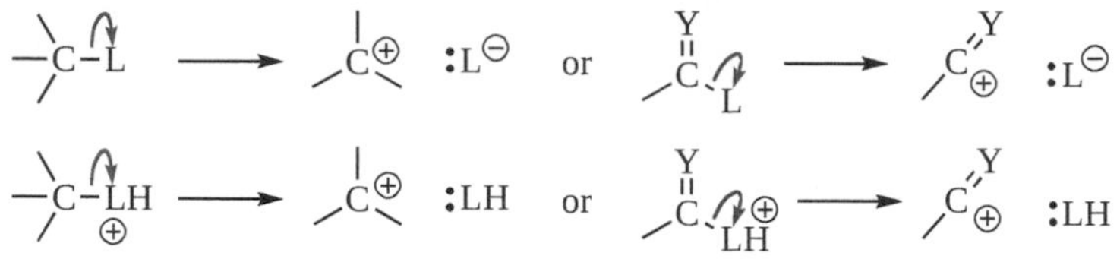

A_N, Trapping of an Electron-Deficient Species (Section 7.1.3)

Cross-check: none, usually OK.

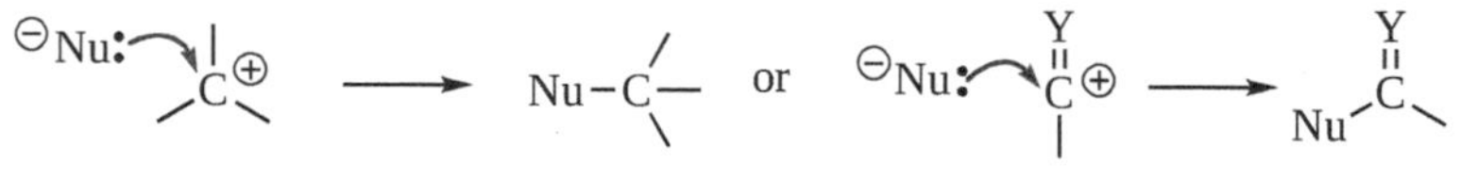

A_E, Electrophile Addition to a Multiple Bond (Section 7.1.4)

Cross-check: Markovnikov's rule — make the most stable carbocation.

D_E, Electrofuge Loss from a Cation To Form a Pi Bond (Section 7.1.5)

Cross-check: Electrofuge lost should be reasonably stable.

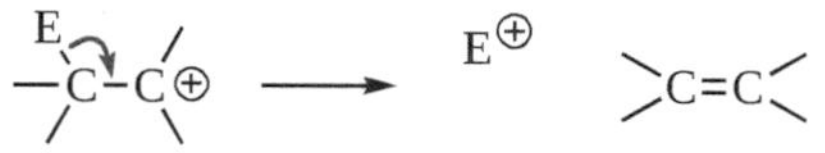

S_N2, The S_N2 Substitution (Section 7.1.6)

Cross-check: ΔpK_a rule on Nu and L, $\Delta pK_{aHL} \leq 12$, good access to C attacked.

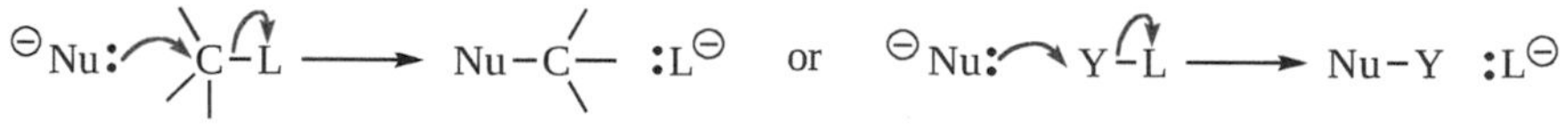

E2, The E2 Elimination (Section 7.1.7)

Cross-checks: ΔpK_a rule on incoming base and outgoing L, C–H bond and C–L need to be nearly coplanar.

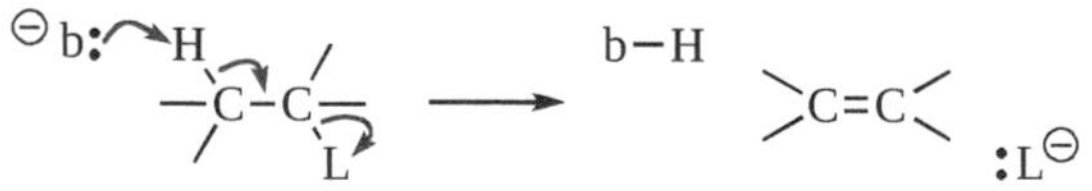

Ad_E3, The Ad_E3 Addition (Section 7.1.8)

Cross-check: Markovnikov's is rule followed because of polarization of pi-adduct.

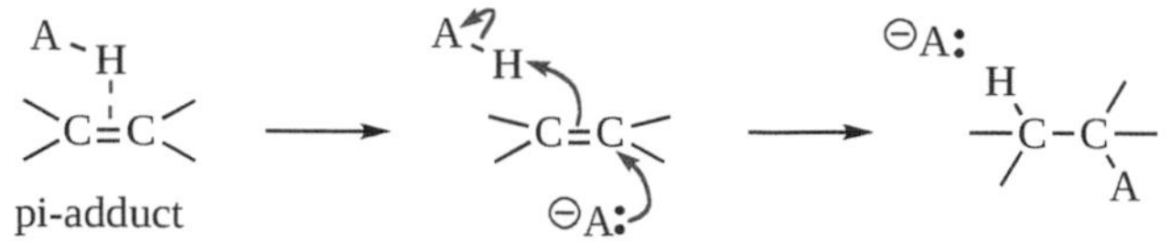

Hetero Ad_E3 (See 7.4.3) Cross-check: ΔpK_a rule on incoming Nu and anion formed.

Ad_N, Nucleophilic Addition to a Polarized Multiple Bond (Section 7.1.9)

Cross-check: ΔpK_a rule on incoming Nu and outgoing anion formed.

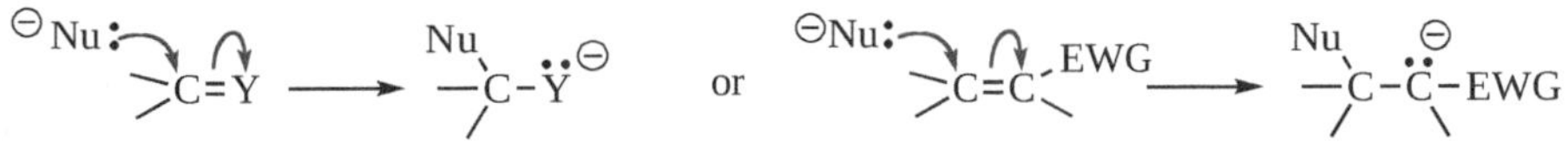

E_β, Beta Elimination from an Anion or Lone Pair (Section 7.1.10)

Cross-check: ΔpK_a rule on L and anion reactant.

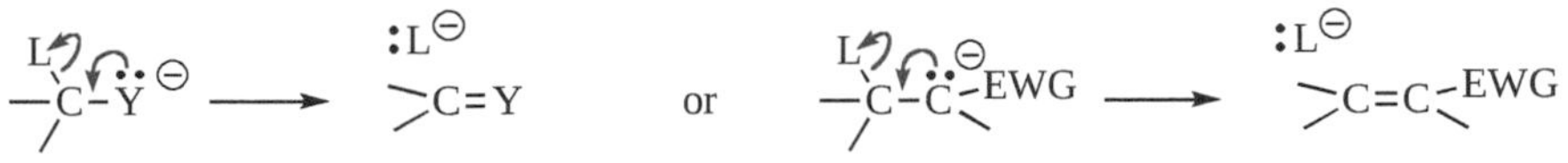

1,2R, Rearrangement of a Carbocation (Section 7.1.11)

Cross-check: Usually forms carbocation of equal or greater stability.

Path 1,2RL, Rearrangement with Loss of Leaving Group (Section 7.1.12)

Cross-check (if no pushing Y): good L, cation usually secondary or better, polar solvent.

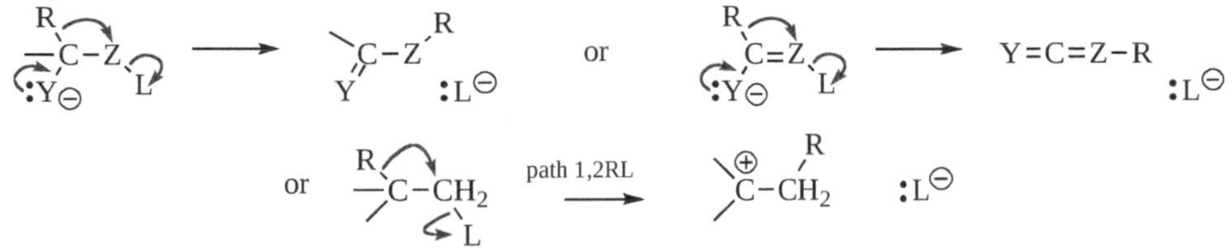

MINOR PATHS

Path pent., Substitution via a Pentacovalent Intermediate (Section 7.2.1)

(Not for carbon, only for third-row and higher elements)

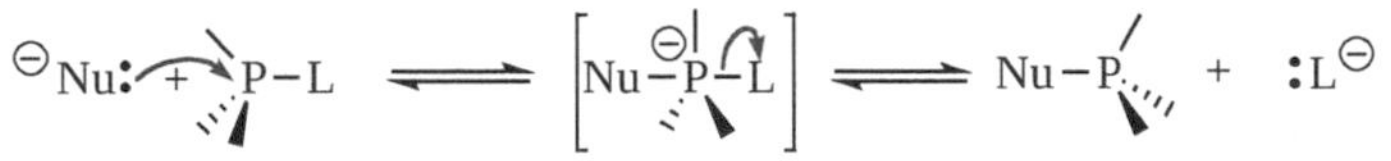

Path 6e, Concerted Six-Electron Pericyclic Reactions (Section 7.2.2)

Thermal rearrangements:

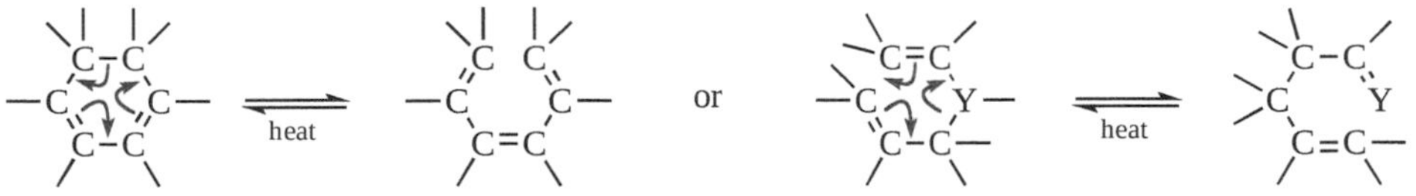

Thermal cycloadditions or cycloreversions (the reverse reaction):

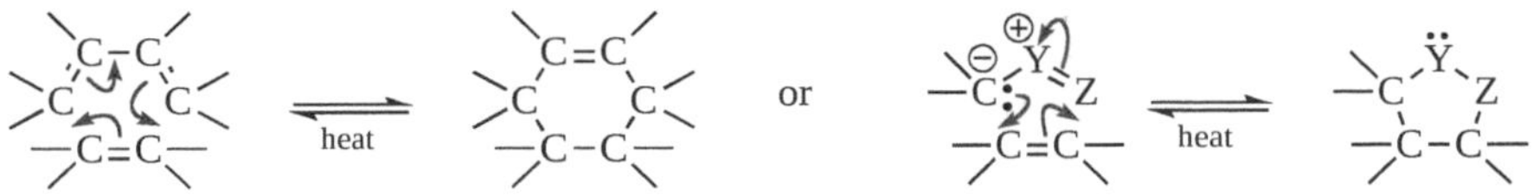

Metal-chelate-catalyzed additions:

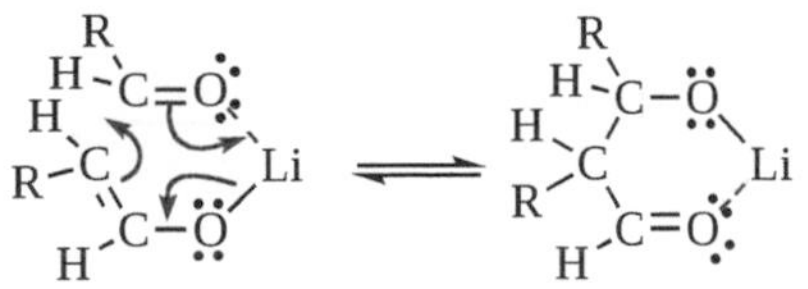

Path Ei, Thermal Internal *Syn* Elimination (Section 7.2.3)

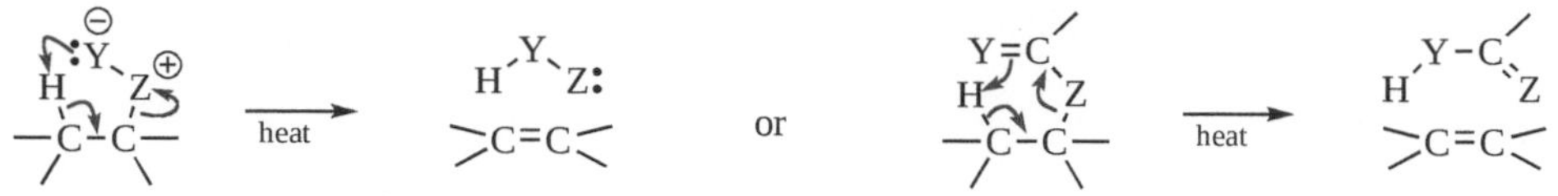

Path NuL, Nu−L Additions (Section 7.2.4) (three-membered ring formation)

Path 4e, Four-Center, Four-Electron (Section 7.2.5) (three cases only)

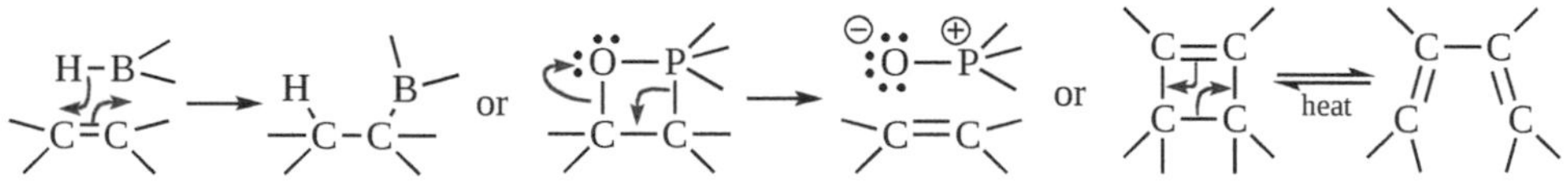

Path H⁻ t., Hydride Transfer to a Cationic Center (Section 7.2.6)

(forms a carbocation of equal or greater stability than the starting carbocation.)

COMMON PATH COMBINATIONS

S_N1 (Substitution, Nucleophilic, Unimolecular), $D_N + A_N$ (Section 7.3.1)

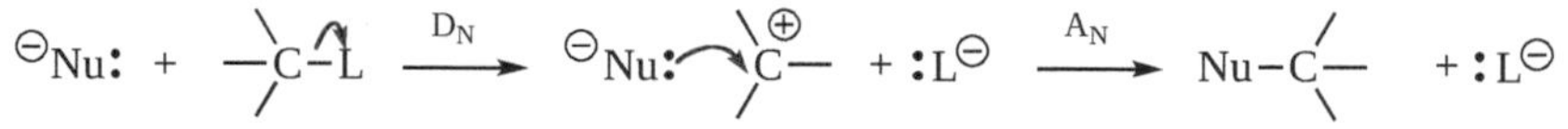

Lone-Pair-Assisted S_N1, $E_\beta + A_N$

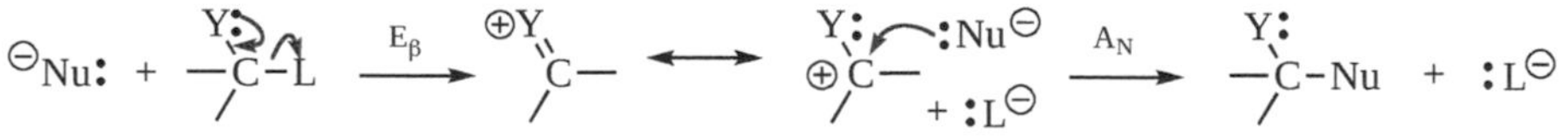

Ad_E2 (Addition, Electrophilic, Bimolecular), $A_E + A_N$ (Section 7.3.2)

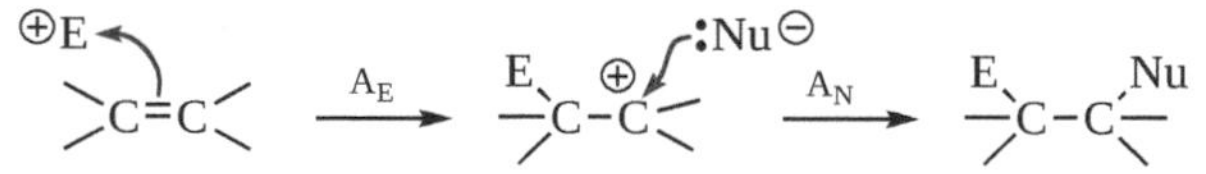

Hetero-Ad_E2, p.t. + Ad_N

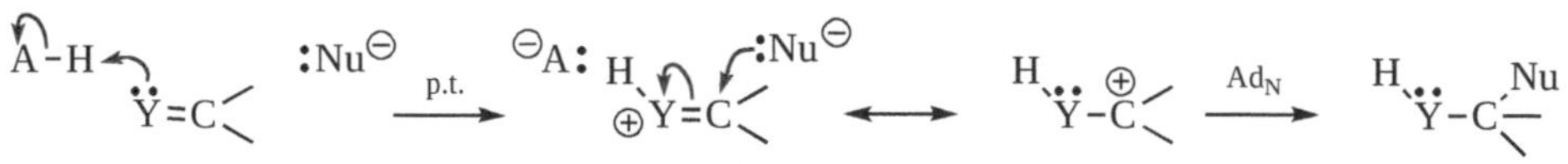

E1 (Elimination, Unimolecular), $D_N + D_E$ (Section 7.3.3)

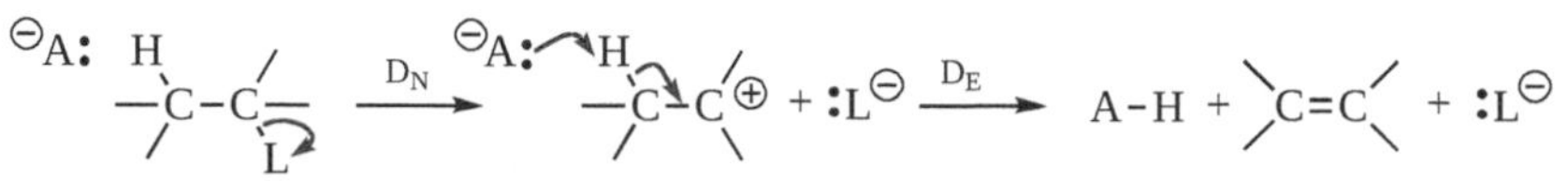

Lone-Pair-Assisted E1, E_β + p.t.

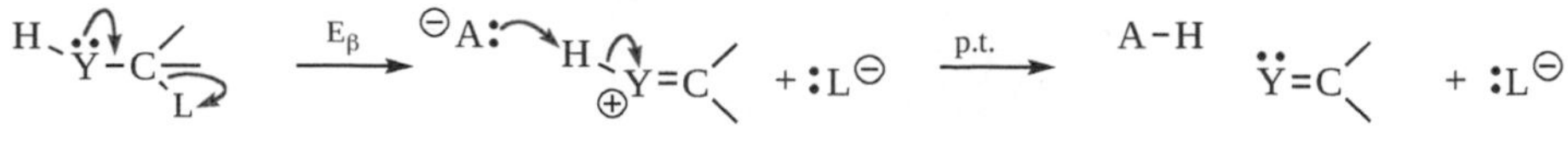

S_E2Ar Electrophilic Aromatic Substitution, $A_E + D_E$ (Section 7.3.4)

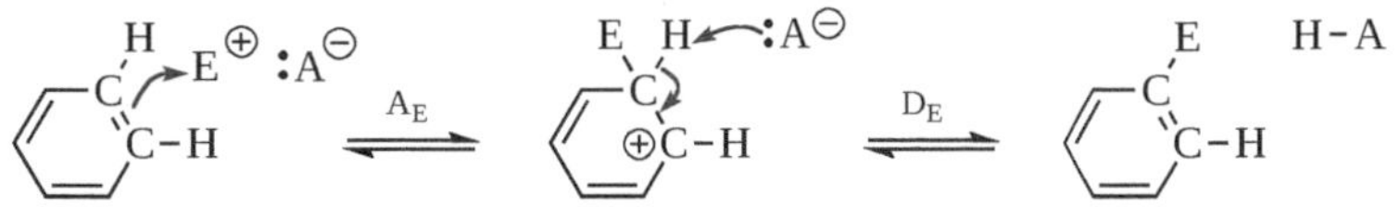

E1cB (Elimination, Unimolecular, Conjugate Base), p.t. + E_β (Section 7.3.5)

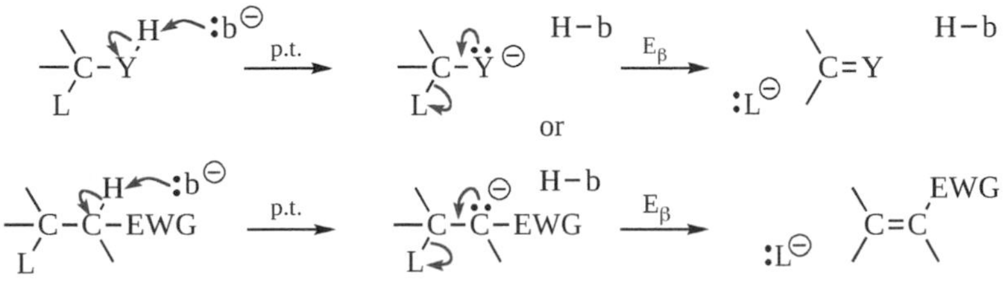

Ad_N2 (Addition, Nucleophilic, Bimolecular), Ad_N + p.t. (Section 7.3.6)

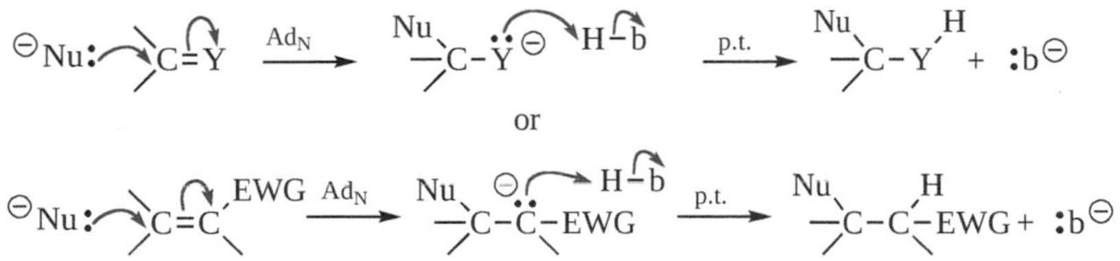

Addition–Elimination, $Ad_N + E_\beta$ (Section 7.3.7)

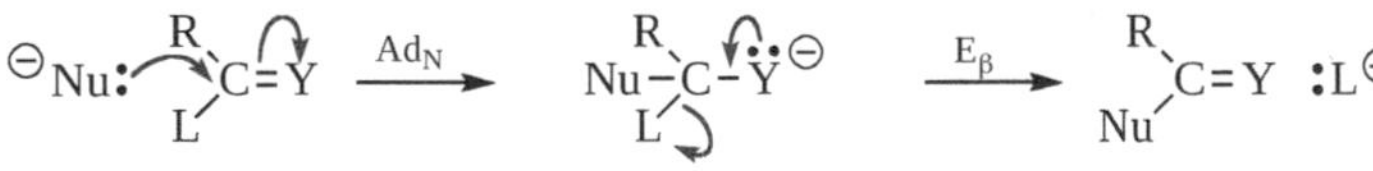

Tautomerization, taut. H–C–C=Z ⇌ C=C–Z–H (Section 7.3.8)

Tautomerization is the shift of an H from a carbon adjacent to a carbon–heteroatom double bond to the heteroatom itself (and the reverse). It is an acid- or base-catalyzed equilibrium. Two examples are the keto/enol pair (Z = oxygen) and the imine/enamine pair (Z = nitrogen). Base catalysis goes via the enolate anion.

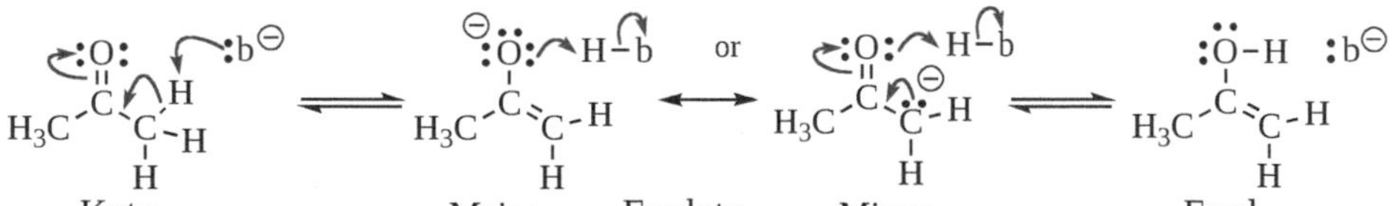

Keto Major Enolate Minor Enol

Acid catalysis goes via the lone-pair-stabilized carbocation.

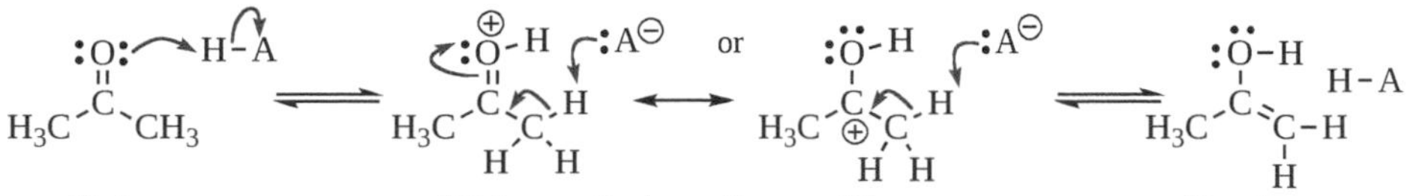

Keto Major Carbocation Minor Enol

Variations on a Theme are Common (Section 7.4)

Atom variations swap out one atom for another, but the electron flow can remain the same. **Vinylogous variations** add an extra arrow but do not significantly alter the electron flow. **Proton-transfer variations** range from hydrogen bonding (partial protonation) to full protonation or deprotonation. General-acid and -base processes are common in enzymes.

Trends Guide

Carbocation Stability Ranking (Table 4.1)

Carbocation Name	Structure	Comments on Stabilization
Stable carbocations		
Guanidinium cation	$(H_2\ddot{N})_3C^{\oplus}$	Three N lone pairs delocalize plus
Tropylium cation		Stable by aromaticity
Protonated amide	$H_3C{-}C^{\oplus}(\ddot{O}H)(\ddot{N}H_2)$	Delocalized by N and O lone pairs
Moderately stable		
Protonated carboxylic acid	$H_3C{-}C^{\oplus}(\ddot{O}H)(\ddot{O}H)$	Delocalized by two O lone pairs
Triphenylmethyl cation	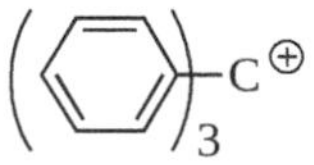	Delocalized by three phenyl groups
Protonated ketone	$H_3C{-}C^{\oplus}(\ddot{O}H)CH_3$	Delocalized by one O lone pair
Diphenylmethyl cation	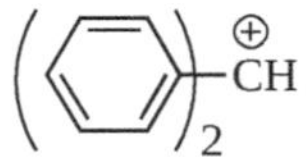	Delocalized by two phenyl groups
Average stability		
Tertiary alkyl cation (3°)	$(CH_3)_3C^{\oplus}$	Stabilized by three alkyl groups
Benzyl cation	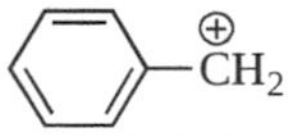	Delocalized by one phenyl group
Primary: allyl cation	$H_2C{=}HC{-}\overset{\oplus}{C}H_2$	Delocalized by one pi bond
Acylium cation	$H_3C{-}\overset{\oplus}{C}{=}\ddot{\underset{\cdot\cdot}{O}}$	Lone-pair-delocalized vinyl cation
Moderately unstable		**No resonance delocalization**
Secondary: alkyl cation (2°)	$(CH_3)_2\overset{\oplus}{C}H$	Stabilized by two alkyl groups
Secondary: vinyl cation	$H_3C{-}\overset{\oplus}{C}{=}CH_2$	Plus on more electronegative *sp* C
Unstable		**(Rarely formed)**
Primary alkyl cation (1°)	$CH_3\overset{\oplus}{C}H_2$	Stabilized by one alkyl group
Primary vinyl cation	$H{-}\overset{\oplus}{C}{=}CH_2$	Plus on more electronegative *sp* C
Phenyl cation	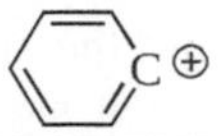	Bent vinyl cation and not delocalized
Methyl cation	$^{\oplus}CH_3$	No stabilization

Carbocations stabilized by resonance with a lone pair are more stable than those stabilized only by resonance with a carbon–carbon double bond, which in turn are more stable than those stabilized only by alkyl group substitution. The effects are additive; three of the lesser type of stabilization are as good or better than one of the better type.

Electron-Releasing Group (ERG) Ranking (Table 4.2)

Excellent donors	**Features of group**
$-CH_2^{\ominus}$	Anionic and least electronegative
$-NH^{\ominus}$	Anionic and more electronegative than C
$-O^{\ominus}$	Anionic and more electronegative than N
Good donors	
$-N(CH_3)_2$	Neutral and less electronegative than O
$-NH_2$	Neutral and less electronegative than O
$-OH$	Neutral and more electronegative than N
$-OCH_3$	Neutral and more electronegative than N
$-NH-C(=O)-CH_3$	Amide carbonyl EWG decreases N lone pair availability
$-SCH_3$	Sulfur 3*p* orbital has poor overlap with a 2*p* carbon orbital
Poor donors	
—Ph	Delocalization through resonance with phenyl
—R	Hyperconjugation only, no lone pair or pi resonance
—H	No substituent at all
Very poor donors	
$-O-C(=O)-CH_3$	Electronegative oxygen, carbonyl EWG decreases O lone pair availability
$-Cl$	Electronegative and poor orbital overlap (a 3*p* with a 2*p*)

Allylic Electron Source Reactivity Trends
The better the donor is, the better the allylic electron source is.

Alkene Stability Trends
The more substituted, the more stable.

Carbanion Stability Trends
The most stable carbanion will be the weakest base and have the lowest pK_{abH}.

Electron-Withdrawing Group Trends

CH_3–EWG has a low pK_a for a good EWG. The acidity of CH_3–EWG increases as groups better at withdrawing electrons more effectively stabilize $^{-}CH_2$–EWG, the anionic conjugate base. If the pK_as of CH_3–EWGs are compared, the more acidic (lower pK_a) the CH_3–EWG is, the better the ewg group is at withdrawing electrons. The following is a list of electron-withdrawing groups ranked by the pK_a of CH_3–EWG.

Better EWG

CH_3–EWG pK_a

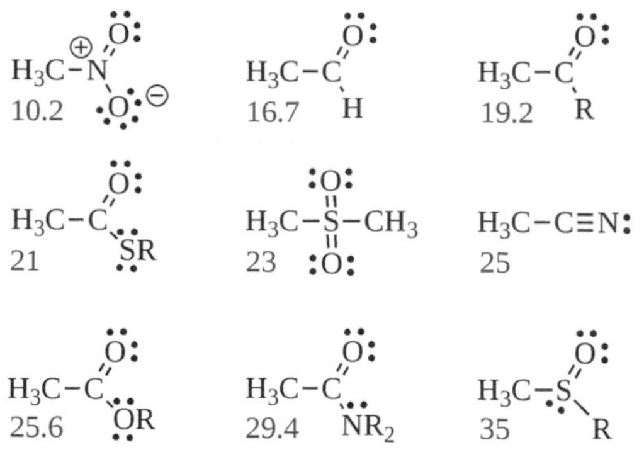

Poorer EWG

Leaving Group Trends

A good leaving group, L, has a negative pK_{aHL}. When using a pK_a chart to rank leaving groups, always look up the conjugate acid of the leaving group on the chart. Note that groups with a pK_{aHL} above 30, are not leaving groups. Both F^- and CN^- are worse than expected.

Excellent	pK_{aHL}
N_2	<−10
$CF_3SO_3^-$	−15
$ArSO_3^-$	−6.5
$CH_3SO_3^-$	−6

Good	pK_{aHL}
I^-	−10
Br^-	−9
Cl^-	−6
EtOH	−2.4
H_2O	0.0
CF_3COO^-	+0.5

Fair	pK_{aHL}
$O_2NC_6H_4COO^-$	+3.4
$RCOO^-$	+4.8
$O_2NC_6H_4O^-$	+7.2
NH_3	+9.2
RS^-	+10.6
NR_3	+10.7

Poor	pK_{aHL}
HO^-	+14.0
EtO^-	+16
$RCOCH_2^-$	+19
$ROCOCH_2^-$	+26

Aromatic Ring Reactivity Trends

The better the donor, the more reactive the ring is to electrophilic attack. Donors direct electrophiles *ortho–para*. The better the electron-withdrawing group, the less reactive the ring is to electrophilic attack. Electron-withdrawing groups direct electrophiles *meta*.

Carboxyl Derivative Trends

Electron releasing groups deactivate carboxyl derivatives toward nucleophilic attack.

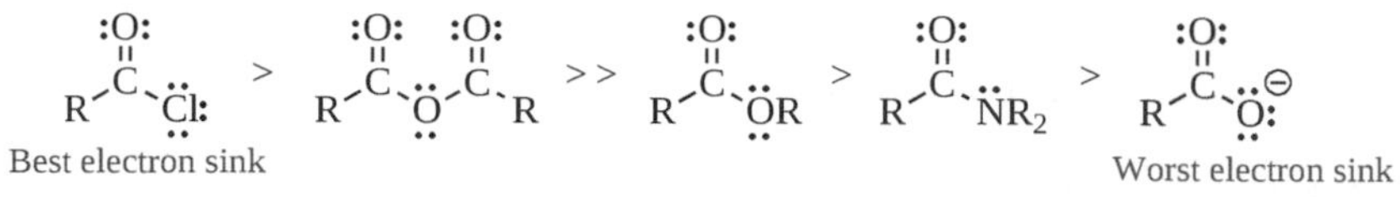

Lone Pair Nucleophilicity Trends

A pK_a chart can be used as a reference for nucleophilicity only if the difference in softness is considered. A partially plus carbon atom is a much softer electrophile than a proton. Soft ions are more nucleophilic in protic solvents because tighter solvation greatly decreases the nucleophilicity of the hard ions. Also, steric hindrance decreases nucleophilicity. To rank lone pair nucleophilicity in protic solvents with soft electrophiles such as R–X, first rank by softness, then by basicity (within the same attacking atom). However, for lone pair nucleophiles reacting with harder electrophiles like a proton or carbonyl, rank by basicity. Very reactive electrophiles like carbocations are not selective and react with the most abundant nucleophile (commonly the solvent).

Relative nucleophilicity toward CH_3I	
CH_3OH (solvent)	1
F^-	500
CH_3COO^-	20,000
Cl^-	23,000
Et_2S	220,000
NH_3	320,000
PhO^-	560,000
Br^-	620,000
CH_3O^-	1,900,000
Et_3N	4,600,000
CN^-	5,000,000
I^-	26,000,000
Et_3P	520,000,000
PhS^-	8,300,000,000
$PhSe^-$	50,000,000,000

Organometallic Reactivity Trends

The greater the electronegativity difference between R and M, the more reactive.

Electronegativity difference	**Most reactive**
1.62	R–Na
1.57	R–Li
1.24	R–MgX
0.90	R_2Zn
0.86	R_2Cd
0.65	$R_2Cu^-Li^+$
0.55	R_2Hg
0.22	R_4Pb
	Least reactive

Metal Hydride Reactivity Trends

The electronegativity difference also governs the reactivity of the metal hydrides. The metal hydride will become less reactive as the metal becomes more electronegative by attaching an inductive withdrawing group. As the hydride source becomes less reactive, it also becomes more selective. The complex metal hydride reactivity trend is:

$$\mathbf{AlH_4^- > HAl(OR)_3^- > BH_4^- > H_3BCN^-}.$$

Conjugate Acceptor Reactivity Trends

Trading C–EWG for Y is simply exchanging an electronegative carbon atom for an electronegative heteroatom. The better the electron-withdrawing group is, the better the electron sink will be. More than one good electron-withdrawing group on a double bond, $C{=}C(EWG)_2$, gives a very reactive electron sink because the resulting anion from the conjugate addition is stabilized to a much greater extent.

MAJOR ROUTES SUMMARY

Substitution Summary

• S_N2 must pass three tests: good access (primary or secondary), a leaving group with a pK_{aHL} of 12 or less, and a decent nucleophile.

• S_N2 does not occur on leaving groups attached to a trigonal planar center or on a highly hindered site like a tertiary or neopentyl center.

• Substitution version of the ΔpK_a rule cross-check: never kick out a leaving group more than 10 pK_a units more basic than the incoming nucleophile.

• S_N1 must pass three tests: forms a carbocation that is usually more stable than simple secondary, a polar solvent, and a leaving group with a pK_{aHL} of zero or less.

• The pentacovalent path is for third-row or higher elements and must pass a ΔpK_a rule cross-check.

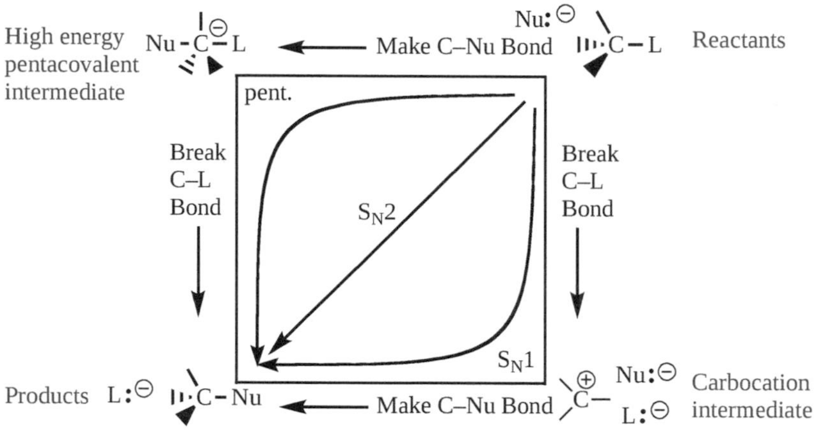

Figure 4.2 Top view of a simplified energy surface for substitution at a tetrahedral center.

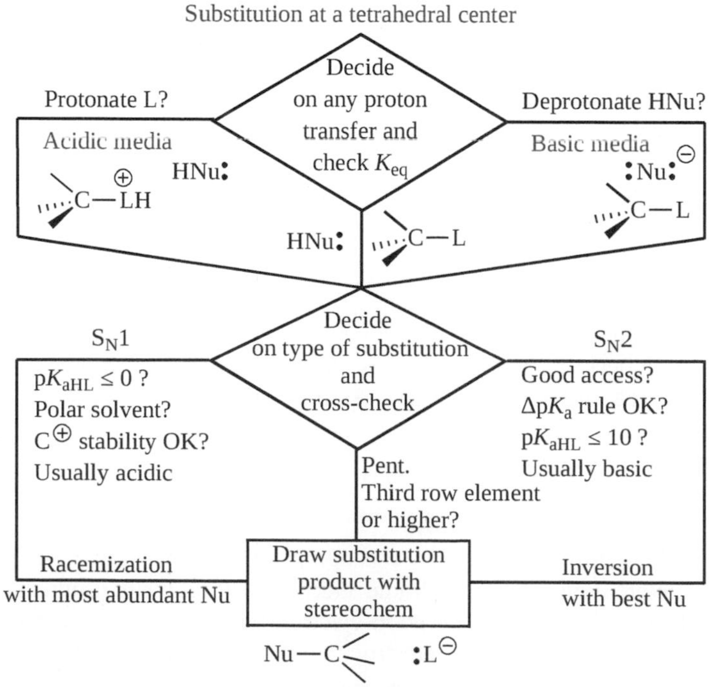

Figure 4.15 The flowchart for substitution at a tetrahedral center.

Elimination Summary

- Elimination version of the ΔpK_a rule cross-check: never kick out a leaving group more than 10 pK_a units more basic than the incoming base.
- E1 must pass three tests: forms a carbocation with a stability usually better than simple secondary, a polar solvent, and a leaving group with a pK_{aHL} of zero or less.
- E2 occurs when the carbanion and carbocation are not very stable. The energy surface folds down the middle. The C–H and C–L eliminated must be close to coplanar.
- E1cB requires an acidic hydrogen and a good base. Check that the proton-transfer K_{eq} is at least 10^{-10}. E1cB tolerates poorer leaving groups, but check the ΔpK_a rule cross-check.

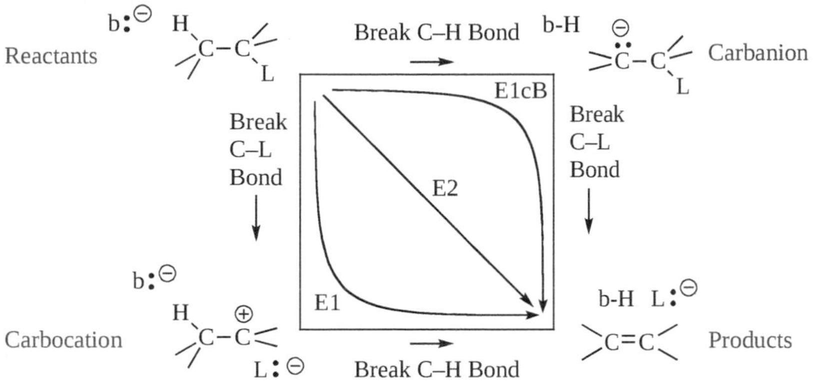

Figure 4.16 Top view of the simplified energy surface for elimination.

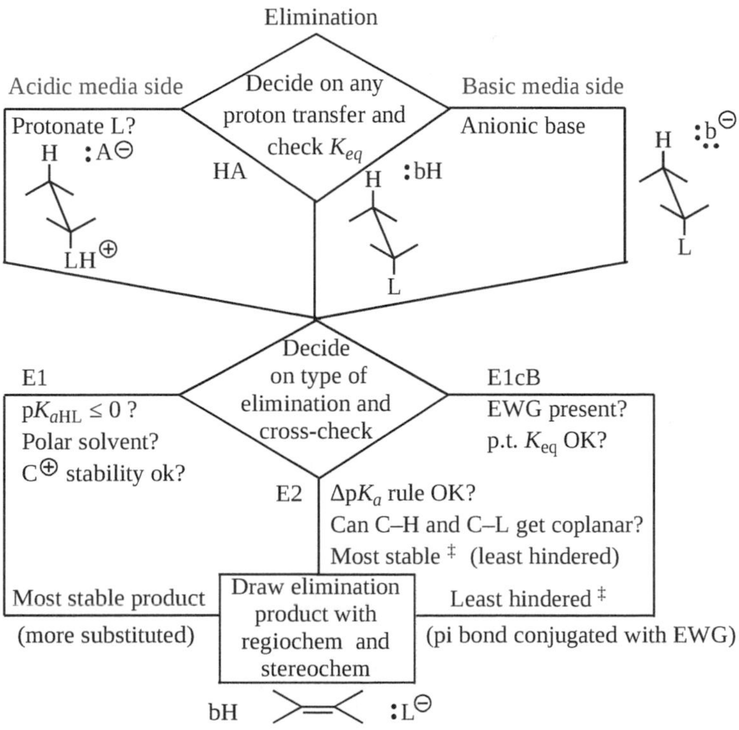

Figure 4.30 The flowchart for elimination.

Addition Summary

• Ad_E2 must form a reasonably stable carbocation usually better than secondary, and tends to produce a mix of *syn* and *anti* addition. Markovnikov's rule is followed: The electrophile adds to form the most stable carbocation.

• Ad_E3 occurs when the carbanion and carbocation are not very stable: The energy surface folds down the middle. The electrophile and nucleophile end up on opposite faces of the pi bond (*anti* addition). Markovnikov's rule is followed.

• Ad_N2 follows the ΔpK_a rule cross-check: Never form an anion more than 10 pK_a units more basic than the incoming nucleophile. For this to happen, the anion formed must usually be on an electronegative atom or stabilized by an electron-withdrawing group.

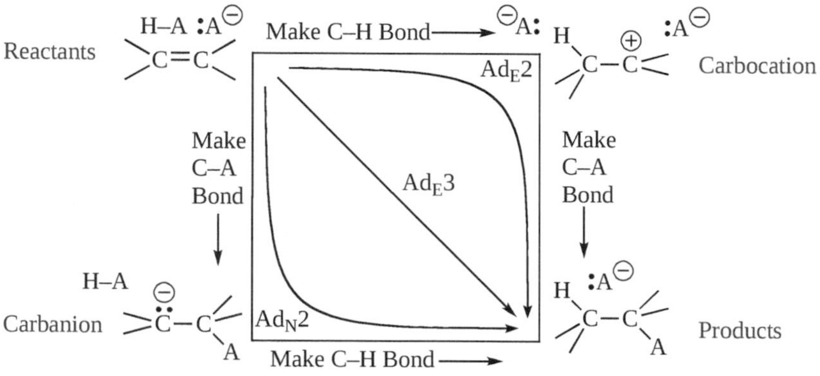

Figure 4.32 Top view of the simplified energy surface for addition to a carbon–carbon pi bond.

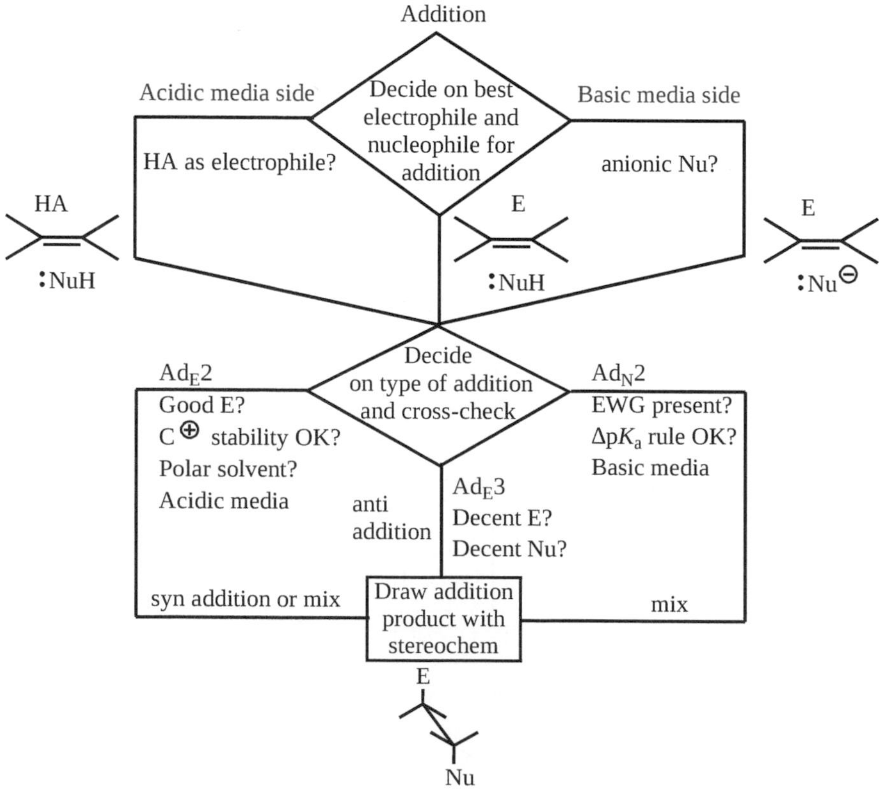

Figure 4.40 The general flowchart for addition. A proton-transfer step may follow.

Addition–Elimination of Carboxylic Acid Derivatives Summary

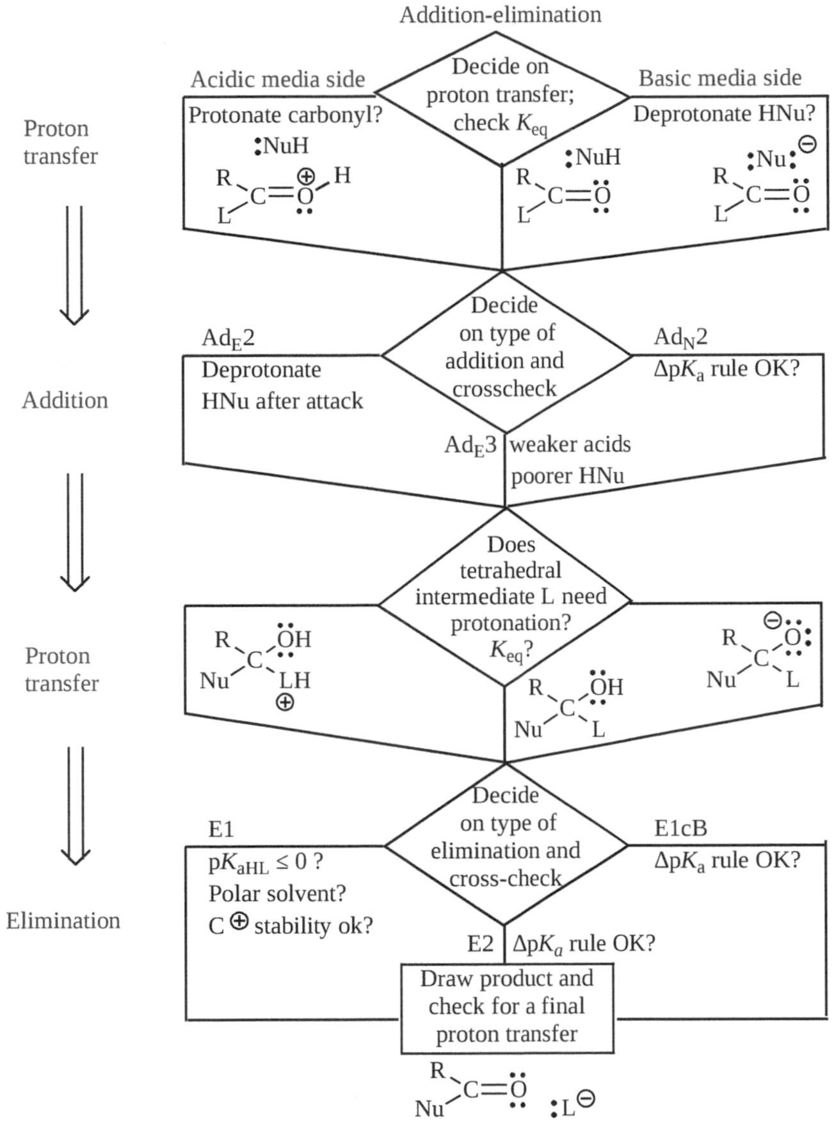

Figure 4.45 The flowchart for addition–elimination on carboxylic acid derivatives.

- Ad_E2 usually occurs in acid because it forms a very acidic protonated carbonyl.
- Ad_E3 occurs in buffered media, where the medium is not acidic enough to protonate the carbonyl and not basic enough to support strong nucleophiles.
- Ad_N2 occurs in base with good nucleophiles and follows the ΔpK_a rule cross-check — never form an anion more than 10 pK_a units more basic than the incoming nucleophile.
- Elimination follows the ΔpK_a rule cross-check: never kick out a leaving group more than 10 pK_a units more basic than the tetrahedral intermediate.
- E1 tends to occur in acid and must pass three tests: a reasonably stable carbocation (usually better than secondary), a polar solvent, and a leaving group pK_{aHL} of zero or less.
- E2 usually occurs buffered media when the energy surface folds down the middle.
- E1cB usually requires basic media and an acidic hydrogen. Check that the proton-transfer K_{eq} is more than 10^{-10}. It tolerates poorer leaving groups, but check the ΔpK_a rule.

Major Decisions Guide

Substitution vs. Elimination Decision Matrices.

The substitution/elimination decision becomes a function of three major variables: nucleophilicity, basicity, and steric hindrance. In acidic media, the S_N1 competes with the E1 process; the first step of both is to lose the leaving group to form the carbocation. There normally is competition in basic media between the S_N2 and E2.

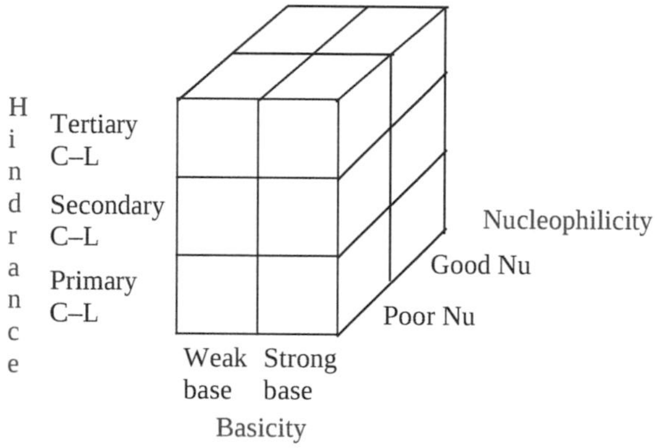

Figure 9.5 A three-dimensional correlation matrix for substitution vs. elimination.

Substitution vs. Elimination Decision

Example Reagents

	Weak base	Strong base
Good nucleophile	I⊖, RS⊖	HO⊖, EtO⊖
Poor nucleophile	H_2O, EtOH	$(CH_3)_3CO^⊖$, $R_2N^⊖$

Unhindered Primary

	Weak base	Strong base
Good nucleophile	Subst. by S_N2	Subst. by fast S_N2
Poor nucleophile	No rxn. or very slow S_N2	Mixture (depends on L)

Secondary

	Weak base	Strong base
Good nucleophile	Subst. by S_N2	Mixture (depends on temp.)
Poor nucleophile	No rxn. or slow S_N2 or S_N1	Elimin. by fast E2

Tertiary

	Weak base	Strong base
Good nucleophile	Mixture (depends on basic.)	Elimin. by fast E2 or E1
Poor nucleophile	Subst. by S_N1 (polar solvent)	Elimin. by fast E2 or E1

Figure 9.6 Substitution/elimination correlation matrix layers from Figure 9.5 sorted by hindrance. The matrix in the upper left gives some common examples of each type of reagents.

The unhindered primary layer has the strong base, poor nucleophile box as its mixture quadrant with all others as substitution. Notice how the mixture quadrant in the primary matrix moves around to the strong base, good nucleophile quadrant in the secondary matrix, then again to the weak base, good nucleophile quadrant in the tertiary matrix, each time leaving behind an elimination quadrant.

Multiple Addition Decision to Carboxylic Acid Derivatives

There are four ways that the reaction may stop after the initial $Ad_N + E_\beta$

1. **Unfavorable Equilibrium:** The Nu is such a weak base that the second addition easily reverses and the equilibrium favors the more stable carbonyl. Heteroatom lone pair sources almost never add twice.

2. **Selective Nucleophile:** The initial electron sink must be more reactive than the product of the addition, and the electron source, although reactive enough to add once, is not reactive enough to make a second addition to the less reactive compound. Organometallics in the reactivity range of organocadmiums, organozincs, and organocoppers (metal electronegativities are 1.69, 1.65, 1.90, respectively) add quickly to the more reactive acyl chlorides but slowly to the ketone products.

3. **Poor Leaving Group:** The leaving group is so poor (in amides, pK_{aHL} = 36, for example) that it cannot be ejected easily from the tetrahedral intermediate. Since the carbonyl does not reform, there is no site for nucleophilic attack. The carbonyl forms in the acidic workup (no Nu left then) by protonating the nitrogen, making it a better L.

4. **Low Temperature:** One equivalent or less of the electron source was added to the sink at a low enough temperature (usually −78°C) so that the loss of the leaving group from the tetrahedral intermediate by path E_β does not occur to a significant extent before the electron source is used up.

Ambident Sink: 1, 2 vs. 1, 4 Addition Decision to Conjugate Acceptors

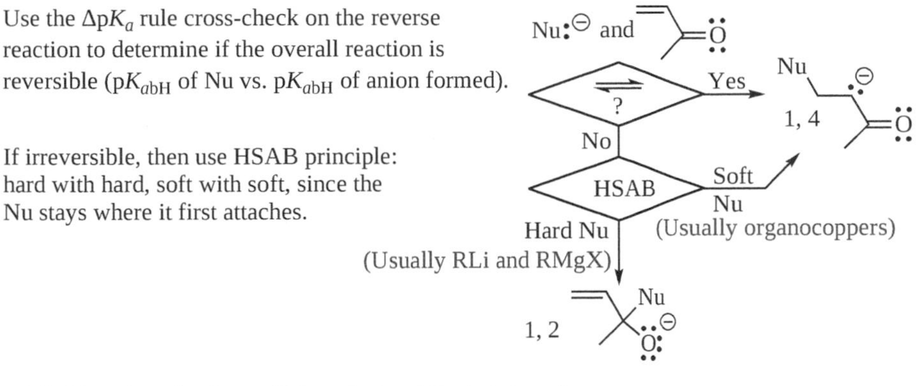

Figure 9.7 Flowchart for the ambident electrophile decision of an enone.

Ambident Source: C vs. O Attack Decision for Allylic Sources

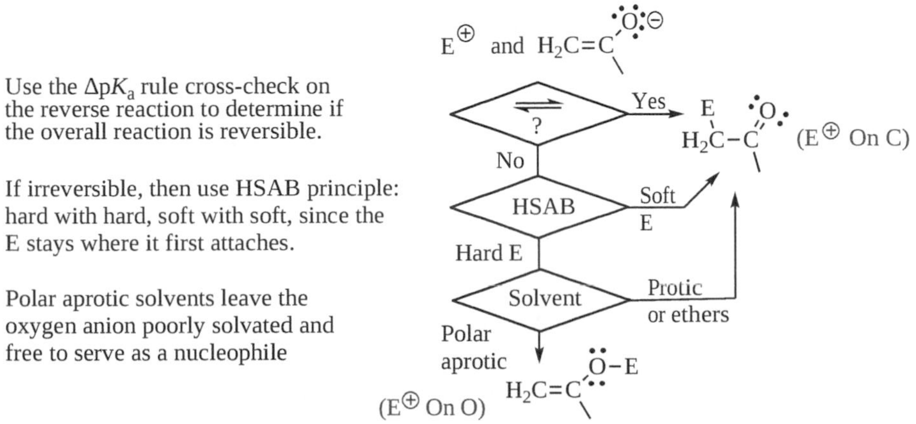

Figure 9.1 Flowchart for the ambident nucleophile decision of an enolate.

Thermodynamics and Kinetics

Free Energy, $\Delta G°$, and K_{eq}

$$\Delta G° = -RT \ln K_{eq}$$

$\Delta G°$ is the free energy difference between the reactants and the products at standard conditions. R is 1.987×10^{-3} kcal/mol-K (8.314×10^{-3} kJ/mol-K), and T is temperature in K. At room temperature, every 1.364 kcal/mol (5.706 kJ/mol) changes the equilibrium constant by a factor of ten.

Table 2.1 $\Delta G°$ and K_{eq} Values for 25.0°C (room temperature)

$\Delta G°$ kcal/mol	K_{eq}	Reactant	Product	$\Delta G°$ (kJ/mol)
+5.45	10^{-4}	99.99	0.01	+22.8
+4.09	10^{-3}	99.9	0.1	+17.1
+2.72	10^{-2}	99	1	+11.4
+1.37	0.10	91	9	+5.73
+1.03	0.18	85	15	+4.30
+0.50	0.43	70	30	+2.10
0.00	1	50	50	0.00
−0.50	2.3	30	70	−2.10
−1.03	5.7	15	85	−4.30
−1.37	10	9	91	−5.73
−2.72	10^{2}	1	99	−11.4
−4.09	10^{3}	0.1	99.9	−17.1
−5.45	10^{4}	0.01	99.99	−22.8
−9.55	10^{7}	Essentially complete		−39.9

The Relationship of Free Energy to Enthalpy and Entropy

$$\Delta G° = \Delta H° - T\Delta S°$$

Enthalpy, Heat of Reaction, $\Delta H°$

$$\Delta H° = \Delta H(\text{bonds broken}) - \Delta H(\text{bonds made})$$

Kinetics, $\Delta G^‡$

The $\Delta G^‡$ of a reaction is the height of the free energy barrier to reaction measured from the reactants. The $\Delta G^‡$ of a reaction that proceeds at a reasonable rate at room temperature is 20 kcal/mol (84 kJ/mol). The diffusion-controlled limit of 10^{10} liters/mol-s corresponds to a reaction upon every collision. At room temperature, dropping the $\Delta G^‡$ by 1.364 kcal/mol (5.706 kJ/mol) increases the rate tenfold. The more reactive a species is, the less selective it is. The more stable a compound is, the less reactive it is.

Generation of Alternate Paths, Reaction Cubes

One of the more difficult tasks in working mechanism or product prediction problems is deciding whether the set of alternate reaction paths is complete or not. Keeping track of those paths and putting them in some logical order is also very important. Quite often reactants have four charge types, but rarely do more than three occur in the same medium; one is usually an acidic medium charge type, one a basic medium type, and two can often occur in either medium. Products likewise can have four charge types, again with the same media preferences. Frequently, these charge types are interrelated by proton transfer (Figure A.1).

It would be nice to have a graph theory way of describing all the possible interactions of each of these four reactant charge types with the four product charge types so that a more complete set of reaction alternatives could be generated. A data structure for describing the interrelation of eight variables—these charge types of both reactants and products—is a cube (Figure A.2). Each charge type can then be related to any of the other seven by an edge, face diagonal, or body diagonal. The horizontal axis is the reaction coordinate for bond formation between the nucleophile and the electrophile. A side of the cube that includes this axis forms an energy surface relating two charge types of reactants with two charge types of products. These two dimensional energy surfaces have been used throughout this text.

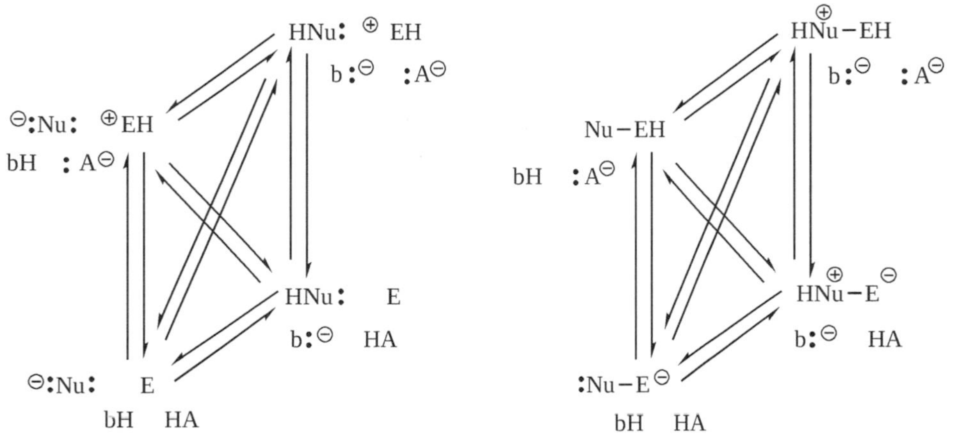

Figure A.1 Conversions of charge types of reactants (left) and products (right) by proton transfer.

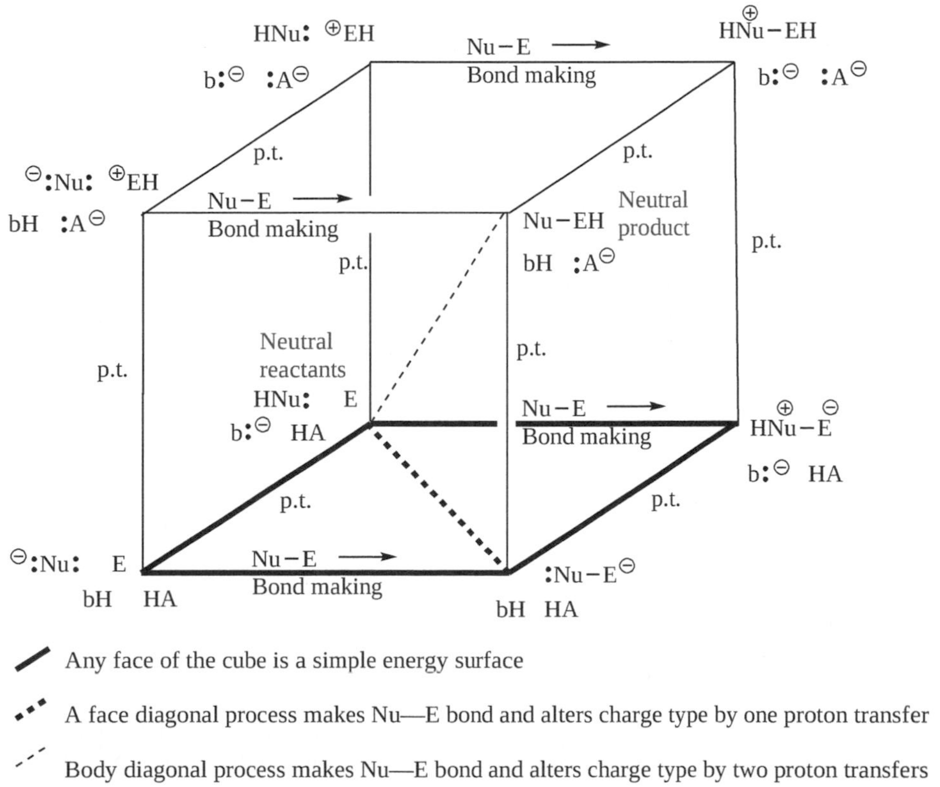

Figure A.2 A generic reaction cube illustrating the interaction of various charge types of reactants and products. p.t. = proton transfer.

Figure A.3 shows the usefulness of the reaction cube as a data structure. Additions to carbonyls often occur between different charge types, and frequently three-dimensional energy surfaces are used to clarify the various equilibria. We have seen two faces of this cube before as individual energy surfaces. The bottom face of the cube is Figure 7.16, polarized multiple bond addition/elimination mechanisms in basic media. The back face of the cube is Figure 7.17, polarized multiple bond addition/elimination mechanisms in acidic media.

One or more corners of this cube can be of higher energy because of a wide pK_a span (corner 3). When we add the dimension of energy to the reaction cube, in essence, it becomes a four-dimensional energy surface. Each corner can be ranked according to its probability of occurrence under the particular reaction conditions, and the most probable lowest-energy route would then be expected to avoid those high-energy corners. Another way to view a reaction cube is that each face is the planar projection of a three-dimensional energy surface. Conditions that raise the energy of a particular corner in the three-dimensional energy surface likewise affect that corner of the reaction cube.

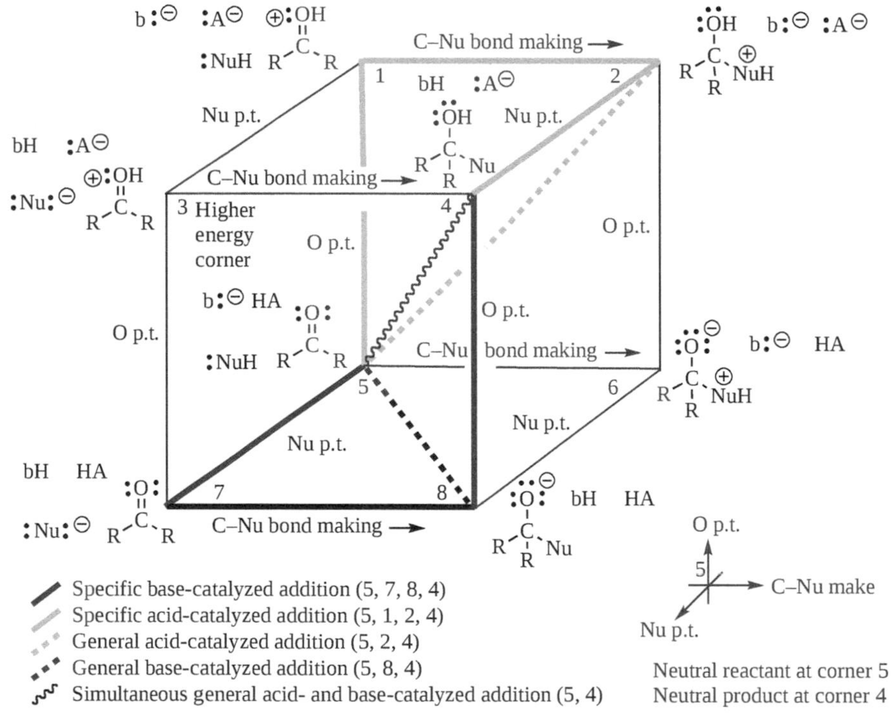

Figure A.3 The reaction cube for the addition of a nucleophile to a carbonyl in protic media (for hemiketal/hemiacetal, NuH = ROH) where O p.t. is proton transfer to/from oxygen and Nu p.t. is proton transfer to/from the nucleophile.

There is a need for interrelating the various mechanisms in such a way that all alternatives are easily visible, so that none are left unexplored. With 12 edges, 12 face diagonals, and 4 body diagonals, there are 28 possible interactions that require checking. Specific acid or base processes lie along the edges of the cube. The general acid-catalyzed, Ad_E3, or general base-catalyzed processes constitute the face diagonals. The body diagonal is the simultaneous general acid, general base catalysis (push–pull catalysis). Reactions proceeding through an intermediate can be represented by cubes sharing a common face, edge, or vertex corresponding to the intermediate charge type(s).

NOTES ON NOMENCLATURE

This section is meant to be a brief compilation of the important aspects of organic nomenclature and is by no means complete. The most important organic nomenclature to know is the names of functional groups, since they are sites of reactivity. Since structure determines reactivity, knowing (for instance) that the structure contains a ketone, and therefore is in the polarized multiple bond class, is more important than being able to name all the grease hanging off a complex structure. The most common functional groups are in Table 1.3; a much larger set is in the functional group glossary in this appendix.

Common Names

Some frequently used compounds do not have a systematic name or have an infrequently used systematic name. Many compounds were named before their structure was known, so the common name has little or no structural information. Some important examples are shown below.

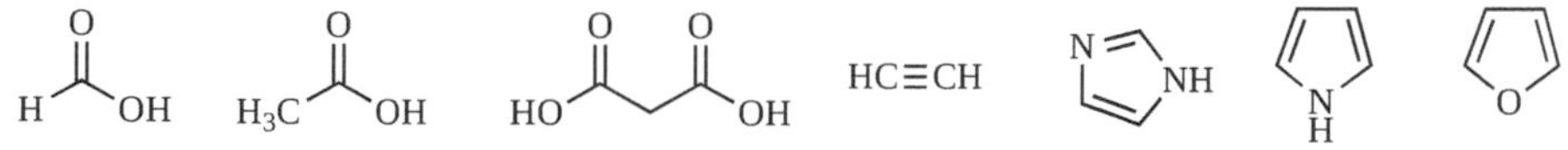

formic acid   acetic acid   malonic acid   acetylene   imidazole   pyrrole   furan

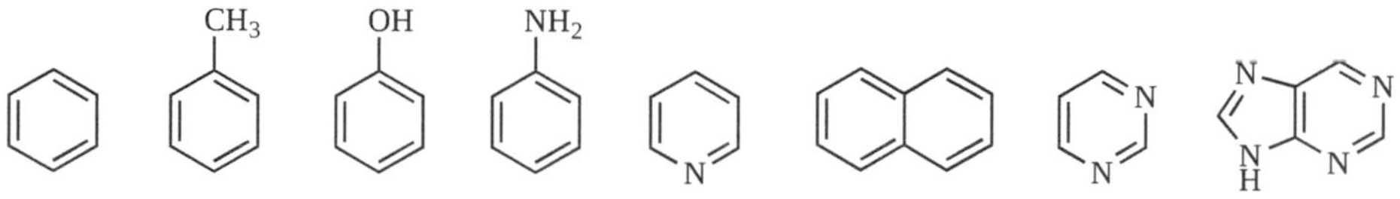

benzene   toluene   phenol   aniline   pyridine   naphthalene   pyrimidine   purine

Group Names

Sometimes a group attached to a compound has a special name that has to be learned separately; listed below are the most common groups.

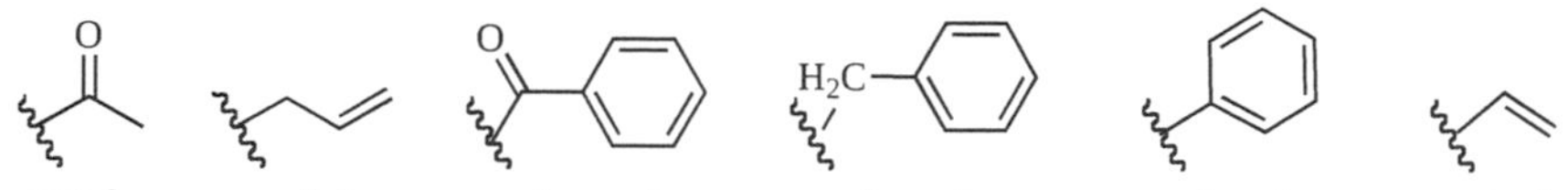

acetyl   allyl   benzoyl   benzyl   phenyl   vinyl

Oxidation State of Carbon

To determine oxidation state, we assign the shared bonding electrons to the more electronegative atom. For organic molecules this means: Carbon-carbon bonds don't count; each carbon to hydrogen bond gives the carbon a −1, and each carbon to heteroatom bond gives the carbon a +1, so the total for any carbon atom can vary from −4 in methane to +4 in carbon dioxide. Therefore, in comparing two organic compounds, the more hydrogen atoms bonded to a carbon atom, the more reduced it is. The more heteroatoms bonded to the carbon, the more oxidized it is. Usually, we just want to know if a redox reaction has occurred and not the actual oxidation states, because oxidation and reduction use special redox reagents. Common reducing agents are complex metal hydrides, MH_4^-, as shown below. Common oxidizing agents vary, but can be high oxidation state transition metals like Cr(VI) as below.

Oxidation example: 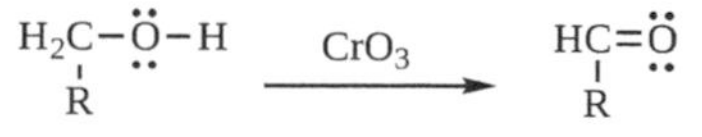 (carbon loses hydrogen)

Reduction example:  (carbon gains hydrogen)

Prefix-Root-Suffix Naming Convention

This is section is just the basics of naming to allow us to understand simple structural names. The longest chain that includes the primary functional group determines the length of the root chain. Knowing the names of the first ten alkanes, shown below, is usually sufficient to name the root for small molecules.

List of the First Ten Alkanes

# of C	Alkane name	Group name	Alkane
1	Methane	Methyl	CH_4
2	Ethane	Ethyl	CH_3CH_3
3	Propane	Propyl	$CH_3CH_2CH_3$
4	Butane	Butyl	$CH_3CH_2CH_2CH_3$
5	Pentane	Pentyl	$CH_3CH_2CH_2CH_2CH_3$
6	Hexane	Hexyl	$CH_3CH_2CH_2CH_2CH_2CH_3$
7	Heptane	Heptyl	$CH_3CH_2CH_2CH_2CH_2CH_2CH_3$
8	Octane	Octyl	$CH_3CH_2CH_2CH_2CH_2CH_2CH_2CH_3$
9	Nonane	Nonyl	$CH_3CH_2CH_2CH_2CH_2CH_2CH_2CH_2CH_3$
10	Decane	Decyl	$CH_3CH_2CH_2CH_2CH_2CH_2CH_2CH_2CH_2CH_3$

The naming precedence for the functional groups is based on the most oxidized, as shown in the table below. The prefix identifies the position and groups attached to the root chain. The suffix identifies the functional group. Numbering is from the end of the root chain that gives the functional group the lowest number.

Precedence of Functional Groups and Examples

	Functional Group	Suffix	Example	Example Name
1	Carboxylic acid	-oic acid	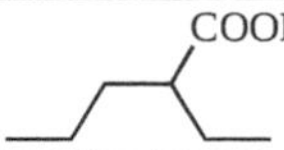	2-Ethylpentanoic acid
2	Esters	-oate	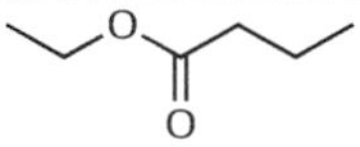	Ethyl butanoate
3	Amides	-amide	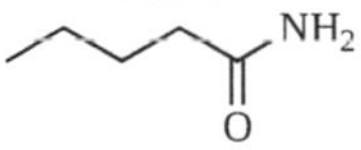	Pentanamide
4	Nitriles	-nitrile		Hexanenitrile
5	Aldehydes	-al	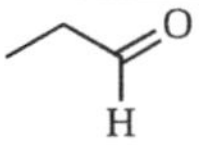	Propanal
6	Ketones	-one	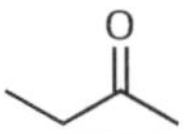	Butan-2-one
7	Alcohols	-ol	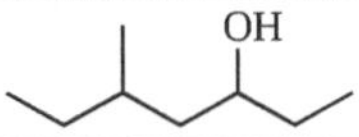	5-Methylheptan-3-ol
8	Amines	-amine		2-Methylpropan-2-amine
9	Halogen	As prefix	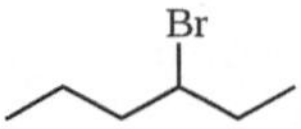	3-Bromohexane

Compounds with two functional groups are named using the higher-precedence group for determining the root numbering and for the group suffix.

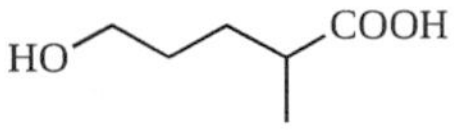

5-hydroxy-2-methylpentanoic acid

Primary, 1°, Secondary, 2°, Tertiary, 3°, and Quaternary, 4°, Notation

For carbocations, the 1°, 2°, 3° designation is based on how many R groups (not counting hydrogen atoms) are attached to the positive carbon atom, as shown below. For alcohols, it is the number of R groups attached to the carbon bearing the OH group. For amines, it is the number of R groups attached to the nitrogen atom.

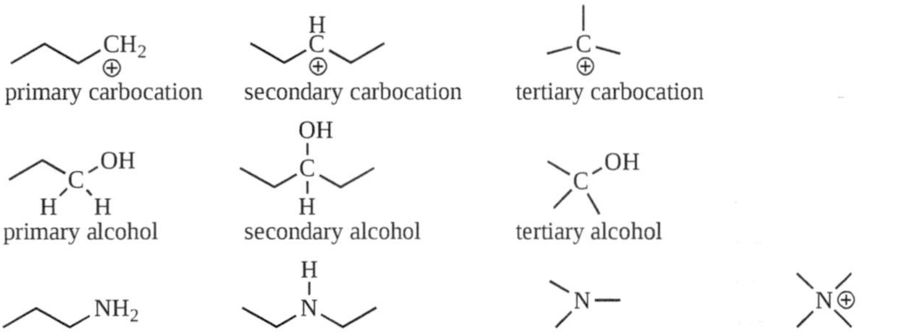

primary carbocation, secondary carbocation, tertiary carbocation

primary alcohol, secondary alcohol, tertiary alcohol

primary amine, secondary amine, tertiary amine, quaternary amine

Projections for Stereochemistry

We also need to be able to describe stereochemistry, the different arrangements of atoms in space, since reaction mechanisms often have stereochemical consequences. We need to have a unambiguous convention for drawing a three-dimensional molecule flat on a page. Shown below are several projections of the same molecule, D-glyceraldehyde. A wedged bond comes out of the page and a dashed bond goes into the page. Molecular models help in visualizing these projections as three-dimensional.

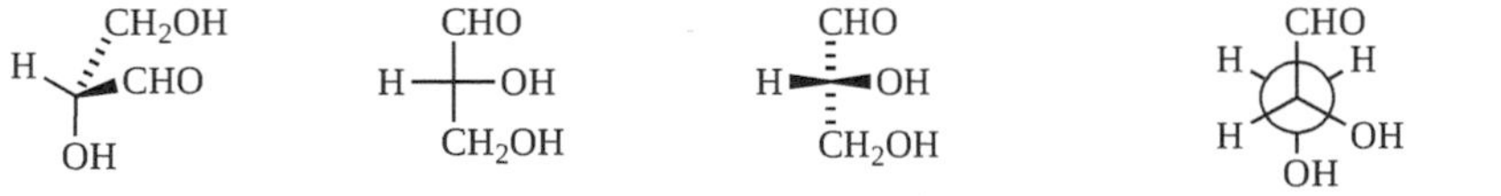

Dash is back and wedge is out; Fischer projection; Translation of Fischer into dash and wedge; Newman projection down the C2–C3 bond

***Cis* and *Trans* on Alkenes**

Double bonds with two attached groups can be designated as *trans*, *cis*, or *geminal*, shown below. Double bonds are named by the lower number of the two double bond carbons; so if it is called a 2-alkene, then the double bond is between carbons 2 and 3.

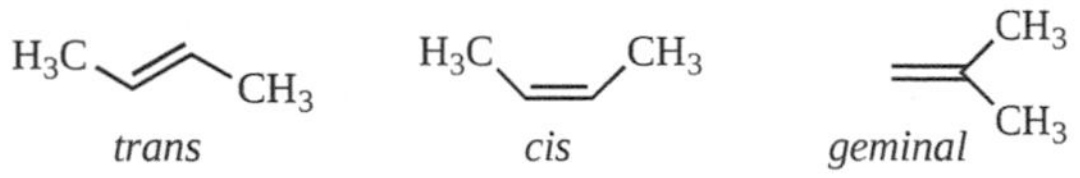

trans-2-butene; *cis*-2-butene; 2-methyl-1-propene

***Cis* and *Trans* on Cycloalkanes**

Cis and *trans* can be used to describe the relationships between groups on a cycloalkane. Groups on the same side of a ring (viewed from the top) are also called *cis*. If the one group is below the ring and the other above they are called *trans* (see below).

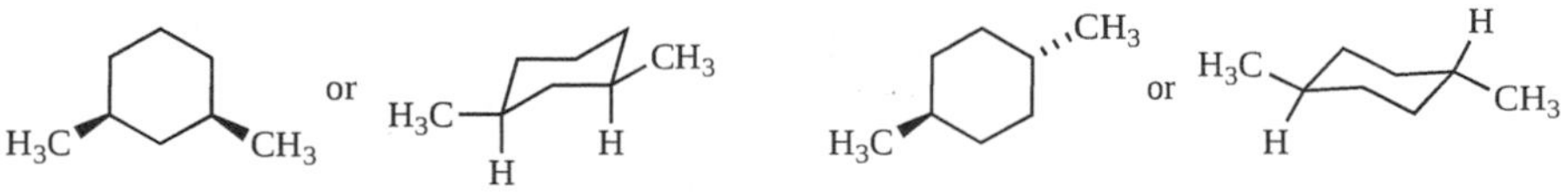

cis-1,3-dimethylcyclohexane; *trans*-1,4--dimethylcyclohexane

Cahn-Ingold-Prelog Priority Assignment

For the next set of stereochemical labels, we need to be able to assign priorities to groups. The Cahn-Ingold-Prelog sequence rules give highest atomic number the top priority. For example, in assigning priorities to the groups on the carbon of bromofluoroiodomethane, CHBrFI, top priority goes to iodine, second to bromine, third to fluorine, and lowest to hydrogen. When there is a tie, the atoms next out from the tying atoms are considered; the highest-priority atoms directly bonded to the tied atoms are compared until a difference is found. See examples below. Double bonds are counted as if they were two single bonds. Let us assign the stereochemistry of the simplest naturally occurring sugar, glyceraldehyde.

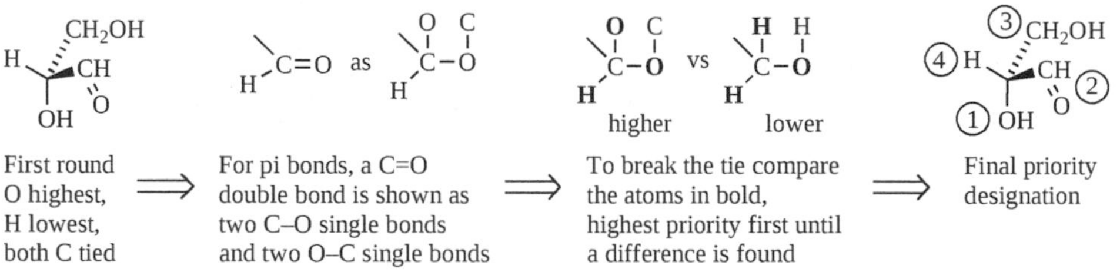

R and S Absolute Configuration

R and *S* are used to designate the absolute configuration about a chiral center, an atom with four different groups. With the lowest-priority group facing away to the back, if the remaining priorities go in a clockwise manner it is *R*; if they go counterclockwise it is *S*. See example below. Molecular models help in assigning *R* and *S*.

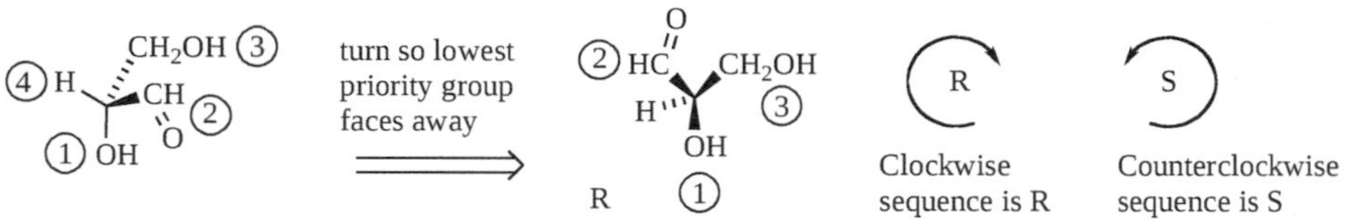

E and Z Nomenclature of Alkenes

Tri- and tetrasubstituted alkenes need their own nomenclature, not cis or trans. If the two highest-priority groups on each end of the double bond are adjacent to each other, then the stereochemistry is called *Z*; if they are across the pi bond, then the stereochemistry is called *E*.

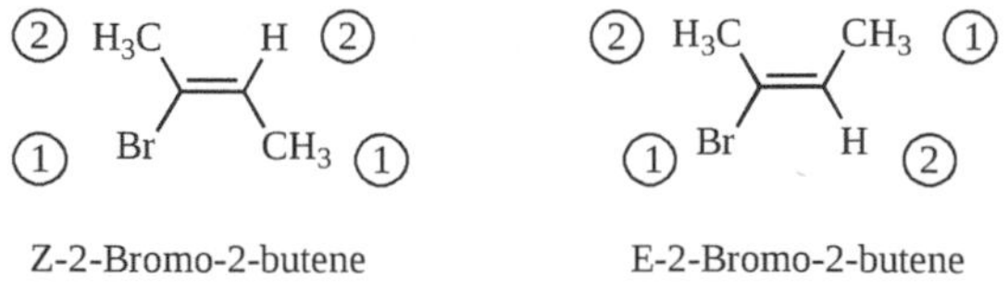

Z-2-Bromo-2-butene E-2-Bromo-2-butene

Multiple Chiral Centers

If a compound has two different chiral centers, then there are four possible combinations, *RR*, *RS*, *SR*, and *SS*. The *RR* and *SS* pair are mirror images (enantiomers), and likewise so are the *RS* and *SR* pair. The *SS* and *SR* compounds are stereoisomers but not mirror images, and are called diastereomers.

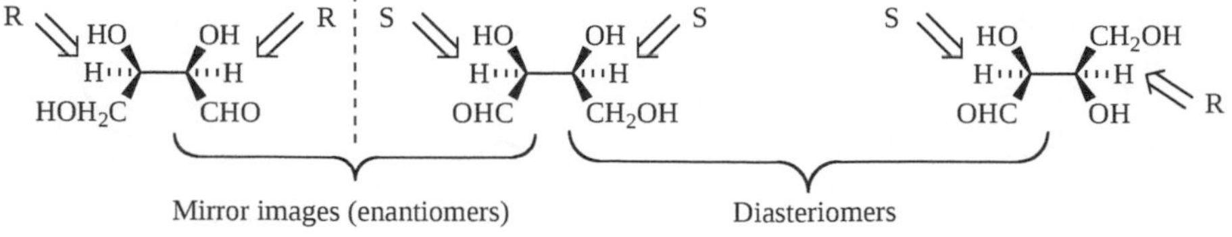

If a compound has two identical chiral centers, then the *RS* and *SR* compounds can be superimposed; they are identical and are called a *meso* compound. *Meso* compounds can usually be recognized by having an internal mirror plane, and are not chiral.

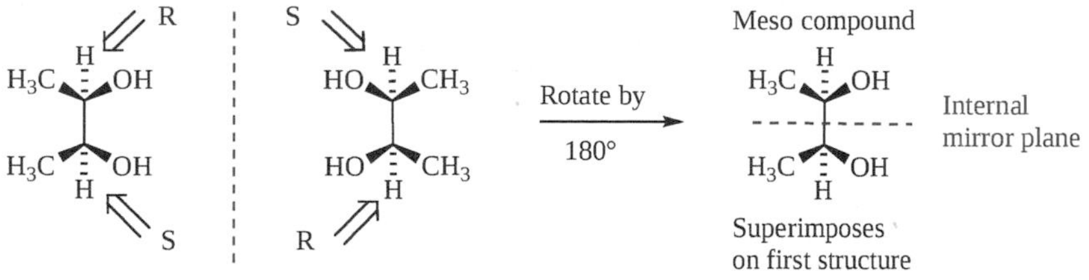

D and L Relative Configuration

Relative configuration was developed before the age of crystallography and is still in use today for sugars and amino acids. The examples below are drawn with their lowest numbered carbon at the top. If the highest numbered chiral center in a sugar has the same stereochemistry as D-glyceraldehyde, then it is a D sugar. With amino acids, if the chiral carbon bearing the amine has the same stereochemistry as L-glyceraldehyde then it is an L amino acid. If the Fischer projection is drawn with the most oxidized end to the top, then the D sugars have the OH of the highest numbered chiral center (arrow below) drawn on the right and L amino acids have the alpha carbon amine drawn to the left (arrow).

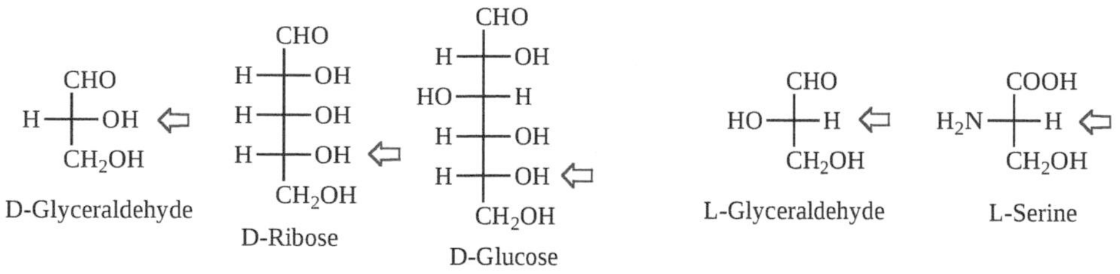

Prochiral Relationships

When, for example, an enzyme creates a chiral compound from an achiral reactant, we need a way to distinguish between the two identical groups in the reactant that will become different in the chiral product. To do this we "promote" one of the two identical groups slightly and then assign *R* or *S* to the "new" chiral center. If the group we promoted would create an *R* chiral center, then we call that promoted group pro-*R*; if it would create an *S* center, then that promoted group is pro-*S*.

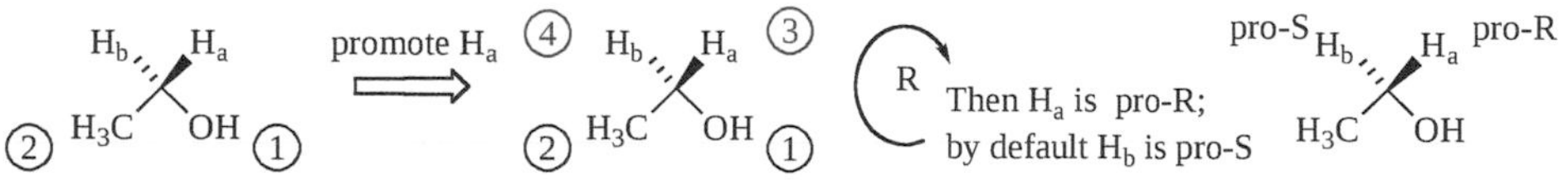

When an enzyme shows selectivity for a face of a reactant, a carbonyl addition, for example, we need a way to label each face of the reactant. If the priorities go in a clockwise manner, then the face showing is the *re* face; if they go in a counterclockwise manner, that face is the *si* face.

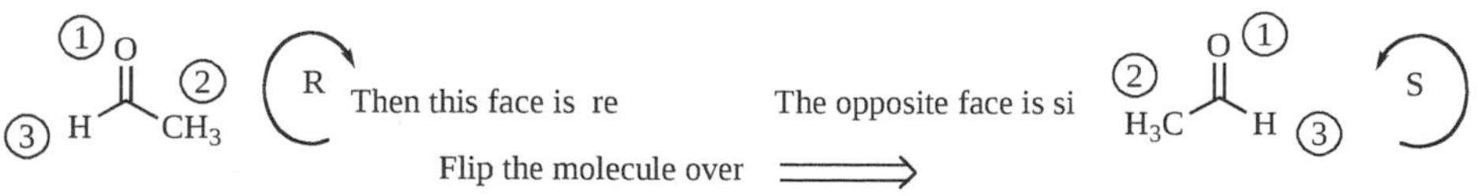

A Bridge to Transition Metal Organometallics

This section is designed to bridge the gap between organic mechanisms and the field of transition metal organometallics by retaining a few electron-pushing concepts to aid in the understanding of organometallic mechanisms.

Statement of the Problem

When students who have gotten really good at organic electron pushing take a course on transition metal organometallics, they are often surprised that the curly arrow way of mechanistic thinking is seldom used in that course. Instead, the metal's oxidation state, the metal's *d* electron count, and the total electron count around the metal are most important. Transition metal organometallic chemists use the names of their elemental mechanistic processes to indicate mechanism without utilizing the organic chemist's curly arrows. For example, a transition metal catalyzed hydrogenation is a synthetically very important reaction. Its mechanism is shown below from Section 8.10.3, using a dotted line to symbolize complexation to the alkene. The processes were named, but no arrows were drawn. This section will attempt to bridge that gap.

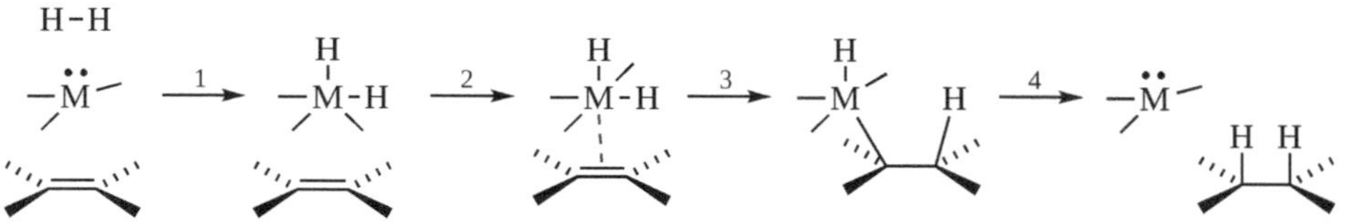

Step 1 is the *oxidative addition* of hydrogen; the metal loses two electrons to form a bond to each of the two hydrogen atoms. Step 2 is the *complexation* of the alkene to the metal, activating the pi bond. Complexation usually occurs from the least hindered face of the pi bond. Step 3 is the *insertion* of the metal hydride bond into the pi bond, forming a metal-carbon bond. Step 4 is the *reductive elimination* (the reverse of oxidative addition), which returns the metal to its original oxidation state and releases the alkane.

Structure Conventions for Inorganic Chemistry

Organic line structure is a fast way to draw organic molecules but can cause confusion because it leaves out most of the hydrogen atoms and lone pairs. To not make errors of omission in our mechanisms, we need to display hydrogen atoms, show lone pair electrons, and keep track of charges near the reactive center. That is why we switch to using a Lewis structure for any part of the molecule near electron flow arrows and do not use line structure.

The inorganic structure drawing convention also contains assumptions that can likewise trigger misunderstandings. Simply put, inorganic structural representations are not Lewis structures. The normal way of drawing organometallic structures emphasizes the complexation of groups (ligands) around the metal rather than the formal charges on the ligands or on the metal. This convention for drawing the structures of metal complexes can be confusing, because sometimes a drawn line can mean a pair of electrons in a bond (as in this book). At other times a drawn line can mean complexation, but the pair of electrons being complexed may be also drawn, seemingly showing the same pair of electrons twice (this book uses a dotted line for complexation). Organometallic structures often use "L" to symbolize a generic ligand, not to be confused with this book's identical abbreviation for a leaving group. When the ligand provides both electrons for the ligand–metal bond, it is called dative bond and they are common with lone-pair donor ligands like CO, PPh_3, H_2O, and with pi bonds. Dative bonds shown as L→M can be confused with curly arrows, so will not be used here.

Inorganic Structure Modifications for Electron Pushing

Like we do for organic line structures, we will need to convert the part of the inorganic structural representation near our curly arrows into Lewis structures in order to facilitate electron counting. To avoid the problem of having an electron pair appear twice, all solid lines drawn will be electron pairs, but complexation of electrons already drawn elsewhere on the structure will be indicated by a dotted line to the metal. Therefore, a simple Lewis base complexed to a metal via a lone pair will have a solid line drawn from the ligand to the metal indicating the pair of electrons involved in the complexation, and that lone pair will not be drawn. However, a sigma or pi bond complexed to the metal will still be drawn as a bond but with a dotted line to the metal to indicate complexation. In this section, a simple dash drawn from metal symbolizes a generic two electron donating ligand, rather than using the letter L for such a ligand. Below in Figure A.4 are several examples of the inorganic structure to Lewis structure conversion used for arrow pushing mechanisms in this section.

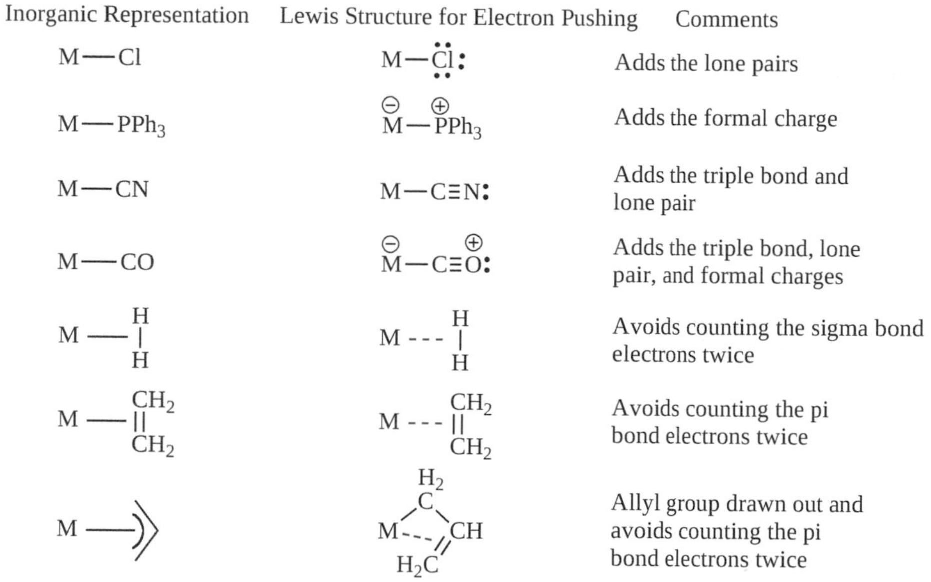

Figure A.4 Conversion of the standard inorganic representation to a Lewis structure equivalent to aid in electron-pushing mechanisms in this section.

Metal Oxidation State, *d* Electron Count, and Number of Valence Electrons

Oxidation state is determined with two rules. Give the metal–ligand bonding electrons to the more electronegative partner and do not count dative bonds to the metal from neutral ligands. With metal–metal bonds, split the electrons equally between the two metals. The oxidation state is the charge left on the metal after adding the total charge of the complex. For example, $Fe(CN)_6^{4-}$, each of the six cyanide ligands would be given the pair of electrons used to bind to the metal. This would leave the iron with a +6 formal charge, but adding in the –4 charge of the overall complex, the iron must have a +2 oxidation state, or Fe(II).

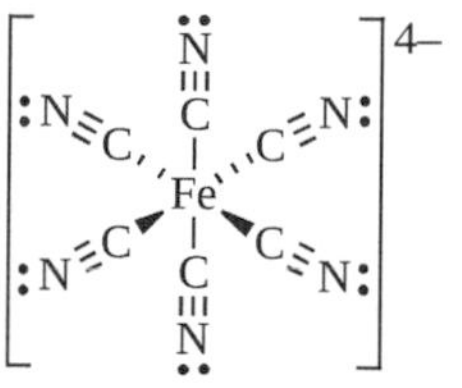

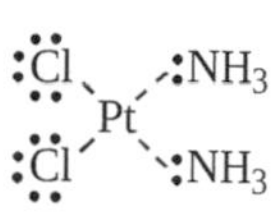

Another example would be the anticancer drug cisplatin, $Pt(NH_3)_2Cl_2$, in which both chlorides get the metal–ligand bonding electrons, and the dative bonds to the neutral ammonia ligands are not counted, leaving the metal as Pt^{2+}. Since the complex has no overall charge, this gives Pt(II) as the oxidation state.

The *d* electron count is calculated by subtracting the metal's oxidation state from its group number in a modern periodic table. For example, iron is in Group 8, so Fe(II) would be 8 – 2 or d^6. The Pt(II) cisplatin example would be 10 – 2 or d^8.

The metal's number of valence electrons or total electron count often determines the stability of the complex. The valence shell of a transition metal requires nine orbitals (the *s*, three *p*, and five *d* orbitals) to be filled for a total of 18 electrons, $18e^-$. We can liken this filled shell stability of an $18e^-$ complex to be similar to completing a valence shell octet in most organic structures. The d^6 complex $Fe(CN)_6{}^{4-}$ example would have 6 electrons from the d^6 iron(II) and two from each of six complexed cyanides, 6 + 6x2 for $18e^-$. The d^8 cisplatin example would have 8 electrons from the d^8 Pt(II), 2 from each of two ammonias, and 2 from each of two chlorides, for a total of 8 + 2x2 + 2x2 = $16e^-$. An $18e^-$ complex is called coordinatively saturated because it is not easy to add another ligand. Cisplatin, at $16e^-$ is coordinatively unsaturated and can accept a lone pair ligand to reach an $18e^-$ count.

Typical Transition Metal Organometallic Elemental Mechanistic Processes

Ligand association (coordination) and **ligand dissociation** are the reverse of each other. **Ligand exchange** or substitution is a combination of both processes. The association then dissociation combination is characteristic of complexes with less than a full $18e^-$ count; the cisplatin example was able to accept a ligand because it was only $16e^-$. If the complex has $18e^-$ to start, then the most common ligand exchange process is to dissociate a ligand to become a $16e^-$ complex then associate a new ligand to return to an $18e^-$ species. The metal's oxidation state and the *d* electron count do not change with these processes. Both the coordination number (the number of ligands around the metal center) and the total electron count increase with coordination and decrease with dissociation.

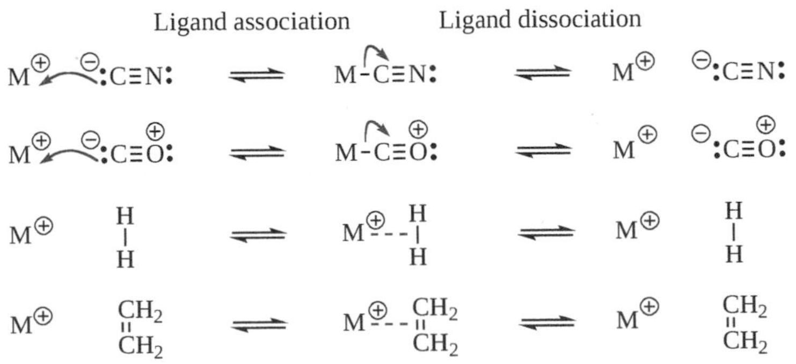

Transmetallation is a ligand exchange between two different metals. From the HSAB principle (Section 2.4, hard with hard, soft with soft), the equilibrium favors formation of a more covalent organometallic from the softer pair and a more ionic salt from the harder pair.

$$2\ RMgCl \quad + \quad CdCl_2 \quad \rightarrow \quad R_2Cd \quad + \quad 2\ MgCl_2$$

R soft–Mg hard + Cd soft–Cl hard → R soft–Cd soft + Mg hard–Cl hard

Oxidative addition and **reductive elimination** are the reverse of each other. With oxidative addition the metal's oxidation state increases by two and the *d* electron count decreases by two. Because of this, the metal must be at least d^2 so there are two electrons to lose in the oxidative addition process. The metal's total electron count increases by two.

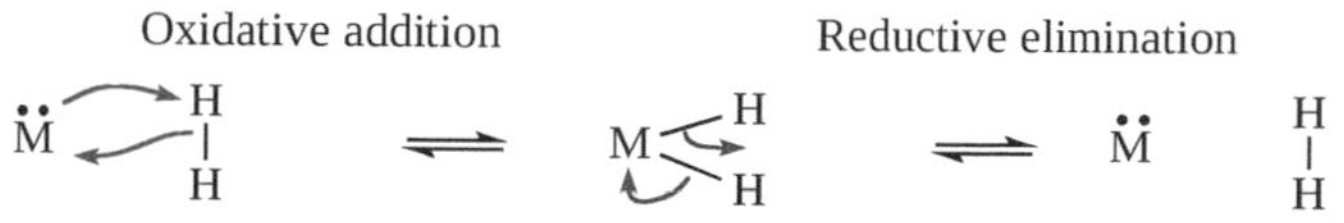

Oxidative addition can also occur in a two-step process of nucleophilic attack by the metal followed by ligand association by the leaving group, as shown below.

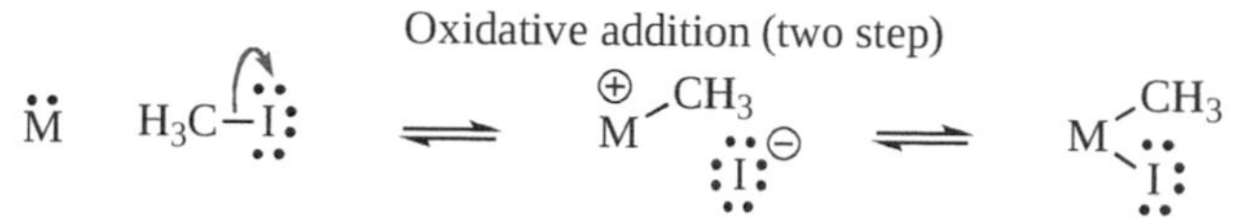

Organometallic formation from magnesium metal and an organic halide (Section 12.8) is believed to follow a similar stepwise oxidative addition involving an initial single electron transfer (s.e.t.) that triggers the loss of the leaving group in the first step.

Protonation and **deprotonation** are similar to the oxidative addition and reductive elimination pair. On protonation, the oxidation state of the metal increases by two and the *d* electron count decreases by two, but since a proton does not bring any electrons to the complex, the total electron count does not change. Deprotonation is the microscopic reverse.

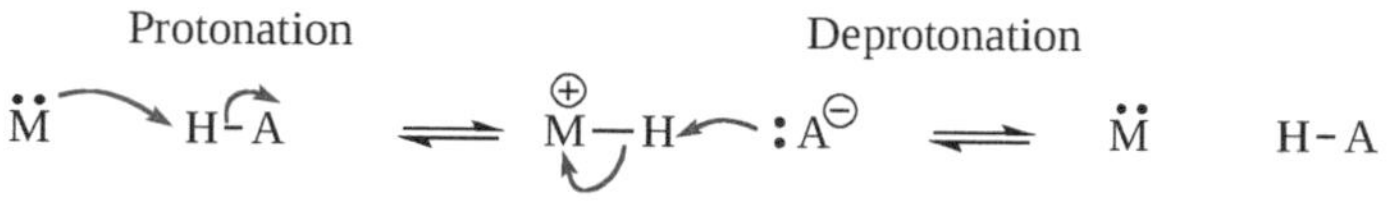

The 1, 1 insertion and **deinsertion** reactions are the reverse of each other. On insertion, the metal's oxidation state and *d* electron count do not change, but the total electron count of the complex drops by two because there is one less electron pair complexed to the metal. This process is similar to the 1,2 rearrangement of a carbocation discussed in Section 9.8.

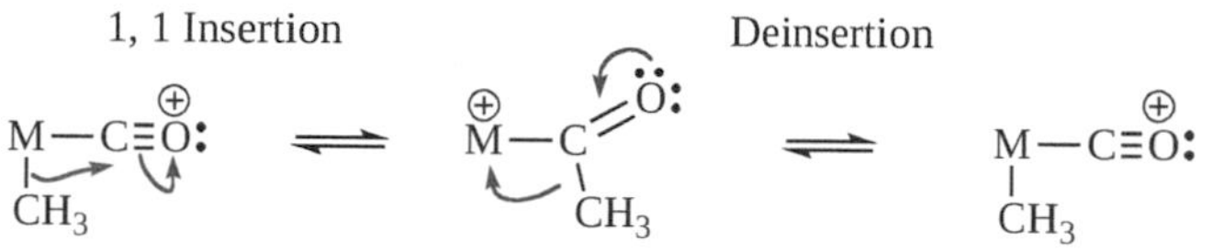

The 1, 2 insertion and **beta elimination** reactions are the reverse of each other. Similar to the 1, 1 insertion, the metal's oxidation state and *d* electron count do not change, but the total electron count of the complex drops by two because there is one less electron pair complexed to the metal. The migrating group and the migration destination must be very close. Similar to the migrations we saw in Section 9.8, the orbital of the migrating group should retain good overlap with the orbital it is leaving while forming a new overlap to the destination orbital. In the transition state, the migrating group maintains as much bonding overlap as possible in what is described as a bridging two-electron three-atom bond. It just slides across to a nearby orbital and does not hop through hyperspace. This 1, 2 insertion is similar to the four-center, four-electron addition of borane to alkenes (Section 7.2.5). The most common beta elimination is the β-hydride elimination shown below.

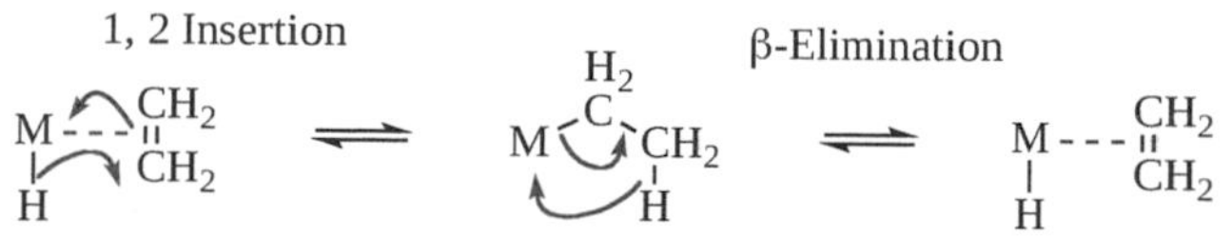

The 2+2 cycloaddition and cycloreversion reactions are the reverse of each other. The metal's oxidation state, *d* electron count, and total electron count of the complex do not change. Note that the cycloreversion can produce two products as shown below, reversion to the original reactant on the left and the alternative break product on the right.

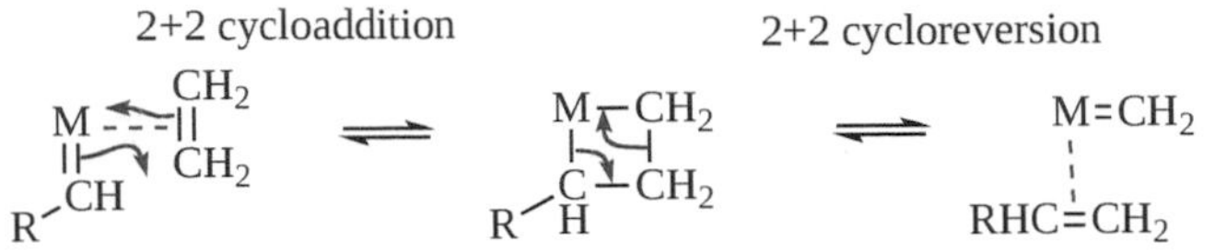

Catalytic Transition Metal Organometallic Processes Are Cyclic

Often the transition metal catalyst is expensive, so ideally it is used in a catalytic cycle in very small quantities. The following are just a few of the synthetically important reactions illustrating the use of the common transition metal organometallic elemental mechanistic processes. The very important hydrogenation reaction at the start of this section can also be written as a cycle.

Pd-catalyzed C–C coupling reactions are very important Nobel Prize winning construction reactions for synthetic organic chemists. It is often the only way to couple two aryl groups. Shown below in Figure A.5 is the general catalytic cycle of oxidative addition, transmetallation, and reductive elimination that produces the coupled product and regenerates the palladium catalyst. A variety of organometallic reagents can be utilized in this cycle.

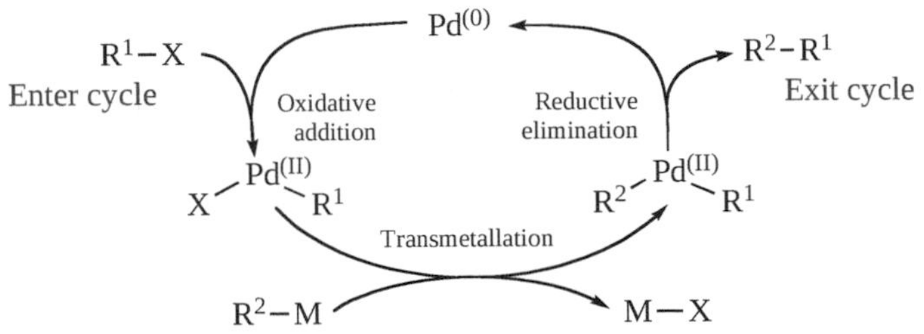

Suzuki coupling: M is $B(OH)_2$
Negishi coupling: M is Zn , Al, or Zr.
Stille coupling: M is Sn

Figure A.5 Common mechanistic cycle for palladium catalyzed carbon–carbon coupling reactions.

Olefin metathesis is another Nobel Prize winning reaction that swaps the ends of two double bonds in a unique and useful reaction. A few transition metal complexes of Ti, W, Mo, and Ru, can catalyze this reaction. Some olefin metathesis reaction examples follow.
Joining two terminal alkenes to make a much longer alkene:

This reaction can also be used to close a large ring in a ring closing metathesis reaction.

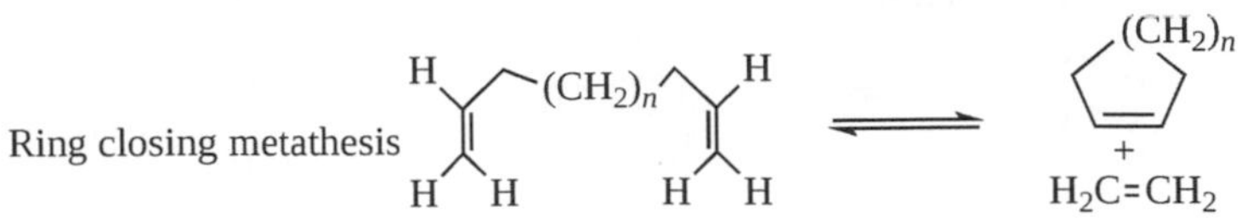

Figure A.6 shows the mechanistic catalytic cycle for the olefin metathesis reaction.

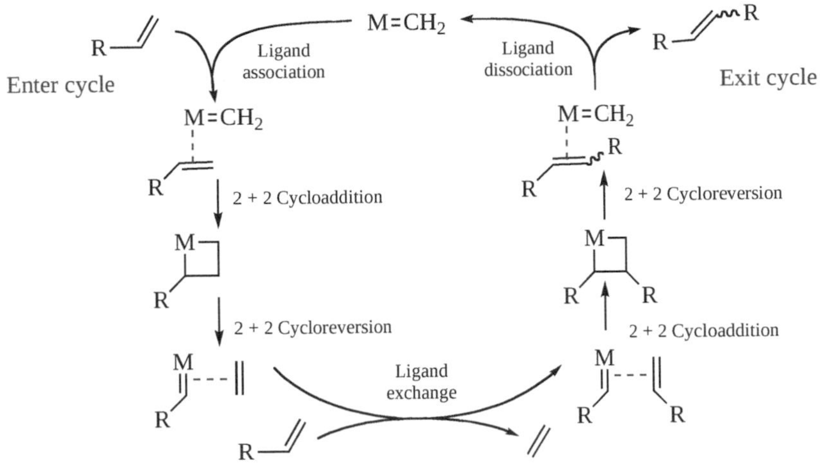

Figure A.6 Common mechanistic cycle for the olefin metathesis reactions.

Hydroformylation adds both hydrogen and carbon monoxide to inexpensive terminal alkenes to give valuable aldehydes for synthesis. Figure A.7 shows the mechanism of the very useful industrial catalytic hydroformylation process.

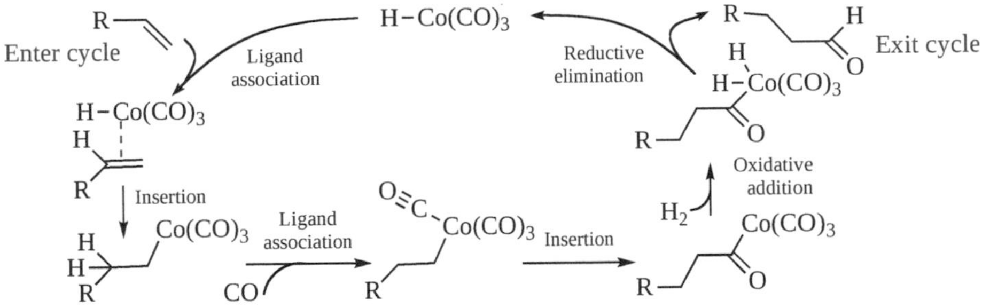

Figure A.7 Common mechanistic cycle for a hydroformylation reaction.

Biochemical Example: Vitamin B_{12} is the structurally most complex of all the vitamins, having a cobalt atom in the center of large ring system, a carbon–cobalt organometallic bond, and a tethered ligand (see below). B_{12} is a cofactor in several important enzymatic reactions. Lack of vitamin B_{12} in the diet can cause pernicious anemia and several other ailments.

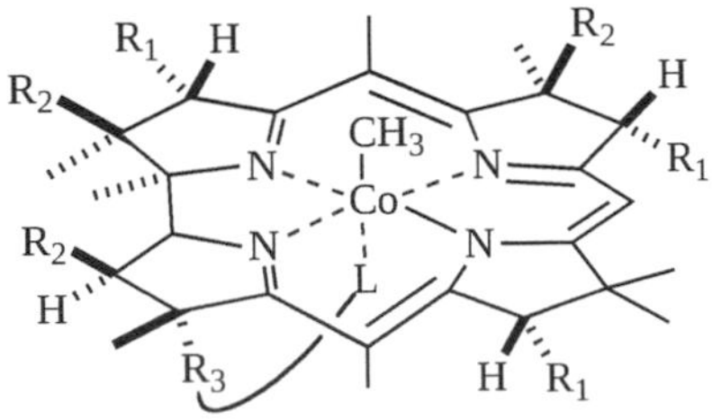

Transition metal organometallic chemistry is a rich and exciting field that continues to have a major impact on organic and biochemistry. We can see that these transition metal organometallic elemental mechanistic processes open up a vast array of previously unattainable synthetic reactions. For further study, see Chapter 12 in Anslyn, E. & Dougherty, D., *Modern Physical Organic Chemistry*, University Science Books, 2006.

HINTS TO SELECTED PROBLEMS FROM CHAPTERS 8, 9, 10, AND 11

The most important point to remember in classifying into sources and sinks is to **stay flexible**. Almost all electron sinks have lone pairs that can serve as sources. Lone pairs on sinks can complex with Lewis acids or be protonated. A polarized multiple bond can serve as an electron-withdrawing group, making protons on adjacent atoms acidic. Looking at both reactive partners can help you decide. If one partner is clearly a sink, like sulfuric acid, then the other partner must serve as a source, even though it may be a polarized multiple bond sink like a carbonyl; protonate the carbonyl lone pair as the source.

Chapter 8

Problem	First Step Source	First Step Sink	Chapter 8 section
8.1	R–M	Y=C–L	8.8.4
8.2	C=C	H–A	8.2.6
8.3	Z (O lone pair)	Y=C–L	8.8.1
8.4	Z–C=C	*sp*3 C–L	8.4.5
8.5	Z or RM (C lone pair)	C=C– EWG	8.7.1
8.6	R–M	Z=C=Y	8.9.3
8.7	R–M	*sp*3 C–L	8.4.4
8.8	Z–C=C	C=C– EWG	8.7.5
8.9	M–H (as base)	Y=C–L (and adjacent CH)	8.8.2
8.10	Z–C=C	Y=C–L	8.8.5
8.11	C=C	H–A	8.2.6
8.12	R–M	Y–L	8.3.4
8.13	Z–C=C	Y–L	8.3.5
8.14	Z (O lone pair)	*sp*3 C–L	8.4.1 and 8.4.2
8.15	R–M	H–A	8.2.4
8.16	Z (N lone pair)	H–A	8.2.1
8.17	Z (O lone pair)	Y–L	8.3.1
8.18	Z (O lone pair)	H–A	8.2.1

Chapter 9

Problem	Site	Base	Nucleophile	Minor Variable
9.1a	Primary	Weak	Good	
9.1b	Primary	Strong	Poor	
9.1c	Primary	Strong	Poor	
9.1d	Hindered primary	Weak	Good	
9.2a	Secondary	Weak	Poor	
9.2b	Secondary	Strong	Good	Heated
9.2c	Primary	Strong	Poor	
9.2d	Secondary	Weak	Good	
9.3a	Secondary	Weak	Good	
9.3b	Hindered secondary	Weak	Good	Heated
9.3c	Tertiary	Weak	Poor	
9.3d	Tertiary	Strong	Good	

Problem	First Step Source	First Step Sink	Decision Type
9.4a	R–M	Y=C–L	Multiple addition
9.4b	Z–C=C	C=C– EWG	Ambident electrophile

Problem	First Step Source	First Step Sink	Decision Type
9.4c	Z–C=C	Y=C–L	Ambident nucleophile
9.4d	Z (O lone pair)	sp^3 C–L	Intermolecular versus intramolecular
9.5a	Z (N lone pair)	C=Y	Thermodynamic versus kinetic enolate
9.5b	Z–C=C	sp^3 C–L	Ambident nucleophile
9.5c	R–M	C=C–EWG	Ambident electrophile
9.5d	Z (O lone pair)	H–A	To migrate or not
9.8a	R–M	Y=C–L	Multiple addition
9.8b	R–M	C=C–EWG	Ambident electrophile
9.8c	R–M	C=C–EWG	Ambident electrophile
9.9a	Z (O lone pair)	sp^3 C–L	Substitution/elimination
9.9b	R–M	Y=C–L	Multiple addition
9.9c	Z–C=C	Y–L	Ambident nucleophile
9.16a	Z (O lone pair)	sp^3 C–L	Intermolecular versus intramolecular
9.16b	R–M	C=C– EWG	Ambident electrophile
9.17a	Y=C	HA	Thermodynamic versus kinetic enol
9.17b	Z–C=C	C=C– EWG	Ambident electrophile and nucleophile

Chapter 10

Problem	First Step Source	First Step Sink	1st Step Archetype
10.1a	Z (O lone pair)	Y–L	substitution
10.1b	C≡C	H–A	addition
10.1c	Z (N lone pair)	C=Y (and adjacent CH)	proton-transfer
10.2a	M–H	C=Y (and adjacent CH)	proton-transfer
10.2b	Z ($^-$OH lone pair)	C=C– EWG	addition
10.2c	Z ($^-$OEt lone pair)	Y=C–L (and adjacent CH)	proton-transfer
10.3a	Z (O lone pair)	H–A	proton-transfer
10.3b	R–M	C≡Y	addition
10.3c	Z–C=C	Y=C–L	addition
10.3d	Z (N lone pair)	C=Y (and adjacent CH)	proton-transfer
10.4a	Z (carbonyl lone pair)	H–A	proton-transfer
10.4b	C=C	H–A	addition
10.4c	Z (O lone pair)	Y=C–L	addition
10.4d	Z (carbonyl lone pair)	H–A	proton-transfer
10.5a	Z (S lone pair)	sp^3 C–L	substitution
10.5b	Z (carbonyl lone pair)	A (Lewis acid)	addition
10.5c	R–M	H–A	proton-transfer
10.5d	Z (O lone pair)	H–A	proton-transfer
10.6a	C=C	Nu–L	addition
10.6b	Z (P lone pair)	sp^3 C–L	substitution
10.6c	Z–C=C	H–A	addition
10.6d	Z (O lone pair)	sp^3 C–L	substitution
10.7a	Z (N lone pair)	Y=C–L	addition
10.7b	Z (N lone pair)	H–A	proton-transfer
10.7c	R–M	Y=C–L	addition
10.8a	Z (N lone pair)	C=Y and H–A	addition
10.8b	Z (O lone pair)	H–A	proton-transfer

Problem	First Step Source	First Step Sink	1st Step Archetype
10.8c	MH_4^-	Y=C–L	addition
10.9a	Z–C=C–C=C	H–A	addition
10.9b	R–M	C=Y and H–A	addition
10.9c	C=C	H–A	addition
10.10a	ArH	*sp*3 C–L	substitution
10.10b	C=C–Z	C=Y	addition
10.10c	Z (carbonyl lone pair)	*sp*3 C–L	substitution
10.11a	Z (O lone pair)	H–A	proton-transfer
10.11b	Z (carbonyl lone pair)	H–A	proton-transfer
10.11c	Z (⁻OEt lone pair)	C=Y (and adjacent CH)	proton-transfer
10.12a	C=C	H–A	addition
10.12b	Z (⁻OEt lone pair)	Y=C–L (and adjacent CH)	proton-transfer
10.12c	Z (⁻OH lone pair)	Y=C (and adjacent CH)	proton-transfer
10.13	Z (C lone pair)	C=Y and H–A	addition
10.14	none	*sp*3 C–L	substitution

Chapter 11

11.1a Failed one of the D_N cross-checks.
11.1b Failed the S_N2 cross-check.
11.2a Failed the p.t. K_{eq} cross-check.
11.2b Failed the media pH cross-check.
11.3a Failed the ΔpK_a cross-check.
11.3b Failed the ΔpK_a cross-check.
11.4a Failed the S_N2 cross-check.
11.4b Failed the ΔpK_a cross-check.
11.5a Failed the ΔpK_a cross-check.
11.5b Failed the ΔpK_a cross-check.
11.6a Failed the ΔpK_a cross-check.
11.6b Failed the p.t. K_{eq} cross-check.
11.7 Failed the cation stability cross-check.
11.8 Failed the media pH cross-check.
11.9 Failed the step order cross-check.
11.10 Failed the cation stability cross-check.
11.11 Failed the ΔpK_a cross-check.
11.12 Missed best proton-transfer.
11.13 Failed the ΔpK_a cross-check and one other.
11.14 Line structure error.
11.15 Missed direct vs. conjugate addition decision.
11.16 Missed direct vs. conjugate addition decision.
11.17 Failed charge balance cross-check.
11.18 Failed Lewis resonance form cross-check.
11.19 Failed the S_N2 cross-check.
11.20 Failed one of the D_N cross-checks.

SOLUTIONS TO ODD-NUMBERED PROBLEMS

Chapter 1 Odd-Numbered Exercise Answers

1.1 Draw Lewis structures with resonance for the following neutral compounds. Decide the major resonance form, if any.

O–O–O H_2C–N–N   O–C–O   H–O–N–O   H–O–NO_2

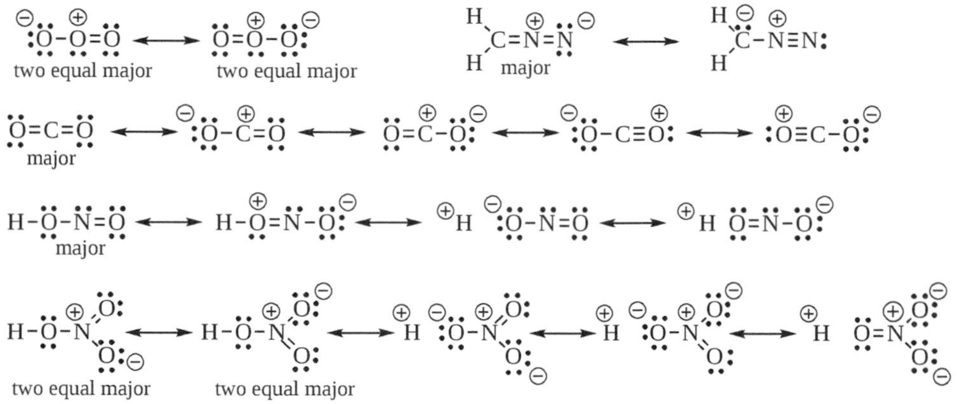

(The acids, H–A, have extra resonance forms that corresponds to H^+ A^-)

1.3 The polarization of the indicated bond is decided by allowing the more electronegative atom to win the electron tug-of war.

Br–Br	HO–Cl	H_2B–H	H_2C=O	I–Cl
None	δ– δ+	δ+ δ–	δ+ δ–	δ+ δ–

1.5 Draw the Lewis structure(s) that would be the product of the arrows.

(a)

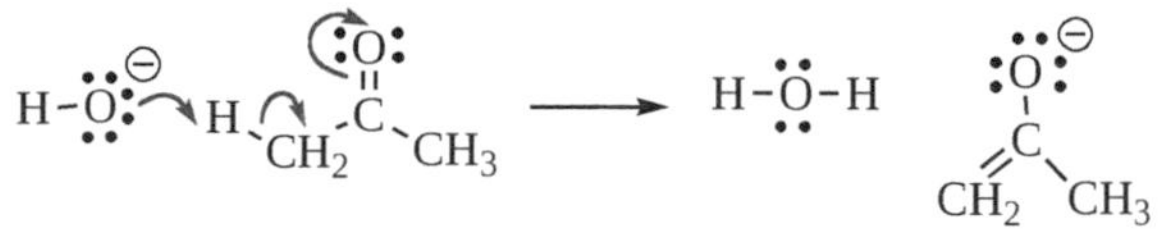

(b)

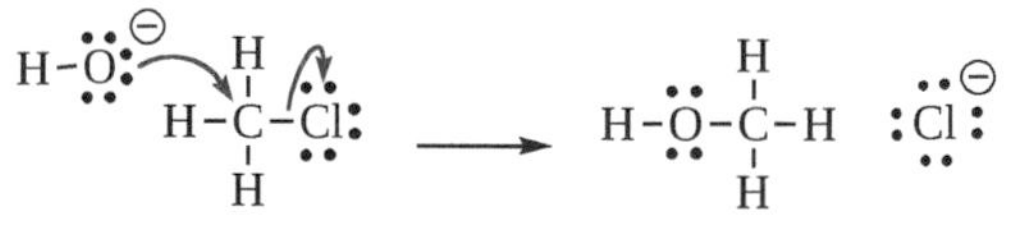

(c)

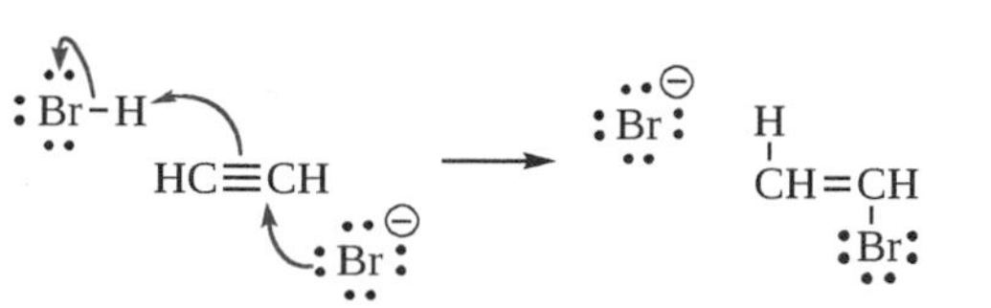

1.7 Give the curved arrows necessary for the following reactions. Every arrow must start from either a lone pair or bond.

(a)

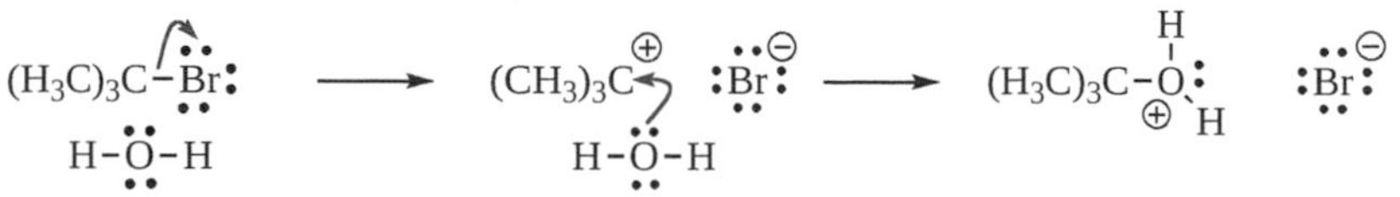

Electron Flow in Organic Chemistry: A Decision-Based Guide to Organic Mechanisms,
Third Edition. Paul H. Scudder.

(b)

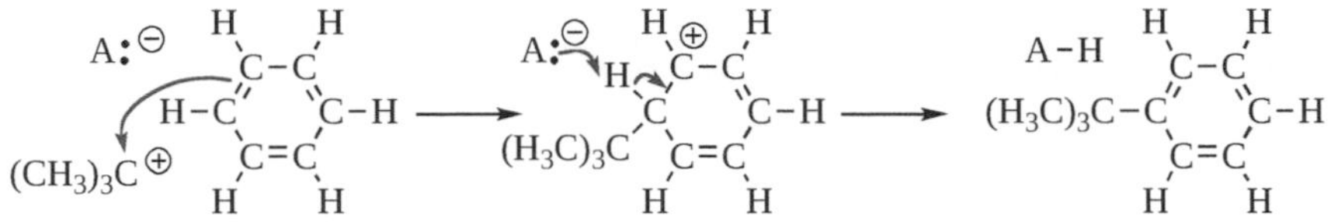

(c)

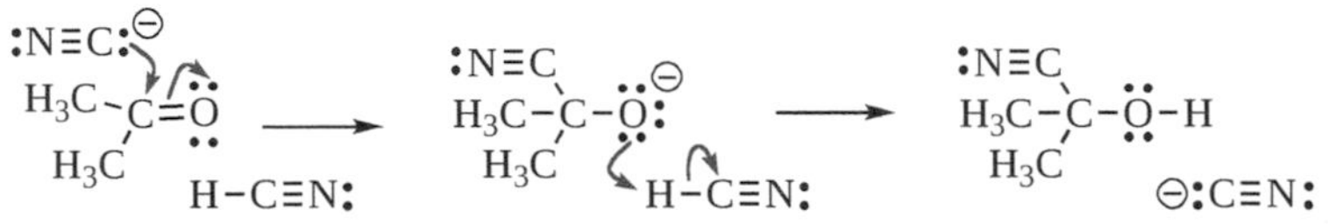

1.9 Draw full Lewis structures for the following line structure drawings. Common errors in this problem are H's missing or five bonded carbons.

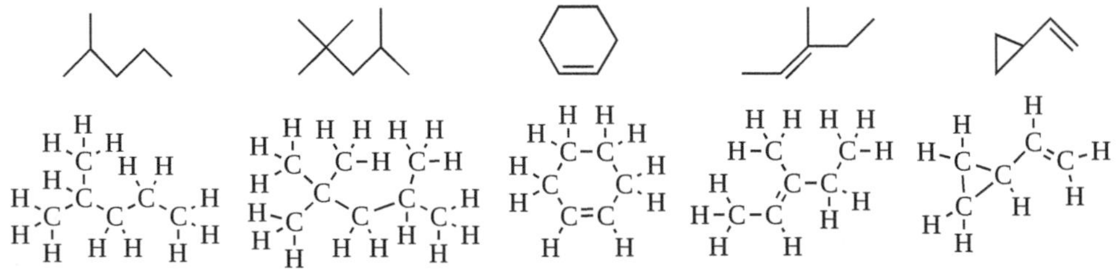

1.11 In the following structures circle any carbon atom that bears a negative or significant partial negative charge. (Hint: draw some resonance forms.) The carbon atom that bears a negative or significant partial negative charge is shown on the following resonance structures.

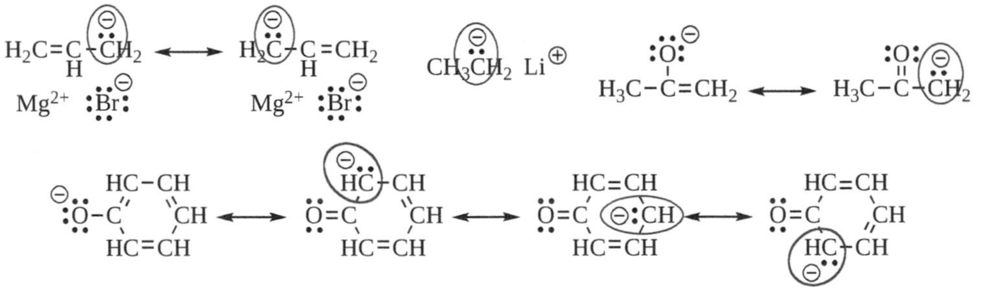

1.13 For each of the resonance pairs below, determine which is the major contributor.

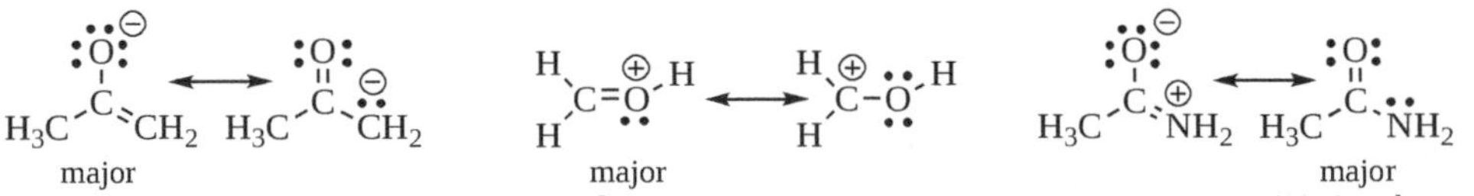

major
minus on O

major
complete octets

major
minimize charge

1.15 The aromatic compounds in the following list are circled; they have $4n + 2$ electrons in a conjugated pi loop. Section 1.9.3 tells when to count lone pairs as part of the loop and when to ignore them.

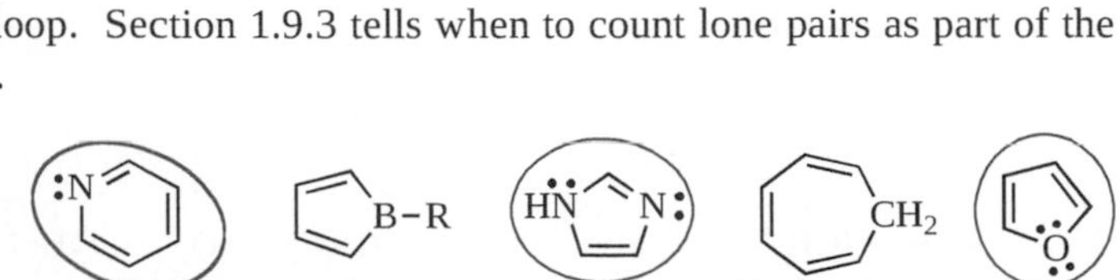

$4n$

Broken loop

1.17 First draw a good Lewis structure, then decide the structure of the following compounds, using VSEPR. For connectivity on these, put the unique atom in the center.

SF_4	CO_2	BF_3	H_2O	NH_3

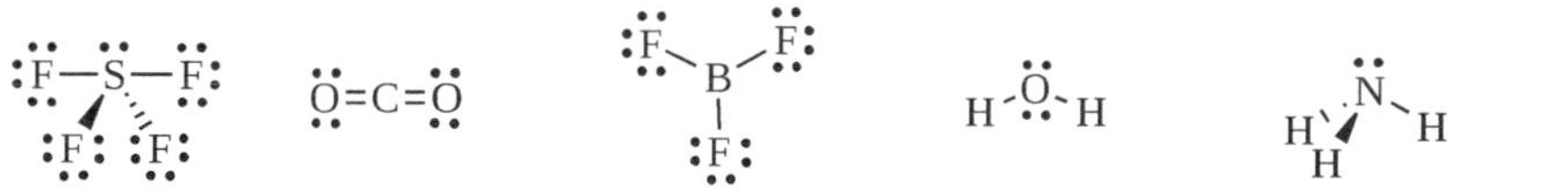

See–saw	Linear	Trigonal planar	Bent	Trigonal pyramidal

1.19 Rank the following on the size of the atom or ion (use #1 for largest). All have the neon electronic configuration of ten electrons; increasing nuclear charge pulls the electrons in.

Ne	Na^+	F^-	Mg^{2+}	Al^{3+}
2	3	1	4	5

1.21 Circle the polar covalent and underline the mostly ionic compounds below.

Br_2 LiBr (underlined)   HCl (circled)   KBr (underlined) H_2O (circled)

1.23 Decide which of the following anions is the most stable, the least stable. #1 is most stable.

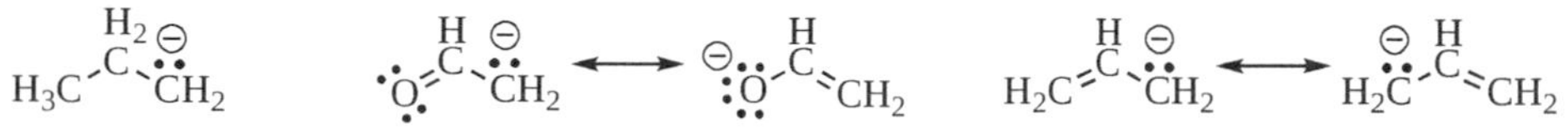

	no resonance	resonance with an electronegative atom	resonance with pi bond
Rank	3	1	2

Chapter 2 Odd-Numbered Exercise Answers

2.1 There are two ways that a compound can be stable: one due to $\Delta G°$ and the other due to $\Delta G^‡$. Draw energy diagrams to explain.

Thermodynamic stability, the $\Delta G°$ reason: A is the most stable of all possibilities; there is nowhere to go but up. Kinetic stability, the $\Delta G^‡$ reason: B is **not** the most stable of all possibilities, but the barriers are too high to traverse with the energy present.

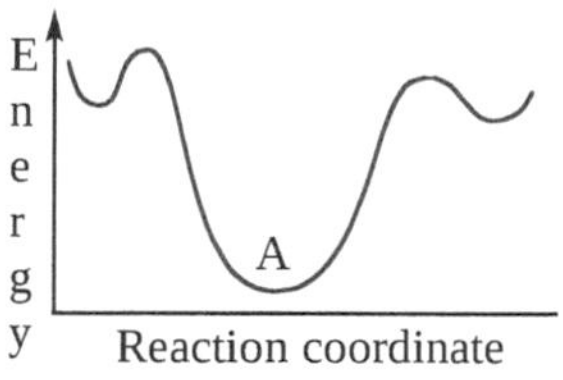

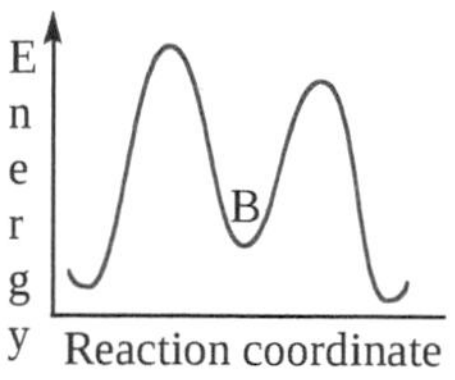

2.3 Using this energy diagram:

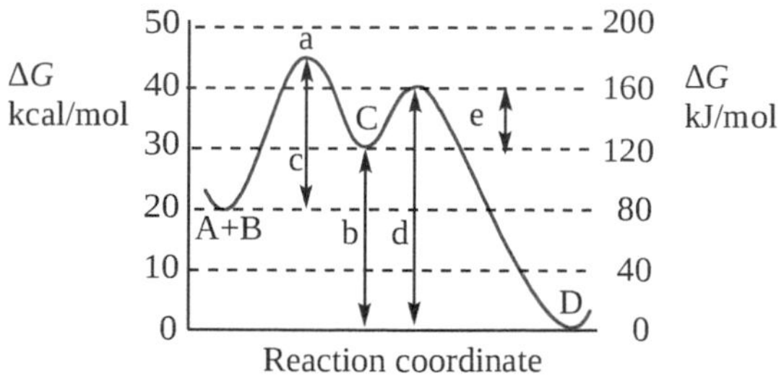

For a different temperature, we need a more detailed equation. If the relative rates (let k_1 be 10,000 and $k_2 = 1$) of two related unimolecular reactions under identical conditions at 25°C (convert °C into K by adding 273), then the difference in barrier heights is related to the ratio of the rates by the formula $\Delta\Delta G^\ddagger = -RT\ln(k_1/k_2) = -RT2.303\ \log(k_1/k_2)$. Plugging in those values gives $-(1.99 \times 10^{-3})(298)(2.303)\ \log(10{,}000) = -5.5$ kcal/mol (–23 kJ/mol). It checks.

2.19 Using the bond-strength table in the Appendix, calculate the $\Delta H°$ of reaction for of the following enolization process. Would an equilibrium favor reactants or products?

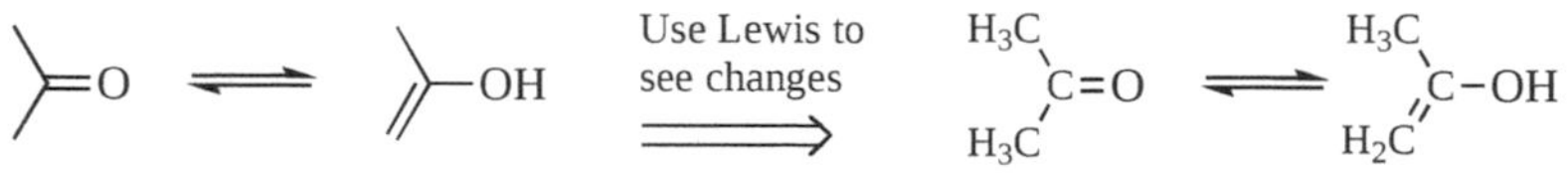

$\Delta H°$ = bonds broken - bonds made. Broken (C=O, C–C, C–H) - made (C–O, O–H, C=C) = (177 + 83 + 99) – (86 + 111 + 146) = +16 kcal/mol (+67 kJ/mol). Because $\Delta H°$ is positive, it's endothermic. Since the change in disorder, $\Delta S°$, is expected to be small for an isomerization, $\Delta H°$ dominates $\Delta G°$ making it also positive, and so the equilibrium favors reactants.

2.21 In each molecule, the red carbon atom is softer because it bears a lesser partial positive charge. In the first, iodine is less electronegative than oxygen; in the second, the CH_2 is farther away from the inductively withdrawing electronegative oxygen.

I–$CH_2CH_2CH_2CH_2$–OSO_2Ph $\quad$ H_2C=CH–CHO

2.23 For each of the following pairs of compounds, tell whether they are enantiomers, diastereomers, conformational isomers, or identical. Making a model helps greatly on problems like these. Interpreting two-dimensional drawings of three-dimensional objects is an acquired skill that takes time to master.

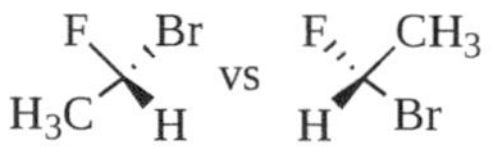

Enantiomers
non-superimposable
mirror images

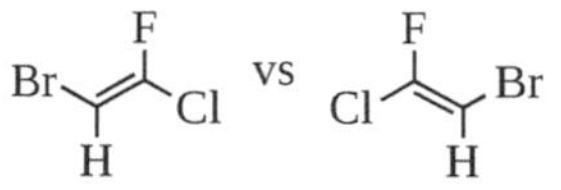

Identical (and flat)
flip over to see

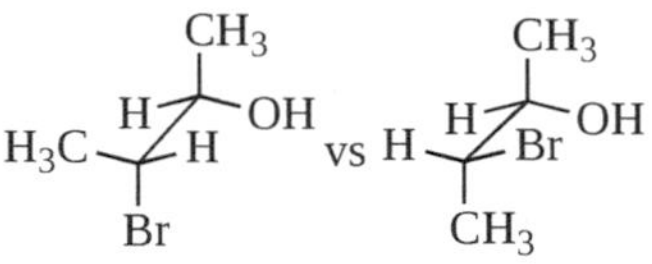

Conformational isomers
rotate the front part to
make same

Chapter 3 Odd-Numbered Exercise Answers

3.1 Using the pK_a chart in the Appendix, calculate the numerical value of K_{eq} for these reactions. Predict whether it favors reactants or products and estimate $\Delta G°$ at 25°C.

The numerical value of K_{eq} is calculated by $\log K_{eq} = pK_{abH} - pK_{aHA}$, and $\Delta G°$ at 25°C by $\Delta G° = -1.364 \log K_{eq}$ for $\Delta G°$ in kcal/mol or $-5.706 \log K_{eq}$ for $\Delta G°$ in kJ/mol.

(a) PhONa + PhSH ⇌ PhSNa + PhOH

$\log K_{eq} = pK_{abH} - pK_{aHA} = 10 - 6.5 = +3.5$, so $K_{eq} = 10^{+3.5}$ favors products.

$\Delta G° = -1.364 \log K_{eq} = -1.364*3.5 = -4.77$ kcal/mol (–19.9 kJ/mol).

(b)

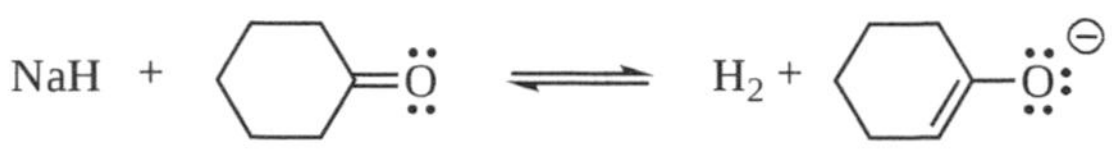

$\log K_{eq} = pK_{abH} - pK_{aHA} = 35 - 19.2 = +15.8$, thus $K_{eq} = 10^{+15.8}$ favors products.

$\Delta G° = -1.364 \log K_{eq} = -1.364*15.8 = -21.6$ kcal/mol (–90.1 kJ/mol).

3.3 Trends: all species are ranked beginning with the numeral 1.

Rank the most stable carbanion by lowest pK_{abH}

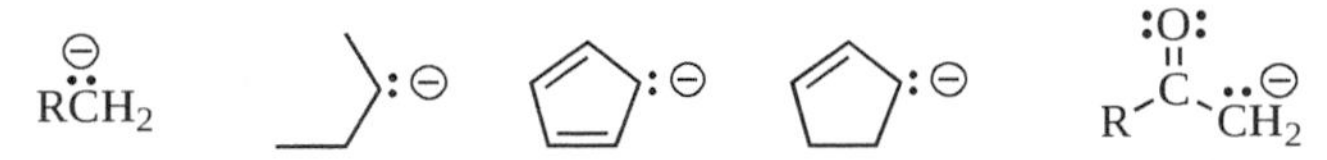

pK_{abH}:	50	51	16	43	19.2
Rank:	4	5	1	3	2

The best electron-withdrawing group will make CH_3–EWG acidic (lower the pK_a)

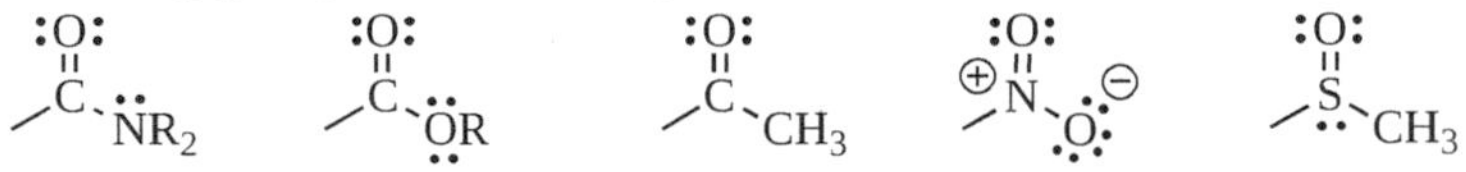

CH_3-EWG pK_a	29.4	25.6	19.2	10.2	35
Rank:	4	3	2	1	5

3.5 For the diprotic malonic acid $CH_2(COOH)_2$ (pK_{a1} = 2.8 and pK_{a2} = 5.7), using crossing and break points, draw a species vs. pH graph for malonic acid, like Figures 3.2 and 3.3. What is the major species in solution at pH 4.3, at pH 7.0?

We need only a qualitative figure. Since pH = pK_a + log([A^-]/[HA]), then one pH unit acidic from the pK_{a1} the [A^-]/[HA]ratio is 1:10. The curves cross at each pK_a since pH = pK_a when [HA] = [A^-]. Since there are two pK_as, 2.8 and 5.7, there are two crossing points. At one pH unit away from the pK_a, the curves are at 9% near the bottom of the graph, or 91% if near the top. The neutral dicarboxylic acid exists in an acidic medium, the carboxylate dianion in a basic medium, and with the monoanion in between. At pH 4.3, the almost exclusive species is the monoanion. At pH 7.0, the almost exclusive species is the dianion.

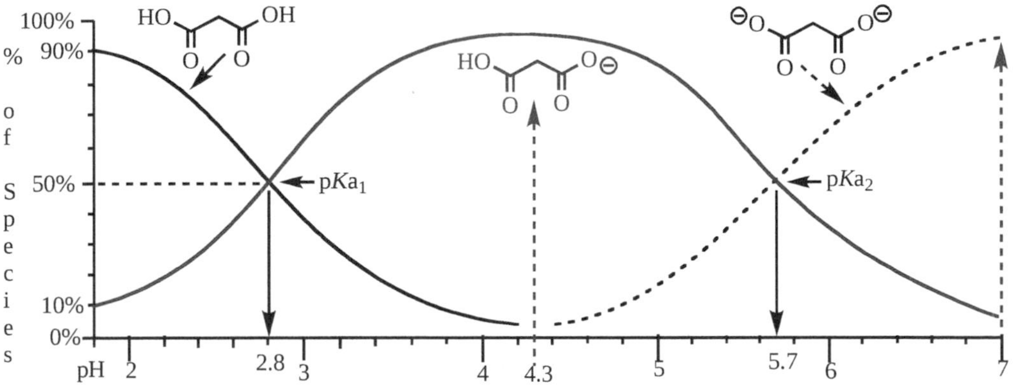

3.7 With the help of a pK_a chart, find the most acidic H in each of the following compounds.
The most acidic H is circled in each of the following compounds, with the pK_a shown below.

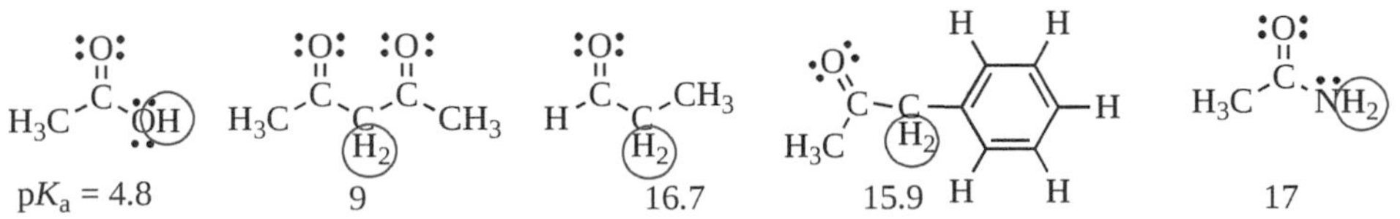

pK_a = 4.8 | 9 | 16.7 | 15.9 | 17

These are all on this book's pK_a chart. Sometimes a resonance form might be drawn instead.

3.9 Ascorbic acid, vitamin C, has one very acidic H (pK_{a1} = 4.2), and all the others are not very acidic (pK_{a2} = 11.6). The circled acidic H is as acidic as a carboxylic acid H (pK_a = 4.8) because it is a vinylogous carboxylic acid and has similar resonance forms. The minus charge in this conjugate base ends up on an electronegative oxygen in two of the resonance forms, just like in the carboxylate anion. For review, see Section 1.9.2 Vinylogy.

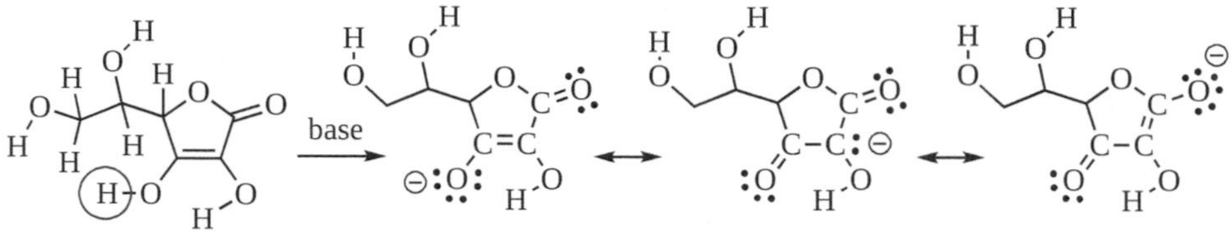

3.11 Decide whether these conditions are acidic or basic and give an approximate pH.

(a) A water solution of 0.1 M PhONa and 1.0 M PhOH.

Since pH = pK_a + log([A^-]/[HA]), and because there is 1:10 ratio of the quantities of conjugate base to acid, the solution pH is 9, basic, one pH unit more acidic than the pK_a of 10.

(b) A water solution of 1 M KF and 1 M HF.

Because there are equal quantities of acid and conjugate base the solution pH is 3.2, acidic, the same as the pK_a = 3.2

(c) A water solution of 1.0 M KF and 0.1 M HF.

Because there is 10:1 ratio of the quantities of conjugate base to acid, the solution pH is 4.2, acidic, one pH unit more basic than the pK_a of 3.2.

3.13 What is the most acidic hydrogen of the following compound?

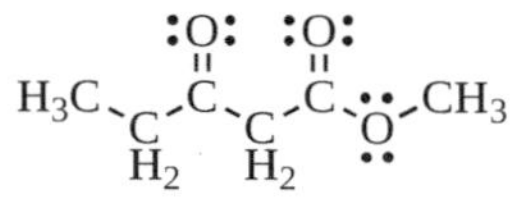

The functional groups are a ketone and an ester. Both CH_2s are adjacent to the δ+ end of a polarized multiple bond and are acidic. The pK_as of each set of hydrogens are:

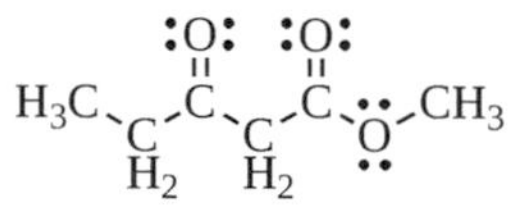

50 19.2 10.7 ≈ 50

Neither methyl group is acidic, for both lack any sort of stabilization of their conjugate base. The CH_2 between the two carbonyls is the most acidic; it has the lowest pK_a because the anion is delocalized onto two electronegative oxygens.

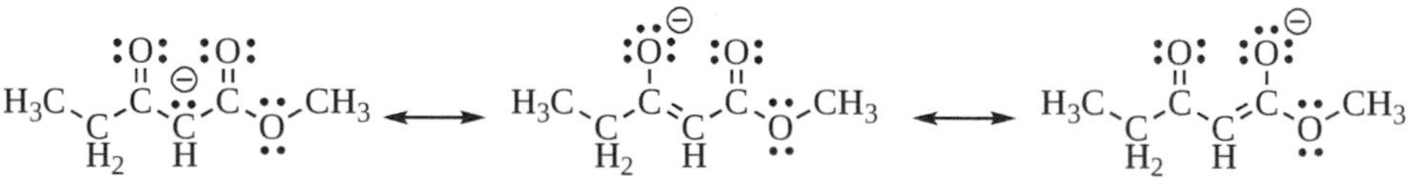

3.15 Fill in the blank for the missing partners in these proton-transfer reactions. Give K_{eq}, $\Delta G°$, and equilibrium direction of each problem.

(a)

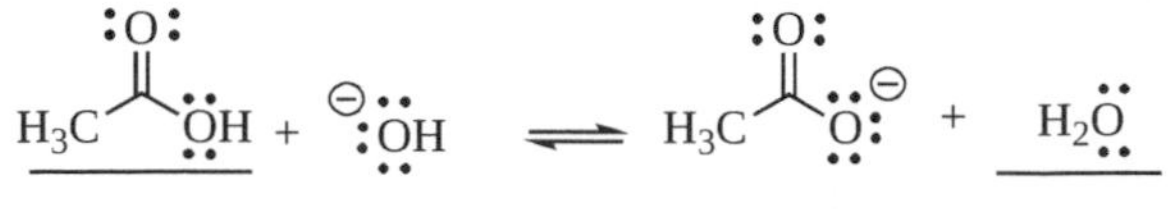

log K_{eq} = pK_{abH} – pK_{aHA} = 14.0 – 4.8 = 9.2 and thus $K_{eq} = 10^{+9.2}$, and $\Delta G°$ = –1.364 log K_{eq} = –1.364*(9.2) = –12.5 kcal/mol (–52.5 kJ/mol), therefore the equilibrium favors products.

(b)

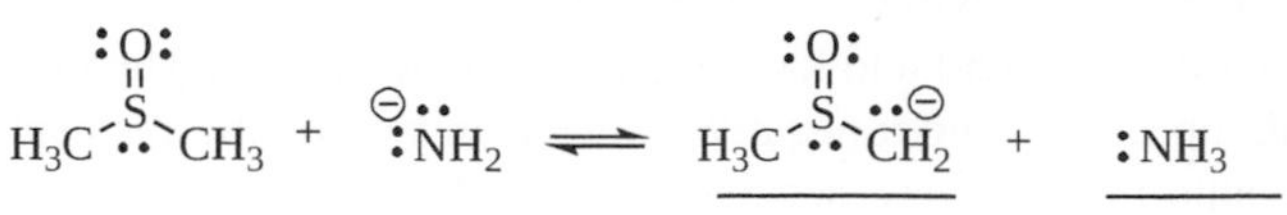

$\log K_{eq} = pK_{abH} - pK_{aHA} = 35 - 35 = 0$, $K_{eq} = 10^0 = 1$, $\Delta G° = -1.364 \log K_{eq} = -1.364*(0) =$ 0 kcal/mol (or 0 kJ/mol), therefore the equilibrium direction is balanced equally between reactants and products.

3.17 Trends: With the help of the pK_a chart when needed, rank all species, beginning with the numeral 1 to designate:

(a) The weakest base has the lowest pK_{abH}

Species	^{-}CN	$PhNH_2$	$^{-}NH_2$	HO^-	$(CH_3)_3CO^-$
pK_{abH}	9.2	4.6	35	14.0	19
Rank	2	1	5	3	4

(b) The weakest acid has the highest pK_a

Species	H_3O^+	HF	MeOH	HCN	NH_3
pK_a	0.0	3.2	15.2	9.2	35
Rank	5	4	2	3	1

3.19 Draw a reasonable proton transfer for the reaction below, and calculate the value of K_{eq}. A cyanoketone pK_a is not listed; it is about halfway between the dicyano and the diketone.

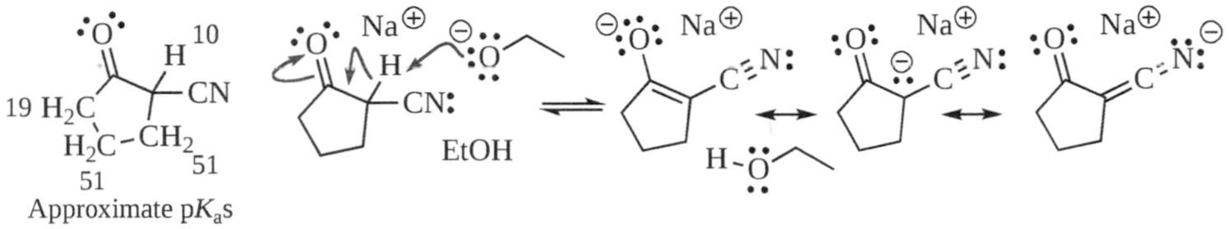

The most acidic H is between the two EWGs, the nitrile and the ketone, because the conjugate base is delocalized onto both giving good stability. However, the pK_a for that H has to be estimated as halfway between the diketone at 9 and the dinitrile at 11.2 or equal to about 10.1. The ethoxide base has a pK_{abH} of 16 giving a $\log K_{eq} = pK_{abH} - pK_{aHA}$ or $16 - 10.1 = 5.9$ or $K_{eq} = 10^{+5.9}$ highly favoring products.

3.21 Draw the most reasonable proton-transfer product given the following conditions.

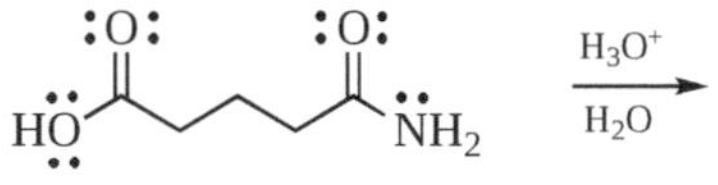

There are four possible lone pairs to protonate: The two left structures below put a positive charge next to the partial plus of the carbonyl, destabilizing the charge. The middle and right two structures delocalize the positive charge, and that is better. The right structure puts a plus on a less electronegative atom and is best.

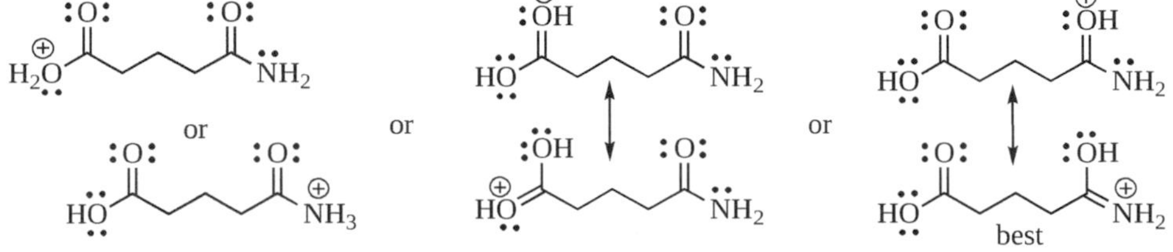

3.23 Draw the proton-transfer products for the following reaction and using the pK_a chart in the Appendix, calculate the value of K_{eq} and estimate $\Delta G°$ at room temperature. Draw a qualitative energy diagram.

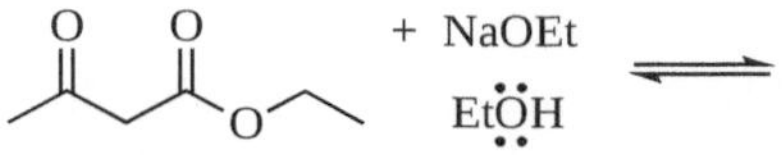

The most acidic group is the CH_2 between the two EWGs. log K_{eq} is $pK_{abH} - pK_{aHA}$ = 16 – 10.7 = 5.3, so K_{eq} = $10^{+5.3}$ and $\Delta G°$ = –1.364 log K_{eq} = –1.364*5.3 = –7.2 kcal/mol (–30.2 kJ/mol). The reaction favors the weaker acid and base of products.

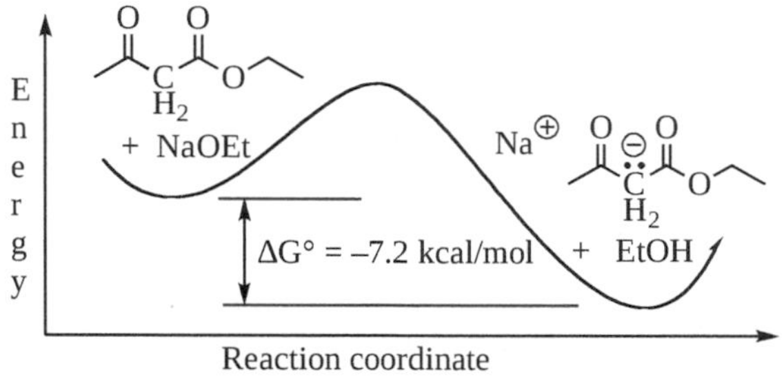

3.25 Draw the proton-transfer products and calculate the value of K_{eq} and $\Delta G°$ at room temperature. The cyanoester is not on the chart, but is easily estimated as halfway between the dicyano and the diester.

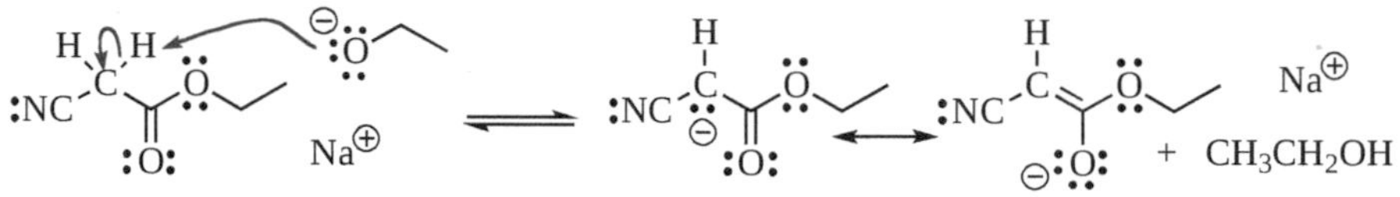

The pK_a of diester is 13 and the pK_a of dicyano is 11.2 therefore the pK_a of the cyanoester is (13 + 11.2)/2 = 12.1. The ethoxide base has a pK_{abH} of 16 giving a log K_{eq} = $pK_{abH} - pK_{aHA}$ or 16 – 12.1 = 3.9 therefore, K_{eq} = $10^{+3.9}$ highly favoring products. $\Delta G°$ = –1.364 log K_{eq} = –1.364*(3.9) = –5.3 kcal/mol (–22.2 kJ/mol).

3.27 In cells, enolpyruvic acid is in equilibrium with pyruvic acid. In the lab in acidic media, the first step most likely is a protonation of enolpyruvic acid to give a resonance-delocalized cation intermediate, followed by a deprotonation to give pyruvic acid. With those hints, propose a reasonable mechanism. Draw both resonance forms of the intermediate cation and decide which form is major. Use H_3O^+ to protonate and H_2O to deprotonate.

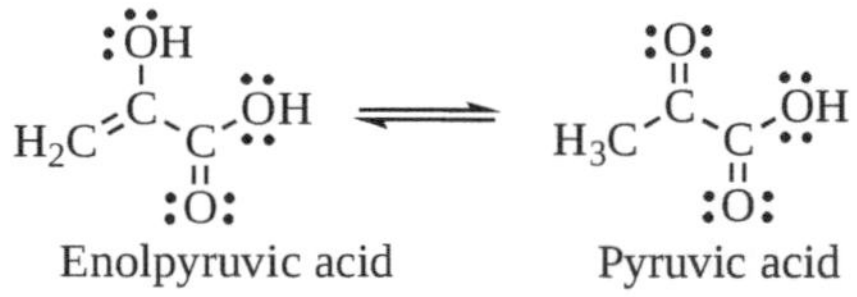

Comparing the two structures, we can see a proton was added to the end methylene of the pi bond and a proton has been lost from the oxygen. Given the hints, we can draw a mechanism.

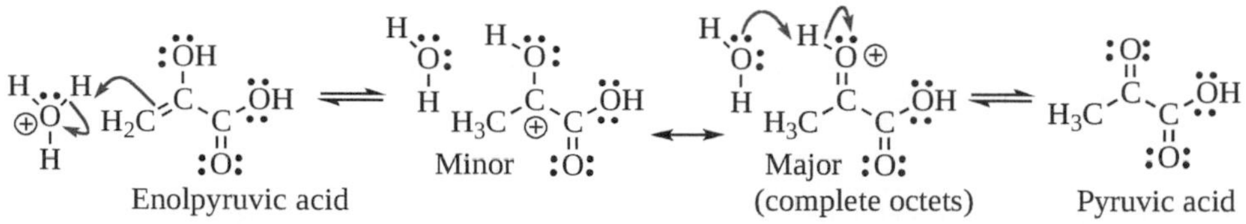

3.29 For the simplest amino acid, glycine, NH_2CH_2COOH (pK_{a1} = 2.31 and pK_{a2} = 9.24), using crossing and break points, draw a species vs. pH graph for glycine, like Figures 3.2 and 3.3. In acid, the nitrogen is protonated. What is the major species in solution at pH 6.0?

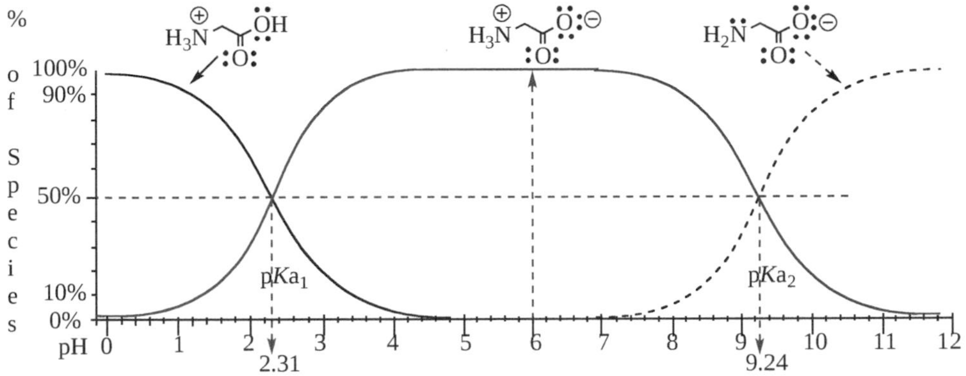

We need only a qualitative figure. Since pH = $pK_a + \log([A^-]/[HA])$, then one pH unit acidic from the pK_{a1} the $[A^-]/[HA]$ratio is 1:10. The curves cross at each pK_a since pH = pK_a when $[HA] = [A^-]$. Since there are two pK_as, 2.31 and 9.24, there are two crossing points. At one pH unit away from the pK_a, the curves are at 9% near the bottom of the graph, or 91% if near the top. At two pH units away from the pK_a, the curves are at 1% near the bottom of the graph, or 99% if near the top. The vertical dashed arrow on the chart at pH 6 points to the zwitterionic species present in solution at that pH.

3.31 Sugars isomerize between aldehyde and ketone forms via an enediolate HO–C=C–O⁻ or enediol intermediate. Give a reasonable proton-transfer mechanism in basic water for the reaction below. Since the alcohol and the H adjacent to the C=O have almost the same acidity, there are two possible first steps. Explore both so you recognize the dead end.

The dead end is deprotonating the OH, because this albeit good process does not have an easy way to go on to product.

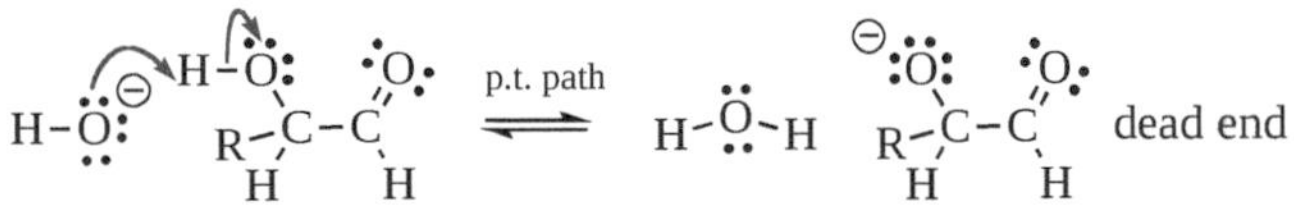

Deprotonating the H next to the C=O gets us to the enediolate or its resonance form. A proton transfer gets us to the enediol or another enediolate, which then protonates to give the product.

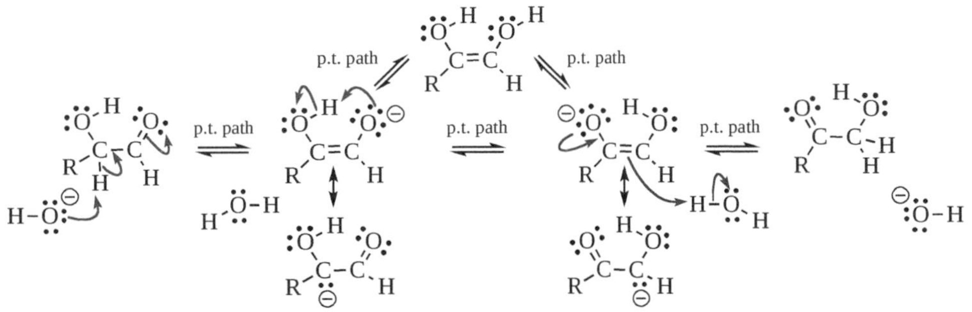

Chapter 4 Odd-Numbered Exercise Answers

4.1 Trends: All species are ranked beginning with the numeral 1.

(a) The most stable carbocations are lone pair stabilized. Lone pairs (Lp) on less electronegative atoms are better donors. Pi bond resonance (Pi) stabilizes to a lesser degree. Alkyl group (R) substitution stabilizes the least. Three of a lesser stabilization is better than one of the next better stabilization.

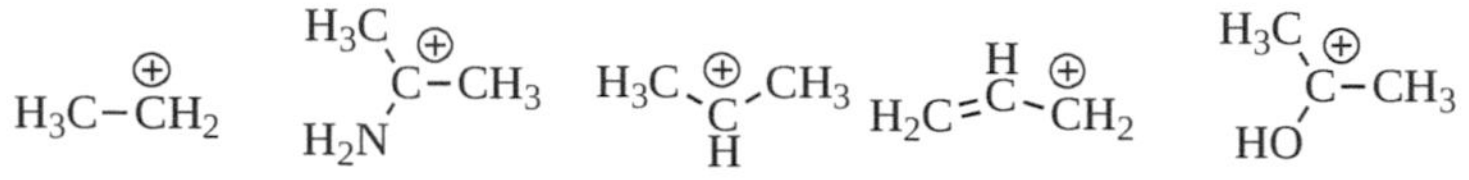

Rank:	5	1	4	3	2
Stabilized by	1R	1Lp, 2R	2R	1Pi	1Lp, 2R

(b) The most stable carbanion has the lowest pK_{abH}

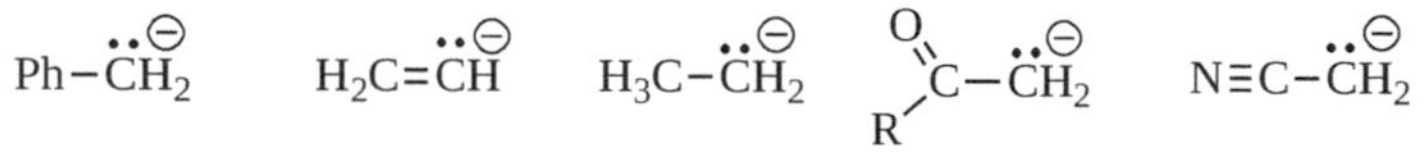

pK_{abH}:	40	44	50	19.2	25
Rank:	3	4	5	1	2

(c) The best EWG has the lowest pK_a for CH_3–EWG. The better the EWG, the more it stabilizes the conjugate base, $^{-}CH_2$–EWG, making the acid more acidic.

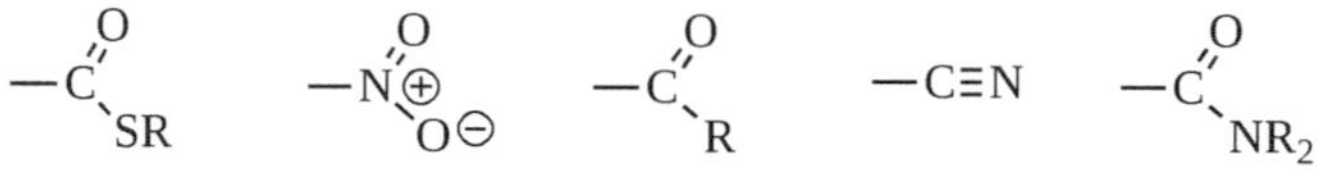

CH_3-EWG pK_a:	21	10.2	19.2	25	29.4
Rank:	3	1	2	4	5

4.3 In the reaction below, find the leaving group then identify the nucleophile. With that information, draw the **substitution** product and decide which path is more likely, S_N1 or S_N2.

S_N1 is most likely: good carbocation, bromide is good L, ethanol is weak Nu and a polar solvent. The carbon bearing the leaving group is too hindered for the S_N2.

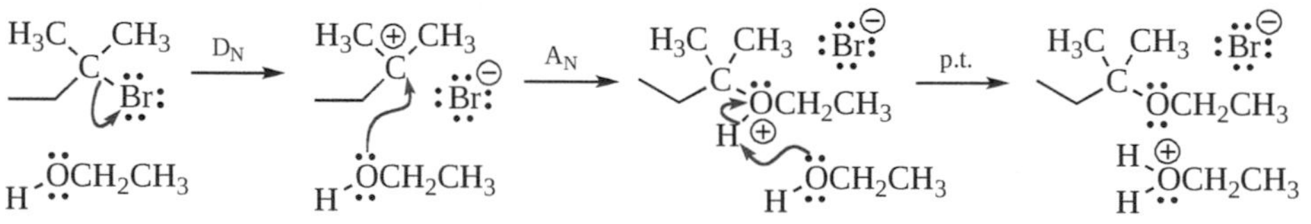

4.5 In the reaction below, find the leaving group. Draw Lewis structures for any carbon(s) next to the leaving group to see which H atoms could be lost. With that information, draw the elimination product and decide which path is more likely, E1, E2, or E1cB.

E2 is most likely: bromide is a good L, and *tert*-butoxide is a strong base. E1 would form a highly unstable 1° carbocation. There is no EWG to make the CH acidic for the E1cB.

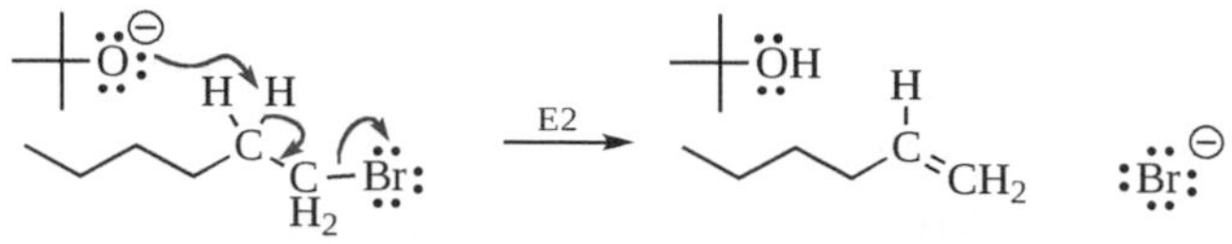

4.7 Draw the addition product. Decide which path is more likely, Ad_E2, Ad_E3, or Ad_N2. The addition will go via the most stable carbocation (Markovnikov's Rule). The secondary benzylic cation intermediate is moderately stable, so the surface tilts toward the cation corner. The Ad_E2 addition is most reasonable. An Ad_E3 addition usually occurs when the carbocation is less stable. The Ad_N2 addition fails the ΔpK_a rule cross-check, climbing over 50 units.

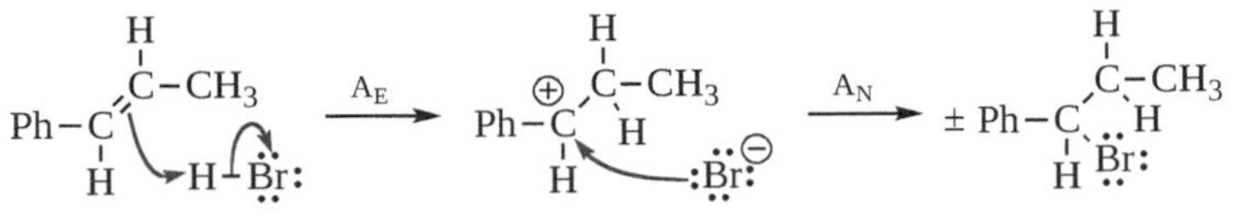

4.9 Draw the product of addition–elimination. What is the nucleophile? Decide whether the route is basic or acidic.

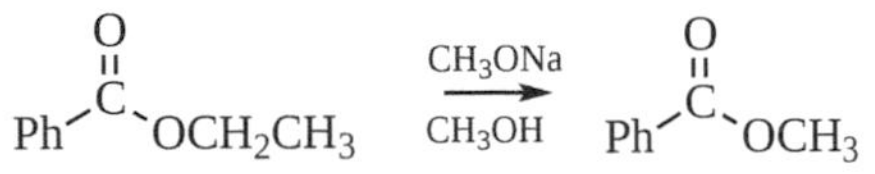

It is in a basic medium with a good nucleophile, methoxide pK_{abH} 15.2. The addition–elimination reaction, Ad_N then E_β, forms a slightly stronger base pK_{abH} 16. This slightly favors reactants but would be easily biased by mass balance. Since the solvent is usually in excess, a reasonable addition–elimination mechanism would form the methyl ester:

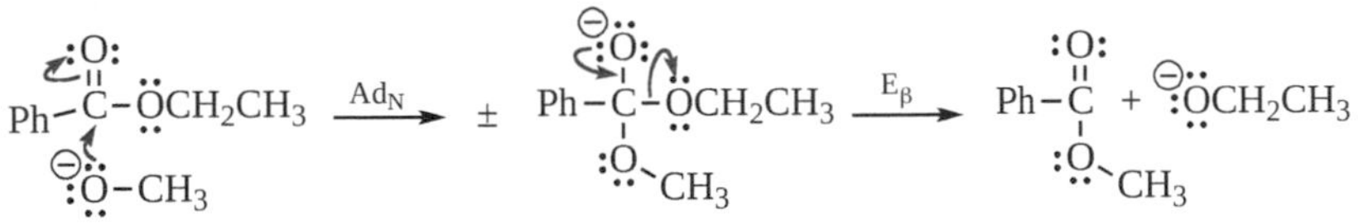

4.11 Draw the product of electrophilic aromatic substitution. The E adds just once.

The large *tert*-butyl group on the ring is an ERG and directs para because the ortho site is too hindered. The *tert*-butyl carbocation can do an electrophilic aromatic substitution by an expected A_E + D_E mechanism:

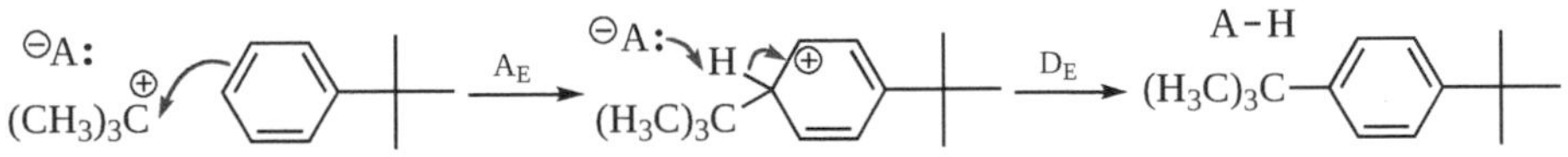

4.13 Use the appropriate electron flow pathway(s) to rationalize the product of these reactions. First, decide which of the four reaction archetypes has occurred. These mechanisms have only one electron source and one sink present in the reaction mixture.

(a) Addition by Ad_E2 is most likely. Simple pi bond source, good carbocation formed, powerful electrophile, Markovnikov's rule followed.

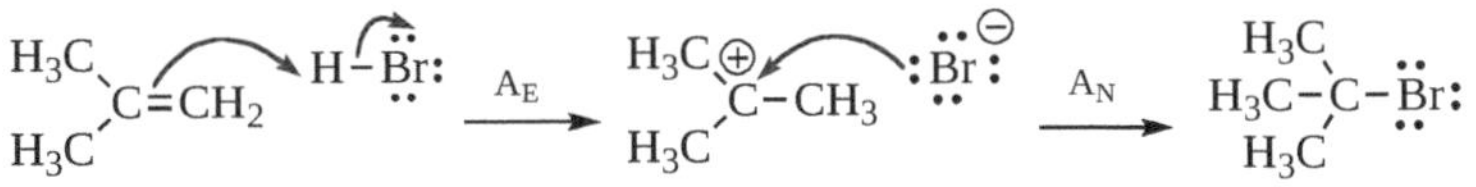

(b) Substitution by S_N2 is most likely. Good Nu, good access to C–L, excellent L. A highly unstable methyl carbocation would form by D_N so S_N1 is out.

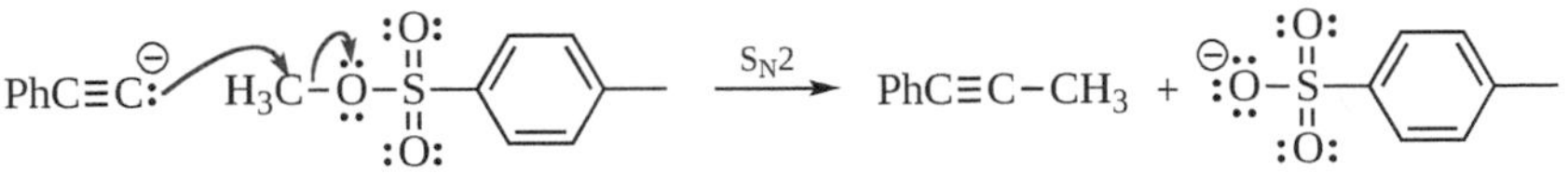

4.15 First decide which of the four reaction archetypes has occurred. Use the appropriate electron flow pathways to rationalize the product of these **two-step** reactions. The first step is a favorable proton transfer that generates the electron source for the reaction; there is only one electron sink.

(a) The proton transfer is favorable; log K_{eq} = pK_{abH} – pK_{aHA} = 16 – 13 = 3, so p.t. K_{eq} = 10^{+3}. Substitution by S_N2 is most likely: a good Nu, good access to C–L, a and good L. The S_N1 is not viable, for a poor 1° carbocation would form by D_N.

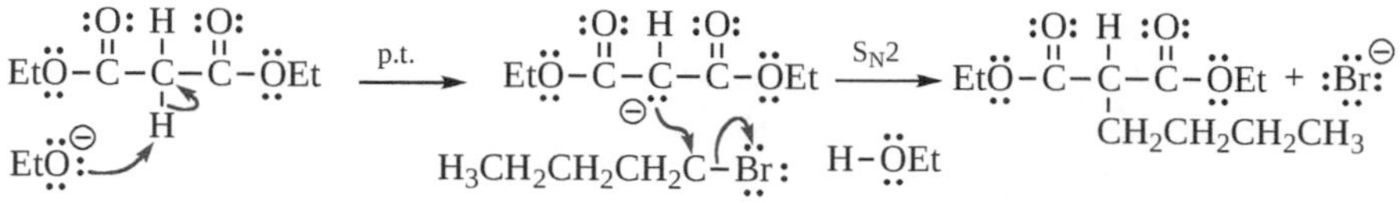

(b) Calculated cyanohydrin pK_a = 12.9, so log K_{eq} = pK_{abH} – pK_{aHA} = 14.0 – 12.9 = 1.1, K_{eq} = $10^{+1.1}$, thus favorable. Elimination by E1cB is most likely. CN is a relatively poor L. The E1cb passes the ΔpK_a rule, incoming pK_{abH} 12.9, outgoing 9.2, so downhill 3.7 units. The E2 elimination is also possible, but the E1 is not, for it requires a better leaving group.

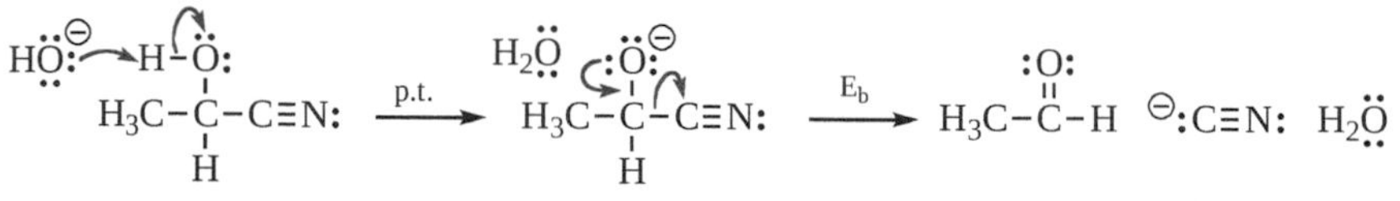

4.17 Use the ΔpK_a rule to predict whether each of these reactions favors reactants or products.

(a) Incoming pK_{aHNu} is +19.2, outgoing pK_{aHL} is +6.5 for a drop of 12.7 pK_a units, and therefore products are favored.

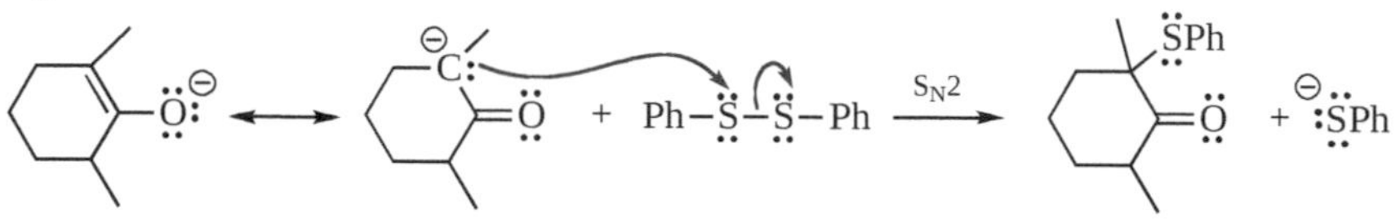

(b) Incoming pK_{aHNu} is –10, outgoing pK_{aHL} is +7 for a climb of 17 pK_a units, and therefore reactants are favored.

$$H_3C{-}SH + I^{\ominus} \rightleftharpoons HS^{\ominus} + H_3C{-}I$$

4.19 Use the ΔpK_a rule to determine which of the two following alternatives is the lower-energy process. The reaction would favor the formation of the weaker base.

CH_3COOEt + HO^- reacting to form H_2O + $^-CH_2COOEt$
pK_{abH} = 14.0going up 11.6 units to 25.6 is less favored than

CH_3COOEt + HO^- reacting to form EtOH + CH_3COO^-
pK_{abH} = 14.0 going down 9.2 units to 4.8, which is more favored.

4.21 Use the carbocation stability trend to decide which of the following two alternatives is the lower-energy process.

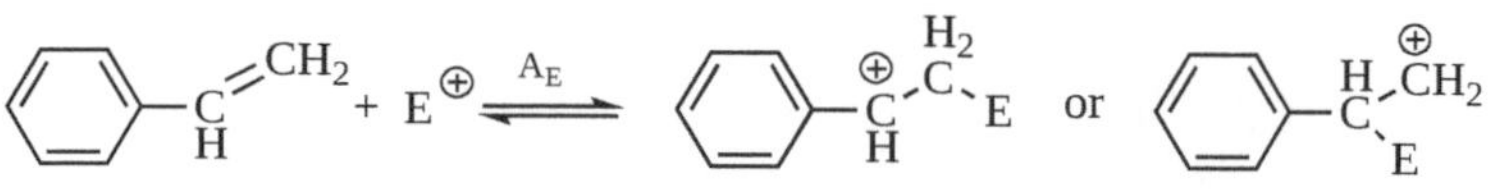

Resonance delocalized Unstable primary

The secondary benzylic carbocation is delocalized, and is much more stable than the primary carbocation. The secondary benzylic carbocation would be highly favored in this addition.

4.23 Decide which of the following two alternatives is the lower-energy process.

The second of the following two alternatives is the lower energy process. The first alternative is the formation of a more basic anion of pK_{abH} of about 10–13 from a nucleophile with a pK_{abH} of 4.8. The second alternative forms an anion of pK_{abH} –9 from a nucleophile with a pK_{abH} of 4.8 and so is favored by the ΔpK_a rule.

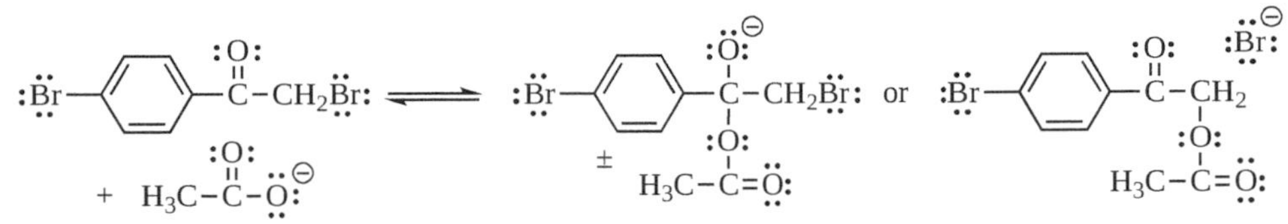

4.25 Explain the fact that thiocyanates react at sulfur whereas cyanates react at nitrogen. Draw Lewis structure for both products. These reactions demonstrate the HSAB principle: Hard with hard, and soft with soft. Alkyl iodides are soft electrophiles. With SCN^- the softest end is the sulfur because it is more polarizable and less electronegative than nitrogen. With OCN^- the softest end is the nitrogen because it is less electronegative than oxygen.

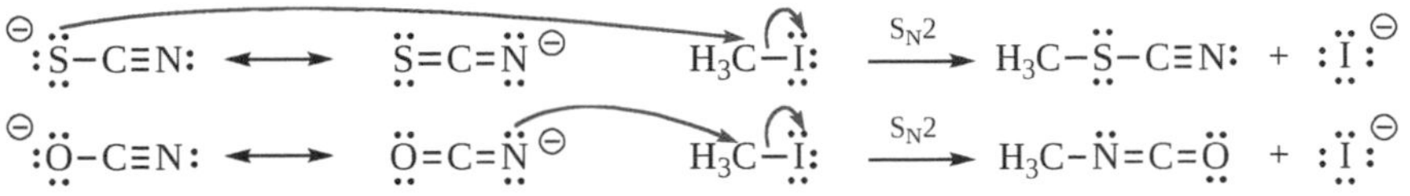

Chapter 5 Odd-Numbered Exercise Answers

5.1 Give the generic class of each of the following electron sources.

$LiAlH_4$	CH_3Li	NH_3	$CH_3OCH{=}CH_2$	Et_2NLi
MH_4^-	R–M	:Z (weak b/Nu)	:Z–C=C	:Z (good base)

5.3 Draw Lewis structures for the species in problem 5.1 and indicate on the structures which atom of the electron source would be attacked by the electrophile in a reaction. The atom on each molecule that is best for electrophilic attack is circled; it gives the most stable species upon E attack. With $LiAlH_4$ all four hydrogen atoms are equivalent.

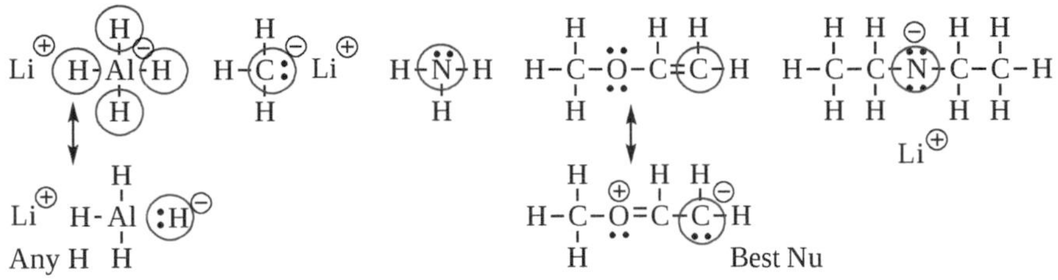

5.5 Give the generic class of each of the following electron sources.

NaH	$CH_3C{\equiv}CH$	$(CH_3)_2C{=}CH_2$	*t*-BuO^-	$PhCH_3$
MH (strong b)	C≡C	C=C	:Z (good base)	ArH

5.7 Draw Lewis structures for the species in problem 5.5 and indicate on the structures which atom of the electron source would be attacked by the electrophile in a reaction. The atom on each molecule that is best for electrophilic attack is circled; it gives the most stable species upon E attack. With $PhCH_3$ the two carbons ortho to the methyl are more hindered than the para carbon, but all react with small electrophiles.

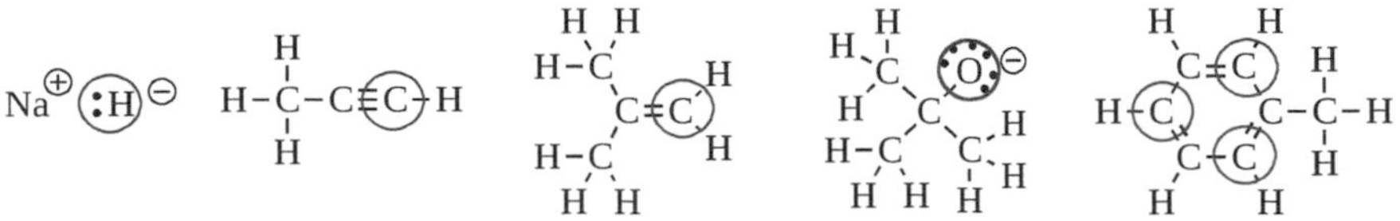

5.9 Give the generic class of each of the following electron sources.

$NaBH_4$	$^-O–CH=CH_2$	HO^-	$PhC(CH_3)_3$	NaOEt
MH_4^-	:Z–C=C	:Z (Nu or b)	ArH	:Z (Nu or b)

5.11 Draw Lewis structures for the species in problem 5.9 and indicate on the structures which atom of the electron source would be attacked by the electrophile in a reaction. The one atom on each molecule that is best for electrophilic attack is circled; it gives the most stable species upon E attack.

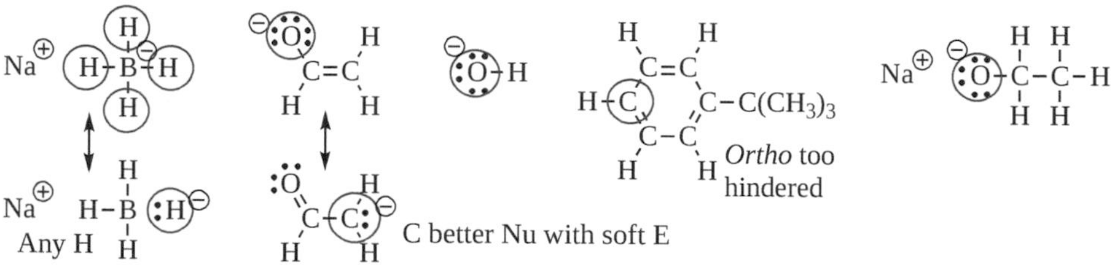

5.13 The one atom on each molecule that is best for electrophilic attack is circled; it gives the most stable species upon E attack. The first two have resonance delocalization of the positive charge; the third attacks the softer and more basic nitrogen atom, the last attacks the negative oxygen and maintains the aromaticity of the ring.

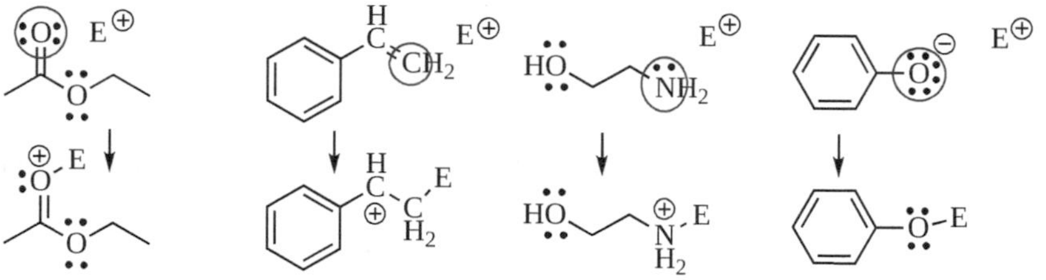

5.15 All species are ranked beginning with the numeral 1.

(a) The best allylic electron source bears the best electron-releasing group.

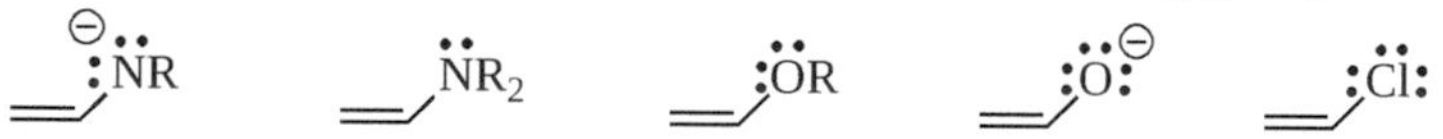

Rank: 1 3 4 2 5

(b) The most electron-rich aromatic ring is the most reactive toward electrophilic attack. Therefore electron-releasing groups activate, and electron-withdrawing groups deactivate.

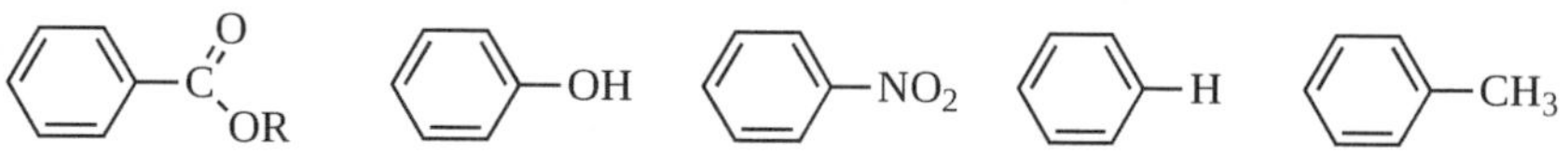

Rank: 4 1 5 3 2

5.17 Use Markovnikov's rule to predict the site of electrophilic attack on each of the pi bonds in the compounds in problem 5.16.

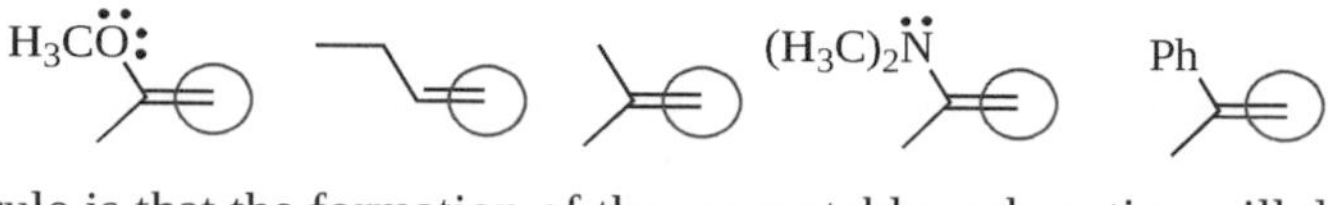

Markovnikov's rule is that the formation of the more stable carbocation will determine the site of electrophilic attack. Electrophiles attack all of the above on the pi bond end CH_2.

5.19 Use the nucleophilicity trend to decide which of the following two alternatives is the faster process in methanol solvent.

$CH_3CH_2CH_2Br + CH_3O^-$ displacing bromide to form $CH_3CH_2CH_2OCH_3$ or
$CH_3CH_2CH_2Br + CH_3S^-$ displacing bromide to form $CH_3CH_2CH_2SCH_3$

In a protic solvent like methanol, the softer CH_3S^- is a better nucleophile than CH_3O^- for reaction with a soft electrophile like an alkyl halide. Tight solvation of methoxide decreases that anion's ability to act as a nucleophile (Section 4.2.2). Therefore $CH_3CH_2CH_2Br$ + CH_3S^- displacing bromide to form $CH_3CH_2CH_2SCH_3$ is the faster process in methanol.

5.21 Rank all species, beginning with the numeral 1 to designate:

(a) The strongest base has the highest pK_{abH}

	NH_2^-	PhO^-	CH_3COO^-	CH_3O^-
pK_{abH}:	35	10	4.8	15.2
Rank:	1	3	4	2

(b) The most reactive complex metal hydride.

The less electronegative the metal, the more negative charge on the hydrogen atoms making them more reactive. The most reactive complex metal hydride is AlH_4^- with the greatest electronegativity difference. Borohydride is next. Cyanoborohydride is the least reactive because the cyano group inductively makes boron more electronegative reducing the negative charge on hydrogen.

	$NaBH_3CN$	$LiAlH_4$	$NaBH_4$
Rank:	3	1	2

5.23 Decide whether the following aromatic rings would be attacked *ortho/para* or *meta*.

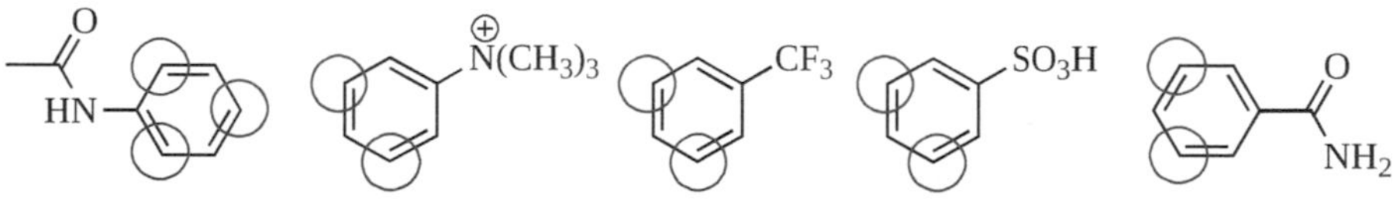

ortho/para	*meta*	*meta*	*meta*	*meta*
ERG	EWG	EWG	EWG	EWG

Electron-releasing groups direct ortho/para and electron-withdrawing groups direct meta. The amide ERG has a lone pair on N that can donate into the ring. All the others have a full or partial plus on the atom bound to the ring and act as an EWG.

Chapter 6 Odd-Numbered Exercise Answers

6.1 Give the generic class of each of the following sinks.

PhNCO	BF_3	HOBr	$(CH_3)_2CHBr$	$(CH_3)_2CO$
Y=C=Z	—A<	Y–L	—C–L	C=Y
$(CH_3CO)_2O$	CH_3Br	$AlCl_3$	CH_2=$CHCOCH_3$	CO_2
Y=C–L	—C–L	—A<	C=C–EWG	Y=C=Z

6.3 Draw Lewis structures for all the species in problem 6.1 and designate which atom(s) on each would be attacked by the electron source in a reaction. With HOBr, the more electronegative oxygen atom makes the bromine atom partially positive and prone to nucleophilic attack.

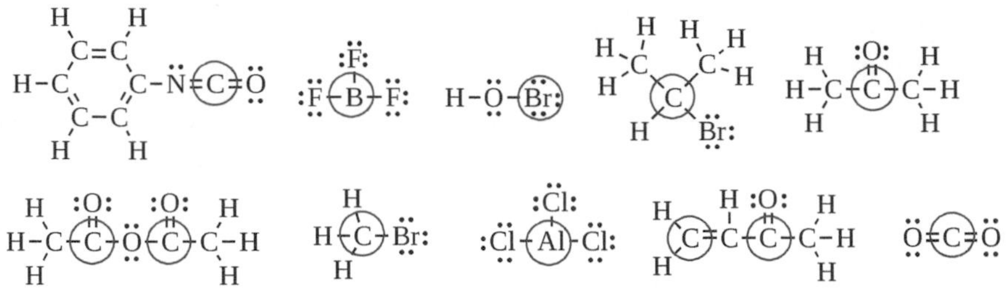

6.5 Give the generic class of each of the following sinks.

$COCl_2$	CH_3COOH	H_2O_2	CH_3COOEt	$CH_3OSO_3CH_3$
Y=C–L	HA or Y=C–L	Y–L	Y=C–L	C–L (tetrahedral)
PhCHO	EtOCOCl	Cl_2	Cl–C≡N	H_2C=C=O
C=Y	Y=C–L	Y–L	L–C≡Y	C=C=Z

6.7 Draw Lewis structures for all the species in problem 6.5 and designate which atom(s) on each would be attacked by the electron source in a reaction. Acetic acid, CH_3COOH, tends to be deprotonated in base as an HA and attacked at the carbonyl in acid as a Y=C–L sink.

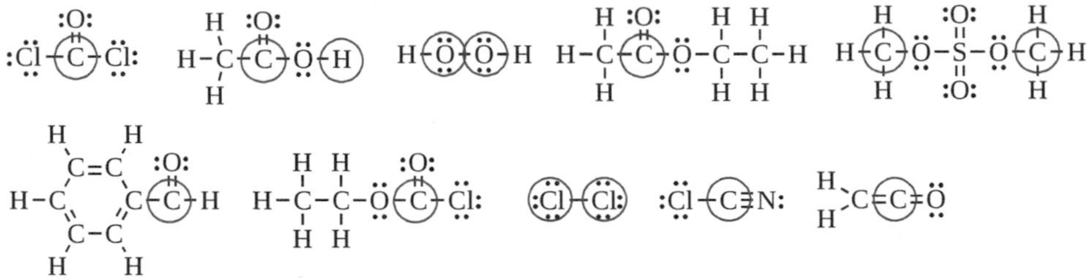

6.9 All species are ranked beginning with the numeral 1.

(a) The most reactive toward nucleophilic attack bears the poorest donor. The leaving group trend can also be used for this set.

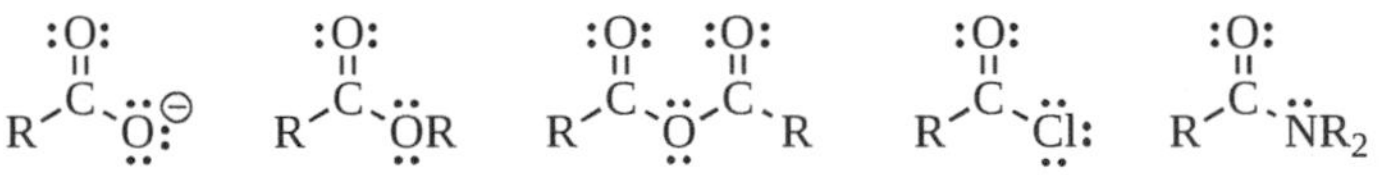

Rank: 5 3 2 1 4

(b) The most reactive conjugate acceptor bears the best electron-withdrawing group.

Rank:	1	2	3	4	5
CH_3EWG pK_a =	10.2	19.2	25.6	29.4	35

6.11 Sodium borohydride reduces ketones quickly but reacts very slowly to reduce esters. Since an aldehyde is more reactive than a ketone, an aldehyde will react very quickly with sodium borohydride.

6.13 Circle the best atom on each molecule for nucleophilic attack.

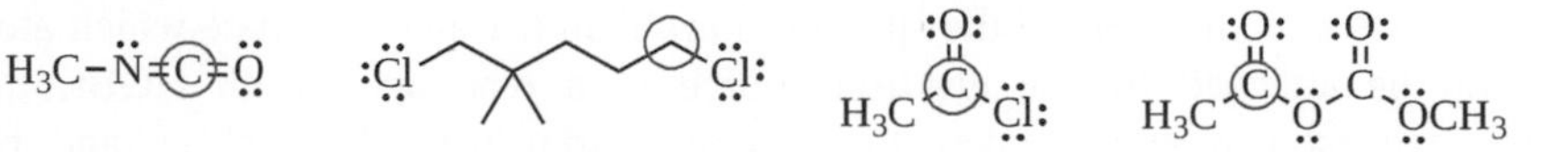

C most δ+ | least hindered C–Cl | C most δ+ | fewer ERG, C most δ+

6.15 Trends: rank all species; use the numeral 1 to designate:

(a) Most reactive toward nucleophilic attack has the poorest electron-releasing group.

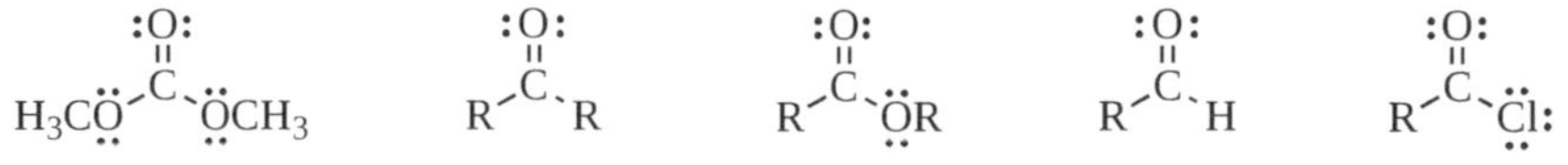

Rank:	5	3	4	2	1

(b) The strongest acid has the lowest pK_a

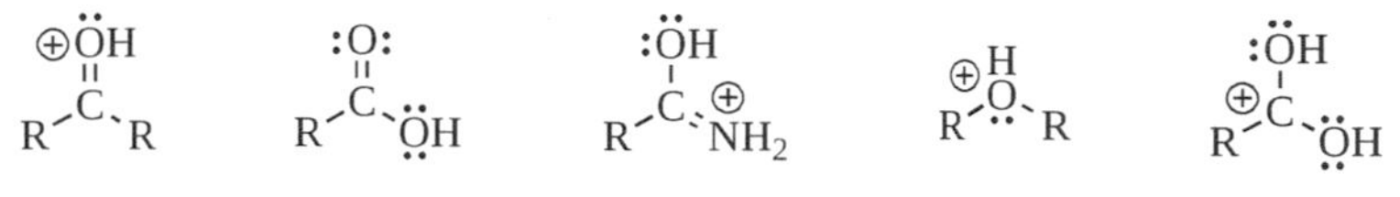

Rank:	1	5	4	3	2
pK_a =	–7	4.8	–0.5	–3.5	–6

6.17 Some carbocations are stable enough to appear on the pK_a chart. How does their reactivity correspond to their acidity? Rank the following, use numeral 1 to designate the most reactive, and explain the trend. The pK_a is listed below the structures.

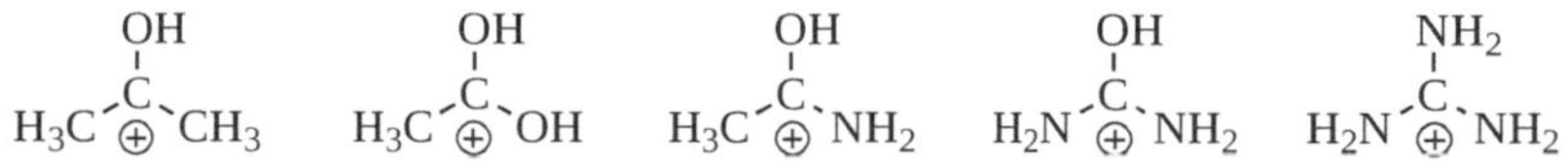

pK_a	–7	–6	–0.5	0.18	13.6
Reactivity rank:	1	2	3	4	5

The more acidic the acid, the less stable it is compared to its conjugate base. As we go from left to right, more lone pair donors are added, and the carbocation becomes more resonance stabilized, thus more stable and less acidic. The pK_a –6 species compared to the pK_a –0.5 species, shows that the less electronegative nitrogen donates its lone pair better than the more electronegative oxygen. The more stable a species is, the less reactive it is, both as an acid and an electrophile.

Chapter 7 Odd-Numbered Exercise Answers

7.1 Go back to Chapter 1, problem 1.5, and place the name of the path over each transformation arrow.

(a)

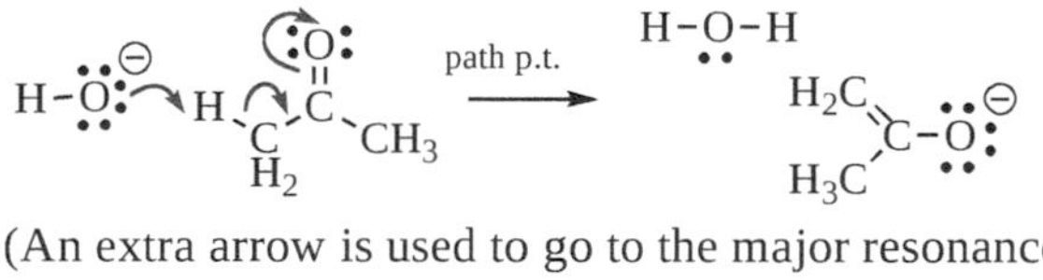

(An extra arrow is used to go to the major resonance form with the charge on O.)

(b)

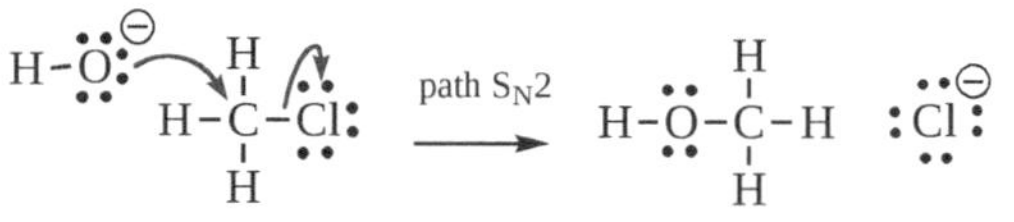

(c)

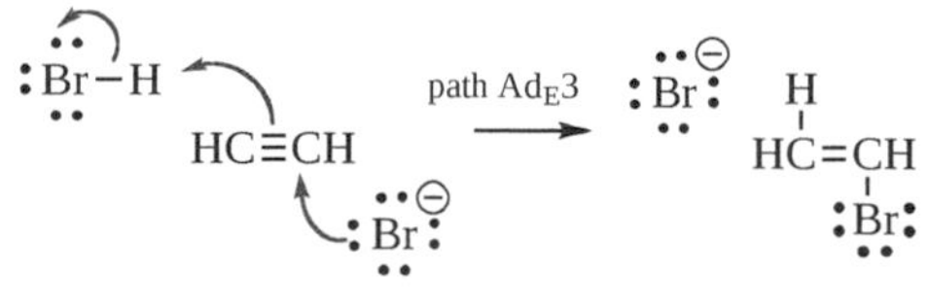

since there is an acid catalyst present, that should be the first step. This gives water as a good leaving group ($pK_{aHL} = 0.0$).

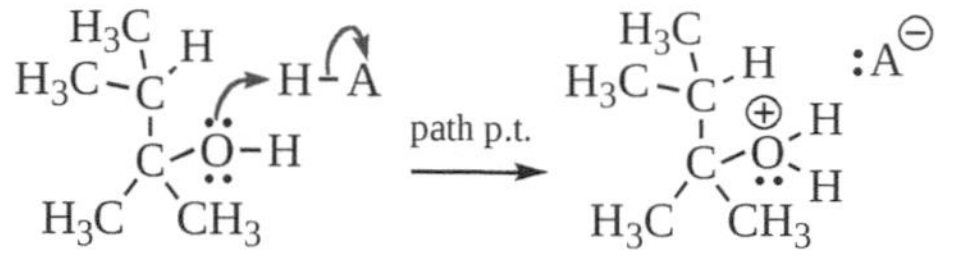

With the sink as a guide, the possible acidic paths are the E1 and E2. The E1cB needs base. The E1 passes the three tests for the slow D_N step: polar solvent, good leaving group ($pK_{aHL} = 0.0$), and a tertiary carbocation which is more stable than secondary. The reaction is most likely an E1, path D_N followed by D_E. Water is the best base present. E2 is also possible.

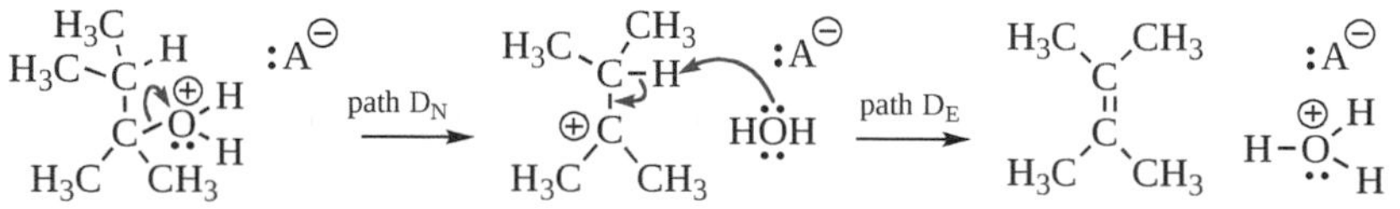

Chapter 8 Odd-Numbered Exercise Answers

On <u>all</u> of these problems, make sure the missing structure balances the reaction equation.

8.1 In the following problem, identify the electron sink and look at the product to identify the electron source and predict the pathway(s) to the product.

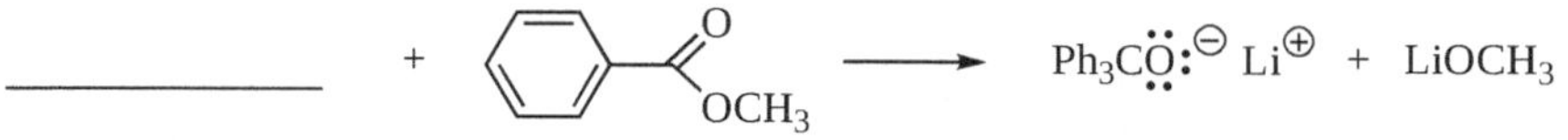

The sink is Y=C–L and two phenyl nucleophiles have been added. Looking at the products, the counterion for the Nu was Li, so the Nu was PhLi, an organometallic (Section 8.8.4). A likely path was an addition, then elimination, and then a second addition: paths Ad_N, E_β, Ad_N.

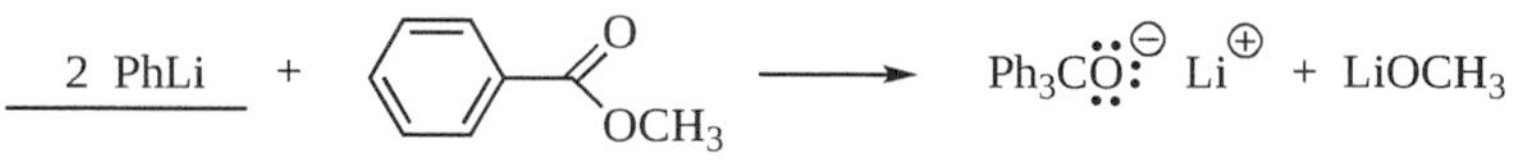

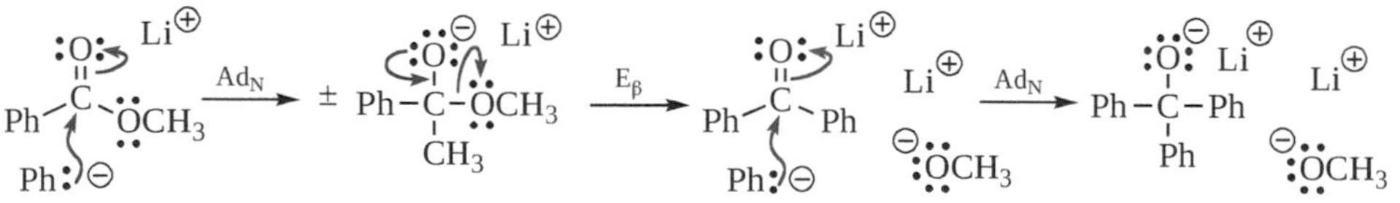

8.3 In the following problem, identify the electron sink and electron source and predict the pathway(s) to the neutral product.

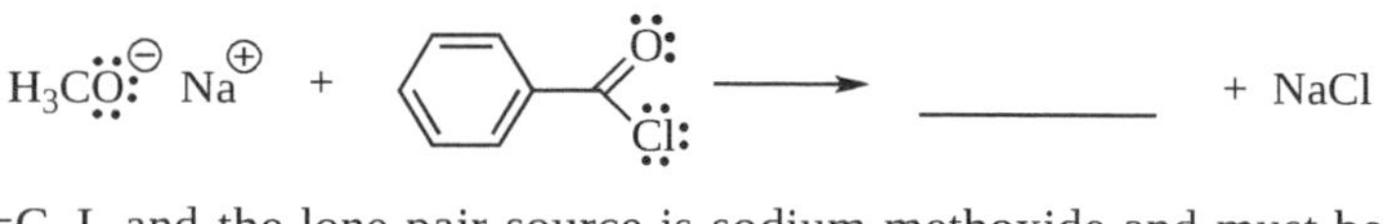

The sink is Y=C–L and the lone pair source is sodium methoxide and must be a nucleophile not a base since the sink does not have any acidic hydrogen atoms. The sink Y=C–L sink usually reacts by an addition, then elimination: paths Ad_N then E_β (Section 8.8.1).

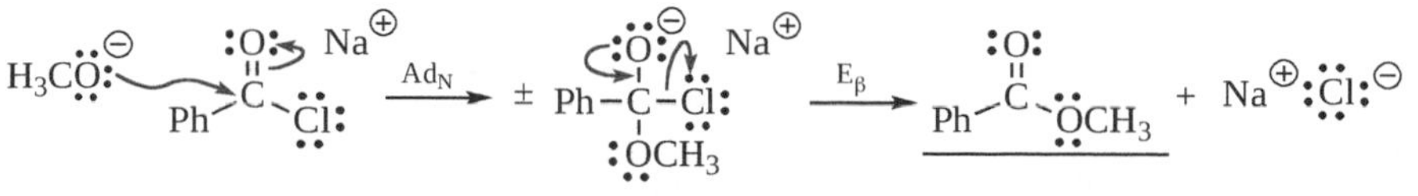

8.5 In the following problem, identify the electron source and look at the product to identify the electron sink and predict the pathway(s) to the product.

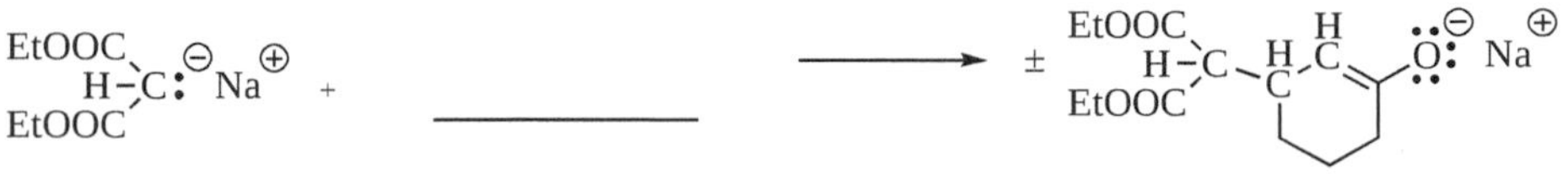

The electron source is the malonate anion, which has added to the electron sink. A way to identify the electron sink is to reverse the reaction by kicking the malonate back out. This gives a conjugate acceptor C=C–EWG as the sink. The Ad_N conjugate addition shown below (Section 8.7.5) is the most likely path in a basic medium.

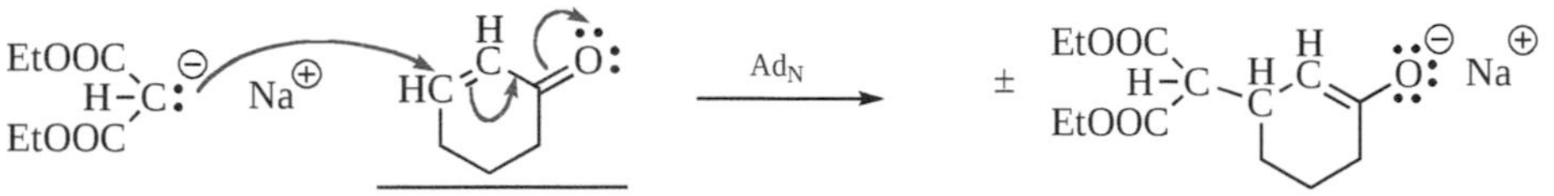

8.7 In the following problem, identify the electron sink and look at the product to identify the electron source and predict the pathway(s) to the product.

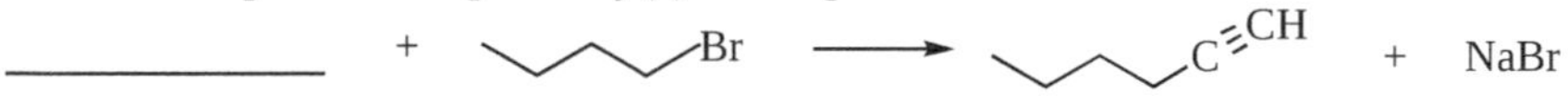

The electron sink is an sp^3 C–L and the source is an organometallic, acetylide anion (Section 8.4.4). Since the leaving group is on carbon, the pent path is out.

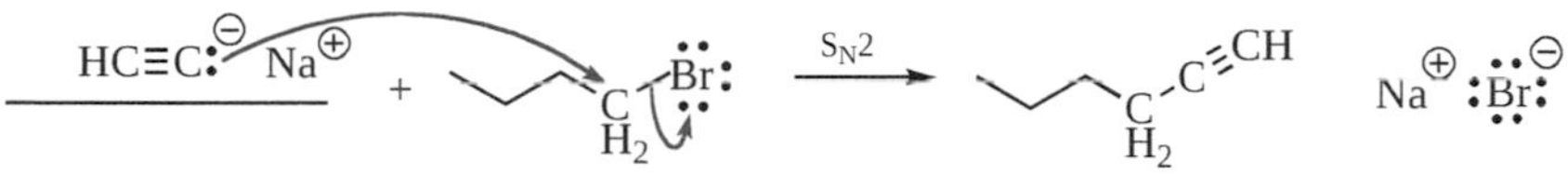

The possible carbocation from loss of the leaving group is primary, so it is very unstable; therefore the S_N1 is out. There is good access and a good leaving group for the S_N2, and the ΔpK_a rule cross-check passes, incoming $pK_{abH} = 25$, outgoing –9. The S_N2 is most likely.

8.9 In the following problem, identify the generic electron sink and electron source and predict the product.

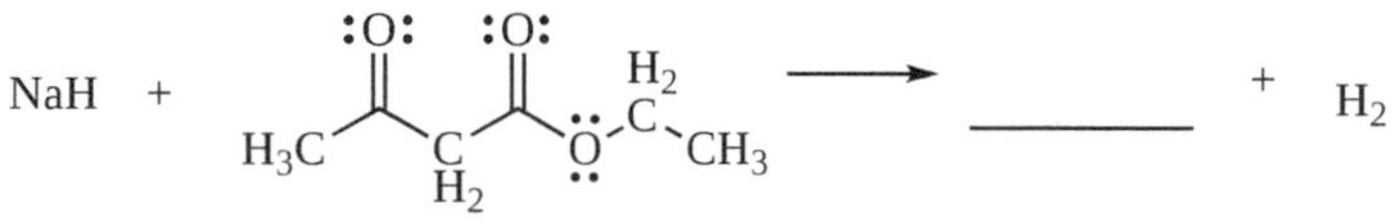

The Group 1A metal hydride sources like sodium hydride serve as bases as and are not nucleophiles (Section 5.3.2). The sink is then an acidic H on carbon, made acidic by two carbonyls as EWGs. The proton transfer, $\log K_{eq} = 24.3$ is very reasonable (Section 8.8.2).

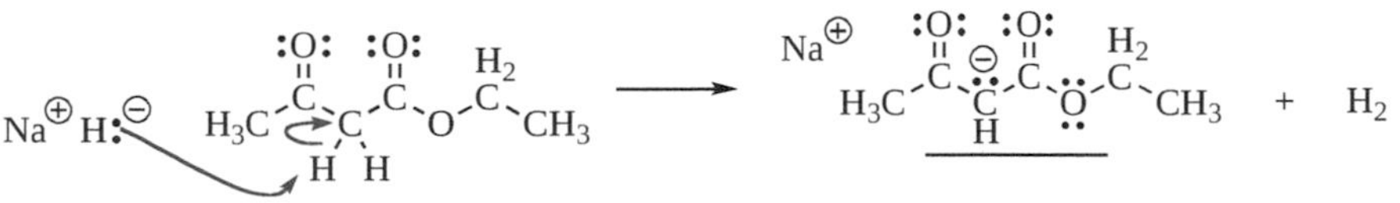

8.11 In the following problem, identify the electron sink and electron source and predict the product and the pathway(s) to the product.

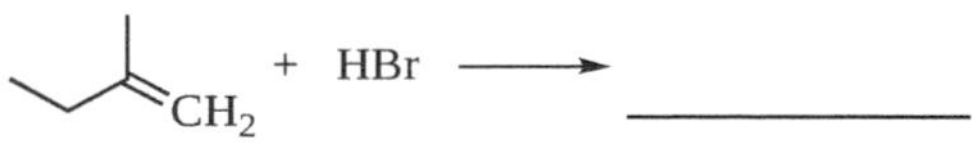

The electron source is a simple pi bond and the sink is an acid (Section 8.2.6).

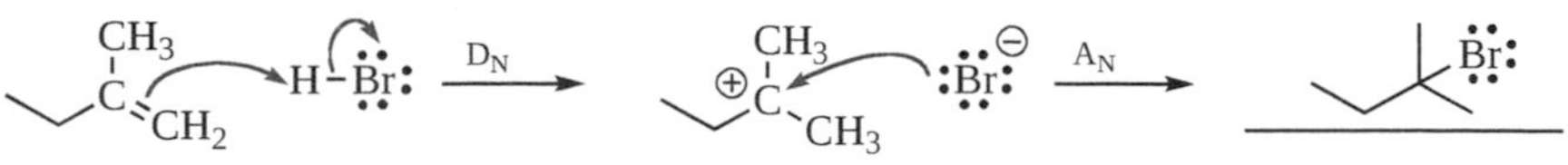

Since it is in an acidic medium and the carbocation that would be formed is tertiary and relatively stable, the most reasonable path is Ad_E2. Markovnikov's rule determines the regiochemistry.

8.13 In the following problem, identify the electron source and look at the product to identify the electron sink and predict the pathway(s) to the product.

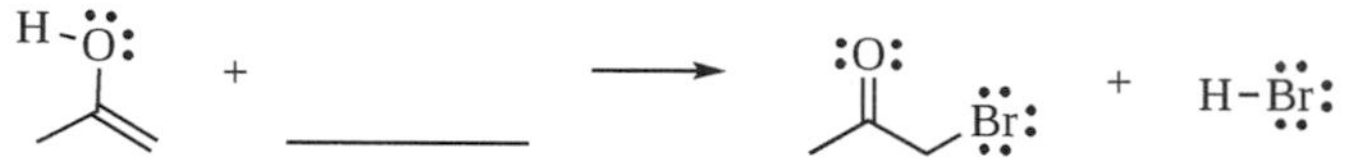

The electron source is the enol, an allylic source, C=C–Z: which has added to the Y–L sink, elemental bromine to make a C–Br bond. A reasonable route is an S_N2 substitution (Section 8.3.5) followed by a slightly uphill proton transfer to the bromide anion (log K_{eq} = –2). Due to the electronegativity of the halogens, an S_N1 is not reasonable.

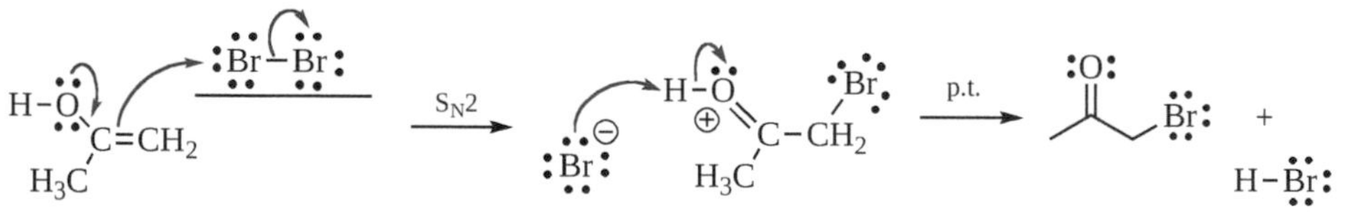

8.15 In the following problem, identify the electron sink and electron source and predict the pathway(s) to the product.

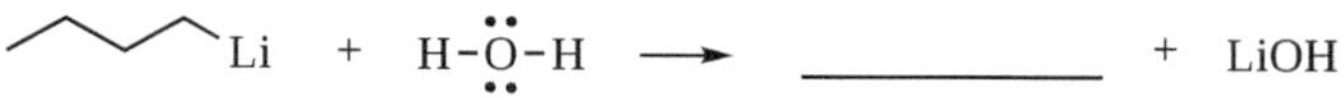

The electron source is the R–M organometallic, which is both a good Nu and very strong base pK_{abH} = 50. The sink is water, which cannot undergo nucleophilic attack but can serve as an acid, pK_a = 14.0. The highly exothermic proton transfer is 36 pK_a units downhill, quenches the organometallic, and is the reason organometallic reactions must be run under very dry conditions (Section 8.2.4).

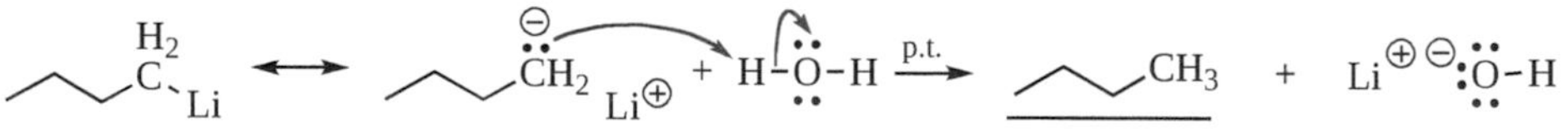

8.17 In the following problem, identify the electron sink and electron source and predict the pathway(s) to the product.

$$2\ NaOH + Cl_2 \longrightarrow \underline{\qquad\qquad} + NaCl + H_2O$$

The source is hydroxide, which is both a good nucleophile and strong base pK_{abH} = 14.0. The sink is elemental chlorine, a Y–L, which is expected to do an S_N2 substitution (Section 8.3.1). To get the water product formed, hydroxide must also serve as a base and deprotonate the hypochlorous acid formed, pK_a = 7.5. A reasonable mechanism is an S_N2 substitution followed by a proton transfer. This is how bleach, sodium hypochlorite, is made. Due to the electronegativity of the halogens, an S_N1 is not reasonable.

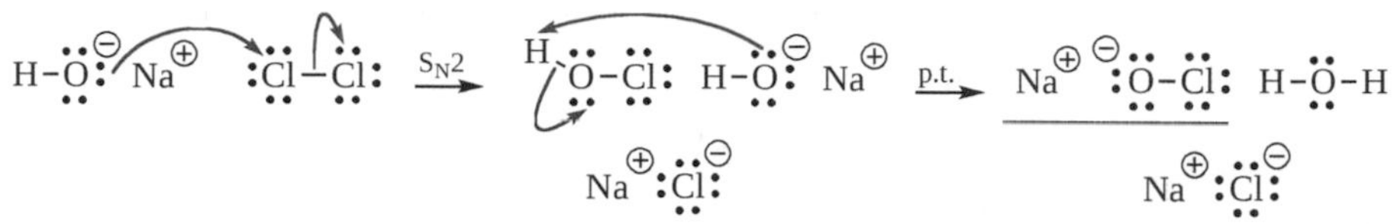

Chapter 9 Odd-Numbered Exercise Answers

9.1 For each of the following reactions decide whether substitution or elimination predominates. First identify the site hindrance, the quality of the nucleophile, and the strength of the base; then use the substitution/elimination decision matrices (Figure 9.6) to predict the main product.

	Substrate, hindrance	Reagent	Base	Nu	Main product
(a)	$CH_3CH_2CH_2CH_2OH$, primary	HBr	Weak	Good	Substitution
	Primary substitution quadrant. Weak base, good Nu. The alcohol protonates to give a good L. Substitution most likely by S_N2.				
(b)	$CH_3CH_2CH_2CH_2OTs$, primary	t-BuO^-	Strong	Poor	Substitution
	Primary mix quadrant. Tosylate leaving group favors substitution even though the Nu is hindered and a good base pK_{abH} = 19. Substitution most likely by S_N2.				
(c)	$CH_3CH_2CH_2CH_2Br$, primary	t-BuO^-	Strong	Poor	Elimination
	Primary mix quadrant. Primary bromide leaving group favors elimination. Poor Nu, hindered base has pK_{abH} = 19. Elimination most likely by E2.				
(d)	$(CH_3)_2CHCH_2Br$,	I^-	Weak	Good	Substitution
	Hindered primary is like secondary. Substitution most likely by S_N2.				

9.3 For each of the following reactions decide whether substitution or elimination predominates. First identify the site hindrance, the quality of the nucleophile, and the strength of the base; then use the substitution/elimination decision matrices (Figure 9.6) to predict the main product.

	Substrate, hindrance	Reagent	Base	Nu	Main product
(a)	$(CH_3)_2CHBr$, secondary	EtS^-	Weak	Good	Substitution
	Secondary substitution quadrant primarily by S_N2. Great Nu, base pK_{abH} = 10.6.				
(b)	$(CH_3)_2CHCHBrCH_3$, 55°C	CH_3COO^-	Weak	Good	Elimination
	Hindered secondary similar to tertiary. pK_{abH} = 4.8, E2 elimination most likely.				
(c)	$CH_3CH_2CBr(CH_3)_2$, tertiary	CH_3OH	Weak	Poor	Substitution
	Tertiary usually substitutes by S_N1. Base too weak for elimination, pK_{abH} = –2.2.				
(d)	$(CH_3CH_2)_3CCl$, tertiary	CH_3O^-	Strong	Good	Elimination
	Tertiary elimination quadrant usually E1 (or E2 with more base). pK_{abH} = 15.2.				

9.5 For each of the following reactions a decision discussed in this chapter is required. Choose which product would be preferred. All of the reactions occur in basic media.

9.5a *Source class* Z *Sink class* C=Y
Decision type Thermodynamic vs. kinetic enolate, Section 9.3

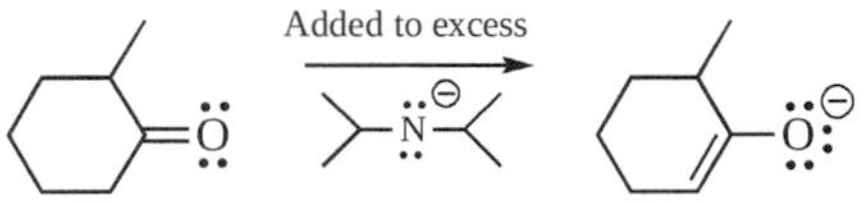

Rationale: The base is very strong (pK_{abH} 36) and hindered and the deprotonation is not at all reversible ($K_{eq} = 10^{+16.8}$). These are kinetic conditions; the most accessible H is removed.

9.5b *Source class* Z–C=C *Sink class* sp^3 C–L
Decision type Ambident Nucleophile, Section 9.4

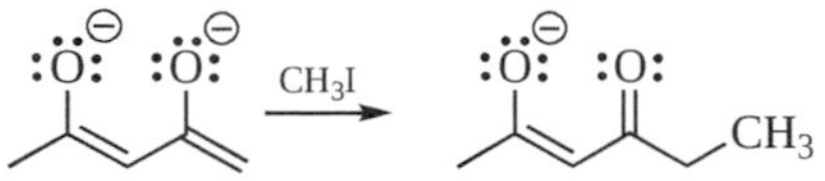

Rationale: These are kinetic conditions; the reaction drops over 40 pK_a units so is not reversible. Considering HSAB is next. The electrophile is soft. The softest nucleophile is carbon not oxygen, and the more reactive nucleophile of the two carbons is the one with the higher pK_{abH} (the end CH_2). This leaves the more stable delocalized anion as the product.

9.5c *Source class* R–M *Sink class* C=C–EWG
Decision type Ambident Electrophile, Section 9.6

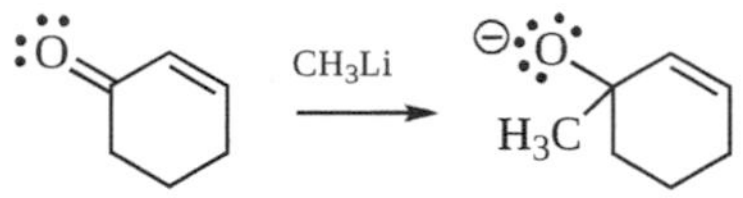

Rationale: The reaction by the ΔpK_a rule is not reversible (incoming pK_{abH} = 48, outgoing = 19) and the organometallic is hard. These conditions favor 1, 2 attack (Figure 9.7).

9.5d *Source class* Z *Sink class* H–A
Decision type To migrate or not, Section 9.8

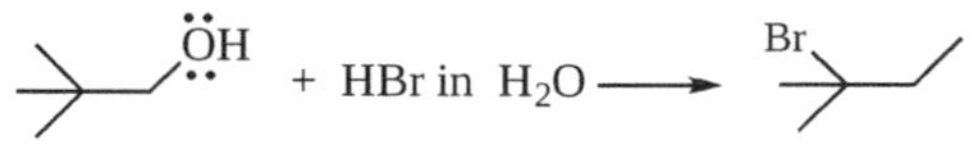

Rationale: An acidic medium can improve the leaving group. The D_N path is not viable because the unstable primary carbocation rarely forms. Therefore S_N1 and E1 routes are out. The $(CH_3)_3C$ group blocks the Nu approach for the S_N2 path. There is no available CH next to the C–L for the E2. This leaves only migration with loss of leaving group (1,2RL path). It would produce a reasonably stable tertiary carbocation, which the bromide ion then traps.

9.7 Provide a mechanism for the formation of the preferred products in problem 9.5. The source and sink for each part was discussed at the beginning of each answer for problem 9.5.

9.7a A reasonable mechanism is deprotonation, path p.t. See the rationale for problem 9.5a.

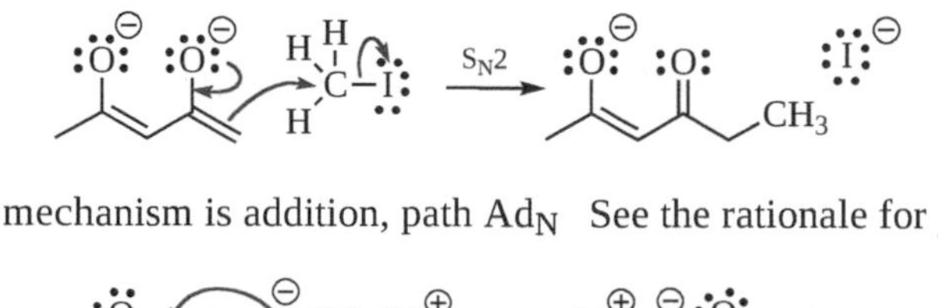

9.7b A reasonable mechanism is substitution, path S_N2. See the rationale for problem 9.5b.

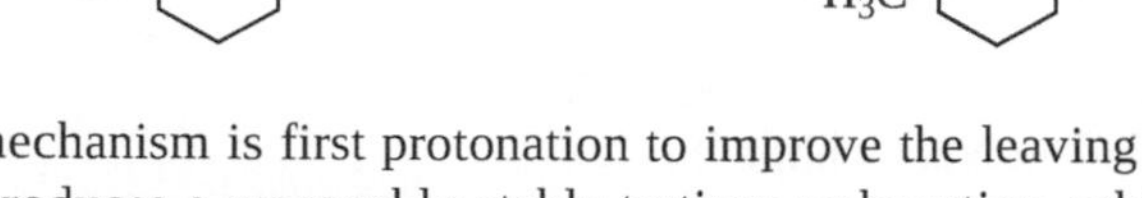

9.7c A reasonable mechanism is addition, path Ad_N See the rationale for problem 9.5c

9.7d A reasonable mechanism is first protonation to improve the leaving group. Then 1,2RL rearrangement path produces a reasonably stable tertiary carbocation, which the bromide ion traps. See the rationale for problem 9.5d.

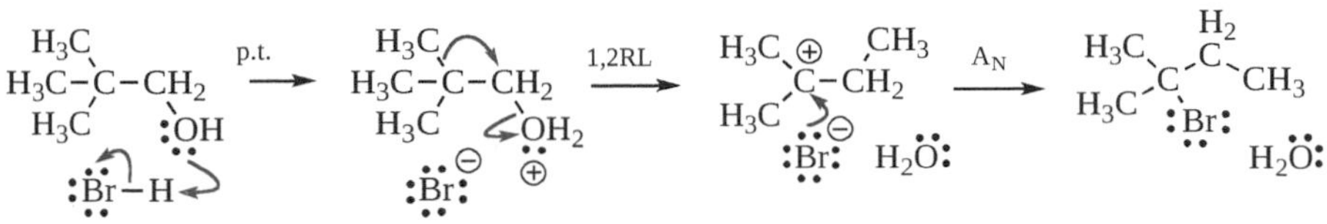

9.9 Each of the following reactions has only one electron source and sink. Give the product of the following reactions; a decision discussed in this chapter is required. Assume a mildly acidic work-up for the organometallic reactions that just protonates any anionic species but goes no further.

9.9a

Source class Z *Sink class* sp^3 C–L
Decision type Substitution/elimination, Section 9.5

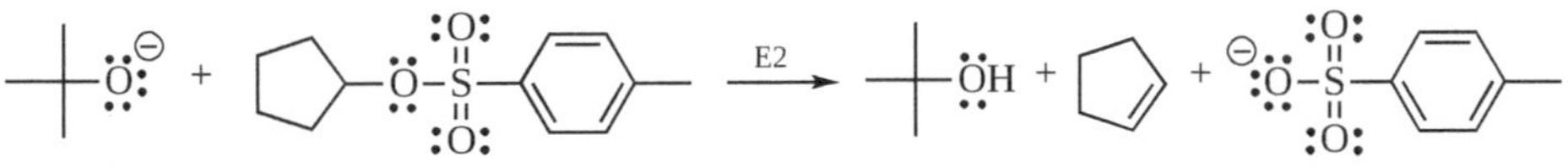

Rationale: The anion is a sterically hindered poor nucleophile and a good base. The secondary site is slightly hindered. Elimination is expected to predominate over substitution.

9.9b

Source class R–M (for first step) *Sink class* Y=C–L (for first step)
Decision type Multiple addition, Section 9.2

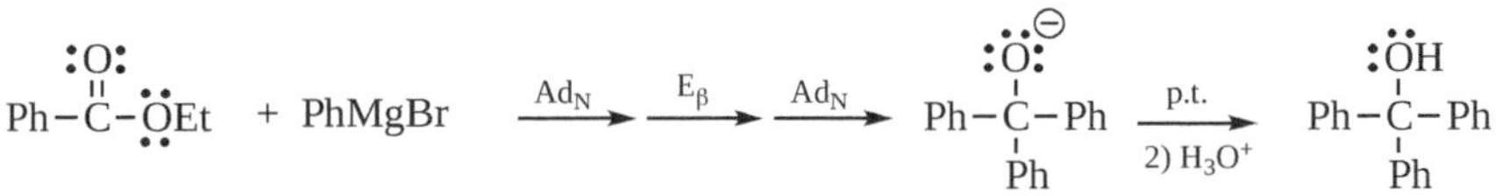

Rationale: Organomagnesium reagents easily add to ketones. Esters are less reactive than ketones, and thus the organometallic reacts faster with the ketone product than with the ester.

9.9c

Source class Z–C=C *Sink class* Y–L
Decision type Ambident nucleophile, Section 9.4

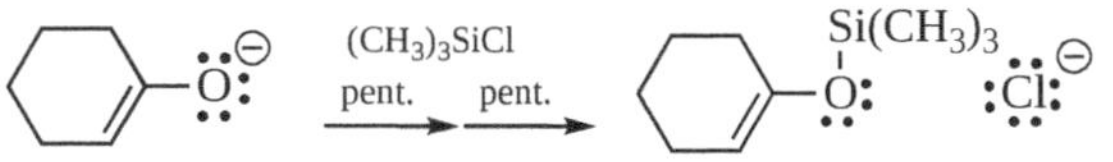

Rationale: The Si–Cl bond is polarized (electronegativity difference is 3.16 – 1.90 = 1.26), and therefore the electrophile is relatively hard. By the ΔpK_a rule, we expect this reaction to be very much downhill in energy (pK_{abH} 19.2 going to pK_{abH} –6) and thus to be irreversible. Irreversible with a hard electrophile is expected to go on the oxygen, the harder end of the enolate anion.

9.11 Provide a mechanism for the formation of the preferred products in problem 9.9. Combine the source and sink with the appropriate pathway(s). See problem 9.9 for rationales.

9.11a The most reasonable route is the E2 elimination.

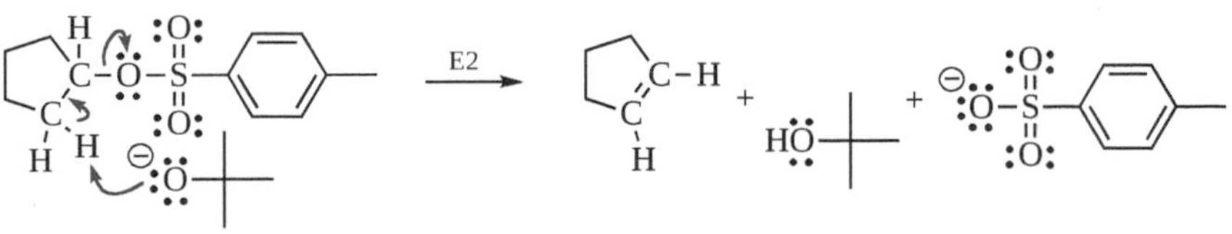

9.11b The most reasonable route is Ad_N + E_β, addition–elimination followed by a second addition, path Ad_N. Proton transfer to the alkoxide in the acidic work up completes the likely Ad_N2 process.

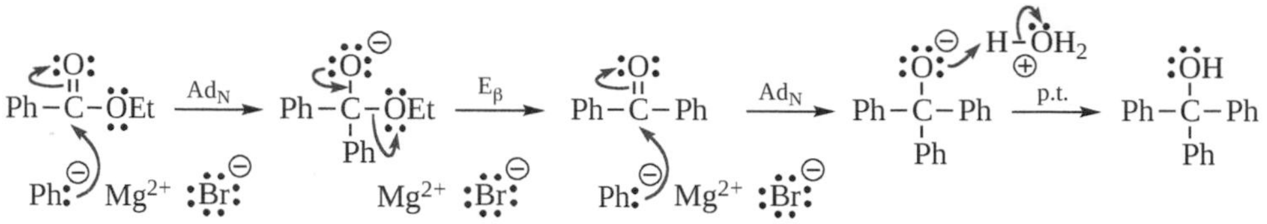

9.11c The most reasonable route substitution via a pentacovalent intermediate

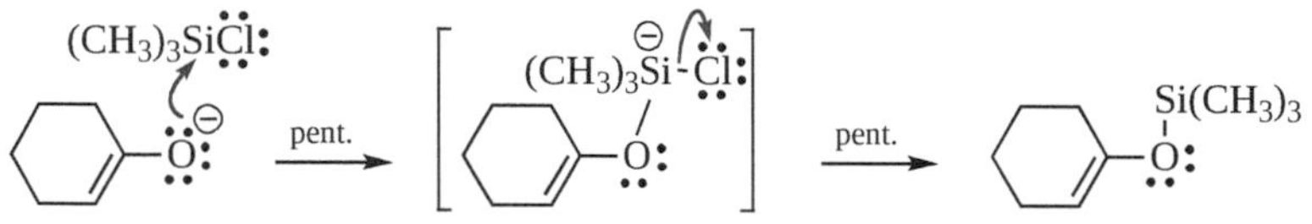

9.13 The following compounds rearrange under the reaction conditions; give the product and a mechanism.

9.13a See Section 4.7.3 for mechanism of this pinacol rearrangement.

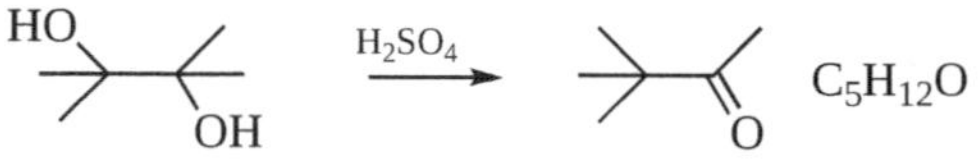

9.13b

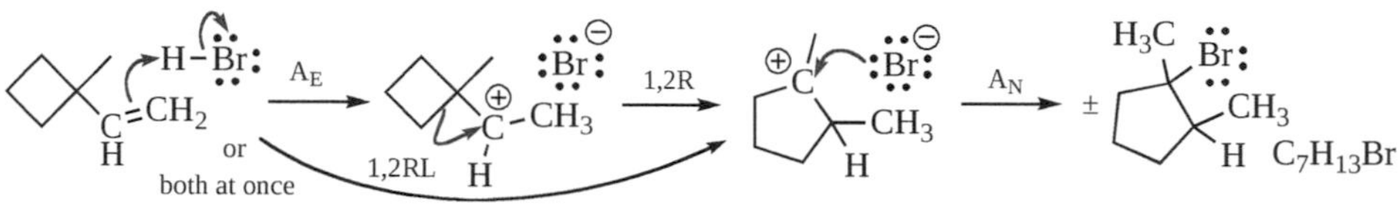

Two possibilities are illustrated by Figure 9.9. The pi bond can be protonated (path A_E) to form a secondary carbocation that then rearranges (path 1,2R) to the more stable tertiary carbocation. The rearrangement may also occur during the initial protonation of the pi bond, avoiding the secondary carbocation altogether and going directly to the more stable tertiary carbocation (path 1,2RL). The ring carbon preferentially migrates since in so doing, it relieves about 27 kcal/mol (113 kJ/mol) of ring strain. Either of the ring carbons next to the carbocation can migrate. Since bromide ion can attack the new carbocation from the top or bottom side, a mixture of stereoisomers are formed.

9.15 The following compound rearranges in acid. Give a reasonable mechanism.

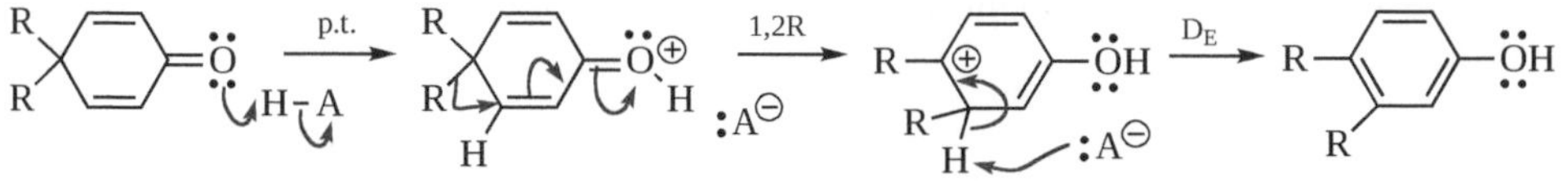

This is called the dienone-phenol rearrangement. Protonation of the carbonyl lone pair gives a resonance-stabilized carbocation that is similar to a sigma complex of electrophilic aromatic substitution, shown below. Migration of the R group allows for aromatization. The driving force for this rearrangement is the formation of the aromatic ring.

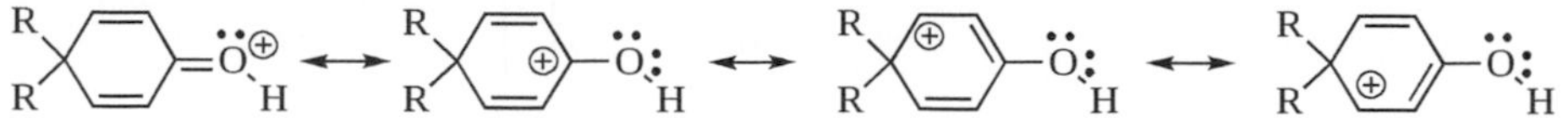

9.17 For each of the following reactions a decision discussed in this chapter is required. Choose which product would be preferred.

9.17a *Source class* C=Y *Sink class* HA *Decision type* Thermodynamic vs. kinetic enol, Section 9.4.3. The conditions are acidic and therefore the enol produced can reverse by protonation of the pi bond. Reversible enolization conditions favor the more stable, more substituted enol. Since protonated acetic anhydride is a hard electrophile, acylation on oxygen of the ambident enol is preferred. Thus the more substituted enol acetate is favored in any equilibrium situation because it is more stable.

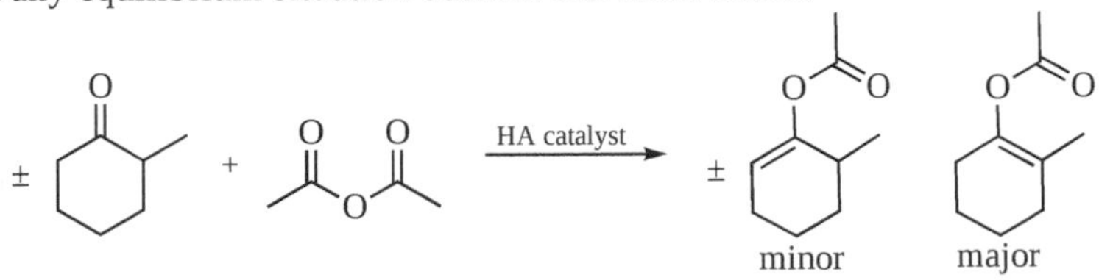

9.17b *Source class* C=C–Z *Sink class* C=C–EWG *Decision type* Ambident electrophile decision, Section 9.6, joined with an ambident nucleophile decision, Section 9.4. The addition is reversible by the ΔpK_a rule, therefore we expect thermodynamic control. This means the ambident enamine nucleophile will attack on carbon not on the very hindered nitrogen. Likewise the ambident enone electrophile will be attacked to give the more stable 1,4-addition product under equilibrium conditions. Both the cis and trans enolate can form.

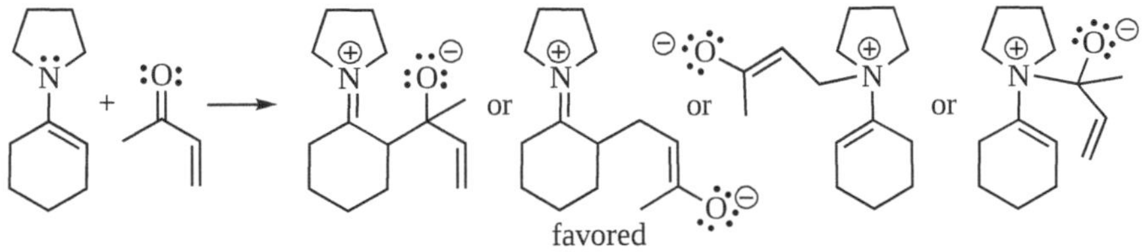

9.19 Provide a reasonable mechanism for the formation of the preferred products in problem 9.17. Combine the source and sink with the appropriate pathway(s).

9.19a A reasonable mechanism begins with a reversible acid catalyzed enolization (Section 7.3.8) to produce the thermodynamically most stable, most substituted enol (Section 9.3). Acylation with the hard electrophile goes on the harder oxygen of the enol (Section 9.4.3). Then the acid catalyzed esterification produces the enol acetate (Section 8.8.1).

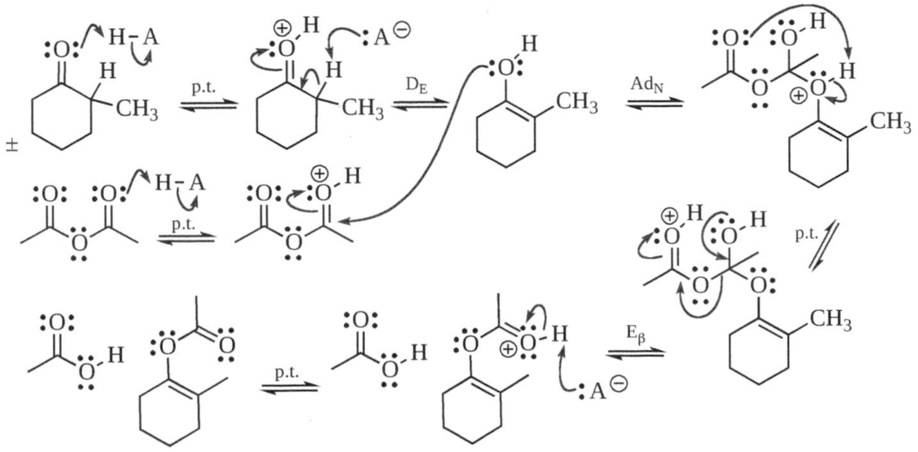

9.19b See the rationale for problem 9.17b. The soft enamine carbon nucleophile can attack the softer methylene end of the enone in an expected reversible Ad_N addition. Since the enone has two rotational isomers, s-cis and s-trans, both the cis and trans enolate can form.

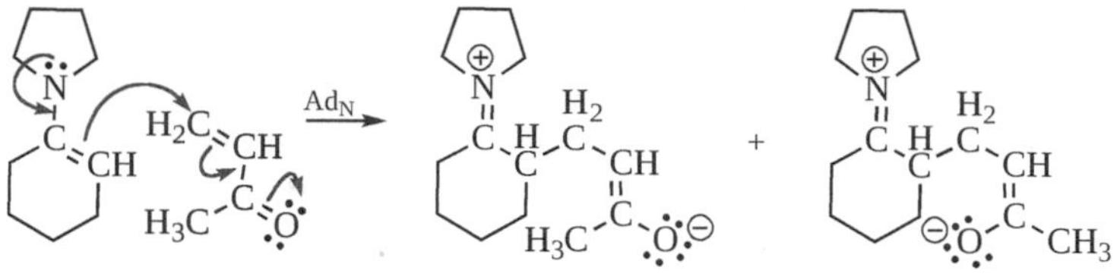

Chapter 10 Odd-Numbered Exercise Answers

Problem 10.1a

Understand the system. A substitution has occurred. The substitution surface alerts us to three possible substitution paths: S_N2, S_N1, and pent. Medium: Acidic, since the reaction produces an acid. Sources: The alcohol oxygen lone pair. Leaving groups: Bromide. Sinks: The phosphorus–bromine bond is a Y–L. Find the pieces of the starting materials in the product and determine what bonds have been made and broken. The new or broken bonds are in red in the following structures.

Find possible routes. Most likely the Y–L sink would be attacked by the oxygen lone pair source. Since the S_N1 would go via a somewhat unstable cation, we can look to other paths first. There are two substitutions that fit the conditions, S_N2 substitution,

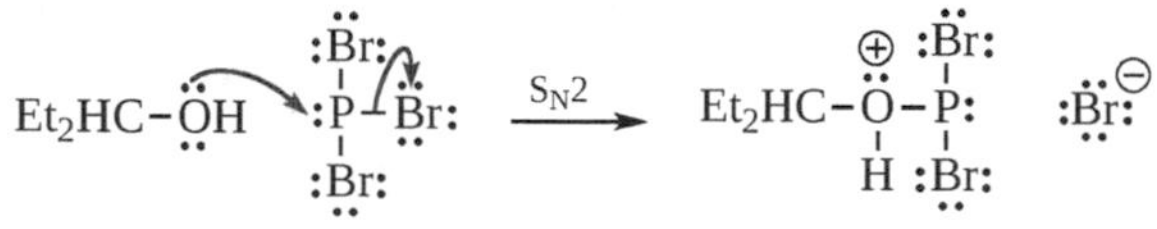

and substitution via a pentacovalent intermediate.

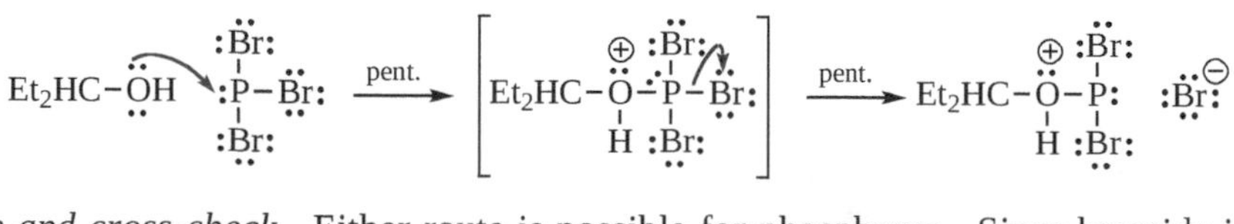

Evaluate and cross-check. Either route is possible for phosphorus. Since bromide is such a good leaving group, the pentacovalent intermediate, if it occurs, will be short lived. The product is expected to be a relatively strong acid (R_2OH^+ has a pK_a of –3.5) and is reasonable for an acidic medium.

Understand the system. We have a new sink, the alcohol hydroxyl has been turned into a good leaving group. In addition, the H on the positive oxygen is now acidic.

Find possible routes. Using the new sink, we find that there could possibly be three paths that fit the medium, sources, and sinks: deprotonation to form a lone pair, path p.t.

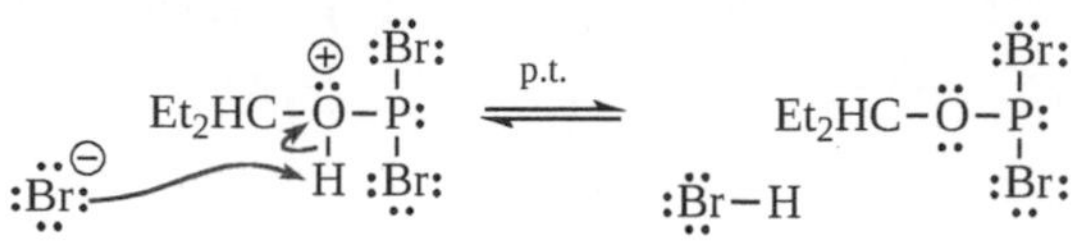

or the S_N1 substitution (path D_N, then A_N),

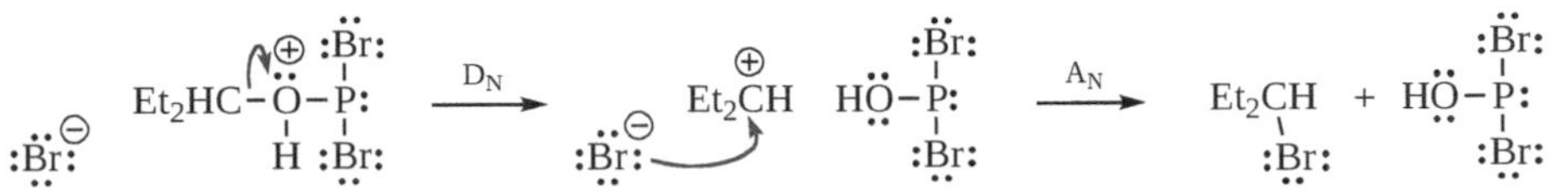

or S_N2 substitution.

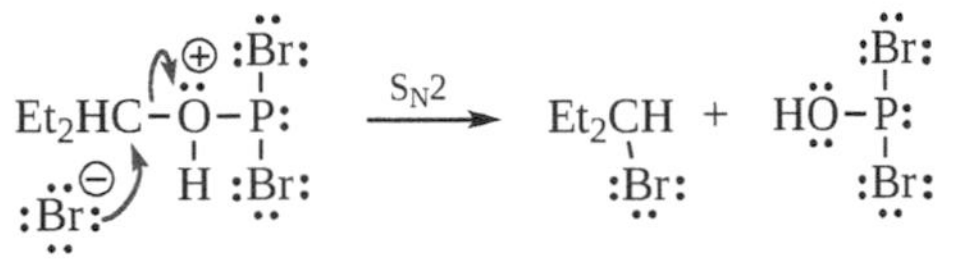

Evaluate and cross-check. Left out of our alternatives is the bromide S_N2 on phosphorus to return to reactants. Since the product is the result of a substitution, we will not explore eliminations in detail, since they do not initially seem to lead to product, and from Section 9.5 we expect substitution. The deprotonation of the oxygen makes the leaving group poorer, and since we wish eventually to do a substitution on carbon, this route is less promising. The S_N1 produces a secondary carbocation, which we know from the carbocation trend is not especially stable. Therefore, this S_N1 process is less likely (as is also the E1). This leaves the bromide S_N2 on the protonated intermediate to be the most reasonable route. The pK_a span of all intermediates is close and fits an acidic medium.

Overview. Our reasonable route is just two S_N2 reactions, but it is not the only way to get to product. The first step could easily go through a pentacovalent intermediate, and our last S_N2 step has an S_N1 side route that might be acceptable. Even if an alternate route's rate-determining barrier were a couple of kcal/mol (10 kJ/mol) higher in energy, many of the more energetic molecules could easily traverse that alternate route.

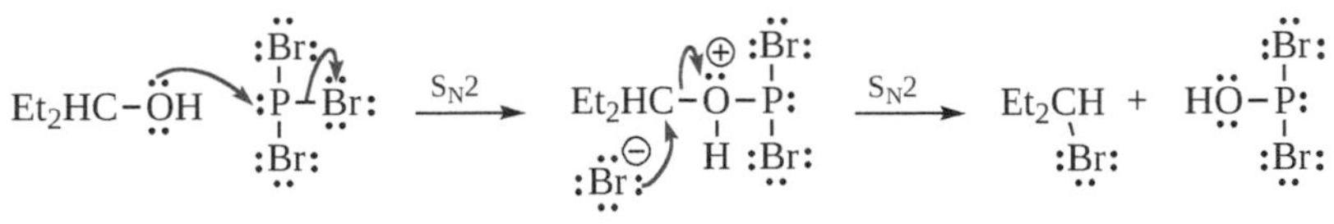

Surface overview. We can look back at the problem space that we just navigated to get an overview, Figure S.1. Red arrows show the most reasonable route of two S_N2 reactions.

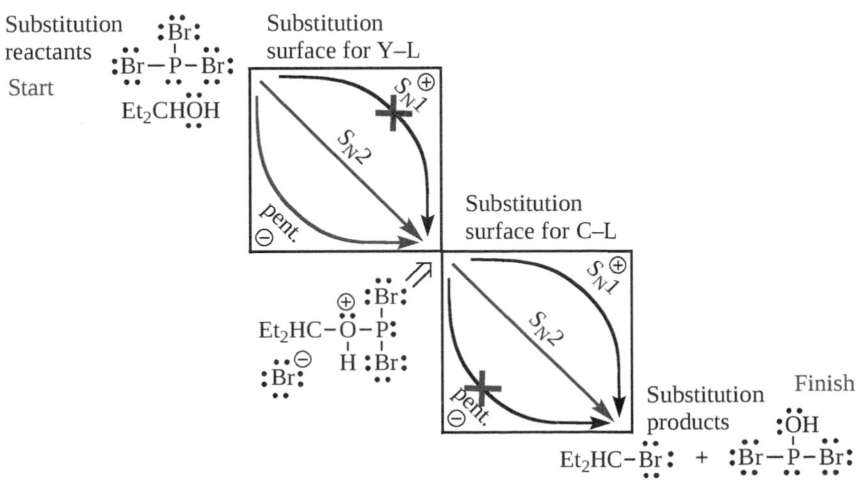

Figure S.1 The surface overview for problem 10.1a

Problem 10.1b

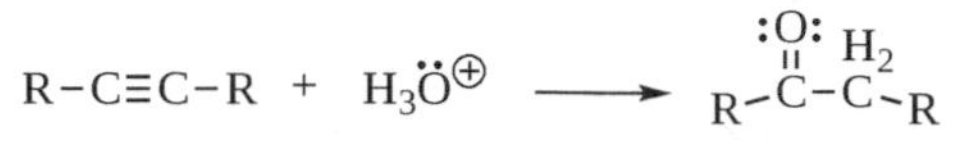

Understand the system. By just comparing the formulas of the reactant and the product, we can see that an addition of water to a pi bond of the alkyne has occurred, but the product is not the expected enol, HC=C–OH, but is instead the ketone. An enol to keto tautomerization must also have occurred. Source class: C≡C. Sink class: HA. Media: acidic. Acids: H_3O^+ at pK_a = 0.0. Find the pieces of the starting materials in the product and determine what bonds have been made and broken. The new or broken bonds are in red in the following structures.

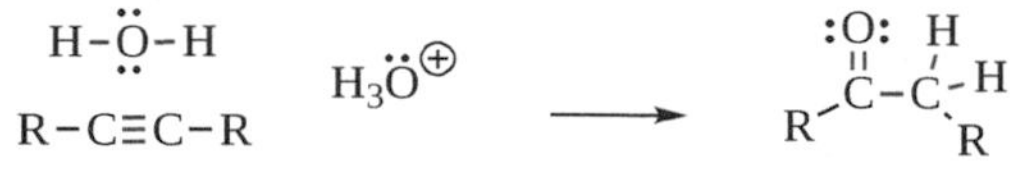

Initial surface overview of problem 10.1b for an addition followed by a tautomerization is shown below:

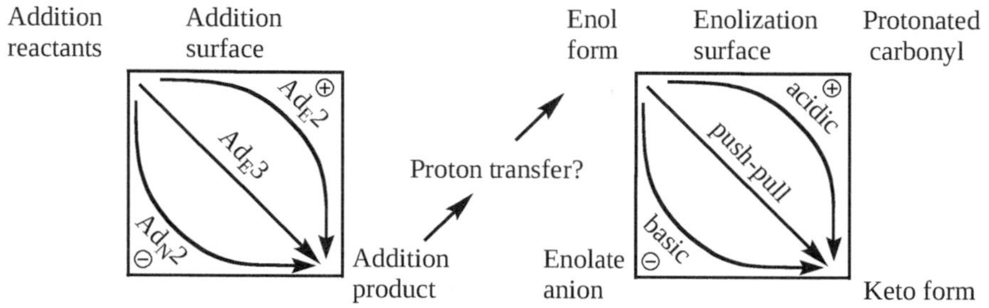

Find possible routes. The Ad$_E$2 and the Ad$_E$3 are reasonable for acidic media. The Ad$_N$2 is not a possibility without an electron-withdrawing group to stabilize the anionic intermediate. We have the Ad$_E$2:

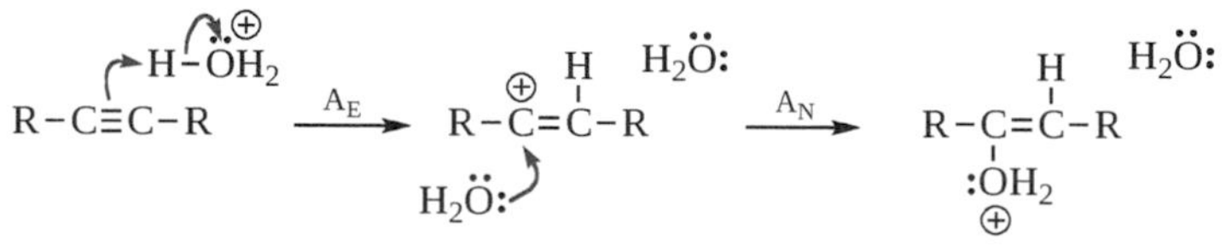

or Ad$_E$3:

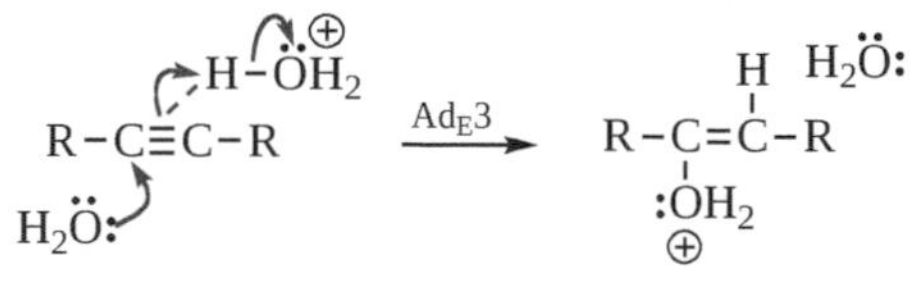

Evaluate and cross-check. The Ad$_E$3 is the most probable path since the Ad$_E$2 goes via the unstable vinyl carbocation. The oxygen protonated enol produced has a pK_a of about –2 to –3 and is consistent with the acidity of the media.

Understand the system. The intermediate protonated oxygen is a good acid. The product has no hydrogens on the oxygen.

Find possible routes. Proton transfer to water produces an enol (path p.t.).

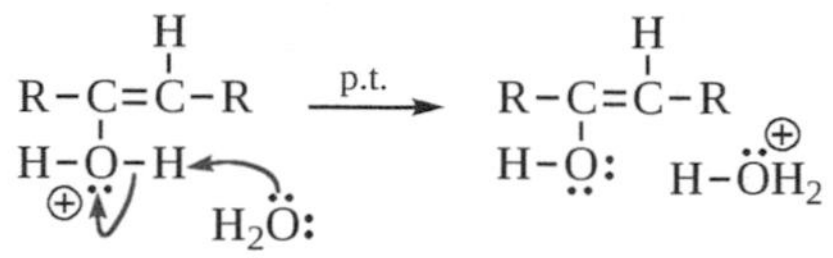

Evaluate and cross-check. The K_{eq} is about 100–1000, so it is very favorable.

Understand the system. Enols are in equilibrium with the keto form via tautomerization (Section 7.3.8). The product ketone is our desired end point.

Find possible routes. Tautomerization can be either base or acid catalyzed. The medium is acidic, so the expected route starts with protonation of the enol followed by either nucleophilic attack by water to give the hydrate or deprotonation to give the keto form.

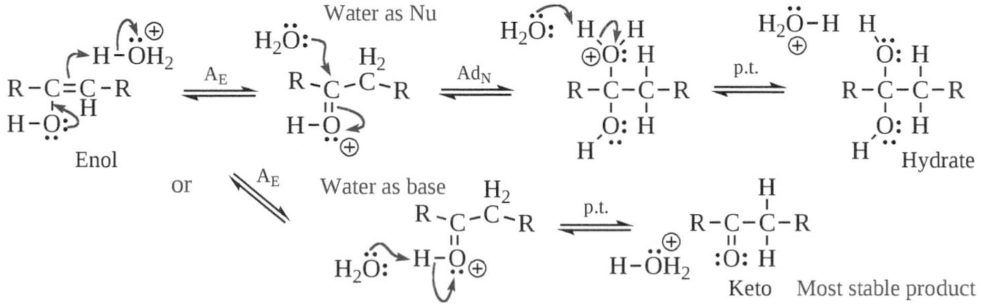

Evaluate and cross-check. The acid-catalyzed route is appropriate for the media. Since a C=O is a stronger bond, 177 kcal/mol (741 kJ/mol), than two C–O bonds, 172 kcal/mol (720 kJ/mol), the equilibrium favors the keto form over the hydrate. The hydrate is a viable minor side product, but the main product given is the ketone.

Overview. A reasonable route is that an Ad_E3 occurs, then a proton transfer to water, giving the enol, which tautomerizes under acidic conditions to the product ketone.

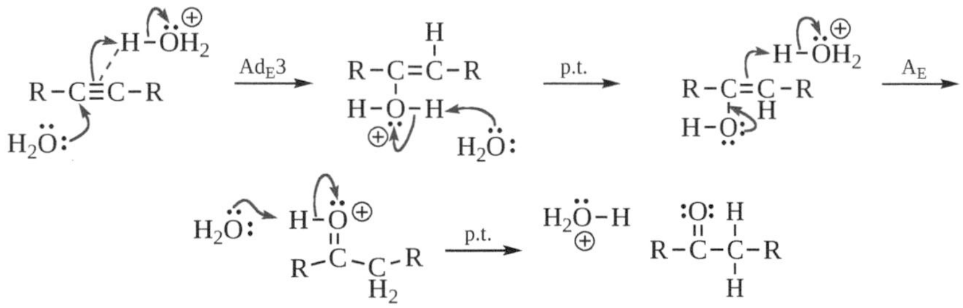

Surface overview. We can look back at the problem space that we just navigated to get an overview, Figure S.2. Red arrows show the most reasonable route of Ad_E3 addition, proton transfer from the nucleophile to give the enol, then finally acidic tautomerization to the ketone. Hydrogen bonding to water also allows the push-pull tautomerization (see Figure 7.18).

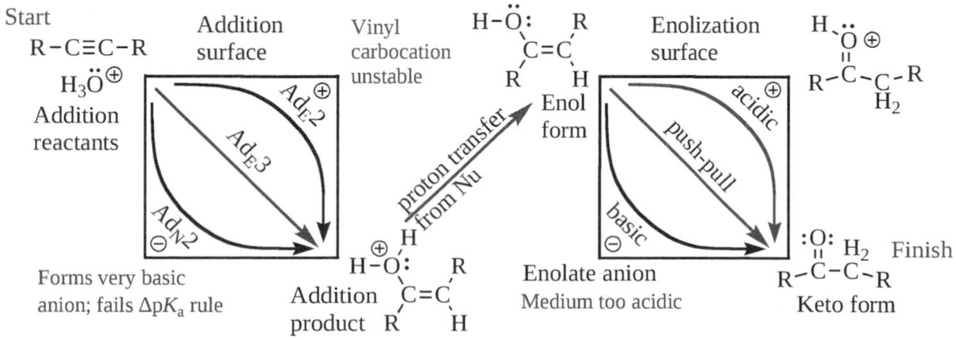

Figure S.2 The surface overview for problem 10.1b.

Problem 10.1c

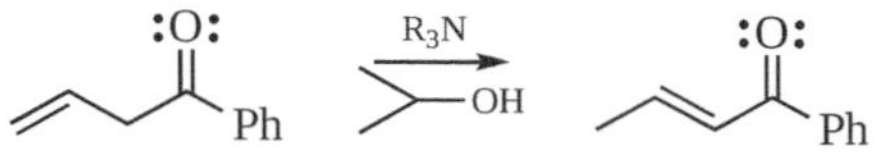

Understand the system. Source class: Z and C=C. Sink class: C=Y. Media: basic. Acidic Hs: the CH_2 next to the ketone, pK_a 19.2, and isopropanol pK_a 17. Bases/nucleophiles: Tertiary amine, pK_{abH} 10.7. Objectives: Line structure is hiding any changes to the hydrogens. To see all the atoms, draw full Lewis structures of the reactant and product. Numbering the carbons, we can see that all that has happened is that an H has been lost from carbon 2 and put back on carbon 4, moving the pi bond. Resonance forms show us that the aldehyde is also an electron-withdrawing group. Find the pieces of the starting materials in the product and determine what bonds have been made and broken. The new or broken bonds are in red in the following structures.

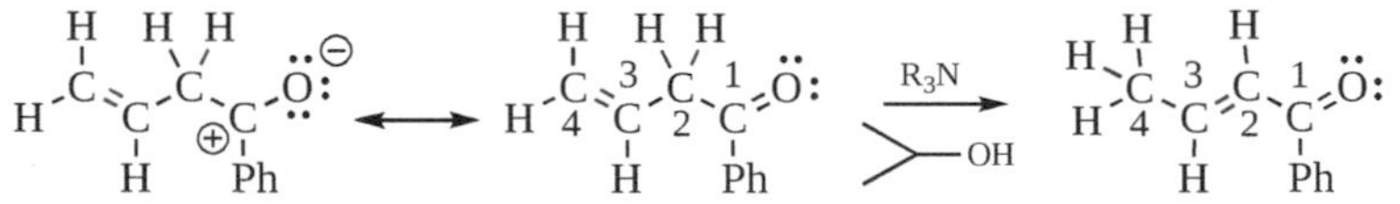

Find possible routes. The tertiary amine is a poor nucleophile and probably would not attack the aldehyde, and the product shows that the aldehyde is still intact, so try something else.

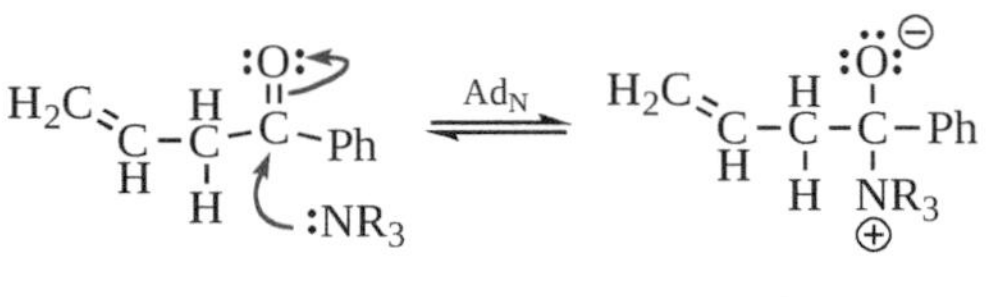

Proton transfer should be all we need here, but the order of steps needs to be decided. The best acid in the reaction is isopropanol pK_a 17. Protonation of carbon 4 by isopropanol is not only extremely uphill but it also forms a secondary carbocation. Secondary carbocations are very strong acids, and isopropoxide is a strong base, thus failing the pK_a span cross-check.

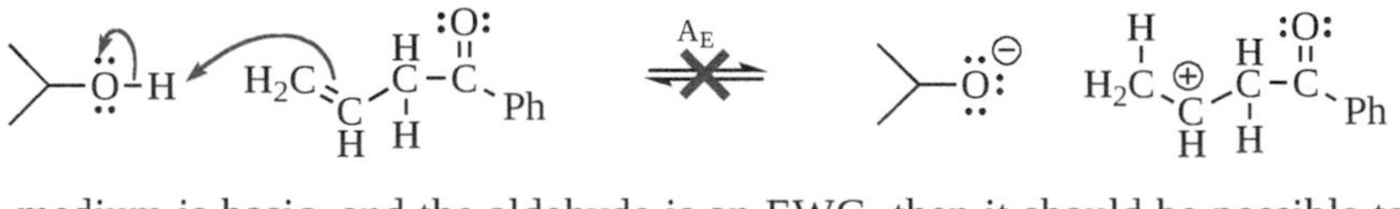

Since the medium is basic, and the aldehyde is an EWG, then it should be possible to use the tertiary amine as a base to remove the H on carbon 2 to generate the enolate.

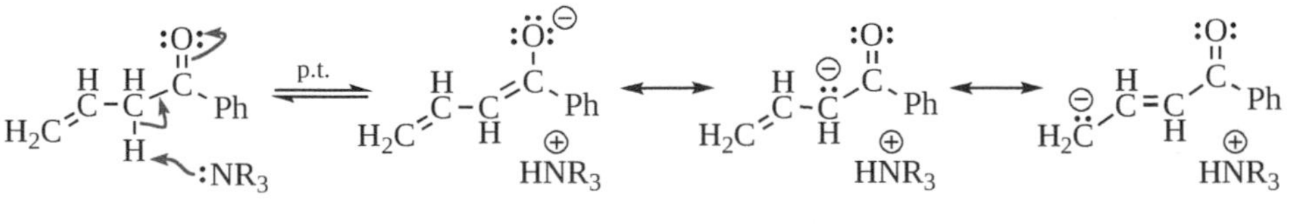

Finally, since the isopropanol medium is protic, the ketone carbonyl is expected to be hydrogen bonded to solvent, setting up a push-pull enolization.

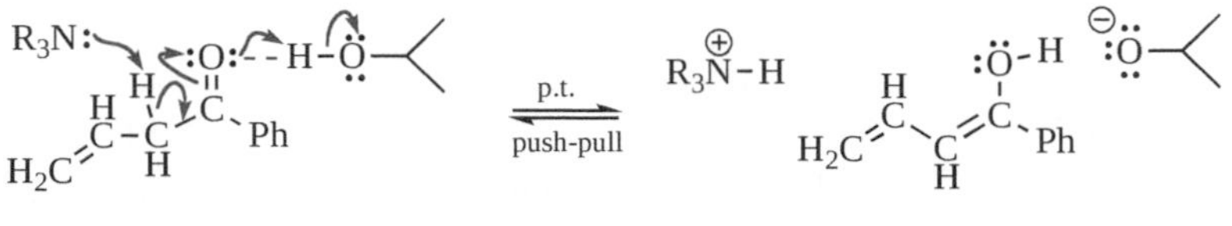

Evaluate and cross-check. The tertiary amine deprotonating the ketone to form the enolate has a proton-transfer K_{eq} of $10^{-8.5}$ and is uphill, reversible, and achievable. The push-pull tautomerization gets directly to the enol without going through the enolate. The push-pull enolization would also be uphill, but how much is harder to guess. The value of K_{eq} needs to include the proton transfer (uphill about 10.7 – 17 = 6.3 units) but also the keto to enol tautomerization. Tautomerization is also uphill since the enol form is less stable than the keto form (acetophenone's keto to enol K_{eq} is 10^{-6}) and this conjugated enol should be more favored. Finally, if the acidic species in the push–pull is the protonated amine instead of isopropanol, then the proton transfer part is a wash (an acid and its conjugate base), so the tautomerization K_{eq} could be around 10^{-6} or better.

Understand the system. The enolate has resonance and is an allylic source. The rightmost resonance form has a minus charge on carbon 4 making that carbon basic. The protonated tertiary amine is the better acid at pK_a 10.7 compared to isopropanol at pK_a 17, but either could protonate the basic enolate, pK_{abH} 19.2.

Find possible routes. Any of the enolate's three negatively-charged sites can be protonated. On oxygen to go to the enol:

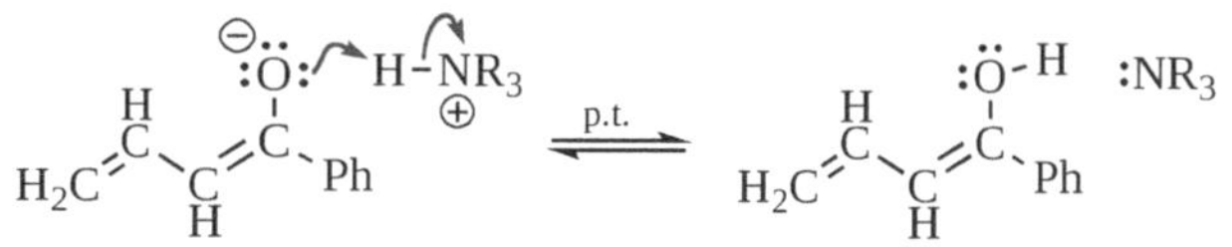

On carbon 2 to return to the starting material:

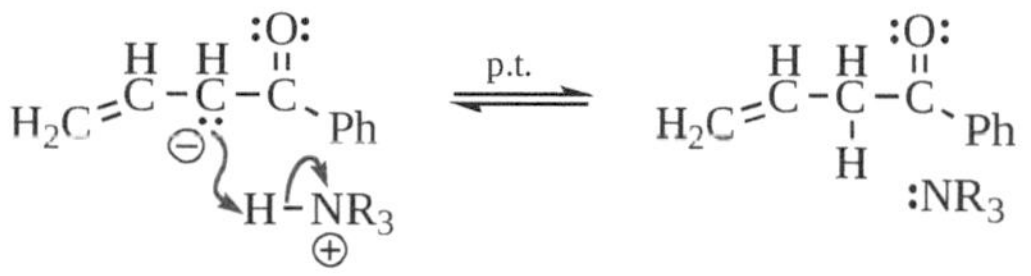

On carbon 4 which yields the desired product:

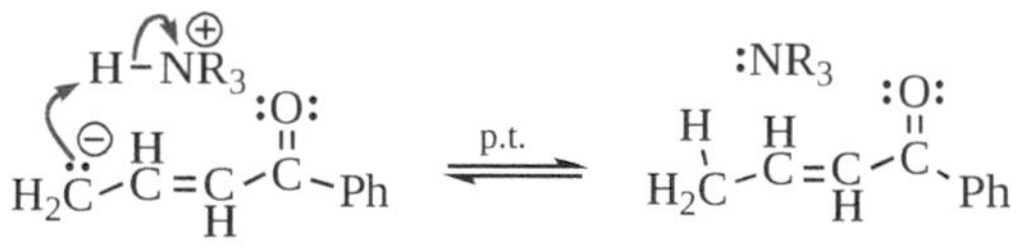

The push-pull would also get us back from the enol form by likewise protonating on carbon 4.

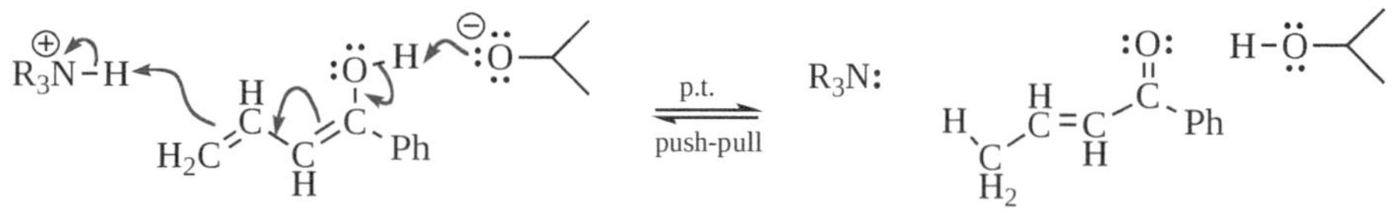

Evaluate and cross-check. The enol is possible but is overall uphill, since the C=O bond of the keto form is stronger than the C=C bond of the enol. Returning to reactant is a good possibility but is not the product. The reversible proton transfer to product has a $K_{eq} = 10^{+8.5}$ and is downhill. The push-pull is likewise downhill in a similar manner, but harder to estimate the exact amount as before. The base in the push-pull could also be the tertiary amine.

Overview. One reasonable route goes via deprotonation on carbon 2 followed by protonation on carbon 4 to give the conjugated product.

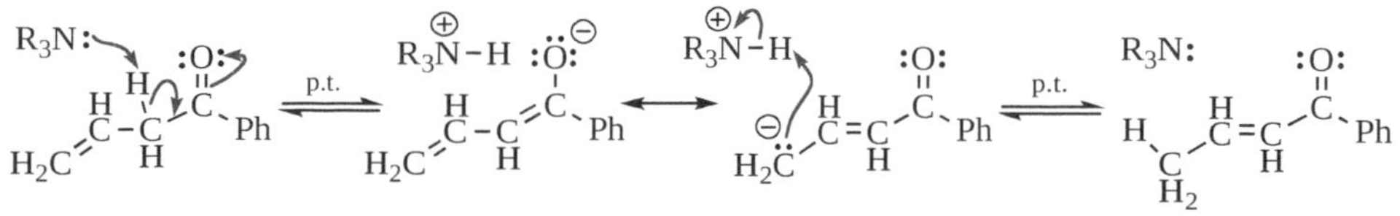

Surface overview. We can look back at the problem space that we just navigated to get an overview, Figure S.3. Red arrows show the most reasonable proton-transfer routes to form the enolate or enol intermediate followed by conversion of the enolate or enol into the conjugated

occurs to give the keto form of the amide. Hydrogen bonding to water also allows the push-pull tautomerization route.

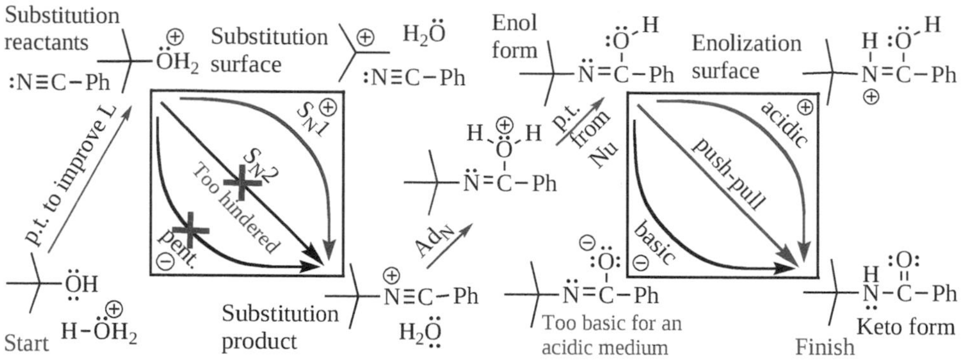

Figure S.4 The surface overview for problem 10.3a.

Problem 10.3b

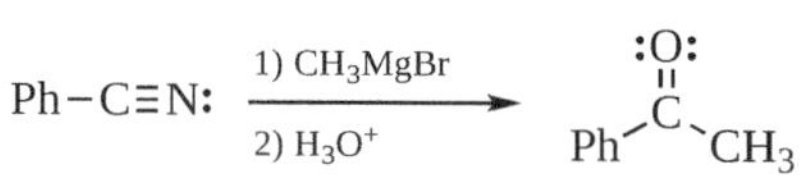

Understand the system. Balanced? No, Mg^{2+}, Br^-, and some nitrogen species are missing from the right-hand side. Generic process? An addition of the organometallic has occurred, but the rest is unclear. Media? Step 1 is highly basic, but step 2 is in acidic water, pK_a 0.0. Sources for step 1? The organometallic. Sinks? The nitrile is a polarized multiple bond. Acidic H's? None. Leaving groups? None. Resonance forms?

Find the pieces of the starting materials in the product and determine what bonds have been made and broken. The new or broken bonds are in red in the following structures. The new ketone oxygen in red came from the water in the acidic workup.

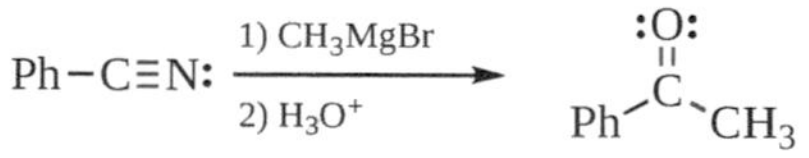

Find possible routes. Only one addition to a polarized multiple bond is possible, path Ad_N, because the other addition paths have proton sources that would quench the organometallic.

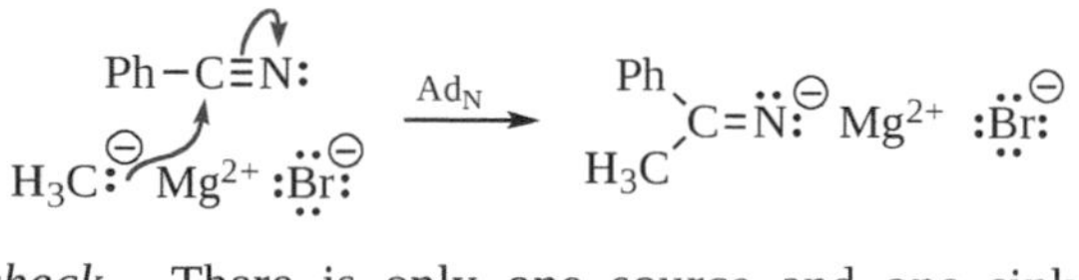

Evaluate and cross-check. There is only one source and one sink and one path. The organometallic addition creates a basic anion whose pK_{abH} is not on the chart. By analogy to the corresponding carbanion, this sp^2 hybridized anion should be more stable than the sp^3 hybridized one but, less stable than similar delocalized anions. Therefore, this anion should be a weaker base than the anion of diethylamine, Et_2N^- pK_{abH} of 36, but should be a stronger base than the delocalized anion of aniline, $PhNH^-$, pK_{abH} 28. The reaction of the organometallic, pK_{abH} 48, to produce this anion is definitely downhill in energy.

Understand the system. Now all the carbons of the product are accounted for. The medium is strongly basic, and the anionic addition product would repel most nucleophiles. The magnesium organometallic, although very reactive, is not a good enough source to attack the anionic addition product. There appear to be no path options left (always finish completely with a step before starting the next one). Now let us explore the second step. The medium is acidic water. The best source is the anionic addition product and the best sink is H_3O^+.

Find possible routes. Proton transfer should be rapid and appears to be the only available path. This neutral intermediate is called an imine.

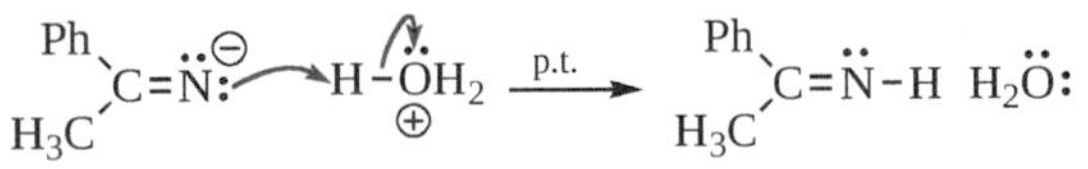

Evaluate and cross-check. The pK_{abH} of the nitrogen anion should be in the low 30's and the pK_a of acidic water is 0.0; therefore the proton transfer is irreversible.

Understand the system. The imine is a polarized multiple bond sink and can also serve as a lone pair source since the nitrogen lone pair is reasonably basic; its pK_{abH} is approximately 5. We can get an overview of the general direction the reaction goes by looking at the final product, which has a double bond to oxygen rather than to nitrogen. We must do an addition to add the oxygen and an elimination to get rid of the nitrogen and form the carbonyl. Initial surface overview for an addition followed by elimination:

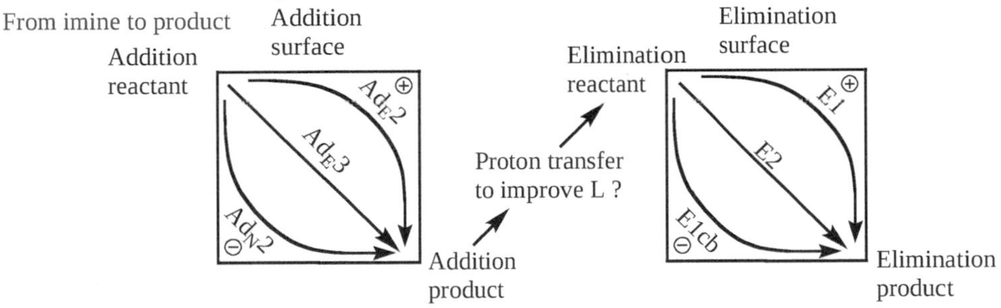

Find possible routes. The only nucleophile present is water. Two addition paths to the imine that would fit the acidic reaction conditions are:
the hetero Ad_E3, (nucleophilic attack by water on a hydrogen-bonded imine).

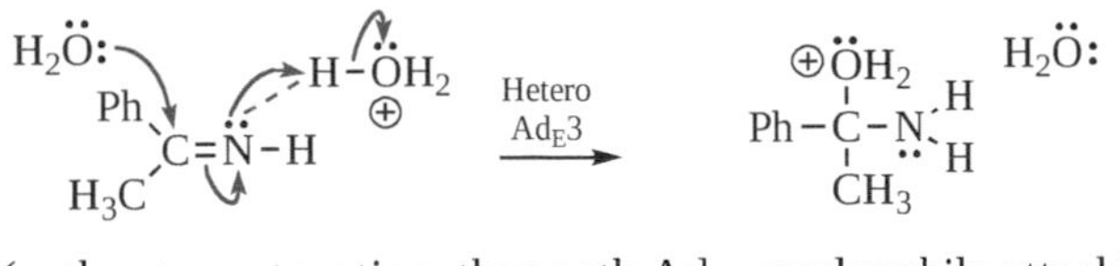

or the hetero Ad_E2 (path p.t., protonation, then path Ad_N, nucleophile attack).

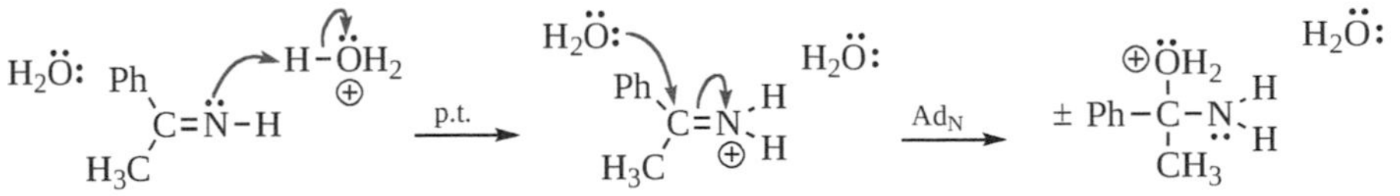

Evaluate and cross-check. Both routes are known to occur in an acidic medium. (A third alternative, the Ad_N2, is ruled out by the ΔpK_a rule, for it would create too basic an anion, pK_{abH} 36, from a nucleophile of pK_{abH} 0.0.) The proton transfer in the hetero Ad_E2 process is very favorable with K_{eq} close to 10^{+5}, and such proton transfers are fast. Both additions form the same product, but the hetero Ad_E2 with its very favorable proton transfer and excellent sink, the protonated imine, may be slightly lower in energy. The overall process begins with H_3O^+, pK_a = 0.0, and ends with an O protonated hemiaminal, pK_a estimated about –3, making a slightly stronger acid. We have broken a C=N pi bond and made two sigma bonds that are stronger. Most likely this addition was downhill in energy.

Understand the system. The O protonated hemiaminal is acidic, and the nitrogen lone pair is reasonably basic, pK_{abH} of approximately 8–9.
Find possible routes. Solvent can mediate the one proton-transfer path.

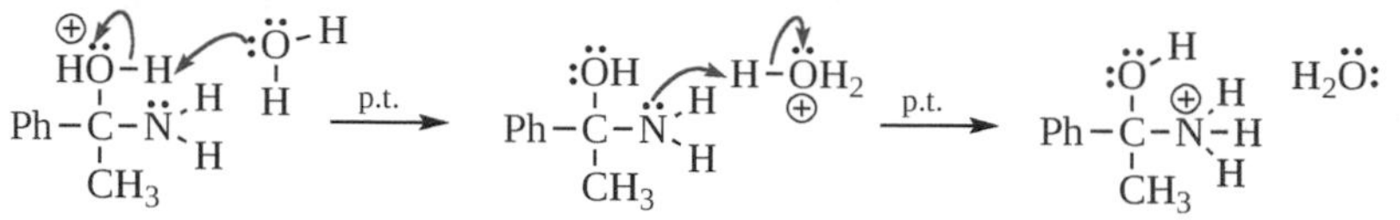

Evaluate and cross-check. An intramolecular proton-transfer process is possible, but since it involves a strained four-membered transition state, such an internal proton-transfer route is probably of higher energy. We can expect proton transfer to be extremely fast between groups that hydrogen bond to the water solvent. Very little needs to happen to turn a hydrogen bond into a "real" bond or vice versa. The overall proton transfer is irreversible with a K_{eq} close to 10^{+12}.
Understand the system. The protonated nitrogen is a fair leaving group, pK_{aHL} 9.2. We need to lose it and a proton to get to the final product. It is time to explore elimination possibilities.
Find possible routes. Three elimination paths are possible, the E2,

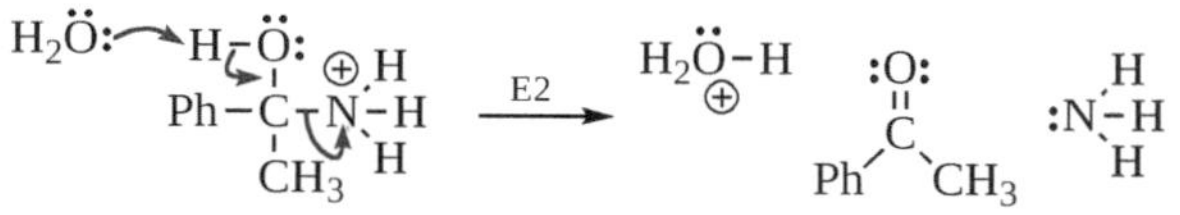

or the lone-pair-assisted E1 (path E_β followed by path p.t.),

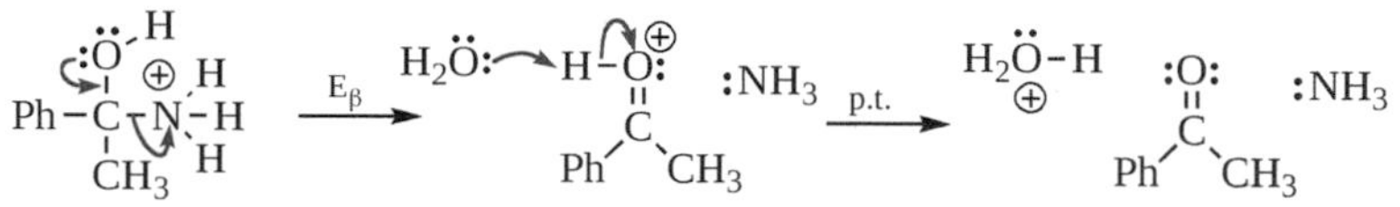

or the E1cB (path p.t. followed by path E_β).

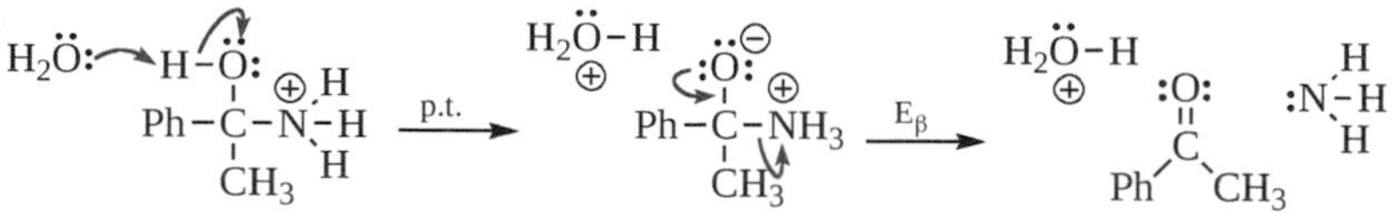

Evaluate and cross-check. We are now at the reactant corner of the elimination surface. The path combination through the upper right corner is the lone-pair-assisted E1, which is favored by a strongly acidic medium and good leaving groups. The diagonal route is the E2 path, with an incoming pK_{abH} of 0.0 and an outgoing pK_{abH} of 9.2, close to our limit. Whereas the E2 produces H_3O^+, pK_a 0.0, the lone-pair-assisted E1 produces a less stable and more acidic protonated carbonyl, approximate pK_a –7. The lone-pair-assisted E1 has a media span cross-check issue, for the L is not quite good enough; it has made products whose pK_as are –7 and 9.2 for a span of 16.2 units, which is a bit past our limit. The E1cB route through the lower left corner has an initial deprotonation step which is difficult to do in strongly acidic media. The E1cB oxyanion is a good enough source to kick out a fair to poor leaving group. In a basic medium the E1cB would be the best route, but here it suffers from a poor proton-transfer K_{eq} (estimated at about 10^{-12}) with water as the base (with a pK_{abH} of 0.0).

All routes have problems; mechanistic studies have shown that the elimination is the slow step in weakly acidic solutions. In all cases a following proton-transfer step is expected. This final proton transfer has a K_{eq} of $10^{+10.9}$ and is almost irreversible. The NH_3 is a much better base than water with a pK_{abH} of 9.2. In fact, if we use the NH_3 generated from a previous elimination as the base in another elimination, both the E2 and E1cB begin to look a lot better.

Overview. We are done, but let us take an overview of the route. The most probable route is a nucleophilic Ad_N2 addition to a polarized multiple bond, protonation upon workup in acidic water to give an imine, the hetero Ad_E2 or Ad_E3 addition, two proton transfers to set up an E2 elimination, and a final irreversible proton transfer. Imine hydrolysis to a ketone may seem familiar, since Section 10.5.2 covered the reverse process, imine formation from a ketone.

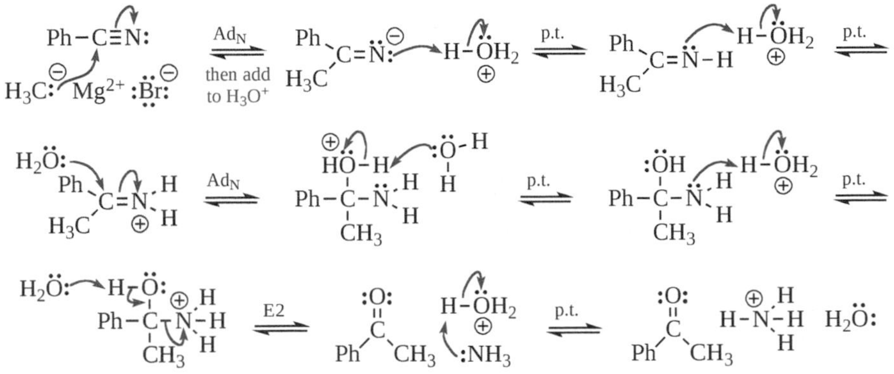

Surface overview. We can look back at the problem space that we just navigated to get an overview, Figure S.5. Red arrows show the most reasonable route(s).

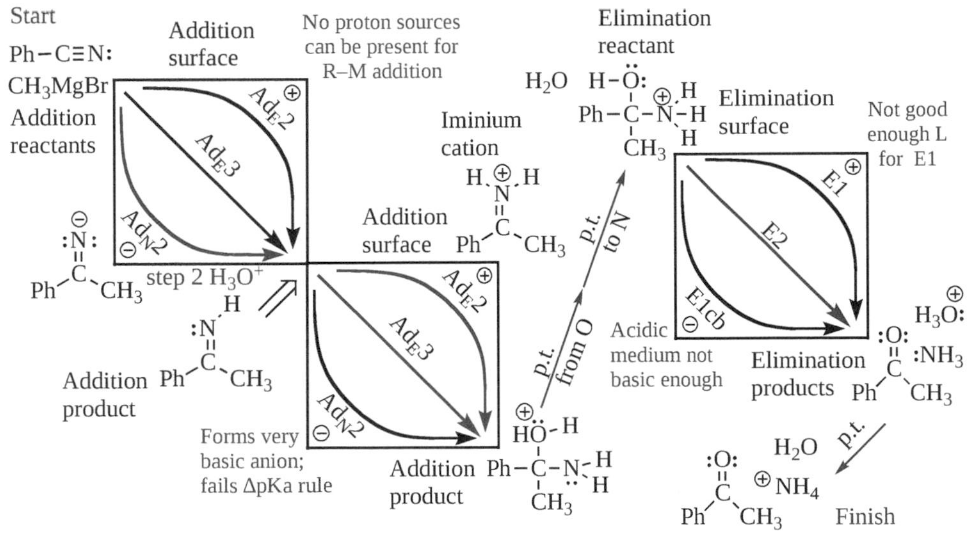

Figure S.5 The surface overview for problem 10.3b.

Problem 10.3c

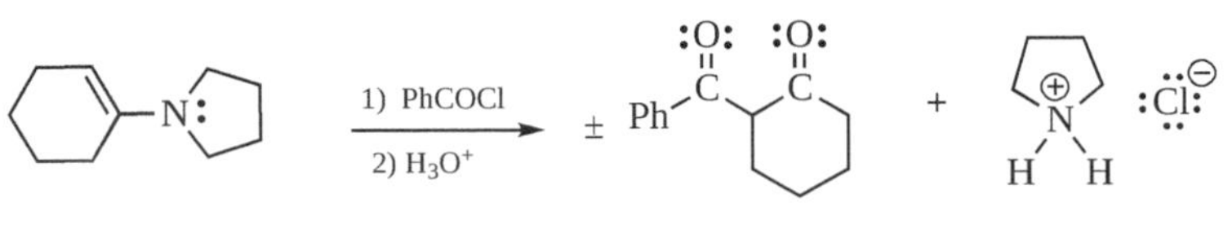

Understand the system. Source class: the enamine is Z–C=C. Sink class: the acyl chloride is Y=C–L. Media: aprotic, because acyl chlorides react with any moisture to produce HCl and a carboxylic acid. The nucleophilic enamine carbon of the source connected to the electrophilic carbon of the acyl chloride sink. Find the pieces of the starting materials in the product and determine what bonds have been made and broken. The new ketone oxygen in red came from the water in the acidic workup. The new or broken bonds are in red in these structures.

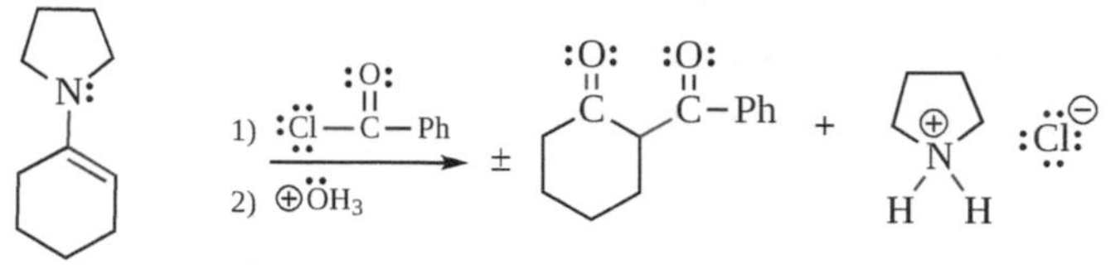

Find possible routes. Y=C–L sinks commonly do addition–elimination, path Ad_N, then E_β.

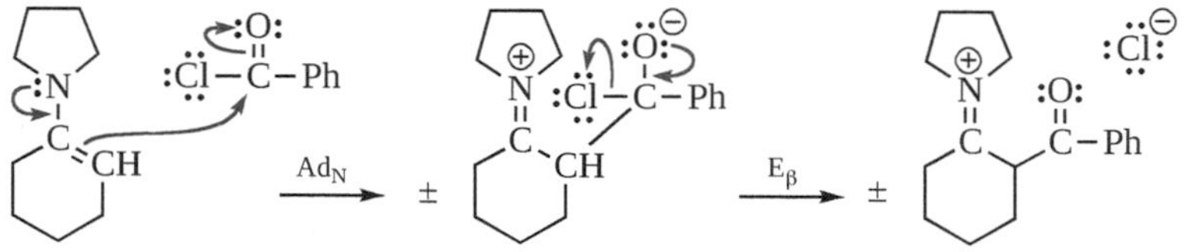

Evaluate and cross–check. The source is ambident, but the heteroatom is hindered and thus a poor nucleophile. Chloride is a good leaving group, pK_{aHL} –6. A look at the product verifies that acylation went on carbon not nitrogen. There's no proton source for other additions types.
Understand the system. The $C{=}N^+$ is an excellent sink, and we need to attack the iminium ion with an oxygen nucleophile to get an oxygen on that carbon. The water in step 2 can serve as the needed nucleophile. By looking at the final product, which has a double bond to oxygen rather than to nitrogen, we must do an addition to add the oxygen and an elimination to get rid of the nitrogen.
Find possible routes. Addition to the polarized multiple bond, path Ad_N.

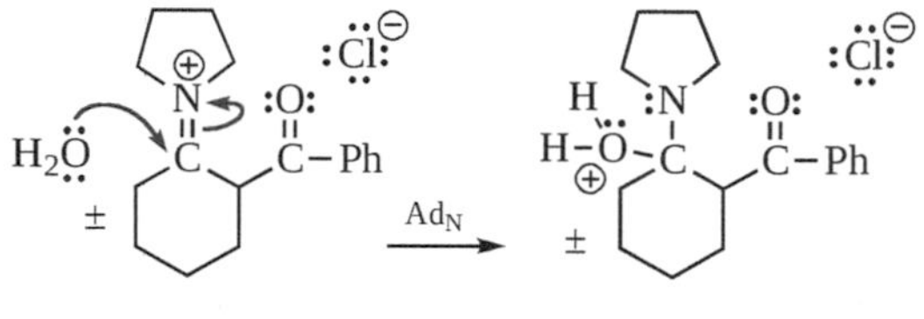

Evaluate and cross-check. This step should seem familiar. We covered the same ground, but in reverse (from the ketone to the imine), in mechanism example 10.5.2
Understand the system. The nitrogen lone pair is basic and the OH_2^+ is acidic.
Find possible routes. Proton transfer is most likely mediated by the water solvent.

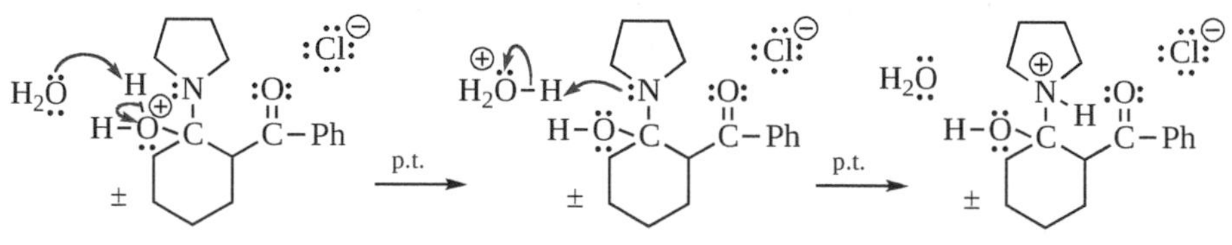

Evaluate and cross-check. The overall K_{eq} should be very good, greater than 10^{+13}.
Understand the system. Elimination of pyrrolidine, pK_{aHL} = 11.3, gives the product.
Find possible routes. We are now at the reactant corner of the elimination surface.
The three choices are E1cB,

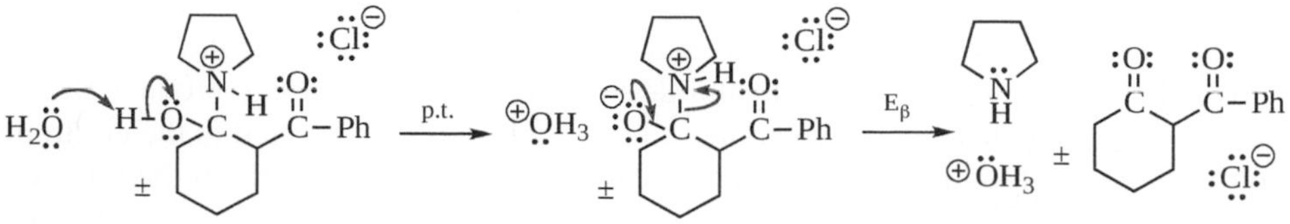

lone-pair-assisted E1

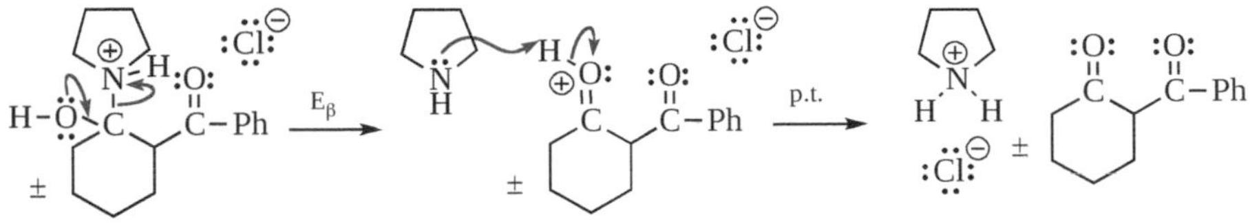

or E2 followed by proton transfer

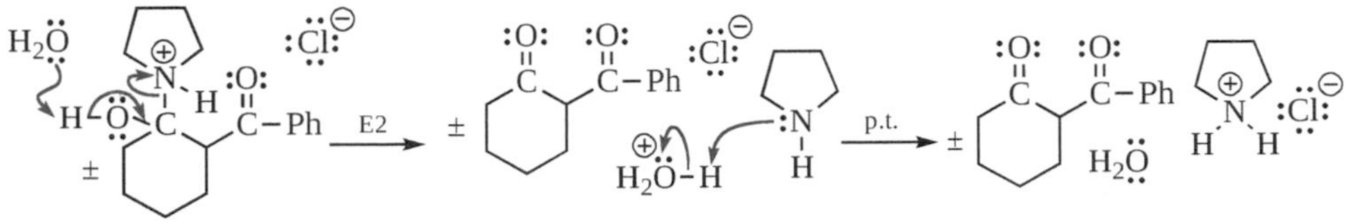

Evaluate and cross-check. The E1cB route's initial deprotonation step seems difficult to do in strongly acidic media. The path combination via the opposite corner of the elimination surface is a lone-pair-assisted E1 that can push out a poorer L than an unassisted E1 ($pK_{aHL} \leq 0$). It would have to pass a ΔpK_a or a media span cross-check instead. A $pK_{abH} = -2.4$ oxygen lone pair pushes out a pK_{aHL} 11.3 pyrrolidine (a ΔpK_a climb up of 13.7 units). The media span cross-check looks at the two products formed; those are a protonated ketone at pK_a –7 and the amine at pK_{aHL} 11.3, for a too wide span of 18.3 pK_a units. The lone-pair-assisted E1 does not seem to have a good enough L this time. We are left with the diagonal route, the E2 path. The E2 with water as a base, $pK_{abH} = 0.0$, may not be quite good enough to kick out a leaving group of pK_{aHL} 11.3. If some reaction proceeds, pyrrolidine forms, and it would be a better base in the E2 or the E1cB. A downhill protonation of N can keep the process from reversing.
Overview. A reasonable process was an Ad_N + E_β, an Ad_N, two proton transfers, an E2 with a follow-up proton transfer.

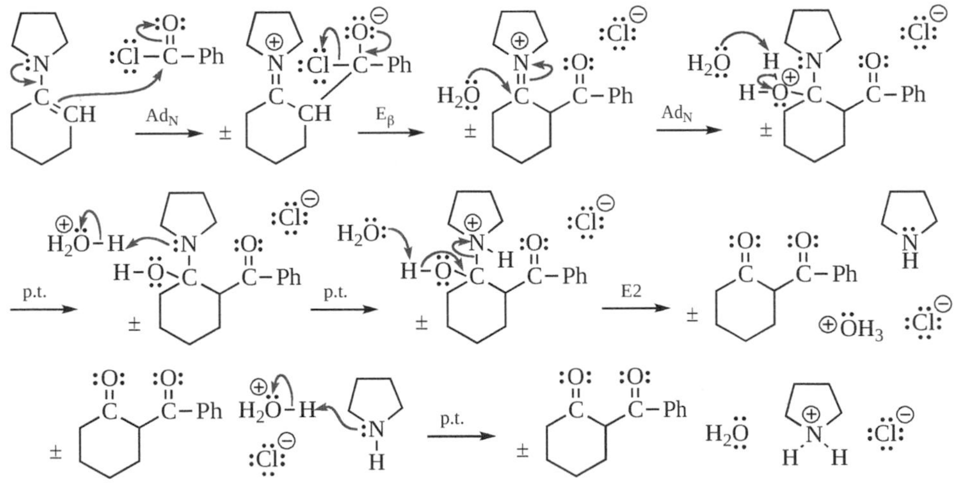

Surface overview. We can look back at the problem space that we just navigated to get an overview, Figure S.6. Red arrows show the most reasonable route. It starts with an addition–elimination on the acyl chloride. The addition–elimination surface shown in Figure 4.41 is central to this surface. The addition–elimination product is the iminium ion, which in step 2 gets attacked by water. After a proton transfer from O to N (probably mediated by solvent) to improve the leaving group, an E2 elimination gives the diketone product. Then an irreversible proton transfer to the nitrogen removes it as a possible nucleophile for the reverse process.

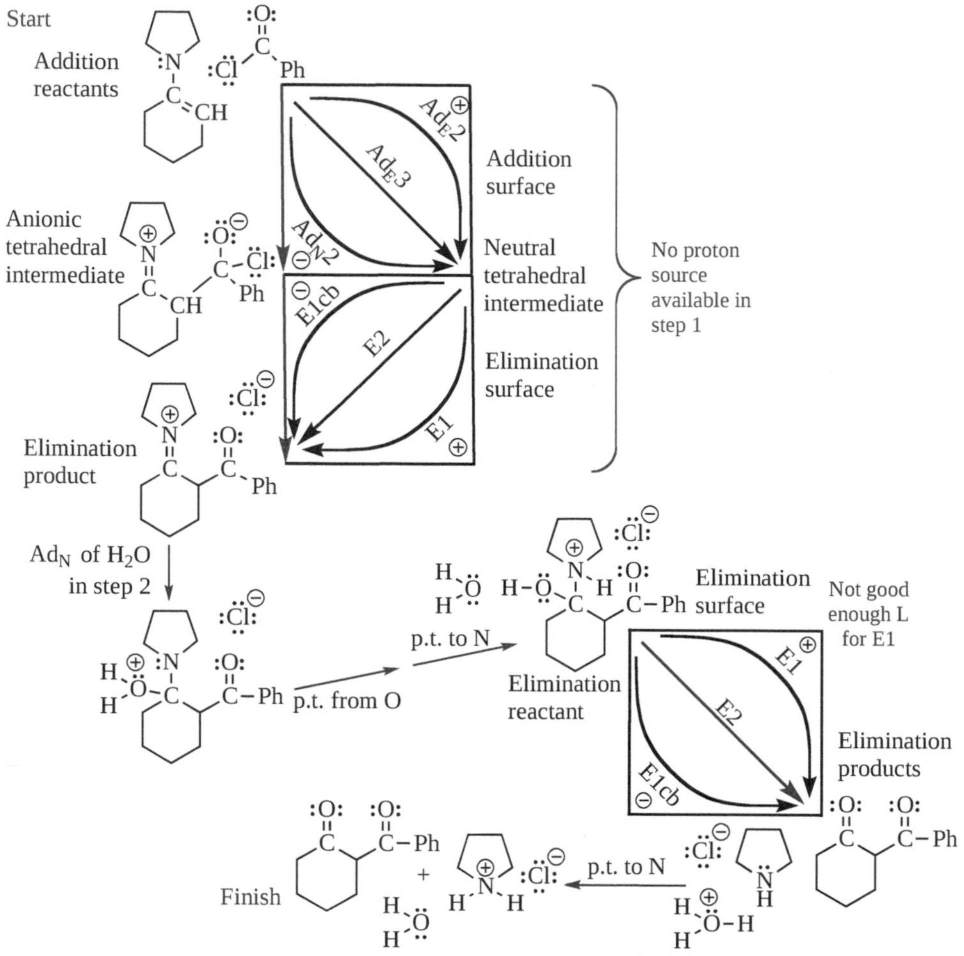

Figure S.6 The surface overview for problem 10.3c.

Problem 10.3d

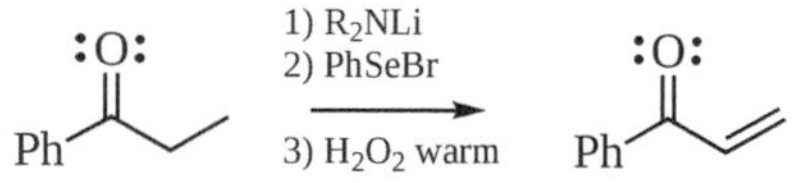

Understand the system. First step source class: Z. Bases: R_2N^-, pK_{abH} 36, is a strong base and a poor Nu. First step sink class: C=Y. Second step sink class: Y–L. Third step sink class: Y–L. Media: basic. The phenyl ketone's CH_2 is the only acidic H ($PhCOCH_3$ pK_a is 18.3). The reaction has formed a pi bond, which almost always happens by an elimination, and that requires a leaving group, but the reactant does not have one! Our intermediate goal then is to attach a leaving group. Find the pieces of the starting materials in the product and determine what bonds have been made and broken. Those bonds are shown in red in the following structures. We have a good Y–L sink, so we need a nucleophile, and the easiest way to get one is to deprotonate the ketone with the strong base.

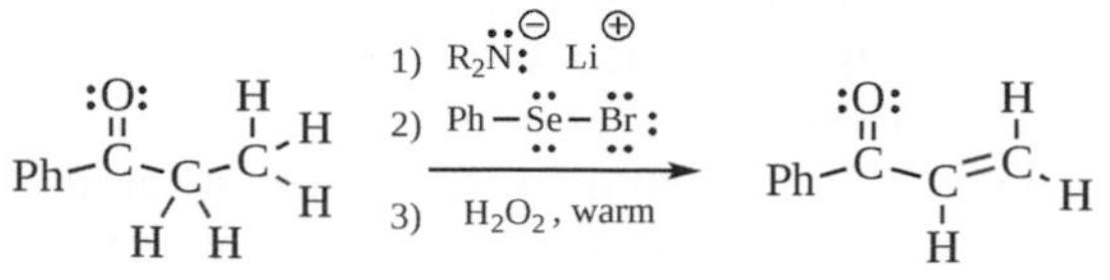

Find possible routes. Proton transfer is about the only reaction a good base but poor nucleophile can do. The methylene next to the phenyl ketone is acidic, about pK_a 18.3.

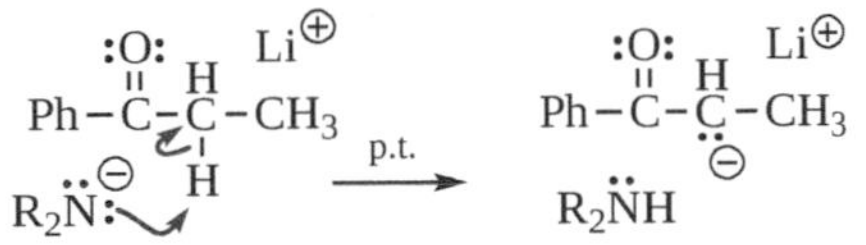

Evaluate and cross-check. The approximate $K_{eq} = 10^{+17.7}$, an irreversible proton transfer.
Understand the system. The enolate is a good allylic source; PhSeBr is a Y–L sink.
Find possible routes. To displace the bromide on selenium use the S_N2 (or pent. path). With bromide as the L, the pentacovalent intermediate would be very short lived. The Y–L sinks usually do not S_N1 because the cation formed would be poor due to the electronegativity of Y.

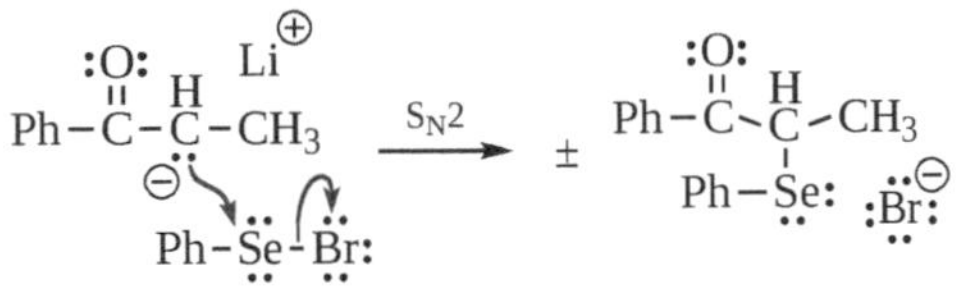

Evaluate and cross-check. By the ΔpK_a rule this should be energetically downhill and irreversible since we went from pK_{abH} of about 18.3 to –9.
Understand the system. The selenium lone pairs are very nucleophilic. H_2O_2 is a Y–L sink.
Find possible routes. The S_N2 oxidizes the selenium. Again the S_N1 is unreasonable.

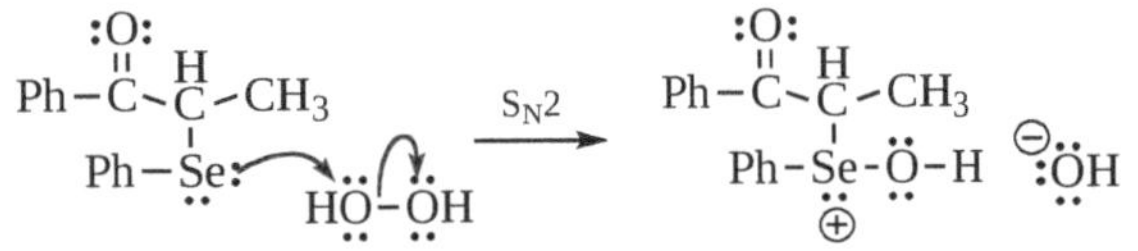

Evaluate and cross-check. The reaction has broken a weak O–O bond.
Understand the system. The SeOH is acidic and the hydroxide is a good base.
Find possible routes. Proton transfer generates water and the selenoxide.

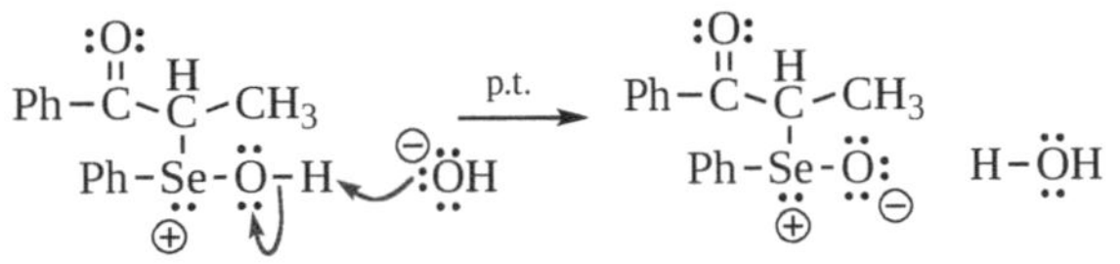

Evaluate and cross-check. The pK_a for the protonated selenoxide is not on the chart. But Se is right below S in Group 16, and they have similar electronegativities, so they should behave similarly. Since a protonated sulfoxide R_2SOH^+ has a pK_a = –1.5, the pK_a of a protonated selenoxide R_2SeOH^+ should be close to that. This proton transfer to hydroxide, pK_{abH} = 14.0, is very downhill and fast since the hydroxide formed from the S_N2 is right nearby.
Understand the system. The oxygen anion is basic and the positive selenium can serve as a leaving group.

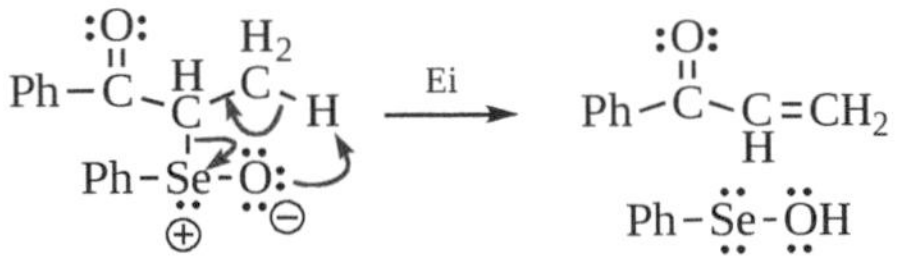

Evaluate and cross-check. This intramolecular elimination reaction that produces our final product is expected to be fast.

Overview. A reasonable process has been a deprotonation to give an enolate, an S_N2 (or pent. path) substitution on selenium, an S_N2 substitution on hydrogen peroxide, a proton transfer to a selenoxide, and finally an Ei elimination.

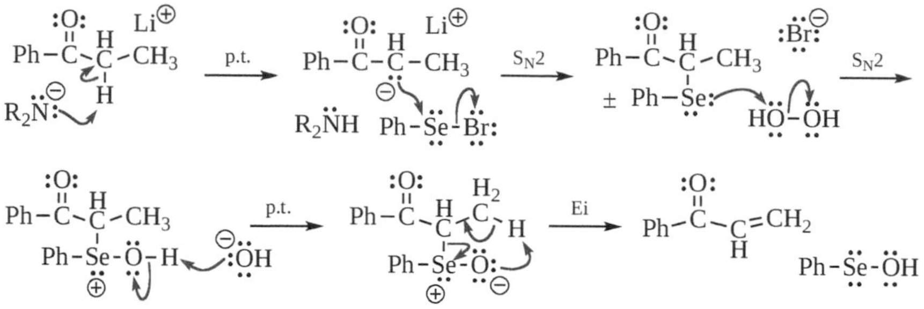

Surface overview. We can look back at the problem space that we just navigated to get an overview, Figure S.7. Red arrows show the most reasonable route.

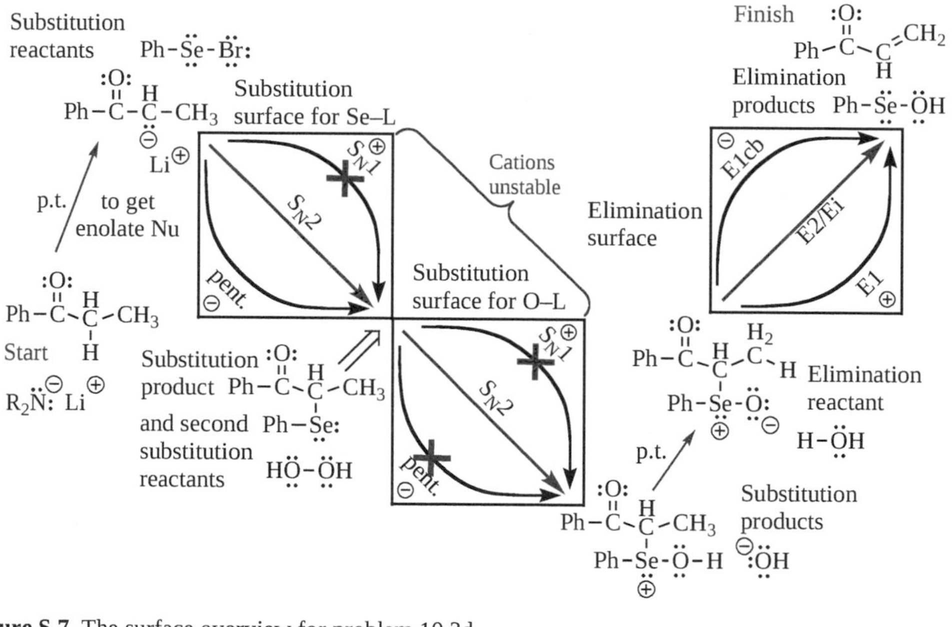

Figure S.7 The surface overview for problem 10.3d.

Problem 10.5a

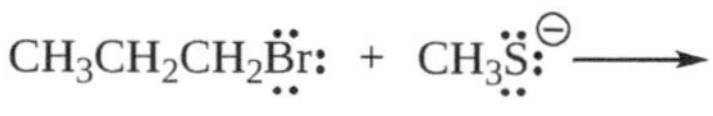

Understand the system. Medium: Basic, CH_3S^- pK_{abH} is 10.6. Sources: CH_3S^-, a lone pair source and an excellent nucleophile. Sinks: The primary sp^3 bound Br leaving group. Acidic H's: There are none. Leaving groups: Bromide is an excellent leaving group.

Find possible routes. Proton transfer is unreasonable because none of the alkyl hydrogens are acidic. Substitution or elimination are the only routes left for this sink; the starting point is on the S_N2/E2 surface, Figure 9.4, not the S_N1/E1 surface, Figure 9.3, since the loss of the leaving group would produce an unstable primary carbocation.

The S_N2 path:

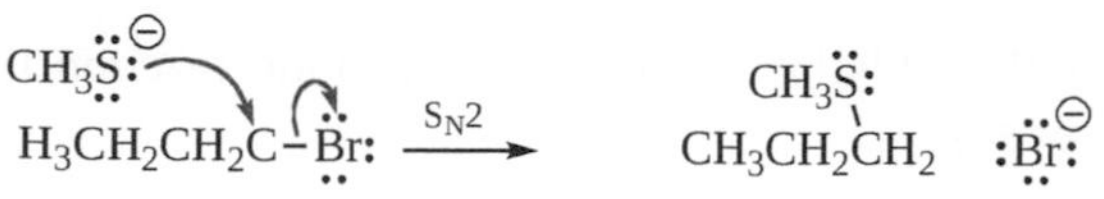

or the E2:

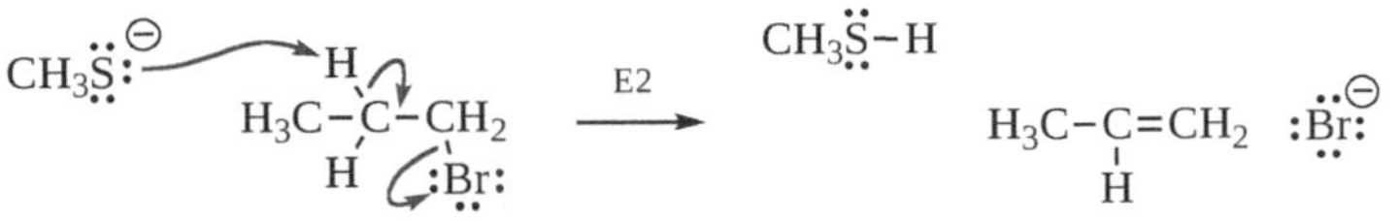

Evaluate and cross-check. We must now make a major decision, substitution vs. elimination, as discussed in Section 9.5. Since the leaving group is on an unhindered primary site and the CH_3S^- is a good nucleophile and a weak base, S_N2 substitution is preferred. The ΔpK_a rule is satisfied. The nucleophile had a pK_{abH} of 10.6 and the leaving group –9; the reaction has definitely formed the weaker base. Since the product appears not to be able to react further, we are done.

Overview. An S_N2 substitution has made a relatively stable thioether and reached a reasonable end point.

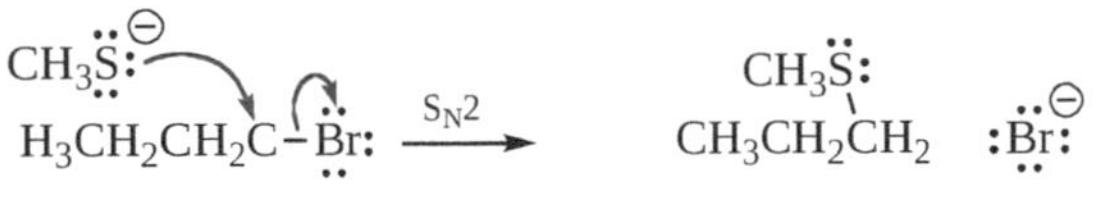

Problem 10.5b

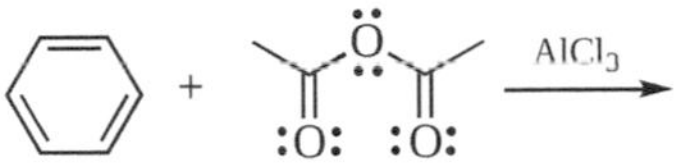

Understand the system. Medium: Acidic; $AlCl_3$ is a strong Lewis acid and will generate HCl with any trace of moisture. Solvent: Benzene, relatively nonpolar (slightly less polar than ether). Sources: The lone pairs of the anhydride and the pi bonds of benzene, an aromatic source. Leaving groups: Acetate is a fair leaving group. Sinks: The anhydride is Y=C–L, a polarized multiple bond with an attached leaving group, and the Lewis acid is excellent. The best sink is the electron-deficient Lewis acid. Resonance forms: The last two resonance forms are very minor, for the second carbonyl's partial plus destabilizes the positive oxygen.

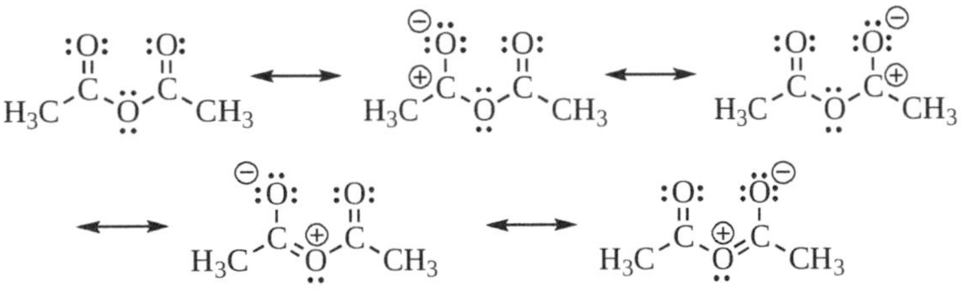

Find possible routes. The Lewis acid is expected to complex with any lone pair source, path A_N. The oxygen lone pairs of the anhydride can complex with the Lewis acid at two different sites:

on the partially positive center oxygen,

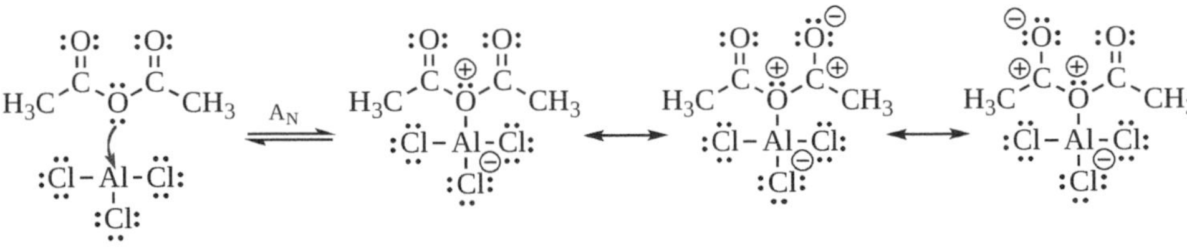

or on a partially negative carbonyl lone pair.

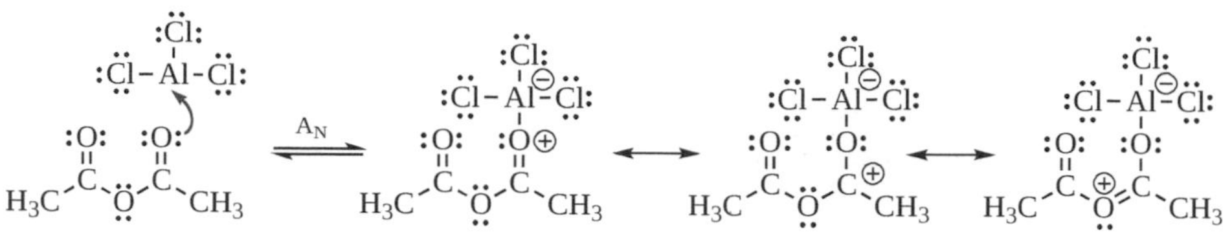

Evaluate and cross-check. The complexation with the ether oxygen is the less favored process since it creates an oxonium ion between two EWGs, the carbonyls. The best site for Lewis acid complexation is the carbonyl lone pair. That Lewis acid complex has a delocalized positive charge and is consistent with a strongly acidic medium.
Understand the system. The complex has several resonance forms. Complexation increases the leaving group's ability to ionize. The leaving group is now good to excellent, equivalent to a protonated acid, pK_{aHL} –6. The only source is the aromatic ring, which needs an excellent electrophile in order to react.
Find possible routes. The Lewis acid complex can be attacked by the nucleophile or the leaving group can be lost. Alternatively, the loss of the Lewis acid, just returns to reactants.

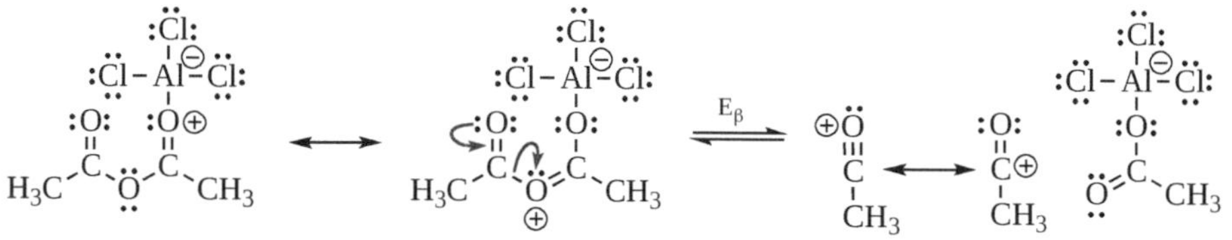

Evaluate and cross-check. The ionization is probably uphill in energy because although the acylium ion is lone pair stabilized, it is also vinyl. A lone-pair-assisted ionization of a leaving group by path E_β passes all cross-checks for cation stability; it has a good leaving group, but the solvent could be more polar.
Understand the system. The acylium ion is resonance delocalized and an excellent electron sink. But, with benzene as the solvent the electrophile may be best considered as an ion pair.
Find possible routes. The aromatic ring can now undergo electrophilic aromatic substitution, path A_E followed by D_E.

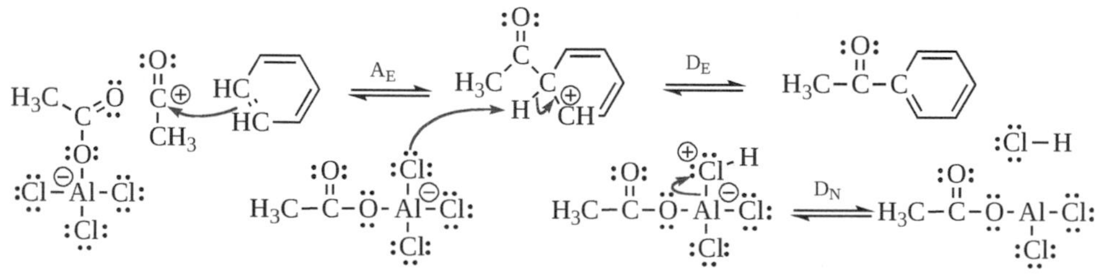

Evaluate and cross-check. The proton is most likely removed by the Lewis acid complex or by chloride to produce HCl, which does not ionize readily in the nonpolar solvent and will escape as a gas. The aluminum-oxygen bond is stronger than the aluminum–chlorine bond.
Overview. The carbonyl is an electron-withdrawing group and deactivates the aromatic ring to further attack. We have reached a good end point. A reasonable mechanism is:

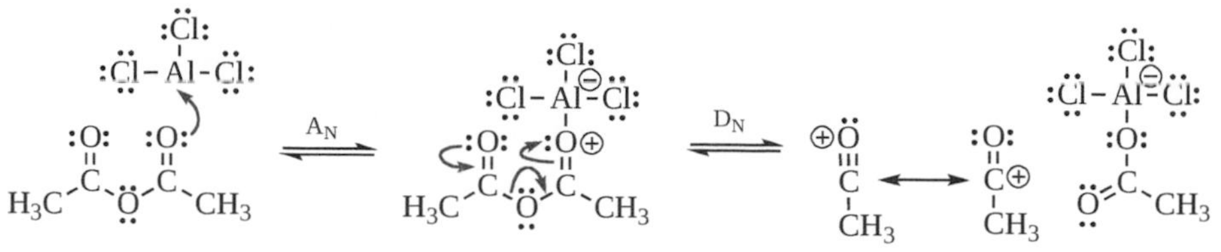

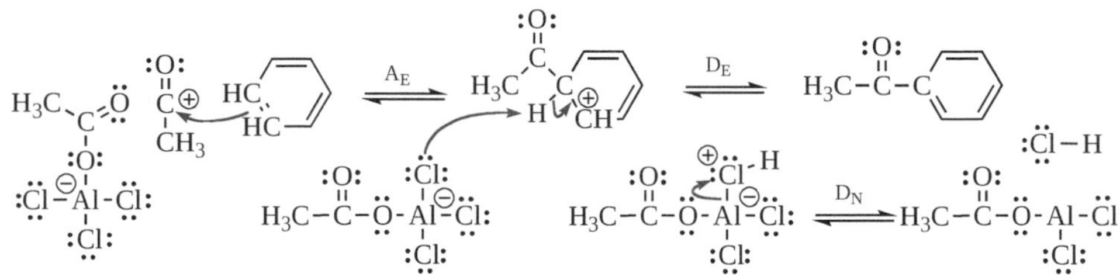

Problem 10.5c

$$\text{PhMgBr} + D_2O \longrightarrow$$

Understand the system. Source class: RM. Sink class: D_2O is an HA, pK_a about 16. Media: very basic due to the organometallic. Most basic site: the anionic carbon atom of the organometallic, pK_{abH} about 43. Most acidic site: the deuterium of D_2O.
Find possible routes. The only likely route is proton transfer.

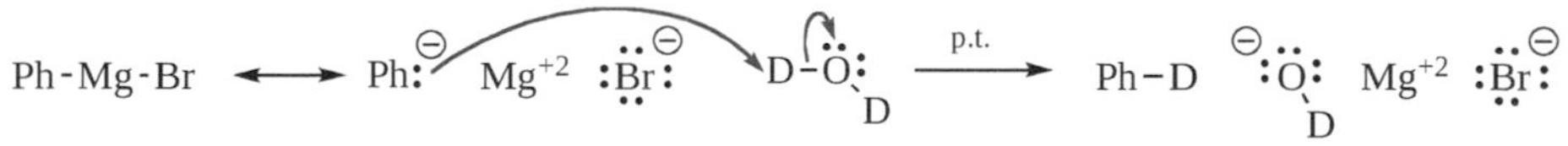

The reaction has formed a much weaker base, and there are no other reactions possible. This is a good way to get deuterium labeled compounds. You could also say it is an easy way to get a PhD.

Problem 10.5d

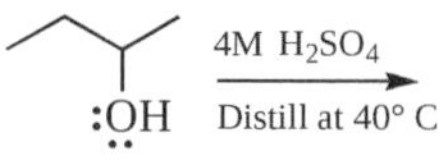

Understand the system. Medium: Strongly acidic. Sources: The alcohol lone pair. Leaving groups: The alcohol OH can get protonated and become a good leaving group. Sinks and acidic H's: Sulfuric acid at a 4 molar concentration dissociates completely in water to hydronium and hydrogen sulfate as shown below (leveling effect).

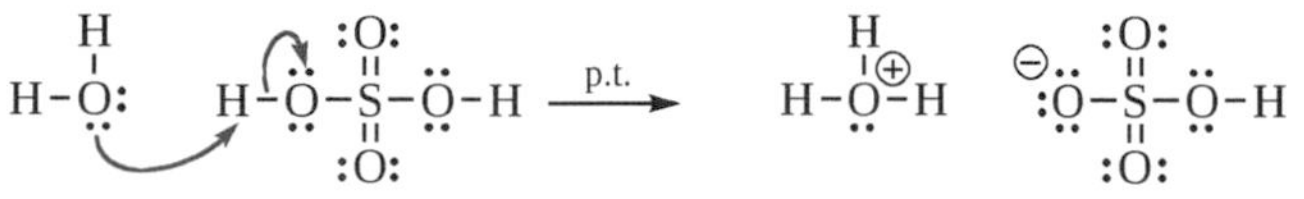

Find possible routes. Since the leaving group is not good enough to ionize and sulfate is a poor nucleophile, the most probable path is hydronium protonating the alcohol lone pair.

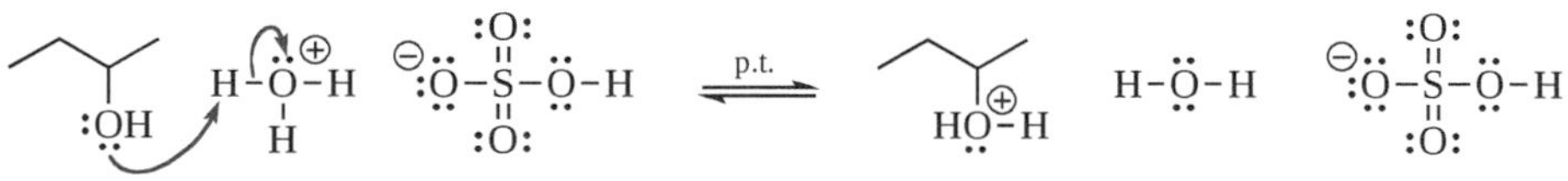

Evaluate and cross-check. Only one path found. The reaction has gone slightly uphill producing an acid of about pK_a = –2.4 from an acid of pK_a = 0.0.
Understand the system. Now we have generated a good leaving group bound to an sp^3 hybridized carbon. Since the reaction is in an acidic medium and there is a good leaving group, we are at a well-known junction, the starting point of the substitution/elimination energy surface, Figure 9.2.
Find possible routes. There are two substitution routes available, the S_N2

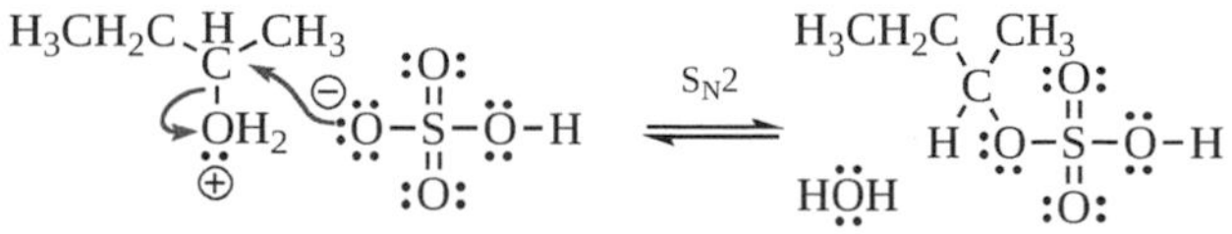

or the S_N1 (path D_N, then A_N),

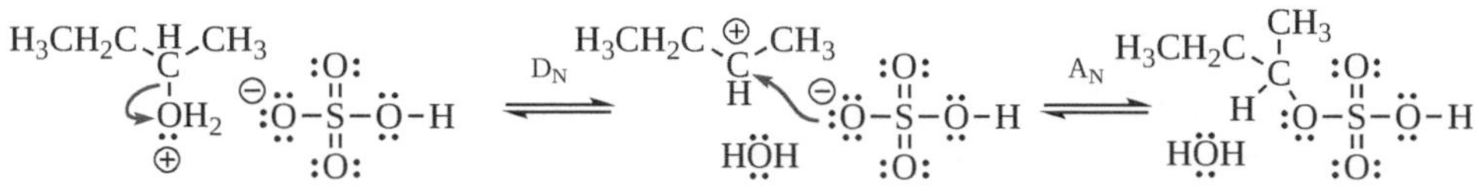

and there are two elimination routes possible the E2

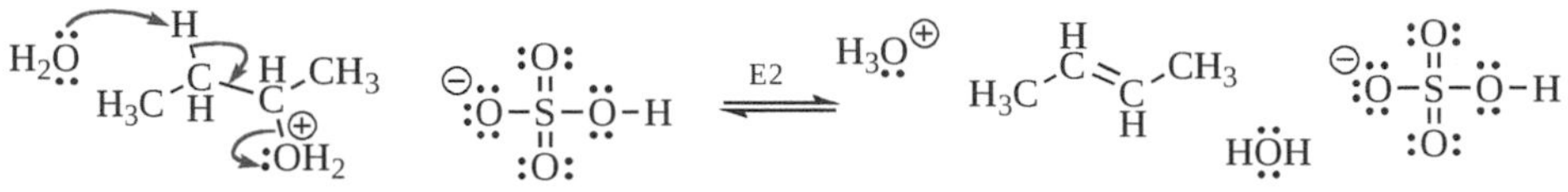

or the E1 (path D_N followed by D_E).

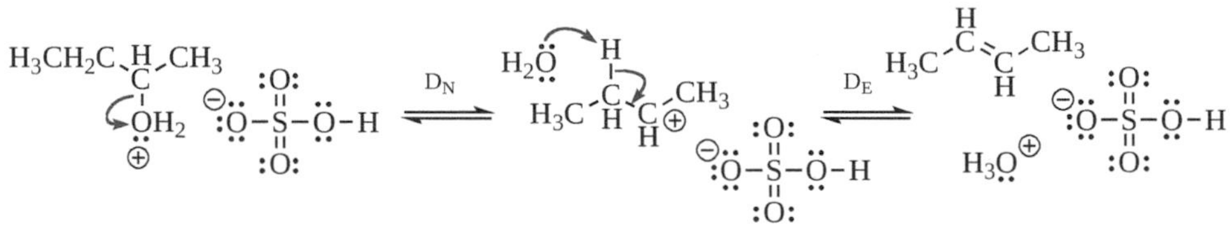

Evaluate and cross-check. As explained in Section 9.5, when the reaction is heated, the nucleophile is poor and also is a good leaving group, then the substitution easily reverses and elimination predominates. The secondary carbocation is the decision boundary between E2 and E1. The alkene can reprotonate and return along these paths, but still the equilibrium will favor the most stable alkene, *trans*-2-butene. The reaction produces a weaker base.
Overview. A reasonable process has been a protonation to give a better leaving group, followed by an E2 or E1 elimination to give the more substituted alkene. The low boiling alkene product is removed by distillation, driving the equilibrium to produce more alkene.

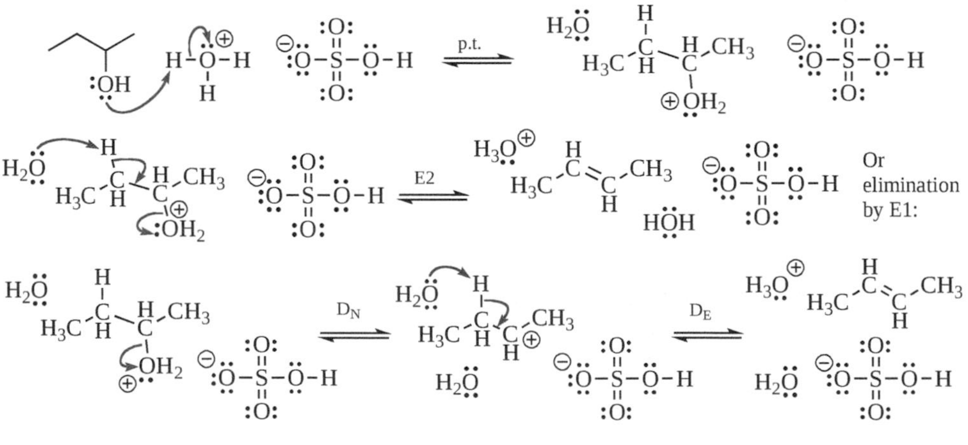

Problem 10.7a

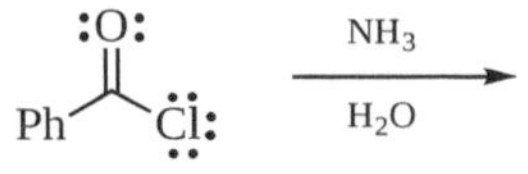

Understand the system. Source class: Z, the water or the ammonia lone pairs. Sink class: Y=C–L. Media: Mildly basic, the pK_a of NH_4^+ is 9.2. Most basic site and best nucleophile: the ammonia lone pair.
Find possible routes. Nucleophilic addition, path Ad_N by the ammonia lone pair.

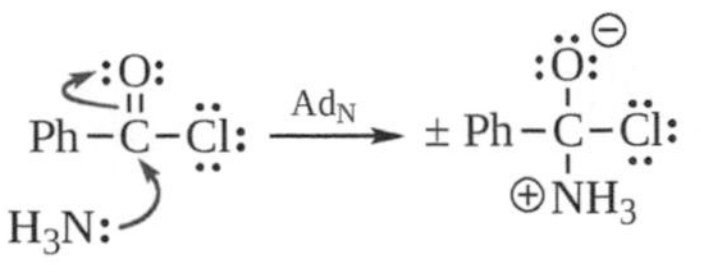

or by the water lone pair.

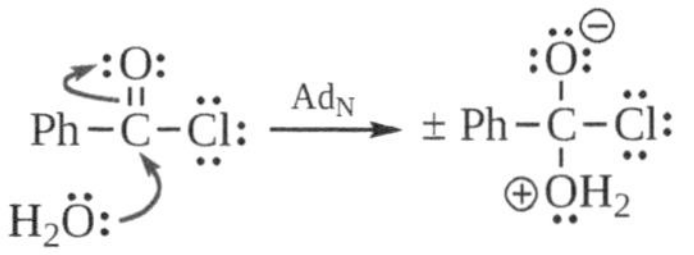

Evaluate and cross-check. The ammonia nitrogen is softer and more nucleophilic than the water oxygen. The ammonia addition tetrahedral intermediate should be a relatively stable anion due to inductive withdrawal of the positive nitrogen and the electronegative chlorine.
Understand the system. The chlorine is an excellent leaving group, and the protons on the positive nitrogen are acidic. Our best base is the ammonia reactant.
Find possible routes. Path E_β, then deprotonation path p.t.,

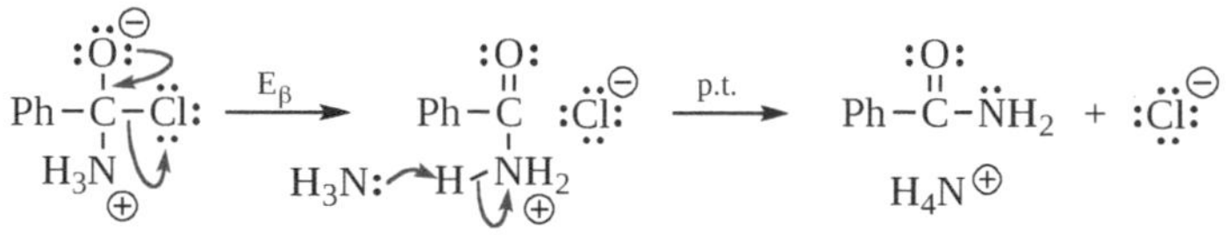

or deprotonation path p.t., then path E_β.

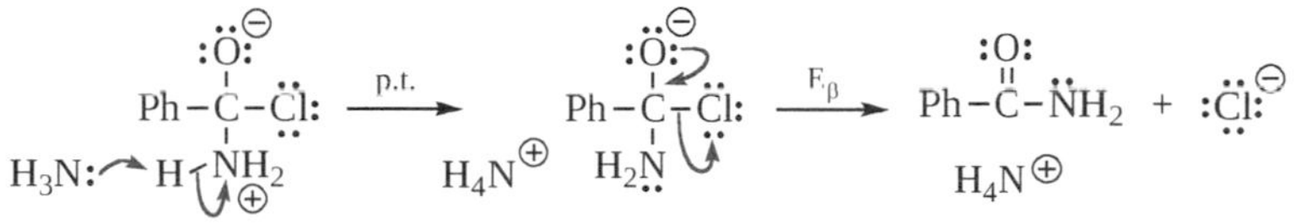

A third possibility since the reaction is in a strongly hydrogen bonding medium is the E2 path.

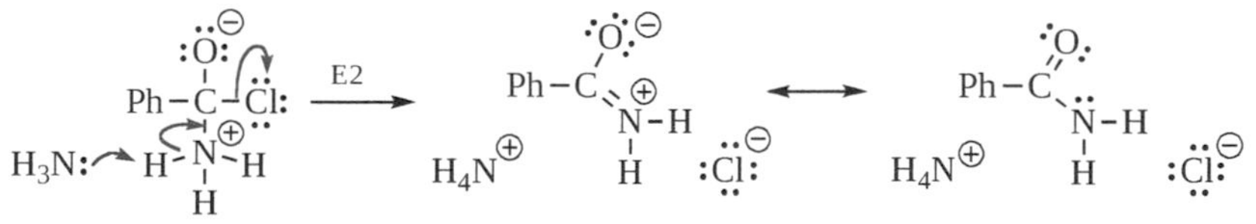

Evaluate and cross-check. In the first route, the loss of an excellent leaving group from an anionic tetrahedral intermediate should be rapid, but this process produces the somewhat destabilized nitrogen-protonated amide. The positive charge on nitrogen is destabilized being next to an electron-withdrawing carbonyl. Proton transfer in water is often at the diffusion-controlled rate so the second route is possibly better. The third route also makes use of the extensive hydrogen bonding expected in this medium. Both of the last two routes are good and produce the same product, the amide.
Understand the system. Amides are one of the least reactive of the class Y=C–L; without any powerful electrophiles or nucleophiles in the reaction mixture, this is a good place to stop.
Overview. A logical process is an addition then (in any order) elimination and proton transfer.

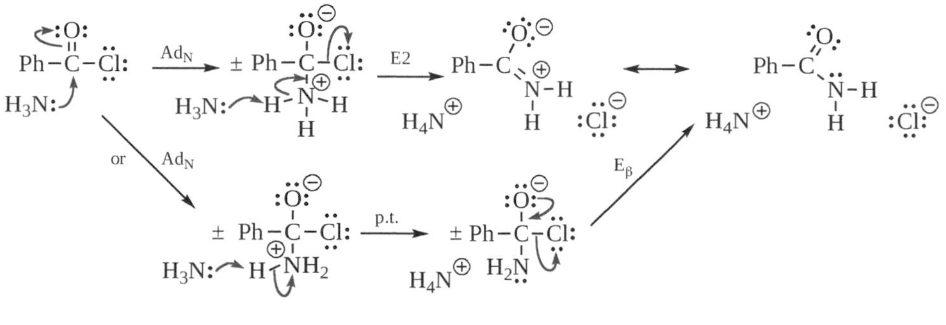

Understand the system. The protonated amine is a fair leaving group, $pK_{aHL} = 10.7$. The center amino-alcohol (hemiaminal) above may seem like familiar ground, for that is halfway between an C=N and a C=O (worked example 10.5.2). Elimination of water takes us to the C=N, and elimination of diethyl amine will take us to the C=O product. A sufficiently acidic medium will drive the reaction toward the formation of the C=O because the amine leaving group will be protonated after departure, removing it as a nucleophile for the reverse process.
Find possible routes. Three elimination routes, the E1cB (path p.t., then path E_β),

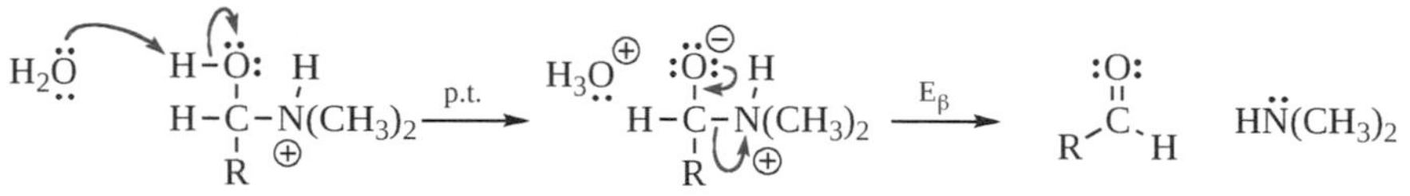

or the E2,

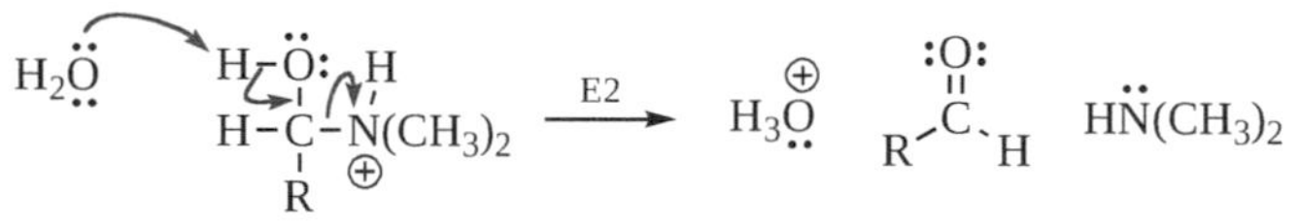

or the lone-pair-assisted E1 (path E_β, then p.t.).

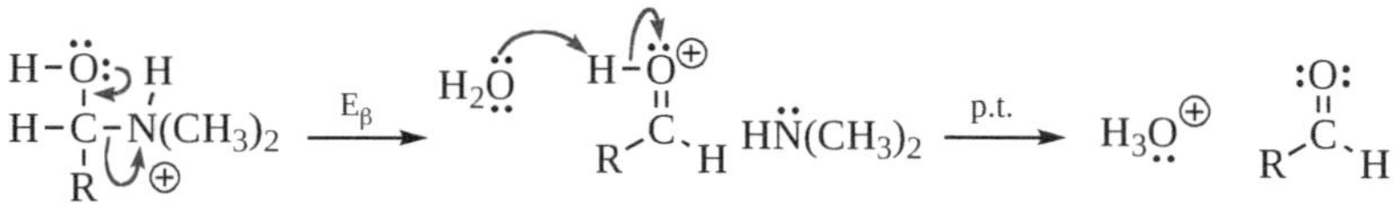

Evaluate and cross-check. The E1cB in acid suffers from a poor proton-transfer K_{eq} (estimated at about 10^{-12}), but the anion formed is a good enough source to kick out a fair to poor leaving group. The lone-pair-assisted E1 produces a very acidic cation, approximate pK_a –10, and usually requires a better leaving group. The E2 must have the O–H and C–N bonds coplanar for proper alignment, but that is not a problem in freely rotating compounds such as this. The E2 may seem to have a slight ΔpK_a rule problem if water is used as the base with incoming pK_{abH} 0.0, outgoing 10.7. However if the leaving amine is used as the base, the ΔpK_a rule is satisfied (in pK_{abH} 10.7 out 10.7). If any of the three routes have enough energy to proceed, then some amine is formed, and the E2 can proceed because it produces more amine. With this adjustment of the identity of the base, the E2 or E1cB is a reasonable route.
Understand the system. The aldehyde will reversibly form the hydrate and the enol, as shown in Section 10.7.3 for a structurally similar ketone. The aldehyde is the most stable, as demonstrated by a $\Delta H°$ calculation.
Overview. A reasonable overall process has been an Ad_N addition, two proton transfers, and an E2 elimination. Finally, the amine is protonated in an irreversible proton transfer with a K_{eq} equal to about $10^{+10.7}$. This reaction is known as the Bouveault aldehyde synthesis.

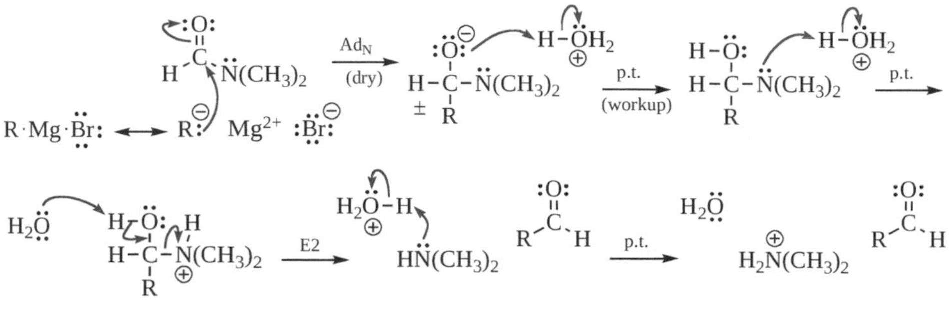

Problem 10.9a

Similar systems run in ^{18}O water show no ^{18}O in the alcohol product.

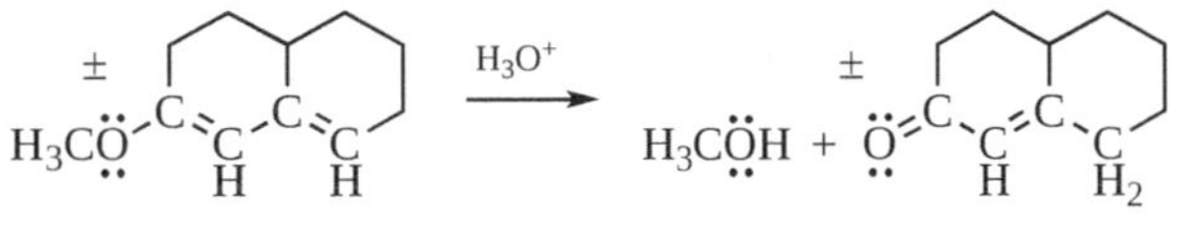

Understand the system. Source class: Z–C=C–C=C (vinylogous allylic source). Sink class: HA. Media: acidic. Best site for protonation: either end or the middle of the vinylogous allylic source, Z–C=C–C=C. If we partially get out of line structure near the reactive part of the molecule, we can see that protonation has occurred on the far C end of the allylic source. Find the pieces of the starting materials in the product and determine what bonds have been made and broken. The new or broken bonds are in red in the following structures.

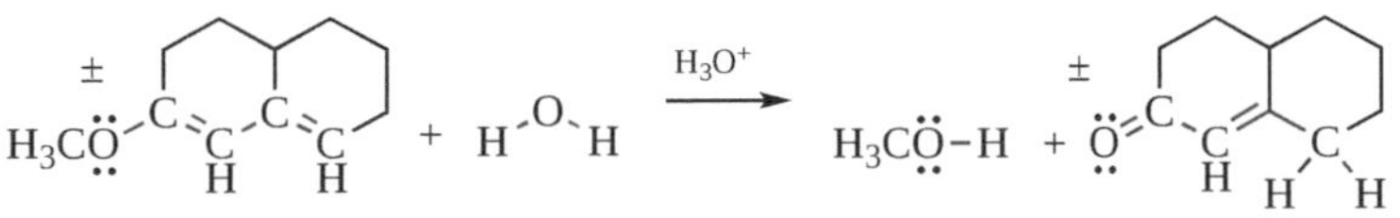

Find possible routes. To simplify the clutter on this mechanism, the ± is not shown since all structures are racemic. Protonation occurring on carbon by path A_E gets us closer to product.

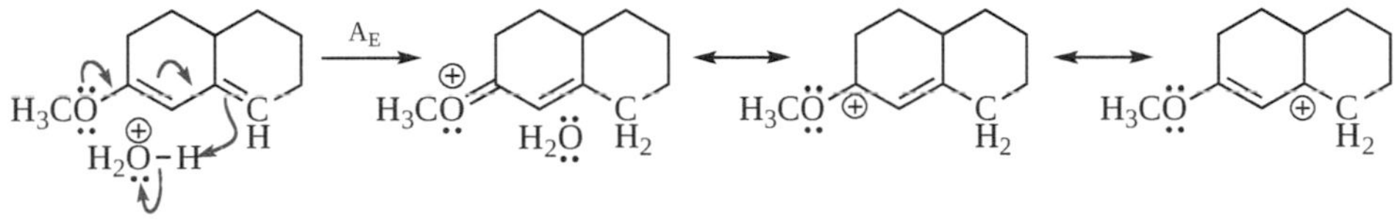

or on oxygen by path p.t. (which sets up the system for an S_N2).

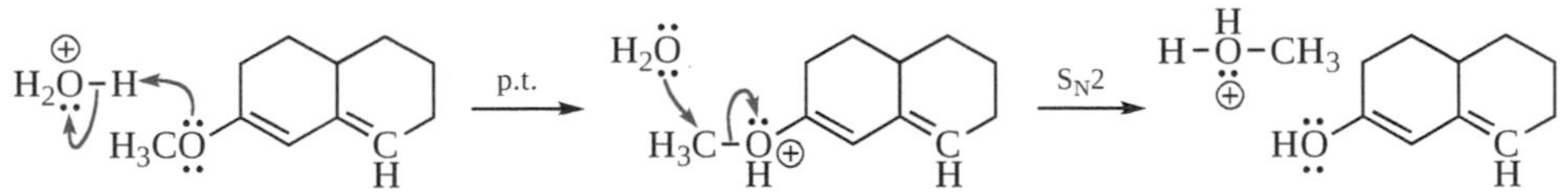

or on oxygen by path p.t. (which sets up the system for an S_N1).

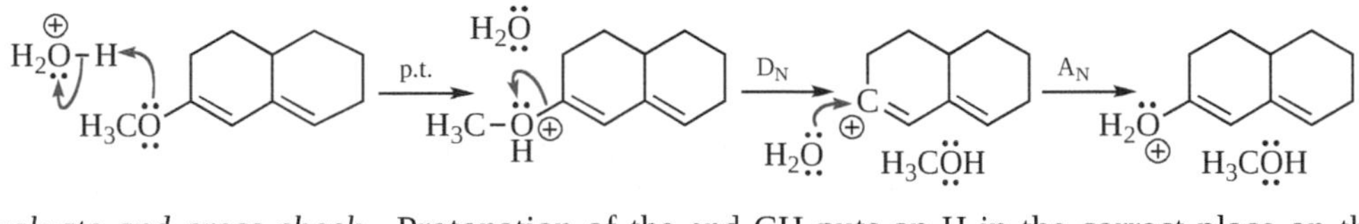

Evaluate and cross-check. Protonation of the end CH puts an H in the correct place on the rightmost ring and creates a very delocalized carbocation. The lone-pair-stabilized allylic carbocation is relatively stable and an excellent electron sink. Protonation in the middle of the pi system does not get us closer to product, and makes a less delocalized carbocation. Proton transfer to the oxygen makes a more stable oxonium ion, which has only one reasonable path, the S_N2, available to it (other than return to reactants). The S_N2 gets us closer to products but is ruled out by the $^{18}OH_2$ study, for the methanol formed would contain ^{18}O if the S_N2 occurred. Finally, the S_N1 is ruled out by the formation of a very unstable vinyl carbocation (not resonance stabilized).

Understand the system. The sink is C=C–EWG, but from the product we can see that no conjugate addition has occurred. Water is our most abundant nucleophile.

Find possible routes. Addition of water to the delocalized carbocation (or polarized pi bond),

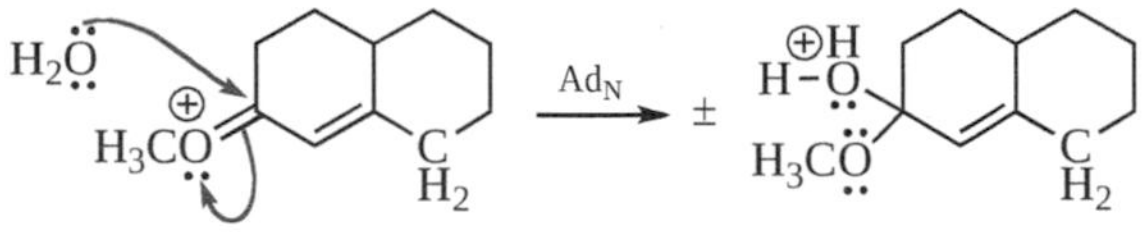

or S_N2 on the methyl.

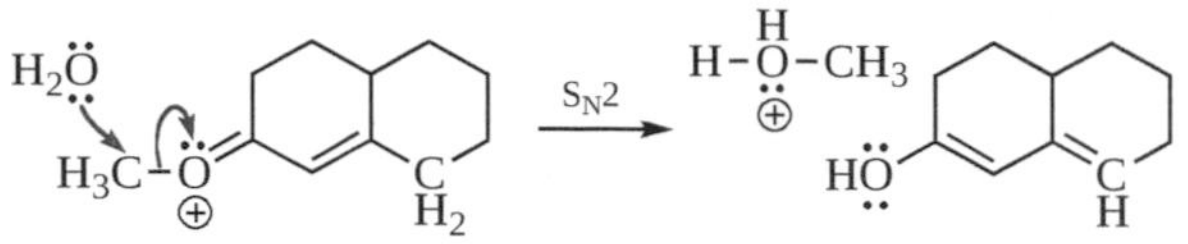

Evaluate and cross-check. This Ad_N could also be considered path A_N, trapping of a cation by a nucleophile. Again the S_N2 can be ruled out by the $^{18}OH_2$ study.

Understand the system. The product is a protonated hemiketal. Proton transfer between the lone pairs should be rapid. We need to make the methyl ether into a leaving group so that we can eliminate it.

Find possible routes. Proton transfer via solvent should be fast.

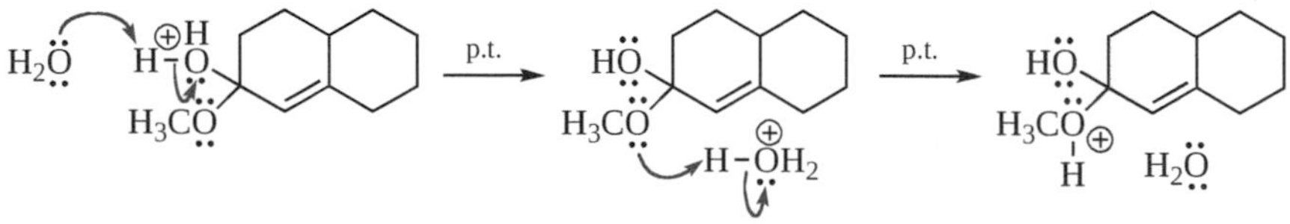

Evaluate and cross-check. The protonated ether is slightly uphill; the K_{eq} is about 10^{-1}.

Understand the system. The protonated ether is a good leaving group, $pK_{aHL} = -2.2$.

Find possible routes. The two elimination possibilities in an acidic medium are the E2,

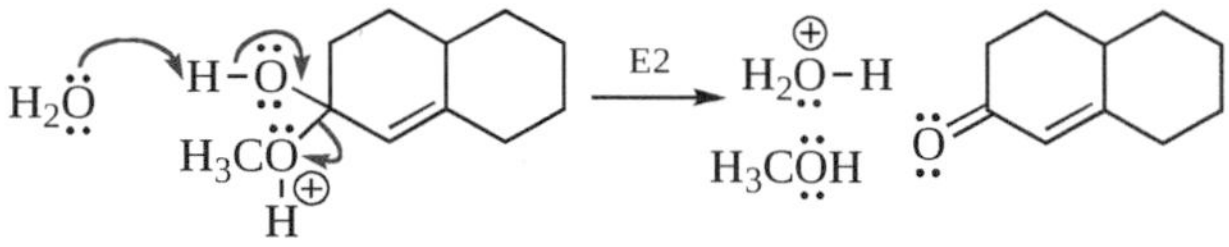

or the lone pair assisted E1.

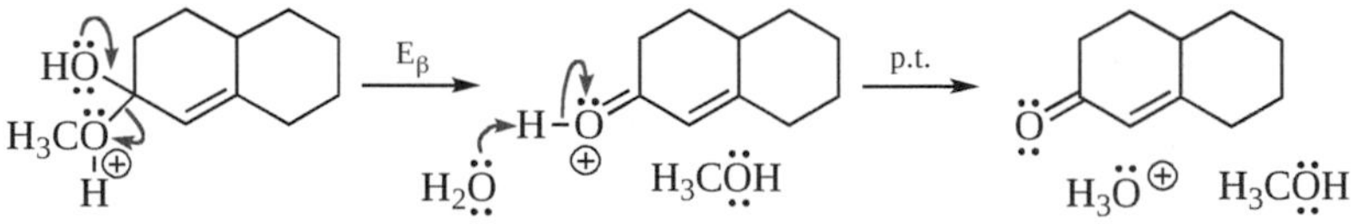

Evaluate and cross-check. Considering the media, the good stability of the carbocation and the good leaving group, the lone-pair-assisted E1 is the best route. However, the E2 is not unreasonable. The lone-pair-stabilized allylic carbocation intermediate is relatively stable. The methanol is not labeled, consistent with the $^{18}OH_2$ study.

Overview. A reasonable mechanism is shown below. Did this problem seem familiar? Look at the overview for the product prediction example in Section 10.7.3. We have gone from the enol ether via the hemiketal to the ketone. The grease has changed but the electron flow remains the same. The only difference was that this system had an extra double bond and started at the enol ether instead of the ketal.

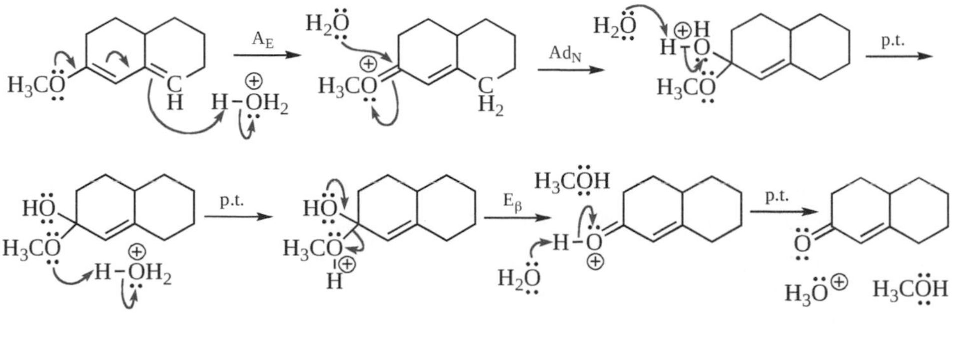

Problem 10.9b

This reaction requires cyanide ion as a catalyst and fails to go if it is absent.

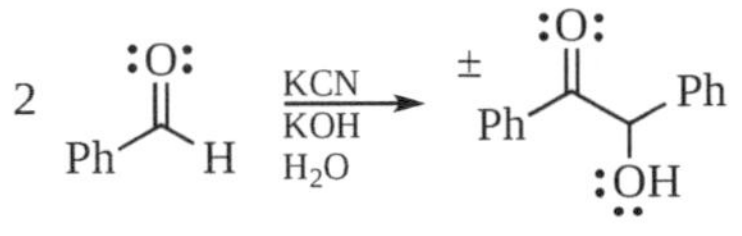

Understand the system. It looks like we have added a benzaldehyde acyl anion equivalent to another benzaldehyde. There are no acidic hydrogens on benzaldehyde. We need to generate a carbon nucleophile for the eventual addition to benzaldehyde. Sources: cyanide and hydroxide. Sinks: benzaldehyde C=Y and water, a weak HA. Media: basic, with bases cyanide at pK_{abH} 9.2 and hydroxide at pK_{abH} 14.0. The best nucleophile is cyanide. Resonance forms:

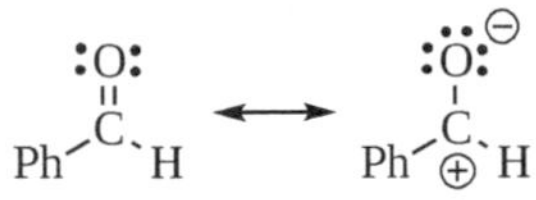

Find possible routes. Since cyanide is required and it is a good nucleophile, then it must add to benzaldehyde since that sink has little else it can do. Only Ad_N2 and the Ad_E3 additions are reasonable for basic media. The hetero Ad_E3 addition uses water as a hydrogen-bonding acid catalyst.

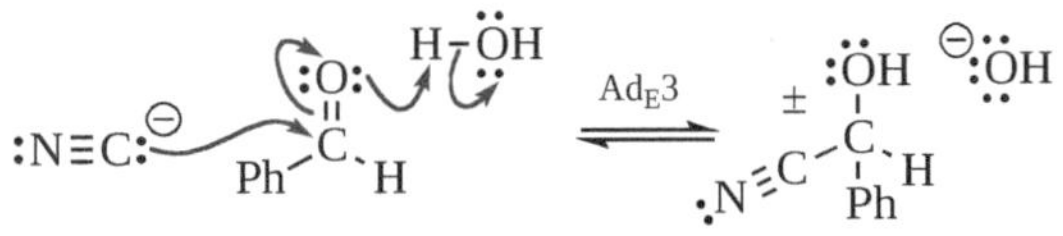

The Ad_N2 does the same transformation in two steps.

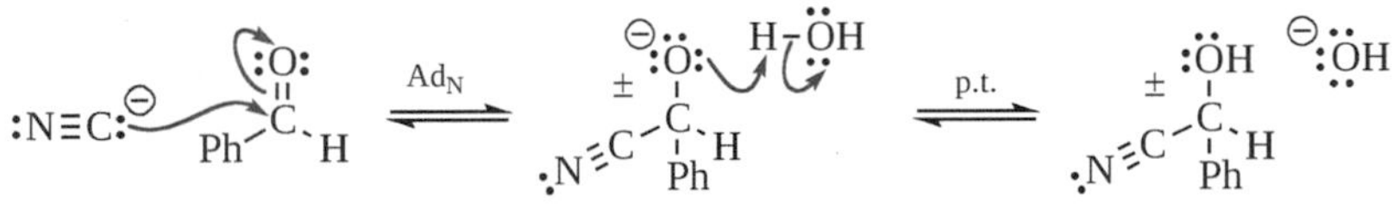

Evaluate and cross-check. The Ad_E3 passes the ΔpK_a cross-check, (incoming Nu is pK_{abH} 9.2 and outgoing hydroxide is pK_{abH} 14.0) for a climb of 4.8 units. Overall, the Ad_N2 addition climbs the about same amount, with the intermediate probably being close in pK_{abH}. The Ad_E3 is the most probable path in water due to extensive hydrogen bonding, but the Ad_N2 is certainly a reasonable route.

Understand the system. Source: Z, hydroxide as a base. Sink: HA, the C–H is now somewhat acidic because the attached cyanide is an EWG. Media: basic, hydroxide pK_{abH} 14.0.

Find possible routes. The electron-withdrawing group makes the C–H acidic because it can delocalize the anion of the conjugate base. The alcohol is also acidic but deprotonating that just returns us to reactants.

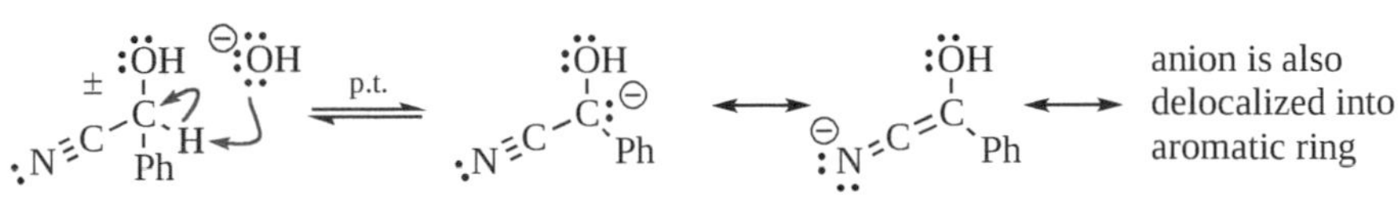

Evaluate and cross-check. The proton transfer passes the ΔpK_a cross-check; the incoming base is pK_{abH} 14.0 and outgoing nitrile enolate is about pK_{abH} 22 (25 of CH_3CN minus about 3 units for the phenyl delocalization) for a climb of about 8 units.

Understand the system. Source: RM, the nitrile enolate. Sink: HA and C=Y. Media: basic.

We now have a nucleophile on the carbon that we wish to attach to benzaldehyde.
Find possible routes. Only Ad_N2 and the Ad_E3 additions are reasonable for basic media. The hetero Ad_E3 addition uses water as a hydrogen-bonding acid catalyst.

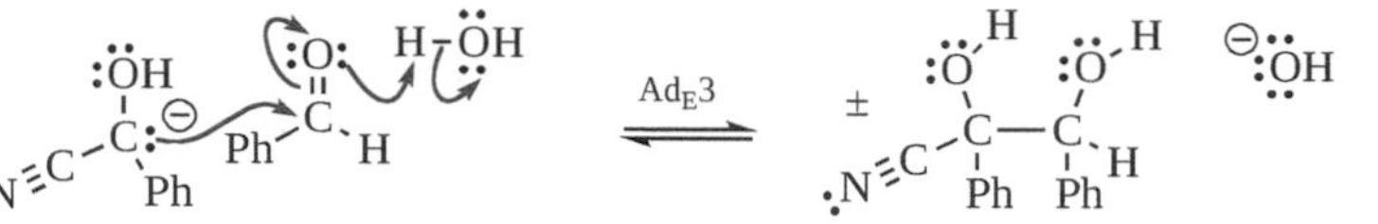

The Ad_N2 does the same transformation in two steps.

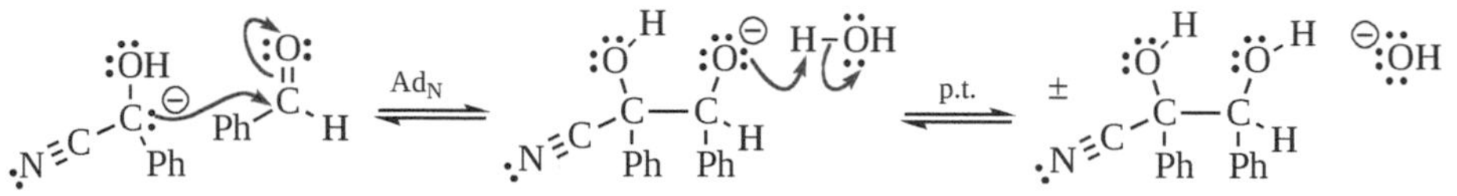

Evaluate and cross-check. The Ad_E3 is downhill by the ΔpK_a rule test, (incoming Nu is pK_{abH} 22 and outgoing hydroxide is pK_{abH} 14.0) for a drop of about 8 units. Overall, the Ad_N2 addition drops the same amount, with the intermediate close to that of $HOCH_2CH_2OH$ that has pK_{abH} of 14.2 due to internal hydrogen bonding. Again, the Ad_E3 is the most probable path in water due to extensive hydrogen bonding, but an Ad_N2 is certainly a reasonable route.
Understand the system. Source: Z, hydroxide as a base. Sink: HA the alcohol hydroxyls. Media: basic.
Find possible routes. The secondary alcohol is acidic and can be deprotonated to form the previous anionic intermediate. That previous anionic intermediate can return us to reactants, or since it is internally hydrogen bonded can easily transfer a proton internally.

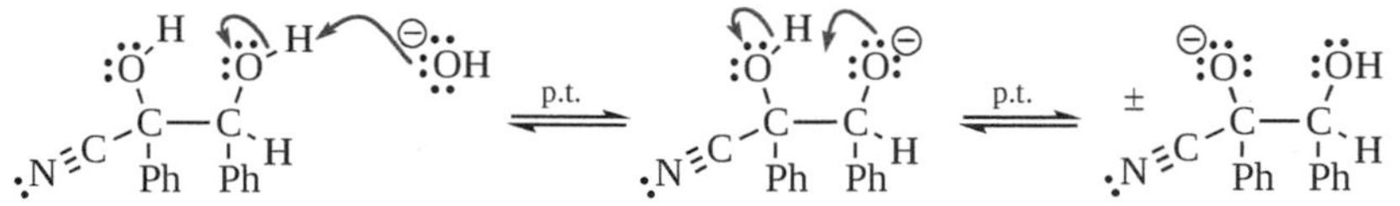

The tertiary alcohol formed in the addition reaction is also acidic and can be part of the first step in an E1cB to kick out our cyanide catalyst.

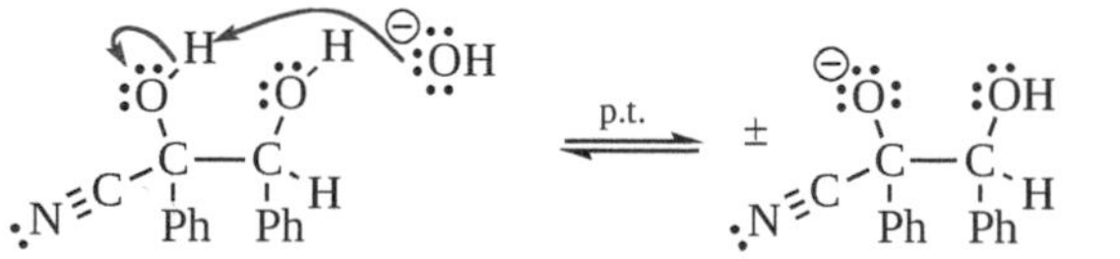

Evaluate and cross-check. Both these proton transfers should have a K_{eq} of about one. Either will work, but the internal proton transfer is expected to be faster.
Understand the system. Source: Z, alkoxide anion. Sink: Cyanide can serve as a sp^3 C–L.
Find possible routes. The deprotonation of the tertiary alcohol is first step in an E1cB to kick out the cyanide catalyst and give us the final product.

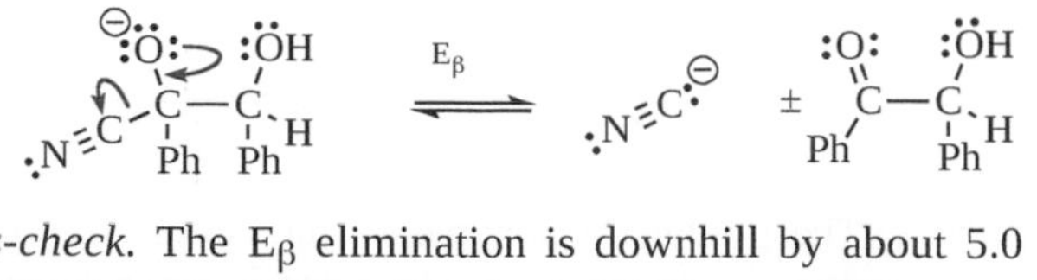

Evaluate and cross-check. The E_β elimination is downhill by about 5.0 units (starting diol anion is $pK_{abH} = 14.2$, and outgoing leaving group is $pK_{aHL} = 9.2$).
Overview. A reasonable process was an Ad_E3, then a proton transfer to generate a carbon nucleophile, an Ad_N2 with an internal proton transfer to set up an E1cB that gives the product.

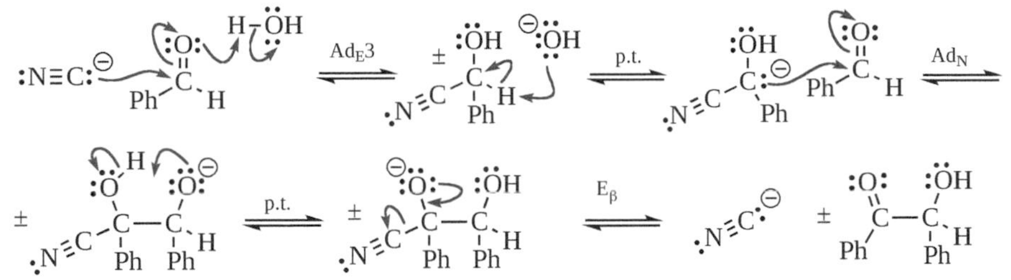

Problem 10.9c Only this stereoisomer of product is formed.

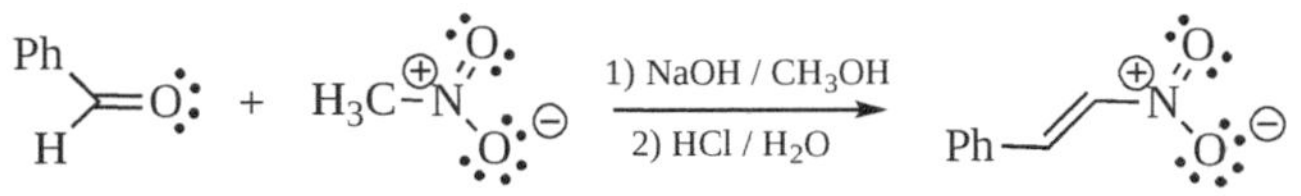

Understand the system. Addition connected the source and sink, and then elimination formed a double bond. The first step is to get a nucleophile on carbon by deprotonating nitromethane.

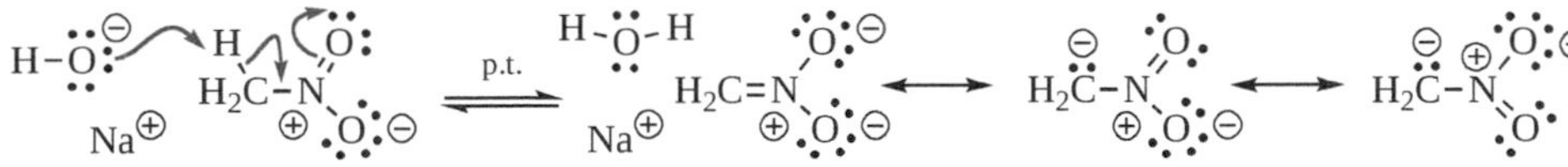

Evaluate and cross-check. Hydroxide, pK_{abH} 14.0, easily deprotonates nitromethane, pK_a 10.2.

Find possible routes. There are three common types of additions, Ad_E2, Ad_N2, or Ad_E3.

Evaluate and cross-check. There is no acid present in the basic medium to make a protonated aldehyde, so the Ad_E2 is out. The Ad_E3 is possible using the methanol solvent as a weak acid. The Ad_N2, shown below, passes the ΔpK_a rule (nucleophile at pK_{abH} 10.2 and the alkoxide intermediate at about pK_{abH} 17). The p.t. step goes downhill to methoxide at pK_{abH} 15.2.

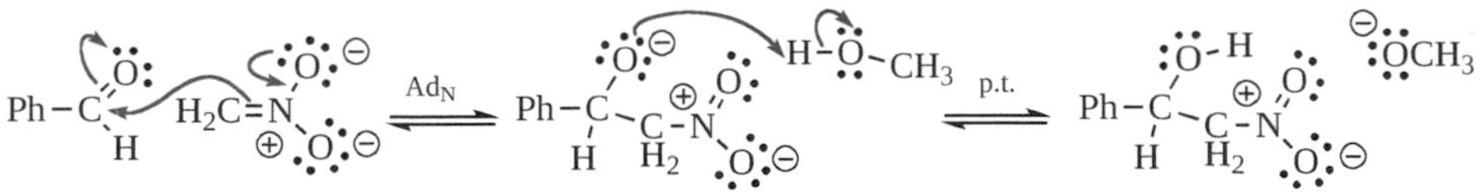

Understand the system. The medium is now changed to acidic, and we need to set up for an acid catalyzed elimination by improving the leaving group by protonation.

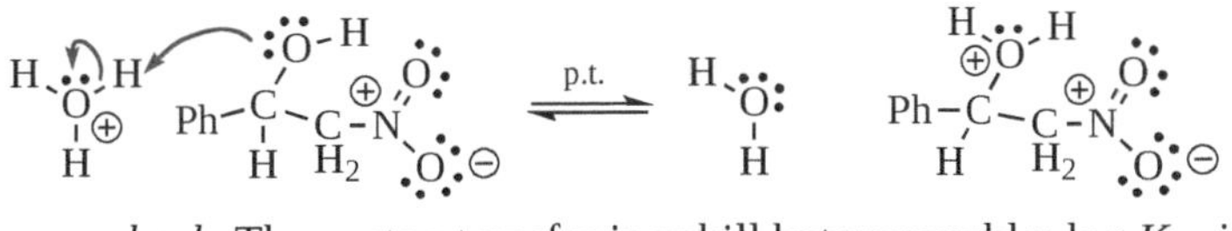

Evaluate and cross-check. The proton transfer is uphill but reasonable; $\log K_{eq}$ is about –2.4.

Find possible routes. There are three types of eliminations, E1, E2, or E1cB.

Evaluate and cross-check. The medium is too acidic for the E1cB. The E1 passes the three tests for the slow D_N step: a polar solvent, a good leaving group (pK_{aHL} = 0.0), and a carbocation more stable than secondary. The E2 is also reasonable; the water base pK_{abH} is 0.0, and the water L pK_{aHL} is 0.0. Shown below is the E1, giving the conjugated product. The most stable product stereochemistry is trans, having much less steric hindrance than cis.

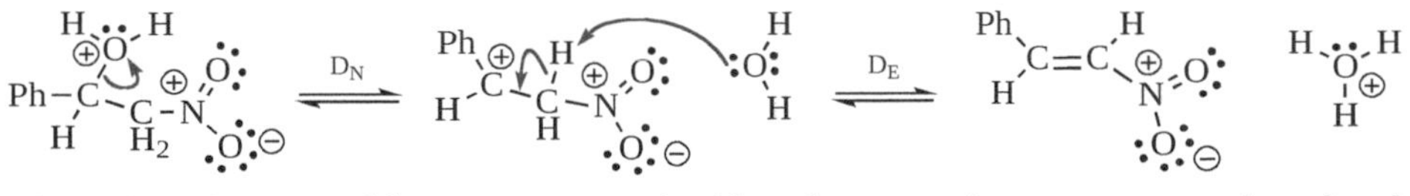

Overview. A reasonable process started with a deprotonation to get a good nucleophile. An Ad_N2 or Ad_E3 addition connected the nucleophile and electrophile. The medium was made acidic so that protonation of the hydroxyl in acid could improve the leaving group.

Finally, an E1 or E2 elimination formed the most stable trans nitro alkene product.

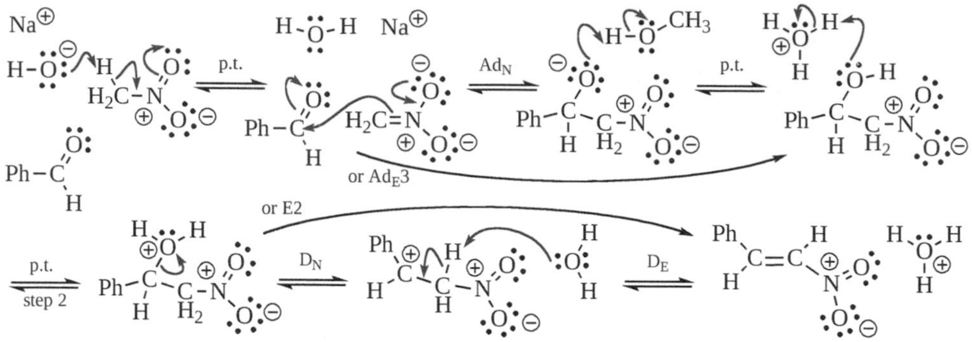

Problem 10.11a Provide a mechanism for the following transformation (more difficult).

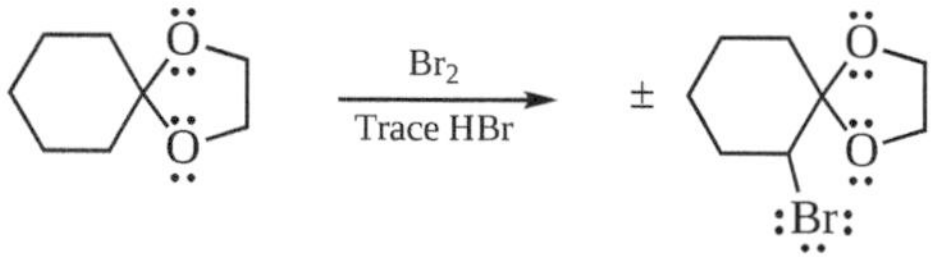

Understand the system. Sinks: Bromine is a Y–L and HBr is an acid. Sources: the oxygen lone pairs of the ketal. Every bond can be considered to be a combination of a nucleophile and an electrophile. The electrophile for the C–Br bond was bromine, the Y–L. A beginner might be tempted to use the CH bond as a nucleophile and get the problem over in one step, but CH bonds, being strong and not very polarized, make rotten electron sources. A better nucleophile must be generated because the CH_2 is not reactive. Before you try to find possible routes, it is critical that you understand what bond changes have occurred. To find the bonds that have changed in this reaction, it is helpful to reposition the reactants and not use line structure near bond changes. Two bonds were broken and two bonds have been made, shown in red below.

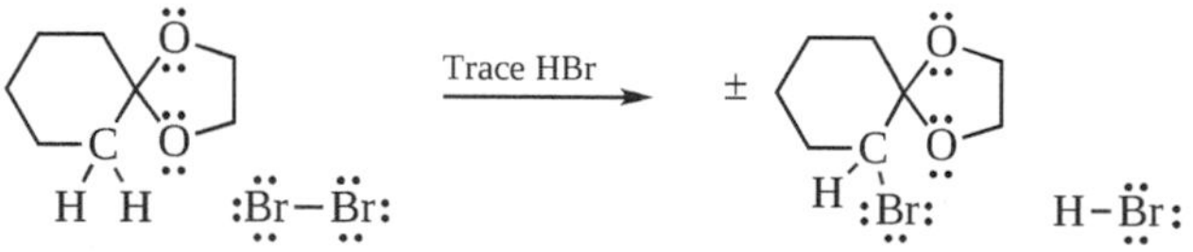

Find possible routes. The trace HBr probably acts as an acid catalyst, and since HBr is a product, we will soon have plenty. There is one path available, proton transfer to the lone pair.

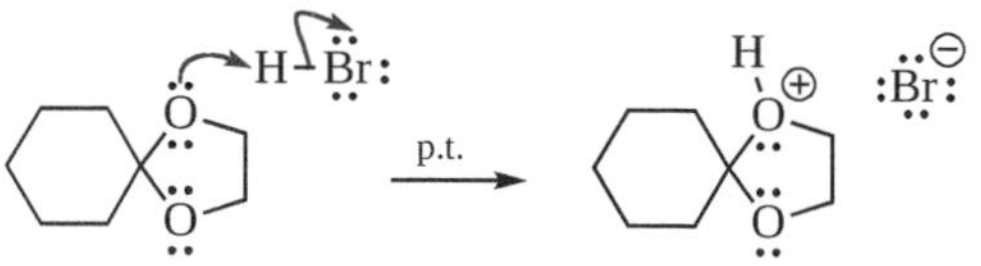

Evaluate and cross-check. The reaction is downhill in energy since the acid catalyst has a pK_a of –9, much more acidic than –3.5, the pK_a of the protonated ether. We made a weaker acid.
Understand the system. The protonated ether is now a good leaving group.
Find possible routes. Elimination would break the correct CH bond and produce a pi bond, which could act as a nucleophile. Two acidic elimination routes are possible, the E2,

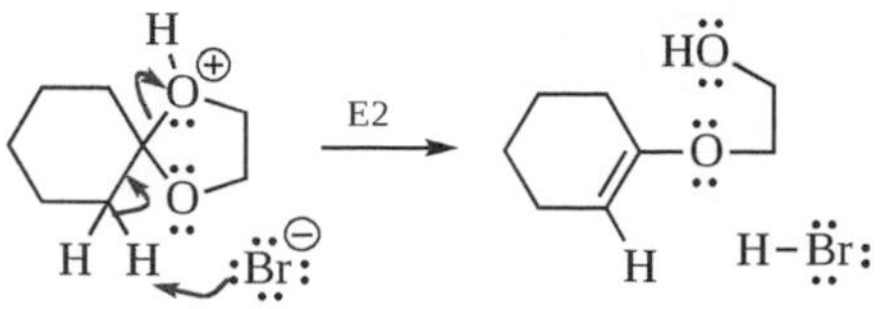

or the lone-pair-assisted E1 (path E_β followed by D_E).

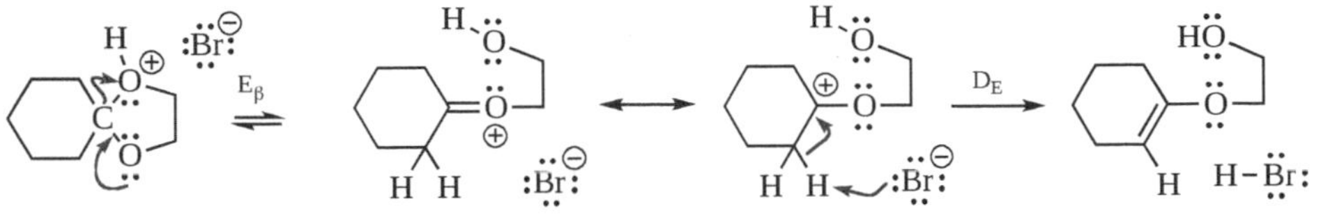

Evaluate and cross-check. Considering the lone pair stabilization of the carbocation formed, and the fact that the leaving group is good, the lone-pair-assisted E1 is a good route. Both are reasonable routes, but the E1 may be lower in energy due to the resonance-stabilized cation lowering that corner of the elimination surface. A $\Delta H°$ calculation shows that the elimination has gone uphill in energy. Bonds broken, C–H, C–O, C–C, minus bonds made, C=C, O–H, gives (99 + 86 + 83) – (146 + 111) = 268 – 257 = +11 kcal/mol (+46 kJ/mol).

Understand the system. The enol ether is a good allylic electron source (Z–C=C), and bromine is a good electron sink (a Y–L). We now have a nucleophile on the correct carbon that can react with bromine.

Find possible routes. Two acidic addition routes are possible, the Ad_E3,

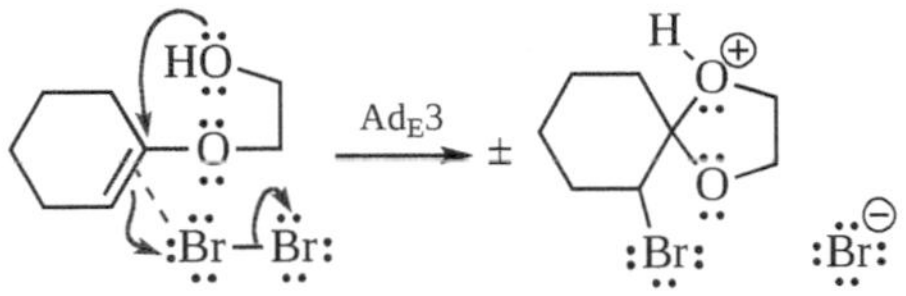

or the Ad_E2, which proceeds via a lone-pair-stabilized carbocation (path A_E then path A_N).

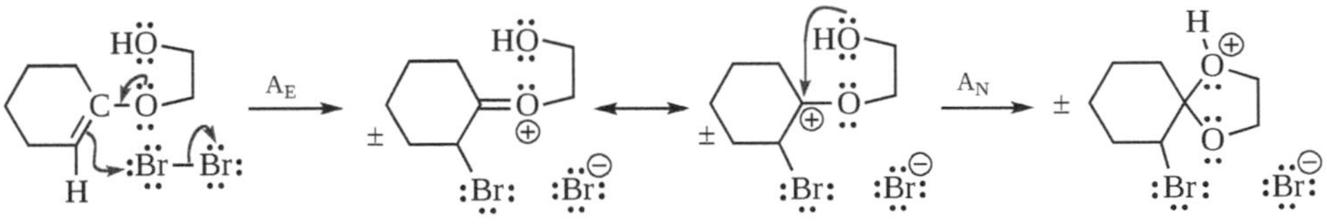

Evaluate and cross-check. The carbocation is relatively good; both routes are reasonable. Five-membered ring formation is expected to be fast. A third possibility is to add bromine by path NuL to form the bromonium ion that would then get attacked by the alcohol lone pair via an internal S_N2. However, the carbocation is stable enough not to require bridging by bromine, thus making this third route essentially the same as the Ad_E2.

Overview A following proton transfer that regenerates the acid catalyst gives us the product. A solvent molecule could pick off the proton, but it is often assumed and not drawn. Using bromide as a base has an uphill proton-transfer log K_{eq} of –9 –(–3.5) = –5.5; so K_{eq} is $10^{-5.5}$.

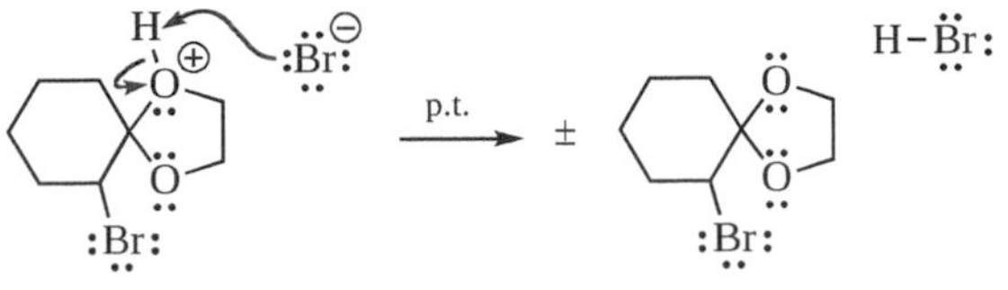

Now let us back off from the problem far enough to see the entire process as a whole. The mechanism for this transformation or "mechanistic sentence" can be made from our "mechanistic phrases" or path combinations. A reasonable mechanism has a proton transfer improving the leaving group, then an E1 (or E2) elimination followed by an Ad_E2 (or Ad_E3) addition, ending with a proton transfer.

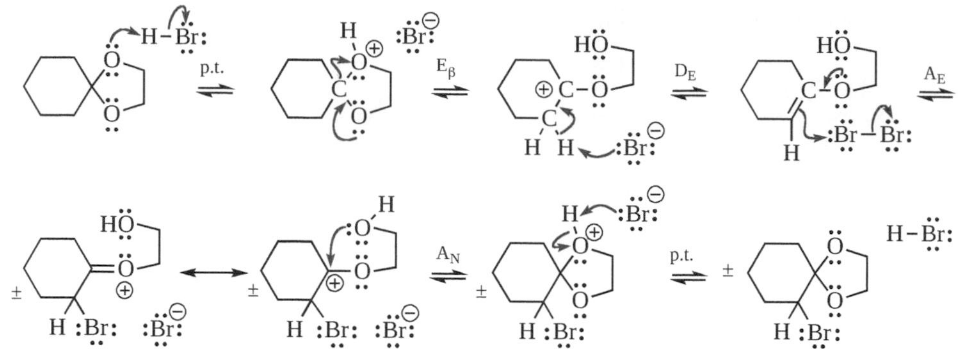

Surface overview. We can look back at the problem space that we just navigated to get an overview, Figure S.8. Red arrows show the most reasonable route(s). A proton transfer from HBr catalyst improves the leaving group, then an E1 (or E2) elimination occurs to give the enol ether, followed by an Ad_E2 (or Ad_E3) addition of bromine to the enol ether, ending with a proton transfer back to bromide to give the neutral product.

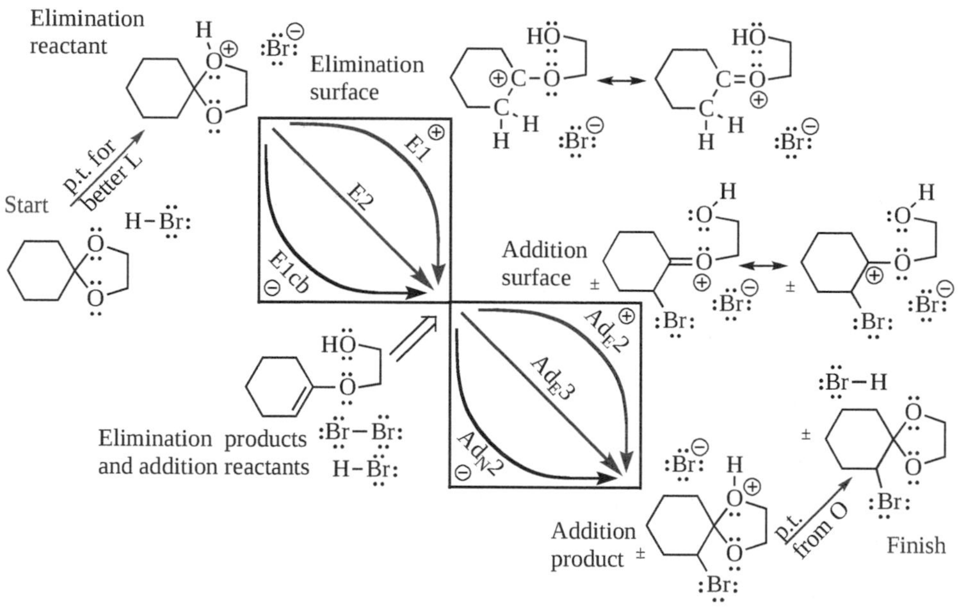

Figure S.8 The surface overview for problem 10.11a.

Problem 10.11b Provide a mechanism for the following transformation (more difficult).

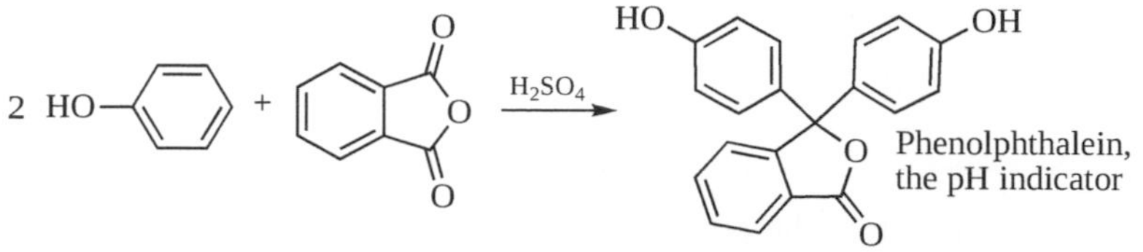

Understand the system. Balanced? No, water must be added to the right-hand side to balance the equation. Generic process? An electrophilic aromatic substitution has occurred twice on the anhydride carbonyl. Medium? Very acidic. Acidic H's? Sulfuric acid. Sources? The oxygen lone pairs. Sinks? The anhydride is a Y=C–L and sulfuric acid is an HA. Leaving groups? The L on the anhydride carbonyl.

Find possible routes. Two possible protonation sites on our sink exist.

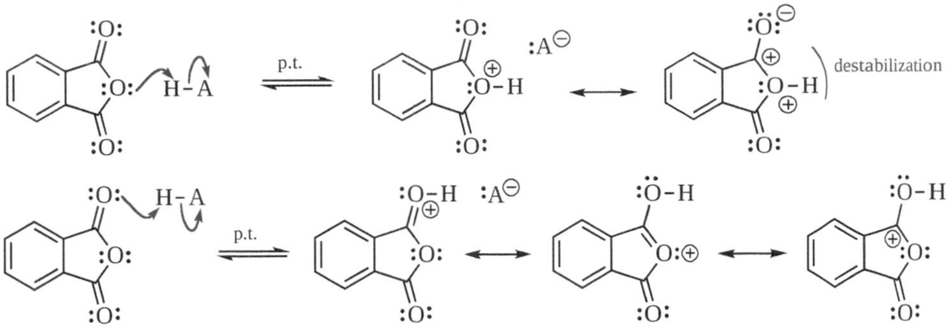

Evaluate and cross-check. Protonating the oxygen between the two carbonyls has no resonance stabilization and puts a positive charge between two electron-withdrawing carbonyls. Protonating on the carbonyl oxygen has lone pair delocalization from the other oxygen, so is of lower energy.

Understand the system. We can view the compound as having a protonated carbonyl or as having a protonated L on a Y=C–L. The phenolic hydroxyl is an electron-releasing group that activates the aromatic ring and directs incoming electrophiles ortho-para.

Find possible routes. Viewing the sink as a carbonyl or protonated carbonyl, we have two routes available in an acidic medium: the Ad_E3 or Ad_E2 addition. Viewing the sink as having a protonated L on a Y=C–L, we can also consider a D_N step to lose the leaving group followed by nucleophilic attack by the aromatic ring. Let us look at all three choices. Attacking the carbonyl as it is being protonated in an Ad_E3 addition, followed by rearomatization, utilizes the mildest electrophile and completes an electrophilic aromatic substitution.

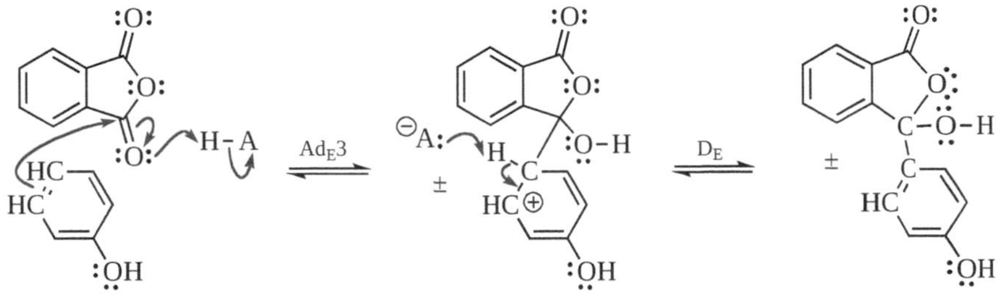

Attacking a more reactive protonated carbonyl as the second step of an Ad_E2 addition, followed by rearomatization, completes the same electrophilic aromatic substitution.

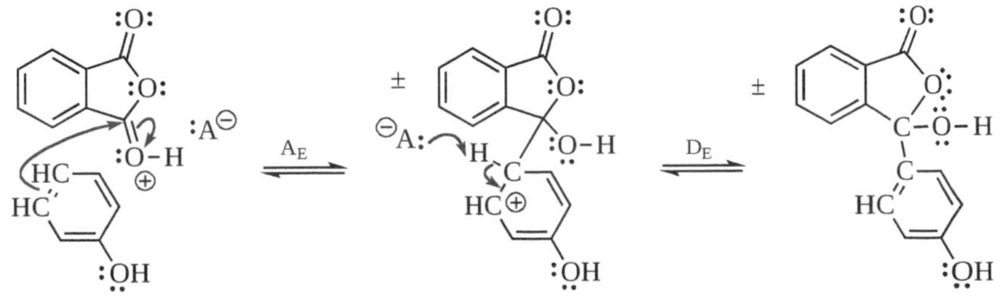

Or we can D_N open the protonated anhydride to the very reactive acylium ion and carboxylic acid pair and hope the aromatic ring attacks the acylium ion before it can reclose back to the protonated anhydride. Leaving group loss followed by electrophilic aromatic substitution:

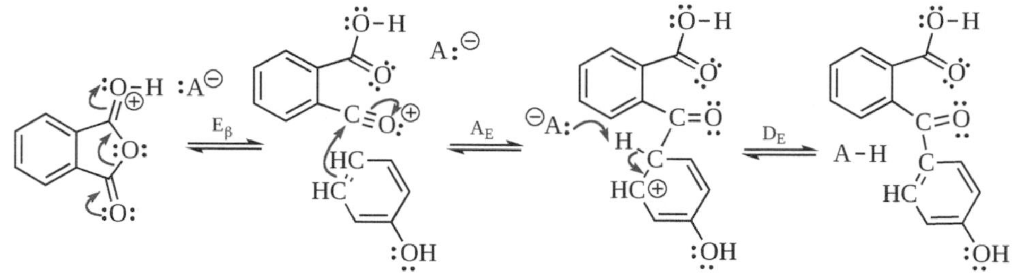

Evaluate and cross-check. The Ad_E3 addition requires the aromatic ring to attack before the carbonyl is fully protonated, so the electrophile is poorer. This aromatic ring is activated by a hydroxyl ERG thus may be good enough to attack at this point. Attack on the protonated carbonyl is better, and only somewhat hindered, because both Nu and E are flat. The ionization of the leaving group passes all cross-checks: the acylium carbocation is about as stable as tertiary, the leaving group has a negative pK_{aHL}, the solvent is polar, and to make it even better, there is a lone pair to push with. Concentrated solutions may have nucleophilic attack by the aromatic ring competitive with the reclosure back to the protonated anhydride.

Understand the system. Although the route seems to have split, we have to set up for a second electrophilic aromatic substitution, and that requires a protonation step. The oxygen lone pairs are the most basic sites.

Find possible routes. We can bring the two routes back together because the protonated ketone can be attacked by the nearby carboxylic acid and close to the hydroxylactone formed earlier.

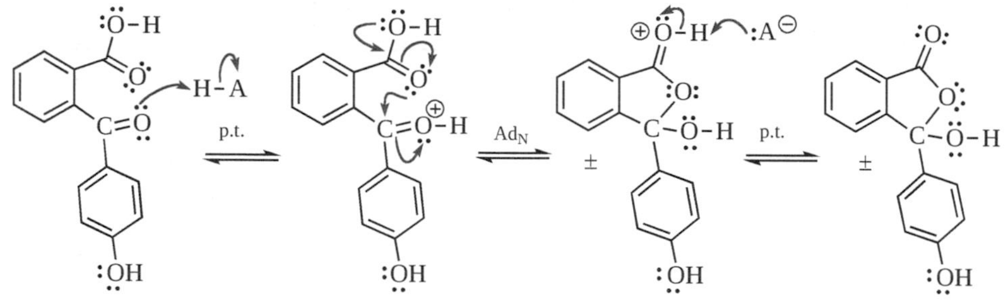

Evaluate and cross-check. Attack on the protonated carbonyl by the carboxylic acid is expected to be fast and reversible because of the large number of effective collisions.

Understand the system. To set up for a second electrophilic aromatic substitution and requires another protonation step. We are still in strong acid and the oxygen lone pairs are the most basic sites. Since the product has a second phenol on the same carbon as the first phenol, let us focus on getting an electrophile on that carbon. Conveniently, the alcohol on that carbon is on a site that would make a very stabilized carbocation if lost.

Find possible routes. Protonation of the hydroxylactone OH sets it up for loss in a D_N step.

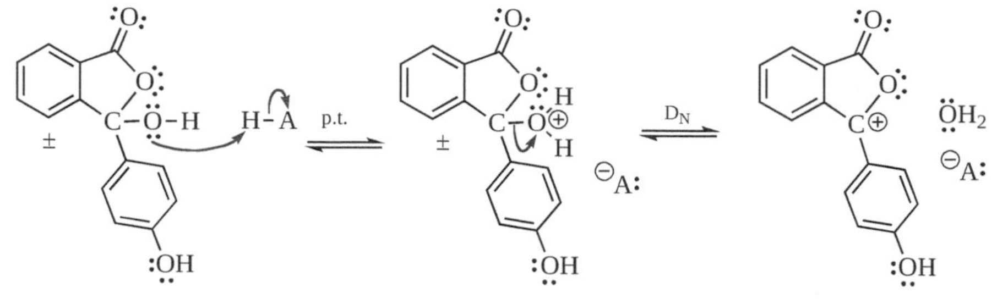

Evaluate and cross-check. Proton transfer is downhill from sulfuric acid and the ionization of the leaving group step passes all three D_N path restrictions: the carbocation is very stable with three resonance donating groups; the leaving group has a $pK_{aHL} \le 0$ and the solvent is polar.
Understand the system. We now have a good electrophile for a second electrophilic aromatic substitution.
Find possible routes. Electrophilic aromatic substitution makes the final C–C bond of the product.

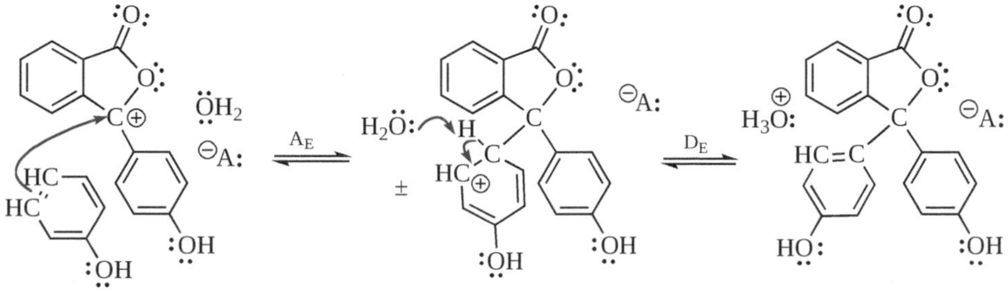

Overview. A very reasonable route starts with a proton transfer to a carbonyl that generates the electrophile for our first electrophilic aromatic substitution. Protonation and loss of a labile alcohol generates the electrophile for our second electrophilic aromatic substitution.

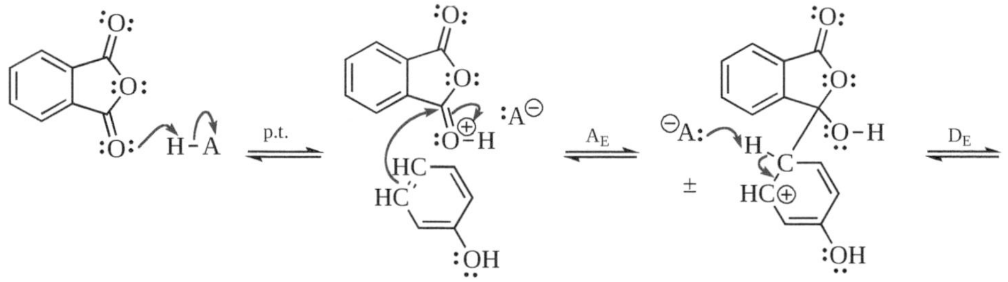

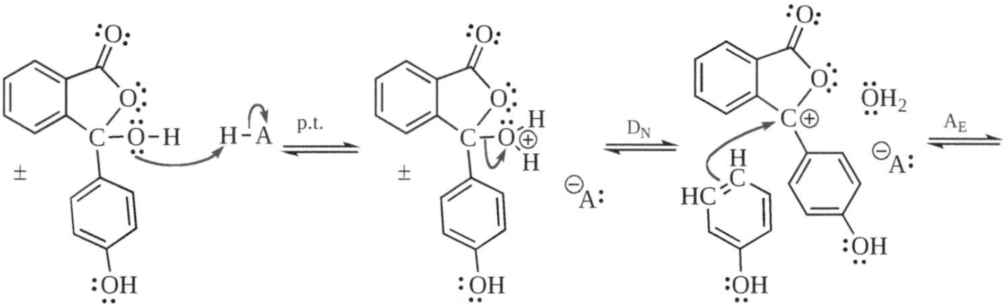

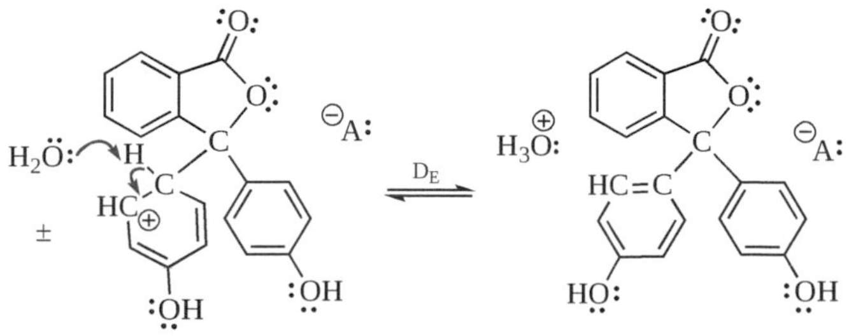

Surface overview. We can look back at the problem space that we just navigated to get an overview, Figure S.9. Red arrows show the most reasonable route. It is shown from the point of view of the electron sink. The first two steps are an Ad_E2 addition to the carbonyl. Next is the rearomatization step of the electrophilic aromatic substitution, followed by protonation of the alcohol to create a good leaving group. An S_N1 process is next (which begins our second electrophilic aromatic substitution), followed by a final rearomatization.

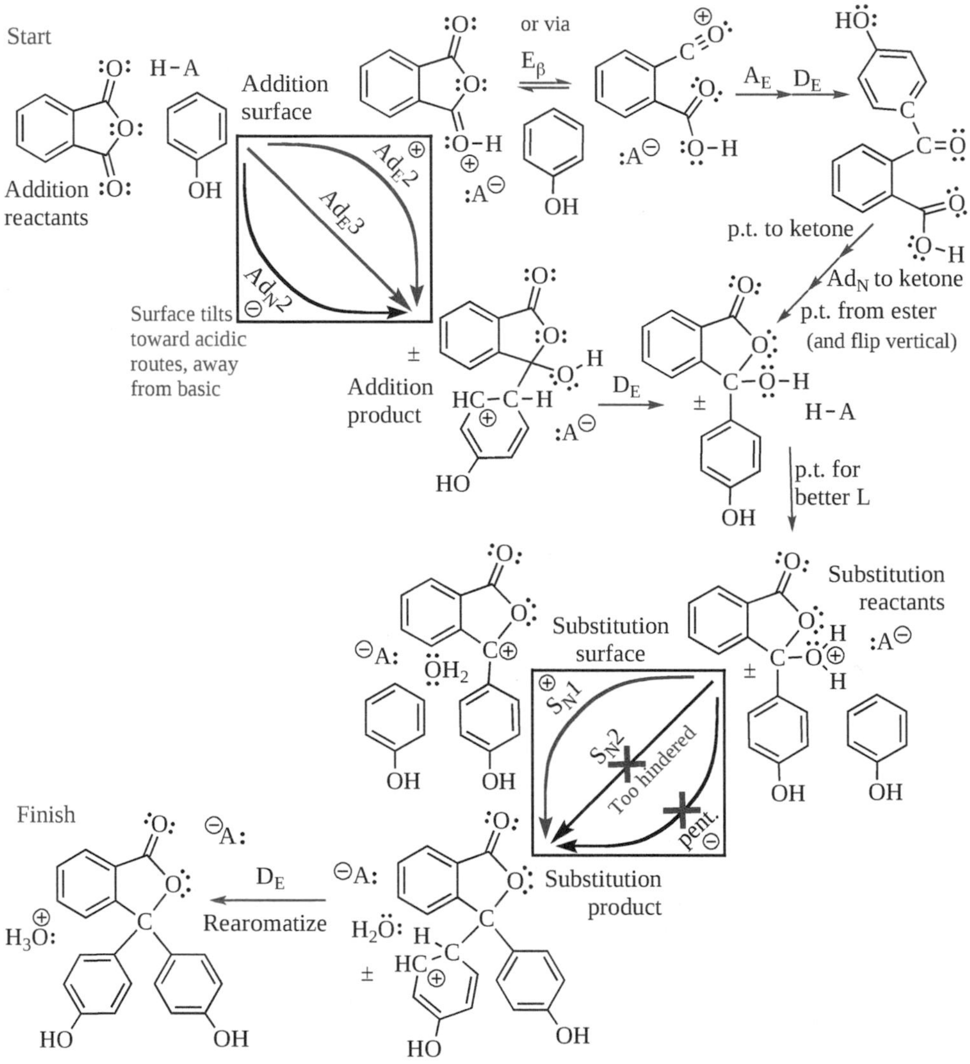

Figure S.9 The surface overview for problem 10.11b.

Problem 10.11c (Robinson annulation)

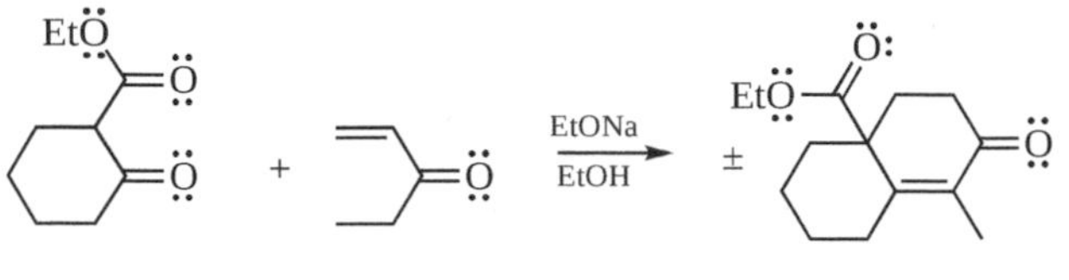

Because of the length of this example, only the more probable paths will be discussed. *Understand the system*. Balanced? No, H_2O is missing from the right hand side. Medium: basic, ethoxide's pK_{abH} is 16. Sources: ethoxide. Sinks: The ketone is a polarized multiple bond, the ester is a polarized multiple bond with leaving group, and the enone is a conjugate acceptor. Acidic H's: The CH between the ester and the ketone is the most acidic H at pK_a

10.7. Leaving groups: No good ones, only the ethoxide on the ester, and that is still attached in the product. Resonance forms of the conjugate acceptor:

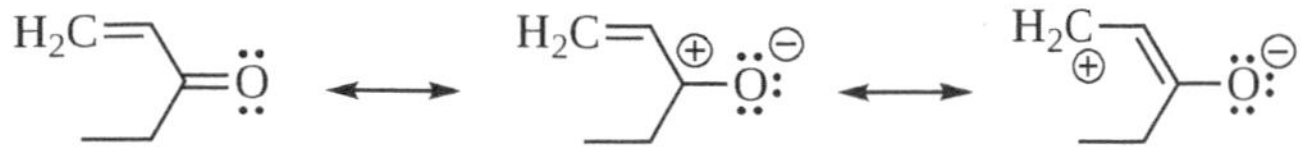

Find the pieces of the starting materials in the product and determine what bonds have been made. The new bonds are in red in the following structure; they connect the two reactants. Reasoning back from adjacent functionality, determine the possible nucleophile and electrophile pairs for each of the new bonds. Atoms next to electron-withdrawing groups can easily become anionic and therefore nucleophilic. Often the first step in a base-catalyzed process is a deprotonation. Deprotonating the most acidic H between the ester and the ketone will give us the needed nucleophile for making the top single bond. The carbonyl and the conjugate acceptor carbon are already electrophilic.

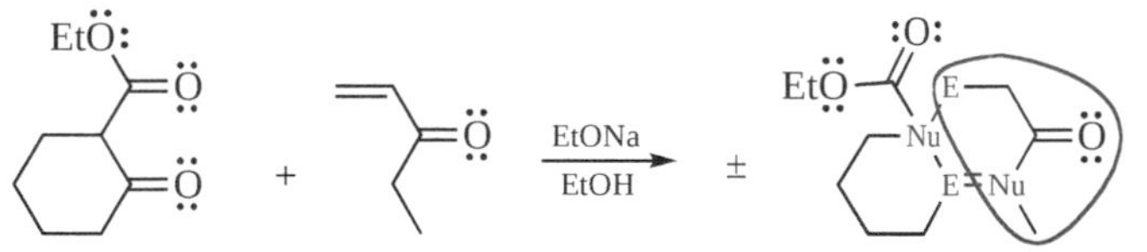

Generic process: Two connections, top and bottom, are needed and a pi bond must be made. A likely sequence is that an addition has occurred to produce the top single bond followed by another addition to then connect the bottom bond, and finally an elimination has occurred to create the bottom double bond. We can use the relevant surfaces as large building blocks to map out a possible problem space. The surface generated problem space might look like:

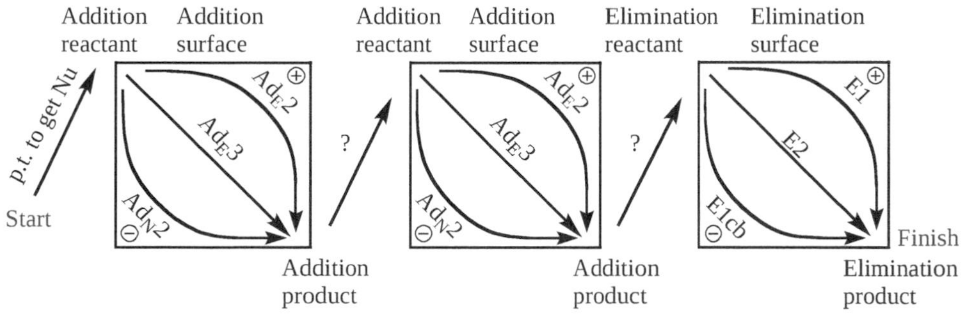

Find possible routes. The lone pair electrons of ethoxide are the best source; but if ethoxide had served as a nucleophile, we would expect a transesterification or an additional ethoxy group in the product. Since neither process connects our two pieces, we explore ethoxide as a base, since there is an acidic H, pK_a 10.7. That deprotonation gives a Nu for the top addition.

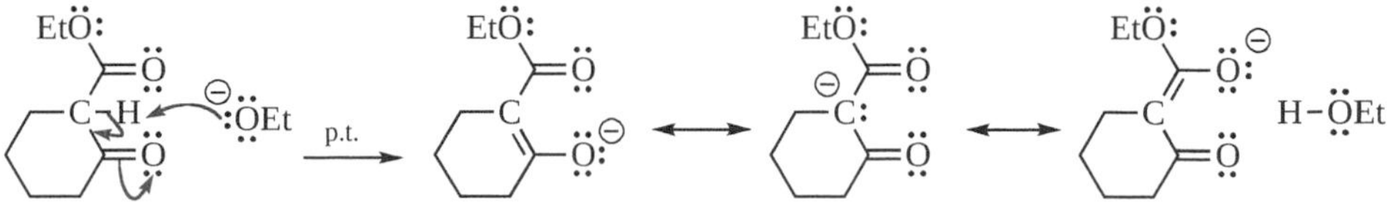

Evaluate and cross-check. The proton transfer has a very good K_{eq} of $10^{(16-10.7)} = 10^{+5.3}$, for the anion is so resonance stabilized. The reaction fits the medium and is downhill in energy.
Understand the system. The nucleophile is now where needed for bonding to the C=C–EWG.
Find possible routes. The delocalized anion adds conjugate to the enone because the addition would be reversible. The Ad_E2 is an acidic medium route, thus not appropriate here. There are two conjugate addition routes consistent with a basic medium. Path Ad_N,

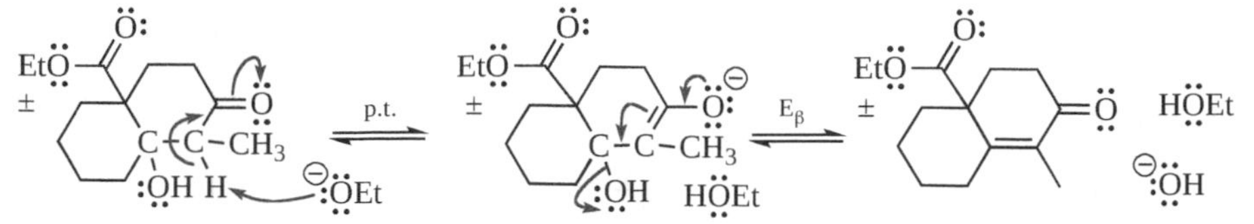

Surface overview. We can look back at the problem space that we just navigated to get an overview, Figure S.10. Red arrows show the most reasonable route(s). A proton transfer created our nucleophile for an Ad_N2 (or Ad_E3) conjugate addition to a C=C–EWG in the first surface, followed by a proton transfer to create the needed nucleophile for another Ad_N2 (or Ad_E3) in the second surface, and finally finishing with an E1cB elimination (or E2) in the third surface. Even though this surface overview may look complex, think of it as assembling just three known blocks, two addition surfaces and an elimination surface.

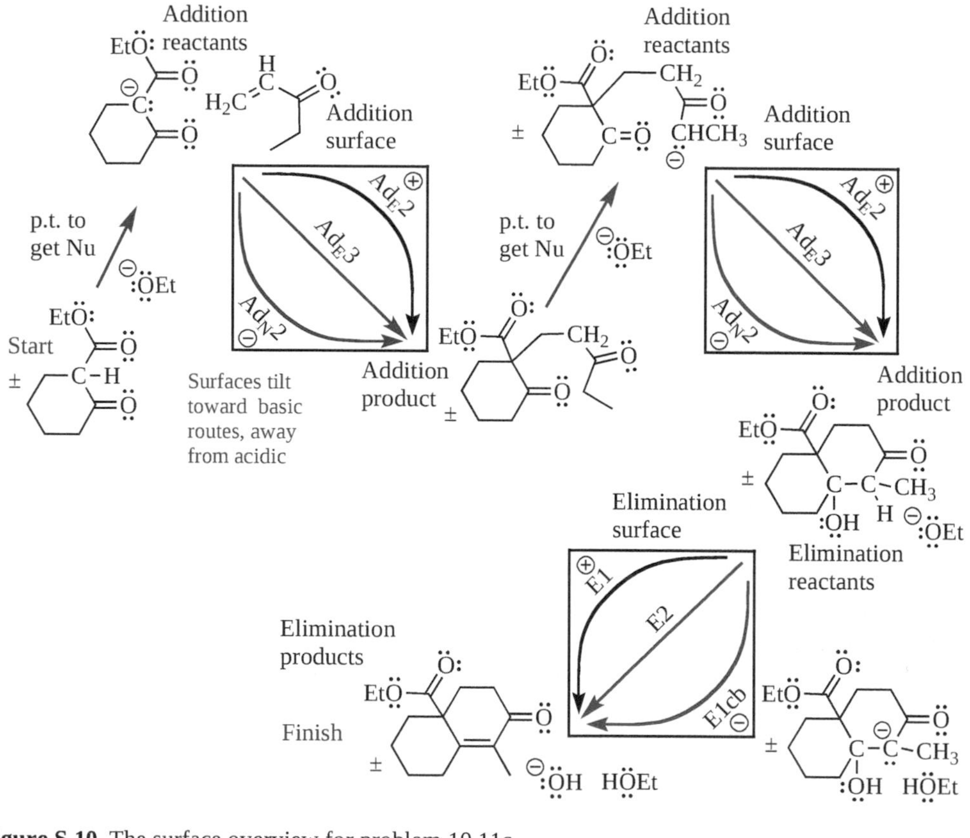

Figure S.10 The surface overview for problem 10.11c.

Problem 10.13 Part of the dark reactions of photosynthesis is interconversion of sugars with an enzyme called transketolase using thiamine pyrophosphate, TPP, as a catalyst (Section 8.12.8). Provide a reasonable mechanism for this enzymatic reaction. In addition to water, there are weak general acids and general bases present in the active site at pH 7.

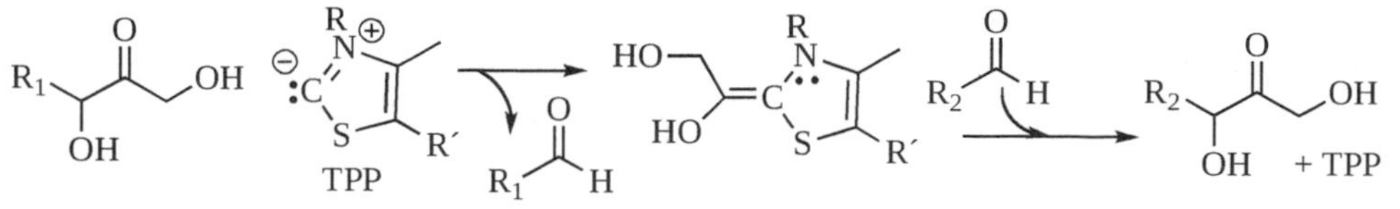

Understand the system. Balanced? For once, yes. Generic process? In the first half, TPP adds to the ketone carbonyl and then an elimination produces the new R_1CHO pi bond. The second half of the reaction looks just like the reverse of the first half. Medium? Neutral. Sources? The TPP lone pair is a soft nucleophile. Sinks? The ketone is a C=Y. Acidic H's? the alcohol hydroxyls are weakly acidic. Leaving groups? No good ones in this neutral medium.

Find possible routes. We can see that just deprotonating the alcohols leads nowhere. The TPP is nucleophilic, and the only site for attack is the ketone. We have three addition paths, Ad_E2, Ad_E3, and Ad_N2, but only the Ad_E3 is reasonable for neutral media.

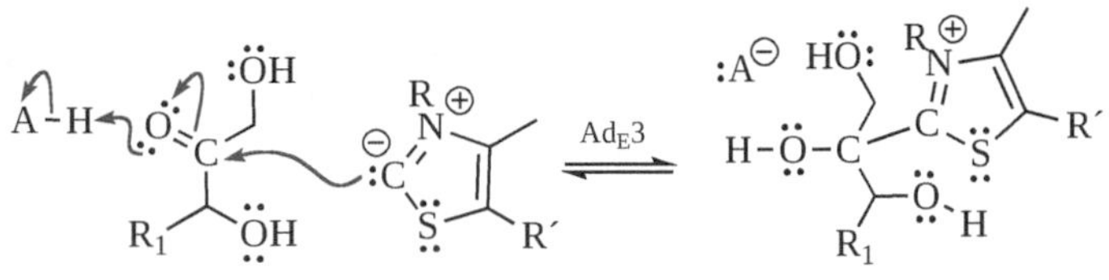

Evaluate and cross-check. The Ad_N2 would produce an alkoxide anion, too basic for neutral media. The Ad_E2 requires the ketone to be protonated first, and that is too acidic for neutral media. Weak acids and bases in the active site of an enzyme can catalyze the Ad_E3.

Understand the system. TPP is also an EWG and will stabilize an adjacent negative charge. There are weak acids and bases present.

Find possible routes. The aldehyde carbonyl can be formed by an elimination, so we have a choice of E1, E2, or E1cB.

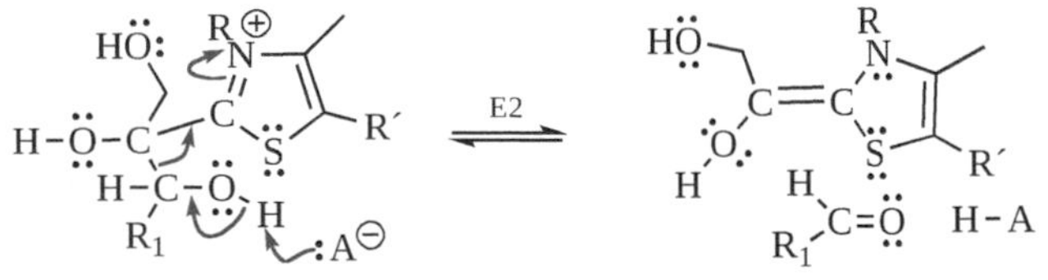

Evaluate and cross-check. E1 would require a better leaving group, and we do not have a strong enough acid present to protonate the leaving group. E1cB would require the deprotonation of the alcohol first, creating a $pK_{abH} = 17$ base, too basic for neutral media. We are left with the E2. The E2 produces a neutral enamine and can be catalyzed by weak acids and bases found in a cell.

Understand the system. We have reached the midpoint in our process and just have to reverse the reaction with a new aldehyde to get to our final product. The R_1CHO diffuses out of the active site and the R_2CHO diffuses in.

Find possible routes. The reverse of the E2 is the Ad_E3.

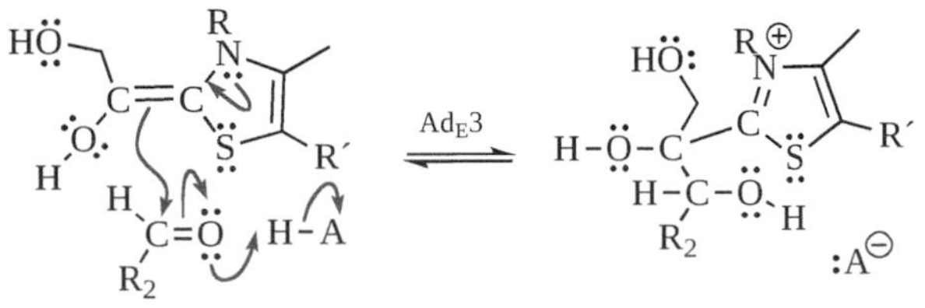

Evaluate and cross-check. As before, weak acids and bases in the active site of an enzyme can catalyze the Ad_E3.

Understand the system. We just have to remove our TPP catalyst and form a new ketone to get to our final product. An elimination is needed.

Find possible routes. The reverse of the Ad_E3 from before is the E2.

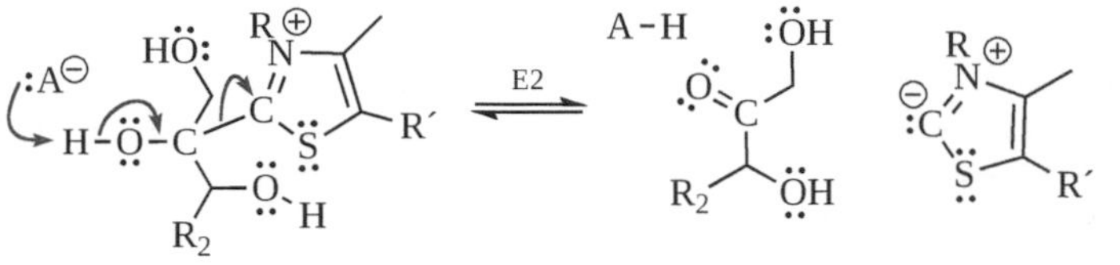

Evaluate and cross-check. As before, weak acids and bases in the active site of an enzyme can catalyze the E2.

Overview. We added our TPP catalyst with an Ad_E3 addition to the carbonyl and created our first aldehyde by an E2 elimination. That got us to our intermediate and halfway. We brought in the second aldehyde and reversed the process (Ad_E3 addition then E2 elimination) to give the final new ketone.

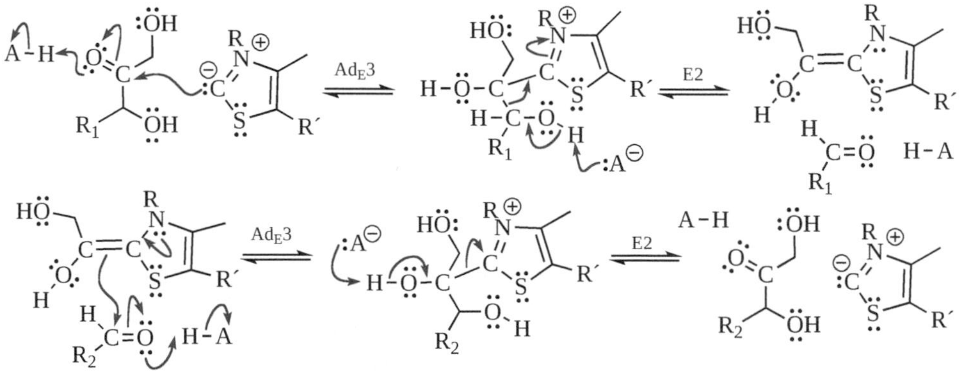

Surface overview. We can look back at the problem space that we just navigated to get an overview, Figure S.11. Red arrows show the most reasonable route. Characteristic of enzymatic reactions is the avoidance of very reactive/less stable anions or cations.

Chapter 11 Odd-Numbered Exercise Answers

11.1 In each of the reactions below, a step is crossed out that failed a cross-check. What cross-check was missed? Use the energy surface for that process to find an alternative that would satisfy the cross-check. Draw a mechanism that passes all cross-checks.

11.1a

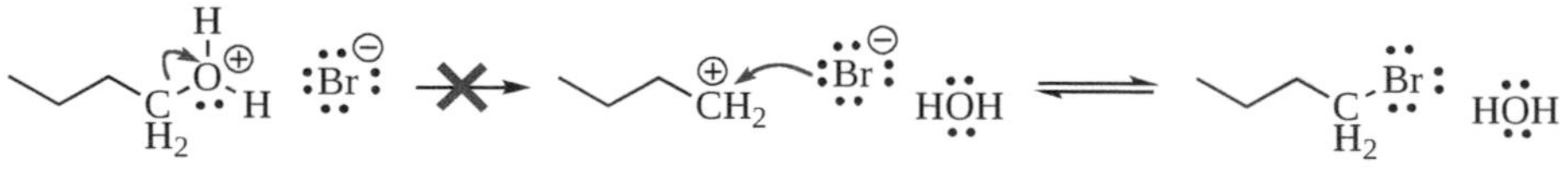

The decision that was missed was the S_N2 vs. S_N1 choice. The reaction drawn failed one of the D_N cross-checks that requires a carbocation usually more stable than secondary. Primary carbocations are very unstable, so an S_N1 is not reasonable. The S_N2 mechanism below passes all cross-checks of having a decent leaving group and good access to the carbon attacked.

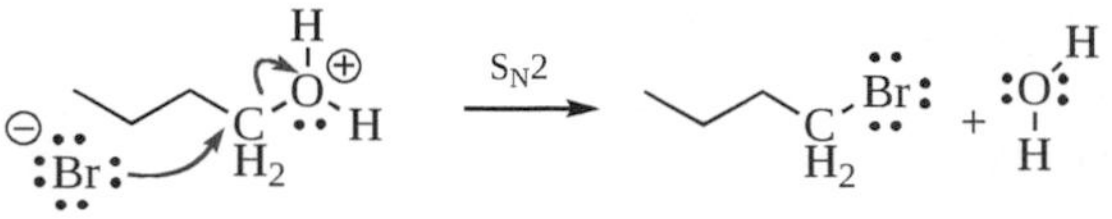

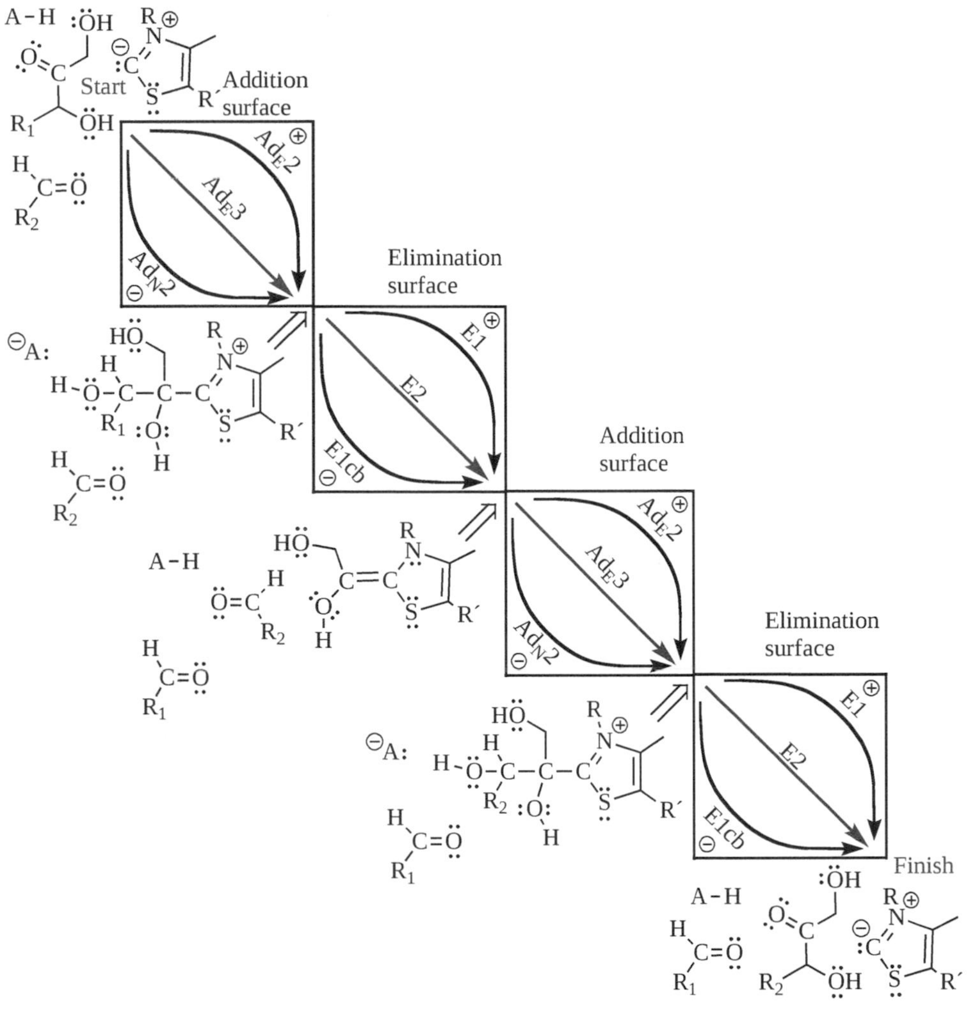

Figure S.11 The surface overview for problem 10.13.

11.1b

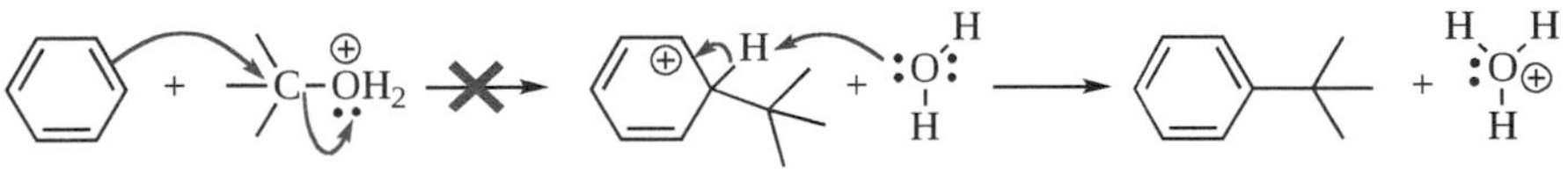

The decision that was missed was the S_N2 vs. S_N1 choice. The reaction drawn failed the S_N2 cross-check by having the nucleophile attack a very hindered center. But this reaction passes all the S_N1 cross-checks: the loss of leaving group (D_N) by having a leaving group $pK_{aHL} \leq 0$, a tertiary carbocation, which is more stable than secondary, and an assumed polar solvent. The reasonable S_N1 mechanism is shown below.

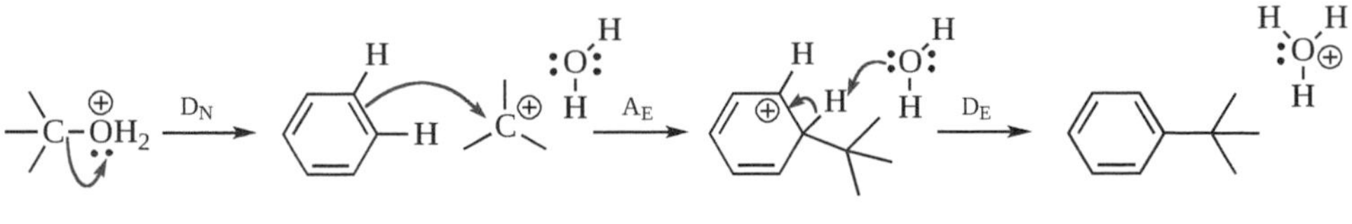

11.3 The reaction below fails a cross-check and the reaction does not happen. Why?

11.3a

The reaction drawn fails the ΔpK_a cross-check. It is an S_N2 with an incoming nucleophile pK_{aHNu} of –6, and kicks out a leaving group pK_{HL} of 14.0 for a climb of 20 pK_a units, well past our decision boundary of keeping to a climb of 10 pK_a units or less. This reaction would run well in the reverse direction but not in the direction written.

11.3b

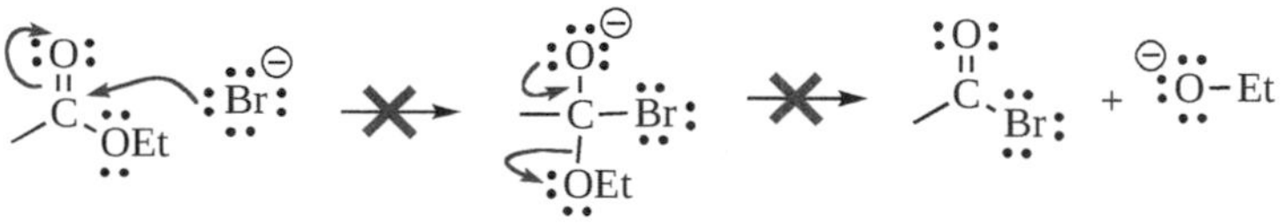

The reaction drawn fails the ΔpK_a cross-check. It is an addition to an ester carbonyl by an incoming nucleophile pK_{aHNu} of –9, and a resulting tetrahedral intermediate anion of approximately pK_{bH} of 13 for a climb of about 22 pK_a units, well past our decision boundary of keeping to a climb of 10 pK_a units or less. The second step kicked out the poorer leaving group, ethoxide pK_{HL} of 16.0, instead of the much better leaving group bromide, pK_{HL} of –9. This overall reaction would run well in the reverse direction, but not in the direction written.

11.5 The reaction on the left is very poor, but a similar reaction on the right works. Why?

11.5a

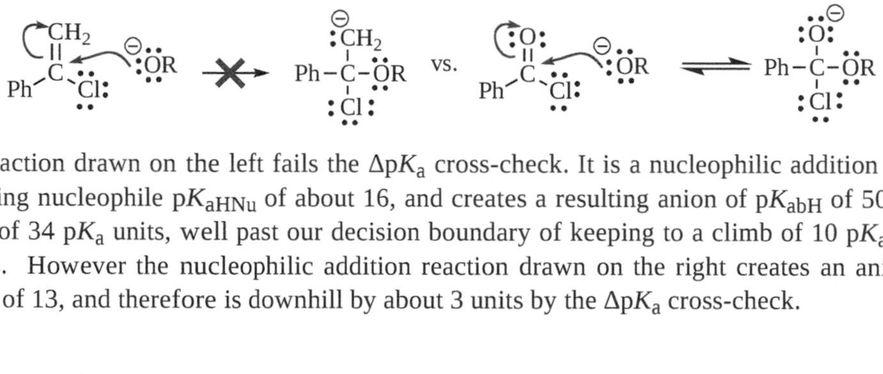

The reaction drawn on the left fails the ΔpK_a cross-check. It is a nucleophilic addition by an incoming nucleophile pK_{aHNu} of about 16, and creates a resulting anion of pK_{abH} of 50 for a climb of 34 pK_a units, well past our decision boundary of keeping to a climb of 10 pK_a units or less. However the nucleophilic addition reaction drawn on the right creates an anion of pK_{abH} of 13, and therefore is downhill by about 3 units by the ΔpK_a cross-check.

11.5b

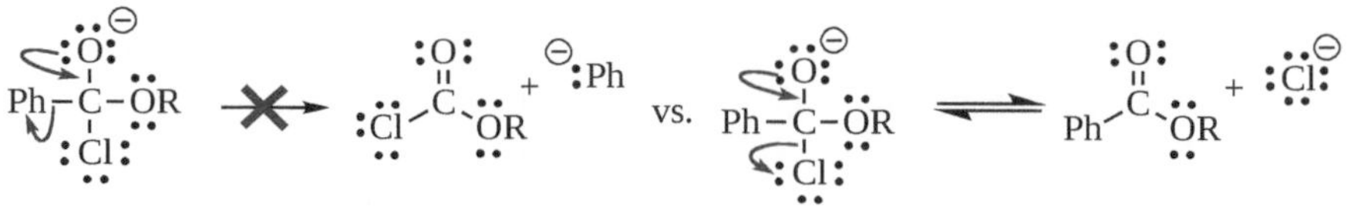

The reaction drawn on the left fails the ΔpK_a cross-check. It is a beta elimination from an oxyanion of about pK_{abH} of 13. Loss of the phenyl anion creates a resulting anion of pK_{abH} of 43 for a climb of 30 pK_a units, well past our decision boundary of keeping to a climb of 10 pK_a units or less. However the elimination reaction drawn on the right loses an anion of pK_{abH} of –6, and therefore is energetically downhill by about 19 units.

11.7 Which steps in the following commonly written mechanism are reasonable and which one is not? What cross-check did it fail? Draw a mechanism that passes all cross-checks.

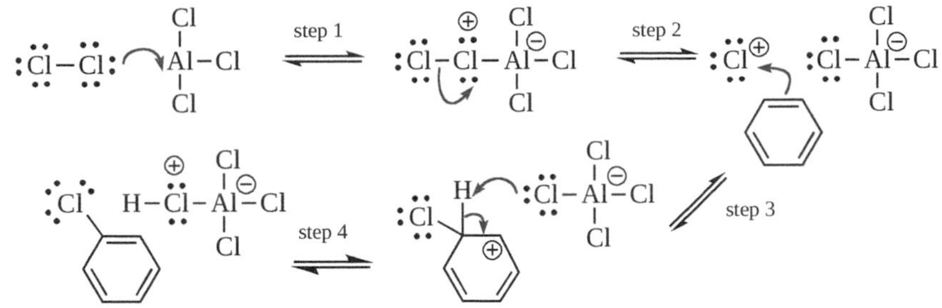

The decision that was missed was the cation stability cross-check. Electronegative atoms with less than a complete octet, like Cl^+, are very unstable. Instead of the unreasonable S_N1 shown in step 2 and step 3, the S_N2 should be our choice since there is good access to the Cl attacked. The more reasonable mechanism is below.

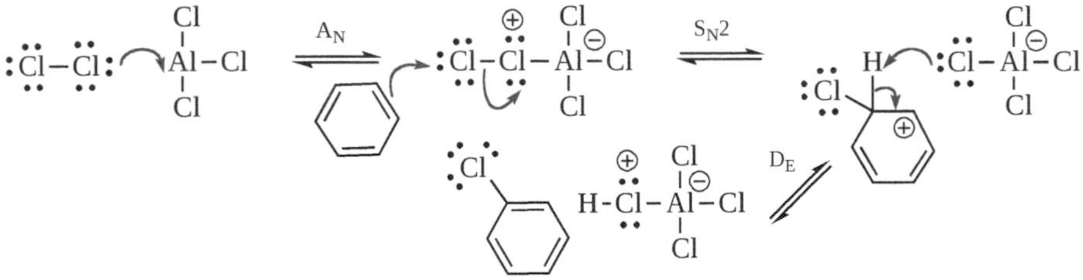

11.9 The student's mechanism written for the reaction below is incorrect.
The problem given was:

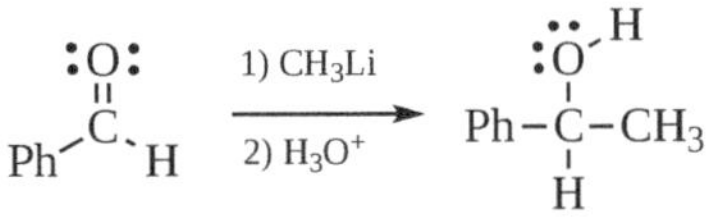

The incorrect mechanism was:

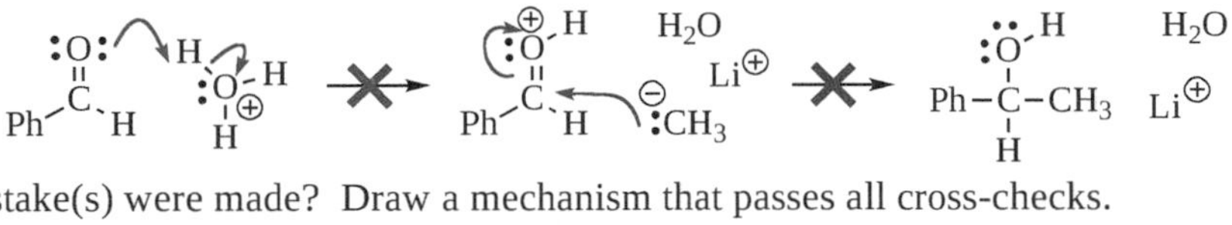

What mistake(s) were made? Draw a mechanism that passes all cross-checks.
There is a temporal anomaly. The problem was that the student used step two before step one. Do not bring in step two until step one is finished. The more reasonable mechanism is below.

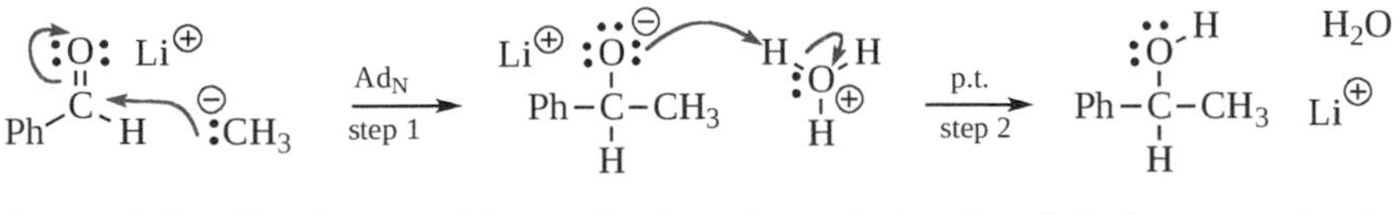

11.11 (More difficult) This possible mechanism shown below has failed a cross-check. What cross-check was missed? Draw a mechanism that passes all cross-checks.

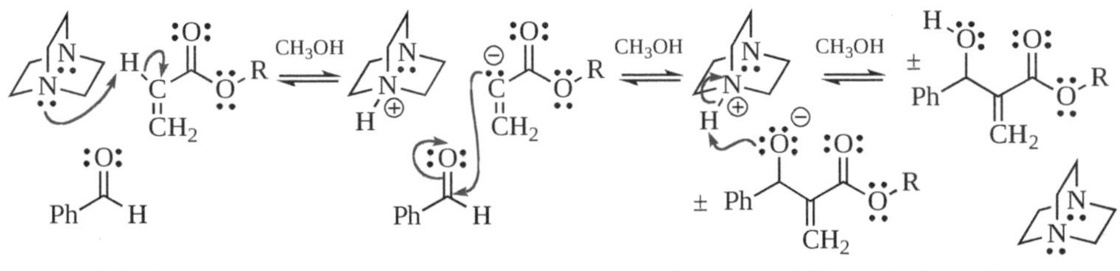

The failed cross-check was the initial proton-transfer step. Although the CH is adjacent to an electron-withdrawing group, the CH is in the plane of the pi system and therefore not aligned to be resonance stabilized by the ester. The diazabicyclooctane (DABCO) has a pK_{abH} = 8.9, so even if the resulting anion were a normal resonance stabilized ester enolate pK_{abH} = 25.6, it would be quite a stretch (16.7 units) by the ΔpK_a cross-check. Without resonance stabilization, this anion is way too far uphill. Another tipoff is that the conjugated ester is a C=C–EWG sink and should behave as a conjugate acceptor. Benzaldehyde is a C=Y with no acidic hydrogens. Since there are no acidic hydrogens in the range of the basicity of DABCO, then the only alternative is for DABCO to be a nucleophile, not a base.

DABCO attacking benzaldehyde is within the possibilities by the ΔpK_a cross-check, but that gives us an intermediate that would easily reverse, since the just attached DABCO pK_{aHL} of 8.9 is also the best leaving group. DABCO would attack 1,4 on the conjugate acceptor since the uphill attack would be easily reversible. The addition surface reminds us that there are three conjugate addition modes to consider, an Ad_E2, Ad_E3, or Ad_N2. The medium is basic with DABCO present, it is not reasonable to expect a protonated carbonyl, ruling out the Ad_E2. The conjugate Ad_N2 of DABCO on the unsaturated ester has an incoming pK_{abH} = 8.9, and an outgoing ester enolate anion of pK_{abH} = 25.6. This climb of 16.7 pK_a units is improbable by the ΔpK_a cross-check. That leaves the Ad_E3, and since the solvent is methanol, we expect hydrogen bonding to the lone pairs of the electron sink. A reasonable mechanism is the hetero Ad_E3, followed by a base catalyzed Ad_N2 or Ad_E3 addition on benzaldehyde, ending with an E2 elimination (the reverse of the Ad_E3 addition) of DABCO to give the final product. The hetero Ad_E3, base catalyzed Ad_N2, and E2 route is shown below.

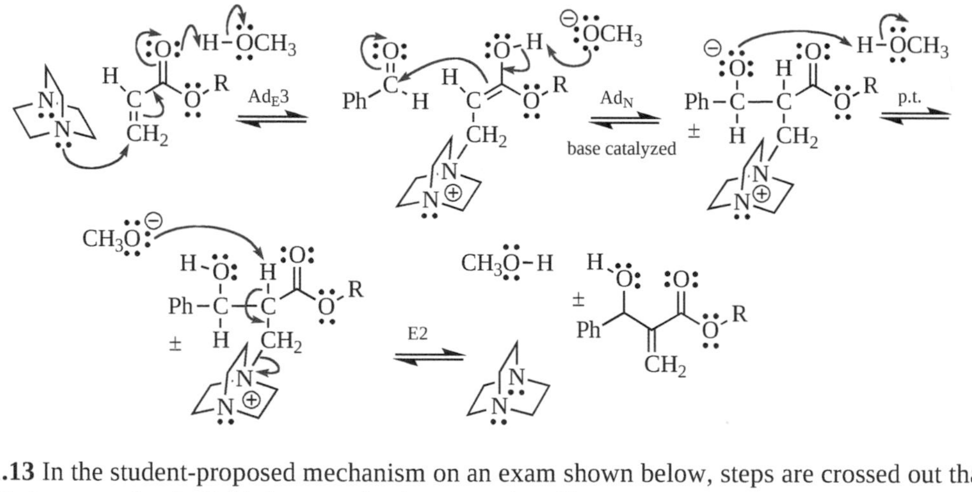

11.13 In the student-proposed mechanism on an exam shown below, steps are crossed out that failed a cross-check. Which cross-checks were missed?

The test question:

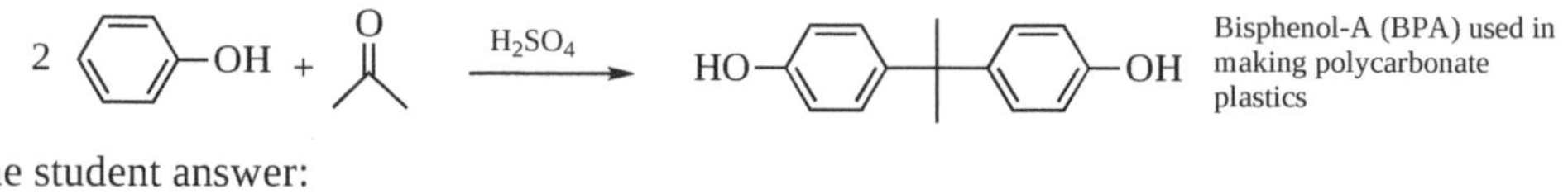

The student answer:

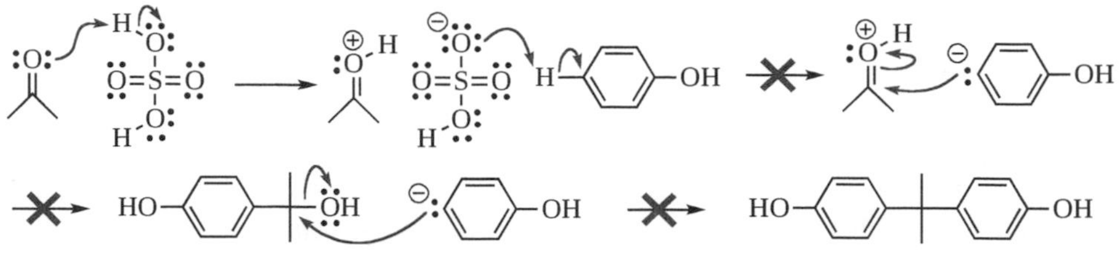

Draw a mechanism that passes all cross-checks.

The first reaction is the correct first step. We need to protonate the carbonyl to create a great electrophile for attacking the not-so-reactive aromatic ring. The conjugate base in this proton transfer is bisulfate anion. Although this anion seems like it might be a base, it is really a very poor base, pK_{abH} of about –9. The second step is a crossed out proton transfer with bisulfate removing a phenyl H (pK_a of about 43), giving an approximate $K_{eq} = 10^{(-9 - 43)} = 10^{-52}$. You cannot get a much poorer K_{eq} than that. In addition, that phenyl H is not the most acidic H on phenol; the OH is. Even the more acidic phenolic OH with pK_a of 10, is beyond the ability of bisulfate to deprotonate, $K_{eq} = 10^{(-9 - 10)} = 10^{-19}$. The nucleophile has been over-optimized, so now the next crossed out step fails a media pH span cross-check. The protonated carbonyl is a strong acid with a pK_a of –7 and the phenyl anion is a strong base at about pK_a 43, spanning 50 pK_a units, and so they cannot exist together without neutralization. The third crossed out step is an impossible S_N2 on a highly hindered tertiary center.

This reaction is definitely in an acidic medium with sulfuric acid present. The first step is to again protonate the carbonyl. Since we have a great electrophile in the protonated carbonyl, the aromatic ring is a good enough nucleophile for electrophilic aromatic substitution, A_E then D_E, shown below. A good leaving group is needed next, and protonation of the OH by sulfuric acid (HA) is downhill.

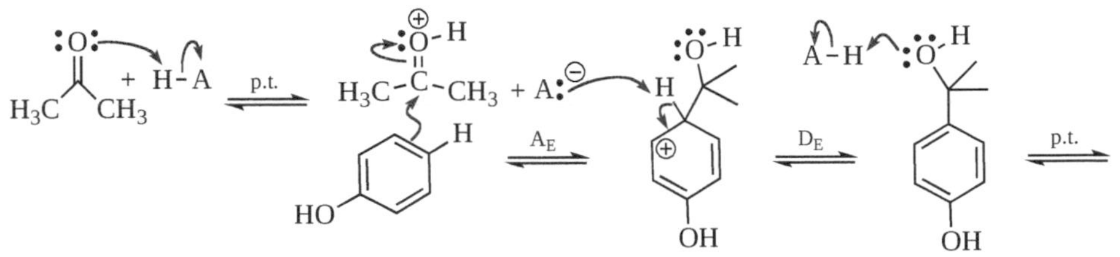

The second half of a reasonable mechanism is an S_N1 on the protonated tertiary alcohol, D_N followed by A_N, nucleophilic attack on the tertiary carbocation by the aromatic ring. Part of this process is also an electrophilic aromatic substitution, A_E then D_E, shown below.

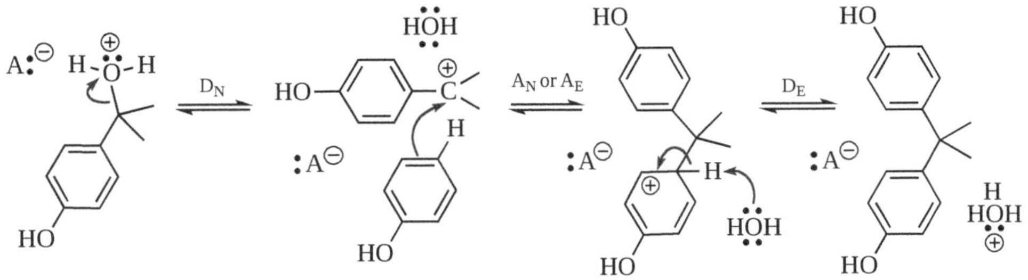

11.15 In a mechanism question on an exam, the student got stuck in the dead end shown below. What decision was missed? The test question:

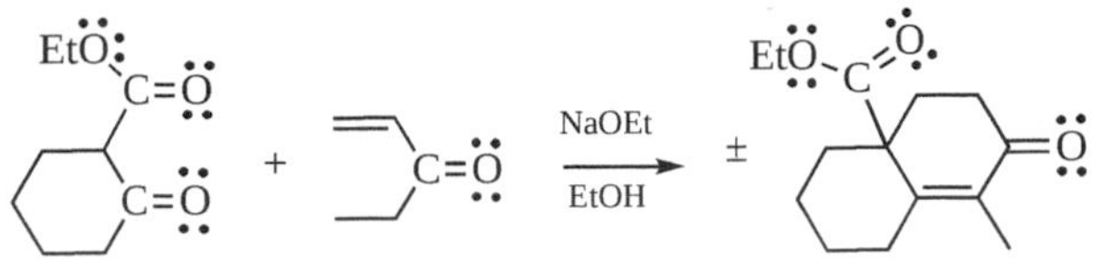

The student answer:

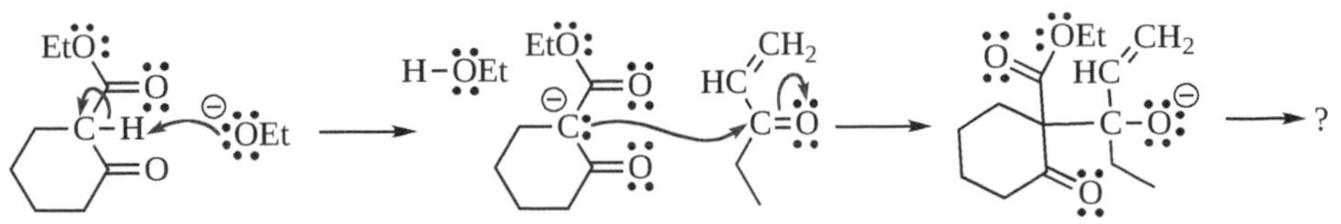

Draw a mechanism that passes all cross-checks.

The first reaction is reasonable; deprotonate the most acidic H between the two carbonyls, pK_a 10.7 with ethoxide pK_{abH} 16, for a $K_{eq} = 10^{(16 - 10.7)} = 10^{+5.3}$, well downhill. The student probably did not notice that the electron sink is a conjugate acceptor, C=C–EWG, and attacked the carbonyl of the electron sink. This is a reasonable thing to do, but is a dead end. Luckily, this reaction is reversible. An anion of pK_{abH} 19 can easily kick out a leaving group of pK_{abH} 10.7 to return to reactants. This error could also have been avoided by a close examination of the connectivity of the product, which shows a conjugate addition has occurred. The forgotten decision point is direct vs. conjugate addition, Section 9.6 Ambident Electrophiles.

A more reasonable answer starts with the same first step, deprotonate the most acidic H between the two carbonyls. However, the next step after enolate formation is an Ad_N2 conjugate addition (paths Ad_N then proton transfer) that connects the two pieces at the correct position. The ΔpK_a cross-checks on both are fine, with the first step going up from pK_{abH} 10.7 to 19.2, and the second step going back down from pK_{abH} 19.2 to 16.

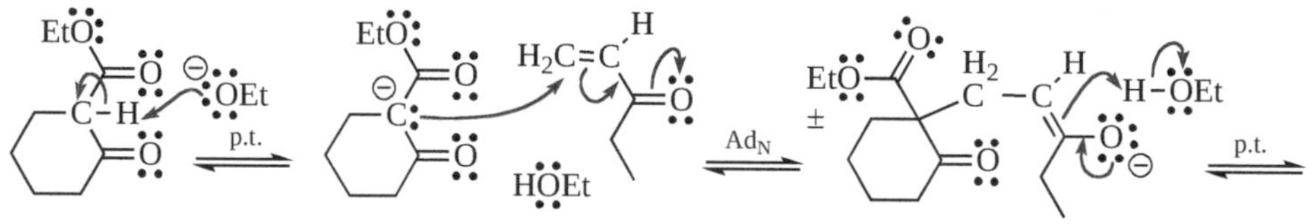

Now the other side of the ring must be connected. A proton transfer that makes an enolate anion on the correct carbon sets the system up for another Ad_N2 addition. The ΔpK_a cross-checks on both are fine, with the first proton-transfer step going up from pK_{abH} 16 to 19.2, the Ad_N step going from pK_{abH} 19.2 to 19, and the last proton-transfer step going from pK_{abH} 19 back down to 16.

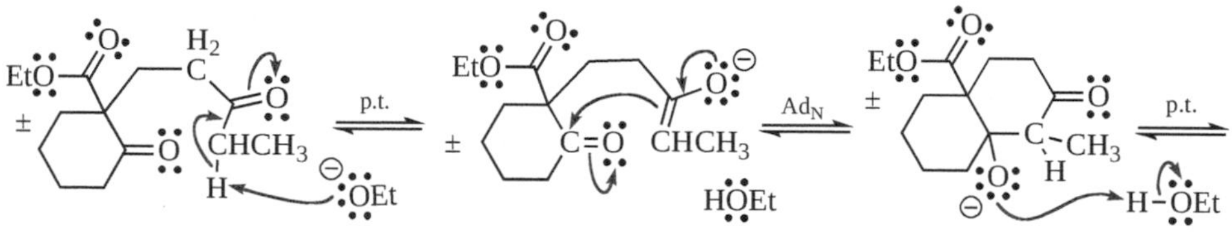

Finally an E1cB elimination (paths proton transfer then beta elimination) shown below completes the route to the product by eliminating hydroxide. The ΔpK_a cross-checks on both are fine, with the proton-transfer step going up from pK_{abH} 16 to 19.2, and the elimination step going from 19.2 to 14.0.

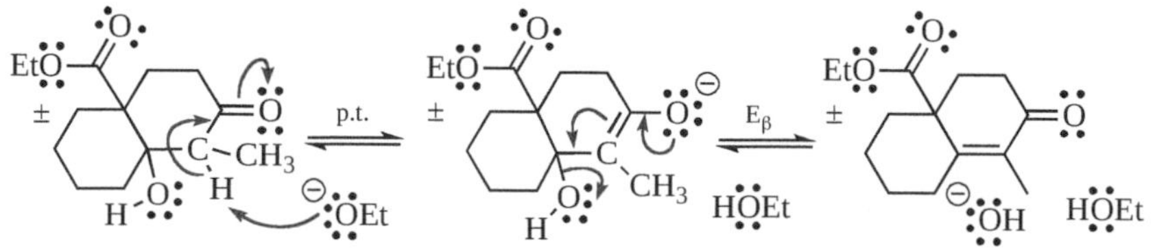

Overview. This ring formation, known as the Robinson Annulation, is summarized below.

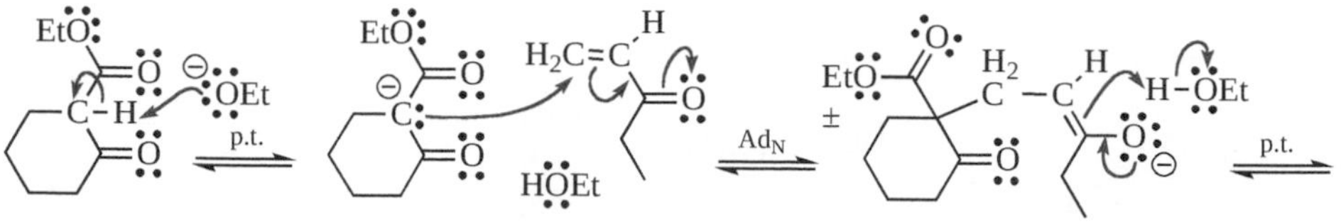

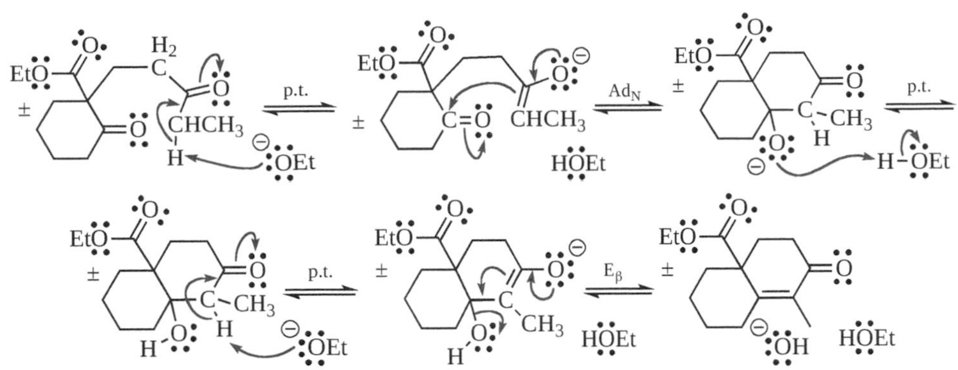

11.17 In the mechanism on an exam, the student got stuck in the dead end shown below. What decision was missed? The test question:

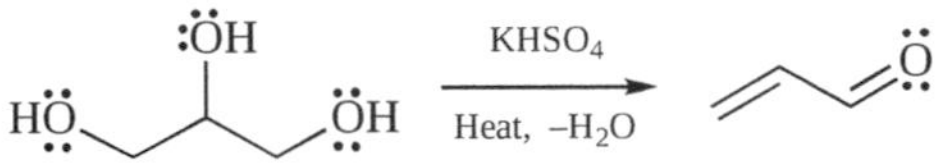

The student answer used HA for the acid catalyst:

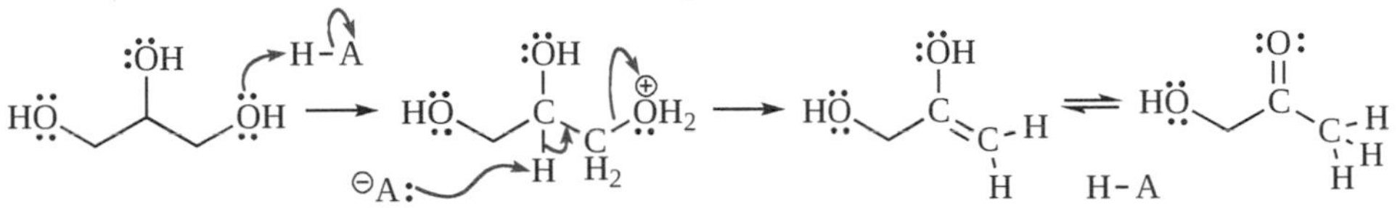

Draw a mechanism that passes all cross-checks.

A branch in the search tree was missed. There are two alcohols that could eliminate first. If the secondary alcohol goes first by E2, then the allylic alcohol has a good E1 because of the stability of the resonance-stabilized carbocation formed (which is just the protonated product).

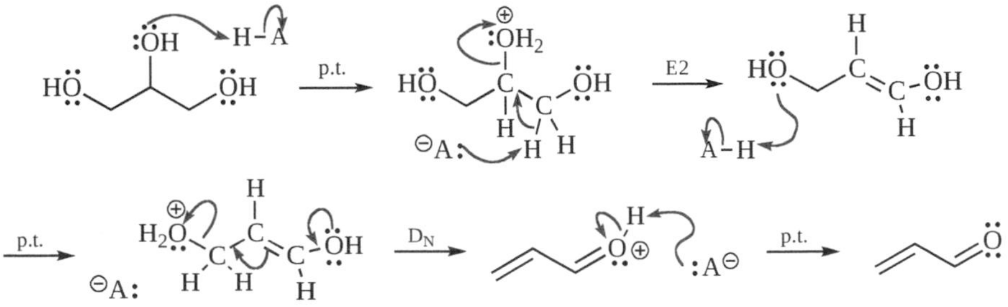

11.19 In the student-proposed mechanism below, a step is crossed out that failed a cross-check. What cross-check was missed? Use the energy surface for that process to find a reasonable alternative. Draw a mechanism that passes all cross-checks.

The cross-check that was missed was that the S_N2 reaction is not known to occur on leaving groups connected to an *sp*2 center, only *sp*3 centers. Substitution on an sp2 bound leaving group goes by addition–elimination, Section 4.5 Nucleophilic Substitution at a Trigonal Planar Center. The more reasonable mechanism is shown below. The Ad_N nucleophilic addition creates an anion of about pK_{abH} of 13, from an incoming nucleophile of pK_{abH} 16, and therefore is downhill by about 3 units by the ΔpK_a cross-check. The beta elimination that follows is downhill 19 pK_a units. Overall, this reaction is very favorable.

11.21 (More difficult) Of the two alternatives, which is the favored product and explain why?

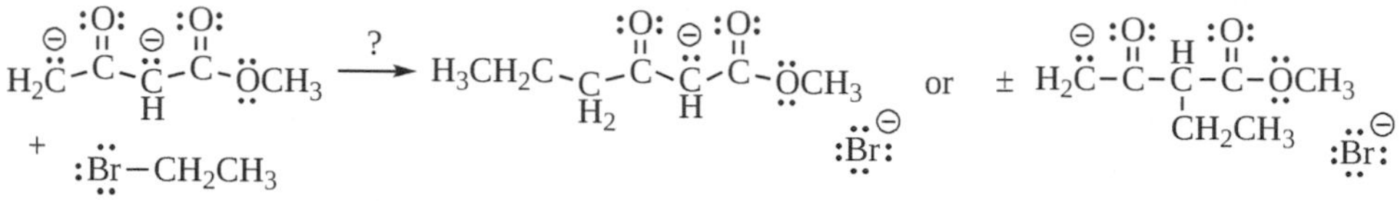

This problem is easy to get the correct answer for the wrong reason. The first of the two alternatives is the product of the reaction. The first product is the weaker base with an anion of pK_{abH} 10.7, and the second product is a more basic anion of pK_{abH} 19.2. This seems to solve the issue by forming the more stable product, and it would be the correct explanation if the reactions were reversible, but they are not. The ethyl acetoacetate dianion reactant is more basic than pK_{abH} 35 (Section 11.7), and the bromide leaving group has a pK_{abH} of –9, for a drop of well over 44 pK_a units. The reaction is clearly irreversible. We cannot use a thermodynamic argument to explain the product; it must be a kinetic rationale (Section 9.9).

So which route has the lowest barrier to reaction and would proceed the fastest? Figure 2.17 shows two energy diagrams for thermodynamic and kinetic control. The fastest reaction joins the most reactive nucleophile with the most reactive electrophile. This is basically a trends question. Since the electrophile is common to both, we need to rank nucleophiles (section 4.2.2). Both nucleophiles are on carbon, so with the same attacking atom, the most basic site is the most nucleophilic. Although the first pK_a of ethyl acetoacetate is just 10.7, it takes a much stronger base than NaH (pK_{abH} 35) to remove a second proton to form the less stable dianion shown below (Section 11.7). Therefore the deprotonated methyl is the most basic and therefore the most nucleophilic. It will react the fastest with ethyl bromide to produce the first product.

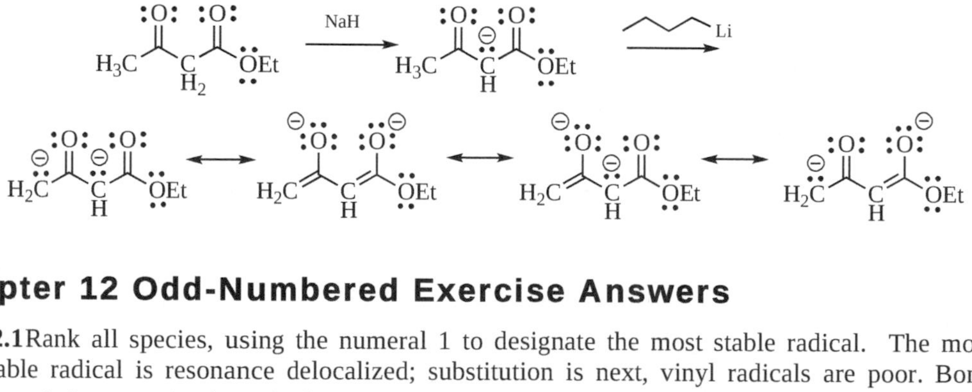

Chapter 12 Odd-Numbered Exercise Answers

12.1 Rank all species, using the numeral 1 to designate the most stable radical. The most stable radical is resonance delocalized; substitution is next, vinyl radicals are poor. Bond dissociation energies (BDE) of R–H → R· + ·H show the stability differences (Table 12.1).

	·CH=CH$_2$	·CH$_2$–CH=CH$_2$	·CH$_2$–CH$_3$	·CH(CH$_3$)$_2$	·CH$_3$	
Rank:	5	1	3	2	4	
CH BDE	110.7	88.8	101.1	98.6	105.0	kcal/mol
	(463)	(372)	(423)	(413)	(439)	kJ/mol

12.3 Calculate the $\Delta H°$ for the abstraction of an H from ethane by a ·OOH radical. Use Table 12.1 for more accurate values. $\Delta H°$ = bonds broken – bonds made = [bonds broken: ethane C–H for 101.1 kcal/mol (423 kJ/mol)] – [bonds made: H–OOH for 87.8 kcal/mol (367 kJ/mol)] = +13.3 kcal/mol (56 kJ/mol).

12.5 Check the $\Delta H°$ of the propagation steps for the radical chain addition of HF to ethene to see whether it is an energetically reasonable process. The two propagation steps are:

Step 1	$CH_2{=}CH_2 + \cdot F \rightarrow \cdot CH_2{-}CH_2{-}F$	$\Delta H° = -53(-222)$
Step 2	$\cdot CH_2{-}CH_2{-}F + H{-}F \rightarrow H{-}CH_2{-}CH_2{-}F + \cdot F$	$\Delta H° = +37(+155)$
Total:	$CH_2{=}CH_2 + H{-}F \rightarrow H{-}CH_2{-}CH_2{-}F$	$\Delta H° = -16\ (-67)$

Step 1: $\Delta H°$ = bonds broken – bonds made = [bonds broken ethene C=C for 146 kcal/mol (611 kJ/mol)] – [bonds made C–C for 83 kcal/mol (347 kJ/mol) + C–F for 116 kcal/mol (485 kJ/mol)] = –53 kcal/mol (–222 kJ/mol). Step 2: $\Delta H°$ = bonds broken – bonds made = [bonds broken H–F for 136 kcal/mol (569 kJ/mol)] – [bonds made C–H for 99 kcal/mol (414 kJ/mol)] = +37 kcal/mol (+155 kJ/mol). The radical chain addition of HF to an alkene is not an energetically reasonable process because the second propagation step is more than 20 kcal/mol (84 kJ/mol) endothermic, even though the entire process is exothermic.

12.7 Predict which you would expect to be more selective: radical chain chlorination or bromination of an alkane. Explain.

Which H is abstracted determines selectivity. Radical chain bromination of an alkane would be more selective because the abstraction reaction is much more endothermic (S_H2 in Section 12.6). Product radical stabilities would make much more of a difference in endothermic abstraction transition states (Section 2.6.1).

12.9 Give the product and mechanism of the following radical reaction (initiated by light).

$$Br_2 + (CH_3)_2CHCH_2CH_2CH_3 \rightarrow$$

The product distribution of this S_H2 radical substitution is dominated by the abstraction of H by bromine radical to form the more stable tertiary radical.

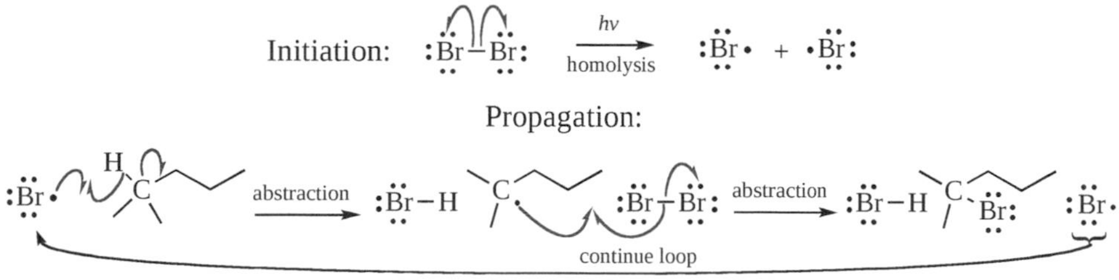

12.11 Write out an $S_{RN}1$ process for phenyl diazonium ion reacting with iodide anion to produce nitrogen gas and phenyl iodide. Use s.e.t. from iodide anion to start the process.
Initiation is by single electron transfer:

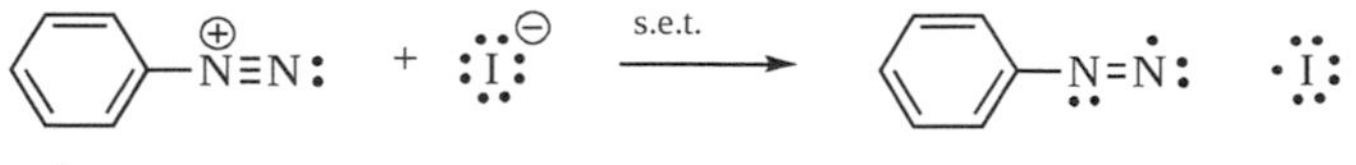

$S_{RN}1$ Propagation loop:

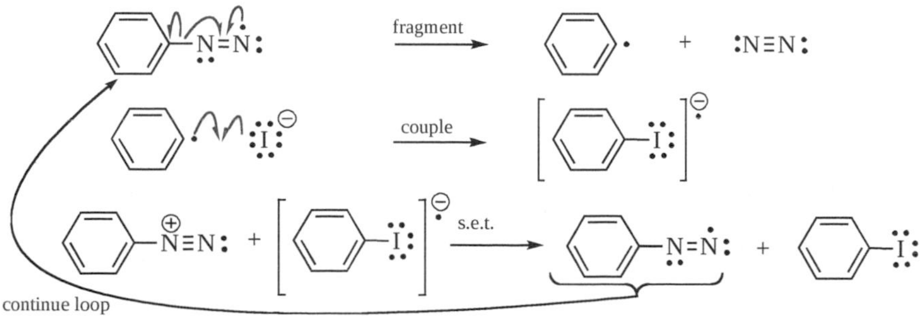

There also is a related solvent cage process that is quite reasonable if the lifetime of the first s.e.t. species is short. Brackets indicate the solvent cage. This is not a radical chain loop.

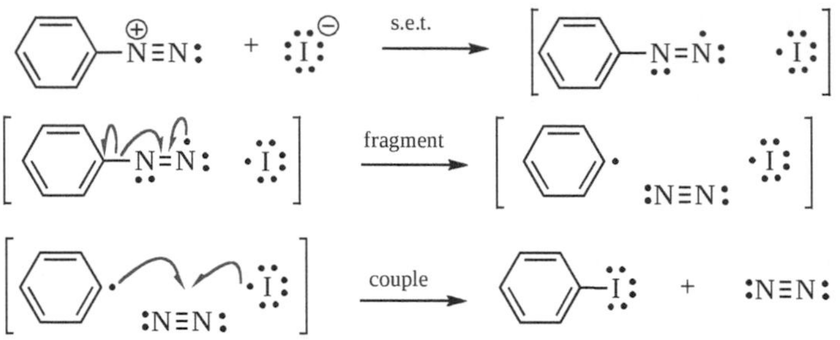

12.13 Give the product and a mechanism for this radical chain addition initiated by light.

$$BrCCl_3 + H_2C{=}CEt_2 \xrightarrow{hv}$$

Homolysis of the weak Br–C bond by light starts the process.

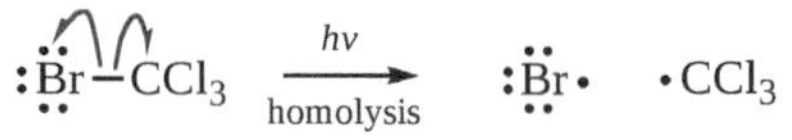

The fact that there are two possible radicals to start the chain may seem confusing at first. Let us explore both. The most likely next step is radical addition to the alkene to form the more stable tertiary radical.

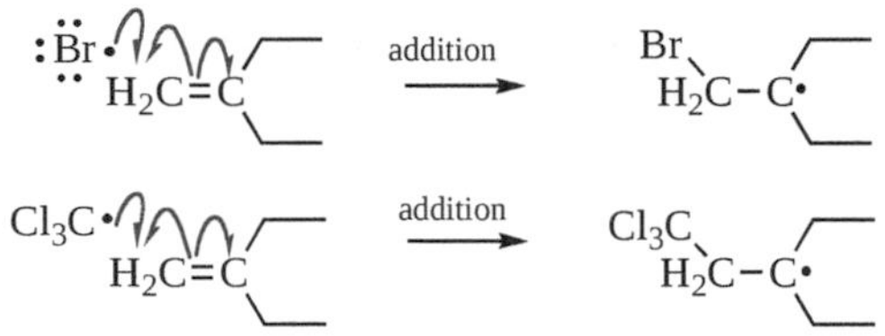

The next most likely step is for the carbon radical to abstract a bromine atom from $BrCCl_3$ as the Br–C bond is the weakest, therefore the lowest energy abstraction.

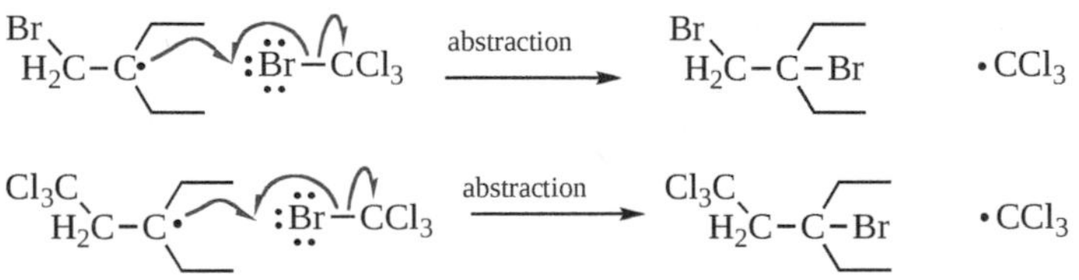

This leaves the trichloromethyl radical as the only radical continuing the chain regardless of which radical we started with. The propagation steps are therefore an addition of trichloromethyl radical to the alkene followed by an abstraction of a bromine atom:

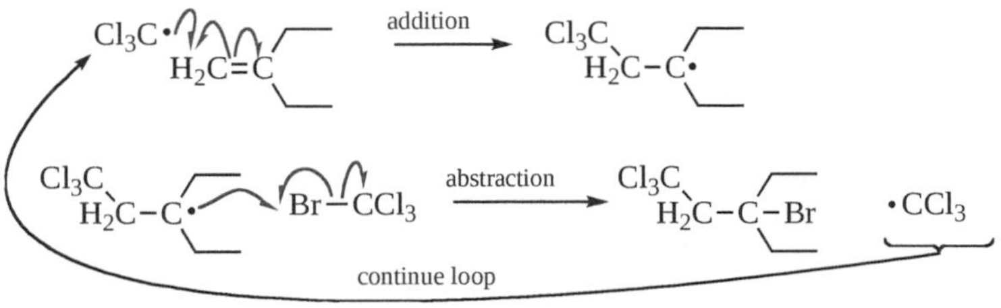

Chapter 13 Odd-Numbered Exercise Answers

13.1 Draw the pi MOs of pentadienyl cation. What is the HOMO and LUMO?

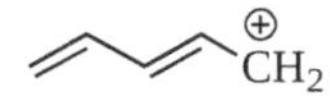

The pi system contains four pi electrons and five p orbitals. Use the wave analogy to generate the molecular orbitals. Systems having an odd number of p orbitals have a non-bonding MO where every other p orbital contributes, Figure 13.27.

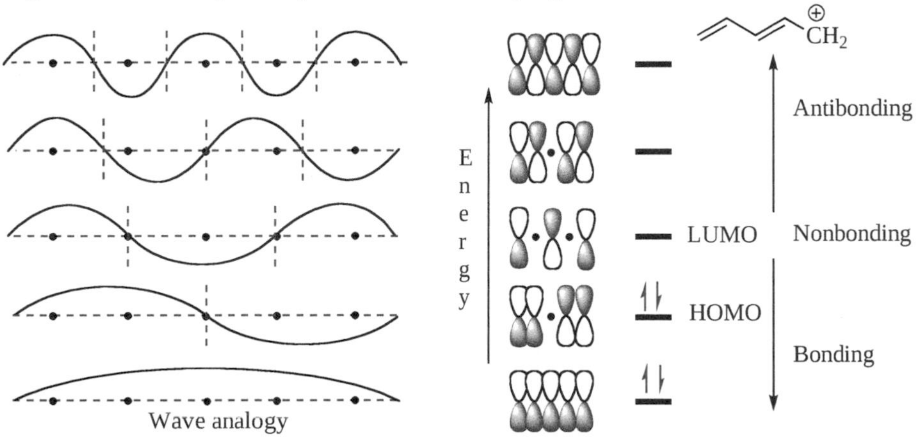

13.3 Would the thermal electrocyclic closure of 1,3,5-cycloheptatriene go con or dis? Draw the product.

The electrocyclic ring closure uses three arrows therefore six electrons, so will thermally go disrotatory (Section 13.8).

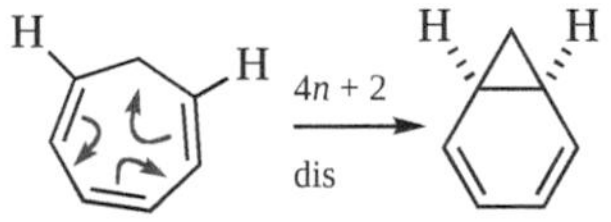

13.5 Is the following sigmatropic rearrangement thermally allowed?

Yes, because this Cope rearrangement is a three arrow, six electron, suprafacial, retention sigmatropic process (Section 13.9).

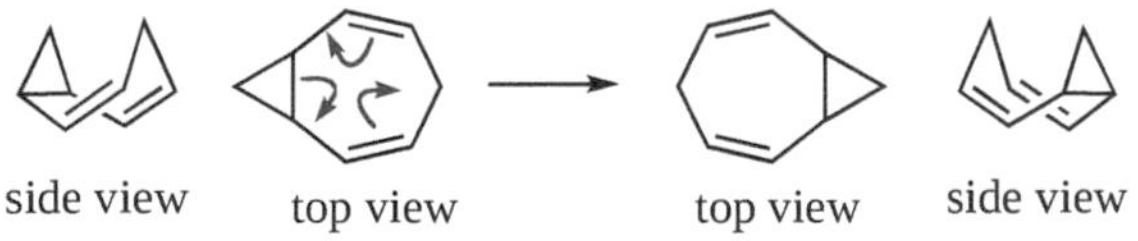

13.7 Give the product of this electrocyclic ring opening showing stereochemistry.

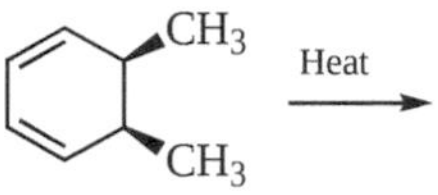

The electrocyclic ring opening uses three arrows therefore six electrons, so will thermally go disrotatory. The methyls will rotate away from each other to avoid colliding, Section 13.8.

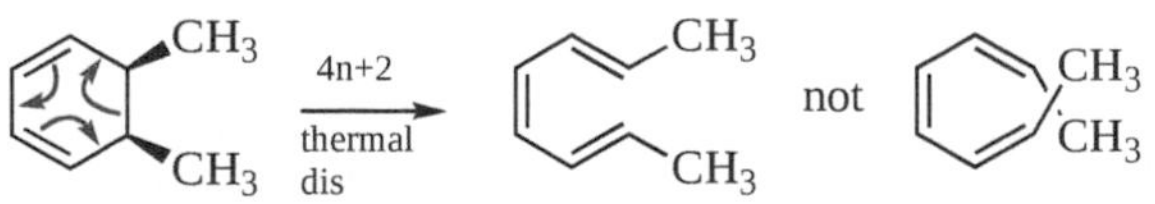

13.9 The compound below does not react with Diels–Alder dienophiles. Explain.

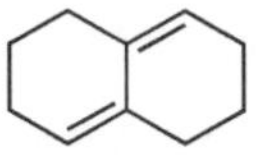

The two rings hold the ends of the diene far apart. The dienophile must be able to overlap with the ends of the diene, and in this case the ends of the diene are just too far apart. See Figure 13.39 for the needed transition state for the Diels–Alder.

13.11 Draw the photochemical ring closure product of the compound below.

The electrocyclic ring closure uses two arrows therefore four electrons. Since the thermal 4n electrocyclic reaction goes con, the photochemical 4*n* reaction goes disrotatory, Section 13.8.

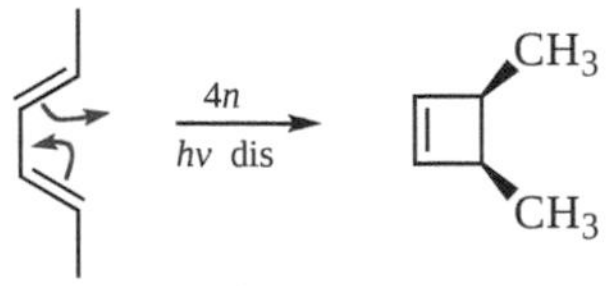

13.13 Ozone is a 1,3 dipole and adds to alkenes to give a molozonide by the 6e path, then breaks by the reverse process, and adds together again to form the ozonide by the 6e path. The ozonide is cleaved by dimethyl sulfide to give two carbonyls. Show the mechanism.

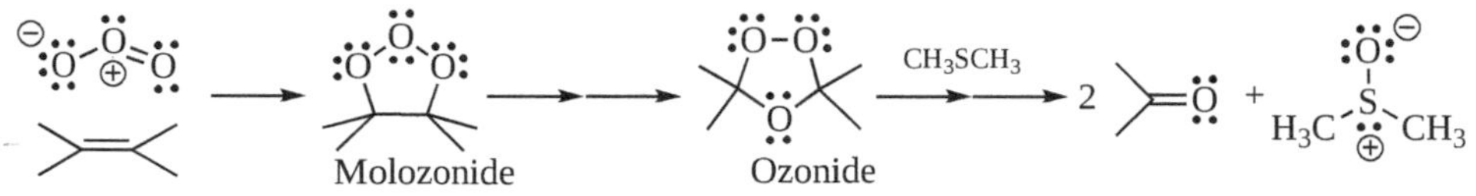

The first 6e path forms the molozonide, which can then break either weak O–O, bond by another 6e path to give a ketone and a carbonyl oxide. The nucleophilic oxygen of the carbonyl oxide attacks the carbonyl carbon in the second 6e path to form the ozonide. The ozonide has a weak O–O bond that the nucleophilic sulfur of dimethyl sulfide can break in an S_N2 reaction. This leaves a hemiketal oxygen anion that can kick out the other oxygen on that carbon to form a carbonyl by path E_β. The newly formed hemiketal-like oxygen anion of the leaving group kicks out dimethyl sulfoxide by path E_β to form the second carbonyl.

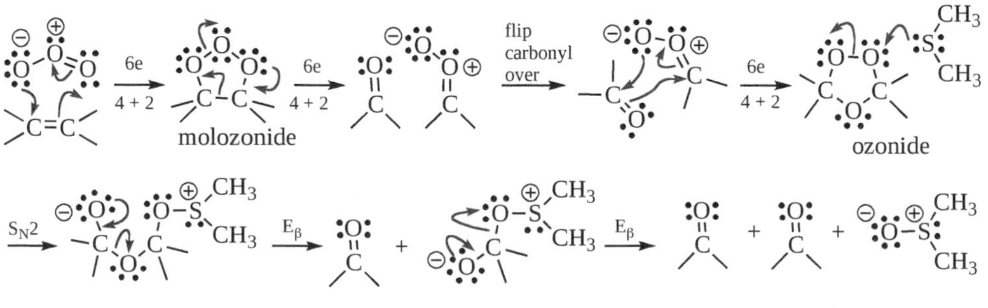

13.15 Predict the stereochemistry of the Diels–Alder addition of these two reactants.

The dienophile comes in endo, which puts its hydrogens on the same side as the CH_2 of the cyclopentadiene (Section 13.7).

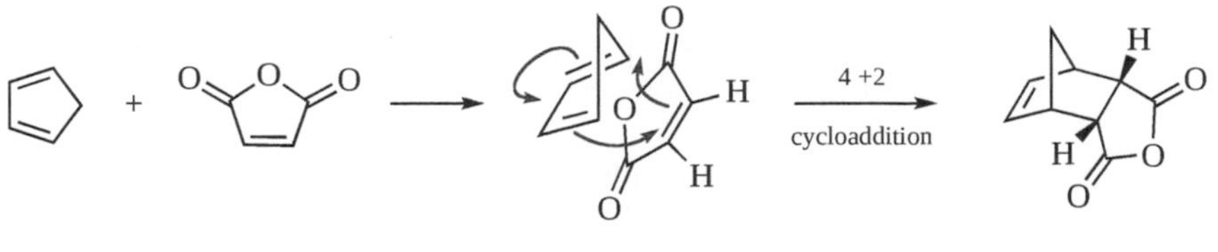

13.17 Give the product of this thermally allowed rearrangement:

This Claisen sigmatropic rearrangement is a three arrow, six electron, suprafacial, retention sigmatropic process (Section 13.9).

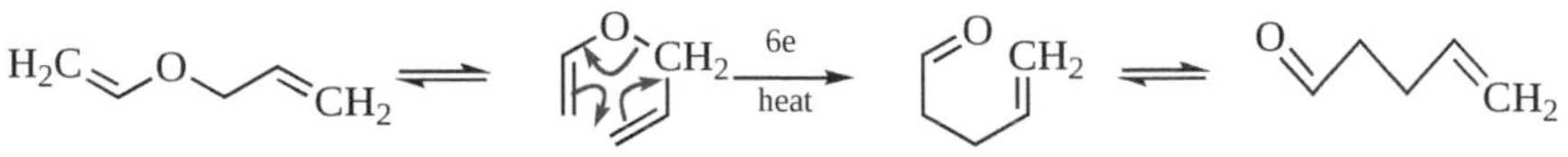

13.19 Vitamin D is synthesized in the skin by UVB light. The steps are a photochemical electrocyclic ring opening followed by a thermal sigmatropic hydrogen shift. Given those hints, propose a reasonable mechanism for this very important biochemical process.

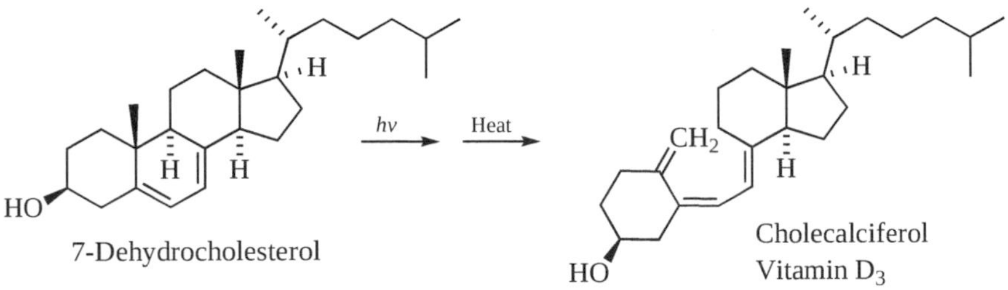

The cyclohexadiene ring opens photochemically in a three arrow, six electron ($4n + 2$), conrotatory electrocyclic process (Section 13.8).

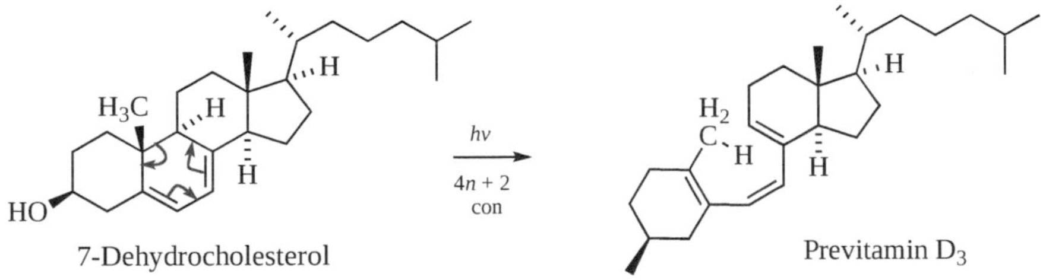

The hydrogen on the methyl is passed antarafacially in a four arrow, eight electron ($4n$), thermal 1, 7 sigmatropic shift (Section 13.9).

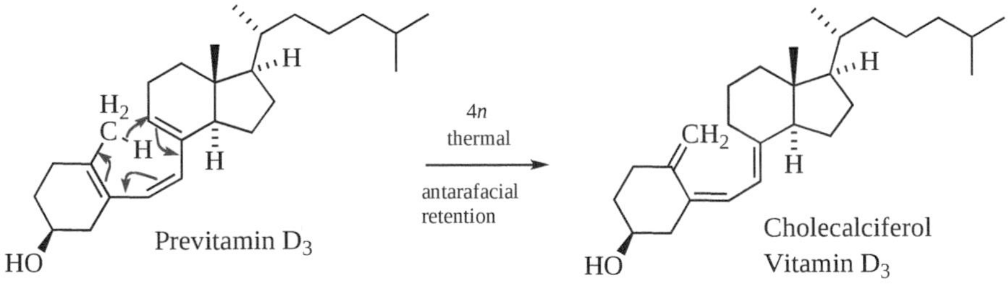

Chapter 14 Odd-Numbered Exercise Answers

14.1 Find the molecular formula and the DU given the MS molecular ion of 88.1 amu and combustion analysis data of 54.53 %C and 9.15 %H.

Refer to Section 14.4. The percentages do not add to 100%, so oxygen is present. If we subtract what we have from 100%, it has 36.32% O.
For C: MS molecular weight * % C / atomic weight of C. (88.1)(0.5453)/12 = 4 carbons
For H: MS molecular weight * % H / atomic weight of H. (88.1)(0.0915)/1 = 8 hydrogens.
For O: MS molecular weight * % O / atomic weight of O. (88.1)(0.3632)/16 = 2 oxygens.
This gives a formula of $C_4H_8O_2$
To get the DU use (saturated H count – actual H count)/2 or (4*2 +2 – 8)/2 = (10 – 8)/2 = 1
This compound has one ring or pi bond.

14.3 The IR of $C_5H_{10}O$ has no peaks from 3600–3100 cm^{-1} but a strong peak at 1715 cm^{-1}. Find the DU and give possible functional group(s) for this molecule.

Refer to Section 14.5. The formula has a DU = (2*5 +2 – 10)/2 = 1, so has one ring or pi bond. No peaks from 3600–3100 cm^{-1} means no O–H stretch, therefore it is not an alcohol. A strong band at 1715 cm^{-1} is probably a C=O stretch, and that is our DU of one. Possible functional groups for this molecule are a ketone or an aldehyde.

14.5 How many C–NMR peaks would each of these compounds have?

Refer to Section 14.6. The first compound has no symmetry, thus has five different carbon environments. The second has a 180° rotational axis of symmetry, so has only two different carbon environments. The third has resonance, which gives it a plane of symmetry, shown below, so has only three different carbon environments. The last has a mirror plane and just three different carbon environments.

C–NMR peaks 5 2 3 3

14.7 List the C–NMR peaks with the DEPT assignment and propose a structure for C_4H_8O from the spectra:

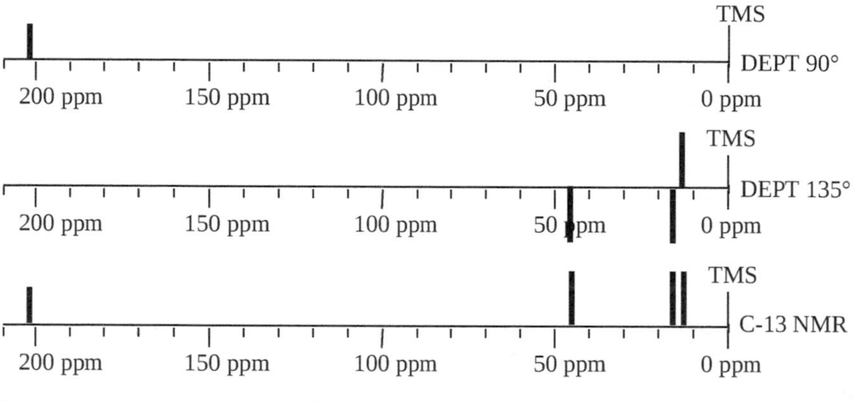

Refer to Section 14.6. Since there are four carbons in the formula and four C–NMR peaks, the carbons are all accounted for, and the molecule has no symmetry. The formula has a DU of one, so has one ring or pi bond. The C–NMR peak at 202 ppm is the correct shift for a C=O and since the DEPT 90° indicates it is a CH then it must be an aldehyde. The C–NMR peaks at 46 and 16 ppm point down in the DEPT 135° so must be methylenes. Finally, the C–NMR peak at 13 ppm that is up in the DEPT 135° but missing in the DEPT 90° must be a methyl. The C–NMR would be listed as: 13 ppm (CH_3), 16 ppm (CH_2), 46 ppm (CH_2), and 202 ppm (CH). With a 202 ppm CHO, the DEPT pieces sum to the molecular formula, so we are not missing any pieces. The molecule has an aldehyde for one end and a methyl for the other, with two methylenes in between. The aldehyde is electronegative so put the most deshielded methylene at 46 ppm next to it. The structure and C–NMR assignment is:

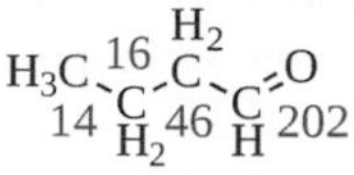

14.9 The C–NMR of $C_6H_{14}O$ is 23 and 68 ppm. Propose a consistent structure.

Refer to Section 14.6. With six carbons and only two C–NMR environments, this molecule has a lot of symmetry. It has a DU of zero, so no rings or pi bonds. The 68 ppm peak is correct for a C attached to O. The only unique atom is oxygen, so put that on our symmetry line with two 68 ppm carbons attached to it. The other four carbons must be in identical environments, so we need two mirror planes. This gives the structure:

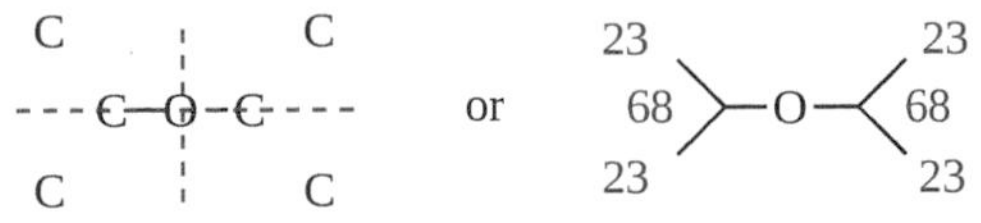

14.11 Make an NMR table, give a structure, and assign all the NMR peaks for the following compound. The MS molecular ion is at m/z 148. The combustion analysis is 81.04% C and 8.16% H. The H–NMR is 1.4 ppm 3H triplet (7Hz), 4.0 ppm 2H quartet (7Hz), 5.2 ppm 1H doublet (7Hz), 6.2 ppm 1H doublet (7Hz), and a 5H multiplet from 7.1 to 7.6 ppm.

Refer to Sections 14.4 and 14.7. For 100% it must have 10.80% O.
For C: MS molecular weight * % C / atomic weight of C. (148)(0.8104)/12 = 10 carbons
For H: MS molecular weight * % H / atomic weight of H. (148)(0.0816)/1 = 12 hydrogens.
For O: MS molecular weight * % O / atomic weight of O. (148)(0.1080/16 = 1 oxygen.
This $C_{10}H_{12}O$ formula has a DU of (2*10 + 2 – 12)/2 = 5 rings and/or pi bonds.

NMR Table

ppm	# of Hs	Coupling	Adjacent H's	Pieces
1.4	3H	Triplet, 7Hz	2H	CH_3 (2H)
4.0	2H	Quartet, 7Hz	3H	OCH_2 (3H)
5.2	1H	Doublet, 7Hz	1H	=CH (1H)
6.2	1H	Doublet, 7Hz	1H	=CH (1H)
7.1–7.6	5H	Multiplet	?	–Ph
	Total 12H			

The H–NMR integration checks at 12 Hs total. The 1.4 ppm 3H triplet (7Hz) next to 2H and 4.0 ppm 2H quartet (7Hz) next to 3H make up a CH_3CH_2 group. The ethyl's CH_2 is very deshielded, most likely attached to oxygen. So we have an end piece of CH_3CH_2O. The 5.2 ppm 1H doublet (7Hz), 6.2 ppm 1H doublet (7Hz), each have one neighbor, share a coupling constant, and have a shift in the double bond region. This gives a middle piece of –CH=CH– and the size of the coupling constant tells us that it is cis. Subtracting these two pieces from the molecular formula gives us C_6H_5 and four DU. The last piece is a 5H multiplet from 7.1 to 7.6 ppm in the aromatic region and can be explained by a monosubstituted phenyl group, which is an end piece. With two ends and one middle piece, it goes together just one way. Resonance with the oxygen lone pair shields the 5.2 CH. The electronegativity of oxygen deshields the 6.2 CH.

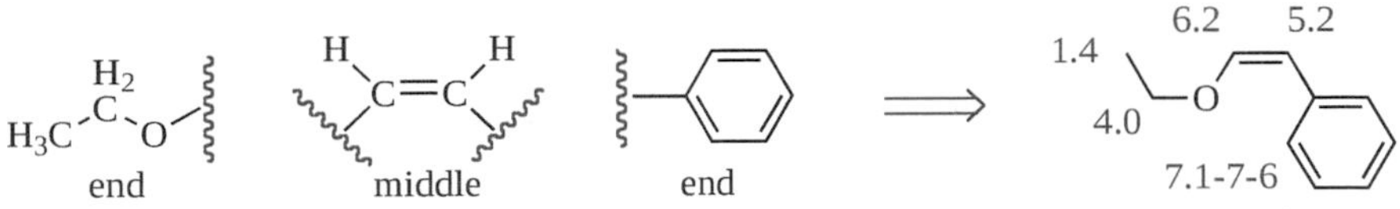

14.13 Make an NMR table, give a structure, and assign all the H–NMR peaks. The MS molecular ion is at m/z 134. The combustion analysis is 89.5% C and 10.5% H. The H–NMR is 0.8 ppm 3H triplet, 1.2 ppm 3H doublet, 1.6 ppm 2H quintet, 2.5 ppm 1H hexet, and a broad 5H singlet at 7.1 ppm. The C–NMR (DEPT) is 12.3 (CH_3), 21.9 (CH_3), 31.3 (CH_2), 41.8 (CH), 125.8 (CH), 127.1 (CH), 128.3 (CH), 147.7 ppm (C).

Refer to Sections 14.4, 14.6 and 14.7. The analysis sums to 100%. It's just C and H.
For C: MS molecular weight * % C / atomic weight of C. (134)(0.895)/12 = 10 carbons

For H: MS molecular weight * % H / atomic weight of H. (134)(0.105)/1 = 14 hydrogens. This $C_{10}H_{14}$ formula has a DU of (2*10 + 2 - 14)/2 = 4 rings and/or pi bonds.

The NMR table follows. Subtracting all the pieces from the molecular formula tells us that nothing was forgotten.

NMR Table

ppm	# of Hs	Coupling	Adjacent H's	Pieces w DEPT
0.8	3H	triplet	2H	CH_3 (2H)
1.2	3H	doublet	1H	CH_3 (1H)
1.6	2H	quintet	4H	CH_2 (4H)
2.5	1H	hexet	5H	←CH (5H)
7.1	5H	broad singlet	?	–Ph
	Total 12H			

The easy way to assemble larger pieces is to take the two end methyls and move inward.

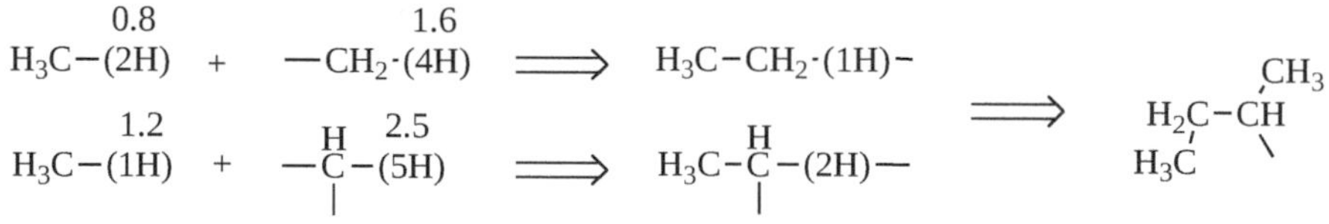

We have two ends and no middles, so just connect those two ends to get the structure.

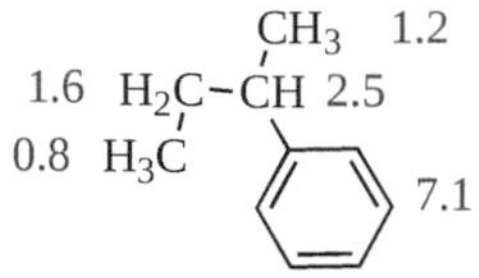

14.15 Fill in the NMR table, and give a structure for $C_{13}H_{16}O_2$. The IR spectrum has nothing 4000–3000 cm^{-1}, a strong peak at 1720 cm^{-1}, and one at 1180 cm^{-1}. The C–NMR (DEPT) is 19 (CH_3), 28 (CH), 71 (CH_2), 118 (CH), 127 (CH), 128 (CH), 131 (CH), 135 (C), 145 (CH), 167 ppm (C). The H–NMR is tabulated below. The calculated HMQC (nmrdb.org) below allows for assignment of which H is on which C (except the 7.3 ppm multiplet).

Refer to Sections 14.5–14.9. Completed NMR Table:

ppm	# of Hs	Coupling	Adjacent H's	Pieces w DEPT
1.0	6H	doublet, 7Hz	1	Two equal CH_3 (1H)
2.0	1H	nonet, 7Hz	8	CH (8H)
3.9	2H	doublet, 7Hz	1	←CH_2 (1H)
6.6	1H	doublet, 18Hz	1	=CH (1H)
7.3	5H	multiplet	?	–Ph
7.6	1H	doublet, 18Hz	1	=CH (1H)
	Total 16H			C=O

$C_{13}H_{16}O_2$ has a DU of (2*13 + 2 – 16)/2 = 6 rings and/or pi bonds. The IR spectrum has nothing 4000–3000 cm^{-1} therefore there is no O–H stretch. The IR peak at 1720 cm^{-1} is most likely a C=O stretch, and the one at 1180 cm^{-1} is likely a C–O stretch.

The compound has symmetry since there are fewer C–NMR peaks than there are carbons in the molecular formula, so the next task is to reconcile the DEPT with the H–NMR integration. Which peaks integrate for more than expected from the DEPT? The H–NMR peak at 1.0 integrates for 6H must be a doubled C–NMR peak, two equal methyls. The other place to look for symmetry is with a phenyl group. There is DU for a phenyl, an H–NMR peak at 7.3 ppm for 5H, and C–NMR peaks in the aromatic region.

A sum of the pieces shows we have them all, with two ends, the phenyl and the pair of methyls next to 1H that should be placed on the same carbon, the 2.0 CH, to get the needed symmetry. When the 6H for the methyls are subtracted, this new piece is next to 2H. There is a 3.9 ppm 2H doublet whose H–NMR and C–NMR shifts fit for being a CH_2O that also has the same 7 Hz coupling constant. This fits as a neighbor to our new piece as shown below.

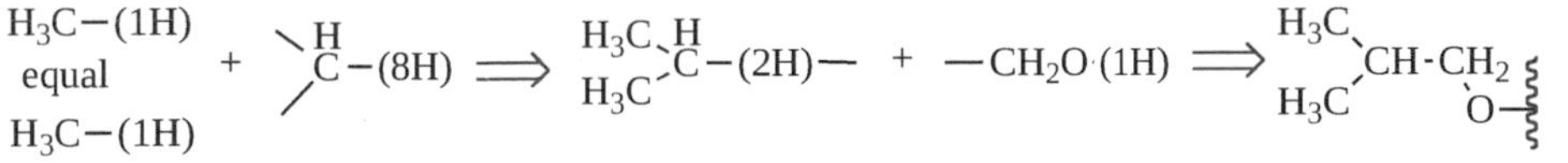

We then put the two vinylic CH doublets (1H) at 6.6 and 7.6 ppm that share an 18 Hz coupling constant together to give a trans –CH=CH– piece. With the C=O from the IR that fits for the 167 ppm C–NMR peak, plus our 5H phenyl, we have all the pieces.

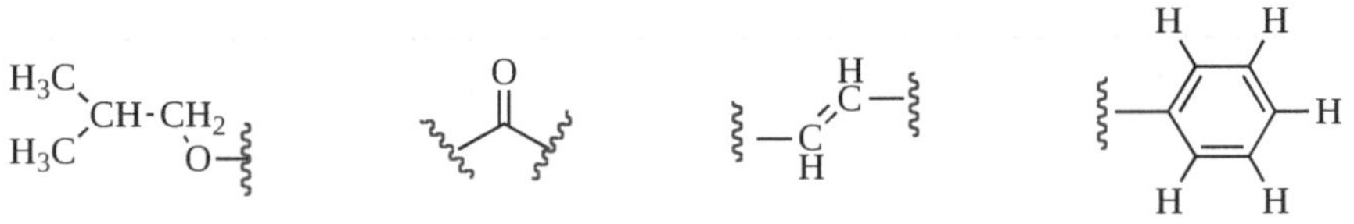

There are two ends and a two middle pieces, so there are two ways to assemble it. We need to use chemical shifts to decide between them. There are on-line programs to calculate spectra from drawn structures. The calculated values from the nmrdb.org website shown below have some error, so don't expect exact correspondence; they are guidelines.

The calculated values for structure on the left definitely fit the actual spectrum better. To assign the carbons, first draw lines from the sides to the peaks within the HMQC rectangle.

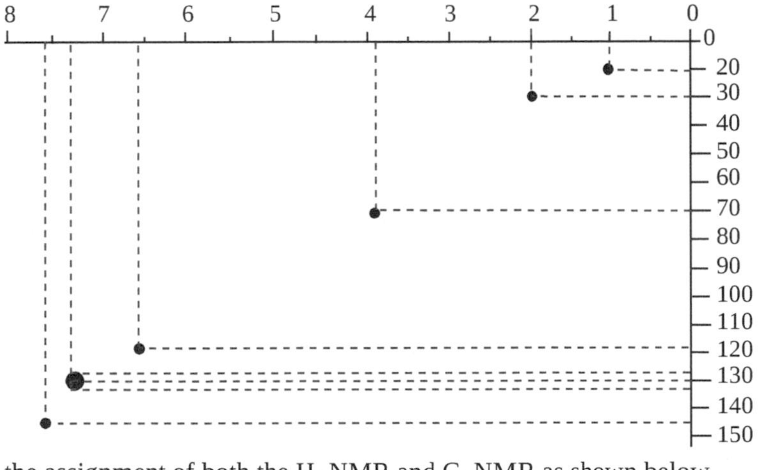

This allows the assignment of both the H–NMR and C–NMR as shown below.

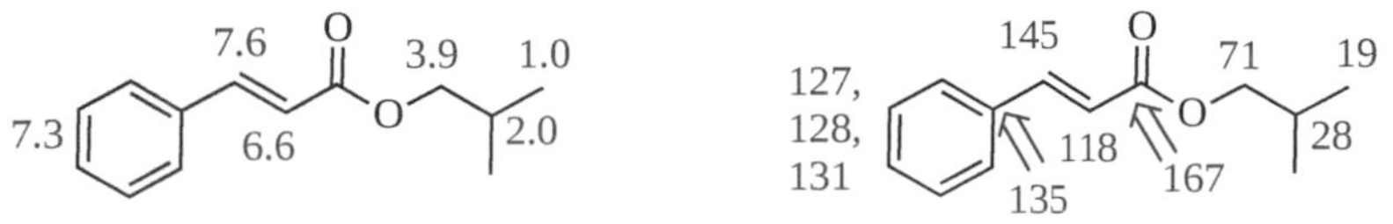

14.17 Make an NMR table and give a structure for $C_8H_{14}O_3$, whose H–NMR is shown below. The C–NMR (DEPT) is 13 (CH_3), 14 (CH_3), 17 (CH_2), 45 (CH_2), 49 (CH_2), 61 (CH_2), 167 (C), 203 (C) ppm. The IR has nothing 4000–3000 cm^{-1}, major bands at 1742 cm^{-1} and 1717 cm^{-1}, and a band at 1200 cm^{-1}. The COSY below allows a definite H assignment.

Refer to Sections 14.5–14.9. $C_8H_{14}O_3$ has a DU of (2*8 + 2 – 14)/2 = 2 rings and/or pi bonds. The IR spectrum has nothing 4000–3000 cm^{-1} thus there is no O–H stretch. The IR major bands at 1742 and 1717 cm^{-1} are likely two different C=O stretches, and a band at 1200 cm^{-1} is likely a C–O stretch. Use an NMR table to gather the H–NMR data together and to integrate it with the pieces from the DEPT spectrum.

NMR Table

ppm	# of Hs	Coupling	Adjacent H's	Pieces w DEPT
0.9	3H	triplet	2H	CH_3 (2H)
1.3	3H	triplet	2H	CH_3 (2H)
1.6	2H	hexet	5H	CH_2 (5H)
2.5	2H	triplet	2H	CH_2 (2H)
3.4	2H	singlet	0H	CH_2 (0H)
4.2	2H	quartet	3H	←CH_2 (3H)
	Total 14H			two C=O

All the Cs and Hs are accounted for, but another O is needed for $C_8H_{14}O_3$. Let us walk the COSY to get big pieces. First draw in the linking boxes on the COSY as done below:

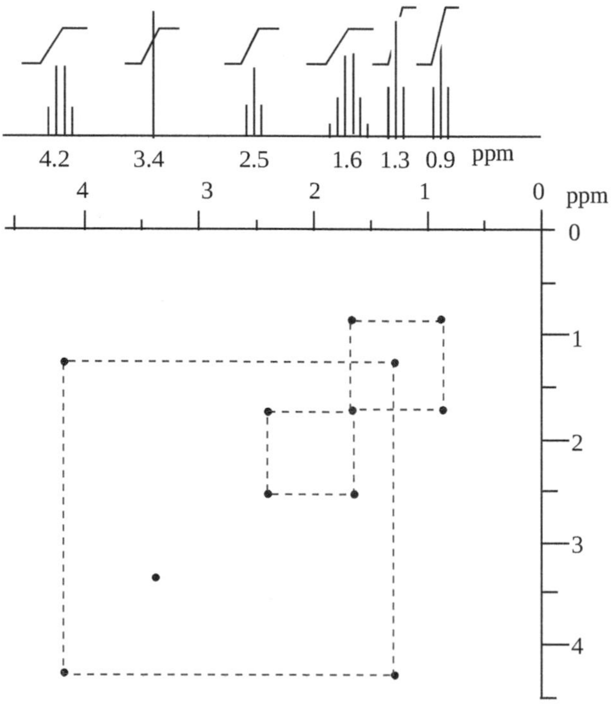

The first thing to notice is that there are two separate sets, and one diagonal peak with no box from the singlet. The 1.3 ppm CH_3 (2H) is linked to the 4.2 ppm CH_2 (3H) and that is all. This 4.2 ppm shift must be a CH_2–O, so we have a CH_3CH_2–O piece. The 0.9 ppm CH_3 (2H) is linked to the 1.6 ppm CH_2 (5H) that is linked to the 2.5 ppm CH_2 (2H) giving us a $CH_3CH_2CH_2$ piece. The remaining pieces are an isolated 3.4 ppm CH_2 and two different C=O pieces, at C–NMR shifts 167.4 (an ester shift) and 202.7 (a ketone shift). Our pieces are:

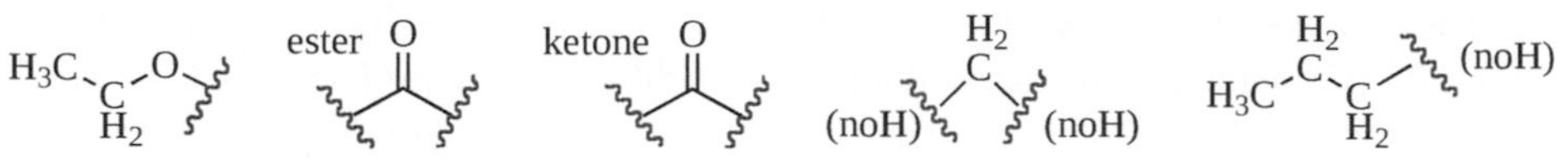

We can put the left two together to make an ethyl ester, but the isolated 3.4 ppm CH_2 must be between the ester and ketone carbonyls to separate it from coupling with the propyl unit on the far right. There is just one way to assemble the molecule with the COSY. We can assign the C–NMR using the similar order of shifts from the H–NMR.

These guessed C–NMR shifts fit those using the on-line C–NMR shift calculation tool. Usually those groups that deshield the H–NMR similarly deshield the C–NMR.

14.19 Give an NMR table and a structure for $C_{10}H_{12}O_2$, whose H–NMR is shown below. The C–NMR (DEPT) is 10 (CH_3), 22 (CH_2), 67 (CH_2), 128 (CH), 130 (CH), 132 (C), 134 (CH), 167 (C) ppm. The IR has nothing 3800–3100 cm^{-1} and a major band at 1687 cm^{-1}.

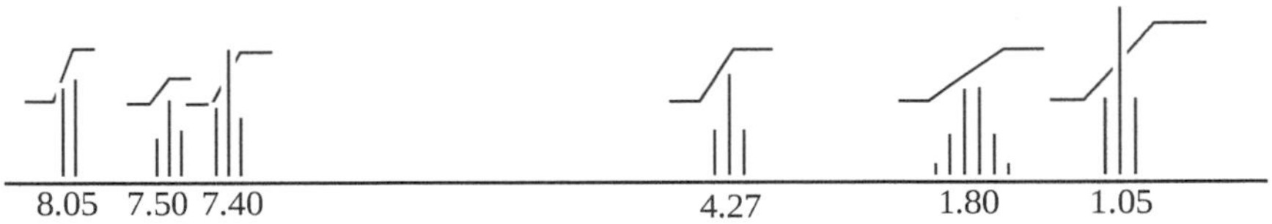

Refer to Sections 14.5–14.7. $C_{10}H_{12}O_2$ has a DU of (2*10 + 2 – 12)/2 = 5 rings and/or pi bonds. The IR spectrum has nothing 3800–3100 cm^{-1} thus there is no O–H stretch. The IR band at 1687 cm^{-1} is likely a carbonyl, and fits with the 167 ppm C–NMR peak.

NMR Table

ppm	# of Hs	Coupling	Adjacent H's	Pieces w DEPT
1.05	3H	triplet	2H	CH_3 (2H)
1.80	2H	hexet	5H	CH_2 (5H)
4.27	2H	triplet	2H	←CH_2 (2H)
7.40	2H	triplet	2H	two =CH (2H)
7.50	1H	triplet	2H	=CH (2H)
8.05	2H	doublet	1H	two =CH (1H)
	Total 12H			C=O

The compound has symmetry since there are two fewer C–NMR peaks than there are carbons in the molecular formula, so the next task is to reconcile the DEPT with the H–NMR integration. The peaks at 7.40 and 8.05 ppm each integrate for twice that expected from the DEPT, so they must be doubled by symmetry. The 2H doublet, 2H triplet, and 1H triplet in the aromatic region is a pattern for a mono-substituted phenyl group. We now have an end. The methyl at 1.05 ppm next to 2H is another end. It would split its neighbor into at least a quartet so the triplet at 4.27 cannot be its neighbor; it must be the hexet at 1.80. That means the 1.80 ppm 2H peak must be adjacent to an additional 2H. A good candidate for that partner is the 4.27 ppm ←CH_2 (2H) peak whose H–NMR and C–NMR shifts fit for being a CH_2O next to 2H. The second end piece goes together thus:

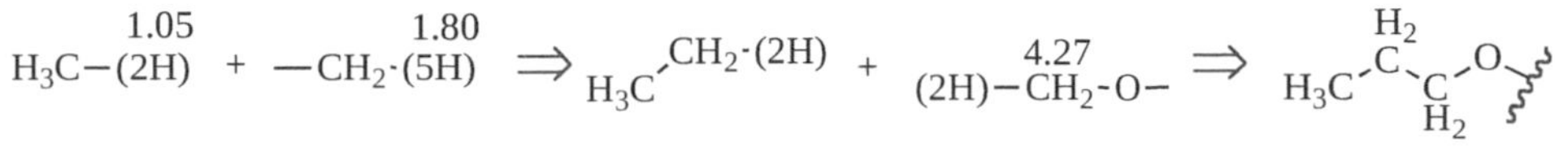

We now have two ends and our C=O for a middle, so it can only go together one way:

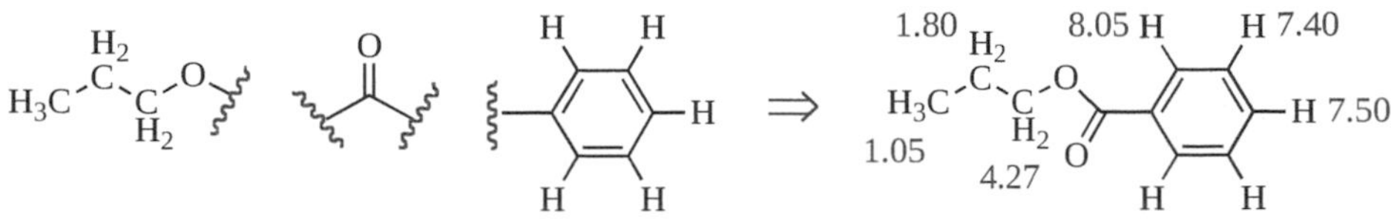

INDEX

A

Electron Flow in Organic Chemistry: A Decision-Based Guide to Organic Mechanisms,
Third Edition, Paul H. Scudder.

B

C

D

E

H

I

J

K

L

M

N

O

S

U

V

W

X

Y

Z